ASSE Professional Development Conference (Safety 2017)

Denver, Colorado, USA
19 - 22 June 2017

Volume 1 of 2

ISBN: 978-1-5108-5001-9

Printed from e-media with permission by:

Curran Associates, Inc.
57 Morehouse Lane
Red Hook, NY 12571

Some format issues inherent in the e-media version may also appear in this print version.

Printed by Curran Associates, Inc. (2017)

For permission requests, please contact American Society of Safety Engineers (ASSE) at the address below.

American Society of Safety Engineers (ASSE)
520 N. Northwest Hwy
Park Ridge, IL 60068
USA

Phone: +1 847 699.2929

www.asse.org

Additional copies of this publication are available from:

Curran Associates, Inc.
57 Morehouse Lane
Red Hook, NY 12571 USA
Phone: 845-758-0400
Fax: 845-758-2633
Email: curran@proceedings.com
Web: www.proceedings.com

TABLE OF CONTENTS

VOLUME 1

VOLUME 2

From Safety Leader to Business Leader – Bridging the Gap

John McBride, SPHR, SHRM-SCP
National Recruiting Team Leader
Consentium Search LLC
Short Hills, NJ

Introduction

Today's most successful companies look for their leaders to be active business partners, regardless of their actual title. However, moving beyond a focus on your area of expertise to a more comprehensive *business* leadership role can be daunting.

Additionally, when it comes to careers, people never plan to fail, but many fail to plan. Their careers seem to go on autopilot or they don't have a specific destination as a target. This interactive presentation will guide the attendees through the process of planning for and becoming a safety business partner.
Although instructional to all levels, this paper will focus on the professional who is already in a role that is responsible for the entirety of a safety or EHS function on a site, region or corporate level and is seeking to move their career to a full business impact position. We will cover: What is a business partner and why do companies seek them over functional managers for leadership roles? What makes a business partner different than a manager/director? And finally: How to get there?

Why Do Companies Seek Business Partners over Managers for Leadership Roles?

EHS professionals who can demonstrate business acumen can help design a strategic plan that balances the needs of the organization, its employees, and other stakeholders with that of Safety. It can help align the efforts of the various functions in the organization with the plan's strategic goals. The key phrase at the beginning is "business acumen." "Business Acumen," in its most basic definition, means that you comprehend and see beyond the confines of your fiefdom and are focused on the success of the business as a whole.

What Makes a Business Partner Different than a Manager or Director?

A business partner is expected to understand how the business is run, how to read the company financials, and most importantly, how their specific expertise, in our case the role of EHS, could improve profitability and cash flow.

The role of the business partner is about shared responsibility for the overall results of the business in all areas. Coincidentally, the internal clients share the responsibility for achieving the EHS goals as well and they should not forget about it. When the internal client wants EHS to share his or her goals, he or she has to accept the EHS goals as a part of the managerial goals in his or her area.

The role of the business partner is about balancing the interests of EHS and the company to honor the company's mission statement and achieve the strategic goals. Many times, EHS represents the interests of other stakeholders inside and outside the organization, including government regulators and shareholders. The role of an environmental health and safety business partner (EHSBP) is in balancing the EHS goals and the long-term stability and sustainable profitability of the organization.

What Do Companies Require to Be Considered for Business Partner Roles?

Simply put, companies need EHSBPs need functional expertise, business knowledge, as well as communication and leadership skills.

Having the ability to see beyond the company's specific sphere of influence and realize the interrelated impact that its function has on all areas of the organization is the end goal of the business partner.

Do You Want to Be a Business Partner?

To fulfill a strategic business partner role, EHS leaders understand the organization's business. Additionally, they must understand the mission of the company (why it exists), the environment in which it operates, the competition, and the circumstances that could influence the progress of the organization. They must see that EHS can no longer focus on its own internal tasks. EHS must be responsible for ensuring that EHS strategy, goals, and priorities are driven by and aligned with the overall business needs. It must establish key business relationships with senior management, as well as key figures in other functions within the organization. Although the operational role of EHS, the day-to-day tasks required to run an organization are not strategic in nature, the responsibilities must mirror the goals of the organization. There needs to be a more integrated global, company-wide process that considers how each of the EHS programs can help move the entire organization in the right direction.

Is this a role you wish to take on?

If Yes, How Do You Get There?

Interview your current business leaders about what they expect EHS to deliver in terms of business success.

Educate yourself.

Be pushy, but not rude.

Too often, EHS still receives a bad rap in organizations as simply being overhead or the necessary evil rather than a business partner. Who is responsible for ending this perception?

Bibliography

Cawley, M. "Lessons Learned." *Risk Management.* Nov 2010: 32-34.

Carillo, R. "Positive Safety Culture, How to Create, Lead and Maintain" *Professional Safety,* May 2010: 47-54.

Combs, D. "Selling Safety, Enhancing Professional Sales Skills." *Professional Safety,* July 2012: 50-52.

Duncan, C. "It's time to take 'magical thinking' out of safety leadership." *Safety + Health,* December 2014: 76.

Dunlap, E. "Safety Leadership, Finding Common Ground." *Professional Safety,* September 2011: 42-49.

Forck, M. "Courageous Leadership." *Professional Safety,* March 2011: 34-35.

Forck, M. "Leading in Accountability." *Professional Safety,* December 2011: 50-51.

Forck, M. "Safety & The Corner Office." *Professional Safety,* August 2012: 24-26.

Forck, M. "Why Safety Leadership Is Hard." *Professional Safety,* October 2012: 34-35.

Fox, A. "Reach New Heights, The best HR business partners connect human resources with business strategies." *HR Magazine,* June 2012: 34-39.

Galloway, S. "Standing Side by Side." *Professional Safety,* July 2012: 26-27.

Gray, D. "Targeting Zero, What Senior Leaders Think." *Professional Safety,* July 2013: 18-19.

Grossman, L. "Why You Need An Entourage." *HR Magazine,* November 2013: 70-71.

Hansen, M. "Know What Makes Them Tick." *Professional Safety,* October 2012: 38.

Hoover, T, "Collaboration builds ownership, better outcomes." *Safety + Health,* August 2012: 52.

Janove, J. "To Know the Business, Start in the Trenches." *HR Magazine,* April 2012: 57-60.

Johnson, A. "2012 Job Outlook." *Safety + Health,* August 2012: 36-41.

Journal of Cognitive Neuroscience, Vol 13(6), Aug 2001, pp. 744-753.

Kendrick, J. "Transformational Leadership." *Professional Safety,* November 2011: 14.

Martin, D. "I talk about safety all the time – but does anyone really hear me?" *Safety + Health,* June 2012: 70.

McKelvey, J. "Lean principles: The connection to safety." *Safety + Health*, November 2012: 50.

Morrison, K. "The ROI of Safety" *Safety + Health,* June 2014 (www.safetyandhealthmagazine.com/articles/print/10414-the-roi-of -safety).

Pater, R. "Two Types of Leaders – Which One Are You?" *Professional Safety*, Jan 2011: 25.

Pater, R. "Overcoming Leadership ADD." *Professional Safety,* November 2013: 30-32.

Raines, M. "Engaging Employees, Another Step in Improving Safety." *Professional Safety,* April 2011: 36-43.

Tyler, K. "Gain Experience Beyond HR." *HR Magazine,* August 2010: 71-73.

Ulrich, D. *Human Resource Champions: The Next Agenda for Adding Value and Delivering Results.* Harvard Business Press, 1996.

Wade, J. "Risk Management's Next Move." *Risk Management.* Sep. 2010: 23-26.

Weiss, K. & Gonser B. "Aligning SH&E With Core Business Initiatives." *Professional Safety,* November 2013: 69-70.

Session No. 502

Partnering: A Key to Building Trust and Safety Excellence

Jay Mullis
Acting Manager
Oak Ridge Office of Environmental Management
U.S. Department of Energy
Oak Ridge, TN

Kenneth J. Rueter
President & Project Manager
URS | CH2M Oak Ridge LLC (UCOR)
Oak Ridge, TN

Figure 1. Gaseous Diffusion Plants at the 2,200 acre East Tennessee Technology Park before Demolition

Introduction

URS | CH2M Oak Ridge LLC (UCOR) combines the talents of two global engineering, design, construction, project management and environmental services companies, AECOM and CH2M, and small business partner Restoration Services, Inc. (RSI). With AECOM as lead partner, UCOR manages the cleanup of 2,200-acre East Tennessee Technology Park for its client and partner, the U.S. Department of Energy, in Oak Ridge, Tennessee.

The site was contaminated with radioactive, hazardous and industrial wastes generated by more than 40 years of national defense and energy missions.

The East Tennessee Technology Park, once called the Oak Ridge Gaseous Diffusion Plant, was built as part of the super-secret Manhattan Project in the 1940s to enrich uranium for the atomic bombs that would end World War II. The site later produced enriched uranium for commercial and defense purposes. Operations ceased in 1985, and the site was permanently shut down in 1987.

That same year, the Oak Ridge Reservation was placed on the National Priorities List as a Superfund site, designating it to be cleaned up under the provisions of the Comprehensive Environmental Response, Compensation, and Liability Act (CERCLA). The Department of Energy (DOE) then began cleanup operations, which include demolition of many of the buildings at the site.

This massive, multi-decade cleanup mission is executed through close partnerships among the DOE, UCOR, community stakeholders, and the represented workforce. These partnerships work to advance environmental cleanup, enable new and enduring missions, and support the shared vision to reindustrialize the East Tennessee Technology Park, turning it into a vibrant private-sector industrial park.

Figure 2. Demolition begins at the K-27 Uranium Enrichment Building, February 2016.

Partnerships Are Integral to Success

Cleanup is paving the way to achieve DOE's goals of creating a safer environment and making land available for future economic development, and partnerships have been integral to the project's progress and success. UCOR and DOE have entered into a formal partnering agreement in which the parties commit to work in a collaborative and cooperative manner through open communication and coordination to achieve the collective vision. More than five years into the contract, partnering has led to a win-win relationship, and established a collective culture of safe sustained performance, as cleanup work has progressed in a safe, cost-effective way.

This productive relationship starts with the mutual understanding that worker safety is a prerequisite and is built on the pillars of trust, open-mindedness, recognition of strengths and weaknesses, and compromise for the greater good. The results have been impressive and, in some ways, unprecedented. In August 2016, the partnership completed DOE's Vision 2016 for ETTP to achieve the first-ever successful demolition of a former uranium enrichment complex's processing buildings. This historic achievement, which was safely accomplished well ahead of schedule and under budget, was made possible by the committed workforce.

Although UCOR's partnering agreement is formal and signed by senior managers, the real strength of partnership is not what is written on the paper but what is unspoken but understood between the lines. That begins with mutual respect between the parties, shared mutual goals, and pursuit of outcomes and, most of all, a passion for results. This means delivering more than the minimum called for in the contract. It means taking it to the next level, and delivering more than what is expected.

Our partnership is not driven by edicts or unilateral mandates. Mutual respect calls for two-way conversations between entities that are both recognized as competent and capable. In our case, DOE is not built to deliver the work themselves. They depend on groups like ours; contractors who have broad capabilities and experience to do the work at hand. It is our job to deliver results. In the process, as DOE's partner, we must also communicate and inform in order to encourage them to give us what we need to get the job done.

For example, UCOR has established a program called Electrical Safety Pursuit of Excellence as a means to resolve long-standing, site-legacy safety issues. *The plan, which grew out of DOE reviews of UCOR's work performance, is designed to improve the UCOR electrical guidance and upgrade worker adherence to the safeguards and controls necessary to attain flawless electrical work performance.* While the visible achievement is the Pursuit of Excellence initiative, between the lines you will find all the elements that make us truly exceptional working together. We excel at stakeholder relations, communicating with regulators, keeping elected officials informed and making sure the media and the public stay in the loop. We also work safely and efficiently, delivering our projects ahead of schedule and under budget.

We participate in regular partnering sessions – meetings with DOE – where we check ourselves on how we're doing. These are candid discussions about our full scope of work where we air our successes, identify problems, discuss solutions and analyze failures. In every case, we leave the room as members of one team.

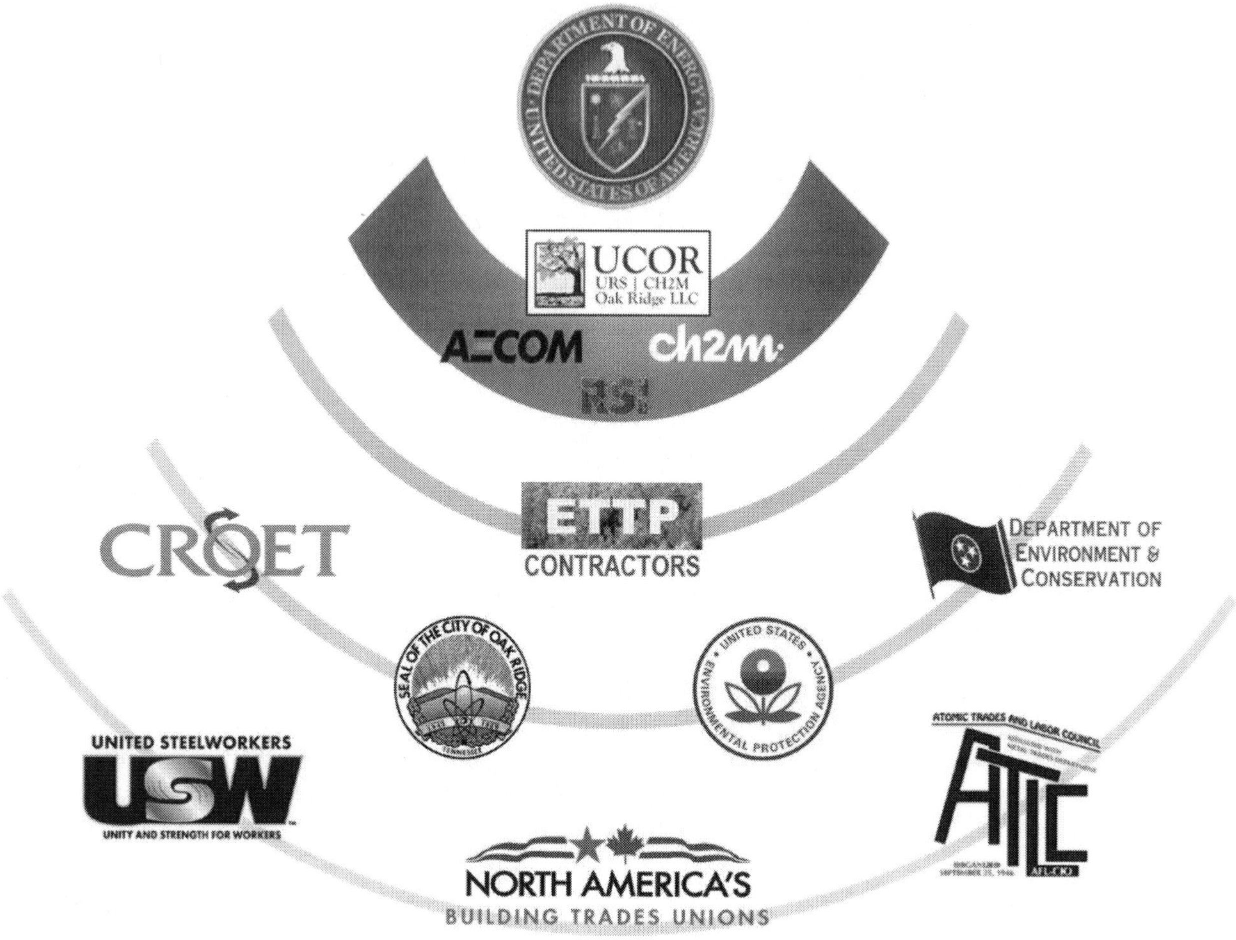

Figure 3. This graphic illustrates the myriad partnerships that contribute to the successful cleanup of the East Tennessee Technology Park.

UCOR's partnerships extend beyond DOE to encompass a host of activities designed to promote safety and the well-being of employees. For example, UCOR is a sponsor of Safety Fest Tennessee, which brings together many groups to promote safety and health at work, at home and in the community. The event attracts hundreds of citizens each year.

In addition, UCOR partnered with all three unions that represent the bargaining employees at ETTP at both the local and national levels to achieve Vision 2016, the first-ever demolition of a complete gaseous diffusion complex. These unions are the Atomic Trades & Labor Council (ATLC), the Knoxville Building and Construction Trades Council (KBCTC), and the United Steel Workers (USW).

On the local level, UCOR's President and Project Manager, along with senior leadership, promote and attend the Safety Advocate and Stewards meeting, where all the unions work to resolve safety issues. During regular meetings of the presidents of UCOR, the ATLC, the KBCTC and the USW, union leaders provide insight into upcoming work, budget and challenges, and are encouraged to work jointly to improve processes, create efficiencies, and resolve challenges together.

On the national level, UCOR's President and Project Manager meets quarterly with senior leadership from the National Building Trades Union, AFL-CIO, and the Metal Trades Department, AFL-CIO, to provide similar updates. These partnerships have enabled UCOR and the unions to develop active safety programs with management and union leadership commitment and encouragement.

Figure 4. Workers completed demolition of the K-25 uranium enrichment building in 2013. It was once the largest building under one roof in the world.

Also, on a national level, the UCOR President has been named to the Board of Directors of the Center for Construction Research and Training, also known as CPWR, further strengthening the company's safety culture, its ties with labor and partnership with DOE. Both DOE and the UCOR workforce benefit from CPWR's considerable training and education resources, and its focus on reducing occupational injuries and fatalities in the construction industry.

The CPWR's mission is consistent with UCOR's Safety Conscious Work Environment and its status as a VPP Star site. CPWR activities include facilitating research on initiatives that improve construction work conditions and prevent injuries, enhancing construction data gathering, and hosting advisory groups that develop best practices and guidelines.

In a related and equally important partnership, DOE and UCOR have aggressively embarked on "Vision 2020," which has, as its goal, the completion of cleanup of the East Tennessee Technology Park. Vision 2020 calls for an accelerated cleanup schedule that translates into significant cost savings for DOE and the taxpayer while making available a multi-use community asset sooner.

Along with DOE, the stakeholders that comprise the greater Oak Ridge community have been, and will continue to be, valuable partners in reindustrialization and economic development, national historic preservation, and conservation efforts on the site. A network of highly experienced government contractors, led by seasoned management, also forms a vital partnership that enables DOE to realize its vision for the Oak Ridge site. Across the Oak Ridge Reservation, contractors support missions from multiple DOE offices.

Several joint projects, including the Secretary of Energy's Excess Contaminated Facilities initiative, depend on coordination and mutual support among these contractors. This partnership will become even more critical as cleanup of the East Tennessee Technology Park is completed and remediation work begins to shift to the Y-12 National Security Complex, a National Nuclear Security Administration facility, and Oak Ridge National Laboratory, which operates under DOE's Office of Science.

In partnership with DOE, UCOR has implemented a safety culture training course that DOE developed for its employees and contractors. Both DOE and UCOR employees have attended the training, which is designed to raise awareness of safety in all aspects of operations.

UCOR has collaborated closely with our DOE client and workforce in implementing substantial improvements to our industrial hygiene (IH) program. This includes remote physiological monitoring for heat stress, state-of-the-art, real-time personal monitors for hazardous workplace gases, a new IH database for all employees, and a novel industrial hygiene work permit process to ensure all hazards are identified and controlled prior to performing any work. These actions were only possible through collaboration among UCOR, the DOE, and our labor workforce. Both DOE and UCOR have shared the details of IH improvements with other DOE sites, such as the Waste Isolation Pilot Plant, to aid in driving improvements.

In another example of client/contractor partnership, UCOR is in the second year of a two-year program to implement human performance improvement (HPI) project wide. With DOE support, all employees will receive training on HPI tools and how to apply them to their jobs. Thus far, nearly 1000, or two-thirds of all UCOR employees along with DOE employees, have completed this training. In addition, 60 UCOR employees have gone to comprehensive three-day HPI training, earning college credit, and serve as UCOR's HPI practitioners in addition to their normal job function. These individuals are institutionalizing HPI tools and methods into UCOR's operating policies and procedures.

Benefits of Partnerships

Our partnerships are benefitting our mission in many ways. UCOR has amassed an impressive safety record of more than 8 1/2 million hours, extending over the past three years, without a lost-work, day-away accident. This achievement has been recognized by multiple awards from the National Safety Council (NSC). We also received the 2016 Medgate EHS Impact Award. We have been recognized as a Tennessee Healthier Work Place. We have been awarded VPP Star status in the DOE Voluntary Protection Program. We support initiatives on Conservative

Decision Making, Safety Culture Sustainment and Human Performance Improvement, all aimed at improving UCOR's safety and conduct of operations.

Over 190 UCOR employees, many of whom are in the represented workforce, have achieved Safety Trained Supervisor formal certification from the Board of Certified Safety Professionals (BCSP). Over 60 percent of all line managers at UCOR hold this certification. UCOR has more employees who hold STS than any other company in the State of Tennessee. UCOR works closely with our employees to enable them to achieve this certification, and each new STS receives a watch as an acknowledgement of their accomplishment.

In 2016, UCOR implemented Mission Possible: Zero for all employees. The objective of this initiative is to encourage all employees to become actively engaged and participate in safety programs. This program has been highly successful. Prior to 2016, employees performed approximately 100 behavior-based safety observations per month. During 2016, on average, over 300 observations were performed. Employees were engaged in numerous other safety programs, such as leading a safety discussion within their workgroup, participating in wellness activities, applying lessons learned to their work, or using a HPI tool to reduce the potential for error in their work. A similar initiative called Target Zero is underway in 2017.

In partnership with DOE as part of a safety culture improvement plan initiative, UCOR performs a comprehensive and independent safety culture survey every two years. The survey was conducted by Oak Ridge Associated Universities. In addition to the survey, management and workers are interviewed individually and in roundtable forums. A report of the conclusions from these three inputs is shared with the workforce. The UCOR Safety Culture Monitoring Panel, composed of employees from all groups and all job classifications, prepares a safety culture improvement plan, and monitors the implementation of the actions and their influence on the safety culture of the UCOR Project.

The UCOR Questioning Attitude Program has been recognized as a good practice by DOE at the headquarters level. Since the inception of this program two years ago, 580 employees have been nominated for exhibiting a questioning attitude that resulted in an improvement to the safety or efficiency of our operations.

Examples of safety events averted include:

- Carpenters were preparing to build scaffolding when they noticed some liquid in the area where the scaffolding was to be erected. They alerted IH specialists, who determined the liquid needed to be removed from the area prior to work resuming.

- A worker notified her supervisor that an identification tag was missing from a canister. The area was barricaded and posted to keep employees at a safe distance until a safe work plan could be developed to remove, package, and dispose of the cylinder.

- A worker noticed a steam line he was preparing to repair had only a single point of lockout when a double lockout was needed. The worker's questioning attitude made sure the process was corrected before beginning work, preventing possible personal injury.

Local Safety Improvement Teams (LSITS) provide yet another avenue for partnering with employees. UCOR supports the operation of seven LSITs, led by craft employees in each of the primary physical areas where work is conducted onsite. These LSITs are empowered to evaluate the safe conditions and behaviors in their workplace, and management supports them by providing the support and resources necessary to address safety issues. A logbook is kept in each area to track the completion of all issues and to maintain accountability.

As a result of the trust that has developed between the partners during this contract, and the collaborative safety initiatives pursued, the safety performance record of UCOR has improved dramatically. In the past three years, the severity of occupational injuries and illnesses at UCOR has been reduced by over two-thirds, as measured by our DART injury rate, which has steadily declined from 0.72 to 0.20. This is a true indication that injury severity levels are being reduced significantly.

At the same time, the frequency of all recordable injuries has declined dramatically as well, as our total recordable case (TRC) rate dropped from 0.96 to 0.34 during this same time period. UCOR's Zero Injury philosophy has established a goal to achieve over one million hours without a recordable injury. Thus far, UCOR has achieved runs of over 800,000 hours and intends to achieve the million-hour goal in 2017.

Conclusion

In summary, the partnership between DOE, UCOR, labor and the community supports a vision of safe, ongoing missions in Oak Ridge, alongside diverse industry, all enabled by strong community support and a highly skilled workforce.

The Elements of an Effective Dust Hazards Analysis (DHA)

Jason Reason, CIH, CSP, CHMM
Senior Vice President of Safety and Health Services
Consultant-Combustible Dust, Safety and Health
Lewellyn Technology, LLC
Indianapolis, Indiana

The issue of combustible dust and its associated hazards (i.e., fire, deflagration and explosion) has existed for hundreds of years, throughout numerous industries in the United States and abroad. When combustible dust fires and explosions occur, they tend to be catastrophic. However, due to their complexity, combustible dust hazards are frequently overlooked and incorrectly perceived as low priority and low risk. After several high-profile accidents and fatalities, as well as increased enforcement by the Occupational Safety and Health Administration (OSHA) and other Authorities Having Jurisdiction (AHJs) (e.g., fire marshals/departments, insurance firms/carriers, etc.), each facility has been forced to reexamine their perception of the combustible dust hazards.

As can be observed from so many of the investigation findings of previous combustible dust incidents, employers and employees appear to be unaware of the hazards posed by combustible particulate solids that have the potential to form combustible dusts when processed, stored or handled. Thus, the Dust Hazards Analysis (DHA) was recently created to identify and combat the potential hazards associated with combustible dusts and combustible particulate solids. Although all the National Fire Protection Association (NFPA) combustible dust standards now retroactively require a DHA to be performed, the importance of a DHA goes far beyond just a requirement. When performed correctly, a DHA not only thoroughly identifies and assesses complex combustible dust hazards, but also provides specific techniques to mitigate these hazards. Unfortunately, all DHAs are not created equal, and all too often, DHAs misidentify or ignore potentially serious fire and explosion hazards (especially DHAs performed by unqualified individuals or entities).

The Creation of the Dust Hazards Analysis (DHA)

Currently, the following five industry or commodity-specific NFPA standards are primarily used to address the combustible dust hazards inherently present in most of the industries located throughout the United States:

- *Standard for the Prevention of Fires and Dust Explosions in Agricultural and Food Processing Facilities* (NFPA 61)
- *Standard for Combustible Metals* (NFPA 484)

- *Standard for the Prevention of Fire and Dust Explosions from the Manufacturing, Processing, and Handling of Combustible Particulate Solids* (NFPA 654)
- *Standard for Prevention of Sulfur Fires and Explosions* (NFPA 655)
- *Standard for the Prevention of Fires and Explosions in Wood Processing and Woodworking Facilities* (NFPA 664)

These five standards broadly apply to varied facilities, processes, equipment types, and dust types to protect against the hazards from combustible dust fires and explosions. A basis of safety is embedded in each of these standards that requires the fuel (dust) to be managed, ignition sources to be controlled, and impact from an explosion to be limited through construction, protection, isolation, and housekeeping. Although each of these standards is similar in that they address the fire and explosion hazards associated with combustible dusts and combustible particulate solids, they are substantially different in their approach as to how these hazards are assessed and mitigated. These differences have caused the requirements prescribed in each of these standards to be inconsistent between the various industry sectors and the dust types, which has lead to confusion in determining which standard applies and how to protect similar hazards within a given process.

To alleviate this confusion and establish consistency across all the industry or commodity-specific combustible dust standards, NFPA recently developed the *Standard on the Fundamentals of Combustible Dust* (NFPA 652). NFPA 652 was developed to establish the relationship and hierarchy between it and any of the other NFPA industry- or commodity-specific combustible dust standards, ensuring that the fundamental requirements are addressed consistently across industries, processes and dust types. Although NFPA 652 was primarily passed to establish consistency across NFPA combustible dust standards, one of its most important contributions was the creation of the Dust Hazards Analysis (DHA). An example of the importance of the DHA is evident in that all the requirements related to a DHA are retroactive (i.e., "grandfathering" does not apply).

The Scope and Requirements of a DHA

NFPA 652 defines a DHA as, "a systematic review to identify and evaluate the potential fire, flash fire, and explosion hazards associated with the presence of one or more combustible particulate solids in a process or facility." Although the term DHA may be new to some, the requirement to effectively assess all the potential combustible dust hazards is not. For at least ten years (i.e., multiple editions), NFPA 484, NFPA 654, and NFPA 664 have all required the fire and explosion hazards associated with combustible dust processes to be examined through a hazard analysis. Essentially, the DHA just builds upon this requirement in that the DHA is a Process Hazard Analysis (PHA) that is solely focused on assessing and mitigating the fire, deflagration, and explosion hazards associated with combustible dusts and combustible particulate solids. Because the DHA only focuses on these specific hazards, there might be other hazards associated with a process (e.g., chemical exposures (industrial hygiene), flammable liquids/vapors, etc.) that are not covered by the DHA.

Any facility, regardless of size, where any quantity of materials (dusts, particulates, etc.) that have been determined to be combustible or explosible are present in an enclosure (i.e., room, building, vessel, silo, bin, pipe, duct, etc.) is required to complete a DHA within three years from the effective date of NFPA 652-2016 (i.e., September 7, 2015). While this duration may seem excessive, the history of fatalities, injuries and loss of business continuity associated with combustible dust incidents reinforces the need for each facility to perform a DHA as quickly as possible. It should also be noted that the absence of previous incidents (e.g., "no fire or explosions have ever occurred at the facility") cannot be used as a basis for not performing a DHA.

How a Sampling Strategy (or Lack Thereof) Can Affect a DHA

To properly assess and mitigate potential fire, deflagration, and/or explosion hazards, one must first know the specific physical characteristics (e.g., particle size, moisture content, etc.) and properties (e.g., combustibility, ignition, etc.) associated with the combustible dust(s). Thus, precisely determining the combustibility and explosibility of a dust or particulate is vital to support the DHA and ensure that any combustible dust hazards are correctly assessed and mitigated. However, the cost associated with dust testing methodologies often causes facilities to question the need for dust testing and rely on incomplete historical or published data and/or the absence of previous incidents. Neither of these concepts should be used as the sole basis for deeming a dust or particulate as noncombustible or nonexplosible. Relying on these or similar concepts is dangerous because it can often lead to missing or incorrectly identifying potentially serious combustible dust hazards.

While attempting to determine the combustibility and explosibility of dusts, it is common to analyze an overabundance of dust samples and perform unnecessary tests. Thus, the importance of an effective combustible dust sampling strategy cannot be overstated. An inadequately designed and/or executed combustible dust sampling strategy will incorrectly comprehend the specific physical characteristics and properties associated with the specified combustible dust(s). At best, this will most likely lead to more than the minimum number of samples being analyzed (i.e., over-sampling), as well as insufficient or unnecessary tests being performed on dust samples. At worst, this could diminish the hazard mitigation strategies proposed by the DHA and allow fire, deflagration, and explosion hazards to remain virtually unchecked.

The Important Role of the Qualified Person

The expertise and experience of the individual(s) who leads or performs the DHA can greatly affect the overall quality of any DHA. Thus, NFPA 652 and the other NFPA industry- or commodity-specific combustible dust standards require any DHA to be performed or led by a qualified person. NFPA 652 defines a qualified person as a "person who, by possession of a recognized degree, certificate, professional standing, or skill, and who, by knowledge, training, and experience, has demonstrated the ability to deal with problems related to the subject matter, the work, or the project." Besides this simplistic definition, NFPA 652 sheds little light onto what qualifications are truly required to perform or lead a DHA.

While NFPA 652 tries to define the essence of being qualified, what makes an individual qualified to perform or lead a DHA is much more difficult that the definition implies. All too often, individuals examine this definition and incorrectly assess their own abilities and expertise to perform or lead a DHA. These inexperienced and unqualified individuals perform DHAs that do not adequately identify and address all the potential combustible dust hazards associated with process equipment and building compartments. The conclusions reached during these DHAs may create additional hazards, and provide a false sense of security that the fire, deflagration or explosion hazards are effectively identified and mitigated, when in fact they are not.

While the fire, deflagration and explosion hazards posed by combustible dusts fall under the umbrella of safety and health, these hazards are much more complex than they appear. Individuals must realize that simply being an engineer, consultant, or environmental, safety and health (EHS) professional does not necessarily mean one has the necessary skills and unique knowledge to perform a DHA. An advanced degree, license and/or certification (e.g., PE, CIH, CSP, CFPS, etc.) also does not automatically qualify an individual to perform or lead a DHA. Despite assertions to the contrary, there is also no one training class that an individual can take to instantly become qualified to perform or lead a DHA.

So, the real question is who is qualified to perform or lead a DHA? Unfortunately, the answer to this simple question is extremely convoluted. NFPA 652 suggests that the qualified person who performs or leads the DHA should be, "familiar with conducting a DHA and with the hazards of combustible dusts." However, these statements are overly broad and underestimate how much an individual's expertise can affect the overall quality and outcome of a DHA. Someone who is truly qualified to perform a DHA simultaneously has intimate knowledge of a wide range of dust handling processes, as well as specialized knowledge of the complex hazards associated with the various types of combustible dusts. This individual has not only performed all the aspects of several DHAs associated with the affected industry or type of combustible dust, but has also honed their expertise by examining and mitigating a variety of fire, deflagration, and explosion scenarios over several months or years.

Although determining the qualifications of an individual is relatively simple, it is frequently not performed and taken for granted. To their own detriment, facilities instead improperly rely on an individual's word or reputation. Thus, it is vital that the facility perform *all* the tasks listed below to ascertain if the selected individual(s) is qualified to perform the upcoming DHA:

- Obtain the resume and other applicable information (e.g., published articles, copies of certifications, etc.) to gauge the individual's specific qualifications and selected role in the upcoming DHA.
- Obtain and contact several references (preferably at least one in a similar industry) to shed light on not only how the individual(s) previously assessed and mitigated combustible dust hazards, but also the overall structure and quality of the DHA.
- Determine what actions, if any, the individual(s) will perform prior to commencing the DHA. Prior to performing the DHA, the qualified person(s) should thoroughly discuss the process equipment, building compartments, safety and health programs, and fire and explosion scenarios with all affected parties.
- Obtain and thoroughly review one or more reports from previous DHAs performed or lead by the selected individual(s) to assess how the vast amount of information gathered during the DHA will be compiled, the thoroughness and quality of the DHA, and effectiveness of the hazard mitigation strategies previously employed by the selected individual(s).

It is vital to remember that unqualified individuals often cannot or do not wish to have their "qualifications" questioned. However, no one who performs any aspects of a DHA should consider themselves above being vetted. Thus, failure to provide any of the above information should be taken as a definitive sign to begin to question the individual's expertise and experience in assessing and mitigating the serious hazards posed by combustible dusts and combustible particulate solids.

DHA Methodology

While the significant role of the qualified person(s) cannot be understated, an effective DHA requires additional organizational involvement beyond that of the qualified person(s). The most effective DHAs are not "one-man shows," but rather are a collaborative effort between multiple layers of an organization and other interrelated entities (e.g., insurance carriers, contractors, etc.). Due to their intimate knowledge of the complexities and underlying factors of the specific process equipment and building compartments, representatives of the EHS, engineering, and maintenance departments should all be heavily involved in any DHA. Due to their vast knowledge of all the normal and abnormal production aspects, equipment operators and mid-level managers (e.g., supervisors, shift leaders, etc.) will also play a crucial role in any DHA. While their involvement in the DHA may not be as substantial, outside entities (e.g., contractors, equipment manufacturers, insurance carriers, etc.) are a vital source of information on the design, operation, and maintenance of process equipment located throughout the facility.

The purpose of a DHA is to identify the fire, deflagration, and explosion hazards associated with combustible dusts and combustible particulate solids, and document how these hazards are currently being controlled or managed. Specifically, the DHA should thoroughly analyze not only any process equipment that handles, generates, stores, etc., combustible dust (e.g., dust collectors, dryers, bins, silos, etc.), but also any building compartments (i.e., rooms, areas, buildings, etc.) where combustible dust could be present in any quantity. The DHA should first identify what are the normal operating conditions (e.g., temperature, speed, pressure, material/product flow, etc.) associated with the process equipment and building compartments. The DHA should then thoroughly identify and evaluate specific scenarios to determine the consequences of what could go wrong in the event of a fire, deflagration and/or explosion. The DHA should next examine the effectiveness of the safeguards that are currently implemented to control or eliminate these consequences. Where the current safeguards are inadequate or nonexistent, recommendations should then be issued that precisely describe what administrative or engineering controls can be implemented to effectively mitigate the fire, deflagration, and/or explosion hazards associated with the process equipment or building compartments.

The Anatomy of an Effective DHA

A DHA is complex in every sense of the word, and as such should not be generalized in its approach. However, facilities often dangerously assume every DHA is created equal, and do not pay enough attention to the actual composition of their own DHA. For a DHA to be effective, it must analyze all the potential fire, deflagration, and explosion hazards associated with the affected process equipment and building compartments. While this approach may seem simple, many of the factors that can contribute to, or be affected by, these hazards are not always adequately analyzed during many DHAs. Thus, for any DHA to be truly called effective, each of the following factors should be painstakingly analyzed for potential deficiencies:

- All potential sources of ignition
- Industrial ventilation systems
- Environmental, safety and/or health (EHS) programs and policies
- Housekeeping frequencies
- Cleaning methods used to remove combustible dusts
- Use of outside contractors (especially cleaning contactors)
- Electrical classification

- Reactivity of combustible dust(s)
- Design, operation and maintenance of explosion protection and prevention systems
- Design, operation and maintenance of fire protection systems
- Preventative maintenance programs
- Construction of building compartments
- Emergency egress
- Testing of representative dust samples
- Use of flame-resistant clothing (FRC)

Unfortunately, all DHAs are not created equal and, for a myriad of reasons, several of these factors are frequently ignored or overlooked during many DHAs. This calls into question a qualified person's expertise in not only combustible dust, but also safety and health in general. This lack of expertise can cause several problems, including falsely believing that any DHA will ensure compliance with all applicable OSHA regulations, state Fire Codes, and/or NFPA standards.

EHS programs such as, but not limited to, hazard communication (HazCom), confined spaces, emergency response, contractor safety, hot work, and electrical safety are almost never examined during most DHAs. In fact, most individuals who perform DHAs will go out of their way to not even consider the idea of examining these programs because their expertise lies in combustible dust and not general safety and health. These individuals don't understand the importance of how these programs individually address all safety and health hazards (including combustible dust), which causes them to label these programs as unnecessary when they most assuredly are not.

Bibliography

National Fire Protection Association (NFPA). 2016. NFPA 652, *Standard on the Fundamentals of Combustible Dust*. Quincy, MA: NFPA.

_____. 2017. NFPA 61, *Standard for the Prevention of Fires and Dust Explosions in Agricultural and Food Processing Facilities*. Quincy, MA: NFPA.

_____. 2015. NFPA 484, *Standard for Combustible Metals*. Quincy, MA: NFPA.

_____. 2017. NFPA 654, *Standard for the Prevention of Fire and Dust Explosions from the Manufacturing, Processing, and Handling of Combustible Particulate Solids*. Quincy, MA: NFPA.

_____. 2017. NFPA 664, *Standard for the Prevention of Fires and Explosions in Wood Processing and Woodworking Facilities*. Quincy, MA: NFPA.

Occupational Safety and Health Administration (OSHA). *Combustible Dust National Emphasis Program (CPL 03-00-008)* (retrieved March 3, 2017) (http://www.osha.gov/pls/oshaweb/owadisp.show_document?p_table=DIRECTIVES&p_id=3830).

Session No. 506

Your Physical Job Does Not Make You Physically Fit, Unless You're a Ninja

Lori Frederic, CSCS, CMT
Biomechanics Educator, Owner
Balance Bio - Movement Mechanics
Littleton, CO

Physical Fitness

Physical Fitness, as Defined by Wikipedia:
"Physical fitness is a general state of health and well-being and, more specifically, the ability to perform aspects of sports, occupations and daily activities. Physical fitness is generally achieved through proper nutrition, moderate-vigorous physical exercise, and sufficient rest."

What is interesting is that before the industrial revolution, fitness was defined as "the capacity to carry out the day's activities without undue fatigue." Although with the recent huge leaps in technology and automation, we have created changes in lifestyles. In present times, physical fitness is considered more of a measure of the body's ability to function efficiently and effectively in work and leisure activities rather than more industrial daily activities. Physical fitness means to be healthy, to resist diseases, and to be functional in emergency situations.

When it comes to physical jobs of today, I believe the previous definition still rings true and workers should be able to carry out the day's activities without undue fatigue. As safety professionals, we work hard to ensure our employees health and safety and send them home in one piece at the end of the day. However, as muscle strains and sprains continue to be at the top of the list for injuries, it is evident that many employees experience undue fatigue.

What's your definition of physical fitness?

As much as we want employees to resist diseases, the external factors of fitness and health are often the easiest to improve. Assessments of posture and functional movement are helpful as external indicators of physical fitness. These assessments can be incorporated in Post-offer Employment Testing (POET) or Physical Capacity Evaluations (PCE). Movement assessments are most helpful to the employee when results are shared for their knowledge and benefit.

Examples of movement assessments are:

- Functional Movement Screen (FMS), by Gray Cook [Exhibit 1]
- 3D Movement Analysis & Performance System (3DMAPS), by Gray Institute [Exhibit 2]

Exhibit 1. These are the 7 Functional Movement Screens

Exhibit 2. This is 1 of 6 Movement Analysis of 3DMAPS

Physical Jobs

Physical jobs are the fabric of our country. Trade skills are underappreciated, and those who have them are often overworked. Many who perform physical work for a living understand that their bodies are much like a train engine fed coal; the more fuel they give it, the faster it will, go so to speak.

Now by "fuel," I mean many things. Fuel can be actual fuel like nutrition, or it can be sleep for recovery, or it can be daily stretching or hydration. I believe it's possible to fuel a body mentally and physically. How do your workers "fuel" their bodies?

Unfortunately, fueling the body is the last thing on a worker's mind. There is work to be done and often in a hurry. Focus is often on production and deadlines…and not getting killed. Valid focuses indeed; however, imagine a world where workers are indeed physically fit. Imagine teaching the value of the human machine, where a worker maintains and fuels it as such.

Change Your Mind

Attitude

Physical tasks can break bodies down if work is not performed with the right attitude. The attitude your workers have will direct the actions of their shift, especially when working in a physical job that requires long hours where the work scenarios and environmental factors are ever-changing. The mental and physical environment creates great potential for soft-tissue injuries from repetitive motions, awkward positions, and worker fatigue.

When a worker has the right attitude and understanding of their physical condition, they are able to take steps that will improve their effectiveness in the field and go home without injury. This is a game changer for most organizations, especially those physical jobs where the external environment cannot be easily manipulated to reduce risk of injury.

Act Like a Ninja

Seriously. Ninjas are agile, focused, and aware of their environment. Long shifts with awkward lifting and repetitive tasks create muscle imbalances in the body. When a body is imbalanced, it is a greater risk for dysfunction and injury.

Workers who are aware of the risks of improper working posture will make greater effort to correct them. Simple corrective exercises and stretches help balance the body and make proper working postures second nature that, in turn, affect productivity, overall safety, and attitude.

Safe lifting and proper body mechanics are all about viewing your body as your premiere tool. The technique of how you use your body properly hasn't changed, but the technique of getting your workers to care sure has. Educating your employees on the importance of using proper mechanics and core strength while lifting, along with appropriate stretching techniques, leads to an improvement in physical fitness, along with a reduction in soft-tissue injuries.

Giving the employees tools and capabilities to understand how their actions can in fact contribute to safe work. By teaching habits of awareness, attitude, and taking action, then conditioning the employee through practice (much like a ninja) so that these become a habit, their ability to call on that knowledge and execute a proper response given the circumstances will, in turn, lead to a reduction of injuries and improved operational effectiveness of the workforce.

Qualify the People

Encouraging retention of knowledge is possible by changing the tactic from 'training' to 'qualifying.' For this modern workforce, the term 'trained' is so one-dimensional. Whereas emphasizing a good attitude, optimal health, fitness for duty, *and* occupational skills builds motivation to be a part of the company culture. Inspiring workers to take action for themselves and for each other attracts and keeps good people.

Being healthier, happier and more physically fit naturally brings increased work quality and productivity. As a supervisor, the way you 'qualify' your workers makes a huge difference. Your job is to train essential safety skills, but it is also to inspire workers to create their own culture. A good leader will help cultivate the collective mind in a way that, in turn, creates more leaders.

Making Work Cool Again

As a huge fan of Mike Rowe and his Mike Rowe Works Foundation, I must agree that now is a great time to evolve the perception of trade workers. I believe the idea of making work "cool" again won't solve every problem. However, when a worker has more pride in the work that they perform, then they are motivated to continue education and training, along with developing an understanding for maintaining their premiere tool; that is, their body.

"Working hard for something we don't care about is called stress. Working hard for something we love is called passion." - Simon Sinek

My best example of making work cool is in my experience working with oilfield employees. When I traveled to different shale plays and ventured onsite, I was able to get a window into the world of roughnecks and frac hands. Unfortunately, I kept seeing the same theme of window decal on most of the trucks. [Exhibit 3]

Exhibit 3. Oilfield Trash Truck Decal

I saw "Oilfield Trash" or "Oilfield Trash and Proud of It," "Oilfield Trash Making Oilfield Cash." I could tell that there was an underlying pride for the industry, but the fact that they identified with being "trash" really got under my skin. These folks were busting their rears 12 or more hours a day for 2-3 week hitches. This industry heats our homes, cooks out food, and supplies us with fossil fuels that make technology possible. I wanted to design a better option, a better identity, thus the "Oilfield Ninja" was born.

Exhibit 4. Oilfield Ninja

Engage and Inspire

As supervisors and safety professionals, it is so important to relate to the worker and help change behaviors. Use your creativity and knowledge to develop your own theme to inspire workers to take action. It is common knowledge that hands-on learning should be used whenever possible when teaching physical tasks. The good news is that it is easy to incorporate an emphasis on physical fitness while you are teaching. For example, when you are training tractor-trailer chain-up, you can speak about body postures, back safety, core strength, and flexibility. The same is true for ladder safety, emergency rescue, and others.

Be authentic, and deliver the message that you care, and you will create more buy in. You care about getting the job done right, but you also want your employees to be able to perform their jobs years down the road without incident or injury. Workers want to have strength and energy at the end of the shift, at the end of the week to head home to their families and play, hunt, etc.

In conclusion, when you authentically care, you get buy in. When you make physical work cool again, you develop a following of workers who strive to qualify to the high standards of a premiere safety culture. With a small shift in perspective, it becomes possible to engage and inspire to create physical fitness from all physical jobs.

Bibliography

Cook, Gray, et al. 2011. *Movement Functional Movement Systems: Screening, Assessment, Corrective Strategies.* Aptos, CA: On Target Publications.

Gray, Gary, 2016. "Three-Dimensional Movement Analysis and Performance System (3DMAPS)." *Body analysis online training program.* Gray Institute.

Rowe, Mike. 2017. "Mike Rowe: If you want to make America great again, you have to make work cool again." Tucker Carlson Interview. (http://mikerowe.com/videos/interviews/).

Wikipedia. *Physical Fitness.* (https://en.wikipedia.org/wiki/Physical_fitness).

Complacency: The Silent Killer

Larry Wilson
Author of SafeStart
Belleville, Ontario, Canada

Introduction

Driving home from the rec. center last week, I saw a familiar scene starting to build on the side road next to the highway. It was the usual suspects: an ambulance, more than one police car, a fire truck, another on the way—sirens blaring, the street being closed off.

In a few blocks, I was home. The neighbors were asking if I knew what happened.

"No, I don't," I said. "But whatever it was, it didn't look good—too many police cars and fire trucks for something minor."

"Was it on the highway?"

"Are they going to close the highway down?"

Exhibit 1. Accident Site on Mons Road by Trail Exit

I answered both people at once. "No—it was on the Mons side road where the bike trail exits (Exhibit 1). They're closing it down. Or, at least it looked like they were going to close it down."

My seven- and nine-year-old girls were on their bikes. They asked if they could ride over and see what was going on. I said I didn't think that was a good idea. Whatever happened—one thing I was sure of—it wasn't going to be pretty.

Several hours later, we eventually heard what happened: a flatbed transport truck (18-wheeler) hit a man on a mountain bike crossing the road—in broad daylight. The truck wasn't speeding.

"How can you get hit by a transport?" one of my friends asked.

"Must not have seen it," I offered.

"Yeah, I know, but that's what I mean—how can you not see a transport truck?"

"Not easy," I said in agreement. But in my head I'm thinking complacency. You've got to be pretty complacent to not see a transport truck. He must not have even looked. Heck, just a quick glance probably would have been enough. If he had just slowed down enough to get that quick look, he'd still be alive.

With violent crime, quite often you hear yelling and screaming before the gun shots. In the movies, the music changes, but with complacency, there's usually no warning; nobody's worried about anything, and then "wham"—somebody's dead—every day, thousands (and thousands) of times a day, all over the world. When you add it all up, complacency contributes to more unintentional deaths than anything else, especially when you combine it with rushing, frustration or fatigue.

How do people get so complacent that they don't look for transport trucks or that they fall asleep at the wheel? Why do people get so complacent that they don't even think about the risk anymore? (When was the last time you "worried" about your safety as you got behind the wheel?) How do people get complacent enough that they will do something that they know contributes to making a "mind not on task" error, like texting while driving? Or to not use a safety device like a fall arrest harness, which reduces the risk if they do happen to make an error with balance, traction or grip? And finally, since it happens to all of us—we all do get complacent—what can we do to fight it?

Complacency and "Mind Not on Task"

As mentioned above, complacency causes many problems, but the biggest or the worst is that it leads to "mind not on task." Once the fear is no longer pre-occupying, your mind can wander (Exhibit 2). And when you're thinking about something else, other than what you're doing at the moment, your most important asset—your star player—is sitting on the bench. Think about all of the times you've been hurt (not including sports). Can you even think of one time in your life when you've been hurt when you were thinking about what you were doing, and the risk of what you were doing, at the exact instant when you got hurt? Note: If you're like most people, you can't even think of one—let alone ten. And yet, we've all been hurt thousands of times if you count all of the cuts, burns, bruises, and scrapes we've had.

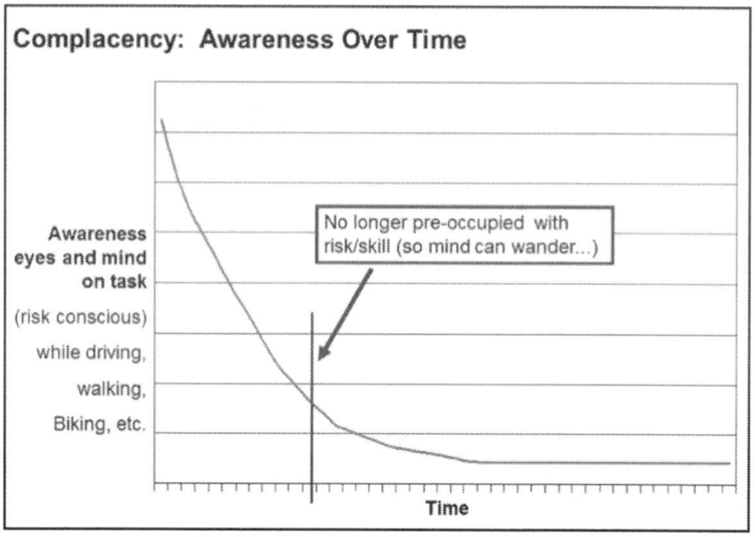

Exhibit 2. Complacency: Awareness Over Time Graph

Case in point: I asked the question to over 1200 linemen of a major electrical utility. Out of 1200 lifetimes—on and off the job—I got two examples. The average age of the group was about 45. That's two examples in 1200 × 45 years.

So, as mentioned before, your mind is indeed your most important safety "device." But here's the catch: we don't always need to be thinking about what we're doing—from a risk perspective—in order to do many things (like drive) without getting hurt. We can all do lots of things on auto-pilot. So the first thing people need to do is recognize or accept that their mind will wander if they've done the same thing many times before. It's going to happen. It happens to everybody. It doesn't make you a bad person—just a dead one, or a disabled one or a lucky one. But it's not hopeless. There are techniques you can use to fight complacency, and there are techniques you can use to compensate for complacency leading to mind not on task—and mind not on task leading to line-of-fire or balance, traction or grip errors. They don't require expensive equipment or yellow and black hashmarks. But they do take a bit of personal effort. And finally, besides teaching people these techniques, we also have to teach them to develop a deep level of respect for complacency, and what it can do, so that it doesn't start creeping in to their decision making as well.

Critical Error Reduction Techniques

Although everyone does get complacent with things they have done over and over, it's not hopeless, because, if you're not thinking about what you're doing, your behavior will be what it normally is.

And it will only change if you make a conscious effort to change it. So we need to get people (including ourselves) to work on their safety-related habits: like moving your eyes before you move your hands, feet, body or car; testing your footing or grip before you commit your weight to it; looking at your "second foot" as you step over a cord or something that you could trip on (it's usually the second foot that you're not paying attention to that gets caught or hung up); looking twice when the sun is in your eyes to make sure you didn't miss something (the sun does play tricks on you); lightly touching something before you grab it—if it might be hot (I learned this one at a steel mill); and finally—sadly in the case of the mountain biker—habitually looking for line-of-fire or what might be coming at you at blind intersections (or where bike trails through the woods exit onto a side road, as in Exhibit 3). Because, if you don't see what might be coming at you, it could easily be the last mistake you'll ever make.

Exhibit 3. Trail Ends at Blind Intersection

Of course, it takes a bit of effort to develop new habits or to change old ones, but the real bonus is that once you do, you will do it automatically. It won't take any effort after that. And then you won't have to worry nearly as much about complacency leading to mind not on task, and mind not on task leading to other critical errors like line-of-fire and a loss of balance, traction or grip. However, even though improving your safety-related habits will help to compensate for your mind going off task, it would also be good if we had a way of helping people "pull their mind back into the ballgame" or to bring them back to the moment. This is because your habits and reflexes alone don't give you the ability to anticipate a dangerous situation and then take yourself completely out of harm's way. For that, you need your mind.

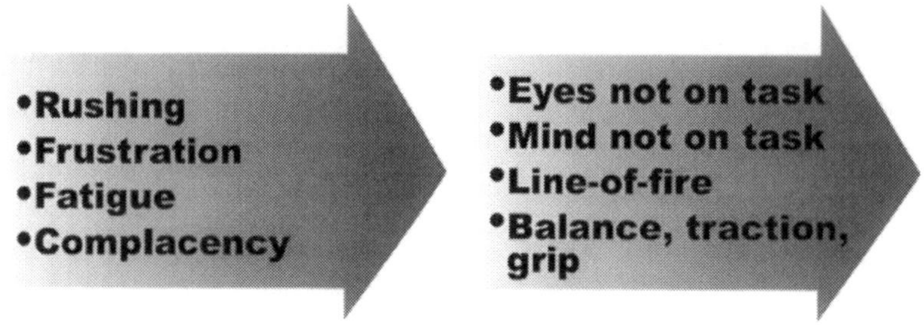

Exhibit 4. State-to-Error Risk Pattern

However, if you watch other people for "state-to-error" risk patterns (see Exhibit 4), every time you see one, it will automatically make you think more about what you're doing. And, if what you see is sensational enough, you'll do more than think about it; you'll actually react to it. For instance, instead of just driving on auto-pilot, you will look around a bit more. As you do, you will notice the driver beside you trying to look at a piece of paper inside a file folder while talking to someone on the phone.

Chances are, you're going to either speed up, slow down, or move over another lane to get out of the way.

But if you didn't look or you don't get in the habit of looking, then you wouldn't necessarily take yourself out of the line-of-fire.

So, working on your safety-related habits helps to compensate for complacency leading to mind not on task, and looking at others for state-to-error risk patterns helps to pull your mind back into the ballgame and make you think about the risk of what you're doing at the moment.

Complacency and Decision Making

While these two critical error-reduction techniques will reduce injury-causing errors like eyes not on task, moving into the line-of-fire, and losing your balance, traction or grip, we also need to help people realize the problems complacency can cause with decision making.

One of these problems is with trusting something important—especially something that is important or critical from a safety perspective—to your memory. This can be as simple as forgetting to bring a lifejacket for everyone in the boat, to something more complicated, like remembering a change to a well-established routine (driving to and from work). In some cases, the consequences are just wasted time or wasted money. In others, if there's enough hazardous energy around, the consequences can be deadly. (Approximately 100 children/infants a year die in Canada and the United States of hyperthermia because they fell asleep in their car seat and their parents forgot they were in the vehicle.)

So, when you say to yourself, "I've got to remember this, or I can't forget to do that", you need to realize that "right now" is your last clear chance to do something to aid your memory (note in line of vision, alarm on PDA/Blackberry, etc.). But once again, complacency can get in the way. Since we don't always forget, or worse—hardly ever forget—it's very easy for people to get complacent about doing something else to aid their memory, especially if whatever they need to do takes a bit of effort or if it seems silly (like putting a post-it note on the dashboard saying "daughter in car seat" if you don't normally take the child to daycare).

Another problem complacency causes is with recognizing change. We can all talk and drive at the same time. And we all know it can be a distraction. Sometimes it's not too distracting (not too dangerous), and sometimes the conversation can be very pre-occupying (very dangerous). In situations like these, it's easy to become complacent and rationalize what we're doing when it's not dangerous, because the risk is low. It only becomes a problem if the conversation starts to get more involved. But this can be difficult to recognize because you're now thinking about whatever it is that's preoccupying your attention, not on what you're doing (driving). So, even though we all know that driving when you're really preoccupied is dangerous, complacency can lead us to do things that may easily become very dangerous without us always recognizing it.

Another fairly obvious problem complacency causes is with overconfidence. Many safety devices, procedures or protocols are redundant if nobody makes a mistake or "nothing goes wrong." We all know that you don't need a life jacket unless you fall in the water. But if someone is a good swimmer, they might be less inclined to wear a life jacket; or, if they were an experienced iron worker, less inclined to wear a fall arrest harness—especially if they were only 6-10 feet above ground.

However, people can also get complacent about using checklists and permits. They can also become complacent about things like telling someone where they're going (hike in the woods) and when they're planning on coming back—sort of like filing a flight plan. People can even get complacent about holding the handrail, even though we've all fallen going down stairs before, and some of us have even fallen going up stairs.

So, as mentioned before, we need to teach people more than just the critical error-reduction techniques they need to learn to fight complacency, and to compensate for complacency leading to mind not on task; we also have to teach them to develop a deep respect for complacency, and what it can also do to our decision making.

As it turns out, the man killed on the mountain bike at the beginning of this paper was not an Australian looking the wrong way, as was originally speculated. He was a ski instructor who had been living in the area for 30 years. Over 300 people came to his wake.

Every day, thousands (and thousands) of times a day, complacency claims another victim.

A Real and Present Danger: The Underestimated Impact of Workplace Carcinogens

Connie Muncy, CIH, CUSP, REM, MS
Senior Health and Safety Administrator
The AES Corporation
Dayton, Ohio

Introduction

Worldwide, workplace cancer prevention has a much lower profile than workplace injury prevention, despite a real and present need to elevate the profile of workplace cancer prevention around the globe. Numerous organizations around the globe attest to the high number of annual work-related cancers and cancer deaths, but then invariably go on to say workplace cancer statistics are underestimated, the problem is even worse than statistics bear out, and the profile of workplace cancer prevention needs to be elevated. This apparent consensus begs several questions. This paper will explore, with the help of reputable resources from around the globe, the following questions:

- What is occupational cancer, how prevalent is it, and what are its causes?
- Why does cancer prevention have a much lower profile than workplace injury prevention?
- Are current occupational exposure limits (OELs) for carcinogens adequate?
- What are the problems associated with cancer cluster investigations, how reliable are they, and what needs to be done to improve them?
- What must be done to advance the cause of workplace cancer prevention?
- What are some valuable resources available to those wishing to help advance the cause of workplace cancer prevention?

What Is Cancer, and How Prevalent Is It?

According to the U.S. Department of Health and Human Services *14th Report on Carcinogens, 2016,* cancer affects almost everyone's life, either directly or indirectly; about 1 out of 2 men and 1 out of 3 women living in the United States will develop cancer at some point in his or her lifetime.[1] According to the American Cancer Society (ACS), cancer is the second most common

[1] National Toxicology Program (NTP). 2016. *Report on Carcinogens, Fourteenth Edition*. Research Triangle Park, NC: U.S. Department of Health and Human Services, Public Health Service (retrieved February 26, 2017) (http://ntp.niehs.nih.gov/go/roc14/).

cause of death in the U.S. and accounts for nearly 1 of every 4 deaths.[2] The World Health Organization (WHO) estimates that, worldwide, there were 14 million new cancer cases and 8.2 million cancer-related deaths in 2012 (their most recent data), and the number of new cancer cases is expected to rise by about 70% over the next 20 years.[3]

According to the American Cancer Society,[4] cancer is defined as, "a group of diseases which cause cells in the body to change and grow out of control." This class of diseases is comprised of more than 200 different types of cancer, and each is classified by the type of cell that is initially affected. The 10 most common cancers in the U.S. and their prevalence as listed by the National Toxicology Program (NTP), Department of Health and Human Services *Report on Carcinogens, Fourteenth Edition* are depicted in Figure 1 below.

Figure 1: The Ten Most Common Cancers in the U.S.[5]

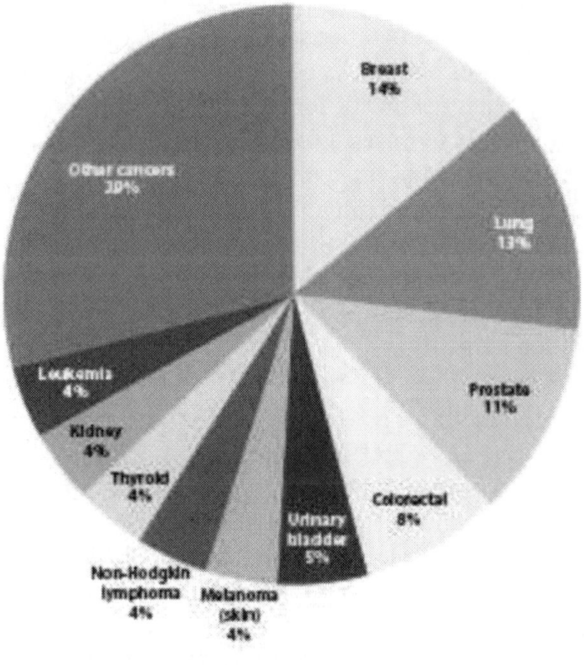

Medical News Today[6] goes on to clarify the damaging nature of cancer:

"Cancer is ultimately the result of cells that uncontrollably grow and do not die. Normal cells in the body follow an orderly path of growth, division, and death. Programmed cell death is called apoptosis, and when this process breaks down, cancer begins to form. Unlike regular cells, cancer cells do not experience programmatic death and instead continue to grow and divide. This leads

---

[2] American Cancer Society (ACS). 2017. *Cancer Basics: Economic Impact of Cancer* (retrieved February 26, 2017) (https://www.cancer.org/cancer/cancer-basics/economic-impact-of-cancer.html).

[3] World Health Organization (WHO). 2017. *Media Center Cancer Fact Sheet* (retrieved February 26, 2017) (http://www.who.int/mediacentre/factsheets/fs297/en/).

[4] American Cancer Society (ACS). 2017. *Glossary: Definitions & Phonetic Pronunciations* (retrieved February 26, 2017)(http://www.cancer.org/cancer/cancerglossary/index).

[5] National Toxicology Program (NTP) (ibid).

[6] MediLexicon International Ltd. 2017. *Medical News Today: Cancer: Facts, Causes, Symptoms and Research* (retrieved February 26, 2017) (http://www.medicalnewstoday.com/info/cancer-oncology).

to a mass of abnormal cells that grows out of control. Cancer harms the body when altered cells divide uncontrollably to form lumps or masses of tissue called tumors (except in the case of leukemia where cancer prohibits normal blood function by abnormal cell division in the blood stream). Tumors can grow and interfere with the digestive, nervous, and circulatory systems, and they can release hormones that alter body function…Dangerous, or malignant, tumors form when two things occur:

- a cancerous cell manages to move throughout the body using the blood or lymphatic systems, destroying healthy tissue in a process called invasion
- that cell manages to divide and grow, making new blood vessels to feed itself in a process called angiogenesis."

So what causes cells to uncontrollably grow and become cancerous? Per *Medical News Today*, cells can experience uncontrolled growth if there are mutations to DNA. Cancer occurs when a cell's gene mutations make the cell unable to correct DNA damage and unable to "commit suicide." Some causes of cancer include:

- Being born with certain genetic mutations.
- An increase in the number of possible cancer-causing mutations in our DNA that occur with age (i.e., age is an important risk factor for cancer).
- Eight viruses (known as "oncoviruses") (e.g., Epstein-Barr Virus, Hepatitis B and C, HIV, Kaposi sarcoma-associated herpesvirus, and more)[7, 8]
- Certain bacteria (e.g., h pylori) and funguses (e.g., aflatoxins).[9]
- Carcinogens (a class of substances that are directly responsible for damaging DNA, promoting or aiding cancer. When a body is exposed to carcinogens, free radicals are formed that try to steal electrons from other molecules in the body. These free radicals damage cells and affect their ability to function normally).
- Ionizing radiation (e.g., gamma and x-rays).
- Ultraviolet radiation (A, B, C).
- Tobacco (smoke, smoking and smokeless).
- Alcoholic beverage consumption (increases cancer risk in multiple ways):
  - metabolizing (breaking down) ethanol in alcoholic drinks to acetaldehyde, which is a toxic chemical and a probable human carcinogen; acetaldehyde can damage both DNA (the genetic material that makes up genes) and proteins;
  - generating reactive oxygen species (chemically reactive molecules that contain oxygen), which can damage DNA, proteins, and lipids (fats) through a process called oxidation;
  - impairing the body's ability to break down and absorb a variety of nutrients that may be associated with cancer risk, including vitamin A; nutrients in the vitamin B complex, such as folate; vitamin C; vitamin D; vitamin E; and carotenoids;

---

[7] National Toxicology Program (NTP). 2016. *Report on Carcinogens, Fourteenth Edition*. Research Triangle Park, NC: U.S. Department of Health and Human Services, Public Health Service (retrieved February 26, 2017) (https://ntp.niehs.nih.gov/ntp/roc/content/listed_substances_508.pdf).

[8] American Industrial Hygiene Association (AIHA). 2016. *TCE, HIV Virus Added to Report on Carcinogens* (retrieved February 26, 2017) (https://www.aiha.org/publications-and-resources/TheSynergist/Industry%20News/Pages/TCE,-HIV-Virus-Added-to-Report-on-Carcinogens.aspx).

[9] National Toxicology Program (NTP). 2016. *Report on Carcinogens, Fourteenth Edition*. Research Triangle Park, NC: U.S. Department of Health and Human Services, Public Health Service (retrieved February 26, 2017) (https://ntp.niehs.nih.gov/ntp/roc/content/listed_substances_508.pdf).

- increasing blood levels of estrogen, a sex hormone linked to the risk of breast cancer; and
- alcoholic beverages may also contain a variety of carcinogenic contaminants that are introduced during fermentation and production, such as nitrosamines, asbestos fibers, phenols, and hydrocarbons.[10]

- Shift work. In 2007, the International Agency for Research on Cancer (IARC) concluded that shift work is "probably carcinogenic to humans." The evidence is strongest for breast cancer,[11] although the risk of prostate and colorectal cancer may also be increased by shift work. Recently, 38 women with breast cancer who had previously worked night shifts for at least 20 years were compensated by the Danish national board of industrial injuries.[12] Possible mechanisms by which shift work increases the risk of cancer under study include:
  - Light at night suppresses the production of melatonin, which has direct and indirect anticancer effects;
  - Sleep disruption stimulates the hypothalamic-pituitary axis to release glucocorticoids, which results in depression of immune function;
  - Phase shift, in which the peripheral rhythms of functions, such as digestion, are out of phase with central sleep and wake rhythms. This may result in changes in the control of cell and tissue proliferation;
  - Shift work may result in changes in lifestyle factors such as smoking, diet, alcohol use, or exercise; and
  - Decreased production of vitamin D.
- Anything else that suppresses or weakens the immune system, i.e., inhibits the body's ability to fight infections and increases the chance of developing cancer.

Principal Carcinogenic Occupational Exposures

Occupational cancer is caused wholly or partly by exposure to a cancer-causing agent (carcinogen) at work, or by a particular set of circumstances at work.[13] These agents may be chemical, biological or physical in nature.

Many factors play a role in the development of cancer. The importance of these factors varies depending on the type of cancer. A person's risk of developing a particular cancer is influenced by a combination of factors that interact in ways that are not fully understood. Some of the factors include:

- Personal characteristics, such as age, sex, and race;
- Family history of cancer;
- An individual's susceptibility to a substance;
- Diet and personal habits, such as cigarette smoking and alcohol consumption;
- The presence of certain medical conditions or past medical treatments, including chemotherapy, radiation treatment, or some immune-system suppressing drugs;

---

[10] National Institutes of Health (NIH), National Cancer Institute (NCI). 2017. *Alcohol and Cancer Risk* (retrieved February 26, 2017) (https://www.cancer.gov/about-cancer/causes-prevention/risk/alcohol/alcohol-fact-sheet#q3).

[11] Straif, K., Baan, R., Grosse. Y., Secretan, B., El Ghissassi, F., Bouvard, V., et al. "Carcinogenicity of shift-work, painting, and fire-fighting." *Lancet Oncol* 2007: 8:1065-6.

[12] Moukangoe, P. I. J., Jansen van Rensburg, M. S. "The Association of Night Shift Work with the Development of Breast Cancer in Women." *Open Journal of Epidemiology*. February, 2015: 14-21.

[13] Institution of Occupational Safety and Health (IOSH). 2017. *Occupational Cancer* (retrieved February 25, 2017) (https://www.iosh.co.uk/Books-and-resources/Our-OH-toolkit/Occupational-cancer.aspx).

• Exposure to cancer-causing agents in the environment (for example, sunlight, radon gas, air pollution, and infectious agents); and
• The amount and duration of an exposure to cancer-causing agents in the workplace.[14]

There are many lists of occupational carcinogens published by a variety of agencies around the world. As one example, the National Institute for Occupational Safety and Health (NIOSH) has published a quite lengthy list of substances considered to be potential occupational carcinogens [15] and the International Agency for Research on Cancer (IARC) lists over 50 substances that are known or probable causes of workplace cancer, and over 100 other possible substances.[16]

Canada's Victorian Trades Hall Council has compiled a list of examples of principal carcinogenic occupational exposures, together with associated conditions; this is presented below as Table 1.

Table 1: Examples of Principal Carcinogenic Occupational Exposures[17]

Cancer	*Examples of principal carcinogenic occupational exposures*
Lung cancer	Asbestos; silica; nickel; indoor radon; diesel fumes; environmental tobacco smoke (ETS) at the workplace; production and refining of: arsenic, beryllium, cadmium, aluminum and chromium; mining of uranium; copper smelting; iron and steel founding; vineyard workers; roofers; asphalt workers; painters; welders
Bladder cancer	Diesel fumes; 2-naphtylamine; benzidine; 4-aminobipheyl; manufacturing of: magenta, auramine, p-chloro-o-toluidine, pigment chromate, and dyes; synthetic latex production; tire curing; calendar operatives; reclaim; cable makers; gas-retort workers; painters
Mesothelioma	Asbestos
Leukaemia	External ionizing radiation; benzene; ethylene oxide; rubber industry; boot and shoe manufacturing and repair
Laryngeal cancer	Sulfuric acid; mineral oils and asbestos; pickling operations
Skin cancer	Intensive solar radiation; coal-tar pitches; coal tar; shale oils; arsenic; mineral oils; polycyclic-aromatic hydrocarbons (PAH); production of coke; vineyard workers; fishermen
Sinonasal and nasopharyngeal cancer	Wood dust; nickel compounds; hexavalent chromium; boot and shoe manufacturing and repair; manufacturing of isopropanol using strong acid process; furniture and cabinet making; carpenters; formaldehyde
Kidney cancer	Coke production

[14] National Institute for Occupational Safety and Health (NIOSH). 2017. *Occupational Cancer* (retrieved February 26, 2017) (https://www.cdc.gov/niosh/topics/cancer/default.html).
[15] National Institute for Occupational Safety and Health (NIOSH). 2017. *Carcinogen List* (retrieved February 26, 2017) (https://www.cdc.gov/niosh/topics/cancer/npotocca.html).
[16] Institution of Occupational Safety and Health (IOSH). 2017. *Occupational Cancer* (retrieved February 25, 2017) (https://www.iosh.co.uk/Books-and-resources/Our-OH-toolkit/Occupational-cancer.aspx).
[17] Victoria Trades Hall Council Australia. 2015. *Cancer-What Causes It?* (retrieved February 25, 2017)(http://www.ohsrep.org.au/hazards/chemicals/cancer-what-causes-it).

Liver cancer	Vinyl chloride; occupational infections with hepatitis B and C; health care workers

Canada reports the most common types of occupational cancer are lung cancer, bladder cancer and mesothelioma.[18]

The Canadian Center of Occupational Health and Safety (CCOHS) has published an extensive list of examples of occupations and occupational groups that are more likely to have been exposed to carcinogens; this is included as Table 2 below.

Table 2: Some Occupations or Occupational Groups Associated with Carcinogen Exposure[19]

Occupations and Occupational Groups	*Suspect Substance*
Aircraft and aerospace industries	Asbestos, beryllium and beryllium compounds; ionizing radiation
Aluminum production	Aromatic amines; pitch volatiles
Asbestos cement industry	Asbestos
Auramine manufacture	Auramine; 2-Naphthylamine; pigments
Battery production workers	Cadmium and cadmium compounds
Beryllium extraction and processing	Beryllium and beryllium compounds
Boot and shoe manufacture/repair	Leather dust, benzene and other solvents
Bus and truck drivers; Dock workers; Filling station attendants; Mechanics; Operators of excavating machines; Professional drivers; Railroad workers; Transport industry	Diesel engine exhaust
Cadmium-copper alloy workers; Cadmium-smelter workers	Cadmium and cadmium compounds
Carpentry and joinery; Furniture and cabinet making	Wood dust
Ceramic production	Cobalt and cobalt compounds
Chemical and rubber industries	Aromatic amines; 1,3-Butadiene; isoprene

[18]Canadian Centre for Occupational Health & Safety (CCOHS). 2017. *OSHAnswers Fact Sheets: Occupational Cancer* (retrieved February 26, 2017) (http://www.ccohs.ca/oshanswers/diseases/occupational_cancer.html).

[19] Canadian Centre for Occupational Health & Safety (CCOHS). 2017. *OSHAnswers Fact Sheets: Occupations or Occupational Groups Associated with Carcinogen Exposures* ((retrieved February 26, 2017) (http://www.ccohs.ca/oshanswers/diseases/carcinogen_occupation.html).

Chemical industry	Acetamide; acrylamide; benzene
Chromate production plants; Chromium ferro-alloy production	Chromium (VI) compounds
Coal gasification, coke production	Coal tar, coal-tar fumes; polycyclic aromatic hydrocarbons (PAHs)
Construction; Insulation and maintenance workers	Asbestos; glass wool; silica (crystalline); toluene diisocyanates
Dry cleaning	Carbon tetrachloride; tetrachloroethylene; trichloroethylene
Dye and pigment production	Aromatic amines (e.g., 2-naphthylamine, 4-aminobiphenyl); benzidine; cadmium and cadmium compounds; chromium (VI) compounds
Electrical capacitor manufacturing	Polychlorinated biphenyls (PCBs)
Electronic production/industries	Beryllium and beryllium compounds; dichloromethane (methylene chloride)
Electroplating processes	Cadmium and cadmium compounds
Fabric manufacture (heat-resistant)	Ceramic fibres (refractory; respirable)
Firefighters	Polycyclic aromatic hydrocarbons (PAHs)
Furnace insulators	Ceramic fibres (refractory; respirable)
Furniture restorers	Dichloromethane (methylene chloride)
Glass production	Cobalt and cobalt compounds
Hairdressers & barbers	Aerosols, dyes (aromatic amines, amino-phenols with hydrogen peroxide); propellants; solvents
Hematite mining; Uranium miners	Radon daughters; silica (crystalline)
Herbicide production	Polychlorophenols and their sodium salts
Hospitals	Ethylene oxide
Iron and steel founding	Formaldehyde; metal fumes; PAHs; silica (crystalline)
Isopropanol manufacture, strong-acid process	Diisopropyl sulfate; isopropyl oils; sulfuric

	acid
Jewellers	Beryllium and beryllium compounds
Leather manufacturing including tanning	Polychlorophenols and their sodium salts; chromium (VI) compounds
Magenta manufacture	Magenta; 4,4′-Methylene bis(2-methylaniline); ortho-Nitrotoluene; ortho-Toluidine
Manufacture of pottery, paper, paint and cosmetics	Talc containing asbestiform fibres
Metal degreasing	Tetrachloroethylene; trichloroethylene
Metals industry	Strong-inorganic mists containing sulfuric acid
Mineral processing	Acrylamide
Miners (including underground)	Cobalt and cobalt compounds; X-radiation, gamma-radiation
Mining and milling	Asbestos
Mining of ores containing arsenic	Arsenic and inorganic arsenic compounds
Nickel refining and smelting; Welding	Nickel and nickel compounds; welding fumes
Nonferrous metal smelting	Arsenic and inorganic arsenic compounds
Nuclear industry; Clean-up workers following nuclear accidents	Beryllium and beryllium compounds; X-radiation, gamma-radiation
Outdoor workers	Solar radiation
Paint stripping; Cleaning and degreasing	Dichloromethane (methylene chloride); 1,2,3-Trichloropropane
Perfume preparation; Epoxy resin formulations; Styrene glycol production; Manufacture of cosmetics, surface coatings, agricultural and biological chemicals	Styrene-7,8-oxide
Petroleum refining and distribution	Acrylamide; PAHs; benzene; diesel fuel, marine; fuel oils, residual (heavy); gasoline
Pharmaceutical production	Dichloromethane (methylene chloride)
Plastics industries	Acetaldehyde; acetamide; acrylonitrile; ethyl acrylate; isoprene; special purpose glass fibres (respirable); styrene; vinyl acetate

Plating and engraving; Lithography; Photography	Chromium (VI) compounds
Plutonium workers	X-radiation, gamma-radiation
Polyester resin manufacture; Production of packaging materials and fibreglass-reinforced polyester	Styrene
Printing processes	Inks; solvents
Processing of copper and nickel ore	Cobalt and cobalt compounds
Production and use of resins, glycerin and propylene-based rubbers	Epichlorohydrin
Production of art glass, glass containers, and pressed ware	Arsenic; antimony oxides; asbestos; lead; PAHs; silica (crystalline)
Production of polyvinyl chloride and co-polymers	Vinyl chloride
Production, packaging, and use of arsenic-containing pesticides	Arsenic and inorganic arsenic compounds
Radiologists and technologists; Radium-dial painters	X-radiation, gamma-radiation
Sheep dip manufacture	Arsenic and inorganic arsenic compounds
Sheet-metal workers	Asbestos
Shiftwork that involves circadian disruption	--
Ship builders	Ceramic fibres (refractory; respirable); toluene diisocyanates
Shipyard workers	Asbestos
Stainless-steel welding	Chromium (VI) compounds
Steel and lumber industries	Acrylamide
Sugar production	Acrylamide
Textile manufacturing/industries	Acrylonitrile; textile dust in manufacturing process; dyes and solvents in dyeing and printing operations; formaldehyde
Water and wastewater treatment	Acrylamide; chromium (VI) compounds
Wood manufacturing	Pentachlorophenol; polychlorophenols and

	their sodium salts
Wood preservation	Chromium (VI) compounds; pentachlorophenol
Wool fibre production	Arsenic and inorganic arsenic compounds
Workers in bars and restaurants	Tobacco smoke

Great Britain reports that almost half (48 per cent) of occupational cancer deaths are in construction workers with the main carcinogens associated with those industries being asbestos, respirable crystalline silica, solar radiation (sun exposure), and diesel exhaust emissions.[20]

Occupational Cancer Statistics from Around the Globe

There is no disputing that statistics from around the globe bear out the fact that occupational cancer is a prevalent and serious problem, and that, more importantly, the total numbers are significantly underestimated.

The National Institute of Occupational Safety and Health (NIOSH) asserts that 3-6% of all cancers worldwide are caused by exposures to carcinogens in the workplace, with estimated annual medical costs associated with occupational cancers to be 4.3 billion dollars in the U.S. alone.[21] Using cancer incidence numbers in the U.S., this means that in 2012 (the most recent year available), there may have been as many as 91,745 new cancer cases that were caused by past exposure in the workplace. NIOSH then continues on to say, "This is probably an underestimate."[22]

According to Jukka Takala, President, International Commission on Occupational Health Workplace Safety and Health Institute in Singapore, the proportion of cancer deaths attributable to occupational causes in Finland was 8.3% (13.2% percent among males), and in the UK it was 5.3% (eight percent among males).[23]

In Australia, the scientific consensus is that, on average, 8% of cancer deaths are work related. For some, like bladder and lung cancer, the figure is even well above 10%.[24] In fact, Australian researchers, Dr. Tim Driscoll, senior lecturer at the University's School of Public Health and Dr. Lin Fritschi from Queensland Cancer, found in a 2006 study that one in ten male workers and one in 50 female workers developed cancer every year due to workplace exposure to carcinogens.

---

[20] Institution of Occupational Safety and Health (IOSH). 2017. *Occupational Cancer: Specific Occupational Causes* (retrieved February 25, 2017) (https://www.iosh.co.uk/Books-and-resources/Our-OH-toolkit/Occupational-cancer/Specific-occupational-causes.aspx).

[21] National Institute of Occupational Safety and Health (NIOSH). 2017. *Cancer, Reproductive, and Cardiovascular Diseases* (retrieved February 26, 2017). (http://www.cdc.gov/niosh/programs/crcd/risks.html)

[22] National Institute of Occupational Safety and Health (NIOSH). 2017. *Occupational Cancer* (retrieved February 26, 2017) (http://www.cdc.gov/niosh/topics/cancer/default.html).

[23] Takala, J. "Eliminating Occupational Cancer." *Industrial Health*. July, 2015: 307-9. (Retrieved February 26, 2017) (https://www.ncbi.nlm.nih.gov/pmc/articles/PMC4551060/).

[24] Victoria Trades Hall Council Australia. 2015. *Cancer-What Causes It?* (retrieved February 25, 2017)(http://www.ohsrep.org.au/hazards/chemicals/cancer-what-causes-it).

The International Labour Organisation (ILO) estimates the human toll at over 600,000 deaths a year – one death every 52 seconds. The ILO calculates that approximately 13% of all cancers in developed countries are the result of preventable, predictable workplace exposure.[25]

In a striking 2013 report, the Occupational Cancer Research Centre, based in Cancer Care Ontario, found that in Canada, occupational cancer has become the leading cause of compensated work-related deaths, a phenomenon which appears to be the most notable in the province of Ontario, where occupational cancer deaths surpass those for traumatic injury by two to one.[26]

Author Jukka Takala asserts "Previous global estimates on occupational cancers by the ILO established that 32% of the deaths in the world related to work are associated with cancers. However, occupational cancers are quite rapidly globalized and in many industrializing countries, the percentage of occupational cancer deaths among all work-related deaths approaches that of the high-income countries; e.g. in the European Union occupational cancer deaths is already at 53% of all work-related deaths."[27]

This phenomenon is further supported by a January 2007 paper in the *American Journal of Industrial Medicine* gives a breakdown of the top workplace killers as: #1 – Cancers (32%), followed by #2 Circulatory Disease (26%), and #3 Work Accidents (17%). The paper further concludes that occupational cancer "represents over half" of all occupational disease cases in established market economies.[28]

Work-Related Cancer Cases Are Often Underestimated

Many reputable sources, including some cited above, openly acknowledge that occupational cancer statistics are, in reality, underestimated. By illustration, in the U.S., the BLS workplace fatality statistics are obtained by cross-referencing source documents, such as death certificates, workers' compensation records, and reports to federal and state agencies. This method assures counts are as complete as possible, but BLS acknowledges that long-term latent illnesses caused by exposure to carcinogens are often are difficult to relate to the workplace and are not adequately recognized and reported.[29] Likewise, according to the UK's Trades Union Congress (TUC), it is almost always impossible to link a specific instance of cancer with a specific exposure to a cancer-causing substance.[30] When a fatality occurs in the workplace, it is very visible; but most people who are killed by cancer will die either at home or in a hospital. Furthermore, many cancers develop decades after the initial exposure, and often the person has retired from work before they develop any signs of cancer.

Author Jukka Takala states, "There is an urgent need to harmonize the estimation methods by various bodies and to resolve these issues. However, experience shows that the more we'll study

---

[25] Victoria Trades Hall Council Australia. 2015. *Cancer-What Causes It?* (retrieved February 25, 2017)(http://www.ohsrep.org.au/hazards/chemicals/cancer-what-causes-it).
[26] Medscape Medical News/Oncology. 2013. *Cancer is Now Top Cause of Job Related Death in Canada* (retrieved February 26, 2017) (http://www.medscape.com/viewarticle/809376).
[27] Takala, J. "Eliminating Occupational Cancer." *Industrial Health*. July, 2015: 307-9. (Retrieved February 26, 2017) (https://www.ncbi.nlm.nih.gov/pmc/articles/PMC4551060/).
[28] Hamalainen P., Takala J., 2 and Leena Saarela, K. "Global estimates of fatal work-related diseases." *American Journal of Industrial Medicine (AJIM)*. Volume 50, 2007: 28-41.
[29] Bureau of Labor Statistics (BLS). (https://www.bls.gov/news.release/history/osh_121595.txt).
[30] Trades Union Congress (TUC). 2017. *Occupational Cancer, A Workplace Guide* (retrieved February 26, 2017)(https://www.tuc.org.uk/sites/default/files/extras/occupationalcancer.pdf).

occupational carcinogens, mutagens and reprotoxic substances, the higher will be the estimates of negative outcomes. Further, research cannot be an excuse for doing nothing, with today's solutions, most or all of such deaths and lost life years can be eliminated."

The Notable Lack of Action

While smoking cessation has become a major public health priority and has spurred an entire prevention industry, no similar campaign has been waged to address the carcinogens encountered and inhaled by millions at work. A staggering number of factors appear to be to blame for the lack of action in preventing workplace cancers. Below is a sampling of postulated factors from a wide variety of sources (with citations for each):

- It has been postulated that because today's exposures cause cancers at least two decades later, it is a problem that doesn't cause alarm bells for corporate executives who are answerable to shareholders from annual general meeting to annual general meeting.[31]
- Approaching 100,000 chemicals are used in workplaces worldwide. Barely 1 in a 100 has been thoroughly tested for health risks.[32]
- Tens of thousands of workers generally have to die before scientific studies identify a workplace problem. It can take a generation of exposures and a generation of deaths for traditional studies to spot an occupational cancer risk, unless the cancer is very rare in the general population. With the exception of the asbestos cancer mesothelioma, however, the most common occupational cancers are also common in the wider community.[33]
- A persistent misconception exists that better regulation has relegated occupational cancer to a fast decline with a curtailment of a dangerous legacy of past practices and exposures. In fact, regulation has not been a cancer cure. Unlike the case of infectious diseases, where a response is frequently swift and draconian, there are typically long delays between the identification of a carcinogenic agent and adoption of adequate measures of prevention. Even then, measures are usually late and incomplete, and will leave a generation to await their fate as a result of prior exposures.[34]
- The reaction of manufacturers that produce or employ products that might be deemed to be carcinogenic has at times been to suppress the damning research rather than to take steps to prevent harm to the exposed populations.
- According to Dr Lorenzo Tomatis, who headed the IARC program until 1993, industry is now dictating terms. "The prevailing assumption, also used as an improper justification, was that the production of certain goods is necessary and vital, even when it was only aimed at increasing consumption of inessential goods, and that the risks involved in their production are an unavoidable price that society must pay."
- This unequal risk of occupational cancer means a minority of the population are facing an enormously elevated, serious and preventable risk. That risk is not being taken seriously and those cancer cases are not, on the whole, being prevented. Work-related cancer is far more common in blue-collar workers; there is an undeniable correlation between employment in lower status jobs and an increased risk. According to the CAREX report for Great Britain, all

---

[31] *Hazards Magazine/Worker's Health International News*. 2017. "Occupational Cancer/Zero Cancer: Worker Cancer Prevention Kit, Part 1" (retrieved February 26, 2017) (http://www.hazards.org/cancer/preventionkit/part1.htm)

[32] *Hazards Magazine/Worker's Health International News*, ibid.

[33] *Hazards Magazine/Worker's Health International News,* ibid.

[34] *Hazards Magazine/Worker's Health International News,* ibid.

the workplace exposures to carcinogens were restricted to about one-fifth of the working population.[35]

- The effective identification of workplace cancer risks has slowed. There's good reason to believe this could be the result of a well-coordinated industry campaign to influence decisions of bodies, including the International Agency for Research on Cancer (IARC) and the World Health Organization (WHO), rather than any actual improvements at work. And as public funding for independent occupational health research is eroded, industry-funded research is swamping the literature, with occupational and environmental risks going underestimated or undetected as a result. A report in the October-December 2005 issue of the *International Journal of Occupational and Environmental Health* (IJOEH), examining "business bias" in workplace studies, concludes, "in spite of claiming primary prevention as their aim, studies of potential occupational and environmental health hazards that are funded either directly or indirectly by industry are likely to have negative results."

The Inadequacy of Current OELs for Carcinogens

One problems with setting occupational exposure limits (OELs) for carcinogens is that there is not always a threshold below which there is no adverse health effect. For example, carcinogens always cause a risk no matter how low the dose is. So, setting OELs is not as straightforward a matter as one might think; since, obviously, the cost of avoidance or treatment goes up as the risk is being reduced, there is a need by society to accept *some* level of risk.[36] This is where the process of risk management comes in to play.

Risk management is the process of selecting the most appropriate guidance or regulatory actions by integrating the results of risk assessment with engineering data and with social, economic, and political concerns to reach a decision.[37] But the OELs reached through the risk management process are not always deemed acceptable by many, and even those organizations setting the OELs reflect this sentiment, as illustrated in cases below discussing the setting of OELs for carcinogens by the American Congress of Governmental Industrial Hygienists (ACGIH), NIOSH, and the U.S. Occupational Safety and Health Administration (OSHA).

P. Valente, MD, of Health Local Unit Roma G, Department of Prevention (Italy), and F. Cavariani, PhD, Health Local Unit Viterbo, Industrial Occupational Laboratory (Italy), speak to the adequacy of current OELs for carcinogens in their paper, "OEL's To Carcinogenic Substances: The Additional Risk At Current TLVs." They state, "At the present there is no consensus on the levels of exposure to carcinogenic agents which can be considered 'acceptable.' Different countries and international agencies have proposed or have adopted different limits, on the basis of different criteria, for those exposed in the occupational setting and for the general population." They continue on in their paper to mathematically illustrate the magnitude of the additional risk to which occupationally exposed groups could undergo on the basis of the current TLVs for a number of carcinogenic substances. Calculations were carried out for all substances in the R45, R49 and R40 category for which U.R. (by the USA-EPA) and C (by ACGIH-TLV-

---

[35] *Hazards Magazine/Worker's Health International News*, ibid.

[36] Dartmouth College. 2017. *Risk Assessment* (retrieved February 26, 2017) (https://engineering.dartmouth.edu/~d30345d/courses/engs37/RiskAssessment.pdf).

[37] Niemi, C. "Acceptable" Risk Levels for Carcinogens: Their History, Current Use, And How They Affect Surface Water Quality Criteria. *SWQS Policy Forum #3* (Retrieved February 26, 2017) (http://www.tmw-law.com/news-pdf/SWQSPolicyForumRiskLevel%2002-08-213.pdf).

TWA) were available. The results showed that the additional risk for the occupationally exposed groups at the current ACGIH TLVs-TWA is always higher than the risk considered "acceptable" for the general population (set by USA-EPA, at 1/100,000). The paper concluded that despite the many uncertainties involved in risk assessment and in risk management, the excess of cancer risk to workers estimated in this study shows that present day prevention policies are inequitable from the ethical point of view and necessitate further study and new solutions.[38]

According to the 2013 Public Comments on NIOSH Draft Intelligence on Carcinogen Classification and Target Risk Level Policy for Chemical Hazards in the Workplace, historically, NIOSH has used a lifetime cancer risk increase of 1 in 1,000 as the acceptable regulatory threshold, while stating that "controlling exposure to lower concentrations is always warranted." NIOSH admitted that "an excess risk of 1 in 1,000 is one or more orders of magnitude higher than what the United States permits for the general public." NIOSH justified this questionable threshold with two arguments: The first is the historic "benzene decision" made by the U.S. Supreme Court in 1980, where a 1 in 1,000 risk was used in a seemingly rhetorical example. The second justification is that workers are a very small subset of the general population, and higher exposures for small numbers of people may be considered acceptable if they are comparable to the overall risks of employment itself. Parties commenting disagreed with these nonscientific justifications making the following observations: "There is increasing evidence that occupational carcinogens spread into the greater environment. For example, trichloroethylene (TCE), an industrial solvent, is now present in approximately one-third of the U.S. water supply. The maximum risk threshold acceptable to the EPA is 10-fold less than the NIOSH threshold – and given the overlapping exposures, that does not make sense. The EPA considers 1 in 1,000,000 to be the target threshold for as many people as possible, but that is 1000 times lower than the NIOSH threshold. The bottom line is that there is no scientific basis for these differential safety standards, and we now know that occupational and environmental exposures frequently become indistinguishable. For that reason, the workforce should be afforded the same level of protection as the general public."[39]

Perhaps recognizing this concern, on December 27, 2016, NIOSH published *Current Intelligence Bulletin 68: NIOSH Chemical Carcinogen Policy*. Underlying this policy is the recognition that there is no safe level of exposure to a carcinogen, and therefore that reduction of worker exposure to chemical carcinogens as much as possible through elimination or substitution and engineering controls is the primary way to prevent occupational cancer. Accordingly, this policy no longer uses the term *recommended exposure limit (REL)* for chemical carcinogens; rather, NIOSH will only recommend an initial starting point for control, called the risk management limit for carcinogens (RML-CA). For each chemical identified as a carcinogen, this level corresponds to the 95-percent lower confidence limit of the risk estimate of one excess cancer case in 10,000 workers in a 45-year working lifetime. Keeping exposures within the risk level of 1 in 10,000 is the minimum level of protection; striving for lower levels of exposure is recommended. When

---

[38] Valente, P., Cavariani, F. 2015. *OELs to Carcinogenic Substances: The Additional Risk at Current TLVs (1999)* (Retrieved February 26, 2017) (https://www.researchgate.net/publication/277952165_OELs_to_carcinogenic_substances_the_additional_risk_at_current_TLVs_1999).

[39] National Center for Health Research. 2013. *Public Comments on NIOSH Draft Intelligence on Carcinogen Classification and Target Risk Level Policy for Chemical Hazards in the Workplace, By Anna E. Mazzucco, Ph.D December 16, 2013* (Retrieved February 26, 2017)(http://center4research.org/public-policy/testimony-briefings-statements/nioshpubliccomments/).

measurement of the occupational carcinogen at the RML-CA is not analytically feasible at the 1 in 10,000 risk estimate, NIOSH will set the RML-CA at the limit of quantification (LOQ) of the analytical method. In addition, NIOSH will continue to evaluate available information on existing engineering controls and also make that information available when publishing the RML-CA.[40]

Consider also OSHA's permissible exposure limit (PEL) for the carcinogen asbestos. In their discussion, entitled "Despite Limits, No Safe Level of Asbestos Exposure," asbestosnetwork.com opines, "Asbestos is extremely hazardous." According to the National Institute for Occupational Health and Safety (NIOSH), "all levels of asbestos exposure studied to date have demonstrated asbestos-related disease," and "there is no level of [asbestos] exposure below which clinical effects do not occur." Therefore, all avoidable exposures to asbestos should be prevented whenever possible. The Occupational Safety and Health Commission (OSHA) set a permissible asbestos exposure limit (PEL) of 0.1 fiber per cubic centimeter (f/cc) for work in all industries, including construction, shipyards, and asbestos abatement work. OSHA is quick to add, however, that the asbestos PEL is a target guideline for regulatory purposes only, and does not establish any level of "safe" asbestos exposure. As OSHA states its Asbestos Final Rule, "The 0.1 f/cc level leaves a remaining significant risk."[41]

Closer examination of the asbestos final rule at FR 59:40964-41162 raises some significant questions on the adequacy of the asbestos PEL. The following excerpt from the Federal Register entry of the Asbestos Final Rule[42] states:

"At the time of the proposal in 1990, the question of whether the proposed PEL reduction would reduce a still significant risk had already been given a tentative answer by the Court. The D.C. Circuit Court of Appeals, in remanding the issue of lowering the PEL to the Agency, noted that based on the 1984 risk assessment, the excess risk stemming from average exposures of 0.1 f/cc "could well be found significant." *BCTD v. Brock*, 838 F.2nd at 1266." (55 FR at 29714). In the proposal, OSHA stated that it believes "that compliance with proposed amendments to reduce the PEL to 0.1 f/cc as a time-weighted average measured over 8 hours would further reduce a significant health risk which exists after imposing a 0.2 f/cc PEL" (55 FR 29714, July 20, 1990). OSHA's 1984 risk assessment showed that lowering the TWA PEL from 2 f/cc to 0.2 f/cc reduced the asbestos cancer mortality risk from lifetime exposure from 64 to 6.7 deaths per 1,000 workers. OSHA estimated that the incidence of asbestosis would be 5 cases per 1,000 workers exposed for a working lifetime under the TWA PEL of 0.2 f/cc. Counterpart risk figures for 20 years of exposure are excess cancer risks of 4.5 per 1,000 workers, and an estimated asbestosis incidence of 2 cases per 1,000 workers. OSHA's risk assessment also showed that reducing exposure to 0.1 f/ cc would further reduce, but not eliminate, significant risk. The excess cancer risk at that level would be reduced to a lifetime risk of 3.4 per 1,000 workers and a 20-year exposure risk of 2.3 per 1,000 workers. Consequently significant risk would be reduced substantially. However,

---

[40] National Institute of Occupational Safety and Health (NIOSH). 2017. "NIOSH Evaluation of Its Cancer and REL Policies – Current Intelligence Bulletin #68: NIOSH Chemical Carcinogen Policy" (retrieved February 26, 2017) (https://www.cdc.gov/niosh/topics/cancer/policy.html).

[41] AsbestosNetwork.com. 2017. *Worker Safety: Despite Limits, No Safe Level of Exposure for Asbestos* Retrieved February 26, 2017) (http://www.asbestosnetwork.com/Worker-Safety/).

[42] Occupational Safety and Health Administration (OSHA). 1994. *Federal Register 59:40964-41162, Occupational Exposure to Asbestos: Final Rule* (retrieved February 26, 2017)(https://www.osha.gov/pls/oshaweb/owadisp.show_document?p_table=FEDERAL_REGISTER&p_id=13404).

OSHA concluded therefore that continued exposure to asbestos at the TWA permitted level and action level would still present residual risks to employees which are significant."

The likely inadequacy of present-day OELs begs the question of whether this might lead to the presence of various cancer clusters in industry.

Workplace Cancer Clusters

A cancer cluster is the occurrence of a greater than expected number of cancer cases among a group of people in a defined geographic area over a specific time period. A workplace cancer cluster may be suspected when people report that several coworkers have been diagnosed with the same or related types of cancer.[43]

The Connecticut Department of Health opines that there is no doubt that occupational cancer clusters are a real phenomenon. One of the first cancer cluster reports involved an epidemic of scrotal skin cancer among 18th century London chimney sweeps with high exposures to PAH chemicals in chimney soot. More recent cancer cluster studies also have been successful in explaining higher cancer rates for certain occupations, such as mesothelioma among asbestos workers and hepatic angiosarcoma (a rare liver cancer) among workers exposed to vinyl chloride during PVC plastics manufacturing. But, that having been said, they further state that, while there are indeed some examples of cancer cluster studies that have identified a common cause, the vast majority have not.[44] The National Cancer Institute (NCI), on its Cancer Clusters webpage, states these very same examples, and also goes on to state that most (but not all) suspected cancer clusters turn out, on detailed investigation, not to be true cancer clusters; that is, no cause can be identified, and the clustering of cases turns out to be a random occurrence.[45] This leads to the question of the effectiveness of the cancer-cluster vetting process itself.

NCI explains how this process typically begins. Concerned individuals contact their local or state health department to report a suspected cancer cluster or to find out if one is being investigated. Health departments provide the first response to questions about cancer clusters because they, together with state cancer registries, will have the most up-to-date data on cancer incidence in the area. If additional resources are needed to investigate a suspected cancer cluster, the state health department may request assistance from federal agencies, including the Centers for Disease Control and Prevention (CDC) and the Agency for Toxic Substances and Disease Registry (ATSDR), which is part of the CDC. Health departments use established criteria to investigate reports of cancer clusters.[46]

According to the Connecticut Department of Health, further investigation should only be considered by the health departments when a cancer cluster report involves: (1) cancers of the same type (particularly if they are rare cancers); (2) cancers in numbers likely to reach statistical

---

[43] National Institutes of Health (NIH), National Cancer Institute (NCI). 2017. *Cancer Clusters* (retrieved February 26, 2017) (https://www.cancer.gov/about-cancer/causes-prevention/risk/substances/cancer-clusters-fact-sheet).

[44] Connecticut Department of Public Health (CTDPH). 2017. *Environmental Health Technical Brief: Cancer Clusters* (retrieved February 26, 2017) (http://www.ct.gov/dph/lib/dph/environmental_health/eoha/pdf/cancer_cluster_tech_brief_final.pdf).

[45] National Institutes of Health (NIH), National Cancer Institute (NCI). 2017. *Cancer Clusters* (retrieved February 26, 2017) (https://www.cancer.gov/about-cancer/causes-prevention/risk/substances/cancer-clusters-fact-sheet).

[46] https://www.cancer.gov/about-cancer/causes-prevention/risk/substances/cancer-clusters-fact-sheet.

significance; (3) cancers diagnosed within a short time period; and (4) the presence of a possible common environmental exposure.[47]

According to the CDC, because cancer is a common disease, cancer can be found among people at any workplace. In the United States, over their lifetime, approximately one in two men and one in three women will develop or die from cancer. These statistics show the unfortunate reality that cancer occurs more often than many people realize. Cancer clusters related to a workplace exposure usually consist of the same types of cancer. When several cases of the same type of cancer occur and that type is not common in the general population, it is more likely that an occupational exposure is involved. When the cluster consists of multiple types of cancer, without one type predominating, an occupational cause of the cluster is less likely. When cancer in a workplace is described, learning whether the type of cancer is a primary cancer or a metastasis (spread of the primary cancer into other organs) is important. Only primary cancers are used to investigate a cancer cluster.[48]

The CDC also states that we should also look to see whether cancer is occurring among employees in particular jobs or areas of the workplace. This can help to identify potential exposures that had been previously unrecognized. The time between first exposure to a cancer-causing agent and clinical recognition of the disease is called the latency period. Latency periods vary by cancer type, but, for some cancers, may be 15 to 20 years or longer. Because of this, past exposures are more relevant than current exposures as potential causes of cancers occurring in workers today. Often, these exposures are hard to document. [49]

A number of other problems are encountered in the study of clusters, based on their diverse and complicated nature. To name just a few more:

- The health events being investigated (often morbidity or mortality) are usually rare, and increases of these events tend to be small and may occur over a long period.
- Some clusters occur by chance, much in the same way that five consecutive tosses of a coin, repeated several times, will eventually result in five tosses that come up all heads.
- Information on the population at risk or on the expected rates often is not available.
- A complicating factor for methods using aggregated data is that health events occur in space and time continua, thus yielding optional and suboptimal units for displaying a pattern. The choice of a geographic area that is too small or too large, or of a time period that is too short or too long, may result in insufficient statistical power to indicate a cluster.
- Most of the cancer clusters brought by citizens to the attention of public health agencies consist of many different cancers, with different etiologies that are unlikely to have a common cause.
- There is our natural tendency to notice cases first, and then define a cluster boundary around the known cases, creating a potentially artificial cluster.
- Cancer has a long latency period. Unlike infections or acute toxic reactions, the effect of a carcinogen in a community will not usually be seen for many years.

---

[47] Connecticut Department of Public Health (CTDPH). 2017. *Environmental Health Technical Brief: Cancer Clusters* (retrieved February 26, 2017) (http://www.ct.gov/dph/lib/dph/environmental_health/eoha/pdf/cancer_cluster_tech_brief_final.pdf).
[48] National Institute of Occupational Safety and Health (NIOSH). 2017. *Cancer Clusters* (retrieved February 26, 2017) (https://www.cdc.gov/niosh/topics/cancer/clusters.html).
[49] National Institute of Occupational Safety and Health (NIOSH). 2017. *Cancer Clusters* (retrieved February 26, 2017) (https://www.cdc.gov/niosh/topics/cancer/clusters.html).

- In our highly mobile society, cancer patients who appear to be clustered geographically may not have lived in the area long enough for their cancers to have a common cause.
- The development of cancer is a highly complex process. To produce a cancer cluster, a carcinogen must affect a huge number of cells in a huge number of people, over a long period of time. And even when people have received a heavy dose of a carcinogen and many cells have been damaged, not all will get cancer. After all, for reasons we do not fully understand, the majority of smokers do not get lung cancer.[50]
- Because of the diverse and complicated nature of clusters, there is no omnibus test for assessing them. Investigators must perform several related tests and report the results that are most consistent with validated assumptions.[51]

In contemplating the many hurdles in evaluating potential occupational cancer clusters, one might naturally wonder how effective any evaluation can actually be at this point in time. Surely there is much work left to be done in the area of cancer cluster evaluation, as well as in other areas.

What Should Be Done?

In addition to improving cancer cluster evaluation processes, additional action is urgently needed. The July 2015 article, "Eliminating Occupational Cancer," which appears in NIOSH Journal *Industrial Health* succinctly sums up the reasons that action must be taken:

"There is a need for clear priorities to prevent major work-related health risks and to identify solutions for action. Obviously, cancer at work is the biggest individual threat when looking at the number of deaths in the developed world and this serious and preventable disease is rapidly becoming the biggest killer at places of work in most countries in the world. We can and should have a more ambitious target: to eliminate occupational cancer."[52]

To this end, below is a sampling of actions recommended by various individuals and organizations to battle the occurrence of occupational cancers. These recommendations are targeted at physicians, researchers, epidemiologists, employers, governments, unions, and individuals. It is hoped that each of these groups will take up the mantle and pursue many of these recommendations in the short term to stem the tide of human suffering and loss of life and the impact and cost these occupational cancer cases and deaths have on our societies today.
- Author Jukka Takala plainly states: "It is clear that cancers caused by work can be prevented by reducing or eliminating the exposures leading to the disease. In fact these cancers are the easiest ones to tackle, such risks can be usually reduced or even eliminated, and ethically this is the right way to go ahead."[53] Takala urges the avoidance of exporting the risks from the developed countries to developing countries, and recommends the following actions be taken:

---

[50] Connecticut Department of Public Health (CTDPH). 2017. *Environmental Health Technical Brief: Cancer Clusters* (retrieved February 26, 2017) (http://www.ct.gov/dph/lib/dph/environmental_health/eoha/pdf/cancer_cluster_tech_brief_final.pdf).
[51] Centers for Disease Control (CDC). 2017. *Morbidity and Mortality Weekly Report (MMWR): Guidelines for Investigating Clusters of Health Events* (Retrieved February 26, 2017) (https://www.cdc.gov/mmwr/preview/mmwrhtml/00001797.htm).
[52] Takala, J. "Eliminating Occupational Cancer." *Industrial Health*. July, 2015: 307-9. (Retrieved February 26, 2017) (https://www.ncbi.nlm.nih.gov/pmc/articles/PMC4551060/).
[53] Takala, J. "Eliminating Occupational Cancer." *Industrial Health*. July, 2015: 307-9. (Retrieved February 26, 2017) (https://www.ncbi.nlm.nih.gov/pmc/articles/PMC4551060/).

- Present unified scientific evidence of the magnitude of the problems, levels of exposures, number of workers exposed, and produce credible data on predicted negative outcomes.
- Provide recommendations on evidence-based solutions that are adaptable to different kinds of circumstances, cultures, countries, sectors, and sizes of workplaces.
- Share findings via well-prepared reports and articles published in a high-impact journals
- Mobilize institutions globally to act on elimination of work-related cancer and expand this action gradually through the International Labour Organization, ILO and its Occupational Safety and Health Flagship program being established, and through WHO and its 'collaboration centers' network, supporting IARC efforts in this field, and mobilizing global action through the national stakeholders.

- A May 2005 report by the Canadian Strategy for Cancer Control (CSCC)'s National Committee on Environmental and Occupational Exposures (NCEOE) identified the following seven priority areas for improving primary prevention of occupational cancers:[54]
  - improved surveillance;
  - better information disclosure and labeling;
  - community education and action;
  - worker education and action;
  - non-governmental organizations' involvement in cancer prevention;
  - employer/industry reductions in carcinogen use; and
  - government intervention in the form of new regulations and policy.
- The Canadian Centre for Occupational Health and Safety (CCOHS) stresses that physicians can play a key role in identifying individual cases and assisting their patients with compensation by being aware of their patients' hazardous exposures and vigilant for early signs of work-related disease. By adopting a model in which physicians are more involved with the active surveillance of occupational cancers, and in which the current and projected burden of occupational cancer has been more systematically assessed, (countries) can be in a position where efforts are aimed at eliminating or minimizing occupational exposures and preventing these cancers from occurring.[55]
- The Victoria Trades Hall Council in Australia stresses that media involvement is needed. Popular media frequently cover such topics as: cancer, the tragedy for the individual; cancer, the challenge for the medical profession; and cancer, the result of smoking and bad diet, but they do not cover cancer from an OHS perspective. Yet occupational cancer tops the International Labour Organisation's (ILO's) workplace diseases and accidents table, with over 600,000 dying of occupational cancers ever year. More than one in five workers face a cancer risk from their work. Between 8 and 16 percent of all cancers result from occupational exposures.[56]
- European Agency for Safety and Health at Work (EU-OSHA) opines that efforts are required at all levels:

---

[54] Hazards Magazine/Worker's Health International News. 2017. *Occupational Cancer/Zero Cancer: Worker Cancer Prevention Kit, Part 1* (retrieved February 26, 2017) (http://www.hazards.org/cancer/preventionkit/part1.htm).

[55] Medscape Medical News/Oncology. 2013. *Cancer is Now Top Cause of Job Related Death in Canada* (retrieved February 26, 2017) (http://www.medscape.com/viewarticle/809376).

[56] Victoria Trades Hall Australia. 2015. *Occupational Cancer and Workers Memorial Day* (retrieved February 26, 2017) (http://www.ohsrep.org.au/hazards/chemicals/occupational-cancer-and-workers-memorial-day).

o improved application of legislation (especially concerning process-generated factors and non-chemical factors),
o awareness-raising strategies to improve the risk perception of all stakeholders,
o specifications of comprehensive preventive measures for all work processes that involve such risk factors,
o improved implementation and enforcement, and
o lowering barriers to compensation.[57]

- The Canadian Strategy for Cancer Control (CSCC), a coalition of cancer prevention, health service and other bodies, has made a public stand in favor of "primary prevention" of occupational cancer. A letter of October 2005 from CSCC's National Committee on Environmental and Occupational Exposures (NCEOE) called on a House of Commons standing committee to back changes in Canadian law to better promote preventive measures. It called for the Canadian government to strengthen the Canadian Environmental Protection Act as it is applied "in particular to IARC 1 and 2a designed human carcinogens." The letter also called for
  o information bulletins to be developed to address cancer prevention and toxic use exposure reduction,
  o an investigation of the possibilities for introducing toxics use reduction legislation, and
  o possible incentives for toxic use reduction programs.[58]

There are a surprising number of resources available to be used by persons who wish to get involved to help solve the workplace carcinogen exposure problem. The following is just a brief sampling of the available materials.

- UK's Trade Union Congress (TUC) Occupational Cancer/Zero Cancer – A Union Guide to Prevention.[59]
- A very useful publication has been developed by a group of UK organizations: *Cancer and Working - Guidelines for Employers, HR and Line Managers.* The publication includes a template policy and provides useful and practical advice for workplaces.[60]
- The Health and Safety Department of the European Trade Union Institute - Research, Education, Health and Safety provides infographics and campaign resources.[61]
- UNIFOR (a large Canadian manufacturing union) offers several resources for initiating Prevent Cancer Campaigns, including a Cancer Campaign book and a Personal Chemical Exposure Journal.[62]

---

[57] European Agency for Safety and Health at Work. 2017. *Exposure to Carcinogens and Work Related Cancer: A Review of Assessment Methods* (retrieved February 26, 2017) (https://osha.europa.eu/en/tools-and-publications/publications/reports/summary-on-cancer).
[58] Hazards Magazine/Worker's Health International News. 2017. *Occupational Cancer/Zero Cancer: Worker Cancer Prevention Kit, Part 1* (retrieved February 26, 2017).
[59] International Metalworkers Federation. 2007. *Occupational Cancer/Zero Cancer: A Union Guide to Prevention* (retrieved February 26, 2017)(http://www.bwint.org/pdfs/ZeroCanceENr.pdf).
[60] http://www2.cipd.co.uk/NR/rdonlyres/62EACA31-115A-4060-83B6-AA4199459DF2/0/3906cancerguideweb.pdf.
[61] European Trade Union Institute (ETUI). 2017. *Union Campaigning* (retrieved February 26, 2017) (http://www.etui.org/Topics/Health-Safety/Occupational-cancers/Union-campaigning).
[62] Unifor/The Union/LeSyndicat. 2017. *Prevent Cancer Campaign* (retrieved February 26, 2017) (http://www.unifor.org/en/take-action/campaigns/prevent-cancer-campaign).

- CCOHS Free e-Course Occupational Cancer Prevention and Recognition.[63]
- Exposure to carcinogens and work-related cancer: A review of assessment methods by European Agency for Safety and Health at Work, EU-OSHA, Table 3: Findings and recommendations offers extensive recommendations.[64]
- *Reducing Environmental Cancer Risk: What We Can Do Now*, U.S. Department of Health And Human Services, National Institutes of Health, National Cancer Institute[65] provides recommendations on what individuals can do to reduce their own cancer risks.

Summary

What is important to know about the occupational carcinogen problem is that it is taking a high toll on society; that there may be no safe limit of exposure to an occupational carcinogen; that all occupational cancers are avoidable; and that experts have proposed a variety of steps be taken to prevent occupational cancers. It is up to the reader to decide how they can get involved to help solve this serious problem.

[63] Canadian Center for Occupational Health and Safety (CCOHS). 2017. *Occupational and Environmental Cancer: Recognition and Prevention* (February 26, 2017) (http://www.ccohs.ca/products/courses/prevent_cancer/).

[64] European Agency for Safety and Health at Work. 2017. *Exposure to Carcinogens and Work Related Cancer: A Review of Assessment Methods* (retrieved February 26, 2017) (https://osha.europa.eu/en/tools-and-publications/publications/reports/summary-on-cancer).

[65] U.S. Department Of Health And Human Services, National Institutes of Health, National Cancer Institute. 2009. *Reducing Environmental Cancer Risk: What We Can Do Now* (retrieved February 26, 2017)(https://deainfo.nci.nih.gov/advisory/pcp/annualReports/pcp08-09rpt/PCP_Report_08-09_508.pdf).

A Global Framework for Practice: INSHPO OHS Professional Capability Framework

Terrie Norris, CSP, ARM, CPSI
President Elect
International Network of Safety and Health Practitioner Organizations

Eldeen E. Pozniak, CHSC, CRSP, CHSMSA, FIIRSM, CMIOSH
President
International Network of Safety and Health Practitioner Organizations

Introduction

Exhibit 1. OSH Professional Capability Framework

The OHS Capability Framework: A Global Framework for Practice **document is available on the INSHPO website – http://www.inshpo.org/work.php.**

The International Network of Safety Health Professional Organizations (INSHPO) developed *The OHS Professional Capability Framework* to provide greater clarity around the generalist Occupational Health and Safety (OHS) professional role.

The *Framework* defines the role of the OHS professional, as well as the requisite knowledge and skills. The framework promotes a high standard of capability among OHS professionals and, in turn, informs employers and regulators as to the capabilities of OHS professionals. The sections on knowledge and skills provide benchmarks for education and training bodies and OHS professional associations in their development of the detail of certification schemes, educational programs, and continuing professional development.

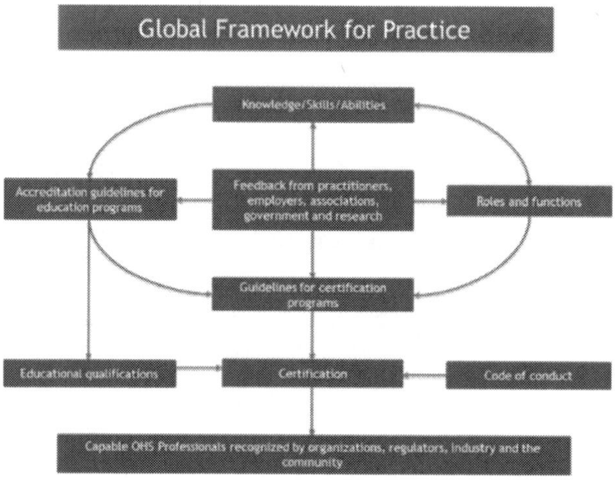

Exhibit 2. ***Global Framework for OHS Practice***

INSHPO has developed this framework to:

- Facilitate a shared understanding of the difference in roles of the OHS professional and the OHS practitioner;
- Position the OHS professional as a key advisor, strategist, and leader in fully integrating the management of OHS risk into sustainable business practice;
- Position the OHS practitioner as a skilled implementer of OHS activities and an effective OHS supporter and communicator at the site level.

"Professional" Versus "Practitioner" Concepts

OHS Practitioner	OHS Professional
Implementer/executor of strategy	Designer of strategy
Communicates mainly with middle management, supervisor an shop floor	Influential with senior managers
Understands local work and industry environment	Considers wider context of business and external regulatory and societal influences
Provides technical advice, mentors and coaches	Provides strategic and technical advice, mentors and coaches
Oversee and drive compliance, supporting change	Problem solvers, involved in organisational review and change
Advice/action based on technical knowledge, experience	Advice/action based on conceptual and technical knowledge mediated by analysis of evidence, experience and critical thought
Focus is on organisations primary processes operating in known contexts within established parameters	Able to extend their understanding and control to novel and complex situations
Usually work under direct or indirect supervision or mentorship with substantial responsibility for own work	Value professional collaboration
Usually educated through vocational or technical sector	Usually educated through university or Higher Education sector

Table 1. Professional versus Practitioner

OHS Capability Framework: Components

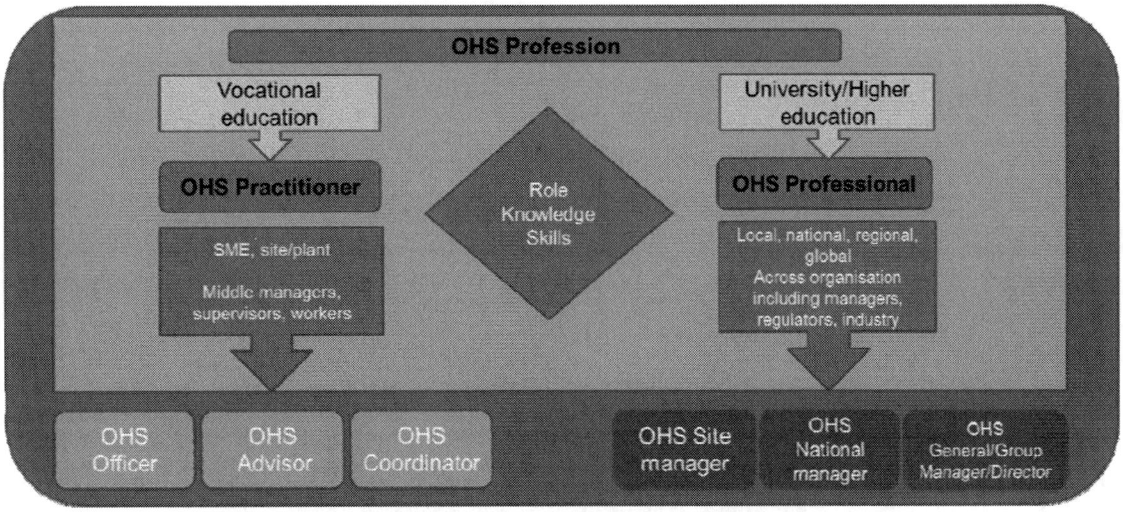

Exhibit 3. Components of the OHS Capabilities Framework

Position Profiles: Position profiles are described for OHS professionals and for OHS practitioners. As there are gradations within each role, the profiles are at three levels for each role.

	Level 1	Level 2	Level 3
Position details			
Representative title			
Purpose of role			
Typical reporting			
Professional parameters			
Autonomy			
Influence/leadership			
Complexity			
Business & organisational skills			
Knowledge and skills			
Knowledge			
Analyse & evaluate information			
Problem solving			
Transmit knowledge, skills & ideas			
Soft skills			
Qualifications			
Qualification level			
Example qualifications			
Experience			

Table 2. Position Profile

The *Capability Framework* describes the **activities, knowledge and skills** for OHS specialists:

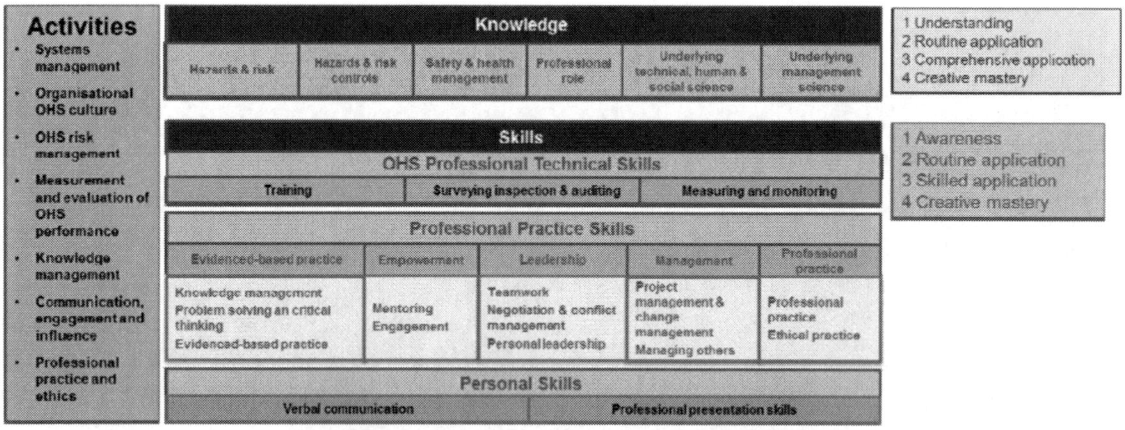

Exhibit 4. Activities, Knowledge, and Skills

<u>Activities</u>

The *OHS Capability Framework* defines the role of the OHS professional and the OHS practitioner in terms of OHS-related activities. These activities are described at two levels of detail:

- Dimensions: Providing the scope of the distinguishing boundaries of the roles.
- Domains: Describing fields of activity within the dimensions.

Seven dimensions are used to outline the roles of the OHS professional and OHS practitioner:

1. Systems management approach
2. Organizational culture and its impact on OHS
3. OHS risk management processes
4. Measurement and evaluation of OHS performance
5. Knowledge management
6. Communication, engagement and influence
7. Professional and ethical practice

The activity statements may be used in a number of contexts:

- As a mapping tool to confirm key OHS activities are addressed by one or more OHS specialists within the organization;
- As a detailed OHS duty statement as part of a position description;
- To create a shared understanding of the role by incumbents, line and senior managers, and others;
- As a basis for performance appraisals; and
- To identify areas for role expansions and further development of an incumbent.

OHS Practitioner				OHS Professional	
	Dimensions		Domain	Dimension	Domain
1. Systems Management approach	Support the implementation of a system approach to OHS.	1.1	Support implementation of and monitor compliance with defined OHS management system policy and procedures. Builds relationships as a basis for influence with managers to understand the limitations of written safety rules and procedures and to favour control through workplace and process design, critical controls and proven competence.	Lead and support the development and implementation of a system approach to OHS.	Lead the development of OHS management systems, policies, procedures. Understand management systems, policies, and procedures in the broader context of the business as a sociotechnical system and develop effective, non-bureaucratic management systems that will be accepted by all stakeholders.
		1.2	Contribute to identification of required resources and work with resource framework.		Advise on and facilitate commitment of appropriate resources for sustainably managing OHS.
		1.3	Support and motivate line management and supervisors to provide OHS leadership and through them to influence workers to give appropriate priority to OHS in relation to other organizational objectives.		Support and motivate senior management, and through them, all people in the organisation, to provide OHS leadership and to give appropriate priority to OHS in relation to other business objectives.

Table 3. OHS Practitioner Versus OHS Professional Activities

Knowledge

The knowledge matrix of the *Framework* is described under six areas, with each area having a number of categories, with illustrative generic topics indicating the intended scope of the knowledge category. This is described at a high, generic level to allow flexibility in the way it is applied to suit the legal and OHS context in individual countries.

The knowledge areas are:

A. Hazards and risks
B. Hazard and risk controls
C. Safety and health management
D. Professional role and functioning
E. Underlying technical and behavioral disciplines
F. Underlying management science

Code	Knowledge category	Illustrative generic topics	OHS Practitioner	OHS Professional
A.	**Hazards and Risks**			
1.	Causation – Health & Psychosocial	• Chronic and cumulative impacts	1	3
		• Multifactorial nature of health determinants	1	3
		• Work related impacts on health	2	3
		• Concept and models of "healthy work" and "wellness"	1	3
		• Models of causation of fatigue and stress	1	3
		• Mental illness in the workplace	1	3
2.	Causation – Safety	• Models of accident causation (linear to complex)	2	4
3.	Causation – Environmental	• Models of environmental harm (air, water, soil)		3
4.	Risk	• Difference between hazard and risk	3	4
		• Risk as a complex concept (uncertainty)	2	4
		• Prioritisation of critical risk	2	4
		• Qualitative/quantitative aspects of risk	1	3
5.	Hazards	• Process and task safety analysis methods (e.g. Job Safety Analysis)	3	4
		• Complex hazard analysis methods (i.e. FMEA, HAZOP, Fault Tree, Bow-tie, etc.)	1	3
		• Knowledge of exposure standards and their application	2	3
		• For each specific hazard o Basic underpinning science to understand the behaviour of the hazard, how it causes damage and how it can be controlled o Relevant definitions, units and methods of measurement	2	3

Table 4. Knowledge Matrix

Skills

Personal and professional skills are vital attributes for effective practice as an OHS professional or practitioner. Such skills have been identified as a priority in recruitment and a key area of professional development for OHS personnel.

A 'Bloom'[1] style taxonomy approach has been used to describe skills in the *Framework*, which are presented in three sections: personal skills, professional practice skills, and professional technical skills.

As with the knowledge requirements, the skills have been annotated with a coding to indicate the expected skill levels for the OHS professional as compared to the OHS practitioner. The skill coding is based on the g four **skill levels**, which also reflect Bloom's taxonomy. However, while definitions of the skill levels are similar to those for the knowledge requirements, there are important subtle differences to reflect the nature of application of skills.

Level	Skill
1	**Awareness:** Understands the need for and general principles of application of the skill.
2	**Routine application:** applies the skill independently to well-known, routine tasks and to non-routine tasks under supervision.
3	**Skilled application:** adapts and applies the skill independently and effectively, also to non-routine tasks.
4	**Creative mastery:** applies the theoretical concepts and the practiced skill critically and creatively to new situations.

Table 5. Application of Skill

As with the activities and knowledge, when considering required skills, it should be remembered that the application of skills is as follows for each of the two roles:

OHS Professional: Across the organization, including senior management and external agencies, including regulators and industry bodies

OHS Practitioner: Within the organization, with a focus on middle management, supervisor and workers.

The skills are structured to support self-assessment and peer or manager assessment. They may also provide a basis for training and development, either formal or informal.

There are three categories with a number of skills under each category, including a number of sub-skills in the professional practice category.

Personal skills:

- Verbal communication
- Professional presentation

[1] Anderson, L.W., Krathwohl, D.R., Airasian, P.W., Cruikshank, K.A., Meyer, R.E., Pintrich, P.R., Raths, J., & Whittrock, M.C. (2001). *A Taxonomy for Learning, Teaching and Assessing: a revision of Bloom's taxonomy of educational objectives.* New York: Longman.

Professional practice:

- Evidenced-based practice
- Influence
- Leadership
- Management
- Professional and ethical practice

Professional technical skills:

- Training
- Surveying, inspecting and auditing
- Investigating
- Measuring and monitoring

B. Professional skills				
B1	**Evidence-based practice**			
B1.1	**Knowledge management**	**Accesses** information from a range of workplace sources using digital skills and a variety of strategies.	3	3
		Uses a mix of strategies to access information from a range of external sources.	2	3
		Reviews uses literacy skills to read and interpret OHS legislation, codes of practice, guidance material, policies and procedures.	3	4
		Investigates and assesses the credibility of sources and reliability and validity of information.	3	3
		Collates information to identify common themes.	2	4
		Critically **evaluates** and validates results through challenging information, concepts, and theories.	2	3
		Synthesises information to identify implications for practice.	2	3
		Applies information, concepts and theories to inform practice	3	3

Table 6. Professional Skills

Target Audiences

This document is intended for six target audiences:

- OHS professional associations and related certification bodies: To inform their certification and other assessment processes.
- OHS professionals and OHS practitioners (subsequently, collectively referred to as 'OHS specialists'): To act as a reference and basis for gap analysis in relation to their professional practice and career development, to aid the development of continuing professional development plans and to assist in promoting the OHS professional and OHS practitioner roles in relation to each other.
- OHS educators: To use while developing and reviewing OHS education programs.
- Employers and recruiters: To assist in developing position descriptions for OHS roles, in recruiting OHS personnel, and in performance evaluation as a basis for professional development.
- OHS regulators: To aid in understanding the range of OHS specialist roles in order to clarify the requirements for certification/registration/licensing of OHS specialists, as well as to inform the professional development of government and other regulatory inspectors.

- The community: To assist in creating a better understanding of the scope of the OHS specialist roles.

Target Group: OHS professional associations and related certification bodies
Purpose: To inform their certification and other assessment processes.
An example would be from Board of Canadian Registered Safety Professionals (BCRSP).

The BCRSP sets certification standards for occupational health and safety professionals, operating out of Canada and is a participant in development of *Capability Framework.*

It defines certification as the "voluntary process by which a non-governmental entity grants a time–limited recognition and use of a credential to an individual after verifying that he or she has met predetermined and standardized criteria."

Standard Setting and Examination Development

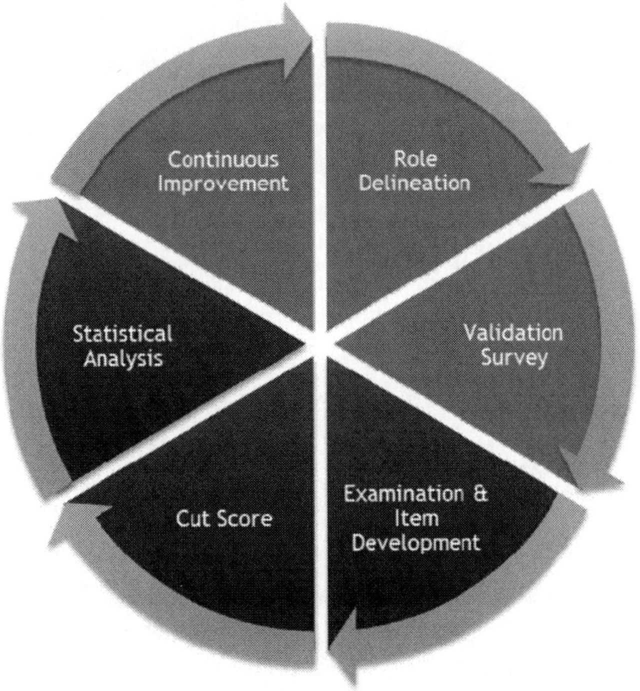

Exhibit 5. Standard Setting and Examination Development Cycle

The CRSP *Examination Blueprint* is updated every five (5) years to ensure the examination reflects current occupational health and safety profession.

The INSHPO *Capability Framework* is used as a resource to inform the competency development/review task force on global practice.

Certification Development:

- 2012: Strategy identified to consider expansion of certification offerings
- 2013-2015: Research, Governing Board discussions / ongoing review
- 2013 & 2015: Input from National Education Symposium attendees (OHS educators)

- 2016: Steering Committee struck to draft competencies for a technician certification
    - INSHPO *Capability Framework* used as one of the resources to inform the competency development/review task force on global practice.
    - Focus group (13 participants) to review draft competencies for technician certification
    - Survey on the Competencies (July/August): input from safety professionals (validation)

Opportunities with utilization of the *Framework* seen by the BCRSP:

- Transportability of credentials (mapping)
- Recertification/learning opportunities
- Competency-based training/development
- Harmonization/Alignment
- HR application

Target Group: Educators

Purpose: To use while developing and reviewing OHS education programs.

Program design:

- Provides a vision for the graduate role
- Provides a benchmark for content (not all content expected to be included in education)
- Encourages inclusion of skills in education

Program delivery:

- Structure for assessment activities

Program review:

- Benchmark for review of content

Accreditation:

- Basis for development of content criteria

Target Group: Industry, mining

The following is a summary the joint International Council on Mining and Metals (ICMM)-INSHPO project where the INSHPO framework was used as a benchmark in an industry-wide project on OHS capability. Not only did the INSHPO framework inform the industry analysis and the outcome results also fed back into refining the INSHPO document and application tools.

Project parameters:

- The project came about in response to a decision by the ICMM Council of CEOs. The ICMM Council of CEOs identified that a review and new approach to OHS capability was required to meet current and future challenges.
- The scope of the project was to redefine the role of their health and safety professionals and establish associated development pathways.
- The focus of the project was full-time safety roles at the corporate, regional/commodity, and site levels.
- The ICMM OHS Forum agreed to undertake a joint project with the INSHPO to develop an *OHS Capability Framework* for the mining industry. This framework would be developed

based on the INSHPO *Occupational Health and Safety (OHS) Professional Capability Framework – A Global Framework for Practice* (hereafter referred to as the *INSHPO Framework*), and informed by input from the mining industry.

Proposed outcomes:

- The defined outcomes were intended as guidance or a reference that individual organisations could use to inform the development of their frameworks and structures to suit their situations. While suggestions were provided as to how the outcomes might be used, it is up to the various organisations to take what suits them.
- A capability framework for the OHS function in the mining industry consisting of:
    1. Example OHS role statements
    2. Roles, knowledge, and skills:
        a. Tasks and activities associated with these roles, including the safety professionals' role in the verification of critical controls for material risk
        b. Underpinning knowledge required to carry out the task and activities
        c. Personal, professional practice, and technical skills required to carry out the roles
    3. Development of a self-assessment tool for use in performance appraisal/professional development planning
    4. Educational, experiential, and professional development pathways for acquiring the knowledge and skills.

The project comprised five phases:

Phase		Activity
1	**Current status of OHS role**	Analysis of PDs against INSHPO framework Questionnaire: demographics, training and activities (with reference to INSHPO activities)
2	**Managers' vision**	Interviews with managers Workshop with OHS GMs
3	**Learning from others**	Literature review on frameworks Review of frameworks from a range of professions, industries and organisations
4	**Framework**	Based on modified INSHPO framework
5	**Pathways**	Self-assessment tool Multi-pathway model for knowledge and skill development

Table 7. The Phases of the Joint ICMM-INSHPO Project

Outcomes: Need for change

- Position descriptions: It was found that often the position descriptions did not reflect core OHS functions or the knowledge and skills specific to the role. It was also determined that activities actually carried out were not reflected in the priority areas specified in position descriptions.

- Role: The current roles were seen as good at maintaining the status quo, while there was a lack of clarity between the site, regional, and corporate perceptions. Those who progressed from site to corporate often were not able to effectively make the transition. The emphasis of the role was compliance, rules and paper based.
- Knowledge and skills: A lack of soft skills was determined, along with a lack of recognition of the need for systematic body of knowledge resources. Additionally, the professional were not able to enunciate the science of OHS.
- Education/training/qualifications: There were variations and conflicts in the advice from OHS personnel. Mining experience was often preferred over OHS qualifications and experience. The training/education did not reflect tasks undertaken or position description positions, especially at corporate levels.

Analysis and development process with INSHPO *Global Capability Framework* as a benchmark

INSHPO Professional Level	ICMM Professional Profile
OHS Professional Level 3	Senior corporate
	Corporate
OHS Professional Level 2	Regional
OHS Professional Level 1	
	Senior site
OHS Practitioner Level 3	
OHS Practitioner Level 2	
OHS Practitioner Level 1	Site support

Table 8. INSHPO Levels Versus ICMM Profiles

	Senior Corp	Corp	Regional	Senior site	Site support
[illegible]	OHSMS	Planning OHSMS	Planning OHSMS Risk Ass Knowledge & skill Address issues	Audits Knowledge & skill Provide information Monitor performance Planning	Provide information Training Monitor performance
Medium focus		Performance monitoring processes Audit processes Knowledge & skills Criteria for monitoring perf Implementation of controls	Performance monitoring processes Provide advice on compliance Monitor controls Address issues Manage team	OHSMS Investigations Risk Assess Emergency prep Monitor compliance Develop team	Inspections Audits PPE Fire equipment Reporting non conformances
Major gaps	Advise on resourcing & priorities Hazard ID& RA Monitoring controls Monitoring performance OHS information management Communication & consultation	Advise on resourcing OHS in change management Provide information for RA Monitor controls Communication & consultation	OHS in change management Innovation	Provide information for RAs Monitor effectiveness of RM	RM

Table 9. Example of the Report's Gap Analysis

The ICMM survey responses on roles and responsibilities were compared to the INSHPO *Global Capability Framework* to address the mining and metals industry requirements and identify deficiencies. The results were that a capability framework was developed for the mining and metals industry and the INSHPO *Global Capability Framework* was updated. Additionally, self-assessment and pathways documents were developed for the mining and metals industry, and a template was developed for the future INSHPO online tools.

Target group: OHS professional and practitioners: Self-assessment

INSHPO has developed online tools to make it easier for OHS professionals and practitioners to conduct a self-evaluation and identify gaps that need to be filled in order for them to progress along a desired pathway. The OHS user first enters the country in which they practice, and then they can either enter their current level as practitioner or professional, or the level to which they aspire.

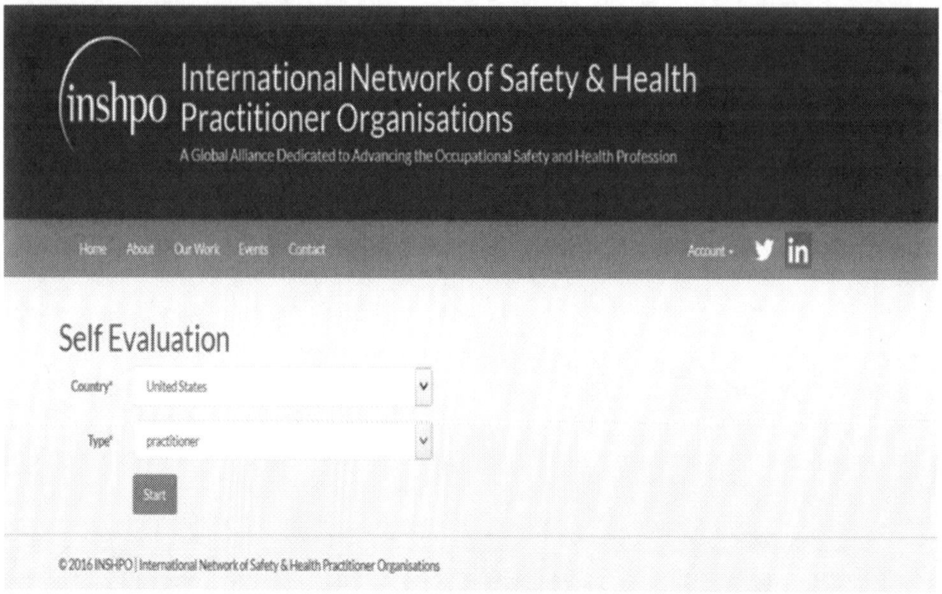

Exhibit 6. Opening Page of INSHPO Self-Evaluation Online Tool

On the next page, the user will work through all or some of the skills and knowledge levels to rate themselves.

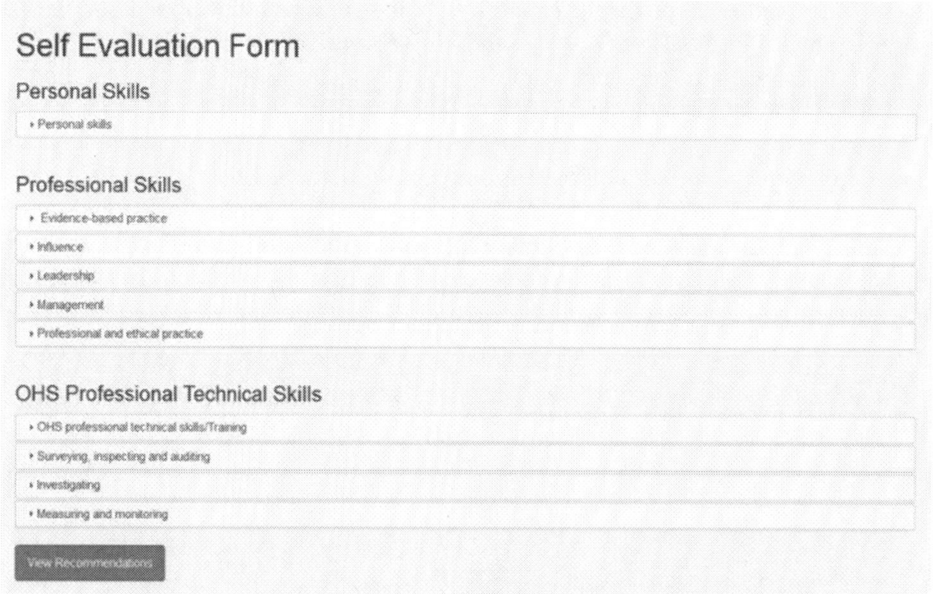

Exhibit 7. Second Page of the INSHPO Self-Evaluation Tool

When the user is finished, they can establish an "account" in which to save the self-evaluation, so that they can come back to it and update it at a later date. This is important due to the depth of the evaluation; it will take some time to complete the entire evaluation.

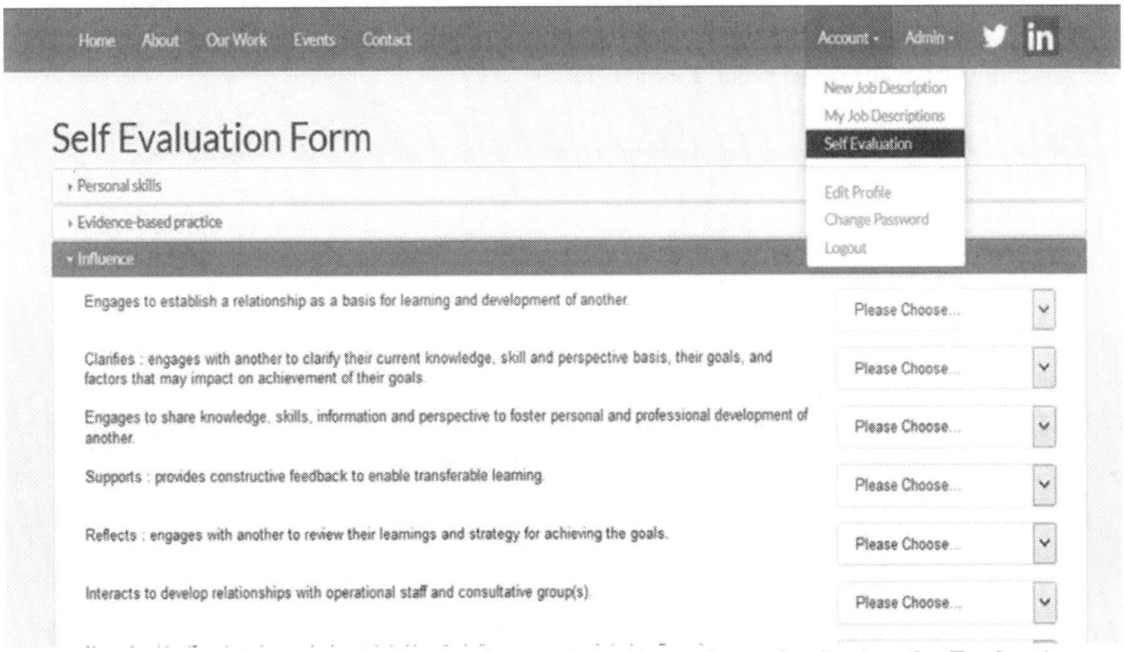

Exhibit 8. Example the Self-Evaluation and the Drop-down for Saving the Evaluation

Once completed, the user can print out the recommendations for filling the "gaps" identified by their self-assessment.

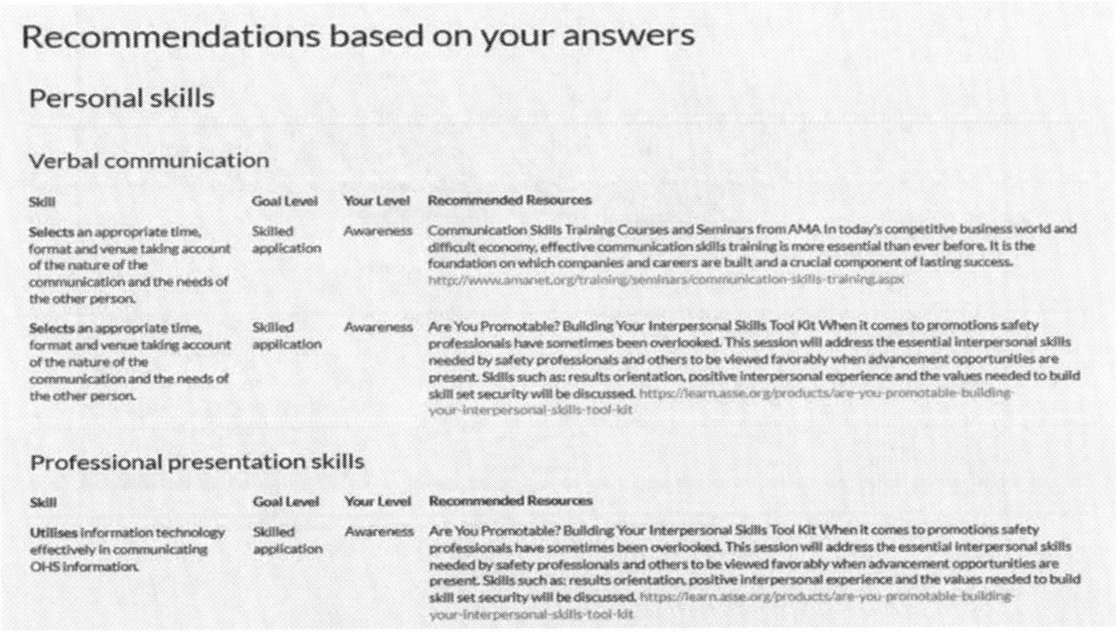

Recommendations based on your answers

Personal skills

Verbal communication

Skill	Goal Level	Your Level	Recommended Resources
Selects an appropriate time, format and venue taking account of the nature of the communication and the needs of the other person.	Skilled application	Awareness	Communication Skills Training Courses and Seminars from AMA In today's competitive business world and difficult economy, effective communication skills training is more essential than ever before. It is the foundation on which companies and careers are built and a crucial component of lasting success. http://www.amanet.org/training/seminars/communication-skills-training.aspx
Selects an appropriate time, format and venue taking account of the nature of the communication and the needs of the other person.	Skilled application	Awareness	Are You Promotable? Building Your Interpersonal Skills Tool Kit When it comes to promotions safety professionals have sometimes been overlooked. This session will address the essential interpersonal skills needed by safety professionals and others to be viewed favorably when advancement opportunities are present. Skills such as: results orientation, positive interpersonal experience and the values needed to build skill set security will be discussed. https://learn.asse.org/products/are-you-promotable-building-your-interpersonal-skills-tool-kit

Professional presentation skills

Skill	Goal Level	Your Level	Recommended Resources
Utilises information technology effectively in communicating OHS information.	Skilled application	Awareness	Are You Promotable? Building Your Interpersonal Skills Tool Kit When it comes to promotions safety professionals have sometimes been overlooked. This session will address the essential interpersonal skills needed by safety professionals and others to be viewed favorably when advancement opportunities are present. Skills such as: results orientation, positive interpersonal experience and the values needed to build skill set security will be discussed. https://learn.asse.org/products/are-you-promotable-building-your-interpersonal-skills-tool-kit

Exhibit 9. Example Recommendation Report

Target group: Associations
The following is an example from the American Society of Safety Engineers (ASSE) of how professional associations can use the *Framework.*

The ASSE audience is their members: OHS professionals. The *Framework* helps them assist their members develop their careers while promoting the ASSE educational offerings. It also assists the ASSE to identify gaps in their educational offerings.

ASSE sees the INSHPO *Capabilities Framework* as the keystone to meeting what learners want and need. ASSE believes that learners of the future will want:

- Variety
- Flexibility
- Choices in learning options
- To control their own learning
- Frequent feedback
- Access to learn from mentors and coaches
- Learning from peers
- Micro-learning (short, digestible)
- Learning to be experiential
- Learning to be On Demand

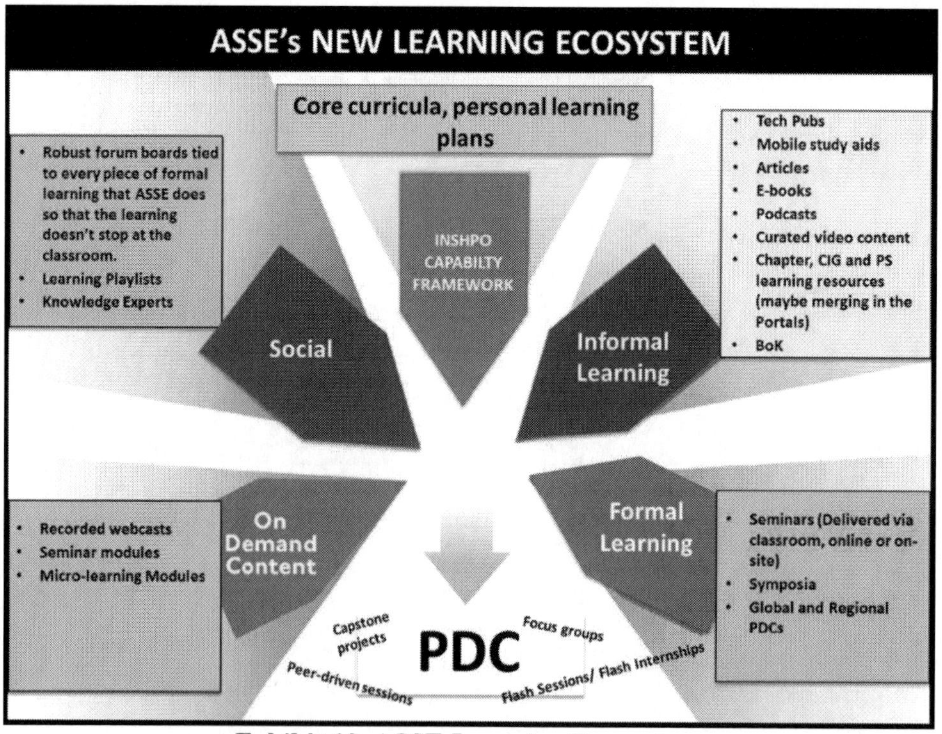

Exhibit 10. ASSE Learning Ecosystem

The INSHPO *Capabilities Framework*'s knowledge matrix provides the means for developing a personalized learning plan that allows the ASSE member to track their progress and CEUs. The ASSE can constantly provide formal and informal learning opportunities. The ASSE system also allows their members to track their progress, including CEUs.

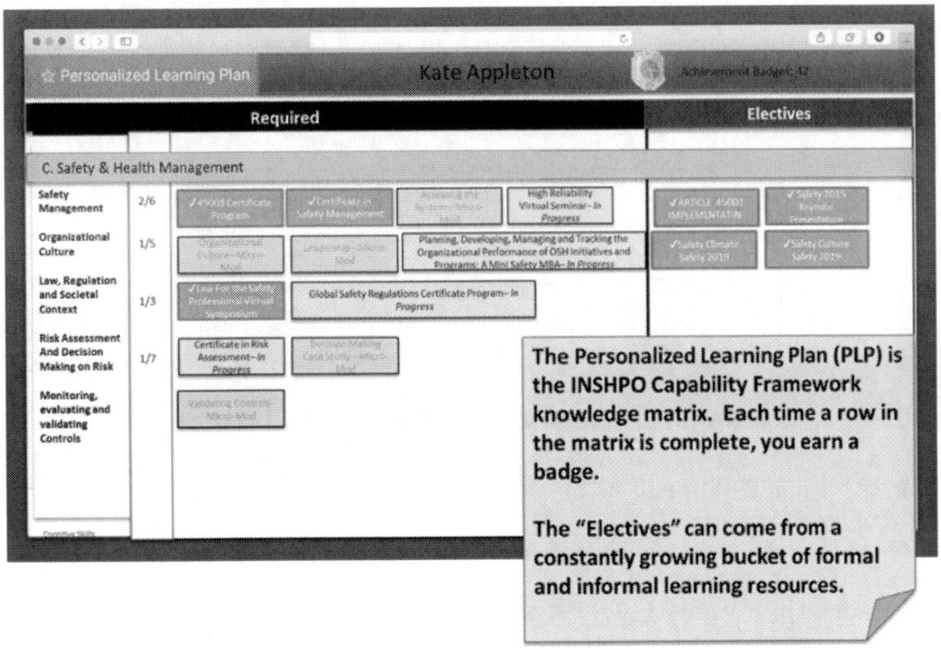

Exhibit 11. Example of ASSE Personalized Learning Plan

Target group: Business community: human resource professionals, employers, recruiters, position description builder

Human resources-related objectives include a means to provide a globally recognized online source for employers/human resources professionals to obtain all relevant information needed for hiring, developing, and retaining qualified occupational health and safety professionals, including recommended competencies, educational qualifications, and credentials. An additional objective is to educate the business and human resources community on the competencies, capabilities and scope of OHS professionals, and how to hire qualified professionals.

- Tools and resources provided by the online *Framework* tools include:
  - Online position description builder
  - Country-by-country credentials guide
  - Salary surveys
  - Career-pathing guide
  - Interview guide
  - Job-description bank

The current online tool that is available has the:

1. Ability to create multiple job descriptions.

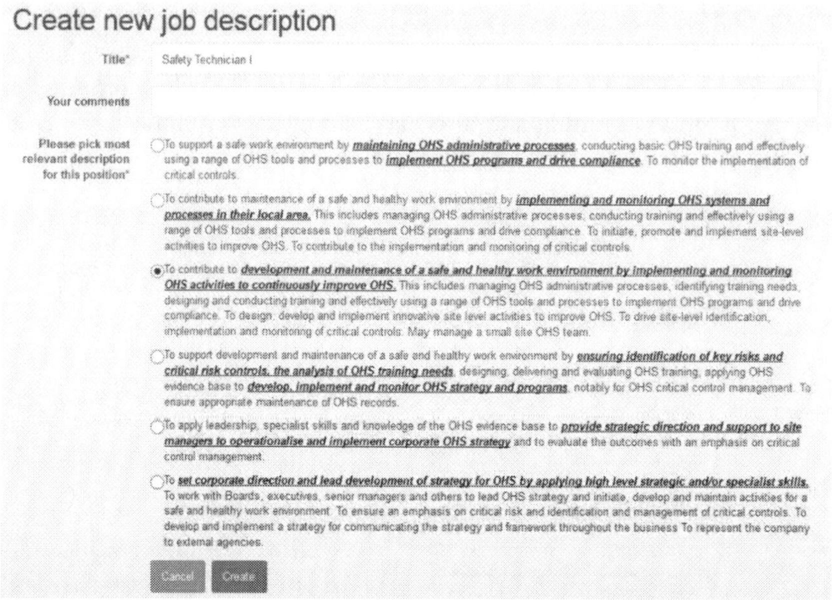

Exhibit 12. Example Job-description First Step

2. Ability to select predefined descriptions from a matrix of either professional or practitioner.

3. Ability to choose different levels within professional or practitioner groups.

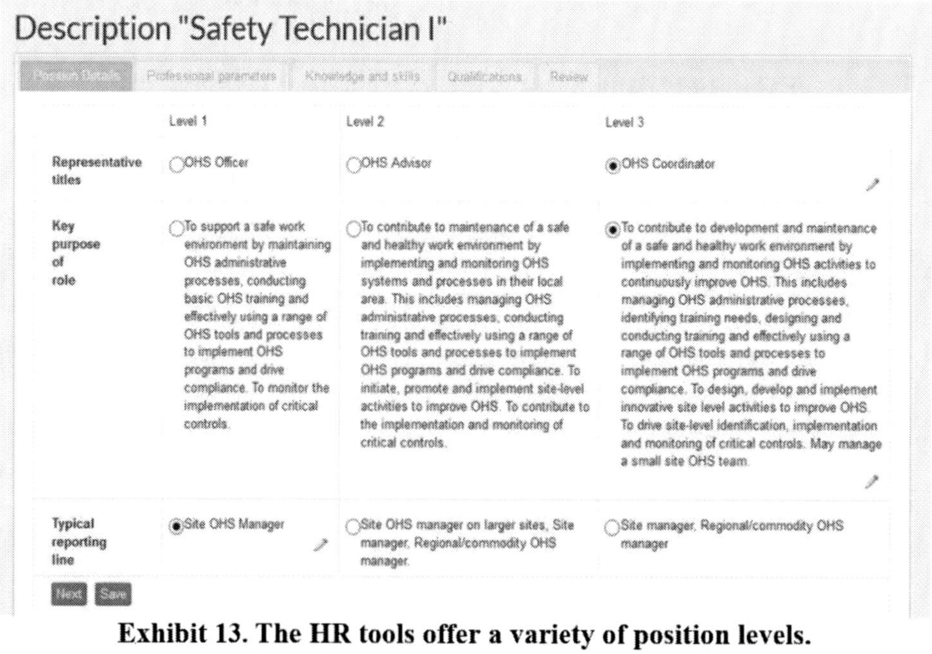

Exhibit 13. The HR tools offer a variety of position levels.

4. Ability to add, modify and save descriptions.

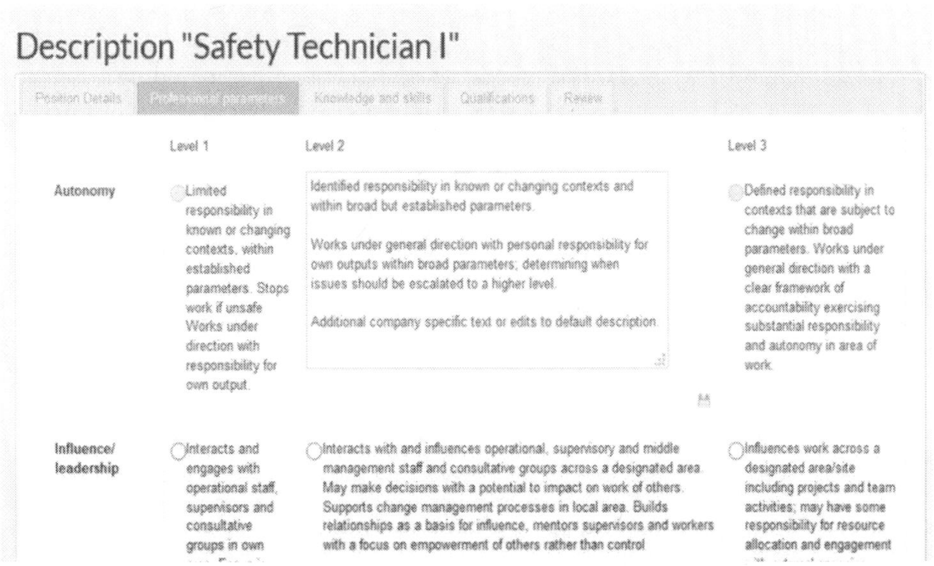

Exhibit 14. Example of Professional Parameters

5. Ability to print generated job descriptions.

Exhibit 15. Sample of Completed Job Description

In Summary: INSHPO Framework Utilization

Establishing an International standard of practice has many benefits, including:

- Better recognition of the value of the safety profession. This, in turn, will provide greater job growth within the profession; better promotion of the profession; higher compensation levels, and increased movements to senior management positions.
- Improved definition of qualifications for voluntary standards, such as ANSI Z10 and ANSI 590, as well as in contracts and bid documents published by government and private corporations.
- Facilitation of cross-border practice as employment becomes more mobile.
- Greater recognition for practitioners in countries that currently lack certification schemes.
- Enabling corporations with global reach to make better OHS hiring decisions.

Session No. 510

Short Service Employee Programs (Onshore Operations)

Wayne Vanderhoof, CSP
President
RJR Safety Inc.
Claysville, PA

With the national downturn, beginning mid-2014, of the oil & gas industry and the associated reduction in workforce by 50% to 60% (anecdotally), and now in 2017 and 2018, the slow resurgence of the industry, we will now require a build-up of the workforce. Many of those workers will be workers new to the industry as those that were in the industry will not return to the oil & gas industry. The general feeling is that many workers will not return due to the volatility of the industry. The story is the workers came to the industry in the boom of 2009 to 2014 and then were laid off. They found jobs or returned to jobs in a more stable industry, such as construction or manufacturing, so now they are not interested in coming back to a volatile job nor are their families interested in the return to a volatile industry where afamily member could lose their source of income because of many factors that the worker cannot control, such as the price of oil and the factors that affect the price of oil.

With the influx of new employees to the oil & gas industry, there needs to be a robust program in which to train the new workers to control and keep the number of workplace injuries low, since new employees historically make up as a significant number of the total number of workplace injuries. This may be done with a well-developed Short Service Employee (SSE) Program that is adhered to.

Short Service Employee (SSE) Definition

A Short Service Employee is defined, generally, in one of two methods. The first definition is an employee with less than six months of continuous employment with their present employer (API RP 76, Appendix C, 2007). The second definition is that of an employee with less than six months' experience in their assigned job with this definition (*Texas Oil & Gas Association (TxOGA) Oil & Gas Roundtable*, March 2015).

Statistics

Published statistics bear out the fact that 65 percent of fatalities occurred within first year of service (employment) and, of those, 34 percent of fatalities occurred within first three months of service (employment). Besides fatalities, the statistics show that "...30 to 40% of injured workers have been on the job less than a year." (*BLR Workplace Safety News,* May 2015).

A review of statistics from the BLS database reveals that employees with less than one year of experience had a significant number of injuries that were considered as Days Away cases as shown below:

- 2013 - 29% of Days Away injuries
- 2014 – 30.2% of Days Away injuries
- 2015 – 31.7% of Days Away injuries

Possible Reasons for High Fatalities and Injuries

Some of the potential reasons for these high percentages of workplace injuries in employees with less than one year of experience include: New employees do not know the specific job tasks or hazards of the job; new employees do not know the specific hazards of the work environment; employers assume new employees know more than they do, when asked if the new employee understands the task they blindly say "Yes!"; new employees may not ask questions; or employers hire the "wrong" people.

Elements of a Short Service Employee Program

In API RP 76, Contractor Safety Management for Oil and Gas Drilling and Production Operations (Nov. 2007), Section 4.3.1, Training Requirements, the need for a Short Service Program is identified and in Annex C, Sample Short Service Employee Program Development Guideline, the elements of a Short Service Employee Program are identified and explained.

The first element is that of identifying the short service employee with either a sticker on their hardhat or with a contrasting colored hardhat different from the other workers. Some possible reasons for identifying the short service employee include that SSEs are a hazard to themselves, their coworkers, and others. Another reason is that we cannot assume everyone is an experienced and well-trained worker.

The next element is for the short service employee receiving an HSE Orientation, which includes a basic safety orientation such as, yet not limited to, SafeLand USA or OSHA Outreach training. The HSE Orientation also is recommended to include training on employer-specific safety procedures/policies and jobsite orientations.

The final element of a Short Service Program is that of identifying and assigning a mentor to the short service employee. The purpose of the mentor is "...to provide guidance, development, and transfer skills & knowledge to SSE..." (API RP 76, 2007), in an organized and documented manner, using peer to peer involvement. The basic qualifications for a mentor is that they must have a positive safety attitude, have the ability and willingness to instruct and coach, have specific knowledge & experience in the tasks being trained, can coach and provide constructive feedback to the short service employee without criticizing or berating the short service employee, which will build up confidence of that short service employee.

Those employees volunteering and being assigned as a mentor have many responsibilities that include leading by example, meaning they are not taking shortcuts or exhibiting unsafe actions or unsafe behaviors. The mentor is to ensure the SSE understands the day's tasks. The mentor reviews hazards and safe work practices for each task with the short service employee, while following the job safety analysis or equivalent document. The mentor must always be open and available for questions, even encouraging questions and discussion from the SSE. The mentor must provide close supervision of the SSE and observe the SSE while performing tasks.

Responsibilities of SSE

The Short Service Employee is responsible to seek assistance and guidance from their mentor, be open to coaching and feedback from the mentor, follow all policies/procedures as taught/shown, and to work safe and in an environmentally sound manner.

Other Elements in SSE Program

A Short Service Program should also include a specific process for an employee who is released or "graduated" from SSE status with the training and competency evaluations specifically documented. There should be a specific recommendation for release from SSE status from each of those involved including the mentor, supervisor, and potentially, an operator representative.

Elements within ANSI Z490.1 - 2016

There is a framework that has been established in the voluntary standard ANSI Z490.1-2016, Criteria for Accepted Practice in Safety, Health & Environmental Training, that may be of use when developing a Short Service Employee Program. The elements of defining training, establishing a Training Program, conducting a Job Analysis, and performing Trainee Evaluations will be discussed below.

The definition of a Training Program is "…an established system of managing, developing, delivering, evaluating, and documenting [SHE] training" (section 2.12), with the Program administered by an individual or individuals with assigned responsibilities and accountabilities. The elements in a Training Program are listed below:

- Conducting a Needs Assessment (section 4.2)
- Developing Learning Objectives (section 4.3)
- Designing specific training course (section 4.4)
- Trainee evaluation & criteria for successful completion (sections 4.5 & 4.6)
- Course evaluation and improvements (sections 4.7 & 6)
- Documentation and Recordkeeping (section 7.0)

The definition of training in the standard is "…any activity provided to trainees to gain, improve, or retain specified knowledge, skills, or abilities." (section 2.9). Training is developed to "…improve the occupational safety, health, or environmental knowledge, skills, or abilities used by the trainees in the performance of their jobs." (section 4.0). The short service employee needs to receive information on the theory of the task, which is basically the classroom portion or reading of the Standard Operating Procedure (SOP) or Work Instruction. It is vitally important for the SSE receive skills instruction and be shown exactly how to do the task per the SOP. Then the short service employee needs to be closely supervised while doing the task to ensure the SSE is doing the task per the SOP or Work Instruction, gaining the correct skills necessary to complete each task while applying the theory within the SOP or Work Instruction. The SSE needs to practice performing each task for which they are being trained so that they develop the skills and understanding to safely and efficiently and correctly perform the task when they are released from the SSE Program.

Conducting a Job Analysis (JA) (section B.6) of each task that a short service employee is expected to be trained in includes identifying specifically how a task is performed by an experienced and qualified person. The JA should list environmental factors, such as indoor/outdoor task, temperature, work area/space, and lighting requirements. It should identify specific skills, knowledge, and abilities required to perform the task, which will be translated onto the Trainee Evaluation and includes specific acceptable standards of performance of each task so that the SSE knows what is expected of them. The JA should contain a hazard analysis and risk assessment of each task. All this information can be used to develop the Standard Operating Procedure or Work Instruction used to train the short service employee.

In the standard ANSI Z490.1 2016 is an element for Trainee Evaluation that can be used to define "success" of the SSE by evaluating the short service employee's (trainee's) achievement of the learning objectives (section 4.5). The evaluation will determine if the SSE has gained the knowledge, skills, abilities, and attitude necessary to be released from the SSE Program. Information that is needed for the Training Evaluation includes determining the specific criteria for completing the training on each task using qualitative measures, criteria that can be measured and quantified using items such as, but not limited to, a test score, the requirement to complete a specific number of steps correctly performed in order per task, or the successful completion of a specific task (project). Once the Evaluation is performed on the SSE for each task for which they are being trained, the results should be discussed with and provided to the SSE, with the mentor providing feedback and coaching on the level of success that the SSE obtained on the specific task. If the SSE does not obtain a high enough level of success on the evaluation based upon the established qualitative measures, the SSE may need to review course materials to reinforce areas they are unsuccessful, and receive continued training and coaching from the mentor as well as additional opportunities to practice such skills.

There are many tools that can be used to evaluate the short service employee, which include oral or written tests, or demonstrating the skill or task in a simulated work setting. A better indicator of success is having the SSE perform the task in the actual work setting under controlled conditions in the presence of the mentor. This allows the mentor to observe the actual performance of the task by the SSE to verify the SSE can demonstrate the targeted skills and or abilities under actual work conditions.

Developing a Short Service Employee Program

Some of the key areas of the SSE program are: defining the company-specific definition of an SSE; defining the responsibilities of the SSE, mentor, supervisors, and management; identifying the anticipated length of training; and setting out the basic process of SSE training. There should be a section that explains an exception to the SSE program whereby an experienced worker comes to the current employer with the requisite skills to perform the tasks and just needs to be indoctrinated with the specifics of the company, such as using the specific tools, learning the specific Standard Operating Procedures, and learning specific customer requirements. In this case, there may need to be an abbreviated training program for the SSE. Every element of the Short Service Employee Program should be documented, and include checklists for the mentor to follow indicating: the specific safety training that is required; the specific SOPs or Work Instructions the SSE is to be trained on; the specific tasks that the SSE will receive training on; and the specific qualitative evaluation criteria the SSE will be evaluated upon. There should be periodic evaluations of the SSE that do not necessarily include specific tasks; such items include their attitude toward the work, their attitude toward safety, their ability to work within the culture of the company, their willingness to accept feedback and coaching, their overall development as an employee, their ability to identify hazards, general compliance with safety procedures, and similar items. This type of qualitative evaluation should be conducted at specific intervals such as two weeks, a month, 60 days, etc., to determine if the SSE is showing improvement as perceived by the mentor. Coaching and feedback should be provided to the SSE on this evaluation as well.

Once the Short Service Employee has successfully completed the Short Service Employee Program, the mentor, supervisor and other members of management will need to specifically sign off on the employee being released from the SSE program, and now a regular employee. It should be communicated to the employee what the responsibilities are now that they are no longer in the SSE Program. This release from the SSE program should be documented and communicated to the operator's representative, if required.

Summary

When a Short Service Employee Program is developed, using applicable elements of ANSI/ASSE Z490.1-2016, a robust program is in place to effectively train new workers on the tasks they will be performing in efforts to keep the number of workplace injuries and fatalities that new employees historically experience low.

Bibliography

American National Standards Institute/American Society of Safety Engineers. 2016. *ANSI/ASSE Z490.1-2016, Criteria for Accepted Practice in Safety, Health & Environmental Training*. Park Ridge, IL: ASSE.

American Petroleum Institute. 2007. *API RP 76, Contractor Safety Management for Oil and Gas Drilling and Production Operations*, Appendix C, 2007. Washington, DC: API.

BLR Workplace Safety News, May 2015.

Texas Oil & Gas Association (TxOGA). *Texas Oil & Gas Association (TxOGA) Oil & Gas Roundtable*, March 2015.

Key Safety and Health Documents: Shields or Swords?

Adele L. Abrams, Esq., CMSP
Law Office of Adele L. Abrams PC
Beltsville, MD

Overview

There are a multitude of reasons why employers create safety and health-related documentation. These include compliance with mandatory paperwork requirements promulgated by the Occupational Safety & Health Administration (OSHA) and the Mine Safety & Health Administration (MSHA), which range from injury/illness records, to documentation of training provided to workers (both formal courses and less formal "toolbox talks"), as well as workplace examinations, equipment inspections, industrial hygiene monitoring, and exposure control plans to guard against health hazards associated with asbestos, lead, silica, and other toxic chemicals.

There is another universe of documents, which are routinely created and maintained in the course of business voluntarily or as a matter of "best practice." These include: employee handbooks, safety and health audits, safety and health programs, incentive and disciplinary programs, safety and health committee minutes, job hazard analysis (JHA) or other standard operating procedures, job descriptions, safety and health management systems, environmental compliance handbooks, occupational health programs, medical surveillance programs, and internal training materials.

Properly prepared and maintained, these documents can be your best line of defense (a shield) against unwarranted citations or other enforcement actions. But failure to invoke legal privilege where available, carelessly written documents, or materials that contain information constituting an admission against interest by the employer or its agents of management, can be used as a sword against you by OSHA and MSHA. Even documents viewed as proactive, or helpful, by an employer can be used to show "employer recognition" of hazardous conditions, which can support enforcement actions under OSHA's General Duty Clause, or to heighten the level of negligence assigned to a citation by showing a pattern or practice of recurring safety and health problems. Documents that can inadvertently be produced voluntarily and then used as the government's "Exhibit A" against the employer include safety and health audits, safety and health committee meeting minutes, near-miss reports, and job hazard analysis forms. Moreover, improper destruction of documents, particularly those where a litigation "hold" has been placed under notice by OSHA or MSHA, can even give rise to criminal prosecution for obstruction of justice or conspiracy!

What Does the Government Want from Us, Anyway?

OSHA often asks to examine complex construction project agreements, staffing agency contracts, and similar documents between an employer and agencies, union hiring halls, or different tier subcontractors, to see if they specify each party's responsibilities for aspects of safety training, hazard controls, provision of personal protective equipment (PPE), and more. Employers must remember that all of this non-mandatory documentation is discoverable by OSHA, even prior to the issuance of citations, by the use of a subpoena *duces tecum*, to obtain such materials. OSHA has even used these subpoenas to gain access to audits conducted by third-party insurance companies.

The wealth of information contained in committee minutes, disciplinary records, near-miss records, safety audits, and even JHAs can be used against the employer to show prior awareness of non-compliant conditions, or to show hazard recognition under OSHA's General Duty Clause (Section 5(a)(1) of the OSH Act). Even disciplinary reports can be incriminating, such as where the narrative admits a violation (e.g., where the worker was counseled for not wearing mandatory fall protection equipment, or for failing to lockout/tagout a system before performing maintenance work), and can be the basis for a citation if the violation occurred within the previous six months.

In addition to records created by the employer directly, there will also other correspondence and other memoranda and materials created by third parties including outside counsel, safety and health consultants, general contractors and subcontractors, equipment manufacturers and other vendors. All of these records can play critical roles in both incrimination during an initial OSHA inspection, or in the ensuing litigation, depending upon what is at issue.

A big game changer occurred on December 1, 2016, when OSHA put into effect new anti-retaliation rules that require "reasonable" procedures for reporting injuries/illnesses (should be documented), and workers must be trained on those procedures (more documentation). The same rule also bars employers from disciplining workers who are injured more harshly than non-injured workers for safety infractions, and from drug testing injured workers in many situations (absent legal requirement to test or where the employer has reasonable suspicion that the injured worker's impairment was a causal factor in the accident).

Therefore, incident reports and worker's comp forms documenting an occupational injury in an employee's personnel file, coupled with a decision to select that worker for layoff, could give rise to a whistleblower action under Section 11(c) of the OSH Act, plus civil penalties of up to $126,749. The safety department and the HR department must have clear communication about a worker's potential protected activities before making any adverse employment decision.

<u>MSHA's Powers to Compel Document Production</u>

MSHA has powers that far exceed those of most federal agencies when it comes to document discovery, including the right to conduct warrantless searches. It is unusual, but possible, for MSHA to seize computers or other files containing documentation during the course of its inspections and investigations, particularly if the agency believes documents might be destroyed before litigation (and formal discovery) can ensue. MSHA does have more limited subpoena power, as subpoenas are available to the agency only when it convenes a public hearing or is in litigation over citations, orders or penalties. But the agency has broader powers overall than does OSHA when it comes to its ability to obtain a host of potentially incriminating materials. Section 103(a) of the 1977 Mine Act, as interpreted by the courts, gives MSHA the ability to demand (without a subpoena) any documents or other information necessary to carry out its activities under the Act. This was construed in the recent *Warrior Coal* case as permitting MSHA to obtain the names, addresses, and phone numbers of every employee, so that offsite interviews can be conducted without management knowledge. If, after issuing a Section 103(a) citation, the employer still does not release the requested documentation, MSHA can seek injunctive relief from the U.S. District Court to compel production of documents, pursuant to Section 108(a)(1)(E) of the Mine Act.

Personnel files, emails, purchase orders, and disciplinary notices are all among the written materials that MSHA has used against mine operators and contractors at mine sites in court. MSHA can also compel employers to obtain and produce third-party documents (such as insurance company audits) under threat of Section 103(a) sanctions. Penalties can reach in excess of $250,000 per violation, and managers who refuse to cooperate can be personally fined up to $70,000 and can even be criminally prosecuted.

Audits: Protected … or Not?

OSHA has a long-standing, formal policy that it will not use safety and health audits against an employer in most circumstances, as long as the violations identified in the report have been corrected prior to the commencement of the OSHA inspection. *See* https://www.osha.gov/pls/oshaweb/owadisp.show_document?p_table=FEDERAL_REGISTER&p_id=16434&p_search_type=CLOBTEXTPOLICY&p_search_str=self-audit. However, this safe harbor is not absolute, and OSHA reserves the right to use these documents against employers to show prior knowledge or patterns of violations to show "what the employer knew and when he knew it" for purposes of sustaining willful violations. Therefore, to the extent possible, it is helpful to privilege audits through counsel if they can be directed to be performed "in anticipation of litigation."

For other classes of documents, the nature and purpose of the material being created will dictate, to a large extent, whether it is eligible to be protected from disclosure in litigation under a legal privilege.

Legal Privilege Issues for Safety and Health Documents

Documents that must be maintained by law and shown to investigators or government agencies (e.g., OSHA Form 300/301 logs, training and workplace examination reports) cannot be privileged because they are mandatory documents. Other voluntarily prepared documents may be covered by attorney-client privilege or the attorney work-doctrine privilege, depending on the circumstances and timing of its creation. The issue of privileging correspondence and other communications can be critical, given an increasing trend toward criminal prosecution of CEOs and other agents of management under the OSH Act, environmental laws, and various other federal statutes.

The attorney-client privilege protects communications between a client and the client's attorney that are intended to be, and were actually, kept confidential for the purpose of obtaining or providing legal assistance. (*Brennan Center for Justice at New York University School of Law v. U.S. Department of Justice*, 697 F.3d 184 (2^{d} Cir. 2012)). The party asserting the privilege has the burden of demonstrating its applicability, and any ambiguities are construed against it. (*Shaffer v. American Medical Ass'n*, 662 F.3d 439 (7th Cir. 2011)).

The attorney-client privilege is usually applied where outside counsel are retained in litigation, but it was found to be applicable as well to communication between managers and in-house counsel by the U.S. Supreme Court in 1981; but, to be protected, the information must be provided by employees to counsel for the purpose of the company obtaining legal advice. The fact that an investigative report was prepared by or for company counsel does not mean that the attorney-client privilege necessarily applies. The primary or dominant purpose of the communication must be to seek legal advice, and not simply to prepare records created and maintained in the normal course of business, absent litigation threat.

When properly invoked, the privilege can also extend to documents created by non-attorney consultants who conduct investigations at direction of in-house or outside counsel for the purpose of representing the employer in future litigation with OSHA. However, the privilege will not apply to communications with the client that are in furtherance of fraud or other criminal acts. Moreover, "fact-based" materials, including photographs, videotapes, audiotapes, and measurements or drawings, cannot be privileged, even if they are created by an attorney.

Even if the attorney-client privilege is inapplicable, documents may still be protected as counsel's work product, prepared by or under the direction of counsel in anticipation of litigation. (*Upjohn v. United States*, 449 U.S. 383 (1981)). There are two type of work product: ordinary, "fact" work product and "core" work product.

To obtain disclosure of "fact" work product, a party has to show that it has a substantial need for the information and cannot obtain the equivalent without undue hardship. (Fed. R. Civ. P 26(b)(3)(A); *FTC v. Boehringer Ingelheim Pharms., Inc.*, 778 F.3d 142 (DC Cir. 2015)). "Core" work product contains an attorney's mental impressions, conclusions, opinions or legal theories. This remains privileged unless the opposing party demonstrates a "highly persuasive showing of need."

Significantly, work product protection can extend to materials created even before the events giving rise to litigation, if the documents were created with an eye toward anticipated litigation. Normally, however, an attorney must direct the person creating the document to privilege it in advance of its preparation.

Regardless of which type of privilege is involved, employers and counsel must guard against inadvertent waiver or the need for disclosure of otherwise privileged statements or reports to make the affirmative defenses of a thorough investigation. Documents intended to be privileged that are written by or to counsel from a client will generally be bullet-proof.

Documents that fall under the "work product" doctrine of privilege must have a controlled circulation (to senior management level personnel) and cannot be left unprotected on a server that can be accessed by rank-and-file workers or union personnel. Moreover, if a collective bargaining agreement, for example, requires that copies of all safety and health audit reports must be shared with the union, then these documents would lose their privilege once disseminated, even if they were originally prepared by or initially distributed through an attorney.

Internal Investigation Issues

Perhaps the most potentially incriminating documents are those arising from internal investigations, whether pertaining to accident and near-miss incidents, to hazard complaint evaluations, or vetting of discrimination complaints (either of the EEOC variety or arising from claims of discrimination under Section 11(c) for safety whistleblower actions). Where possible, before any documents are drafted, counsel should be consulted to determine which, if any, privileges may be applicable and to set up the infrastructure for controlled circulation of key materials prepared under legal protection.

Complaints of possible safety or health hazards or OSHA violations, or infringement of whistleblower protections under the OSH Act or environmental or DOT hours-of-service laws (all of which are enforced by federal OSHA, – sometimes in concurrence with state occupational safety agencies) must be handled with care. Those also mandate immediate internal investigations to ensure that workers are protected and to avoid having the complainant file a formal hazard complaint with OSHA (which will bring inspectors to your door). The investigative actions and abatement of hazards must be documented in order to avoid OSHA prosecutions.

Personnel who create documents germane to these investigations will become witnesses in any litigation as the records custodian or developer of policies and procedures. Supervisors who conduct safety and health evaluations, or write up disciplinary reports, must be clear on how to consistently and effectively document the good, bad and ugly, as these records may be the only rebuttal evidence of "business necessity" or "due diligence" to defend the company against OSHA/MSHA action. Determining the appropriate person(s) to conduct and receive information about internal investigations will largely depend on the nature of the matter under scrutiny, and will also impact to some extent whether attorney-client privilege or attorney work-product privilege will apply to all, or some, documents relevant to litigation matters.

The information that should be recorded during internal investigations typically includes:

- Memos and letters sent to employees,
- Written summary of the date/time/place/people involved in any alleged incident,
- Any work documents involved,
- Notes from meetings,
- Witness statements,
- Investigation interviews,
- Performance evaluations, and
- Disciplinary actions of any kind, including termination.

Aside from the need to demonstrate a thorough internal investigation and remedial measures (where warranted) for purposes of any follow-up OSHA/MSHA investigation or litigation, timely documentation of any incidents or complaint investigations will make it easier to recall the facts if the complaint is filed long after an incident occurred, or if witnesses subsequently leave the company, and may prevent a lawsuit entirely.

Commencement of any hazard or whistleblower complaint-related actions, OSHA/MSHA inspections or accident investigations should be a signal to alert managers and counsel that a litigation hold should be considered on all relevant documents until the complaint is resolved internally or through litigation, arbitration or mediation. Absent a litigation hold, following a written document retention/destruction program is normally helpful in defending against claims of spoliation of evidence.

Once the investigation is concluded, all witnesses have been interviewed, and all documents examined that support and refute the allegations, the investigator should prepare a privileged summary report and this ideally should be sent to corporate counsel (in-house or outside). If the matter is resolved at this stage, make sure to document the end result, and any discipline imposed or training provided, so that there is institutional memory of this in case of future infractions and for use under a progressive discipline system. Written reports can also be valuable to help management develop corrective procedures to address questionable conduct.

You also must document situations where a hazard or whistleblower complaint was received, it was investigated, and it was closed without further action. Make sure all know that the employer is committed to preventing any discrimination, harassment, abuse or retaliation in the workplace for protected safety activity, and reaffirm the procedures for reporting any future problems.

If a formal report is needed, when preparing the final report and any recommendations that might be requested in terms of follow up action or discipline, the investigator should include:

- The nature of the report and issue;
- A summary of the facts gathered and a chronology of events;
- The people interviewed and documents reviewed;
- Any credibility assessments reached;
- Whether the report/complaint was substantiated, unsubstantiated, or inconclusive;
- Specific conclusions on each issue;
- Identification of issues that could not be resolved in the investigation;
- Whether a breakdown in an internal control was implicated; and
- The financial impact (if any) on the employer or third parties

Normally, the final investigative report will not include recommendations on how an employer should respond, specific corrective actions needed, or any discipline to be meted out. However, if the employer has already decided on a course of action prior to conclusion of the investigation, include that in the report.

Conclusion

Documentation of safety and health training and related activities, investigations, and audits can be useful, but also has the potential to be damning. Documents, whether formal written policies, disciplinary reports, or even informal emails, that concern safety and health practices or actions, must be written with an eye toward being read in court by the trier of fact. Where the threat of litigation exists, privileging documents to the extent possible, as soon as possible, can be key to blocking unwarranted citations or defending against personal injury or wrongful death tort actions.

Session No. 513

Improving General Industry Qualitative Risk Assessments Using LOPA Concepts

John Mulhausen, PhD, CIH, CSP
Director
Corporate Safety and Industrial Hygiene
3M Company

Overview

Effective and efficient safety programs rely upon accurate and consistent risk assessments. Underestimate risk, and our safety and health programs become ineffective, as employees whose risk was underestimated continue to perform their duties with the risks unabated. Overestimate risk, and our programs become inefficient, as limited resources are wasted on risk controls where none were needed.

In general industry, formal risk assessments are often made qualitatively, using a matrix approach with severity and likelihood of harm as key inputs (Figure 1). Accurate and consistent risk assessments require selection of the correct severity and likelihood inputs. In particular, a clear understanding and assessment of the adequacy of controls is critical for selection of the correct likelihood input parameter and is therefore a critical determinant for the accuracy and consistency of risk assessments.

Techniques developed to support formal layer of protection analysis (LOPA) in the process industries hold promise for improving our understanding of controls used in general industry. In particular, LOPA concepts may be able to increase our understanding of the strength and reliability of various options in the hierarchy of controls.

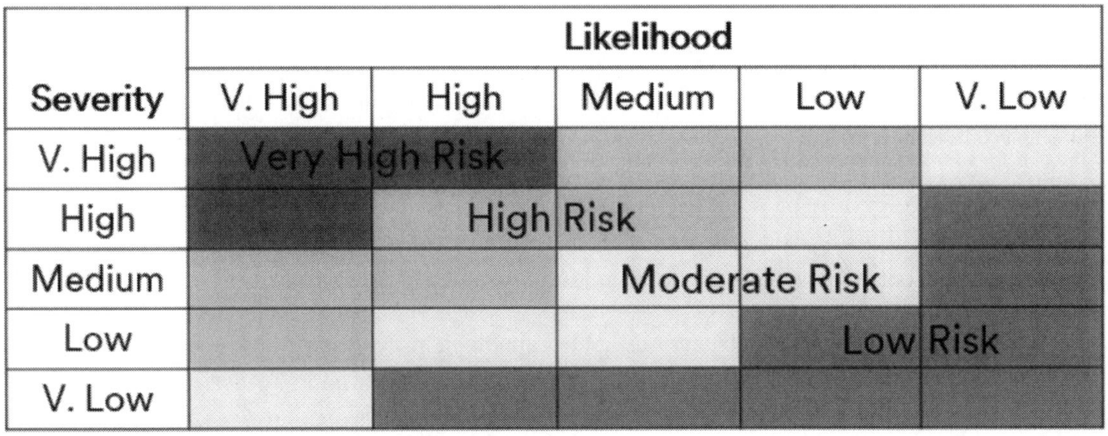

Figure 1. Example of a qualitative risk matrix.

LOPA is a semi-quantitative approach to risk assessment. It uses order-of-magnitude control layers as basic building blocks to systematically understand the risk controls that are in place, and whether adequate protection is provided for a particular scenario outcome.

The LOPA control-layer building blocks are called independent protection layers (IPLs). IPLs are very similar in concept to layers of protection in the commonly used "Swiss Cheese" hazard control model (Figure 2), and their relative strength corresponds roughly with the traditional hierarchy of controls. However, there are rigorous rules for their proper validation and use. Understanding IPLs and the rules for their use may provide promising opportunities for improving the accuracy and consistency of general-industry, matrix-based qualitative risk assessments.

Figure 2. Typical layers of protection in a "Swiss Cheese" hazard control model.

Promising LOPA Concepts

Several key LOPA concepts may offer promise for improving the consistency and accuracy of qualitative risk assessments.

Independent Protection Layer (IPL)

The authors of *Layer of Protection Analysis: Simplified Process Risk Assessment*, one of the key texts on LOPA, define an independent protection layer (IPL) as follows:

> An IPL is a device, system, or action that is capable of preventing a scenario from proceeding to its undesired consequence independent of the (scenario) initiating event or the action of any other layer of protection associated with the scenario. The effectiveness and independence of the IPL must be auditable.

They further make the important distinction between an IPL and other safeguards that may be capable of preventing an undesired consequence but are not IPLs because they may not be independent or because the effectiveness of their performance may be unknown or un-auditable.

Independence

A critical aspect of an IPL is its independence. An IPL must be able to properly function, regardless of the actions of other protection layers associated with the scenario and the event that initiated it. If the same condition that initiated the event scenario renders the safeguard ineffective, then that safeguard is not an IPL. If the condition that caused one safeguard to fail also renders a second safeguard ineffective, then the second safeguard is not an IPL.

A shared reliance on human initiation or intervention from the same person is one common cause, resulting in two or more safeguards not meeting the requirements for independence. Even if redundancy is built in with separate individuals, if both rely on the same information source to initiate action, or if they have separate information sources, but both sources could be incapacitated by a common cause, then the safeguards are not independent.

Probability of Failure on Demand (PFD)

Probability of failure on demand (PFD) is a measure of performance effectiveness for an IPL. It is a dimensionless value between zero and one that quantifies the number of times the IPL fails, divided by the number of times it is needed. For example, if a pressure relief valve fails once in 100 trials, it will have a PFD value of $1/100 = 10^{-2}$. Because LOPA IPLs are expressed as order of magnitude, PFD is typically summarized by orders of ten (Table 1). The lower the PFD value, the more reliable the IPL.

Values for PFD can be determined using tables of generic values, literature from device manufacturers, or from actual performance data in a facility.

Independent Protection Layer (IPL)	Generic Probability of Failure on Demand (PFD) *
Personal Protective Equipment (PPE)	10^{-1}
Human Response to an Abnormal Condition	10^{-1}
Automatic Fire Suppression System for a Room	10^{-1}
Safety Interlock	10^{-1}
Permanent Mechanical Stop That Limits Travel	10^{-2}
Spring Operated Pressure Relief Valve	10^{-2}
Safety Instrumented System Control Loop – SIL1	10^{-1}
Safety Instrumented System Control Loop – SIL2	10^{-2}
Safety Instrumented System Control Loop – SIL3	10^{-3}

Table 1. Examples of PFD values.*

*Source: Center for Chemical Process Safety. 2015. AIChE *Guidelines for Initiating Events and Independent Protection Layers in Layer of Protection Analysis*. New York, NY.

Auditability

A third critical feature of an IPL is its auditability. One must be able to audit the performance of the IPL and the systems that support it to ensure that assumptions about independence and PFD that were used in the risk assessment hold true in the actual work environment.

Of course, it is not only important that the IPL be auditable; it is also important that the IPL be audited. Inherent in the LOPA process is the expectation that there be a program in place to systematically and regularly audit IPL performance.

Conclusion

LOPA concepts may offer promise for improving the consistency and accuracy of qualitative risk assessments performed in general industry. In particular, the concepts of independence, probability of failure on demand, and auditability are worth exploring as mechanisms for clarifying discussions regarding the strength and reliability of controls; both are crucial inputs into making accurate likelihood decisions when performing qualitative risk assessments.

Bibliography

Center for Chemical Process Safety. 2009. *Guidelines for Developing Quantitative Safety Risk Criteria*, New York, NY: AIChE.

Center for Chemical Process Safety. 2013. *Guidelines for Enabling Conditions and Conditional Modifiers in Layer of Protection Analysis*. New York, NY: AIChE.

Center for Chemical Process Safety. 2015. *Guidelines for Initiating Events and Independent Protection Layers in Layer of Protection Analysis*. New York, NY: AIChE,

Center for Chemical Process Safety. 2001. *Layer of Protection Analysis: Simplified Process Risk Assessment*. New York, NY: AIChE.

Session No. 514

Hazards of Homeless Encampments and Working in Urban Environments

Cynthia E. Braun, CSP, CHMM, CET
Braun Safety Associates, LLC
Celebrate Safety, Inc.
Littleton, Colorado

Introduction

Working in urban environments presents a wide variety of hazards and risks for employees and organizations across many industries. Hazards and risks associated with this work are often not anticipated and can be overlooked in the preplanning stages of projects and tasks. The unique, non-traditional hazards of working in urban environments come from many sources, such as homeless encampments, drug labs, criminal activity, human interaction, and animals, just to name a few. Working in urban environments can be complicated by other factors, such as needle-sharing programs, encountering stolen property, working in areas of high crime and gang activity, and the use of temporary workers by employers.

Health and safety hazards in urban environments can be encountered by workers on construction sites, city streets and parking lots, vacant or abandoned property, environmental cleanup projects, surveying and locating projects, urban campuses, right-of-ways, in retail operations, in and around commercial establishments and healthcare facilities, and in many other work locations and industries. Traditional risk assessment methods provide a starting point for evaluating potential worker exposures. However, non-traditional health and safety hazards must also be considered to fully assess worker health and safety hazards in urban environments.

This paper provides an introduction to the identification of non-traditional health and safety hazards that can arise in urban settings, and suggestions for control methods to protect workers from potential harm. This paper is not a sociopolitical study or statement regarding causes, reasons, or remedies for the hazards and conditions that confront workers in urban environments.

Urban Environments

The Census Bureau's urban-rural classification is a delineation of geographical areas:

- Urbanized Areas (UAs) have 50,000 or more people;
- Urban Clusters (UCs) have at least 2500 and less than 50,000 people; and

- Rural areas encompass population, housing, and territory not included within an urban area.[1]

For the purposes of this paper, the phrase "urban environments" is broadly used to represent densely developed metropolitan and rural areas, including residential, commercial, and other non-residential urban land uses.

Working in Urban Environments

Millions of workers are exposed to urban hazards in various work capacities. Examples of industries and workforces that are affected are listed below, although this list is far from exhaustive:

- Building inspectors
- Charitable organizations
- Construction and trades workers
- Directional drilling companies
- Emergency responders
- Engineers, Project managers
- Environmental consultants
- Facilities maintenance companies
- Healthcare and community care workers
- Law enforcement
- Lawn- and tree-care industry workers
- Municipal workers – city, county, state
- Oil and gas operations
- Property management personnel
- Right-of-way workers
- Retail store workers
- Sanitation workers
- Security personnel
- Surveyors and locators
- Transportation workers
- Urban planners
- Utility crews

Hazards of Working in Urban Environments

When assessing the health and safety hazards of working in urban environments, "traditional" and "non-traditional" hazards must be considered. These terms are discussed below.

Traditional Hazards

"Traditional" hazards of working in urban environments are those commonly known and used in workplace safety programs for hazard assessments, safety training, and safety procedures. Examples are presented below; this list is not exhaustive:

	Hazard Category	Examples of Traditional Hazards
1.	Physical Hazards	Fall from heights, slip/trip/fall, electrical, mechanical, thermal, noise, pinch points, struck-by, struck against, caught in-between, pressure, engulfment, sharp objects and edges, radiation
2.	Chemical Hazards	Liquid, gas, vapor, particulates, solid, semi-solid, sludge
3.	Biological Hazards	Animals, insects, plants, bloodborne pathogens
4.	Ergonomic Hazards	Awkward postures, overexertion, repetitive motion, vibration, contact stress
5.	Human Factors	Fatigue, stress

Non-traditional Hazards

"Non-traditional" hazards of working in urban environments may not be as obvious to workers, safety professionals, and project managers. They include hazards created by urban conditions, persons, and activities. They can be easily overlooked during project/task risk assessments, and workers must be trained to identify them. Examples of "non-traditional" hazards of working in urban environments are presented in Table 1; this table is not exhaustive:

	Hazard Category	**"Non-traditional" Hazards of Working in Urban Environments**
1.	Physical Hazards	Assault: street violence, crime, gangs Electrical: exposed underground utilities, missing electrical panel covers, live electrical connections from exposed wiring, connections to transformers Explosion: compressed cylinders, chemical reactions Fire: camp fires, wood stoves, grills, hazardous materials, sterno Cuts/lacerations from sharp objects sticking out of bags/containers Weak/deteriorating structures Slips/trips/falls from poor lighting, ground features, debris and trash Unstable walking/working surfaces
2.	Chemical Hazards	Methamphetamine (meth) labs Propane gas cylinders "Death bags:" Garbage or plastic bags with offgas from meth production Hazardous materials Alcohol Universal waste: mercury bulbs, asbestos, batteries
3.	Biological Hazards	Human waste: urine, feces, vomit, blood, body fluids Human contact: persons with lesions and infections Human corpses Methicillin-resistant Staphylococcus aureus (MRSA) Needles and syringes Food waste, food-borne diseases Animal feces, carcasses Animal bites (e.g., pets, stray animals, rats) Bedbugs, lice Old landfill waste
4.	Ergonomic Hazards	Awkward postures, reaching Heavy lifting of odd-shaped and odd-sized objects Holding objects away from body Quick movements and reactions
5.	Human Factors	Human interaction - Aggressive persons - Persons under the influence of substances - Persons with mental illness - Unpredictable behavior - Startled persons Gang activity Panhandlers Shoplifters, employee theft Counterfeiters

		Intruders Stolen property Weapons Odors Crime (broad range)

Table 1. In addition to traditional hazards, non-traditional hazards of working in urban environments must be considered during project and task pre-planning. This list is an introduction to hazards, and is not exhaustive.

Exhibits 1 and 2. Weak structures and ground debris present physical hazards to workers.

Exhibit 3. Workers at charitable organizations, such as this thrift-store donation station, encounter unique hazards when storing, unloading, and sorting donated goods. Hazards, such as human body fluids and waste, hazardous materials, needles/sharp objects, and homeless persons seeking shelter, are some of the hazards that affect these workers.

Homeless Encampments

The U.S. Department of Housing and Urban Development (HUD) classifies homeless persons in two categories: sheltered and unsheltered. A "sheltered" homeless person lives in an emergency shelter or transitional housing, such as domestic violence shelters; residential programs for homeless or runaway youth; or a hotel, motel or apartment paid for with a voucher provided by a governmental or private agency. An unsheltered homeless person lives in "a place not meant for human habitation, such as cars, parks, sidewalks, abandoned buildings, or on the street."[2] About 44 percent of homeless people are unsheltered.[3]

Locations of Homeless Encampments

Homeless encampments are commonplace in most U.S. cities. Homeless encampments are typically located where individuals can seek personal shelter, be out of sight, and be protected from the weather. Examples of locations where homeless encampments are found include: campgrounds, parks and alleys, vacant property, vacant and abandoned buildings, under bridges and overpasses, inside viaducts, on right-of-ways, around heat sources and vents, in empty trucks and train cars, near overgrown brush and vegetation, along waterways, and adjacent to pedestrian paths. Some individuals use facilities and structures for short or long-term residence, such as dumpsters, public transportation systems, public restrooms, vehicles, trash compactors, balers, stairwells, rooftops, parking ramps, make-shift shantytowns and tent cities, and commercial establishments.[4]

Encampments are usually located close to goods, services, and resources that people need, such as food, alcohol, transportation, walking paths, and employment opportunities. These temporary "homes" can have from a few to several dozen residents and their pets. Often, encampment residents are convicted felons or persons with outstanding warrants.

Exhibit 4. Homeless encampments are typically located where individuals can seek personal shelter, be out of sight, and be protected from the weather. Encampments are often located close to goods, services, and resources that individuals need to live.

Hazards of Homeless Encampments

Homeless encampments bring a distinct set of environmental and safety hazards to the areas inhabited. Table 1 contains examples of the health and safety hazards associated with homeless encampments.

Dismantling and Cleanup of Homeless Encampments

Workers who dismantle and cleanup encampments are directly in the path of safety and health dangers. Cleanup efforts must be executed in an organized manner, and planned well in advance of the event. Because of the hidden dangers and unique hazards associated with encampments, persons involved in cleanups must be trained to recognize health and safety hazards, and methods to avoid personal injury and illness.

Pre-planning for worker safety must take place because of the logistics involved. Establishing a team approach and chain of command are essential for a successful cleanup at these sites, starting with local law enforcement sweeps of the encampment. During sweeps, police officers check for persons who may still inhabit the encampment, stolen property, drug labs, residents' personal possessions, and other dangers they may recognize.

Encampment cleanups are a team effort between the police department, other agencies, and private sector services, based on needs of the project. For example, homeless encampment cleanups in Denver, CO, such as on a right-of-way along a major metropolitan highway, can include some or all of the following participants:

1. Denver Police Department,
2. Colorado Department of Transportation (CDOT),
3. City and County of Denver Homeless Outreach Team (HOT),
4. City of Denver Public Works Solid Waste Department,
5. City of Denver Parks Rangers (Parks & Recreation),
6. City of Denver Fire Department, including hazardous materials specialists,
7. Colorado Department of Public Health and Environment (CDPHE),
8. Colorado Department of Corrections prison crews,
9. Hazardous waste management contractor,
10. Hazardous waste cleanup contractor,
11. Xcel Energy (public utilities services), and
12. Tow-truck operators.

Safety Management Strategies

Commonly, workplace safety programs do not sufficiently anticipate, recognize, evaluate, or mitigate the hazards and risks posed by working in urban areas. As a minimum, employers should have written policies, programs, and procedures that specifically address traditional and non-traditional hazards (in either field or facility work environments). Employee training is a must to assure that workers can anticipate and identify the types of hazards that can arise. Training must include protective measures to minimize employee potential for injury and illness, and explicit instructions for emergency readiness and response. During project/task work, risk assessment methods should be used, such as job safety analysis (JSA) or task hazard analysis (THA), plan-of-the-day, project risk assessments, pre-task safety plans, and site safety plans.

General recommendations are presented below for managing and mitigating the non-traditional health and safety hazards and risks of working in urban environments. These recommendations serve as an introduction to hazard management; they are general/minimal. Each project site, facility, and work/task situation must be evaluated for the appropriateness of these recommendations, and to determine the engineering, administrative, work practice, and equipment controls needed to protect workers.

	Hazard Category	Hazards of Working in Urban Environments	Safety Management Strategies
1.	Physical Hazards	Assault: street violence, crime, gangs Electrical: exposed underground utilities, missing electrical panel covers, live electrical connections from exposed wiring, connections to transformers Explosion: compressed cylinders, chemical reactions Fire: camp fires, wood stoves, grills, sterno, hazardous materials Weak/deteriorating structures Unstable walking/working surfaces Slips/trips/falls from poor lighting, ground features, building debris	Automation of tasks: garbage trucks, sidewalk sweepers Buddy system, team approach Clearing brush, excess foliage Use of heavy equipment: loaders, bobcats Locks on electrical panels and other equipment Security: - lighting of work and storage areas - security cameras - fencing, walls - locking of fences, vacant lots, buildings, storage areas - hired security staff - security patrols - signage and warnings - posting of personnel Fire extinguishers Designated staging areas Established paths/walkways Personal protective equipment (PPE)
2.	Chemical Hazards	Methamphetamine (meth) labs Propane gas cylinders "Death bags:" garbage or plastic bags with offgas from meth production Hazardous materials Alcohol Universal waste: mercury bulbs, asbestos, batteries	Dust suppression Securing hazardous materials Spill kits Segregation of materials Specially trained teams Hazardous waste management company (experts) Designated staging areas Personal protective equipment (PPE)
3.	Biological Hazards	Human waste: urine, feces, vomit, blood, body fluids Human contact: persons with lesions and infections Human corpses Methicillin-resistant Staphylococcus aureus	Hands off approach: "do not touch," hold away from body Use of tools, shovels, rakes Use of heavy equipment: loaders, bobcats First aid kits

		(MRSA) Needles and syringes Food waste, food-borne diseases Animal feces, carcasses Animal bites (e.g., pets, stray animals, rats) Bedbugs, lice Old landfill waste	Biohazards kit Sharps containers Dust suppression Use of sturdy bags and containers Scheduling by seasons, time of day Clearing brush, excess foliage Use of dust mask or respirator HBV vaccination Personal protective equipment (PPE)
4.	Ergonomic Hazards	Awkward postures, reaching Heavy lifting of odd-shaped and odd-sized objects Holding objects away from body Quick movements and reactions	Weight limits of bags/containers Automation of tasks Mechanical lifting devices Use of sturdy bags and containers Deliberate footing/steps
5.	Human Factors	Human interaction - Aggressive persons - Persons under the influence of substances - Persons with mental illness - Unpredictable behavior - Startled persons Gang activity Panhandlers Shoplifters Counterfeiters Intruders Stolen property Weapons Odors Crime (broad range)	Call police Call supervisor Buddy system, team approach Stay close to vehicle Verbal approach, loud voice when announcing oneself Reliable communication devices Automation of tasks Clearing brush, excess foliage Stay in close proximity to escape routes, vehicles Park away from questionable areas Do not engage with people Avoid personal encounters Do not pursue or "give chase" to shoplifters, violators, etc. Security - lighting of work and storage areas - security cameras - fencing, walls - locking of fences, vacant lots, buildings, storage areas - hire security staff - security patrols - signage - posting of personnel

Table 2. Management of non-traditional hazards facing workers in urban environments can include the methods above. This list provides an introduction to managing these types of hazards. These ideas are basic, general, and not exhaustive. Control methods must be specific to the job/task and the working environment.

In addition to hazard management methods in Table 2, the following general precautions are offered for field and facility work. This list is not exhaustive, and project/task-specific controls must be selected, based on thorough risk assessments and the hazards anticipated.

1. Prior to field/facility work:
   a. Facility plan or site-specific safety plan preparation (can encompass any of the elements below)
   b. Emergency planning: contact names/agencies, phone numbers, response actions, shooter/intruder plans, lockdown procedures, drills, plans with nearby resources
   c. Worker identification system: e.g., check-in/check-out policy, personnel scan codes on hard hats
   d. Communications: reliable devices, contacts, primary and backup plans
   e. Research and review of historical data, current area/ neighborhood data, crime rate
   f. Research of legal and liability issues, local laws
   g. Law enforcement actions required in advance, e.g., eviction preparation
   h. Permits required in advance, e.g., demo, right-of-way, asbestos abatement
   i. Site access considerations
   j. Worker orientation, training/briefing (including temporary and volunteer workers)
   k. First aid kits, bloodborne pathogens response kits
   l. Hazardous materials spill kits
   m. Decontamination supplies
2. During facility or project/task site operations:
   a. Provide designated and/or secured parking for workers
   b. Real time review of the site-specific safety plan for the expected shift/work activities
   c. JSA/THA review with crews, contractors, agency reps, volunteers
   d. Tailgate talks with crews, contractors, agency reps, volunteers
   e. Use of cell phones, radios; testing of communication plans
   f. Use of law enforcement and security personnel
   g. Use of GPS personnel locator apps
   h. PPE: e.g., high-visibility vest, steel-toed boots, Kevlar® gloves, safety glasses, hard hat, heavy jeans, respirator, face shield, Tyvek® suit/booties
   i. Use skilled supervisors and project managers
   j. Assign full-time safety professionals for oversight
   k. Schedule the work to minimize exposure: e.g., work in daylight hours, work in winter for fewer possible encounters with homeless persons
   l. Ample lighting for conditions
   m. Post signage and warnings
3. Debrief project/task management of health and safety hazards:

Exhibit 5. Workers involved with homeless encampment cleanups should wear personal protective equipment (PPE), including high-visibility vest, steel-toed boots, Kevlar® gloves, safety glasses, hard hat, heavy jeans, respirator (as needed), face shield (as needed), and Tyvek® suit/booties.

Conclusion

Employees who work in urban settings face traditional and non-traditional hazards and risks. The non-traditional safety and health hazards facing employees may not be sufficiently addressed by an organization's health and safety program or management system. Even companies who systematically use hazard and risk analysis techniques may not be adequately addressing worker exposure to unknown ("surprise") hazards. Prevention of worker harm takes place before, during, and after urban projects/tasks are conducted during field work or in facility settings. Research, pre-planning, worker training, and facility preparedness are crucial to minimize worker exposure to the hazards of working in urban environments.

Endnotes

[1] The United States Department of Commerce, Bureau of the Census. "Urban Area Criteria for the 2010 Census." *Federal Register*, Vol. 76, No. 164, Wednesday, August 24, 2011.

[2] Chamard, Sharon. 2010. "Homeless Encampments, Guide No. 56: Center for Problem-Oriented Policing." State University of New York at Albany.

[3] Ibid.

[4] U.S. Housing and Urban Development (HUD). 2008. "A Guide to Counting Unsheltered Homeless People." Second Revision.

Session No. 515

Intersection of Wearable Safety Devices, Data Collection and Cyber Policy Risk

Rachel Michael, CPE, CHSP
Ergonomics Thought Leadership
Sr. Consultant
Aon Global Risk Consulting

Introduction

Wearable technology is currently defined as: Electronics that can be worn on the body, either as an accessory or as part of material used in clothing, and are connected to a data collection device that monitors personal metrics or environmental conditions.

While the lure of these evolving technologies is undeniable, the costs of purchasing and implementing wearables can be substantial. Without careful planning, wearables might not deliver either the anticipated results or a meaningful return on investment (ROI).

Wearables are hardly a panacea for solving the spectrum of workplace safety and productivity issues. Thus, their acquisition and deployment should include an assessment of potential issues and a framework for addressing them.

Potential pitfalls with wearables include: information overload, integrating wearables into employee-focused safety systems, corporate exposures with personal data collection and protection, potential liabilities with health-related monitoring, and specific logistical issues in providing meaningful feedback.

Definitions

For the purpose of this paper, the following definitions will be used:

- Ergonomics is a body of knowledge around human ability, human limitation, and human characteristics relevant to design.

- Ergonomic design is the application of that knowledge to the design of tools, systems, and environments for safe and efficient use.

- A safety device is a device which seeks to prevent death, injury or illness or is designed to reduce the potential severity of injury (i.e., a seatbelt).

- A safety tool may be a safety device, but will require human decision-making to be effected (i.e., pulling a cord to activate a parachute).

- Wearable technology is defined here as: Electronics that can be worn on the body, either as an accessory or as part of material used in clothing, and are connected to a data collection device that monitors personal metrics or environmental conditions.

Wearable Technology Applications for Safety

A 2017 survey of wearable technology marketed as "safety" in web-based advertising descriptions found over 100 devices. There is currently no standard requiring a company to define how a device provides "safety" or interacts with an established safety program. Advertised devices ranged from ingested technology designed to monitor body temperatures to lighting worn on the body.

Many of these devices do have applications within an established safety program. Let's consider wearable technology such as gas monitors and GPS location trackers. If an employee wears this technology as part of an established safety program, what is the expected behavior from the employee? Likely, we expect the employee to remove themselves from an area that is defined as high risk. We may expect only to be able to track the location of an employee without expecting the employee to undertake any action.

Another popular category of "safety" wearables is a communication item, such a phone/radio/applications and "smart" glasses. In this case, we expect that communication of safety protocols would be the outcome of use of this type of technology. Again, a change in employee behavior is not, necessarily, the expected outcome so long as a standard protocol is followed.

A large share of the marketplace for wearable "safety" devices is dedicated to data recording and transmission accessories and jewelry. These devices are heavily marketed towards women and advertise that, in the event of a physical attack, the device will make environmental recordings and/or record location, and potentially send a distress message to designated contacts or local law enforcement. While the initial marketing push for these devices has been individuals, employers may consider these types of devices for employees who may be involved in tasks that take place in a private home (i.e., home health care) or employees that are in remote locations alone (i.e., sales representatives). It is important to note that these devices should not be used independent of a comprehensive safety program.

The final category we are considering in this paper is made up of devices that are commonly advertised as personal data monitors. These devices may advertise the following types of features:

- Monitor goals related to fitness activities
    - steps taken, stairs climbed, calories burned, etc.
- Provide status of current health
    - heart rate, blood pressure, sitting times, etc.

  - Provide status of current physical work
    - heat rate, distance covered, elevation gains, power output, etc.
  - Provide biofeedback for posture, muscle force activation, and training.

When considering these types of devices, there is an important distinction to be made between a device that facilitates direct biometric measurement, and a device that provides algorithm-based feedback to a user.

Many wearable devices measuring employee biometrics provide an algorithm-based output. This algorithm is likely proprietary to a manufacturer and, as such, will differ between device manufacturers offering the same outputs. It is also likely that a company employing wearable devices would not be privy to the underlying program thresholds or variances. While algorithm-based feedback is a viable data source, it may not be as appropriate as direct measurement for some aspects of health and safety programs.

Considerations for Wearable Technology as Part of EHS and Ergonomics Programs

Prior to implementing wearable technology, consider the potential outcomes and goals of the initiative, such as:

1. Is the goal of data collection to compare objective data gathered to a threshold?
2. Is the goal of data collection to change *individual* behavior?
3. Is the goal of data collection to change *organizational* behavior?

If the first statement is the goal, consider the aforementioned implications of direct measurement versus algorithm-based feedback. Has the risk or action threshold been validated by a party independent of the device manufacturer?

If the second statement is the goal, consider that EHS has a long history of failure with this type of implementation. As we learned with the widespread implementation of back belts, leaving a poorly designed system to be fixed by individual behavior is a recipe for a failed initiative.

If the third statement is the goal, the study should be short term with well-defined limits, and data streams mapping to supported potential future state outcomes.

Potential Pitfalls

Potential pitfalls associated with implementing wearable safety devices as part of your employee health and safety program can include:

- Information Overload
  - Consider the potential different outcomes an employer can realistically expect to achieve by investing in wearable technology. Are there other data sources that could provide the level of detail needed for decisions to be made? For example, if an employer has injury data associated with palletizing at floor level, and the

EHS team has identified risk potential in floor palletizing tasks, does measurement of employee trunk angle using a wearable device measurably affect the potential solution? A careful review of your firm's work processes might yield all the information you need to reduce injuries and improve productivity.

  - Most employers aren't set up to be research labs. Ensuring appropriate data collection, study design, and statistical analysis can add significant but necessary costs to this initiative.

- Employee-reliant Systems
  - The fundamental idea behind many wearables ultimately is to change behavior. Evidence gathered from years of progress in this area is that behavior is driven predominantly by organizational culture and systems to reinforce desired behaviors and discourage deviations.
  - If a tool to measure behavior is used, then the feedback loop–positive reinforcement, habit-making time measurement, and further reinforcement–must already be in place.
  - If the right systems aren't in place, you can demoralize employees by erroneously holding them solely responsible for accidents, incidents, or poor performance resulting from bad design.

- Privacy and Data Protection
  - There are numerous potential privacy issues associated with gathering and maintaining personal information of employees. Check with your risk management department to find out if your firm has insurance that covers the personal data on individual employees that you will be collecting and storing.
  - If you're using wearables to record vital statistics, such as heart rate, temperature, pain response, etc., that may trigger HIPAA (Health Insurance Portability and Accountability Act) compliance issues, as well as problems associated with tracking employees on anything other than a voluntary basis. In some instances, you may not be able to require employees to participate in data collection if that was not explicitly disclosed at hiring.
  - Consider data storage and access as well. How can data be stored securely? For how long? Are you obligated to share data with employees? Are employees required to share their data with employers?

- Liability Exposures
  - Some data collected through wearables may create other potentially significant exposures for employers.
  - Consider a potential expectation of live monitoring of health data and the OSHA General Duty Clause exposure potential.
  - Depending upon the nature of a data breach, criminal, cyber or professional liability insurance policies may be triggered. Ensure the Risk Management department of the organization has reviewed any wearable program for potential insurance coverage needs.

- Feedback Delays
    - Wearables provide feedback for individual employees or the system. However, if employers only review data collection at the end of the day, they may miss opportunities to provide reinforcement exactly at the time of positive behavior.
    - While real-time feedback elements can be built into any system, they typically require additional planning and resources.

Summary

Using wearable technology to collect broad data assessments has the potential to overload an organization, and may introduce additional risk. Using wearable technology to gather information on the nuance of risk to an individual, however, may allow for targeted solutions that reduce specific risk.

References

Michael, Rachel. (8/10/16) "Getting the Most from Wearables: 6 Tips. How choosing the right technology can improve safety and work processes." *EHS Today*. (http://ehstoday.com/health/getting-most-wearables-6-tips)

National Institute for Occupational Safety and Health (NIOSH). 1997. Publication 91-117, *Elements of an ergonomics program*. (http://www.cdc.gov/niosh/docs/97-117.pdf.)

National Institute for Occupational Safety and Health (NIOSH). 1997. Publication 97-141, *Musculoskeletal Disorders (MSDs) and Workplace Factors: A Critical Review of Epidemiological Evidence for Work-Related Musculoskeletal Disorders of the Neck, Upper Extremities, and Low Back.* (http://www.cdc.gov/niosh/docs/97-141/default.html.)

Punnett, L. and Bergqvist, U. "Visual Display Unit Work and Upper Extremity Musculoskeletal Disorders: A Review of Epidemiological Findings." *National Institute for Working Life* – Ergonomic Expert Committee Document No 1, 1997:16

"Wearables for workplace safety" (video interview). (1/18/16) *Business Insurance*. (http://bcove.me/rpkzif9l and http://www.businessinsurance.com/article/20160117/VIDEO/160119858/business-insurance-in-focus-video-wearable-technology-being-used-to)

Using Exposure-Based Scenarios to Enhance Hazard Recognition and Response

Jim Spigener, Senior Vice President
DEKRA Insight
Oxnard, California

Introduction

Over the last several years, organizations across the world have made great strides in safety performance. Technological advancements, holistic safety systems, and better understanding of the elements of exposure reduction have all contributed to a safer work environment. Even with all the progress, people continue to get hurt, and rates for the most serious injuries have yet to decline at the same pace as other injury types.

Some organizations are taking a proactive approach to addressing hazards. For many, their performance is good, sometimes very good. But they know to become great, they must take safety to the next level. It is not enough to establish rules and procedures around safe performance. Even when employees follow them to the letter, there are situations that can't be covered in advance, leaving individuals exposed to hazards that can arise unexpectedly.

This paper outlines a mechanism for developing a robust hazard recognition ability into the organization's toolkit. The process uses real-life, exposure-based scenarios to assist supervisors and their teams in recognizing when exposure is changing and enhancing their response to it as it arises. The process builds upon practical exposure-reduction practices to enhance teamwork and communications.

Learning Through Scenarios

There is a wealth of educational and organizational research that shows learning through scenarios or stories can have a profound and long-term impact on people. This is especially true when the learning is centered on things that are significant to learners and administered by people that have the most experience with the work. Enhancing hazard recognition and response using exposure-based scenarios allows supervisors and workers to train in risk mitigation practices that directly address the hazards they face every day.

Scenario-based learning uses practical examples to development employee and supervisory skills for addressing exposure. Safety scenarios set up a job situation, provide facts, and ask workers to identify exposures based on typical tasks. Supervisors then introduce changes to the scenario, have participants identify the new exposures, and discuss ways to mitigate and remove the barriers to working safely.

Safety scenarios can be integrated into daily discussions: safety briefings, tailgate meetings, pre-task risk assessments, and walk-arounds. This hands-on approach helps supervisors and workers partner to identify exposures and take mitigating action when exposure changes.

Supervisors are critical to the process. They are the closest representatives of management to the work that is being done, making them uniquely qualified to align leadership's values with those of the workforce. Because of this position, supervisors have the most immediate impact on improving the culture. Whether it be a shift from a production-first mentality to a view that safety is integral to the success of production or a maturation from a culture of compliance to one of that is committed to safety, mid-level leaders can drive change through how they communicate concerns, listen to subordinates, and encourage discretionary efforts.

Enhancing hazard recognition and response with scenarios should not be thought of as a safety initiative; it is a supervisor initiative related to safety. The framework it provides creates the space for supervisors to develop leadership skills while they engage workers in reducing exposures. Because they know the work the best, supervisors design and implement scenarios based on data gathered about injuries and potential incidents. This ensures that the scenarios are both relatable and meaningful to employees.

The safety scenarios set up a job situation, provide facts, and ask workers to identify exposures based on typical tasks. Supervisors then introduce changes to the scenario and have participants identify the new exposures and discuss ways to remove the barriers to safe work. The exercise brings key organizational development opportunities into focus and trains workers to be active observers looking out for increased risks.

Safety scenarios can be done anytime and anywhere. Supervisors and workers can talk through scenarios during before they begin work, when they are pausing to address changes in exposure, or when debriefing after the work is done. This hands-on approach helps supervisors uncover barriers to safe work, while targeting those employees that aren't yet fully engaged in safety. Scenario-based learning is an effective way to develop adaptive workforces that can identify and manage changes in exposure, while enhancing problem solving and team building.

When people are fluent at identifying exposures and detecting changes, they are more effective at protecting themselves, others, and the organization itself. Creating a workforce that is dedicated to safety and aligned with organizational goals around exposure reduction takes more than a training seminar or solitary initiative. Becoming great at safety is an iterative process of learning and shaping responses to exposures. Through a collective focus, open communication, and safety-building exercises, the process helps people build a culture that is proactive, engaged, and effective at problem solving. The process is transferable; it provides a model for developing a learning organization that can be applied beyond safety—to any problem the company encounters.

Why Exposure-based Scenarios?

For some leaders, there comes a point in their effort to improve safety where the on-the-ground facts are not lining up with the goals and values of the organization. They may have found initial success with a behavior-based safety program that engaged employees and involved site leaders in the safety process. Perhaps they saw a dramatic improvement in injury rates and performance. But as time went on, the gains plateaued, people continued to be exposed, and injuries didn't stop. Their safety effort had slowed to a crawl.

The problem, in many cases, is not in the efforts of the people on the ground or in their leaders who are truly dedicated to the safety of their teams. The problem is that exposure is never static. It changes and shifts with the introduction of new equipment or technology, new hires, and new tasks. Exposure, like a live workplace, is always changing—and that means people need to be able to change with it and recognize when new risks emerge.

To navigate the complexities of a shifting and fluid work environment, where safety issues don't always arise as experts predict, takes leadership—both at the executive and supervisory levels—and a workforce that is adaptive and actively engaged in eliminating risks. For many organizations, this means implementing an exposure reduction process that helps supervisors involve employees in identifying exposures and addressing them as they change. Such a process incorporates scenario-based exercises that help supervisors be better leaders, cultivates a stronger safety culture, and trains workers to know how to recognize, speak up about, and adjust to changes in exposure.

Employees often do not recognize changes in exposure or fail to respond to developments that might lead to an incident because they are not trained to. In some cases, frontline workers have become desensitized to the risks they work with every day. While workers understand the job hazards and dangers of the chemicals, substances, and equipment they are handling, they can still become overly comfortable with the risks. Tasks and activities are often routine and so familiar that they require little of the workers' attention, leading them to believe nothing is going to go wrong; after all, they've done things the same way for years without incident. When employees become comfortable with risks, they miss changes in exposure that can lead to serious injuries or worse.

Even when workers recognize the signals that indicate a change in exposure, the legacy of how a job is performed or the culture of the organization can drive them to "push ahead" and work through the issue without pausing to assess emerging dangers. This type of single-minded approach is usually the outcome of a culture that promotes production over safety. In an exposure reduction process, supervisors play a leading role because they are the best positioned to align workers with the organization's values. When executives, supervisors, and workers are all on the same page with regard to safety, employees will be more willing to speak up about a minor problem before it can become a major incident.

A Culture of Exposure Reduction

The goal of any safety improvement effort is to build a culture that has a persistent focus on reducing exposures in order to minimize the number of incidents and injuries. This type of culture engages employees, has supervisors who are effective at leading safety, and is driven by executives who truly value the welfare of everyone in the organization. Specifically, a workplace culture embraces reducing exposures by:

- Respecting and acknowledging current hazards
- Recognizing when exposure changes
- Responding in a consistent and safety-focused manner
- Removing barriers (personal, equipment, conditions) to hazard recognition and safe work

Respecting and Acknowledging Hazards

Although it may seem like common sense, it cannot be stressed enough: Potential dangers should never be ignored. Unfortunately, many people fail to maintain sufficient vigilance in the workplace because they believe nothing will happen to them. Sometimes, they are in a rush to get the work done and take shortcuts that expose them to serious injury. It is vital that employees recognize, acknowledge, and respect the hazards that exist in the line of work. Leaders should ensure that employees are trained to be aware of workplace risks and that they treat them with the highest degree of caution at all times. Respecting and acknowledging hazards is the foundation of effective exposure-reduction practice.

Recognize When Exposure Changes

Rules and procedures can go a long way to encourage safe behaviors and ensure the security of facilities and equipment. But even when employees follow the rules to the letter, they may struggle to recognize the subtle changes in exposure. Being good at exposure reduction takes something more than rules alone can accomplish. It requires every person on the shop floor to maintain vigilance consistently during their work. It also takes a culture that supports open communication and the free exchange of ideas about ways to reduced exposure. There are three key areas vigilant employees pay attention to with regard to recognizing changes in exposure:

- Equipment: Employees should be alert for pumps running louder than usual, temperature rising too quickly, plugged equipment, and so on.
- Condition: Workers should keep an eye out for new water or liquid on the floor, daylight to darkness changes that affect vision, and emergency situations.
- Personnel: Veteran employees know that a new employee may be unaware of some the risks in the workplace or may be unfamiliar with all safe work practices. Sometimes, during the middle of a job, an experienced employee is called away. This removes an experienced set of eyes from the job, leaving those remaining susceptible to new exposure.

Responding to Exposures

When workers identify a change in exposure that suggests heightened risk, the first thing they should do is pause the job, and take a time out to evaluate current conditions. Stop-work authority is one of the most important powers given to employees. Leaders should encourage its practice by supporting workers who pause work to evaluate risk. After the job is paused, employees can discuss concerns with fellow workers, and develop a plan to mitigate potential risks. Supervisors support this practice by having two-way discussions with workers, listening to their concerns, and being open to the solutions they suggest.

Removing Barriers

Removing barriers is about addressing changes in equipment, conditions, and personnel. This could include altering the way the work is done, using personal protection equipment, or acquiring outside help if the exposure is beyond the workers' control. Leaders support this practice by first ensuring that workers know their concerns will be listened to. If a piece of equipment is damaged and using it puts people at risk, workers should never feel hesitant to inform their supervisor, and the supervisor should never balk at doing what needs to be done to position his team to perform their tasks safely, regardless of production schedules or time constraints.

Leadership Practices that Advance the Process

Enhancing hazard recognition and response with exposure-based scenarios fosters a working environment where safety is at the forefront of people's minds. They routinely talk about safety and proactively root out exposures. In many industries, these exposures include serious risks that can lead to catastrophic events and life-altering injuries. There are three practices both executives and mid-level leaders can adopt to bolster their exposure reduction process.

Anticipation

Staying out in front of exposures requires a culture that is sensitive to the weak signals that indicate potential for catastrophic events. Leaders must ensure they have data-collection methods in place that capture information from a variety of sources. The data should be integrated with other information to provide a truer picture of on-the-ground facts. Leaders should also encourage employees to report any indication of increased risk. Supervisors need to foster an environment where people readily report near misses and other "non-events" that potentially could have been a serious incident. Supervisors who practice anticipation might ask some of the following questions:

- There's been a sound coming from a pump for some time; is there a risk present we haven't investigated?
- Given this crew's experience, what gaps in knowledge are there that could increase the risk of something going wrong?
- Given the amount of time it will take to complete this job, are we better off starting it with a fresh crew in the morning?

Inquiry

Inquiry measures perceptions of how leaders mitigate bias by seeking a wide range of ideas and opinions and by listening to and showing respect for dissenting opinions. Inquiry minimizes the chances cognitive bias and "group think" will lead to poor safety decisions. Promoting open dialogue and seriously considering dissenting opinions helps avoid normalization of deviance and decisions based on flawed and unstated premises. Supervisors who practice inquiry develop sensitivity to subtle changes and learn how to ask the right questions in the right ways to avoid biases. For example, a supervisor might ask a worker:

- Before I share what I believe our issues are, what do you think about the safety of this plan?
- What do you think might cause us problems?

- If I ask you to play the devil's advocate, what would you say to challenge my thinking?

Resilience

Resilience refers to the organization's and individual supervisor's capacity to remain levelheaded during a crisis to prevent it from escalating into a catastrophe. This can be done by emphasizing employee readiness, preparation, flexibility, decision-making ability, and authority to take action. These behaviors relate to the expectations established around how decisions are made in non-routine situations and the way the organization reacts in the aftermath of those decisions. To cultivate resilience, supervisors might ask:

- What will you do if it starts to rain and gets dark?
- How will you protect workers if the job requires them to be in the line-of-fire of a crane or in a cramped space?
- If the second pump fails, what do you do?

Conclusion

Using exposure-based scenarios to enhance hazard recognition and response helps leaders build a knowledgeable and flexible workforce that can identify and address risks far upstream of injuries. Through scenario building, training, and guidance coaching, supervisors learn to get the most out of their teams while becoming better leaders and creating a stronger culture. Through this training, organizations develop enhanced communications among employees and between employees and supervisors. They build better teamwork and a more engaged workforce. Scenario-based learning increases employee capabilities and skills that strengthen their ability to uncover mechanisms for identifying and dealing with exposure. This approach contributes to the development of a learning organization, where employee, supervisor, and organizational values are aligned across all levels.

Bibliography

Duncan, C. 2012. *The Zero Index: A Path to Sustainable Safety Excellence.* Ojai, CA: Safety in Action Press.

Krause, T.R. 2005. *Leading with Safety.* Hoboken: Wiley Interscience.

Krause, Thomas. 2004. "Influencing the Behavior of Senior Leaders in EHS." *Professional Safety.* June edition.

Stricoff, S. and Groover, D. 2012. *The Manager's Guide to Workplace Safety.* Ojai, CA: Safety in Action Press.

Session No. 518

How to Understand and Interpret ANSI Z359.2 to Develop a Fall Protection Program

Kevin Denis
Gravitec Systems, Inc.
Poulsbo, WA

Since 2007, ANSI Z359.2 has been a guidance document for employers to establish a fall protection program. Changes to the document in 2017 are intended to make the standard more manageable and complete, hopefully increasing the safety of people who work at height.

Developing and implementing a comprehensive fall protection program is the most effective tool for employers to deal with fall hazards. Most organizations that have workers at height have some type of fall protection program already. Usually, equipment has been purchased and training has been conducted. Unfortunately, fall-related accidents continue to hold steady, and in some jurisdictions, increase. Z359.2 is another tool in the battle against fall-related accidents.

The purpose of ANSI Z359.2 is to establish criteria and requirements for an employer's fall protection program. Included in Z359.2 are the policies, duties and responsibilities, training, survey and identification of fall hazards, fall protection procedures, eliminating or controlling fall hazards, rescue procedures, program implementation, incident investigation, and evaluating program effectiveness.

If you have already been using or are familiar with Z359.2, the changes in the 2017 version are subtle. The greatest change is a reorganization of the content so the document is more user-friendly and assembled in a logical order. Another change is the inclusion of clear and succinct language as to what an employer must do to be in compliance with the standard. Training requirements have been reorganized to be clear, but also flexible to the needs of employers.

The standard has been written to reflect the order of steps that an employer uses to establish a fall protection program. This illustration is representative of how the standard is organized.

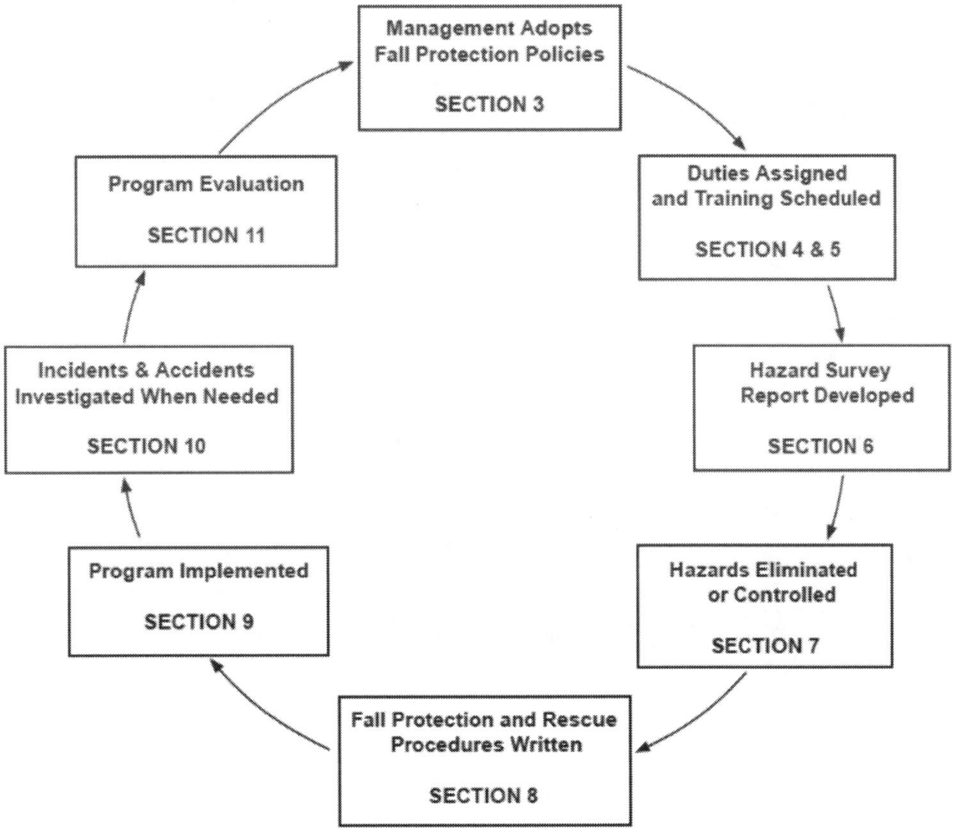

Figure 1. Excerpt from ANSI Z359.2-2017 (Reproduced with permission from ASSE.)

Reorganization of the document should make the standard easier to understand and read. Employers will find the standard establishes what you need to include in your fall protection program in an easy-to-understand format and order. Understanding how to comply with the standard is straightforward.

The revision of Z359.2 has also taken into account the unique training needs of various employers. Program administrators are now able to teach content based on their own site's needs. Basic training that is applicable to all is still required, but once completed, program administrators can pick from multiple topics and skills catered to their jobsite needs. This allows program administrators flexibility in the education they develop for their training. Program administrators will develop training programs based on a needs assessment, not prescriptive listings from the standard.

Users will also find there is clarification on the selection of anchorages. When selecting an anchor, users should seek out ones that have been proven through physical testing or engineering analysis, but when this isn't possible, the standard provides guidance on how to use authorized, competent, and qualified people to select anchorages.

The hierarchy of fall protection has also been slightly modified. There was a significant amount of confusion and discussion surrounding the term "administrative controls." Generally speaking, when this term is used in fall protection circles, it refers to a fall protection control that is compliant, but does not supply a physical secondary system, such as a safety monitor, controlled access zones, qualified climbers, and fall protection plans, which means exposure to a fall hazard still exists. These control measures are outside the scope of this standard, and the terminology used to describe the hierarchy of fall hazard controls has been modified slightly to reflect these changes.

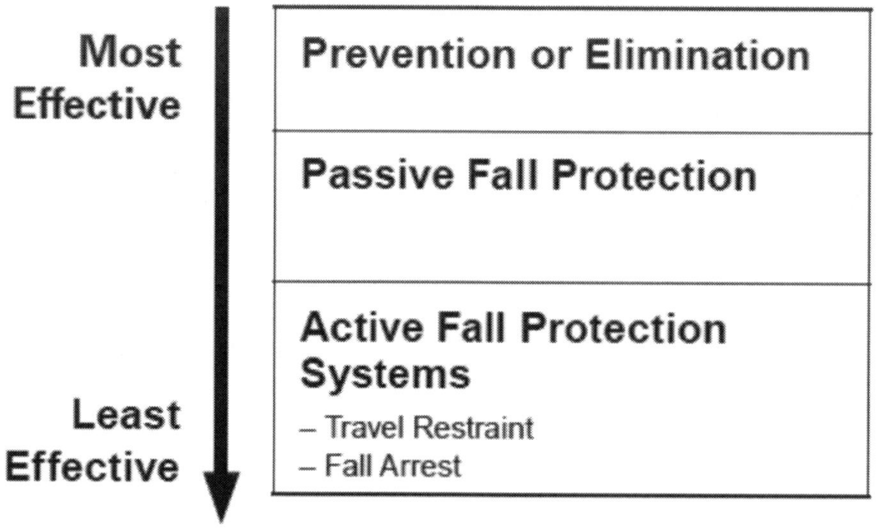

Figure 2. Excerpt from ANSI Z359.2-2017. (Reproduced with permission from ASSE)

The standard has been broken into sections that are representative of how employers establish safety programs. Those sections are:

Section 1: Scope, Purpose, Applications, Exceptions and Interpretations
Section 2: Definitions
Section 3: Program Requirements
Section 4: Duties and Responsibilities
Section 5: Training and Evaluations
Section 6: Fall Hazard Survey
Section 7: Eliminating and Controlling Fall Hazards
Section 8: Minimum Requirements for Fall Protection and Rescue Procedures
Section 9: Fall Protection Program Implementation and Management
Section 10: Incident Investigations
Section 11: Evaluating Program Effectiveness

Sections 3, 4 and 5 have had the most significant changes to them. An excerpt from Section 3 is as follows;

> *3.1.3 The fall protection program shall establish requirements and assign responsibilities for the identification, evaluation, elimination or control of every existing or foreseeable fall hazard in the employees' workplace environment, including:*
>
> *– Establishing employer responsibilities, in accordance with section 3.2.*
> *– Surveying and documenting fall hazards, in accordance with section 6.*
> *– Preparing written fall protection and rescue procedures in accordance with section 8.*
> *– Implementation and management of requirements in accordance with section 9.*
>
> *These requirements shall conform to the fall hazard hierarchy of controls specified in Section 7.*

The intent of the standard is every employer has an objective to reduce fall-related accidents of employees (safety policy). Once this objective is established, employers will assign the appropriate resources (time and people) to mitigate fall hazards encountered (identification of personnel). Once the infrastructure is in place, the workplace conditions can be surveyed, and all existing and potential fall hazards can be identified (fall hazard survey). From the survey, solutions to every fall hazard can be determined by using the hierarchy of fall hazard controls. Once the best solution is determined, written procedures and rescue plans can be made and the last element is to maintain the program.

Every location and employer will have unique needs, but ANSI Z359.2 is structured in such a manner that all employers can take advantage of its guidance and create or maintain an effective fall protection program.

Highlights of the New ANSI/ASSE Z244.1 Standard, Control of Hazardous Energy Lockout/Tagout and Alternative Methods

Todd C. Grover
Global Sr. Manager
Applied Safety Solutions
The Master Lock Company
Milwaukee, WI

Introduction

In the cumulative experience of author is that, to date, the majority of U.S. employers still struggle to maintain a consistently effective lockout/tagout protocol with their workers, contractors, and vendors. Driven by the limited guidance provided by regulatory requirements only, compliance is very difficult to routinely achieve.

The latest version of the ANSI/ASSE Control of Hazardous Energy Lockout, Tagout and Alternative Methods standard, released in December 2016, is a well-resourced and progressive look at how to include well-described energy control practices into daily productive operations. This appeals to employers who seek to understand how to improve their energy-related protective practices and resonates with people whose work exposes them to the hazards of sudden machine startup. The newly revised Z244 standard speaks to these needs by offering comprehensive information on the latest methodology and how to accomplish across all industries, and especially in your workplace.

How the New ANSI Z244 Standard influences the Practice of Lockout/Tagout and Alternative Methods

Certainly, the methods of protecting workers against the sudden startup of machinery have evolved greatly over the years. The most often referenced source of lockout/tagout information is OSHA's 29 CFR1910.147 regulation, which came out in 1989. It was based heavily on ANSI's original Z244.1 Lockout Standard first published in 1982. We have come a long way since then in terms of technology and new methods, but there certainly is a long way to go. Each year OSHA publishes its Top 10 Most Cited Violations and again, for Fiscal 2016, lockout was ranked fifth (with very similar outcomes as 2015) in terms of the particular rules that were cited and value of the citations issued. Heightened self-reporting requirements for serious injuries and fatalities are bringing more violations to OSHA's attention, and it seems that many U.S. employers are coming to an understanding that these types of accidents continue to occur with significant frequency and often with great severity.

ANSI/ASSE Z244.1 (2016), The Control of Hazardous Energy–Lockout, Tagout and Alternative Methods, seeks to address these trends with the most significant expansion of best practice guidance since the 2003 Z244 Standard revision was published. Although OSHA has clarified their position on elements of the 1910.147 regulation through letters of interpretation, no definitive updates or substantial changes have been issued in the more than 25 years the law has been in effect. The American National Standards Institute (ANSI), on the other hand, has looked closely at the progression of technology and solutions, as represented by a wide range of industries, and has reaffirmed or revised Z244.1 every five years to continuously improve the application of this important safety practice. This serves as an effort to better protect people doing potentially hazardous work on machinery, equipment or processes.

The Z244 committee agrees with OSHA: Workers should be protected from injury due to unexpected equipment startup or release of potentially hazardous energy. However, the committee did not try to align fully with every historic OSHA compliance requirement. Instead, they carefully took into consideration the improvements made in safety-rated technology and other advancements in energy control methods proven by experience to effectively provide protection beyond basic lockout regulatory requirements. This offers enhanced opportunities to meet the needs of technology-driven and fast-paced modern workplaces without compromising the safety of personnel working hard to keep the wheels of industry turning.

The committee has made deliberate efforts not to make major changes in recommended practice that would create completely new obligations for employers and participating personnel. Instead, they focused on addressing the right thing to do to prevent sudden startup accidents based on the current knowledge available to us. The intent has been to offer a plain-language approach that describes how a range of well-proven energy control practices can be accomplished and adapted to a wide range of industries. The 2016 standard was written to be scalable to meet the needs of both large and small companies with varying levels of technology present in their machinery and processes. Sections 1-3 respectively address Scope and Purpose, References and Definitions. They are not commented on here for reasons of brevity. Important new and expanded information provided in the 2016 revision of Z244.1 is covered in the following sections.

Section 4: Responsibilities

A reference is given to **supplier,** which is defined in the 2016 standard as: "An entity that designs, redesigns, fabricates, modifies, integrates, assembles or supplies machines, equipment or processes." The section directs suppliers to provide the users of machinery provided to them, with equipment meeting the requirements of this section, addressed below. It further compels users to obtain machinery equipment and processes that complies with Section 5.

Section 5: Design of Machinery/Equipment for the Control of Hazardous Energy

The advancement of technology and engineering options in new machinery construction has resulted in a heightened capability for machinery to be dependably controlled during servicing operations. This section provides guidance on roles and responsibilities of suppliers and builders to design for integral lockability, tamper resistance, and the use of control-reliable safeguarding technology when applicable. Suppliers are compelled by the standard to document safety systems, provide risk assessments, and offer procedures that can justify alternative methods of protection if necessary, and how risk exposures can be minimized during routine and repetitive servicing, repair, and maintenance activities.

This results in end users being better prepared to finalize their own safe work procedures for personnel operating the equipment. Machinery manufacturers, integrators, modifiers, and remanufacturers are encouraged to use the tenants of Section 5 to enhance the value of their product offerings to customers who are looking for engineering safeguards and information to be part of their purchase considerations. End users purchasing new or modified equipment can also use the guidance in section 5 in their specifications to make sure equipment is designed to be readily isolated, securable, and ready to be serviced safely and efficiently.

There is significant detail provided on the location and identification requirements for energy isolating devices. This includes providing the capability to be locked or otherwise secured in an isolating position. And there are documentation requirements for suppliers to provide users with relevant procedural information and instructions along with posting any special warnings. When stored or residual energy presents a risk, the ability to effectively release or dissipate these energy hazards must be incorporated into the design of the machinery.

Section 6: The Hazardous Energy Control Program

The importance of a well thought-out and documented lockout program is broadened by the additional information provided in this section. Section 6.3 provides detail on the required elements of the control program, including identifying responsibilities, tasks, energy sources and isolation devices. Operational procedures and implementation requirements for hardware, as well as communications and training, are covered in terms of content expectations. And a review process is provided to periodically determine the effectivity of the Hazardous Energy Control Program and the need for continuous improvement where indicated. Figure 1, as shown below, is provided as a decision tree for implementing a fully resourced program.

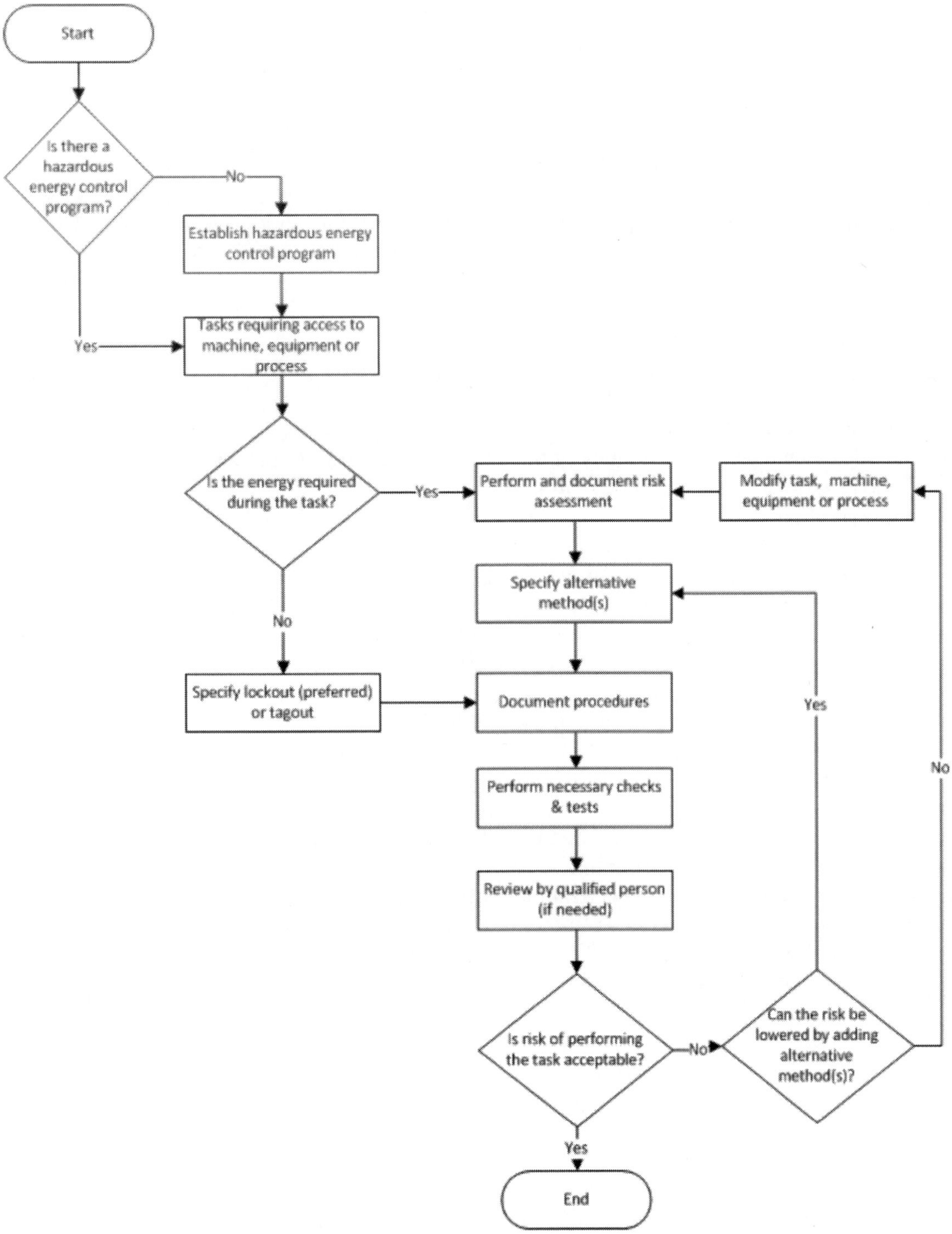

Figure 1. Flowchart for Controlling Hazardous Energy

(Source: ANSI/ASSE Z244.1-2016 – The Control of Hazardous Energy. Lockout, Tagout and Alternative Methods Section 6.3 Page 28)

Methods for managing change for new or modified equipment in the workplace and ways of timely updating and document control are noteworthy additions in Section 6. This assures analysis is performed prior to startup to allow for proper engineering and personnel preparation. Also benefitting are new or revised operating procedures, maintenance requirements, lockout instructions, training information, and tasks that may require alternative procedures.

Section 7: Control of Hazardous Energy

This section has been expanded significantly to discuss improved hazardous energy control procedures, the identification of energy-isolating devices, verification by testing of the effectiveness of de-energization, as well as new provisions for interrupting energy isolation for positioning and testing requirements. Lockout and tagout practices are clearly laid out in 7.1.

Lockout is required in the following situations:

- When no risk assessment has been completed.
- When the tasks and/or hazards are unknown.

Lockout or tagout is required:

- During assembly and disassembly of the machine, equipment or process.
- When energy is not required to perform the task.
- For major repairs.

Section 7.2 addresses the detailed requirements of hazardous energy control procedures. A clear explanation of when equipment-specific procedures are required is provided, along with the elements that make up a well prepared lockout or tagout procedure. Procedure management is also discussed as a means of providing timely access to the most up-to-date information. Information is also offered on protective hardware and tags, as well as transfer of responsibility between work shifts, including the use of transfer locks.

Section 7.6 continues with step-by-step guidance on the elements to apply lockout/tagout, including the use of warning placards where applicable. Details regarding partial energization of machinery as required for operational and maintenance tasks are allowed, provided no danger is posed to the individuals performing work on the equipment. The section continues with information regarding partial energization when some energy is required to be present in machinery, equipment or processes in order to complete the work assignment (this is also further addressed in Section 8 of the standard). Section 7.6 also covers the processes of verifying isolation has been successfully achieved, and returning equipment back to service.

Section 7.9 contains an expanded discussion of working with outside services or contractors. Details regarding notification of hazards, and coordinating contractor programs with host programs, are given for integrated tasks. Group lockout techniques are covered in Section 7.10, and complex group lockout guidance is covered in 7.11. Both reference extensive technical discussion in the ANSI/ASSE Z244.1 annexes I and J, which draw information from the OSHA instruction to their compliance officers, titled CPL 02-00-147, *The Control of Hazardous Energy – Enforcement Policy and Inspection Procedures*.

Section 8: Alternative Methods

This section expands guidance beyond OSHA's regulatory limitation to tasks that are routine, repetitive, and integral to production operations. The Z244 Committee recognized that there are also many maintenance, repair or service work assignments where energy may need to be present to perform the job properly that do not allow for the full application of isolation and lockout. The outcome was that competent evaluation methods completed specific to the task being performed are the best method of determining the feasibility of applying full energy isolation techniques or justifying dependable alternative methods to lockout or tagout.

The Z244 Committee's position is made clear, lockout (or tagout) shall be used unless the user can demonstrate that a well-established alternative method will provide effective protection. In situations where the task is not well understood or the risk not assessed, lockout shall be the default protective measure applied to control machinery or processes. Section 8.2.1 specifies that alternative methods shall only be used after hazards have been assessed and documented through the application of a practicability (or justification) study. This results in a first-step decision as to whether a risk assessment is reasonable to perform as the next step in the determination process.

Once a decision to move forward has been justified, Section 8.2.2 requires that a risk assessment process be completed to determine that the techniques used will result in a negligible risk or no risk for sudden start up. A description of multiple methods of risk reduction methods is found in this section as follows:

- Industry best practices/methods (8.2.3);
- Architecture/structure (8.2.4);
- Using well-tried components (8.2.5);
- Using well-tried designs (8.2.6);
- Common cause failure (8.2.7);
- Fault tolerance (8.2.8);
- Exclusivity of control (8.2.9);
- Tamper resistance (8.2.10);
- Program to support (8.2.11);
- Procedures in place (8.2.12);
- Periodic checking and testing (8.2.13);
- Review by a qualified person (8.2.14).

These assessment methods are intended to provide supportive details on suitable control measures that confirm whether or not reliance can be placed on technically reliable safeguarding methods to engineer-out hazard exposures. These engineering controls are considered to potentially be an improvement over the procedural measures provided by the behavior-reliant application of lockout techniques to significantly reduce the likelihood of human error. Following the hierarchy of control model, ANSI/ASSE Z244.1 (2016) provides detailed guidance on if, when, and how a range of alternative control methods can be applied to result in equal or improved protection for people performing specific tasks.

Additionally, the new standard offers over two dozen informational annexes that provide examples of risk assessment techniques, sample policies and lockout/tagout procedures, group lockout guidance, and inspection information. Alternative risk-reduction methodology is covered in detail specific to a number of new technologies including the packaging, pharmaceutical, plastics, printing, and steel industries; semiconductor and robotic applications, and others challenged by the current regulatory limitations.

The author reflects the position of the Z244 Committee that injuries and fatalities related to unexpected startup of machinery or release of hazardous energy are preventable while recognizing that zero risk is only a theoretical possibility. The better we understand the tasks that are associated with potentially hazardous energy exposures and that can be protected by the use of hazard elimination, engineered controls, procedures and training, the better we can more reliably move risk to the lowest possible levels. The newly revised ANSI/ASSE Z244.1 Standard provides the latest information on achieving acceptable levels of risk by knowledgeably applying conventional lockout, tagout or well-determined, alternative methods to prevent these avoidable accidents from occurring.

Endnotes

[1] The ideas, opinions and summary of content of the ANSI/ASSE Z244.1 (2016) Standard expressed in this technical paper reflect the opinion and experience of the author and cannot be implied to represent a consensus of the thoughts of the many organizations and individuals that also participated in the Z244.1 Standard revision process. Best-practice guidance is subject to interpretation, and the author encourages those interested in this standard to obtain a copy and explore the possibilities and applications they believe will best suit the needs of their working environment. Every ANSI committee encourages feedback from those who put these standards into practice and encourages participation in the continuous improvement process via committee membership and other valuable contributions.

[2] Permission granted from ASSE, as secretariat of ANSI Z244.1, to use Figure 1 from Section 6.3 of the Standard.

Bibliography

American National Standards Institute/American Society of Safety Engineers (ANSI/ASSE).2016. *The Control of Hazardous Energy. Lockout, Tagout and Alternative Methods.* ANSI/ASSE Z244.1-2016. Park Ridge, IL: ASSE.

American National Standards Institute (ANSI). *Lockout/Tagout of Energy Sources – Minimum Safety Requirements.* ANSI Z244.1-1982.

National Safety Council (NSC). "OSHA's Top 10 Most Cited Violations for 2016." *Safety and Health Magazine* (http://www.safetyandhealthmagazine.com /articles/14927-2016-oshas-top-10-most-cited-violations). December 2016 Issue.

Occupational Safety and Health Administration. *The Control of Hazardous Energy – Enforcement Policy and Inspection Procedure.* CPL 02-00-147, Chapters 4-7 through 4-14.

Occupational Safety and Health Administration. Title 29 CFR 1910.147. *The control of hazardous energy (lockout/tagout).* December 2016.

Session No. 522

Auto Safety and DOT Compliance

Charlie Halfen
Transportation Safety Practice Leader
Chubb ESIS
Houston, TX

Introduction

The safety of employees and the general public should always be the highest priority of any organization. Workplace injuries, illnesses and auto crashes take a tremendous toll on any organization. The prevention of injuries and auto accidents has been an ongoing battle for years. Auto crashes still account for the highest number of severe injuries and fatalities in the workplace and are the leading cause of death from the ages of 5 to 35 throughout the world. Without a comprehensive safety process, injuries, accidents, and fatality trends will go up and down over time with no consistent improvement or solution.

According to the "2015 Motor Vehicle Crashes: Overview," published by the U.S. Department of Transportation, our nation lost 35,092 people in crashes on U.S. roadways during 2015, an increase from 32,744 in 2014. The 7.2 percent increase is the largest percentage increase in nearly 50 years. The estimated number of people injured on U.S. roads increased in 2015 to 2.44 million injured people. The estimated number of police-reported crashes increased by 3.8 percent, from 6.0 to 6.3 million.

In the Bureau of Labor Statistics' "Census of Fatal Occupational Injuries Summary, 2015," it was noted that roadway incident fatalities were up nine percent from 2014 totals, accounting for over one quarter of the fatal occupational injuries in 2015. Heavy and tractor-trailer truck drivers recorded the most of any occupation.

Many consider auto accidents to be based on chance or bad luck. Nothing is further from the truth. Accidents can be reduced by the elimination of unsafe driving behaviors, and fatal accidents can only be decreased by the reduction of all accidents. While changing these long-entrenched mindsets is not easy, business results will be sustainable once a methods-based safety culture is implemented.

A world-class organization builds and maintains safety excellence by incorporating a methods-based safety culture throughout all levels of the organization. For a safety culture to flow smoothly, **management commitment** is established with a diagnosis and assessment of the current structure and the formulation of policies, procedures, and methods. **Communication and training** will ensure company-wide knowledge and education. **Employee involvement** must be encouraged through company-sponsored activities and employee recognition. **Tracking and analysis** is needed to monitor the effectiveness of the process. Workers' compensation, auto and general liability must be closely monitored. The flow of the safety process leads to **accountability** at each level of the organization. Taken in its entirety, the safety culture will ultimately lead back to a review of management's commitment to the prevention of injuries and auto crashes. Workplace safety is essential to any organization's success.

While the reduction of injuries and crashes will always begin with management's commitment, the organization must communicate the policies, procedures and methods so that everyone understands the expectations. Training must follow immediately to ensure that all employees possess the adequate skills and knowledge to perform their assigned tasks.

Everyone learns at a different pace. There is no correct amount of training for all employees. To be sure that training is effective and begins to develop into a habit, let the following quote be your guide:

> What I hear, I forget;
> What I see, I remember;
> But what I do I understand.

The basic management skill of demonstrating the method and accountability for its performance will always produce the greatest comprehension. Awareness and education are also important ingredients in the prevention of injuries and auto accidents.

A comprehensive communication plan should be created utilizing daily, weekly, and monthly publications. This standardized safe driving material should be sent to all driver locations and posted to an internal web address. Topics would include safe driving methods and regulatory compliance, as well as current issues.

Safe Driving Methods

The goal of safe driver training and education is to have everyone understand what is needed to prevent an automotive crash. While all driver training, whether computer-based, classroom or hands-on, works to some degree, drivers need on-road training. Mental and physical habits, such as driving, cannot be taught in a classroom. Much like any athletic sport, safe driving calls for continuous improvement, so even the most experienced drivers need retraining on-road in the safest driving methods.

Careless drivers take more chances and are more likely to be involved in auto crashes. They have learned their driving habits through their own experiences and will continue to take those same chances until they have a significant emotional event (a crash where they accept responsibility), or they receive instructional training to correct their bad methods.

Companies can no longer accept the liability that comes with a poor driver. Trainers must have the tools to be successful in training safe driving methods. Driver communications with classroom and on-road training should be developed for the introduction and acceptance of a Record of Safety Ride.

While on-road methods training can be started in a classroom, the safe driving methods that the driver will use will be set behind the wheel. Records of safety rides are used to observe and correct the driver's ability to recognize the hazards faced while driving. A good trainer should take the wheel during this ride and show the student how to keep their eyes scanning, recognizing those hazards that may affect their travel path, and how to react accordingly. It is important that constructive feedback on the mental and physical driving skills demonstrated by the driver should be given, since there will always be areas for commitments to improve.

The behind-the-wheel Record of Safety Ride should concentrate on the hazards the driver will face on road. First, distractions are the number one cause of all crashes. Driving requires your undivided attention from start to finish. In other words, if we just pay attention and look for hazards while driving, we cut our chance of having a crash in half. Second, the most common crashes are backing and hitting stationary objects. Learn how to handle the vehicle. Pay attention even when doing the simplest maneuver such as parking. The same is true with backing. No one feels comfortable moving backwards but it's something we do every time we drive. Learn to use the mirrors. Third, the most serious injury-producing crashes are intersection, head-on, hitting another vehicle in the rear, pedestrians, and cyclists. While any crash can be deadly, these crashes cause the most injuries. With limited resources, organizations should concentrate training in these most dangerous and most common areas.

When reviewed with the driver, the Record of Safety Ride can be the first step in accountability for correct methods use before a crash occurs. Using this easy-to-use form, drivers can be observed at any time of day by any company employee.

Like injury prevention methods, not using safe driving methods will result in more auto crashes. Managers, supervisors and trainers should monitor and track the unsafe driving behaviors of their drivers. Management must also "Walk the Talk" by regularly communicating and demonstrating their commitment to safe driving. An employee will never be accountable to follow the safe methods by an immediate supervisor who does not follow those same methods. The key for management is to never overlook an unsafe act, whether in the building or behind the wheel. Silence is acceptance.

DOT Compliance

Every commercial motor vehicle must be operated in accordance with the laws and regulations of the jurisdiction in which it is operated. The Department of Transportation (DOT), through the Federal Motor Carrier Safety Administration (FMCSA), continues to hold motor carriers responsible for driver behavior and job performance.

The Compliance, Safety, and Accountability (CSA) initiative is the FMCSA enforcement and compliance program used to reduce vehicle crashes, injuries, and fatalities. The program leverages the use of technology by using safety performance data collected during roadside stops

and through crash reports to identify unsafe carriers and drivers and to identify safety performance problems. Within the CSA, the safety measurement system (SMS) quantifies the on-road performance of carriers and drivers to identify candidates for intervention, determine the specific safety problems exhibited, and to monitor whether safety concerns are improving.
The SMS uses a motor carrier's data from roadside inspections, including all safety-based violations, State-reported crashes, and the federal motor carrier census to quantify performance in the following seven **B**ehavior **A**nalysis and **S**afety **I**mprovement **C**ategories (BASICs):

- **Unsafe Driving:** Employers are held responsible for the poor driving habits of their drivers.

- **Hours of Service (HOS):** HOS regulations were developed to get tired drivers off the road before serious accidents could occur. Both employers and drivers have an obligation to comply. The roles and responsibilities of all personnel as they pertain to HOS policies and procedures should be reviewed in new-hire, initial and refresher training.

- **Driver Fitness:** Motor carriers are required by law to use qualified drivers to move passengers and goods. It is also in the best interest of the carrier, since a safe, dependable driver can be a top asset to an organization.

- **Controlled Substances/Alcohol**: Carriers must develop written company policies incorporating all regulations on controlled substances and alcohol use, testing, required training, and record retention for all employees.

- **Vehicle Maintenance**: A carrier's inspection and maintenance programs are critical to a carrier's on-road safety process. Vehicle maintenance helps to prevent crashes caused by vehicle deficiencies. Worn, failed or incorrectly adjusted components can cause or contribute to an incident. Preventive maintenance and periodic inspection procedures can help prevent mechanical failures from occurring while vehicles are being operated. Required inspection procedures and documentation assist the process.

- **Hazardous Material Compliance**: Transporting hazardous materials (HazMats) can be extremely dangerous, not only to the driver but the general public. The general public is subject to a greater safety risk if HazMats are involved in a motor carrier crash, and any unmarked or poorly marked cargo can result in a less effective emergency response, leading to injuries or fatalities for emergency responders.

- **Crash Indicator**: Maintain high standards when it comes to crash avoidance. An auto crash is any occurrence in which a vehicle is in any way involved and which results in personal injury, property damage, or collision. While the DOT has its own definition of what's recordable, a carrier should track and demand accountability on any incident no matter how slight and regardless of fault.

Accountability for results can be the impetus for change to any organization. Auto crashes and the absence of regulatory compliance are no longer a cost of doing business; they are a failure of business. While all safety improvement is seen in the bottom line, gains are a direct result of the increased accountability and the use of safe driving methods. Organizations should not tolerate unsafe actions in the work area or behind the wheel.

By meeting high safety standards, the organization contributes to the well-being of its employees, its company, and the communities it serves. As the safety culture takes hold, consumers will respect the brand as much for on-road safety as they do now for their excellent service and reliability.

Responding to Safety Interventions in a Positive Way: Changing Safety Cultures

John W. Drebinger Jr., C.Ht.
Certified Speaking Professional
Owner: John Drebinger Presentations
Galt, California

How to Read This Paper

This paper is written in two different styles. The first portion is written in story form. All of the concepts are contained in the story. The main characters of the story are Jeff, a safety team member, and Joe, who comes to Jeff during a dream he has on an airplane flight to a major safety conference.

The second part of the paper addresses the "why and how to" aspects of the subject. The same principles are explained in each section, so if you like a good story, start at the beginning. Either way, it is said repetition is the mother of learning, and that is designed into this practical lesson.

Thank you for taking the time to care enough about yourself, your family, your co-workers, and your friends to read this paper.

Responding When Others Help You: A Story

Joe shares a new idea with Jeff, "Up until now, we have been focused on how to share safety with someone else, and yet all communication has two sides. One of the most important things you can do to help yourself and others be safe is to respond in the best way possible.

"Jeff, let me show you a situation where this happened. Interestingly enough, it was at a safety meeting for a company that cared about safety. They have special safety meetings for their employees where they bring in a professional safety speaker to remind them of how important it is to watch out for each other's safety."

Transported to a new location, in the audience of this safety meeting, Jeff watches as the safety speaker asks someone in the audience for a dollar bill. [1] Several people offer, and the safety speaker takes the first one available. He smiles and says, "Thank you," as he put the dollar bill into his pocket.

With that little joke, he takes the dollar back out and brings a volunteer on the stage, hands them the dollar bill, and asks them to read the serial number. The safety speaker writes the number down on a large Post It® Note flip chart. He steps out of the way and asks the volunteer to read the number once again so everyone can verify he had written it down exactly as read. He takes the large paper and hands it to the person who gave him the dollar bill and says, "Here is your receipt." The audience laughs as the speaker returns to the stage.

He then instructs the volunteer to tear the dollar bill in half. The volunteer looks at him as if he thinks the safety speaker is kidding. "No, I'm serious. Tear up the dollar." Reluctantly, the volunteer tears up the dollar bill. The safety speaker then announces it is a federal offense to tear up United States currency. Once again, the audience chuckles. He then takes the bill, saying he wouldn't ask the volunteer to do anything he wouldn't do himself and he tears the dollar again into four pieces.

The speaker quickly comments to the person who gave him the dollar, "Now you have four quarters, just don't try and use them in a vending machine." He holds the pieces in plain sight and with a wave of his hand they vanish. He exclaims, "Now you have a problem. That guy in the audience is going to want his dollar back, and I have just the solution."

"As you may have noticed, this lemon has been sitting on the stage since the beginning of my presentation. Watch!" The safety speaker picks up the lemon and a knife and he slices the lemon in half. Leaving it intact, he hands it to the volunteer on stage and instructs them to twist it apart. They do so and, to everyone's amazement, there is what appears to be a dollar bill rolled up in the center of the lemon. The volunteer unrolls it and reads off the serial number, which matches the number on the receipt for the original bill that had been destroyed. The audience breaks into applause.

Jeff watches after the presentation and notices someone approach the safety speaker. The person introduces himself as Jerry and asks the speaker if he would like him to watch out for his safety. As the speaker responds yes, he looks around the stage and on his prop table to see if he could notice what the person has observed.

Jerry then says, "I noticed you cut the lemon during that last magic trick and I would hate to see you get cut doing it."

The safety speaker responds, "Yeah, that would be very embarrassing to get cut during a safety presentation."

Jerry then offers, "We have some special leather gloves that are Kevlar-lined. Would you like me to get you a pair?"

"Sure!" is the safety speaker's response. Jerry brings a pair of gloves back, and gives them to the speaker, who is grateful and says, "Thanks, I will be sure to use them in all my future presentations. Also, thank you for watching out for my safety!"

A couple weeks later, the safety speaker is doing the "dollar-bill-in-the-lemon trick" again, and afterwards; another gentleman comes up to him, introduces himself as Jack, and says, "Would you like me to watch out for your safety?"

The speaker replies, "You bet."

Jack then says, "I noticed when you are cutting the lemon you wear leather gloves and those are good. We have gloves that are Kevlar and are more cut-resistant than leather. You could cut through the leather with a sharp knife."

Joe chimes in, "Jeff, do you realize what is happening? Initially, what Jack said seems to only affect the speaker's safety. However, how the safety speaker responds to Jack will shape both Jack's actions with other people in the future and the future safety of those individuals. The safety speaker's response will definitely influence whether Jack will choose to help other people in the future."

Joe continues, "Once somebody points out a hazard to you, you're now aware of the hazard. You have benefitted from the advice, and I'm sure you would want others to continue benefitting from Jack's observations. It's the very next person he comes in contact with that the speaker's response is going to affect."

Now, remember Jack was telling the safety speaker about the Kevlar cut-resistant gloves. Remember, also, the leather gloves the safety speaker had were Kevlar-lined. No one knew that except the safety speaker himself. The speaker could have said to him, "Oh, hey, don't sweat it, these have a Kevlar lining." Yet he chose not to. I wonder why."

Jeff begins to understand this whole concept more and more. He says, "If the safety speaker had done that, it is possible Jack would have felt like he was being put down. Jack may have felt foolish that he didn't realize the safety speaker would have had the best gloves. Jack could have walked away thinking he shouldn't have said anything. It would have made him feel uncomfortable about watching out for other people's safety in the future. Because he was helping the safety speaker, it was important for the speaker to encourage the correct behavior."

Jeff continues, "We know people will do things with which they are comfortable and we always want them to feel comfortable watching out for someone else's safety. Imagine if the next person Jack sees near a hazard is doing something much more hazardous than cutting a lemon. He thinks, 'Oh, should I say something?' How horrible would it be if he'd gotten ridiculed the last time he decided to watch out for someone's safety and, as a result, he keeps his mouth shut and the next person gets hurt?"

Joe says, "You are really getting good at this, and the best part is you understand the principles behind why it works."

Jeff replies, "So, what else do you want to teach me?

Joe responds, "Actually, nothing. You now have all you need to go out there and help other people work safely. In fact, you have more than that; you have the tools to help them watch out for others' safety."

Jeff says, "Ok, but are you sure I don't need more? It seems like it should take a long time to be able to do this."

Joe smiles and said, "Remember what I told you about how people want to make things more complicated? Well, that is an easy trap in which to fall. Sure, you could get lots of safety training, yet even an untrained person can spot hazards out there in the world. That magician safety speaker once told me a saying his dad would often say, 'The better is the enemy of the good.' What you have now, Jeff is good. The most important thing now is to put it to work and to teach it to others."

"Will I be seeing you again?" Jeff asks.

"Sure, I will probably be cheering you on now and then. I will drop by to remind you of the impact you are having. Maybe in the future, I can take you around and show you how to help people avoid shortcuts. Have fun, Jeff, and may God bless you and every person who watches out for the safety of God's children."

With that, the wheels of the plane hit the ground and Jeff awakens to find himself back home in Sacramento.

The How to's of How to Respond When Someone Watches Out for Your Safety

When you want to improve your safety culture, teaching people how to respond when someone watches out for them is critically important. I got an outstanding view of this when I was performing a trick where a dollar bill from an audience member is destroyed and then ends up restored inside a lemon. To finish this trick, I borrow a knife from someone in the audience and cut the lemon in half. One day, after doing this trick, a gentleman walked up to me afterwards and said, "John, would you like me to watch out for your safety?" I said, "Sure." He then said, "When you cut the lemon in half, it would be much safer if you had gloves on. Would you like a pair of leather gloves that are Kevlar-lined?" I said, "That would be great." He went and got me a pair, and I started using them during my next presentation. I was now safer because he watched out for my safety, and I thanked him for that.

A couple weeks later, I was doing the same trick, and another gentleman came up to me afterwards and said, "John, would you like me to watch out for your safety?" I said, "You bet." He said, "I noticed when you are cutting the lemon, you wore leather gloves and those are good. We have gloves that are Kevlar and are more cut-resistant than leather. You could cut through the leather with a sharp knife."

At that point, he's watching out for my safety and protecting me. How I respond to his concern goes beyond protecting just me. It can impact other people this gentleman observes in the future.

How I respond has everything to do with whether he will choose to help others in the future. Remember, the gentleman was telling me about the Kevlar cut-resistant gloves, and recall also that the leather gloves I was using were Kevlar-lined. No one knew that except me. I could have said to him, "Oh, hey, don't sweat it, these have a Kevlar lining."

What effect would that response have on the second gentleman in our story? One possibility is he might have felt I was putting him down because he didn't know. He might have felt foolish that he didn't realize I would have had the best gloves. He was trying to help me, so it was important for me to let him do so. Otherwise, he could have walked away thinking he shouldn't have said anything. I would have made him feel uncomfortable about watching out for other people's safety. If I made him feel uncomfortable, he might choose not to help the next person he sees doing something more hazardous. His internal dialogue might be, "Should I say something? No, I got beaten up the last time I tried to help." He fails to say anything, and someone gets injured.

Once the hazards been pointed out to me, how I respond is not about me; it's about the next person. So, I said to him, "Hey, that's great." He went to get me a pair, and they are the ones I use now. To this day, he believes he helped me out, and he did. So did the other guy. The secret to the interaction I had with the second person is I didn't steal his moment of helping from him. That moment is about him and the next person he's going to help.

Whenever somebody points out safety to you, please do whatever you can to make it a big a deal. Say, "Thanks very much, I really appreciate that. Thank you for actively caring!" This response will generate the continued behavior of watching out for others' safety.

Now Take Action

You now know how to watch out for the safety of others. Knowledge without action is useless.

I challenge you to use one of these techniques in the **next 24 hours** to point out a hazard or unsafe behavior to someone on or off the job. When you have done this, write the date in the space below

I used these techniques today.

Bibliography

Drebinger, J. 2011. *Would You Watch Out For My Safety?®: Helping Others Avoid Personal Injury*. Galt, CA: Wulamoc Publishing.

Session No. 526

Getting the OSH Professional's Voice Heard: Communicating Clearly about Acceptable Risk

Graham Parker, Msc CFIOSH, FIIRSM, CBIFM
President
Institution of Occupational Safety and Health (IOSH)
Leicester UK

Jo Frape, BSc (Hons), Grad IOSH
Technical Publications Officer
Institution of Occupational Safety and Health (IOSH)
Leicester UK

OSH professionals have a major role in communicating risks to employers and employees at all levels. They need to be technically competent, but that isn't enough; they also need to be good listeners and communicators. Health and safety can be regarded as complex, and it is the job of the OSH practitioners to simplify the messages and make them relevant to the people they advise, using language that will fit into the business culture. Messages need to be tailored to different audiences.

For example, migrant workers may not speak the language of the country fluently and, therefore, may have difficulty in understanding instructions, asserting their rights at work, and seeking help. As a result, they are less likely to report accidents. They may be unable to communicate effectively with other employees, and their understanding of risk may be impaired. In these cases, different ways of communicating need to be found to get the message understood. On the other hand, senior managers or human resource professionals need to be able to hear the business case for health and safety to persuade them to invest both financially and emotionally in health and safety. Workers need to understand how they can integrate safety into their tasks so that any risks are 'acceptable'. Building trust and credibility is key to successful communication and forming functional alliances helps with this process (Joyce, Hand 2012).

This paper discusses:

- Why the OSH practitioner needs to get heard
- What the OSH practitioner needs to communicate regarding acceptable risk and control
- Who the OSH practitioner needs to communicate with

- How to choose the best communication channel
- The need for planning a communication strategy
- How to get the message across; the skills needed for good communication.

Communication case studies from a public sector utility in the UK and a construction contractor are outlined.

Participants will gain an understanding of the seven Cs of effective communication to appreciate the need to make functional alliances and learn about practical examples of how communication has been improved in companies.

Introduction

Senior managers are ultimately responsible for ensuring the health, safety and welfare of their employees. In the UK, The Management of Health and Safety at work regulations 1999 requires every employer to provide employees with information on, the potential risks to their health and safety, preventative and protective measures for those risks, emergency procedures and the identities of those who have a role within the organisation's health and safety management system (Pearson and NAS 2013). This includes giving employees information on any process or tasks that might involve significant risk; this information must be comprehensive, and it must make sense to the people it affects.

As life can never be risk free, and zero risk is unachievable, there are some levels of risk that everyone will find 'acceptable' (Hunter and Fewtrell, 2001). Understanding how to communicate the acceptable risks with employees will help to engage and gain trust within the organisation.

To communicate clearly to employees regarding risks, senior managers have not only to find the communication method that will work for their audience but also have to be technically competent, a good listener and a good communicator.

Why the OSH Practitioner Needs to Get Heard

OSH practitioners have many different titles, from safety manager and loss prevention engineer, to name a few. However, the core duty of the profession is the prevention of events that cause harm to people, property, or the environment. Traditionally, the safety professional has been viewed as a technical expert who must be proficient in a wide range of methods, controls, and administrative tasks in order to drive safety, and is occasionally seen as an inconvenience to the running of any commercial activity.

Safety professionals have the skills and ability to help implement processes and technology with reliability and sustainability (Groover, et al, 2008). In this way, safety professionals position themselves as consultants to the organisation and trusted advisors to the line organisation and its leaders.

Health and safety can be regarded as complex, and it is the job of the safety professionals to simplify the messages and make them relevant to the people they advise, using language that will fit into the business culture (Ludhra and Joyce 2013). Messages need to be tailored to different audiences; for example, migrant workers may not speak the language of the country fluently and, therefore, will have difficulty in understanding instructions, assessing their rights at work, and seeking help. As a result, they are less likely to report accidents. They may be unable to communicate effectively with other employees, and their understanding of risk maybe impaired.

Communicating safety risks clearly can ensure that the employees are engaged with senior management in safety activities, cooperation, and support, and to also maintain a positive safety culture. Communicating the specific safety risks in forms such as policies, procedures, performance statistics, hazards and incident reports, workplace inductions, risk assessments and training will all be vital, but having these forms in itself will not be enough to communicate the risks; it is how they are communicated that is vital, and knowing and understanding the target audience with clear and constructing safety communication will improve knowledge and understanding that prevents at-risk behaviours and enhances safe work practices (Angelica M. Vecchio-Sadus 2007).

What the OSH Practitioner Needs to Communicate Regarding 'Acceptable' Risk and Control

'Openness and inclusiveness,' the principles that underpin effective communication, are key demographic principles and are important in their own right (Gov.UK 2011). The UK government information and communication service (2011) defines communication as follows: "building relationships with others, listening and understanding them, and conveying thoughts and messages clearly and congruently; expressing things coherently and simply, in ways that others can understand, and showing genuine knowledge, interest and concern; brining these aspects together to make change happen."

Communication with the workforce can help departments handle risk more effectively; it can help to prevent crisis from developing; it can lead to better decisions about how to handle risks; it can help ensure smoother implementation of policies to tackle risks; and it can help to empower and reassure employees. Over time, it can help to build trust within the organisation from the information it provides (Gov.UK 2011)

The inability to communicate is a problem if a migrant worker has an accident. McKay et al. (2006) reported that HSE (Health and Safety Executive) inspectors in the UK often have problems investigating accidents involving migrant workers because the injured person is not able to give an account of what happened. Also, some migrants are worried about reporting accidents in case they lose their jobs (Joyce, Hand 2012).

IOSH sponsored research, carried out by Cameron et all (2011), looked at the impact of pictorial occupational safety and health training on migrant worker behaviour and competence. This study looked at whether pictorial materials, compared to training materials using just text, resulted in improved knowledge and understanding in a sample of migrant workers in the construction sector. Topics covered included exclusion zones, storage of materials, safe use of portable tools and personal protective equipment (PPE). They measured subsequent knowledge with a multiple choice pictorial test and by behavioural observations. The results showed that the pictorial materials improved knowledge and understanding but the behavioural measures were more viable. Improved behaviour was observed where managers had posted pictorial materials near work areas. The report concludes that using pictorial tool-box talks, alongside poster campaigns, might improve the overall impact and effectiveness of pictorial aids to communicate health and safety information.

Investigators determined that NASA's space shuttle Columbia explosion on Feb 1, 2003, which claimed seven lives, was partially due to lack of effective communication of critical safety information. They concluded that organisational causes, including lack of communication, contributed to the incident (Lyon and Hollcroft 2012).

One of the core skills of being a good communicator is listening; listening requires attending to the person talking and being curious enough and interested in the other person and taking what the other person is talking about seriously (Ludhra & Joyce). According to the HSE (December 2011), we only actually remember 25-50% of what we hear; many people when listening tend to start concentrating on something else, which prevents them from really understanding what the other person is trying to say, which may not only affect the way jobs are done but could also lead to health and safety risks. Good listening is important and is likely to gain the respect, cooperation and trust needed from the other person to make change happen. Not only encouraging good listening from senior managers but between employees with each other, for example, will help find ways on how best to accomplish tasks, which improves the production of work and will help improve clarity and maintain focus on the task, thereby reducing risks.

For example, in the modern world of flexible working arrangements and remote workforces, an individual's personal safety and health, and how they manage risk in their working environment, is increasingly in their own hands, as for many health and safety professionals cannot always manage risks by being 'on the ground' 24/7 to ensure controls are in place. The way to make this work is by education, entrusting and empowering employees; giving them the training and skills is the first step and putting trust in their capacity to apply common sense and sound reasoning is the second; then, by empowering them to make decisions, not removing responsibility or accountability from the health and safety professional but recognising that, as reasonable people, they can work with the health and safety teams to maintain safe and health workplaces. Relying on the three topics, education, entrusting, and empowering the employees understanding of 'acceptable risk', is essential (Parker, 2017).

Organisations face a wide range of risks each day that can affect their ability to achieve certain business objectives and stay in business. Risk management is an important and sophisticated process used to assess organisations risks so that it mitigates and reduces risks to an 'acceptable' level (Lyon and Hollocroft 2012).

Since risk assessment is fundamental to the risk assessment process, OSH professionals must be clear on what a risk assessment entails and how it is performed and to also communicate the risks and controls effectively to all staff within the organisation, taking into account their target audiences and using communication methods that will work for all staff within the organisation, whether it's verbal, written or visual communication. This will enable the OSH practitioner and the organisation to ensure that acceptable levels of risk are communicated effectively. Lyon and Hollocroft (2012) explain that two of the reasons organisations tend to fail to perform good risk assessments are (1) failing to understand an organisations acceptable risk level, and (2) failing to communicate, before, during and after the risk assessment.

Manuele (2010) suggests that safety professionals have not yet fully embraced the concept of acceptable risk. Some organisations promote zero risk as their primary goal; however, some residual risk will always remain. As described by ASSE previously, safety and health management goals should be specific, measurable, actionable, realistic, and time-oriented. A zero-risk goal may not be realistic, and should be redefined to an achievable and acceptable level for the organisation.

How to Choose the Best Communication Channel

IOSH's technical guide, *Getting the Message* (2015), explains communication within an organisation typically comprises of five crucial interrelated types:

- written
- verbal
- visual
- non-verbal
- grapevine

An organisation cannot survive without written communication as its core method, though increasingly this is transmitted electronically rather than as hard copy. Types of written communication are memos, letters reports, newsletters, and email. Although email, as a type of written communication, is a consistent message about an issue that can be easily and quickly shared with many staff; however, this can also be perceived as 'too much information' or spam, and is not always appropriate for the audience.

Effective verbal communication is also valuable as part of an overall plan when rolling out a new policy, strategy or operational system as it is proactive, i.e., it does not rely on others receiving and reading a written communication. Verbal forms of communication can be meetings, presentations, lectures, conversations, i.e., face to face or via phone. The benefits to verbal communication are that face-to-face communication builds rapport and trust and generates high levels of understanding and transparency. However, verbal communication can't be relied upon for most critical health, safety and environmental communications and, without written backup, no proof of a conversation can lead to 'your word against mine' disputes.

Visual communication, in the forms of charts, diagrams, drawings and graphs, can be ways of demonstrating complex information in a quick-to-digest way and is an asset to effective communication. For example, the IOSH research mentioned earlier describes how pictures were used in safety training to improve migrant worker behaviour and found that using pictorial tool-box talks, alongside poster campaigns, might improve the overall impact and effectiveness of pictorial aids to communicate health and safety information. Other benefits of using visuals in communications are that a combination of words and images has greater impact than words alone and information in a table, chart or graph increases audience understanding and encourages participant questions. The downside to this is that not all topics can be presented visually.

Non-verbal communication/body language is an important aspect of face-to-face and one-to-one communication involving non-verbal elements. Being aware of what to look for in others, while making sure body language gives the right signals, is a major factor in managing workplace situations. Body language can back up the words used and how those words are said are those words, but can also betray true feelings. Experts believe that upwards of 75 percent of all human communication is non-verbal. So, while body language statistics vary according to each situation, it's generally accepted that non-verbal communication is crucial to how we understand, each other, especially when communication involves emotional or attitudinal aspects. Key body language characteristics include, posture, gestures, eye movement/contact, facial expressions, and mirroring.

Grapevine communication is an informal channel of business communication, so called because it extends throughout an organisation in all directions, regardless of authority levels. Typically, it's more significant at lower levels and/or remote parts of an organisation, as employees use the grapevine when other communication channels are weak or non-existent.

In today's fast-paced work environments, management depends on technology to communicate. Yet most organisations fail to provide training on how to choose the most appropriate and productive communication channel.

The more complicated the message is, the more personal and interactive the channel should allow for human elements such as voice tone, facial expression, and physical presence. For example, bad (and good) news is better delivered in face-to-face or one-to-one situations, depending on the group or individual content. Routine, easy-to-understand messages, on the other hand, can be presented via a more 'static' channel.

The Need for Planning a Communication Strategy

The IOSH *Guide* also explains whether there is the need to communicate general day-to-day information or news about major changes in the organisation or work practices, the best communication starts with thorough planning. It's a good idea to set a plan in writing. As part of the planning process, considering the organisation's health and safety mission statement can underpin the communication strategy.

As part of the planning process, consider how the organisation's safety and health mission statement can underpin the communications strategy. A mission statement's purpose is to get employees to comprehend what the organisation stands for, its purpose, and how they can work towards common strategic goals.

Be clear about the overall communication objectives: What is the objective aiming to be achieved, why and by when? Carefully address;

- the target audience;
- what the person or group wants or is expecting;
- the messages that are going to be conveyed;
- the communication channel(s) and method(s) to be used;
- the timescale, overall and for each step;
- any integral or external constraints on budget or resources;
- potential language barriers; and
- those in the target audience who have special requirements, e.g., personnel with disabilities.

The subject matter to communicate maybe simple, but the process can be complex. To begin changing behaviours and attitudes directly can be extremely challenging. Your communication should aim to inform and persuade the audience, while giving them a sense of control in any decision process. It's much better to target behaviour; attitudinal changes may follow if the communication is successful.

Don't forget to incorporate any relevant legislative, regulatory, technological or organisational requirements as part of the communication management. However, bear in mind though that the 'threat' of legislation as the primary reason for any safety and health communication should be carefully considered to avoid audience switch-off. Try to identify good internal business reasons for any proposed changes, rather than implying, 'we have to because a third party says so.' Care must also be taken to keep communication 'document control' up to date in this scenario.

When thinking about what to communicate, develop key messages to meet both the objectives and the different communication needs of each target audience. Be concise, and don't overload each message with irrelevant material or detail. Structure when and how to disseminate key messages as part of your communication plan.

To avoid information overload or confusion, keep the number of key messages to a minimum, ideally between three and five per communication.

An example of a communication strategy that has worked and made a considerable difference to an organisation is a leading UK Gas and electric company E.ON, which has more than 50,000 employees at over 50 UK sites. In a determined strategy to address the increase in occupational health issues, estimated to be costing around £30m a year, an occupational health team was set up. The team's communication strategy was literally to go desk-to-desk, talking to thousands of staff about health issues. Online self-help support materials were then developed to tackle stress and other mental health issues, and resulted in over 7000 hits.

The number of new cases and days lost were reduced by 25%. The programme was delivered to a diverse range of employees, including office workers, remote lone workers and engineers in power stations. E.ON's innovative communication strategy won the mental health and stress award in the occupational health awards 2010. The company estimates the programme saved around £11.8m a year.

How to Get the Message Across: The Skills Needed for Good Communication

Effective communication mechanisms are critical to engage employees in safety activities and to gain cooperation and support to maintain a positive safety culture. These mechanisms need to complement the practical and technical safety strategies. Employees with effective communication skills are better able to provide correct feedback for risky behaviours, thereby decreasing the probability of an injury, as well as rewarding feedback for safe behaviours, thereby increasing the likelihood of future behaviours being performed safely and enhancing safe work practices.

Employers have a legal duty to consult employees on safety and health risks. The management of health and safety with any organisation is more likely to succeed if it encourages active participation of workers and sets up a dialogue between employees and management.

Employees who work in the organisation are not health and safety professionals so will need to be trained or instructed on the organisation's health and safety practices and legislations. There are ways to communicate this effectively, which will not only lead to improved incident rates. The HSE (2017) suggests getting employees talking, listening, and cooperating with each other and can help the organisation in many ways, including;

- identifying joint solutions to problems;
- developing a positive health and safety culture where risks are managed sensibly;
- reducing accident and ill health, plus their related costs to the organisation;
- bringing about improvements in overall efficiency, quality and productivity; and
- meeting customer demands and maintaining credibility and complying with legal requirements.

Workplaces where employees are involved in taking decisions about health and safety are safer and healthier.

Stronger employee involvement means better control of common workplace risks, such as slips and trips. Accident rates are lower where employees genuinely feel they have a say in health and safety matters, compared with workplaces where employees do not get involved.

Communication and consultation are the most basic forms of employee involvement in health and safety management (HSE, 2015). It is crucial that employers get them right, not only because the law requires this but because they form the foundation for fully engaging the workforce to achieve the type of changes in behaviour in managers and employees that will lead to safer and healthier workplaces.

Key actions in ensuring senior managers are communicating effectively involve leaders ensuring that enough time is allocated to ensure communications can take place. Senior managers should think about formulating plans for cascading information, remembering to plan how to get the message across to the target audience, including anyone with low levels of literacy, migrant workers, vulnerable workers. They also should ensure that what is being communicated is easily understood by everyone, and that any written communications are clear and up to date, and lay out clear communications procedures for safety-critical tasks, and that any safety-critical messages have been given attention and are understood by employees.

If employees have difficulties understanding English, or employees have low literacy levels, there are a number of ways employers can communicate with them to encourage their involvement. Employees should ensure that there is adequate time to consult with employees where language and/or literacy may be issues, so they can absorb the information and respond to you. Encourage employees to express their views in their preferred language by using interpreters, pictorial information, and internationally understood pictorial signs where appropriate (HSE, 2015).

Evidence from the IOSH-funded research looking at lessons learned from the 2012 Olympic park construction project indicated that key factors in the high standard of health and safety achieved there included effective communication and good leadership. Along with worker involvement, this helps to develop a climate of trust. Communication was multi-directional and included a formal network. Success was due to their coordinated approach to communication (Joyce and Ludhra 2013).

"Hold points" were included in method statements to allow regular review of work carried out. Otherwise, there were many opportunities created for communication, such as a regular project safety leadership team meetings, safety forum meetings, risk assessment briefings, workforce representative meetings, safety circle meetings, and one-to-one interviews between workers and supervisors. Good practice was also shared between contractors and clients. Specialist software packages were used to help the exchange of health, safety and environmental information between workers and supervisors. Good practice was also shared between contractors and clients. Specialist software packages were used to help the exchange of health, safety and environmental information between the Olympic Delivery Authority (ODA) and contractors. This information assisted monitoring of the safety climate. This illustrates the need to make opportunities available to talk about health and safety.

The strategy went beyond paperwork and the health and safety management system and engaged people in the process. Cheyne et al. (2012) also highlighted the importance of helping people to understand why a health and safety measure was necessary, not just what they needed to do.

Presenting health and safety information in different formats was also important. For example, Wusteman (2012) explained e-learning packages translated into many different languages were used for health and safety inductions.

'Visual standards' were developed in consultation with contractors which identified 'what good looks like' and 'what bad looks like.' This involved taking 74 photographs of key-risk areas. These were put into a manual made of waterproof material that could be used by supervisors to explain the standards required. This helped ensure consistency across the site and provided an easier method of communicating with those on the site who were migrant workers. This will help establish the positive health and safety culture, which has resulted in the safest project to build an Olympic park ever. Further details can be found at http://learninglegacy.london2012.com

Further to this, this organisation found ways to communicate effectively to their employees by understanding the target audience and selecting the right method for them. The Tamdown Group, a construction contractor with 600 employees, introduced a new work safety strategy to motivate the workforce to improve standards in an industry that claims the life of roughly one worker and 27 serious injures a week. The visually based communication programme majored around Power Point presentations, supported by interactive technical sessions, workshops, and daily 10-minute, forum-style worker engagement sessions on every operational site. Group Health and Safety Manager, Andrew Denby, said, "successful communication is about getting people to take key messages on board. There's a lot more trust and cooperation, and workers are far more open and likely to come up with suggestions for change." Crucially, the Tamdown Group saw employers' liability insurance premiums cut by £50,000 a year. Accident rates dropped from eight in a typical year to just one. The drop in accidents has also resulted in further savings of between £14,000-£25,000 a year, based on reduced investigation costs and replacement labour charges alone (IOSH 2015).

European agency for safety and health at work (EU-OSHA) (2012) explains that risk communication manages how a potential hazard is represented to multiple audiences. Good risk communication requires dedicated effort, training and resources to built-in sites and will not compensate for failings for shortcomings elsewhere, such as a poor underlying scientific evidence base. It demands consistency and clear focus throughout, and requires both expertise and experience. There are a wide range of approaches to risk communication with different characteristics serving different requirements and audiences. Risk communication is not simply the provision of information; it is a two-way process by which all sides can learn something from one another.

Risk communication will not necessarily resolve conflict, guarantee understanding or cause people to behave in certain ways, but it can clarify the nature of disagreements, restrict their scope, help people make more informed choices and minimise resentment caused by people feeling excluded from decisions that affect them. Risk communication can help: build trust, allow better decisions to be made on how to address risks, ensure smoother implementation of risk management policies, empower and reassure the public, help prevent crisis developing, and manage them when they do. Risk communication is likely to be most effective if it is done proactively; accurately, it is relevant and getting the timing right is also essential because attitudes and beliefs, once formed, are resistant to change, and these will shape how new information is perceived, making it difficult, if not impossible, to take any real dialogue forward.

Before any communication exercise, the evidence first needs to be assembled, acknowledging and considering any possible uncertainty to demonstrate a credible basis for the position being taken. It is then necessary to identify the audience by finding out what different groups maybe be affected by the risk, i.e., benefit or suffer loss, have relevant expertise and whose cooperation is crucial. The communication method, approach and content needs to be tailored to targeted groups, taking into account the diversity of information needs, expectations, perceptions, concerns, and possible underlying reasons such as value systems.

The 7 Cs of Communication

According to Murphy, Hildebradt and Thomas, by selecting the message content and style that best suits the purpose and recipient of a message. The seven Cs of communication are shown in Table 1, below and discussed below:

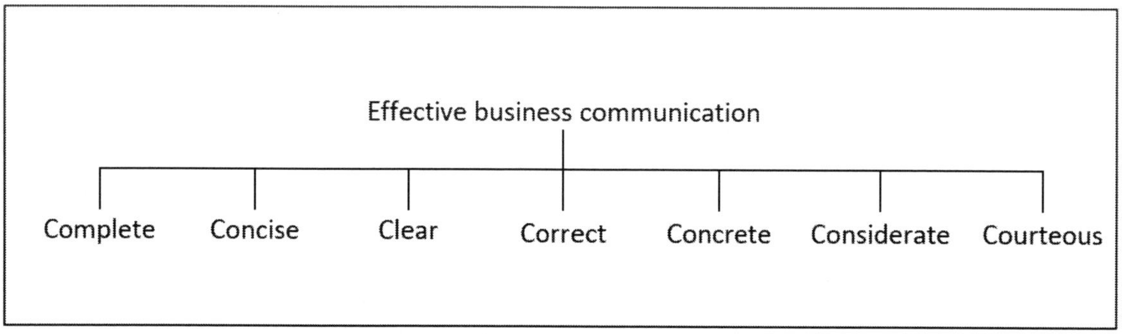

Table 1. 7 Cs of effective business communication

The following seven Cs are applicable to both written and verbal messages. The sender (speaker or author) should plan and prepare to ensure that each communication is;

Complete: A complete safety and health communication should answer six questions: who, what, when, where, why, and how. It should include all the facts, and the target audience needs to fully understand and, if necessary, respond. It should aim to leave no questions in the mind of the receiver. As a result, it:

- Aids better decision making, as the recipients get all the desired and critical information that's relevant to them
- Saves the time and costs of additional 'fill in' communication
- Develops and enhances the reputation of the sender/department/organisation
- Ensures shared meaning

Concise: Each message should communicate relevant information to the receiver clearly, without unnecessary wordiness, ambiguity or irrelevant detail. Concise information:

- Focuses on the key messages (s)
- Is more appealing and comprehensible to the audience

Clear: Don't use technical or business jargon unless confident that the audience is familiar with the terms. A clear communication focuses on a specific message or goal rather than trying to achieve too much at once. For clarity, use exact and appropriate language. Clear communication:

- Enables audiences to understand a message effortlessly

- Enhances the meaning of the message

Correct: All information in a communication should be checked for accuracy. Correctness also refers to the use of proper punctuation, grammar and spelling. Correct communication:

- Has a greater impact on the audience
- Increases confidence levels in the communicator

Concrete: Supporting a message with concrete facts and figures, as appropriate, makes the communication more solid and convincing. A concrete message:

- Strengthens confidence in the overall message
- Minimises the potential for misinterpretation

Considerate: Effective communication should consider the target audience. Strive to empathise with the audience's viewpoints, background, mindset and comprehension levels, and make the message personal. Emphasise what's impossible. By showing an interest and fostering mutual respect, communication becomes more valuable. Consideration:

- Stimulates a positive reaction from the audience
- Conveys optimism

Courteous: Courteous communication is where the message sender is sincere, polite and judicious, reflective and enthusiastic. Courteous communication:

- Takes other viewpoints into consideration
- Is positive and audience-focused
- Is unbiased

Conclusion

This paper has discussed why OSH practitioners need to get heard, explaining that by communicating safety risks clearly, they can ensure that employees are engaged with senior management in safety activities, cooperation and support and also maintain a positive safety culture. Also, simplifying messages to ensure they are relevant to the audience is also important, especially if the audience includes migrant workers and that do not speak the language of the country. It has covered basic principles, including 'openness and inclusiveness,' which underpin effective communication and by communicating with the workforce, departments can handle risk more effectively, preventing crisis from developing. The paper also includes how to choose the best communication channel by using various methods of communication and the need for planning a communication strategy by explaining how to go about the communication plan and areas the address. It has also included ways to get the message across; the need for effective communication mechanisms to engage employees; and finally the seven 'Cs of communication show how to select the communication message content and style that bests suits the purpose and recipient of a message.

Bibliography

Brink, Kyle E., and Robert D. Constigan. 2015. "Oral communication skills: are the priorities of the workplace and AACSB-Accredited Business Programmes Aligned?" *Academy of management learning and education,* Vol. 14 No. 2, 205-221.

Cameron, I., Hare, B., Duff, R., and McNairney, F. 2011. "Using pictures in training. The impact of pictorial OSH training on migrant worker behaviour and competence." IOSH.

Cheyne, A., Gibb, A., Hartley, R., and Finneran, A. 2012. "Talk the talk – walk the walk." An evaluation Olympic Park safety and communication initiatives. *Report submitted to the IOSH Research Committee.*

Christensen, W.C. and Manuele, F.A. 2000. *Safety through design.* Itasca, IL: National Safety Council.

European Agency for Safety and Health at Work (EU-OSHA). 2012. *Risk perception and risk communication with regard to Nano materials in the workplace.*

Groover, Donald R., and Jim Spigener. 2008. *The emerging role of the safety professional, part 1.* (https://ohsonline.com/Articles/2008/04/The-Emerging-Role-of-the-Safety-Professional-Part-1.aspx?Page=1)

Harris, Grant T., Christopher T. Lowenkamp, and N. Zoe Hilton. 2015. *Evidence for Risk Estimate Precision: Implications for individual risk communication.*

Health and Safety Executive (HSE). 2015. *Involving your workforce in health and safety, guidance for all workplaces.* London: HSE.

Health and Safety Executive (HSE). *Key actions in communicating effectively* (retrieved February 2017) (http://www.hse.gov.uk/managing/delivering/key-actions/communication.htm).

Health and Safety Executive (HSE). *Leadership and worker involvement tool kit, effective communication and gaining co-operation.* December 2011. (http://www.hse.gov.uk/construction/lwit/assets/downloads/communication-skills.pdf).

Health and Safety Executive (HSE). *Why talk to your employees about health and safety?* (Retrieved February 2017) (http://www.hse.gov.uk/involvement/talktoyouremployees.htm).

Hunter and Fewtrell. "Acceptable risk." In World Health Organisation (WHO*). Water Quality: Guidelines, Standards and Health.* Edited by Lorna Fewtrell and Jamie Bartam. London: IWA publishing.

Institution of Occupational Safety & Health (IOSH). 2015. "Getting the Message, Guidance on communication." Leicestershire, UK: IOSH.

Joyce, J. and Hand, G. 2012. *Migrant workers–sustainable not disposable- a challenge for safety and society.*

Ludhra. S and Joyce.J. 2013 *Delivering messages to your board and organisation so they will listen.*

Lyon, Bruce K and Bruce Hollocroft. *Risk assessments Top 10 pit falls and tips for improvement. Professional Safety.* www.asse.org (http://www.oshrisk.org/docs/Risk_Assessments_Top_10_Pitfalls.pdf)

McKay, S., Craw, M., Chopra, D. 2006. *Migrant workers in England and Wales - An assessment of migrant worker health and safety risks.* HSE research report RR 502. Working Lives Research Institute. London Metropolitan University (retrieved February 2, 2012) (http://www.hse.gov.uk/research/rrhtm/RR502.htm).

Murphy, H. A., Hildebrandt, H. W. and Thomas, J. P. 1997. *Effective business communications.* 7th edition. New York: McGraw-Hill.

Parker, G. IOSH President. "Empowering non-OSH specialists." *IOSH Magazine*, February 2017. (httos://www.ioshmagazine/article/empowering-non-osh-specialists).

Pearson and NAS. 2013. *6.0 Maintaining systems for health and safety, welfare and environmental protection in the workplace. Part 6.2 communications of health and safety in the workplace.* (http://www.contentextra.com/eet/files/topicguides/Topic-guide-6.2-Communication-of-health-and-safety-in-the-workplace.pdf).

Shen, Fuyuan, Vivian C. Sheer, and Li, Ruobing. 2015. *Impact of narratives on persuasion in health communication: A meta- analysis,*

UK Government, Cabinet office, emergency planning. January 2011. *Communicating risk guidance.* (Retrieved February 2017) https://www.gov.uk/government/publications/communicating-risk-guidance

Vecchio-Sadus, Angelica M. 2007. *Enhancing safety culture through effective communication.* CSIRO process science and engineering, Australia.

Wusteman L. The final straight - *HSW - Health and Safety at Work* June 2012 http://www.healthandsafetyatwork.com/hsw/2012-final-straight accessed 8 March 2013.

Session No. 528

Construction SH&E Best Practices: Guidance for Young Professionals

Timothy Bates, BS, CSP
Hensel Phelps
Boydton, VA

Andrew Mahoney, BS, GSP
PCL Construction
Shreveport, LA

Molly O'Brien, MS
FDR Safety
Franklin, TN

S. Alex Sewing, MS, CSP
AECOM
Gallatin, TN

Krystal Zwinggi, MPH, ASP
Zachry Industrial, Inc.
San Antonio, TX

Introduction

Established in March of 1986, the National Construction Safety Executives (NCSE) is a group of Executive Safety Professionals from 30 of the top design, engineering and construction firms in the nation. These individuals meet regularly to discuss and share information, ideas, and construction industry best practices.

The NCSE assembled the Future Leaders of Construction Safety (FLOCS) in the fall of 2015. The FLOCS is a group of safety, health & environmental (SH&E) young professionals under the age of 35 who exhibit high potential and an inclination to expand their industry expertise. The NCSE tasked the FLOCS with collecting, reviewing, and sharing SH&E best practices. The project allowed the FLOCS to network with experienced SH&E professionals, and further develop their understanding of SH&E best practices currently deployed throughout the construction industry.

This paper reviews the Future Leaders of Construction Safety SH&E Best Practices Project. The definition of best practice, challenges faced when implementing a best practice, effective persuasion strategies, and advice and recommendations for SH&E young professionals are examined herein.

What Is a Best Practice?

Best practice is a term or expression that has long been used throughout various industries. Organizations often develop, implement, and utilize best practices with the goal of increasing production, reducing costs, and separating themselves from the competition. But what exactly is a best practice? More specifically, what is an SH&E best practice with regard to the construction industry?

The FLOCS interviewed NCSE members in an effort to uncover the true connotation of a best practice. During the interview process, the FLOCS asked, "What is an SH&E best practice?" When this question was asked, many of the members' responses, though insightful, were not immediate. The FLOCS believe this is because commonly used terms, such as best practice, tend to take on an identity of their own, thus the true definition is forgotten.

<u>Findings</u>
The FLOCS compiled all individual definitions and identified six (6) key elements of an SH&E best practice according to the NCSE members. These elements, in no particular order, are as follows:

1. Specific
2. Innovative
3. Exceeds regulation(s)
4. Impactful
5. Beneficial
6. Shareable

1. Specific
An SH&E best practice seeks to reduce, eliminate, or improve a specific hazard, task, activity, or trend.

2. Innovative
An SH&E best practice is new, different, or an alteration to an existing initiative, policy, program, or procedure.

3. Exceeds regulation(s)
An SH&E best practice is an initiative, policy, program, or procedure that is *not* enforced by a government entity. Therefore, a best practice is not simply adhering to an OSHA regulation. Rather, it is an organizational initiative utilized in order to exceed accepted standards.

4. Impactful
An SH&E best practice directly and positively impacts workers' health and well-being.

5. Beneficial
Along with positively impacting workers, an SH&E best practice benefits the organization as a whole. This may be in the form of increased production, reduced costs, or an improvement in quality.

6. Shareable
An SH&E best practice is something that produces an outcome so favorable it is worthy of being shared with other companies on a project, other projects within the organization, or other organizations throughout industry.

Based on this feedback, the FLOCS defines a construction industry best practice, with regard to safety, health & environment as, "a unique or innovative endeavor to improve the condition or well-being of individuals, organizations, and industries by exceeding a specific standard that is traditionally accepted by society."

Implementation Barriers

As part of the FLOCS SH&E Best Practices Project, the NCSE members shared the best practices currently expanded throughout their organizations. These ranged from various forms of risk assessments, hazard analyses, training, employee development programs, management committees, and technology. The NCSE members were asked to describe some of the common challenges they experienced when expanding these best practices.

Findings
The FLOCS identified five (5) common obstacles encountered by NCSE members. These obstacles, in no particular order, are as follows:

1. Resistance to change
2. Geographical differences
3. Obtaining buy-in
4. Lack of information
5. Language barriers

1. Resistance to change
Change can be difficult to accept. There is often kickback from both field personnel and executive management. Many employees have been set in their ways for countless years and do not see the need for change. Organizations must show employees how a new practice can improve safety on their job. Once resistance is overcome, the implementation process becomes less difficult.

2. Geographical differences
Geographical differences can play a significant role in executing a best practice. A best practice may have different outcomes, results, or effects when implemented in the Gulf Coast versus other parts of the country. This may be attributed to cultural differences or attitudes towards safety in certain demographics of the world. Environmental and climatic conditions can also play a role in a best practice being effective, and adjustments may have to be made to fit the conditions. For

example, a construction company requiring the use of spoggles at jobsites across the company had issues with lenses fogging in areas with higher temperatures and humidity.

3. Obtaining buy-in
Like other safety initiatives, the buy-in of upper management is essential to the effective implementation of a best practice. In addition to upper management, middle management must also show commitment to a best practice through leading their employees by example and providing the appropriate resources. The support of middle and upper management reinforces the importance of a best practice and helps to ensure people are held accountable.

4. Lack of information
Upper management will quickly reject a proposed best practice if SH&E professionals fail to present data, statistics, facts, or supporting information. SH&E professionals often fail to include cost-benefit data in conjunction with SH&E-related statistics when presenting and selling a best practice.

5. Language barriers
Construction is a diverse industry with employees who speak different languages and come from various backgrounds. These factors can present potential barriers to the successful implementation of a best practice. Statistics show that Hispanic workers are two times more likely to be injured on the job. A construction company recently participating in a Spanish language course pilot study found that 65% of course participants indicated feeling more effective when they spoke Spanish on the job, and over half experienced a victory utilizing the skills acquired during the pilot study. Sixty-three percent reported having improved relationships with Spanish speakers on the job when they made the effort to communicate in Spanish to Spanish-speaking workers. These findings confirm the importance of communicating a best practice to employees who speak other languages so they understand the importance of the best practice and how it will keep them safe in the workplace.

Persuasion Strategies

Identifying and recognizing potential implementation barriers is imperative. Equally important is an SH&E professional's ability to persuade and convince upper management, workers, and other stakeholders to accept and support a best practice.

The NCSE members were asked to share their strategic advice for persuading and selling a best practice to stakeholders.

Findings
The FLOCS identified four (4) common persuasive strategies recommended by the NCSE. These strategies, in no particular order, are as follows:

1. Make it personal
2. Provide facts
3. Understand the risk and consequences
4. Educate

1. Make it personal
In order to implement a best practice effectively, it needs to be made personal. This means that employees need to understand how the best practice will benefit them. For example, a best practice that reduces hazards will help keep workers safe so they are able to return home to their families at the end of every day. Another way to make a best practice "personal" is to encourage feedback or suggestions from employees throughout the implementation. By involving employees in the process, they are more likely to have a sense of ownership in the success or failure of the best practice.

2. Provide facts
Facts and data are used to support the implementation of a best practice. The implementation process becomes easier when there is factual data to support the new best practice. Facts come in many forms, including studies and proven techniques from other companies. When facts are given to management and field supervision, and they are proven to have an impact on safety, they can make obtaining support easier.

3. Understand the risk and consequences
Understanding the risk and consequences when executing a best practice can cause varying actions relating to the success of its implementation. In broad spectrum, risk vs. reward must be thoroughly understood from an organizational standpoint. Consider the business fundamentals first when evaluating ways to improve safety. The "business" still must be profitable to sustain jobs. How can this best practice increase the safety performance, while also enhancing the profitability?

4. Educate
A key to the successful implementation of a best practice includes a well-organized, properly executed rollout strategy. A rollout strategy should involve communicating details of the new best practice throughout the company and providing training on how to integrate the best practice into programs at the jobsite level. Employees will feel a sense of ownership and accountability to honoring the best practice when they have a strong understanding of the reasons behind the best practice.

Advice & Recommendations for SH&E Young Professionals

Implementing an SH&E best practice may be increasingly problematic for SH&E young professionals. Minimal experience and a lack of familiarity in the construction industry place young professionals in a difficult position.

How can you improve your persuasion skills in order to sell your best practice? The FLOCS asked the NCSE members what their advice was for SH&E young professionals.

Findings
The FLOCS identified six (6) common recommendations made by the NCSE. These recommendations, in no particular order, are as follows:

1. Improve soft skills
2. Improve leadership skills
3. Become involved

4. Know more than safety
5. Know that you don't know
6. Seek professional development

1. Improve soft skills
Soft skills are critical and often overlooked. Much of what you learn in school is focused on the technical side of SH&E. Until you craft your ability to communicate, only then will you be able to put your technical skills to good use. As one executive told the FLOCS, "You can be the smartest guy in the world but if you can't explain it to people, you won't prevent injuries. Go out and learn people's names and shake their hands" (Bob Johnson, Granite Construction Inc.).

2. Improve leadership skills
An SH&E professional's span of influence can at times reach many different areas and have a profound effect on the direction of the company. Strong leadership skills evolve over time through individual experiences and personal growth. The greatest leaders are said to be those who can put others' success above their own. In his online article, "Great Leadership," best-selling author Jon Gordon, explained, "Good leaders get people to believe in them. Great leaders inspire people to believe in themselves" (Gordon, n.d.).

3. Become involved
Do not limit yourself to your current organization or project. Join professional safety groups and expand your network. These connections will provide you with invaluable knowledge and information. Likewise, give back to your profession. One executive shared, "Get involved with ASSE or National Safety Council, for example. Go back to your university and share with future SH&E Young Professionals" (Brad Giles, FDR Safety).

4. Know more than safety
Understanding the business as a whole is vital. Cindy DePrater, Turner Construction Co., advised, "Do as much as you can. Don't silo yourself within just pure safety. You can't truly understand the operations or business side unless you have done it."

5. Know that you don't know
Regardless of what industry you work in, no one expects you to know everything on day one. It is important is that you know your limits: "Know what you know and know what you don't know" (Page Warde, Weeks Marine).

6. Seek professional development
Seek out opportunities to grow personally and professionally. Certifications and continuing education are becoming increasingly important. Obtain nationally recognized certifications, and pursue a Master's or PhD degree. By doing so, you will ensure that you remain highly competent and knowledgeable in your field. The learning process should never stop.

Final Reflection

In conclusion, the NCSE's support and sponsorship of the SH&E Best Practices Project exposed the FLOCS to a unique and valuable opportunity to establish relationships with highly experienced safety professionals in the construction industry and draw from the expertise they have gained through the years. This project also provided the FLOCS with a great deal of insight as to different interpretations of the term "best practice," common barriers in the successful implementation of best practices, and how these barriers can be overcome.

The FLOCS strongly recommend becoming involved in other young professional groups. Similarly, the FLOCS encourage organizations to promote and assist young professionals in pursuing various professional growth opportunities.

The Future Leaders of Construction Safety (FLOCS) would like to thank the National Construction Safety Executives (NCSE) members for their continuous support and devotion to the development and betterment of SH&E Young Professionals.

Bibliography

Gordon, J. (n.d.). *Great Leadership*. (Retrieved from http://www.jongordon.com/articles-greatleadership.html).

Session No. 529

911, What's Your Emergency? We Have an Active Shooter; Hurry, Please Hurry!

Scott B. Lassila, CSP, ASP, CSM, APS
Managing Consultant
Aon Global Risk Consulting
Houston, Texas

Missy Bertrand
Senior Manager of Enterprise Safety
Best Buy, Inc.
Richfield, Minnesota

The Definition and the Issue

The United States Department of Homeland Security defines *active shooter* as, "an individual actively engaged in killing or attempting to kill people in a confined and populated area; in most cases, active shooters use firearms and there is no pattern or method to their selection of victims. Most incidents occur at locations in which the killers find little impediment in pressing their attack. Locations are generally described as *soft targets*, that is, they carry limited security measures to protect members of the public. In most instances, shooters commit suicide, are shot by police, or surrender when confrontation with responding law enforcement becomes unavoidable." According to New York City Police Department (NYPD) statistics, 46 percent of active shooter incidents are ended by the application of force by police or security, 40 percent end in the shooter's suicide, and in only 14 percent of the time the shooter surrenders.

An *active shooter incident* is a slightly narrower definition of a *mass shooting* in which an individual begins shooting at people "in a confined and populated area," according to the FBI and in which law enforcement or a citizen response can intervene.

There is a considerable amount of overlap between mass shootings and active shooter incidents. The number of people killed or injured generally defines mass shootings. Active shooter incidents, on the other hand, are defined only by *attempts* to kill or wound people.

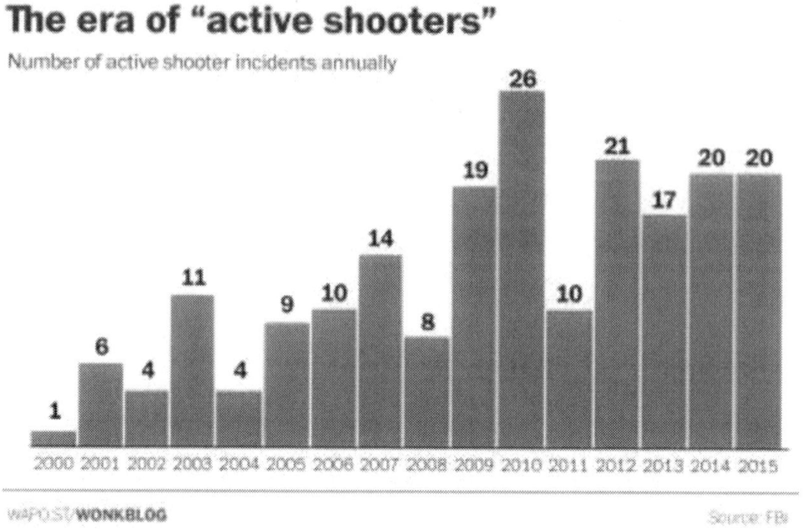

Figure 1. Annual number of active shooters

The FBI has recently released data that shows active shooter incidents are on the rise and have actually reported, "We now live in the age of public shootings." As shown in the graph in Figure 1, the annual number of "active shooter" incidents has continued to rise since 2000, when there was only one incident listed in the FBI database. However, active shooter incidents date back as far back as August 1, 1966, when ex-Marine sniper Charles Whitman killed his wife and mother, then climbed a 27-story tower at the University of Texas and killed 16 more people before police shot him to death.

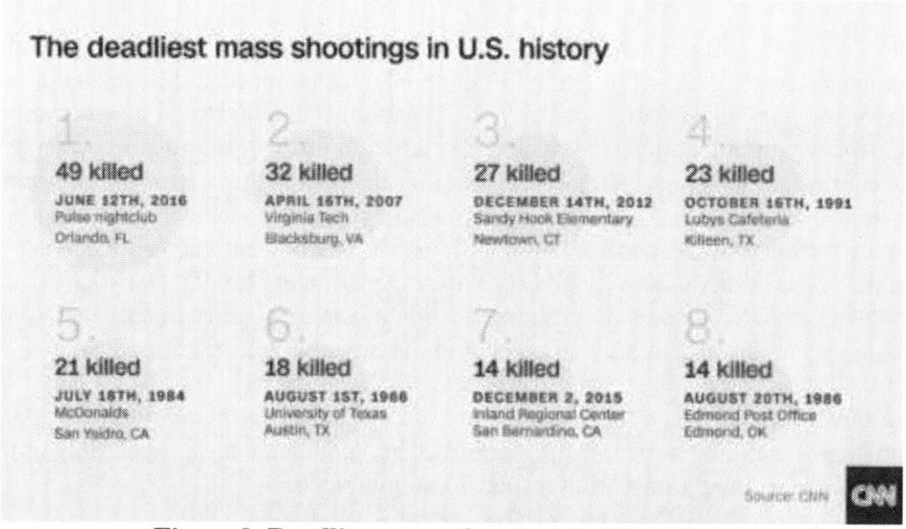

Figure 2. Deadliest mass shootings in U.S. history

Introduction to Run-Hide-Fight Approach

Active shooters may have a target in mind, but often have no pattern or method in their selection of victims. They often spend time planning and rehearsing their attack and, in most cases, they do not expect to survive. Active shooter situations are unpredictable and evolve quickly. For this reason, individuals must prepare both mentally and physically to deal with an active shooter situation.

Most organizations have adopted the Run, Hide, and Fight approach, developed by law enforcement and vetted by the Department of Homeland Security. It is becoming the core concept that most large entities are following as a best practice. In the event of an active shooter event, Run, Hide, Fight could save your life.

Employer's Responsibility

As we have seen an increase in active shooting occurrences in the work place, employers have the responsibility to ensure employees are trained on how to respond in these types of events. Regardless of the work environment, all employees should know how to respond, and some employees may actually need to be given permission to protect themselves, because of the strict training they have received around locked doors, a non-violence policy in the workplace, or a signed code of conduct. Training employees on how to respond to an active shooter incident and immediately after could potentially limit the number of people impacted by the event.

Best Buy's Case for Change

We all know that this topic is not easy, but it is necessary and, as safety professionals, it is our responsibility to influence leadership on the importance of this training.

When approaching leadership, know your facts when looking to get buy in and for funding. Here are some examples to consider when discussing with leadership:

- There have been 12 active shooting incidents with two miles of our locations during a three-month period.
- Our locations have had to go into lock down three times during a three-month period because of an active shooting in the mall.
- The workplace is the number 1 location for an active shooter event.
- Employees have requested training.
- There is civil unrest throughout the county; let's be prepared.

Take partners along in this conversation, such as legal, employee relations, finance, human resources, risk management, and others. This will help with the conversations with the senior leadership.

Best Buy's Approach

Best Buy has retail space and large warehouses, so they evaluated the needed training based on the location.

Retail: Public locations

Best Buy has 1400 retail locations, so there is an online training option, with a soft opening message by a high-ranking leader (SVP of Retail) introducing the topic and reinforcing an internal HR support contact line for further support if needed. Next, a video was rolled out with a with the Run-Hide-Fight methodology. This video targets all employees, is required annually, and is reinforced with a store drill.

Warehouse: Mostly in remote locations

With 35 facilities in remote locations and a large footprint, we worked with the local Police Departments and created a Run-Hide-Fight, hands-on training for location leadership. This is a classroom setting, and a tour of the facility is led by the local safety team for the Police Department. This gives the Police Department a chance to visit and view the facility and get familiar with the leadership. The leadership then delivers the training to all employees without the local Police Department present, as this can be intimidating.

Learnings

Some items Best Buy learned as they went through this process, as well as remaining decisions include:

1. Holding training during non-peak hours so you do not scare employees or start rumors with the Police Departments presence. Communicate the reason for the police presence before they arrive.

2. Ensuring that you communicate to the location that this training is *not* in response to a threat.

3. Using rally points. Some Police Departments do not recommend rally points, and some do. An advantage is that an employee head count can be taken; a disadvantage is if a current or former employee is the perpetrator, they may know the rally point and put employees in greater danger. Another disadvantage is the use of a second device, meaning that the shooter may have placed explosives at the rally point.

4. Using intercoms. Since our facilities are large, and the intercom does not come though clearly, Best Buy uses radios.

5. Providing a "blueprint" of the building to the local Police Department. Most departments can download the document into their system and, if a call comes, in witnesses and those on the scene can articulate the location of the shooter. The dispatcher or responders will be better prepared to enter.

6. Having an Active Shooter Code. We have fire alarms and severe weather alerts, but Best Buy did not have a standard for active shooters. During the planning, it was discovered

that most locations, the only announcement over the intercom or radio is SHOOTER, SHOOTER, SHOOTER.

7. Pulling the fire alarm is *not* recommended as you may be sending employee directly into the line of fire.

8. Most employees are not allowed cellphones on the working floor, so how can they call 911? Best Buy has landlines placed throughout the location.

9. Allow time for questions and allow people to share their stories.

10. You are better off when more people call 911, as they can gather facts over the phone from several people.

11. Best Buy has third-party security guards at their front doors so they were included the training.

The Survivalist Mindset

As we learn more from the people that survive these types of situations, we are hearing that they have had training or just conversations about what they would do such if an event occurred (i.e., Run-Hide-Fight). The bottom line is to take direct responsibility for your personal safety by having a survival mindset that is comprised of three components:

Awareness: Be aware of your surroundings, regardless of where you are at, work, mall or restaurant.

Preparation: Training and talking (watch videos on line, talk to your family).

Rehearsal: By yourself or with others (with co-workers and family).

Let's face it; an active shooter incident is most likely to be over in two-to-three minutes, so that means you are on your own. Law enforcement will be first on the scene to take down the shooter, not to help the wounded.

What Now

When an active shooter event is occurring in your vicinity, follow these steps which are spelled out in the FBI's active shooter pocket card.

Run

Always have an escape route, and plan in your mind on where you are going to go. Always get out of the building if you can. Do not run into unfamiliar areas within your facility because you could become trapped. Help others to escape, if possible. Always leave your belongings behind. Active shooter incidents are not robberies. Get out; you will always be able to go back and get your personal items. Prevent others from entering the area and call 911 when you are out and safe. Keep your hands up and visible at all times. Remember that the police will be responding, and they do not know who the active shooter is, or how many there are, and where exactly they are located in your building. You don't want to be mistaken as a perpetrator. Remember that, under the new guidelines for law enforcement training, is Advanced Law Enforcement Rapid Response Training (ALERRT). When law enforcement enters a building, they are not going to help you if you are wounded; they are not going to drag out the dead; they are there to neutralize the threat *first*. Do not run to the officers unless told to do so. Do not make a threatening move to officers, and always keep your hands visible.

Hide
If you cannot get out of the facility, you have to hide in an area out of the shooter's view. Block entry to your hiding place, and lock the doors. Push file cabinets, desks, and chairs in front of the doors if possible. Turn off the lights and hide behind a large object to provide cover in the event they fire in your direction, and stay low to the floor. Silence your cell phone and/or pager, and be quiet. Fear causes many different reactions in people. Now is the time to be as quiet as possible. Try to refrain from crying is possible. Any noise could give away your hiding place, and make you the target.

Fight
As a last resort, and only when your life is in imminent danger, you are going to have to attempt to incapacitate the shooter. If you are with a group, you have to act as one. You will need to be physical aggressive toward the shooter. This includes improvising weapons, throwing items at the active shooter, charging them, and taking them "out" if need be. Anything can be a weapon. You have to understand that it is you or them. You have to remember that you can wait to die or have a *fighting* chance. Always remember that an active shooter is not expecting a confrontation from potential victims. In their mind, they are in control of the situation, and you are simply a pawn. By fighting back, you have turned the tables on them.

Information to Provide to Law Enforcement or 911 Operators

When it is safe to do so, you should call 911, tell them the location of the active shooter, the number of shooters, and physical description of shooter(s), if you know. Height, size, race, hair color, color of their clothing, shoe color, number of perpetrators, and type of weapons held by shooters will be beneficial to the first responders to help identify the shooter if they engage him. Telling them the number of potential victims at the location will also be helpful. If you can, don't hang up; they can listen in on your phone, and this may help during the investigation.

How to Respond When Law Enforcement Arrives

Remain calm and follow instructions of law enforcement. Put down any items in your hands (i.e., bags, phones, jackets, etc.), raise your hands, and spread your fingers. Keep your hands visible at all times, and avoid quick movements toward officers, such as holding on to them for safety. Remember, they are there to confront the shooter. You must avoid pointing, screaming or yelling. Do not stop to ask officers for help or direction when evacuating unless they tell you to stop. Follow their directions when told to do so.

The Crime Scene

Your place of business has now become a crime scene, and you should expect to be closed for several days/weeks as the officers work to complete the investigation. In some situations, you may not reopen. You will not know how long you will be required to be closed, but here are the most common drivers of the investigation:

- The more people that have access to the scene, the more difficult of a job the investigators have, such as first responders.
- Do you have witnesses, cameras or security guards?
- How many injured or deaths?
- Types of weapons used.

The Employees

Once the Police Department is on the scene, there are several things you can be doing to assist them and support your employees:

- Provide a list of all the employees scheduled to work; try not to let them leave.
- Provide a list of missing employees.
- Gather the employees who witnessed what occurred in a general area; however, segregate those from the others because, as people talk, they could change their stories based on what other say; have them write it down.
- Follow your company Emergency Response Plan.
- Do not talk to the media.
- Share cellphones for employees to call family members.
- Offer support by keeping them updated on returning to work.

The Feedback

Although Best Buy has completed the training, there continue to be incidents near our locations. However, what is important is that our employees are empowered to Run-Hide-Fight. Best Buy received a lot of positive feedback from employees and local Police Departments. Below is a sampling of some of the hundreds of comments received on the training provided:

"After being in a live active shooter situation at the mall where I worked, I realized you hear those words 'active shooter' in your headset, you won't know what you will do. Our locations did an excellent job of securing our store and our customers last November. It was an experience that I hope to never deal with again, but I am prepared for it should it happen again. It is the reality of the world we live in. Kudos to my Best Buy team for such a great job that fateful day!"

"Good quick overview of complex situations that impact many of us every day."

"Thank you Best Buy, for keeping me safe."

"Good to keep up with relevant strategies."

"Good stuff to know for our everyday life."

"Thanks for caring and keeping us safe."

References

Federal Bureau of Investigation (FBI). 2000 to 2016. *Active Shooter Incidents.* (https://www.fbi.gov/file-repository/activeshooter_incidents_2001-2016.pdf/view).

______. 2014. *Active Shooter Study: Quick Reference Guide.* (https://www.fbi.gov/file-repository/as-study-quick-reference-guide-updated1.pdf/view).

Department of Homeland Security. n.d, *Active Shooter Safety Card.* https://www.dhs.gov/sites/default/files/publications/active_shooter_pocket_card_508.pdf

Ready Houston. 2012. *RUN. HIDE. FIGHT.® Surviving an Active Shooter Video.* https://www.youtube.com/watch?v=5VcSwejU2D0

Environmental Regulations: What Safety Professionals Need to Know

Salvatore Caccavale, CPEA
Global HSES Manager
Air Liquide Advanced Materials
Mount Bethel, PA

Introduction

For years, industries claimed they didn't realize what they were emitting into our air, into our waterways, and into the ground, and what effect it would have on our environment. Since the advent of the environmental disciplines (mid-1960s), our society now realizes what a mistake that approach was. Even today, abandoned industrial sites and misused landfills are still being cleaned up through the Environmental Protection Agencies (EPA) "Superfund" program. Industry was not the only culprits; however, mom-and-pop shops, farmers, dry cleaning operations, and other types of small businesses, had just as much effect on polluting the environment.

In the early 1970s, the story of Love Canal, a housing development built over a capped industrial dumpsite in suburban Niagara Falls, New York, ran rampant in the media and took the public by surprise. Concern for this type of event happening in "our backyard" led to public demand for our legislative body to develop regulations that would ensure the proper management of industrial wastes.

You Say You Want an Evolution

Management of environmental program elements has increasingly become a major focus for safety professionals over the past few decades. Many companies had separate environmental departments; however, as companies streamline and reengineer their professional support staffs there has been a marked trend towards an increase in consolidation of the safety, health, and environmental functions. The traditional safety professional has found the knowledge of environmental affairs critical to their well being. For companies involved in international business, the ISO 14000 standards represent a revolution in corporate environmental management. The safety professional must become familiar with the ISO approach to the environmental affairs.

There are approximately 30 federal laws pertaining to environmental regulations based on air, water, and land. Some of the most important laws are as follows:

- **National Environmental Policy Act (NEPA):** One of the first laws written, NEPA established a national framework for protecting the environment, using a basic policy to ensure all branches of government give proper consideration to the environment prior to undertaking any major federal action that significantly affects the environment. This included projects such as building airports, military complexes, highways, parkland purchases, and other similar activities covered by NEPA.

- **Comprehensive Environmental Response, Compensation and Liability Act (CERCLA):** The law provides a "Superfund" to clean up uncontrolled or abandoned hazardous waste sites, accidents, spills, and other emergency releases of pollutants or contaminants into the environment. Through CERCLA, the EPA imposes liability on owners/operators of a facility from which there is a release of hazardous substances. Transfer of a property does not relieve the owner/operator of liability; in addition, the current owner may be held strictly liable. The last reauthorization of CERCLA was the Superfund Amendment and Reauthorization Act (1986), otherwise known as SARA. These amendments covered emergency planning, community right-to-know (Tiers I and II), and toxic release reporting (Form Rs).

- **Resource Conservation and Recovery Act (RCRA):** This law regulates the generation, transportation, treatment, storage, and disposal of hazardous waste. RCRA requires the owner/operator of a facility to undertake corrective actions to clean up a facility used for the treatment, storage or disposal of hazardous waste. The entire premise of this regulation is hazardous waste is covered from time of inception to time of final disposal (known as "cradle to grave"). RCRA went through it latest revision in May 2017.

- **Clean Air Act (CAA) of 1990:** This act is a complex, multifaceted statute that is designed to regulate air emissions from both stationary and mobile sources. The act is composed of 11 different titles, such as the national ambient air quality standards (Title I), hazardous air pollutants (Title III), acid deposition control (Title IV), operating permits (Title V), and ozone protection (Title VI). Another important topic under the CAA is the Chemical Safety Information, Site Security and Fuels Recovery Relief Act, an amendment to Section 112(r) of the CAA, which addresses reporting and disseminating information on flammable fuels and public access to off-site consequences. This section also includes the Risk Management Plan (RMP), which almost parallels the OSHA Process Safety Management (PSM) 1910.119 elements.

- **Clean Water Act (CWA):** This act prohibits the discharge of pollutants from point sources and storm water into the navigable waters of the United States without a permit. The act imposes liability on any person who is responsible for the operation and/or equipment that results in a discharge. The basic requirement of the act is to force compliance with both uniform technology-based effluent limitations, regardless of the receiving waters, and with more stringent limitations necessary to meet state-imposed water quality standards.
- **Safe Drinking Water Act (SDWA):** This act imposes federal drinking water standards on virtually all public water systems. The act requires the establishment of drinking water standards for maximum concentration levels for organic and inorganic chemicals, turbidity, coliform bacteria, and various measures of radioactivity.

- **Toxic Substance Control Act (TSCA):** This act governs the manufacture and use of chemical products. The import and export of chemicals are regulated under the act. In addition, specific regulations control the use, management, storage, and disposal of polychlorinated biphenyl (PCB) materials.

- **Oil Pollution Act of 1990:** This act imposes strict liability on responsible parties for removal costs and damages resulting from discharges of oil into navigable waters of the United States. An owner/operator of an onshore facility that resulted in a discharge would be considered a responsible party.

Organizing an Environmental Management Program

Establishing an environmental compliance program can be overwhelming. Any company that is regulated by the Environmental Protection Agency (EPA), the Occupational Health and Safety Administration (OSHA), and the Department of Transportation (DOT), among others, cannot afford to be without an environmental management program.

Whether a company is large or small, senior management must communicate a strong commitment to responsible environmental practices through the ranks down to the front-line workers. Unfortunately, a sincere environmental management commitment that alters old practices often occurs only after a major disaster or stringent penalty. This does not have to be the case.

Deciding where to begin can be a big obstacle to organizing an environmental management program. The ways to comply with governmental regulations can seem endless.

Preventing Common Violations

Mislabeling of chemical and hazardous waste containers is probably the most commonly cited violation, and can most easily result in environmental problems. It is also the easiest violation to avoid. Companies should attach proper labels to all chemical containers. In the case of hazardous waste, labels must meet both RCRA regulations and DOT requirements prior to shipping.

Regulations require storage containers to be in good condition. Leaking containers are a common source of environmental fines. Inspect tanks daily, and check drums and similar containers weekly. Ensure aisle space is adequate to allow inspectors easy access to the facility. When waste is stored in tanks, regulations require secondary containment to prevent leaks or spills.

Other common violations include the following:

- Improper waste disposal
- Oil spills
- Destruction of wetlands
- Dumping into oceans, streams, lakes or rivers
- Improperly hauling pesticides or other toxic chemicals
- Improperly removing and disposing of asbestos
- Falsifying lab data pertaining to environmental regulations
- Committing fraud related to environmental crime

These violations are considered white-collar crimes. They can result in criminal fines, probation, jail time, injunctions, or any combination of these punishments. Regular audits can identify areas where potential violations exist.

Recordkeeping

Environmental regulations have mandated that organizations document many of their environmental practices during transportation and disposal of hazardous waste. During an inspection by a regulatory agency, records are usually the first target for an inspector. Often, the mere existence of the records and easy accessibility may be all an inspector needs.

It is also very important to document all environmental decisions. For example, whenever a facility has their waste streams analyzed and found to be non-hazardous, records of the information supporting that determination should be maintained and made readily available to an inspector. Relying on a sketchy explanation from a compliance manager can lead to an inspection report listing possible violations. Accurate, available, and complete records save companies time, money, and legal headaches. It is also important to give the inspector only what is requested. Often, it is a good idea to take copies of the records to the inspector at a location away from where the records are kept.

Create a Spill-reporting Plan

Under numerous statues and regulations (notably Superfund), companies must notify the United States National Response Center immediately after the release of a reportable quantity (RQ) of any listed hazardous substance or pollutant. As regulators decrease the time allowed between an incident and the reporting time, environmental authorities strictly enforce existing laws. In some cases, such as air pollution discharges, the permitted lag for tag for reporting may only be minutes.

An unintentional spill or release of a toxic material can occur during a facility emergency, when personnel are literally putting out fires or responding to explosions or other emergencies. For that reason every facility must have written instructions for employees to follow whenever a release or spill occurs. The reporting plan must detail the facility's procedures, telephone numbers of all agencies that should be notified, and who has the authority to report spills. Employees should be aware of these instructions and where they can be found at various locations. Not only does a spill-reporting plan permit a company to respond immediately to an emergency, but its existence is also evidence of a firm's environmental responsibility to regulatory agencies that may question the company's ability to respond to an emergency. It is important to conduct regular training sessions, and to practice incident responses, so that personnel feel comfortable dealing with emergency situations.

Summary

This session is designed for professionals who have assumed environmental responsibilities within their organization or who just want to learn more about the major federal environmental regulations, including: the Clean Air Act; Clean Water Act; Resource Conservation and Recovery Act; Comprehensive Emergency Response, Compensation and Liability Act; and Federal insecticide, Fungicide and Rodenticide Act. Session attendees will gain a better understanding of the basic requirements of each of these regulations, and how they apply in industry and the public sector.

Environmental managers should suggest realistic compliance schedules, and discharge limitations to environmental regulators. Often, in the spirit of cooperation, a manager will offer overly ambitious schedules and limits that ultimately cannot be met. So do not over commit. Once limits are set in discharge permits and orders, it is difficult to get them changed.

When reporting a release that exceeds set limits, always include the reason and try to solve the problem. Companies have reported pollutant releases above limits, received no response from the regulatory agency, and over time, assumed the increased rate is acceptable. The company's own reports then establish its lack of compliance, which can prompt an enforcement action or serve as evidence in a citizen suit against the company.

Focus will be on five of the major environmental regulations that touch almost every industry in one way or another, and provide brief real life scenarios. The information provided will help the novice to intermediate professional attain a better understanding of how these key regulations work and provide some helpful tips as to how to keep oneself out of regulatory hot water.

Efforts to reduce and minimize waste can save companies money, reduce potential environmental liabilities, protect public health and worker health and safety, and protect the environment. Waste can be minimized through source reduction and recycling. Source reduction generally is more cost-effective and has less impact on the environment than reuse/recycling.

Competitive advantages can be realized through the alignment of critical safety, health, and environmental factors, along with a company's value chain: the process by which products (or services) are conceived, produced, marketed, distributed, used and recycled, or disposed.

Steps in an environmental management program include preventing common violations, maintaining accurate records, creating a spill reporting plan, setting realistic limits and schedules for meeting the environmental regulations, and training employees in your program objectives and procedures. Companies must also establish a relationship with an attorney skilled in environmental law.

The safety professional must be familiar with the most current environmental regulations covering such areas as the release and recovery of hazardous substances, air, water, soil pollution, and toxic substance control.

Bibliography

Caccavale, Salvatore. 2012. *A Basic Guide to RCRA: Understanding Solid and Hazardous Waste Management.* Des Plaines, IL: ASSE.

Environmental Protection Agency (EPA). 2015. *Guidance to Identify Waters protected by Clean Water Act.* Washington, D.C.: EPA/U.S. Army Corps of Engineers.

Environmental Protection Agency (EPA). 2015. *Title V Policy and Guidance – Basic Information about Operating Permits.* Washington, D.C.: EPA.

Environmental Protection Agency (EPA). 2015. *Waste Analysis at Facilities that Generate, Treat, Store, and Dispose of Hazardous Waste – A Guidance Manual.* Washington, D.C.: Office of Solid Waste and Emergency Response.

Hagan, Philip. 2009. *Administration and Programs.* Itasca, IL: National Safety Council.

Keegan, Robert. 2016-2017. *Hazardous Materials, Substances, and Waste Compliance Guide.* Kutztown, PA: Hazardous Materials Publishing Company, Inc.

Session No. 532

Identifying Sprinkler System Inadequacy at Your Facility

Walter S. Beattie, CSP, CFPS, CSHM
Beattie Fire Protection & Risk Consulting, LLC
Pottstown, PA

Introduction

Many safety professionals do not re-evaluate the fire sprinkler systems at their facilities on a frequent basis. If there is a major change in operations, the sprinkler systems may be evaluated to verify that they are adequate for the new occupancy. However, they are seldom evaluated to ensure they are still adequate for the existing occupancy. Sometimes, changes creep into the facility and no consideration is given to the fire sprinkler systems, which may have been installed many years ago.

Identifying sprinkler system inadequacy seems to be a backwards concept. Unfortunately, facility hazard analysis reviews that incorporate fire protection adequacy reviews are not updated on a regular basis, if they exist at all. Hazard analysis reviews should be performed on a regular basis, but, unfortunately, many times they are not. Fire sprinkler systems are many times considered just another building system that is designed and installed when the building is built and not evaluated on an ongoing basis. This paper will help the safety professional to spot red flags that should make them take a more detailed look at the adequacy of the fire sprinkler system. The items identified in the paper should not be considered an all-encompassing list, but a starting point. The items identified in this paper may help identify conditions that should be further investigated. The presence of a red flag item does not necessarily mean the system is inadequate. It should be used by the safety professional as an investigative tool.

This paper will specifically address potential sprinkler system inadequacy. Sprinkler system inadequacy may be as a result of physical issues with the sprinkler system or its appurtenances, misapplication, occupancy beyond the capability of the system, or special hazard(s) being introduced which are beyond the capabilities of the system design.

Fire Prevention Versus Fire Protection

It is important to make the distinction between fire prevention and fire protection. Fire prevention is eliminating the possibility of a fire being started. This may be accomplished by preventing heat sources, maintaining good housekeeping, following good recommended practices, and ensuring standards and codes are upheld, and maintaining an active and vigilant self inspection program.

For the purposes of this article, fire protection involves fire suppression and mitigating the loss that is not prevented. If and when prevention activities fail and a fire erupts, proper protection must be implemented. There are many aspects and means of fire protection.

Human element programs address plans of action to implement. Some of the plans include:

- Emergency evacuation procedures to ensure all employees are outside of the hazard area and are counted;
- Plans of action to be implemented by management;
- Plans of action to be implemented by fire emergency response teams;
- Safe use of portable and manual fire extinguishing equipment, such as fire extinguishers and hose stations;
- Plans of action to be implemented by the public fire department and other public responders; and.
- Safe shutdown procedures for operators prior to leaving their post, if safe to implement.

Physical fire protection to limit the loss:

- Fire alarm and signaling systems,
- Fire sprinkler systems,
- Adequate fire protection water supply,
- Gaseous fire protection systems,
- Dry chemical extinguishing systems,
- Foam fire extinguishing systems.

Management of Change

Companies and their operations constantly change. The changes may be subtle, but over time, the subtle changes build upon one another, may develop, and have a significant impact on the ability of the fire protection sprinkler system to control or extinguish a fire. When making changes, building owners or occupants should consider the impact on or to the existing fire sprinkler system.

Companies build, purchase or lease buildings as their operations grow or contract. When relocating into a new or existing facility, a complete and thorough design review of the sprinkler system should be performed by a competent fire protection professional.

When existing spaces within buildings are renovated, the renovation design should incorporate appropriate fire protection design improvements when needed. Existing sprinkler systems should be updated, modified, or replaced as needed.

When new production lines are added they may be updated or changed. The new production line may use different technologies or be composed of different materials than older production lines. For instance, the amount of plastics used in the new production line may be significantly more than in older production lines. The new production line may use different materials with more hazardous properties than older production lines.

Companies change warehouse storage configurations. Floor space is a valuable commodity in a factory or warehouse. Sometimes, it is necessary to change the storage configuration of a storage area. The 20-ft. high, single-row, rack storage may be changed to double-row rack storage. An eight-foot aisle may be reduced to 4 ft. because new style, material-moving equipment uses less space. Multiple row racks may replace a lower hazard-storage arrangement. The storage height may be increased by installing higher racking systems or by double stacking the top level of storage.

Manufacturing changes may have a significant impact on the commodity classification of materials being used in the production of products. When metal and wood parts are replaced with plastic parts, the commodity classification of the product will also change. What was once a Class III or IV commodity may now be a Group A Plastic commodity. The change in commodity will severely overtax a sprinkler system designed and installed to protect the product when it was of a lower commodity classification.

Even packaging materials have an impact on the storage commodity of a product. When cardboard fillers are replaced with plastic foam blocks, the commodity classification may increase from a prior commodity to a Group A Plastic, or even an expanded foam commodity, increasing the required sprinkler density immensely.

When you evaluate your sprinkler systems and water supply adequacy, there are many ways to make the existing protection adequate. Reinforcing the fire protection systems is one way. You may also lessen the hazard so the existing sprinkler protection is adequate. For instance, idle pallets may be stored outdoors, storage heights may be reduced, solid shelves may be removed, or other elimination techniques may be used so the protection is adequate for the hazards present. The ultimate goal is to have an adequately protected facility.

Why Water?

In the average facility, water is the extinguishing agent of choice. It is readily available, inexpensive, abundant in large quantities, easy to store, and an effective extinguishing agent. Water is a stable material. It does not decompose under fire conditions. It can absorb large amounts of heat. When converted to steam, it expands almost 1700 times its liquid volume and can cool and displace air in the area of the fire.

Water extinguishes fire in several ways. It can cool the surface of the burning material. It will soak the burning material and displace available oxygen on the surface of the material. It can turn to steam and displace air around the burning material. It can dilute water soluble materials, such as alcohols.

Water may be used to extinguish many different types of materials. Ordinary combustible materials, such as wood and cellulose fibers, plastics, electrical equipment, flammable and combustible liquids, many various chemicals, and other special hazard applications may use water as an extinguishing agent.

Water may be adapted to meet the demands of a particular environment. Nonfreeze additives may be added to the water so it may be used in subfreezing temperatures. Foam concentrates may be added to the water to produce firefighting foam, which may be applied with specially designed foam application systems, hoses, or sprinkler systems. Surfactants may be used to reduce the surface tension of the water.

Fire Sprinkler Systems

Automatic sprinkler systems are often considered the most significant component of a building fire protection system. Fire sprinkler systems have a stellar record in mitigating major loss from fires. A building's sprinkler system must be properly designed, installed, and maintained throughout its life, which may be the life of the building. However, sprinkler systems have limitations. Their performance may be impacted by factors that are not related to the sprinkler system itself.

It should go without saying, but I will say it anyway; if a sprinkler system is not maintained in proper operational status, it will not work as intended! Proper inspection, testing, and maintenance are imperative to maintaining sprinkler systems. Your inspection, testing, and maintenance program extends beyond the annual service provided by your sprinkler contractor. Your sprinkler contractor is servicing the sprinkler system in accordance with NFPA 25, Standard for the Inspection, Testing, and Maintenance of Water-Based Fire Protection Systems. NFPA 25 does not require that the sprinkler contractor to identify inadequate design for a particular hazard or occupancy. Identifying inadequate sprinkler system design is the responsibility of the local facility management or building owner.

Types of Sprinkler Systems

Wet Pipe Fire Sprinkler Systems

In wet fire sprinkler systems, the overhead sprinkler piping is filled with water that is under pressure. When a fire occurs, one or more sprinklers activate, and water immediately discharges from the sprinkler head. Wet pipe systems are the most common type of fire sprinkler system installed in commercial and industrial facilities. They are also the simplest systems to install and maintain while being very reliable. Wet pipe sprinklers must be maintained at or above 40 °F (4 °C).

Dry Pipe Fire Sprinkler Systems

Buildings or areas that cannot be reliably maintained at or above 40 °F require nonfreeze-type systems. Dry pipe sprinkler systems are one of the most common nonfreeze sprinkler systems and use a dry pipe sprinkler valve as the actuator. A dry pipe sprinkler valve is a special valve that prevents the pressurized water in the fire mains from entering the sprinkler system piping. Normally, this is accomplished by filling the system with air. Most valves use a differential method of keeping the valve shut; generally, the differential is 5:1 or 6:1. This means that 5 or 6 psi of water pressure is held back by 1 psi of air pressure. In the event of a fire in which a sprinkler head actuates, the air pressure in the system decreases until the valve trips.

Dry pipe sprinkler systems have a higher likelihood of internal corrosion in the piping due to moisture-laden air. Sometimes, air drying systems are installed or an alternate gas, such as nitrogen, is used. Another disadvantage of dry pipe sprinkler systems is the time delay between the operation of the sprinkler head and the water actually discharging from the sprinkler head.

Pre-action Fire Sprinkler Systems

Pre-action fire sprinkler systems are similar to dry pipe sprinkler systems in that the sprinkler pipes do not contain water. Rather than using a dry pipe sprinkler valve, a pre-action valve is installed. A pre-action sprinkler valve is actuated with an electrical or pneumatic releasing system that operates after a fire is detected with a detection system.

Two separate events must occur to initiate sprinkler discharge. One or more sprinkler heads must activate from heat operating the fusible element. Plus, a fire detection system must detect a fire and actuate the releasing system which allows the pre-action valve to open and allow water to flow into the pipes.

If a sprinkler head opens or a system pipe is broken, a supervisory signal operates from pressure release, but no water comes out of the sprinklers or pipes. On the other hand, if a pre-action valve actuates, but there is not enough heat to activate a sprinkler head, the water fills the sprinkler system pipes, but no water flows into the protected area because a sprinkler head has not actuated. This type system is common in occupancies where water discharge is undesirable due to the operation being performed or equipment installed.

Deluge Fire Sprinkler Systems

Deluge fire sprinkler systems are similar to pre-action fire sprinkler systems in that the sprinkler pipes do not contain water. However, all the sprinkler heads are open. Deluge systems use a deluge sprinkler valve, which is similar to the pre-action valve. A deluge sprinkler valve is actuated with an electrical or pneumatic releasing system that operates after a fire is detected with a detection system. A fire detection system must detect a fire and actuate the releasing system, which allows the deluge valve to open and water to flow into the pipes. Water flows from every sprinkler open head in the system, after a brief delay during the time when the water fills the sprinkler system pipes.

Deluge fire sprinkler systems are used in high-hazard industrial and commercial applications where the hazards require quick response to prevent a fire from growing out of control or to prevent a fire from spreading to other areas, occupancies, or hazards. Deluge systems provide a simultaneous application of water over the entire hazard to suppress a fire quickly and effectively.

Foam-water Fire Sprinkler Systems

Foam-water fire sprinkler systems are a special type of sprinkler system designed for a specific hazard. The foam-water sprinkler system is designed to meet a desired performance. They may have open or closed heads, and the sprinkler heads may be ordinary style sprinklers or be specially designed air-aspirating heads. The system pipes may have water or foam concentrate in the pipes, or the pipes may be dry. A foam concentrate delivery system, such as a bladder tank or venturi delivery system is used to introduce the foam concentrate into the water flow. The foam discharge continues for a period of time, after which water flows until it is manually shut off. The operation of the sprinkler system is determined when it is designed and may or may not require a releasing system.

Foam-water fire sprinkler systems are used in high hazard industrial and commercial applications where the hazards require quick response to prevent a fire from growing out of control or to prevent a fire from spreading to other areas, occupancies, or hazards. Foam-water systems provide foam application to smother a fire and coat support structures. They may use less water than comparable water only sprinkler systems. They are usually used to protect high-challenge fires, such as flammable liquids, chemical processes, and airport hangars.

Types of Sprinkler Heads

Automatic sprinkler heads come in many types, sizes, and shapes. It is important that you identify the particular sprinkler head installed in the sprinkler systems at your facility. Sprinklers may be designed to be installed in the upright, pendant, sidewall, or concealed configuration. Sprinkler heads may be designed as a general use spray-type sprinkler head, or it may be a special purpose sprinkler. Special purpose sprinklers may be designated for storage, foam application, or other use, such as extended pattern head used in attics. Some special heads may be coated to resist corrosion from specific chemicals or atmospheres. Some special heads may be designed to be used in cold environments. Temperature ratings of sprinkler heads may range from 165 degrees to over 500 degrees Fahrenheit.

Certain features will be apparent in every sprinkler head. Every sprinkler head will have a deflector, and a frame. Most will have a water seal cap and a fusible link. Fusible links may be thermal, solder-type link or a glass-bulb style. Sprinkler heads in a deluge system will not have fusible links or a water seal cap because the deluge systems are designed to flow water from every head at once. The characteristics of each sprinkler head will determine the usage application. Understanding the application for the sprinkler head will help you know if the system has been designed for the application it is protecting.

To determine the sprinkler head installed in your sprinkler systems, you must look at the individual sprinkler heads. A good indication of what is installed is to look in the spare sprinkler head box. The installer is required to stock the spare sprinkler head box with an appropriate number of each sprinkler head used in the sprinkler system. You can also identify heads by physically looking at them as installed. Every sprinkler head will be labeled so you can identify the manufacturer, model, year of manufacture, and temperature rating. The information may be stamped on the frame or the deflector. Temperature ratings will be identified by the color of the liquid in the sealed bulb or stamped on the fusible link.

Once the sprinkler head manufacturer and model is acquired, the best way to identify the usage of a particular sprinkler head is to refer to the manufacturer's data and usage sheets. The manufacturer of every listed or approved sprinkler head will produce a datasheet, which describes the installation and care required for the particular sprinkler head. The data sheet will describe important items, such as how the sprinkler head must be installed. It will identify the type of sprinkler systems in which it may be installed. Some sprinklers may only be installed in wet pipe systems. It may identify the occupancy and fire hazard the sprinkler is permitted to protect. For instance, it may be capable of being used in only ordinary hazard occupancies and not to exceed a given maximum storage height which will be identified in the data sheet. The data sheet may identify spacing and obstruction requirements, minimum and maximum working pressures, and minimum water supply duration. It will also identify usage limitations such as not suitable for open-top storage containers or ceilings with joists, deep beams, or other significant obstructions. By reading and understanding the manufacturers' data sheet, you will be able to identify sprinkler heads that are not designed to protect the occupancy in which they are installed. If the permitted sprinkler head usage does not match the occupancy in which it is installed, then it may be prudent to further investigate the sprinkler system to verify that it is adequate for the present occupancy.

Design of Sprinkler Systems

From the beginning of sprinkler systems' existence until the 1960s, fire sprinkler systems were designed based upon a given pipe schedule. The schedule identified the maximum number of sprinklers that could be fed by a given size pipe. Hydraulic friction loss is created as water flows through a pipe. Friction loss decreases the available pressure of the water as it travels. If the amount of friction loss is not controlled, there may be insufficient pressure at the sprinkler head to produce a properly formed water spray with sufficient volume to control a fire below it. A number of fire sprinkler pipe schedules were developed, with the final pipe schedule being adopted in 1953.

Hydraulically designed sprinkler systems were installed from the early 1970s. Hydraulically designed and pipe schedule systems coexisted. A sprinkler designer could use either design method. Also, any pipe schedule system could be hydraulically analyzed and be proven to provide a given hydraulic density and area of operation. In the later 1980s, there was a change in NFPA 13, Standard for the Installation of Sprinkler Systems, that required hydraulically designed sprinkler systems in all but small systems. From a practical perspective, any sprinkler system installed in the last 30 years was probably designed with a designed hydraulic density and area of operation.

Even though a design placard is not installed on the riser, the sprinkler system may possibly have been hydraulically designed. Other criteria which may be a clue that the sprinkler system was hydraulically designed include:

- A sprinkler system that is from the mid-1980s or before that does not adhere to the pipe schedule criteria of the time period when the system was installed.
- The sprinkler system has a looped feed main system.
- The sprinkler system is gridded.
- The sprinkler system uses branch lines of all one pipe size.
- The branch lines use large pipe sizes.
- The head spacing is irregular or less than 100 sq. ft. per head.
- High-temperature sprinkler heads are used.
- Quick-response sprinkler heads are installed.
- Special-application sprinkler heads are installed.
- Large-orifice sprinkler heads are used.
- The ceiling appears to have two sprinkler systems, one atop the other. This is referred to as an overlay system, and was used to upgrade existing sprinkler systems, usually over a storage area, in the early days of hydraulically designed systems.

Hydraulically Designed Sprinkler Systems

Hydraulically designed sprinkler systems are designed so the sprinkler system will deliver a minimum sprinkler density, identified in gpm/ft^2 (L/min/m2), operating over a designated design area of application, identified in sq. ft. (sq. m). From the design calculations, a base of riser demand is identified as total flow in gpm (liters) at an identified pressure in psi (bars). NFPA 13 has K-factor specifications when densities reach various thresholds. The increase in K-factor for the sprinklers is to help prevent excessive flow through a single sprinkler head, maintain reasonable end head pressures, and reduce overage across an entire sprinkler system design.

With the development of computer software to perform sprinkler hydraulic analysis, looped and gridded sprinkler systems were calculated more quickly than by hand methods, and became more cost effective. Also, sprinkler designers were able to minimize material and installation costs by using small pipes, which cost less to purchase and required fewer installers.

When a sprinkler system is hydraulically designed, NFPA 13 requires that a permanent placard be installed on the sprinkler riser. The placard will provide at least the following information:

- The location of the design area or areas of the sprinkler system,
- The system discharge density or densities over the design area or areas,
- The flow and residual pressure demand required at the base of the riser,
- The occupancy classification, or for storage areas, the commodity classification, maximum permitted storage height, and storage configuration,
- The hose stream allowance included in addition to the sprinkler demand, and
- The name of the installing contractor.

An example of how the design of the sprinkler system may be expressed is, "The system should be capable of supplying at least 0.25 gpm/sq. ft. over the most hydraulically remote 2,000 sq. ft. of floor area with a base of riser demand of 658.0 gpm at 97.4 psi using 165 degree F. sprinklers with a K-factor of 8.0" Note the primary components of the design:

- The density, which in this case is 0.25 gpm/sq. ft. of floor area.
- The design area of the system, which in this case is 2,000 sq. ft.
- The total minimum flow required at the base of the riser, which in this case is 658.0 gpm.
- The minimum pressure required at the base of the riser, which in this case is 97.4 psi.
- The temperature rating of the sprinklers is 165 degrees F.
- The sprinkler orifice size has a K-factor or 8.0.

When you evaluate your occupancy and determine a design from NFPA 13, you should have each of the components identified above. Each required parameter must be met or exceeded by the existing sprinkler system for it to be considered adequate in accordance with NFPA 13. If any single parameter is not met, the system may be considered deficient, and a more detailed analysis should be performed.

Where to Find Occupancy and Design Criteria

There are several sources you may use to identify the design criteria for your occupancy or storage. The most common fire protection codes and standards are National Fire Protection Association (NFPA), Factory Mutual Data Sheets (FM), or your local building code.

NFPA 13 provides the requirements for design, installation, and maintenance of sprinkler systems. NFPA 13 is probably the most used fire standard in use in the United States. To view the latest edition of NFPA 13, you can visit NFPA's website at nfpa.org. NFPA standards may be viewed for free after a registration. For the purposes of this paper, we will limit our discussion to

NFPA 13. Keep in mind that other standards may also apply. For instance, protection for combustible and flammable liquids may be found in NFPA 30, Flammable and Combustible Liquids Code.

What Design Do I Need for My Occupancy?

NFPA 13, Chapter 5, Classification of Occupancies and Commodities, identifies the occupancy classifications related to sprinkler system and water supply requirements. Occupancy Classification identifies the parameters for Light Hazard, Ordinary Hazard (Group 1), Ordinary Hazard (Group 2), Extra Hazard (Group 1), and Extra Hazard (Group 2). The occupancies increase in hazard as the classification increases from light hazard to extra hazard. Typically, warehouses and storage occupancies will be more adequately addressed by other portions of the standard.

Examples of each occupancy class are provided in the Annex. The Annex provides examples of occupancies having uses and conditions similar to the occupancy definition. It is important to realize the examples provided in the annex are general, and may not accurately reflect the hazard exposure in your facility. Your occupancy, or portion thereof, must be classified according to the quantity and combustibility of contents. The materials in your facility must be evaluated based on expected rates of heat release, heights of stockpiles, and presence of flammable and combustible liquids. If you have an occupancy that is listed as one of the examples in the Annex for a particular occupancy classification, but your facility possesses higher hazards than the occupancy listed as an example, then you should choose an occupancy classification that more accurately reflects the hazards in your facility.

Once you have determined that your facility falls into one of the occupancy classifications, you may define the level of protection needed for your occupancy. Chapter 11, Design Approaches, identifies the minimum sprinkler density and area of operation for sprinkler systems. It also addresses water supply criteria and other factors that may affect sprinkler system adequacy. One of the most commonly used sources of density/area for occupancy group is NFPA 13, Fig ure11.2.3.1.1, Density/Area Curves. This figure shows curves that show a minimum density and a corresponding minimum area of operation for each occupancy class.

For example, the portion of your facility in question is used for metal working, and we have determined the proper occupancy is Ordinary Hazard (Group 2). By looking at Figure 11.2.3.1.1, we determine one of the points on the curve is 0.20 gpm/sq. ft. over the most hydraulically most remote 1,500 sq. ft. of floor area. Other points on the curve that may be used include 0.19 gpm/sq. ft. over the most hydraulically most remote 2000 sq. ft. of floor area or 0.17 gpm/sq. ft. over the most hydraulically most remote 3000 sq. ft. of floor area. Each of these points is an acceptable density area.

In a perfectly designed sprinkler system, the volume of water required to feed the sprinklers would be the density times the area of operation. A sprinkler system designed to provide 0.20 gpm/sq. ft. over the most hydraulically most remote 1500 sq. ft. of floor area would require 0.20 gpm/sq. ft. × 1500 sq. ft. = 300 gpm. Unfortunately, sprinkler systems have friction loss, elevation loss, and other considerations that make the volume of water flowing from each sprinkler head a little different than from other sprinkler heads in the system. The most hydraulically remote sprinkler head will flow the minimum allowed flow, and other heads will flow slightly more water. If the sprinkler system has a base of riser demand of 366 gpm, then the extra amount of water required to meet the system demand is called "overage." Overage is expressed as a percentage above 100% and, in this case, the overage would be 366 gpm / 300 gpm = 1.20, or 20 % overage.

The overage is unique for each sprinkler system, is dependent upon the design of the system, and is dependent on how much friction loss the designer incorporates into the system. For practical and simplistic purposes, the smaller the pipe diameters used, the greater the friction loss and the greater the overage. For example, friction loss and overage may be reduced by designing the system with a large diameter pipe, for instance using 1.25 inch pipe rather than 1 inch pipe. A sprinkler system designer may use smaller diameter pipe when the density and area of operation is low or the water supply is strong with good volume and high pressure available. When the density and area of operation is high or when the water supply is marginal, larger pipe is needed to reduce friction loss to design the system within the available water supply.

Any sprinkler system may be analyzed for another density and area of operation. A sprinkler system designed to provide 0.30 gpm/sq. ft. over the most hydraulically most remote 2500 sq. ft. of floor area may be recalculated to find a new density, area, and base of riser demand. If the new occupancy requires 0.19 gpm/sq. ft. over the most hydraulically most remote 2000 sq. ft. of floor area, the resulting base of riser demand would be lower than the original demand. Likewise, if the new demand required a higher density, the system could be recalculated at the higher design and another (more demanding) base of riser demand would be identified. The new base of riser demands would then be compared to the water supply to verify that the new demand is available. If not, an analysis of the actual density and area of operation could be calculated, based on the available water supply. As a general rule, when the overage is extremely high, the flexibility to recalculate a sprinkler system to a higher density and area of operation substantially decreases. This is because as flow through a given pipe increases, the corresponding friction loss increases exponentially.

The safety professional would usually rely on a fire protection specialist or sprinkler designer to recalculate the system. In the past, methods were used to approximate an average density for pipe schedule systems. These methods used tables to approximate an average density for a pipe schedule systems and to graphically represent approximate densities and base of riser demands using logarithmic graph paper. These older hand methods have been replaced with more accurate computer calculations. If the original sprinkler prints and calculations are not available, the recalculation cost may be high. Recalculating a sprinkler system involves measuring the pipes, identifying each pipe diameter, redrawing or providing a sketch of the system, and inputting it into a sprinkler calculation program.

Metal-working Facility Example

For example, let's assume you have an area of your facility that you would like to verify for sprinkler protection adequacy. The area in question is used for metal working, and we have determined the proper occupancy is Ordinary Hazard (Group 2). You moved into the building after the original tenant vacated. The work they performed was glass products manufacturing. When we look up the occupancy hazard of their operation, we determine they may have been considered Ordinary Hazard (Group 1). Your occupancy is a higher hazard classification than the prior tenant. This should be a red flag. As hazard increases, so does sprinkler density, and possibly area of operation. When you visit the sprinkler riser, you see a placard that identifies the system was designed to protect Ordinary Hazard (Group 1) occupancy. The design of the system is 0.15 gpm/sq. ft. over the most hydraulically most remote 1500 sq. ft. of floor area. The water demand is 553 gpm @ 82 psi, including 250 gpm for hose streams.

The design placard has very important information. The sprinkler system itself requires 303 gpm @ 82 psi at the base of the riser. From the information provided, the system has an overage of approximately 35%.

The base of riser is where sprinkler system riser pipe comes out of the floor. The hose demand for ordinary hazard is 250 gpm and may require 50 or 100 gpm for inside hoses that may have been installed as part of the sprinkler system. If there are no hose stations installed from the sprinkler system in the protected area, the entire demand is for outside hose streams that the fire department may use.

The base of riser water supply feeding the sprinkler system in this example is: Static pressure = 130 psi; Residual pressure = 55; flow = 700 psi. The available flow at 20 psi is 860 gpm. Figure 1 shows the water supply graph and the sprinkler base of riser demand which includes hose demand. The sprinkler demand cannot exceed the available water supply, and any attempt to increase the density and/or area would exceed the available water supply.

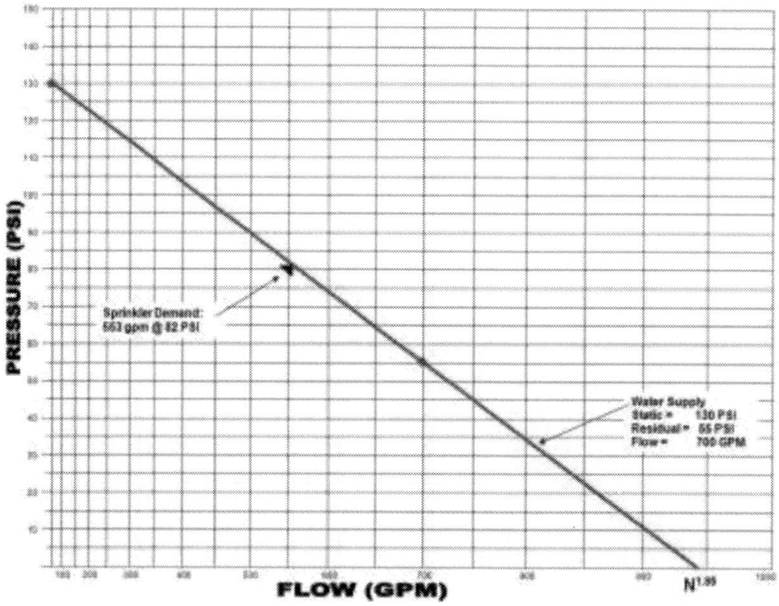

Figure 1. This is logarithmic graph paper with the sprinkler base of riser demand and available water supply.

Your metal working operation is considered Ordinary Hazard (Group 2) occupancy. The minimum density design over the most hydraulically remote 1500 sq. ft. of floor area is 0.20 gpm/sq. ft. Based on the available information, the existing sprinkler system is suspected to be deficient for your new occupancy.

To make the existing sprinkler system adequate, it would probably need to be modified by a sprinkler contractor. The contractor could replace some of the existing sprinkler pipes or feed mains with large diameter pipes. Other methods may also be available, such as replacing one-inch pipes on a tree-type sprinkler system, gridding the system if it is a tree-type system, increasing the water supply with a booster pump or a pump and tanks, or by making other changes.

Modifying a sprinkler system to obtain a greater density and/or area of operation may be an expensive exercise. Older pipe schedule sprinkler systems generally have more options available to improve their performance characteristics than hydraulically calculated or gridded systems.

Storage Commodities

Adequately designed and maintained automatic sprinkler protection is very effective in controlling fire in storage occupancies. Storage occupancy fires may be some of the most challenging fires for sprinkler systems to control. There are many variables that may impact the effectiveness of the sprinkler system. Materials flow in and out of the storage area on a regular basis, and new materials may be introduced without being evaluated for the effect they may have on the ability of the sprinkler system to control a fire after the new materials are introduced into the storage. Storage arrangements may change, and subtle changes in storage methods can severely impact the effectiveness of the sprinkler system.

Solid Shelves in Rack Storage

The use of solid shelving in racks can prevent sprinkler discharge water from reaching lower levels of storage. The solid shelves shield the lower levels of storage and can allow a fire in lower levels to grow, extending the length of the racks. The buildup of heat can open additional sprinkler heads, but the fire may continue to extend along the rack. At some point, the sprinkler system may become overtaxed, allowing the fire to extend from rack to rack.

For this reason, the use of solid shelves in racks is not permitted unless special precautions are implemented. The solid shelves must be limited in area and flue spaces of sufficient width must be provided at prescribed intervals. In general, properly designed in-rack sprinklers must be installed below each solid shelf to provide adequate protection.

If your facility's rack storage area has solid shelves, an engineering study should be completed to determine that the areas of the storage with solid shelves meet the provisions of NFPA 13. Requirements for solid shelving without in-rack sprinkler protection are very specific. If the criteria are not applied appropriately, the sprinkler system may be overtaxed, and a total loss in the storage area may ensue.

Block Flues

In storage arrangements, either piled or rack storage, there are flues. Flues are spaces between units, usually pallet loads, of storage. Longitudinal flue space is the space between rows of storage that is perpendicular to the direction of loading. When looking at the end of a row of racks, the longitudinal flue space is the space between rows of storage that travels the length of the rack. Transverse flue space is the space between rows of storage parallel to the direction of loading. When looking at the face of rack storage, the transverse flue is the space at the uprights and between pallet loads as you look at the front face of a rack. The flues are important because they help to allow water discharge from the sprinkler system to penetrate into the racks and reach the lower levels of storage. Blocking the flues may prevent sprinkler discharge water from reaching the seat of the fire.

Size of Sprinkler Orifice in Storage Areas

Automatic sprinkler protection in storage areas has been refined through research and testing. Older sprinkler heads have a nominal orifice diameter of 1/2 inch (K = 5.6) and a usual water flow of 25 gpm or less. Today, sprinkler heads have evolved to being capable of discharging more than 125 gpm for each head, as much water as a fire service initial attack hose line. ESFR sprinkler systems can have a sprinkler water flow demand of 2,200 gpm plus another 500 gpm for hose streams. In order to flow the large volume of water from each sprinkler head, the orifice size must be larger than 1/2 inch spray sprinklers.

NFPA has defined the smallest orifice size permitted to protect storage areas based on the required density. In storage areas, spray sprinklers should be larger than 1/2 inch, K=5.6. If your storage requires a density of 0.20 gpm/sq. ft. to 0.34 gpm/sq. ft., the smallest standard-response sprinklers should have a nominal K-factor of K-8.0 or larger. For densities greater than 0.34 gpm/sq. ft. standard-response spray sprinklers with a K-factor of K-11.2 should be installed. Other applications may require other minimum orifice sizes. If a review of a sprinkler system identifies a required density that is greater than allowed for the sprinkler heads installed in the sprinkler system, then a thorough review should be performed. The sprinkler system may not be designed to provide the needed density for the existing storage arrangement.

Multiple Hazard Classifications within a Given Area

Areas that have multiple hazard classifications should be protected for the highest hazard classification. In addition, the water supply requirements and hose stream allowance should be in accordance with the highest hazard classification. For instance, storage areas that store an assortment of Class I, II, III, and IV commodities, the sprinkler system should be designed to protect the highest classification of storage, in this case, Class IV commodity.

Sometimes, a warehouse may be designed for a given commodity, and a few pallet loads of higher commodity goods are introduced into the storage area. Incidental pallet loads of goods of a higher commodity class may seem to be inconsequential, but in reality are not. More and more pallets may be introduced, and before long, the entire system should be modified to protect the higher commodity goods. NFPA 13 addresses this situation in Chapter 5, Classification of Occupancies and Commodities. Up to 10 pallet loads of a higher hazard commodity are permitted to be present in an area not exceeding 40,000 sq. ft. The pallet loads must be randomly dispersed in the storage, and no two pallet loads may be adjacent to one another. This means beside, above, or diagonally. Also, when the storage area is designed to protect Class I or Class II commodity, only five pallets of Class IV or Group A plastics may be stored in the 40,000 sq. ft. area.

The effect of having multiple hazard classifications within a given storage area that was not designed for the higher storage commodity may result in an inadequately protected area. The higher commodity materials may be segregated into a specified area that is protected adequately. This may be accomplished by relocating the pallet loads closer to the riser where the density may be greater than at the far end of the sprinkler system, modifying the sprinkler system so it provides adequate design density and area of operation, reducing the amount of storage of the higher commodity materials, or eliminating their storage all together.

Encapsulated Storage

Encapsulated storage presents a significant increased challenge to sprinkler systems when introduced without upgrading the sprinkler system. NFPA 13 identifies encapsulation as "a method of packaging that either consists of a plastic sheet completely enclosing the sides and top of a pallet load containing a combustible commodity, a combustible package, or a group of combustible commodities or combustible packages, or consists of combustible commodities individually wrapped in plastic sheeting and stored exposed in a pallet load." Encapsulation is used by shippers to stabilize pallet loads and protect goods in transit. This encapsulated unit load is stored in a warehouse or storage area without removing the encapsulation. Encapsulated materials are a higher hazard than unencapsulated materials and require a significantly increased density discharge from the sprinkler system.

Changing Pallet Types

NFPA 13 protection levels are established based on the use of wood or metal pallets. If your original storage method used wood or metal pallets and the pallets have been changed to another pallet style, the sprinkler protection may no longer be adequate. The sprinkler protection should be revaluated to determine if the change adversely affects the level of protection.

If unreinforced polypropylene or unreinforced high-density polyethylene plastic pallets replace wood or metal pallets, the commodity classification will increase by one commodity classification. This means that commodities Class I through Class IV are increased by one commodity level. A Class III commodity storage area is now a Class IV storage area.

If reinforced polypropylene or reinforced high-density polyethylene plastic pallets replace the wood or metal pallets, the commodity classification will increase by two commodity classifications. This means that commodities Class I through Class IV are increased by two commodity levels. A Class II commodity storage area is now a Class IV storage area. Class IV commodity goods are increased to a cartoned, unexpanded Group A plastic commodity.

The change from wood or metal pallets to plastic pallets significantly increases the hazard level, and may render the sprinkler system inadequate to protect the storage configuration. In the event of a fire, the sprinkler system may be severely overtaxed and fail to control a fire.

Idle Pallets

Idle pallets can present a high challenge to sprinkler systems. Protection requirements for idle pallets have increased over the last several years. If the sprinkler design is adequate for Ordinary Hazard Group II occupancy, and there is a hose stream demand of at least 250 gpm for 60 minutes, a limited number of wooden pallets can be stored inside the production or storage area, provided that the height of storage is no more than 6 ft., each pallet pile of no more than four (4) stacks of pallets is separated from other pallet piles by at least 8 ft. of clear space or 25 ft. of stored commodity. The maximum clearance to the ceiling is 20 ft.

The inside storage of idle plastic pallets should be limited to a height of 4 ft., and each pallet pile of no more than two (2) stacks of pallets is separated from other pallet piles by at least 8 ft. of clear space or 25 ft. of stored commodity. The sprinklers in the ceiling sprinkler system must also be high temperature-rated heads.

If idle pallets are stored in storage areas, the sprinkler system should be reviewed to verify the sprinklers are adequate for the high challenge. Plastic pallets have an extremely high rate of heat release and an overabundance of idle plastic pallets can quickly overtax a sprinkler system.

NFPA 13 lists very specific maximum storage height, maximum ceiling/roof height, sprinkler density, and area of operation based on the temperature rating and K-factor of the sprinkler heads. In addition, idle pallets may be stored on the lowest level of storage on racks or shelves, but only when additional sprinkler protection criteria are met. If these criteria are not met, your sprinkler system may be inadequate to protect the hazard. The sprinkler system may require reinforcement or the idle pallets should be removed.

Open-top Containers

NFPA 13, paragraph 3.9.1.19 defines open-top containers as having any shape that is entirely or partially open on the top and arranged so as to allow for the collection of discharging sprinkler water cascading through the storage array. When sprinkler heads discharge over a fire in a storage area, the sprinkler discharge water flows over the boxes or materials being stored and continues to flow downward through the storage array, whether it is piled or rack storage. This flow of water helps to quickly control the fire in lower levels of storage. When open-top containers are used in the storage, the sprinkler water will collect in the bins, delaying water flow to burning storage materials on lower levels. The fire may rapidly grow, overtaxing the sprinkler system. Once the sprinkler system is overtaxed, the fire cannot be adequately controlled, and total loss may ensue. Protection of storage with open-top containers is outside the scope of NFPA 13.

Types of Water Supplies

Sprinkler systems must be fed from a reliable water supply. The main types of water supplies can be grouped into those which are gravity fed or pumped. Some fire protection water supplies may have both gravity tanks and pumps.

Gravity-fed systems may have lower system pressures than pumped systems because the pressure in a gravity-fed system is directly related to the height of the water supply storage above the usage point. A column of water from a gravity tank will produce approximately 0.433 psi per foot of elevation. If a gravity tank is 100-feet high, the resultant static pressure from the tank with no water flowing is 43.3 psi. For fire systems, the design of the sprinkler system should be based on the bottom of the tank because as water is used by the sprinkler system, the pressure drops until the water is at the bottom of the tank.

Water-supply systems fed with pumps will be designed to meet the system demands, and every system using pumped supplies is designed for that specific application. More research is required to determine the design parameters of a water-supply system fed with pumps.

Fire protection systems may be fed from public water supplies, private water supply systems, or a combination of private and public. Sometimes, both the public and private water supplies may be capable of supplying the total demand of the fire protection systems. In other systems, both the private and public systems are required to meet the total fire protection demand.

When the public water supply has sufficient volume, but lacks adequate pressure to adequately feed the sprinkler systems, a fire booster pump may be installed. If the fire pump fails to start in the event of a fire, the sprinkler system may not be adequately supplied and may not control a fire.

Fire pumps may be fed from a dedicated water storage tank. In this situation, the fire pump is fed directly from the tank, and may have no other source of supply. When the tank is drained, the available water supply is then depleted. Depending on the values at risk at the facility, a second or even a third fire pump and tank may be provided. Fire protection systems with fire pumps and tanks are usually designed to meet specific hazard demands. You should be able to go back to the original design correspondence and blueprints to identify the system design criteria.

Understanding the water supply that is feeding the fire sprinkler systems at your facility is very important. By understanding the water supply and the available volume and pressure supplied, you may be able to identify water systems that are not capable of adequately supplying the sprinkler system demands.

If you know the sprinkler system needs a density of 0.35 gpm over the most hydraulically remote 3,000 sq. ft., the water volume required will probably be in excess of 1700 gpm when both sprinkler system and hose demands are considered. If the sprinkler system requires a base of riser pressure of 80 psi, the water supply must be capable of supplying at least this amount. If a hydrant test at your facility shows that the water supply will supply less than this volume and pressure, your system is possibly deficient and additional investigation is needed.

Water Supply Adequacy

Water supply adequacy should be a focus point when determining sprinkler system adequacy. Fire sprinkler systems must be adequately fed by a water supply of sufficient flow and pressure, and both the flow and pressure must be adequate to meet the sprinkler system demands. A water supply system capable of flowing large volumes of water, but at an insufficient pressure will be inadequate, and sprinkler systems that are not adequately supplied should be considered inadequate. This is not to say that a small fire which opens only one or two heads will be ineffective, but in the event of a major fire, the sprinkler system may be quickly overtaxed. Water supplies feeding sprinkler systems must be capable of supplying the largest sprinkler demand. This should include both sprinkler and hose demands. Hydraulically designed sprinkler systems should have a placard identifying the needed sprinkler flow, pressure, and hose demand and, if the water supply is not capable of providing the required demand, the sprinkler system will deliver a lesser density over the demand area. Depending on the severity of water supply inadequacy, the sprinkler system may be totally inadequate.

Water supplies fed from a public water system may become inadequate over time. Deterioration of the pipes in the public system may decrease the available flow. As a community grows, there may be more housing units and industrial plants built. As the community grows, the water demand increases, and the available water supply at your facility may decrease. The water supply available 20 years ago may not be available today. During times of draught, the water supply may be reduced. If the sprinkler system was built with little or no cushion between the sprinkler demand point and the available water supply, that sprinkler system may be inadequately supplied today.

Private fire protection pipe systems may deteriorate over time due to obstructions, corrosion, or tuberculation forming within the pipes. Water leakage in the public water system may increase as the system ages, reducing the available water supply at the sprinkler system. Water authorities may reduce their preventative maintenance programs in an effort to balance their operating budgets. Regular water supply testing should help identify degradations in the water supply system.

Other Red Flags

Many times, the easiest safety system evaluation is to identify systems that are definitely inadequate. Have you ever looked at a situation, and emphatically stated, "That's not going to work!" or "That's not strong enough!" or "That will never hold!" This technique to identify potential weaknesses is many times called "Red Flag" identification. Red flags are hints that you should look deeper into an arrangement, situation, or critical component. Red flags are items of concern or identification. Additional study regarding red flag items may help identify the design of a sprinkler system.

For instance, 1/2 inch sprinkler heads will not be used in a ceiling sprinkler system designed to protect storage occupancy with 20-foot high rack storage of plastic commodity. A 1/2 inch head simply will not produce adequate flows at a reasonable pressure to properly protect such a hazard. The sprinkler heads should be a red flag to look closer at the sprinkler system design.

When fire protection systems are installed, they are designed to protect a specific occupancy. Normally (but not always), the correct design criteria is used in the design of the sprinkler system. The original design criteria may not be a life-long design. There are many changes that may occur in a building or in an occupancy. Some of these changes may include:

- The use of the area may change.
- The materials used in the production of products may change.
- The storage configuration may change.
- The storage height may change.
- A different material is stored in an existing storage area.
- The water supply has a material change
- As business park develops, new factories are built, water demand increases, resulting in a reduced available water supply for the sprinkler systems.
- Deterioration of fire pumps or water mains that supply the sprinkler system.
- Structural changes to the building, walls installed or removed, change in building height, mezzanines constructed without updating the sprinkler systems.

<u>Obstructions to Prevent Proper Water Distribution from Sprinklers</u>

Obstructions are a reason for concern. NFPA 13 devotes a significant amount of interpretation to obstructions. Some obstructions affect spray pattern development from the sprinkler head and others prevent or interrupt sprinkler water discharge from reaching hazards. Regardless of the reason, if the water cannot reach the fire, the fire cannot be properly controlled.

Obstructions affecting spray pattern may be from columns, girders, beams, bar joists, light fixtures, soffits, and HVAC ducts. Sprinklers are designed as high technology devices. They may flow large amounts of water and distribute the water in specific, precise patterns. Each type of sprinkler is designed by the manufacturer, and tested by a listing or approval organization. The listing requirements specify the installation parameters, including obstruction requirements. The obstruction requirements must be followed or the system may be rendered partially or wholly ineffective in controlling or extinguishing a fire.

Sprinklers must be installed with adequate distance to beams or building members. The listing for each sprinkler head and NFPA 13 identify criteria for positioning sprinklers to avoid obstructions. These instructions include the maximum allowable distance of the sprinkler deflector above the bottom of the obstruction. Usually, both horizontal and vertical distances are provided and must be followed.

Other obstructions may be large horizontal obstructions, such as esthetic panels in large areas used to visually enhance the space. They may also be vertical room dividers, privacy curtains, or free-standing, privacy-partition dividers. These items may inhibit proper sprinkler coverage and can render the sprinkler system inadequate.

Inappropriate Building and Decorating Materials

Buildings evolve over time. As buildings change owners and occupancies, they may be renovated to meet the new needs. Sometimes, the changes introduce hazards that were not anticipated during the original sprinkler system design. Many engineered construction systems use synthetic materials that are very dependent upon their orientation when installed. For instance, some insulated panel assemblies used in industrial buildings are made with rigid and foam plastics and have very high flame-spread characteristics.

Spray-on foam plastics used for insulation and esthetics may have extremely high rates of heat release under fire conditions, especially when installed in a vertical configuration. The Station nightclub fire, which occurred on February 20, 2003 in West Warwick, RI, involved plastic foam in the walls and ceilings. The fire engulfed the nightclub within several minutes, killing 100 and injuring 230 others.

Introducing new construction materials such as these may exceed the capability of the existing sprinkler system. While plastic materials are the most obvious materials, other materials, such as fabric on the walls or used as drapes, may also introduce a fuel load beyond the capability of the existing sprinkler system. In schools, hanging posters or artwork in classrooms or hallways may introduce a potential for a flash fire or add significant fuel loads that may make the sprinkler system inadequate.

Impaired Fire Protection

Any impairment to the sprinkler system or water supply may render the sprinklers inadequate in the event of a fire. A good impairment handling program should be implemented to help prevent fire system impairments. Impairments may occur from a number of reasons, including fire pumps or water supply out of service for any reason, shut valves, or freezing. It is critical that sprinkler systems be maintained in operational order; when they are not, they should be quickly repaired and returned to service. NFPA has codes and standards covering these issues, including NFPA 25, Standard for the Inspection, Testing and Maintenance of Water-Based Fire Protection Systems. Any sprinkler system that is impaired in any way, whether the impairment is known or not, may be inadequate to protect the occupancy.

Conclusion

It is easy to feel protected from a fire event simply by the presence of an automatic sprinkler system within your facility. As safety professionals, we know that most safety measures are more complex than what actually meets the eye. Sprinkler systems are no exception. Regular evaluation and remediation of the many components of these systems, including water supplies, sprinkler heads, sprinkler design, commodity of storage, and regular maintenance help to ensure that your facility will be able to withstand the effects of a fire.

Bibliography

Beattie, Walter S. "Evolution of the Fire Sprinkler." *Fireline* 2011: 11-21. Beattie Fire Protection & Risk Consulting, LLC. Web. <https://waltbeattie.files.wordpress.com/2010/08/evolution-of-as-beattie.pdf>.

Beattie, Walter S. *Fire Sprinkler Systems*. 2009. Des Planes, IL: American Society of Safety Engineers. Beattie Fire Protection & Risk Consulting, LLC. Beattie Fire Protection & Risk Consulting, LLC. Web. <https://waltbeattie.files.wordpress.com/2010/08/fire_sprinkler_systems1.pdf>.

Beattie, Walter S. *Fire Pumps*. Vol. 2. 2010.Des Planes, IL: American Society of Safety Engineers, Ser. 2. Beattie Fire Protection & Risk Consulting, LLC. Beattie Fire Protection & Risk Consulting, LLC. Web. <https://waltbeattie.files.wordpress.com/2010/08/firepump.pdf>.

Beattie, Walter S. 2008. "Handling Impairments to Fire Protection Systems." *Professional Safety*. Beattie Fire Protection & Risk Consulting, LLC. Web. <https://waltbeattie.files.wordpress.com/2010/08/handling_impairments_to_fire_protection_systems.pdf>.

Ferguson, Lon H., and Christopher A. 2005. Janicak. *Fundamentals of Fire Protection for the Safety Professional.* Lanham, MD: Government Institutes.

National Fire Protection Association. 2013. NFPA 13, *Standard for the Installation of Sprinkler Systems.* Quincy, MA: National Fire Protection Association.

National Fire Protection Association. 2014. NFPA 25, *Standard for the Inspection, Testing, and Maintenance of Water-Based Fire Protection Systems*. Quincy, MA: National Fire Protection Association.

Session No. 533

An Athletic Approach to Heat Illness: Beyond Water, Rest, Shade

Allie Thunstrom
Alsie Nelson
Ergodyne
St. Paul, MN

Scholars in the athletic field have studied heat-related illnesses and their impact on athletes of all ages and levels. Their findings can be applied to industrial settings to aid safety professionals in preventing heat-related illnesses in the workplace.

Heat-related illnesses, injuries, and death are entirely preventable in the workforce. OSHA's national heat stress campaign centers around prevention, water, rest, and shade best practices, and has been very successful at raising awareness to the safety concern around heat stress. Yet heat-related deaths doubled from 2014 to 2015, and the number of non-fatal cases resulting from excessive heat exposure rose to 2830 (Bureau of Labor Statistics, 2015). With weather patterns becoming increasingly unpredictable, it is more important than ever for safety professionals to be informed on the latest research to prevent future heat-related illness and injuries. We will explore the science behind heat stress, how to recognize symptoms in the field, and how research from athletics can be used to prevent occupational heat illnesses, injuries, and most importantly, deaths.

Performance and Physiology

The Science Behind Heat Stress

Heat-related illness (HRI) is an umbrella term used to describe a variety of health issues that result from prolonged exposure to environmental heat and humidity, including heat rash, cramps, heat syncope, heat exhaustion, and heat stroke. They are studied at length by professionals in various fields, including occupational health and athletics. To better understand how these illnesses and injuries occur, it is important to understand the underlying body processes that are taking place, and how they contribute to the onset of symptoms.

Thermoregulation is the complex interaction of the body's central nervous system, the cardiovascular system, and the skin that helps maintain a core temperature of approximately 98.6° Fahrenheit. Thermoregulation operates like a home thermostat system in that the body (the hypothalamus) receives information from skin receptors and circulating blood and initiates heat transfer responses as needed. When core body temperature rises, the blood vessels near the surface of the skin widen, and sweating begins to occur in an attempt to dissipate heat (Casa et. al, 2012). As the thermoregulation system is overwhelmed by prolonged heat exposure, hyperthermia – a body temperature greater than normal – can occur. Moderate hyperthermia is normal and protective, but if it continues for a prolonged period of time and develops into an HRI, harmful long-term neurological effects can result. It is not necessarily the degree of hyperthermia that is important, but rather the duration of the increased body temperature, which is why it is critical to drop body temperature as soon as possible. Studies done by Dr. Doug Casa and the Korey Stringer Institute (2012) have shown that if a body is cooled within 30 minutes – the "golden half hour" – there is a 100-percent survival rate.

In general, the body gains or loses heat in one of four ways:

1. **Radiation:** Heat is transferred to or from an object or body via electromagnetic radiation from higher to lower energy surfaces (i.e., sunlight).
2. **Conduction:** Heat is transferred from warmer to cooler objects through direct physical contact (i.e., ice packs).
3. **Convection:** Heat is transferred to or from the body to surrounding fluid or air (i.e., fan or immersion in water).
4. **Evaporation:** Heat is transferred via vaporization of sweat.

Any of these heat-loss methods can help overheating workers lower body temperature and prevent HRIs.

Ice packs and cooling towels are a good example of conduction, while a cold bath or misting system are examples of viable convection methods.

Evaporation is the most effective means of heat transfer. Efficient evaporation depends on water saturation of the surrounding air and velocity of the air, but its effectiveness is greatly reduced when humidity is high, as water cannot easily vaporize. In some cases, those experiencing severe heat exhaustion or heat stroke will stop sweating because their body can no longer react to an elevated internal temperature.

Mental Function

Performance and physiology are key areas of study for exercise physiologists. The information obtained during athletic and military studies is valuable to those outside the aforementioned fields and can also be applied to those in the safety industry. Decreased mental performance may be linked to the discomfort of a high skin temperature, high skin wettedness, and strain to their cardiovascular system.

Military research has found that mental performance degrades most when a person is performing repetitive tasks (Department of Army and Air Force, 2003). This is often the case in industrial and construction settings, when routine tasks are completed more slowly when a worker is suffering from heat stress. Vigilant task performance worsens significantly after two to three hours of heat stress. Furthermore, heat stress can slow a worker's reaction time and decision-making ability, leading to errors.

Dehydration intensifies a worker's decreased mental function during heat exposure. Even simple tasks such as addition, response time, and recognition of words and phrases decrease as dehydration is prolonged.

Staying Off the Injured List

Safety professionals can use research from athletics to help workers affected by heat stress return to work or to prevent heat stress from occurring in the first place.

Korey Stringer Institute

After professional football player Korey Stringer died from complications of heat stroke in August of 2001, his wife Kelci teamed with exertional heat stroke expert Douglas Casa, Ph.D., ATC to form the Korey Stringer Institute at the University of Connecticut in April 2010. The institute has identified four "big-ticket" items that, if followed, will almost certainly eliminate HRIs and fatalities:

1. Hydration
2. Heat acclimatization
3. Work-to-rest ratios
4. Body cooling throughout the session

Hydration is an obvious prevention method; however, it is only one piece of the HRI puzzle. All four items must be considered and implemented to prevent HRIs.

Hydration

Water is an integral component of body processes as it carries nutrients and oxygen to cells, helps convert food to energy, and regulates body temperature. It accounts for 60 percent of body weight, 75 percent of muscles, and 80 percent of the brain. When an individual is adequately hydrated, called euhydration, body processes such as thermoregulation are able to occur. When an individual is dehydrated, however, these processes are unable to successfully happen. As discussed earlier, dehydration is a major contributing factor to HRIs. To prevent dehydration, and thus minimize the risk of contracting an HRI, workers should have free access to readily available fluids at all times. People should aim to prevent weight loss of more than two percent and keep urine light in color. One way to determine the amount of fluid an individual lost is to weigh themselves in the morning before their shift and immediately after. Every pound lost should be replaced with eight ounces of water (Casa et. al, 2012). Furthermore, electrolytes will be lost during the perspiration process and should be replaced accordingly.

Acclimatization

Acclimatization is a key prevention method to which states with existing heat illness standards refer. The California Heat Illness Prevention Standard, for example, defines acclimatization as a "temporary adaptation of the body to work in the heat that occurs gradually when a person is exposed to it. Acclimatization peaks in most people within four to fourteen days of regular work for at least two hours per day in the heat."

Military research agrees that optimal acclimatization requires specific and deliberate exposure to heat conditions. Though the effects of acclimatization to heat on mental performance have not been determined, acclimatization should lead to better mental performance for workers because their thermal comfort would be improved.

During the first few days of acclimatization, physiologic strain will be the highest and will be manifested by elevated core temperature and heart rate. This strain decreases with each subsequent day, and dramatically improves comfort and physical work capabilities as time goes on. Successful acclimatization increases stroke volume, which is the amount of blood pumped per heartbeat, and sweat rate, and decreases heart rate, core body temperature, skin temperature, and sweat salt loss.

It is important to note that dehydration negates many of the thermoregulatory advantages of acclimatization (Department of Army and Air Force, 2003). The military is one of the best organizations at helping their employees get accustomed to heat. Some of their heat acclimatization strategies that might apply to an industrial setting include:

- Mimic the work environment climate.
- Use work and rest to modify the heat strain.
- Ensure four to fourteen days of heat exposure.
- Maintain the daily duration of at least 100 minutes.
- Begin the acclimatization process up to one month *before* work in hot conditions begins.
- Upon arrival to the hot work environment, start slowly and reduce work intensity and duration and limit heat exposure.
- Acclimatize in the heat of the day, work in the coolest part of the day.

Work-to-Rest Ratio

Once temperatures reach 75 degrees Fahrenheit, begin adjusting work-to-rest ratios. There are many variables, including temperature, work intensity, and clothing/equipment that come into play, but the general idea is that, as the heat increases, so to should the frequency and length of breaks. OSHA has separate, temperature-dependent guidelines in place for workers doing light, moderate, and heavy work. For light work, the defined temperature is a Wet Bulb Globe Temperature (WBGT) of 86° Fahrenheit, 80° F for moderate work, and 77° F for heavy work. These guidelines suggest workers should not be subjected to the aforementioned temperature situations for more than two hours without a break (Minnesota Department of Labor and Industry, 2016).

Body Cooling

Preventing body temperatures from reaching dangerous levels while working in high-heat environments is key to heat-related illness prevention. One way to combat rising temperatures is to utilize body cooling techniques throughout the workday or training session. These techniques include removing outer layers of clothing and/or equipment to allow the body to cool off and receive air flow. Place cooling towels, wet sheets, or ice packs over major arteries or areas with a conglomeration of blood vessels to facilitate cooling (Department of Army and Air Force, 2003). Wetting the skin surface via a hose, mist, or other means, and applying a fan can accelerate the evaporation process.

Psychology of Toughness

HRI is prevalent in athletes during the preseason and among new workers in industrial settings. In both populations, fighting the notion of, "I don't need a break," is a challenge. The new worker may set out to prove their worth to an organization in the same way an athlete aims to impress their coach early in the season. This overzealousness may obviate a worker's training on HRI symptom recognition, causing them to bypass normal behavioral adaptations to heat and ignore the early warning signs of an HRI.

On the athletic field, recognition of early HRI symptoms is supplemented by the presence of an athletic trainer or team doctor who is well versed in HRIs and takes prompt action when symptoms develop in an athlete. It is very unlikely to have such oversight in an industrial setting, leaving recognition of HRI symptoms to safety managers, supervisors, and the workers themselves. Often, those supervising athletes have incorrectly assumed that an HRI condition is not serious and that the athlete will recover on their own with rest (Casa et al, 2012). In the workplace, this attitude can be addressed in frequent training when workers will be exposed to hot working environments.

William M. Adams et. al, in conjunction with the Korey Stringer Institute, conducted a study of high school football coaches and their perceived confidence in dealing with heat stress. What they *thought* they knew versus what they *actually* knew was worlds apart. Despite being self-assured, most coaches were very deficient in their knowledge of the signs, symptoms, and emergency treatment of heat stroke (Adams et al, 2014). Studies show that, if a heat stroke victim is not treated and cooled within a half hour of collapse, survival rate drastically decreases. So it is essential that authority figures recognize the signs of HRIs swiftly and address them properly.

Treatment and Return to Work

HRIs are cumulative in nature and can manifest themselves over the course of hours or days (Casa et. al, 2012). To counter the accumulation of heat stress, individuals should aim to incorporate some proactive measures into their daily routine while working in hot environments:

- Get at least seven hours of sleep per night.
- Eat a balanced diet.
- Hydrate properly *every day*, not just on work days.
- Rest in a cool environment and resist the urge to spend additional time in the heat.
- Decrease or eliminate the use of dietary supplements or substances that have a dehydrating effect, increase metabolism, or affect thermoregulation processes (i.e., diuretics, coffee, and energy drinks).
- Be cognizant of any underlying health conditions that could affect the body's natural response to high heat exposure.

Treatments depend on the type of HRI the worker is experiencing (National Athletic Trainers' Association, 2002).

1. **Heat Rash:** Make sure skin is dry, move to a shaded area and rehydrate.
2. **Heat Cramps and Muscle Cramps:** Get rest and do some light, static stretching. A massage may help. Hydrate and replace electrolytes.

3. **Heat Syncope:** Move to a shaded area and elevate the legs above the heart, cool the skin, and rehydrate.
4. **Heat Exhaustion:** Use the same treatments as described for heat syncope, and also remove excess clothing and equipment to facilitate evaporation. Fluid replacement via IV may be necessary and should be done in a healthcare facility.
5. **Heat Stroke:** The worker's core body temp should be lowered to under 102° F within 30 minutes of collapse. This will return blood flow from the skin to the heart and lower the core body temperature by reducing the hypermetabolic state of the organs. The worker's trunk and extremities should be quickly immersed in a pool or tub of cold water (35° F-59° F) or wet towels should be placed over the entire body or doused with cold water. Consult a medical professional before submerging a person into a tub of water.

If a worker experiences any of the HRIs described above, returning to work on the same day is not recommended. According to the National Athletic Trainers' Association, the fastest way to decrease core body temperature is by immersing the trunk and extremities into a pool or tub filled with cold water (as described above). Other forms of cooling commonly used, including water spray, ice packs, or blowing air, decrease the body at a slower rate than cold-water immersion. If immersion is not possible, cooling with ice bags or cooling towels on as much of the body as possible, at major vessels in the armpit, groin, and neck regions as well as the hands and feet, and cold towels on the head and trunk can assure the most rapid heat loss.

The Industrial Athlete

The Korey Stringer Institute's unofficial fifth big-ticket item is education. The industrial athlete faces similar extreme environments, pressure, and exertion as a world-class athlete in their field of competition. Both can benefit from repeated HRI education on prevention, recognition, and treatment. HRIs can be prevented by adequate hydration, acclimatization, implementing work-to-rest ratios, and body cooling as temperatures soar. Having the ability to recognize and treat heat stress victims early can be the difference between life and death. By understanding an athletic approach to HRI prevention, the industrial worker can get a new, illuminating perspective on the matter.

Bibliography

Bureau of Labor Statistics (BLS). 2015. *Fatal Occupational injuries by industry and event or exposure, All U.S.* (www.bls.gov/iif/).

California Department of Industrial Relations. 2017. *Heat Illness Prevention, Section 3395.* (https://www.dir.ca.gov/title8/3395.html)

Department of the Army and Air Force. 2003. *Heat Stress Control and Heat Casualty Management.* (http://armypubs.army.mil/med/dr_pubs/dr_a/pdf/tbmed507.pdf)

Minnesota Department of Labor and Industry, Occupational and Health Division. 2016. *MNOSHA Heat Stress Guide.* (http://www.dli.mn.gov/OSHA/PDF/heat_stress_guide.pdf)

Korey Stringer Institute, William M. Adams, MS et al. 2014. “The Secondary School Football Coach’s Relationship With the Athletic Trainer and Perspectives on Exertional Heat Stroke.” *Journal of Athletic Training.*

Korey Stringer Institute, Douglas J. Casa, PhD et al. 2012. “Exertional Heat Stroke: New Concepts Regarding Cause and Care.” *Current Sports Medicine Reports.*

National Athletic Trainers’ Association. 2002. “Position Statement: Exertional Heat Illnesses.” *Journal of Athletic Training.*

Medical Marijuana: Miracle Drug or Safety Nightmare?

Fred H. Kohanna, MD, MBA, FACOEM
V.P. of Occupational Health & Chief Medical Officer
AllOne Health
Woburn, MA

Historical Perspective on Medical Marijuana

The Chinese Emperor Shen Nung, in 2737 B.C., described the therapeutic use of marijuana for joint pain, constipation, malaria, and childbirth. Therapeutic and religious use achieved great popularity in India around 1000 B.C. Medicinal use of marijuana continued, and spread to Africa, the Middle East, and the Arabian Peninsula into the 18th century.

Marijuana was introduced into western medicine by a physician, W.B. O'Shaughnessy who, in 1839, described its use for pain control, muscle relaxation, appetite stimulation, and as a treatment for nausea, and for seizures. The psychiatrist Moreau, in 1845, published an article describing the use of marijuana in his patients. In 1850, marijuana was added to the U.S. Pharmacopoeia. Information regarding medical uses of marijuana disseminated throughout North America and Europe. By 1900, more than 100 scientific articles had been published on medical marijuana. Marijuana extracts were being marketed by prominent pharmaceutical companies.

Use of marijuana declined in the U.S. from 1900 through the 1930s due to difficulty in standardizing preparations, development of alternative "mainstream" pharmaceuticals, and taxation by the Federal Marijuana Tax Act of 1937. By the 1930s, there were at least 2000 medicines, with over 280 manufacturers. Marijuana was removed from the U.S. Pharmacopoeia in 1942. In 1970, The Controlled Substances Act classified marijuana as a "Schedule 1" drug, in the same category as heroin.[1]

Mechanism of Action of Marijuana

Marijuana is a synonym (slang) for cannabis, which is derived from the marijuana plant, Cannabis sativa. The primary active constituents in the marijuana plant are Delta9-tetrahydrocannabinol (THC) and Cannabidiol (CBD).[2] The human body has a receptor-signaling system for marijuana known as the Endogenous Cannabinoid System (ECS). The ECS has two types of receptors that activate very different bodily functions: CB1 receptors are very abundant in the brain, and solely mediate the behavioral, analgesic, and euphoric effects of marijuana (THC) (e.g., memory, cognition, perception, and pain). CB2 receptors are not very abundant in the brain, but are highly expressed in the gut and immune cells where they regulate inflammation and immune function. Cannabidiol (CBD) activates these receptors (e.g., anti-inflammatory, anti-seizure, anti-nausea, and anti-anxiety).

Potency of cultivated marijuana has steadily increased. In the 1980s, THC content averaged 3%. By 2012, THC content had increased to 12%.[3] The duration of action varies with the route of administration. Inhaled marijuana peaks in 15 to 30 minutes, and effects taper in 2-3 hours. Ingested marijuana peaks in 2-3 hours, and effects taper in 4-12 hours. Marijuana is metabolized predominantly in the liver and is excreted into the bile and urine. Marijuana is stored in body fat and can be excreted in the urine for up to 1-2 months in heavy users.

Federal Versus State Laws

Marijuana remains illegal at the federal level. As of January, 2017, 28 states and the District of Columbia have legalized medical marijuana.[4] Eight states and the District of Columbia have legalized marijuana for recreational use. Several synthetic marijuana preparations have been FDA approved. Dronabinol (Marinol) is synthetic THC and is approved for chemotherapy-induced nausea/vomiting, and for appetite stimulation in patients with AIDS. Nabilone (Cesamet) is a synthetic analogue of THC with similar indications for use.

The FDA is investigating two additional marijuana preparations for "fast-track" approval. Nabiximols (Sativex) is an oral spray with THC and CBD in a 1:1 ratio, indicated for symptomatic relief in multiple sclerosis, and for pain relief in cancer patients. Epidiolex is a concentrated CBD oil, indicated as an anti-seizure medication for intractable epilepsy.

Indications for Medical Use

Marijuana is *not* a first-line drug of choice for any medical condition. Marijuana is typically recommended when available therapies have been ineffective, or when side effects of those therapies have been unacceptable. In 2014, 94% of applicants for medical marijuana in Arizona reported "severe pain" as their reason for requesting marijuana use. Marijuana is not typically helpful for acute pain. Marijuana may act synergistically with opioids to enhance analgesia and to allow lower doses of opioids.

Appropriate indications for medical marijuana have been promulgated by organizations such as the National Academy of Sciences, Health Canada, the State of New York, and the Office of Medicinal Cannabis (OMC) in the Netherlands. These indications only apply to patients for whom standard therapies have been ineffective, or who have had intolerable side effects. Indications for medical marijuana include the following:

- Poor appetite, weight loss, and wasting in cancer or HIV/AIDs
- Autoimmune disorders, such as rheumatoid arthritis, inflammatory bowel disease, and lupus erythematosis
- Pain from HIV/AIDs or cancer
- Chronic neuropathic pain from multiple causes, including diabetes, nerve trauma, post-shingles, complex regional pain syndrome (CRPS), etc.
- Nausea and vomiting from chemotherapy, radiation, or other medications
- Neuropsychiatric conditions, including epilepsy, Parkinson's disease, Tourette's syndrome, and PTSD

- Intractable glaucoma
- Pain and spasticity from multiple sclerosis or spinal cord injury

Contraindications and Side Effects

Medical marijuana should not be used in the following situations:

- Pregnancy or planning to get pregnant
- Breastfeeding
- Severe heart, lung, liver, or kidney disease
- Allergy to cannabinoids or smoke
- Schizophrenia or other psychotic disorders

Relative contraindications to medical marijuana include:

- Substance abuse history
- Use of psychoactive medications
- Other psychiatric illnesses (e.g., anxiety/panic attacks)
- Asthma/Chronic obstructive pulmonary disease (COPD)

Side effects include:

- There are no cases of fatal overdoses from marijuana alone
- Panic attacks
- Cognitive impairment
- Seizures
- Hallucinations/psychosis
- Sedation
- Dry mouth
- Impaired reaction time
- Palpitations/Arrythmias
- Addiction (approximately 9% of recreational marijuana users will develop addiction)
- Withdrawal syndrome (e.g., cravings, restlessness, irritability, anger, insomnia, and loss of appetite)
- No definitive evidence at this time that marijuana is a "gateway" to other drugs

The Risk of Impairment and Impact on Safety

Marijuana use has a negative effect on learning, memory, attention, reasoning, and concentration.[5] In a flight simulator study, pilots were impaired after marijuana use at 1, 4, and even 24 hours after consumption.[6] Marijuana use (acutely) is associated with an increased risk of motor vehicle collisions, especially fatal crashes.[7] Marijuana use results in delayed braking reaction time, impaired lane tracking, aversion to lane changing, and inattention to speed.[8] Simulator studies indicate that use of marijuana adversely affects psychomotor skills required for safe driving in a dose-response fashion.[9] A meta-analysis of 21 studies in 13 countries. with 239,739 participants, found that marijuana use (self-reported or found in blood, urine, or saliva tests) was associated with a 20% to 30% greater risk of a motor vehicle crash.[10] Marijuana use could impair human performance up to 24 hours.[11] Postal workers who tested positive for marijuana on a pre-placement drug test had 55% more industrial accidents, and their absentee rate was 75% higher.[12]

In contrast, some studies have shown no association between marijuana use and workplace safety. In 1994, a survey of 9097 employees aged 18 and older were surveyed on drug use. There was no association between marijuana use (recent or remote) and the risk of work-related accidents.[13]

In 2014, employees who had a work-related injury were tested for marijuana and compared to a control group. The workers who tested positive for marijuana were not more likely to have had a work-related injury than the control group.[14] Studies suggest that tolerance to the effects of marijuana develop in long-term users, and that impairment for short-term users is significantly greater by comparison.

Implications for Drug Testing Programs

Should Employers Maintain the Status Quo?

Medical marijuana cannot be allowed by any employer who is governed by federal or state regulations that specifically prohibit its use (e.g., FMCSA, NRC). Employer prohibition of marijuana use by employees (medical or recreational) is still supported by federal law and does not violate the Americans with Disabilities Act (ADA). Levels of marijuana in the urine or saliva do not correlate with the level of impairment (unlike blood or breath alcohol levels), and so there is no "safe or acceptable" level. Dronabinol and Nabilone (both FDA-approved drugs) would still constitute legitimate use, and would result in a Medical Review Officer (MRO) negative drug test, but safety concerns would still need to be considered by the employer.

Should Employers Consider Allowing Employees to Use Medical Marijuana?

Employers will probably face increasing pressure from employees to allow medical marijuana use for non-federally regulated jobs.[15] Ongoing legal challenges to workplace drug testing policies may force changes to employer drug testing protocols. Consideration for medical marijuana use by workers must address critical considerations such as:

- Safety sensitive vs. non-safety sensitive jobs
- Underlying diagnoses
- Frequency of use and route of administration (inhalation vs. ingestion)
- Duration of use (short-term vs. long-term)

- Necessary work accommodations or restrictions
- Worker would have to notify the employer of any changes to dose, frequency, or route of administration
- In most cases, medical marijuana should not be administered while the employee is at work
- Fitness-for-duty evaluations, including consultation with an Occupational Health physician, and neuro-cognitive testing should be available as needed

Conclusions

Medical marijuana has been in use throughout the centuries. Although marijuana is not a first-line therapy for any medical condition, it does have proven efficacy for a number of difficult to treat chronic medical illnesses. Medical marijuana has both benefits and risks. The increasing use of medical marijuana by employees will pose a threat with regard to workplace safety, and will create challenges for current drug-testing program policies. Should a company decide to allow the use of medical marijuana by an employee, strict safeguards will need to be implemented to address safety sensitive work, any necessary job accommodations or restrictions, as well as ongoing fitness-for-duty issues. The bottom line is that medical marijuana is *neither* a miracle drug nor a safety nightmare.

Endnotes

1. Goldsmith, RS, et al. "Medical Marijuana in the Workplace: Challenges and Management Options for Occupational Physicians." *JOEM* 2015; 57:5:518-525.

2. Schatman, ME. *Medical Marijuana: The State of the Science.* February 6, 2015. Medscape (retrieved February 11, 2017) (http://www.medscape.com/viewarticle/839_print).

3. Volkow, ND, Baler, RD, Compton, WM, and Weiss, SRB. "Adverse Health Effects of Marijuana Use." *N Engl J Med* 2014; 370: 23:2219-2227.

4. The National Academies Press. 2017. *The Health Effects of Cannabis and Cannabinoids: The Current State of Evidence and Recommendations for Research* (retrieved February 11, 2017) (http://www.nap.edu/24625)

5. Crane NA, Schuster RM, Fusar-Poli P, Gonzalez R. "Effects of cannabis on neurocognitive functioning: recent advances, neurodevelopmental influences and sex differences." *Neuropsychol Rev.* 2013; 23:117-137.

6. Yesavage JA, Leirer VO, Denari M, Hollister LE. "Carry-over effects of marijuana intoxication on aircraft pilot performance: a preliminary report." *Am J Psychiatry.* 1985; 142: 1325-1329.

7. Asbridge M, Hayden JA, Cartwright JL. "Acute cannabis consumption and motor vehicle collision risk: systematic review of observational studies and meta-analysis." *Brit Med J.* 2012: 344.

8. Liquori A, Gatto CP, Robinson JH. "Effects of marijuana on equilibrium, psychomotor performance and simulated driving." *Behav Pharmacol.* 1998; 9: 599-609.

9. Sewell, RA, J. Poling, and M. Sofuoglu. 2009. "The effect of cannabis compared with alcohol on driving." *American Journal on Addictions.* 18(3):185-193.

10. Rogeberg, O, and R. Elvik. 2016. "The effects of cannabis intoxication on motor vehicle collision revisited and revised." *Addiction.* 111(8):1348-1359.

11. Heishman SJ, Huestis, MA, Henningfield JE, Cone EJ. "Acute and residual effects of Marijuana: profiles of plasma THC levels, Physiological, subjective, and performance measures.: *Pharmacol Biochem Behav.* 1990; 73:561-565.

12. Normand J, Salyards S. "An Empirical Evaluation of Pre-employment Drug Testing in the United States Postal Service: Interim Report of Findings. Drugs in the Workplace: Research and Evaluation Data. National Institute on Drug Abuse." 1989: *NIDA research monograph* 91:111-138.

13. Hoffman, J. and C. Larison. 1999. "Drug use, workplace accidents and employee turnover." *Journal of Drug Issues.* 29(2):341-364.

14. Price, JW. 2014. "Marijuana and workplace safety: An examination of urine drug tests.: *Journal of Addictive Diseases.* 33(1):24-27.

15. Phillips, JA, et al. "Marijuana in the Workplace: Guidance for Occupational Health Professionals and Employers." *JOEM* 2015, 57:4:459-475.

Session No. 537

Practical Tools for Gender Equal Protection in OSH Programs

Morgan M. Bliss, M.S., CIH, CSP
Assistant Professor
Central Washington University

Jacob Krzystowczyk, M.S., CIH, CSP
Corporate Industrial Hygienist
The Albemarle Corporation

Introduction

In the past year, there have been significant changes to regulatory and legal aspects of gender equal protection in the United States (U.S.). These changes have been enacted at both the U.S. Supreme Court and state level. After the 2016 Supreme Court Decision in *Peggy Young v. UPS,* occupational safety and health (OSH) professionals may notice an increased interest in, and concerns about, gender equal protection from management, occupational medicine providers, and human resource partners. Additionally, several states have passed, or are in the process of deliberating on, workplace laws about pregnancy accommodations and protection. These national conversations have re-energized the discussion about gender equal protection in the workplace.

State-based laws have generally been termed a Pregnant Workers Fairness Act (PWFA) or similarly named acts. As of June 2016, there are 18 states, the District of Columbia, and 4 municipalities that have enacted PWFA legislation, including: Alaska, California, Colorado, Connecticut, Delaware, Hawaii, Illinois, Louisiana, Maryland, Minnesota, Nebraska, New Jersey, New York, North Dakota, Rhode Island, Texas, Utah, and West Virginia (National Partnership for Women and Families, 2016). There is also a proposed Pregnant Workers Fairness Act (S.1512) at the federal level, but progress in passing the act is slow. Congress has "attempted multiple times to pass" the federal PWFA, but the law has been opposed by multiple sessions of Congress (Pisko, 2016).

Also relevant are the upcoming regulatory changes associated with the revisions to the U.S. Environmental Protection Agency (EPA) Toxic Substances Control Act. As part of the Frank R. Lautenberg Chemical Safety for the 21st Century Act, a definition for "potentially exposed or susceptible subpopulation" has been established, to include a group of individuals with greater risk than the general population for adverse health effects relating to chemical exposure. This greater risk is further explained to be either from greater susceptibility or greater exposure, and includes infants, children, pregnant women, workers, and the elderly (Frank R. Lautenberg Chemical Safety for the 21st Century Act, 2016).

In a June/July 2016 article from *The Synergist,* entitled "Windows of Susceptibility," authors Lewandowski and Dodge provide a sobering statistic: "[…] 66% of women who gave birth to their first child between 2006 and 2008 worked during their pregnancy." In a section of their article discussing occupational exposure limits (OELs) and evaluating developmental and reproductive toxicity (DART) effects, the authors state that it is "important to encourage workers to inform health and safety personnel of their pregnancy (or intended pregnancy) as early as possible" so that the "windows of susceptibility" for the fetus are sufficiently protected (Lewandowski & Dodge, 2016).

The challenges associated with this topic and the recent regulatory developments can create a problem for an OSH professional. If a worker notified their company safety professional or industrial hygienist about their pregnancy or intent to conceive, would the OSH professional know what to do? Where can they go for this information?

Based on a review of common safety and industrial hygiene texts that would be accessible to a U.S.-based OSH professional, relevant guidance on managing the safety and health of pregnant or breastfeeding workers may be insufficient or difficult to understand. The indexes of texts from the American Society of Safety Engineers (ASSE), American Industrial Hygiene Association (AIHA), American Conference of Governmental Industrial Hygienists (ACGIH), and other commonly used OSH texts were reviewed for terms such as "pregnant," "reproductive," "teratogen," "breastfeeding," and other similar terms. The findings of this review are in Table 1.

Table 1: Available U.S. Guidance on Safety and Health for New or Expectant Mothers

Name of Text/Document	OSH Guidance Provided
The Safety Professional's Handbook, 2nd Edition (American Society of Safety Engineers, 2012)	• The index does not contain the terms pregnant, breastfeeding, or reproductive. • Page 364 mentions concerns about chemicals in breast milk and whether there are "safe" levels of chemicals in breast milk; this is the hazard communication section of the text. A recommendation from the authors is to develop a "set of flow charts to help evaluate the level of concern in worker populations and workplaces and a series of proactive steps for protecting both workers and company resources." • Page 718 mentions OELs and teratogenic materials, with a goal of minimizing or eliminating, to the extent possible, any worker exposure.
Safety and Health for Engineers, 2nd Edition (Brauer, 2006)	• Pregnancy and reproductive hazards are not discussed, only a brief mention of teratogens. • Page 447 discusses teratogens and states that: "Typically, there is no exposure effect on the mother."

Name of Text/Document	OSH Guidance Provided
Hazard Recognition and Control in Institutional Settings (Byrns, 2013)	• When referencing healthcare workers, describes reproductive hazards associated with ethylene oxide, lead, solvents, waste anesthetic gases (WAGs), radiation, and polychlorinated biphenyls. • Page 38 recommends complete medical histories for workers exposed to WAGs, including the "outcomes of all pregnancies of female workers or the wives of male workers."
Essentials of Industrial Hygiene, 1st Edition (Fuller, 2015)	• This text only contains a list of reproductive and developmental toxins and basic toxicology information; no specifics on OEL considerations or concerns.
Fundamentals of Industrial Hygiene, 6th Edition (Plog & Quinlan, 2012)	• Page 136 notes the "very considerable pressure to address the topic of pregnant women in the workplace." • Page 136-137 recognizes the concern of women and teratogen exposure; notes difficulty of setting OELs. • Page 136-137 states: "The workplace should be safe enough that fertile men and women are able to work without likelihood of harm." • Page 188 states: "In a comprehensive industrial hygiene program, all women who become pregnant or are trying to become pregnant should have their workplace evaluated for potential exposure to teratogenic or fetotoxic exposures."
Toxicology Principles for the Industrial Hygienist (Luttrell, Jederberg, & Still, 2008)	• Page 116 states: "Only a few chemicals can be incriminated as being reproductive toxic agents after conception in humans." • Page 117 states: "The impact of occupational agents on the fertility of male workers has been explored only to a limited extent. Infertility or unsuccessful pregnancy outcome [...] assumed to be linked with female health problems." • Page 123 notes that more data is needed to evaluate reproductive and developmental toxicities of occupational exposures.
The Occupational Environment, 3rd Edition (Anna, 2011)	• Page 67 notes that women of reproductive age are now commonly employed in the workplace, so evaluation of teratogenic chemicals is needed. • Page 67 notes that the agents of concern for reproductive and developmental toxicity are "readily encountered" outside the workplace, which "makes the evaluation of workplace exposures and their contribution to normal pregnancy outcome an extremely difficult task."

Name of Text/Document	OSH Guidance Provided
A Strategy for Assessing and Managing Occupational Exposures, 4th Edition (Jahn, Bullock, & Ignacio, 2015)	• On page 293, when explaining similar exposure groups and their use in hazard and control banding, the text states: "The [similar exposure group] priority rank may be raised or lowered based on the following data: […] presence of workers whose personal health condition might put them at increased risk (e.g., pregnant workers […]."
2016 TLVs and BEIs (American Conference of Governmental Industrial Hygienists, 2016)	• This mentions the German MAK guidelines for pregnancy. • Page 4: "Susceptibility to the effects of chemical substances may be altered during different periods of fetal development and throughout an individual's reproductive lifetime."

As the research-based component of OSH, the U.S. National Institute for Occupational Safety and Health (NIOSH) has a website dedicated to reproductive health and the workplace. The website is written for the everyday worker, and explains that many chemicals have not been tested to determine the hazards to reproductive health. NIOSH also states that the existing U.S. laws are not sufficiently protective for reproductive health or the health of a worker's family. The website appears to place the burden on the worker and their healthcare professional, rather than the employer. The section entitled "What Workers Should Know" simply tells workers to ask their organization's safety professional how to "stay safe" while doing their job, ask for any air monitoring results, and review Safety Data Sheets (SDS). In the section entitled "What Employers Should Know," NIOSH recommends that organizations "make a plan for pregnant and breastfeeding workers" but does not provide any guidance for OSH professionals (National Institute for Occupational Safety and Health, 2015).

There is substantial information about reproductive toxicants in toxicology, occupational health, and occupational medicine textbooks, but the focus is primarily on known reproductive toxicants. Much of the available U.S. literature focuses on theory, toxicology, and reproductive science, but does not provide concrete, practical methods to implement necessary elements of gender equal protection in OSH programs.

In this proceedings paper and corresponding presentation at Safety 2017, implementation strategies for improving gender equal protection within their OSH programs are provided, including:

- Risk assessments for new and expectant mothers
- OELs and the pregnant or breastfeeding worker
- Model/tool to evaluate special exposure groups and how to take appropriate action

As recommended by ASSE's *The Safety Professional's Handbook* (2012), a model flow chart risk assessment tool has been developed for evaluating the risk to susceptible workers. This model risk assessment tool is accompanied by a table explaining the susceptible worker categories and recommended exposure control measures. The following is practical information about the available scientific recommendations and regulatory requirements for protecting new and expectant mothers in the workplace.

Risk Assessments for New and Expectant Mothers

Risk assessments are an important tool for OSH professionals when evaluating the hazards present for new and expectant mothers. By definition, a new or expectant mother is "a woman who is pregnant, has given birth within the last six months or is breastfeeding" (Health and Safety Executive, 2016). Workplace risk assessments are recommended to consider risks to all female workers of childbearing age, with enhanced focus on risks for new and expectant mothers. These risks can include working conditions, physical hazards, chemical hazards, biological hazards, and others. It is commonly recognized that risk assessment can be a fundamental part of job hazard/safety analysis, where an OSH professional identifies jobs or tasks with hazards that present an unnecessary risk to the worker, and eliminate or reduce the risk.

Risk assessments, per ANSI/ASSE Z690.3-2011, Risk Assessment Techniques, consist of a three-part process: risk identification, risk analysis, and risk evaluation. In risk identification, OSH professionals should anticipate, recognize, and record hazards. As part of risk analysis, OSH professionals should apply the data obtained in risk identification to an analysis that helps them understand consequences of the hazard, probabilities of the hazard, and existing controls in place. Through risk evaluation, an OSH professional will compare the risk levels and consider additional needed controls to keep exposures as low as reasonably practicable (ALARP).

Some organizations may not be comfortable conducting risk assessments for new and expectant mothers, either because they use more of a hazards-based or compliance-focused approach, or due to a lack of knowledge and resources on the part of the OSH professional. Organizations may also have a "fear of discovering and documenting certain risks that may be difficult to address or mitigate" (Lyon & Hollcroft, 2012). Risk assessments by qualified, trained OSH professionals are needed when risks pose serious consequences for the worker or the organization, and when control measures are unclear. Workplace exposures to new and expectant mothers can pose serious consequences for the mother and the developing fetus, and guidance on control measures may be unclear for OSH professionals.

Safety Management Systems and Risk Assessments

Organizations that have developed their OSH programs in compliance with voluntary safety management system standards, such as ANSI/AIHA Z10-2012, Occupational Health and Safety Management Systems (OHSMS), should already have processes in place for workplace risk assessments. Section 4.1 of the Z10 standard identifies risk assessments and evaluations to be part of the OHSMS review process. Section 5.1 requires an organization to "establish and implement a risk assessment process(es) appropriate to the nature of hazards and level of risk." Additionally, Section 6.1 requires an organization to "establish a process to monitor and evaluate hazards, risks, and their controls [...]" including exposure assessments and occupational health assessments. Within informative Appendix F of the Z10 standard, AIHA provides an explanation of risk assessment and a general risk assessment example with a risk matrix. The Z10 standard notes that risk assessments should be done in a team approach whenever possible, with members providing "knowledge gained through education and experience" (American Industrial Hygiene Association, 2012).

It is also anticipated that the upcoming ISO 45001, Occupational Health and Safety Management Systems standard, will include requirements for risk assessments of the workplace (NSF-ISR, 2016). These risk assessments, like those required by regulation and industry standards, should be conducted by qualified and trained OSH professionals.

Existing Risk Assessment Templates

There are several available new or expectant mother risk assessments available for use by an OSH professional; most were developed by companies in the United Kingdom (UK) or the European Union (EU). In the UK, Health and Safety Executive guidance for OSH professionals on new and expectant mothers, a general risk assessment is provided, including a two-phase flowchart for OSH professionals to follow (Health and Safety Executive, 2016). The applicability of this flow chart to U.S.-based OSH professionals is currently limited, due to the differing availability of parental leave between the U.S. and the UK or EU.

For example, the University of Bristol has a policy for new and expectant mothers, which also contains a thorough hazard checklist (University of Bristol, 2012). The University of Glasgow provides guidance for students and staff on the occupational health risks associated with pregnancy, with a simple risk assessment form (University of Glasgow, 2016). Guidance from the International Labour Organization (ILO) about safe maternity in the workplace is also available, compliant with the international labor standards and conventions, which integrates new and expectant mother risk assessments as a fundamental element of workplace risk assessments.

Applicability of OELs to Pregnant and Breastfeeding Workers

As part of the risk assessment process, during the risk evaluation phase, OSH professionals may consider reducing their organization's OELs for specific chemicals that pose risks to new or expectant mothers. When evaluating the applicability of regulatory or voluntary OELs to pregnant and breastfeeding workers, OSH professionals may encounter resistance from management about proposing reduced OELs for the duration of the pregnancy and breastfeeding. However, a precedent already exists for reduced OELs in the OSHA regulation for lead.

The OSHA lead standard at 29 CFR 1910.1025 has different exposure limits for workers of reproductive age, both male and female. The standard provides workers with the ability to ask their occupational health provider for "medical advice concerning the effects of current or past exposure to lead on the employee's ability to procreate a healthy child" in 1910.1025(j)(3)(i)(C). In Appendix A of the lead standard, the lower OELs are explained: "The blood lead levels of workers (both male and female workers) who intend to have children should be maintained below 30 μg/100g to minimize adverse reproductive health effects to the parents and to the developing fetus." These blood lead levels for male and female workers who intend to have children are only 50% lower than the regulatory blood lead level of ≥ 60 μg/100g, at which medical removal of the worker from their job role is required.

Similar language exists for the OSHA ethylene oxide (EtO) standard at 29 CFR 1910.1047. The standard provides workers the right to ask their occupational health provider for "medical advice concerning the effects of current or past exposure to EtO on the employee's ability to produce a healthy child" per 1910.1047(i)(2)(i)(E). The standard does not have recommended exposure levels that differ for females and males of reproductive age. Interestingly, the potential reproductive effects of other chemicals that have OSHA standards, such as arsenic, benzene, beryllium, cadmium, and others are not addressed in the standard or health effects presented in the Appendices of the standards.

In a 2013 article from *The Synergist*, Frank Mirer discussed the OSHA lead standard and other hazards to reproductive health in the workplace. Mirer explained that the "most recent authoritative recommendations for [blood lead] levels of concern for both children and adults vary between 5 and 10 μg/dl depending on the health effect." Further in his article, Mirer noted that risk assessment for reproductive and developmental toxicity is not well developed; however, "evolving information on key chemicals shows high risks and the need for action" (Mirer, 2013).

Lewandowski and Dodge explain a process for developing OELs that consider DART effects. The authors state that "OELs that are not documented to have considered DART effects should be viewed with suspicion." Recommendations for selecting or developing OELs for new and expectant mothers included a review of existing DART data from available databases and hazard banding. The other strategies recommended by the authors (read across, structural activity relationships, and high-throughput screens) would likely require the expertise of a toxicologist or other scientist to interpret the results (Lewandowski & Dodge, 2016).

Referencing an occupational health textbook can provide helpful risk assessment and OEL guidance that can be interpreted by OSH professionals, although these textbooks are generally written for occupational health providers. In *Occupational and Environmental Health: Recognizing and Preventing Disease and Injury*, 6th Edition, the assistance of industrial hygienists is recommended. When discussing the evaluation and control of risk, this text notes that "[p]recise assessment of exposures is difficult" due to the incompleteness of available research and databases. A four-question approach is recommended. First, what is the agent(s) the employee is exposed to? Second, is exposure to the agent(s) actually occurring, and at what dose rate/timing? Third, what is the evidence that this agent(s) causes adverse DART effects? This third question requires a literature review and may involve the input of a reproductive hazard assessment expert. Fourth, reviewing the available data, does the agent(s) pose a significant DART risk? The fourth question involves an estimate of other concurrent exposures, in order to place the potential workplace exposure in context. Finally, from an occupational medicine perspective, it is recommended that a "precautionary approach" be used, to emphasize the importance of "exposure avoidance in the workplace and at home, beginning in the preconception period for both men and women" (Levy, Wegman, Baron, & Sokas, 2011).

Additionally, one of the most comprehensive and referenced sources for reproductive health in the U.S. workplace is the U.S. Navy and Marine Corps Public Health Center's 2010 technical manual entitled *Reproductive and Developmental Hazards: A Guide for Occupational Health Professionals.* This document is also written for occupational health providers serving U.S. Navy and Marine Corps personnel. In Section V of the technical manual, recommendations for evaluating and managing a pregnant worker are made. The manual explains that the ACGIH Threshold Limit Values® (TLVs) "rarely provide adequate guidance" for pregnant workers, and if the TLV or other OEL does not mention hazards to reproductive health, it "does not imply there is no hazard or that any substance with a given TLV or [OEL] is 'OK' in regards to [reproductive and developmental health] issues." The first step recommended in this technical manual is to have the occupational health provider take a detailed history that encompasses personal, social, occupational, and environmental considerations, and include a spousal history. When considering modifications to the work environment, the assistance of industrial hygienists is recommended. Modifications to the work environment follow the hierarchy of controls, and include product substitution, engineering controls, administrative controls, and personal protective equipment (PPE). In relation to OELs, one of the most helpful parts of this technical manual is Table 8 – Chemical Reproductive and Developmental Hazards List, which contains 11 pages of listed agents, with Chemical Abstract Service (CAS) numbers and notes referencing the reproductive and developmental concerns with available OELs (Navy and Marine Corps Public Health Center, 2010).

In Table 2, recommended modifications of OELs for specific types of substances are provided. These were developed based on a risk evaluation of new and expectant mothers and scientific precedence in establishing action levels and OELs.

Table 2: Example Modifications of Occupational Exposure Limits

Substance	Example	Modified OELs in ppm
Hydrocarbons, PAHs, solvents	Toluene	25 ppm (50% of Action Level)
	Acetone	500 ppm (50% of OEL)
	Hexane	125 (50% of Action Level)
	Benzene	0.05 ppm (5% of OEL)
Acids	Hydrogen Chloride, Hydrogen Bromide, Phosphoric, Phosphinic, Hydrogen Fluoride, Sulfuric	No adjustment of OEL. No physical contact permitted.
Amines	Dimethylaniline	No open handling.
High Potency Pharmaceuticals	Succinyl Choline Chloride	No contact and/or access to areas.

ACGIH does not freely provide information on whether the TLVs and biological exposure indices® (BEIs) are established with consideration of hazards to reproductive health; this information may be present in the *Documentation* for each TLV, when available, but each *Documentation* must be purchased. ACGIH explains that TLVs and BEIs are health-based values for the "typical worker" or "nearly all workers" (American Conference of Governmental Industrial Hygienists, 2016). However, questions were raised as early as 1986 on whether the TLVs and BEIs considered pregnant women; an article by Pamela de Silva noted that susceptible workers, such as pregnant women, may not be adequately protected by the TLVs (De Silva, 1986).

Due to the lack of guidance from AIHA and ACGIH for OEL adjustments for new and expectant mothers, a model for evaluating special exposure groups, such as new and expectant mothers, is provided in the next section. This model is based on the concept of hazard banding, which is also called occupational exposure banding (OEB).

A Model for Evaluating Special Exposure Groups

Effective implementation of a susceptible worker assessment program (SWAP) requires communication at all levels. Detailed, accurate information about specific job hazards, such as chemical and physical hazards, must be elevated to a worker's supervisor, primary safety contact (i.e., the union safety representative, safety advocate, lead operator, safety coordinator, or safety manager), and industrial hygienist or toxicologist, if available. The program must be developed with controlled documents including, but not limited to: SWAP worker self-evaluation form, work-specific risk assessments, sign-off forms for communication and worker buy-in, and detailed plans for continuing hazard evaluations, which can take the form of an industrial hygiene sampling plan. As soon as a worker indicates their desire to be included into an assessment program, a chain of events should be initiated, whereby the supervisor and OSH team go to extra lengths to ensure that an at-risk worker (such as a new or expectant mother) is adequately protected.

Completion of a self-evaluation form by the at-risk worker is the first and most critical step in the process. An effective SWAP Health History Questionnaire should capture basic worker demographic information, basic medical information that is relevant, physical job requirements, potential exposures (if known) for that worker, any control measures for those exposures, and any relevant reproductive health history if the worker is to be placed into a SWAP with fetal embryo protection requirements. This information provides the baseline risk data for an industrial hygienist or occupational health provider to begin their assessment.

Management is then required to provide the worker-completed questionnaire and any other known information to the OSH team. A qualified safety professional must then perform a qualitative risk assessment in which the worker is placed into an appropriate protection category, and a plan is developed to mitigate potential risks. Placing a worker into a protection category should be based on two primary factors: the underlying health concern of the worker, and the relative risk of the work that must be performed. One approach to this is to follow a banding strategy for risk and add into the banding a consideration for the vulnerability of the worker.

Table 3 provides an example categorization for at-risk workers that considers the vulnerability (pregnant or trying to become pregnant) and job risk, based on the type of work conducted by the worker. In Appendix 1: Risk Assessment Tool for SWAP, a flowchart is provided to show the risk assessment process in assigning a worker to a particular work safety category.

For example, a worker who is pregnant, or plans to become pregnant, will be notified of her ability to be placed in the SWAP. When a pregnant worker expresses her desire to participate in the SWAP, a qualified OSH professional conducts a total risk assessment. If the work or work environment has teratogen, mutagen, carcinogen, or reproductive hazards, the worker is placed into Work Safety Category 4. This category is for workers who are pregnant or trying to become pregnant that conduct high-risk work with high-, moderate-, or low-hazard materials. Similarly, for a male worker working in an area with reproductive or teratogen hazards, whose partner is trying to conceive, the male worker can be placed in Work Safety Category 4.

Table 3: Example Work Safety Categorization of Worker Vulnerability and Job Risk

Work Safety Category	**4**	**3**	**2**	**1**	**0**
Work Type	High-risk work with high, moderate- or low-hazard materials	Moderately high-risk work with moderate- or low-hazard materials	Moderate-risk work with moderate- or low-hazard materials	Low-risk work with moderately low-risk materials	Low-risk work with low-hazard materials
Worker Example	*Those in a SWAP who are pregnant or trying to be pregnant who handle moderate/high-risk materials*	*Those in a SWAP who are pregnant or trying to be pregnant who handle low-risk materials*	*Those in a SWAP who are not pregnant or trying to be who handle moderate/high-risk materials*	*Those in a SWAP who are not pregnant or trying to be who handle low-risk materials*	*Those not in a SWAP who are not pregnant or trying to be pregnant who handle-low risk materials*

Categorizing workers and risk with a banding approach allows OSH programs to provide increasing levels of protection with increasing levels of risk. This follows industry-wide accepted programs for banding occupational hazards.

After proper categorization has been performed, documented actions must be taken to address all risks. Special precautions, such as limiting access to risk areas, restricting worker tasks to ventilation-protected environments, providing additional training, and labeling high-risk areas are required. The risks of a task and the vulnerability of a worker may require that all work be performed inside of a fume hood or with a respirator. If the nature of a hazard, such as toluene, includes skin sensitization or absorption through the skin, contact with the material must be restricted or effective PPE (such as a chemical protection suit and gloves) must be used. The control strategy must be approved by a qualified OSH professional. First-line supervisors are not qualified to perform a thorough hazard assessment for at-risk workers.

Exposure assessment, communication of exposure, and periodic re-evaluation are the steps that close the cycle of a SWAP. If exposure monitoring is possible, it must be performed by a qualified industrial hygienist. An exposure sampling plan that considers the risk of each chemical hazard and the potential for exposure should be developed and followed. Exposure monitoring performed without an objective, simply to monitor and gather data, is useless at best, and can be misinterpreted.

Timely analysis and issuing of reports is critical. Workers must be informed of their personal exposure, and the results of exposure assessments are required to evaluate the effectiveness of a SWAP. Results must be communicated to the worker in a meaningful way. Documentation of the worker's reception of the results is important. If the results show unacceptably high exposure potential, the worker should be moved to the next highest work-safety category. If the worker is already in the most protective work-safety category, then that worker must be removed from all work that contains that risk. Such requirements for medical removal already have precedence, as discussed previously in the OSHA standard for lead.

The removal of the worker from high-risk work cannot be held against them in any way with regard to career performance or promotion. The failure to eliminate or control the hazard rests squarely on the employer, not the worker. The final step is to periodically re-evaluate either placement into, or removal from, a SWAP. This re-evaluation should be conducted monthly and requires all available information and input from first-line supervisors, safety professionals, industrial hygienists, occupational health providers, and the at-risk worker. The most conservative measure is to keep a worker in a SWAP once they have been placed into it until the risk can be eliminated from the workplace.

In Appendix 2: Work Safety Categories and Exposure Control Measures, each work safety category is explained (from 0 to 4), including an example worker description for each work-safety category and recommended controls to reduce exposure to the susceptible workers. The controls are separated into the following categories: general controls, containment, PPE, housekeeping and maintenance, labeling and transport, waste, and training. The tabular format used is intended to simplify the risk assessment and evaluation process for practicing safety professionals. This is intended to be a proactive assessment of worker health and safety in a variety of work environments. The definitions of low-, moderate-, and high-exposure risk work practices are provided in Table 4 below.

Table 4: Exposure Risk Definitions

Exposure Risk Level	Definition/Examples
Low-exposure Risk	• Wetted, controlled material handling. • Closed systems: ventilation present. • Work with compounds that are not teratogens, mutagens, reproductive toxins, or carcinogens.
Moderate- to High-exposure Risk	• Open handling of dry material. • Work with compounds that are teratogens, mutagens, reproductive toxins, or carcinogens. • Handling of material outside of process equipment or controlled systems (i.e., charging and open handling of dry solids, working outside of containment).

In Appendix 3: SWAP Health History Questionnaire, a sample worker questionnaire is provided to show OSH professionals the type of information gathered by an occupational health provider that can be used in a risk assessment for new and expectant mothers. ***Please note*** that the authors reserve the copyright for the SWAP risk assessment tool and work safety categories assessments in Appendix 1 and Appendix 2.

Conclusion

Recent research on OELs has shown that OELs may not be sufficiently protective against the effects of developmental and reproductive toxicity. Workers may express their concerns about reproductive health effects, prenatal exposure, or exposure during breastfeeding to their company's OSH professional. The available recommendations for risk assessments, OEL corrections, and PPE provision to pregnant or breastfeeding workers is scattered over many disciplines, and can be difficult to understand. This paper aimed to provide practical methods of implementing gender equal protection within OSH programs. This has become especially relevant, given the number of female and at-risk workers who continue to work during pregnancy. Understanding OELs and their applicability to new and expectant mothers is part of a national conversation about gender equal protection, privacy, and reproductive hazards in the workplace that has been heightened by state-level regulations relating to pregnancy accommodations and fairness in the workplace.

Appendix 1: Risk Assessment Tool for SWAP

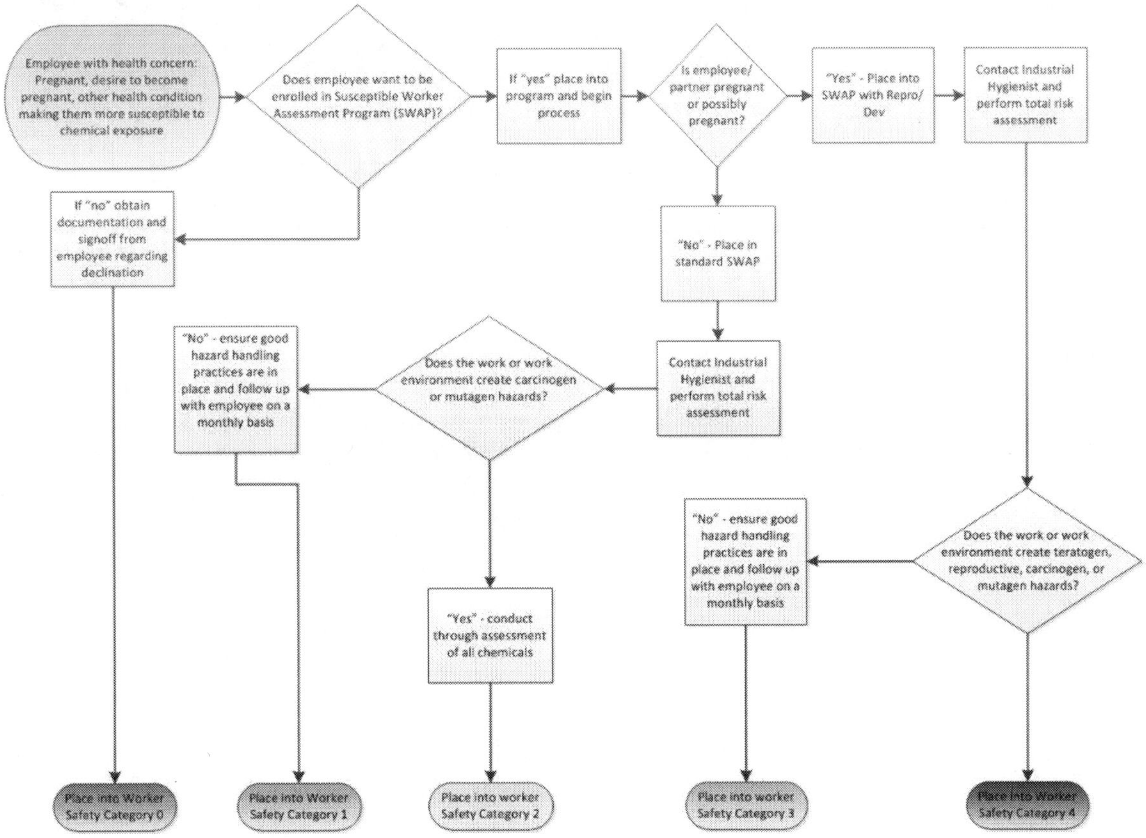

Appendix 2: Work Safety Categories & Exposure Control Measures

Work Safety Category	4	3	2	1	0
Work Type	High-risk work with high-, moderate-, or low-hazard materials	Moderately high-risk work with moderate- or low-hazard materials	Moderate-risk work with moderate- or low-hazard materials	Low-risk work with moderate- or low-risk materials	Low-risk work with low-hazard materials
Worker Example	*Those in a SWAP who are pregnant, or trying to be, who handle moderate- or high-risk materials*	*Those in a SWAP who are pregnant or trying to be who handle low-risk materials*	*Those in a SWAP who are **not** pregnant or trying to be who handle moderate- or high-risk materials*	*Those in a SWAP who are **not** pregnant or trying to be who handle low-risk materials*	*Those not in a SWAP who are **not** pregnant or trying to be who handle low-risk materials*
General Controls: Access	Establish regulated or designated areas for the use of high-risk materials	Establish regulated or designated areas for the use of product and intermediates	Establish limited access areas	Regular access with area owner permission	Regular access
General Control: Surfaces and Cleaning	All surfaces in the work area should be smooth and non-porous for easy cleaning; should be resistant to decon. agents that will be used	All surfaces in the work area should be smooth and non-porous for easy cleaning; should be resistant to decon. agents that will be used	Surfaces should be cleanable and provisions for decon., such as wetting and washing, should be considered	No special cleaning requirements	No special cleaning requirements
General Controls: Area Labeling	Label the process building where high-risk materials are used, present or	Label the process building where moderate- or low-risk materials are	Label the process area where moderate- to low-risk materials are used, present	Label the process area low-risk materials are used, present or stored	No special labeling requirements

Work Safety Category	4	3	2	1	0
	stored, example: "High-Risk Materials Present - No Entrance by Workers in SWAP"	used, present or stored	or stored		
General Controls: Equipment Labeling	Fume hoods, ovens, or glove boxes where product or intermediates are used, present or stored should have prominently displayed warning signs or labels both in the areas and access to the areas	Fume hoods, ovens, or glove boxes where product or intermediates are used, present or stored should have prominently displayed warning signs or labels both in the areas and access to the areas	Equipment expected to contain the material should have prominently displayed sign	Equipment expected to contain the material should have prominently displayed sign	Routine signage requirements
	"CAUTION: High-Risk Materials Used, Present or Stored in This Area"	"CAUTION: Moderate-Risk Materials Used, Present or Stored in This Area"	"CAUTION: Moderate-Risk Materials Used, Present or Stored in This Area"	No special signage requirements	No special signage requirements
General Controls: Equipment Staging	Only essential equipment in work area	Only essential equipment in work area	Only essential equipment in work area	No equipment restrictions	No equipment restrictions
General Controls: Work Practice	No smoking, eating, drinking, or application of cosmetics is permitted	No smoking, eating, drinking, or application of cosmetics is permitted	No smoking, eating, drinking, or application of cosmetics is permitted	No smoking, eating, drinking, or application of cosmetics is permitted	No smoking, eating, drinking, or application of cosmetics is permitted

Work Safety Category	4	3	2	1	0
General Controls: Hygiene	Wash hands and face upon exiting containment	Wash hands and face upon exiting containment	Wash hands upon exiting area	Wash hands upon exiting area	Wash hands upon exiting area
Containment	All solids or volatile material handling must be carried out under air pressure that is negative relative to surrounding areas	Closed systems and procedures to eliminate potential for aerosolization	Open handling allowed under a vent or hood	Open handling allowed where exposure monitoring documents exposures below the required fraction of the OEL	Open handling allowed along with good handling practices
Personal Protective Equipment (PPE)	Full protective equipment required and total encapsulation of worker may be required based on risk assessment. **Example:** supplied air respirator and chemical protective suit	Other protective material may be used based on the composition of the material. **Example:** air purifying respirator and protective suit	Other protective equipment may be used based on the composition of the material and risk assessment	Standard PPE	Standard PPE
Housekeeping and Maintenance	Keep all work surfaces and equipment clean at all times	Keep all work surfaces and equipment clean upon completion of task	No dry sweeping of material; wetted sweeping or vacuum clean up preferred; material should be prevented from accumulating	No restrictions on housekeeping/ maintenance beyond good work practices if monitoring shows exposure below 25% of OEL	No restrictions on housekeeping /maintenance beyond good work practices

Work Safety Category	4	3	2	1	0
			on surfaces		
Labeling and Transport	Prior to removal from the containment area, all containers (intermediate storage containers, finished product drums, waste drums, sample bottles, and otherwise) must have a warning label affixed to them indicating their hazard	All containers (intermediate storage containers, finished product drums, waste drums, sample bottles, and otherwise) need to have a warning label affixed to them indicating their hazard	Containers should be labeled according to their hazard	Containers should be labeled according to their hazard	Routine labeling requirements
Waste	Solid waste, including used PPE, must be kept in a waste bin inside the containment	PPE should be disposed of in designated waste bins	Standard disposal	Standard disposal	Standard disposal
Training	Personnel shall be trained on all aspects and hazards of high-risk materials with detailed, job/task	Personnel shall be trained on all aspects and hazards of moderate- and low-risk materials with detailed,	Standard job/hazard training and background information on moderate- and low-risk materials and work and	Standard job/hazard training and background information on moderate- and low-risk materials and hazards	Standard job/hazard training

Work Safety Category	4	3	2	1	0
	specific training	job/task specific training	hazard recognition		

Definitions and Abbreviations:

Low-exposure Risk: Wetted dry material handling. Closed systems, ventilation present. Work with compounds that are not teratogens, mutagens, reproductive toxins, or carcinogens.

Moderate- to High-exposure Risk: Open handling of dry material. Work with compounds that are teratogens, mutagens, reproductive toxins, or carcinogens. Handling of material outside of process equipment or controlled systems (i.e., charging and open handling of dry solids, working outside of containment).

Decon: Decontaminating or decontamination.

Appendix 3: SWAP Health History Questionnaire

Workplace Health History Questionnaire (For Participating SWAP Worker) Confidential Medical Record	
Date:	
Name of Worker:	
Department:	
Location:	
Supervisor:	
Supervisor Phone#:	
Pregnancy Information	
EDD:	
GPTAL:	
LMP:	
OB/GYN:	
PERINATOLOGIST:	

Have you been told this is a high-risk pregnancy?	Yes ________	No ________	
EXPLAIN: *(existing co-morbidity, advanced maternal age, assisted reproductive technology, etc.)*			
Personal Information			
Date of Birth:		Height: ________	Weight: ________
B/P: ________	HR: ________	R: ________	Temp: ________

List Physical Requirements of Job					
Bending:	Yes _______	No _______	Sitting:	Yes _______	No _______
Climbing:	Yes _______	No _______	Standing:	Yes _______	No _______
Lifting:	Yes _______	No _______	Stooping:	Yes _______	No _______
Pulling:	Yes _______	No _______	Twisting:	Yes _______	No _______

List Job/Environmental Exposures (If Unknown, Contact Supervisor or Industrial Hygienist)	
Chemicals:	
Dusts, Fumes, Vapors, Mists:	
Noise:	
Metals and Metal Powders:	
Radiation:	
Vibration:	
Temperature Extremes:	
Sharp Objects:	
Infectious/Biological Agents:	
Do you have any other types of exposures (e.g., second job)?	

Personal Protective Equipment Used					
Ear protection:	Yes ______	No _______	Apron/coat:	Yes _______	No _______
Eyewear:	Yes ______	No _______	Respirator:	Yes _______	No _______
Gloves:	Yes ______	No _______	Facemask:	Yes _______	No _______
Lift devices:	Yes ______	No _______	Ventilation (work hood):	Yes _______	No _______
Other (*specify*):					

Exposures (Past or Present)		
Do you smoke?	Yes ______	No _______
Do you drink more than one glass of alcohol per week?	Yes ______	No _______
Do you use marijuana, cocaine, crack, or heroin? (circle any that apply and list others) ______________	Yes ______	No _______
Do you use saunas or hot whirlpool baths?	Yes ______	No _______
Do you eat raw or very rare meat?	Yes ______	No _______
Do you have a household cat?	Yes ______	No _______
Who changes the cat's litter box?	Self ______	Partner _____
Do you take any prescription or over-the-counter drugs? List : ______________________________	Yes ______	No _______

Reproductive History		
Have you and your spouse/partner had:		
Any problems with becoming pregnant?	Yes ______	No _______
Any miscarriages?	Yes ______	No _______
A baby born prematurely or with low birth weight?	Yes ______	No _______
Any child born with a birth defect?	Yes ______	No _______

Any child diagnosed with cancer?	Yes ______	No ________
Please list any other information you feel relevant to your reproductive planning or to current pregnancy:		

Bibliography

American Conference of Governmental Industrial Hygienists. (2016). *2016 TLVs and BEIs.* Cincinnati, OH: American Conference of Governmental Industrial Hygienists.

American Industrial Hygiene Association. (2012). ANSI/AIHA Z10-2012 Occupational Health and Safety Management Systems. Falls Church, VA: American Industrial Hygiene Association.

American Society of Safety Engineers. (2012). *The Safety Professional's Handbook* (2nd Edition ed.). (J. M. Haight, Ed.) Chicago, IL: American Society of Safety Engineers.

Anna, D. H. (Ed.). (2011). *The Occupational Environment: Its Evaluation, Control, and Management* (3rd Edition ed.). Falls Church, VA: American Industrial Hygiene Association.

Brauer, R. L. (2006). *Safety and Health for Engineers.* Hoboken, NJ: John Wiley & Sons, Inc.

Byrns, G. (2013). *Hazard Recognition and Control in Institutional Settings—A Guide for Hospitals, Universities and Nursing Homes.* Chicago, IL: American Society of Safety Engineers.

De Silva, P. (1986, April). TLVs to Protect 'Nearly All Workers'. *Applied Industrial Hygiene, 1*(1), 49-53.

Frank R. Lautenberg Chemical Safety for the 21st Century Act, Public Law 114-182 (130 Stat. 448 June 22, 2016).

Fuller, T. P. (2015). *Essentials of Industrial Hygiene* (1st Edition ed.). Itasca, IL: National Safety Council.

Health and Safety Executive. (2016, January). *FAQs.* Retrieved January 13, 2016, from New and Expectant Mothers: http://www.hse.gov.uk/mothers/faqs.htm#q3

Health and Safety Executive. (2016). *General Risk Assessment.* Retrieved January 13, 2017, from New and Expectant Mothers: http://www.hse.gov.uk/mothers/docs/pregnant-workers-flow-chart.pdf

Jahn, S. D., Bullock, W. H., & Ignacio, J. S. (Eds.). (2015). *A Strategy for Assessing and Managing Occupational Exposures* (4th Edition ed.). Falls Church, VA: American Industrial Hygiene Association.

Levy, B. S., Wegman, D. H., Baron, S. L., & Sokas, R. K. (Eds.). (2011). *Occupational and Environmental Health: Recognizing and Preventing Disease and Injury* (6th Edition ed.). New York, NY: Oxford University Press.

Lewandowski, T., & Dodge, D. (2016, June/July). Windows of Susceptibility. *The Synergist*, p. web edition.

Luttrell, W. E., Jederberg, W. W., & Still, K. R. (Eds.). (2008). *Toxicology Principles for the Industrial Hygienist* (1st Edition ed.). Falls Church, VA: American Industrial Hygiene Association.

Lyon, B. K., & Hollcroft, B. (2012, December). Risk Assessment: Top 10 Pitfalls and Tips for Improvement. *Professional Safety.*

Mirer, F. (2013, February). Suffer Little Children: Risk Assessment for Developmental and Reproductive Toxicity. *The Synergist*, online.

National Institute for Occupational Safety and Health. (2015, February 27). *Reproductive Health and the Workplace*. Retrieved January 14, 2017, from NIOSH Safety and Health Topics: https://www.cdc.gov/niosh/topics/repro/

National Partnership for Women and Families. (2016, June). *Fact Sheet - Reasonable Accommodations for Pregnant Workers: State and Local Laws.* Retrieved January 13, 2016, from The Pregnant Workers Fairness Act: http://www.nationalpartnership.org/research-library/workplace-fairness/pregnancy-discrimination/reasonable-accommodations-for-pregnant-workers-state-laws.pdf

Navy and Marine Corps Public Health Center. (2010, April). *Reproductive and Developmental Hazards: A Guide for Occupational Health Professionals, Technical Manual NMCPHC-TM-OEM 6260.01C.* Retrieved January 14, 2017, from Bureau of Medicine and Surgery: http://www.med.navy.mil/sites/nmcphc/Documents/policy-and-instruction/oem-reproductive-and-developmental-hazards-a-guide-for-occupational-health-professionals.pdf

NSF-ISR. (2016, September 13). *ISO 45001 Occupational Health and Safety Management Systems Information Guide.* Retrieved January 13, 2017, from NSF Newsroom: https://www.nsf.org/newsroom_pdf/isr_dis45001_guide.pdf

Pisko, S. A. (2016). Towards Reasonable: The Rise of State Pregnancy Accommodation Laws. *Michigan Journal of Gender and Law, 23*(1), 147-185.

Plog, B. A., & Quinlan, P. L. (Eds.). (2012). *Fundamentals of Industrial Hygiene* (6th Edition ed.). Itasca, IL: National Safety Council.

University of Bristol. (2012, July). *New and Expectant Mothers at Work Guidance.* Retrieved January 13, 2017, from Health and Safety Office: http://www.bristol.ac.uk/safety/media/gn/new-expect-mothers-gn.pdf

University of Glasgow. (2016). *Guidance for Staff and Students on the Occupational Health Aspects of Pregnancy.* Retrieved January 13, 2017, from Safety and Environmental Protection Services: http://www.gla.ac.uk/media/media_172820_en.pdf

Understanding Low-Probability/High-Consequence Events

Carol Robinson, CIH, CSP
Specialty Technical Consultants, Inc.
Oakland, California

Debby Shewitz, CSP
Shewitz Consulting, LLC
Cleveland, Ohio

Introduction

Not all high-consequence events make the news and capture national attention, like the Deepwater Horizon explosion or the release of methyl isocyanate in Bhopal, but by definition they cause serious, life-altering damage, injuries and deaths, affecting not only the persons involved, but their families, friends and coworkers, and in severe cases, the local community. At the very least, businesses are likely to suffer damage to their reputations, and the consequences may be much worse. According to the Federal Emergency Management Agency (FEMA), 40 percent of businesses do not reopen after a disaster, and another 25 percent fail within one year of reopening. A large number of these events had an extremely low probability of occurring, and yet they did. Nicolas Taleb refers to these types of occurrences as "black swan" events, and describes them as outliers, with extreme impact, and retrospective (though not prospective) predictability.

Current approaches to risk assessment are based on a determination of the probability and severity of incident outcomes. In addition to evaluating normal operating conditions, more sophisticated risk assessments often also evaluate the outcome of interruptions to normal operations (upset conditions). However, simple probability and severity analysis may not be adequate in evaluating low probability, high-consequence events.

Accident Causation Theories

In order to start understanding low-probability/high-consequence events, we want to look first at theories of how accidents are caused. One of the earliest theories is the Domino Theory, developed by H.W. Heinrich in 1932. As depicted below, it posits that accidents occur when the linkage for a chain reaction lines up like dominoes on end. Each of the factors in the chain is dependent on the preceding factor.

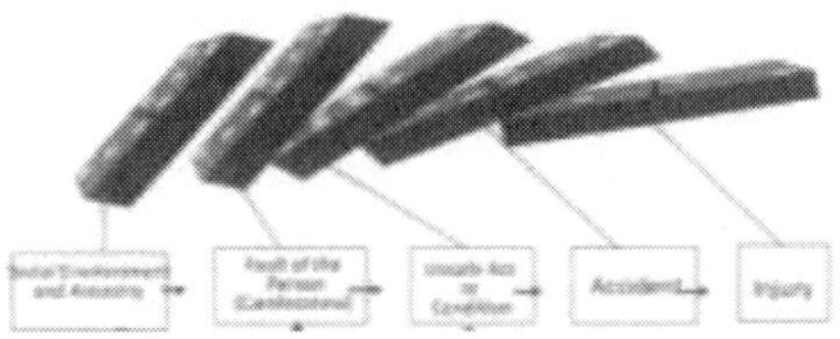

Exhibit 1. The Domino Theory of Accident Causation

There are other event-chain theories of accident causation, such as the human-factors theory, which attributes accidents to a chain of events ultimately caused by human error. It consists of three broad factors that lead to human error: overload, inappropriate response, and inappropriate activities.

The systems-theory model of accidents holds that accidents arise from the interaction between humans, machines, and the environment. These are not simply chains of events or linear causality, but more complex of causal connections. This theory also assumes that, under normal conditions, a "steady state," or harmony, exists between individuals and the work environment, as opposed to looking at an environment full of hazards and people prone to errors, and therefore, the probability of accidents under normal conditions is low.

The theory that we focus on in this presentation is the cumulative-acts theory, commonly referred to as "the Swiss Cheese Effect." It compares human systems to multiple slices of Swiss cheese, stacked together side by side. It was originally put forward by British psychologist James T. Reason of the University of Manchester in 1990. If we think of the slices of cheese as our accident prevention activities, we would all love to have programs and systems with absolutely no holes in them that might allow an accident to occur. But we also all know that perfection cannot be attained. So we build in layers of prevention, so that if something slips through a hole in one activity, it will be blocked by a solid part of the next preventive activity, and no single hole can lead to a major event. Now, the challenge becomes trying to ensure that the holes don't all line up, allowing a series of deviations to lead to a catastrophic result.

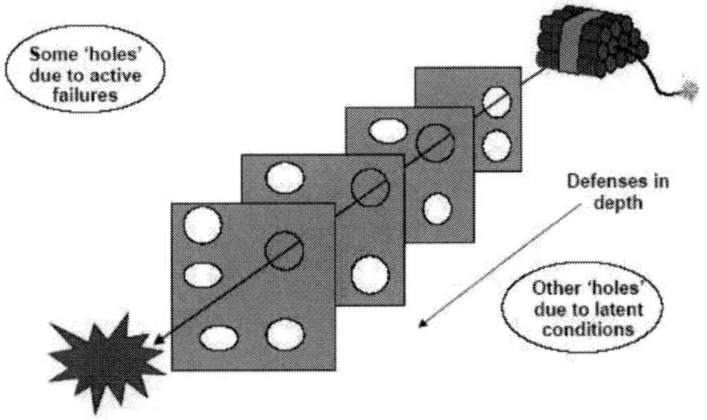

Exhibit 2. The Cumulative Acts (or Swiss Cheese) Theory

Risk Assessment Techniques

There are numerous risk assessment techniques in use in various industries. As previously mentioned, the simplest techniques assess risk based on likelihood/probability and impact/severity, and use a risk matrix to rank risk and the need for additional controls (using a table such as the Table 1 below). Some also include a step to identify the risk of upset conditions. These techniques are well known to the safety generalist due to their wide use, but are not adequate in evaluating low probability, high-consequence events due to their characteristics.

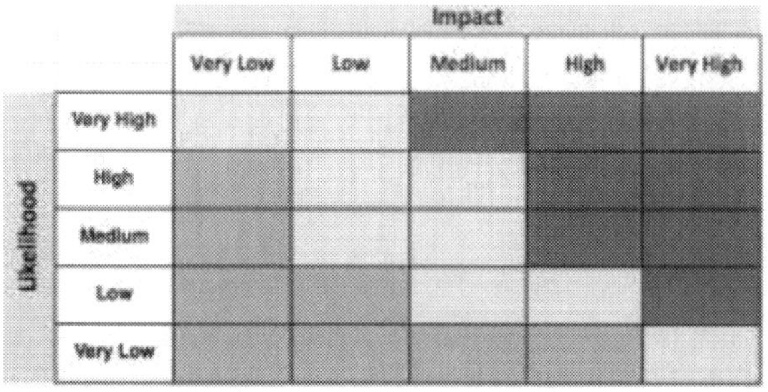

Table 1. Risk Assessment Based on Likelihood/Probability and Impact/Severity

High-hazard industries, such as the nuclear, petrochemical, and space industries, have led in the use of more specialized tools, such as fault tree analysis and failure mode and effects analysis (FMEA). Process safety management is an example of a regulatory requirement to perform process hazard analysis (PHA), which is a systematic approach for identifying, evaluating, and controlling the hazards of processes involving highly hazardous chemicals. A newer tool, bowtie risk assessments, is becoming more prominent. This method visualizes the risk assessment in a single diagram that includes the hazard/top event, threats (causes), consequences (outcomes), and barriers (controls). The details of these methods are beyond the scope of this paper, due to the specialized training and knowledge needed to use them. So what can we generalists do to help ourselves sleep better at night and not have to worry that the holes in the cheese are not going to line up?

Possible Approaches

Each of us needs to evaluate what additional activities targeted towards preventing low-probability/high-consequence events might work in our own operations, based on available resources, the existing systems we might be able to build on, and the overall culture. The following are few things that a generalist can look at.

<u>Information Sources</u>

Even the smallest of operations will have informal sources of information, and if you work someplace that's big enough to have a safety specialist, there are likely some formal ones, even if they're pretty basic. These might include accident investigations, inspection and audit programs, and safety committees. Use the information from these sources to identify the holes that have the potential to line up to contribute to a high-consequence incident, and work to ensure that appropriate action is taken to keep that from happening.

Think about other systems that can be implemented to collect additional information. Some fairly common ones include suggestion programs, confidential reporting systems, and near-miss reporting systems. These might even be things that you could set up largely on your own, since they have minimal cost. If nothing else, getting a voicemail box, a dedicated e-mail address, or a physical "safety concern" box outside your office shouldn't be that hard to do.

Preventive Maintenance

In the simplest terms, if protections are designed into systems, they have to be kept in reliable condition or they're useless. The mechanical integrity element in OSHA's PSM regulation (which is basically about preventive/predictive maintenance) is identified by the agency as one of the most critical. Even if there's no facilities or maintenance department in your operations, you can identify the components (e.g., vessels, piping, interlocks, controls, monitors, alarms) that are most critical to safety, and ensure that they are inspected and maintained. The minimum requirement is always to follow the manufacturer's recommendations.

Change Control

This is another aspect of making sure that safety controls/features that were designed into a system aren't lost; ensure that there's some sort of EHS review before the system gets changed (including brand new processes/systems). In addition to trying to make sure that changes don't add any hazards, this might be an opportunity to get some safety improvements worked in that were hard to justify by themselves. Reviews should happen as early as possible in the change process so that equipment that would be expensive to alter isn't already ordered.

The new amendments to the Environmental Protection Agency's (EPA's) RMP Rule include a requirement for using the "safer technology and alternatives" (STAA) hierarchy of controls. The hierarchy consists of inherently safe design/technology, passive controls, active controls, and procedural controls. Inherently safer strategies include:

- Minimization: Using smaller quantities of hazardous substances;
- Substitution: Replacing a material with a less hazardous substance;
- Moderation: Using less hazardous conditions or a less hazardous form, or designing facilities that minimize the impact of a release of hazardous material or energy; and
- Simplification: Designing facilities to eliminate unnecessary complexity and make operating errors less likely.

If nothing else, the simplification technique can be used in any type of operation to make human error less likely.

Know Industry Best Practices

In the PSM Standard, OSHA has for a number of years been putting a focus a phrase that's seemingly buried in a couple of the elements: "Recognized and Generally Accepted Good Engineering Practices," called by its acronym RAGAEP. In a 5/11/16 Enforcement Memorandum to Regional Administrators, the agency clarifies the types of published standards that they would consider a covered site would need to know about and implement to the extent applicable (NFPA, ASME, ANSI, Chlorine Institute, etc.). It also makes clear that facilities need to keep up with the most current versions/amendments to these standards. If it's common practice for your industry to follow a standard, the certainly makes it "recognized and generally accepted."

Some of the places where you can connect with others in your industry to find out what they're doing to deal with concerns, and to get some help addressing concerns that might be new, include trade associations, local safety councils, local ASSE chapters and ASSE practice specialties, and the Chemical Safety Board (CSB). The CSB posts great analyses/videos of high-consequence incidents that have actually happened, along with recommendations for how to prevent similar occurrences. Even if these incidents weren't in your exact industry, there may be information that you can use.

Modification of Standard Risk Process

Another simple approach that could be as effective at organizations already conducting risk assessment is the implementation of a modified risk assessment process specifically to identify potential low-probability/high-consequence events, using some of the simpler PHA techniques identified in OSHA's Process Safety Management Standard. This would involve identifying a cross-functional group of site staff knowledgeable about site activities. It could be an existing group (such as the safety committee, depending on their expertise) or an ad hoc group. This group would conduct a structured brainstorming session (such as "what-if," and maybe using checklists) directed at identifying potential low-probability/high-impact events. To minimize the risk of missing events, the process should consider potential events over the entirety of the facility, and then potential events outside the sites, e.g. events at neighboring businesses. It is important to consider potentially applicable historical events, such as those that have occurred at similar organizations. If this process identifies events that hadn't been thought of before, or that had been underrated using traditional techniques, then of course the critical next step is to identify potential controls to minimize the risk. At this point, it may be necessary to learn to use some of the specialized risk assessment methods and tools mentioned above, and/or bring in specialized expertise to help quantify the risk and identify additional controls.

Impediments

Culture of Silence

In many cases, people may already know about holes in the safety systems that are just waiting to line up, but don't say anything about it. This can be for fear of reprisal, or a softer fear of standing out from the group. Maybe they don't think it's their job to say anything, and that somebody else must already know about it and will handle it. And there's the universal reason: Few of us really believe that anything seriously bad will really happen to us (especially if we've overlooked the problem before and nothing happened).

Solving a culture of silence, like any cultural change, may be difficult. Two elements that are required in the offshore oil and gas industry safety and environmental management system are stop-work authority and ultimate-work authority. The first establishes procedures whereby any employee can stop work when witnessing an activity that creates a threat of danger. The second involves clearly defined requirements establishing who has the ultimate authority on the facility for operational safety and decision making at any given time. With appropriate fine tuning of details to obtain management support, these two elements can be implemented in any operation.

Resources

This paper would not be complete without a discussion of resources. Major impediments to low-probability/high-consequence event risk assessment are the constraints of personnel, time and money. There may be a perception in the organization that this is not an important activity precisely because the probability is low and there are other more compelling ways to direct resources. The lower the perception of risk of site activities, the more likely resources are to be an issue. There is no one easy answer to resource constraints, but one method that may be useful would be to start small, build on the existing risk assessment processes and culture, and build on successes.

Conclusion

We hope we have convinced you that low-probability/high-consequence events are an important consideration for all industries, and something safety generalists should understand. These events are easy to overlook as because they are so rare, and it can be difficult to argue for the resources to identify and control them.

The standard risk assessment process does not work well for low-probability/high-consequence events. Instead, a separate, dedicated process with a cross functional expertise is needed. With some modifications, the elements of OSHA can be used.

Both identification and control is essential. Unidentified hazards leave the organization vulnerable to costly disasters. The control process is equally important because careful use of available resources is necessary to implement effective controls to avoid the situation where holes in the cheese line up, and costly disasters occur.

Session No. 539

The Intersection of Risk Management and Fleet Management

Sheila Palmer Coppinger, BS, MS, CSP, CFPS
Chubb Inc.
New York, NY

Jennifer Guerrini MS, CSP, ASP, CDS, CSS
Chubb Inc.
Whitehouse, NJ

The Impact of Fleet Management on Your Company's Risk Management Program

Employers should consider that a collision can impact numerous policies. For commercial auto policies, there can be first-party bodily injury and property damage and third-party bodily injury and property damage. When employees are involved in a collision, their injuries are typically considered to be covered under workers' compensation. This will impact the employer's workers' compensation modification and, ultimately, the premium. If negligence is found, the claim cost can impact the generally liability and/or excess liability policies.

Highlights of the 2015 Census of Fatal Occupational Injuries tell us that:

- Roadway-related fatalities climbed 9 percent to 1264, accounting for 26 percent of all fatal work-related injuries.
- Heavy and tractor-trailer truck drivers experienced 745 fatal injuries, the most of any occupation

Costs of Motor Vehicle Crashes to Employers Worksheet

Direct Costs to the Organization	
Workers' compensation benefits	$__________
Healthcare costs	$__________
Increases in medical insurance premiums	$__________
Auto insurance and liability claims and settlements	$__________
Physical and vocational rehabilitation costs	$__________
Life insurance and survivor benefits	$__________
Group health insurance dependent coverage	$__________
Property damage (equipment, products, etc.)	$__________

Motor vehicle repair and replacement	$______________
EMS costs (ambulance or medivac helicopter)	$______________
Vehicle towing, impoundment, and inspection fees	$______________
Municipality or utility fees for damage to roads, signs or poles	$______________
Direct Total	$______________

Indirect Costs

Supervisor's time (rescheduling, making special arrangements)	$______________
Fleet manager's time to coordinate vehicle repair, replacement, etc.	$______________
Reassignment of personnel to cover for missing employees (less efficient)	$______________
Overtime pay (to cover work of missing employees)	$______________
Employee replacement	$______________
Re-entry and retraining of injured employees	$______________
Administrative costs (documentation of injuries, treatment, absences, crash investigation)	$______________
Inspection costs	$______________
Failure to meet customer requirements resulting in loss of business	$______________
Bad publicity, loss of business	$______________
Indirect Total	$______________
TOTAL	$______________

Types of Fleets

Regulated Fleets

A motor vehicle or combination of motor vehicles used in commerce to transport passengers or property if the motor vehicle: (1) has a gross combination weight rating of 26,001 or more pounds inclusive of a towed unit with a gross vehicle weight rating of more than 10,000 pounds; (2) has a gross vehicle weight rating of 26,001 or more pounds; (3) is designed to transport 16 or more passengers, including the driver; or (4) is of any size and is used in the transportation of materials found to be hazardous for the purposes of the Hazardous Materials Transportation Act and which require the motor vehicle to be placarded under the Hazardous Materials Regulations (49 CFR 172, subpart F) (49CFR383). These are vehicles that are regulated under the Federal Motor Carrier Safety Administration (FMCSA). These include vehicles that transport hazardous materials.

Non-regulated Fleets

This typically includes a fleet of private passenger vehicles, light trucks and SUVs. The vehicles are used daily to transport personnel, tools, equipment, goods, and supplies to and from different locations. Vehicles may be used to deliver or pick up products for a variety of operations. These vehicles are not regulated by DOT. ANSI/ASSE Z15.1, *Safe Practices of Motor Vehicle Operations,* provides guidance on managing these fleets.

Non-owned/Hired Fleets

A non-owned auto is a vehicle that is used for company business is not owned, leased, rented or hired or borrowed by the company. Essentially, this is a vehicle owned by someone else used for company business. Typically, this is an employee who uses their own vehicle and is reimbursed through a stipend or submits mileage-based expenses.

A hired fleet are vehicles the company rents, hires, leases or borrows, other than vehicles owned by the employees or partners of the policyholder.

Basis of liability is the common law doctrine of respondeat superior, which means an employer can be held responsible for the wrongful acts of an employee who is acting on the behalf of the employer

Fleet Management

Many companies will have a combination of the different types of fleets. Successful management of the risk they create is a management control program. The typical elements of a program will include the following:

1. Program ownership and accountability
2. Driver selection
3. Driver management
4. Driver training
5. Vehicle selection
6. Vehicle maintenance
7. Vehicle inspection
8. Vehicle replacement
9. Accident management

Best practices to reduce risk through better selection of vehicles and drivers include:

1. MVR review
2. Drug testing
3. Driver files
4. Best practices for accident management

The employer should strive for a program of continuous improvement based on the latest technologies. The speakers will provide a brief review of the new technologies and Benchmarking Tools from Chubb and other sources.

References

American National Standards Institute/American Society of Safety Engineers. ANSI/ASSE Z15.1-2012, Safe Practices of Motor Vehicle Operations. Park Ridge, IL: American Society of Safety Engineers.

Bureau of Labor Statistics. 2015. *Census of Fatal Occupational Injuries.* (http://www.bls.gov/news.release/cfoi.nr0.htm).

Occupational Safety and Health Administration (OSHA). *Guidelines for Employers to Reduce Motor Vehicle Crashes.* (www.osha.gov/Publications/motor_vehicle_guide.pdf).

To The Point
Fleet Safety Management

CHUBB®

Managing the risks associated with the use of vehicles for business

It is difficult to imagine a company without some businessrelated vehicle use. The scope of vehicle use may range from large, interstate-transport companies with a fleet of thousands of large trucks to a small company that relies on the use of an employee's car to run occasional business errands. The use of vehicles at some level exists in almost all companies.

The use of vehicles is so common that it is often overlooked by those companies that are not in the transportation business. Although occasional use of vehicles such as light trucks or passenger cars does not present the same level of risk as an interstate trucking company, the risk of crashes and the resulting damage and injury can pose a severe threat to a company.

Motor-vehicle related incidents are the leading cause of work-related fatalities. Based on Bureau of Labor Statistics, more than 35% of occupational fatalities are motor-vehicle related. In addition to the potential for fatalities, vehicle crashes also produce severe debilitating injuries to drivers, occupants, and pedestrians. Vehicle crashes can also cause severe business disruptions when critical equipment or material is damaged or lost. The media attention that often follows such events can also damage the image of the company and attract legal and regulatory attention.

When you considered all forms of business-related driving your employees are exposed to during the year, you will realize the significant risk to your company. It is easy to identify the regular business driver, but employees who rent vehicles when they travel or employees who use their personal vehicle for company business are not as easy to control. For many companies, the exposure to vehicle crashes and the resulting injuries is greater than initially believed.

Risk Engineering Services

Your company can more effectively manage the risk of vehicle-related incidents by applying a process to determine:

- The total exposure to vehicle operations
- The risk associated with the exposure
- The effectiveness of controls

Once a critical review has been conducted in these areas, a formal plan can be developed to address the concerns.

If your company operates vehicles that qualify as "commercial motor vehicles" as defined by the Federal Motor Carrier Safety Administration (FMCSA), you are likely very familiar with the management of vehicle safety. The FMCSA regulations do not apply to the operation of passenger vehicles, light to medium trucks or vans, but they do apply to vehicles that are used on a highway in interstate commerce to transport passengers or property. Examples include any vehicle that:

- Has a gross vehicle weight rating or gross combination weight rating, or gross vehicle weight or gross combination weight, of 4,536 kg (10,001 pounds) or more, whichever is greater;
- Is designed or used to transport more than 8 passengers (including the driver) for compensation;
- Is designed or used to transport more than 15 passengers, including the driver, and is not used to transport passengers for compensation; or
- Is used in transporting hazardous material requiring placarding by the Department of Transportation (DOT).

The FMSCA regulates how companies manage their commercial motor vehicles as defined above. There is no current Federal or State regulation that requires companies to formally manage the rest of their vehicle operations.

Still, companies not regulated by the FMCSA are legally responsible for the safe operation of vehicles used for business purposes and the safety of their employees. Regardless of the size or complexity of your vehicle operations, the following approach can result in a heightened understanding of the risk to crashes and injuries:

- Identify the nature and extent of your company's vehicle use, including vehicles owned, leased or borrowed on behalf of the company. Also include non-owned vehicles used for company business as well as company-owned vehicles used by others for personal reasons.
- Quantify the exposure to crashes and injuries posed by the use of vehicles as noted above. How many occupants are typically in these vehicles? How many miles are driven? What types of vehicles are used and what is their condition? Where are the vehicles used? The risk of crashes is different on local roads, expressways and interstate highways.
- Review the frequency and severity of vehicle-related claims or incidents over the past five years. Are accurate records kept of vehicle-related claims or incidents? Are these incidents properly reported? Are the total costs of crashes determined? Are vehicle incidents properly reported?
- Consider how you currently control your vehicle risk. Are the vehicles you use appropriate for the job, inspected and maintained? Are employees provided driver training? Are vehicle crashes and incidents reported and investigated? Is seatbelt use required? Are employees prohibited from cell-phone/smart-phone use or other distractions?

Assessing, and then managing, your company's vehicular risk can seem a daunting task. Fortunately, there are standards companies can use to develop their vehicle risk management practices. These standards provide a framework companies can use to implement, review and further develop their management practices for vehicle operations.

Resources

The Newly Revised ANSI/ASSE Z15.1 Standard - Safe Practices for Motor Vehicle Operations
www.asse.org/assets/1/7/Z15_1_Tech_Brief_4_2012.pdf

This standard applies to all types of companies and vehicles. It addresses the administration of vehicle management, various vehicle operation issues, driver exposures and controls, vehicle exposures and controls, and incident reporting and analysis. The standard is available from the American Society of Safety Engineers (ASSE) website. Summaries of the standard can also be found online.

National Institute for Occupational Safety and Health (NIOSH) Center for Motor Vehicle Safety
www.cdc.gov/niosh/topics/motorvehicle/NCMVS.html

This internet-based knowledge center provides employers with data, research, and resources related to the identification and control of vehicle hazards.

Occupational Safety and Health Administration (OSHA) Motor Vehicle Safety Topics
www.osha.gov/SLTC/motorvehiclesafety/index.html

This page of the OSHA website is devoted to motor vehicle safety. It provides easy access to motor vehiclerelated regulations, hazard and control references, and other related resources.

Federal Motor Carrier Safety Administration (FMCSA)
www.fmcsa.dot.gov/

This agency governs the operation of large trucks and buses and establishes safety requirements for drivers and business operators.

Network of Employers for Traffic Safety (NETS)
www.trafficsafety.org/

NETS offers employers access to fleet safety promotional materials. Members of NETS can also share "best practices" through a benchmarking process.

Connect With Us

For more information about protecting your business, contact your local Chubb risk engineer or visit us at www.chubb.com/engineering.

Chubb. Insured.℠

To The Point
Non-Owned Auto Risk Management

A Commonly Overlooked Risk

In an effort to reduce operating costs, more and more organizations are reducing or eliminating the number of owned or leased cars they provide employees. As a result, the company car goes away but the need to use automobiles for business purposes usually doesn't. To meet the need, companies sometimes reimburse their employees for using their personal vehicles on company business. Employees usually don't give much thought to this, especially since they are reimbursed for mileage and/or receive a monthly stipend for the use of their car. Additionally, employers with few- or no-owned autos often don't give much consideration to managing their non-owned auto risk. When combined, these factors can lead to unexpected auto claims for the company as a result of this new arrangement.

A Non-Owned Auto includes any automobile leased, rented, hired or borrowed by the employer that is not owned by the employer, but is nonetheless used in connection with the business of the employer. Some examples of this business usage could include when employees use their personal autos to pick up office supplies, make sales calls, run to the post office to get the company's mail, or go to the bank to make a deposit for the business.

It is imperative that you identify, understand and manage your non-owned auto risk to minimize the potential for losses. To properly manage this risk, your company should make a concerted effort to implement controls similar to those you would for a fleet of owned or leased vehicles.

To accomplish these controls, a business-use policy for non- owned auto best practices should be developed. This could include specifics as to who is approved to drive for business purposes, how they will be qualified by the company (e.g., driver license verification, MVR review), and the type of vehicles that can be used. It should establish policies regarding seat-belt usage, as well as distracted driving that can result from cell phones, smart phones, texting, and inputting data in GPS units while driving. Accident reporting procedures should also be detailed.

Risk Engineering Services

Use of Personal Autos is a Business Risk

The practice of allowing employees to regularly use their personal auto for business purposes creates increased liability for the company. The company has little or no control over who the legal owner is, or the type and condition of the vehicle being driven. The type of auto-liability insurance carried by the employee may include low limits of liability, high deductibles and/or adverse terms. The company may also have increased exposure if the employee has a poor driving record or an invalid driving license.

In the event of an accident, the driver's personal-auto insurance is the first layer of insurance, but depending on the circumstances, the company's commercial- or business- auto policy may also respond once personal auto policy limits are exhausted.

Non-Owned Auto Controls

Because the use of personal autos for business purposes presents a significant loss potential for the company, it is recommended that the following controls be established:

- Establish a driver-authorization program that requires each individual driving a non-owned vehicle for company business to be authorized to do so by management or a designated responsible person. Authorized drivers should be required to:
  - Provide a valid driver's license for the type of vehicle to be operated, and proof of registration in the employee's name
  - Be 21 years of age or older and have a minimum of three years licensed driving experience
  - Have an acceptable driving record as determined through an initial review of the driver's Motor Vehicle Record (MVR) by the company, and an annual review thereafter
  - Be versed in the safe-driving policies of the company
  - Provide proof of insurance by supplying the company a current certificate of auto-liability insurance, or a copy of their current auto-insurance policy Declarations page for the vehicle being authorized for use (with established minimum-acceptable levels of auto-liability coverage)
  - Report all accidents to the employer when they occur while the vehicle is used for company purposes
- The driver-authorization program should also address:
  - The type of vehicle that is acceptable to be driven for business purposes
  - Regular defensive-driver training
  - Rules addressing impaired driving, seat-belt use, etc.
  - Procedures to follow when an accident occurs
  - Rules regarding the transport of non-business- related passengers while driving for business purposes
  - A distracted-driving policy that addresses common distractions such as phone calls, texting, computer or tablet use, GPS use, etc.
  - Providing the driver a company vehicle for business use or renting an appropriate vehicle instead

Prepare Now

Don't let this often hidden, overlooked, or seemingly inconsequential issue impact your bottom line. Evaluate your non-owned auto exposure based on your operations and the expectation of employees using their personal vehicles on company business. Develop a plan to address the risks you identify and implement a formal policy to mitigate these risks. Train affected employees and get their acknowledgement so that they understand what is expected of them as a prerequisite for driving their vehicles on company business.

Resources

ANSI / ASSE Z-15.1 - 2012
Safe Practices for Motor Vehicle Operations
www.asse.org/shoponline/products/Z15_1_2012.php

National Safety Council - Safety on the Road
www.nsc.org/safety_road/Pages/safety_on_the_road.aspx

Connect With Us

For more information about protecting your business, contact your local Chubb risk engineer or visit us at www.chubb.com/engineering.

Chubb. Insured.℠

Session No. 540

What is Best-in-Class in Contractor Management

Dag Yemenu, M.B. A., M.A, B.S.
Senior Vice President

Richard Cerenzio
Corporate HSE Director
ISN
Dallas, Texas

What is Best-in-Class Contractor Management?

When we identify key issues facing safety health & environmental professionals, one that stands out among many as an emerging issue for the safety profession, is the focus on global industry trends, indicating a continued growth of outsourcing non-core business activities to third-party contractors, ranging from construction and maintenance activities to security and catering functions. This "contractorization" of operations has become a key strategy pursued by both private and public sectors and is even embraced by the military in some countries. Decades of data compiled by the International Association of Oil & Gas Producers on the Upstream Oil and Gas industry,[1] for instance, indicates that while employee work hours have doubled over the past 25 years, contractor work hours have more than quintupled.

Contractorization allows for more competent, well-equipped and specialized contractors to undertake tasks or projects in a much safer and sustainable manner than could be achieved when simply relying on in-house resources. The challenge and opportunity lies in establishing a robust contractor management strategy to effectively mitigate risk exposures from contractorization, while allowing the owner to streamline its operations and focus on its core competencies.

An effective contractor management system goes beyond a one-time pre-qualification step. It levels the playing field by holding all contractors to established safety and compliance standards and promotes a culture of transparency, accountability, and continuous improvement. In other words, it adds value to the hiring client's overall operations by going beyond simply meeting internal or external compliance requirements to establishing and sustaining a mutually beneficial partnership between a hiring client and its contractors.

In this paper, we outline 10 essential attributes we find in best-in-class contractor management systems. This list of attributes is not intended to be exhaustive, but it presents key items considered to be of strategic significance for establishing effective contractor management processes. Whether you are a hiring client who uses contractors or a general contractor who hires subcontractors, you will find reference points for benchmarking your practices and identifying opportunities for reducing your risk exposure. We focus on the non-commercial and non-technical aspects of contractor management, though most of our points would also apply to those considerations.

This paper demonstrates how these elements can foster an improved culture of prevention in the ever-growing sector of outsourced operations. These findings highlight the importance of consistent expectations and requirements tailored to local operating conditions in order to drive safety performance. Through use of a system for managing and evaluating contractor safety data, client companies have seen significant improvement in their contractors' performance. For companies that hire contractors, the elements discussed will assist by providing ideas for benchmarking and give useful and actionable takeaways for enhancing contractor management processes.

The presentation that usually accompanies this paper primarily focuses on addressing key considerations under each of these ten strategic elements as well as prepare SH&E professionals for the new challenges introduced by contractorization. By drawing from more than 15 years of global experience working along 500+ owners and 63,000+ contracting companies on contractor management, the best practices discussion will incorporate case studies, anecdotes, and lessons learned for practical insight. Attendees should obtain useful and actionable takeaways from the discussions helpful for expanding their knowledge and professional skills surrounding contractor management and enhancing their contractor management and health and safety improvement efforts for their jobsites.

The Value of Benchmarking Contractor Performance: Leading and Lagging Indicators

What gets measured gets improved. Measuring and benchmarking contractor performance is a key driver of continual health and safety improvement efforts in the workplace. The presentation will also address some of the key leading and lagging (historical) performance indicators used in the industry, and how the data can be used for benchmarking and continual improvement purposes. Without capturing and analyzing contractor work hours, quarterly and annual incident rates, the number of open and closed action items, post-job evaluation feedback or the status of certificates of insurance in place for all contractors for instance, the owner's system will struggle to proactively identify unfavorable trends and implement timely and objective corrective actions.

In addition to basic data trending and analysis methods, the presentation will share leading edge methods, including regression and correlation statistical analysis approaches, being used in the industry to identify factors that are positively or negatively associated with increased or decreased health and safety performance. Such methods are useful in identifying the vital few key performance indicators (KPIs) out of the trivial many that perhaps need closer attention by health and safety professionals. Real world analytical examples will be presented and discussed.

Best-in-Class Contractor Management Attributes

Effective and best-in-class contractor management systems are characterized by the following key attributes.

1. Establish a Clear Communication Strategy

Contractor management processes are often bureaucratic, inefficient, and a source of confusion to contractors due to poorly designed or executed communication strategies. At their core, effective contractor management systems must have a well-thought-out plan for communicating with internal stakeholders as well as external contractors. The plan should establish what needs to be communicated to contractors, how to accomplish it, and how to proceed to the next step of engaging contractors on a personal level.

2. Designed to Measure Performance
Processes designed with a mind-set of simply completing the administrative task of pre-qualifying contractors, with documents and data on completing the contractor onboarding process, then locked away in filing cabinets, will not drive continual improvement. Without capturing and analyzing contractor work hours, quarterly and annual incident rates, the number of open and closed action items, post-job evaluation feedback or the status of certificates of insurance in place for all contractors, for instance, the system will struggle to proactively identify unfavorable trends and implement timely and objective corrective actions.

3. Incorporate a Risk-ranking Process
Contracting companies come in different sizes, structures, capabilities, and risk profiles. Some might be present at the hiring client's job site only a few days a year, but may be involved with high-risk activities that could have consequences for the safety, integrity or continuity of operations. Others might be 'nested' at the client's jobsite year round. The question is, how should contractor requirements and expectations (including health, safety, insurance, training, competency, and the like) be tailored to best fit the appropriate risk exposure?

4. Follow a Three-step Due-Diligence Process
For best-in-class systems, due diligence involves collecting self-reported information from contractors, reviewing and verifying contractor information and undertaking on-the-ground inspections and audits of contractors' worksites, manufacturing facilities, or home offices. These three levels are synergistic and enable a holistic and sustainable due diligence process.

5. Maintain an Approved Contractor List
Hiring clients sometimes find themselves under pressure to approve a contractor in an "urgent" manner, even when the situation itself doesn't meet the definition of a true emergency. How many times have you had to play catch up to approve a contractor who has already been procured, or even worse, is already on your jobsite providing services? A process that allows for recurring last-minute procurement of services, usually due to a poorly maintained list of approved contractors to tap into or an inconsistently enforced contractor procurement policy, is risky.

6. Establish Consequences and an Exit Strategy
If you have an approved contractor list, do you remember the last time you, your procurement group or your HSE department decided to "clean it up?" If your list has grown significantly without good commercial and operational reasons, and the list includes numerous dormant, redundant or consistently non-conforming service providers, it is a good time to re-evaluate your contractor exit strategy.

7. Address Subcontractor Management
Have you ever run into a situation where a non-vetted subcontractor working for one of your general contractors is on your jobsite driving a vehicle with logos you don't recognize? Such instances are not new to veterans in contractor management. No matter how robust your contractor management system is, your subcontractor management process may be a weak link.

8. Follow Standardized and Transparent Guidelines
Contractors are often frustrated by multiple requests for information from various departments within a hiring client's organization. These usually stem from a lack of consistency among a hiring client's sites, plants or operational departments. Best-in-class contractor management systems are built around a centralized and standardized set of guidelines for managing

contractors. They keep contractor data in a single, well-organized platform, as opposed to maintaining multiple database silos that do not communicate with each other.

9. Integrated with Internal Business Processes
Best-in-class systems consider their contractor management process to be an integral part of their extended internal business processes. For instance, key contractor scorecards and compliance data may be electronically integrated with the hiring client's purchasing system to ensure only contractors meeting its standards receive purchase orders. The achievement of a well-integrated contractor management approach requires the support of multiple stakeholders within the hiring client organization (such as HSE, procurement, security, risk management, and operations) and the full commitment of senior leadership.

10. Take Safety Culture to the Next Level
Do you have issues with building and maintaining a strong site safety culture at your site? Best-in-class cultures help to ensure the message you're trying to convey to your contractors is not hit or miss. Building this culture takes time, resources and engagement by both client and contractors. Having a best-in-class culture helps you to communicate and share best practices, lessons learned and promotes peer engagement. Thus, engagement can be realized by site workers taking responsibility to recognize, prevent and when necessary, mitigate hazards on site works effectively.

Summary

Contractors have become an increasingly essential part of the wider business ecosystem. Competent contractors enable owners to streamline their operations, optimize their cost structure, and sharpen their focus on core competencies. Contractorization is a trend that is here to stay and will continue to grow. Safe and sustainable contractorization of operations, however, depends on a robust process for managing contractors, especially when hundreds and even thousands are involved.

Many will find the ultimate success of a contractor management process is the mutually beneficial partnerships it helps build between an owner and its contractors. It also has a direct human value. As Juan Somavia, the Director General of the International Labor Organization (ILO), has stated, "Occupational safety and health is first of all a matter of human right and respect for human dignity."

Establishing and sustaining a robust contractor management process is therefore a clear demonstration of the owner client's commitment to protecting both the human and environmental dignity involved in its operations. While even the most comprehensive contractor management system can't eliminate risk altogether, it will go far in bringing contractorization to the forefront by both reducing what could be left to chance and in mitigating overall risk exposure. For companies that hire contractors as either an owner client or as a prime/general contractor, the ten contractor management attributes, along with the statistical performance analysis methods presented, will assist such organizations by providing them with ideas for benchmarking and continually improving their contractor management processes.

Endnotes

[1] International Association of Oil and Gas Producers. 2010. *Safety performance indicators–2010 data*. (http://www.ogp.org.uk/pubs/455.pdf).

Safe Decision Making in Organizations: New Research and Improvement Strategies

Thomas R. Krause, Ph.D.
Kristen J. Bell, M.A., M.B.A.
Krause Bell Group
Ojai, CA

Introduction

Thought leaders in the safety community have known for many years that safe decision making is an essential element in the complex of factors that go into making a safe workplace. But until the last ten years or so, we hadn't realized two critically important things: (1) The people making the most important safety decisions are not frontline workers. Workers have a role, but the most critical role in safe decision making is the one played by leadership. (2) Leadership decision making is pivotal to the understanding and prevention of serious injuries and fatalities (SIFs).

In this paper, we will examine the evidence for these statements, as well as the implications that follow, for policy, strategy and the role of senior leaders in SIF prevention.

The Role of Decision Making in Safety

New Research in Brain Science and Implications for Safety

There has been extensive research in psychology, and more specifically in brain science, over the last 20 years that has deepened our understanding of decision making. An entirely new field has developed that examines the cognitive processes involved, and identifies natural failures in the process that undermine sound decision making. Based on this body of research, Tom Krause and Kristen Bell began to study its implications on safety performance improvement. Questions began to arise about how this science could be applied, particularly to the prevention of serious injuries and fatalities. How important is decision making? Who makes safety-related decisions? Which decisions are most important? Could safety be improved through development of decision-making skills and knowledge?

The Decision-making Study: Who, What, When and How

In 2015, Krause and Bell conducted a study to answer these questions. Along with their partners at ORCHSE Strategies (formerly Organization Resource Counselors), they developed a research project that examined decision making that led to serious injury and fatality (SIF) events. Decisions were analyzed across 60 SIF events across multiple industries, and the aggregate data showed promising insights. The study was coined the safe decision making (SDM) study.

To support the study, a group of five large corporations came together to form the "Safe Decision Making Innovation Group." This innovative and collaborative approach was key in connecting academic research with actual, real-life application. The group members were from a variety of industries, including oil and gas, metal manufacturing, and health services. A key safety leader represented each organization, some well known worldwide for their contribution to safety. These representatives submitted data on past serious injury events that occurred at their organizations. They supported the analysis by providing access to internal personnel (who were personally familiar with the incidents) for interviews. The group continued to meet periodically, about once a quarter, to review emerging findings, discuss, and provide valuable insight based on each member's unique experience. This was an important element of the method; it is what differentiates this innovation work from academic research, and ensured that the focus remained on the practical and relevant use of the findings.

The Safe Decision Making (SDM) study included analysis of 60 SIF events, based on data extracted from investigation reports, interview transcripts, root-cause report outs, live interviews, and any other incident-related sources. This effort uncovered 612 individual decisions made that influenced the outcome. In addition, each incident was evaluated against 58 attributes, and each decision was evaluated then against 37 attributes, including who (level and function), what (type of decision), why (situation), when (relative to incident, and which stage of the decision-making process), and the biases or other decision-process breakdowns that potentially undermined the decision.

Whose Decisions Matter?

Most organizations recognize that the ability of leaders to make sound decisions is directly related to overall business performance. It is widely understood that, "Ultimately, a company's value is no more (and no less) than the sum of the decisions it makes and executes."[i] Yet, this view does not always carry over into the safety arena. The relationship between sound decision making and safety performance is not as widely known, and is generally not incorporated into the approaches and methodology many organizations apply to safety. At most, efforts to combat faulty safety decisions are focused at the worker level, embedded in the typical response of, "What was the worker thinking?" Influenced by a shallow understanding of behavior-based safety principles, many organizations are blind to the significance of managerial decisions in safety performance.

One of the most pronounced findings of the study was that decisions made above the front line have a major role in injury prevention. Of the decisions found to have influenced the 60 SIF events, 67% were above the front line (department managers through corporate executive leadership).

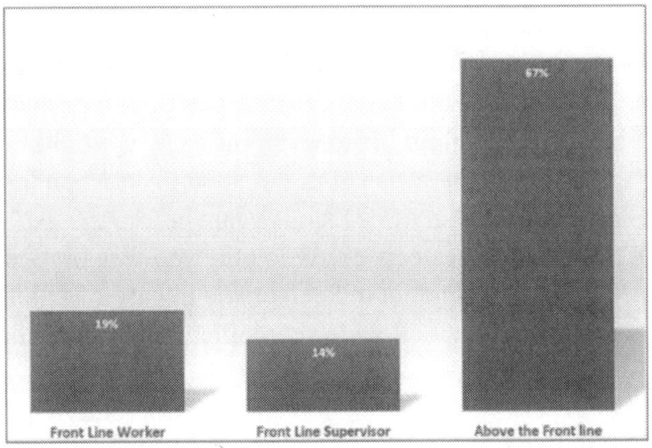

Exhibit 1. 67% of the decisions were made above the front line.

The implications of this are far reaching. It seems common knowledge that managers play a role in incident prevention and causation, but to suggest that they have a larger role than those at the front line could revolutionize the way we approach safety. It is critical that this finding is not misconstrued as blaming the manager, any more than it would make sense to blame the worker. What this suggests instead is that leaders are positioned to have a greater influence over safety than previously thought, and by equipping them with the appropriate knowledge and skills in decision making, they could provide the greatest opportunity in safety improvement.

In addition, the study found that of the 67% of decisions made above the front line, 79% of those were made at the site level. This was an unanticipated, yet incredibly meaningful, finding. What this says to the safety world is that the shortest distance to a zero-injury rate is through a direct focus on decision making within site management. These are the leaders that make critical, operational decisions every day that have direct impact on safety. Some examples of this are scheduling, over-time allocation, purchasing of new equipment, hiring, and so on.

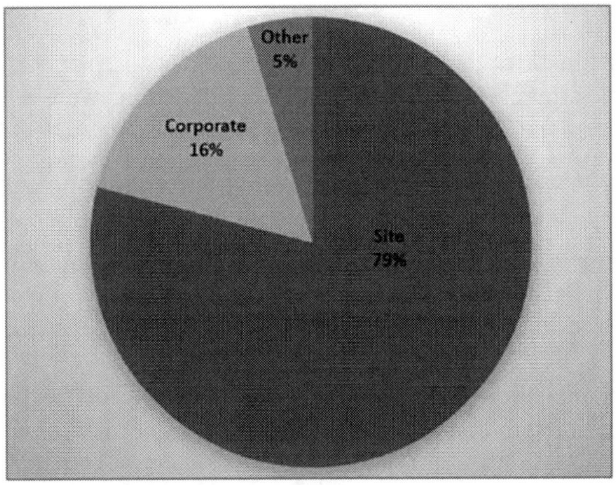

Exhibit 2. Of the decisions made above the front line, 79% of them were made at the site level.

To illustrate this point, we can use an example from a case that was included in the study. In 2010, a metal dust explosion caused three fatalities and one serious injury.[ii] An explosion had occurred inside the metal blender, resulting in three people fatally injured, and one critically injured. One of the blender blades had struck the side of the blender, causing a spark that ignited the extremely explosive metal dust. During the investigation, it was found that the workers had repaired the blade multiple times. When it continued to function improperly, they decided to replace it with a used blade from an older blender. In this case, it would be completely reasonable to point to the worker's decision as the cause of the incident. But when investigated deeper, we find that management had made a chain of decisions years before the explosion that set the stage for the worker to make that final, fatal decision. When the site was first opened, corporate decided not to purchase back-up equipment to keep onsite in case of failure. They also decided to work on a 24/7 schedule, which did not allow for equipment downtime and exerted time pressure on the operation. The site manager decided not to enforce a known rule for closing the lids of the blenders while they were operating. When first informed of the faulty equipment months before, site management decided to defer maintenance due to cost. Although the maintenance worker had the option of making a different choice, it becomes evident in cases such as this that decisions made by management had the greatest influence, and created the conditions under which such a disaster could occur.

Which Decisions Matter?

When we think about the sheer numbers of decisions made by leaders that could potentially have safety impact, the task of improving them all seems quite daunting. Although the case could easily be made that all decisions are indirectly related to safety, the study found a subset that could be viewed and having the greatest impact, both linearly within a specific case and laterally across the organization.

To better understand this quantitatively, all decisions found were categorized by situation and type. The situation was defined as the circumstance that required a decision to be made, such as "design of a hazard mitigation" or "installing equipment." Within these situations, there are many types of decisions that need to be made. This is best understood as, "What is the question that needs to be addressed?" For instance, decision types could be, "How fast?" or "Go or no-go?" or "What resources?"

Out of the decisions made at the site level (above the front-line supervisor but below corporate), the top three situations where critical safety decisions were made that impact a SIF event were:

- Hazard Mitigation (28%): Defining need, designing, implementing and addressing compromises
- Planning (20%): Project planning, scheduling, starting, addressing delays, and designing work processes
- Safety Management Activities (11%): Conducting audits, investigations, and inspections

The types of decisions seen most in the data were defining specifications, identifying the information needed, and how to communicate. Therefore, if a cross-section of these types of decisions made within the example situations listed above were identified and tracked, a new indicator could be developed that signals a moment in time where safety should be the highest priority.

From this, we can conclude that there are certain situations where leadership decisions should be examined more carefully. Even if the focus is only on certain types of decisions made in certain situations, great opportunities could arise for safety improvement.

How Are Decisions Compromised?

In psychology, it is widely accepted that a complex set of cognitive processes occur prior to making a decision. This is how we evaluate given information to establish the best course of action. The process can take days or a millisecond, and can be conscious or unconscious. For the most part, these processes help us, but often they lead us astray, stealing our objectiveness and replacing it with biased, unfounded, or even distorted subjectivity. We like to call this a "breakdown" in the decision-making process that can often lead to less than desirable outcomes.

In our study, we identified a breakdown in the decision-making process in 85% of the decisions. Sometimes, it broke down because they were overconfident in their own abilities, or that of their team. At other times, the incorrect information was used to inform the decision, (looking at the wrong report). There were also times that the organizational pressures of continuing the status quo were given a higher priority than safety. There are infinite ways a decision can be compromised. Krause and Bell have already written extensively about cognitive bias and its effect on safety, and this study confirmed what was already said.

Another finding, based on input from the Safe Decision Making Innovation Group members, was that the academic concept of cognitive biases doesn't mean much to organizational safety leaders. Organizations need more practical tools to help managers make decisions. They care less about the academic label or its constructs, and more about how to make that next decision that comes across their desk. In a perfect world, leaders would make all decisions objectively, based on nothing more than factual evidence and statistical probabilities. But we know that is not the way it works. All humans, including the greatest leaders, are biologically predisposed to "systematic errors in thinking"[iii] that can derail the most intelligent decision maker from making a sound choice. The SDM study confirmed this, and identified the need for further study.

The New View: Focusing on Decision Making at the Leadership Level Is a Critical Component to Safety Improvement

What this all says is that we can get better at decision making. A new perspective must be taken in regards to incident causation: If you want to protect your workers, you must create the safest context for them to work in. As Krause and Bell have stated in their previous writings, leadership creates culture.[iv] Every decision made at the management level speaks to the cultural values and beliefs of the organization. Therefore, proper decision making is critical in creating a safe workplace. If a leader wants to develop a culture of safety, the decisions they make must reflect that value. The next step is to find out how to improve decisions by identifying the best practices for decision making as it applies to organizational safety. This will be the next study for Krause Bell Group, coming in 2017.

Acknowledgement

Special thanks to Jessica Adams who collaborated with the authors on the writing and editing of this paper.

Endnotes

[i] Blenko, M. W., Mankins, M., & Rogers, P. (2010, June). *Harvard Business Review.* (Retrieved from https://hbr.org/2010/06/the-decision-driven-organization.)

[ii] U.S Chemical Safety and Hazard Investigation Board. (7/16/2014). *AL Solutions Fatal Dust Explosion Report.* (Retrieved from CBS.gov: http://www.csb.gov/al-solutions-fatal-dust-explosion/).

[iii] Khaneman, D. 2011. *Thinking, fast and slow.* New York, NY: Farrar, Straus and Giroux.

[iv] Krause, T.R., Bell, K. 2015. *7 insights into safety leadership.* Ojai, CA: The Safety Leadership Institute.

Session No. 542

Opening Our Eyes to Fatigue in the Workplace

Susan Sawatzky, B. Ed, CRSP
In-Scope Solutions

Introduction

"Eye on Fatigue" is a unique approach for communicating a comprehensive fatigue risk management strategy (FRMS) to organizations. The model visually demonstrates the need for integrating organizational strategies and worker engagement to create a strong fatigue management safety culture. It further demonstrates the value of this approach by highlighting the financial drivers for implementing the strategy. This approach addresses fatigue as a complex issue that requires comprehensive strategies to fully mitigate. Most importantly, it recognizes the need for both management and workers to share responsibility in adequately managing the risks of fatigue.

This model addresses the fact that many organizations today still underestimate and under-address existing fatigue risks. These organizations have achieved compliance, but not comprehensive and fully effective fatigue risk management. Many lack the capacity to formally identify fatigue risk exposures and lack the best practice knowledge needed to create mitigation strategies specifically targeted to known fatigue risks.

The Eye on Fatigue Model strives to communicate a more effective and systematic approach for managing fatigue. It incorporates the learnings from other fatigue strategies existing in North America and abroad, and is reflective of the key components of a generic safety management system. The model seeks to provide a more holistic framework for fatigue mitigation. This comprehensive strategy is based on the recognition that effective mitigation is reliant on organizational strategies inclusive of leadership support and embedded operational processes and procedures, as well as worker engagement strategies developed foundationally through education and awareness training.

These organizational and worker engagement strategies work synergistically in a comprehensive approach that, if implemented effectively, should demonstrate positive business drivers for the organization. The Eye on Fatigue Model uses the recognition of these business drivers to incentivize organizations toward controlling their fatigue-risk exposures. It does this by specifying the financial values that can align with comprehensive FRMS implementation. It uses common business language by detailing positive key performance indicators (KPIs) that may be influenced by the successful implementation of a FRMS program.

This paper explores the following:

- Current fatigue research and modern science-based understandings to increase recognition of how fatigue impacts worker health and safety.
- The Eye on Fatigue Model that demonstrates the integrated elements of organizational strategies and employee empowerment needed to create demonstrated positive business drivers.
- How this model reflects practical and implementable fatigue risk management strategies that can be used to assist an organization in proactively managing fatigue to reduce fatigue related risk exposures.

Fatigue Impairment Background Information

Fatigue Impacts on Safety and Health

There is a plethora of research demonstrating that fatigue can impair a worker's abilities. This impairment then creates a demonstrated impact on worker safety.

Fatigue can impair workplace performance in four key areas:

1. **Alertness**: to notice or react to changes in a work environment, to remain vigilant
2. **Emotional Stability:** to remain emotionally calm and stable when dealing with critical situations
3. **Mental Ability:** to remember, to think clearly, to make decisions, to communicate
4. **Physical Ability:** reaction times and overall coordination

Short-term effects of fatigue can lead to significant safety hazards as the worker's ability to safely and effectively perform his/her work is impacted. Examples of physical, mental, and emotional effects of short-term fatigue are detailed in Figure 1.

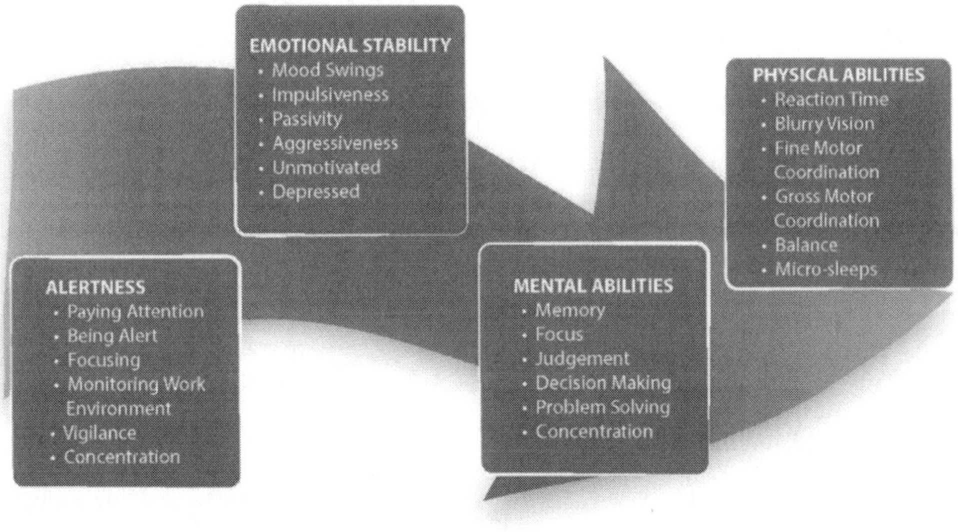

Figure 1. Short Term Effects of Fatigue

Findings from occupational research support a strong business case for the need to address worker fatigue impairment, as it has been shown to increase safety risks. For example, fatigue has been shown to be four times more likely to contribute to workplace impairment than drugs or alcohol. However, unlike drugs and alcohol, which can be measured to assess impairment, measuring and combating fatigue in the workplace is often more difficult.[i] This impairment can lead to higher frequency and severity of incidents. In fact, many sources list fatigue as one of the top five causal factors in workplace incidents.[ii]

An 88% excess risk has been demonstrated for individuals working more than 64 hours a week.[iii] Even working more than 50 hours per week has been shown to increases the risk of making an error by almost two times.[iv]

Shift work can lead to significant incident risks, particularly when sleep debt is allowed to accumulate. Compared with the day shift, incidents have been shown to increase by 15% during an evening shift and 28% during a night shift.[v] Overall, the excess risk of work injury attributed to shift work was 14.4% for women and 8.2% for men.[vi]

Fatigue risks are higher in tasks that can be boring or monotonous and that require constant vigilance. A common task that involves both these elements is driving; as a result, driving is often a key task highlighted when assessing fatigue risks. It has been estimated that 20% of all fatal collisions involve driver fatigue.[vii] In fact, time of day was the single best predictor of decreased driving performance, as a driver is 4 times more likely to have a fatal crash if they are driving between 10 p.m. and dawn.[viii] The majority of single-vehicle accidents occur during our circadian trough (midnight to 7:00 a.m., with 1:00 to 4:00 a.m. being the highest), the second most occur during our circadian dip during 1:00 to 3:00 p.m.[ix]

A ground-breaking study raising awareness on fatigue safety issues was conducted in 1997 by Dawson and Reid. Their study, "Fatigue, Alcohol and Performance Impairment," compared the effects of fatigue with alcohol in terms of performance impairment.

In this often-referenced study, researchers took 40 participants and conducted two counterbalanced experiments. The first group they kept awake for 28 hours; the second were asked to consume 10-15g of alcohol at 30-minute intervals. They then tested the levels of both cognitive and physical impairment at half-hour intervals, using a computer-administered test of hand-eye coordination. The results clearly show how impaired individuals were by fatigue.

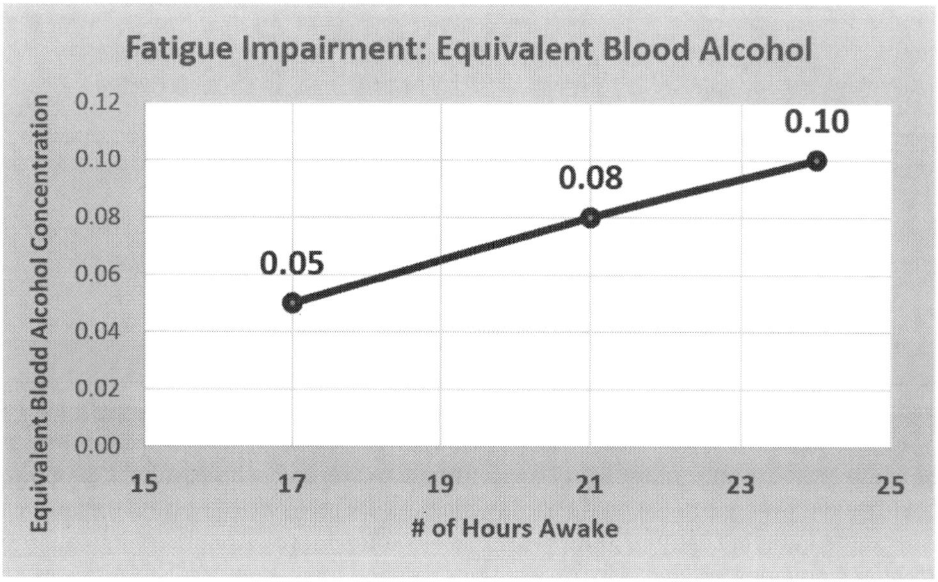

Exhibit 2. Fatigue Impairment: Equivalent Blood Alcohol (Created by In-Scope Solutions, based on Dawson & Reid 1997)

Individuals who had been awake for 17 hours performed with similar impairments to those with a blood alcohol content (BAC) of 0.05%. Based on the study, one could extrapolate that those awake for approximately 21 hours performed similarly to the legal impairment level of 0.08% BAC. Individuals awake for 24 hours were impaired equivalent to those well over the legal limit at 0.10% BAC.[x]

When we add to this study the knowledge that many shift workers often obtain minimal or no sleep the day they begin shift, we can infer many night workers are highly impaired in their physical and mental abilities, often having been awake for over 24 hours by the end of their first night shift.

In addition to occupational safety risks, there are human health effects relating to fatigue. Fatigue has been shown to correlate to numerous health impacts including: risk of stroke, cardiovascular disease, hypertension, gastrointestinal issues, obesity, mental health issues and more. The International Agency for Research on Cancer of the World Health Organization (WHO) identified fatigue and interrupted sleep, when associated with circadian disruption, as a type II carcinogen, or a probable cause of cancer.[xi]

It should be noted that the majority of statistics on health effects are based on shift workers who regularly have their sleep habits outside of normal circadian rhythms. Sleeping and being awake outside of normal daily rhythms appear to have the largest impact on human health.[xii]

The following diagram (Figure 3) illustrates findings from several studies on human health and fatigue.

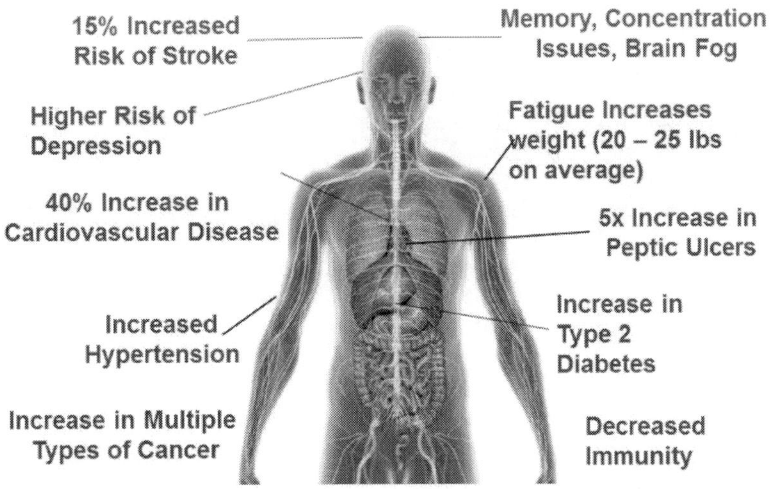

Figure 3. Effects of Fatigue on Human Health (Copyright In-Scope Solutions[xiii])

Fatigue is an unusual risk and safety hazard for employers, in that organizational influences combined with worker tolerance for fatigue introduce workplace risk and the potential for long-term health issues. This is why it is imperative that organizations and workers share the responsibility for addressing fatigue and implementing strategies to reduce fatigue-related risks.

Fatigue in the Workplace

It is often not recognized that fatigue risks can be impacted, due to a variety of factors in a workplace setting. When addressing workplace factors that impact fatigue, typically only work scheduling and worker-related, off-duty factors are recognized. There are a multiple factors that influence the likelihood of fatigue-creating risk within an organization, and by reviewing and recognizing these key factors, an organization can begin to accurately assess fatigue risks inherent to their current operations. Key areas likely to influence fatigue (Figure 4) include the following:

Figure 2. Key Factors Influencing Fatigue

- **Work scheduling**. This includes shift scheduling, shift frequency, regular and modified rotations, overtime, call-out and on-call practices. Controls in this area should be focused on managing operational needs while providing optimum opportunities for restorative sleep for workers. It should be based on modern scientific understandings, such as recognizing the importance of circadian rhythms and sleep cycles.

- **Work environments** also impact alertness and fatigue. Non-optimal work environments, even in controlled work settings, can add to worker fatigue. For example, working in extreme temperatures can be tiring and can dehydrate workers. Also, chaotic or stressful work environments will increase mental fatigue.

- **Work tasks** are another area to examine when assessing fatigue risks. It is important to recognize there are many tasks that increase the likelihood of inducing fatigue and reducing alertness. Tasks most likely to induce drowsiness are boring and monotonous tasks, where a worker's mind tends to wander, alertness is reduced, and drowsiness symptoms are more likely to occur, especially if the worker is already somewhat fatigued.

- **Other factors** that may influence fatigue within a work environment can include travel or working in a circumpolar region that has significant daylight hour variations based on seasonal changes.

- **Worker factors** illustrate that not all fatigue risks come from within the work environment. Many risks are created through the workers themselves. Many individuals are more susceptible to drowsiness or fatigue. Factors that influence worker susceptibility to fatigue include age, overall health, lifestyle, sleep disorders, chronotypes, habits, and medications.

Recognizing and addressing the multitude of factors that influence fatigue risk in a work environment is an essential part of comprehensive fatigue risk management strategies. It is the reason a Formal Fatigue Risk Assessment is an integral part of the recommended organizational strategies. It is also the reason that worker engagement is such a critical part of the overall strategy.

Current Management of Fatigue in North America

There is a tendency for many organizations in North America to under-address fatigue risks, and this persists for a multitude of reasons. In some industries, such as transportation, it persists because existing regulatory approaches, such as "hours of service" regulations, create the illusion that comprehensive fatigue management is occurring. Although effective in managing many elements of fatigue, regulations alone cannot fully mitigate an issue as complex as fatigue. A combination of approaches and strategies is needed. The National Transportation Safety Board (NTSB) recommends that, "fatigue-related accidents can be avoided with a combination of science-based regulations, comprehensive fatigue risk management programs, and individual responsibility."[xiv]

Transportation is not alone in its focus on regulatory compliance to address fatigue risk; in many other industries (e.g., construction, manufacturing), fatigue is not addressed because the industries are focusing primarily on physical risks, like confined space and fall protection. The safety management systems in these industries tend not to be complex or robust, and focused on strict compliance and risks that are easy to identify. Less conspicuous risks, such as psychosocial risks and fatigue, rarely enter into the consciousness or the conversations of these organizations.

Other reasons given for not addressing the issue of fatigue include a demonstrated lack of knowledge regarding the multitude of strategies available to manage the issue. Some organizations will state that they are not able to manage fatigue because operational requirements or lack of human capital necessitate the need for long working hours. It is ironic that these are exactly the circumstances that create the highest fatigue hazards, and require the most mitigation to effectively reduce risk to personnel and equipment.

The Eye of Fatigue Model

The Eye of Fatigue Model (Figure 3) strives to address the issues of lack of awareness, complex influencing factors and current compliance focus by increasing the awareness of both the methods for best managing fatigue and the reasoning for managing it. In essence, the model strives to communicate to organizations both the 'how' and the 'why' for implementing fatigue risk mitigation strategies.

Eye on Fatigue FRMS Initiative

"Fatigue should be supported from the top but driven by the bottom" NAFMP

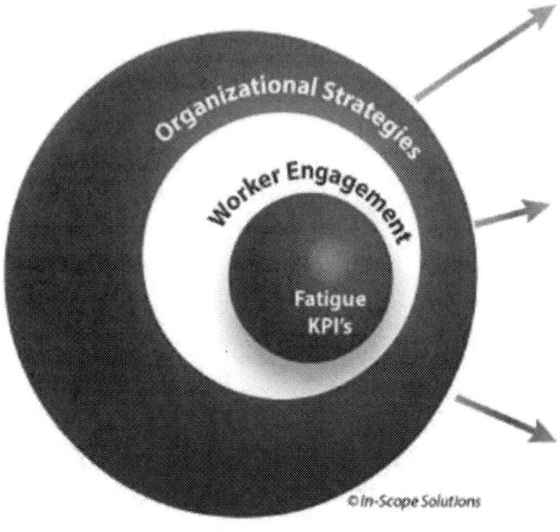

ORGANIZATIONAL STRATEGIES

I. Create Corporate Fatigue Risk Management Strategy
II. Formal Fatigue Risk Management
III. Fatigue Training and Education
IV. Fatigue Monitoring Systems
V. Sleep Disorders Management

EMPLOYEE ENGAGEMENT/WORKER RESPONSIBILITIES

I. Attend Fatigue Management Training and Education
II. Participate in Sleep Disorder Screening (and if required, Treatment and Compliance Monitoring)
III. Report Fatigue Related Incidents, Near Misses, Issues, or Concerns
IV. Prioritize Sleep and Report Fit for Duty
V. Utilize Fatigue Monitoring Devices, if supplied

KPI'S (KEY PERFORMANCE INDICATORS)

I. Reduced Fatigue Related Incidents
II. Increased Productivity
III. Reduced Overtime Costs
IV. Reduced Absenteeism
V. Increased Employee Retention
VI. Reduction in Some Health Issues and Associated Costs

Figure 3. The Eye of Fatigue Model

The model first begins by focusing on the 'how' of fatigue management. Fatigue is somewhat unique from many other work-related health and safety issues in that certain factors, such as the workers attitude, lifestyle factors, existing circumstances and habitual behaviours, can have a significant impact on workplace safety. This is one of the reasons that regulatory compliance alone cannot fully mitigate workplace risks. The National Academy of Sciences has noted the inability of regulatory compliance alone to manage the issues in the transportation industry. It stated that current hours of service regulations are limited and cannot fully mitigate the issue of driver fatigue due to their inability to mandate drivers rest in off-duty times.[xv]

The model visually depicts the idea that fatigue management is an integrated responsibility between workers and the organization. It notes the key strategies that need to be employed by the organization and also details how these are related to the key responsibilities of workers. The model implies that worker engagement can be best obtained by non-punitive environments that encourage workers and organizations working together to manage the issue.

The strategy recognizes that fatigue is somewhat unique from many other workplace hazards in that many of the factors influencing it occur outside of work environment. It recognizes that in order to effectively incomprehensively manage fatigue, organizational strategies must be implemented in combination with worker empowerment strategies.

Engaging workers is a critical component of comprehensive fatigue risk management strategies. Although most safety strategies involve some form of worker engagement, fatigue is unique in that it must fully engage workers in order to encompass both on- and off-site attitudes and behaviors. The Eye of Fatigue recognizes the foundational necessity of engaging the workforce and having designated worker responsibilities. Worker responsibilities components for the strategy include fatigue management training and education, sleep disorder screening, reporting forward of fatigue-related issues and prioritizing sleep to be able to report fit for duty.

Fundamentally, the Eye of Fatigue Model recognizes that worker engagement is a critical component to achieve sustainable fatigue mitigation. Some of the risk of fatigue occurring in the workplace develops not just because of workplace factors; it comes from the worker themselves. This is different than most work-related risk or from other safety management elements; worker health disorders, worker lifestyle choices, and worker values around sleep can introduce risk into the workplace. Having workers involved as a partner to reduce fatigue-related hazards is critical, because their off-duty choices and personal health issues can contribute directly to increased risks.

Knowledge and education changes the work climate, and climate is what influences the workplace culture. Educated and aware workers and organizations, working in symbiosis, are the only effective and sustainable method to openly addressing the risk of fatigue in the workplace.

Organizational strategies reflect typical safety management systems, including things like garnering leadership support, developing organizational policies and procedures, formally assessing fatigue risks, developing controls based on assessed risks, and ensuring these controls align with existing safety management systems. Organizational strategies also include assessing strategy effectiveness through audit processes, incident reporting an investigation, and regular review. Unique to fatigue management, however, is the need for organizational strategies that specifically target sleep disorder identification and management. These are a critical component for any work settings involving driving or safety sensitive work requiring consistent concentration.

The transportation industry and others have begun to recognize that to fully manage the issue, organizations need to employ an approach that is "supported from top (by organizational strategies) but driven by the bottom (through worker engagement)." [xvi]

Targeted fatigue risk management strategies based on assessed risk acknowledges that different strategies need to be employed for low-, moderate- and high-risk environments. A proven strategy for addressing risks in higher risk environments is the Five Level Fatigue Hazard Control Model.[xvii] This strategy employs multiple levels of predictive, proactive and reactive controls to create a comprehensive risk management strategy. Most key components of this and other comprehensive strategies have been integrated into the Eye of Fatigue, either at the model overview stage or during the implementation process.

The Eye of Fatigue Model endeavors to recognize fatigue risk exposures and take corrective action in advance to eliminate potential fatigue errors or incidents prior to their occurrence. Like most hazards in an occupational environment, identification of the hazard is a crucial first step in devising controls. Being able to understand and communicate fatigue as a safety issue is critical to influencing the development and implementation of Fatigue Risk Management Strategies

(FRMS). Predictive controls include a gap analysis process based on a formal fatigue risk assessment of existing operations.

Comprehensive proactive strategies involve the creation of a corporate fatigue risk management strategy that incorporates and multi-levelled fatigue training and education across the organization. Reactive controls include organizationally embedded fatigue monitoring systems and incident investigation procedures designed to identify or rule out fatigue as a causal factor during investigations.

The inner circle of the model begins to address the 'why' of fatigue management.

It can be argued that there are really three key drivers that incentivize organizations to implement any safety measure. The first key driver is moral obligation. This implies that companies are morally obligated to protect the workers they employee. The second key driver is societal. Societal obligations are reinforced through various systems, such as regulations, societal pressures that impact brand popularity and buying decisions, and the need for companies to create sustainability in the social environments in which they operate. Financial drivers are those that recognize that many safety systems influence other business values.

Financial incentives were used to discuss the 'why' of fatigue management for a variety of reasons. Moral values apply to any safety initiative in varying degrees and societal values often have other systems that encourage employers to align. Financial drivers are those frequently discussed by senior leaders and are also the least intuitively recognized when discussing fatigue management. There is a variety of occupational research that supports the likelihood of successful alignment between financial drivers and FRMS integration. As well, financial incentives were chosen because these are often the ones given for delaying or not implementing a FRMS.

As discussed earlier, fatigue is proven to have a negative impact on worker health and safety and this, in turn, impacts multiple business drivers. Fatigue issues directly impact other business goals through increased absenteeism, reduced productivity, short- and long-term benefit costs, as well as reduced employee retention, resulting in increased onboarding and training costs. Discussion of these financial drivers is facilitated by focusing on Key Performance Identifiers (KPIs).

KPIs that should reflect a fully integrated Fatigue Risk Management Strategy include a demonstrated reduction in the risk of fatigue-related incidents, reduced fatigue-related worker errors, and increased employee productivity. Other long-term benefits can include improved employee health and wellness, typically resulting in reduced absenteeism, improved employee morale and retention, and employees equipped with knowledge and tools to better understand and manage fatigue in their work and personal lives.

These anticipated KPIs are supported by various occupational research recognizing business costs relating to worker safety, worker health, productivity and overall operational costs. The relationship of fatigue impairment relating to increased workplace incident is well researched, with some suggesting that highly fatigued workers are 70% more likely to be involved in accidents.[xviii] Other research has shown a sleep-deprived employee is 2.9 times more likely to cause a workplace accident and 3.6 times more likely to engage in safety-compromising behaviors.[xix] Incidents impact an employer's bottom line in multiple ways, and this is further acknowledged by some industrial surveys that indicate that workers' compensation costs per employee per year were almost five times higher in facilities with severe fatigue problems.[xx]

The relationship of fatigue and worker health has been detailed above. Industrial surveys reveal that absenteeism rates are nearly double in facilities with severe fatigue problems.[xxi] The impact on worker health can also increase overall employer costs related to health care, especially in the United States, where the costs of private insurance usually rests directly with the employer.

Other KPIs relate to productivity. A 2010 study estimates a loss of $1967 per employee per year in lost productivity due to fatigue.[xxii] According to some studies, the amount of productive work performed per week per employee caps out at as little as about 50 hours. Consequently, more than 10 hours of overtime per week may not provide much additional productive work.[xxiii]

A 2012 NAOSH study further detailed KPIs, noting that shift work, long work hours, and fatigue cost employers an estimated $116 billion or $2000 per employee per year, owing to a variety of factors: reduced productivity; increased errors; absenteeism; increased healthcare and worker compensation costs; and worker attrition due to disability, death, and moving to jobs with less demanding work schedules.[xxiv]

A current flaw does exist with the model. Although KPIs are the main indicator used to rationalize financial incentives to implement a FRMS, these KPIs have not yet been formally demonstrated through a significant quantity of peer-reviewed research. More research is needed to fully relate the integration of the strategies with realized business values. The author of this paper is currently embarking on a research project that will strive to more fully demonstrate this alignment. It is hoped that other projects will also be conducted to further demonstrate this incentive for FRMS program integration.

Fatigue Management Strategies

Once the Eye of Fatigue Model has been used to explain the overarching process of a Fatigue Risk Management Strategy and incentivize an organization to move forward with it, the only remaining step is the full integration of the program into existing Safety Management Systems (SMS).

A Fatigue Risk Management Strategy (FRMS) is an integrated set of management practices, beliefs, and procedures for monitoring and managing the risks posed to health and safety by fatigue. It is based in safety management system theory, with an emphasis on risk management. The following graphic (Figure 4) outlines the five key steps needed to implement a comprehensive FRMS.

Figure 4: Five Steps to FRMS Implementation

Three critical components of this integration process need to be expanded upon. First, there must be a strategy that will focus on providing senior leaders the information they need to fully understand fatigue as a risk factor in the workplace. This engagement step is necessary to empower senior leaders to understand both the risks and suggested management strategies to mitigate this risk. Additionally, it should help senior leaders to be engaged in actively leading, supporting and/or participating in the FRMS change management process.

The second is the importance of key fatigue information to be disseminated across the organization. ***The importance of training and education with regards to fatigue management cannot be overstated.*** For any FRMS to succeed, there needs to be a climate of awareness and understanding regarding the hazard and mitigation of fatigue. Training must occur at all levels of the organization. As mentioned, senior leadership should be aware of the risks and current strategies being employed. Managers and supervisors should understand the risk, be able to recognize signs of fatigue in workers, and be empowered to intervene. All levels, including workers, should be aware of best practices in terms of managing the risks of fatigue.

A third important element of the implementation is the importance of a non-punitive system of mitigation. Checks and balances can be in place to ensure systems are not abused, but workers cannot be penalized for identifying fatigue-related risks. Typically, only work climates that identify and manage the issue of fatigue, without laying fault, succeed in implementing controls that allow for successful mitigation.

Conclusion

Despite extensive occupational research illustrating the health and safety impact of fatigue, many industries in North America do not currently acknowledge, assess or control fatigue risks.

To push past the existing inertia of lacking fatigue management, current cultural and work-related barriers must be overcome. These barriers are propagated by lack of knowledge and awareness by many industries and organizations and by many safety professionals. These barriers are further supported by existing workplace practices and existing safety cultures, which must be effectively influenced to properly implement fatigue controls. Using the Eye of Fatigue Model helps organizations and at-risk workers to understand their roles in the effective management of fatigue.

The model further supports the concept that fatigue controls are best implemented in a Fatigue Risk Management Strategy approach. This approach focuses on identifying fatigue risk exposures and addressing risks in a targeted and comprehensive strategic approach that aligns with existing safety management systems.

The Eye of Fatigue Model, although simplistic in appearance, incorporates the learnings from previous models and integrates predictive, proactive, and reactive tools and methodology to provide a comprehensive framework of mitigation strategies designed to deal with the issue.

Finally, the model seeks to align safety strategies with senior leadership thinking. It does this by translating the need for a FRMS into the language of business and trying to demonstrate likely KPIs. Demonstrating that successful FMRS implementation aligns with positive business values is thought to increase the likelihood an organization will move forward in implementing the program.

As North American industries progress in recognizing and controlling nonphysical hazards within their work environments, fatigue risk assessment and strategies will become a more commonly integrated part of safety management systems. In the same way general health and safety has evolved over the years, fatigue management and other worker-related factors will need to evolve and improve. Key stakeholders within organizations will need be trained to recognize and manage fatigue hazards when they arise. The hope is that fatigue management will become as common here as it is in other countries, and we will then begin to see a reduction in fatigue-related incidents and injuries.

Endnotes

[i] Chan, M., "Fatigue: the most critical accident risk in oil and gas construction." *Construction Management and Economics* 29, 341–353.April 2011.

[ii] Various sources, including (Chan, D. *News | The University of Sydney. Sydney.edu.au.* 2010. (Retrieved from http://sydney.edu.au/news/84.html?newsstoryid=4436).

[iii] Vegso, S., Cantley, L., Slade, M., Taiwo, O., Sircar, K., & Rabinowitz, P. et al. 2007. "Extended work hours and risk of acute occupational injury: A case-crossover study of workers in manufacturing." *Am. J. Ind. Med.*, 50(8), 597-603. doi:10.1002/ajim.20486.

[iv] Rogers, A., Hwang, W., Scott, L., Aiken, L., & Dinges, D. 2004. *The Working Hours of Hospital Staff Nurses and Patient Safety* (1st ed.). (Retrieved from http://www.protectmasspatients.org/docs/Rogers.pdf).

[v] Caruso, C. "Workplace Safety and Health, Running on Empty: Fatigue and Healthcare Professionals." NIOSH. Aug 2, 2012.

[vi] Wong, I., McLeod, C., & Demers, P. "Shift work trends and risk of work injury among Canadian workers." *Scandinavian Journal of Work, Environment & Health*, 37(1), 54-61. doi:10.5271/sjweh.3124, 2010.

[vii] U.S. Federal Motor Carrier Safety Administration and Transport Canada. *Commercial Motor Vehicle Driver Fatigue and Alertness Study*. 1996.; see also "Fatigue." *Center for Accident Research & Road Safety: State Of The Road.* 2011. (Retrieved Dec 15, 2013) (http://www.police.qld.gov.au/EventsandAlerts/campaigns/Documents/fatigue_fs.pdf).

[viii] Ibid.

[ix] Mitler, M., Carskadon, M., Czeisler, C., Dement, W., Dinges, D., & Graeber, R. (1988). "Catastrophes, sleep and public policy: consensus report." *Sleep*, 11(1):100-9. (Retrieved March 29, 2014) (http://www.ncbi.nlm.nih.gov/pmc/articles/PMC2517096/).

[x] Dawson, D., & Reid, K. 1997. "Fatigue alcohol and performance impairment." *Nature (388:235).* (Retrieved Dec 15, 2013) (http://www.nature.com/nature/journal/v388/n6639/full/388235a0.html).

[xi] Straif, K., Baan, R., Grosse, Y., Secretan, B., Ghissassi, F., & Bouvard, V. et al. "Carcinogenicity of shift-work, painting, and fire-fighting." 2007. *The Lancet Oncology*, *8*(12), 1065-1066. doi:10.1016/s1470-2045(07)70373-x).

[xii] National Institute of Neurological Disorders and Strokes. (Accessed March 1, 2016) (http://www.ninds.nih.gov/disorders/brain_basics/understanding_sleep.htm).

[xiii] Created by In-Scope Solutions: Body Image from Thinkstock, *15% Increased Risk of Stroke:* Colten, H. and Altevogt, B., "Sleep Disorders and Sleep Deprivation: An Unmet Public Health Problem," 2006. *40% increased risk of cardiovascular disease:* Fujino, Y. "A prospective cohort study of shift work and risk of ischemic heart disease in Japanese male workers," *American Journal of Epidemiology*, volume 164, pages 128-135, 2006. *Increased hypertension:* Gangwisch JE, Heymsfield SB, Boden-Albala B, et al. "Short sleep duration as a risk factor for hypertension: analyses of the first National Health and Nutrition Examination Survey." *Hypertension.* 2006; 47:833–839. (Accessed September 21, 2011). (http://hyper.ahajournals.org/content/47/5/833.long). *Increase in multiple types of cancer:* Megdala, S. et al. "Night work and breast cancer risk: A systematic review and meta-analysis." *European Journal of Cancer*, volume 41, issue 13, pages 2023-2032, 2005. Tatsuhiko Kubo et al. "Prospective cohort study of the risk of prostate

cancer among rotating-shift workers: Findings from the Japan Collaborative Cohort Study," *American Journal of Epidemiology*, volume 164, pages 549-555, 2006. T Lahti et al. "Night-time work predisposes to non-Hodgkin lymphoma," *International Journal of Cancer*, volume 123, pages 2148–2151, 2008. Fortson, K. "The diurnal pattern of on-the-job injuries," *Monthly Labor Review*, pages 18-25, September 2004. *Memory, Concentration Issues, Brain Fog:* Xie, L., Kang,H., et al. "Sleep Drives Metabolite Clearance from the Adult Brain," *Science* 18 Vol. 342, Issue 6156, pp. 373-377. Oct 2013. *Fatigue Increases weight*: Buxton, C., O'Connor, P., Duffy, W., Czeisler, S. "Adverse Metabolic Consequences in Humans of Prolonged Sleep Restriction Combined with Circadian Disruption." April 2012. *Increase in peptic ulcers:* Pietroiusti, A., Forlini, A., Magrini, A., Galante, A., Coppeta, L., Gemma, G., Romeo, L., Bergamaschi, A., "Shift work increases the frequency of duodenal ulcer in H pylori infected workers." *Occ & Envir Med*, 63, 773-775. 2006. *Increase in Type 2 Diabetes.* Buxton, C., O'Connor, P., Duffy, W., Czeisler, S. "Adverse Metabolic Consequences in Humans of Prolonged Sleep Restriction Combined with Circadian Disruption." April 2012. *Decreased Immunity:* ibid.

[xiv] 2016 Annual Report, A Report to Congress and the National Transportation Safety Board - The U.S. Department of Transportation's Status of Actions Addressing the Safety Issue Areas on the NTSB's Most Wanted List. April 2016.

[xv]Commercial motor vehicle driver fatigue, long-term health and Highway safety: research needs, national Academy of sciences, 2016.

[xvi]NAFMP. Accessed from http://www.nafmp.com/en/ March 2013.

[xvii]Taken from Development and Implementation of Fatigue Risk Management Systems in the Canadian Aviation Industry, 2011.

[xviii] Citation accessed from www.sleepfoundation.org March 2016.

[xix]Sleep quality and the risk of work injury: a Swiss case–control study, Katrin Uehli1,2, et al, http://onlinelibrary.wiley.com/doi/10.1111/jsr.12146/fullSummaryJUN 2014

[xx] Circadian. Myths & Realities of Fatigue: Reducing the Costs, Risks, and Liabilities of Fatigue in 24-Hour Operations. William G. Sirois

[xxi]Circadian. Myths & Realities of Fatigue: Reducing the Costs, Risks, and Liabilities of Fatigue in 24-Hour Operations. William G. Sirois.

[xxii]Rosekind M, Gregory KB, Mallis M, Brandt S, Seal B, Lerner D. The cost of poor sleep: workplace productivity loss and associated costs. J Occup Environ Med. 2010;52:91–98.

[xxiii]ACOEM Guidance Statement, 2012. Fatigue Risk Management in the Workplace, ACOEM Presidential Task Force on Fatigue Risk Management: Lerman, S. et al, American College of Occupational and Environmental Medicine

[xxiv]NIOSH: Workplace Safety and Health, Running on Empty: Fatigue and Healthcare Professionals, Claire C. Caruso, PhD, RN, Aug 02, 2012

Case Study Research: Safety: Implications of Information Overload in a Hospital

Joel M. Haight, Ph.D., P.E
Associate Professor of Industrial Engineering
University of Pittsburgh
Pittsburgh, PA

Introduction

The results of an analysis of the cognitive information-processing workload of patient monitoring technicians working in central monitoring units (CMU) of five hospitals in Oklahoma are presented. The analyses were conducted in June 2016 during a visit to the four of the hospitals. The analysis included conversations with the management team, nurses and monitoring technicians at all five hospitals. The objective of the study was to determine whether the information processing (IP) load exceeds the processing capacity of monitoring technicians, and whether any overload is great enough to necessitate consideration for IP load-reducing strategies and interventions. This determination was accomplished by identifying whether an IP load gap exists between the current load and an industry-accepted IP load limit. The healthcare literature appears to be sparse in this area, making difficult any determination that changes in error rates are a function of changes in IP load. However, it is assumed for this study that an increased IP load or length of exposure to the current load will result in an increase in error rates among technicians. Errors are defined (albeit forensically) in this study as events, such as missed alarms, late response to alarms, misinterpretation of alarms and/or EKG strips, missed alarm notifications of nursing staff, etc.

The alarm processing load for a technician is partially derived from a measured number of alarms collected at one of the hospitals between 12/23/2015–1/19/2016. During that time, there were 748,150 alarms noted. The load may be smaller at the other hospitals; however, fewer technicians on duty make the processing load similar across all five hospitals. There is a probability of 31% that an alarm will activate in any one operational second (0.31 alarms/second). A measure of the rest of the IP load was derived from the amount of additional data analyzed and/or managed by the technicians, such as EKG interpretations, hospital admissions and discharges, phone calls made to nursing staff, etc. The visual IP load per technician was calculated and determined to be approximately **seven to eight bits/second** (a "bit" is the amount of information necessary to decide between multiple alternatives (Freivalds, 2014)). This is compared to a recommended IP load limit of **three to four bits/second** for visual information processing (Miller, 1955). While the information load for each technician is high, it appears that the screen information density is reasonable, and that this may be able to be safely and effectively managed, but only in the short term (~two to four hours). Two-to-four hour rotations should be considered (Chewning and Harrell, 1990; Goldstein and Gigerenzer, 2002; Marois and Ivanoff, 2005; and Nadav-Greenberg and Joslyn, 2009). This rotation is suggested due to the fact that even mindful-task (meaningful) vigilance decrement occurs in all people without exception (Grier, Warm, Dember, Mathews, Galinsky, Szalma and Parasuraman, 2003). The ability of technicians to manage that much information without error, consistently over the long term is of concern, and the probability of error increases with time (lower probability of error for two-hour rotation and a higher probability of error for a four-hour rotation) (Murayama, K., Blake, A.B., Kerr, T., Castel, A.D., 2016).

Monitoring System Description

The subject hospitals monitor the health status of many of their cardiac patients via telemetry and bedside portable monitoring units. An example of a staffed monitoring work station is shown in Figure 1.

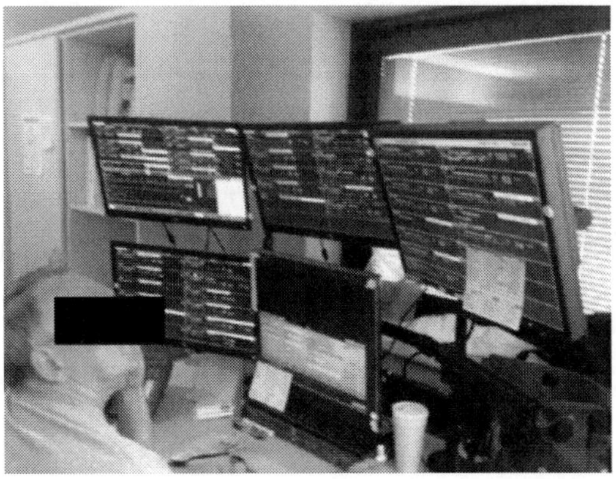

Figure 1. A photo of one CMU workstation.

These telemetry and/or bedside monitoring devices transmit from the patients to the CMUs where technicians monitor their health status (see Figures 2 and 3, below). The main responsibility of the technicians is to monitor the data screens for alarms triggered by patient conditions that would require attention from the attending nurses. This alarm condition occurs when an out-of-acceptable-range for a measured parameter (sometimes referred to as vital signs) is detected by the system. These acceptable ranges are medically determined and for each variable and each patient, may be different. The biometric variables monitored include: heart rate (from electrocardiogram (EKG) reading), respiration rate, pulse oxygen level, mean arteriole blood pressure, central venous pressure, premature ventricular contraction, and continuous real time blood pressure, as well as several others (all parameters are not monitored on all patients throughout their entire stay in the hospital). The parameters that are measured are monitored by technicians 24 hours a day at the CMU's computer stations; each working eight-hour shifts (technicians at one of the hospitals work 12-hour shifts).

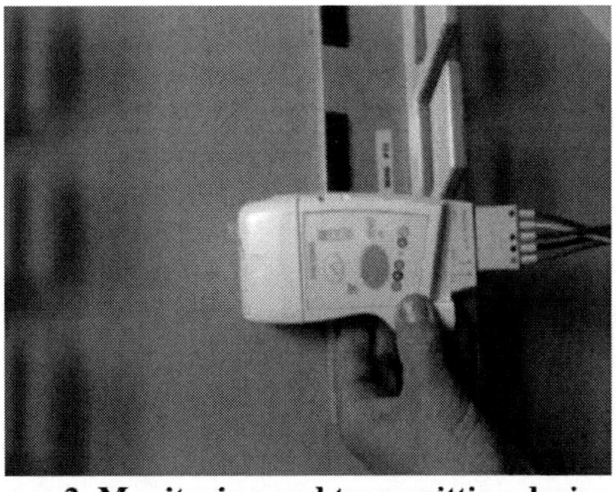

Figure 2. Monitoring and transmitting device.

An example of workstation layout is shown below in Figure 3. There are five data screens and one reading screen where electrocardiogram (EKG) wave signals are read.

Figure 3. An example of a work station showing five data screens and one reading screen for viewing and managing electrocardiogram results.

The system is programmed to recognize acceptable ranges for each biometric variable monitored. Upon development of a potentially adverse health condition, during which one or more of the monitored variables dips below or exceeds the predetermined acceptable limit, an alarm will be generated. The resulting alarm indication appears on a designated screen at one of the stations in the CMU. Depending upon the magnitude of the out-of-range parameter, the alarm will be either "yellow," indicating the development of a critical condition (categorized as either "yellow arrhythmia" or "yellow parameter") or "red" indicating a potentially life-threatening condition ("red arrhythmia" or "red parameter"). Each is recognized by its own distinctive tone and yellow- or red-colored light activating on the applicable screen. When an alarm condition that requires a response develops, the technician calls a nurse on the floor where the patient experiencing the condition is notified so that they can then deliver the necessary medical response.

Cognition

The brain has significant processing power. Even though few would argue against the existence of this power, there is a research supported claim that the brain is still limited in the amount of information it can process at any one time (Marois and Ivanoff, 2005). According to Marois and Ivanoff (2005), the brain can store billions of bits of information over a lifetime, but its processing power is limited nonetheless. There are three main bottlenecks in information processing; attentional blink (inability to process a second visual target when presented within a half of a second of a first target; in this case, a second alarm coming in within a half second of a first alarm), visual short-term memory limitations, and psychological refractory-period phenomenon (processing a second piece of visual information is slowed due to the fact that the brain is still processing the first). Even though we claim that we can function effectively in our "multi-tasking" world, it is difficult for us to effectively and accurately do more than one thing at a time (Marois and Ivanoff, 2005).

The literature is sparse on the subject of quantifying information processing capacity, but there is support for some generalized quantification. In general, it is accepted that increasing the volume of information can be detrimental to both the technicians' performance in decision-making and their situational awareness. Several studies support this generalization (Chewning and Harrell, 1990; Goldstein and Gigerenzer, 2002; and Nadav-Greenberg and Joslyn, 2009). Even if there are essentially no demands placed on short-term memory (as visual data are maintained on computer screens and do not necessarily have to be remembered), our processing capacity is limited (Norman, 2013). Endsley (1995a) suggests that the processing limitations affect "our ability to perceive, comprehend and make projections," and this can also erode our situational awareness.

To quantify information being processed, the term "bit" is used. A bit is the amount of information necessary to allow a person to decide between equally likely alternatives (in some cases, the alternatives are not equally likely a different calculation is used). It is not an absolute amount of data or information, but includes all the information necessary to make up the decision alternative. In a 1955 paper, Miller suggests that we are only capable of processing three to four bits of visual information per second. He suggests, however, that in reality, we are seemingly able to manage much more information than that due to our ability to identify and select salient information and only process information that we consider "important." Therefore, even though we can manage (i.e., remain aware of) much more information than we can actively process, we are still limited in the amount of information that we actually process (Miller, 1955). The downside of this is that interpretations and decisions don't readily come from this "managed' information. Information determined to be salient is then further processed and therefore, it better informs interpretations and decisions. So while an interpretation of the meaning noted by Miller (1955) suggests that more than 3 or 4 bits of information may be manageable, there are still concerns associated with the processing an information load greater than 3 to 4 bits/second.

Experimental Methodology

During this study, the hardware, process and operations were evaluated and compared with Recognized and Generally Accepted Good Engineering Practice (RAGAGEP), at similar types of operations for layout, displays, lighting, activity level, task and information processing and distraction loads. The analysis was performed through facility observations, technician and management interviews and extensive calculations using alarm and information data. The results of these analyses were compared with findings in and impressions of similar studies in the published literature.

This cognitive processing workload is addressed in the analysis by quantifying the amount of data to be monitored by each technician, the number of data points they are responsible for monitoring, the number of alarms to which each technician has to respond, and the additional duties and distractions to which the technicians are exposed. The cognition process involves the input of information, through visual and auditory channels (as well as other channels, but visual and auditory are the primary ones in this case); the interpretation of what that information is and what it means (i.e., Is it critical? Does it require action? If so, what type of action? etc.); and selection of an appropriate response (e.g., acknowledge the alarm, call floor nurse, etc.). This process takes time, and this time is often determined and identified as simply "response time" or "information processing time (IP time)" (see Eq. 3 below). IP load, by definition, increases as the amount of information to process increases and/or the number of decision alternatives to be considered increases. In this case, an effort was made to capture and quantify this "amount of information." Granted, while these data were collected over just two days of observation and one month of alarm data, each technician interviewed explained that the amount of information was typical of any one day or one month. Therefore, the analysis, interpretation, conclusions, and recommendations are based on the alarm data collected over one month and that which was determined by two days of observation and interviews.

In addition to IP, target-search time and decision-making time (or "choice reaction time") were also determined. Decisions must be made about whether a critical alarm has activated and if so, where and what type of alarm. Once these determinations and decisions have been made, the technician must then decide what actions to take (another decision). It is this information processing workload that the analyst quantifies here, along with determining its impact on decision time, reaction time, choice reaction time, target search time and the need for prospective memory (the need to remember to go back to readdress an alarm that had not yet established its full meaning when first detected).

Analytical Methodology

The first method to quantify information and its flow starts with defining "bits." Again, a bit is an amount of information that is necessary to allow a person to decide between alternatives. If the alternatives are equally likely to occur, the amount of information needed or used can be calculated as the $\log_2$ of the number of alternatives there are; as shown in Equation 1 (Freivalds, 2014 and Wickens, Lee, Liu, Gordon-Becker, 2004):

$$\mathbf{H = \log_2 n} \qquad \textbf{(Eq. 1)}$$

Where H = the number of bits of information and n = the number of decision alternatives.

If the alternatives are not equally likely, the calculation is a little more complex. The amount of information is calculated by:

$$\mathbf{H = \sum p_i \times \log_2 (1/p_i)} \qquad \textbf{(Eq. 2)}$$

Where H = the number of bits, and p_i = the probability of the ith alternative being selected or likelihood of occurring.

The concept of redundancy can be considered a measure of reduction in the amount of information that must be processed in order to make the decision when decision alternatives are not equally likely to occur. Essentially, more redundancy means less attention capacity is needed. It is calculated by:

$$\mathbf{R = (1 - H/H_{max}) \times 100} \qquad \textbf{(Eq. 3)}$$

Where H = the number of bits processed when there are different occurrence likelihoods and H_{max} is the maximum information to process if all alternatives are equally likely. R = % redundancy

From this information, we are able to determine decision-making or information processing time. This is calculated by equation 4:

$$\mathbf{IP = -2.328 + 0.7325\ H} \qquad \textbf{(Eq. 4)}$$

Where H = the number of bits of information and IP is information processing time (seconds). The numerical values are constants derived from research data distribution plots (Freivalds, 2014 and Wickens, Lee, Liu, Gordon-Becker, 2004).

An analysis of a previous internal alarm study conducted over one month to determine the various aspects of alarms registered in that period (December 2015/January 2016) formed a significant basis for the quantification of the current decision-making load.

In the first step, the analyst determined that each technician can be expected to monitor as many as 130 data points every second (each monitored value can change from one second to the next). While each technician does not have to make a decision about every data point every second (an impossible task), they do have to continually monitor those data; this consumes attention capacity and must be considered in decision-making time determinations. All 130 data points are not exclusively or explicitly used in the calculations because that amount of information, and the attention capacity that it consumes, is reduced by an alarm system. The alarm system allows quick dismissal of any change in the monitored data from one second to the next if the change does not indicate that the monitored parameter exceeded or dipped below the range of values considered to be medically acceptable. Without the alarm system, the technicians would not be able to process the information necessary to make decisions. It would be difficult for anyone to process without error, the number of bits/sec introduced by 130 data points over even one hour, let alone eight or 12 hours ~ the bit value of 130 alternatives nonetheless is $H = Log_2\ 130 = 7.022$ bits/second). Normal conversational information processing (auditory information) is roughly 50 bits/second, which is a significant amount of material to process, but this processing is much different than visual, numerical information processing. Even though some may be able to effectively process this much information, it would not be possible for long periods of time. Technicians would begin to be drop information and miss alarms after even perhaps just 30 minutes. The vigilance research also shows performance drop off after less than one hour when monitoring data this way (Freivalds, 2014 and Wickens, Lee, Liu, Gordon-Becker, 2004). Alarm system reliability is critical. Although those 130 data points must be considered, processing the alarms in categories and the number of response alternatives reduce the required IP load. The alarm choices (category, type, location, and criticality, etc.) become the decision alternatives. From the January 2016 alarm report, there are over 240 alarm types, impossible to process every second. Management of this phenomenon is made more possible by the fact that only the top 20 alarm types (all greater than 10,000 alarm activations in a month) would consume the most significant amount of attention capacity. Therefore, for consideration of alarm type, we can assume 20 alternatives.

Even with the alarm limitation, there are processing demand offsets that can increase the information processing load. Several types of distractions (distractions consume attention capacity) occur in the work day of the technicians. Attention capacity is drained from the monitoring task by the admission and discharge process. While it was not the case at all of the hospitals evaluated, leaving the work station to print out and fax EKG charts to the floor nurses consumes significant attention capacity. The need to answer and make telephone calls about the status of a patient or to ask questions about a particular alarm also consumes attention capacity. Interpreting each EKG requires significant information processing as well, especially given that there are 15 or more conditions that the technicians determine from the EKG wave form. Even though there are more possible interpretations, a "bit" calculation is done based on an average of 15 alternatives.

One other criterion that was explored was the concept of target search time. This is the time that it takes someone to search a scene (in this case, six computer screens) for target information. It takes time to find important information on a computer screen and if all 130 data points would have to be searched every time there is a change, it would not be possible to find the critical information (parameter exceedance) in anything less than about 16 minutes. This calculation is determined by:

$$\mathbf{T = N \times I/2} \qquad \textbf{(Eq. 5)}$$

Where T = target search time in seconds, N = number of data points that must be searched in the target, I = typical screen inspection time (seconds). This quantity is divided by 2 because it is most often the case that the correct information will be found within the time it takes to search half of the data points. (Freivalds, 2014 and Wickens, Lee, Liu, Gordon-Becker, 2004)

The cognitive processing load on monitoring technicians; i.e., the amount of information they monitor, the number of alarms they experience, the number of interpretations they have to make for each alarm, their additional duties (calls to nurses, printing out and sending strips to nurse stations, etc.) and how the they integrate these into their primary objective of keeping patients medically cared for were analyzed. Additionally, shift length and fatigue were also considered in the development of recommendations. The primary objective of the technicians is to determine incoming critical alarms, correctly interpret them, and communicate the interpretation results to the appropriate nurses on the units being monitored. Anything that does not contribute to that objective is considered a distraction. All of the technicians' daily activities are important to the function, but from a value-stream perspective, consideration is given to the importance of each activity in terms of its direct contribution to the primary objective, i.e., monitoring patient signals. Calling the attending nurse to ascertain the criticality of an equipment failure alarm consumes the technician's attention; however, it does not directly contribute to alarm monitoring of real and critical signals. Therefore, this activity should be considered a distraction, as it reduces attention capacity available for the more critical cognitive processing of monitoring data and alarms.

Results

Each technician manages as many as 130-180 data points and that can change from one second to the next, and from these data points the technicians determine if and what response is needed. Any job requiring full attention to all of those data points would be impossible. It is made more possible by the alarm categorizations and the existing alarm system that activates (colored light and an audible tone) for five different conditions. It activates when a device becomes inoperable or when a measured parameter goes outside of its medically defined acceptable range in four different categories. These categories are labeled: "Red-Arrhythmia," "Red-Parameter," "Yellow-Arrhythmia," "Yellow-Parameter," and "Inoperable." This reduces the load to a decision about the category, and type of alarm, the alarm location, and then a decision about the alarm's criticality. There is still significant information to process, but it is more possible because a technician processes only the salient information related to the alarms, minimizing the demand on attention capacity.

Between December 2015 and January 2016, 748,150 alarms were registered at one of the hospital's Central Monitoring Unit. An alarm registers on any one parameter at a rate of 0.31 alarms per operational second, or identifies an estimated 0.31 or 31% probability that in any one second, an alarm will activate. Throughout the day, roughly 4-5 alarms were observed at any one workstation at any point during the observation. Because there are 12 monitored units, five different alarm categories, and over 200 types of alarms, there is a significant amount of information to process in order to successfully make appropriate patient care decisions. In the current decision-making process, the following were considered and determined:

Information Processing load: approximately **7-8 bits/second** for each technician (research literature indicates (Miller, 1955) a more manageable load of **3-4 bits/second**.)

Fatigue: Eight hours in three hospitals and 12 hours in one of the hospitals. Information fatigue and vigilance (a rapidly deteriorating commodity) is a concern.

Prospective Memory: There are several times in the shift that technicians recognize an alarm condition associated with artifacts (movement-generated signals that are not associated with a heart condition). At the time of the alarm, it is not known whether it is a meaningful alarm. In order to establish alarm importance, the technicians give the signal time to stabilize, clear or become definitive. The problem is that they have to go back to the signal in a few seconds or a few minutes to ascertain the alarm's true meaning. This uses a cognitive capacity called prospective memory. It is similar to turning on the water to fill a bucket and then leaving to attend to other issues and having to remember to return later to turn off the water without overfilling the bucket. This form of memory unreliable, and task design engineers should strive to minimize the need to use it. This problem needs further study.

Recommendations

While the information load for each technician is high, screen density seems reasonable. The technicians should be able to be safely managed for short periods of time. However, this much information processing limits the technicians ability to accurately perform their duties for the full shift without error. A lower likelihood of error can be achieved if changes are made to the task, the environment, the schedule and the patient load. The following can be considered for error probability reduction:

1. Reduce patient load per technician to a maximum of 25 patients per technician (will reduce average data points to monitor to roughly 100 or less, which will reduce target search time and response time).

2. Reduce or eliminate the need to print out and fax EKG strips (significant distraction). The existing electronic systems at two of the hospitals are much improved (more efficient and less distracting) over the largest hospital's printing out and faxing paper copies.

3. Consider a rotation schedule that allows technicians to rotate to other non-monitoring functions every 2-4 hours (this is especially important for the 12-hour shift technicians).

4. Maximize the use of headphones with one-button, call activation for calls to floor nurses.

5. Automate, as much as possible, the interpretation of EKG results to reduce the interpretation load for the technicians.

6. Automate the admission and discharge process as much as possible, so technician attention is not taken away from alarm monitoring.

7. There is a high percentage of "inoperative" alarms. Reduction of these alarms would help to reduce the information processing load on each technician.

Limitation and Future Research

This study was limited by the fact that only one month's worth of alarm data was available. Results could be strengthened by using a year's worth of monthly data so a frequency distribution could be developed. The study was done in only one hospital system. It is not known if that one hospital system is representative of all hospitals, but it can be a starting point. Another limitation is that it is difficult to accurately measure information processing in an applied setting without disrupting the work that the technicians do and still have it be representative of what each technician is processing. Generalization had to be used to establish reasonable estimates of the IP load on which to based staffing and control strategies.

References

Chewning, E.G. and Harrell, A.M. 1990. "The Effect of Information Overload on Decision Makers' Cue Utilization Levels and Decision Quality in a Financial Distress Decision Task." *Accounting, Organizations and Society*, 15: 527-542.

Clemens, P.L., Simmons, R.J. 1998. *NIOSH Instructional Manual, System Safety and Risk Management, A Guide for Engineering Educators*. U.S. Department of Health and Human Services, Public Health Service, Centers for Disease Control.

Endsley, M.R. 1995. "Toward a Theory of Situational Awareness in Dynamic Systems." *Human Factors*, 37 (1): 32-64.

Freivalds, A. 2014. *Niebel's Methods, Standards, and Work Design*, 13th ed. New York: McGraw Hill.

Goldstein, D.G., Gigerenzer, G. 2002. "Models of Ecological Rationality; The Recognition Heuristic." *Psychological Review*, Vol. 109, No. 1, 75-90.

Greir, R.A., Warm, J.S., Dember, W.N., Mathews, G., Galinsky, T.L., Szalma, J.L., Parasuraman, R. 2003. "The Vigilance Decrement Reflects Limitations in Effortful Attention, Not Mindlessness." *Human Factors*, Vol45 N3 Fall 2003, pp. 349-359.

Marois, R., Ivanoff, J. 2005. "Capacity Limits of Information Processing in the Brain." *Trends in Cognitive Science*, 9(6): 296-305.

Miller, G.A. 1955. "The Magical Number Seven, plus or minus Two - Some limits on Our Capacity for Processing Information." *Psychological Review*, Vol 101, No 2, 343-352.

Murayama, K., Blake, A.B., Kerr, T., Castel, A.D. 2016. "When Enough Is Not Enough: Information Overload and Metacognitive Decisions to Stop Studying Information." *Journal of Experimental Psychology: Learning, Memory, and Cognition*, Vol. 42, No. 6, 914–924.

Nadav-Greenberg, L. and Joslyn, S. 2009. "Uncertainty Forecasts Improve Decision-Making Among Non-Expert." *Journal of Cognitive Engineering and Decision-Making*, 2(1): 24-47.

Norman, D.A. 2013. *The Design of Everyday Things*. New York: Basic Books, A Member of the Perseus Books Group.

Petersen, D. 1996. *Human Error Reduction and Safety Management*. New York: Van Nostrand Reinhold.

Salvendy, G. ed. 1987. *Handbook of Human Factors*. New York: Wiley Interscience Publication.

Wickens, C.D., Lee, J.D., Liu, Y., Gordon-Becker, S.E. 2004. *An Introduction to Human Factors Engineering*. 2nd ed. Saddle River, NJ: Pearson Education Incorporated.

American Society of Safety Engineers

Professional Development Conference June 19-22, 2017

SAFETY 2017 ASSE

Session: 546

Fundamentals of SH&E: Hazard Identification and Control

David F. Coble, CSP

CTJ Safety Associates

Overview

SAFETY 2017 ASSE

- Definitions
- Hazard Identification and Control
- Accidents
- Work Planning
- Workplace Hazards
- System Safety

2

Hazard Identification and Control

SAFETY 2017 ASSE

Hazards and Risk

Hazard:

- a dangerous condition, either potential or inherent, which can interfere with the expected, orderly progress of a given activity

Risk:

- a measured or calculated chance of exposure to hazard(s) which may or may not result in loss

<u>Both</u> should be managed!

3

Hazard Identification and Control

SAFETY 2017 ASSE

Accidents and Incidents

Accident:

- an *unplanned* and therefore unwanted or undesired event resulting in physical harm and/or property damage

Incident:

- an unplanned event that did not *but could have* resulted in physical harm and/or property damage

<u>Both</u> should be investigated!

4

Hazard Identification and Control

Hazard Identification

5

Hazard Identification

- Acceleration
- Corrosion
- Chemical Replacement
- Fire
- Electrical
  - Shock
  - Thermal
  - Inadvertent Activation
  - Power Source Failure
  - Electromagnetic Radiation
- Contamination
- Chemical Dissociation
- Oxidation (other than air)
- Explosion
- Heat & Temperature
  - High Temperature
  - Low Temperature
  - Temperature Variation
- Leakage

6

Hazard Identification

- Pressure
  - High
  - Low
  - Rapid Changes
- Moisture
  - High Humidity
  - Low Humidity
- Stress Reversal
- Vibration and Noise
- Toxicity
- Radiation
  - Thermal (IR)
  - Electromagnetic (RF)
  - Ionizing
  - Ultraviolet (UV)
- Weather / Environment
- Shock
- Stress Concentration
- Structural Damage or Failure

7

Hazard Identification

Where to Look for Hazards

- Machines
- Lifting Equipment
- Facility Design
- Confined Spaces
- Fire Considerations
- Biological
- Ergonomic
- Energy
- Chemicals
- Pressure Vessels
- Electrical

HAZARDS ARE EVERYWHERE!

Who should be looking for them?

8

Hazard Identification and Control

SAFETY 2017 ASSE

Hazard Reduction Measures

- Design Feature
- Safety Device or Feature
- Warning Device
- Procedure and Training

It may require some serious thinking before the best approach is determined

9

Hazard Identification and Control

SAFETY 2017 ASSE

Hazard Identification and Closed-Loop Tracking System

- Hazard Identification
  - Know your requirements!
  - Inspection, Audit & Identification Program
    - Checklists
  - Procedures and Training

10

Hazard Identification and Control

Hazard Identification and Closed-Loop Tracking System

- Corrective Action Plan
  - Individual and completion date must be assigned
- Must be Closed-Loop
  - Make Closure Rate one of your leading indicators

11

Hazard Identification and Control

Accidents

12

Accidents

Look at past accidents for your first step in Accident Reduction

- Were prior hazards adequately resolved?
- What work controls were implemented to prevent recurrence, and are they working?

13

Accidents

Five Sources of Accidents

- People
- Equipment
- Material
- Procedures
- Work Environment

Note: There are typically multiple sources; rarely does a single source result in an accident

14

Accidents

SAFETY 2017 ASSE

The Domino Effect

Failure "here" may cause failure there"

- *Cause* and *effect* relationship

Five Phases of accident process

- Like dominos falling against one another
  - Management
  - Origins
  - Symptoms
  - Contact

} **LOSS**

15

Accidents

SAFETY 2017 ASSE

Accident Investigation
What is it?

- A systematic look at the event
- Examination of the risks
- Evaluation of the losses incurred
- An inquiry (who, what, when, where, why, how)

If done right

It will result in the control of future loss

16

Hazard Identification and Control

Work Planning

17

Work Planning

Accidents don't usually happen when people are doing what they usually do

- It is usually something they don't do everyday
  - Different job
  - Different place
  - Different equipment

18

Work Planning

SAFETY 2017 ASSE

Look at how new work is planned

It takes two people not doing their jobs to lead to an accident

- Supervisor – Inadequate planning and oversight
- Worker – Doesn't stop when the work doesn't look like the plan

19

20

21

Work Planning

Incorporate checks for work planning

- PPE
- Required equipment
- Required training
- Other work controls
  - LOTO
  - Confined space
  - Etc.

22

Work Planning

SAFETY 2017 ASSE

Job Risk Assessment

- Identify the task
- Break-down the steps
- Assess the risk
  - Too risky – implement controls
- Reassess after controls to be sure adequate risk reduction is achieved
  - Include the people who have done the work or similar work before

23

Hazard Identification and Control

SAFETY 2017 ASSE

Workplace Hazards

24

Workplace Hazards

SAFETY 2017 ASSE

Facility Risk Assessment

- List the activities, processes and elements
- Identify the hazards and controls
- Assess the risk
  - Too risky – implement more controls
- Reassess after controls to be sure adequate risk reduction is achieved

25

Workplace Hazards

SAFETY 2017 ASSE

Facility Layout and Maintenance

- Work area design and layout
  - Corners, aisle-ways, work stations, laboratory setup, shop areas, location of hazardous operations
- Maintenance and housekeeping
  - Trip/slip hazards, clutter, damaged walking or working surfaces
- Changes in original design or intent
  - Layout changes may create new hazards to workers in the area

26

Workplace Hazards

SAFETY 2017 ASSE

Facility Layout and Maintenance

- Illumination
  - Improper lighting for the type of work
- Ventilation
  - Insufficient air exchange or air flow
- Noise
  - Interfere with other work, stress
- Life safety considerations
  - Means of egress
- Access and Egress

27

Workplace Hazards

SAFETY 2017 ASSE

Office Safety Considerations

- Furniture, equipment, electrical, bookcases, etc.
- Walking / working surfaces
- Illumination levels
- Sanitation and refrigeration
- Emergency egress considerations

Housekeeping

- Trash, Stock, Up-keep, Fire Concerns, etc.

28

Workplace Hazards

SAFETY 2017 ASSE

Machines – Anything Power-Assisted

- Controls
  - Lockout Provisions
- Guards & Interlocks
- Point-of-Operation
- Operator's Manual & Training
  - Beware of "Customary" Operations

29

Workplace Hazards

SAFETY 2017 ASSE

Lifting and Elevating Equipment - General

- Cables; chains; ropes; hooks; hasps; slings; shackles; rails; limit switches; lifting beams; related equipment
- Oil systems and hydraulics; fuel systems; electrical systems; warning devices; fail-safe mechanisms
- Maintenance records; proof testing; operator training and certification; procedures; rated loads

30

Workplace Hazards

SAFETY 2017 ASSE

Elevated Work Platforms

- Exceeding manufacturer's load limit
- Improper (or no) railings or toe boards
- Slippery floor surfaces
- Out-riggers not functioning / not used
- Defective / poorly maintained equipment
- Fall protection devices not available or not being used

31

Workplace Hazards

SAFETY 2017 ASSE

Forklifts

- Exceeding manufacturer's load limit
- Poor or improper maintenance practices
- Personnel riding on forks or load
- Operator not properly trained/authorized
- Exceeding posted speed limits or speeds appropriate for the conditions
- Backing with people in the area
- Carrying loads improperly
- Narrow aisle-ways/not marked
- Platform / ramp edges

32

Workplace Hazards

SAFETY 2017 ASSE

Cranes and Hoists

- Exceeding rated capacity
- Contacting high voltage lines and cables
- Excessive winds
- Insufficient guide ropes/tag lines
- Working under suspended loads
- Leaving elevated load unattended
- Personnel riding on loads, slings, hooks, etc.
- Dangerous motion, direction, or use of crane

33

Workplace Hazards

SAFETY 2017 ASSE

Working in Confined Spaces - General

- Any area, space, or enclosure not designed for continuous occupancy
- Large enough to enter and perform work
- Has limited or restricted openings for entry and exit
- May lack adequate ventilation
- May contain a hazardous atmosphere
- May not be safe to enter

34

Workplace Hazards

SAFETY 2017 ASSE

Working in Confined Spaces

- Boilers
- Furnaces
- Degreasers
- Storage tanks
- Test chambers
- Vessels
- Tunnels
- Pits
- Vats
- Sewers
- Underground utility vaults
- Utility access holes
- Aircraft work areas
- Around some spacecraft

35

Working in Confined Spaces

SAFETY 2017 ASSE

Classifications

- Class A Permit-Required Confined Space
  - Immediately dangerous to life and health (IDLH)
- Class B Permit-Required Confined Space
  - Potential for injury or illness, but not IDLH
- Class C Confined Space
  - Remote potential for injury or illness; no permit required

36

Workplace Hazards

SAFETY 2017 ASSE

Fire Protection and Prevention

- General
  - Reporting and response procedures
  - Construction contracts
  - Emergency vehicles, fire lanes
  - Fire fighting equipment, sprinklers, alarms, hydrants, extinguishers
  - Personnel training and awareness

37

Fire Protection and Prevention

SAFETY 2017 ASSE

Fire Hazards - General

- Appliances
- Holiday decorations
- Carpets, curtains, upholstery
- Housekeeping
- Storage cabinets
- Supply and bulk storage
- Smoking policies
- Extension cords

Fire Hazards - Specific

- Welding equipment, other open flame devices
- Flammable and combustible materials
- Outdoor inspections
- Vehicle fueling areas
- Pyrotechnics, explosives
- Life safety issues

38

Fire Protection and Prevention

Pyrotechnics and Explosives

- Procurement procedures or property transfers
- Shipping and receiving
- Transportation on- and off-site
- Inventory and control - use of logbooks
- Occupational exposure monitoring
- Training and certification of personnel

39

Fire Protection and Prevention

Welding and Other Open Flame Devices

- Operator training and certification
- Welding (or "Hot Work") permits
- Personnel protection
- Fire prevention measures
- Ventilation concerns
- Working in confined spaces
- Electrical hazards (arc welding)
- Open flame (gas welding)

40

Fire Protection and Prevention

SAFETY 2017 ASSE

Outdoor Inspections for Fire Hazards

- Excess vegetation
- Debris and refuse collection
- Storage areas and yards

Vehicle Fueling Areas

- Proper spacing
- Fire protection measures
- Procedures enforced

41

Fire Protection and Prevention

SAFETY 2017 ASSE

Life Safety Issues

- Means of egress
  - Fire doors
  - Stairwells
  - Corridors
  - Emergency exits
  - Storage areas
- Inspections
  - Building codes
  - NFPA 101

42

Workplace Hazards

SAFETY 2017 ASSE

Electrical Safety - General

- National Electrical Code, NFPA 70, 70B, 70E
- Other standards and regulations (local, state, etc.)
- Use of "Listed" or "Labeled" equipment
- Use of Lock-out/Tag-out procedures
- Personnel training and awareness
- Ground Fault Interrupter (GFI) usage
- Arc-flash and Shock Hazards

43

Electrical Hazards

SAFETY 2017 ASSE

What to Look For:

- Shock hazards
- Fixed equipment
- Safety disconnect
- Dangerous conditions
- Clear working space
- Equipment mounting
- Automatic start equipment
- Extension and power cords
- Arc-flash hazards
- Portable power strips
- Portable tools
- Computers
- Portable power systems
- Temporary wiring
- Openings in enclosures
- Unusual or special conditions

44

Electrical Hazards

Common Appliances

- Radios
- Desk lamps
- Hot plates
- Refrigerators
- Coffee makers
- Television Sets
- Toasters and toaster ovens
- Microwave ovens
- Fans
- Typewriters
- Space heaters
- Adding machines
- Paper shredders, copy machines, fax machines
- Personal computers and peripheral devices

45

Lockout / Tagout

General

- OSHA 29 CFR 1910.147
- Applies to the control of ALL types of hazardous energy sources
- Lockout:
  - *To positively hold an energy isolating device in a safe position*
- Tagout:
  - *The use of a tag on an energy isolating device to warn others*

46

Lockout / Tagout

SAFETY 2017 ASSE

Requirements

- Applies to all employees performing work around machinery where they could be injured by:
  - *Unexpected start-up of the equipment*
  - *Release of stored energy*
- Examples when "LOTO" should be used include:
  - *A guard must be removed or a safety interlock by-passed*
  - *Repairing electrical circuits*
  - *Cleaning or oiling machinery that normally has moving parts*
  - *Clearing a jammed mechanism*

47

Lockout / Tagout

SAFETY 2017 ASSE

Requirements

- Only the person who applied the lock/tag can remove it
  - *Supervisor, with concurrence from Safety & Health representative*
- Training
  - *Proper methods of applying LOTO*
  - *Periodic re-training (semi-annual, annual, etc.)*
  - *Only those trained can use locks and tags*
- Employee Responsibilities
  - *Must understand the LOTO program*
  - *NEVER interfere with a lock or tag*

48

Workplace Hazards

SAFETY 2017 ASSE

Pressure Vessel Hazards - General

- ASME design specifications
  - Includes piping
- ANSI requirements
- Compressed Gas Association
- Local or State codes and standards
- Labels and placards
- Piping Identification

49

Workplace Hazards

SAFETY 2017 ASSE

Pressure Vessel Hazards

- Design
  - Safety factors; MAWP; Supports (for horizontal and vertical vessels); Hydraulic systems (hydraulic shock)
- Components
  - Relief valves; Redundancies; Identification plates; Testing; Gauges; Piping and tubing; Flexible hoses

50

Workplace Hazards

SAFETY 2017 ASSE

Pressure Vessel Hazards

- Compressed Gas Cylinders
  - Proper handling, use, and storage practices
  - Maintenance and certifications
  - Leaking and defective containers
- Cryogenic Systems
  - Hazards of cold temperatures
  - Pressure relief devices
  - System and materials compatibility requirements

51

Hazard Identification and Control

SAFETY 2017 ASSE

System Safety Techniques

52

System Safety Techniques

SAFETY 2017 ASSE

Definition of System Safety:

"A sub-discipline of systems engineering that applies scientific, engineering, and management principles to ensure adequate safety, the timely identification of hazard risk, and initiation of actions to prevent or control those hazards throughout the life cycle and within constraints of operational effectiveness, time, and cost."

53

System Safety Techniques

SAFETY 2017 ASSE

THE SAFETY PROFESSIONAL MUST ASSESS THE RISK OF HAZARD EXPOSURE AND DETERMINE THE PRACTICAL METHODS TO BEST ELIMINATE OR REDUCE THAT RISK TO THE LOWEST POSSIBLE LEVELS

THIS IS THE ESSENCE OF SYSTEM SAFETY

54

System Safety Techniques

SAFETY 2017 ASSE

System Safety Concepts

- *A forward-looking process that considers how:*
  - Human resources *(people)*
  - Operating equipment *(parts or hardware/tools)*
  - Facilities *(positions or location in plant or factory)*
  - Operating procedures *(paper or written instructions)*

Affect the performance of the job/task in the

WORK ENVIRONMENT

55

System Safety Concepts

SAFETY 2017 ASSE

System Safety Criteria

- Hazard severity:
  - A categorical description of hazard level
  - Based on real/perceived potential for causing harm/damage
- Hazard probability:
  - The calculated likelihood that a condition(s) will exist
- Can be Qualitative or Quantitative

56

System Safety Concepts

SAFETY 2017 ASSE

DESCRIPTION	SEVERITY CATEGORY	MISHAP DEFINITION
CATASTROPHIC	I	DEATH OR SYSTEM LOSS
CRITICAL	II	SEVERE INJURY, ILLNESS, OR SYSTEM DAMAGE
MARGINAL	III	MINOR INJURY, ILLNESS, OR SYSTEM DAMAGE
NEGLIGIBLE	IV	LESS THAN MINOR INJURY, ILLNESS, OR SYSTEM DAMAGE

57

System Safety Concepts

SAFETY 2017 ASSE

DESCRIPTION	PROBABILITY LEVEL	MISHAP DEFINITION
FREQUENT	A	LIKELY TO OCCUR FREQUENTLY
PROBABLE	B	WILL OCCUR SEVERAL TIMES DURING ITEM LIFE
OCCASIONAL	C	LIKELY TO OCCUR SOME-TIME DURING ITEM LIFE
REMOTE	D	UNLIKELY, MAY POSSIBLY OCCUR DURING ITEM LIFE
IMPROBABLE	E	SO UNLIKELY, IT CAN BE ASSUMED THAT HAZARD WILL NOT OCCUR

58

System Safety Concepts

SAFETY 2017 ASSE

System Safety Techniques

The Hazard Matrix

- A recommended tool to assess & understand hazard risk
- Combines severity and probability into single rating
- Establishes Risk Assessment Code ("RAC")

59

	HAZARD CATEGORIES			
PROBABILITY LEVEL	I Catastrophic	II **Critical**	III Marginal	IV Negligible
A - Frequent	1A	2A	3A	4A
B - Probable	1B	2B	3B	4B
C - Occasional	1C	2C	3C	4C
D - Remote	1D	2D	3D	4D
E - Improbable	1E	2E	3E	4E

HAZARD RISK INDEX

Risk Classification	**Risk Criteria**
1A, 1B, 1C, 2A, 2B, 3A	Unacceptable
1D, 2C, 2D, 3B, 3C	Undesirable
1E, 2E, 3D, 3E, 4A, 4B	Acceptable, with review
4C, 4D, 4E	Acceptable, without review

60

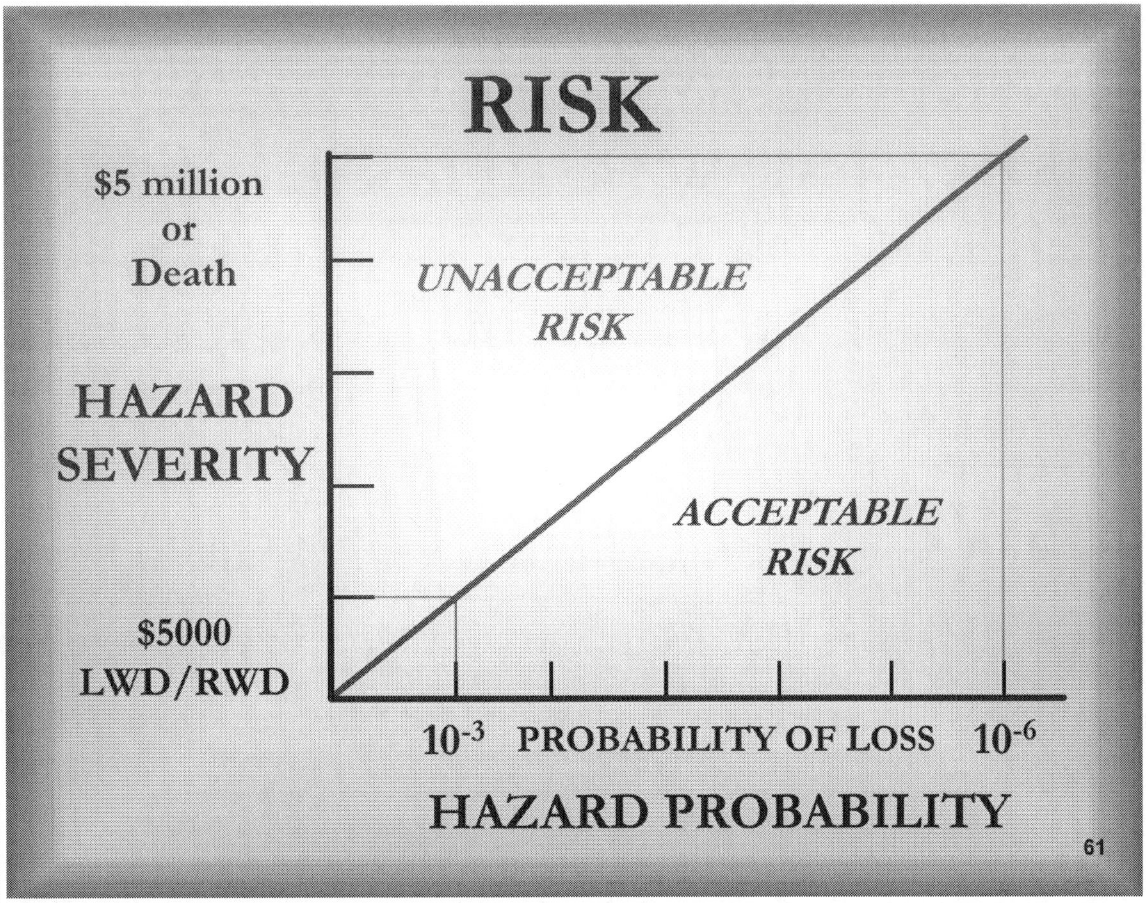

System Safety Concepts

SAFETY 2017 ASSE

Precedence for Reduction of a Hazard

What do you do if a risk is unacceptable?

1. Design feature
2. Safety device
3. Warning devices
4. Procedures and training

62

System Safety Concepts

1. Line Safety Program
2. Management Commitment and Buy-in
3. Hazard Identification, Reporting, and Tracking
4. Life-Cycle Phases and Hazard Analyses
5. Hazard Risk Management and Acceptance

63

Hazard Identification and Control

Analytical Techniques

64

Analytical Techniques

- Checklist
- Job Safety Analysis/Job Risk Assessment
- Facility Risk Assessment
- Management Oversight & Risk Tree (MORT)
- HAZOP

65

Analytical Techniques

- Fault Tree
- Failure Modes & Effects or Fault Hazard
- Operating & Support
- Preliminary Hazard Analysis
- Sneak Circuit
- Energy Analysis
- What-if

66

Session No. 550

Delivering Effective Safety Communications in the Boardroom and the Break Room

Randall C. Butler, MS, CSP
Consumers Energy
Jackson, MI

Introduction

This paper will focus on how to emphasize the importance of safety-based on the needs of the audience. Safety professionals who use traditional media to interact may have to reassess the personal/professional driver of their audiences to align their safety and health message with the evolution of communication. While technology has contributed useful methods of delivering workplace safety data, new challenges arise in maintaining effective workplace communications specific to safety and health. New strategies to develop safety and health communications target optimizing communication by increasing conversation amongst top management and line employees.

Understanding Your Audience (The Setup)

In the workplace, the first objective to effectively communicate safety is to engage the audience. In the workplace, mass communication can be efficiently through a website, share point, personal email, text message or even a safety stand-down; however, delivering the content does not mean the message was received. Individuals that do not feel connected to the topic(s) or material will ignore, disengage, or simply 'go with the flow' to satisfy a safety and health objective.

Most individuals relate best to information that could results in a direct benefit. Drivers that influence management may be different than the interests of the line workers. For a safety professional to communicate one message that can reach all, researching common themes of each workgroup and its knowledge as it relates to each topic can assist in developing a plan of action. This action plan must work to meet a clearly defined objective you want to reach with the audience. Safety communications generally target to inform, excite, and/or persuade the audience to engage in a safety and health protocol.

Inform

The best informative communications are concise, factual, and unbiased. With regard to safety communications, the basis of this type of communication has traditionally been to advise about a best practice, rule, or regulation. This approach shifted during the information age.

The information age is described as a period that began in the late 20th century, which allowed for instant transfer of information, leading to the rise of information-based industries and the world-wide web. Technology has made most information formally delivered through the specialty skills offered by a safety professional immediately available to all. Due to this innovation, communications that targets to 'inform' an audience must be developed in a way that fulfills a specific knowledge gap, interprets a rule, or reports the status of an event, for the information to be useful for the end user.

As information is received, verbally or in writing, the audience will subconsciously filter the message to determine what is relevant based on its perspective. Therefore, it may be best to build interest before attempting to supply the information necessary for safety success.

Excite

An aroused audience is one that is interested in the topic, making them more receptive to the information that is communicated. When communicating to a large or diverse group, it may be necessary to use methods that 'excite' an audience to enable active listening, or otherwise, accept a non-verbal written message. Examples of exciting an audience includes: asking a thought-provoking question, making a provocative statement or a joke, or including clickbait, taglines, or a photo that could be used as a openers in gaining the initial attention of an audience.

After introducing the audience to a safety topic, it may be necessary to re-excite interest if the presentation or publication is not brief. In such cases, the safety professional should use simultaneous brain-mind-body stimuli to create an experience for each member of the audience. This is an area where communication to management and line workers may be different, based on the content of the message.

One of the best ways to stimulate multiple senses is through storytelling. A good storyteller can connect with the audience by depiction of an event using the end users' current knowledge base in a way that guides their perception of an event. This technique works in the boardroom and in the break room to effectively communicate almost any message. For management, the safety professional may tell a story to connect the dots of the organization's past, present, and forecasted trends. In the break room, similar methods can be used to communicate a scenario related to the worksite that could result in an incident if safety rules are not followed. Visualizations can be used with either audience to capture attention before influencing behaviors.

Persuade

An English idiom states that, "a picture is worth a thousand words." Barriers that exist between languages, cultures, and within a society are now bridged by memes on social media. If used appropriately in the workplace, safety visualizations can be a powerful and preferred tool to communicate a means to an end.

A connotation adds direction to the visualization, which would otherwise be, at best, a conversation piece, rather than a directed communication. Gaining the audience's attention provides an opportunity to gain support, provide direction, solicit requests, or push forward an idea by generating a new perspective. As a support unit for your organization, the safety professional should continually work to motivate all constituents to implement 'best practice' in resolving each hazard.

Motivation is attributed to each individual's desire, which can be influenced through a logical assertion that will work to prevent a realistic consequence. A successful safety and health communication can be measured by its outcome relative to those impacted. Informing, exciting, and persuading provide methods to connect; however, the effectiveness of the communication is dependent on the level of engagement of the audience resulting from the communication. Making information available is not the same as delivering a message. Also, delivering a message is not the same as communicating effectively. To deliver an effective safety communication, the message must be communicated, received, and proven to work in inciting action, thus engaging the target audience to improve safety and health. Engagement should focus on how the message was used by the audience.

It is not possible to please everybody; however, by using multiple communication techniques to deliver each message, it is possible to reach and engage each person in the workplace. To engage an audience, there must be a want or a need. Safety conscious companies that have: robust training programs that provide workers with all of the information necessary to keep them safe; automated reporting systems to track workplace safety metrics in real time. As with most other information readily available online, these organizations can find it challenging to make safety personal through non-repetitive communications. Also, this can make it difficult for a safety professional to find a point of engagement.

Rather than being a single source that generates information, motivates the workforce, and encourages safety action, the safety professional should seek opportunities to create and develop sub-channels of communication that naturally push a message throughout an organization.

The Alignment

When processing information, adults tend to compare new information with their previous knowledge or experience, filtering each message to determine what is useful, objectionable, or not applicable. The massive amount of information available through technology leads to people in each audience rejecting more information than they accept because they are already familiar with a topic, do not need the information communicated, or perceive that the message does not add value. To overcome these barriers, the safety professional must develop a way to intrigue an audience to want to know more. This can create discussion 'peer to peer' versus the tradition 'one-to-many' seminar approach.

Similar to the strategies that are used in social media, the safety professional should target to engage a portion of the audience in a way that motivates them to share the communication. This method can bridge a message throughout an organization naturally, from peer to peer, with the safety professional needing to reach only a few influential or outspoken stakeholders for their communication.

Our workforce is intelligent; therefore, to create curiosity that engages an audience, it may be more appropriate to speak as a subject-matter expert, rather than directly to your audience. Formally, it was heavily frowned upon to speak in this manner because the presenter could not effectively communicate detailed information, due to a lack of understanding by the audience or a lack of their knowledge in the subject area. This is no longer true, as audiences continually seek new information; however, it very important to include them in the message to make it meaningful.

Technology provides the ability for motivated recipient(s) to immediately fact check information, look up foreign information, and/or photograph slides that can be studied at a later time. Traditionally interpreted as distractions, these actions are now how the audience works to experience communication. The key tactic for the safety professional is linking the information to an actual event that the audience is interested in or has experienced.

Rather than seeking to understand the entire audience, the safety professional may be able to speak to everyone while targeting only a few influential person(s) that will understand, interpret, and redistribute the communication. These management, or field, influences can connect the message more closely with their peers to influence safety by redistributing the message in their own way. People are shifting away from wanting to receive information; now they want to experience knowledge. This era is now referred to as the 'experience age.'

Experience will focus beyond the message to determine the look, feel, smell, and sound interpreted in a communication. Consider the difference in feel when a person types in capital letters versus when they type in all small letters. Or, consider the feeling of a quiet lobby versus a lobby with soft music playing. The safety professional must embody the message to align with how the audience should feel before, during, and after it is delivered.

The Delivery

A decade ago, safety professionals were challenged with trying to get a break room of workers to listen. Today, they are challenged by this same workforce to have them open an attachment on an email before deleting the communication. No longer are taglines that contain bold language such as: IMPORTANT, URGENT, or HURRY considered an influence to make a message a priority. In verbal communications, shock value, elevated enthusiasm, and extreme emotion are seen as gimmicks. The audience may be interested in 'who' is delivering the message, rather than the content of the message itself.

Whether the audience is familiar with the topic or the presenter, an introduction is always necessary to prepare the audience for what it will receive. In the delivery, an introduction is not the same as a start. An introduction can communicate what will be explored; be a catch phrase to build interest; or be a summary of the presenter's experience/association with the topic. Beginning a communication the right way sets the tone for everything else that follows. Consider how a non-traditional salutation, such as "We Made It," will make you feel, compared to "Dear Sir/Ma'am." The rule to a good introduction is 'keep it simple.' The message should be more complex.

After understanding your audience and developing targeted objectives for your message, an execution plan should consider the best way to align with the motivated recipient(s) without losing the rest of the audience. To do this, consider a dynamic structure in communicating.

Similar to writing a story, a dynamic structure does not have to follow the start, middle, end format. When communicating to management, consider fewer words and more visuals to depict a message as you explain the details of each chart or graph. Use similar methods when posting break room communications. A meme, rather than a lengthy notice or publication, can generate more conversation while communicating the same message. If the detail offered by a newsletter or publication is preferred, consider color, style, and formatting to make it attractive and stylish for the audience. Safety professionals must embrace a less utilitarian and more decorative style of communicating. The goal is to depend on social factors and critical thinking to deliver each message.

Conclusion

To deliver an effective safety communication, a safety professional must understand what specifically the audience needs in regards to safety. Then, determine clear objectives that can be creatively communicated to the audience to fulfill this need. In each organization, it is the safety professional's responsibility to align with the workforce. However, aligning may not be possible with every individual; therefore, it may be less cumbersome to follow up with those that express an interest in what they have to offer.

To identify your target audience, notice who is asking questions, challenging a perspective or those known as the office go-to person; these individuals will more than likely be the most motivated and influential. Allow these individuals to take on the responsibility, rather than delegating responsibility. Use positive motivators to encourage action. To deliver effective communications, develop a peer-to peer liaison team that champions the message of safety throughout an organization from the boardroom to the break room.

Bibliography

Bale, L. 2017. "How to Measure Effective Communication." *Houston Chronicle*. (Retrieved 2017) (http://smallbusiness.chron.com/measure-effective-communication-3180.html).

Brounstein, M. 2001.*Communicating Effectively for Dummies*. New York: Wiley Publishing.

Bruno, P. "How People Learn: An Evidence-Based Approach." *George Lucas Educational Foundation*. October 9, 2015.

Castells, M. 2009. *The power of identity: The Information Age: Economy, society and culture. Volume II (The information age)*. Oxford: Wiley-Blackwell.

Collinge, R. "Designing for the Experience Age." *Usabilla*. February 9, 2017.

Livingstone, S. 2003. "The Changing Nature of Audiences: From the Mass Audience to the Interactive Media User," pp. 337–359 in A. Valdivia (ed.), *The Blackwell Companion to Media Research*. Oxford: Blackwell.

Livingstone, S. 2004. *The Challenge of Changing Audiences or, What is the Audience Researcher to Do in the Age of the Internet?* (Retrieved from: http://journals.sagepub.com/doi/abs/).

Meier, D. 2000. *The Accelerated Learning Handbook: A Creative Guide to Designing and Delivering Faster, More Effective Training Programs*. Burr Ridge, IL: McGraw-Hill Education.

Wadhera, M. 2016. *The Information Age is over; welcome to the Experience Age*. Crunch Network. (Retrieved from: https://techcrunch.com/2016/05/09/the-information-age-is-over-welcome-to-the-experience-age/).

Williams, J. and Geller, S.G. "Communication Strategies for Achieving a Total Safety Culture." *EHS Today*. July 7, 2008.

Williams, J. H. "Improving Safety Communication Skills: Becoming an Empathic Communicator." *Proceedings of the Annual Professional Development Conference for the American Society of Safety Engineers*, Seattle, WA, 2006.

Session No. 551

Executive Animation of EH&S^3 at L'Oréal

Jay R. Harf
Kristin K. Herman
Rema T. Bianco

Successful animation is the right message at the right time in your organization. L'Oréal Environmental Health Safety, Security, and Sustainability (EH&S^3) executives will share how they animate various EH&S^3 data, programs, and actions to organizational leaders in an effort to inform and influence desired outcomes and share beauty with all.

Introduction

At L'Oréal, we animate and present the safety process through metrics and actions. We are also trying to highlight key themes during regular interactions with each of our senior executives. These key themes are:

1. Objectives (key messages)
2. Corrective action plans
3. Organizational issues
4. Performance (metrics/adherence to goals and objectives)
5. Communication
6. Animation
7. Special projects
8. SWOT analysis

Although these are the key points of our paper, we are trying to deliver a message of expertise and competence to our teams. This competency arises out of leadership, executive communications, and presentation skills.

Leadership

Recently, the roles of leadership and culture as critical components to achieving high-performing safety systems have become more visible. Top executives set the tone for safety culture. Their personal commitment and attitude toward safety excellence becomes interwoven in the fabric of organizational norms and expectations. Through value-driven safety, transactional, and transformational leadership determine safety culture and impact safety excellence. When addressing senior executives in overall EH&S^3 Performance, all aspects of the aforementioned leadership styles must be referenced and utilized. For example, we must determine the level of goal obtainment we are achieving (transactional leadership), the ability to continuously improve the organization (transformational leadership), and finally, prove to senior management how we are supporting the organization's mission (servant leadership). The process of presenting this information to senior executives is critical from a leadership perspective, and safety professionals must deliver on this topic.

Safety professionals, more often than not, do not have the positional power or financial power to simply make decisions for the organization. Consequently, their actions and leadership are a matter of influence through technical and organizational servant leadership and by influencing the followership of an organization. Additionally, the safety professional serves as a role model of the followership of an organization through optimism and trust. As safety professionals work at all levels of the organization (in the boardroom and on the factory floor), they can "instill optimism, confidence and faith in their followers by suggesting that although their challenges appear formidable, they be can be successful by working together to create a better future" (Berson et al. 2001, p.56). Furthermore, optimists "see adversity as a challenge, transform problems into opportunities, put in the hours to refine skills, persevere in finding solutions to difficult problems, maintain confidence, rebound quickly after setbacks and persist" (Schulman, 1999, p.33). "Positive expectation is embracing the inevitability of change as both expected and positive. It is the attitude that whatever the circumstances or situation, some good can result. It is also having the attitude of 100% responsibility toward our ability to respond to and affect the circumstances in some manner" (Gore, 2002, p.49). In summary, the most important and frequently identified theme of a EH&S^3 leader is the ability to foster a culture of interdependence through service to the organization and the followership. One of the ways this is accomplished is through executive communication and the animation of the current state of the EH&S^3 process.

Executive Communication

Effective communication with a group or an individual requires information gathering and research. You need to ask yourself the following questions: Who is my audience? What are their goals and priorities? How do you gain trust with the group? How much time do I have? Who is presenting before and after me, and what priority are those presentations to the group or individuals in the room? What is the preferred method of presenting (i.e., using slides, handouts, etc.)? Should I have animations? Should I utilize a flash drive? The keys to successful executive-level presentations revolve around establishing relationships with senior executives, meeting preparation, and performance management.

Establishing Relationships

Establishing individual relationships with the executives prior to delivering presentations builds trust and respect. Transparency in your approach improves the level of trust one has with their management team. As a courtesy, you should inform an individual if you are presenting information that may be seen as negative or controversial to their part of the business. You never want to present information that was not communicated with an individual prior to a bigger group meeting. There is a strong effort to ensure there are no surprises both before and after an executive-level presentation. Now that you have managed your relationships prior to any presentation, it is time to prepare for the meeting.

Meeting Preparation

Meeting preparation is critical to the success of any executive interaction. First scenario: You are asked to present an EH&S^3 update to an executive group once per quarter. How do you start? First, find out who is in the audience. Research their names, and determine what businesses and/or functions are within their scope. If possible, obtain an organizational chart to understand the reporting structures. Next, determine if you currently possess a relationship with any of the individuals in this audience. If you do not have an established relationship, consider organizing a one-on-one meeting with the executive prior to your meeting. Based on the organizational chart, bring all the relevant business data. This meeting should be no more than 30 minutes, and you should be prepared and ready to answer any and all anticipated questions. Also ask questions specific to goals, actions, and priorities. This will help to determine what information would be useful during future meetings. After this initial meeting, suggest meeting on a regular basis and calling it a 'touch base.' Proactively schedule the touch-base, bi-monthly meetings via calendar invites.

Ask for the agenda to determine what subjects will be discussed, and who is presenting before and after your presentation. What will be the tone of the meeting? Will the presentation prior to your time going to be numbers-oriented and about business objectives? Is the presentation prior to your presentation going to be numbers-oriented or tone of the meeting and business objectives? Prepare your presentation with a limited number of slides, just in case your allotted time is decreased. You want to ensure that all the important information is able to be presented, even if in a short period of time. Create an appendix that you can refer to if someone has a specific question (i.e., specifics on an incident report, statistics etc.). If someone asks a question that you cannot answer at that moment, let the group know that you do not have the answer, but you will make sure that they are provided with follow up after the meeting.

Performance Management

Now let's get into the specifics of the content that should be covered in an executive meeting regarding EH&S[3]. The content should remain consistent with the same outline for each quarterly meeting. You should include the following subjects in each presentation:

1. Objectives (key messages)
2. Corrective action plans
3. Organizational issues
4. Performance (metrics/adherence to goals and objectives)
5. Communication
6. Animation
7. Special projects
8. SWOT analysis

Review the status of the key objectives and goals for the company. For those objectives that are performance-related, manipulate the data into graphs and/or tables that will translate well for the audience. Always compare the current data with the data from the previous year. Consider comparing the data over a span of a few years, if this data is consistently reported and available. Analyze the information, and present the key messages for the objectives. If the data is favorable, commend the group; however, if the data is not favorable, offer potential solutions or an action plan for each attendee to own.

Presentation Skills and Animation

For the L'Oréal Americas Operations Group, PowerPoint presentations can be an effective means of communication to executive leaders when appropriately presented. Poorly organized or irrelevant presentations can often lead to confusion or misunderstandings between colleagues. To avoid this situation, the Environment, Health, Safety, and Sustainability teams of L'Oréal Operations Americas follow PowerPoint and presentation practices that allow communication to be simple yet efficient.

One of the most important pieces to being an impressive presenter is to know who you are presenting to. Prior to creating your presentation, think about your audience. What do they already know about the topic? What you want them to know about your topic? These answers will help structure the content included. The material should be no more than six words per line and six lines per slide. The bullets should be key points to serve as reminders to speak to, rather than long notes. If bullets on slides are too long, the audience could easily become distracted from reading the slide and not pay attention to the speaker. More concise bullet points also allow the audience to understand and remember the information better.

The aesthetics of a PowerPoint presentation are also vital to a compelling presentation. Images and diagrams are also effective to demonstrate relationships instead of text. They can help make a presentation interactive and engaging. The template and font should be consistent throughout all slides. Light-colored fonts should be used against darker backgrounds, and vice versa. Text should be large enough to be read from the back of a room, but not so big that the information cannot fit on the slide. Animation on slides can be helpful to not let the audience read too far ahead of pace but should be simple enough to keep readers focused. Transitions between slides should also be kept uniform throughout the presentation, making for smooth transitions from slide to slide.

In addition to a visually pleasing presentation, the flow of the content is important. An opening agenda outlines what will be presented, and can help keep an audience's attention. The audience will know what is next to come, but will have to wait until that slide or section. As always, any presentation should be given with the utmost confidence. Speaking clearly and succinctly will prove your knowledge of your content to executive leaders.

Interdependent Cultures

Interdependent cultures are seen as collaborative. "They are broadly characterized by the assumption that leadership is a collective activity that requires mutual inquiry and learning. This assumption may lead to the widespread use of dialogue, collaboration, horizontal networks, valuing of differences, and focus on learning. Other characteristics associated with interdependent cultures include: the ability to work effectively across organizational boundaries, openness and candor, multi-faceted standards of success and synergies being sought across the whole enterprise" (Paulus, 2009, p. 6). From an industrial and occupational safety perspective, the culture of interdependency is the vision that must be achieved. The interdependent culture model has been made famous by DuPont, which published what is widely known in the safety profession as the Bradley curve safety culture model (See Figure 1). Teams of employees feel ownership for safety, and take responsibility for themselves and others. People do not accept low standards and risk-taking. They actively converse with others to understand their point of view. They believe true improvement can only be achieved as a group, and that zero injuries is attainable. The model portrays the four states of safety culture: reactive, dependent, independent, and interdependent, and describes that, as safety cultures mature into the different levels, accidents decline. Ideally, the model focuses on the culture of interdependence, where the collective group is able to achieve the highest reduction in accidents based on key components of conforming, contributing, values, and looking out for each other (my other's keeper). Figure 1 illustrates the Bradley Curve by DuPont Inc.

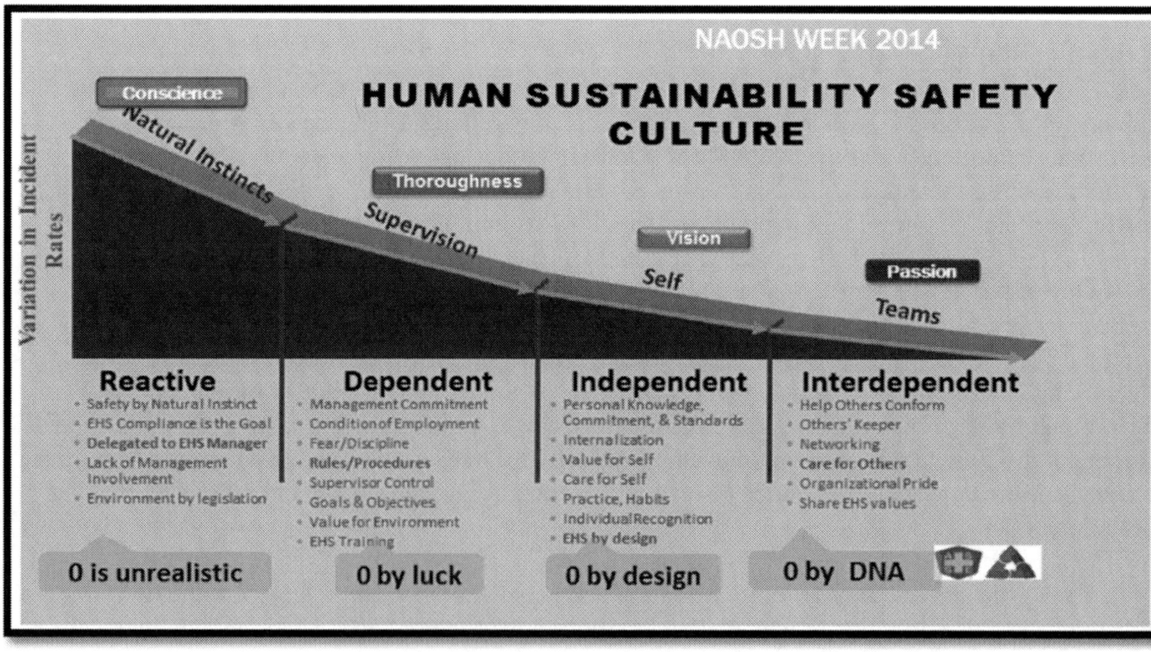

Figure 1. Safety Culture Model of Interdependence

Conclusion

According to the Center for Creative Leadership (2016), there are three ways to influence in the theory of leadership. They are logical appeals, emotional appeals, and cooperative appeals. Logical appeals "tap into people's rational and intellectual positions. You present and argument for the best choice of action based on organizational benefits, personal benefits, or both." Emotional appeals "connect your message, goal, or project to individual goals and values. An idea that promotes a person's feelings of well-being, service, or sense of belonging has a good chance of gaining support." Lastly, cooperative appeals "involve collaboration, and alliances." Working together to accomplish a mutually important goal extends a hand to others in the organization and is an extremely effective way of influencing. We are trying to maximize the environmental health safety, security, and sustainability visibility process to the executive leaders, in short, regularly schedule timeframes and convey the topics of most importance in that time period.

The vision of the future is an ideal safety culture where people are guaranteed not to be injured in the course of their employment. The roadmap to achieve this vision is a collective approach by industry and society to systematically reduce or eliminate all possible risks of industrial accidents, and to continuously improve our standing until our colleagues, friends, and family members are free from all hazards. If the vision is not attainable, the next step must be to revisit progress and statistics in the quest for an interdependent safety culture. This process will ultimately lead to the ideal culture of interdependence that the occupational health and safety profession should aspire to achieve.

References

Berson, Y., Shamair, B., Avolio, B.J., & Popper, M. 2001. "The relationship between vision strength, leadership style and context." *The Leadership Quarterly*, 12, 53-73.

Gore, Gary. 2002. *Navigating Change*. Memphis: Team Trek.

Harvard Business Review. 2016. "Three Ways to Influence."

Knapp, D. 2008. "Eight Time-Proven Presentation Tips." *Program Manager*.

Linggi, D. 2009, November 20. *Key Steps to Effective Executive Communications*. (Retrieved April 26, 2017) (http://apps.prsa.org/Intelligence/TheStrategist/Articles/view/8437/1004/Key_Steps_to_Effective_Executive_Communications#.WP5C1fnyuM8).

United States Naval Academy. (n.d.). *Microsoft PowerPoint: Best Practices* (Rep.). (Retrieved April 26, 2017) website: https://www.usna.edu/IDSC/_files/documents/PowerPoint_BestPractices.pdf

Schulman, P. 1999. "Applying learned optimism to increase sales productivity." *Journal of Personal Selling and Sales Management, 19* (1), 31-37

Session No. 552

Enhancing Construction Safety Using Wearable Technology

David B. Korman, CSP
Skanska USA
New York, NY

Albert Zulps, AIA
Skanska USA
Boston, MA

Introduction

Construction is still one of the most dangerous occupations in the world, with jobsites being complex networks of people, equipment, and activity that are risky and ever-changing. With continued pressure to drive more value to clients, how do project teams use technology to improve productivity while maintaining safe working conditions? Skanska sought to answer this question and piloted the use of an Internet of Things (IoT)-based technology called real-time location systems (RTLS) to dynamically track people and assets in order to make our jobsites smarter, safer, and more productive. This paper will focus on how IoT, including wearable technology and RTLS, can enhance environmental awareness and real-time visibility of worker's safety and productivity on active jobsites, as well as into building operations.

The IoT network created for the RTLS allows for information from wearable technology to be used in real time and then archived for future reference and analysis. Information from accelerometers indicating falls and trips can be recorded and acted upon. Additionally, the possibility exists to monitor a worker's vital signs to warn against fatigue, heat stroke and other conditions.

We will demonstrate a live RTLS during this presentation, courtesy of RedPoint Positioning.

Wearable Technology for Construction Safety

Wearables for construction safety are an emerging technology. There are still some technical challenges that are being addressed, such as battery life, administration of wearables on a daily basis, and creating robust analytics. Social acceptance is also an area of study. The University of Michigan has begun studying acceptance in diverse worker groups, where age, role, and common perceptions factor into the desire to use wearables on construction sites. Individuals may have concerns on how information is used, if it is anonymized, and may need reassurance that it does not operate beyond the confines of the construction site.

Many wearable IoT devices in use on construction sites are often temporary badge-like monitors that can be worn on the body or within a safety vest. For instance, the Triax Spot-R solution is intended to be worn around the waist, easily clipping onto a belt, much like a wireless pager.

With virtually all uses, the wireless mesh infrastructure created by an IoT network is required to allow wearables to communicate to a remote dashboard, for monitoring and analytics. This, in turn, allows for warnings and alarms to be disseminated to the appropriate parties, depending on the incident. For example, a fall from a ladder may create a signature in the accelerometer worn on a worker's vest, wristband or badge that is recognized as unusual. This information is recorded and relayed to the worker's supervisor or person monitoring the system automatically to help mitigate a proper safety response. A worker can also manually trigger a message to alert others of danger, or to report that they have been hurt. Leading IoT companies, such as Triax, RedPoint

Positioning, Human Condition Safety, and other companies, incorporate these basic features into their systems.

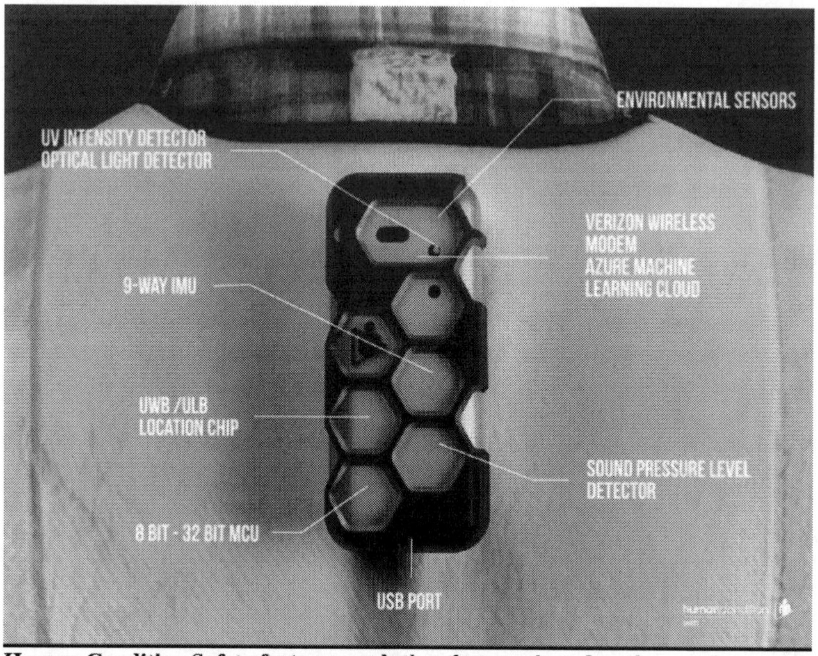

Human Condition Safety features a solution that monitors location, environmental exposure, and biometric markers of an individual worker through a proprietary, sensor-laden safety vest.

Caterpillar Prevent system detecting personnel in machinery danger zones

IoT Network: What Are Real Time Location Systems (RTLS)?

Definition

Real-time location systems (RTLS) represent a class of technologies that encompass hardware and software used collectively to provide location information relative to a coordinate system. A global positioning system (GPS), which is a car navigation system, is perhaps the most familiar equivalent to RTLS, combining complex software with a sophisticated network of satellites that orbit the earth to tell you precisely where you are in relation to the streets and buildings around you. In short, an RTLS provides location information for people or things on the move inside of a building or within an active construction site, something that most satellite-based GPS solutions are not able to do accurately.

Indoor GPS Isn't Really GPS

Instead of RTLS, the term "indoor GPS" is used in the construction industry because it's a short and convenient means for conveying a familiar concept. The U.S.'s Global Positioning System (GPS), along with competing Global Navigation Satellite Systems (GNSS) built by other countries, has infiltrated our lives, from robotic tractors to Google Maps running on smart phones. But to be clear, the whole reason RTLS is needed indoors is because GPS satellite signals do not effectively penetrate buildings, leaving commercial GPS receivers searching aimlessly for direction. Indoors, other systems are needed.

Global Positioning System

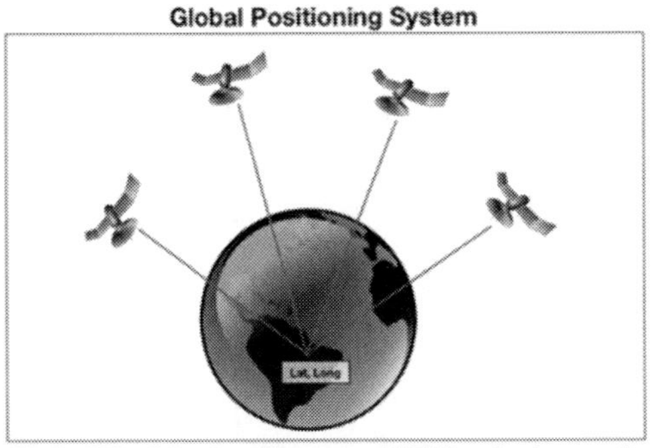

Redpoint "Indoor GPS"

Global Positioning Systems (GPS) solutions rely upon access to satellites to provide orientation, while real time location system s (RTLSs)establish a network of beacons and access points inside of a building not reliant on satellites.

Common Uses of RTLS

Over the last decade, RTLS systems have been deployed indoors to solve a variety of problems. The following are some commonly discussed applications for RTLS, as well as some more unique uses.

Hospitals

Hospitals can be large, complex, and ever-changing facilities. Staff, patients and equipment are constantly moving across medical campuses, between building floors, and from room to room. Locating any one particular person or object can be particularly challenging. Perhaps even more interesting, RTLS can be used to track exposure to infections or even document doctors-patient… interactions.

Factories

RTLS can be used to identify bottlenecks and workflow inefficiencies in industrial factories. Complex assemblies often require many parts and are touched by many individuals. Location information enables visibility into component delays, assembly dwell times, and worker interactions.

Retail (Consumer Behavior)

Retailers want to sell more stuff. They can do that by increasing traffic to their stores, or by convincing their existing customers to purchase more. RTLS helps retailers with both. By tracking consumer behavior (in stores and out!), vendors can individualize marketing data, incentivizing visits to the store. Once there, the vendors can use customer tracking data to refine product placement, down to shelf level.

Indoor Navigation (Malls, Museums etc.)

As technology increasingly permeates our lives, GPS for navigating unfamiliar roads is no longer a novelty, but rather something we simply expect to work. For the most part today, however, those turn-by-turn directions cease as soon as we step foot indoors. Future RTLS systems will help all of us navigate complex buildings like museums, hospitals, and shopping malls with the same ease we have become accustomed to in the streets. But do not be surprised when your smartphone alerts you to let you know that that new jacket you have been wanting is on sale in the very store you are standing nearby.

Technology

Positioning Concepts

Any RTLS system seeks to answer a simple question: Given that I know the location of objects X and Y, what can I infer about the location of object Z? There are a lot of different ways to arrive at an answer, and sophisticated tools for getting more and more precise solutions. Broadly speaking, they can be separated into a handful of different techniques.

Presence Detection

Presence detectors can measure light, sound, and radio waves. When you walk into a room and the lights turn on automatically, there's a motion sensor that's detecting your presence. When you wander into a coffee shop and your phone senses the Bluetooth radio signal from a beacon located in the store, the application running on the phone informs you of the latte special of the day, while the store owner learns about your caffeine addiction.

Triangulation

Explorers of all sorts use triangulation to find themselves using visual cues, and have been for thousands of years. A wanderer in the eastern plains of Colorado might look to the West and use a compass to take bearings to Pikes Peak to the south and Longs Peak to the north. Drawing those lines on a map, the point where they intersect should precisely line up with the explorer's own location. Modern RTLS systems use the same technique, but measure radio signals instead of light. The angle-of-arrival (AOA), as perceived by three or more radio receivers, can precisely identify the location of a transmitting radio beacon.

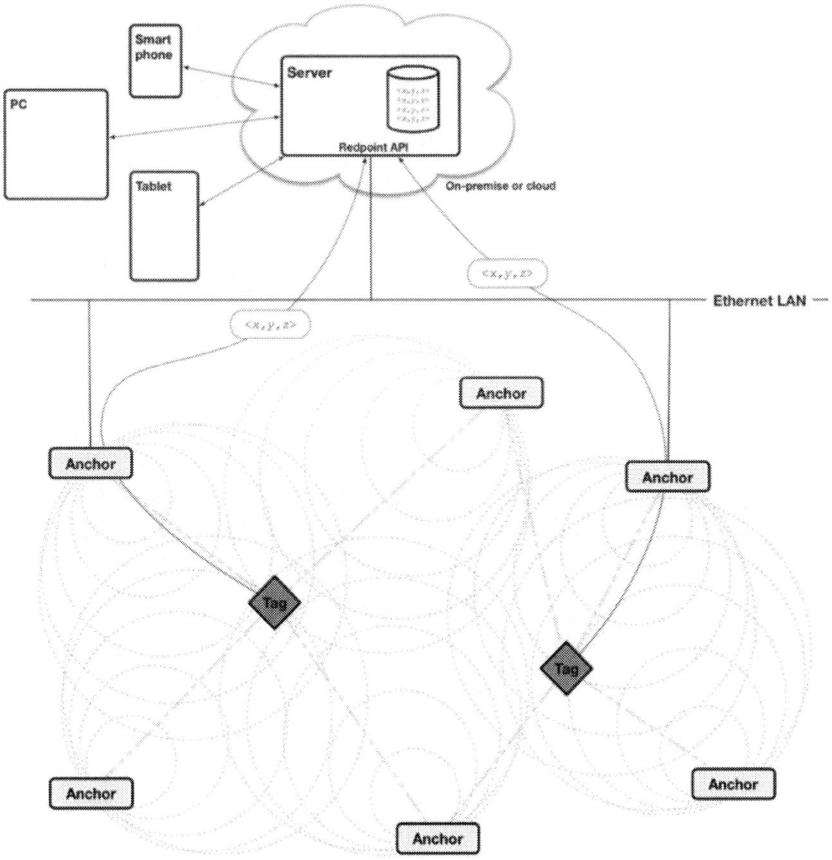

RTLS IoT network

Applying "Indoor GPS" to Industrial Construction

Safety

Overview

Real-time location monitoring of crews and equipment provides many options for improving the safety and security of everyone on the jobsite.

Static Geo-fences for Objective Hazards

Jobsites contain many areas that should be off-limits to most workers under any condition or, at the very least, require careful attention and perhaps specialized equipment before entering the area. Examples include working below a lifting operation, high-voltage electrical sites, exposed ledges, and chemical storage areas.

Traditional safety procedures require erecting barriers and signs that are frequently overlooked or ignored. These measures can be supplemented with electronic detection and alarming that alerts not only the individual at risk, but appropriate supervisors as needed.

Project managers can use tablet-based software to visualize and identify static risk across a project, mark hazardous areas on the site plan, and document a reason for the hazard. For example, consider a geo-fenced zone marked "core drilling above." A worker wearing a tracking tag who enters a geo-fenced area is immediately alerted through an audible alarm and visible strobe, together with a brief text display that reads "DANGER! Core drilling above." The worker's foreman or the project manager may receive an alert as well to expedite corrective action. In all cases, events are logged for later analysis.

Dynamic Geo-fences for a Jobsite in Motion

Dynamic geo-fences, similar to their static counterparts, generate alerts whenever a worker enters a hazardous area. In this case, the location of the hazard area is tied to that of another object. For example, it is possible to define a hazard area as being anywhere within 30 ft of a forklift or within 70 ft of a bulldozer. With struck-by incidents second only to falls as leading injury causes, dynamic geo-fencing has the potential to prevent many incidents, and even save lives.

Credentialed Access

Some locations should be accessible only to workers with appropriate authority, credentials, and skills. Unlike physical barriers and signs, indoor GPS systems can allow a worker trained in confined spaces to walk into a crawlspace, while a coworker without the same training immediately triggers an alarm upon entering the space. With 8 in. accuracy, the protected area can be as small as an electrical closet accessible only to electricians, or as large as an entire jobsite (e.g., prohibit anyone without OSHA-10 training).

Last Known Location and Mustering

During incidents that may affect several crew members (e.g., explosions, gas leaks), one key to providing workers with the best chance for survival is to know either where they are or what their last-known location was right before the incident. During an emergency, tracking solutions can provide rescue personnel with the exact locations of missing workers and, thereby, save valuable search time.

Indoor GPS eliminates the need for manual accounting during a muster event as well. Through a tablet-based visualization tool, supervisors can identify at a glance anyone not at a designated muster point, and pinpoint their location anywhere on the site.

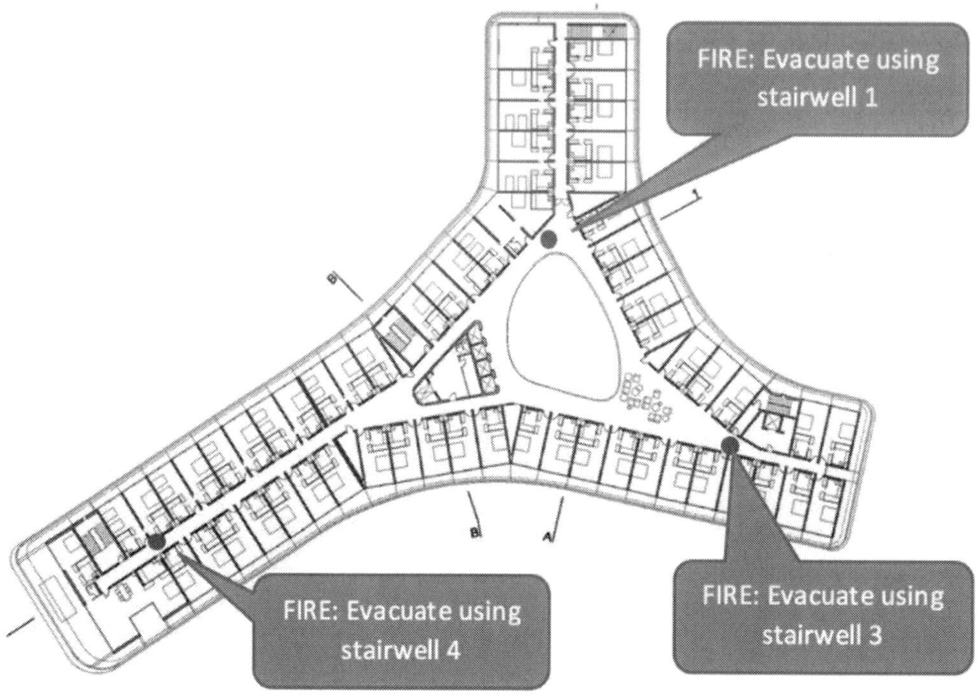

Identifying fire evacuation location on RTLS plan

Protecting the Lone Worker

Location-aware badges can also provide tracking of lone workers in an emergency, in confined spaces or during the night shift to help manage efficient emergency response. The call button on a wearable safety badge enables the worker to manually trigger an emergency help request that indicates his/her exact location. The message can be sent to the closest colleagues, based on their proximity to the incident.

The technology can also detect "man-down" events by monitoring workers' movements with an accelerometer inside the badge. If the person is not moving, an automated emergency request and location will be dispatched.

High-worker and Occupancy Areas

When workers, equipment, and materials are tracked across a site, high-traffic areas are easily identified and rerouted. High-occupancy areas shared by multiple crews become apparent, and work can be rescheduled to mitigate spatial conflict and attendant hazards. Without real-time location records, project managers lack the visibility to understand the flow of people and machines throughout a site; with the aid of indoor GPS, collisions and occupancy-related accidents can be averted before they happen.

Visibility and Interaction

Real-time vs. Delayed

The first time someone sees a precision RTLS system at work, it can be mesmerizing. A crew of painters shows up on the second floor as yellow dots moving in unison. The blue electricians are more sparse, and spread out across the floor. And then there's that lone plumber up on floor seven. What's he up to? This aquarium-style perspective into the real-time workings of the project affords supervisors unprecedented understanding. More than just moving dots, activities can be broken down and reported on a dashboard. How many workers are there on the site? How many from each trade? What's the ratio of workers to supervisors?

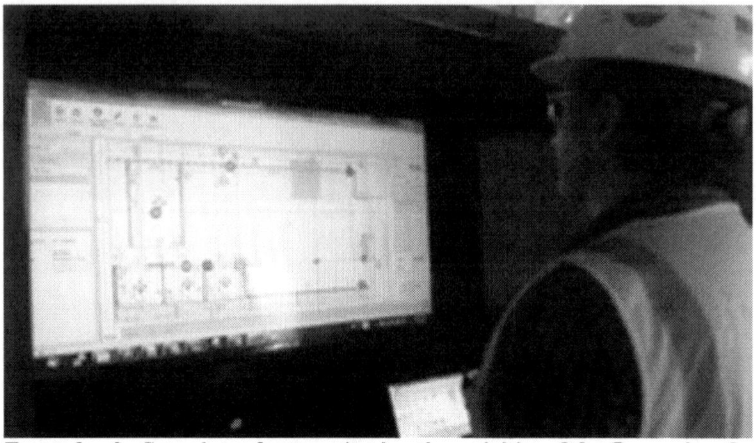

Example of a Superintendent monitoring the activities of the floor using the RTLS dashboard on the jobsite

Powerful Metrics for Continuous Improvement

Researchers have studied the link between lean construction and worker safety and shared empirical results. The benefit is derived from the continuous improvement feedback loop, where best practices are regularly reviewed and adapted to incorporate new findings. While this is a step in the right direction, current practices operate on limited information about what is really happening on a project. Jobsite injuries are readily documented, but near-hits may never be reported or even recognized.

Real-time tracking on the jobsite provides a new level of granularity and comprehensiveness to incident awareness. A worker walking dangerously below a lifting operation will be immediately flagged (and alerted) by the automated tool, while no human may ever notice that it happened, unless that action results in an incident. With indoor GPS and connected reporting tools, project managers are given powerful metrics and site-wide visibility that they can use to adapt the company and project procedures to improve safety.

Efficiency Through Safety

According to Liberty Mutual's 2016 *Workplace Safety Index*, more than $62 billion in direct workers' compensation payments were generated in 2013 for all industries, including construction, for workplace injuries and illnesses. In an earlier Liberty Mutual survey, 60% of financial officers stated that a $1 investment in injury prevention generated returns of more than $2. Other surveys indicate a savings of at least $3 for every $1 invested.

Investing in safety technology has a clear financial return on investment. This is particularly true when many of these new solutions, such as jobsite tracking technologies, also positively affect project workflows, practices, and efficiency. Let's call it "efficiency through safety."

Workflow Analysis

Traditionally, we might look at historical data and see that it took subcontractor X two weeks to install the flooring. Therefore, for the next bid, we'll estimate the same amount of time for similar space and materials. We may have similar experience with contractor Y and choose accordingly between them. But how do we shorten the installation time as it happens? Enter RTLS. If you can track the flooring crews and their materials, you might find that all of the tile is located in the NW corner of the project, while the work is being done in the atrium in the SE corner, and 20% of the worker's time is spent moving materials around, often the same materials more than once! Indoor GPS can highlight this and suggest time-saving actions that can be implemented immediately.

Context-sensitive Documentation

The age of building information modeling (BIM) has ushered in an overabundance of information, where a single model can explode into thousands of individual PDFs. As the project progresses, the sheer volume of information can quickly become overwhelming. Smart filters are needed! And none is perhaps more powerful than location. A PM walking the floors of a building can open his tablet, and with the aid of location information, can access:

- Geo-referenced punchlists
- Automated time and attendance
- Distance traveled by workers
- Site-wide search for people, materials, equipment, tools

The Human Aspect of Tracking

Privacy Concerns

Workers may question whether tracking technologies represent an invasion of their privacy. However, these concerns can usually be allayed if the deployed system is clearly designed to reduce risk and improve safety. Further, badge identification can be made anonymous to the degree desired. Nameless badges may have only a group name or a color code, per trade or subcontractor. With this method, a worker's identity is not disclosed, but his/her location, used for safety purposes, is still known. Helping a worker in an emergency does not require checking the individual's ID.

A typical RTLS wearable device can be about the size of a standard access badge

Considerations for Selecting an RTLS for Construction

Accuracy

Not all connected and wearable safety solutions are created equal. The cost, accuracy, and reliability can differ greatly based on the vendor and underlying technology. While radio-frequency identification (RFID), Wi-Fi, Bluetooth and cellular-based solutions can provide some helpful data, the biggest problem typically encountered is inaccuracy, prohibitive cost, or the high complexity of system deployment. The most accurate and affordable tracking solutions today are based on the new radio frequency standard called ultra-wideband (UWB), which can support up to 8 in. (20 cm) real-time tracking capability due to its RF characteristics. Location accuracy and reliability are highly important in enabling reliable hazard zone identification and proximity alerts.

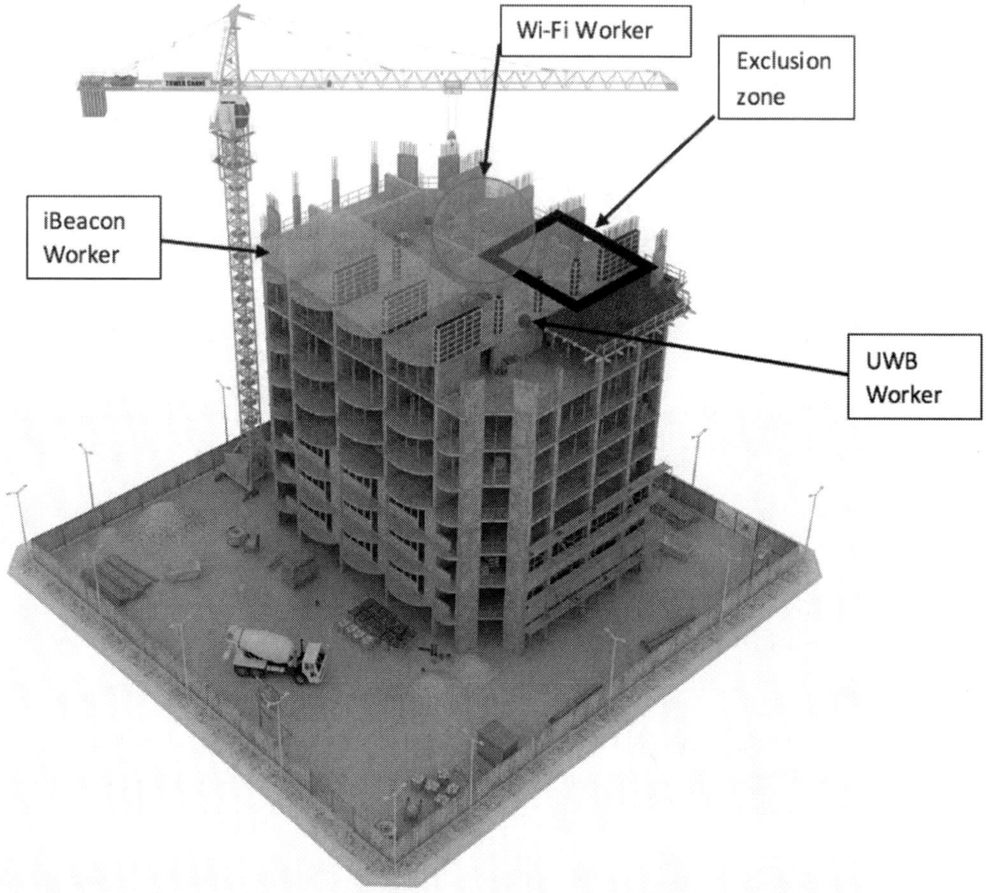

The Future

IoT networks are the physical backbone behind the current and soon-to-be-realized construction wearable market. The infrastructure and hardware are improving, and the devices worn by workers, such as smart vests, clip-on devices and tags are improving in functionality. With a mature infrastructure and instrumenting, the power and promise of wearable IoT will be in the data analytics. Being able to turn the lagging indicators (warnings, near misses, falls and accidents, etc.) into leading indicators will provide predictive and meaningful adjustments to our jobsites and work patterns that will make our projects safer and more efficient.

We can then add in artificial intelligence/machine learning as a part of the data analytic process.

Artificial intelligence
Machine-learning and artificial intelligence (AI) are poised to evolve typical design and construction processes in ways we can't even imagine. Research is being conducted to use AI to predict injuries, improve human-machine interfaces, surveying, and generative design. In a BIM environment, AI will be able identify missing elements like waterproofing or detect real clashes versus false ones. Skanska is currently taking part in research with Smartvid.io to deploy machine-learning to organize and analyze photo, video, and voice data collected from drones, smartphones, and other internet-connected devices to identify safety, productivity, QA, and efficiency improvements on the jobsite.

Daqri headset: Augmented reality. Daqri headsets, for example, are beginning to be used to project useful safety and maintenance information, supported by 3D models. The promise is to enable workers to access information and overlay construction and safety sequences in real time, while allowing free use of their hands, as the information can be accessed without the use of a keyboard.

Session No. 554

Preparedness: Are You Ready to Deal with a Crisis?

Sharon Kemerer, MSN, COHN-S/CM, FAAOHN
Corporate Director, Occupational Health & Safety
Baxter International
Deerfield, IL

Deborah R. Roy, MPH, COHN-S, CET, CSP, FAAOHN
Corporate Director, Health Safety and Wellness
L.L.Bean, Inc.
Freeport, ME

Introduction

In the past year alone, a number of significant events have occurred that have had implications and impact on global and local business. When the bombing in Brussels occurred, the airport was shut down for a significant period of time. What if you had business travelers in Brussels, or needed to ship product through that airport? Zika virus seemed to emerge from nowhere, and suddenly the fear of mosquito-borne illness took on mammoth proportions. Do you have employees and business operations in Brazil or Puerto Rico? What if business travelers to affected areas contract Zika virus? What if that business traveler is pregnant? How will you deal with this? How will you prevent or minimize exposure? Do you have the communication systems and resources in place for your company to anticipate, respond, and recover from the threats of a constantly changing landscape?

Business today must not only be able to respond to an emergency, but must also be aware of the situations that can cause business interruption and take steps to mitigate them. What does it mean to be "prepared" for all hazards and to be resilient as an enterprise? What process should be in place on an ongoing basis to decrease risk from external threats to the business? Who should be on the "team," and what is the role of the health and safety professional in the process? This paper will address the basic elements of an all-threats preparedness strategy, the steps to achieve sustainable internal systems to anticipate and respond, and the essential role of the health and safety professional. The difference between preparedness, emergency response, and business continuity will be clarified. Businesses must have plans and teams in place that can flexibly respond to whatever event occurs, and base their process on previously agreed-upon principles. Stakeholder communication, roles and responsibilities, and tools to protect employees and company assets, must be in place *before* an event is active.

This paper outlines the basic elements of a sound preparedness management system and describes how two global manufacturing companies have structured their preparedness and response organizations.

Basic Elements for a Preparedness Strategy

When asked about preparedness and business continuity, it's not unusual for a company to respond with, "No problem. I have emergency response plans at all my sites. We're covered!" But knowing how to exit the building during a fire is only a small part of what you must be able to do to protect a business and let it continue through crisis. The framework of business continuity can be depicted in this way:

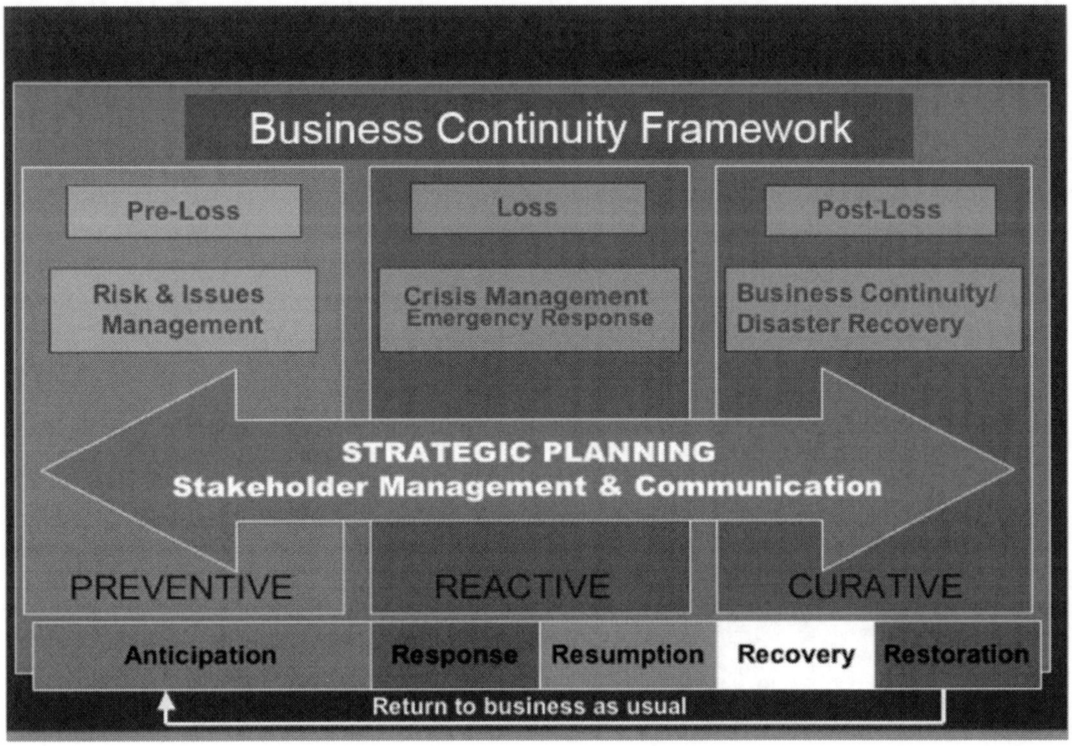

Exhibit 1. Business Continuity Framework

Preparedness focuses on anticipation and preparation for a variety of threats to a business. Threats can come in the form of political unrest, natural disasters, terrorist actions or public health issues. Likely business interruptions can vary by location, but when the basic elements of preparedness are in place, they serve to address a wide variety of situations.

Emergency response is what needs to happen during an active threat: a fire, earthquake, active shooter on campus, impending hurricane or loss of power. Clear procedures are essential, along with a recognized chain of command and communication. This aspect of preparedness is the one that most EHS professionals are familiar with and have in place.

Business recovery is what needs to take place *after* the event has occurred. What manufacturing lines must come back up to operation first? What personnel are needed for those operations? What resources are pre-arranged to support your critical recovery efforts?

For these elements to function well, basic program components must be in place.

Organizational Structure and Reporting Lines

Standing groups or committees that meet regularly and are comfortable working with each other are a key element for success. Expectations for communication flow and timeliness enable early intervention and help minimize the impact of many threats.

Defined Roles and Responsibilities

Clearly defined roles and responsibilities are critical. The time to decide who is in charge is NOT when a crisis is emerging. In deciding what is best for your organization, the RACI model can be a helpful framework. Going through this model can also help define expectations for communication.

THE "RACI" MODEL	Who	Activities & Responsibilities
Responsible Individuals who perform an activity. Responsible for action/ implementation. The degree of responsibility can vary.		
Accountable The TMT leader who is ultimately accountable for day-to-day decisions and for making recommendations to the executive who has ultimate veto power on policy and spending decisions.		
Consulted Individuals to be consulted before a final decision or action is taken in order to assess impact of the decision on various stakeholders. Two-way communication.		
Informed Individuals who need to be informed after a decision or action is taken. One-way communication.		

Exhibit 2. RACI Model.

When defining roles, it's best to take a simple, direct approach. Stating responsibilities in a bullet format provides clear direction and visibility to all team members. In any case, the role of the environment, health and safety (EHS) professional is key. With knowledge of injury potential, safety during recovery efforts, and potential for chemical and other hazards to be involved, the threat management team, at all levels of the organization, should include a strong EHS team member.

Communication Plan and Resources

Who needs to be informed when a threat becomes active? Taking the time to clearly identify key stakeholders, both internal and external, will reap enormous benefits when a crisis occurs.

Exhibit 3. Stakeholder Communication

Clear Expectations and Verification Process

Business units or sites need to have a clear sense of what needs to be in place, access to identified resources and support for questions. A layered approach to verification can provide visibility at the corporate level to what sites feel is in place (through a periodic checklist, for example), and should be coupled with some form of external verification (for example, incorporated into an internal audit).

Business Recovery Planning

Operations must have a sense of what events have the potential to cause business interruption and how they will recover if they occur. Critical products and services must be identified so thay recovery can take place in order of priority. The chain of command with alternates identified, along with identified resources for recovery/repair, is critical to secure prior to an interruption.

Protection for International and Domestic Travelers

Business travel is a frequent expectation in today's workforce, and unexpected events can occur anywhere in the world. Organizations that expect their employees to travel have an obligation to protect and support them during the process.

Notification Systems

A variety of systems are available to push communications to both travelers and groups of employees that may be impacted by an emergency. Being able to locate and communicate with traveling employees during an emergency is a key component to look for in a service.

Global Health and Security Support

It is not uncommon for employees to experience a health issue or emergency during business travel. How will they identify and access reliable sources of care? How will they be transported to their home base in the event a health issue is more serious? Services are available to support businesses from a variety of vendors and should be identified and communicated to employees well in advance of travel.

Examples from Two Global Companies

Preparedness at a Global Healthcare Manufacturing Company

At a large, global healthcare product manufacturing company, threat management teams are organized at the global/corporate, regional/business, and site levels with clear scope of responsibilities defined. Annual preparedness checklists help to verify that essential components are in place at individual sites. A variety of support tools are used to locate and communicate with employees during emergencies, support health and security needs occurring during travel, and track emerging issues on a global basis.

Exhibit 4. Example of Global Threat Management Team

Preparedness at an Outdoor Retailer with Global Operations

At an outdoor retailer with global operations but a mixture of manufacturing in the U.S., and vendors throughout the world, the business continuity team is at the corporate level, and each country with company employees or an operations area such as distribution, manufacturing, retail, contact centers, or sourcing has a business unit plan with a contact person in charge and backup person. The corporate plan includes clear responsibilities and the business unit plans flow from the corporate one. An annual table top is used to exercise the process. Each table top encompasses a different business unit and/or external threat. Lessons learned from the drills are used to upgrade systems and processes to prevent incidents whenever possible. A variety of support tools are also available, as above.

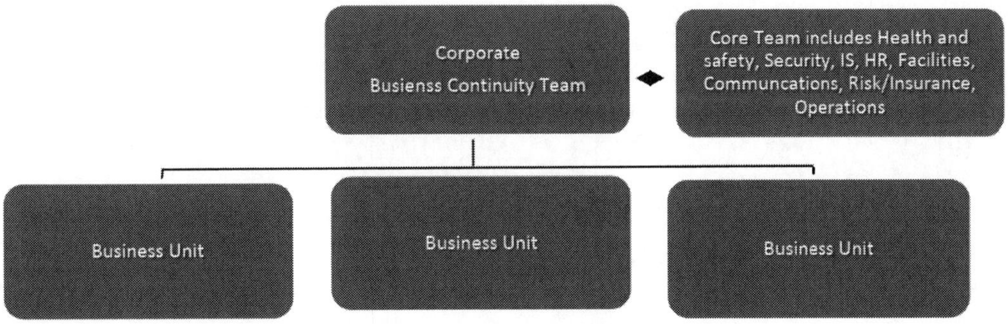

Exhibit 5. Example of Corporate Business Continuity Team

Test the System

Just as you would test to see if employees can exit the building effectively during a fire by conducting periodic fire drills, a business continuity system should be periodically tested through a simulation, table-top exercise or a drill. This can be done through the assistance of outside resources (who can provide unexpected scenarios) or by internal team members posing group challenges. A good test is one in which something fails; then learning can take place. Both following a "drill" and when threats do emerge that are handled by the team, and after-action evaluation should be conducted for maximum learning and continuous improvement.

Summary

Although many of us are quite expert at emergency response, businesses today have a need to understand external threats and how to be resilient as an enterprise. Threats may occur in another part of the world that could impact raw materials flow, travel, or the health of employees at one of your company's manufacturing facilities. Preparedness involves having communication plans, clear roles and responsibilities, and tools in place, even before an event occurs.

Bibliography

Bennett, Brian. "Effective emergency management: A closer look at the Incident Command System." *Professional Safety*. Nov. 2011.

Iske, Jr., S.D. Allen and Linda G. Lengfellner. "Fire, water & books: Disaster preparedness for academic libraries." *Professional Safety*. Oct. 2015.

Nicoll, Scott R. and Owens, Russell W. "Emergency response & business continuity: the next generation in planning." *Professional Safety*. Sept. 2013.

Roy, Deborah R. "Preparing for a pandemic: Lessons from H1N1." *Professional Safety*. June 2011.

U.S. Department of Homeland Security. *National Preparedness Guidelines*. (https://www.dhs.gov/national-preparedness-guidelines)

Session No. 557

BBS: The Good (Ice Cream), The Bad (Green Beans), and the Ugly

Bill Sims, President
The Bill Sims Company

BBS Processes and Their Potential

Behavior-based safety (BBS) methods entered their "heyday" in the 1980s and 1990s, largely driven by successful implementations in the military, petrochemical, and manufacturing industries.

Widely touted and praised, they were adopted by many firms. Notably, many firms stayed "on the sidelines," to determine whether the methods would have sustainability and staying power.

Properly done, these systems provide a new way to engage workers in safety, building on what we have learned about participative management over the last 100 years.

Behavior-based methods providing positive feedback for desired behaviors, and those which encourage worker input, will continue to provide solid injury reductions.

However, many BBS processes "miss the mark" and become a "paper chase."

Does BBS Really Work?

In a three-year study conducted by the American Society of Safety Engineers (ASSE) that was published in *Professional Safety* magazine in 2004, 300 construction firms were tracked. One half of the companies chose to implement a behavior-based, positive-reinforcement program. The other 150 companies did not implement a system.

At the end of the three-year study, the firms who chose to implement a behavior-based, positive-reinforcement system had injury rates that were *50 percent lower* than the firms that refused to try behavior-based recognition programs.

There are numerous case studies that support these same findings, proving that these programs work. But to understand why they work, we must also explore why they fail.

Why Companies Avoid BBS

- Too cumbersome & too many resources needed. After the "right sizing" that swept U.S. industry in the 1990s, many safety leaders found the new "lean" environment had eliminated the extra positions that were used to run BBS processes. They watched employee engagement in these systems wither on the vine. Some of symptoms include these:

  - Pencil whipping

  - Inability to get accurate data, and unable to "crunch the numbers" so as to make meaningful decisions.

  - **A belief that "quantity is king." One Board of Directors of a major U.S. firm mandated two observations per 15 work hours. Pencil whipping was rampant. Data quality was questionable. Lost production time required to do the observations was estimated at $10,000,000 per year.**

  - **One worker stated:**

  I work for a large firm and we have BBS from one of the top "big name" bbs consultants. I am a agreement employee who has coordinated the process For over two years. It has flat lined and we are looking to go back and motivate our teams or try. I am not convinced the Same ole materials will do the trick, looking for fresh ideas. I am afraid we have to many so called experts, within, who Don't understand the design of BBS and I am but one voice, not that I have all the answers, but I understand I know just Enough to get me into problems. Now because management has invested big $$$ and they did put up some good numbers, which was More the result of luck than the process, they want to move in and manage in hopes of a repeat. WE made all the classic mistakes, didn't get union buy in, no management involvement starting up, etc. They put the process in OUR, rank & file, hands Which was a big mistake, I am afraid we are on a crash course.

How to Fix BBS
Some ways to "fix" BBS include:

- Add R+. Ask this question about your BBS process: "Where's the R+?" If you can't easily spot it, there is a good chance your process uses R-.

- Make it fun.

- Integrate incentives with it. Many firms espouse a devotion to leading indicators, and yet their KPIs and incentives are still tied to lagging metrics. OSHA has raised the fine to in excess of $120,000 per employee denied a lagging-indicator incentive.

- Focus on finding what "could happen" and less of "what has happened."

- Ask people how to make the process safer; and then, be sure to empower people to act on it and fix the problem.

Use the Proactive Approach
Using a proactive approach will give companies more of the results they want: a work environment with a higher awareness of safety and fewer or no injuries.

Before an employee experiences an injury, he has "near misses" that involve him taking a chance or risk but somehow escaping injury. These near misses reinforce his belief that it is "ok" to take chances. Then, lightning strikes, and a recordable injury occurs.

The proactive approach rewards things employees do right and alerts them to things they are doing wrong. Though consistently successful, this proactive approach is viewed as "new school," and is inherently far more complex than the old school approach to recognition.

With the old school approach, all a manager has to do is count the safe work hours, and hand out T-shirts accordingly. Working under the old school system, the all the employee has to do to be "recognized" is show up at work, and hide injuries (or somehow escape them).

With the new school approach, managers have more to do. They must decide in advance what behaviors to reward, and do so on a daily basis, without abuse or favoritism.

You want lasting behavior change. Affecting behavior change in employees requires that you to *train* them and *reward* them through recognition and thanks.

Watch Out for Speed Bumps
Once you have made the decision to switch from an "entitlement" program to one that requires action, you can expect to hear plenty of grumbling. Middle managers will complain they have too much to do to reward employees for safe behavior. CAVE employees (Citizens Against Virtually Everything) will complain that they do not want to "fool with that stuff."

Overcoming these complaints can be as simple as rewarding middle managers and safety committee members for driving the behavior-based process. The people who do not want to "fool with safety" are exactly the people who should not receive an award. In fact, employees who are willing to put extra effort into safety should pick up a greater percentage of the rewards available. That's what behavior-based safety is about: being rewarded for doing something.

Why Middle Managers Must Be Rewarded
Supervisors and middle managers represent where the rubber meets the road. Squeezed from both ends and pulled in many competing directions, they can easily come to view quality, production, and safety as the red-headed stepchild.

Most behavior change programs fail to reward the supervisor, justifying this mistake with the attitude, "Why should I reward my supervisors? Isn't a paycheck enough?" As with any employee, middle managers who do not feel appreciated will not perform as well.

Cash Incentives Versus Recognition
Cash is not king. Findings from studies conducted by Dr. Frederick Herzberg in 1959, and confirmed more recently by Bob Nelson, indicate that employees ranked *achievement* and *recognition* as the top contributors to job satisfaction. Money ranked a statistically insignificant 5th.

Recognition will produce superior job satisfaction, while cash will only produce average satisfaction. Many studies have shown that satisfied employees perform at far higher levels than unhappy ones.

In a study conducted by the University of Waterloo, employees were assigned a task and asked what type of reward they preferred: cash/gift cards or a tangible gift item. Eight-two percent of the employees picked cash. When the results were completed however, the group that was given a gift item instead of cash *outperformed* the cash group by a margin of two to one.

Why Does Recognition Work Better?
In the 1920s, Hawthorne Labs conducted a study to survey the effects of different lighting levels on worker productivity. A group of employees from the plant who knew they were part of a study were placed in a special room to perform their normal jobs. Output was measured each day as lights were dimmed and brightened.

As the lighting was changed, productivity went up, ultimately increasing by more than 30 percent. The engineers concluded that there was no correlation between productivity levels and a particular light setting. However, the sense of recognition and achievement the workers experienced by "being a part of things" translated into better job performance.

The Hawthorne Effect explains why new initiatives that solicit input and ideas from employees often improve performance.

Common Mistakes Made by Incentive Planners

Mistake #1: Using the Word Incentive Instead of Recognition
Remember, an incentive is akin to placing a giant hoop in front of your employees and saying, "Jump through and you get the treat!" What does this really communicate to our employees? Recognition says, "I like the way did your job. You are a valuable employee and here's proof…"

Mistake #2: No Top Management Support
Trickle-down theory is hard at work in this scenario. If top management doesn't support the program, middle managers will follow suit, and the whole program will collapse like a house of cards.

Mistake #3: No Middle Manager Buy-in
Employees can spot indifference in a heartbeat. Theories of positive reinforcement tell us that rewarded behaviors are repeated. Negative reinforcement–giving people the opposite response from what they want–will extinguish a behavior. So, if top management is giving opposite

responses to the program, employee zeal will be extinguished.

Mistake #4: No Way to Measure Return on Investment
When writing an annual budget, does your company set out to make "a good profit," or do they put an exact number next to the dollar sign? They use an exact number; that's how they know if profits are on target from one quarter to the next. The same is true when designing an employee recognition program. Establish measurable objectives that will clearly indicate behavior changes: the return on your investment. For example, "Decrease lost time injuries by 50% during the first 12 months of the recognition program."

Mistake #5: Under-funding the Program
In the business world, you reap what you sow. Sow the bare minimum in employee recognition, and you'll reap bare minimum results in changed employee behaviors, if you're lucky.

Tax and Legal Consequences of Cash and Gift Cards

Many well-intentioned employers have rewarded their employees with gift cards and prizes, not realizing the full ramifications of the tax issues they are placing on their most-valued employees (not to mention their own payroll departments). Under various IRS codes, "same as cash" gifts are subject to federal and state income taxes.

Tower Automotive in Traverse City, MI, thought they were giving a grand reward to their hourly employees in the form of $15 gift cards. Imagine the employees' surprise (and that's putting it mildly) when $5.51 was withheld from their next paycheck for federal and state income taxes due on the gift.

According to IRS Code section 62(a)(1), the term "wages" for FICA purposes is defined as all remuneration for employment, including the cash value of all remuneration (including benefits) paid in any medium other than cash (i.e., gift cards and prizes.)

But is there a parachute? According to IRS Code section 132(a)(4), gross income does not include any fringe benefit that qualifies as a "*de minimis*" fringe benefit. The IRS goes on to define a "*de minimis*" fringe benefit as any property or service with a value so small as to make accounting for it unreasonable or administratively impracticable. How small is so small? Better ask your corporate accountants. We all know what happens when you plead ignorance to the IRS.

Here's a great example of "worst-case scenario" that can be applied to employee incentive programs. Oprah Winfrey's studio audience went wild when she gave away 276 brand-spanking-new Pontiac G-6 sedans on one of her shows in September 2004. However, the recipients soon learned they had to pay taxes on the winnings–a prize with a sticker value of $28,500! The so-called winners had three choices: Keep the lovely car, and pay $7,000 in taxes. Sell the lovely car and pay $7,000 in taxes. Or forfeit the lovely car.

As if tax troubles aren't bad enough, enter the lawyers! Ralph's Grocery Stores in California designed an unpopular bonus program for store managers. When calculating the bonus, Ralph's deducted general business losses–like store theft and damaged merchandise–from the final figure. Feeling that employees should not have to shoulder these "costs of doing business," one former store manager believed employees should not have to shoulder these "costs of doing business." This employee filed a class action lawsuit against Ralph's … and won.

The state court found that Ralph's "bonuses" actually qualified as wages, as defined by the California Labor Code. The store managers did not qualify as exempt employees, as Ralph's maintained. Thus, Ralph's could not lawfully deduct costs of doing business from the employees' wages. The jury found in favor of the employees in the class action, and Ralph's was slapped with a hefty settlement.

The lesson is clear, when it comes to rewards and recognition: What you don't know can hurt you!

Conclusion

After all is said and done, the evidence is clear that recognizing people for upstream behaviors is a highly effective behavior change tool, one that every successful manager must master.

Seventy-three percent of all employees report that they are not thanked or praised enough. In work environments where employees are thanked and praised regularly for positive behaviors, their need for support and recognition is being met. These environments don't just "happen;" they are created, with the help of a well-planned recognition program.

Neanderthal companies that cling to the misconception that, "a paycheck is recognition enough" have created a culture where employees will hide injuries to be a "hero" or "help" the team.

Finally, regardless of where you stand on the issue of injury hiding, I would caution all of us to accept the humbling truth that nobody is the master of human behavior. There are no easy answers or quick fixes. Anyone who supposes otherwise will always find himself proven wrong.

Acknowledgements

Thank you to Dr. Jitu C. Patel, CPEA, for reviewing and preparing this paper, and for his valued comments and counsel. Sincere gratitude is also extended to various ASSE Chapters globally for inviting the Bill Sims Company, USA, to present the keynote address.

References

Goodrum, Paul M. and Manish Gangwar. 2004. "Safety Incentives: A Study of Their Effectiveness in Construction." (http://www.asse.org/professional_safety_feature_index.htm).

"The Hawthorne Experiments: Management Takes a New Direction." http://www.mgmtguru.com/mgt301/301_Lecture1Page10.htm

Herzberg, Frederick and Bernard Mausner. 1959. *The Motivation to Work.* New York: Wiley Press.

Jeffrey, Scott, Ph.D. "Right Answer, Wrong Questions." (http://www.salesforcexp.com).

Ralphs Grocery Company, Petitioner, v. The Superior Court of Los Angeles County, Respondent; David Swanson, Real Party in Interest. Oct. 23, 2003. Court of Appeal of California, Second Appellate District, Division Seven. 112 cal. App. 4th 1090; 2003 Cal. App. LEXIS 1592.

Sims, Bill Jr. "Effective Motivation." (http://www.billsims.com/oshmag2.php)

"Winners of Oprah Car Give-Away Must Pay Up to $7,000 Tax." Sept. 22, 2004. http://money.cnn.com/2004/09/22news/newsmakers/oprah_car_tax/index.htm?cnn=yes

"You Call This a Bonus?" The Associated Press.

Respiratory Protection Programs
Let the Standard Work for You

Dennis A. Terpin, PhD, O.H.S.T.
ANSI/ASSE Z88. Respiratory Protection and
Fit-Testing Committee Member
Senior Industrial Hygienist
University of Illinois at Chicago
Chicago, Illinois

Introduction

In the control of occupational diseases caused by breathing air contaminated with harmful dusts, fogs, fumes, mists, gases, smokes, sprays, or vapors, the primary objective shall be to prevent atmospheric contamination. This shall be accomplished as far as feasible by accepted engineering control measures (for example, enclosure or confinement of the operation, general and local ventilation, and substitution of less toxic materials). When effective engineering controls are not feasible, or while they are being instituted, appropriate respirators shall be used pursuant to this section.

Respiratory Protection

Respirators are designed to **protect** individuals from harmful airborne contaminants or low oxygen levels.

Respiratory Protection Programs

Respiratory protection programs establish acceptable practices for respirator use, responsibilities, and provides guidance on medical surveillance, proper selection, fit testing, use and care of respirators in accordance with the OSHA regulatory requirements.

Permissible Practice

OSHA has established hierarchy of controls requiring the use of feasible engineering controls as the primary means to control air contaminants. Respirators are required when "effective engineering controls are not feasible, or while they are being instituted." Paragraph (a)(2) requires employers to provide employees with respirators that are "applicable and suitable" for the purpose intended "when such equipment is necessary to protect the health of the employee."

Qualified Program Administrator

- Maintaining the respiratory protection program
- Evaluating for potential workplace respiratory hazards
- Assessing worker exposure for these hazards
- Selecting the appropriate respirators
- Medical evaluations are conducted PRIOR to fit testing
- Respirators are appropriately cleaned, disinfected, stored, inspected, repaired, discarded, and maintained
- Proper air quality air is supplied if supplied air respirators are used
- Respirator users are trained in respiratory hazards, and the proper use and maintenance of respirators
- Annual evaluation of the respiratory protection program implementation
- Workers who voluntarily wear comply with Appendix D of the standard

Selection of Respirators

We must select a respirator certified by the National Institute for Occupational Safety and Health (NIOSH), which must be used in compliance with the conditions of its certification.
We must identify and evaluate the respiratory hazards in the workplace, including a reasonable estimate of employee exposures, and identification of the contaminant's chemical state and physical form.

Selection of Respirators

- Loose-fitting
- Tight-fitting
- Air-purifying respirator types
- Atmosphere-supplying respirator

Loose-fitting respirators do not depend on a tight seal with the face to provide protection. Typically, it is a hood draped over the head and shoulders, used in a PAPR or supplied-air configuration.

Tight-fitting respirators require a tight seal between the respirator and the wearer's face in order to work properly. If the respirator's seal leaks, contaminated air will be pulled into the facepiece, and can be breathed in.

Air-purifying respirators have filters, cartridges, canisters, or a combination of filters and cartridges/canisters that remove contaminants from the air by passing the ambient air through the air-purifying element before it reaches the user.

Supplying respirator supplies clean breathing air from another source.

Medical Evaluation

- Must provide a medical evaluation to determine employee's ability to use a respirator.
- Must identify a physician or other licensed health care professional (PLHCP) to perform medical evaluations.
- Must obtain a written recommendation regarding the employee's ability to use the respirator from the PLHCP.
- Additional medical evaluations are required under certain circumstances.

Fit Testing

All employees using a negative or positive pressure, tight-fitting facepiece respirator must pass an appropriate qualitative fit test (QLFT) or quantitative fit test (QNFT).

Fit Testing

- Qualitative fit test (QLFT)
- Quantitative fit test (QNFT).

Qualitative fit test procedures rely on the subjective sensory (taste, irritation, smell) response of the respirator wearer to a particular test agent.

Quantitative fit test (QNFT) measures real-time fit while the user simultaneously performs a series of moving, breathing, and talking exercises designed to simulate the same movements made in the field.

Use of Respirators

Personal protective equipment shall be worn in such a manner that does not interfere with the seal of the facepiece to the face of the user.

Maintenance and Care of Respirators

Appendix B-2 to §1910.134: Respirator Cleaning Procedures (Mandatory), Employees must clean and disinfect respirators using the procedures in Appendix B-2, or equally effective manufacturer's procedures.

Breathing Air Quality and Use

Breathing air for atmosphere-supplying respirators must be of high purity, meet quality levels for content, and not exceed certain contaminant levels and moisture requirements.

Program Evaluation

- The type and extent of hazards in your workplace.
- The types of respirators used by your employees.
- The number of your employees who use respirators.
- The amount of experience your respirator-wearing employees have in using respirators.

Recordkeeping

- Records of medical evaluations must be retained and made available per 29 CFR 1910.1020.
- A record of fit tests must be established and retained until the next fit test.
- A written copy of the current program must be retained.

Bibliography

Occupational Safety and Health Administration. 1998. 29 CFR 1910.134, *Respiratory Protection.*

Your Most Costly Workers' Comp Claim?
Hint: It's Auto Accidents – Avoid Mod Abuse!

Gerard P. Condello, BS, ARM, ALCM
Nationwide Insurance
New York, NY

Introduction

Over the course of my more than 30-year career in the health and safety profession, I have seen often dramatic changes in workplace function and exposures, perhaps none more so than the rise of the Internet and the connected world. This innovation led to dramatic changes in the way that business is conducted. Gone was the familiar row after row of work desks in the centralized office workplace. Today, telecommuting employees have become increasingly more common. According to a recent Gallup Poll, in 1995, only 9% of the workforce telecommuted; today, 37% of workers' say they telecommute, a more than 400% increase. Among college graduates, 55% say they telecommute. This change brought new exposures, and elevated existing exposures to new levels.

Despite this workplace change, many businesses still view workers' compensation and business auto coverages as separate, stand-alone coverages from a risk management prospective. Fleet safety programming is often delegated to a fleet manager in larger organizations or as an additional duty to another management level employee in smaller companies. Workers' compensation, on the other hand, is usually handled by a human resources department or a benefits manager.

Few organizations, except the savviest, grasp the powerful interplay these two seemingly unrelated coverages have on a risk management and insurance program. Fewer still have a cohesive risk-management approach to address this dynamic. This paper will discuss the altered insurance and workplace safety landscape for businesses, and offer some insight into practical fleet safety solutions

The Scope of the Issue

Motor Vehicle Crashes and the Mobile Workforce

Motor vehicle crashes are a leading cause of death in the U.S. (CDC 2015). More than 2.5 million drivers and passengers were treated in emergency departments as the result of being injured in motor vehicle crashes in 2012 (CDC 2015). The economic impact is also notable: In 2012, a one-year period, the cost of medical care and productivity losses associated with injuries from motor vehicle crashes exceeded $80 billion (CDC 2015). But the cost actually extends well beyond the monetary impact; there are the hidden costs, such as post-traumatic stress syndrome, physical scars, and the emotional scars of families that have lost loved ones.

Employees operating outside of the traditional office increased. Between 2005 and 2012, telecommuting grew by 79%, and now makes up 2.6 percent of the American work force, or 3.2 million workers (Tugend 2014). This trend seems to be accelerating as companies look to save on real estate costs. Real estate costs are reported to be an organization's second largest expense item after salaries and benefits. By having employees telecommute, an organization has access to a larger available talent pool, and also produces benefits in the form of cost savings from the reduced real estate footprint.

While telecommuting was enabled by the arrival of the Internet, the new mobile workforce that often operates from vehicles was enabled by a second technological innovation – the cell phone. The arrival of the cell phone as a business tool has been both a blessing and a bane. It has allowed for great enhancements in productivity, calls between locations, text and email messages, and GPS navigation. This, however, came with a significant cost in property damage and physical injury for the employee and society.

The National Safety Council (NSC) reports, in the 2015 edition of *Injury Facts*, that 26% of all crashes involved the use of a cell phone (CDC 2015). The issue of distracted driving has had a chilling impact on the annual number of people killed in automotive crashes. In 2015, there was a 7.2% increase in the number of deaths (35,092) over 2014. A large year-to-year increase such as this has not been seen since 1966 (NPR 2016). Distracted driving was cited by the National Highway Traffic Safety Administration (NHTSA) in approximately 10% of the crashes that occurred in 2015, and is responsible, at least in part, for the deaths of 3509 people (NPR 2016).

According to a 2012 National Council on Compensation Insurance (NCCI) study, traffic accidents are the leading cause of high-severity workers' compensation injuries. Almost 6% of motor vehicle events are multiple-claim events, compared to 2.5% on average for all causes. More than 12% of motor vehicle claims are from multiple-claim events versus 6% for all causes. In the study where the primary medical diagnosis was cervicalgia, the average cost per claim was $20,148 (Restrepo & Shuford 2012).

Some facts to consider on the relationship between auto crashes and workers' comp claims:

- More likely to result in multiple claims
- More severe than average workers' comp (WC) claims
- Impacts a diverse range of occupations
- 33% longer duration
- Three times more likely to involve an attorney

Fleet Safety

Practical Resources and Solutions

To be effective, a fleet safety program must be comprehensive, become part of the corporate culture, and be reinforced frequently with company drivers:

- Audit your existing fleet safety program against a national consensus standard, such as ANSI Z15.1-2012, *Safe Practices for Motor Vehicle Operations,* or The NETS (Network of Employers for Traffic Safety) *Comprehensive Guide to Road Safety*, as both are excellent resources.
- Talk to managers and employees to discover any company policy disconnects where company policy does not match company practices. This will help identify policies that need to be modified or clarified.
- Put an expiration date on your fleet safety program, and plan to update it every three years so it will properly address any exposure changes presented by the changes in technology, the company, or regulatory agencies.
- Communicate regularly about your fleet safety program. An interactive dialog will reinforce the safety culture, and help to make sure drivers understand not only what the policy is but why the policy exists as well.
- Conduct driver training on hazards that could be encountered, discuss actual losses, what lessons were learned, and what to do/not do in the event of a crash.
- Consider making changes to your vehicle purchasing or leasing program to incorporate technological advances such as anti-collision systems. According to an Insurance Institute for Highway Safety (IIHS) study, systems with automatic braking reduce rear-end crashes by about 40 percent on average, while forward collision warning alone cuts them by 23 percent. The study also showed that autobrake reduces injuries. The rate of rear-end crashes with injuries decreased by 42 percent with forward-collision warning with autobrake and 47 percent with City Safety (Volvo system) (IIHS 2016).
- Have a specific policy banning texting while driving.

Summary

Safety professionals understand that business auto crashes affect their organizations, and that the costs can be significant. However, recognizing the crossover impact on their workers' compensation programs, and taking into account the return on investment potential from the combined effect of fleet safety enhancements on both the business auto and workers' compensation insurance programs can be significant. Understanding this, and applying practical adjustments to the vehicle safety and vehicle purchase programs, will maximize the success potential of their safety program overall.

Bibliography

Centers for Disease Control and Prevention (CDC). 29 Dec. 2015."Cost Data and Prevention Policies." (Accessed March 10, 2017) (www.cdc.gov.motorvehiclesafety/costs.)

Insurance Institute for Highway Safety, Highway Loss Data Institute (IIHS/ILDS). January 28, 2016. "Crashes Avoided: Front crash prevention slashes police-reported rear-end crashes." (Accessed March 10, 2017) (www.iihs.org/iihs/news/desktopnews/crashes-avoided-front-crash-prevention-slashes-police-reported-rear-end-crashes.).

Gallup, Inc. "In U.S., Telecommuting for Work Climbs to 37%." 19 Aug. 2015. (Accessed March 10, 2017) (www.gallup.com/poll/184649/telecommuting&g_medium=searching&g_campaign=tiles).

Gonzales, Richard. August 29, 2016. "Traffic Deaths in 2015 Climb By Largest Increase In Decades." *NPR*. (Accessed March 10, 2017).

National Safety Council (NSC). 2015. *Injury facts*. Itasca, IL: National Safety Council. (Accessed March 10, 2017)(www.nsc.org /Membership%20Site%20Facts/NSC_InjuryFacts2015.pdf).

Restrepo, Tanya, and Harry Shuford. 2012. *The Role of Traffic Accidents in Workers Compensation - An Update*. (Accessed March 10, 2017) (www.ncci.com/Articles/Pages/II_Role-of-Accidents-in-Workers-Compensation–An-Update.aspx).

Tugend, Alina. March 7, 2014. "It's Unclearly Defined, but Telecommuting Is Fast on the Rise." *The New York Times*. (Accessed March 10, 2017) (www.nytimes.com/2014/03/08/your-money/when-working-in-your-pajamas-is-more-productive.htm?_r=0).

Session No. 563

Setting Frontline Supervisors Up for Success in Safety

Judy Agnew, Ph.D.
Sr. Vice President
Safety Solutions
Aubrey Daniels International
Atlanta, GA

Introduction

A safe work place takes a coordinated effort on the part of all employees. Senior leaders establish safety-oriented vision and values, middle managers create safety management systems, safety professionals develop procedures, and frontline employees complete the work as safely as possible. Frontline supervisors, however, have perhaps the most crucial role. Frontline supervision is where safety comes together: where the vision and values are executed, where the procedures are complied with, where decisions are implemented, and where safety can be seen. It is the moment-by-moment decisions and actions of the frontline supervisor that ensure the workplace is safe, and the work is done safely, or not.

Improved safety performance requires organizations do a better job of setting frontline supervisors up for success. Too often, supervisors are told that safety is of primary importance, but their role in making that happen is poorly defined, their priorities are competing, and they aren't trained in the skills needed to be successful.

Engagement Is Key

It's instructive to explore how supervisors view their role in safety. Too often, frontline supervisors see their job as primarily ensuring compliance with safety rules and regulations. But compliance only goes so far. To create the safest possible work environment, organizations need more than just compliance. They need workers to do the right thing all the time, even when supervisors aren't watching. They need engaged employees. So how can frontline supervisors go beyond compliance and create an environment where the workforce is truly engaged in safety?

The first step is to define employee engagement. While there are many definitions of engagement, it is helpful to define engagement from a behavioral perspective. By focusing on what engaged employees do, it is easier for supervisors to understand their role in encouraging engagement. Below are examples of the behaviors that are characteristic of engaged workers (and exemplary safety cultures).

Participate in Safety Discussions

Engaged employees actively participate in all discussions about safety and share their perspective. No one knows the challenges of performing the work safely better than the people who perform the work. Therefore, it is important that frontline employees talk about what gets in their way. Active participation from the front line helps to avoid the creation of "solutions" that won't work because they don't get at the true root causes.

Plan for and Anticipate Hazards

Engaged employees are constantly scanning for hazards, thinking through their work thoroughly, anticipating what might go wrong and planning carefully how to complete the work safely. Engaged employees go beyond basic requirements of pre-job hazard assessments and do continuous assessments. When all eyes, ears, and brains are engaged, more hazards are identified, and risk is minimized.

Actively Work to Keep Peers Safe

Engaged employees watch out for each other. It is clear that frequent observation and feedback are an essential part of minimizing incidents. Realistically, supervisors don't have the time to do enough observations and provide enough feedback to ensure consistent safe behavior. When frontline employees actively participate in peer feedback, safe habits can be built and sustained.

Report Hazards

Engaged employees report hazards; not just the physical hazards (e.g., an old piece of equipment that shows signs of failure) but behavioral and procedural hazards as well (e.g., decisions that have resulted in unanticipated risks, or procedure changes that are causing dangerous shortcuts). Hearing about hazards and problems before they become incidents allows for truly preventative safety management.

Report Near Misses

Engaged employees report near misses. Near misses provide rare opportunities to learn about what is not working as planned. Truly engaged employees go beyond "softball" events, like close encounters with a wasps' nest and paper cuts. Any event that could have resulted in injury, illness or damage is seen as a learning opportunity and is reported.

Challenge Decisions

Engaged employees will challenge decisions or directions when they have safety concerns. A new procedure, a new piece of equipment, a new vendor may seem like a good idea to improve products, reduce waste, protect the environment, etc., but those making such decisions sometimes don't or can't thoroughly assess the impact on safety. When those who do the work are willing to challenge decisions, minor and major safety issues can be avoided.

Understanding what true engagement looks like is an important first step, but engagement is not possible with a compliance focus, the typical way in which supervisors view their role. All of the engagement behaviors listed above are discretionary; that is, they are not requirements of the job, but they make all the difference in creating the optimal safety environment. Getting discretionary effort takes a different approach to managing than most frontline supervisors use. What follows are seven steps to help supervisors build engagement and be more successful in all their safety leadership efforts.

Seven Steps to Improved Supervisor Safety Leadership

Here are seven supervisory practices that will foster engagement and enhance safety performance:

1. *Build relationships.* Relationships are important in all aspects of life and business, and they are particularly important in safety. It's no coincidence that supervisors who have strong relationships with their crews tend to have safer crews. So how are relationships built?

    a. The first step is to treat direct reports like people, not just employees. Supervisors must demonstrate they truly care about their direct reports, and in particular about their health and safety.

    b. The second step is to ask more, and tell less. Supervisors too often believe that because they are the "boss" they are supposed to have all the answers. By *asking* more than *telling*, leaders learn more about direct reports, leave them feeling valued and respected, which leads to more optimal safety solutions.

    c. A third key to relationships is building trust. Trust is important in safety because it is a key ingredient for engagement. Engagement is risky business for frontline employees. Giving constructive feedback to a peer, stopping a job because it isn't safe, challenging a decision; these are risky. Not the kind of risk that might lead to injuries; the kind that might lead to other negative consequences, such as getting in trouble, flak from peers, getting a reputation as being difficult, not getting promoted, etc. Most of us take risks only when we trust the person we are taking a risk with or for. We must trust that they understand our good intentions; we must trust that they won't react negatively; we must trust they won't use our words or actions against us. Building trust is not easy but the recipe for trust is simple. We trust people who do what they say they will do. Supervisors must be vigilant about doing what they say they will do, no matter how small. Getting back to a direct report with an answer to a question, acquiring a new tool the team requested, or fixing the coffee machine; each one is important. Every time a supervisor does what s/he says s/he will do, trust is built. Every time a supervisor fails to do what s/he says s/he will do, trust is eroded.

2. *Relentlessly address hazards.* It's impossible to have the safest work environment if hazards don't get reported and remediated. Hazard remediation relies on frontline employees to identify and report hazards, and on management to fix them. The frontline supervisor is often the point person, the person to whom the hazards are reported. Thus, a critical role of frontline supervisors is to shepherd hazards through whatever organizational systems are in place to remediate them. While it isn't always the role of the frontline supervisor to do this, it behooves them to make it their responsibility. Why? Because frontline employees gauge how truly important safety is in an organization by management's willingness to invest in keeping the workplace as free of hazards as possible. Hazard remediation is the litmus test of management support for safety. Since frontline supervisors are the face of management (and often the person to whom the hazard was reported), it is essential they be proactive around hazard remediation.

3. *Build critical safe habits*. Frontline supervisors have opportunities every day to observe safe and at-risk behavior, and provide feedback that helps build safe habits. Too often, that feedback is focused on less critical behaviors. Unless they deliberately plan otherwise, supervisors are likely to provide feedback on what they happen to see when in a work area, most often PPE, housekeeping, and other easy-to-observe safety behaviors. Given that supervisors have limited time and opportunity to give feedback, it is important to ask if those are the behaviors that will protect the workforce from the most frequent and the most catastrophic injuries. More deliberate and effective feedback starts with using data to target behaviors that prevent both high-frequency incidents (e.g., slips, trips, and falls) and less likely catastrophic events (e.g., explosions, chemical releases). Taking the time to ensure that safe habit development is focused on the most important behaviors ensures the best impact of supervisor efforts.

4. *Conduct daily safety interactions*. Nothing is more important than interacting with those doing the work on a daily basis. Such interactions provide opportunities to ask questions, to learn about hazards, to address concerns, and importantly, to influence behavior. Unfortunately, too often supervisors only initiate safety interactions when there are problems. This sets supervisors up to have largely negative interactions. While these are a necessary part of the job, supervisors who are most successful initiate safety interactions when things are going well just as often (if not more often). When supervisors purposely recognize the safe behaviors they want more of, they strengthen those safe behaviors, make them more consistent, and at the same time build relationships, trust, morale, and engagement. But it isn't easy. It requires fighting the temptation to manage only exceptions (the default approach for most supervisors), which leads to the use of more negative consequences. While it can be challenging, the positive side effects and direct impact on safety are significant. Below are some tips for making safety interactions positive and effective.

    a. <u>Ask More Than Tell</u>. As noted above, asking more than telling enables supervisors to truly learn what obstacles are in the way of safety, understand what direct reports are concerned with, and build trust and better relationships.
    b. <u>Be Specific.</u> Vague statements like "good job" or questions like "how's it going" don't lead to improvement. Asking specific questions and pinpointing the specific safe behaviors ensures meaningful conversations that lead to improvement.
    c. <u>Be Sincere</u>. Everyone can see through disingenuous attempts at reinforcement, and superficial interest in safety, so sincerity is essential.
    d. <u>Interact Frequently</u>. The more frequently supervisors interact with direct reports, the more direct reports believe safety is a priority and the more trust is built. Furthermore, being out in the workplace frequently enables supervisors to catch people doing things that promote safety and provide the necessary positive reinforcement.

5. *Respond positively to reporting*. While hearing about incidents, accidents and near misses isn't a positive experience, shooting the messenger only reduces reporting and leaves supervisors with a false sense of how safe the workplace is. Undesirable safety events provide insight into how safety is really working, as opposed to how we think it is working. They uncover weaknesses in safety systems and processes that, in turn, enable changes to be made to prevent future incidents. But most supervisors (and managers)

inadvertently discourage reporting of minor incidents and near misses by how they react. Signs of frustration, disbelief or anger are only the start. Reporting often leads to unpleasant paperwork, investigations that feel like inquisitions, and sometimes discipline for those who report. While these situations can be complicated, supervisory reaction to the reporting of such events should be some form of, "Thanks for telling me, let's figure out how to prevent this from happening again." To be clear, this isn't about positive reinforcement for incidents, near misses or unsafe actions; it is positive reinforcement for the honest reporting of those events. Deliberately encouraging reporting is an essential skill for frontline supervisors, but a difficult one. It also requires support from above because, if supervisors experience negative consequences when such safety events are reported, that response will trickle down through the organization.

6. *Focus on prevention.* While supervisors are often held accountable for lagging safety indicators, it is most productive to hold the front line accountable for safety-related behaviors that prevent incidents and accidents. Telling workers to "not get hurt" has proven ineffective time and again. Supervisors need to actively manage behaviors that prevent incidents, not just react when incidents occur. But reactive safety management is hard to resist because it is often how safety has been managed, how supervisors are managed, and it is easier than proactive management, which requires daily effort. When supervisors change their focus to preventative behaviors and change their strategy to purposely recognizing the safe behaviors they want more of, they will strengthen those safe behaviors, make them more consistent, and at the same time build relationships, trust, morale, and discretionary effort.
7. *Consider safety in every decision.* Organizations are interconnected systems. A change in one area inevitably has impact in other areas, often unanticipated. While it is impossible to anticipate all the ways a decision will influence safety, it is important to try. Senior leaders are always concerned about safety; however, they are responsible for keeping their organizations profitable. It is easy to see how well-intended leaders make decisions that are good for the business but end up having negative implications for safety. Frontline supervisor decisions likely have smaller effects than those of senior leaders whose decisions impact large parts of the organization, but there is clearly an impact. Rather than influencing safety through organizational systems, supervisor decisions are likely to have a more direct impact on conditions and on the day-to-day safe and at-risk behavior of their team. Examples of such decisions include which employees to pair up for a job, when to order replacement parts, whether to work the crew overtime, how many people to assign to a job, how often to check in during a task, and time estimates given to management regarding completion of work. While most supervisors believe they already do consider safety in every decision, doing it well requires a good understanding of human behavior, and thus the ability to predict how decisions will affect behavior at the front line.

In Summary

Frontline supervisors have challenging jobs. They have pressures from above, pressures from below, and they are ultimately the most responsible for the success of safety efforts. Developing truly effective frontline supervisors is not a simple task; however, it can begin with some simple changes. The tips above can help build engagement, which makes the job of frontline supervision much easier. But supervisors often need skills training and ongoing coaching to be maximally effective. If becoming an impactful safety leader was as simple as reading a book or attending a seminar, there would be many more effective supervisors. What is needed is better coaching for supervisors to help them improve their skills in real time, in the midst of all the pressures of daily work. Ideally, managers would provide that coaching, but most organizations fail to provide managers the time required to do this coaching. If organizations want to improve safety, making an investment in ongoing supervisor development is an important next step. There is no job more difficult nor one more important to safety.

NOTE: Portions of this article are reprinted with permission from *A Supervisor's Guide to Safety Leadership* by Judy Agnew, and with permission of Performance Management Publications (PMP).

Session No. 564

Strategies for Preventing Fatal and Life-changing Injury Events

Frank J. Baker, CSP, ARM, CFPS, ALCM
Eastern Alliance Insurance Group
Carmel, IN

Gregory A. Clone, CSP
Eastern Alliance Insurance Group
Carmel, IN

Introduction

Preventing fatal and life-changing injury events moving forward will require a different approach than what the safety profession has been doing for most of the last 80 years. Heinrich's safety pyramid approach could only take us so far with its frequency-of-incidents approach. To get beyond that, we must look deeper into an organization's climate (or culture) to address the precursor events that lead to those situations that have the potential to result in a fatal or life-changing injury. Eastern Alliance Insurance Group, a worker's compensation insurance carrier, has developed an approach to address the critical cultural elements that must be in place to help an organization identify and address their risk factors, and implement proper controls to reduce the likelihood of fatal and life-changing injuries.

Our approach to preventing fatal and life-changing injury events is called N3L3®, an acronym we use to identify the "culture" that needs to be in place to impact the potential for bad outcomes. N3L3 is derived from "*the **Next Three** Seconds – Protects Your **Life**, Your **Loved** Ones and Your **Livelihood***"®. It combines elements like: changing the perception of an employer as to what risk is; engaging them to be actively involved in identification of their potential exposures and managing them through various deeply ingrained practices that are intended to ensure every worker is aware of what needs to be done to accomplish each task safely; and empowering them without fear of reprisal to speak up when they have concerns.

The N3L3 acronym was chosen, because it would be easy for workers to remember. N3L3 is intended to define a way of thinking and a mindset to help protect workers both at and away from work. Getting workers to think through a task and consider the potential consequences before they act, because in addition to protecting their ***Life**, their **Loved** Ones and their **Livelihood***; *the Next **Three** Seconds, could be your Last **Three** seconds* ®.

Prevention of fatal and life-changing injuries requires changing how we think about risk: We must understand there is risk in everything we do; it can be negative, positive, minimal or great, but it is always there. Several key elements are incorporated into the N3L3 culture, including: understanding that the past does not always accurately predict the future when dealing with those events that seem to occur at random; conscious vs. intuitive thinking; actively assessing risks; pre-work planning; stop-work authority; and root cause analysis by teams. Each of these elements must be in place to support the culture and ensure that those high-risk activities (precursors) are identified in the workplace and are treated appropriately when the situations are encountered. The maturity of the safety culture is measured for each of those precursors based on the organization's knowledge of its potential risks, competency of the training for its workforce, resources dedicated to managing its exposures, auditing to ensure adherence to policies, and feedback systems to close the loop on corrective measures and integrate into the day to day operations.

Why the Focus on Fatal and Life-changing Injuries?

While the Bureau of Labor Statistics (BLS) separates occupational injuries and illnesses as to whether or not they are fatal, we place the same level of importance on those injuries severe enough to be "life changing" for the injured worker, and as such treat them both as catastrophic in nature and deserving of the same degree of prevention effort. From this point forward, we will use the term F&LC to indicate what we term a "Fatal and/or Life-changing" injury event.

Multiple sources for safety related data such as the BLS, National Safety Council (NSC), etc., all bear out the fact that we are not seeing the same level of downward impact on F&LC injury events that we are with less serious incidents. In 1970, the fatality rate was about 18 per 100,000 workers employed, the result of approximately 13,800 workplace deaths (NSC 1994). In 2014, that rate was down to 3.3 per 100,000 employed involving 4679 fatal workplace accidents (OSHA 2016). While this is an impressive drop in workplace fatal injuries over that time period, the reductions became less significant from year to year after 1992, while non-fatal injury rates (although measured on a basis of 100 workers) continued to decline steadily.

One critical aspect that makes F&LC events different from less catastrophic injuries is that of their predictability. If we have employees using box cutters, are suffering laceration injuries as a result of improper use, but do nothing to change the outcome, we should expect to continue to have laceration injuries. Because of what appears to be the random nature of F&LC events, the ability to predict when they might occur is significantly hampered. We can however, focus our efforts on managing those activities that have a high potential for injury and actually have an impact on whether or not all the causal elements come together to result in an F&LC event.

It is the devastating injury or fatality events that have the most impact on the worker, their family, the rest of the workforce, and the employer; from not only the physical and emotional toll, but the financial impact such events leave in their wake. Because of the potential financial implications, the security of the worker's family might also be at risk: What would the potential impact be to your life, your loved ones, and your livelihood? The tragedy of it all is that many of these F&LC events can be prevented by implementation of the right strategies. We have to actively manage our exposure to high-risk activities with the same level of intensity as being efficient, producing a quality product, and delivering it on time. Our hope is that ultimately the strategies become part of the safety culture and are deeply woven into the fabric of the organization, and that safety migrates from being another business priority to become an intrinsic value of the organization.

In analyzing our own loss data, we elected to classify any injury of $100,000 or greater, as well as any fatality, as an F&LC event. This does not mean that a lesser injury is not "life changing." Take, for instance, the maintenance employee who inserts his left hand into a woodworking machine to point out to a co-worker a problem that needs to be corrected during an upcoming shut down. Because he failed to ensure the machine was properly shut down and locked out, he severs the ring finger. For many individuals, losing a single digit might not be life changing; but in this case, the injured worker has been playing guitar on weekends in a local band for over 20 years. His future as a guitar player was now in jeopardy because of this life-changing injury.

Changing the Perception of Risk

Every business organization and individual has a different perception of what poses a risk to them and how they are going to deal with it. If an organization or individual is on the far end of the spectrum and does not recognize the fact there is the potential for negative (downside) risk in almost any activity, a large problem exists. That level of risk increases with specific activities that have the potential for high-energy transfer to a person, or are unforgiving in nature (no second chances). That level of risk rises even higher when we engage in activities that are high risk, but occur infrequently.

Predicting when and where an F&LC event might occur is difficult. We can analyze the activity involved, the potential consequence of a bad outcome, how many employees might be exposed to that activity, and how often that occurs to conclude that there is an elevated risk. Because these events tend to be less predictable as to when they occur, they are often called "black swan" events or random anomalies over which we have not control; however, this is far from truth. We do, in fact, have control over the level of risk involved by making sure that those high-potential risk activities are handled properly every time!

Since risk (Fig. 1) is a mathematical product of the "likelihood ,(or probability)" of an event, and the "potential or actual consequence" we can apply a numerical value to the chances of the event occurring. However, when dealing with F&LC events, we cannot rely on past history, because there may not have been a past adverse event within the timeframe that organization has been in existence. This tends to cause a false sense of security for organizations, because they tend to think that it cannot happen here because it has not happened before. Due to the unpredictability of these events, we have to calculate our risk absent any historical data as to whether or not any past events have occurred.

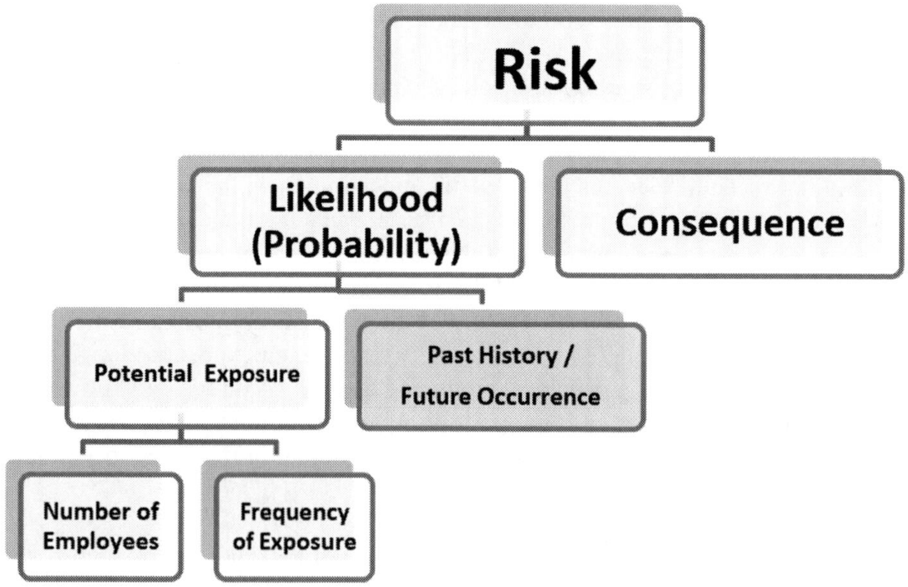

(Fig. 1)

Using a simple four-quadrant risk/frequency matrix (Fig. 2), we can visually see which area might cause the highest potential for bad outcomes with any event. Those activities, as previously mentioned with the potential for high levels of energy transfer (e.g., work from heights, electrical work, motor vehicle accidents, etc.), have inherent risks. However, if we do these things on an infrequent basis, our risk of human error actually increases because our proficiency level for the task may not be as good as if it were something we do on a frequent basis. This is particularly evident when dealing with plant shutdown work, some infrequent general maintenance activities and special projects. Other types of situations, such as confined space entry or trenching operations, also have the potential for high-impact consequences if not handled properly, and may require extensive preparation, equipment and training to be done safely by in-house personnel. Shortcutting or failing to follow prescribed procedures (either by intent or) due to lack of familiarity can have catastrophic outcomes.

	FREQUENCY	
RISK	High Risk Low Freq.	High Risk High Freq.
	Low Risk Low Freq.	Low Risk High Freq.

(Fig. 2)

While activities that are both high risk and high frequency may still present a significant exposure, there is an advantage that can help to manage the likelihood of an adverse outcome through the competency gained by performing the activity regularly (Graham 2016). While doing a specific high-risk activity more often increases the number of potential exposure events, those involved would in theory have more experience at doing those activities well (e.g., a well-experienced driver who has been trained in defensive driving practices vs. an inexperienced driver operating a motor vehicle). If you don't perform the activity often, then you need to train more frequently to maintain the necessary competency.

Those primary precursors that we have identified include the following:

- Motor vehicle operations
- Work at heights
- Slip / fall exposures
- Manual material handling
- High-voltage electrical work
- Mobile equipment use
- Non-routine work
- Lockout/Tagout
- Confined spaces
- Trenching /excavation / underground work
- Crane / hoist work
- In-plant construction projects or machinery rigging
- Occupational illness or disease potentials
- Infectious disease exposures
- Noise
- Workplace violence

While this is not intended to be an exhaustive list of potential high-impact precursors, it does represent a large percentage of what we might encounter within the scope of the customers we work with. Each organization needs to evaluate their own operations to see if these are potential exposures for them, and if there are others not listed that could adversely impact them.

Conscious Thinking (System 2) vs. Intuitive Thinking (System 1)

It has been commonly accepted in the field of psychology there are two modes of thinking for the average person. Psychologists Keith Stanovich and Richard West originally used the terms System 1 and System 2 to describe these modes of thinking, System 1 being intuitively based, while System 2 involves conscious thought for dealing with more complex issues. The importance of these two modes is that each has its strengths and weaknesses when it comes to decision making. As human beings, we make many decisions every day. The outcome of any given situation that requires a decision is going to be largely dependent on which mode of thinking we use at the moment. In many cases, we make decisions which provide the easiest pathway (or the least resistance) in attempting to accomplish a task or goal, whether right or wrong. We are often influenced by various motivating factors to come up with the fastest way to complete a task, usually for the reason of getting to something else we may tend to like more. If all goes well with a given decision, we continue to make similar decisions, as long as the desired results continue to be achieved. There is a belief that, "it has always worked for me in the past, so why won't it work again for me this time?" But remember that the past does not always predict the future.

Our world tends to revolve around intuitive thinking (System 1), and we spend the majority of our time utilizing this type of thought process because it takes the least amount of effort. This represents an intuitive decision making process requiring little or no conscious thought or effort, but rather more of a reactionary response based upon input from prior experiences. We are not really conscious of how this is happening, but more conscious of the results. This is a reflection of our associative memory, working automatically in more of a stimulus-Response relationship. System 1 thinking generally continues to automatically address problems (with System 2 running in a low-effort mode), until its limits are reached, and the brain is forced to fully engage System 2 to solve the problem. Noble prize-winning author Daniel Kahneman, in his book *Thinking Fast and Slow,* also points out the fact that System 1 cannot usually be turned off unless you are completely focused on something else.

Conversely, conscious thinking (System 2) arrives at conclusions based upon critical or analytical decision making, where facts or data are applied and considered. Thinking consciously pays much more attention to potential outcomes during the decision-making process, and requires and investment of time and effort. Using System 2 (or conscious) thinking to address problems usually requires the individual to make additional effort and think the situation through to the potential consequences of their ultimate action; it requires more work or effort! If the decision maker does not remove the "auto-pilot" of System 1 (intuitive) thinking to solve the problem and stops at the first available solution, they never get the chance to fully use System 2 to find alternative solutions to the problem.

We intuitively tend not to see risk if we have not had a prior negative experience. This also brings about complacency and defines yet another challenge for safety professionals when examining the workplace. We must remove the barriers to good decision making by removing the opportunity for failure. We do this by encouraging employees to think the process or action through to the potential consequences in order to maximize their ability to use System 2 thinking, and as we will discuss later; empowering them to speak up if things are not right, not to be complacent or accept the status quo without thinking things through. Every decision is an opportunity to either win or fail. By removing the opportunities to make the wrong decision, either through complacency and taking the path of least resistance, or due to lack of knowledge regarding the potential risks, we remove the opportunities to fail and produce bad outcomes, such as F&LC events.

In industrial settings, it is common for certain organizational drivers to have a bearing on this way of thinking. Corporations place incentives that promote this type of intuitive behavior and way of thinking. Such things as piece-rate systems, supervisory pressures, demands for the highest output in the shortest amount of time, lack of proper training, and even lack of proper tools to get the job done; all of these factors tend to create a culture or climate within the workplace that places production at a premium and puts safety somewhere in the rear-view mirror. Most workplace accidents are the result of a failure of management systems, so instead of placing all of the blame on the injured worker following an F&LC event, we need to focus more deeply to determine whether or not the organizational systems have failed the employee and are promoting or reinforcing bad behaviors and influencing poor decision making. Instead of being a just another *priority* (along with quality, production, timeliness of delivery, etc.), safety must become a *value* that is woven into the fabric of an organization.

Adjusting Our Focus

For many years, organizations have relied on traditional injury-based metrics to define and measure our safety success. In actuality, our incidence rates do not measure the level of safety, but rather the number of people who were injured. So we are measuring our safety success based upon our failures, which seems like a very odd way to gauge our success. The problem stems from the fact that it is extremely difficult to measure something that does not occur. It is hard to quantify the numerous decisions that are made each day that do not lead to some type of failure. With these numbers seemingly impossible to count, we settle for the next best tool of measurement, injury-based metrics. Measuring our injuries is not measuring the presence of safety; it is measuring the absence of safety.

Safety has been traditionally defined as a condition where the number of things that go wrong is acceptably small. Therefore, achieving zero accidents becomes the mission for all organizations to strive for. But if we are measuring how "safe" an organization is based upon the number of people who were injured, do we truly know how well that organization is performing if the goal of zero accidents is achieved? How close to the edge did they come to having an accident, or possibly an F&LC event? The truth is we wouldn't know. Without some type of true barometer to tell us how we are doing, organizations will continue to react to events. Instead, organizations should be proactively seeking out opportunities for potential failures to increase the probability for good decision making. Safety is not just the elimination of hazards and the prevention of injuries. Safety should also focus on how to develop an organization's capacity to sustainably respond, monitor, understand, and anticipate failure.

Risk Assessment

To determine if there is risk, an active process of assessing exposures needs to be conducted. This should be a systematic approach to determine where these exposures lie to what extent they can impact the organization. Here, the adage "you don't know, what you don't know" clearly applies.

ISO 31010: 2009 (ANSI Z690.3-2011) recognizes 31 formal methods for conducting risk assessments. These can range from simple methods, such as brainstorming and checklists, to the highly complex systems, such as Delphi Technique, Preliminary Hazard Analysis, or Failure Mode and Effect Analysis. Each method has its own advantages and disadvantages, based on the elements of risk that can be identified by the particular process (e.g., consequences, probability, level of risk) and would involve quantitative or just qualitative results. Some are focused on identifying the causes, while others are oriented toward the end result of the exposure. This requires the organization to make its own decision as to which method(s) will work best for its particular applications. Ideally, it should review multiple methods and determine which one(s) would work best for the organizational needs and standardize the internal processes in order to develop internal staff proficiency.

The important issue is that there be some organizational mechanism for identifying where risks are present before anything can be done to control them.

Pre-work Planning

By now, the phrase "failing to plan is planning to fail" is something that we've undoubtedly heard. And while there may be some discrepancy as to whether this quote was originated by either Benjamin Franklin or Winston Churchill, the truth lies within the words of the quote. Failing to pre-plan for safety in the workplace is a recipe for almost certain failure.

As with any type of planning, the decision making must occur *before* the actual job or task is to take place. We must actively engage in pre-job, pre-task, pre-shift huddles or meetings to discuss the steps in the process before the work begins. This is particularly important when the task is engaged in very infrequently such as: start-up tasks, plant shut-down work, and every other non-routine task. The need to discuss safety ahead of time arises before *any* task is conducted. We must take time to think about the potential consequences, and take three seconds to think about what could possibly go wrong.

One of the most common methods of pre-work planning is to conduct a Job Hazard Analysis (JHA). Some may still refer to this process as a Job Safety Analysis (JSA). Regardless of the name, the purpose of the process remains the same; to break down each individual job task as a way to identify hazards before they occur. The goal is to identify any potential hazards, address those that are uncontrolled, and take steps to either eliminate or reduce them to an acceptable level of risk. Spending the time to analyze work processes can help to identify such things as the qualifications of the personnel involved in the process, the resources that will be needed, the hazards that could be anticipated, and the proper control measures to be used. Not only can the results of a JHA help to eliminate any potentially dangerous health hazards, but are also likely to result in fewer worker injuries and illnesses, safer and more effective work methods, increased worker productivity, and reduced workers' compensation costs. Also, one often overlooked benefit of a JHA is its usage as a training tool for new employees to show them the proper steps required to perform their jobs safely.

Crew Resource Management

The concept of Crew Resource Management incorporates several critical aspects of the cultural change an organization must adopt to be successful: bringing the concepts of open communication between affected personnel and empowerment to stop work without fear of reprisal for safety concerns, with the end goal of fostering teamwork and breaking down barriers that may exist within the organization.

Crew Resource Management arose in the 1970s following several commercial aircraft accidents with experienced professional flight crews; the most notorious was the 1978 crash of United Airlines flight 173. In this accident, the flight crew was troubleshooting a landing gear problem, but ran out of fuel while doing so and subsequently crashed.

The National Aeronautical and Space Administration (NASA) and the National Transportation Safety Board (NTSB) convened a workshop in 1979 to investigate commonalities between several recent commercial aircraft disasters. It was found that there were deficiencies in the ability of the flight crews to effectively communicate due to barriers created by the authoritarian culture in place within the crew hierarchy. This led to a change in the aircraft cockpit with respect to making it possible for crew members to speak up and challenge situations they felt might be detrimental to the mission. Up until that time, subordinate crew members did not have the permission to question the senior officer's decisions or actions when it came to safety issues.

This empowerment to step up and say something in the event of a questionable safety situation is carried one step further with "Stop Work Authority" being granted to all the affected personnel. In many work environments, there is a fear (whether valid or not) of potential reprisal directed toward an employee who speaks up, or questions whether an activity is safe. It must be clearly communicated through a written policy, reinforced by management action, that there will be no bullying or reprisals for speaking up, and that management will respond to those questions with the appropriate action to investigate, evaluate and change the processes or equipment to eliminate the concern.

Root Cause Analysis by Teams

It is a commonly accepted fact that accidents represent a failure of our management systems, based on the writings of Dan Petersen, one of the most recognized safety theory authors of all time. Regardless of what theory of accident causation you subscribe to (Heinrich's Domino Theory, Haddon's Energy Release Theory, etc.), the fact is that when they occur, we must do everything possible to prevent them from recurring. Performing a true Root Cause Investigation is critical to preventing recurrence, as we do not stop at the low-hanging fruit, which are actually direct and surface causes, but are mistaken by inexperienced investigators as the "root causes,"

Often, those direct or surface causes that are identified by the direct supervisor of the injured employee result in recommended corrective actions, such as "told the employee to be more careful" or "pay more attention to the task," focusing on the failure of the employee. When in fact it was the management systems in place that failed the employee by not ensuring they had the knowledge to recognize the risk in the work, were not well trained, did not have the proper resources, no one monitored them to ensure they were following the rules, and there were no feedback mechanisms in place to close the loop when a deficiency was found in the system and make sure proper corrective actions were in place.

If, in fact, most of the responsibility usually lies with failure within the management systems, is the employee's direct supervisor likely to point fingers back at himself when determining the true cause of an accident? We recommend a multi-disciplinary team that does not include the direct supervisor to investigate accidents. This removes the risk of not identifying the true management system shortcomings by eliminating the self-serving (or self-preserving) bias of the direct supervisor from the system; but it also brings in fresh eyes that would be more likely to question the methods or processes used because they are not already deeply invested in the status quo and might not see other opportunities.

Once the investigation is completed and appropriate corrective actions are developed, there must be a mechanism or process to ensure consistent accountability for implementation, monitoring of the future outcomes to ensure they are successful, and communication throughout the organization regarding the improvements to involve employees at all levels in improving safety.

Evaluating the Maturity of Your Safety Culture

The maturity of the safety culture is based on how well integrated into the fabric of the organization the processes for actively managing risk have become. We use five criteria to separately evaluate each exposure potential from the previously discussed list to determine how well the organization is managing their exposure to risk.

- Knowledge: Is there the technical ability within the organization to recognize that risk is present as a result of their activities and evaluate the exposure potentials? Often equated to: "We don't know, what we don't know."
- Competency: Does the workforce have the knowledge, skills, and abilities necessary to effectively handle the situations they might encounter? This can only be accomplished through an effective program that incorporates active training and measures performance outcomes.
- Resources: Has management made the necessary investments in workplace safety efforts to ensure that the safety and risk management programs are effective and sustainable? This includes a wide variety of aspects, including management engagement, the necessary time for personnel to train and audit, funding of engineering changes, and the commitment to process changes and improvement, etc.
- Audit: Does the organization have a system in place for ensuring they do what they say they do? Is there a Plan-Do-Check-Act (PDCA) cycle in place, are they following similar auditing processes for monitoring safety as they employ with their quality management?
- Feedback: How extensive is the capability to perform effective Root Cause Analysis to correct deficiencies in processes or following accidents? Do they perform planned observations that include mentoring/coaching to monitor and change behavior? Are there systems in place to ensure accountability for meeting the improvement objectives?

The five elements listed above, while independent, tend to have a cascading effect as we move down the list, each building upon the strength of the previous item. If any element is not sufficiently strong, the successive elements tend to be progressively weaker (e.g., if there is not a strong rating under Knowledge, there is usually a lesser degree of commitment for the subsequent items of Competency, Resources, Audit and Feedback, etc.).

Summary

Prevention of Fatal and Life-changing injury events will require a different approach than we have been using in the past. Merely addressing those risks that are easily identifiable from a frequency of occurrence standpoint will have little if any impact on preventing F&LC events. The new approach needs to focus on changing the culture, which will inevitably change behaviors within the organization by empowering and involving the workforce in their own safety future. While there are several key initiatives to integrate into an organization, they must become values within the organization and not just priorities that can bend to the ever-changing business cycles. Those steps include:

- Recognizing that there is risk present in many of the endeavors the organization undertakes and doing an effective job at evaluating the potential impact. We cannot rely of the absence of F&LC events in our past as an accurate predictor of what might be in our future.
- Instilling the practice of engaging in Conscious Thought (System 2 Thinking) when performing tasks; taking the time to think the activity through and realize the potential consequences if things do not go as planned.

- Shifting our focus to identifying potential weaknesses in management systems to make failure less likely, and the consequences of failure less severe.
- Systematically identifying, evaluating and treating risks to employees and organizations using consistent formalized risk assessments.
- Utilizing Pre-work Planning as a methodology to ensure safety is integrated into every task, particularly those done infrequently.
- Incorporating the principles under Crew Resource Management to empower all employees to speak up and take an active role in managing exposures, by stopping work to reassess the procedure if necessary.
- Conducting effective Root Cause Analysis investigations by a team to take advantage of multiple viewpoints and ask the necessary questions to get past the obvious direct causes; then close the loop and integrate the needed changes into the organization.

Every day, managers, supervisors, team leaders and co-workers need to challenge each other to think things through; *before* they act! With the harsh reality that: *The Next* ***three*** *seconds could your Last* ***three*** *seconds*; following through with the N3L3 concepts can help each of them *Protect their* ***Li****fe, their* ***Lo****ved Ones, and their* ***Li****velihood*!

Bibliography

Bureau of Labor Statistics (BLS). 2016. "Incident rates of non-fatal occupational injuries and illnesses by industry and case types 2015" (http://www.bls.gov/iif/oshsum.htm#14Summary_News_Release).

Crew Resource Management.org. 2016. "Crew Resource Management, Introduction". Retrieved from URL March 7, 2016. (http://www.crewresourcemanagement.net/introduction)

Elliot, Michael W (ed.). 2012. *Risk Assessment and Treatment (ARM 55)*. Malvern, PA: The Institutes.

Federal Aviation Administration (FAA). 2012. FAA TV – *The History of CRM*. Retrieved from URL March 7, 2016. (http://www.faa.gov/tv/?mediaId=447)

Graham, Gordon. 2016. *Some thoughts on Real Risk Management: A presentation at the Carmel, (IN) Fire Department (January 12, 2016)*. Aliso Viejo, CA: Lexipol, LLC.

Kahneman, Daniel. 2011. *Thinking Fast and Slow*. New York: Farrar, Straus and Giroux.

National Safety Council Statistics Department. 1994. *Accident Facts*. Itasca, IL: National Safety Council.

Occupational Safety & Health Administration (OSHA). 2016. Commonly used statistics. (www.osha.gov//oshstats/commonstats.html)

Moderating the Use of Discipline to Achieve a Positive Safety Culture

Mike McCarroll, CSP
President & CEO
PROSAFE Solutions, Inc.
Villa Rica, GA

Michael Belcher, CSP
VP, Risk and Safety
DS Services of America, Inc.
Atlanta, GA

Organizations benefit from having clear and concise ways to address common issues like tardiness, absenteeism, theft, substance abuse, violence, harassment, and other violations of company standards. Discipline is often used to accomplish this. In a document published by the Society for Human Resources Management (SHRM), discipline is described as a way to "address conduct issues such as poor work performance or misconduct to encourage workers to become more productive workers and to adapt their behavior to company standards and expectations." Also, if discipline is applied consistently and documented accordingly, this can help defend organizations against grievances, EEOC charges, OSHA whistleblower allegations, and wrongful termination lawsuits. That's all good. And it might seem intuitively obvious to include unsafe acts as behaviors worthy of discipline. But, after carefully analyzing numerous safety-related disciplinary events in light of current research about human performance, it's become increasingly apparent that using discipline as a reaction to safety infractions often does more harm than good.

The rationale for disciplining workers for safety infractions often follows this thought process: (1) the worker broke a safety rule or did something unsafe, (2) the worker was aware of the safety rule, (3) the worker's actions had, or could have had, serious consequences, and (4) other workers have been disciplined for similar infractions. This approach isn't much different than disciplining for other fundamental workplace rules. But before rushing to judgment, there are some critical issues that must be understood in order to use discipline effectively in a positive safety culture.

Issue # 1: Discipline Is Punitive

Despite assurances from some well-meaning Human Resource professionals that the purpose of discipline is not to punish or embarrass workers, traditional discipline is still punitive in nature. Common forms of discipline, typically doled out in a progressive fashion, include counseling, written warnings, suspensions, and termination. Depending on the circumstances, discipline may also include a demotion or job transfer. Do any of these consequences sound positive to you?

The word discipline has been closely associated with punishment since the 13th century, when it was used to describe chastisement of a religious nature. The *Oxford English Dictionary* describes the verb discipline as follows: "To punish, chastise, or reprimand (a person), esp. with the intention of correcting or altering subsequent behavior." This is closely related to the notion that crime must be followed by punishment. After all, without punishment, there would be chaos and disorder.

Unfortunately, far too many workers end up being punished for safety "crimes" that were either a result of simple human error and/or system induced. The reason for this is rooted in our reactions to failure. When an unsafe act or outcome occurs, a natural reaction for the observer is to focus on what the worker should have done or could have done, rather than trying to understand why he/she acted in the way they did. And, when trying to understand why somebody did something unsafe, research shows that observers are more likely to ascribe (and emphasize) causes to internal or personal factors, rather than external or situational factors. This is especially true in Western culture, where we place great emphasis on personal responsibility. These tendencies perpetuate a superficial understanding of human error and can lead us to blame the worker simply because he/she did something unsafe.

Issue # 2: Disciplining Human Error Is an Exercise in Futility

Human error is inadvertent. It's where a "planned sequence of mental or physical activities fails to achieve its intended outcome" (Reason). To understand why punishment doesn't work, let's begin by describing three basic types of human errors: slips, lapses and mistakes.

A slip can be thought of as an unintended or unplanned action; for example, over-tightening a bolt or turning a switch the wrong direction. Slips are considered "errors of commission" that usually occur when a simple, routine physical action goes awry. Even the most skilled and experienced people are susceptible to these inadvertent errors. A lapse is an "error of omission;" for example, missing a step in a lockout/tagout procedure or forgetting to unplug a heater. As tasks become more repetitive and familiar, they can be performed with less conscious attention, and the easier it is for the mind to be occupied with other thoughts. Slips and lapses are execution errors, and they are quite common. In your everyday activities, you can expect to make four to five inadvertent execution errors per hour.

Consider delivery drivers who are required to operate commercial trucks and make timely delivery of goods to customers on specific routes. For some drivers in the industry, it's not unusual to make 50-75 stops a day. Every time the truck is parked, the driver is responsible for securing the vehicle and applying the parking brake so the vehicle doesn't roll away. Is it conceivable that the driver might forget to put the brake on? Of course it is. And it has led to numerous vehicle accidents all over the country. This kind of mental lapse has nothing to do with skills, knowledge, judgment or motivation. You simply can't prevent lapses by disciplining drivers. The only proven way to minimize these types of errors is through controls like task and fatigue management, reminders, checklists, alarms or warning devices.

Another type of human error is the mistake, when somebody does the wrong thing believing it to be right. Mistakes result from failures in thinking, where the worker takes action as planned using deliberate thought processes, but it doesn't pan out; for example, when a worker tries to apply a rule that was successfully used before in a similar situation. Or when a worker lacks expertise for a given situation and has to analyze and plan based on limited time and resources. Humans are prone to mistakes when tasks become more difficult and complex.

Disciplining workers for committing errors will not prevent the recurrence of slips, lapses and mistakes. Human error is not a violation. So don't treat it as such.

Issue # 3: All Violations Are Not the Same

Violations occur when a worker knowingly deviates from safe operating procedures. Violations are certainly the most disturbing of all human failures, due to their very nature as deliberate actions that contradict established protocols; for example, taking shortcuts or failing to follow safety procedures. On the surface, a violation might seem deserving of discipline. After all it was intentional! And the worker knew better! But in most cases, very few people are culpable or malicious in their intent. Most are blameless, and were actually trying to do their best in an imperfect system. As it turns out, the majority of safety violations are system-induced.

Do you remember Domino's famous "30 Minutes or It's Free" guarantee? Introduced in 1979, this slogan was familiar to millions of people looking to have hot, fresh pizza delivered to their doorstep. It became a popular slogan, and a cornerstone of success for Domino's. But by the summer of 1989, after a 17-year old delivery driver lost his life in a fatal crash after skidding off a wet road, the press picked up on the story, and many periodicals began publishing stories about their drivers' accidents and questioning the safety implications of this policy. After dozens of lawsuits were filed by relatives of workers killed while delivering pizzas, many critics, including the National Safe Workplace Institute, urged Domino's to abandon their 30-minute promise. An underlying presumption was that franchise operators and their drivers faced pressure to deliver pizzas, thus inducing them to take risks while driving, like speeding or running stop signs.

The reality is that modern workers face enormous challenges in today's workplace. Factors like time pressure, peer pressure, fatigue, workload, task complexity, suitability of tools and equipment, and organizational influences play a role in shaping worker behaviors. It's shortsighted to discipline workers for safety infractions without first considering how their decisions and actions made sense to them, given the circumstances that surrounded them.

For example, let's say we have an hourly worker who has been instructed to change a high-bay light bulb over a production line. The production line is going to start up in 30 minutes, and the light bulb is critical for conducting quality inspections. In this case, the proper procedure calls for the use of a man-lift to reach the light bulb. But unfortunately, the man-lift is broken (and won't be fixed anytime soon, due to budgetary concerns). To avoid delaying the production process, the worker decides to use a forklift and a makeshift platform. This is an example of a *situational violation*, where the worker took actions deemed necessary to get the job done. It wasn't done for personal gain. The lack of personal gain is a key difference between safety violations and other types of violations.

Issue #4: Discipline Creates Disengagement and Mistrust

Using cookbook disciplinary responses to unsafe acts and outcomes without adequately diagnosing the behavior and accounting for system influences will undermine efforts to produce workers who are engaged and committed to the goals of the organization. From the results of thousands of employee surveys and personals interviews, we've also found that that using discipline in this fashion fosters mistrust. When discipline is used as punishment for safety errors, this can lead to a "cycle of blame," which is characterized by reduced trust, less communication, and less managerial awareness of jobsite conditions. As a result, latent organizational weaknesses persist, which perpetuate more errors.

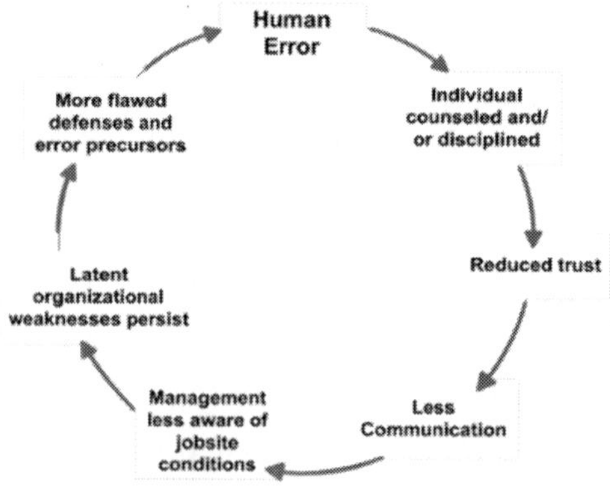

Using discipline as punishment for safety violations or accidents isn't much different. When a serious accident occurs, it is almost always attributed to an unsafe act or condition that immediately precedes the failure. However, these events are generally not the result of a single cause or individual action. The cause is usually rooted in underlying latent conditions or organizational weaknesses that allow the failure to occur. Latent conditions are often not identified in traditional accident investigations. As a result, actions are taken against a culprit or "safety violator," while allowing latent conditions to continue to exist. Punishing people for system-induced violations is both unjust and retributive.

The mere fact that an accident can be behaviorally prevented is not sufficient grounds to conclude that behavior is the root cause of the accident.

Issue # 5: Discipline May Be Considered Discriminatory

According to OSHA, reporting a work-related injury or illness is a core worker right, and retaliating against a worker for reporting an injury or illness is considered illegal discrimination. OSHA has reiterated their position that incentive programs and disciplinary programs may discourage workers from reporting injuries, and that this can constitute unlawful retaliation under Section 11(c) of the OSH Act. Below are some of the practices that OSHA lists as potentially discriminatory:

- Disciplinary actions against workers who are injured on the job, regardless of the circumstances surrounding the injury.
- Disciplinary actions for violations of a safety rule, if it is used as a pretext for discrimination against a worker for reporting an injury or illness. Compliance officers and whistleblower investigators are directed to focus on whether the employer consistently monitors compliance with the safety rule in the absence of an injury, and whether it uniformly disciplines for violation of the safety rule in the absence of injury. If discipline is issued only when a worker is injured, such disciplinary decisions will be considered suspect.
- Programs that intentionally or unintentionally provide workers with an incentive not to report injuries or illnesses, such as rewarding workers with prizes or bonuses if the workers have not been injured over a given period of time (e.g., safety bingo). OSHA's position is that if the effect of an incentive program is great enough to dissuade a reasonable worker from reporting a work-related injury, the employer may be found to have unlawfully discriminated against a worker. Instead, OSHA encourages employers to develop incentives that promote worker participation in safety-related activities, such as identifying hazards, participating in the investigation of accidents, or serving on a safety committee.

Issue # 6: Terminating Experienced Workers Can Decrease Safety

The most extreme form of discipline is termination. This can have a toxic effect on an organization, especially when severing experienced workers. Studies show that injury rates decline with job tenure and that newly hired workers are more likely to be injured than those who have been on the job for longer periods of time. In a major study involving 58,000 workers over a period of six years, this was shown to be true in all age groups and for all job tenures. So when an experienced worker is fired and has to be replaced with a new worker, chances are that new worker will be more likely to be injured, especially during the first six months.

Several years ago, one of our clients in the transportation industry terminated a driver who had experienced a minor collision due to a mental lapse. The collision could have been disastrous, but fortunately only minor property damage resulted. The driver, an experienced veteran, had an excellent safety record with only the one blemish. But he was determined to have "violated a safety rule," and the management team upheld his termination. He was then replaced with a new, less-experienced driver, a driver that, according to the organization's own data, was four times more likely to be in an accident than his predecessor.

Using Discipline in a Positive Fashion

By reflecting the social and cultural underpinnings of an organization, discipline defines the relationship between workers and management. If this relationship is not built on trust, mutual respect, and fairness, then the safety system will be adversely affected.

1. Become a learning organization. Before administering discipline for unsafe acts and/or accidents, seek first to understand. Take the time to diagnose what happened before reacting. This may require a shift in thinking in order to overcome our natural tendency to attribute personal factors as the reason for unsafe acts which often leads to blame. Learn from mistakes; don't try to punish them out of the system by blaming your workers. As Sydney Dekker says, you can learn or you can blame. You can't do both!

2. Use a methodical process to determine culpability. A good starting point is to use James Reason's "substitution test," which asks the question, "Would other workers with similar experience and in a similar situation and environment have behaved in the same manner as the worker being evaluated?" If the answer is most likely "yes," then punishment is not the appropriate response. Another way to administer the substitution test is to ask the worker's peers "given the circumstances that prevailed at the time, can you be sure that you would not have committed the same or similar type of unsafe act." If the answer is most likely "no," then punishment is inappropriate. Other questions you can ask to determine culpability are as follows:

- Have work rules been effectively communicated?
- Were the rules and procedures workable and correct?
- Was the worker adequately trained and coached?
- Did the worker knowingly violate rules or procedures?
- Was the worker's action for personal gain or organizational gain?
- Was the associate experienced in this work?

3. Make your disciplinary program "pro-worker." If you've determined discipline is the appropriate response, consider a "pro-worker," non-punitive approach. Organizations such as Frito Lay, GTE/Verizon, and Tampa Electric have utilized these non-punitive approaches with positive outcomes. Namely, problems get resolved favorably, without all of the headaches often associated with traditional discipline like stress, reduced trust, reduced communication, absenteeism, and turnover. The premise behind this approach usually begins by helping the worker identify a performance gap and coach and counsel them in a positive, constructive fashion. If that fails, the worker is suspended, with pay, for one day to give him/her an opportunity to reflect on their behavior and respond accordingly. At that point, either they make the necessary changes to improve their performance or leave the company.

References

Bena, A., Giraudo, M., Leombruni, R. & Costa, G. Job tenure and work injuries: a multivariate analysis of the relation with previous experience and differences by age.

Berg, Eric N. 1989. *Fight on Quick Pizza Delivery Grows.* (Retrieved from http://www.nytimes.com/1989/08/29/business/fight-on-quick-pizza-delivery-grows.html?pagewanted=all.)

Goldberg, Allan T. 2012. "Is Firing an Effective Safety Tool?" *Professional Safety.*

Grote, Dick. 2006. *Discipline Without Punishment; The Proven Strategy that Turns Problem Employees into Superior Performers*. 2nd ed. AMACOM.

Holden, Richard J. 2009 "People or systems? To blame is human. The fix is to engineer." *Professional Safety*.

McDonald, Chris. 2011. "Death by Pizza Delivery: Domino's Korea." (Retrieved from https://businessethicsblog.com/2011/02/10/death-by-pizza-delivery-dominos-korea/.)

McCarroll, Mike. 2016 "Catch the Drift–How Organizational Drift Leads to Accidents."

Reason, James. 1990. *Human Error*. Cambridge, UK: Cambridge University Press.

Reason, James. 1997. *Managing the Risks of Organizational Accidents*. Burlington, VT: Ashgate.

Society for Human Resources Management (SHRM). 2014. *Progressive Discipline Policy: Varyied Disciplinary Process Based on Type of Conduct*. Alexandria, VA: SHRM.

Scaling an Occupational Safety Management System for Small- and Medium-sized Organizations

Pamela Walaski, CSP, CHMM
GAI Consultants, Inc.
Pittsburgh, Pennsylvania

Introduction

Most occupational safety and health (OSH) professionals would agree that there is an evolution underway in how we practice. Going back ten years or so, the predominance of thought leaders postulating that our compliance-based approach to developing workplace safety programs had run out of steam, so to speak, grew. This thinking was based upon a review of statistical data that showed a slowing in the reduction of fatalities and, in some cases, an increase in the number of serious injuries; those which result in permanent disability or lengthy periods of time off work.[1] It also was not unusual for large organizations with deep pockets for OSH resources to end up with low or nonexistent incident rates of less serious injuries, but still experience multiple fatalities or catastrophic events in the same time frame. These awakenings resulted in an introspective review of the of the basic tenets upon which most OSH programs were based, and many OSH professionals concluded that the profession has been mistaken in believing that focusing efforts on the reducing in frequency of minor incidents would lead to a simultaneous reduction in the severity of them.[2]

It is important to note that movement away from a reliance on compliance-based approaches does not mean ignoring the requirements of applicable laws and regulations. In fact, all OHSMS mention the need to comply. As noted below, OSH policies must mention compliance as a primary expectation. However, compliance is not intended to be the only focus nor is it intended

---

[1] Bureau of Labor Statistics (BLS). 2016. "Census of Fatal Occupational Injuries Charts, 1992-2014 (preliminary data)." (http://www.bls.gov/iif/oshwc/cfoi/cfch0013.pdf).

[2] *Industrial Safety and Hygiene News* (ISHN). 2012. "Preventing serious injuries and fatalities." (http://www.ishn.com/articles/93471-preventing-serious-injuries-and-fatalities).

to be the type of focus that obscures the systemic foundation needed for achieving continuous improvement.

As discussions and debate have worked their way through the ranks of OSH professionals, what has emerged is a fairly broad consensus that approaches relying on compliance with mandatory regulations are not effective enough at managing the risks of the workplace, however well-intentioned the underlying theme of the regulatory framework. Large multi-national organizations have seen an effectiveness with the use of a multi-layered safety management system, based on published consensus standards, such as the American National Standards Institute/American Society of Safety Engineers' Occupational Health and Safety Management Systems (OHSMS) Standard Z10-2012 (ANSI/ASSE Z10), first published in 2008. The OHSMS approach has substantially changed the way large organizations approach OSH, many of whom have invested enormous resources (both money and time) to train their staff to understand how an OHSMS works and implement this approach.

But where does that leave the majority of organizations that are much smaller and may believe they are without the resources needed to develop and implement an OHSMS? Can they hope to accomplish the same success as their much larger counterparts? Based upon this author's experience and that of others consulted for this paper, many smaller organizations do not even attempt to implement an OHSMS. The reasons vary, but resources, internal expertise, and an awareness of safety management systems in general are the most likely culprits.

This paper and the presentation that will follow during Safety 2017 will demonstrate that organizations of just about any size can develop and implement an effective OHSMS by understanding the larger picture of why a systemic approach is effective, identify its essential concepts, and then "right-size" as needed.

What Is an SME, and How Many Are There?

The vast majority of American companies fit into the definition of a small or medium enterprise (SME). The U.S. Small Business Administration (SBA) establishes guidelines for SMEs, based upon the number of employees (no more than 1500), or average receipts (no more than $38.5 million), based upon the organization's North American Industry System Classification (NAISC) code. The number of SMEs in the U.S. alone is roughly 28 million, and while many of those are home-based businesses employing only one person, larger SMEs employ approximately 54.5 million people, or about 54% of the private workforce.[3] The U.S. is not alone in this regard. Canadian SMEs employ 10 million workers, or 90% of the workforce[4] and, in the European Union, there are roughly 10 million SMEs, or 99% of all enterprises, employing 66% of all workers.[5]

---

[3] Retrieved from http://www.sba.gov on March 10, 2016.

[4] Retrieved from http://www.ic.gc.ca/eic/site/061.nsf/eng/02805.html on February 28, 2016.

[5] Retrieved from http://siteresources.worldbank.org/CGCSRLP/Resources/SME_statistics.pdf on February 28, 2016.

SMEs, while more exposed to the negative aspects of risk, stemming from the lack of an effective OSHMS, are actually more able to adopt an OSHMS framework due to the flexibility, stemming primarily from their size. By learning and applying the foundational elements of an OSHMS, their ability to effectively manage their OSH risks increases.

OHSMS Components

As noted above, the three most common frameworks for OHSMSs have many elements that are similar and some differences. Because this paper is based upon ANSI/ASSE Z10, this is the only one that will be discussed in full.

ANSI/ASSE Z10 is divided into seven sections and eleven appendices. It should be noted that the sections provide mandatory elements, and the appendices are intended to be informative only. They are as follows:

- **1.0 Scope, Purpose, and Application**: This section identifies the minimum requirements for an OHSMS, and explains the purpose of the standard. Most importantly, this section categorically states that this standard applies to organizations of all sizes and types.
- **2.0 Definitions:** This section provides definitions for critical terms used in the standard, including conformance, hazard, and top management.
- **3.0 Management Leadership and Employee Participation**: This section establishes the overarching responsibility of the organization's top management to "establish, implement and maintain an OHSMS" that conforms to the requirements of the standard. It also mandates the establishment of a process to endure "effective" participation by employees at all levels.
- **4.0 Planning**: This section describes the components of a planning process with the goal of identifying and prioritizing management system deficiencies and opportunities for continuous improvement. It also requires the establishment of objectives upon which to measure the success of the improvement process.
- **5.0 Implementation and Operation**: This section includes definitions of the operational elements of the OHSMS, and the specific means by which they will be implemented within the organization. It is also important to note that the continuous improvement foundation of an OHSMS requires that the application of these elements is fed back to the planning processes.
- **6.0 Evaluation and Corrective Action:** This section defines requirements and processes of how the performance of the OHSMS is regularly evaluated, how corrective actions are identified and implementation when conformance problems are identified, and how the results of both activities are looped both back the planning process and forward to management review (Section 7.0).
- **7.0 Management Review**: Provides a mandate for ongoing management review of the OHSMS and outlines the expected processes and outcomes.
- **Appendix A:** Provides sample policy statements that can be adapted for use by an organization.
- **Appendix B:** Provides sample roles for many of the top management positions within an organization as they related to the commitment to the OHSMS.
- **Appendix C:** Provides guidance and ideas on the methods to best assure effective employee participation. It also identifies many of the common barriers and means to overcome them.

- **Appendix D:** Provides an excellent process for conducting a gap analysis, which is discussed in greater detail below.
- **Appendix E:** Provides guidance on setting organizational objectives of the OHSMS, and uses the SMART method for effective goal and objective writing.
- **Appendix F:** Contains a lengthy discussion of risk management, and how it fits into an OHSMS. It also provides examples of basic risk assessment methods.
- **Appendix G:** The hierarchy of controls is a well-known framework for managing organizational risks. In this appendix, six different methods are control are detailed, ranging from the highest and most effective (avoidance, elimination) to the least (personal protective equipment and safe work practices).
- **Appendix H:** This appendix describes the elements of a management-of-change process, which addresses the means of reducing the introduction of new hazards and risks into the work environment when changes are made in operations, tasks, technology, equipment, and facilities, among other things.
- **Appendix I:** Identifies method that should be incorporated into the procurement process to prevent the introduction of new hazards and risks when purchasing such items as products, tools, and equipment.
- **Appendix J:** This appendix provides guidance on the management of contractors, who often expose organizations to risks.
- **Appendix K:** A critical component of any OHSMS is incident investigation, which is addressed in this appendix, along with a sample document.
- **Appendix L:** Section 7.0 mandates management review of the OHSMS. One means to do so is an audit, both internal and third-party. This appendix provides a detailed review of the types of documents that can be used to verify compliance, but it also provides a secondary benefit of the types of documents that can be used when conducting the gap analysis.
- **Appendix M:** Describes the management review process, along with a sample "scorecard" for evaluative purposes.
- **Appendix N:** Provides a comparison of the most common OHSMS standards.
- **Appendix O:** Bibliography and references.

How to Right-size an OHSMS

Becoming the Expert

Going from a complete absence of an OHSMS to a fully functioning one can be daunting, even more so when considering many SMEs likely do not employ a dedicated OSH professional. But as the adage about how to eat an elephant goes, the recommendation for how to get to the end point is to take one bite of the process at a time. For many SMEs, the first bite is an OSH professional who believes the process is important and takes up the cause.

OSH professionals wishing to lead this type of an effort in their organizations must first start by becoming the internal expert on OHSMSs if they do not have prior experience or knowledge. This means starting out by reading and understanding the current published OHSMS standards, including the following three documents:

- ANSI/ASSE Z10-2012, Occupational health and safety management systems[6]
- BS OHSAS 18001, Occupational health and safety management systems[7]
- International Standards Organization(ISO)/Draft International Standard (DIS) 45001, Occupational health and safety management systems[8]

Although the focus of this paper, and the presentation associated with it, will be ANSI/ASSE Z10, it is helpful to be familiar with all three documents because they have many similarities in their approach to management systems, but also because all three feature some differences that are important to understand. (The reader is referred to Appendix N of ANSI/ASSE Z10-2012 for a comprehensive comparison of the OHSMS standards. It should also be noted that, as of this writing, ISO 45001 remains in the draft stage. As this paper is being written, DIS2 is expected to be released, and it is almost certain to remain in the draft stage by the time this presentation occurs in June 2017. However, it is recommended reading now because much of its essential framework is not expected to be changed once finalized.)

Conducting a Gap Analysis

Following the first step of preparation, the OSH professional should next undertake a thorough gap analysis of their organization to ascertain the current status with regard to the framework of ANSI/ASSE Z10-2012. As discussed above, this is the fundamental part of the planning process, noted in Section 4.0 of ANSI/ASSE Z10-2012. (The reader is referred to Appendix D of the standard for a comprehensive discussion of how to conduct a gap analysis.)

This analysis should subsequently be used to develop a customized plan that addresses the means and methods required to make the transition from what exists to an OHSMS. In addition, the gap analysis can be used to assure that the OHSMS aligns with the SME's strategic plan (if one has been developed), a connection that will aid tremendously in selling the approach to management. The OSH professional should also closely review any organizational governance procedures because of their influence in the organization's activities. Since OHSMSs are essentially based upon how an organization controls its activities with regard to OSH, further alignment with governance is necessary. In addition, the SME's governance framework also helps clarify lines of management responsibility, crucial for going beyond assessing into actual implementation and continuous monitoring of an OHSMS.

Engaging Senior Management

Once the above preparatory steps have been completed, the OSH professional is now ready to engage senior leadership in the effort. This management commitment is essential for moving forward; without it, the successful implementation of an OHSMS is likely to fail to achieve its objectives. Some might argue that a implementing a partial system is better than none at all, and

---

[6] American Society of Safety Engineers (ASSE). 2012. ANSI/ASSE Z10-2012, *American National Standard for Occupational Health and Safety Management Systems.*

[7] British Standards Institute (BSI). 2007. BS OHSAS 18001, *Occupational Health and Safety Management Systems.*

[8] International Standards Organization (ISO). 2017. ISO/DIS 45001, *Occupational Health and Safety Management Systems.*

taking the time to achieve full commitment is not required. But in order to have an effective OHSMS, a shift in organizational culture and values is needed, and since management commitment is the fundamental pillar of the process, a partial implementation is, at the least, a waste of time and, at worst, it communicates a message to the workforce that may hinder and even derail full implementation.

OSH professionals wishing to develop a method of engaging senior management buy-in would be helped by reading some of the writings of current thought leaders in OSH, such as Dr. Tom Krause, founder of BST; Ron Gantt, Vice President of SCM Safety; or look at some of the postings on Safety Differently.[9] This review will provide plenty of backup for the need to look OHSMS, as well as provide documentation of the success of the efforts of other organizations.

At the end of this process, an engaged management will have adopted a formal OHSMS policy that will then guide the efforts moving forward.[10] The critical elements of this policy include a commitment to the following:

- Protection and continual improvement of employee health and safety;
- Effective employee participation;
- Conformance with the organization's health and safety requirements; and
- Compliance with applicable occupational health and safety laws and regulations.

Find Your OSH Champions

The two pillars of all OHSMSs are management commitment (discussed above), and employee participation. The OSH SME leader must be skilled at navigating both at the same time and integrating them into the system.

Every organization has OSH champions. They exist in all staffing levels, and come from all backgrounds with regard to their passion for workplace safety. Most OSH professionals know who they are (or should be), and many of them can be educated about the value of an OHSMS with some time and engagement. Some of those employees may even have OHSMS experience from previous employment or intuitively understand the goals of the process. Regardless, finding them and developing them creates a contagion for the process that can spread throughout the organization, leading to the culture transformation that will be essential to achieving the overall goals.

Conclusion

OHSMSs have increasingly demonstrated their value in recent years, as singularly relying on compliance has fallen out of favor with most OSH professionals. They have been the norm in larger organizations, particularly those with a global footprint. OSH professionals in those organizations are typically skilled at the process, including understanding the essential

---

[9] Retrieved from http://www.safetydifferently.com/ on March 1, 2016.

[10] American Society of Safety Engineers (ASSE). 2012. ANSI/ASSE Z10-2012, *American National Standard for Occupational Health and Safety Management Systems.*

components of an OHSMS, knowledge of how to implement many of the more common OHSMS strategies, and in using the OHSMS to drive continuous improvement, along with a reduction in the incident of injuries, particularly those with more severe outcomes.

OSH professionals in most SMEs may aspire to replicate an OHSMS in their organization, but resist due to beliefs about the intensity of the process and the need for extensive resources that may not be available. The result is an exposure to higher levels of risk that may not be necessary if the OHSMS can be right-sized. This paper has provided some general reference points to methods that can be utilized by OSH professionals in SMEs, allowing them to adapt an effective OHSMS in their organization.

Bibliography

American Society of Safety Engineers (ASSE). 2012. ANSI/ASSE Z10-2012, *American National Standard for Occupational Health and Safety Management Systems.* Des Plaines, IL: ASSE.

British Standards Institute. 2007. BS OHSAS 18001, *Occupational Health and Safety Management Systems.* London: BSI.

Bureau of Labor Statistics. 2016. "Census of Fatal Occupational Injuries Charts, 1992-2014." (http://www.bls.gov/iif/oshwc/cfoi/cfch0013.pdf.)

Innovation, Science and Economic Development Canada. (http://www.ic.gc.ca/eic/site/061.nsf/eng/02805.html.)

International Organization for Standardization (ISO). 2017. ISO/DIS 45001, *Occupational Health and Safety Management Systems.* Geneva, Switzerland: ISO.

Industrial Safety and Hygiene News (ISHN). 2012. "Preventing serious injuries and fatalities." (http://www.ishn.com/articles/93471-preventing-serious-injuries-and-fatalities.)

United States Small Business Administration. (http://www.sba.gov.)

World Bank. "SME Statistics." (http://siteresources.worldbank.org/CGCSRLP/Resources/SME_statistics.pdf.)

Session No. 567

Green Light! Advancing Understanding of Sustainable Safety Innovation Benefits

Louise E. Vallee, MS, CIH, CSP, CPE
Crum and Forster
Morristown, NJ

Green Jobs and Sustainable Safety Culture

The expansion of green jobs has promoted inspiration and hope for positive impacts on the economy and the environment. Green jobs are traditionally defined as jobs supporting environmental sustainability. Globally, countries have set "green" objective goals to reduce carbon footprint, expanding the green economy and job growth. The U.S. Bureau of Labor Statistics (BLS) defines green jobs as "either:

- Jobs in businesses that produce goods or provide services that benefit the environment or conserve natural resources.
- Jobs in which workers' duties involve making their establishment's production processes more environmentally friendly or use fewer natural resources."[1]

These jobs include renewable energy, energy-efficiency products and services, pollution and greenhouse gas reduction, recycling, natural resource conservation, and environmental compliance. A perception disconnect has arisen about the safety of green jobs. These jobs are not necessarily safer just because they are green. Some green jobs are associated with brand new technology, introducing new hazards and control strategies, while others may be an evolution of existing jobs, like construction. Often, the potential increased hazards due to new or different conditions and materials that have not yet been fully assessed. This leads to a lack of identified control measures to incorporate into existing safety programs and training. Our challenge is to address safety on the front end of these projects, not lag behind.

Simply put, we need a sustainable culture change in our approach to green job safety. Worker safety should receive as much consideration as environmental impacts. These interests do not compete; both are important. In order to promote culture change, leadership engagement should drive and support continuous safety context awareness and impact. We all need to be on the same page. Green jobs should be safe jobs. Companies making facilities, equipment, and materials acquisition decisions should be able to review the full range of corporate social responsibility impacts, including safety, not just environmental. Job safety hazard analysis, a traditional safety skill, should be a standard process for new jobs or variations. Don Ellenberger, Director of Hazardous Waste Training at The Center for Construction Research and Training, has said, "It is essential that conserving our human resources receive the same consideration as conserving our natural resources."[2]

---

[1] Bureau of Labor Statistics (BLS). 2013. "Measuring Green Jobs." (http://www.bls.gov/green/#definition).
[2] Ellenberger, Don. 2010. "Green Jobs: A Safety and Health Outlook for Workers." CPWR Center for Construction Research and Training. (http://www.cpwr.com/research/cern-meeting-october-2010.)

Social equity, including worker safety, is one of the three fundamental pillars of sustainability. The solidification of worker safety into sustainability efforts is in process, as evidenced by G4 Sustainability Reporting Guidelines and Centers for Disease Control (CDC/NIOSH) and Occupational Safety and Health Administration (OSHA) green jobs websites. The Center for Safety and Health Sustainability (CSHS) was formed as a global collaborative effort among the American Society of Safety Engineers (ASSE), the American Industrial Hygiene Association (AIHA), and the Institution of Occupational Safety and Health (IOSH). The CSHS objective is, "[f]or all organizations to recognize their responsibility to ensure that the protection of human life and the safety, health, and well-being of workers, customers, and neighboring communities is among the primary considerations in any business practices, operations, or development."[3]

LEED Green Buildings, Prevention through Design, and Life Cycle Assessment

The U.S. Green Building Council (USBGC) has developed the Leadership in Energy and Environmental Design (LEED) Green Building rating system. LEED is a voluntary and consensus-based rating system that provides a distinct business marketing advantage and drives green practices. Key LEED criteria include sustainable siting, energy efficiency, water conservation, pollution and waste reduction, material and resource efficiency, and indoor air quality. The quality of LEED certification is a recognized and highly marketable brand. LEED offers materials and resources credits for use of building products from manufacturers that detail environmental performance results across the life cycle. Life Cycle Assessment (LCA) is an emerging "cradle-to-grave" products hazard assessment that is now becoming integrated into environmental impact analysis for products used by the construction industry. LCA has been incorporated into ISO 14000, Environmental Management standard, which provides guidance to corporations on measurement and management of their environmental impacts.[4]

The U.S. Green Building Council and Centers for Disease Control (CDC) have teamed up to accelerate safety and health accident prevention efforts throughout green building planning, construction, and operations. Christine Branche is leading CDC efforts. They have jointly developed an inexpensive online .5 CE course that raises awareness of the benefits of Prevention through Design (PtD) and associated LEED pilot credit for inclusion of worker safety throughout the building construction and operations process. These continuous building construction and operational safety efforts, incorporating PtD, are now known as "Life Cycle Safety."[5]

Building occupant safety and health has long been a LEED objective. Green buildings offer significant occupant benefits over past practices in this regard, in addition to expected resource efficiency. Life Cycle Safety's fundamental positional is that worker safety efforts should extend beyond occupants alone, including appropriate attention throughout the construction and operational time frame. It is ironic that green building construction and maintenance employees may have disproportionate risk working on or in green buildings due to the moniker's safety implications. Green buildings are perceived as safer and the

---

[3] Cecich, Tom. 2013. "The Role of Worker Safety and Health in Sustainability." (www.esafetyline.com/eei/conference pdf files/eeiFall2013).
[4] United States Green Buildings Council. 2016. "Life Cycle Safety: Basics and Connections to Sustainability." http://www.usgbc.org/education/sessions/life-cycle-safety-basics-and-connections-sustainability-6679047).
[5] National Institute of Occupational Safety and Health. 2015. "Life Cycle Safety – What does it mean and why is it important?" (www.cdc.gov/niosh/topics/greenconstruction/pdfs/lifecyclesafety2.pdf).

elevated performance criteria should also encompass continuous safety. LEED, by giving credits for occupant safety, presented a platform for safety expansion recognition and credits.

Together, the USGBC and CDC identified potential credit areas where green practices could improve construction, operations, and maintenance staff safety and health. In February of 2015, the USGBC developed a new LEED "pilot credit" for Prevention through Design. The credit supports review of continuous operational and maintenance safety design, as well as a construction safety throughout the building life cycle. The pilot phase provides an opportunity for user feedback.[6]

Green Construction

Construction remains a high-hazard industry, due to worker frequency and severity exposures to heights, heavy equipment, electrical hazards, power tools, and excavation, as well as industrial hygiene hazards like silica and asbestos. The term "green construction" refers to the segment of energy-efficient and environmentally responsible buildings, whether new construction or retrofits. In 2011, 71% of the construction industry reported using some element of green technology.[7] Green buildings are also known as high-performance buildings, giving design consideration to occupant health. Even with all these attributes, construction worker safety advances remain a "missing link." The LEED pilot presents a potential safety equity solution.

Green building designs are often open and airy, with lots of natural light from atriums and skylights. Roof solar-panel installation, extra interior insulation, and exterior weatherization are common green building construction tasks. These elements increase the risk of falls from heights, electrical hazards, and chemical exposures. Sadly, reports of workers falling through skylights and off roof edges are regularly in the news. Updated skylights are designed and installed to meet OSHA's fall protection standard, but older skylights typically are not. A skylight and roof safety assessment before job initiation is a critical safety task. Older skylights are not "grandfathered" out of OSHA fall protection requirements.[8]

Workers are often unfamiliar with green construction job hazards, and the additional risks introduced. Some of the new construction materials have a limited performance and safety track record. An initial building Life Cycle Analysis review should incorporate a review of tasks and materials with an exposure assessment and control implementation strategy. Continuous training, including classroom sessions and toolbox talks, has long been recognized as a critical component of construction safety programs.

E-waste Recycling and Repurposing

E-waste recycling businesses have expanded with the exponential growth of discarded electronics arising from continual technological advances, the advent of cheap "disposables," and the surge of new consumer products. In the U.S., about 400 million pieces of consumer electronic equipment are recycled annually, yet about 40% of e-waste is discarded.[9] The U.S. Environmental Protection Agency (EPA) indicates that

---

[6] American Industrial Hygiene Association (AIHA). "USGBC Introduces Prevention through Design Pilot Credit for LEED." *Industry News*, March 2015. (https://www.aiha.org/publications-and-resources/TheSynergist/Industry%20News/Pages/USGBC-Introduces-Prevention-through-DesignPilot-Credit-for-LEED.aspx)

[7] Centers for Disease Control. 2015. "Safe, Green, and Sustainable Construction." (http://www.cdc.gov/niosh/topics/greenconstruction/default.html).

[8] Fairfax, Richard. 2006. "Letter of Interpretation." OSHA, Occupational Safety and Health Administration, 1926.501 Fall Protection. (https://www.osha.gov/pls/oshaweb/owadisp.show_document?p_table=INTERPRETATIONS&p_id=25542),

[9] Electronics Takeback Coalition. 2016. "E-Waste Problem Overview." (http://www.electronicstakeback.com/resources/problem-overview/).

electronic waste is the fastest growing municipal waste stream. Recycling is preferred over landfill disposal due to toxic metals released during waste decomposition. Large volumes of electronic waste are shipped overseas, especially to developing countries, generating toxic pollution and hazardous unprotected worker exposures. Residents of these countries may acquire and burn electronics for scrap, which generates clouds of toxic smoke.[10]

It is not easy to recycle electronics due to the diverse components and packaging. Many e-waste recycling startups make very little money due to the resource requirements to break up electronics, and separate out the very small amounts of valuable metals and other recyclables. Many start ups' workers used hammers to break up electronics cases and tools to drill out and unscrew components. Protective eyewear, hearing protection, gloves, and clothing are worn by these employees. Recently, some computer companies have begun to factor in "end-of-life" LCA into computer design to facilitate ease of dismantling and demanufacturing, a positive advancement.

Manual product sorting, dismantling, and separation can result in a variety of injuries including:

- Back injuries from lifting, twisting, and bending
- Repetitive motion and upper extremity injuries from hand- and power-tool use
- Cuts from sharp product pieces, glass, and tools
- Eye injuries from flying debris and dust
- Noise
- Toxic metal dust, such as lead, mercury, and cadmium from CRTs and LCDs[11]

Technological improvements have driven automation in the e-waste recycling industry. Large volumes of materials are now moved via conveyors, machine processed, and automatically sorted, reducing hazards to workers and increasing profitability. Typically, toxic products, such as fluorescent bulbs and batteries are removed from the waste stream for separate handling. Discarded electronics are shredded and glass, metals, and plastics are separated by magnets and optical scanners.

"Repurposing" electronics has become the new fourth "R" (after reduce, reuse, and recycle). Repurposed electronics are restored and repaired back to a working condition and sold. Profitability is increased from the sales. Repaired electronics provide a true social equity value, giving people access to inexpensive or donated electronics so they can benefit from the technology. Developing countries are an additional market for repurposed electronics.

Renewable and Alternative Energy

Wind turbines are very large structures, up to 300' tall, which generate wind-driven electricity. Wind turbine project installations are growing at 36% per year. Wind energy turbine construction and maintenance workers are exposed to a number of potentially severe hazards that can result in death or serious injuries. Exposures include falls from heights, electrical shock and arc flash, fire and egress, confined spaces, struck by moving equipment, and bad weather, including lightning. Falls from heights are a recurring catastrophic injury, due to absence of or improper use of fall protection equipment and worker fitness issues.[12] There are more than 500 U.S. facilities, manufacturing over 800 wind turbine components. Chemical exposures during manufacturing include epoxy resins, styrene, various solvents, hardeners, and fiberglass. Wind turbine manufacturing and transportation involve special crane and

---

[10] IfixIt.org. 2016. "E-waste is the Toxic Legacy of Our Digital Age." (http://ifixit.org/ewaste).

[11] Hazard Evaluation System & Information Service, California Department of Public Health. 2012. "Electronic Waste Recycling: Working Safely." (www.cdph.ca.gov/programs/hesis/Documents/eWaste.pdf).

[12] OSHA. 2016. "Green Job Hazards: Wind Energy." (https://www.osha.gov/dep/greenjobs/windenergy.html).

rigging risks, due to the extremely large size of structural components and turbines. Vibroacoustic complaints have arisen from residents located near wind farms.

Solar energy continues to be a green-jobs growth segment. Currently, over one million homes and businesses have been equipped with solar panels, as solar energy has become both more efficient and less expensive. Multiple worker health and safety hazards exist in the manufacture, installation, and servicing of solar panels. Worker installation exposures include electrical shock, arc flashes, falls from roofs and heights or through skylights, and burns. Solar-panel manufacturing entails worker exposures to toxic and/or flammable chemicals, metals, and gasses.[13]

Geothermal energy is rapidly expanding, and is considered to be one of the largest energy-related natural resources, second only to solar. Geothermal systems create electricity by tapping natural heat from water and steam reservoirs deep within the earth. Some systems drill into the earth's geothermal reservoirs, using the hot water and steam to create electricity. Other systems pump recirculating water through underground piping, to be heated by the earth and used to heat and cool buildings. Traditional drilling and excavation hazards exist, as do exposure to brine solutions or glycols.[14]

The demand for biofuels is expanding due to a low carbon impact. Biofuels are manufactured from renewable resources, including plants and grains, vegetable oils, and municipal and industrial wastes that have undergone treatment. Many of these chemicals or intermediates are flammable, and some are capable of creating hazardous chemical reactions. The two major types of biofuels produced are ethanol and biodiesel, which are typically blended with petroleum biodiesel.[15]

Reforestation

Reforestation refers to the replenishment of forests and woodlands that have been depleted by deforestation. Reforestation work occurs seasonally and in remote locations. Workers typically work long days clearing areas and planting seedlings, especially due to payment incentives. Workers reside at the jobsites in tents or cabins. Expected hazards include bears and wildlife, insects, falls from heights, power tool use, tree clearing or pruning injuries, landslides, slips and falls, back injuries, repetitive motion injuries, burns, forest fires, severe weather, and vehicle accidents. Some seasonal workers travel and camp with their families who then share risks. The remote work makes emergency response difficult; each camp should have medical emergency response trained staff. Regular communication with workers is important early on to identify potential problems and respond, requiring additional non-cell-tower based equipment.

Sustainable Product Issues: Tin Whiskers, Plastics, and Drones

Tin whiskers, conductive crystalline filaments, have long been recognized as an electronics failure problem that develops from internal and external tin-plating stresses. Tin plating enhances solderability. Whiskers can sprout quickly, or grow over many years, to lengths ranging from 7 to 10 millimeters. The potential for whiskers to develop on circuit boards and components exists for any tin/high-tin content or cadmium plating, but especially bright tin plating. Electronics plating, in the past, incorporated lead into the tin plate to reduce whisker formation. With the advent of anti-toxics legislation worldwide, such as European Restriction of Hazardous Chemicals (RoHS), lead had to be removed, resulting in an increased incidence of tin whiskers. Many electronics failures have been attributed to the tin-whiskers phenomenon, including satellite deaths and failures involving the space shuttle, medical devices, weapon systems, power plants, and consumer products. These failures arise from transient and stable short circuits, metal

---

[13] OSHA, Green Job Hazards: Solar Energy. 2016. (https://www.osha.gov/dep/greenjobs/solar_electrical.html).
[14] OSHA, Green Job Hazards: Geothermal Energy. 2016. (https://www.osha.gov/dep/greenjobs/geothermal.html)
[15] OSHA, Green Job Hazards: Biodiesels. 2016. (https://www.osha.gov/dep/greenjobs/biofuels.html.)

vapor arcs, and whisker debris spread and contamination.[16] Supplier mitigation strategies have varying success and are application dependent. Conformal coatings reduce whisker potential but may be penetrated by growth long term (>10 years). Original equipment manufacturers should integrate viable tin-whisker prevention strategies, particularly for critical risk products.[17] The International Electronics Manufacturing Committee (iNEMI) identified environmental test conditions for tin-whisker prediction, and three test methods to characterize and qualify tin finishes.[18]

The ocean currents of the earth form five natural, extremely large, whirlpools called gyres. These massive, slowly moving whirlpools collect highly durable plastic debris. Ten percent of plastics manufactured in the world end up in the ocean, while 50 percent end up in landfills, and about ten percent is recycled. Gyre plastic debris comes from Asia and the United States, taking one year and six years, respectively, to arrive. Sunlight and wave action, over time, cause the plastics to break down into tiny particles, which remain in the ocean and are joined by fresh plastic debris. The center of the gyre is stable, which is why debris collects there. Debris also sinks to the ocean floor and accumulates. This debris is harmful to marine life. These gyres are becoming ocean deserts, devoid of ecosystems. The North Pacific Gyre is twice the size of Texas, and is known as the "Great Garbage Patch."[19] Multiple organizations have initiated research and awareness efforts to initiate action on this vast problem.

Drones are here to stay, and usage is expanding globally at an exponential rate, even while the United States Federal Aviation Administration (FAA) struggles with regulatory decisions and frameworks. While privacy and safety concerns are not yet fully resolved, clearly drones have the potential to contribute significantly to economically beneficial sustainability in agriculture, construction, petrochemical, energy, transportation/delivery, and telecommunications. Environmental, crop condition, and remote property surveillance have demonstrated positive results.[20]

Conclusion

Green jobs' sustainability presents a "horizon of opportunity" to achieve environmental health and safety objectives in alignment with both business profitability and corporate social responsibility goals. Safety culture development is key to continuous improvement and success. EHS professionals are often tapped for initiative leadership. In the June 2013 *Professional Safety* article, "Safety and Sustainability; Understanding the Business Value," Kathy Seabrook and Daryl Hill referenced an Aberdeen study that found that EHS professionals led sustainability efforts for the majority of companies reporting. EHS expertise can be found leading safety metrics development and benchmarking, resource conservation, waste reduction and zero landfill success, "green" project safety, "green' buildings and fire protection, and sustainable culture development. Communication and development of common context understanding are key for strategic initiative promotion. Collaborative sustainability leadership presents welcome opportunities to advance the environmental health and safety agenda and innovatively contribute to sustainable profitability.

---

[16] NASA. 2009. "Basic Information Regarding Tin Whiskers." (http://nepp.nasa.gov/Whisker/background/index.htm).

[17] Osterman, Michael. 2015. "Tin Whisker Mitigation." Venter for Advanced Life Cycle Engineering. (http://www.calce.umd.edu/tin-whiskers/presentations/TinWhiskerMitigations.pdf).

[18] International Electronics Manufacturing Committee. 2004. "Tin Whisker Acceptance test requirements." (http://thor.inemi.org/webdownload/projects/ese/tin_whiskers/Tin_Whisker_Accept_paper.pdf).

[19] National Geographic Society. 2016. "Great Garbage Patch." (http://education.nationalgeographic.org/encyclopedia/greapacific-garbage-patch/).

[20] Hering, Garret. 2014. Green Biz, "Sky's the limit: Drone Makers Target Sustainability." (https://www.greenbiz.com/blog/2014/07/11/8-drone-developers-target-sustainability.)

Bibliography

American Industrial Hygiene Association. "USGBC Introduces Prevention through Design Pilot Credit for LEED." *Industry News*. March 2015. (https://www.aiha.org/publications-and-resources/TheSynergist/Industry%20News/Pages/USGBC-Introduces-Prevention-through-DesignPilot-Credit-for-LEED.aspx).

Bureau of Labor Statistics (BLS). 2013. "Measuring Green Jobs." (http://www.bls.gov/green/#definition).

Cecich, Tom. "The Role of Worker Safety and Health in Sustainability." (www.esafetyline.com/eei/conference pdf files/eeiFall2013).

Electronics Takeback Coalition. "E-Waste Problem Overview." (http://www.electronicstakeback.com/resources/problem-overview/).

Ellenberger, Don. 2010. "Green Jobs: A Safety and Health Outlook for Workers." CPWR Center for Construction Research and Training. (http://www.cpwr.com/research/cern-meeting-october-2010).

Fairfax, Richard. 2006. "Letter of Interpretation." OSHA, Occupational Safety and Health Administration, 1926.501 Fall Protection. (https://www.osha.gov/pls/oshaweb/owadisp.show_document?p_table=INTERPRETATIONS&p_id=25542).

Hering, Garret. 2014. Green Biz. "Sky's the limit: Drone Makers Target Sustainability." (https://www.greenbiz.com/blog/2014/07/11/8-drone-developers-target-sustainability).

IfixIt.org. 2016. "E-waste is the Toxic Legacy of Our Digital Age." (http://ifixit.org/ewaste).

International Electronics Manufacturing Committee. 2004. "Tin Whisker Acceptance test requirements." (http://thor.inemi.org/webdownload/projects/ese/tin_whiskers/Tin_Whisker_Accept_paper.pdf).

National Aeronautics and Space Administration (NASA). 2009. "Basic Information Regarding Tin Whiskers." (http://nepp.nasa.gov/Whisker/background/index.htm).

National Geographic Society. 2016. "Great Garbage Patch." (http://education.nationalgeographic.org/encyclopedia/great-pacific-garbage-patch/).

National Institute of Occupational Safety and Health (NIOSH). "Life Cycle Safety – What does it mean and why is it important?" (www.cdc.gov/niosh/topics/greenconstruction/pdfs/lifecyclesafety2.pdf).

Occupational Safety and Health Administration (OSHA). 2016. "Green Job Hazards: Wind Energy," (https://www.osha.gov/dep/greenjobs/windenergy.html).

Occupational Safety and Health Administration (OSHA). 2016. "Green Job Hazards: Solar Energy." (https://www.osha.gov/dep/greenjobs/solar_electrical.html.)

Occupational Safety and Health Administration (OSHA). 2016. "Green Job Hazards: Geothermal Energy." (https://www.osha.gov/dep/greenjobs/geothermal.html."

Occupational Safety and Health Administration (OSHA). 2016. "Green Job Hazards: Biodiesels." (https://www.osha.gov/dep/greenjobs/biofuels.html).

Osterman, Michael. 2015. "Tin Whisker Mitigation." Venter for Advanced Life Cycle Engineering. (http://www.calce.umd.edu/tin-whiskers/presentations/TinWhiskerMitigations.pdf).

United States Green Buildings Council, 2016. "Life Cycle Safety: Basics and Connections to Sustainability." (http://www.usgbc.org/education/sessions/life-cycle-safety-basics-and-connections-sustainability-6679047).

The Lies We Tell Ourselves and Other Challenges to Reducing Fall Risk

Thomas E. Kramer, P.E., C.S.P.
LJB Inc.
Miamisburg, OH

Introduction

Falls are one of the leading causes of workplace fatalities and violations related to fall protection have been at the top of the Occupational Health & Safety Administration's (OSHA) list for the past six years. Unfortunately, fall injuries and fatalities continue to increase, even with the safety industry working hard to reverse this trend.

To be clear, I do not think building better and stronger personal protective equipment (PPE) for fall protection will change these results. Despite significant efforts by OSHA, the American National Standards Institute (ANSI) and other groups, the struggle against knowingly and unknowingly risky behaviors is our focus, and it is an uphill battle. A good safety record, coupled with entrenched methods of information processing, training, and social influences, can lull organizations into a false sense of security, leading many to believe they are good, rather than just lucky. While self-talk and mental shortcuts can lead to risky decision-making for workers, there are many practical and evolving methods for reducing falls.

By recognizing the powerful social and cognitive factors that can impact behavior—and by implementing better systems through proven fall prevention and protection techniques—organizations can improve safety and reduce risk.

Social and Cognitive Factors

It is common for people to be shocked when they witness or view documentation of a worker actively overcoming safety precautions and putting themselves in harm's way. But it is understandable due to information-processing shortcuts—or heuristics—that are a commonly used cognitive tool. Heuristics are mental shortcuts that ease the load of decision making, although they limit a person's ability to reason intuitively. The failure in judgment and decision making is especially poor when related to issues with greater orders of magnitude (i.e., life and death) (Tversky & Kahneman, 1974).

Specifically related to fall risk, heuristics can lead a person to reason that falls do not happen often or that they certainly are not planning on falling. If a person has never witnessed, experienced or nearly missed a fall, it is mentally easier to justify risky behavior. Anecdotally, this is exactly why, in my experience, those who have experienced or witnessed falls or near misses are the primary advocates for fall protection, since they know the tragedy.

Still, several issues can lead to and perpetuate the cognitive biases that minimize a person's ability to reason and react to fall risk. The brain's overall limited capacity, mental noise, as well as emotional and moral motivations, can further minimize an individual's ability to reason properly. One of the most problematic issues for establishing a positive and effective safety culture is social influences.

As the saying goes, "Repeat the lie loudly and frequently and it will become truth." I have found the following statements to be progressive in some organizations or, in some cases, it is simply part of the worker's self-talk. In these cases, it can be challenging for workers to go against the norm and execute appropriate safety policies:

- "I'll be careful."
- "I'm not that high."
- "It will only take a minute."
- "I've got a harness on."
- "I'm tied off."

Even when confronted by data that compares fall speed and human response factors, some workers will still fall back on excuses and rationales for why they will not fall. In reality, the human body can fall up to six feet before it even responds. In most scenarios, this response time is much too late to avoid an accident.

Evolving Regulations and Standards

One of the most significant ways the industry is working to reduce fall risk is through the establishment of updated regulations and standards. It is an unprecedented time related to fall protection standards, as both OSHA and ANSI have just updated fall protection requirements and guidelines. However, regulations and standards should never be the primary focus to protect workers. It is critical to take this information and put it in terms of systems built to protect workers and the behaviors necessary to keep them safe.

OSHA regulations are the law, and they establish the bare minimum requirements of compliance. Since these regulations are created through the legislative process—and must address needs for all sizes and types of organizations—it can be a significant challenge to finalize updates. Fortunately, regulations that reflect today's equipment and general principles just went into effect in January 2017 (OSHA 29 CFR 1910 Subparts D and I).

Some of the most significant and impactful changes that have resulted from the new regulations are listed here, with references to the actual regulation location:

- Requirement to provide training for exposed workers (1910.130)
- Requirements for fall hazard risk assessment (1910.132 (d))
- Load rating of walking-working surfaces (1910.22 (b) and (d)(1))
- Establishment of safe distances from a roof edge (1910.28 (13))

Similarly, the ANSI/ASSE Z359 Standards Committee has been working on updating fall protection standards for the past decade to address specific equipment components, such as harnesses, connectors, and lanyards. The ANSI/ASSE Z359.2 standard is also the only place to find vetted guidance on a managed fall protection program, including basic safety principles like hazard surveys, hazard elimination and training. While the guidelines in the committee's Fall Protection Code are not legally binding, they are a higher standard of care, and represent industry best practices established by experts from all facets of the industry.

In 2016, the Committee undertook a landmark revision to the Z359.1 standard that requires manufacturers to meet all provisions of the Fall Protection Code to claim compliance with ANSI. These requirements are now more stringent and protective than the previous Z359.1 standard, which was published in 2007 (the original standard was published in 1992). The new Z359.1 standard takes effect on August 14, 2017.

This breakthrough revision most directly impacts equipment manufacturers. But it is critical for all buyers of fall protection equipment to learn about the new standards so that they can be better consumers, with a full understanding of the changes that come with the updated requirements.

Prevention through Design

One effective way to reduce fall risk is to eliminate hazards before they are ever created. Although it may be easier to recognize safety hazards in an existing structure, it is safer and more cost effective to implement protection measures before structures or processes are built. Prevention through Design (PtD) ensures that safety measures are evaluated and implemented during the programming and design phases of a project. This allows organizations to create systems where a worker's bias is less important relative to their exposure.

In addition to increased safety, incorporating safety at this stage saves money because designers will not even need to erase lines on their drawings; the safety aspects are simply programmed into the design. Organizations that have applied PtD programs have proven meaningful results: Life-threatening work hazards are reduced, productivity is improved, and costs are lowered.

Save Money and Lives

When safety professionals—and front-line employees who know the work best—are brought in at the beginning of a project, safety concerns can be addressed within the project's original timeline and budget. Evaluating and implementing safety measures early in the project ensures optimal safety for workers, from construction into operations, and throughout ongoing maintenance.

Besides these safety benefits, applying PtD significantly reduces the cost of addressing fall protection issues. The impact to bottom-line cost can be dramatic. For example, eliminating the need to erect, use, and dismantle scaffolding to solve fall protection issues after project completion—even in just one area—can result in significant savings and reduced risk over the life of a building. PtD programs can also reduce overall operating costs. When fall hazards are eliminated or managed through engineering control solutions, there are fewer fall protection issues to contend with through the life of a facility. This translates to fewer equipment purchases, less training, and a reduced number of systems to manage.

Fall Protection Equipment Advances

Fall protection equipment is the most visible aspect of a fall protection program and meaningful advances have been made in recent years to improve its functionality and application. Proper use of modern equipment can significantly reduce risk for workers at height. However, it's important to note that an organization's investment in equipment can be nullified–or even magnified–due to equipment misuse or "cheating." It is critical to always use equipment in the manner in which it was intended and tested, which is the challenge when it comes to equipment standards. Most standards specify design requirements and test methods, while workers need to be trained on use and limitations. Training helps organizations create better systems while also addressing safer behaviors for workers.

The following key advances in equipment can have a positive impact on worker safety when applied correctly.

Body-support Device

Unlike other forms of personal protective equipment (PPE), such as hard hats, boots and safety glasses, fall protection equipment is not intuitive. Proper harness fit is not easy to achieve, but it is critical in order for a body-support device to work as intended in the event of a fall. When workers wear harnesses that are not properly fitted and hang loose on their bodies, serious injury, or worse, can result during a fall. As I often share in our competent person training, "Either you secure the strap where you want it, or it will go where it chooses!"

All organizations should train each authorized person in the proper fit of their harness. Another recommendation is to suspend authorized persons in their harnesses to show them how it feels to be suspended. This exercise has proven to impact worker behavior, as they have personally felt the effects of misuse.

When considering personnel assignment to tasks requiring use of a harness, there are several factors to consider, including weight, height, underlying medical conditions, previous injuries, vertigo, side effects of medication, etc. To create better systems, medical evaluations and harness-fit testing could be required in the next 5-10 years within the ANSI standards.

Twin-leg, Self-retracting Devices

One of the fastest growing segments of the fall protection industry is the personal mounted twin-leg, self-retracting devices. In some cases, manufacturers boast that self-retracting devices are safer alternatives to energy-absorbing lanyards. However, consider the following items when evaluating this issue:

- Self-retracting devices often have a much lower tensile strength relative to an energy-absorbing lanyard (5000 pounds vs. 3000 pounds). Therefore, in a misuse application, an energy-absorbing lanyard provides additional residual strength.

- The testing of the twin-leg, energy-absorbing lanyards is much more rigorous than twin-leg self-retracting devices.
- Self-retracting devices, per the ANSI Z359.14 standard, are not tested in a dorsal-mounted position.
- There have been cases where the lack of the snaphook on the dorsal d-ring for the twin-leg, self-retracting device has prevented workers from disconnecting when another health and safety incident, such as a fire, occurs, resulting in the need to rapidly disconnect.
- At least one manufacturer refuses to sell its leading-edge, self-retracting devices if the consumer shares that it will be used in an application where concrete is the edge. This is because the leading-edge test in ANSI Z359.14 is a steel edge, which may not be the most aggressive edge in causing line failure.

Horizontal Lifelines (HLL)

HLL systems are perceived as low-cost, simple, aesthetically pleasing solutions, and they are touted for their ability to span long distances without overhead structure. However, they always require a qualified person's involvement due to the energy equations involved to properly design them. If the system is not designed properly, the end anchorage structure may be overloaded or cause an individual who has fallen to hit the ground, due to the increase in anchorage system displacement.

End-anchorages forces can easily exceed 5000 pounds, and fall clearance requirements can also be very high with these systems. Changing parameters, such as number of users, span length, intermediate anchorages, and presence of in-line energy absorber, can greatly affect end-anchorage forces and fall clearances. Because of this, it is critical not to make adjustments in the field without oversight by a qualified person.

These systems can also be misused due to poor installation (e.g., wire rope clips installed incorrectly) and improper intermediate anchorages that could result in point loading and failure of the cable.

The primary way to mitigate issues with horizontal lifelines is to involve a qualified person in design, inspection, and use. This involvement is required by OSHA regulations, and is truly critical for such complex systems. In addition to initial design, qualified persons should also:

- Specify all equipment components used in an HLL, including intermediate anchorages, pass-through devices and anchorage connectors.
- Develop detailed use and rescue procedures that provide design basis (number of users, maximum weight, clearance requirements, and inspection requirements) to ensure systems are not used outside the manner in which they were designed.

Conclusion

Significant advancements have been made in the fall protection industry in recent years, especially related to standards, regulations, advanced equipment, and application of Prevention through Design. Still, in many cases, organizations have not truly addressed the real risks and exposures for workers, and the number of fatalities continues to rise. It is critical to minimize negative social influence and mental shortcuts that can lead workers to making risky decisions.

When organizations create better systems, and workers stop telling themselves (and their co-workers) the lies that lead to accidents, the application of new standards, techniques and equipment can have a truly meaningful impact on worker safety.

Bibliography

Tversky, A. & Kahneman, D. 1974. "Judgement under uncertainty: Heuristics and biases." *Science.* 185 (4157): 1124–1131. doi:10.1126/science.185.4157.1124. PMID 17835457.

Humor in Safety: From Blah Blah Blah to Ha Ha AHA!

Tim Page-Bottorff, CSP, CET
Senior Consultant
SafeStart
Belleville, Ontario, Canada

Introduction

Wah wah wah wah wah—the trombone sound of the teacher talking in the Charlie Brown cartoon. Do you know what I am talking about? Another mind-numbing, boring regurgitation of uninteresting information presented by a droning monotone voice, "death by PowerPoint" or the dreadful "persistent slide reader." Have you ever attended this type of lecture, meeting, or conference session before? Of course you have. I have. Everyone has. Time and again, more often than not, this is the type of presentation we're exposed to.

Let's face it, you would rather be playing sports, cards, exercising, or even working, than listening to this drivel. More importantly, you would rather be doing your favorite things with your family. Now, imagine you're not in this session by choice but your job requires you to attend. This can't get much worse! Oh wait, it can—today's topic, "Safety Training."

Maslow's hierarchy, common sense, and those "favorite things with your family" would lead you to believe that staying alive would be of interest to most people. But sadly, safety is not everyone's favorite subject. No surprise though, considering most adults feel they are safe enough already, coupled with years of rules, procedures, negative enforcement (as opposed to positive reinforcement) and uncomfortable behavior change. I mean, think about it. When was the last time a police officer pulled you over and said, "Thanks for driving the speed limit?"

As a safety instructor, this is what you're up against. Before you even begin, your students have a predisposed expectation that you will deliver yet another boring or repetitive class on a subject that they don't want or need.

But what if you had an important message to deliver; a critical piece of information that would help them make it home to their family and their favorite things? And safety is always an important message, even if it is the umpteenth time through a regulatory refresher. Always. That's the real risk here; when an important message is missed because the instructor didn't get their students' attention, because of the "blah, blah, blahs."

I teach a class called HAZWOPER. If you're a safety professional, chances are you know what HAZWOPER means, for those of you that don't, it's an acronym that stands for Hazardous Waste Operations and Emergency Response. This class requires 40 hours of instruction in order to work at a hazardous waste site. It has to be one of the most boring classes I have ever attended, or taught for that matter. But, as with any training course, the instructor can make or break the class. HAZWOPER, as boring as the material can be, is no different.

If an instructor's presentation capabilities are so crucial to delivering important messages, then we need to understand what makes a good instructor or trainer. Well, it starts with you. First, think about the instructors who have made an impact on you, those that really touched a chord with your life. I have a few, but one is my dad. He really taught me a lot. But what did he do that was so special? What did he do that was so much better than Miss Glentzer in the fifth grade? It was mostly hands-on training and a very stern approach. Those characteristics are great, but that's not what I remember the most. What I remember the most is my dad saying, "Why work if you're not going to have fun doing it?" I never will forget that.

After working with several disgruntled municipal employees, I took my dad's advice and started adding tidbits of humor into my never-ending compliance classes. Once that happened, I received some significant positive feedback. The more humor I used the more positive safety response I received in the field during practical application. Negative energy became positive energy when students remembered what I had told them in a class a year ago. Replacing the "blah, blah, blahs" with a little "ha, ha, ha" went a long way in helping me give these students their "Aha! moment" and solidifying the points I was teaching. So where does humor fit? It fits everywhere and in many ways.

Poking Fun (at Yourself)

First and foremost, you need to set the class at ease, let them know you've been in their shoes and provide self-disclosure to the students. Open yourself up a little. Letting the students know you are as human as they are can go a long way, especially if you can relate some humorous mistakes you've made and the lessons you've learned as a result. I teach a lot of SafeStart® classes, and one of its great successes is this self-disclosure component. When you open the door, what does that do for everybody else? It gets them to open the door. You're going to see automatic participation, whether you want it or not. Self-disclosure is huge, and you can have fun with it to make yourself more relatable.

Finding Humor in Visuals

There are so many funny video clips, photos and cartoons in circulation now, you can fairly easily locate one that will serve as a valuable anecdote in your presentation. These visuals provide laughter, but with appropriate selection can also illustrate or support important teaching points—helping students relate, form more complete thoughts and retain the information better than they would without humor. Don't use random pieces just because they're funny. Make sure they fit, that includes selecting visuals that are appropriate for your audience. It's also important to know a bit about your audience before you present—what may not be offensive to some could potentially ruin your experience (and your rapport) because you didn't do your research to know about a particular incident that happened on site or to an employee of that site.

Additionally, visual aids are extremely helpful when it comes to information retention. Simply listening to a lecture or reading a document doesn't have as much of an impact. People are more likely to remember something presented this way or relate to a piece of information that comes with some form of visual representation, such as a YouTube video. Plus, they're more likely to share this with someone else, which helps retain the information by 90 percent (Exhibit 1).

Exhibit 1. The Learning Pyramid

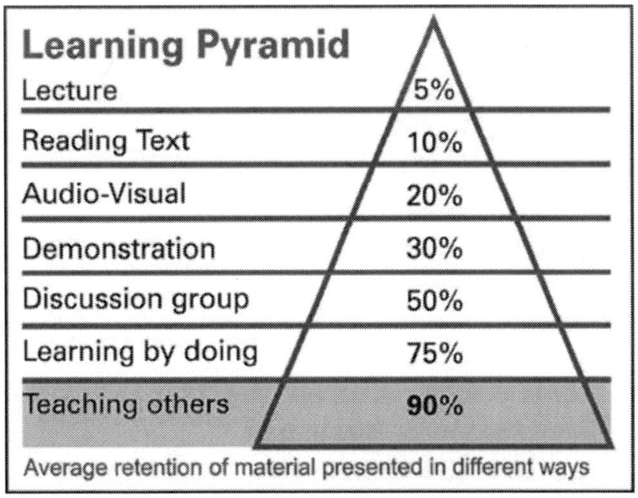

Fun and Games

Back to the HAZWOPER class I mentioned earlier. Other than the typical hands-on component that gets attached to a class of this magnitude, there are several other auxiliary items you can add to make it more upbeat or humorous. For example, play a game for reviewing test questions. There are several free PowerPoint®-based game templates like "Jeopardy," "Who Wants to be a Millionaire," and "Family Feud" that lend themselves to this type of activity. They increase retention levels, participation and fun. It also applies a teaching style that shows employees you've put a lot of thought into teaching for their benefit and not just because you have to.

What sounds more appealing? Being given a series of questions to answer on paper where you silently write the answers (usually on your own) in a limited amount of time, then submit them to be graded for a score based on how many correct answers you gave OR playing a Jeopardy-like quiz game where everyone competes to shout out the right answer, receives points for each correct answer given, and the person that finishes the game with the highest score gets bragging rights.

Introducing a fun way of learning a concept will provide people with a challenge and feed into the competitive nature of the games, making the participant forget that they're learning.

Another tactic I use during the Site Characterization component is to incorporate building toys, like Lego®, Lincoln Logs®, Tinkertoy® or K'nex®. Beyond the childhood memories and fun, these toys give the students an opportunity to "see" the hot, warm and cold zones inside before they have to build them outside. If you don't spend the time to make the learning experience memorable, how can you expect your students to take anything valuable away from your session?

Stories

As I touched on earlier, I teach a lot of SafeStart classes. SafeStart's greatest ingredient is that it is filled with real people telling real stories about their real injuries. It even teaches participants how to tell their own SafeStart stories. And for good reason, these stories stick like glue. After providing consulting services for clients over several years, I can still remember my very first stories involving the State-to-Error Risk Pattern (Exhibit 2). Shoot! I can still remember stories told to me way back in 1976! The point I am trying to make is that stories can linger in a good way, and I often remember them when I'm doing similar activities and will modify my behavior to avoid the injury or close call that happened in the story I was told.

Exhibit 2. The State-to-Error Risk Pattern

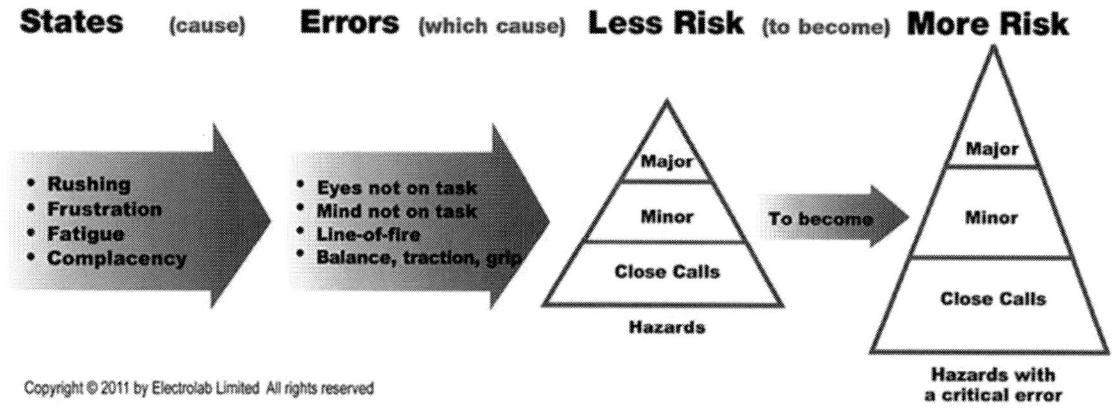

When broken down, a story is simply a connection of cause and effect. The brain works on the existing knowledge of what we know and what we can prove. As a result, the basic human instinct of learning by analyzing cause and effect can be very useful from the teaching perspective. When we listen to a lecture or read text, our brains only decode the meaning of the words, but when we listen to a story almost our entire brain becomes involved, including the areas we would use if we were the ones to whom the story happened. The more visual and descriptive to the senses the story is, the better we'll relate to it and remember it.

Feel the Love

Back to my dad for a minute: He has worked in many different fields after leaving the Air Force. He didn't have one job he really enjoyed doing, but always found enjoyment in them. If you love what you do, you will LOVE what you do. You have a choice, and you can choose to find and focus on those things that you love. In safety it's easy; saving one life and reducing injuries overshadows an awful lot of tiring or thankless work. How fortuitous is it that having fun at work can translate to having fun while training or learning? And if you're passionate about safety, it will show in your presentation, and it will become infectious among your students and your colleagues.

Energy

A speaker without an animated personality might have a lot to share, but they can very quickly turn into that teacher from Charlie Brown. A lack of movement, charisma or expression (on your face or in your voice) are immediate turn-offs, and it won't matter how many years of experience you have or how much you know if you're not conveying it with some energy. Remember, if you're bored with the material, you're boring presenting it. Your audience is more likely to be captivated by someone who moves around, changes pace, makes them laugh, and stimulates their senses. It's nothing personal; it's just how the human brain works.

A Balancing Act

These are just a few ways to make training fun and impactful. Pick and choose the types that work for you, just try to avoid relying too heavily on any one thing. Too many videos will eventually become boring, just like too much lecture will, and too much comedy will almost certainly obscure or obliterate your important message.

Having fun is great, but you must know where to draw the line. I mean, you will always come back to the seriousness of safety—the whole reason why you are there in the first place—but finding a good balance of humorous or interesting activities to intersperse with the important messages you need to communicate will ultimately make for a memorable class. In fact, it has been proven time and again that using a combination of learning activities increases retention.

Start slow and build elements into your class. Especially with humor, practicing your delivery is essential. Like an actor on a stage, you need to hit your marks. You need to keep your audience on a roller coaster of emotions: laughing one minute, drop dead serious the next, followed by anticipation, then through a winding tunnel of self-discovery until they cheer for more when the ride is over.

OK, I might be optimistic about the cheering, but I bet you'd settle for attentiveness and a few smiles. Just remember, it all starts with you. If I can leave you with one point it is this: the minute you get comfortable with your preparation, then it is time to do more work.

Don't forget the reason you are doing what you are doing. Standing in front of people, telling them stories and sharing information might not seem heroic, but what you do can actually save lives and prevent serious injuries. Safety professionals might be perceived as a necessary evil or big bores, but you can change that perception and still get them home safely at the end of the day.

If you don't enjoy it, guess what? Your students will see it. Your employees will see it. It all starts with you. And if you do well by your students in giving them an enjoyable class, you will be rewarded with an enjoyable class. Remember that it isn't about you, it's about them. But it can be beneficial to you both. Life is funny that way.

My Car Does WHAT? Engineering Controls for Common Vehicle Crash Types

Paul H. Farrell, CDS
Nationwide Mutual Insurance Co.
Harleysville, PA

Introduction

Vehicle collisions may result in physical damages, bodily injuries, fatalities and litigation under various theories of liability. The costs of collisions represent a major economic burden on society, and the prevention, mitigation or outright avoidance of these events present an urgent call to action for safety professionals.

Historically, our efforts to improve results fell into one of several categories:

- Driver education and licensing
- Vehicle modification to enhance driver control
- Vehicle modification to increase survivability
- Roadway design and traffic control improvements
- Traffic laws and regulations (creation/enforcement) to limit behaviors

Now, technology has advanced to a point where we may be able to make the vehicle able to drive itself, with predictable rules and reactions to various dynamic situations. The research into autonomous vehicles (AVs) has yielded many technological advances that can be applied to normally piloted vehicles immediately to either enhance driver awareness of their surroundings or to intervene directly on behalf of the human operator to prevent or mitigate a collision event.

Unfortunately, the availability of these innovative enhancements is typically introduced to the automotive marketplace very slowly, and is usually as expensive options on high-end luxury cars. Original equipment manufacturers (OEMs) can view safety enhancements as low priority, since many buyers are unfamiliar with the technology, and so-called "info-tainment" systems tend to captivate buyer's attention.

This paper provides a recap of the most common collision types recorded annually, and highlights technology options currently available for light-duty, medium-duty and heavy-duty commercial vehicles in response to those collision types. Further, non-brand-specific resources are available to help educate car operators, fleet managers and safety professionals on additional technology that is rapidly becoming available on most makes and models to address other common, but less severe, collision types.

Common Collision Types in the USA

The National Highway Traffic Safety Administration (NHTSA), the Federal Motor Carrier Safety Administration (FMCSA), the Insurance Institute for Highway Safety (IIHS) and National Safety Council (NSC) each tally and publish crash data on an annual basis. The research for this paper comes principally from NHTSA resources. Sixty-seven percent of all two-vehicle crashes were categorized as forward collisions (aka "rear-enders"), sideswipes, head-ons, and striking pedestrians or cyclists. By including single-vehicle collisions related to unplanned lane departures, the total increases even further. Because some collisions have multiple causes or contributing factors, the statistics can become confusing.

Without belaboring the statistical point, it is clear that these three types of collisions are highly common. Further, from an insurer's perspective, we find that there are some common characteristics, which make these events especially troublesome.

First, they are likely to include fatalities or multiple, severe injuries. Even at moderate speeds, two-vehicle collisions with a great speed differential release a lot of energy.

Secondly, there are often multiple people injured.

Third, the negligence on the part of the insured driver is typically clear by the fact that if they had been more attentive, they might have avoided the collision or taken steps to mitigate the crash.

Fourth, some element of driver "impairment" may have been involved: drunk, drugged, drowsy, or distracted.

Fifth, the likelihood of litigation is elevated due to these previously mentioned factors.

Sixth, the potential to mount a material, affirmative defense is limited by the circumstances, and we typically do not blame the victim of the collision unless there is clear, well-documented, gross or wanton negligence exhibited by the other parties.

Finding any practical way to avoid these collisions or to mitigate the severity of the impact would be helpful for all parties involved.

Teaching drivers to be more vigilant or making crumple zones more energy absorbing does not have a material effect on reducing the likelihood of future collisions of this sort; however, pieces of autonomous driving platforms can make a difference.

Adding a Hypervigilant Copilot

Autonomous vehicles need to be able to see their surroundings and, even more critically, understand the context of what is being observed. It is insufficient to detect an object in the roadway when that object may not represent a threat to the vehicle. Distinguishing threats from non-threats requires a degree of artificial intelligence (AI) and object libraries that can help classify the object as a pedestrian, cyclist, car, truck, or motorcoach. Measuring the rate of closure and tracking objects that are moving laterally to the roadway lane are additional tasks needed to recognize threats.

With the combination of lasers, radar and cameras, credible detection systems have been combined with AI to:

- Provide warnings of real threats (with less than a 1 in 1000 failure rate).
- Recognize lane markings to alert drivers of lane drift (or make adjustments in some cases).
- Track up to 100 pedestrians and cyclists who may be crossing in front of the vehicle at an intersection or crosswalk.
- Optically read speed limit signs and compare that reading to actual speed recorded via the engine communication bus.
- Leverage facial recognition software to detect drowsy head-drooping and glances away from the road to read or send text messages.

While all of this data is recorded for later analysis by fleet managers and (potentially) insurers, we recognize that the most immediate output of threat detection has two applications in the vehicle:

- Providing an alert to the driver to prompt his/her modified action to avoid the collision; and
- If the alert is unsuccessful in generating a human response, intervening electronically to make minor steering adjustments (for lane drift) or to providing full braking response or partial braking response in forward collision, pedestrian, and cyclist situations.

A human operator who is fully rested, fully sober, and does not possess any electronic or other distracting devices can still become momentarily distracted by a coughing fit, spilled beverage, attention-diverting outside influence (i.e., billboard, witnessing another crash event, etc.). Whether the driver is operating under optimal conditions or not, the addition of a hypervigilant copilot gives them a second chance to avoid a crash, and if connected to the operating systems of the vehicle, a third chance as well.

How Does It Work in Reality?

A recent report, published in January of 2016, cited research by IIHS,

> Vehicles equipped with front crash prevention are much less likely to rear-end other vehicles, IIHS has found in the first study of the feature's effectiveness using U.S. police-reported crash data. Systems with automatic braking [AEB] reduce rear-end crashes by about **40 percent on average**, while forward collision warning [FCW]

alone **cuts them by 23 percent**, the study found. The autobrake systems also greatly reduce injury crashes. If all vehicles had been equipped with autobrake that worked as well as the systems studied, there would have been **at least 700,000 fewer police-reported rear-end crashes in 2013**. That number represents 13 percent of police-reported crashes overall.[1]

In February of 2017, Schneider National CEO Chris Lofgren offered testimony to the Senate Commerce subcommittee on surface transportation that the voluntary inclusion of Automatic Emergency Braking documented "...a 69% decrease in rear-end accidents and a 95% reduction in rear-end accident claims cost since installing the collision mitigation system."[2]

There are a number of videos available that show these systems in action on test tracks, and in staged situations, to show the science behind the practical application. The good news is that these systems can be provided by the OEM, third-party upfitters, and as retrofit kits for vehicles of all sizes, makes, and models. Increasingly, OEMs of light-duty vehicles are offering these safety options across all models for a nominal upcharge, and many have volunteered to make these systems "standard" by 2022 for sedans, and 2025 for light-duty pickups. For light-duty cars and trucks, both consumer reports and IIHS provide portals where fleet managers can look up make, model, and year to determine what advanced systems are available as standard or optional equipment.[3,4]

For heavy-duty trucks and medium-duty vocational trucks, it takes a little more research, but the major brake system suppliers, Bendix and Wabco as two examples, have specialized systems ready to go via OEMs or upfitters.

What Else Is Coming?

There are a whole range of advanced driver assistance systems being deployed at the moment. If you are looking for an excellent resource where you can find short educational videos on each technology and how it works, then navigate your web browser to "mycardoeswhat.org." Here's how they describe their website; "*MyCarDoesWhat.org* is a national campaign to help educate drivers on new vehicle safety technologies designed to help prevent crashes. These technologies range from increasing the stability and control of cars to providing warnings about crash threats to automatically intervening to avoid or reduce the severity of a crash." The site was built and is maintained by the National Safety Council (NSC) and the University of Iowa.

There are almost three dozen distinctive programs or systems that can help drivers avoid crashes, and drive more safely. The website breaks these down by major categories, and provides a great introduction to the benefits and availability of each system. Those categories include:

- Forward Collision Prevention
- Braking, Tire Pressure and Anti-Rollover
- Driver State Monitoring and Communication
- Parking and Backing Assisting
- Lane and Side Assisting

Summary

Fleet managers, insurers, and safety professionals recognize that collisions are usually preventable, and the costs associated with them are a detriment to families, employers, and even our partners in government agencies. Finding new ways to engineer solutions to risk mitigation represents a tremendous leap forward in our ability to preserve life, and avoid needless damages.

While self-driving cars may be many years in the future, the research that is pioneering that vision can help us in valuable ways right now. To make an effectual contribution to safety results, however, we must speed up the adoption of these systems and not be willing to wait for their eventual inclusion as standard equipment.

Endnotes

[1] Insurance Institute for Highway Safety (IIHS). January 28, 2016. "Crashes Avoided: Front crash prevention slashes police-reported rear-end crashes." (http://www.iihs.org/iihs/news/desktopnews/crashes-avoided-front-crash-prevention-slashes-police-reported-rear-end-crashes).

[2] Jones, Kevin. February 16, 2017. "Trucking exec 'stuns' senator with safety stats." (http://fleetowner.com/regulations/trucking-exec-stuns-senator-safety-stats)

[3] Consumer Reports. "Cars with Advanced Safety Systems." April 26, 2016. http://www.consumerreports.org/car-safety/cars-with-advanced-safety-systems/

[4] Insurance Institute for Highway Safety (IIHS). "Crash avoidance in action." http://www.iihs.org/iihs/ratings/crash-avoidance-features

Bibliography

Insurance Institute for Highway Safety. "Automation and Crash Avoidance." (http://www.iihs.org/iihs/topics/t/automation-and-crash-avoidance/topicoverview)

Insurance Institute for Highway Safety. 2017. "Driver trust in five driver assistance technologies following real-world use in four production vehicles" (http://www.iihs.org/frontend/iihs/documents/masterfiledocs.ashx?id=2131)

Insurance Institute for Highway Safety. 2016. "Warning systems neither curb driver distraction nor worsen it" (http://www.iihs.org/iihs/sr/statusreport/article/51/9/3)

Insurance Institute for Highway Safety. 2016. "Drivers who drift from lane and crash often dozing or ill" (http://www.iihs.org/iihs/sr/statusreport/article/51/7/3)

Insurance Institute for Highway Safety. 2016. "Front crash prevention slashes police-reported rear-end crashes" (http://www.iihs.org/iihs/sr/statusreport/article/51/1/1)

National Safety Council and University of Iowa. (https://MyCarDoesWhat.org).

Session No. 577

ANSI/ASSE A10 Construction Safety Forum: Significant Standards Pending

Terry W. Krug, MS, CIH, CSP
EXOSHA, Inc.
Bartlett, IL

Introduction

When an employee attempts to conduct an activity that is covered by OSHA, the employer is required by law to ensure workers are protected by the requirements in the standard that addresses that activity. OSHA's laws and standards are the minimum requirements the employer must abide by when workers are exposed to workplace hazards. Major differences in a particular OSHA standard and ANSI accredited standard committee (ASC) standards are the following:

1. Even though ANSI standards are also minimum requirements for protecting the worker, the standard is developed by a committee of experts in the field of that standard, and addresses more safe work practices and accepted safe industry practices than the OSHA regulations have specified. The OSHA regulations are designed for enforcement, where the ANSI standards are developed with industries' best practices to assist the worker in mind, and provide references, examples, and state-of-the-art procedures.
2. The ANSI standards are required by established rules and procedures to be reviewed every five years, and the committee votes to revise, reaffirm or withdraw each standard. This allows committees and ANSI to ensure the information and recommendations in the standard are relevant and up-to-date. Some of the OSHA standards are more than 40 years old, referencing old technology and practices.
3. Each ANSI standard gives the reader more references, examples; some have been used in the past to be the base language of new OSHA standards.

This panel discussion will discuss the following A10 standards:

- ❑ A10.2 Safety, Health and Environmental Training
- ❑ A10.21 Safe Construction and Demolition of Wind Generation/Turbine Facilities
- ❑ A10.25 Sanitation in Construction
- ❑ A10.29 Aerial Platforms in Construction

This session is designed with enough detail to reach the intermediate level of attendee's knowledge of construction activities; however, it is comprehensive enough to offer technical information to an advanced audience. The attendee will go away with an appreciation of the new A10 standards under development, and a recommended revised language to one standard under revision (A10.25). I will discuss A10.2 and A10.21 for this paper.

The occupational safety and health administration (OSHA) for construction activities in 29 CFR 1926 has, for over 45 years, contained regulations governing activities on construction sites.[1]

The ANSI A10 ASC was founded in 1944 and develops, interprets, and revises voluntary national consensus standards relating to the protection of employees and the public from hazards arising out of, or associated with, construction and demolition operations.[2]

This paper addresses certain activities specific for the construction industry. It also compares the new consensus standards to what OSHA has as regulatory standards if any.

A10.2, Safety, Health and Environmental Training

This A10.2 standard's basic outline is similar to an existing Z490.1 standard, titled *Accepted Practices in Safety Health and Environmental Training.*[3] This Z490.1 standard has undergone many revisions since it was introduced in April 1998, and is geared toward general industry hazards and training, where the A10.2 standard addresses construction hazards and training issues such as competent person training, supervisor training, and site safety orientation.

<u>ANSI A10.2, Training Management, Development, Delivery, and Evaluation</u>

1. General
    - 1.1 Scope
    - 1.2 Purpose
    - 1.3 Application
2. Definitions
3. Types of Training Applicable to Construction
    - 3.1 New Hire Orientation
    - 3.2 Site Safety Orientation
    - 3.3 Regulatory Training
    - 3.4 Pre Task Training
    - 3.5 Competent Person Training
    - 3.6 Safety Leadership
    - 3.7 Supervisor Training
4. Training Program Administration and Management
    - 4.1 Accountability
    - 4.2 Training Program Elements
    - 4.3 Resource Management and Administration
    - 4.4 Program Evaluation
5. Training Development
    - 5.1 Systematic Process
    - 5.2 Needs Assessment
    - 5.3 Learning Objectives and Prerequisites
    - 5.4 Course Design
    - 5.5 Evaluation Strategy
    - 5.6 Criteria for Completion
    - 5.7 Continuous Improvement
6. Training Delivery
    - 6.1 Trainer Criteria
    - 6.2 Training Delivery Methods and Materials
7. Training Evaluation
    - 7.1 General Criteria
    - 7.2 Evaluation Approaches
    - 7.3 Continuous Improvement
8. Documentation and Record Keeping

8.1 Systems and Procedure
8.2 Records
8.3 Record Confidentiality and Availability
8.4 Issuing Certificates

OSHA has compiled a pamphlet, titled *Training Requirements in OSHA Standards* (#2254) in 2015. It is a very comprehensive document, specifically highlighting each standard's training requirements for general industry, maritime, and construction. Here is an example of the training requirements in construction for Subpart K Electrical.

Subpart K – Electrical
1926.416 General requirements (a) Protection of employees (3) Before work is begun the employer shall ascertain by inquiry or direct observation, or by instruments, whether any part of an energized electric power circuit, exposed or concealed, is so located that the performance of the work may bring any person, tool, or machine into physical or electrical contact with the electric power circuit. The employer shall post and maintain proper warning signs where such a circuit exists. The employer shall advise employees of the location of such lines, the hazards involved, and the protective measures to be taken.

The requirements are similar in most OSHA standards in regards to training and that is the employer shall ensure that training is provided:

- To each affected employee
- Before the employee is first assigned duties under this section
- Before there is a change in duties
- Whenever there is a change in operations
- When any new hazards are introduced or employee's knowledge is deficient
- To establish employee proficiencies
- To recommend and ensure PPE or effective signage is used

A10.21, Safe Construction and Demolition of Wind Generation/Turbine Facilities

This A10.21 standard's basic outline follows and discusses the hazards and controls of erecting and dismantling wind generation tower facilities. These are being erected nationwide, and may also be used to support cell phone satellite equipment.

ANSI A10.21, Safe Construction and Demolition of Wind Generation/Turbine Facilities

1. General
    - 1.1 Scope
    - 1.2 Purpose
    - 1.3 Application
    - 1.4 Modifications & Exceptions
    - 1.5 Mandatory & Advisory Rules
2. Definitions

3. General Requirements for Wind Energy Projects
    - 3.1 General Requirements
    - 3.2 Site Hazard Identification & Assessment
    - 3.3 Coordination of Roles, Responsibilities & Activities

Appendices:

It has been brought up in the past that the definition of confined spaces needs to be changed to be consistent with other ANSI standards. For example, the definition in A10.21 defines confined space as the following:

2.5 Confined Space. An enclosed area that is large enough and so configured that an employee can bodily enter and has the following characteristics:

- its primary function/purpose is something other than human occupancy; and
- has restricted entry and exit; **or**
- **contains another known or recognized hazard.**

When you have a bullet point like the last one in the definition, "contains another known or recognized hazard" is so broad that would include all areas that had any recognized hazard, such as an unguarded saw blade, a wall receptacle without a cover, an uncharged or missing portable fire extinguisher, etc. This last bullet needs to be removed along with the preceding "or." Also, use the explanation of restricted entry and exit in Z117.1 R 2016[4] to add clarity.

For example the definition in Z117.1 2016 states the following:

2.5 Confined Space. Enclosed area large enough and configured to allow a person to bodily enter and has the following characteristics:

- Its primary function is other than human occupancy.

- Has restricted entry and exit. (Restricted entry and exit is a physical configuration, which requires the use of the hands for support or contortion of the body to enter into or exit from a confined space.)

This panel of ANSI A10 committee members will discuss the safeguards of each standard under development and why we think these standards, if followed by the employer and employees, are necessary to ensure the specific construction activity is conducted as safe as possible.

Endnotes

[1] Occupational Safety and Health Administration (OSHA). 2017. 29 CFR 1926, *Safety and Health for Construction*. (http://www.osha.gov .)

[2] American National Standards Institute/American Society of Safety Engineers (ANSI/ASSE). 2017. *A10, Safety in Construction and Demolition Operation*. Park Ridge, IL: ASSE.

[3] American National Standards Institute/American Society of Safety Engineers (ANSI/ASSE). *Z490.1 R2015, Accepted Practices in Safety Health and Environmental Training*. Park Ridge, IL: ASSE.

[4] American National Standards Institute/American Society of Safety Engineers (ANSI/ASSE). *Z117.1 (R2016), Safety Requirements for Entering Confined Spaces*. Park Ridge, IL: ASSE.

Exoskeleton Technology: Making Workers Safer and More Productive

Terry Butler, CSP
President
Lean Steps Consulting Inc.
Des Moines, IA

Daniel Wisner, CEA, CBES
Ergonomics Engineer
John Deere
Ankeny, IA

Introduction

"Space, the final frontier. These are the voyages of the Starship Enterprise. Its 5-year mission: to explore strange new worlds, to seek out new life and new civilizations, **to boldly go where no one has gone before.**"[1]

This is probably one of the most recognized quotes, and one that conjures up a sense of exploration, pride, and a can-do attitude. Just like the members of the Starship Enterprise, it is that sense of exploration that motivates the authors of "Exoskeleton Technology – Making Workers Safer and More Productive," to explore unchartered territory of using exoskeleton technology in the workplace. The mission of the authors was to test if upper body exoskeleton technology, in fact, makes a worker safer, and in given applications, more productive through the reduction of muscle activation and fatigue associated with an employee's job. This paper will provide some insight into the depths the authors went on their mission to provide evidence for or to disprove this hypothesis.

Unlike subjective and limited ergonomic assessment tools, such as the Rapid Entire Body Assessment (REBA) and Rapid Upper Limb Assessment (RULA),[3] the authors worked with Jason C. Gillette, Ph.D., Associate Professor, and Mitchell L. Stephenson MS, CSCS, in the Department of Kinesiology at Iowa State University, to gather quantifiable data to test the physical benefits of upper body exoskeleton technology on reducing fatigue. This paper presents those findings, along with the need "to boldly go where no one has gone before," with expanding current ergonomic assessment tool effectiveness by quantifying the physical benefits of applied exoskeleton technology, in addition to subjective risk assessment scores derived from these standard ergonomic risk assessment tools. The authors will discuss the results of measuring muscle activity, using EMG sensors on the shop floor while employees performed their jobs, with and without the aid of an upper body exoskeleton. Findings from measuring productivity with and without an exoskeleton on users performing dynamic work will also be presented.

Wireless electromyography (EMG) sensors were used to record the electrical activity associated with skeletal muscle contraction of all test subjects. "The electrical activity from the muscle cells is conducted through body fluids to the skin, where it can be picked up by surface electrodes, amplified and recorded. The recordings thus obtained are called electromyograms." [9]

Having an understanding of this science will better equip safety and ergonomic professionals when considering deployment of such technology into their own workplace as a solution to an ergonomic risk hazard or in an attempt to aid with improving quality and productivity.

Improved productivity, improved quality, and making workers safer gets almost everyone's attention in manufacturing and other industries whenever a new technology enters the marketplace. However, like fall protection technology, wearable exoskeleton technology needs to be approached by safety and ergonomic professionals with an analytical and technical eye for detail to ensure the proper training, care, use, fit, function, and limitations are communicated to the user. A list of considerations is included within the body of this paper.

This paper will also take a look at the absence of regulations surrounding this technology, and what are the beginnings of such work. The authors have started work with the Musculoskeletal Health Cross-Sector of the National Institute for Occupational Safety and Health (NIOSH) via an Exoskeleton Technical Interchange at the National Institute of Science and Technology (NIST) this past January 2017. The Exoskeleton Technical Interchange was focused on bringing together experts from the industrial, military, academia, and medical communities to discuss potentially applicable standards developed thus far, and others in development that may help in developing exoskeleton standards. The goal is to aid with safely advancing this innovative technology as it is being commercialized and works its way into the work environment.

In Brief

This paper discusses the use of exoskeleton technology, what exoskeleton technology is, and the physical responses to wearing such technology throughout a workday. This paper will review what types of upper body exoskeletons are available and in use in various industries, with a special focus on those used to ergonomically reduce upper body fatigue that can lead to overexertion in employees involved in repetitive, dynamic, and static type of work. From a real life study and its results, readers will learn how to properly assess where and how to add exoskeleton technology to their list of ergonomic solutions as a form of personal protective equipment (PPE). Readers will learn why the absence of regulations on new technology such as this necessitates safety professionals be intimately involved in the assessment and deployment of such standards and devices.

From a Vermeer Corporation maximal fatigue study completed in 2015, those who participated in the welding and painting simulation study were able to produce more welds (86%) and more painted parts (53%) because the use of the exoskeleton delayed their maximal fatigue level from occurring. That was the goal of the study. It was desired to know if the exoskeleton would prevent someone from getting fatigued or just delay it and if so, the impact on quality and productivity.

With the current study, the aim was to determine the impact of fatigue reduction while wearing an exoskeleton over the course of eight- and ten-hour work shifts. Because the benefits employees would experience was not known, and it wasn't understood what kind of productivity improvement could be expected under normal, everyday work conditions, data were collected over a three-month period. What the findings show is the users of upper-body exoskeletons who have control over their work pace will out-produce someone not wearing an exoskeleton. Quality is also improved, with improved hand steadiness where this is a contributing factor to the work they perform.

Exoskeletons Defined

So what is an exoskeleton, and how does it differ from a wearable robot? According to *Merriam-Webster,* an exoskeleton is an artificial external supporting structure.[11] According to the same source, a robot is a machine and performs complicated, often repetitive tasks that vary in complexity, which can be guided by automatic controls. According to an online source by Technopedia, the definition of a wearable robot is a specific type of wearable device that is used to enhance a person's motion and/or physical abilities. Wearable robots are also known as bionic robots or exoskeletons.[12] However, the authors believe it is important to note the distinction between a non-powered, non-tethered exoskeleton and a powered, or tethered, exoskeleton. The latter would be consistent with the definition of a wearable robot, and the former an exoskeleton, as defined above by *Merriam-Webster*. This paper and the research conducted only deals with non-powered, non-tethered upper-body exoskeletons. Exoskeletons of this type use springs, weights, rubber bands, cables, and pulleys to create an upward force supporting the arm. Some of these non-powered, non-tethered upper-body exoskeletons have adjustments for controlling the amount of upward force by turning a screw or knob or changing out a cassette. Exhibit 1 below shows one such exoskeleton in use during the study.

Exhibit 1. Exoskeletons In Use

As this paper is being written, exoskeletons are advancing in sophistication and simplification. As a result, the cost of some are becoming affordable as workplace ergonomic solutions, while others are still priced out of reach. In a July 2015 article titled "The Exoskeletons are Coming," Mr. Yamanaka, a Panasonic spokesperson, said, "We can expect that exoskeletons, or power-assist suits, will be widely used in people's lives in 15 years."[4] The global exoskeleton market is expected to grow 75%, and revenues from commercial and industrial exoskeletons are expected to grow more than 20 times between 2016 & 2021 (Arcluster syndicated study, Aug. 2016;

arcluster.com). Safety professionals have to be on the front of this wave of new technology to ensure it is properly used so workers are not hurt by it, but truly helped by it. Safety professionals have to know how to identify if marketing claims are legitimate, and if adding such a device to the personal protective equipment cabinet makes sense for their business.

Fatigue Defined

According to an article published in May of 2000 in the Brazilian *Journal of Medical and Biological Research*, entitled "Coexistence of Potentiation and Fatigue in Skeletal Muscle," "Fatigue is the force decrease observed after a period of repeated muscle stimulation."[7] According to Dr. Gillette, "[M]uscular fatigue is a complicated condition that is not fully understood and equally complicated to accurately measure."

However, in a prior study completed at Vermeer Corporation and reported in the September 2016 issue of *Professional Safety*, welders and painters were taken to a maximal level of fatigue where they could no longer meet the quality parameters set for the test. They were subjected to the same level of work difficulty while wearing an exoskeleton, and on separate days while not wearing an exoskeleton. The goal was to determine if the use of an exoskeleton could delay fatigue. The researchers also wanted to know the improvement in productivity from such a delay.

During this study, the researchers were able to physically observe the welder's arms and hands shaking, indicating fatigue was happening. Fatigue was also evident in their lack of muscle control, recorded by the weld and paint simulator performance monitors. The welders were able to produce 86% more welds with the use of the exoskeleton, and the painters 53% more painted parts. The welders were able to go twice as long with continuous welding without a break, while wearing the exoskeleton. However, they too eventually became fatigued. This reinforces the results obtained from a study, entitled "Effects of Fatigue on Inter-muscular Coordination During Repetitive Hammering" by Cote et al. (2008). In the study, the researchers stated: "Results suggest that fatigue influences motion at both local and global levels."[13]

Dr. Gillette, who was consulted on the Vermeer test protocol, stated "the results obtained at Vermeer were a good indicator of muscle fatigue." He went on to say, "the 50% decline in performance with the Vermeer welding simulator is a good indicator of functional fatigue in his opinion." However, he noted, "we need to acknowledge the differences," stating "a decline in performance with the welding simulator is based on accuracy rather than purely strength, which likely involves declines in moderate force, larger movements such as maintaining shoulder position and low force, finer movements in the wrist and hand." People typically do not work at their maximal force so this statement by Dr. Gillette makes sense. In fact, this type of fatigue is what we could expect to see in most manufacturing assembly, painting, parts handling, and welding type jobs. Dr. Gillette said, "when looking at the Vermeer simulation data, the performance decline was based on accuracy, which would likely involve a combination of slow twitch and fast twitch muscle fiber fatigue. The results do provide indirect evidence of muscle fatigue."

But what happens over the course of a normal eight- and ten-hour work shift? What kind of fatigue reduction and performance improvements can someone expect if they were wearing an exoskeleton? This question was not answered by the Vermeer study, and is what motivated Dr.

Gillette and the authors of this paper to "boldly go where no one has gone before" and, as Dr. Gillette said, do some "unique things." Unique, because previous studies have restricted themselves to a lab environment, where the researchers typically try to recreate a shop-floor work environment. Instead, the researchers involved in this study decided to "go where no one has gone before," and use wireless EMG sensors with workers on the shop floor. (See Exhibit 5.)

One last thing on fatigue that was interesting with the Vermeer study was the curve of the performance data when plotted over time. (See Exhibit 2.) It followed a similar trajectory as the force over time curve, measured with EMG sensors, that was plotted and noted in a study entitled, "Mechanisms of Fatigue as Studied in Single Muscle Fibers," conducted in 2006 by J. Lännergren, H. Westerblad & O.G. Allen at the University of Sydney in New South Wales Australia.[10] In both studies, the muscles rapidly lost their ability to perform. Although at Vermeer, it was a completely different kind of test, it was interesting to see how the force required to do the weld job, plotted over time, followed a consistent level of performance until fatigue kicked in, and the performance dropped off quickly.

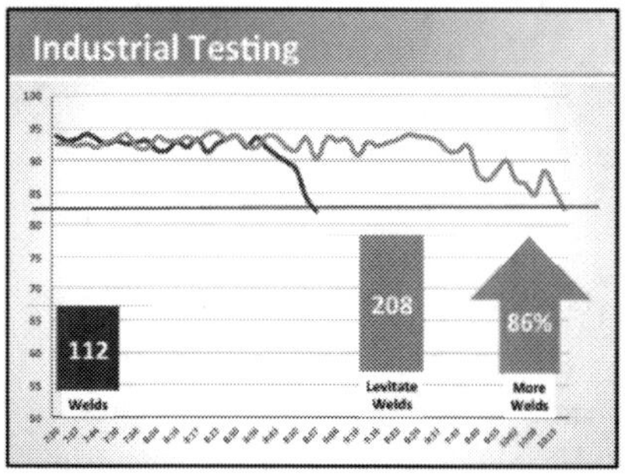

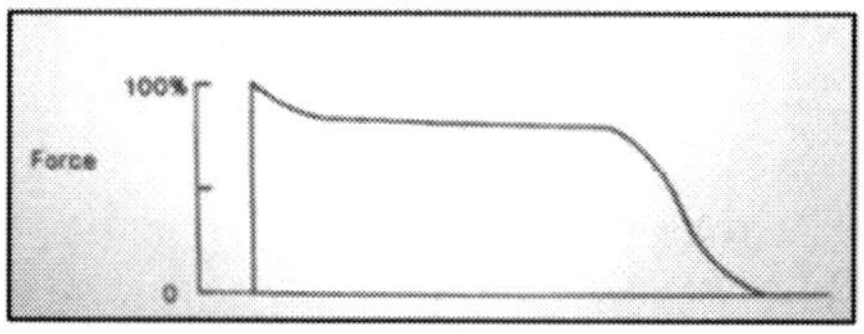

Single Muscle Fibers conducted in 2006 by J. Lännergren

Exhibit 2. Vermeer Study Results & Force Curve

Test Methodology

The team from Lean Steps Consulting, Iowa State University, and John Deere worked to develop a test methodology focused on capturing real-time data from employees while working with and without the aid of an exoskeleton. A total of six volunteers at two John Deere industrial manufacturing sites were selected. The volunteers included three assemblers, one painter, one parts hanger, and one welder. (See Exhibit 4.) EMG data were collected from each employee during a challenging cycle of their job tasks, which ranged from approximately six minutes to ten minutes. EMG data were collected consecutively without and with the exoskeleton at the beginning and end of the work shift to determine fatigue levels. Maximum voluntary isometric contractions (MVIC) were also collected with each employee. EMG signals were collected from eight muscles: right and left Deltoids, Biceps, Trapezius, and Erector Spinae (lower back) (See Exhibit 5). The data were collected using a Delsys Trigno wireless EMG system (see Exhibit 3), and sampled at 1926 Hz, using Delsys EMGworks software. Dr. Gillette at Iowa State University analyzed all EMG data, using a custom ISU written code in Matlab.

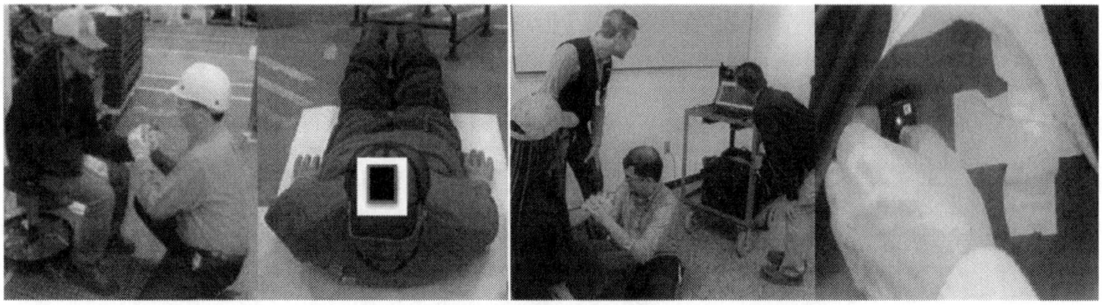

Exhibit 3. MVIC Exercises and EMG Sensor Reading and Placement

Exhibit 4. Some of The Selected Jobs For The Study

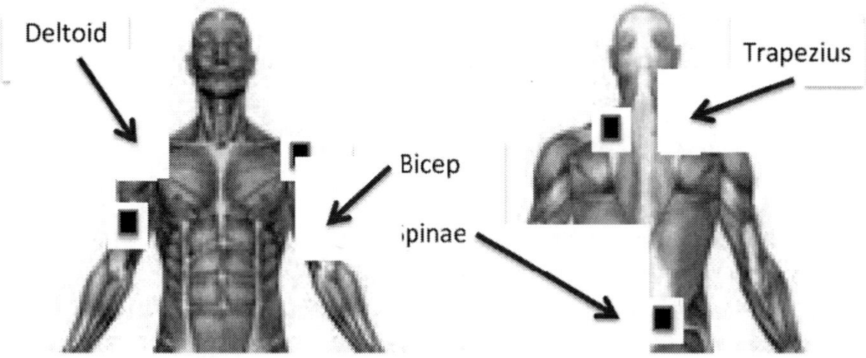

Exhibit 5. Muscles Monitored During Exoskeleton Testing

This paper presents a high level summary of the "big data" collected during the study. For those not familiar with the term, Rich Grant, PE from Fleis & VandenBrink (F&V) engineering and architecture firm in Michigan and Indiana, said "the importance of big data lies in the value of many measurements. Millions and millions of measurements! In lieu of estimating modeling or guessing, the ability to inexpensively collect big data and to powerfully analyze that data allows the analyst to wring at conclusions that were previously too expensive, too time consuming to chase, subjectively derived at, or hidden from the naked eye." The importance of this new analytical methodology adds a new dimension to ergonomic risk assessment tools that currently use subjective inputs to come up with risk scores. According to Claudia Patricia Escobar, "ergonomic assessment tools are used to detect risk factors or unsafe conditions that might result in work-related musculoskeletal disorders (WMSD). Risk factors such as force/load, posture, frequency of movement and vibration, besides individual stressors affecting the worker are thought to directly increase the risk level for occupationally related musculoskeletal disorders (Li & Buckle, 1999)."[5] Notice here the absence of measuring fatigue and its contribution to occupationally related MSDs.

In her thesis study, Claudia concluded that failure to accurately report and train those reporting on those variables considered critical to the resulting risk score could impact the final hazard level scoring. Claudia went on to say, " The results obtained in this study are important and should be used in industrial environments when performing subjective ergonomic assessments. The requirements for training and the selection of the tools to use should consider the critical levels encountered for RULA, REBA, and JSI, because of the direct and significant dependency of the final scores on the critical variables found for them. Practitioners need a basis for preparing effective self-reporting evaluators when using ergonomic assessment tools."[5]

It is the opinion of the authors of this paper that the addition of the use of wireless electromyography sensors in capturing muscle activity as a percent of maximum voluntary contraction, before and after implementation of an ergonomic solution, is an enhanced way to physically measure the benefit to the employee by taking into account the impact of fatigue. Adding this knowledge derived from analysis of EMG data to the way scores are obtained from risk assessment tools that currently use subjective inputs will add a new dimension to ergonomic risk assessments. In fact, this position is further reinforced by the following conclusions from a study entitled, "Shoulder Muscle Fatigue During Repetitive Tasks as Measured by Electromyography and Near-Infrared Spectroscopy" by Ferguson et al. (2013), where the researchers combined EMG with Near-Infrared Spectroscopy (NIRS) to monitor blood hemoglobin levels in various muscles being monitored with EMG sensors. They said, "By measuring both EMG and Near-Infrared Spectroscopy (NIRS), we are gaining a better understanding of the fatigue process."[14]

Process for Analyzing Maximum Voluntary Isometric Contractions (MVIC)

After all the data were collected, Dr. Jason Gillette from Iowa State University set out to perform his analysis. During the analysis process, Dr. Gillette removed any signal noise caused by the EMG sensors being bumped or from lost contact with the skin. Next, EMG signals below 20 Hz were filtered to remove potential noise created by movement of the skin relative to the muscle, and EMG signals above 450 Hz were filtered to remove potential noise created by background electrical interference in the environment. Noise from welding, which showed up as spikes of electrical interference detected at 200 Hz and 400 Hz, were instead filtered for signals above 150 Hz for this job task to remove background electrical noise from the welding equipment.

Dr. Gillette said his next step was to remove all non-zero EMG bias from constant background electrical noise, but he said in general, there was no EMG bias. EMG signals were rectified to make all data positive, rather than alternating positive and negative. EMG signals were then low-pass filtered at 10 Hz filter to create a linear envelope representative of the volume of muscle activity. A moving window was used to determine the maximum one second of sustained EMG amplitude for both the MVICs and job tasks. Dr. Gillette then determined a second EMG amplitude measure by finding the maximum 60 seconds (one minute) of muscle activity distributed throughout the job task.

Table 1 & 2 below show the summary results of the above analysis of the six test participants involved in the study.

Table 1. Maximum sustained one second EMG amplitude. Average values are reported with standard deviations (n = 6 participants). * indicates significantly lower EMG amplitude when using the Levitate Airframe exoskeleton ($p < 0.05$).

Variable	Without Exoskeleton	Levitate Exoskeleton	P-Values
Deltoid EMG (%MVIC)	63.5 ± 15.8	52.1 ± 7.5	0.060
Biceps EMG (%MVIC)	45.3 ± 17.4	37.2 ± 15.1*	0.010
Trapezoid EMG (%MVIC)	48.1 ± 9.5	50.9 ± 5.2	0.510
Spinae EMG (%MVIC)	52.2 ± 21.6	48.4 ± 20.7	0.252

Table 2. Maximum distributed one minute EMG amplitude. Average values are reported with standard deviations (n = 6 participants). * indicates significantly lower EMG amplitude when using the Levitate Airframe exoskeleton ($p < 0.05$).

Variable	Without Exoskeleton	Levitate Exoskeleton	P-Values
Deltoid EMG (%MVIC)	32.0 ± 8.2	24.8 ± 2.8*	0.037
Biceps EMG (%MVIC)	19.2 ± 9.1	14.0 ± 5.9*	0.016
Trapezoid EMG (%MVIC)	25.3 ± 6.4	22.2 ± 4.1	0.261
Spinae EMG (%MVIC)	28.7 ± 15.1	23.3 ± 11.1	0.069

Fatigue Test Findings

Both Tables 1 and 2 above represent the collective summary of all fatigue data collected for all six test volunteers. Using just the deltoid results as an example, let's explore the impact of wearing an exoskeleton device while performing static, dynamic, repetitive, and/or elevated arm type of work. If the maximum distributed one-minute EMG amplitude for the deltoid is at a 32% of the MVIC without the exoskeleton and with it, it is at 24.8% MVIC, there is a net 22% MVIC reduction. This appears to be a substantial benefit to the user, as the amount of work the individual is doing with their deltoid is now reduced by 22%. This, along with the other data collected on the other muscles monitored, explains why workers wearing the exoskeleton did not feel as tired at the end of the day. It further explains why there is an increase in the ability of the employee to maintain a higher level of production. For example, assuming this level of work represents 10% of their job, working with 22% less effort for this work over the course of the day should provide cumulative benefits.

According to Stone & Sands (2007), "During a contraction, the muscle cannot actively develop force or perform work against a resistance until the elastic components are stretched out and the muscle tension and resistance (load) are in equilibrium."[16] Dr. Gillette said that, "the greatest benefits are likely the reductions of greater than 5% of MVIC in the muscles with the highest levels of EMG amplitude." Making jobs easier for employees through the partial augmentation of the arms is a good thing. Muscle atrophy should not occur when using the exoskeleton, as evidenced by EMG amplitudes that have been reduced, but not eliminated. The muscles are still working, but just not as hard as they do when unsupported. This greatly lessens the level of fatigue an employee experiences over an eight- and ten-hour workday.

Even support at lower levels of MVIC reduction are worth capturing, as noted in an article entitled, "The Effect of Limb Support on Muscle Activation," by Wise et al. (2004). Here, the researchers were paying attention to and capturing those benefits from a reduction in the muscle activity greater than 5% of MVIC, while using support of the arm during shoulder rehabilitation exercises. In fact creating low levels of MVIC reductions through the use of support during rehabilitation exercises is encouraged for patients when access to aquatic exercise environments is not available to help strengthen the arm, so as not to let it become reinjured.[15]

In another study, referenced in *Occupational Biomechanics*, Chaffin et al, 1999, "… Sjogaard and coworkers (1986) found that there was a significant increase in perceived fatigue at 5% MVC…" "Other experimenters have placed the endurance limit at 8% of MVC (Bjorksten and Jonsson, 1977), while recommending 5% as a general population limit for continued static work by the upper extremity." "When repetitive isometric contractions are performed, recovery phases become interspersed with periods of fatigue. Very short periods of relaxation (recovery) may not be sufficient to reverse the fatiguing effects of the contraction, although, the time required producing fatigue is prolonged by even short rest periods (Sjogaard et al., 1988; Lee, 979)."[17]

By partially supporting the weight of the arm with an exoskeleton, it is likely that the muscles doing the work spend more time below a fatiguing level of contraction. This partial support of the arm prolongs endurance, thereby reducing the level of fatigue an individual may experience by the end of the day or a given task. This assumption is further supported by the statement that, "The longer the rest period, the longer is the muscle endurance time for a given contraction level, although at the higher MVC levels, very long rest periods are required."[17] The study referenced from *Occupational Biomechanics* goes on to say "… it is clear … that muscle endurance can be maintained in sustained or frequent exertions only by assuring that the contractile force is low or by interjecting adequate rest periods."[17] Although further study is needed, it is believed the results obtained thus far demonstrate that an exoskeleton helps to reduce force and fatigue by partially supporting the muscles doing the work, producing similar effects on fatigue as sufficient and adequate rest periods.

Another finding that surprised the researchers shows up in the results of the %MVIC reductions of the Spinae (lower back). The maximum distributed one-minute EMG results for the six participants working without the exoskeleton was 28.7 %MVIC, and a 23.3 %MVIC when using the exoskeleton, producing a percent reduction in %MVIC of 18.8 %. These finding show the lower back muscles were experiencing less work, which was a surprise to the researchers. The researchers thought with the transfer of the weight of the arms to the hips, there may be an increase in the %MVIC for the Spinae (lower back) and as the results show, we found just the opposite. Understanding that the p value of 0.069 is above the statistical significance threshold of 0.05, it is approaching significance with only six participants.

Productivity Results

Productivity improvements should not be the first reason for wanting to deploy exoskeletons into the work environment. Safety and quality improvements should be the primary motivation for deploying such technology. Exhibits 6 and 7 show over 365 cycle-time observations over the course of three consecutive eight-hour Mondays as a painter worked with and without the aid of an exoskeleton. The results show the painter outperformed his daily production by 8.2% to 19% with the aid of an exoskeleton. Considering the painter said, "he had more energy and felt fresh at the end of his shift" with the use of the exoskeleton provides evidence that he was experiencing less fatigue, even though he produced more painted parts. This was also supported by the results of his EMG analysis, as shown in Table 1 and 2 above. For operations that can benefit from the use of such technology, exoskeletons can be a game changer.

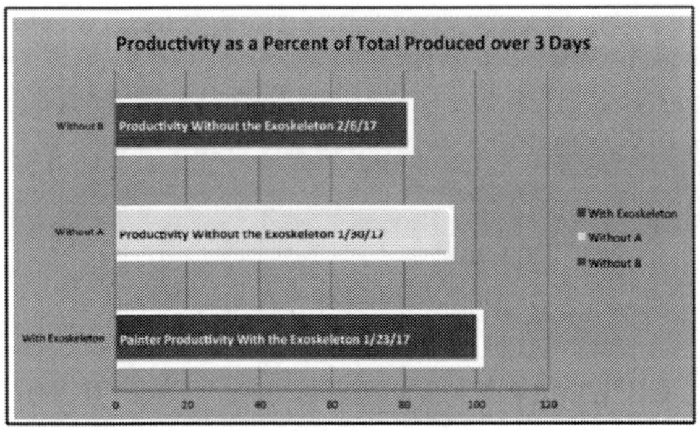

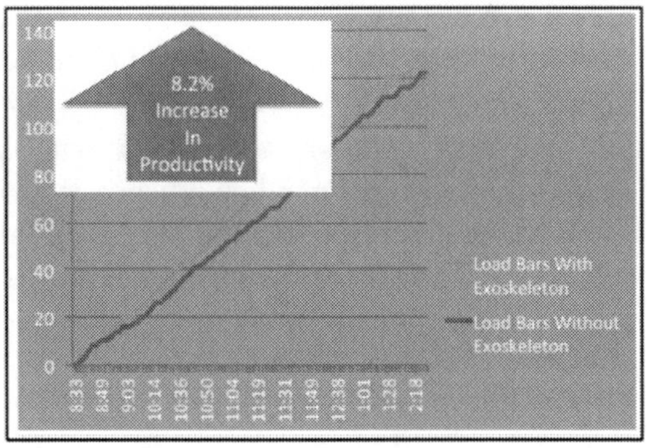

Exhibit 6. Paint Line Productivity **Exhibit 7. Load Bars W & W/O Exoskeleton**

When comparing the level of production over the three days, one with the aid of an exoskeleton and the other two without, it was clear there may have been other factors on 2/6/17 that impacted production measurements, and contributed to such a large decrease in overall productivity for that day compared to 1/23/17 when the painter used the exoskeleton. In interviewing the painter, we found that he had not gotten much sleep due to a family emergency and in his words he was "dragging." It was felt it was important to point out this fact because it's never known how employees are being impacted outside of work, which obviously impacts their ability to produce. The good news is with the aid of an exoskeleton, some of that loss, due to variables out of the control of the company, can be minimized.

Another significant finding from this productivity test was how it relates to the findings from the earlier Vermeer test where painters and welders both experienced significant increase in productivity while using an exoskeleton. From the discussion, it is known that fatigue is not good for the body and, if left unchecked by inadequate periods of rest, injuries and illness can result. That is exactly what happens if you repeatedly take someone to his or her maximum level of fatigue like was done in the Vermeer study. What the results from this study indicate is that when comparing two equivalent days of production (for example, on 1/23/17 where the painter used the exoskeleton, and on 1/30/17 when he did not), it can be expected there will be higher levels of production with the aid of an exoskeleton. This is in cases where the user has the ability to pace their level of activity. In this case, an 8.2% productivity increase was observed because the painter controls the painting process by the push of a button. When pushed, the line will index the work through the paint booth at a pace determined by his physical abilities. The mix of painted parts (large and small) is balanced by the operations feeding and pulling from the paint booth, so this was an "apples-to-apples" comparison. This is why the painter was selected to perform our productivity measurements and cycle-time observations.

Another significant finding that became illuminated while on the shop floor is the relationship between the benefits of fatigue reduction and how staffing levels are calculated in a lean manufacturing environment. Staffing levels are calculated by dividing the sum of the cycle time required to complete all the jobs on a line, by the Takt Time. Takt Time is customer demand divided by available time. This staffing level for the line to meet Takt Time is critical to production synchronization. The product of the sum of the cycle time divided by the Takt time tells the manager the number of people they need to staff the line, as the customer demand goes up or down. Variability in calculating staffing levels is introduced in the singular way cycle time is determined. Cycle time is based on time observations, and is the most repeatable time to do a process. What this does not take into consideration is the age, sex, condition, size, etc., of the various workers that may be assigned to perform a variety of tasks as assigned on a cross-training matrix. Operator "A" may have been observed to establish the cycle time for various parts of the job, and now you have operator "Z,, who is 5 inches shorter, 10 years older, is not in as good of shape, and is weight challenged, assigned to the job. In most cases, Operator "Z" will have to work harder than Operator "A" to try and meet the cycle times for his or her jobs on the line or risk falling behind output standards. This is because his or her physical characteristics were not used to determine the cycle times reflected on standard work combination sheets.

This cycle-time observation to staffing level relationship is a key variable to employee safety and productivity. Without some physical ergonomic intervention, it is impossible to address this mismatch of physical characteristics. Without overstaffing the line to accommodate for these variances, production can suffer, and so can the employee. However, deploying exoskeletons where physical demands, such as repetitive motion of the arms in moderate to elevated ergonomic, static and/or dynamic positions, can help reduce the variances in cycle times between different individuals, as demonstrated in the productivity results noted above. Without implementing solid ergonomic solutions, such as an exoskeleton, individual employees with less physical capabilities than the person observed when calculating cycle times, are at greater risk for overexertion injuries due to the physical demands put upon them being closer to their maximum capability. In such cases, productivity and worker's compensation cost can be negatively impacted.

Fit, Comfort, Weight, and Other Considerations

If someone is asked to wear an external skeleton on their body for potentially eight to ten hours a day, the acceptance by employees to use such exoskeletons will be dependent upon the exoskeleton actually helping, rather than hindering, the employee's ability to get the job done. As shown in the post user survey (Exhibit 9) from the study, several things were learned. The responses to the survey bring to light the fact that it has to fit the employee, be comfortable, and be lightweight enough so as not to add work to the employee just by having to carry the exoskeleton on their body. Ensuring the exoskeleton is body conforming, so it does not catch on something or scratch painted parts, is also a feature that needs to be considered. Having a single point of release is a safety consideration in case of emergency. Smoothness of support activation, as well as the ability for that force to go away when the arms are lowered to a neutral or relaxed position at one's side, is also a critical user feature to consider. If the user has to fight the force or it is jerky, they may become frustrated, and not use it. Consideration as to where the exoskeleton comes into contact with the person and the breathability of the padding material and support structure is critical. As the person sweats, the contact point can become a source of irritation. Contact points should be covered with removable covers for washing. Range of motion should not be restricted while using the exoskeleton. Comfort when bending, squatting, and twisting are also important user considerations.

Shown in Exhibit 8 are the subjective fatigue reduction responses of another independent study. This was a prospective randomized three phase crossover trial approved by the Internal Review Board (IRB) of the University of California at San Diego, involving 18 surgical attendants and residents who participated and agreed to wear a surgical exoskeleton (the same one used for this study). An increase in musculoskeletal disorders and injuries involving the back, shoulder, and neck are among the most common complaints experienced by laparoscopic surgeons. The goal of the study was to see if exoskeletons worn in the operating room can reduce fatigue and provide protection and relief for surgeons exposed to the stressors leading to this increase in MSD's.[18] The results were statistically significant ($P<0.05$) between wearing or not wearing the exoskeleton, furthering support for the deployment of exoskeletons to aid with fatigue reduction in static, repetitive, dynamic, and moderate-to-severe ergonomic positions of the arms.

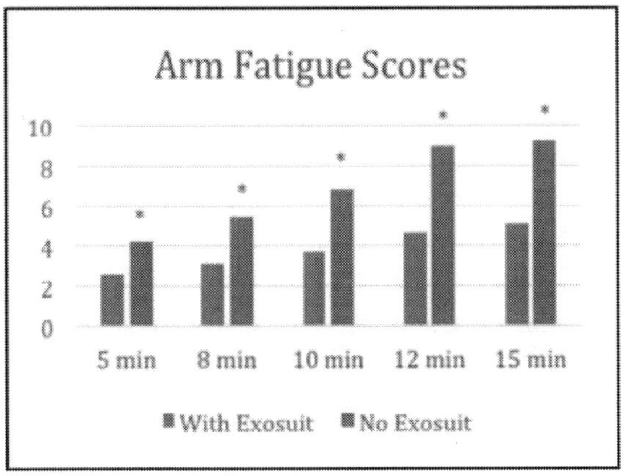

University of Calf San Diageo Solving the Surgeon Ergonomic Crisis with Surgical Exosuit

Exhibit 8. Surgeon Fatigue Response

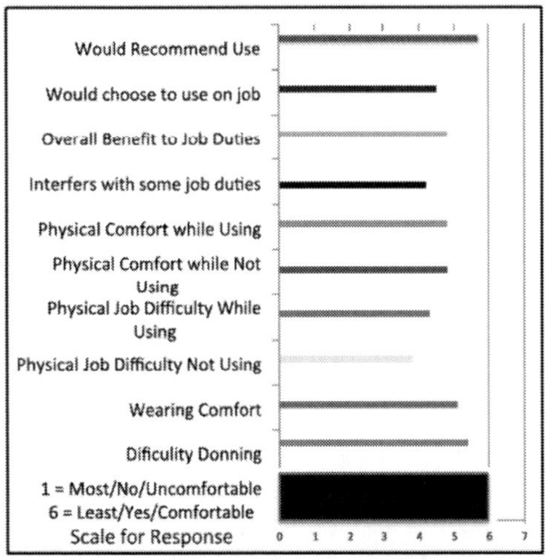

Exhibit 9. Post User Survey

Standards Development Status

While in attendance at the National Institute of Science and Technology (NIST) in January 2017, the authors learned that there are currently no standards that apply specifically to non-powered, non-tethered exoskeleton technology. In fact, the National Institute for Occupational Safety and Health (NIOSH) has already been receiving calls asking if such standards exist. The Exoskeleton Technical Interchange was the first of its kind, and focused on bringing together experts from the industrial, military, academia, and medical communities to discuss potentially applicable standards developed thus far, and others in development that may help in developing exoskeleton standards. (See Exhibit 10). It was clear the medical sector is the furthest along in the deployment of powered lower-body exoskeletons and wearable robots for spinal-cord injury-related needs, as well as development of standards that could potentially be used to develop industrial type of exoskeleton standards. What NIOSH is looking for now are interested parties willing to work on committees to help draft exoskeleton standards. The challenge is huge, due to the multi-faceted use and ranges of complexity of some designs, which includes robotics. The timetable is long and will require a lot of input from the various potential end users. Safety and health care professionals interested in using this technology should consider participating in the standards development process, or at least watch over their development. If you are interested in learning more contact Roger Bostelman at NIST via email roger.bostelman@nist.gov.

Existing Standards & Other Related Work	Military	First Responder	Industrial	Medical	Consumer
ISO TC299 Robotics (e.g. ISO134828)	X	X	X	X	X
ASTM - Response Robots	X	X	X		
NIST Collaborative Robot Tests			X		
RoboMate - Inductrial Robots			X		
Army (NSRDEC & ARL - HRED) - Military Systems	X	X	X		
Navy - Industrial Human Aug. Systems (iHAS)			X		
Department of Energy - Office of Nuclear Energy			X		
Vetrans Health Adminstration Office of Research and Development	X			X	

Exhibit 10. Summary of Existing Standards

It is important for ergonomic and safety professionals to understand the current lack of standards so they can better interpret manufacturers' claims before acquiring and deploying these devices in various applications.

Conclusion

Are the findings shown in Table 1 and 2 above enough to prevent injuries? This is not definitive. However, in the study quoted earlier, entitled "Shoulder Muscle Fatigue During Repetitive Tasks as Measured by Electromyography and Near-Infrared Spectroscopy," the researchers said the following about their findings: "Research has shown that fatigue creates increased muscle coactivation and changes in kinematics that result in increased joint loading, thereby increasing the risk of injury to the joint (Marras et al., 2006; Potvin & O'Brien, 1998; Psek & Cafarelli, 1993). Furthermore, fatigue is a key element in the development and progression of pain (Cote & Hoeger Bement, 2010).[2] Shoulder pain symptoms attributable to muscle fatigue may lead a worker to seek medical attention, potentially resulting in a diagnosed Muscular Skeletal Disorder (MSD)."[14] Combine this with a loss of muscle control and coordination from fatigue, as noted in the above study, entitled "Effects of Fatigue on Inter-muscular Coordination During Repetitive Hammering," where the researchers stated that their "[r]esults suggest that fatigue influences motion at both local and global levels [13] and there is always a greater risk of a acute injury as well." [14]

So to answer the question if the above findings with the use of an exoskeleton are enough relief in fatigue over the course of an eight- and ten-hour work shift to actually prevent injuries, the jury is still out. However, when considering the study quoted above, the authors went on to say, "In theory, decreasing levels of hemoglobin may lead to a deficiency in tissue oxygen levels, in turn resulting in reduced tissue tolerance, which, when combined with tissue load, may lead to cellular damage and fatigue."[11] So to fight fatigue any way possible is a good thing for the employee and the results from this study show that exoskeletons can help reduce the risk of injury, and thereby improve employee safety as well as productivity and quality.

Endnotes

[1] "Star Trek." IMDb: The International Movie Database. (Accessed February 20, 2017.) (http://www.imdb.com/title/tt0060028/quotes.)

[2] Moores, Danielle, and Erica Cirino. "Electromyography." *Healthline.* February 8, 2016. (Accessed February 20, 2017.) (http://www.healthline.com/health/electromyography#Overview1.)

[3] Middlesworth, Mark. "A Step-By-Step Guide to the REBA Assessment Tool." *Ergonomics Plus.* (Accessed February 20, 2017.) (http://ergo-plus.com/reba-assessment-tool-guide/.)

[4] Knight, Will. "An Iron Man Suit for Factory Workers." *MIT Technology Review.* March 21, 2016. (Accessed February 20, 2017.) (https://www.technologyreview.com/s/539251/the-exoskeletons-are-coming/.)

[5] Escobar, Claudia. "Sensitivity Analysis of Subjective Ergonomic Assessment Tools: Impact of Input Information Accuracy on Output (Final Scores) Generation." M.S. Thesis, Auburn University, 2006. May 11, 2006. (Accessed February 20, 2017.) (https://etd.auburn.edu/xmlui/bitstream/handle/10415/386/ESCOBAR_CLAUDIA_15.pdf?sequence=1.)

[6] Lännergren, J., H. Westerblad, and D. G. Allen. "Mechanisms of fatigue as studied in single muscle fibers." *Neuromuscular Fatigue* (1993): 3-9.

[7] Rassier, D. E., and B. R. MacIntosh. "Coexistence of potentiation and fatigue in skeletal muscle." *Brazilian Journal of Medical and Biological Research* 33, no. 5 (May 2000): 499-508. (Accessed February 20, 2017.) (http://www.scielo.br/scielo.php?script=sci_arttext&pid=S0100-879X2000000500003.)

[8] Eberstein, A., and A. Sandow. "Fatigue Mechanisms in Muscle Fibres." In *The Effect of Use and Disuse on Neuromuscular Functions*, 515-26. Amsterdam: Elverstein.

[9] *Electromyography (EMG) Lab.* PDF. (https://sierra.instructure.com/courses/312197/files/25456855/.)

[10] *Collins English Dictionary: Complete and Unabridged.* 12th ed. New York: Harper Collins.

[11] *Merriam Webster Online Dictionary.* (Accessed May 3, 2017.) (https://www.merriam-webster.com)

[12] *Techopedia: Online Encyclopedia.* (Accessed February 14, 2017.) (https://www.techopedia.com/definition/15325/wearable-robot.)

[13] Cote, J. N., A. G. Feldman, P. A. Mathieu, and M. F. Levin. "Effects of fatigue on intermuscular coordination during repetitive hammering." *Motor Control* 12, no. 2 (April 2008): 79-92. (Accessed February 20, 2017.) (https://www.ncbi.nlm.nih.gov/pubmed/18483444.)

[14] Ferguson, S. A., W. G. Allread, P. Le, J. Rose, and W. S. Marras. "Shoulder muscle fatigue during repetitive tasks as measured by electromyography and near-infrared

spectroscopy." *Hum Factors* 55, no. 6 (December 2013): 1077-087. (Accessed February 20, 2017.) (https://www.ncbi.nlm.nih.gov/pubmed/24745200.)

[15] Wise, M. B., T. L. Uhl, C. G. Mattacola, A. J. Nitz, and W. B. Kibler. "The effect of limb support on muscle activation during shoulder exercises." *Journal of Shoulder and Elbow Surgery* 13, no. 6 (Nov. & Dec. 2004): 614-20. (Accessed February 20, 2017.) (https://www.ncbi.nlm.nih.gov/pubmed/15570229.)

[16] Stone, Michael H., Meg Stone, and William A. Sands. 2007. *Principles and Practice of Resistance Training*. Champaign, IL: Human Kinetics.

[17] Chaffin, Don B., Gunnar B J Andersson, and Bernard J. Martin. 2006. *Occupational Biomechanics*. Hoboken, NJ: Wiley.

[18] University of California at San Diego. 2016. "Solving the Surgeon Ergonomic Crisis with Surgical Exosuit." White paper. This was an independent study approved by the Internal Review Board (IRB) for the University of Calf. San Diego conducting the study

Session No. 580

A Healthy Workforce: How Workers' Comp and Wellness Programs Go Together

Matthew J. Condon, JD/MBA
CEO
Bardavon Health Innovations, LLC
Overland Park, KS

Jennifer B. Edwards, MA, LAT, ATC, WCP, CWC
Senior Wellness Coordinator
Saint Luke's Health System
Kansas City, MO

Introduction

Most successful businesses are beginning to understand that the reactive approach to employee health–providing group health insurance to cover employees when they get sick–is far less effective than one that combines preventive efforts with transparent/reactive medical services. Adding a traditional "wellness" program to group health benefits used to be the first step for employers who want to "proactively" reign in their increasing group healthcare costs. This approach is slow, sometimes ineffective, and difficult to measure success. More importantly, this philosophy is becoming obsolete.

By adding a new generation of non-traditional "wellness" programs to their healthy workforce strategy, employers can take their program to a whole new level. Employers will look to a new generation of "comprehensive healthy workforce programs" to address not only nutrition, cessation education, exercise, and disease management, but also disease prevention, functional employment testing, job analysis, ergonomic assessment, and injury prevention. By looking at the total health of our workforce, employers can impact all healthcare costs including:

- Group health insurance premiums
- Workers' compensation insurance premiums and claims
- Absenteeism
- Presenteeism (workers who come to work but underperform due to illness or stress)
- Morale
- Worker satisfaction
- Recruitment and retention

- Cost of rehiring and retraining workers to replace hires that cannot perform the essential functions of a job and therefore, become injured and make workers' comp claims

Many of these programs will be introduced within the context of the workers' compensation and safety, rather than the traditional benefits market. Instead of being an outsider to the process, workers' compensation philosophies will be the foundation of implementation, and the key mechanism to measure the success of the program and an employer's return on investment.

With regard to "physical interventions," the next generation of comprehensive wellness programs will continue to utilize benchmark programs, like biometrics and health risk assessments, but will also analyze additional relevant co-morbidities and other useful data. Most importantly, by analyzing this "expanded" data set, employers can establish new benchmarks that will help them tailor programs specific to their workforce needs; programs including hiring, ergonomics, physical development, provider choice and analysis, functional testing, incentive management, and much, much more. All of these formerly disparate efforts will revolve around the nucleus concept of comprehensive healthy workforce programs. Additionally, comprehensive healthy workforce programs provide employers more direct opportunities to educate employees about how their lifestyle choices impact aggregate healthcare costs.

The world of employer wellness is about to evolve. Those early adopters on the front end of this shift will win in the open market by doing what their competitors fail to do–EFFECTIVELY manage the health and productivity of their workforce, and truly differentiate their workforce, their culture, and their success.

I. Understanding the Problem: The Startling Statistics

To determine a company's total healthcare cost, they must factor in both direct and indirect healthcare costs. Direct costs generally include the cost of healthcare coverage for both group health and workers' compensation segments, including claims costs and legal fees, where appropriate, along with the cost of the company's wellness plan. Indirect costs generally include those related to loss of productivity and declining corporate culture. While these indirect costs have been historically difficult to measure, the implementation of employer-driven healthcare and accountable care organizations (ACOs) will make it easier. Overall, apathetic employers that fail to aggressively address the health of their workforce face staggering costs, and a frightening future on the horizon for healthcare spending.

General Workforce Health Costs

The Centers for Disease Control (CDC) estimates, "if tobacco use, poor diet and physical inactivity were eliminated, 80 percent of heart disease and stroke, 80 percent of Type 2 diabetes and 40 percent of cancer would be prevented."[1] An achievement of that magnitude, and for just those issues alone, would result in over a half a trillion dollars in savings each year. Those savings would go primarily in the pockets of employers, since employers cover nearly 74 percent of the population not eligible for Medicare or Medicaid.[2]

General productivity losses related to personal and family health problems cost U.S. employers $1685 per employee per year, or $225.8 billion annually.[3]

Workforce Smoking Costs

The CDC estimates that smoking alone costs employers a least $170 billion per year in direct medical costs,[4] with 8.7% of annual healthcare spending in the U.S. being attributed to cigarette smoking.[5] Workers' compensation healthcare costs related to employee smokers cost employers $2189 annually per employee, compared to $176 for non-smoking employees.[6] Much of this increased cost is due to the fact that smokers visit healthcare professionals up to six times more than non-smokers,[7] are admitted to the hospital almost twice as often as non-smokers,[8] and average 1.4 additional days in the hospital per admission.[9]

Declines in productivity related to smoking cost employers almost 156 billion per year, including $5.6 billion in lost productivity due to secondhand smoke exposure.[10] These hidden and exposed costs simply cannot be avoided in our existing business environment. Employers must address these costs aggressively and effectively.

Workforce Obesity Costs

Obesity rates in the United States since 1985 have risen at epidemic rates.[11] When correlated to the rise in healthcare costs as a percentage of gross domestic product (GDP) in this same time period, it is clear that obesity alone has had a tremendous impact on overall healthcare costs for our country.[12] The annual cost attributable to obesity among full-time employees is $73.1 billion.[13] Among men, estimates range from $322 for overweight to $6087 for grade III obese men.[14] For women, estimates range from $797 for overweight to $6694 for grade III obese women.[15] Individuals with a body mass index >35 represent 37% of the obese population but are responsible for 61% of excess costs.[16]

Compared with an employee with a body mass index of 25, an employee with a BMI of 35, which is considered obese, has nearly double the risk of filing a short-term disability claim or a workers' compensation claim.[17] Normal-weight employees cost an average of $3830 per year in covered medical claims, sick days, short-term disability, and workers' compensation, while a morbidly obese person costs employers more than double that, at $8067.[18]

Regarding obesity and productivity: Obese workers miss significantly more work days: an average of 1.1 to 1.7 additional absences per year, compared to normal-weight workers. Obesity-attributable absenteeism among American workers costs the nation an estimated $8.65 billion per year.[19]

Clearly, an employer's right to request that their employees become personally accountable for their lifestyle choices can and should be tied more directly to the level the employee shares in the cost of their healthcare insurance. This "employer-centric" approach represents our best chance to impact the health of this country, which is the source of our overwhelming healthcare costs.

II. Understanding the Solutions: A New Approach

Learning from Workers' Compensation Claims

While often forgotten, perhaps the best opportunity for an early and significant return on investment for a comprehensive healthy workforce program is not within the group health insurance spectrum, but rather, within the workers' compensation spectrum. Conveniently enough, the workers' compensation arena also allows some of the most innovative mechanisms for identifying health-risk factors, implementing interventions, tracking and measuring total cost, and for success. That's because, while it's difficult to track co-morbidity data in the group health market because of privacy protections afforded by The Health Insurance Portability and Accountability Act (HIPAA), HIPAA specifically allows an exemption for workers' compensation related matters:

1. If the disclosure is "[a]s authorized and to the extent necessary to comply with laws relating to workers' compensation or similar programs established by law that provide benefits for work-related injuries or illness without regard to fault." 45 C.F.R. § 164.512(l).

2. If the disclosure is required by state or other law, in which case the disclosure is limited to whatever the law requires. 45 C.F.R. § 164.512(a).

3. If the disclosure is for the purpose of obtaining payment for any healthcare provided to an injured or ill employee. 45 C.F.R. § 164.502(a)(1)(ii).

This means employers can, should, and inevitably will demand from their providers the kind of data needed to accurately analyze the actual total cost of an injury, and all the factors that impacted that cost. This also means that several organizations inside their chosen healthcare team have had access to the data they need to analyze the impact of lifestyle choices on the cost of workers' compensation healthcare services and cost of claims. In the new era before us, an organization's chosen healthcare team will be expected to collect, analyze, and share that data, as well as use the data to improve future processes, and increase the likelihood of success for everyone involved.

Perhaps one of the best descriptions of the impact an approach like this can have is in Duke University's study, "Obesity Increases Workers' Compensation Costs," published in 2007.[20] In Duke's study, researchers examined the records of 11,728 employees of Duke University who received health-risk appraisals between 1997 and 2004, to analyze the relationship between body mass index (BMI), and the rate of workers' compensation claims. They found that workers with a BMI over 40 had 11.65 injury claims per 100 workers, compared with 5.8 injury claims per 100 for workers within the recommended BMI range. Obese employees averaged 183.63 lost work days per 100 employees, compared to just 14.19 lost work days per 100 employees for those who had a BMI in the recommended range. Last but not least, the average medical claim cost per 100 employees was $51,019 for the obese, compared to $7503 for those employees with a BMI in the recommended range.

In a subsequent study by Tao, et al, workers' compensation costs and outcomes were compared for obese, overweight, or normal-weight workers. Obesity was defined as a body mass index (BMI) of 30 or higher, and overweight as a BMI between 25 and 30.[21]

After three years, about 11 percent of claims for major injuries were still open—indicating that the worker had not yet returned to work. For workers with major injuries, high BMI was also associated with higher workers' compensation costs. In this group, costs averaged about $470,000 for obese and $270,000 for overweight workers, compared to $180,000 for normal-weight workers.[22] After adjustment for other factors—including high-cost spinal surgeries or injections—obese or overweight workers with major injuries were about twice as likely to incur costs of $100,000 or higher.[23]

These cost and productivity disparities between people with high vs. average BMI are unlikely to be limited to the workers' compensation segment; they are simply more easily benchmarked.

III. Understanding the Future: The Impact of Legislation

PPACA and Defining "Wellness Programs"

The Patient Protection and Affordable Care Act (PPACA) not only set the legislative stage for the importance of workplace wellness; it also sets the debate stage for our legislators regarding the absolute need to approach this problem comprehensively. PPACA provides $200 million in grant funds to assist small employers with the implementation of wellness programs. By putting its money where its mouth is in such a large way, it is clear legislators are beginning to understand both the need for comprehensive employer wellness programming, as well as the need to incorporate occupational health issues into the equation. The preamble to the first Workplace Wellness initiative stated:

> Workplace health promotion programs are more likely to be successful if occupational safety and health is considered in their design and execution. In fact, a growing body of evidence indicates that workplace-based interventions that take coordinated, planned, or integrated approaches to reducing health threats to workers both in and out of work are more effective than traditional isolated programs. Integrating or coordinating occupational safety and health with health promotion may increase program participation and effectiveness and may also benefit the broader context of work organization and environment.[24]

In other words, companies instituting wellness programs that incorporate workers' compensation strategies, including job analysis, valid and legally defensible, post-offer employment testing, and functional testing for return-to-work assessments, alongside more traditional wellness services, such as biometric and health-risk assessments are functionally and financially more successful. Additionally, by including workers' compensation strategies, employers can gather transparent data to tailor interventions that address their employees' specific needs and risk factors.

PPACA and Selecting Quality Providers for Medical Home

Another interesting impact of the PPACA legislation involves the opportunity for employers to select their own group of preferred healthcare providers into a structure referred to as a "medical home." Medical homes have been defined generally as a team-based, healthcare delivery model, led by a physician, P.A., or N.P., that provides comprehensive and continuous medical care to patients with the goal of obtaining maximized health outcomes. When this concept is modified and utilized in the employer-directed healthcare segment, medical homes will be charged to develop and manage plans of care specifically based on an employee's essential job functions. These unique "medical home providers" will be rewarded for savings on total cost of healthcare and maintenance and improvement of function. Therefore, employers can save additional dollars by identifying quality providers who deliver successful, cost-effective outcomes.

These medical provider partnerships will be an integral part of the comprehensive healthy workforce program. Because, in this model, providers will be part of the treatment team, and therefore aware of the employer's specific programming around health, they can consider those programs in their recommendations for return to work. Employers will no longer be limited to choosing providers based on subjective information. In a new generation of comprehensive healthy workforce programs, employers will decide (when allowed by state law), to choose providers, based on a value proposition that goes beyond price, reputation, and/or percentage discount. This new value proposition will include overview of injury types, utilization, days in treatment, functional improvement, co-morbidity factors, and functional outcome. In other words, what is the true total cost to treat this patient, and return the employee to work safely? What unique considerations does this particular patient present? What providers are best equipped to address those unique considerations? Finally, what proactive steps should that employer take now to mitigate future exposure to these cost drivers? In the new era of employer-driven healthcare, all these efforts will be housed within the context of a comprehensive healthy workforce program.

PPACA and Wellness Incentives

While employers are beginning to see the financial and cultural benefits from incorporating effective comprehensive wellness programs, it is equally clear that there is a significant legislative trend aimed at further incenting aggressive employer strategies on health and wellness. A key provision within

PPACA increased the amount an employer can incent improvements in specific health factors, as identified within their wellness program, up to 30 percent of the total cost of benefits; in some circumstances, up to 50 percent. Other relevant legislative initiatives include S. 803/H.R 1897, Healthy Workforce Act; H. Con. Res 405, Resolution Recognizes the First Full Week of April as "Workplace Wellness Week;" S. Res. 673, Resolution Recognizes the Importance of Workplace Wellness as a Strategy to Help Maximize Employee's Health and Well Being.

In addition to the physical and financial benefits of engaging in a comprehensive healthy workforce program, the bottom line is this: In the end, our competitive marketplace itself will inevitably move comprehensive healthy workforce programs from an ancillary program/benefit to a core strategic initiative for successful companies. As more companies learn they can increase their margins by controlling ALL their healthcare costs, group health, workers' compensation, productivity, etc., they will use their improved margin as a competitive advantage. Those aggressive early adopters will be able to reinvest those saved dollars to make similar or higher quality services or goods and sell them at the same or lower prices. They will be able to recruit and retain the most talented in their industry, and they will drive economic development in their respective communities. In this way, the wellness evolution is destined to grow as companies that employ and manage effective healthy workforce programs dominate and overwhelm those that don't.

Endnotes

[1] Mensah, George A., Associate Director for Medical Affairs, CDC. "Global and Domestic Health Priorities: Spotlight on Chronic Disease." *National Business Group on Health Webinar*, May 23, 2006.

[2] National Center for Health Statistics. 2016. *Health, United States, 2015: With Special Feature on Racial and Ethnic Health Disparities.* Hyattsville, MD.

[3] Stewart, W.F., Ricci, J.A., Chee, E., Morganstein, D. 2003. "Lost productive work time costs from health conditions in the United States: results from the American productivity audit." *J Occup Environ Med.*, 45(12):1234-1246.

[4] U.S. Department of Health and Human Services. 2014. "The Health Consequences of Smoking—50 Years of Progress: A Report of the Surgeon General." Atlanta: U.S. Department of Health and Human Services, Centers for Disease Control and Prevention. National Center for Chronic Disease Prevention and Health Promotion, Office on Smoking and Health.

[5] Xu, X., Bishop, E.E., Kennedy, S.M., Simpson, S.A., Pechacek, T.F. 2014. "Annual Healthcare Spending Attributable to Cigarette Smoking: An Update." *American Journal of Preventive Medicine,* 48(3):326–33.

[6] Tobacco Public Policy Center at Capital University Law School. October 6, 2005. "Business Costs Associated with Tobacco Use." (http://www.law.capital.edu/Tobacco/TobaccoInTheWorkplace/BusinessCostsFactSheet.pdf).

[7] Berman, K. October 12, 1987. "Firms hope smoking bans will trim health costs." *Business Insurance.* 21 (41): 16-17.

[8] Halpern, M.T., et al. September 2001 "Impact of smoking status on workplace absenteeism and productivity." *Tobacco Control* 10(3): 233-38.

[9] Stewart, W.F., Ricci, J.A., Chee, E., Morganstein, D. 2003. "Lost productive work time costs from health conditions in the United States: results from the American productivity audit." *J Occup Environ Med.*, 45(12):1234-1246.

[10] U.S. Department of Health and Human Services. 2014. "The Health Consequences of Smoking—50 Years of Progress: A Report of the Surgeon General." Atlanta: U.S. Department of Health and Human Services, Centers for Disease Control and Prevention. National Center for Chronic Disease Prevention and Health Promotion, Office on Smoking and Health.

[11] Centers for Disease Control and Prevention (CDC). "Overweight and Obesity". (https://www.cdc.gov/obesity/data/databases.html)

[12] Centers for Medicare and Medicaid Services. "National Health Expenditure Fact Sheet". (http://www.cms.gov/NationalHealthExpendData/25_NHE_Fact_Sheet.asp.)

[13] Finkelstein, E.A., DiBonaventura, M., Burgess, S.M., Hale, B.C. 2010. "The Costs of obesity in the workplace." *J Occup Environ Med.* 52:971–976.

[14] Ibid.

[15] Ibid.

[16] Ibid.

[17] Van Nuys, K., Globe, D., Ng,-Mak D., et al. 2014. "The association between employee obesity and employer costs: evidence from a panel of U.S. employers." *Am Journal of Health Promotion.* 28(5):277-285.

[18] Ibid.

[19] Finkelstein, E.A., DiBonaventura, M., Burgess, S.M., Hale, B.C. 2010. "The Costs of obesity in the workplace." *J Occup Environ Med.* 52:971–976.

[20] Truls, Ostbye, et al. "Obesity and Workers' Compensation: Results from the Duke Health and Safety Surveillance System," *Archives of Internal Medicine,* April 23, 2007.

[21] Tao, X., Su, P., Yuspeh, L., Lavin, R.A., Kalia-Satwah, N., Bernacki, E.J. 2016."Is obesity associated with adverse workers' compensation claims outcomes?" *J Occup Environ Med.* 58(9):880-4.

[22] Ibid.

[23] Ibid.

[24] GovTribe. 2011. Solicitation Number 2011-N-13420, *Comprehensive Health Programs to Address Physical Activity, Nutrition, and Tobacco Use in the Workplace.*(https://www.govtribe.com/project/comprehensive-health-programs-to-address-physical-activity-nutrition-and-tobacco-use-in-the-workplace)

The Human Side to Change: Make Safety Protocols "Inevitable and Resistance Futile"

David Maxfield
Safety VitalSmarts

Gadgets are smart these days. You can buy a refrigerator that tells you you're out of milk. Your car knows when its tires are low. You can even ask your mobile phone for parenting advice. We owe the increasing power of our toys and tools to the fact that all of them now sport powerful *operating systems.*

Similar to our gadgets, organizations have operating systems. Think of them as cultural operating systems. Both electronic and cultural operating systems provide a set of rules that guide the behavior of the device or the people respectively. For example, when you press the ON button of an iPhone, a whole series of coded rules display an Apple logo, followed by the time of day, your customized wallpaper, and an unlock bar.

Similarly, a cultural operating system engages a set of rules that guide the behavior of employees when they enter a high-stakes meeting, react to abrupt organizational changes, or contemplate what to do with discretionary time. But while electronic intelligence is processed by written coded rules, a cultural operating system (COS) is often driven by a combination of written and unwritten rules. For example, executives may have written cultural mandates in the form of values, mission statements, and corporate creeds that prompt employees to take initiative, speak candidly, and act as a team. Yet when you watch what employees actually do, you realize the unwritten rules are avoid risk, defer to the boss, and stay in your silo.

Executives have known for decades that their COS largely determines organizational success or failure. For example, renowned restaurateur Danny Meyer has created a cultural operating system that influences his 1500 employees to create superlative hospitality experiences for his 100,000 daily guests. Recently, a harried guest forgot her cell phone and purse in a taxi that raced away after depositing her in front of the restaurant. The maître d', prompted by Danny Meyer's COS, welcomed her, assured her all would be well, and seated her with her party. In the meantime, the maître d' began calling the guest's cell phone repeatedly until the cab driver noticed and answered it. Before the guest had even ordered her entrée, an employee had been dispatched to meet the cab driver, retrieve the phone and purse, and deliver it to the overwhelmingly appreciative customer. And, while much credit for the remarkable experience is due to an enterprising team, less dramatic versions of unrivaled customer service happen hundreds of times a day because Danny Meyer has designed crucial skills into his intentionally created COS.

Leaders who invest in fostering a potent cultural operating system that enables their organizational strategy reap daily rewards for their investment. Leaders who fail to make such an investment tend to suffer from dysfunctional, unwritten rules that inevitably fill the void. For instance, the CEO of one company struggling to survive an economic downturn by reducing expenses had a COS in which employees' loyalties aligned with their department first and the enterprise last. When the CEO charged managers to pull together to close the financial gap, they vigorously nodded in agreement while in the room, and then vigorously defended their department budget after leaving. After months of haranguing unresponsive managers to find efficiencies and reduce headcount, the CEO was forced to impose top-down solutions. Losses incurred from managements' dallying amounted to millions of dollars. Unwritten rules always trump written ones.

It's no wonder studies suggest that the culture of an organization can account for one third to one half of its growth and profitability. One extensive study showed that a well-designed COS correlated with 400 to 500 percent increases in revenue, income, and share price growth over organizations with less effective written and unwritten rules.

Two Purposes of a Cultural Operating System

While today's software systems rely on mind-bogglingly complex operating systems that offer hundreds of features and functions, an organization's cultural operating system only needs to enable two functions. The measure of a good COS is the degree to which it enables an organization to execute superbly and innovate consistently.

Gallery Furniture in Houston, Texas provides a stunning example of how a COS is crucial to these two functions. One of Gallery's competitive advantages is its ability to offer same-day delivery of any of its merchandise anywhere in its enormous service area. This is no small promise when you consider that, even purchases made at dinnertime, are promised to arrive before bedtime. Yet Gallery's ability to execute flawlessly on this high-value promise has distinguished the retailer for decades, and enabled the company to grow, even through tough economic times.

Recently, Jim McIngvale, Gallery's CEO, found that 20 percent of the company's signature deliveries had to be reworked due to errors or damages. The rework was costing the firm millions every year. Most any organization has its own version of Gallery's 20-percent rework problem—imperfections in execution that cost time, trust, and money. What's unique about Gallery—and what accounts for its distinct capacity to execute—is how routinely these problems are surfaced, analyzed, and addressed. When the extent of the rework became apparent, McIngvale's team did what it usually does: They candidly discussed deficiencies, explored potential solutions, and developed a robust way of changing behavior to drive better execution. Gallery's written and unwritten rules of initiative, teamwork, accountability, and candor are key to their remarkable competitive advantage.

But Gallery's COS doesn't just enable flawless execution. It is also a powerful impetus of innovation. For example, Gallery makes more from selling mattresses than most any other product line. Yet competition for mattress sales is fierce. Most mattresses you could buy at Gallery can be price-shopped online with a few simple clicks. Gallery, however, is reinventing mattress sales by reframing the customer relationship around the entire sleep experience. Their goal is no longer simply shipping you the mattress you want. Rather, they have realigned their team to enable sales reps to follow a customer's sleep experience for months after the sale. The goal is to ensure the customer's new mattress provides the optimal mix of emotional, physical, and mental ingredients they need to feel rested and well—forever.

Executing on this innovation took much more than a simple brainstorming session. It meant realigning teams, developing new competencies, and perfecting new processes. Most organizations fail to innovate not because they lack ideas but because their unwritten rules stifle or stall them. Gallery stands head and shoulders above a fiercely competitive industry because execution and innovation are hardwired into their COS.

While scholars often confuse the issue by suggesting an organization can't be world class at both execution and innovation, real-world executives know success depends on both. Execute unimpressively, and you risk both capital and customer flight. Innovate lethargically, and you sprint toward irrelevance. Doing both better than others is the only sure path to long life and mission success.

The Four Most Deadly Viruses for a Cultural Operating System

For the past 25 years, we've asked the question, "Are there a small number of skills required in any human system to make it supremely effective at achieving its mission?"

Think about it. It's an audacious question. We've asked, in essence, whether you're the Army, the Peace Corps, a chip manufacturer, an Amish furniture shop, a fast food chain, or a fitness center, etc., are there basic human "operating system" competencies you'll need to execute flawlessly and innovate consistently over time?

Initially, you'd think the answer would be a clear "no." After all, the obvious competencies required for brain surgery, for example, bear no obvious similarity to those required to run an airline.

But dig a little deeper and you find that while the "apps" required to be world class at healthcare or transportation are unique, the human "operating system" supporting those apps is strikingly similar. Just as all electronic devices need basic functions like input, output, data management, and so on, there are foundational skills required for any human system to function effectively. When these crucial skills are present, things run smoothly and improve routinely. When they are absent, the system gets mired in mediocrity, or worse. In most organizations, these skills are in short supply. When organizations do not take intentional action to foster these four competencies, four crippling viruses fill the void. Our research shows that these four viruses produce the greatest drag we've found on any organization's ability to perform, independent of size, industry, or geography.

In fact, leaders who do not intentionally develop the four crucial skills in their organization's operating system in effect pay a tax for the omission. Through inattention, they allow the unwritten rules of their system to lead to cultures of:

- Resistance
- Silence
- Collusion
- Cynicism

The antidote to these cultural viruses includes four skills that define an organization's performance capacity at four levels:

1. Personal: Self-directed change
2. Interpersonal: Open dialogue
3. Team: Universal accountability
4. Organizational: Influential leadership

The best way to appreciate the necessity of these four skills is to consider the viruses that grow in their absence and how each virus constrains execution and innovation.

About the Research

We surveyed a worldwide sample of more than 11,000 managers and employees. The survey measured each of the four cultural competencies (personal, interpersonal, team, and organizational), using multi-item scales that achieved reliability scores of $\alpha >.80$. Innovation and execution were evaluated at the department level, using seven-item scales that achieved reliability scores of $\alpha >.85$.

The 6930 managers rated their region or division leaders along all four competencies, and rated their direct reports on the personal, interpersonal, and team competencies. The 4576 non-supervisory employees rated their immediate managers along all four competencies, and rated their coworkers on the personal, interpersonal, and team competencies. All participants rated the level of innovation and execution at the department or unit level.

We used two calculations to determine the contribution each competency makes to innovation and execution. First, we calculated the tax an organization incurs when it fails at any one of the four competencies. This calculation compared organizations that were strong at all four competencies to those that were strong in three, but weak in the one being examined.

Second, we calculated the force an organization gains when it adds competencies, one on top of another. The simplest way to calculate the force is with a multiple regression. The regression coefficient for innovation is $R=.67$; the regression coefficient for execution is $R=.60$. We illustrate this force by showing the incremental improvement each competency adds, in the following order: personal, interpersonal, team, and organizational.

Virus #1: A Culture of Resistance

The unwritten rule in most organizations is resist change. Change only happens when people are dragged kicking and screaming into it. For example, a hospital that invested tens of millions of dollars in a new electronic medical records system found most doctors were still using paper prescriptions 18 months after it went live. Resistance to the new-fangled system overrode any written rule executives tried to impose.

Most leaders trying to drive improved execution or accelerated innovation encounter cynicism and lethargy. Ultimately, organizational change requires individual change. And the evidence shows we're no better at the latter than we are at the former. One VitalSmarts study revealed that fewer than 10 percent of employees are able to change their own deeply ingrained habits, even when their lives or careers depend on it. For example, the typical employee who is fired for failure to change bad habits is counseled about those habits for years before being shown the door.

Our deficiency in self-directed change explains the almost complete failure of twentieth-century performance management practices to produce any meaningful improvement in organizational performance. Managers who have just finished giving performance feedback to employees believe their employees have less than a one in ten chance of making meaningful change in targeted areas. Our research supports this lack of confidence. Specifically, fewer than one in 20 people succeed when attempting to change longstanding work, health, spending, or relationship habits that undermine their professional and personal goals.

Limited efficacy for personal change robs organizations of a core competence required to drive flawless execution and vital innovation. As a result, key initiatives are led by a small pool of self-starters, or they don't happen at all.

Most organizations attempt to compensate for widespread ineffectiveness at self-directed change through supervision rather than leadership. They install tall hierarchies designed to ensure people behave as needed, assuming that without compulsion, little would happen.

In an ideal world, hierarchies exist so different individuals can focus on different levels of strategic problems. Management should exist to organize work, not control individuals. When individuals are incapable of absorbing performance feedback and developing robust methods for changing their own habits, hierarchies are diverted to micromanaging, compelling, and monitoring individual change rather than coordinating the central work of innovating and executing.

So what's different in a robust COS? The most basic building block of a powerful cultural operating system is the capacity to effectively engage in independent action. Imagine the potential of an organization where continuous personal initiative is a cultural norm. Walk the campus of the tech juggernaut Google, and you'll be immersed in a culture of self-directed change. Everywhere you turn, there is an opportunity or resource to support you in getting fit, learning a new skill, developing a new relationship, or engaging in an innovative task. Walls are plastered with pictures of people who have made diverse contributions to the planet, from Mother Teresa to Steven Hawking. Speakers are invited every week to stimulate thought and enable personal improvement. Discretionary time is available to employees to follow their passion; in fact, employees get up to 20 percent of their workweek to pursue special projects. Google's 20-percent rule codifies the expectation that employees spend a significant part of their time taking independent action to help achieve the mission of organizing all of the world's information.

Consider the message Google is sending its employees. First, it's saying, "We trust you." Employees who test the limits by using Google's time to explore a new risky project hear this message loud and clear. Second, it's saying, "We care about you." Employees who focus their efforts on self-improvement hear this message. However, what sets Google apart is the way it sends these messages. These aren't just words a jaded employee reads in a corporate brochure. These are actual programs that encourage and enable action, the kind of direct experiences that convince even the cynics.

The payoff for the organization comes in employees who are confident that their individual initiative will be rewarded and supported instead of punished and undermined. In this way, Google has created a culture where individuals don't wait to be told. When they see opportunities, they leap, and Google benefits.

But you don't have to have billions in cash and resources to create a culture of self-directed change. Take Gallery Furniture, for example. The small business is no less interested in creating this cultural capacity than Google. McIngvale has spearheaded several efforts to support individual initiative. He has created scholarships that make it easy for employees to further their education in any direction they want. He sponsors employees' volunteer efforts in the community, and gives them time off to make a difference. He has also created a track and gym in the warehouse to support employees who want to improve their fitness. In fact, McIngvale models his own self-improvement efforts by arriving at 7:00 a.m. for calisthenics.

If an organization does not intentionally invest in creating a culture of self-directed change, the virus of resistance will fill the void. Attempts at organizational change will be met by profound resistance.

But think of the bottom-line results organizations achieve when employees are continually engaged in self-directed change. In this kind of culture, employees know they are trusted, so oversight is minimal. Management attention is focused on solving problems, rather than supervising people. As individuals consistently and competently examine their own behaviors and take intentional action to align their habits with their values, they do more than just enhance their own lives. They become part of a human system with unparalleled capacity for unified performance. This core capacity for self-reflection and self-direction is the most fundamental and potent force in any collective human venture. When individuals become skilled at self-directed change, the organization's capacity to change accelerates. The result is unbeatable levels of execution and innovation.

It's this kind of employee, working in this kind of cultural system, who sees a guest who just lost her cell phone as an opportunity for service rather than a distraction from duties.

Results

Tax: Comparing organizations that score low on Self-directed Change, but high on the other three competencies to organizations that score high on all four competencies: Organizations that score low on Self-directed Change are 35 percent less effective at Execution, and 32 percent less effective at Innovation.

Benefit: Comparing organizations that score high on Self-directed Change but low on the other three competencies to organizations that score low on all four competencies: Organizations that score high on Self-directed Change are 32 percent better at Execution and 46 percent better at Innovation.

Virus #2: A Culture of Silence

The greatest impediment to leveraging the potential of a workforce of subject-matter experts is people's innate reluctance to speak honestly about emotionally and politically risky issues. This impediment is a primary cause of poor decisions, half-hearted execution, employee disengagement, and stifled innovation.

When people can't speak the truth to those in power, the organization loses its ability to self-correct and its capacity to handle the unexpected. For example, imagine how many organizations have an equivalent of Gallery Furniture's 20-percent rework problem, but the information is buried rather than discussed. In one organization, the IT department knew for years that maintaining both Macs and PCs for executives was costing hundreds of thousands in support expenses. But everyone stayed mum because they believed the CEO would resent them for questioning his personal preference.

Innovation is stifled less often because a workforce lacks ideas and more often because of a failure to generate spirited debate among differing ideas. Vigorous disagreement is the crucible of innovation, yet human beings are nurtured from childhood to defer to those in authority. If leaders do not make intentional efforts to break down a culture of silence, they will, by default, have a COS that undermines continuous improvement and innovation.

However, our research shows that a culture of tremendous open dialogue is possible. Danny Meyer says that the crucial skill he needs to maintain his world-class COS is helping employees learn that anything can be said if it will help the team offer better hospitality to guests. His most fundamental challenge is ensuring his unwritten rule matches his written one. For example, during one blistering summer, the air conditioner failed at Eleven Madison Park, Danny Meyer's high-end French restaurant. By the time the first hundred lunch guests arrived, the temperature in the restaurant was already at 80 degrees and rising, not to mention the stifling humidity.

Rather than bury the problem or defensively cancel lunch to avoid embarrassment, the Eleven Madison Park team immediately acknowledged the problem and candidly discussed ways of ensuring guests' comfort. Because the team wasted no time finger pointing, they were ready when the guests arrived, offering each the option of a personal fan (purchased by creative team members at a local hardware store) or a rescheduled reservation. The majority of guests weathered the heat in good humor and admired the caring response of the staff.

Individuals yearn for the freedom to express their views. They don't need others to agree; they only want to be listened to respectfully. When leaders take action to develop a COS of candid dialogue, mistakes are caught more quickly, decisions are implemented more effectively, and innovation flows more routinely.

<u>Results</u>

Tax: Comparing organizations that score low on Open Dialogue, but high on the other three competencies to organizations that score high on all four competencies: Organizations that score low on Open Dialogue are 31 percent less effective at Execution and 28 percent less effective at Innovation.

Benefit: Comparing organizations that score high on both Self-directed Change and Open Dialogue, but low on the other two competencies, to organizations that score low on all four competencies: Organizations that score high on both Self-directed Change and Open Dialogue are 54 percent better at Execution and 66 percent better at Innovation.

Virus #3: A Culture of Collusion

One wise healthcare executive discovered that rampant hospital-acquired infections (infections patients contract from healthcare workers!) were not a fact of life in a complex healthcare system. Rather, they were largely a consequence of silence. Many of her employees knew how to avoid spreading infections but were afraid to hold others accountable to following those best practices, even simple practices like washing hands or wearing masks and gowns. For example, we once watched a doctor working with tiny, critically ill, premature babies enter the intensive care unit without donning the required protective clothing. We were stunned not only that he did that, but also that the dozen or so nurses and doctors who saw him said nothing.

When accountability requires position power, very little accountability actually happens. It becomes a line-of-sight problem. Few managers see their employees often enough or understand their performance deeply enough to hold them accountable in a meaningful way.

If team members can't hold each other accountable, then an organization's capacity to carry out unified action on complex interdependent tasks plummets. And nearly all significant results, whether they are related to innovation or to execution, require unified action across multiple teams and functions.

A weak COS is one where those who see problems say nothing because they assume they don't have the power or authority to raise a concern. Time, resources, and morale are wasted through an unwritten rule that "accountability is someone else's job." In most organizations, a culture of collusion prevails; one where individuals see problems but say nothing, assuming someone "up there" will deal with it. In so doing, they unwittingly collude in undermining results. Then, when it comes time to debrief the disaster, the focus is more on blame than on learning, since there is no widely shared sense of accountability. Because mistakes threaten power, the immediate motivation in a power-focused culture is to deflect blame.

Organizations that want to pivot rapidly and execute flawlessly can't afford to wait for accountability to happen top down. Successful organizations depend on creating a culture where anyone can hold anyone else accountable: managers to employees, employees to managers, and peers to peers. The core value in this COS is not power but results.

John Noseworthy, CEO of the Mayo Clinic, knew his efforts to build a cultural operating system of patient safety were paying off when he went to visit an aunt who was recovering from surgery in his Rochester, Minnesota facility. "I had a few minutes during my lunch hour to check in on her so I took the elevator up to her floor," he told us. "As I headed toward her room, I was stopped by a nurse. She immediately recognized me, and greeted me by name, 'Welcome to our floor Dr. Noseworthy. It's nice to have you here!' she said with a broad smile." Noseworthy thanked her politely then attempted to continue forward. She stepped in front of him again, still smiling, and said, "Dr. Noseworthy, we have a rule on our floor to keep you and our patients safe. We require you to wash your hands when you enter the floor and wash them again when you leave it. Would you be so kind as to put your hand under the gel dispenser behind you before proceeding?" After relating this story, this wise leader concluded, "When everyone in our hospital can do what that nurse did, we'll be as safe as we need to be."

When a norm develops that allows people to raise concerns with those beside, above, and below them on the org chart, problems are solved routinely and quickly, and a pervasive sense of empowerment and responsibility for results prevails. Even when problems happen, the sense of personal responsibility that accompanies a culture where anyone can hold anyone else accountable drives a focus on learning rather than blame. This focus accelerates the development of collective competence, dramatically improving execution and facilitating innovation.

Results

Tax: Comparing organizations that score low on Universal Accountability, but high on the other three competencies to organizations that score high on all four competencies: Organizations that score low on Universal Accountability are 24 percent less effective at Execution and 20 percent less effective at Innovation.

Benefit: Comparing organizations that score high on Self-directed Change, Open Dialogue, and Universal Accountability but low on Influential Leadership to organizations that score low on all four competencies: Organizations that score high on Self-directed Change, Open Dialogue, and Universal Accountability are 73 percent better at Execution and 76 percent better at Innovation.

Virus #4: A Culture of Cynicism

We've worked in hundreds of organizations over the past 30 years. The most common cultural characteristic we encounter is deep cynicism about leaders' ability to execute change. When leaders are not skilled influencers, they create "flavor of the month" organizations that rotate through trendy change initiatives, even when their own strategic goals remain constant.

What struck us about Jim McIngvale's COS at Gallery Furniture is that when new behavior was called for in the same-day delivery process, intricate plans to drive that behavior were not only developed, they were also implemented and embraced. The dominant feeling about change in his organization is not cynicism; it is optimism.

Few organizations can drive rapid, comprehensive change, even when an organization finds itself on a burning platform. Leaders aren't able to mobilize their people to execute or innovate in a conscious, strategic way. One change initiative after another fails, not because the strategic ideas were inadequate, but because leaders were incapable of influencing change in the behavior required to execute on the new ideas. Cynicism builds year after year, as bold strategies produce small effects and, ultimately, employees learn it's a waste of time to invest in understanding and supporting change efforts. Instead, they wait them out, realizing that if they wait long enough, the idea of "change" will simply go away.

Ironically, organizations with a weak COS stay that way because their leaders lack this one COS functionality: the ability to execute behavior change. In fact, leaders' unwillingness or inability to think systematically about influencing rapid, profound, and sustainable behavior change is so damaging that it directly leads to cultures of silence, collusion, and resistance to change. And when, in moments of clarity, leaders feel a need for change, their ineffective attempts—which are often delegated to staff functions—fall short of creating real change, which in turn leads to a culture of cynicism.

The picture is profoundly different in organizations where leaders are competent at intentionally influencing system-wide change. Dr. David Dull was part of a team trying to get hundreds of doctors to use a new IT system to fill medication orders at Spectrum Health. Over the years, we've seen large systems like Spectrum struggle to gain this kind of compliance. The challenge is particularly difficult when independently minded doctors are also self-employed, work in multiple hospitals, and only occasionally frequent the facility attempting to implement new systems. For example, one hospital system we worked with found that after three years and 170-percent budget overruns, they still had fewer than half of their doctors consistently using a new IT process that would save time and money.

But at Spectrum, the results were startlingly different. Compliance soared past 90 percent within three days of launching their order-entry system. The key difference at Spectrum was that leaders up and down the chain of command were skilled at creating intentional plans to lead change within their own areas of responsibility. This widely shared competency resulted not just in careful plans to implement technology, but also in complementary efforts to influence the behavior required to make the technology achieve its goals.

Leaders who think deeply and skillfully about the myriad forces that shape behavior and who creatively engage all of these forces to create positive change create cultures of confidence. In these organizations, employees hold leaders in higher regard and feel optimistic that even profound changes will lead to improved future performance.

Results

Tax: Comparing organizations that score low on Influential Leadership, but high on the other three competencies to organizations that score high on all four competencies: Organizations that score low on Influential Leadership are 19 percent less effective at Execution and 18 percent less effective at Innovation.

Benefit: Comparing organizations that score high on all four competencies to organizations that score low on all four competencies: Organizations that score high on all four (Self-directed Change, Open Dialogue, Universal Accountability, and Influential Leadership) are 104 percent better at Execution and 98 percent better at Innovation.

Conclusion

The question is not whether you have a cultural operating system; every organization has one. The question is whether your COS is one that advances or impedes continuous improvement in execution and innovation.

Leaders who lead the way in creating a COS characterized by self-directed change, open dialogue, universal accountability, and influential leadership harness the full potential of their human resources. In so doing, they leverage what research shows is the most potent predictor of sustained value for customers, employees, shareholders, and the world—a high-performance cultural operating system.

Session No. 582

Diacetyl and Other Flavorings: New Health Concerns for Coffee Processors

Sharath F. Smith, MENG, CSP
3M Company
St. Paul, MN

Introduction

Concerns about diacetyl and related flavorings have spread from the microwave popcorn industry to new areas, such as coffee processing. The CDC published a report on the health effects of chemicals produced during coffee processing, that stated, "Roasting coffee beans naturally produces diacetyl and 2,3-pentanedione. Volatile organic compounds, including alpha-diketones (e.g, diacetyl and 2,3-pentanedione), can be released during grinding of coffee."[1] The CDC also reported that exposures to these chemicals at certain levels may result in reduced lung function, severe fixed obstructive lung disease, and an irreversible, and sometimes fatal, lung disease, obliterative bronchiolitis.

After five cases of former workers of a coffee processing facility that roasted, ground, and flavored coffee were identified in 2012, additional studies were undertaken based on worker concerns.[2] In August 2015, the *American Journal of Industrial Medicine* reported several other cases in the coffee processing industry, leading to the need for evaluation of work places beyond the microwave popcorn and flavoring industries, where concerns about these chemicals first arose.[3] Workers may be exposed to diacetyl and 2,3-pentanedione during several phases of coffee processing, including grinding, roasting, flavoring, and packing. One study measured the highest exposures in the flavoring and grinding of unflavored coffee.[4] Other industries with fermentation and pyrolysis products can also generate diacetyl exposures, such as in the manufacture of beer, wine, and dairy products.[5]

As these relationships illustrate, this is clearly an area where medical and scientific knowledge is evolving. Source material includes numerous points made by government agencies or regulators (such as NIOSH, CDC, OSHA, etc.), and the science at this point is likely incomplete or subject to change. Employers facing issues related to flavoring-type exposures during coffee processing, such as diacetyl and 2,3-pentanedione, need to be aware of, and follow, all applicable regulations for their specific circumstances.

The purpose of this paper will be to provide information regarding:

- Occupational exposure limits for diacetyl and 2,3-pentanedione.
- Select exposure assessment strategies.

- Potential engineering controls, such as local ventilation and properly designed room dilution ventilation systems, maintaining negative pressure in and isolating handling areas. Substitution may be challenging, as some exposures are naturally occurring. The health effects of any flavoring substitutes would need further evaluation.
- Work practices to minimize exposures may include covering containers, reduction of spills, closed processing (i.e., avoiding open pouring), good housekeeping, providing proper work clothing, and training. Housekeeping techniques may also include using HEPA-filtered vacuums and wet cleanup techniques to remove spills, and washing hands after exposures.
- Hazard communications to ensure workers have access to appropriate safety data sheets (SDSs), hazards are labeled, and training has been conducted. NIOSH recommends workers should be trained to immediately report eye/skin problems, cough, shortness of breath, or wheezing.
- Personal protective equipment (PPE) options, if monitoring and hazard assessment indicate. Employees may need to wear appropriate, fit-tested respirators until workplace interventions can be put into place. Additionally, eye, face and skin protection should be considered, based on results from the hazard assessment.
- Medical Monitoring,[6] if used, typically should include both questionnaires and breathing tests (e.g., spirometry) before the first exposure, and on a regular basis thereafter, to all workers at risk of hazardous exposure of diacetyl, 2,3-pentanedione, or related flavorings. Workers can then be referred for evaluation by a physician if they have abnormal test results, an accelerated drop in test results over time, or persistent symptoms.

After reading this paper, attendees should be better able to:

- Identify new health hazards associated with naturally occurring diacetyl and other flavorings in food processing, and understand the new proposed recommended exposure limits.
- Understand the latest NIOSH recommendations for engineering controls, administrative controls, employee education and communications, and medical monitoring requirements.
- Understand potential personal protective equipment solutions for respiratory, eye, and dermal hazards.

Health Hazards of Diacetyl and 2,3-Pentanedione

NIOSH identified several potential health effects of diacetyl and 2,3-pentanedione.[7] As the chemical structures of the two chemicals are similar, similar health effects are expected.

Respiratory Hazards

NIOSH reported the following respiratory concerns:

- Dry, persistent cough
- Shortness of breath on exertion
- Wheezing
- Reduced, rapid or permanent reduction in lung function
- Occupational asthma

In rare cases, NIOSH reported the following:

- Flavoring-related lung disease
- Fixed-airways obstructive disease
- Symptoms consistent with obliterative bronchiolitis, a rare disease with airways obstruction

In addition, NIOSH indicated that employees or physicians may report these as asthma or chronic bronchitis.

Other Health Hazards

NIOSH also found that many employees exposed to higher levels of diacetyl also reported eye, nose, and throat irritation, as well as skin rashes in some cases.

Occupational Exposure Limits for Diacetyl and 2,3-Pentanedione

The NIOSH recommended exposure limits do not differentiate between natural and synthetic chemical origin of diacetyl or 2,3-pentanedione. In 2010, California promulgated a regulation for occupational exposure to food flavorings containing diacetyl that requires installation of exposure controls to food flavorings to reduce exposures to the lowest feasible levels. OSHA has not established any permissible exposure limits (PEL.) for these substances as of this date.[8] OSHA has cited coffee companies for exceeding the ACGIH TLV®, using the General Duty clause.

Table 1. Occupational Exposure Limits for Diacetyl and 2,3-Pentanedione

	8-hour Time Weighted Average (TWA)	**15-minute Short-Term Exposure Limit (STEL)**
Diacetyl		
ACGIH TLV®	10 ppb	20 ppb
NIOSH REL	5 ppb	25 ppb
2,3-pentanedione		
NIOSH REL	9.3 ppb	31 ppb

Concern for the Coffee Industry

After five cases in former workers of a coffee processing facility that roasted, ground, and flavored coffee were identified in 2012,[2] additional studies were undertaken, based on worker concerns. In August 2015, the *American Journal of Industrial Medicine* reported several other cases in the coffee processing industry, leading to the need for evaluation of workplaces beyond the microwave popcorn and flavoring industries, where concerns about these chemicals first arose.

Coffee Industry Findings[4,5]

Workers may be exposed to diacetyl and 2,3-pentanedione during several phases of coffee processing, including grinding, roasting, flavoring, and packing. One NIOSH study measured the highest exposures in the flavoring and grinding of unflavored coffee. Exhibits 1 and 2 show examples of mean full-shift area air samples collected by NIOSH for diacetyl with OSHA Method 1012 and for 2,3-pentanedione with OSHA Method 1016. Many measurements exceeded the NIOSH REL of 5 ppb. Fifteen-minute, short-term air samples collected near employees faces, as they frequently and momentarily opened hatches to monitor coffee levels, measured concentrations of 14,300 ppb diacetyl and 13,800 ppb 2,3-pentanedione.

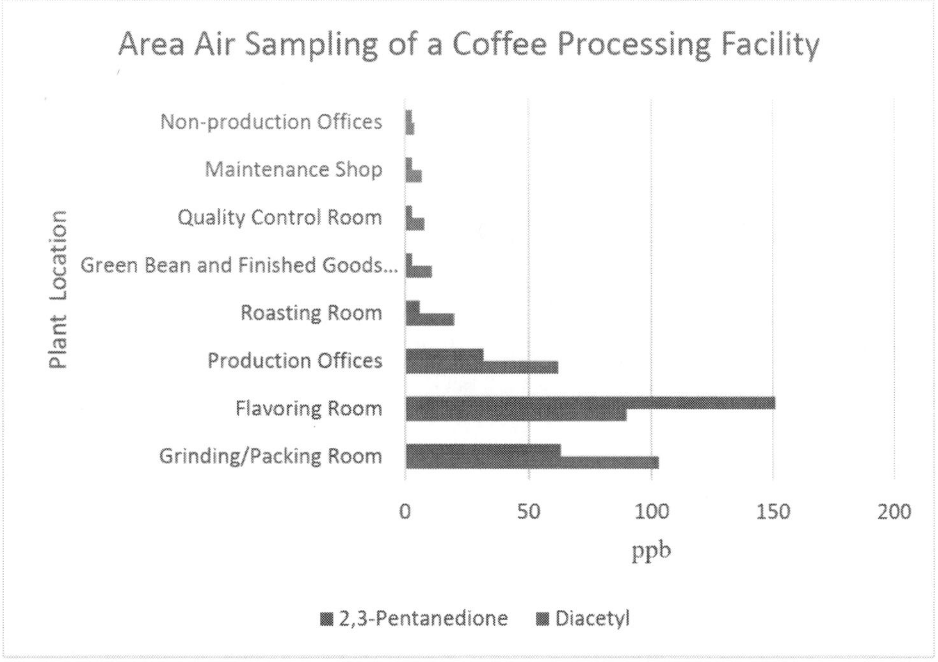

Exhibit 1. NIOSH Reported Levels of Diacetyl and 2,3-Pentanedione measured at a coffee processing facility.[4]

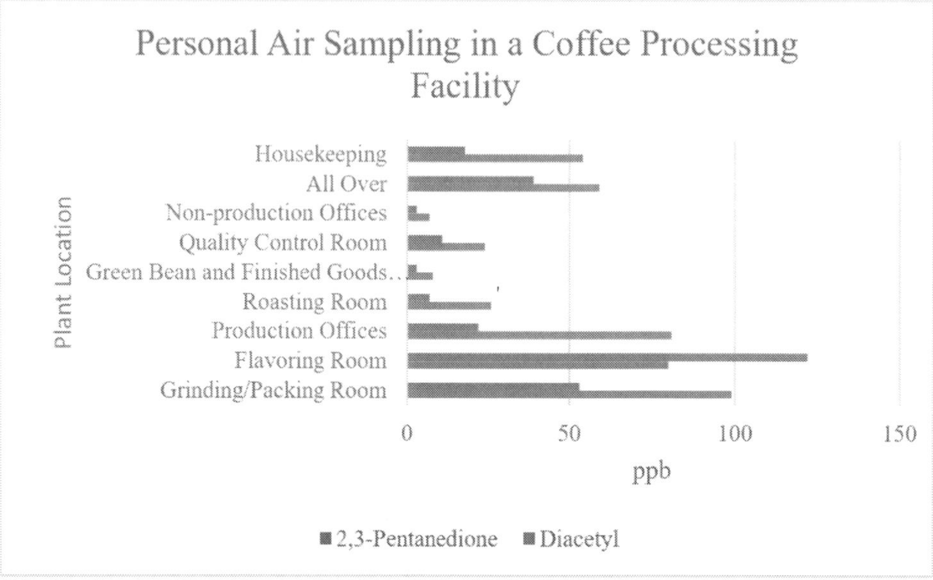

Exhibit 2. NIOSH Reported Levels of Diacetyl and 2,3-Pentanedione measured at a coffee processing facility.[4]

Sampling Methods[9]

Examples of the available sampling methods include:

- Air sampling and analytical techniques for monitoring diacetyl and 2,3-pentanedione vapor
- Techniques for measuring diacetyl and 2,3-pentanedione in airborne dust and bulk materials
- Real-time techniques for measuring relevant airborne analytes and other flavoring compounds

Recommended Air Sampling Techniques for Vapor

NIOSH recommends two validated sampling and analytical methods that have been published by OSHA:

- OSHA Method 1012 for monitoring ppb levels
- OSHA Method 1013 for monitoring low ppm levels

Both of these methods can be used for the simultaneous determination of diacetyl and acetoin.

Per NIOSH recommendations, NIOSH Method 2557 should no longer be used, due to limitations with humidity, which may lead to potentially underestimating the amount of diacetyl.

Recommended Airborne Dust Sampling Techniques

Two validated methods for airborne dust sampling include:

- NIOSH Method 0500 for total dust
- NIOSH Method 0600 for respirable dust

Bulk Sample Testing

In order to reduce diacetyl sources, many companies have requested diacetyl-free flavorings. If higher than expected values are found in the flavoring room, it may also be necessary to have bulk materials tested. Bulk samples should be carefully selected to be representative, and carefully packaged to keep samples cool or protected from light. Follow all required shipping regulations. Recommended analytical methods are under development by NIOSH.

Analysis of Real-time Measurement Techniques

Real-time monitoring instruments may be helpful for screening, to discover leaks or hotspots, and perhaps to help identify appropriate work practices.

However, current real-time measurement techniques available generally lack sufficient sensitivity and specificity for monitoring the ppb levels of diacetyl and 2,3-pentanedione of the REL.

Examples include:

- Photoionization detectors
- Infrared analyzers–IR gas analyzers
- Fourier transform infrared gas analyzer (FTIR) spectroscopy
    - Interferences may pose difficulties
- Photoacoustic analyzers

Control and Prevention of Diacetyl and 2,3-Pentanedione Exposures

The hierarchy of controls for these substances includes the following:

- Elimination/Substitution
- Engineering Controls
- Administrative Controls
- Personal Protective Equipment (PPE)

Substitution of diacetyl may be challenging, as some exposures in coffee processing are naturally occurring. The health effects of any flavoring substitutes would also need further evaluation. If it is not possible to eliminate or substitute, engineering controls may be necessary.

NIOSH Recommended Engineering Controls

Engineering controls in these types of environments commonly include ventilation, production process changes, and general considerations. Common ventilation techniques include proper room dilution ventilation systems, and local exhaust ventilation.

Ventilation[10,11]

Highlights of NIOSH's general ventilation considerations include the following (not exhaustive list):

Local exhaust ventilation

Local exhaust ventilation (LEV) systems, including slot hoods, exhaust hoods, and slotted exhaust hood with workstations, can help remove contaminants at the source:

- Install LEV where alpha-diketones are produced/ and or released, such as:
  - roasters;
  - grinders;
  - hoppers or storage bins of off-gassing roasted coffee;
  - packaging areas;
  - flavoring; and
  - quality control areas.
- Position LEV collection hood a close as possible to source.
- Ensure appropriate supply of make-up air if LEV is installed.
- Install hoods and ensure proper static pressure, face, and capture velocity are maintained:
  - Consider installing a manometer or pressure gauge and monitoring monthly;
  - Inspect hoods and enclosures for signs of damage, leaks or obstructions;
  - Place hoods away from doors, windows, and aisles to avoid cross drafts;
- For small-scale weighing and handling of roasted coffee or flavorings, consider installing benchtop workstations with slotted backdraft ventilation.

Assess general ventilation system

The general ventilation system should also be evaluated to ensure that contaminants do not spread to other areas of the facility. Dilution ventilation can help lower level of contaminants by bringing in uncontaminated air. After installation of ventilation, assess worker exposures to ensure effectiveness of controls. The following considerations may be helpful in implementing general ventilation changes:

- Isolate rooms where flavorings are handled.
- Maintain ventilation systems under negative pressure relative to rest of plant, for production areas where alpha-diketones may be generated or added, such as roasting, grinding, flavoring, off-gassing, packaging, and quality control areas.
- Ensure that air from processing areas (such roasting, grinding, flavoring, off-gassing, packaging, and quality control areas) is not recirculated to other areas of the facility, and is exhausted directly outdoors.
- Consider installing an "on/off" indicator on the exhaust fan.
- Ensure proper design of exhaust stack with respect to fresh air intake, doors and windows:
  - Work with local air pollution control agency to ensure compliance with emission requirements.
  - Guidance on stack design is available from a variety of organizations, including: American Society of Heating, Refrigerating and Air-conditioning Engineers and the American Conference of Governmental Industrial Hygienists.
  - Consult a qualified ventilation engineer to help with proper air flow patterns and system balancing.
  - Consider seasonality; the amount of air may change based on season (e.g., door open in summer, closed in winter).

Examples of Other Engineering Controls Include[10,11]

- For transfer of roasted beans and/or flavorings, where feasible:
  - Use closed transfer processes to help reduce worker exposures to alpha-diketones.

NIOSH Recommended Administrative and Work Practice Controls[10,12]

NIOSH recommends administrative and work-practice controls to help minimize exposures that include using closed processing (i.e., avoiding open pouring), reduction of spills, good housekeeping, and training.

General Recommendations

- Use a closed process wherever possible to transfer flavoring ingredients instead of manual pouring:
  - When poured, liquids should be added through a funnel into a covered mixing vessel to limit spills.
- Develop written procedures for assessing and maintaining ventilation systems.
- Restrict access to all areas where flavorings are handled openly, or any areas for which ventilation may not be sufficient to properly protect workers.

Housekeeping

- Keep containers of roasted grounds or whole bean coffee and flavorings covered when not in use.
- Use vacuums with high-efficiency particulate air (HEPA) filters and wet methods to clean. Avoid compressed air and dry sweeping wherever feasible to avoid re-entraining airborne particulate.
- Establish standard procedures for cleaning containers and mixing tanks for flavorings. Consider cold-water rinse, followed by warm-water cleaning, to reduce the potential for volitization.
- Work areas should be kept clean and spills handled promptly.
- Require workers to wash hands after exposures.

Training

OSHA's Hazard Communications Standard, 29 CFR 1910.1200, requires that employees are informed and trained on the potential hazards, and related safe practices, procedures, and protective measures. NIOSH recommends the training should include at least the following:

- The potential hazards of the coffee roasting industry, including occupational asthma and obliterative bronchiolitis.
- Safe practices.
- Protective measures.
- Procedures.
- Appropriate safety data sheets (SDSs), hazards are labeled, and training has been conducted.
- Workers should be trained to immediately report new, persistent or worsening respiratory symptoms, including cough, shortness of breath, or wheezing, to their personal health care provider, and to the employer-designated individual at their workplace.

NIOSH Recommended Medical Monitoring[13]

If workplaces choose to institute medical monitoring, employees are commonly included who work in or enter these target areas for a total of 40 or more hours per year.

Who does NIOSH recommend direct the monitoring program?

- A licensed physician with training and experience in identifying and preventing occupational lung disease
- Must be able to quickly evaluate clinical data and make medical judgements about appropriate diagnostic and therapeutic measures, including medical removal.
- NIOSH specifically noted that medical monitoring and surveillance was needed due to the potential rapid progression and grave consequences of flavoring-related lung disease, compared to slower progressing, occupational lung diseases.

NIOSH states that the medical program should include:

- good quality spirometry testing for pulmonary function,
- medical evaluation for employees found with abnormal spirometry,
- removal from exposure pending medical evaluation, and
- analysis of group medical surveillance and longitudinal spirometry data to assess work-related risk factors on the basis of job, task, area, and other exposure indices.

NIOSH PPE Suggestions[14]

NIOSH acknowledges that the frequent use of personal protective equipment (PPE), including respirators, may be required for some employees who handle diacetyl, 2,3-pentanedione, diacetyl-containing flavorings or flavored products. Per NIOSH, the frequent use of PPE may be required during job tasks for which:

- Routinely high airborne concentrations of diacetyl or 2,3-pentanedione (e.g., pouring, mixing, packaging) exist.
- The airborne concentration of diacetyl or 2,3-pentanedione is unknown or unpredictable.
- Job tasks are associated with highly variable airborne concentrations because of environmental conditions or the manner in which the job task is performed.

Respiratory Protection

Respiratory protection should be selected based on results of air monitoring, and in compliance with the assigned protection factors (APFs) outlined in the OSHA Respiratory Protection Standard, 29 CFR 1910.134. NIOSH suggests a full-face piece respirator, with a combination organic vapor cartridge/P100 filter or a powered-air purifying respirator with an organic vapor cartridge/HE filter. Per OSHA regulations, a site-specific cartridge change-out schedule must be developed; noticeable odor cannot be relied upon. Supplied air respirators may also be used to control inhalation exposures.

NIOSH-approved respirators must be used in accordance with the NIOSH cautions and limitations specified on the NIOSH approval label, and comply with OSHA's respiratory protection regulations, 29 CFR 1910.134. OSHA requires employers to implement a written respiratory protection program meeting all the requirements of the standard when respirators are used. The respirator manufacturer or a health and safety professional should be consulted if there is any question regarding respirator selection and use. Users must understand the respirator capabilities, as well as limitations, and follow the respirator manufacturer's user instructions in order to receive the assigned level of protection. Misuse of any respirator may result in sickness or death.

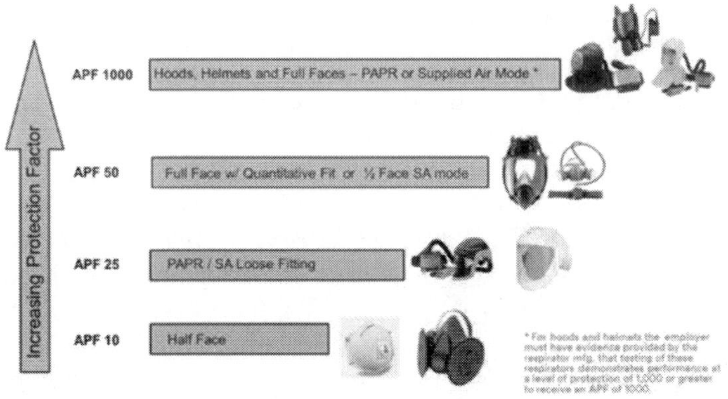

Exhibit 3. OSHA Assigned Protection Factors with Respirator Examples
(*Image source: 3M Company.*)

Key Elements of a Written Respiratory Protection Program
Many coffee processing facilities may not have a written respiratory program. Written worksite-specific procedures for a respiratory program should include the following:

- Procedures for selecting respirators
- Medical evaluations
- Use of respirators
- Maintenance and care
- Assuring breathing air quality and use
- Training (including donning and doffing) and fit testing
- Program evaluations
- Requires a program administrator
- Recordkeeping

Eye Protection

NIOSH recommends eyes be protected from contact with diacetyl or 2,3-pentanedione, or related flavorings. Employers should consider either a full-face-piece respirator, hood, or helmet. If not required to wear a respirator, tight-fitting goggles meeting the ANSI Z87.1 D3 Splash and D4 Dust rating can be considered.

Skin Protection

NIOSH recommends preventing skin contact to reduce the risk of dermal exposure. Protective coveralls, lab coats, aprons, and gloves should be considered. In addition, chemical-resistant gloves (e.g., butyl, nitrile) should be worn to protect the hands. Refer to coverall and glove manufacturers recommendations for specific products.

Conclusion

Workers in the coffee processing industry may be just learning of potential health effects of naturally occurring diacetyl and 2,3-pentanedione. The proper assessment of exposures, and implementation of the needed engineering controls, work practices, and personal protective equipment, will help to keep workers in the coffee processing industry protected.

Endnotes

[1] National Institute of Occupational Safety and Health (NIOSH). 2016. *Flavorings Related Lung Disease- Coffee Processing Facilities.* (Retrieved 9 MAR 2017) (www.cdc.gov/niosh/topics/flavorings/processing.html).

[2] Centers for Disease Control and Prevention (CDC). 2013. *Obliterative Bronchiolitis in Workers in a Coffee-processing Facility - Texas, 2008-2012. Morb Mortal Wkly Rep (MMWR)* 62(16):305-307.

[3] Bailey, Cox-Ganser, Duling, et al., "Respiratory Morbidity in a Coffee Processing Workplace With Sentinel Obliterative Bronchiolitis Cases." *American Journal of Industrial Medicine.* Aug. 2015.

[4] National Institute of Occupational Safety and Health (NIOSH). 2016. *Criteria for a Recommended Standard: Occupational Exposure to Diacetyl and 2,3-Pentanedione. DHHS,* CDC, Publication No. 2016-111:30-31.

[5] Ibid, 7.

[6] Ibid, 225-235.

[7] Ibid, Chapter 3.

[8] Ibid, 4.

[9] Ibid, 15-21.

[10] Ibid, Chapter 8.

[11] CDC. National Institute for Occupational Safety and Health Respiratory Health Division (NIOSHHRHD). 2016. *Workplace Interventions, Flavorings Related Lung Disease, Coffee Processing, Engineering Controls.* (Retrieved 9 MAR 2017) (www.cdc.gov/niosh/topics/flavorings/engineeringcontrols.html).

[12] CDC. NIOSHHRHD. 2016. *Workplace Interventions, Flavorings Related Lung Disease, Coffee Processing, Administrative Controls.* (Retrieved 9 MAR 2017) (www.cdc.gov/niosh/topics/flavorings/administrativecontrols.html).

[13] National Institute of Occupational Safety and Health (NIOSH). 2016. *Criteria for a Recommended Standard: Occupational Exposure to Diacetyl and 2,3-Pentanedione. DHHS,* CDC, Publication No. 2016-111: Chapter 9.

[14] Ibid: 204-207.

Note: This paper is based on current United States federal OSHA requirements 29 CFR 1910, and NIOSH recommendations. State or other country requirements may be different. Requirements can change in the future. This paper contains general information as of the date of this presentation, and should not be relied upon to make specific decisions. No warrantees are made, as this paper only presents suggested procedure(s) that may be applicable for work in these environments. The employer has ultimate responsibility for assessing the worksite and implementing adequate safety precautions.

Session No. 584

Addressing Worker Fatigue Issues in the Oil and Gas Extraction Industry

Ranjana K. Mehta, Ph.D.
Texas A&M University
College Station, TX

Introduction

Fatigue plays a major role in impairing performance, safety, and productivity in the workplace. It is one of the most prevalent symptoms in the workforce, affecting 54% of individuals. Worker fatigue has substantial economic burden, estimated at $18 billion a year in injured workers or workers' compensation and reduced productivity (Caruso, 2014; Lerman et al., 2012). Despite extensive research regarding mechanisms and consequences of fatigue, it still remains a significant concern in various industries, ranging from aviation to healthcare to oil and gas operations, and public safety.

Oil and gas extraction (OGE) workers are exposed to intensive shift patterns and long work durations, coupled with intense physical and mental workload inherent of the OGE environment. From 2003–2014, 1331 OGE workers died while working, resulting in an annual fatality rate seven times higher than that for all U.S. workers (Figure 1).

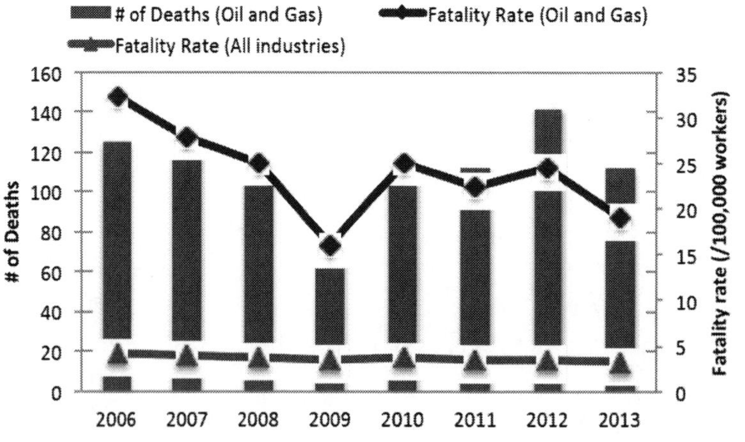

Figure 1. Number and rate of occupational fatalities in the OGE industry from 2006-2013 (BLS, 2013)

In the OGE industry, fatigue is especially a critical safety risk. Operator fatigue has been implicated as a contributing factor in significant safety incidents, such as the BP Texas City refinery, which resulted in 15 deaths, 180 injuries, and significant economic losses. The Chemical Safety Board investigation (CSB, 2007) concluded that the Isomerization Unit operators were likely fatigued from working 12-hour shifts, some working as many as 29 consecutive days during the turnaround of the unit prior to startup, and that, as a result, the operators' judgment and problem-solving skills were likely degraded, hindering their ability to determine that the tower was overfilling with hydrocarbons and to take prompt corrective actions.

Fatigue, generally defined as a physiological state of reduced mental or physical performance capability resulting from sleep loss, circadian phase, and workload, has been implicated as a serious risk factor in a majority of the cases affecting worker safety. Both industry and federal agencies have determined that "*decreasing fatigue-related injuries and fatalities in the OGE industry*" is one of their top strategic research (to practice) priorities. However, one of the major barriers that currently impacts the development of effective fatigue mitigation practices in OGE workers is the assessment of fatigue.

Fatigue is a complex, multidimensional construct, and its definition and assessment differs based on different occupations. Several objective and subjective fatigue assessment methods exist, but none are comprehensive, relevant, and feasible for the OGE workforce. This presents challenges in identifying high-risk workers and developing fatigue management practices that are targeted and effective in reducing fatigue-related incidents in the OGE industry. This is a critical issue, given that in the next few years OGE employment rates will experience dramatic changes. Statistics indicate that new OGE workers are at highest risk of a fatal injury, regardless of the length of time working in the industry, with over half (53.4%) of fatal injuries occurred within one year of service with employer (Mason et al., 2015). This can potentially increase the already high fatality rate in the OGE industry. This is an even bigger challenge with small OGE businesses, which have been shown to contribute disproportionately to the fatality rate in this industry (Figure 2).

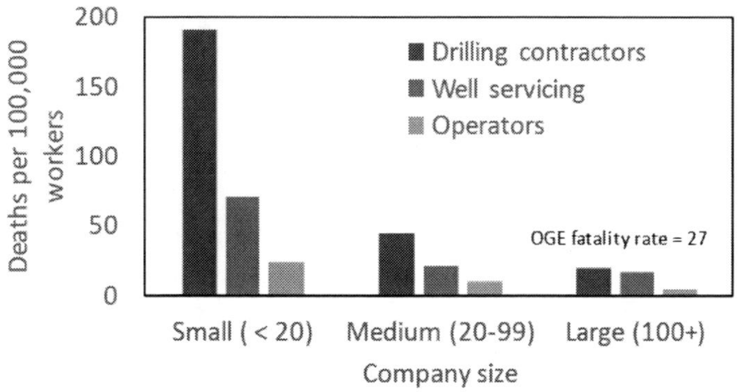

Figure 2. Occupational fatality rate by OGE company size, 2003-2009 (BLS, 2013)

To develop sustainable, evidence-based fatigue mitigation and/or management practices within the framework of OGE operations and constraints, it is first important to understand the nature and extent to which fatigue affects worker health, behavior, and safety. This paper aims to provide a better understanding of:

- OGE-specific fatigue risk factors
- OGE-relevant fatigue assessment methods
- The applicability and limitations of OGE-specific fatigue management guidelines

Fatigue Risk Factors in OGE Operations

Oil and gas extraction is a 24/7 high-risk operation. OGE workers are often continuously exposed to many harsh working environments, including long working hours, shift-swings (i.e., changing from night shift to day shift or vice versa), heavy physical work, intense cognitive load, hazardous work environments, heat and cold stress, noisy environments, psychosocial stressors, a lack of privacy, and isolated work locations. Fatigue is associated with insufficient sleep and disruptions in normal sleep patterns. In OGE operations, it is very common to have swing shifts, where workers rotate between day-night shifts in the middle of their hitch offshore. This introduces circadian variability, which further impacts both sleep duration and quality, in addition to the circadian disruption associated with day/night shift rotation. Coupled with 12-hour shift lengths, often seen in offshore work, workers experience fatigue due to extended wakefulness. These factors can result in impaired sleep and cardiovascular recovery between shifts.

Individual factors that can further exacerbate these relationships are age, physical health status, e.g., chronic conditions, sleep apnea, etc. Compounding these effects, environmental factors, such as heat/cold stress, weather-related visibility, etc., can also contribute to fatigue development in OGE workers. Finally, psychosocial aspects inherent to the OGE environment, such as isolated locations, cramped living/sleeping space, and being away from family, can adversely affect coping mechanisms that otherwise counter worker fatigue development. These effects carry over when the worker is returning home from a long-duration hitch. About ~40% of the worker fatalities in the OGE industry occur from transportation crashes (BLS, 2013), and the rest can be attributed to the work-hour patterns, intensive workload, hazardous chemical exposure or accidents involving machinery.

Within the actual tasks that workers perform in OGE operations, intensive physical and cognitive work demands during the shifts can also have major impact on worker fatigue and recovery. In offshore environments, particularly for drill operators, fatigue may originate from the task itself. Rapid technology adoption, from the well site to the refinery, has led to the introduction of increasingly complex digital displays and operator interfaces, particularly in the driller's cage. In such vigilance-monitoring environments, performance may be influenced by display design elements and multi-display configurations, as well as drill operator experience and knowledge with the interfaces. Cognitive fatigue can also develop with extended time on task and/or heavy cognitive load associated with the task (Gunzelmann et al., 2011). Collectively, these factors contribute to performance failures and costs associated with preventable incidents. This is a critical issue, particularly as task-related cognitive fatigue can potentially interact with fatigue due to sleep loss during shift transitions that can further affect performance (Balkin and Wesensten, 2011).

Fatigue Assessment Methods in OGE Operations

Fatigue is a complex, multi-dimensional construct, and its definition and assessment differs, based on different occupations (Dittner et al., 2004). Objective indicators of fatigue may include operational outcomes, such as declines in rig productivity; performance outcomes, such as decrements in operator performance; or physiological outcomes, such as elevated heart rate and respiration rates. Subjective indicators of fatigue may be obtained using surveys, interviews, sleep logs, etc. While the former method offers high precision and validity, the latter is a feasible assessment method. Figure 3 illustrates potential fatigue assessment measures available that offer some level of feasibility and applicability in the OGE environment.

Subjective Assessment	Objective/Performance Assessment	Physiological Measures
•Swedish Occupation Fatigue Inventory	•Psychomotor Vigilance Test	•Heart Rate
•Occupational Fatigue Recovery	•Iowa Gambling Task	•Heart Rate Variability
•Fatigue-related Symptoms Questionnaire	•Purdue Pegboard Task	•Muscle Activity
•Fatigue Scale	•Simple Reaction Time	•Respiration Rate
•Borg Ratings of Perceived Exertions	•Unstable Tracking	•Skin Temperature
•Karolinska Sleepiness Scale	•Dual Task	•Tympanic Membrane Temperature
•Epworth Sleepiness Scale	•Finger Tapping	•Brain Activity
•Profile of Moods State	•Time Perception	
•State-trait Anxiety Inventory	•Mackworth Clock Test	
•NASA Task Load Index	•Handsteadiness	
	•Force Steadiness	
	•Force Production	

Figure 3. Subjective, performance-, and physiological-based fatigue assessment measures

The selection of a valid indicator to assess fatigue in the complex, multi-component tasks of an offshore environment is challenging. Fatigue may be distinguished not just in relation to the sources, i.e., sleep loss due to shiftwork or intense workload, that aggravate the condition but also to the responses that indicate a fatigued state, i.e., performance decrement or elevated heart rate. For example, decrements in performance metrics may indicate a fatigued state, but these performance changes alone cannot indicate the source of the fatigue, and thus cannot inform fatigue mitigation strategies. To assess operator fatigue in offshore OGE environments, it is important to first distinguish the physiological consequences (or cost to maintain performance) of operator workload (Lorist and Faber, 2011). This, in turn, can facilitate development of fatigue mitigation/intervention strategies that are robust and evidence-driven.

Physiological (and wearable) fatigue monitoring systems are gaining rapid attention, particularly in the transportation and aviation sectors (Borghini et al., 2014; Jung et al., 2014; Lal et al., 2003). However, given the hazardous nature of OGE work conditions, passive or real-time monitoring of physiological indicators of fatigue present safety risks due to sensor materials being explosion-prone in hazardous volatile rig environments. As such, subjective fatigue assessment methods may, in fact, enable viable, safe, and sustainable evaluations of fatigue risks in OGE work environments. The development of a reliable, sensitive, and valid fatigue assessment survey that takes into consideration not only the various OGE-specific sources of fatigue, but also the barriers associated with effective fatigue assessment in OGE operations (identified in this study) are imperative (Mehta et al., 2017). Without access to such methods, developing effective, evidence-based controls or preventive strategies for reducing fatigue-related injuries and fatalities in the OGE industry will likely remain limited.

Existing Fatigue Management Guidelines for the OGE Industry

In high-risk operations, such as oil and gas extraction and petrochemical refineries, worker fatigue is an especially large and complex issue that can have detrimental and catastrophic effects. In these operations, workers are exposed to high levels of cognitive and physical workload, intensive shift patterns, coupled with long work hours and loss of sleep quality. Operator fatigue was cited as a contributing factor to the BP Texas City Incident of 2005 in the Chemical Safety Board (CSB) investigation. Some operators had been working 12-hour shifts for as many as 20 consecutive days.

Fatigue has been recognized as a cause of major accidents in the transportation sector. Fatigue prevention regulations have been developed for aviation and other transportation sectors, but there are no fatigue prevention guidelines that are widely used and accepted in the oil and chemical sector. Based on the CSB investigation, it was recommended that the American Petroleum Institute (API) and the United Steelworkers Union work together to develop a guideline pertaining to fatigue management. The outcome was the Recommended Practice 755: Fatigue Risk Management System (FRMS). FRMS mitigates fatigue through a comprehensive approach and should be informed by sound science. A large component of the FRMS is the Hours of Service Guideline, which recommends maximum shift lengths based on daily work hours (8, 10, or 12 hours) and type of operation (normal vs outage).

The RP 755 FRMS has several limitations. First, there is the assumption that, by implementing a FRMS, it will compensate for the risk from excessive hours and days at work. Additionally, within the RP itself, the guidelines state that "consistently working at the limits shown is not sustainable and may lead to chronic sleep debt." This implies that work schedules should be designed to prevent employees from frequently working near or at the limits over time. Second, current maximum consecutive shift recommendations are widely outside the range of supported by relevant research. In addition, 14 consecutive 12- or 10-hour shifts has never been peer reviewed. Finally, the FRMS was developed for refineries, petrochemical and chemical operations, natural gas liquefaction plants, and other facilities such as those covered by the OSHA Process Safety Management Standard, 29 CFR 1910.119. This document is intended to apply to a workforce that is commuting daily to a job location. As such, these guidelines are not targeted to address and manage fatigue for a majority of OGE onshore and offshore operations.

Bibliography

Balkin, T., and Wesensten, N. 2011. "Differentiation of sleepiness and mental fatigue effects." In Ackerman, P. (Ed.), *Cognitive fatigue: Multidisciplinary perspectives on current research and future applications* (pp. 47-66). Washington, D.C.: American Psychological Association.

Borghini, G., Astolfi, L., Vecchiato, G., Mattia, D., Babiloni, F. 2014. "Measuring neurophysiological signals in aircraft pilots and car drivers for the assessment of mental workload, fatigue and drowsiness." *Neurosci. Biobehav. Rev.* 44: 58–75.

Bureau of Labor Statistics (BLS). 2013. "Fatal Occupational injuries by industry and event or exposure, All U.S." (www.bls.gov/iif/cfoi/cflb0268.pdf).

Caruso, C.C. 2014. "Negative impacts of shiftwork and long work hours." *Rehabilitation Nursing*, *39*(1): 16-25.

Chemical Safety Board (CSB). 2007. "BP America Refinery Explosion." (http://www.csb.gov/assets/1/19/CSBFinalReportBP.pdf).

Dittner, A.J., Wessely, S.C., Brown, R.G. 2004. "The assessment of fatigue: a practical guide for clinicians and researchers." *J. Psychosom. Res.* 56: 157–170.

Gunzelmann, G., Moore, L.R., Gluck, K.A., Van Dongen, H.P. and Dinges, D.F. 2010. "Fatigue in sustained attention: Generalizing mechanisms for time awake to time on task." In Ackerman, P. (Ed.), *Cognitive fatigue: Multidisciplinary perspectives on current research and future application*, pp.83-96. Washington, D.C.: American Psychological Association.

Jung, S.J., Shin, H.S., Chung, W.Y. 2014. "Driver fatigue and drowsiness monitoring system with embedded electrocardiogram sensor on steering wheel." *IET Intell. Transp. Syst.* 8: 43–50.

Lal, S.K.L., Craig, A., Boord, P., Kirkup, L., Nguyen, H. 2003. "Development of an algorithm for an EEG-based driver fatigue countermeasure." *J. Safety Res.* 34: 321–328.

Lerman, S.E., Eskin, E., Flower, D.J., George, E.C., Gerson, B., Hartenbaum, N., Hursh, S.R. and Moore-Ede, M. 2012. "Fatigue risk management in the workplace." *Journal of Occupational and Environmental Medicine*, 54(2), pp.231-258.

Lorist, M.M., Faber, L.G. 2011. "Consideration of the influence of mental fatigue on controlled and automatic cognitive processes and related neuromodulatory effects." In Ackerman, P. (Ed.), *Cognitive Fatigue: Multidisciplinary Perspectives on Current Research and Future Applications*, pp.105–126. Washington, D.C.: American Psychological Association.

Mason, K.L., Retzer, K.D., Hill, R. and Lincoln, J.M. 2015. "Occupational fatalities during the oil and gas boom—United States, 2003-2013." *Morbidity and Mortality Weekly Report*, *64*: 551-554.

Mehta, R.K., Peres, S.C., Kannan, P., Rhee, J., Shortz, A.E. and Mannan, M.S. 2017. "Comparison of objective and subjective operator fatigue assessment methods in offshore shiftwork." *Journal of Loss Prevention in the Process Industries*. http://dx.doi.org/10.1016/j.jlp.2017.02.009

Session No. 587

Beyond the Matrix: Preventing Incidents with a Risk Assessment Process

Francis P. Sehn CSP, ARM
Willis Towers Watson

An Overview of Risk Assessment and the Importance of the Process

This paper will focus on deploying a risk assessment process to reduce risk and effectively minimize incidents. Safety professionals need to move away from purely conducting hazard identification and implementing corrective action to a format of assessing the risks associated with task completion. We have seen many organizations make the move to conduct risk assessments in some fashion. The next step is to effectively deploy these documents into an integrated process to align with the business process. The paper will include a brief case study of two organizations' success in the deploying a risk assessment process to minimize risk.

Importance of this Topic

Over the last ten plus years, we have received many opportunities to learn the risk assessment methodology. There are numerous assessment techniques. Using the techniques will depend upon the organization's goals and objectives, and the acceptance of residual risk. Safety and health professionals are being asked to step out of the compliance box and become partners in the business environment. Deploying a robust risk assessment process beyond regulatory requirements will require a strategic approach in place of the tactical efforts of the past.

What Is a Risk Assessment?

Risk assessments have been used by various disciplines for many years. The terminology is not new to the safety profession, although it may have been somewhat misused. Many safety professionals referred to the hazard identification process or to safety inspections as "risk assessments." To their credit, they were on the right track, but they did not assess as much as identify hazardous conditions or hazardous activities.

There are several definitions of the term "risk assessment" including the following:

> "Risk Assessment is the overall process of risk identification, risk analysis and risk evaluation. Risks can be assessed at an organizational, at a department level for projects, individual activities or specific risks. Different tools and techniques may be appropriate in different

contexts. Risk Assessment provides an understanding of risks, their causes, consequences and their probabilities. This provides input to decisions about:

- Whether an activity should be undertaken;
- How to maximize the opportunities;
- Whether risks need to be treated;
- Choosing between options with different risks;
- Prioritizing risk treatment options;
- The most appropriate selection of risk treatment strategies that will bring adverse risks to a tolerable level."

This definition, from the ANSI/ASSE/IEC/ISO 31010 (Z690.3-2011), *Risk Assessment Techniques,* is a national adoption of IEC/ISO 31010:2009. While it may look as it would be used in a broader risk management initiatives, it should provide guidance for the EHS professional.

Another definition that may be germane to the OSH professional comes from the ANSI/ASSE Z 590.3-2011 standard, *Prevention through Design: Guidelines for Addressing Occupational Hazards and Risks in Design and Redesign Processes.* It states that:

> Risk Assessment is a process that commences with hazard identification and analysis, through which the probable severity of harm or damage is established followed by an estimate of the probability of the incident or exposure occurring, and concluding with a statement of risk.

This definition probably appeals to safety professional due to use of terms more commonly used in the safety vernacular. It is important to note that more standards are including a risk assessment "requirement." A good source for an overview of is *Risk Assessment: A Practical Guide to Assessing Risk* by Popov, Lyon and Hollcroft. The authors are commended for providing the safety professional with a text to enhance the skills and knowledge of the need to deploy risk assessment in assessing operational risk. It should be a worthwhile text for the many schools that offer degrees in safety to provide the future safety professional with knowledge beyond compliance.

Why Should We Deploy a Risk Assessment Process?

The intent of the risk assessment process is to assess all significant occupational safety and health hazards in the organization's sphere of influence. Risk assessments should be completed for all processes and operations having recognized hazards.

In general the risk assessment process emphasizes three primary elements as noted above:

1. Identification of the hazard
2. The evaluation of a hazards potential impact
3. Systematic review of how it can best be controlled

If we agree that this is a viable process to minimize risk in the operations of an organization, why is it not an accepted best practice in most companies? The simple answer is that it is not a regulatory requirement in the U.S. and many other countries. The U.K. and many European countries have adopted the use of risk assessments because it has been a regulatory requirement for companies employing five or more employees.

At a recent American Society of Safety Engineers (ASSE) meeting, sponsored by the Oil and Gas Specialty Practice, the speaker presenting on the topic of hazard analysis and risk assessment asked if anyone has attended the Risk Assessment Institute training. One of the attendees raised his hand indicating he attended the course, and found it to be very beneficial. He was then asked if the risk assessment has been deployed in his organization, and he responded "No." He explained that risk assessment has met with resistance for a variety of reasons, including that it is not a regulatory requirement, and it is a change in how management has approached the safety and health effort.

At this same meeting, two senior students from a local university with a safety management curriculum were asked if they have been introduced to risk assessment. Both young ladies responded that they have discussed it in some of their course work, but there is no course dedicated to risk assessment or risk management. Many of the other attendees at the meeting graduated from "safety" schools, and echoed the comments of the two students.

The late comedian George Carlin once stated, "If you have a solution, you are part of the problem." Does this hold true in the safety profession? We have tried to solve the safety problems of injuries, incidents, and illnesses with regulations, and have been reluctant to use other non-regulatory methods.

Where Is the Problem?

As OHS professionals, we frequently must "prove to management" that various safety initiatives are needed or necessary to improve safety and health or to comply with some aspect of a standard. Compliance with a standard may be required, but showing some type of return on investment (ROI) may be difficult at best. The OHS professional may be required to step out of his/her comfort zone to create a new view of safety and health. At Safety 2016, John Stevens and Lawrence Bamber presented a session, *How ISO 45001 will drive Business Excellence.* Hopefully, we will see this standard in late 2017 or early 2018. It is a much needed safety and health management standard. In their presentation, they discussed the late W. Edwards Deming's 14 point for transforming American industry from Deming's book, *Out of the Crisis,* published in 1982. The 14 points in safety may look like this, with apologies to Dr. Deming:

1. Create a constancy of purpose toward improvement of safety and health to minimize injuries and illnesses in the workplace.
2. Adopt a new philosophy toward safety using risk management tools including risk assessment methodologies.
3. Minimize the need for regulatory inspections by assessing risk and deploying appropriate corrective actions.
4. Provide quality safety devices to employees as needed. Not the low cost devices.
5. Improve constantly the safety management system and decrease the cost of injuries and illnesses.
6. Educate employees in their safety responsibilities and accountabilities.
7. Lead the safety efforts in co-operation with supervisors and managers.
8. Use discipline only as need to drive out fear in the workplace.
9. Be an agent of change and help break down the silos in the organization.
10. Eliminate the slogans and use loss drivers as the target for improvement.
11. Move toward best in class or world class safety metrics.
12. Supervisory and management responsibility must focus on safety as a value.
13. Institute a program of education throughout the organization for safety and health with risk assessment methods.
14. Engage the entire work force in the safety and health effort using risk assessment methods.

Deming's principles apply today as well as they did thirty five years ago. It is the OHS professional's role to move the needle forward. Risk assessment is a tool to help this process take place.

Deploying a Robust Risk Assessment Process

The keys to an effective risk assessment process are that it must be a part of a continuous improvement initiative and it must be formalized. The following are suggestions to deploy a risk assessment process in any organization.

1. The process should be formalized with defined responsibilities for management, supervisors and employees.
2. The risk matrix that is used for estimating probability of occurrence should define acceptable organizational risk and evaluate risk with and without controls.
3. The process should then be used as a basis for new hire training and ongoing training by department managers and supervisors.
4. Risk assessments should be integrated other elements of the safety management "program" including but not limited to accident/incident investigations, inspections, audit results, and health evaluations.

The above elements are easily auditable for any organization that adapts a process of risk assessment.

Putting Risk Assessment to Work: Two Case Studies

The author has had the opportunity to consult with two global organizations that have deployed robust process for risk assessment.

A manufacturer of plastics products began to use risk assessments as an integral component of their safety and health efforts in the 1990's. This is a UK based organization that had facilities in the UK and began a significant expansion on a global basis. Risk assessments are the "heart and soul" of their safety and health efforts. Combined with a robust effort to report hazards and near misses (hits) they have reduced lost work day cases by 90% while more than doubling the size of the work force to 22,000 on a global basis. In less than six years the US operation experienced a similar reduction in LWD cases by implementing similar aspects of safety and health while increasing the size of the facility by 50% and adding 20 employees to the work force.

A global manufacturer of medical devices with a similar foot print but less employee based in the US deployed a safety management system in 2004. The global safety director was in search of a method to reduce injuries and illnesses and the overall cost of risk. She has deployed risk assessment as a key performance indicator. During a recent audit in Mexico the General Manager stated that "safety has been the driving force for the success in all aspects of the operation. She noted that cost, delivery and quality are enhanced by the safety process. One of the supervisors discussed the on- going use of risk assessments as a tool for training and retraining. Risk assessments define the culture for safety improvement at these two organizations.

Risk Assessment in Like the Spaghetti Sauce

Risk assessment has been adopted by many standards as previously noted. The best example of this is the ANSI/AIHA Z10-2012 standard Occupational Safety & Health Management Systems. It clearly states in section 5.1.1 Risk Assessment "The organization shall establish and implement a risk assessment process (es) appropriate to the nature of hazards and level of risk". The committee that wrote this standard is updating and revising the standard as this article is written. It is unlikely this section will be removed. Ironically the standard also refers to Dr. Deming's Plan, Do, Check Act approach as a methodology for implementation. The late Doctor would be proud.

What Is the Future of Risk Assessment?

We are truly at a crossroad in the profession of safety. Some good work has been started but more needs to be done. The following are suggestions to move forward:

- Continued education of safety and health professionals. There should be easily accessible opportunities for the education process to take place.
- Required course work for student studying safety and health in terminal degreed programs.
- Engaging employees in risk assessment strategies. They hold the key to success.
- Deploy a compliance standard for risk assessment.

Change is never easy. Safety and health professional must change to move forward. Let's meet at the crossroads to change the profession.

References

Stevens, John, and Lawrence Bamber. 2016. "How ISO 45001 will drive Business Excellence". Stevens and Bamber. *Safety 2016: Proceedings of the ASSE Professional Development Conference*, June 26-29, 2016, in Atlanta, Georgia.

Deming, W. Edwards. 1982. *Out of the Crisis*. Cambridge, MA: MIT Press.

American National Standards Institute/American Industrial Hygiene Association/ American Society of Safety Engineers. ANSI/AIHA Z10-2012, *Standard Occupational Safety & Health Management Systems.* (www.asse.org/ansiaihaasse-z10-2012-occupational-health-and-safety-management-systems/.)

American National Standards Institute/American Society of Safety Engineers. 2011 (R2016). ANSI/ASSE Z 590.3-2011 (R2016), *Prevention through Design: Guidelines for Addressing Occupational Hazards and Risks in Design and Redesign Processes.* (w/ww.asse.org/ansi/asse-z590-3-2011-r2016/.

American National Standards Institute/American Society of Safety Engineers (ANSI/ASSE/ISO). ANSI/ASSE/IEC/ISO 31010 (Z690.3-2011), *Risk Assessment Techniques; a National Adoption of: IEC/ISO 31010:2009.* (www.asse.org/ansi/sio-risk-management-standards-package-national-adoption-of-iso-31000-31004-31010.)

Does Your RTW/Injury Management Program Need Tweaks? An Injured Employee's View

Angela DeLong Banerjee, CSP, ARM, ALCM
Acuity Insurance
Sheboygan, WI

Introduction

While management may feel they successfully help injured employees through the healing process and ultimate return to work, an injured employee's experience can give useful insights on where your injury management program needs to be tweaked. Enhancing your efforts can improve the employee's experience and morale, reduce lost time, and reduce the company's costs associated with the injury.

Pitfalls and Solutions

Initial Report of Injury or Absence

Injury management can go wrong from the very first step. Each workplace injury has its own unique onset and method of being reported. Acute and severe injuries have urgency and typically present a clear and immediate indication that the injury is work-related. When an injury is cumulative, or involves vague pain, its origins and relation to the workplace are less clear. To complicate matters, the employee may report their injury, pain, or doctor visit via phone. Additional difficulty arises if time has passed since the injury onset or the doctor visit. Management needs to know how to quickly get things back on track.

When Joe called his supervisor, Sam, to say he wouldn't be coming to work because his back hurt, Sam asked no questions. You can't fault a supervisor when his mind immediately jumps to worrying about how he will fill the gap of the absent employee or how production may get behind. In this case, the lack of simple questions led to Joe being off multiple days. Sam thought he was giving Joe time off, but it turned out to be a Workers' Compensation claim, so those were counted as lost-time days. Since it became a lost-time claim, the entire cost of that claim went into the company's Experience Modifier. The delay in identifying the employee injury as work-related also delayed medical care, and prolonged the recovery period.

Had Sam merely asked, "What do you think caused the back problem?" the lost days, delayed treatment, and increased costs could have been avoided. Had he asked, he would've heard the story that came out later: While doing a two-person manual lift of a section of pipe, his coworker dropped his end. The weight and impact of the other end hitting the floor caused Sam immediate back pain. He didn't report it at the time, but later said he "took it easy" the rest of the day. He thought that occasional pain was just part of the job he did every day, and wondered if he'd lose his job if he reported an injury.

Reasons for delayed or lack of injury reporting vary. Some people just tough it out; others don't know how what is expected; and still others fear losing their job if they report injuries. In a study by The Center for Construction Research and Training, carpenter apprentices indicated that accidents were rarely reported, largely because they feared being fired or other retaliation (Lipscomb, 389).

Solutions: Supervisors and others in direct communication with employees must be educated on the need to be gently inquisitive about pain and absences so they don't overlook a work-related injury. Employee education on Workers' Compensation benefits at the time of hire, periodically, and at the time of the injury, is helpful to their understanding of what happens following an injury. While you communicate to employees the need to report incidents and injuries, you must also demonstrate respect and trust for those who do report injuries and avoid any action that could be seen as retaliation. You must show empathy and support of injured employees, knowing that news of poor handling of an injured employee will spread and can prevent reporting of future injuries

Medical Progression

If the supervisor assumes that Human Resources knows the status of the employee, and Human Resources assumes the supervisor took ownership, you may get to a point where you find that doctor appointments have been missed, or changes in restrictions were overlooked. The insurance claims adjuster should track the treatment, but if you don't actively monitor the notes, restrictions, and next steps indicated by the medical providers, the employee is left to passively follow whatever the most recent treating medical provider indicates; or worse yet, *not* follow them. This can lead to the employee exceeding restrictions or, at times, can lead to excess costs from unnecessary diagnostic tests, unneeded visits, excessive treatments, or unnecessary physical therapy tools. A conscientious employee may think that since she "feels better," she can save the company money by not attending the scheduled doctor appointment. She may not realize that it is necessary to have the doctor's release in order to return to normal duties.

When Ann broke her wrist after a fall, surgery was conducted a few days later, and physical therapy began soon after. Ann was allowed back to work with no use of the injured hand. When the physical therapist said, "Here is some therapy putty for you to squeeze for five minutes several times a day to build strength," Ann took it without question. Upon the next visit, the therapist handed Ann a fingerless neoprene glove for support while sleeping. Ann thought it odd when the surgeon, at the next visit, said she didn't need that glove or the putty. She took the doctor's indication and stopped using both. After her restrictions lessened, she tired of fitting the therapy visits into her schedule, so Ann asked the therapist if she thought her strength and range of motion goals had been met. When the therapist measured those things, she found Ann had met and exceeded those targets and was released from treatment. As you might expect, Ann was slightly irritated when she realized the last few visit had not been needed. Even though she felt she had wasted time on recent therapy sessions, she kept the thought to herself so the employer never knew of the wasted time and money.

Solutions: Assign responsibility for monitoring employee progress and medical visits, and provide education on handling the employee and information. Regardless of the employee being at home or back in the workplace, someone besides the insurance company's claims representative must monitor medical appointment outcomes, know when the next visit is, confirm the treating providers are in communication, and help the employee follow restrictions and do assigned therapy. If a supervisor is aware of a medical appointment, they will know when to ask for updated restrictions or a work release from the employee. If there is a relationship with the medical provider, then they can communicate on next steps and needed tools for physical therapy to help avoid unneeded products and unnecessary medical visits.

In *Workplace-based Return-to-Work Interventions: A Systematic Review of the Quantitative and Qualitative Literature Institute for Work & Health,* the findings from the quantitative studies state, "Our best evidence synthesis finds that RTW interventions are effective in reducing the duration of work disability. They also reduce associated wage replacement and healthcare costs." The study goes on to indicate, "There is moderate evidence that three components – early contact with the worker by the workplace; a work accommodation offer; and contact between healthcare providers and the workplace – significantly reduce work disability duration and associated costs." The study also showed there is moderate evidence that, "the involvement of an individual with responsibility for RTW coordination" also reduces work disability duration and associated costs (IWH 2004).

While attending to the employee's progress, those involved must also know how to handle situations that arise. Supervisor and management training are essential, including periodic updates, to share "lessons learned" from experiences. You must ensure the supervisor knows what to do and say when a worker indicates he feels good and wants to "do their normal job," which would exceed their current medical restrictions. Conversely, the supervisor needs to know what to do and say when the employee insists they cannot perform tasks that are within their current restrictions. In both cases, most likely the employee needs to return to the medical provider for evaluation and updated restrictions. No matter how good an employee feels, they must stay within the current restrictions until the medical provider releases them or updates the restrictions. If no one is monitoring this and addressing such situations as they arise, the employee could reinjure themselves and delay their healing.

Planning for Return

If you wait until the doctor releases a lost-time patient or until a surgeon's first post-surgery checkup to hear if and when the employee may return to work, you lose valuable time, and the costs will increase. You may be surprised at times when you expected an employee to be released with restrictions, but the surgeon continues to keep them off work.

When conservative efforts to relieve Joe's back pain failed, surgery was scheduled. The surgeon assumed that the scenario that resulted in the back injury was an essential and ongoing part of his work tasks. In light of that, post-surgery checkups and physical therapy resulted in more weeks with Joe off work completely. The employer did not communicate otherwise or inform the surgeon of availability of modified work tasks, and was surprised when Joe was not released after his post-surgery checkups.

The Washington State Department of Labor & Industry, in its 2016 Spring Vocational Conference, indicated, "For all injured workers, the probability of successful return to work diminishes rapidly with more time spent off work." Their data indicates that the probability of returning to work is only 42.7% if the employee has been off for more than six months, and diminishes further with the passage of more time (Kennedy and Shatto, 16).

Solutions: Work cooperatively with your insurance carrier's claims adjuster and the medical facilities to discuss the expected treatments or surgeries, timing of checkups, and expected restrictions at each phase. All parties need to communicate early and often in order to plan for return to the workplace. If needed, request that the insurance carrier assign a nurse case manager to assist with doctor visits and guiding care. If you have video clips, modified duty task lists, or descriptions of job tasks that you expect may be within the person's restrictions as they heal, provide those to all parties so they have an understanding of your offerings. If providers see your intent to accommodate the employee, they are more likely to collaborate with you about how the employee can incrementally progress and return to work. Consider offering tours of your facility to the medical providers so they are familiar with the actual job tasks.

Employee Frustration

Being restricted or disabled will be a new experience for many people. Often, the injury leaves them unable to perform normal tasks like dressing, brushing their teeth, driving, or preparing meals. If the person lives alone, some tasks are impossible, and minor tasks take much time and effort. Even if the employee has family to help, relying on others for simple tasks or transportation can become frustrating. The frustration can easily lead to anger, resentment towards the company, or depression.

Solutions: As HR or the supervisor interacts with the employee to ensure medical visits occur and restrictions are updated, you should monitor for signs of emotional distress. Not all people are equipped with coping mechanisms to deal with the pain and helplessness that may come with a temporary or permanent disability. Simply asking, "How are you doing with your injury?" and "How's the rest of your life going?" can open the dialogue and give you insights on their frame of mind. Be ready to offer the help of your Employee Assistance Program (EAP), or ask the adjuster what options exist to address any distress you sense in the employee.

Coworker Frustration

An injured employee's coworkers can also become frustrated following the colleague's injury. They may feel they are taking on the work of the injured employee or may question the validity of the injury or restrictions. Sometimes, even when the employee is back, coworkers are frustrated with what they may perceive as the "easy" work the injured employee is doing. While that may be understandable, this can breed detrimental resentment that can impact the company and the injured employee.

In the study on Return-to-Work interventions, it was indicated that, "There are important social aspects to modified work. Modified work can involve difficult social dislocation which produces new sets of relationships and routines. The injured worker may have to deal with co-workers who resent having to take over some of his or her workload, or who may feel that the worker has managed to get an 'easier' job. An awkward fit with a modified work environment or a negative social atmosphere can contribute to a breakdown of the RTW process" (IWH 2004).

Solution: Educate employees at the time of hire and periodically afterward about the reasons for accommodation of workplace injuries. At the time of their colleague's workplace injury, educate coworkers so they understand how and why the employee is out or doing modified tasks. Explain that being at work, even at lower capacity, can help the employee avoid lost wages and keep them connected to their routine and social network. If employees understand the personal impact on the employee and healing process, it can help them have empathy, as well as understand that they would be accommodated if in that situation themselves. Share that the state Workers' Compensation system encourages employers to accommodate restrictions and imposes penalties, in the form of how the injury costs increase your Experience Modifier, for not returning employees with restrictions.

Big Picture Strategies

With so many moments where an employee's progress and claim costs can be derailed, how can employers successfully manage the process? A documented program is critical to smoothing the path, but is only one piece of the solution. To find those hidden pitfalls, you must show empathy, get continuous feedback from injured employees, and tweak your program and training based on what you learn.

Finding Empathy

Empathy is defined as the intellectual identification with, or vicarious experiencing of, the feelings, thoughts or attitudes of another. It's not an easy sentiment if you harbor thoughts that the employee was at fault for their injury, suspect it happened outside of work, or suspect the person is "working the system." Trust and empathy is needed on a much broader scope than just within the realm of injury management, and therefore cannot be explored deeply here. Suffice it to say that if there is not trust in the workplace or employees witness injured employees being treated disrespectfully, you may be more likely to have those points of derailment following injuries. Without trust, a good track record, and employee education, employees might abuse the system out of fear, lack of trust, or simple lack of understanding. Supervisors and others involved must put aside blame and suspicion and consider the employee's perspective. Only then can you best assist and have success with the healing and return-to-work process. The key to being empathetic is listening to the employee. Through listening, you will be better able to meet the employee's needs, help the healing process, and glean information to enhance your program for the future.

In *Workplace-based Return-to-Work Interventions: A Systematic Review of the Quantitative and Qualitative Literature*, it was indicated that, "In the Qualitative review within that study, it was shown that Supervisors can play an important role in the RTW process. They were identified as important to the success of RTW due to their proximity to the worker, their ability to manage the immediate RTW work environment, and their organizational position, which provides a link between the worker and senior workplace decision-makers. These findings are reinforced by moderate evidence from the quantitative literature that educating supervisors and managers leads [sic]to reductions in work disability duration" (IWH 2004).

Communication

In the *Return-to-Work Interventions* study, it was concluded that employees need help during their recovery process. That study indicated that, "the RTW process is laden with potential for miscommunication and misunderstanding." There they state, "Successful return to work after injury or illness is not a straightforward process. The qualitative studies highlight the many difficulties workers face in meeting their 'duty to cooperate' with employers and workers' compensation boards. Their navigation through that system is often arduous, marked by a lack of information about process and procedures at a time when they feel vulnerable and less than self-reliant. Workers are usually not familiar with rules about workers' compensation or the specialized language used by health care and insurance professionals" (IWH 2004).

In light of this, the constant communication mentioned previously is invaluable, as is having a point person in the organization that can explain how Workers' Comp works, understand medical treatment progression, and help the employee navigate the process. This contact can also give them needed empathy, and meet that employee's need for connection to the workplace and feeling valued. In addition, insurance carriers' claims representatives are typically very knowledgeable, and can work with you to guide and educate the employee.

Using Feedback to Support Periodic Training

Formal, documented feedback and data that can be analyzed can have value, but nothing replaces the insights and information you can glean from just chatting with the injured employee. If the employer actively shows empathy, and reaches out to the employee in person or via phone, that feedback will happen along the way. You can gain useful information that you can apply to tweak your program.

Had Ann's employer asked her what was going well and what was not going so well, she would have shared the stories of the therapist's products that the surgeon found unnecessary, and that more physical therapy sessions were scheduled than was necessary. Ann thought it inconvenient, but the employer would also realize there were unnecessary costs incurred from those extra visits. That conversation may also have revealed that Ann was not eating well because she typically cooked for the family and could not achieve much in the kitchen one-handed. They could have brain-stormed to help her think of easier options during that time.

Results of feedback from employees can result in identifying different medical providers that better collaborate with you, establishing a means for assisting the employee to their initial medical appointment and helping them understanding the next steps, looking at options for employee transportation to work and medical appointments, developing a system for getting the right information prior to the employee leaving their medical appointment, or identifying more productive and meaningful modified duty tasks.

Return-to-work Program Documentation

A documented thoughtful Return-to-Work program can be the cornerstone to establish and execute your program. While this paper assumes that such a program is in place, we should be reminded of the key components of a RTW program, below:

1. Program Objectives
2. Assignment of Responsibility
3. Training
    a. Supervisors and Management
    b. Employees
    c. Coworkers
4. Medical Provider Relationship

5. Procedures
6. Transitional Duty Tasks
7. Forms

Conclusion

As with most aspects of life, learning is continual. If you engage the injured employees by communicating and getting feedback throughout their healing process, you will have new lessons to apply to your process. Applying their input can positively affect your employees and your injury costs. Each scenario encountered during employee injury management will have its own unique variables, timing, and challenges. While the documented Return-to-Work program cannot address all points and "lessons learned," you can continuously enhance your program, communication, and training to share the lessons that are learned so they can be applied to future situations.

Bibliography

Asfaw, Abay, PhD; Souza, Kerry, PhD. "Incidence and Cost of Depression After Occupational Injury." *Journal of Occupational & Environmental Medicine.* September 2012: 1086-1091. (http://journals.lww.com/joem/Abstract/2012/09000/Incidence_and_Cost_of_Depression_After.4.aspx.)

Franche, R.L., Cullen, K., Clarke, J., MacEachen E., Frank, J., Sinclair, S., Reardon, R. 2004. *Workplace-based Return-to-Work Interventions: A Systematic Review of the Quantitative and Qualitative Literature Institute for Work & Health.* Institute for Work & Health (IWH). (retrieved March 2017). (https://www.iwh.on.ca/system/files/sbe/summary_rtw_interventions_2004.pdf.)

Kennedy, V. and Shatto, W. "More About the AWA Effort." *Washington Dept. of Labor & Industries Research Services, Spring Vocational Conference.* 2016:16. (http://www.lni.wa.gov/ClaimsIns/Files/Vocational/SpringVocConfPresentation.pdf.)

Lipscomb, H. Ph.D. "Safety, incentives, and the reporting of work-related injuries among union carpenters: 'You're pretty much screwed if you get hurt at work;' Workers Speak Out on Injury Reporting & Safety Incentive Programs*." American Journal of Industrial Medicine.* April 2013: 389-99. (http://www.ncbi.nlm.nih.gov/pubmed/23109103; (http://www.cpwr.com/sites/default/files/publications/LipscombWorkersSpeakSafety_webpost.pdf.)

Session No. 590

Safety Differently: A New View of Safety Excellence

Ron Gantt
SCM Safety
San Ramon, CA

Introduction

According to the United States Bureau of Labor Statistics (BLS), in 2015 the occupational fatal injury rate in the United States was 3.4 (BLS, 2016). This represents a disappointing lack of change over the previous years' occupational fatal injury rates, with the average occupational fatal injury rate in the United States remaining at 3.4 since 2008, according to analysis of data found on the Bureau of Labor Statistics' website. Others within the safety profession have noted this stagnation of fatal incident statistics (Dekker & Pitzer, 2016; Loud, 2016; Manuele, 2013), each noting that progress in safety, as measured by the number of major accidents, appears to have plateaued. In a review of fatal injury rates in the United States conducted by the author, although the fatal injury rate has declined 35% from 1994 to 2015, in the last 10 years (2006-2015), the fatal injury rate has only dropped by 15.8%.

This plateauing of fatal injury rates suggests that progress in preventing major accidents may have the features of an *asymptote*, a line that approaches a curve but never touches it. As Dekker (2015, 28) notes, "asymptotes point to dying strategies." The strategies utilized to achieve the progress in preventing major accidents are providing diminishing returns. As a result, calls for new approaches to safety management have grown (Dekker, 2015; Dekker & Pitzer, 2016; Hollnagel, 2014; Hollnagel, Woods, & Leveson, 2006; Loud, 2016; Manuele, 2013).

Safety Differently grew out of these calls, as safety researchers began to realize that traditional approaches were growing increasingly inadequate to deal with the complex realities of the work processes of today (Dekker, 2015; Hollnagel, 2014; Hollnagel, Woods, & Leveson, 2006; Hummerdal, 2017). Building upon decades of research in social and safety sciences, Safety Differently seeks to develop a more proactive, productive, and inclusive approach to safety management. Traditional safety management approaches tend to focus primarily on the prevention of negatives (Hollnagel, 2014). By contrast, Safety Differently seeks to identify and learn from things that go well as a means to not only prevent or minimize the negatives, but also to achieve more successful outcomes. This paper provides an overview of the Safety Differently approach, the basic tenets of Safety Differently, and recommendations for implementing a Safety Differently approach within an organization.

Safety: The Focus on Negatives

Typical approaches to safety management emphasize the identification and elimination or reduction of negatives. These negatives include accidents, incidents, risks, hazards, "human error," and "unsafe actions" (Hollnagel, 2014). For example, the Board of Certified Professionals (2017) states that:

> Safety professionals identify hazards and evaluate them for the potential to cause injury or illness to people or harm of property and the environment. The safety professional recommends administrative and engineering controls that eliminate or minimize the risk and danger posed by hazards.

The American National Standards Institute's voluntary consensus standard that defines the scope of practice for a safety professional similarly offers:

> The scope and functions of the professional safety position shall be to:
> - Anticipate, identify and evaluate hazardous conditions and practices.
> - Develop hazard control designs, methods, procedures and programs.
> - Implement, administer and advise others on hazard controls and hazard control programs.
> - Measure, audit and evaluate the effectiveness of hazard controls and hazard control programs (American Society of Safety Engineers, 2012).

Although the organization may utilize the outcomes of these processes in a productive way, the scope and function of the safety professional, as defined by each of these organizations, is oriented around the elimination or management of negatives. The American Society of Safety Engineers (ASSE, 2017), in defining what occupational safety and health (OSH) professionals do, explains it this way:

> [OSH professionals] provide employers advice, support and analysis to their employers to help them establish risk controls and management processes that promote sustainable business practice. They work to reduce (and eliminate) fatalities, injuries, occupational diseases, sickness and property damage.

This definition situates the practice of the safety professional toward the goal of promoting "sustainable business practice," but identifies that the mechanisms that the safety professional uses to achieve this goal is through reduction, control, and elimination of negatives.

The goal of eliminating negatives, such as fatalities and injuries, is certainly a laudable one that is not questioned in this paper. However, the intense focus on negatives has analytical consequences. The safety profession is oriented toward identifying, understanding, and managing things that go wrong at the expense of understanding things that go right. Dekker (2006) notes that there currently is not a good understanding of normal work processes in organizations, which may be necessary to enable accident prediction.

Hollnagel (2014) notes that misunderstanding normal work processes may not be trivial. As an illustration, Hollnagel (2014) created a chart (shown in Figure 1) of a system with a failure rate of 10^{-4}, or one failure in 10,000. The traditional focus of understanding and eliminating negatives means we focus intensely on the one failure at the expense of the 9999 times when things went well. This means that most of organizational life is unexamined by traditional safety approaches.

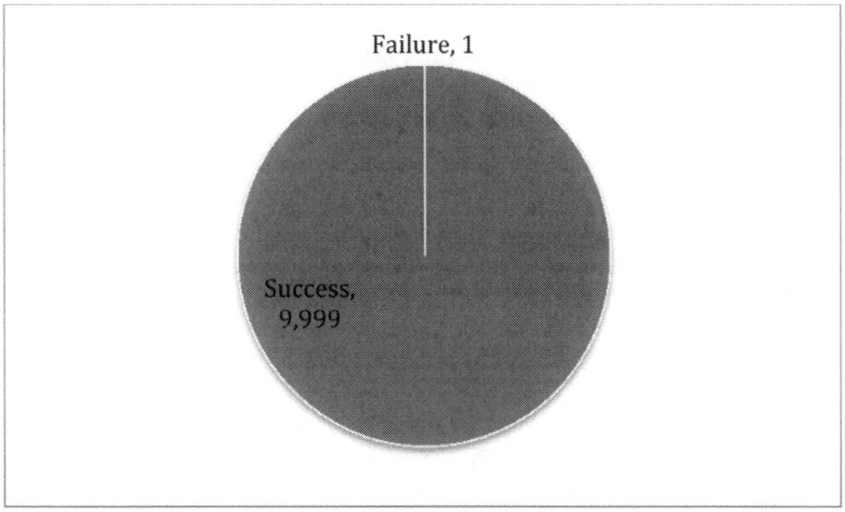

Figure 1. One failure in 10,000 (Adapted from Hollnagel, 2014)

Hollnagel (2014) argues that the focus on negatives not only leads to misunderstanding of what it takes to make an organization successful, it also leads to misunderstandings of the causes of failure. For example, the belief that "human error" or "unsafe acts" are the primary cause of most accidents may stem from this negative focus (Besnard & Hollnagel, 2014). However, the underlying processes that lead to many of the errors and faulty decisionmaking are present because they are adaptive (Hollnagel, 2009; Gigerenzer, 2008). Put another way, when it comes to human performance, the underlying processes are the same for success and failure, and one cannot fully understand how humans fail without understanding how they succeed (Woods et. al., 2010). This calls into question the effectiveness of the current negative focus of safety management.

Safety Differently

By contrast, the Safety Differently model of safety management is based on the idea that understanding and improving normal work processes is central to improving safety performance (Hummerdal, 2017). Rather than seeing safety management as a means to improve work performance, Safety Differently sees the improvement of work performances as a means to improve safety performance. This change in perspective orients the safety professional to interact with the organization in new ways. In the Safety Differently model, the safety professional is a facilitator of work, rather than one that constrains work processes.

In particular, Safety Differently provides an alternative perspective to other safety management practices in three areas:

- How safety is defined
- The role of people
- The focus of the organization

How Safety Is Defined

Although numerous definitions exist in legal and regulatory standards, there currently is no agreed upon definition of "safety" within the safety profession. However, as discussed above, the scope of the safety profession is based heavily on the reduction of negative events, such as accidents. This is consistent with a common way that safety performance is measured, the use of injury statistics. The implicit message is that safety is defined by the absence of negatives.

By contrast though, **Safety Differently defines safety as a capacity to be successful in varying conditions**. This definition inherently sees safety not as a goal unto itself, but as an enabling objective within the organization. Indeed, no organization exists with the sole goal to be safe. Instead, safety enables the organization to achieve its fundamental mission. However, the achievement of a successful outcome itself is not sufficient. Research suggests that organizations can push margins to achieve success in a way that makes failure more likely (Dekker, 2006; Woods, 2003; Patterson & Wears, 2016). Therefore, the goal of a safety management system should be to help the organization achieve success as conditions change. This is consistent with the definition of organizational resilience (Hollnagel, Woods, & Leveson, 2006).

A key threat to the achievement of sustainable success is when resources are insufficient to meet demands (Hummderdal, 2017). In these situations, workers are forced to adapt to fill the resource-demand gap. These adaptations are similar to the "practical actions" described by Snook (2000). Unfortunately, this creates significant variability in work processes as resource availability, work demands, and the abilities of workers to adapt all adjust over time (Hollnagel, 2009). Most of the time this variability goes unnoticed because workers adjust their performance in a way that hide the resource-demand gap (Woods & Hollnagel, 2006). However, sometimes the performance adjustments that workers make fail, and an accident may result (Hollnagel, 2009).

This makes normal work a central area of focus in the Safety Differently model. By studying normal work and what conditions enable or constrain normal work, safety professionals can assist workers in overcoming the constraints and expand the features that enable them to be successful (Hummderdal, 2017). Where work becomes difficult, you will often find adaptation, shortcuts, and 'human error.' Increasing worker capacity to handle these situations will not only decrease the probability that unwanted risk will sneak into the system, but also will increase the ability of your workers to be successful. This naturally will have a net positive effect on production, aligning the goals of safety with the organization (Hollnagel, 2014).

People Are a Solution to Enable

As stated above, the focus on eliminating negative events in safety often led to many misunderstandings about the role of people in both accidents and safety (Besnard & Hollnagel, 2014). As a result, much of safety practice has oriented itself in a way that treats people as if they are a problem or hazard that must be controlled (Dekker, 2015). The assumption appears to be that the system is reasonably safe enough as designed, but the erratic actions of people disrupt the system design, triggering accidents (Dekker, 2014b).

However, as noted above, organizations often work in a degraded system state. There are often never enough resources to meet the demands. Human ability to adapt fills this gap in reliable ways. Unfortunately, these same adaptations that work normally sometimes fail in context-specific ways (Woods et. al., 2011). When this happens we call it human error, but these same adaptations are also responsible for keeping the organization successful (Dekker, 2014b).

For this reason, the Safety Differently model notes that **people are a solution to enable or facilitate**. The unique ability for workers to adapt creatively to changing circumstances is a strength that organizations need to bring to bear on the complex problems they face (e.g., goal conflicts, resource scarcity, competition). Workers at all levels in the organization are constantly facing the realities of these complex problems, often on a daily basis. Therefore, they will be in a unique position to identify innovative solutions, or at least to assist in identifying problems that will be unsuccessful. As Hummerdal (2015) notes, "organizations are filled with people whose capacity goes above and beyond the roles and responsibilities that we have assigned them."

Safety Is an Ethical Responsibility

One of the most disturbing trends in safety management is the explosion of safety as a bureaucracy (Dekker, 2014a). Much of safety practice is dominated by regulatory requirements and international standards. One only needs to peruse the brochures of safety conferences to see this. Heads of regulatory agencies are often keynote speakers, and many breakout sessions are devoted to how navigate complicated regulatory requirements. What are not commonly seen in these brochures are presentations from people who do the work in question (with the curious exception of those who have been involved in accidents). By this anecdotal evidence, it appears that safety professionals are far more interested in what the regulatory agencies believe is 'safe' than what workers believe is 'safe.'

This leads to practices where safety professionals spend much time trying to ensure that workers comply with regulatory standards (Dekker, 2015). These safety practices are oriented to manage the bureaucratic accountability put on organizations and workers. This may lead to the development of the 'compliance heuristic,' where regulatory compliance becomes a substitute for safety management.

In the Safety Differently model, the organization is refocused toward seeing **safety as an ethical responsibility to those who do the organization's risky work**. This focus requires the organization to become far more interested in the realities of normal work and creating a safety management system that facilitates workers rather than one that holds them back. In effect, "safety is a service that [the organization] provides to their employees" (Hummerdal, 2017). This focus does not neglect regulatory compliance, but merely aims to create balance. The organization should be at least as interested in the concerns of their workers as they are of the concerns of the regulatory agency.

Challenges to Safety Differently

In the experience of the author, in discussing Safety Differently with numerous managers, safety professionals, and workers in many industries, many common challenges to the model have arisen. Some of these will be addressed below.

Safety Differently as Flavor of the Month

Many organizations report seeing many "flavors of the month" come and go, promising amazing, "silver-bullet" type results (e.g., behavior-based safety, safety-culture initiatives). Safety Differently is often put into this category. However, Safety Differently is not an individual process or tool that organizations can tack on to their existing organizational programs. Instead, Safety Differently is more of a mental model or a way to see the world. It does not replace what organizations do (Hummerdal, 2017). Rather, it changes the lens by which we view organizational life, work processes, the 'causes' of accidents, and the sources of safety creation within the organization. By change this lens, it naturally changes what the organization sees and does. New practices will emerge, but old practices will continue. They will just be conducted in a different light.

As an example, the organization will still investigate accidents and incidents in the Safety Differently model. However, instead of looking for who the guilty party/parties are, investigations in the Safety Differently model focus on understanding normal work processes, and how those processes led to an unintended and unwanted event in this particular case.

Safety Differently Is Not New

Some have pointed out that some of the philosophies and/or practices in the Safety Differently model have been around for decades. This is, in many respects, correct. Safety Differently is based on decades of research and practice in the social and safety sciences, incorporating elements of systems thinking, complexity theory, cognitive systems engineering, social psychology, sociology, management science, human-factors engineering, high-reliability organization theory, and resilience engineering. In many respects, pointing out that Safety Differently is based on decades of research and practice is not a criticism, but a strength of the model.

However, many organizations have used the point that Safety Differently is not new to make the argument that they are already doing Safety Differently, and, therefore, nothing more is necessary. In the experience of the author, this is often not the case. The organization may have some pieces in place that resemble Safety Differently tenets, but closer examination often yields a focus on preventing negatives, on treating people as a problem to control or fix, and/or on orienting the organization toward satisfying bureaucratic accountabilities at the expense of enabling workers at the sharp end to perform work successfully.

Safety Differently Undermines Accountability

A key concern for many organizations upon hearing of the Safety Differently approach is based on the general idea that organizations need accountability to function effectively, and that Safety Differently undermines that accountability. Typically, what the organization is referring to when they speak of accountability in this context is their ability to discipline individuals in response to errors or violations. However, Safety Differently does not mean to remove the organization's ability to discipline employees. Instead, Safety Differently reminds the organization that the goal is not rote compliance, but improved performance. By reviewing the normal work processes and looking for ways to enable workers to be more successful (instead of merely less prone to fail), the organization has more options for intervention. In effect, Safety Differently gives managers and safety professionals more "tools in the toolbox."

Furthermore, organizations must remember that accountability is one goal amongst many goals within the organization, and some of these goals compete. For example, when the organization places an intense emphasis on accountability, in the form of discipline, this often leads to a reduction in the organization's ability to learn as the reporting of events decreases (Dekker, 2014b). The Safety Differently model allows the organization to balance the goals of accountability with learning and employee engagement more effectively.

Doing Safety Differently

For those organizations seeking to do Safety Differently should begin a discussion in their organizations about what the definition of safety is within the organization. Even if the organization has an explicit definition of safety, often the practices of the organization imply a different definition. For example, many organizations admit that safety is not defined by the absence of accidents, yet still measure how safe they are by the number of accidents they have. Or, an organization may value the creativity and innovation of its workforce, yet have a safety management system based on rote compliance with regulations and procedures. Having discussions with employees at all levels of the organization about how they define safety, and how they perceive the organization defines safety, may yield some interesting data about the values of the organization.

Once this information is gathered, the organization should seek to change or eliminate practices that reinforce the belief that safety is defined as the absence of negatives. In its place, the organization can begin conversations about what it wants out of a safety program. How can the safety program be oriented to enable positives, like innovation, creativity and meaning? One question that can spark interesting discussion is to ask employees at all levels *how would the organization manage safety if the regulator disappeared tomorrow?* Begin to create a safety organization that adapts to the conditions of work, rather than one that does the opposite. As you do so, additional practices will likely emerge, revolving around enabling successful completion of work.

Focusing on Normal Work

Rather than wait for an accident to occur, begin the process of understanding and analyzing normal work. Build in processes that enable the organization to learn more about what happens on a daily basis. This can be as simple as conducting daily debriefing sessions for workers, similar to after-action reviews, commonly used in military contexts. Observations in the form of management site-walks or Gemba walks can also be used for this purpose (Gesinger, 2016). However, managers and safety professionals conducting these observations need to do so in the spirit of curiosity and learning, rather than looking for deviations, unsafe acts, etc. Training on how to interact with workers during these observations, including appreciative questioning skills, may be useful for this.

More focused and resource intensive approaches to understanding and analyzing normal work could be the use of "A Day in the Life of" sessions. In A Day in the Life of session, managers and/or safety professionals conduct work shifts with the workers (Gantt, 2015). This allows the managers the ability more closely to see the world the way the workers see it. To see how things that seem like minor annoyances when discussed in a conference room can be major obstacles in the work environment. Additionally, camaraderie between workers and managers/safety professionals is built up through shared experiences, which enhances trust in the organization (Gantt, 2015).

Learning teams (Conklin, 2016) are another tool that may be used to understand normal work processes. In a learning team, employees work with a facilitator to learn about a particular operation, and then collaborate to identify opportunities for improvement. By engaging those who do the work in this process, the workers get a better understanding of their job tasks, as well as the opportunity to identify improvements. This allows for a sense of ownership in the solutions, which will make the improvements more effective overall (Conklin, 2016).

From all of these tools, data will be gained about how work is normally performed. Safety professionals should treat all data as symptoms, understanding that people adapt their behavior to the conditions they are in (Woods et. al., 2011). By focusing on the conditions that influence behavior, safety professionals can look for sources of variability, such as the frustrations, dependencies, and sensitivities inherent in the work process (Hummderdal, 2017). Opportunities to dampen the variability or creating more capacity for the workers to respond to the variability can be identified and implemented. Examples of this include:

- If work processes are dependent upon certain tools to be available, how can the organization ensure that those tools are readily available in foreseeable conditions? What are the conditions that would cause those tools to become unavailable? How can the organization create the capacity to respond to those conditions so that the necessary tools are available?
- If workers have to deal with frustrating procedures (e.g., complicated, unrealistic, etc.) how can those procedures be changed to make them more workable? Is it possible to have the workers develop their own procedures for the task that might be better?
- If the work processes are sensitive to time pressures, how can we provide slack into the system to allow workers to achieve all necessary goals? How can we create an organizational response to stressors, rather than relying solely on individual responses?

Additionally, good practices and innovations will likely be identified in the evaluation of normal work. These should be shared throughout the organization.

Conclusion

The Safety Differently model provides an alternative to many traditional safety approaches, particularly as related to the focus on the reduction of negatives as the central role of the safety profession. Instead, Safety Differently seeks to not only minimize negatives, but also to expand positives. In particular, Safety Differently defines safety as the presence of a capacity to be successful in varying conditions. As a result of this change in perspective, Safety Differently orients itself to seeing safety as an ethical responsibility from the organization to those doing the organization's risky work. The organization focuses on enabling workers to complete work tasks successful. In doing so, it sees people as a solution to enable or facilitate, rather than merely the problem to control. As discussed in this paper, this is not a "silver bullet" or "flavor of the month." Rather, Safety Differently is a new way of approaching organizational safety that promises more significant alignment between safety, productivity, quality, and other organizational goals.

Bibliography

American Society of Safety Engineers (ASSE), 2012. ANSI/ASSE Z590.2-2003 (R2012): *Criteria for establishing the scope & functions of the professional safety position*. Des Plaines, IL: ASSE.

American Society of Safety Engineers (ASSE). 2017. *Becoming an OSH professional: Careers in safety*. (http://www.asse.org/about/careers/#undefined.)

Board of Certified Safety Professionals (BSCP). 2017. *About BCSP*. (http://www.bcsp.org/About.)

Besnard, D. & Hollnagel, E. 2014. "I want to believe: Some myths about the management of industrial safety." *Cognition, Technology, & Work*. 16(1), 13-23. doi: 10.1007/s10111-012-0237-4.

Conklin, T. 2016. *Pre-Accident Investigation: Better Questions – An Applied Approach to Operational Learning*. Boca Raton, FL: CRC Press.

Dekker, S. 2006. "Resilience Engineering: Chronicling the emergence of a confused consensus." In Hollnagel, E., Woods, D.D., & Leveson, N., eds., *Resilience Engineering: Concepts and Precepts* (77-92). Boca Raton, FL: CRC Press.

Dekker, S. 2014a. "The bureaucratization of safety." *Safety Science*. 70, 348-357. doi: 10.1016/j.ssci.2014.07.015

Dekker, S. 2014b. *The Field Guide to Understanding 'Human Error.'* 3rd ed. Boca Raton, FL: CRC Press.

Dekker, S. 2015. *Safety Differently: Human Factors for a New Era*. 2nd ed. Boca Raton, FL: CRC Press.

Dekker, S. & Pitzer, C. 2016. "Examining the asymptote in safety progress: A literature review." *Journal of Occupational Safety and Ergonomics*. 22(1), 57-65. doi: 10/1080/10803548.2015.1112103.

Gantt, P. & Gantt, R. 2014. "Safety – Not good enough." *The Advisor*. 14(1), 4-7.

Gantt, R. 2015, February 10. *A Day in the Life Of*. (http://www.safetydifferently.com/a-day-in-the-life-of/.)

Gesinger, S. 2016. "Experiental learning: Using Gemba walks to connect with employees." *Professional Safety*. 61(2), 33-37.

Gigerenzer, G. 2008. *Gut Feelings: The Intelligence of the Unconscious*. London, UK: Penguin Books.

Hollnagel, E. 2009. *The ETTO Principle - Efficiency-Thoroughness Trade-Off: Why things that go right sometimes go wrong*. Boca Raton, FL: CRC Press.

Hollnagel, E. 2014. *Safety-I and Safety-II: The Past and Future of Safety Management*. Boca Raton, FL: CRC Press.

Hollnagel, E., Woods, D. D., & Leveson, N. 2006. *Resilience Engineering: Concepts and Precepts*. Boca Raton, FL: CRC Press.

Hummerdal, D. 2015, September 22. *People are the solution*. (http://www.safetydifferently.com/people-are-the-solution/.)

Hummerdal, D. 2017, January 17. *Understanding and managing Safety Differently*. (http://www.artofwork.solutions/blog/2017/1/17/5emivj92zctlo2c2fcseqyf0y9f4lp.)

Loud, J. 2016. "Major risk: Moving from symptoms to systems thinking." *Professional Safety*. 61(10), 50-56.

Manuele, F. A. 2013. "Preventing serious injuries and fatalities: Time for a sociotechnical model for an operational risk management system." *Professional Safety*. 58(5), 51-60.

Patterson, M. D. & Wears, R. L. 2015. "Resilience and precarious success." *Reliability Engineering & System Safety*. 141, 45-53. doi: 10.1016/j.ress.2015.03.014.

Snook, S. 2000. *Friendly Fire: The Accidental Shootdown of U.S. Black Hawks over Northern Iraq*. Princeton, NJ: Princeton University Press.

United States Bureau of Labor Statistics (BLS). 2016, December 16. *Census of Fatal Occupational Injuries Summary, 2015*. (https://www.bls.gov/news.release/cfoi.nr0.htm.)

Woods, D. D. 2003. "Creating foresight: How resilience engineering can transform NASA's approach to risky decision making." *Work*. 4(2), 137-144.

Woods, D. D., Dekker, S., Cook, R. Johannesen, L., & Sarter, N. 2010. *Behind Human Error*. 2nd ed. Boca Raton, FL: CRC Press.

Woods, D. D. & Hollnagel, E. 2006. *Joint Cognitive Systems: Patterns in Cognitive Systems Engineering*. Boca Raton, FL: CRC Press.

Session No. 592

Harnessing the Power of Appreciative Inquiry to Achieve Transformational Results

Chris Ross, CSP, CPLP
Chief Executive Officer
The Engagement Effect
A division of Ross Performance Group, LLC
Anchorage, Alaska

At the heart of it, many safety professionals need to understand, encourage, and enable change. So much of safety is really about influence and change: trying to get leaders to change, trying to get employees to change their behaviors, and helping the organization to change its culture.

Therefore, the more tools and techniques that are available to the safety professional, the greater the chances of success in enabling change. And make no mistake about it; even under the best of circumstances, change can be very difficult and challenging.

A number of safety professionals have embraced the concepts of Human Performance Improvement (HPI), a field of organizational performance tools pioneered by Thomas Gilbert (*Human Competence: Engineering Worthy Performance*). One of the key elements to understanding and using HPI is developing an "HPI mindset." The underlying principles of HPI help to transition from the traditional "training is the answer," to "performance is the answer."

When reviewing the HPI model, it should be noted that the entire model is wrapped in change management, which is really at the crux of nearly every successful organizational endeavor.

There are many tools, models, and processes for driving organizational change, but one of the most exciting tools we can use is called Appreciative Inquiry (AI).

The assumption of AI is simple: Every organization has something that works right; things that give it life when it is most alive, effective, and successful (which also describes what happens when there are high levels of engagement). AI identifies what is positive and connects it to ways that heighten energy and create a vision for change.

There is a term, *positive revolution,* initially coined by GTE to describe the impact of their work that resulted in significant and measurable increases in stock prices, employee morale, quality, customer relations, union-management relations, and more. AI was at the core of those changes, and inspired many other positive approaches to change.

AI is a powerful yet simple process that anyone can use. Through collaborative inquiry and a connection to the positive core, many organizations have created new systems that:

- Create a common ground for vision and strategy.

- Accelerate organizational learning, building innovation, and boosting the power of even the smallest victories.
- Unite labor and management in new co-created partnerships.
- Improve communications.
- Build high-performance teams.
- Create exceptional and lasting positive change.

AI has been used to revitalize many processes and programs, building on the positive core, encouraging people to work together to promote a deeper understanding of the human system, the heart of organizations.

AI has been described by Diana Whitney in this way:

> *Appreciative Inquiry is the cooperative co-evolutionary search for the best in people, their organizations and the world around them. It involves the discovery of what gives "life" to a living system when it is most effective, alive and constructively capable in economic, ecological and human terms. AI involves the art and practice of asking questions that strengthen a system's capacity to apprehend, anticipate and heighten positive potential. The inquiry is mobilized through the crafting of the "unconditional positive question," often involving hundreds or thousands of people. AI interventions focus on the speed of imagination and innovation – instead of the negative, critical and spiraling diagnose commonly used in organizations. The discover, dream, design and destiny model links the energy of the positive core to changes never thought possible.*

Very powerful stuff indeed!

The birthplace and co-founding of AI happened in the doctoral program in Organizational Behavior at Case Western Reserve University in the collaboration between doctoral candidate David Cooperrider and his advisor Suresh Srivastva in 1980. Cooperrider was involved in a standard diagnosis at the Cleveland Clinic of "what's wrong with the human side of the organization?" Along the way, he changed his focus to an analysis of the factors contributing to the highly effective functioning of the Clinic when it was at its best. This laid the foundations for a process of inquiry, focusing on life-giving factors as the basis for organizational analysis.

The traditional approach of what is the problem by focusing on what is wrong or broken can sometimes yield very illuminating causes and is the most common approach: We often strive to fix everything. AI takes a very opposite approach, suggesting that, if we look for what works in an organization, we can appreciate it, build upon it, and achieve success.

As a method of organizational analysis, AI differs from conventional managerial problem solving. The basic premise of traditional problem solving is that, "Organizations are a problem to be solved." The task of improvement is then to remove deficits by identifying the problems, analyzing the causes, exploring solutions, then developing an action plan.

In contrast, the underlying assumption of AI is that organizations are a "solution to be embraced," rather than a "problem to be solved."

The following chart illustrates the two paradigms of organizational change:

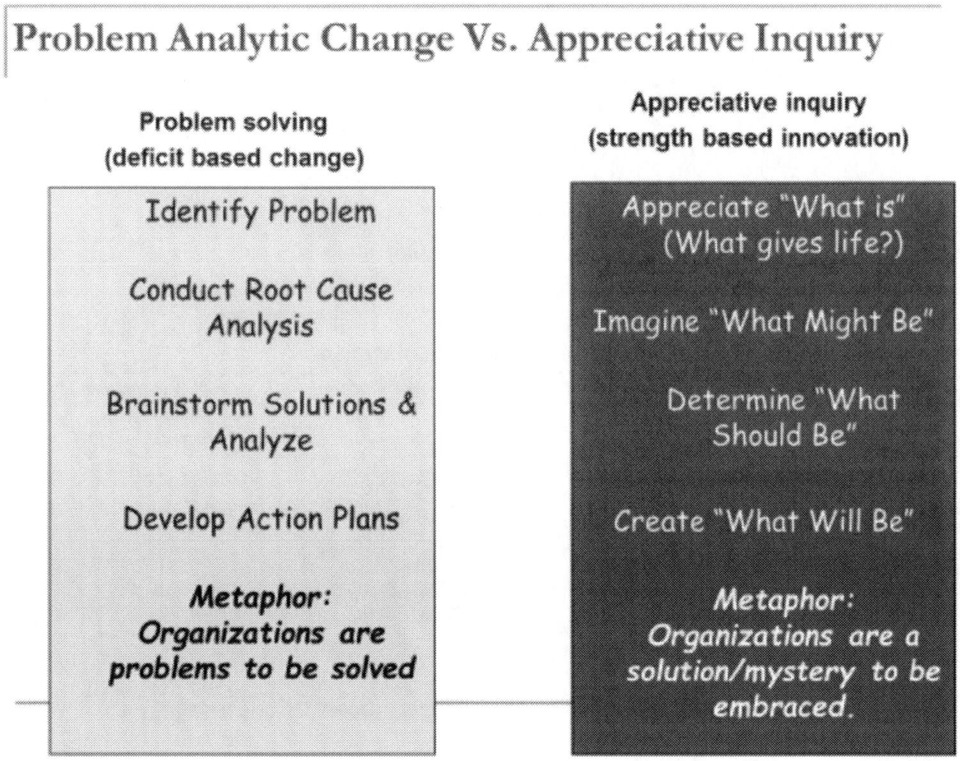

There are some additional assumptions (a set of beliefs shared by a group that causes the group to think and behave in certain ways) that are usually not visible or verbalized, but exist on an unconscious level. They must be made visible and discussed before the group can challenge their beliefs, and create change. And we all know that organizational change can be very difficult at best.

Sue Annis Hammond, in *The Thin Book of Appreciative Inquiry,* lists the eight key assumptions that drive AI:

1. *"In every society, organization, or group something works."*
2. *What we focus on becomes our reality.*
3. *Reality is created in the moment, and there are multiple realities.*
4. *The act of asking questions of an organization or group influences the group in some way.*
5. *People have more confidence and comfort to journey to the future (the unknown) when they carry forward parts of the past (the known).*
6. *If we carry parts of the past forward, they should be what is best about the past.*
7. *It is important to value differences.*
8. *The language we use creates our reality."*

The model for Appreciative Inquiry can be described as the "4-D Cycle," although there is actually a "fifth D," which actually precedes the other four.

Define: Choosing the topic to inquire about. It is critical to pick the right topic of importance, since the very act of asking questions in an organization begins to influence the group in some way.

Discover: Using a series of focused questions, participants begin to identify the best of "what is" in the organization. This phase is not only information gathering, but begins to shape change.

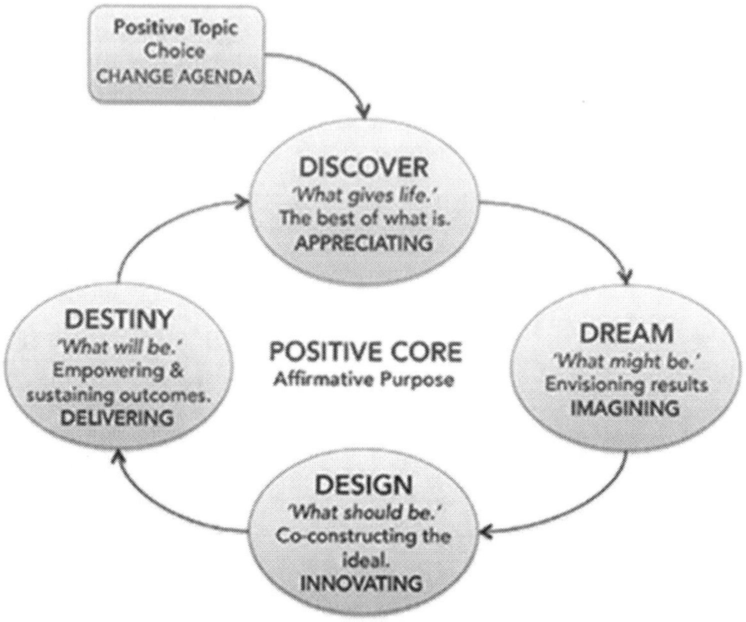

Appreciative Inquiry 4D Model – David L Cooperrider et al

Dream: Dream of what the best could be. This phase is taking a deep dive into what hopes, and aspirations can be used to recreate success factors identified in the Discover phase, and begin forming an idea of what high-level performance can be.

Design: This is the decision-making phase to begin developing concrete steps for success and performance. This phase is often described by using "provocative propositions" that answer the question, "How do we accomplish what we dreamed?"

Deliver/Destiny: This phase is putting all the previous steps into action in order to accomplish the desired results and increased performance.

Using AI can result in transformational organizational results. And it ties in so well with the model for Human Performance Improvement (HPI). The phases of AI dovetail nicely with the phases of HPI.

In HPI, we conduct a performance analysis by using AI; we can explore positive experiences, and also begin to reframe the conversation to strengths analysis. This also begins to create the beginnings of change (see the eight assumptions above).

We can also use the Dream phase of AI to help us understand and quantify desired performance, another aspect of the HPI model.

Then we can use the Design phase of AI to begin our selecting, developing and designing interventions to achieve our desired results.

And because the very process of asking questions begins to influence change, AI, just as HPI, has change management wrapped around every phase, including implementation (or Destiny).

There are times that the problem-solving approach is not sufficient. By paying attention to organizational strengths and shifting to a focus on creating and building a positive future and be a very positive and successful strategy, AI can be a very enriching and exciting tool in the safety practitioner's toolset.

A Deeper Dive into AI by Using Various Applications

Appreciative Inquiry can be used in many ways, including an AI Summit, Appreciative Coaching, and even for a very different twist on investigations.

AI Summit

The Appreciative Inquiry Summit (AI Summit) is a methodology for "whole system positive change." The benefits of the experience of "whole system positive change" range from building relationships and partnerships to building businesses and global organizations.

Appreciative Inquiry Summits have been conducted at medical centers, universities, on consumer products, and high technology companies. They have been used to craft an innovative, union-management partnership to form a strategic business alliance between two high-technology companies, to do annual company-wide strategic planning, and as a forum for culture change.

Whitney and Cooperrider make the following observations about the power of an AI Summit:

> *The Experience of Wholeness*
>
> *What is it about the experience of "Wholeness" that brings out the best in people, teams and organizations?*
>
> *This question has been with us since the inception of The Appreciative Inquiry Summit as we have watched over and over again, tension turn to enthusiasm, cynicism to collaboration and apathy to inspired action. We began wondering and asking people what happens when the whole system is in the room? When posed to clients, workshop participants and colleagues the answers typically include:*
>
> - *It evokes trust. When everyone is there you don't have to feel suspicious about what the others will do – there are no others.*
> - *It lets people see and experience a purpose greater than their own or their departments.*
> - *You get the sense that you are connected to a goodness that comes from the power of the whole. You realize you really need one another.*
> - *It satisfies the human need to be part of a larger community. It taps into our tribal consciousness. You feel like you belong*
> - *It establishes credibility in the outcomes. When everyone is part of the decision you know will stick. Public commitments engender responsibility.*
> - *New norms form quickly. You start to value relationships and getting the "whole story."*
> - *People transcend the "I" and become a "We." What's common becomes apparent.*
> - *It eliminates false assumptions about other people and other groups. When you get to know someone, you realize they aren't exactly what you imagined them to be. You develop compassion for different people instead of judgements.*
>
> *According to Jim Ludema (2000), "having the whole system in the room also brings an ecological perspective: all the pieces of the puzzle come together in one place and everyone can gain an appreciation for the whole. The unique perspective of each person, when*

combined with the perspectives of others, creates new possibilities for action, possibilities that previously lay dormant or undiscovered."

An AI Summit is designed to flow through the Appreciative Inquiry 4-D Process of Discovery, Dream, Design and Destiny. Typically, summits will last for a minimum of one day, though preferably at least two days, and as long as a week for a large or complex organization or issue. Each summit draws people together for a unique purpose and task focus. Each organization has a unique set of relationships within the company, and among stakeholders. Each organization has a unique and diverse cultural and geographical mix of people. And each organization has its unique ways of expressing its most precious values, of honoring relationships, and of making meaning and decisions. Each of these unique ways must be considered in the design of an AI Summit to ensure resonance, engagement, and collaboration among participants.

Appreciative Coaching

Appreciative Coaching is much different than other coaching approaches. Appreciative Coaching is deeply rooted in the principles of Appreciative Inquiry. It focuses on what has and is working in the person's professional and personal lives, and is designed to capitalize on strengths, generating positive changes in performance, professional development, and organizational leadership.

During Appreciative Coaching sessions, problems are acknowledged, but are used only as tools for learning. Appreciative Coaching takes the best and helps make it better by expanding people's capacity to perform, and by channeling their abilities for the greatest impact. This approach to coaching will take the client from good to great because it fosters personal transformation through their essential talent and strengths, increasing the likelihood of producing longer lasting results quickly.

Appreciative Coaching is a shared experience between the coach and the client. Together, the coach and the client collaborate to build a design for the future, develop strategies, and construct supporting actions for success. The relationship formed between the coach and the client is one of discovery. By intensely listening, asking unconditionally positive questions, and offering sincere encouragement, the coach facilitates the process for the client to unearth their own answers, and to take responsibility for acting on them. By increasing the client's self-awareness, they will understand how to better employ their strengths. Appreciative Coaching is the truest form of empowerment.

Appreciative Coaching can be used for any level of leadership in any type of organization. Using AI principles provide powerful and transformational results that can have a huge impact on performance, results and personal/professional development.

Investigations

Using the principles of AI during investigations is a very unique application. Rather than the traditional "fact-finding," "fault-finding," "root cause of the problem" approach, the parties involved develop a set of questions to discover everything in place that is "working" and contributes to the overall good health and function of the organization, its people and its processes.

This approach does not ignore problems or root causes; rather, it minimizes the emphasis on the negative and focuses on the "positive best," with the realization that, if individuals and organizations can find more of what is working, they will thrive and flourish.

Developing a new mindset and skill set can help you turn the corner to a powerful, results-based approach to help you achieve organizational success.

Bibliography

Cooperrider, David L., and Diana Whitney. 2005. *Appreciative Inquiry: A Positive Revolution in Change*. San Francisco: Berrett-Koehler Publishers.

Cooperrider, David L., Diana Whitney, and Jacqueline M. Stavros. 2008. *The Appreciative Inquiry Handbook: For Leaders of Change.* San Francisco: Berrett-Koehler Publishers.

Gilbert, Thomas F. 1978. *Human Competence: Engineering Worthy Performance.* 1st ed. San Francisco: Pfeiffer.

Hammond, Sue Annis. 1998. *The Thin Book of Appreciative Inquiry*. 2nd ed. Bend, OR: Thin Book Publishing Company.

Orem, Sara L., Jacqueline Binkert, and Ann L. Clancy. 2007. *Appreciative Coaching: A Positive Process for Change.* 1st ed. San Francisco: Jossey-Bass.

Whitney, Diana, and Amanda Trosten-Bloom. 2010. *The Power of Appreciative Inquiry: A Practical Guide to Positive Change*. 2nd ed. San Francisco: Berrett-Koehler Publishers.

Whitney, Diana, Amanda Trosten-Bloom, Jay Cherney, and Ron Fry, 2004. *Appreciative Team Building: Positive Questions to Bring Out the Best of Your Team.* (www.iuniverse.com.)

PtD Business Case: Reducing Cr VI Exposure in a Tailpipe Assembly Facility

Georgi Popov, PhD, QEP, CMC
University of Central Missouri
Warrensburg, MO

Elyce Anne Biddle, PhD
University of West Virginia
Morgantown, WV

Introduction

Eliminating or reducing risk of injury or illness in the workplace is critical to maintaining a safe and healthful environment. It is already known that the most beneficial risk-reduction solutions are implemented according to Prevention through Design (PtD) hierarchy of controls (ANSI/ASSE Z590.3-2011). This project examined existing hazards and determined, through visual observations and sampling results at hand-welding stations, that most welding fumes are not captured using the current strategies. While various alternative solutions were identified and evaluated, a new engineering control was implemented, representing a higher level on the PtD hierarchy of control of any current risk-reduction strategies. These engineering controls reduced the risk of Hexavalent Chromium (Cr VI) exposure to welders in an exhaust tailpipe and muffler assembly facility. The ergonomic risks were also addressed. The project ended by determining the business case associated with that solution or, more succinctly, "what was the value to the company for implementing this solution."

Introduction to the PtD Business Case Model

What Is the SH&E Business Case?

A business case in occupational safety and health can be defined as a structured method used to justify a project (intervention or solution) for decision-makers. It captures the effects of implementing programs or activities on employee health (injury or illness), hazards present, risk level, risk management, and the business process. The effects include customary financial business metrics, such as net present value (NPV), internal rate of return (IRR), return on investment (ROI), and payback period (PbP), which are meaningful to business management. The effects also include non-financial measures, those that cannot be directly or easily monetized, such as changes in customer satisfaction, corporate social responsibility, product defects, presenteeism, turnover rate, process cycle efficiency (PCE), and corporate reputation. Finally, the business case presents the health and business risk analysis by explaining opportunities, threats, strengths, and weaknesses associated with identified hazards. In reality, the business case answers the question: What's in it for the company?

Why Develop the Business Case?

A well-developed business case links safety, health, and environmental (SH&E) investments in protecting worker safety and health to business value. Changes in the economic environment and intensified business pressures, such as increased global competition, energy costs, and the recent Great Recession, have some organizations willing to take greater risk under the assumption that profits will increase. "Leaner is better" has been adopted by more and more companies with the thought that less spent on environmental, safety, and health is a wise choice for saving resources. The business case will assist in demonstrating that occupational safety and health activities are more than an expense but can indeed have a positive return on the investment. Perhaps even more importantly, the business case provides support that environmental, safety and health professionals are integral partners in the organization and essential contributors to the bottom line.

When to Develop the Business Case?

If an SH&E protection project will be or has been implemented by business, on business premises, or under the direction of business, then a business case should be developed. Business managers may request a logical and well considered argument—the business case—to justify funding to continue or expand existing solutions. Necessary resources for a new safety, health, or environmental investment under consideration can be identified through the development of the business case. When various competing solutions, including the current solution, to an environmental, safety or health protection problem need to be prioritized, the business case is critically important for making decisions. The business case can also be used to support the overall safety, health, and environmental function when competing for resources against other functions, such as production, within the organization.

How to Use the PtD Business Developer Model?

The Business Case Developer Model uses five steps to determine and illustrate the business value of occupational safety, health, and environmental projects, programs and activities designed to eliminate or mitigate the risks to reduce the associated occupational injuries and illnesses. For each step, the purpose is described, Modules with data collection forms are provided to derive the necessary information, and additional support is available through the following instructions that includes a glossary. It should be noted that throughout the PtD Business Case Developer Model, the terms solution and intervention are used interchangeably. To demonstrate the use of PtD Business Case Developer Model, the authors would like to present the following case study.

Case Synopsis

Implementing new engineering controls reduced the risk of Hexavalent Chromium (Cr VI) exposure to welders in an exhaust tailpipe and muffler assembly facility. Visual observations and sampling results at hand-welding stations indicated that most welding fumes are not captured. Various alternatives were identified and implemented according to PtD hierarchy of controls to reduce Cr VI exposure and the risk of musculoskeletal disorders (MDs). The financial analysis estimated a positive return on investment when adopting these PtD-based solutions.

Case Study Description

This Prevention through Design (PtD) effort was conducted by a mid-size manufacturer of exhaust tailpipe and muffler assemblies for pickup trucks. The 20,000 ft^2 facility receives pre-cut tubing and mufflers, bends and sizes the tubing, and then assembles and welds the completed muffler/exhaust units. The facility runs two 8.5-hour work shifts per day (including 30 minutes for lunch and breaks).

Risk assessment, SH&E intervention, risk reduction, financial and non-financial benefits from the improvements were presented, using a well-established PtD business case methodology (Biddle, E. et al.; Popov, G., Lyon, B.K. & Hollcroft, B. 2016; NIOSH PtD n.d.). The PtD business case model includes five steps, as presented in Figure 1.

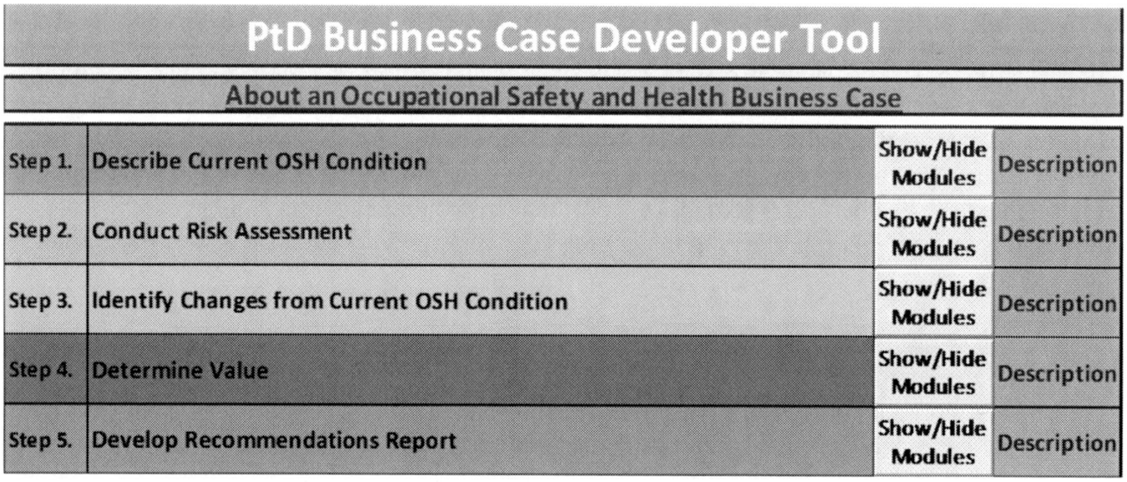

Figure 1. PtD Business case model

Step 1: Describe current OSH condition
Five production welders work on each work shift; four are hand welders, and one operates a Melton welder. During the morning shift, four welders use a handheld, trigger-activated GMAW/MIG gun, equipped with an automatic filler wire feed. The welding stations were not equipped with local exhaust ventilation. Beginning on November 27, 2006, OSHA implemented significantly reduced the Permissible Exposure Limit (PEL) for hexavalent chromium from 52 to 5 $\mu g/m^3$, and an Action Limit (AL) of 2.5 $\mu g/m^3$. Exposures above these values mandated a variety of follow-up and corrective actions.

Because of these standard requirements, exposure assessment was performed as a part of the risk assessment process to evaluate the welders' exposure to insure compliance. Figure 2 provides a chart of Cr VI welders' exposure before PtD intervention.

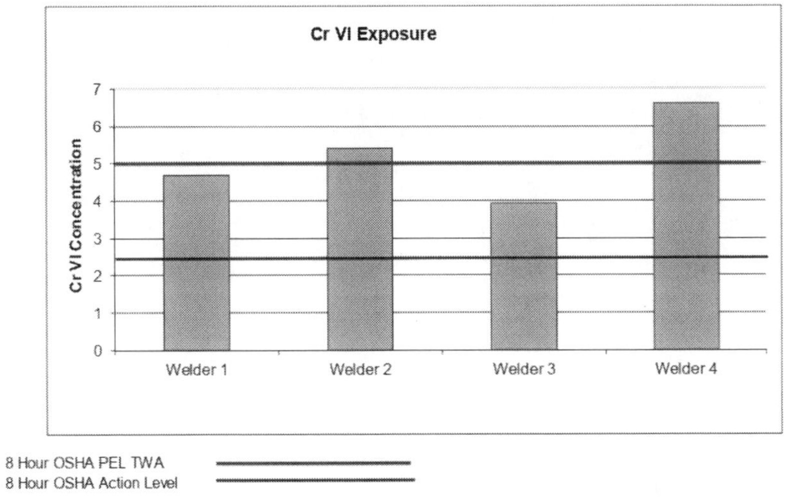

Figure 2. Cr VI welder's exposure before PtD intervention

The results from this exposure assessment bordered or exceeded occupational exposure limits (OELs), requiring the need for periodic air monitoring and initial/annual medical evaluations.

Management wanted to reduce welders' Cr VI exposure below OELs. In addition, the design of the work stations was evaluated to address ergonomic risks. Using multiple risk-assessment methods, the risk of Cr VI exposure and MSDs was determined to be moderate to high, indicating attention was needed. With that knowledge, a team of industrial hygiene consultants, ergonomic consultants, EHS and quality manager, engineers, operations manager and a welder was formed. Project objectives included:

Health and Safety Objectives:

- Cr VI exposure reduction below OSHA PELs
- Reduce lower back injuries (LBI) associated with the welding stations
- Reduce shoulder and neck injuries (S&N Inj.) associated with the welding stations

Operational Objectives:

- Improve company reputation
- Improve process cycle efficiency (PCE)
- Retain talented employees

Step 2: Conduct risk assessment
Risk Assessment–Current State: Current welding operation condition before intervention is presented in Figure 3 below:

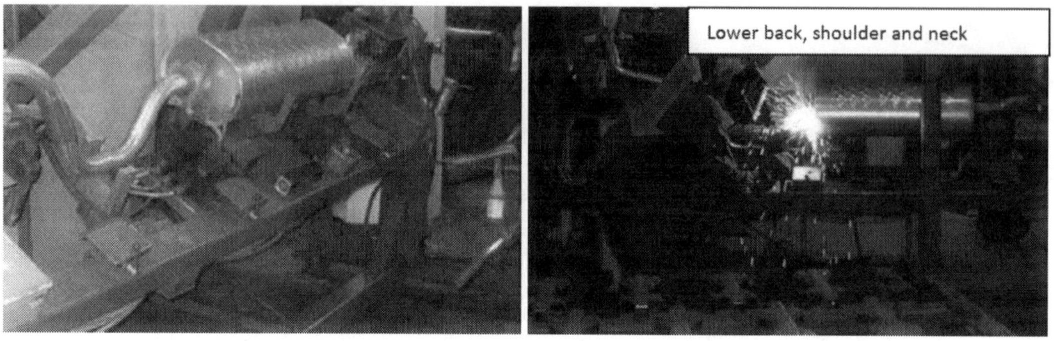

Figure 3. Current welding operation conditions

Initial risk assessment was conducted using a modified preliminary hazard analysis (PHA) method. Three hazards were selected for further analysis:

1. Irritation and damage to respiratory tract due to Cr VI exposure (1 RT Cr VI): Chemical/Physical
2. Lower back injuries (LBI): Ergonomics
3. Shoulder and neck injuries (S&N Inj.): Ergonomics

Current state hazard assessment is presented in Figure 4.

Project Name:	PtD Intervention - Exhaust System Welding						
Hazard Assessment Before Intervention							
Date:							
Prepared by:							
Only the first three hazards entered will be used as the basis for the Business Case Project. However, you may include any hazards that you wish to consider for future evaluation or as a matter of record by inserting and completing additional rows to this table.							
Enter a Hazard Number and Short Name (15 Character Limit)	Hazard Description	Process Description	Business Unit/Department			Description of Current Controls	PtD Hierarchy of Controls
			Name	Number of Employees Exposed	Annual Number of Injuries/Illnesses		
1 RT Cr VI	Irritation and damage to respiratory tract - Chemical/Physical	Welding	Production	18	1	None	None
2 LBI	Lower Back Injuries - Ergonomics	Welding	Production	18	1	None	None
3 S&N Inj.	Shoulder and Neck Injuries - Ergonomics	Welding	Production	18	1	None	None

Figure 4. Current state hazard assessment

Figure 5 provides the risk assessment of the three health outcomes of concern, using risk assessment matrix (RAM).

Risk Assessment Matrix (RAM)					
Hazard Number and Short Name	1 RT Cr VI	2 LBI	3 S&N Inj.		
Severity:	4	4	3		
Probability:	3	2	3		
Total Score	12	8	9		

RAM		Occurrence Probabilities and Values				
		1	2	3	4	5
Severity Level and Values	5	5	10	15	20	25
	4	4	8	12	16	20
	3	3	6	9	12	15
	2	2	4	6	8	10
	1	1	2	3	4	5

ANSI-ASSE Z590.3 2011: PtD Standard Example Risk Assessment Numerical Grading and Scoring	
Incident or Exposure Severity Descriptions	
Catastrophic (5)	One or more fatalities, total system loss, chemical release with lasting environmental or public health impact.
Critical (4)	Disabling injury or illness, major property damage and business downtime, chemical release with temporary environmental or public health impact.
Marginal (3)	Medical treatment or restricted work, minor subsystem loss or damage, chemical release triggering external reporting requirements.
Negligible (2)	First aid or minor medical treatment or minor medical treatment only, non-serious equipment or facility damage, chemical release requiring routine cleanup without reporting.
Insignificant (1)	Inconsequential with respect to injuries or illnesses, system loss or downtime, or environmental chemical release.
Incident or Exposure Probablility Descriptions	
Frequent (5)	Likely to occur repeatedly.
Likely (4)	Probably will occur several times.
Occasional (3)	Could occur intermittently.
Seldom (2)	Could occur, but hardly ever.
Unlikely (1)	Improbable, may assume incident or exposure will not occur.
Very high risk: 15 or greater High risk: 9 – 14 Moderate risk: 4 – 8 Low risk: Under 4	

Risk Assessment Matrix (RAM)			
Health Outcome	1 Respiratory Tract Irritation	2 Lower Back Injuries	3 Shoulder and Neck Injuries
Severity:	4	4	3
Probability:	3	2	3
Total Score	12	8	9

Figure 5. Risk assessment under current conditions

After the initial risk assessment, the health risk of the current operation was considered not acceptable. Two of the welding stations' exposure exceeded the Cr VI PEL. All four of the welding stations exceeded the action limit for Cr VI. In addition, lower back and shoulder/neck injuries were reported. Cr VI and MSDs' health outcomes were included in the risk assessment matrix.

SH&E hazards and risks impact on the business operations are presented in Figure 6 below.

Project Name:	PtD Intervention - Exhaust System Welding				
Business Impact Review Before Intervention					
Date:					
Prepared by:					
***Only the three hazards identified earlier are incuded in this table. Of the five Business Impacts, enter those you wish used in this Business Case Project in the first three cells. However, you may include any Business Impact that you wish to consider for future evaluation or as a matter of record by inserting and completing additional rows to this table. ***					
Hazard Number and Short Name	Enter Business Impact Name	Enter Description of the Impact from an incident or exposure associated with any or ALL hazards combined evaluated in this Business Case	Potential or Existing Business Loss	Description of Current Controls	Hierarchy of Mitigating Business Risk
1 RT Cr VI	Strategic	Hazard resulted in failure to meet business objectives	Existing	None	Accept
2 LBI	ORG	Hazards resulted in the loss of critical personnel	Existing	Temp workers	Reduce
3 S&N Inj.	ORG	Hazards resulted in delay and interruption of operations	Existing	Temp workers and Overtime	Reduce

Figure 6. SH&E hazards and risks impact on the business operations

A business impact risk assessment matrix (BRAM) was used to evaluate the SH&E risk on the business operations. Business impact risk assessment is presented in Figure 7.

Business Impact Risk Assessment Matrix (BRAM)					
Business Impact Name	Strategic	ORG	ORG		
Extent:	4	4	3		
Likelihood:	4	3	3		
Total Score	16	12	9		

BRAM		Business Loss Likelihood and Values				
		1	2	3	4	5
Extent of Impact and Values	5	5	10	15	20	25
	4	4	8	12	16	20
	3	3	6	9	12	15
	2	2	4	6	8	10
	1	1	2	3	4	5

Example Business Impact Risk Assessment Numerical Grading and Scoring	
Extent of Financial Loss	
Severe (5)	Unsustainable losses, total business failure, or inability to continue business operations.
Major (4)	Substantial business downtime, significant business losses
Moderate (3)	Moderate effect on financial losses resulting from business disruption or harm.
Minor (2)	Minor financial losses resulting from the effect on Impacts
Insignificant (1)	Inconsequential with respect to financial losses
Likelihood of Financial Loss	
Almost Certain (5)	Almost certain to occur in most circumstances
Likely (4)	Likely to occur frequently
Possible (3)	Possible and likely to occur at some time
Unlikely (2)	Unlikely to occur but could happen
Rare (1)	May occur but only in rare and exceptional circumstances
Very High Risk: 15 or greater High risk: 9 – 14 Moderate risk: 4 – 8 Low: Under 4	

Figure 7. Business impact risk assessment

To present a 36,000-ft. view of the current state, SH&E risks and business consequences are then presented, utilizing a modified bow-tie risk assessment methodology. SH&E risk levels and business impact numbers are transferred to the modified bow-tie risk assessment diagram. Current state bow-tie risk assessment is presented in Figure 8.

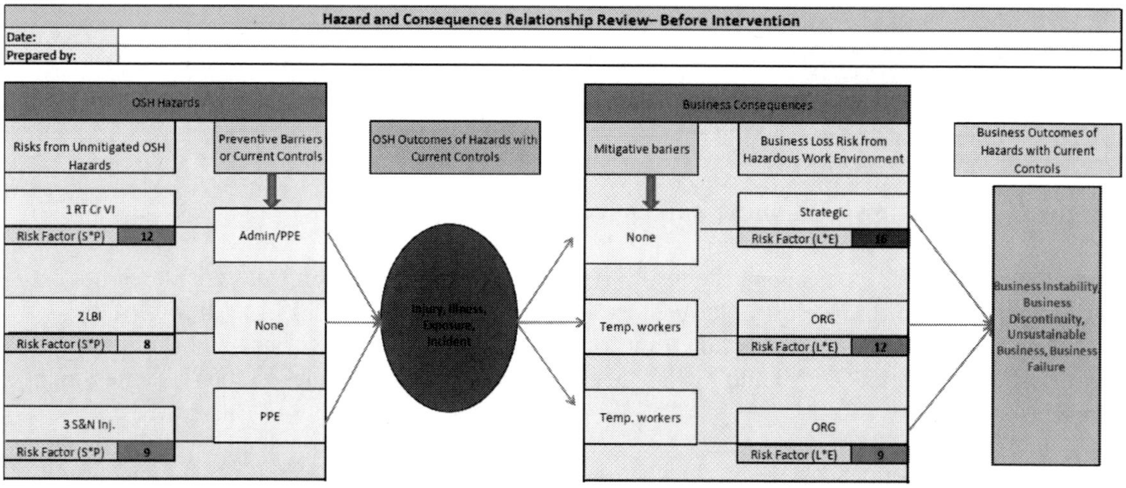

Figure 8. Current state bow-tie risk assessment

Bow-tie risk assessment methodology presents opportunities to safety professionals to present OSH hazards and business consequences in one diagram, as suggested in ANSI Z10 Section 4.2 (ANSI/AIHA Z10-2012).

It was apparent that control measures had to be implemented. The welders experimented with a new welding helmet with supplied air system. However, such controls measures were very unpopular with the welders. The supplied air system was inconvenient for the welders, and productivity was negatively affected. Therefore, other (higher) levels of controls needed to be considered.

Step 3: Identify changes from current OSH conditions
In order to reduce the Cr VI exposure and eliminate the use of respirators, each weld station had to be equipped with unique local exhaust ventilation (LEV). Captured air is transported overhead to a conventional branched and tapered duct system, where it is filtered, and then exhausted. To address musculoskeletal disorders of the back, neck and shoulders, adjustable welding work stations were designed. As a result, the following PtD solutions were agreed upon to reduce risk to welding operators:

1. *PtD solution regarding Cr VI exposure*

As a lower risk alternative, it was agreed that a unique LEV would be developed. It was decided that the supporting frame would be used to capture some of the welding fumes. An inlet pipe with a flare on the end goes into 4×4 inches support frame on the front of the fixture. It also has a few perforated holes on underside of 4×4 to capture some of the welding fumes. The new LEV system is presented in Figure 9 below.

Figure 9. New LEV system–Engineering control

As a result of the PtD intervention, Cr VI levels were reduced below the PEL and even below the action limit. Figure 10 provides a chart of Cr VI welder's exposure after the PtD intervention.

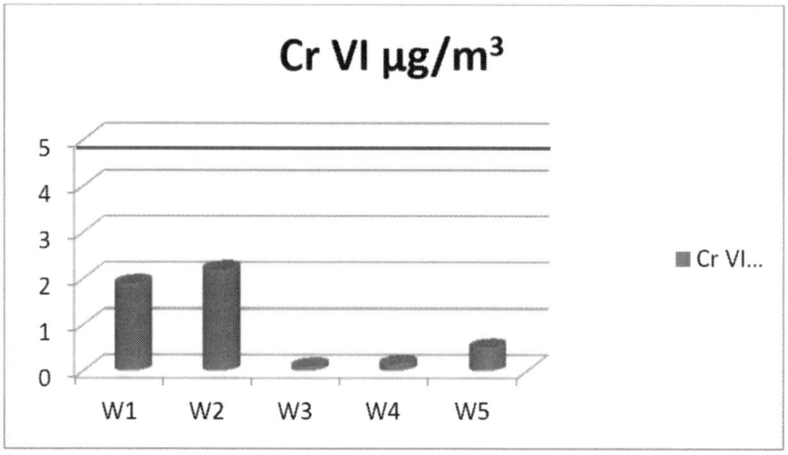

Figure 10. Cr VI welder's exposure with PtD intervention

2. *PtD solution regarding musculoskeletal disorders*

All welding work stations were modified, including height adjustments. Risk assessment was conducted after the intervention as well. The same risk assessment methodology was followed to assess risk after the intervention. The results are presented in a modified bow-tie diagram in Fig. 11.

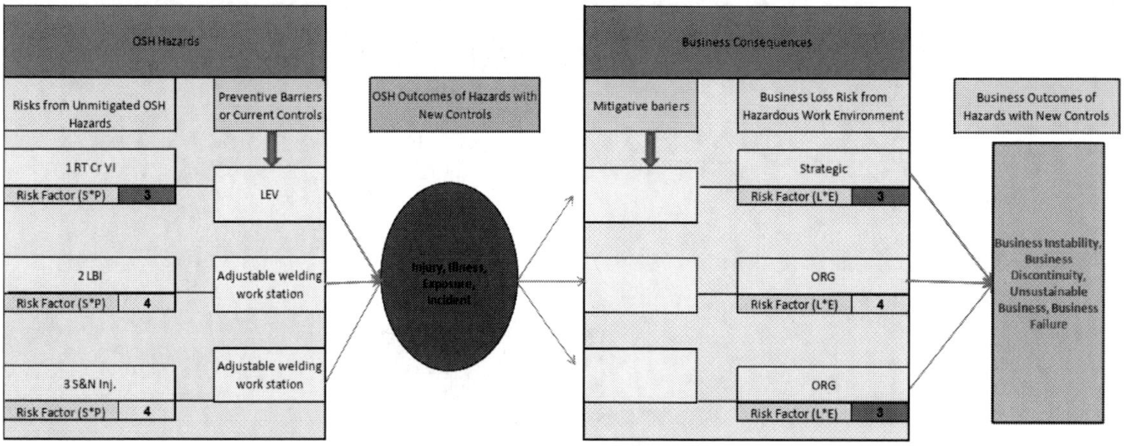

Figure 11. Future state bow-tie risk assessment

Step 4: Determine value

Value of the PtD Solution: The proposed changes led to substantial risk reduction, financial and non-financial benefits to the company. Risk reduction is considered the most important non-financial benefit. SH&E risk levels before and after the interventions are presented in Figure 12 below.

Health and Safety Risk Reduction				
Hazards	Risk Factor Without Intervention	Risk Factor With Intervention	% Risk Reduction	% Residual Risk
1 RT Cr VI	12	3	75.00%	25.00%
2 LBI	8	4	50.00%	50.00%
3 S&N Inj.	9	4	55.56%	44.44%
Total	29	11	62.07%	37.93%
Is this an acceptable level of Health and Safety Risk?		Yes		

Figure 12. Health and safety risk reduction

It was estimated that the SH&E improvements will lead to reductions in business losses. Business risk levels before and after the interventions are presented in Figure 13 below.

Business Loss Risk Reduction				
Business losses from hazardous work environment	Risk Factor Without Intervention	Risk Factor With Intervention	% Risk Reduction	% Residual Risk
Strategic	16	3	81.25%	18.75%
ORG	12	4	66.67%	33.33%
ORG	9	3	66.67%	33.33%
Total	37	10	72.97%	27.03%
Is this an acceptable level of Business Risk?		Yes		

Figure 13. Business loss risk reduction

PtD intervention outcomes are presented in Figure 14 below.

Outcome Measures–Financial

Net Present Value (NPV)	$319,415.14
Internal Rate of Return (IRR)	82.40%
Return on Investment	269.79%
Payback Period (years)	1.95

Outcome Measures–Non-financial

OSH Risk Reduction	62.07%
Business Risk Reduction	72.97%
Company Reputation	EHS intervention leads to reduced Cr VI community exposure
Process Cycle Efficiency	Fewer illnesses/disorders leads to improved employee performance
Employee Turnover	Eliminated unpopular PPE leading to retaining valuable employees

Figure 14. Overall financial and non-financial outcomes of the PtD intervention

Step 5: Develop Recommendations Report

Recommendations: The project team determined that adoption of these PtD solutions would be beneficial to the organization. Redesign of the welding workstations to include LEV and adjustable frame height was approved. The management agreed to adopt the LEV PtD solution, despite a large initial investment to substantially reduce Cr VI exposure without affecting productivity.

Conclusion

Using the PtD Business Case Developer Tool, the proposed changes were shown to lead to substantial risk reduction, financial and non-financial benefits to the company. Risk reduction was considered the most important non-financial benefit.

The project team determined that adoption of this PtD solution would be beneficial to the organization. The management decided to adopt this PtD solution and modify existing welding stations to substantially reduce Cr VI exposure without affecting productivity. The authors demonstrated how to reduce SH&E and business risk by implementing PtD concepts. The financial results were presented to the management chain for their adoption decision using a business case format and business language.

One of the many challenges SH&E professionals are facing is developing skills to implement PtD business case methods. In order to be successful, the SH&E professionals have to learn to use risk assessment tools and incorporate them into PtD interventions and develop business cases to improve workplace conditions. We also have to be able to link SH&E risk, business risk, and develop business cases to support our interventions proposals. PtD business case methodologies, such as those presented, can be successfully utilized to demonstrate the link between SH&E risk and business risk.

Bibliography

American Industrial Hygiene Association. 2009. *Strategy to Demonstrate the Value of Industrial Hygiene.* (http://www.ihvalue.org)

American National Standards Institute/American Society of Safety Engineers (ANSI/ASSE). 2011. *Prevention through Design: Guidelines for Addressing Occupational Hazards & Risks in Design & Redesign Processes* (Z590.3-2011). (http://www.asse.org/shoponline/products/Z590_3_2011.php).

American National Standards Institute/American Society of Safety Engineers/American Industrial Hygiene Association (ANSI/ASSE/AIHA). 2012. *American National Standard—Occupational Health and Safety Management Systems* (Z10-2012). Fairfax, VA: American Industrial Hygiene Association.

Biddle, E., Carande-Kulis, V., Woodhull, D., Newell, S., and Shroff, R. 2011. "The Business Case for Occupational Safety, Health, and Environment and Beyond." In S.Clark, R.J. Burke & C.L. Cooper (eds.). *Occupational Health and Safety.* Famham, England: Ashgate Publishing Ltd.

National Institute for Occupational Health and Safety (NIOSH). 2010. *Prevention Through Design: Plan for the National Initiative*. Publication No. 2011-121. (Retrieved February 21, 2017) (http://www.cdc.gov/niosh/docs/2011-121/pdfs/2011-121.pdf.

Popov, G., Lyon, B.K. and Hollcroft, B. 2016. *Risk assessment: A practical guide to assessing operational risks*. Hoboken, NJ: John Wiley & Sons. (http://centralspace.ucmo.edu/xmlui/handle/123456789/407.)

Funding for this PtD Business Case Brief was partially provided by the National Institute for Occupational Safety and Health. Findings and conclusions in this report are those of the author(s) and do not necessarily represent the views of the National Institute for Occupational Safety and Health. Mention of any company or product does not constitute endorsement by the National Institute for Occupational Safety and Health (NIOSH). For additional details on the Business Case, contact the authors Georgi Popov, PhD, QEP at gpopov@ucmo.edu and Elyce Biddle, PhD at: Elyce.biddle@mail.wvu.edu

Session No. 594

OSHA's Enforcement and Significant Changes in 2015 NFPA 70E

Doug Tellin, ME, Owner
Electrical Safety Specialists LLC
Louisburg, KS

Introduction

The need to assist employers and employees in preventing electrical injuries and deaths has never been greater. With this need, there have been some major changes in the electrical safety standards, such as the National Fire Protection Association (NFPA) 70E, as well as Occupational Safety and Health Administration (OSHA) enforcement of these standards. Assessing electrical hazards and training workers can be a very complex task for most companies.

One of the biggest problems with electrical hazards is that electricity cannot be seen, heard, or even smelled until it's too late. It is very difficult for people to give the proper respect for arguably the most powerful hazard in a facility when it is virtually invisible. In recent years, arc flash has been given a lot of attention. Explosions and burn injuries are easy to visualize compared to the invisible hazards, such as shock and electrocution.

With the unacceptable death and injury rates caused by electrical hazards, OSHA compliance officers and safety professionals are taking another look at what can be done to protect workers and reverse years of complacency in electrical safety.

Following the simple concepts in the hierarchy of safety:

- Electrical Safety Programs and Procedures
  - Train
  - Audit
  - Document
- Risk Management
  - Eliminate
  - Reduce
  - Protect
- Management
  - Plan
  - Implement
  - Evaluate
  - Correct

This paper will help guide employers and employees in the direction of becoming compliant with the latest standards and enforcement of these standards.

OSHA's Mission

OSHA's mission is not to seek and destroy; OSHA's mission is to identify and fix.

It is a common misconception that OSHA is out to fine, penalize, and make life difficult for employers, when, in fact, OSHA's primary mission is to guide and assist companies in maintaining a safe environment for their employees. Companies rarely use this resource as a tool in achieving safe work conditions at their facility.

Never work on live circuits unless it is absolutely necessary. OSHA has strict guidelines for working on live circuits. Production or inconvenience is not justification for working energized. Not having the authority to shut off power cannot be used as an excuse for workers.

NFPA 70E®

NFPA 70E is "the" standard for electrical safety in the workplace.

NFPA 70E 2015 has set the bar high, requiring:

- Written safety programs and procedures
- Arc flash assessments and reviews every five years
- Only qualified persons working within the limited approach boundary of energized parts
- Documentable/demonstrable proficiency in electrical safe work practices for qualified persons
- Retraining every three years
- Annual audits of employee performance
- CPR/AED annually
- Release of victims' training annually

Arc Flash Risk Assessments

How can employees be protected from a hazard if the hazard is unknown?

On August 4, 1991, OSHA included language in its Title 29 CFR, Subpart S, for the electrical safe work practices for general industry that added arc flash as an additional hazard to the shock hazard of electricity

If a piece of electrical equipment equivalent to a stick of dynamite or a fire cracker, one would protect themselves in far different ways when approaching these hazards. Is an arc flash risk assessment necessary? This can be quite the debate. Looking back at the hierarchy of safety, the first steps to follow are to assess, eliminate, reduce, avoid, and protect. In the case of electrical hazards, the questions every EHS professional should have the answer to are:

- Are employees trained in work practices and the proper selection of PPE?
- Was an arc flash hazard analysis performed?
- Were flash protection boundaries established?
- Were incident energies reduced to a safe level where possible?
- Was a shock hazard analyses performed?
- Were shock hazard boundaries established?
- Were all other potential electrical hazards identified?

This can be a very complex task for facilities. It takes skilled individuals with extensive training to properly assess these hazards.

Tables Versus Detailed Assessment

There's an APP for That

There are several different ways companies address arc flash risk assessments. One method commonly seen is for a company to use the tables found in the NFPA 70E®. These are task-based tables in which a qualified person can simply identify the task and condition of equipment found in Table 130.7(C)(15)(A)(a). This will give a yes-or-no answer whether arc flash personal protective equipment (PPE) is required.

The next step is to identify the location in which the task will be performed via Table 130.7(C)(15)(A)(b) for AC systems or 130.7(C)(15)(B) for DC systems. This table will then give you the arc flash PPE category.

Word of Caution

There are certain parameters that must be met before the tables can be used as a viable option. These parameters can be found in Table 130.7(C)(15)(A)(b), in the cell below the type of equipment. The parameters give the maximum limits of available short-circuit current, and the maximum clearing times of the upstream overcurrent device, along with working distance the task will be performed at.

If the short-circuit current is greater than the maximum allowed or the clearing time exceeds the allowable time limit, the table would not be allowed in determining the level of PPE. The incident energy method must be performed. Another shortcoming of the table is that once hazards are identified, OSHA requires hazards to be eliminated or reduced to a manageable level. The only way to mitigate these hazards is to first perform a detailed assessment. The tables do have there place. They can be used as a good "stop gap" measure. In other words, the table will provide some protection for employees, until adequate funding and time is allowed to assess and reduce the hazards correctly by performing a detailed analysis.

Proper Maintenance and Testing of Electrical Equipment

How does proper maintenance and testing procedures affect safety?

> *NFPA 70E 110.1(B) Maintenance. The electrical safety program shall include elements that consider condition of maintenance of electrical equipment and systems.*

As with most electrical hazards, the degree, extent, and frequency are all dependent on how well the equipment functions. Arc flash, for example, is based on the assumption that breakers will operate at manufacturer's specifications. If breakers are not maintained and periodically tested to insure proper working order, the energy levels at the equipment may increase exponentially, thus making PPE virtually ineffective. Simple things like dust, loose connections, and inadequate maintenance can literally mean life or death. It takes highly trained personnel and equipment to analyze and test electrical components.

Training for Employees

It has been a long-standing debate on who is qualified and who is not.

OSHA and the NFPA 70E both require qualified persons when working near energized parts. In order for an employer to deem a person qualified, employees must receive training on safe work practices, and demonstrate their ability to perform their duties in a safe manner:

> *NFPA 70E 110.4(A)(1)*
> *Testing.Only qualified persons shall perform tasks such as testing, troubleshooting, and voltage measuring within the LAB or energized electrical conductors or circuit parts operating at 50 volts or more or where an electrical hazard exists.*

If a person is within 42" of an energized part 50 volts or greater, they would require training and documentation as a qualified person. Tasks such as troubleshooting, voltage measurements, and lockout/tagout (LOTO) are some of the very common tasks that require the use of qualified persons. There has always been debate on the determining factors of what is needed for a person to be qualified. OSHA can be fairly vague on the subject, where the NFPA 70E is more precise. Think of OSHA as the people who tell you what to do, and the NFPA 70 E that tells you how to do it.

<u>The Best Defense Is a Good Offense</u>

Employees who have been properly trained have the skills to avoid and prevent an accident before it occurs. If an event does occur, they have the proper PPE to maintain a survivable level. The question, "Who is a Qualified Person?" is often asked. There is frustration that OSHA does not clearly define what a qualified person is, except for the fact that a person must be deemed qualified to work on or near energized parts. OSHA and the NFPA 70E also have strict guidelines on when you can and can't work near energized parts. Every qualified person should know these regulations or standards, as well as many others.

OSHA's definition of a qualified person (1910.399 8/07) includes the phrase "has demonstrated skills."

This has a big impact on the type of training required for electrical workers. In order to meet this requirement, a person actually has to demonstrate the ability to perform the task in a safe and timely manner. The level of training should match the level of the hazard. The 2015 NFPA 70E also added "demonstrate" to the definition of a qualified person, as well as many of the additional training requirements for qualified persons.

This type of training is seen in the community, and is accepted because there is a proper perception on the hazards and consequences of these hazards.

Fire fighters are an example; if they make a mistake "on the job," the worst-case scenario would be death for themselves or others nearby. The public in general would not feel safe if they knew the only training fire fighters received was watching a short video on how to drive a fire truck and put out fires. The question could then be asked: What is the worst-case scenario if an electrician makes a mistake on the job? It could also be death for themselves or others nearby. If the training an electrician receives does not include individual, hands-on demonstrations and auditing procedures, training would be incomplete. If the worker cannot perform the task with proper PPE in a safe manner, a reasonable person can assume the training has failed. Justification to work, hazard risk assessments, PPE, shock boundaries, CPR/AED, release of victims, and LOTO are but a few of the items a person needs to clearly know and demonstrate their ability to perform before an employer can deem them qualified.

A person with a Master Electrician's license may not be a qualified person if they have not received training in safety, have not demonstrated their ability to perform the task in a safe manner, and the training and demonstration of skills have not been documented.

Bibliography

National Fire Protection Association®. 2015. *Handbook for Electrical Safety in the Workplace.* 4th ed. (NFPA 70E), Article 90, Chapters 1,2,3. (http://www.nfpa.org/).

United States Department of Labor, Occupational Health and Safety Administration. "1910, Occupational Safety and Health Standards, Subpart S, Electrical" (https://www.osha.gov/pls/oshaweb/owadisp.show_document?p_table=STANDARDS&p_id=10135).

Session No. 595

Bag of Tricks: Presentation Design 1

Don Weatherbee
EHS Manager
DuPont
St. Joseph, MO

Introduction

This is intended to be a summary of the presentation as it is given. I'll break down my understanding of, approach to, and general thoughts for presentation design, with a noticeable emphasis on safety training.

When I had first arrived at my prior place of employment, I used whatever I could lay my hands on for training. I found several CDs at the site with a bunch of presentations on them, and went to town. Don't misunderstand me; purchased training presentations can be a good starting point. But once I got past the initial rush, I began to create my own site-specific training.

In my personal protective equipment (PPE) training, for example, I used a football equipment analogy to get the employees to "buy into" my plan to change and upgrade their PPE. So, I showed them a visual history of football helmets. Much like football helmets had changed over the years, our PPE would also be changing. I had to overcome the "if is not broken, don't fix it" syndrome. (Not that that doesn't happen anywhere else, of course.)

Now, besides being site specific, another benefit was realized. The employees knew I was putting in the time to create the training, and this made the training have a greater value to them. After all, perception is often reality, when it comes to effective training.

Of course, I'm not the first person to realize that good training can be made better with a well-designed presentation. There are lots of books on the subject, so I'll only mention three of them here.

Presentation Zen, by Garr Reynolds, is a book, and worth checking out. Its main advise is: "Use pictures." Images are remembered longer by people. There are several examples of before and after slides in this book.

Beyond Bullet Points, by Cliff Atkinson, is another book on the subject. Its' big message is: "Tell a story, have a plot." Think of the presentation as a mini-movie, with a beginning, middle, and end. One of the interesting ideas from this book is the use of hidden slides for organizing a presentation when in "view-all-slides" mode.

Then you have *Slide:ology*, by Nancy Duarte; this book contains lots of insight into slide design. It covers fonts, colors, placements, space, arrangement, backgrounds, text use, and images. It also gets into some fundamental animation, motion and multimedia methods. I've gone back and have reread this one a few times.

Fundamental Considerations for Any Presentation

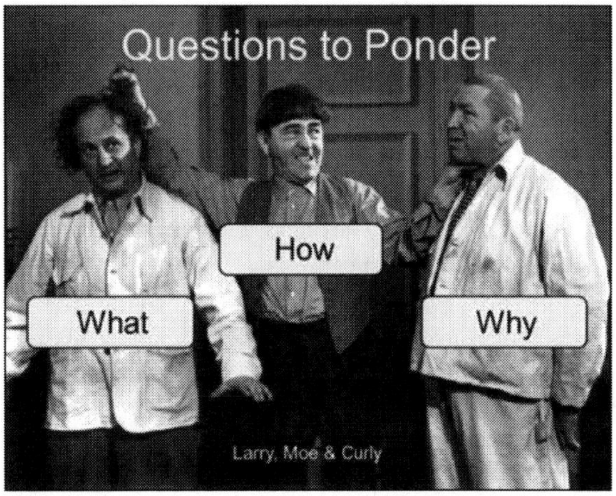

So what is needed for any presentation? Speaker, Audience, Room, and Deck. Hmm, maybe we can ask these guys?

That we should all be as good a speaker as John F Kennedy. But, alas, most of us are not. Certainly, I'm at best an average speaker, so I want something behind me that is going to help. I've seen good speakers, Steve Jobs comes to mind, and I strive to be as good or as effective as he was.

It helps to know the size of your audience; a few people, a few hundred, or a few million. Under 40 to 50 people, you can have good audience interaction. But above 200 or so, you move from being an educator to an entertainer. Your presentation will need to be adjusted accordingly. Is the audience your employees at the plant, your CEO, or your peers? Everyone will have different expectations.

The room is everything; from the projector used, to the physical size of the room, and how the seating is arranged. From the presentation design perspective, will green look green on the screen is a question I ask. That is to say, "Will colors in my presentation show as I intend?"

Those in the "seminar" business tend to call the collection of slides that they use their "deck." The deck, by itself, is not the presentation, but is an important part and warrants due consideration with proper design. The author of *Slide:ology* is the same person who helped Al Gore's "Inconvenient Truth" go from being a somewhat boring to a great presentation. Shucks, it even became a movie, and earned an Academy Award for Best Documentary Feature in 2006.

There are five basic paths or "ways" to give a presentation: live, web pages, webinars, ePresentations, and eLearning. Each has its' own set of pros and cons.

Live: This is what most of us do. The audience gets the full benefit of the speaker: hand gestures, facial expressions, and the ability for the speaker to answer questions as they are asked.

Presentations can be saved as *web pages* (HTMLs). If you have an internal website, this can be uploaded to it. But you, as the speaker, will not be there, so having good notes on each slide is important.

A *webinar* is basically a web version of a presentation, with someone talking behind the slides as they are being shown. The design of slides becomes much more important, as people cannot see the speaker. So the slides need to carry more of the load, and be more interesting.

An *ePresentation* is a Power Point presentation (PPT) that has had narration added it, custom animation applied, and been converted to a YouTube-like format. It basically plays like a movie. In this format, keeping what is on the screen moving is important. (We are a TV generation after all.)

And finally there is *eLearning*, which is basically an ePresentation with some level of interactivity build into it. Once considered the second coming of training, I now see this as just another tool in the toolbox.

Safety presentations tend to fall into one of three main types: topic, task, and story.

A *topic* presentation is the primary kind that most of us are dealing with at work. Many of these are rooted in one regulatory requirement or another. These include most of the OSHA, EPA and DOT training topics. You are basically trying to pass on information about something. (And hoping they'll remember it all!)

Typical training of this type includes: confined space, hot work, hazard communication, fall protection, machine guarding, industrial hygiene, lockout/tagout (LOTO), hazardous waste, heat stress, bloodborne pathogens, back safety, drug and alcohol programs, and physical properties of chemicals.

Presentations of the *task* type will also have some level of hands-on activity associated with them. This type of presentation will be rooted in some direct aspect of the employee's job. In the case of respiratory protection, there are: the fit test, positive and negative flow tests, and a piece of PPE that they use in the plant. For forklift safety, for example, employees will be asked to do an inspection of a forklift and to conduct defined tasks with the forklift to show they can handle one. Electrical safety, NFPA-70E, arc flash, safe use of portable power tools are all examples of job-specific training. Operational training is often of this type, the classic "show me" presentation.

Another type of a task-based presentation is when it used as a visual SOP. For example, I have a presentation on how to do a proper plant restart after a power failure (a not uncommon occurrence in my neck of the woods). In several slides, I show the proper valve settings for a piece of equipment, allowing employees to visually verify the correct arrangement. A copy of this presentation is in the control room and shipping office at my plant, just in case. (After all, with no power, there will be no computers or internet: Hard copy to the rescue).

A *story* type is one that draws upon life experiences, personal beliefs or interests. Since I grew up in Buffalo, NY, I know cold, snow and winter well, perhaps better than I would have liked to have known it. But in either case, I give a cold stress class one year, and a winter safety on the next, based on my experience.

Many off-the-job safety presentations will be like a story. Examples include fire or electrical safety in the home, bugs and bunnies (insects, etc.), Halloween safety, and mower safety. In effect, most of my off-the-job safety topics are my attempt to sell safety as a 24/7 idea. To do this effectively, I should relate work-based safety ideas to day-to-day activities associated with home and play.

TOPIC:

- Least Interesting
- Benefits from plant-specific photos
- Important to change these up every time they are given

I have about 18 topics that the regulations say that I have to cover every year. So, keeping these presentations fresh can become a job into itself.

TASK:

- More interesting due to the hands-on component
- Harder to create due to specific imagery needed
- Site-specific images critical

I'm a decent photographer, but getting good enough lighting inside any plant building can be rough. But not withstanding that, I've gathered a good collection of site-specific photos over the years for use in my plant-specific presentations.

STORY:

- Most interesting
- Lends itself to strong imagery
- More practice than normal needed

I find this often harder to present effectively due to the subject matter. I'll practice three to five times more for this type of presentation. Your timing of your talk with the slides becomes more critical.

Another thing that happens when you do a story type is that any lack of flow or choppiness in presentation will become more obvious.

Ok, so what? Well, it's a matter of perspective, eh. In each year, I'll give 25 to 35 training presentations at my site; about 50% of them are topic, 25% task, and 25% story.

Realizing that there are three kinds of presentations can help as one builds one type or the other. The resources used will change, and the flow and structure of presentation will be different. But my general goal is to make all presentations seem like a story. I go this route, as this tends to be more enjoyable to the audience. The more they like it, the better they'll remember it. And to further complicate things, it is not usual for a presentation to be a blend of all three types. Well, you didn't think it would all be easy, right? Goals are good, eh. (I did mention that I grew up in Buffalo, NY, so just a bit of Canadian influence there).

Do's and Don'ts for Presentations

Borrowing from a famous late-night routine, let's cover the top 10 things to remember about presentation design and delivery.

10. People can read a slide faster than you can speak it. I talk fast, maybe 150 words a minute, but even the average reader is at 250 to 400 words a minute. This means your audience has read a slide before you have gotten part way through it. In fact, the words and image on the slide should support, but not be the same as what you are saying. Or to put another way: Never read a slide to the audience!

9. The bigger the room, with more people, the larger the font with less words on the slide! I suggest a minimum 32 point for text, and 44 point for headers. For myself, I tend to use 48-point headers and 40-point text.

8. Remember that color wheel from your elementary school days? Guess what, it's made a comeback! It is important that the right contrast be used, and that colors complement each other. With the classic color wheel, any two colors opposite each other will have sufficient contrast, but that doesn't mean you want to put that combination on a slide. While red and green go together for Christmas and have great contrast, they do not belong on the same slide. In fact, avoid red as a background or for lettering, as it generally associated with danger. Of course, if you're doing a course on hazard recognition, then some red may be involved.

7. In earlier days with overheads, we used black or very dark backgrounds with white or yellow lettering. Today, we have training rooms with digital projectors, and in these cases, a light or earth color for the background is preferred. Of course notwithstanding all that, if you have a corporate standard, well then you have a corporate standard, don't ya? I tend to use contrasts in the 4.5 to 1 range (http://trace.wisc.edu/contrast-ratio-examples/ColorSets_6x6x6_OnWhite.htm).

As you construct your slide, try to keep text at least a ¼" to ½" from the edge of the slide; this will allow for the variations in computers, printers and projectors that you are likely to run into.

6. Animation in the right place at the right time in a presentation can work; just avoid any that come with Windows or anything that would get your kids to giggling. Everyone remembers the flying toaster screensaver from early Windows. That was cute for about 30 seconds, and then it got old real fast. Same is true for most cartoon-like animations, only faster.

5. Just because I can put the entire Declaration of Independence into a slide, doesn't mean I should. Power Point is not Word; it is not a book or magazine. Think of it as a billboard. The presentation is meant to be a visual representation of the idea or concept you are conveying to the audience via your spoken word.

4. When using images, one wants to be careful to use a high enough resolution picture. As you make a picture bigger, pixilation can start to be a problem. But good imagery can go a long way, but only if the imagery works with your presentation, and what you were talking about. Always try to get the highest resolution image that you can. Generally, the higher the resolution, the sharper the image.

3. *Remember: Rules, not guidelines. I'm from NJ; don't make me come to your house, and pull a Soprano on you.* ☺

2. Use Arial as opposed to Times New Roman. Times New Roman (Serif) has little feet, while Arial (San Serif) has none. Arial is more informal, while Times New Roman is formal. It works well for the *NY Times*, as they use 2-inch columns with really tiny type. It helps your eye to keep on the same line. On a slide with fonts so large, the feet become more of a visual distraction.

1. A well-constructed presentation will help an average speaker, but it takes a very good speaker to overcome a poorly designed one. Why put yourself in the hole at the start? Remember, presentations work dirt cheap. Slides are free, but you can always make more. Use more if needed to break down a complex slide.

Miscellaneous: Avoid all CAPS; it is harder to read, and is akin to shouting, just like in e-mails. A fair percentage of the male population in the Americas is red/green colorblind, about 7 to 10%. (Another good reason not to use red and green on the same slide.) And think about some kind of mental break or diversion every 10 or 15 minutes or so.

Basic Process for Presentation Design

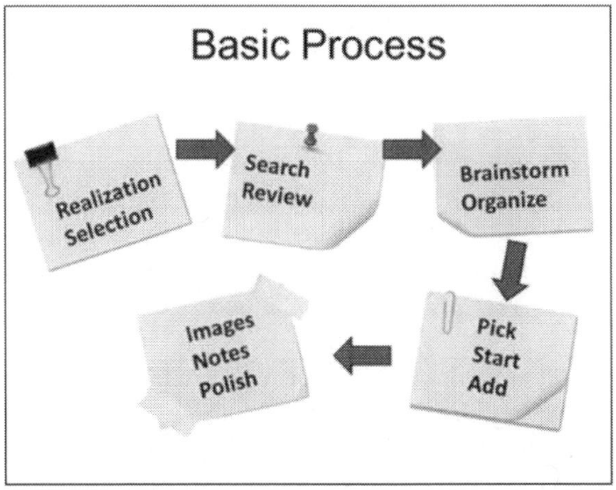

Well, I'm a Chemical Engineer by degree so of course I have a PROCESS! ☺

Realization: Oh no, a training session is needed! To be fair, that's what it was like when I first got to my plant. However, now I lay down the training schedule for the next year in December of the prior year, and put it up on our internal website. When I set up my schedule, I take into account, whether a new or existing presentation will be used to give myself time. No point in setting up a training session with three new presentations, when I can spread those out.

Selection: Pick a focus for the presentation. Sometimes the hardest thing to do is make a choice. If it is standard topic like hot work, or confined space, it's no big deal, but when you get to off-job safety topics, the volume of possibilities can be a bit overwhelming. I tend not to dwell on this; I just pick something and run with it, like winter safety or home-fire safety.

Search: Beg and borrow as needed from the internet. For just about every topic, there are lots and lots of presentations via searches through Google or Yahoo. Most states and universities have safety training available for download. Get it all; good ideas can come from almost anywhere. Only make sure you acknowledge the sources where appropriate.

Review: After you have searched your files and the internet, and you've gotten everything you could lay your hands on, it is time to review your bounty. For example, I found 14 or so presentations for winter safety. I open, review, and close each of these in 30 seconds or so. I do this to basically prime my brain for the next step.

Brainstorm: Time to brainstorm, jot down ideas/concepts, with pen and paper. Yeah really, I actually use a pen and paper for this. Get into a comfortable chair and relax. Write down everything, no matter how silly it may seem to you.

Organize: Then we come to organizing the ideas into subgroups or an outline. I'll use Excel or Word as needed, whichever is easier. For my winter safety training, I ended up with four categories: weather, activities, automobile, and the holidays. This is also the point at which I begin to think about a plot per se. Basically, do I have a beginning, middle and end. "*Beginning: Why; Middle: The Meat; End: Summary, Why Again.*" This was taken from one of son's first grade homework assignments, but I thought it gets the point across.

Pick: Time to select a template (use existing presentation if a good one found). I used to use a group of five or templates, but these days, I actively search for a template that may fit the nature of the presentation, such as one with a wintery background for winter safety. Just like searching the internet for presentations, you can search for templates. If I don't find a template I like, I will default to my basic all white background, no bullets, header-only template.

Start: Construct the presentation with one idea/concept per slide. I suggest that the title block reflect the main idea clearly. Keep in mind you can also combine later if you need to.

Add: Add info to each slide, different from notes or narration. Basically, add whatever ideas or concepts as related to the topic header. You can always remove some of the text later.

Images. Then add the images; use as appropriate. This is also the time at which, if I am thinking of making this a part of an ePresentation or eLearning system, that I give thought to multiple pictures or objects on a slide. Generally, I seem to have better luck with getting images on Google, then Yahoo. Yahoo tends to give you a lot of Flicker images, but not all of those can be copied to your presentation.

NOTE: Always, always, always, cut and paste the link for an image to the notes section of slide. You can construct a photo credit page if needed when the presentation is done then.

I really prefer to use a digital photo that I take at my site over internet pictures. It is easy to get site-specific photos for some topics, harder to do for others. But you have to be careful with the camera you use. Today, you can get cameras for \$50 to \$200, with 8 to 12 megapixels per picture and lots more memory. However, my old camera still takes a far better picture due to the lens construction. The distinction becomes more apparent if your presentation is to be converted to a movie or Flash (i.e., YouTube). The better the picture at the start, the better it will convert later on if needed.

When I decide to use a picture, my pecking order is site-specific photos over web-based photos over illustrations over cartoons over clip art.

Notes: The notes can be used for narration and later reading by others. Yeah, I know no one likes doing the notes. But what I discovered is that, as I write down what I'd like to say, my slides change. I realize that some things are better for me to talk about, rather than to show, and vice versa. Between the images added and notes written, you will discover the words on the slide should drop considerably.

Polish. Once the PPT is functionally done, set it aside for a week, and come back to it. Decide if the PPT flows properly, maybe put in a few slides with a question for the audience. Work in audience interaction in the presentation, almost any of that is a good thing.

You can put in a picture with something that is wrong relative to the training you're giving and query the audience on it.

Just the act of giving your presentation out loud will likely lead you to make subtle changes to it. And if all else fails, give your presentation to any chair you have in house. I often use my young son for this. He often asks very good questions. If you have other trainers that you know, ask if they would be willing to peer review your work.

So you've done everything to the presentation that you wanted to, then it is time to "test" it. I shoot for 10 to 60 seconds per slide.

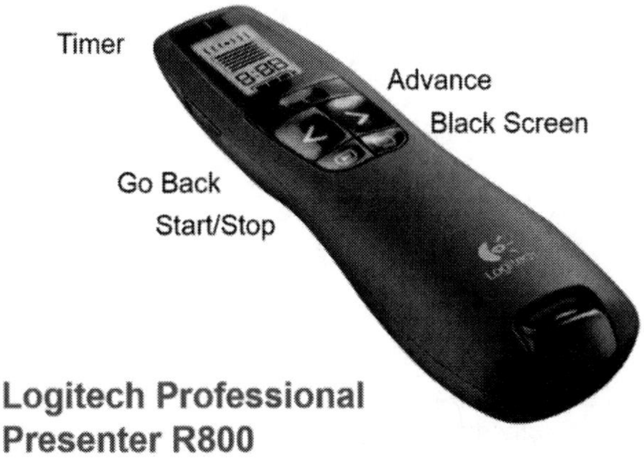

A remote control presenter is something I've come to appreciate having. No longer am I tied to the computer or podium when doing a presentation. What I like on it is the timer, which vibrates when I reach 5 or 10 mins. to go. The Logitech, shown here, is just one example of such a device.

Now let's consider the three versions of the Einstein slides:

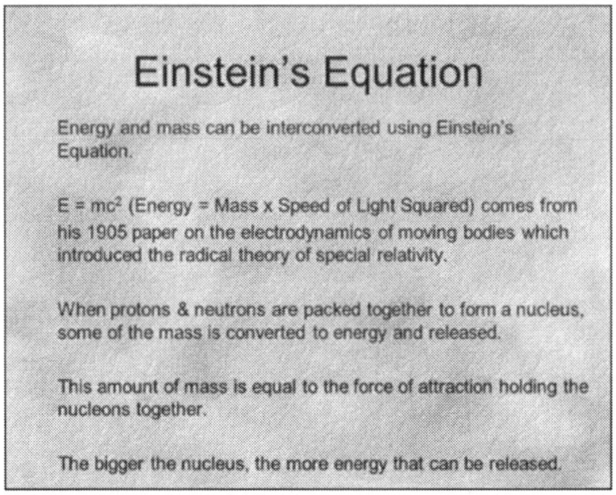

Version 1: This slide is a word document. If I were print out it out, leave it on a table, and go home, you would know 80-90% of what I had intended to say. Basically, the presenter was not needed.

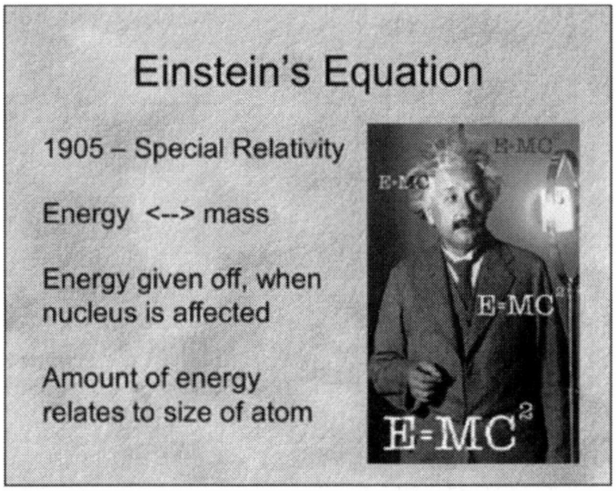

Version 2: This version of the slide is a classic Power Point. If I were print it out, leave it on a table, and go home, you would know 50-60% of what I had intended to say. Basically, the presenter might be needed. Also, consider the words on this typical slide; while fewer than Version 1, you have to ask yourself: Are these words for the audience or crib notes for the presenter?

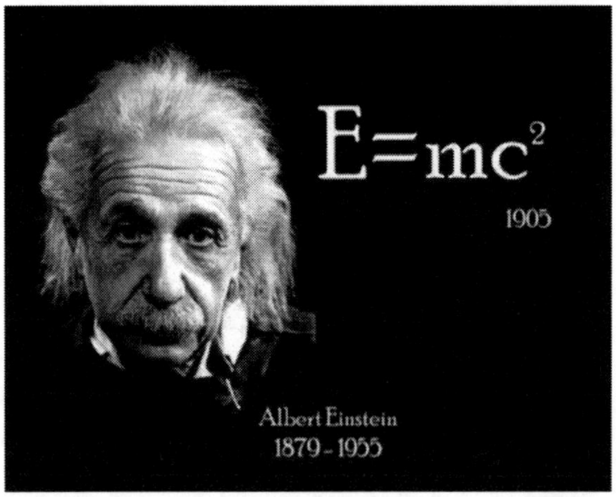

Version 3. This version of the slide is a presentation. If I were print it out, leave it on a table and go home, you would know 10-20% of what I had intended to say. Basically, the presenter is needed.

And that is a key point: The presentation is meant to provide a visual key/anchor to help facilitate your message getting into people's brain and staying there. *Remember: Rules, Not Guidelines. I'm from NJ, don't make me come to your house and pull a Soprano on you.* ☺

Presentation Design Theory

Why presentation design, do's and don'ts, and the process? Well this is why: cognitive load theory, developed by Professor Sweller in 1988. (Well, at least of one of the whys.)

This is the theory behind why the amount and balance of text, color, contrasts, imagery, sound, movies, and animation are important.

There are two sides to your brain; the left side (math, facts, data), and the right side (music, art, emotions). During any presentation, you can hear me (right side) or you read the slide (left side). But if I try to make you do both, your brain goes into "panic" mode, and you miss part of the presentation.

This theory is straightforward; if you want to be effective when doing training, keep in mind the human brain can only take in so much at a time.

You have stuff trying to get into your brain (working memory), and the stuff that has gotten into your brain (long-term memory). The barriers or "filters" on your mind are what I want to get through. If I can do that, I'm golden.

Every person has filters; these filters, in effect, determine what gets into your brain easily, and what has to really work to get in. Everyone has different filters: where you grew up, went to school, got a degree, had kids, who you work with, etc.; all affect your filters. Now imagine you are giving a presentation to room full of people all with different filters: How does one get through to everyone? Effective use of images and cognitive load theory is how.

So, cognitive load theory, as it applies to presentation design, is suggesting to us to limit the info per slide, use images, and give time for the audience to absorb the slide.

Retention

Who knew you were getting advance learning theory when you were two years old?

The cone of learning, developed by Dale Eggar in 1969, tells us that in two weeks, your audience will forget 50% of your presentation. But you got to love the Confucius quote from about 2500 years ago, "I see and I forget; I hear and remember; I do and I understand." This quote seems to say basically the same thing, eh. So what are some of the things that can be done to help improve retention?

Giving proficiency tests after you have presented training can be effective way to gauge whether they understood, or if there is a gap in your training. However, creating tests can be time consuming, or you may think the people you are training will revolt at the idea of a test. I created an access database and threw about 4000 questions into it across 60 categories to make my life easier. I pick a topic, then how many questions, and how many different versions I want, and click print. I sometimes give a test right after the training, or occasionally months later.

I found a good template for *Jeopardy* on the internet, and adapted it to my needs. I did one after NFPA 70E arc flash training with the employees. Basically, I split the room into two groups, and alternated between one side and another. This version had double and final jeopardies in it. Of course, after doing it, I realized, that I would have been better served to have given them the question and asked for the answer, rather than the other way around. It turns out not everyone watches *Jeopardy*.

Near-miss/unsafe condition reporting is another important element. I've used this to focus on a recent training session like lockout/tagout. I'll grab two or three employees, take some near-miss cards, and look specifically for near misses on lockout/tagout. Also, I'll make use of the good/positive check box whenever I see it being done right.

Contests can be fun. I did one for hazard recognition that used five photos with multiple hazards in them. Everyone got five minutes per photo to write down as many things as they could identify, with the best set of answers getting a certificate to a local restaurant.

Why Presentation Design?

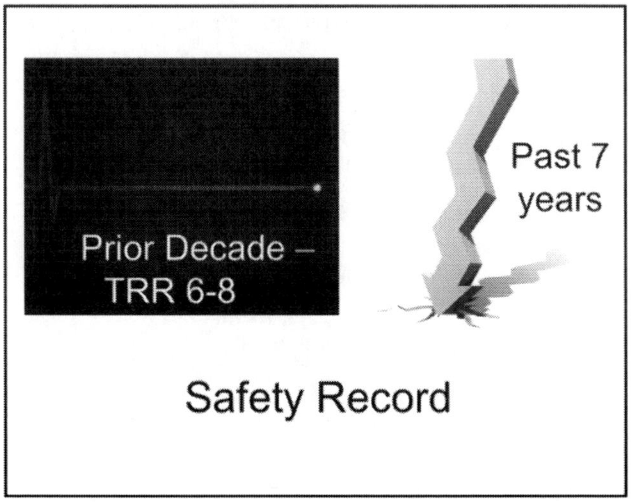

It has worked for me. After all, getting people home in the same general shape as they came to work is the goal. Good training is one of the big ways you can get there.

In the end, a well-designed presentation will be remembered. Better retention on safety training leads to a safer workplace. Presentations will be more fun to sit through. Once you get in this mode of the presentation design, it ends up taking less time to get to a quality presentation. And of course, since you're the one creating it, it hardly costs anything, except your time.

If you have done it right, your audience will never appreciate the time you put into it. Oh, you'll hear if it is bad, but you are not likely to get any compliments if is good. Such is the life of safety professional.

Now, being from Buffalo, NY, originally, I know many of the bands in Canada. One of the more famous is the Barenaked Ladies. They did a remake of a song back in 1992 called "Love in a Dangerous Time." I remember this song for a line that is in it, that in my mind uniquely states what training is all about. "Kick at the darkness, until it bleeds daylight." That's right, if you don't get through to them the first time, try something else.

So, I have two questions for you: How many slides did I use? In general, the average answer for the number of slides is 50; that is about one per minute. And 50 is considered a big (long) presentation and, therefore by default, boring. But it doesn't have to be; that is the mindset we have to get past. I used way more than 50 for this session, and it didn't feel like that at all. In fact, I strongly suspect had I mentioned the actual number of slides at the start, most of the audience would have exited the room at that point.

And how many slides had bullet points on them, not counting the slide that actually had a picture of a bullet on it? Of course, I used no bullets. Just because Microsoft gives them to us in the master slide doesn't mean you have to use them. I create a master slide, with no bullets, white background, and only a title block. This is what I start with. Using bullets tends to cause you to put too much information on a given slide. So try to avoid bullets if you can.

So keep this in mind: We can save the world, one presentation at a time!

Bibliography

Atkinson, C. 2005. *Beyond Bullet Points.* 1st ed. Redmond, WA: Microsoft Press.

Duarte, N. 2008. *Slide:ology.* 1st ed. North Sebastopol, CA: O'Reilly Media.

Moreno, R. & Mayer, R. 1999. "Cognitive principles of multimedia learning: The role of modality and contiguity." *Journal of Educational Psychology,* 91: 358-368.

Reynolds, G. 2008. *Presentation Zen.* 1st ed. Berkeley, CA: New Riders.

Sweller, J. 1988. "Cognitive load during problem solving: Effects on learning." *Cognitive Science* 12, 257-285.

Links and Other Useful References

Duarte, N. 2010. *Resonate.* 1st ed. Hoboken, NJ: Wiley.

Medina, J. 2008. *Brain Rules.* 1st ed. Seattle, WA: Pear Press.

Reynolds, G. 2010. *Presentation Zen–Design.* 1st ed. Berkeley, CA: New Riders.

Williams, R. 2010. *The Non-Designer's Presentation Book.* 1st ed. Berkeley, CA: Peachpit Press.

American Society of Safety Engineers

Professional Development Conference June 19-22, 2017

SAFETY 2017 ASSE

Session: 596

Fundamentals of SH&E: Fire Protection

Stephen J. Musur, CSP, CFPS

Chubb Group of Insurance Companies

Fundamentals of SH&E

SAFETY 2017 ASSE

This presentation is for informational purposes only and has an edition date of June 2017. This presentation is necessarily general in content and intended to provide an overview of certain aspects of fire science. No liability is assumed by reason of the information this document contains.

The views, information and content expressed herein are those of the authors and do not necessarily represent the views of Chubb Insurance. Chubb did not participate in and takes no position on the nature, quality or accuracy of such content. The information provided should not be relied on as legal advice or a definitive statement of the law in any jurisdiction. For such advice, an applicant, insured, listener or reader should consult their own legal counsel.

2

Agenda

SAFETY 2017 ASSE

Introduction

- Science of Fire
- Fire Controls
- Fire Protection
  - Detection / Suppression
- Testing - Maintenance
- Questions

3

What is Fire?

SAFETY 2017 ASSE

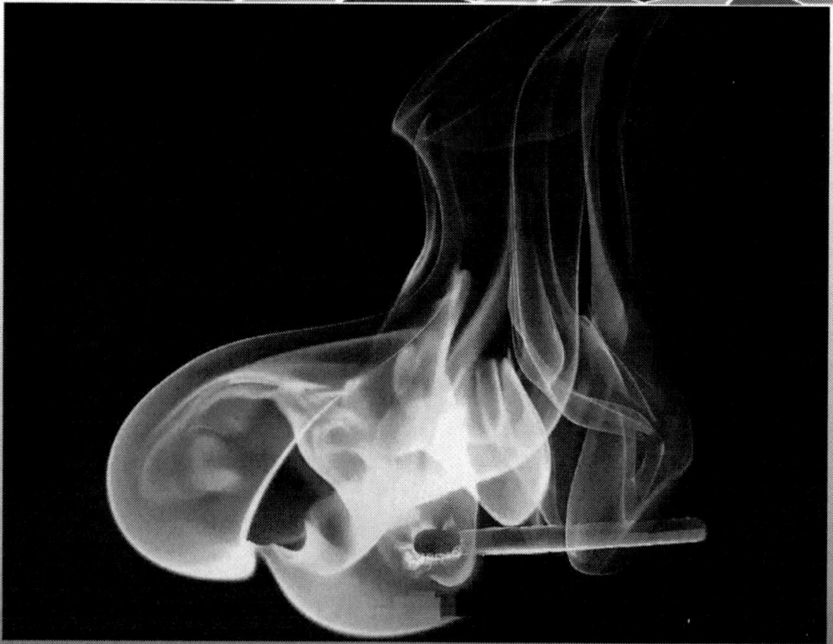

4

Classic Definition

SAFETY 2017 ASSE

Fire is rapid oxidation with the evolution of heat and light

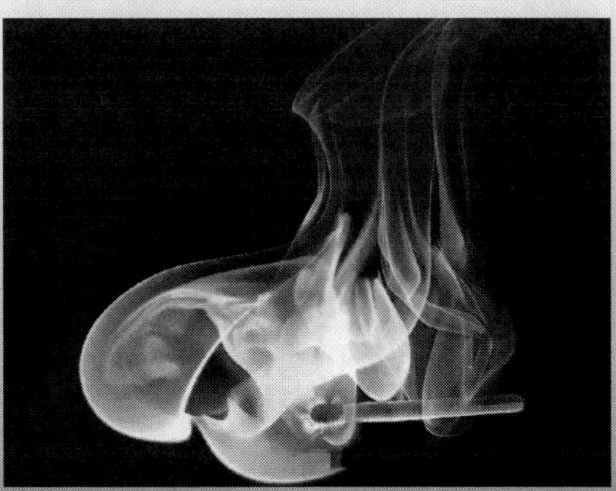

5

The Fire Triangle

SAFETY 2017 ASSE

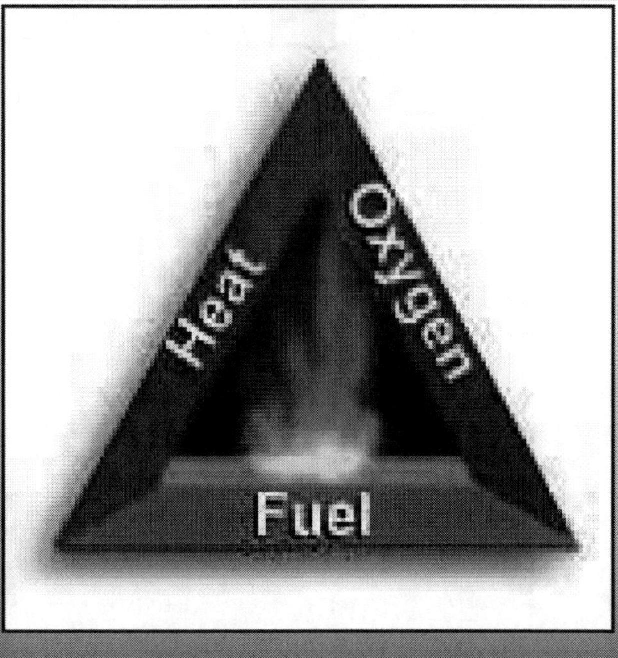

6

Triangle? Not a Tetrahedron?

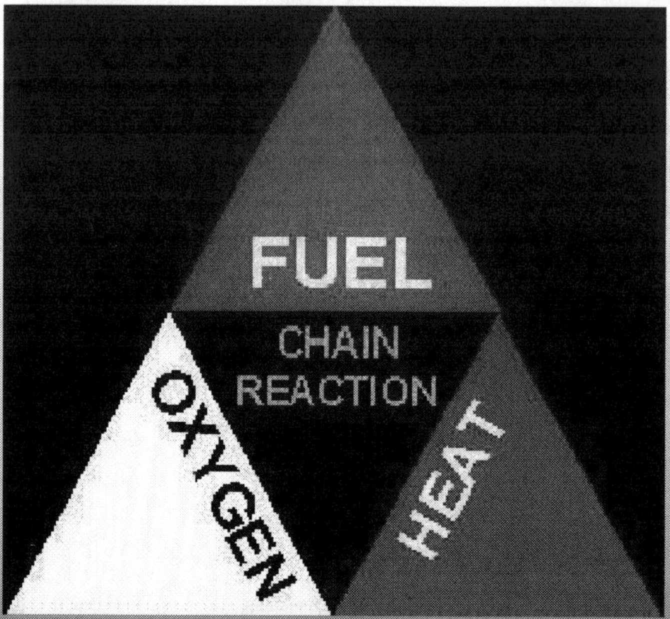

7

What's Burning?

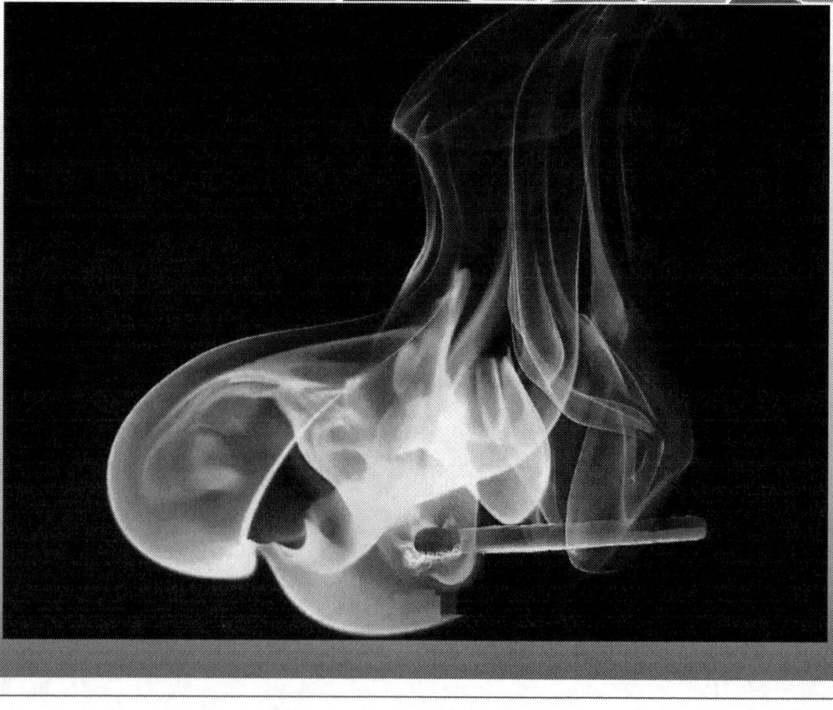

8

Pyrolysis

SAFETY 2017 ASSE

- Some of the heat is lost (convective)
- Some of the heat goes back into the system (conductive)
- Heat produces vapors
- Vapors ignite and propagate

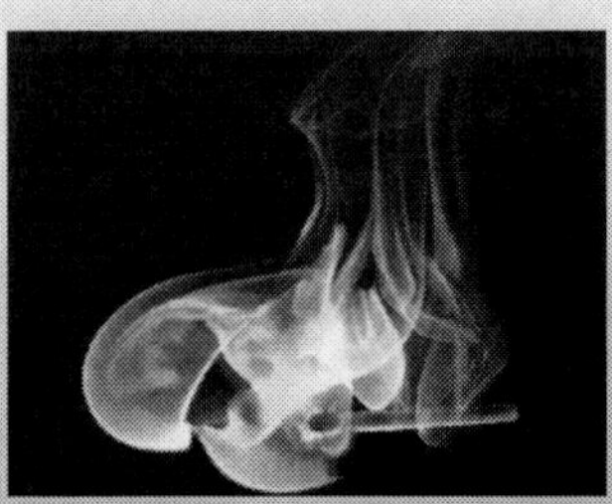

9

Vapors

SAFETY 2017 ASSE

Flammable Range

LFL – UFL, LEL – UEL

- Lean
- Rich

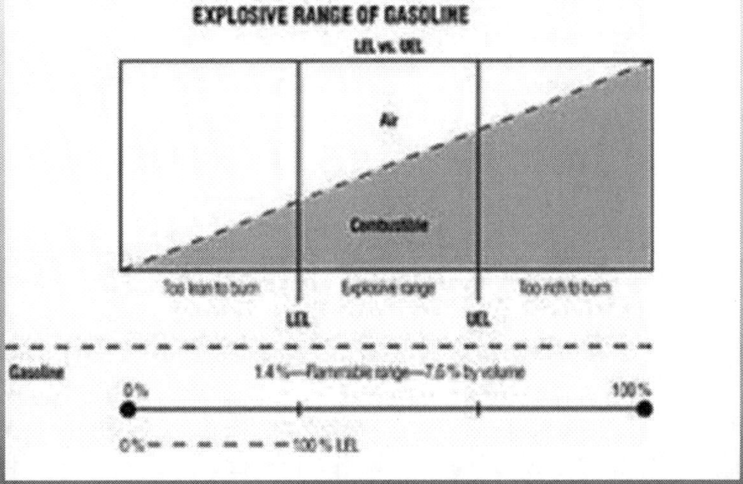

10

So Far

SAFETY 2017 ASSE

- Defined Fire
- Know what is burning
- Concept of Flammability Range

11

Control

SAFETY 2017 ASSE

- Remove one or more legs of the triangle
  - Omit the Fuel
  - Inert the Atmosphere
  - Cool the Reaction
  - Interfere with the combustion process

12

Applied Controls – Fuel

SAFETY 2017 ASSE

- Segregate fuel from processes
- Minimize the amount of fuel
- Use less combustible materials
- Housekeeping - Dusts

13

Applied Controls – Oxygen

SAFETY 2017 ASSE

- Can the process operate rich?
- Can the process operate lean?
- Inert the operation – N_2, CO_2

14

Applied Controls – Heat

SAFETY 2017 ASSE

Exothermic processes

- Frictional heat
- Chemical heat
- Sparks - Electrical

15

Applied Controls – Heat

SAFETY 2017 ASSE

- Sparks - Welding, Hot Work
- Grinding
- Open Flames
- Lightning
- Smoking

16

Applied Controls – Chain Reaction

SAFETY 2017 ASSE

- Less Hazardous Materials
  - Water Soluble vs. Oil Based
    - Paints, Cutting Oils, Lubricants, Inks Etc.
- Fire Resistive Materials
  - Phenols, PVC's vs. Polyethylene, Styrene
  - Inerting Fillers for Plastic
  - Intumescents

17

Applied Controls – Management

SAFETY 2017 ASSE

Written Procedures

- Air Sampling
- Hot Work
- Self Inspections
- Outside Contractors

18

So Far

- Principles of Extinguishment
- Know what is burning
- Concept of Flammability Range
- Applied Controls

19

System Controls – Devices

- Fire Detection
- Fire Suppression

20

System Controls – Devices

Fire Detection

- Smoke
  - Ionization, Photoelectric
- Heat
  - Restorable, Rate of Rise
- Beam
  - Obscuration
- Flame
- IR

21

System Controls – Devices

Four Stages of a Fire

- Incipient
  - Microseconds to days
- Smoldering
  - Microseconds to hours
- Flaming
- High Heat

22

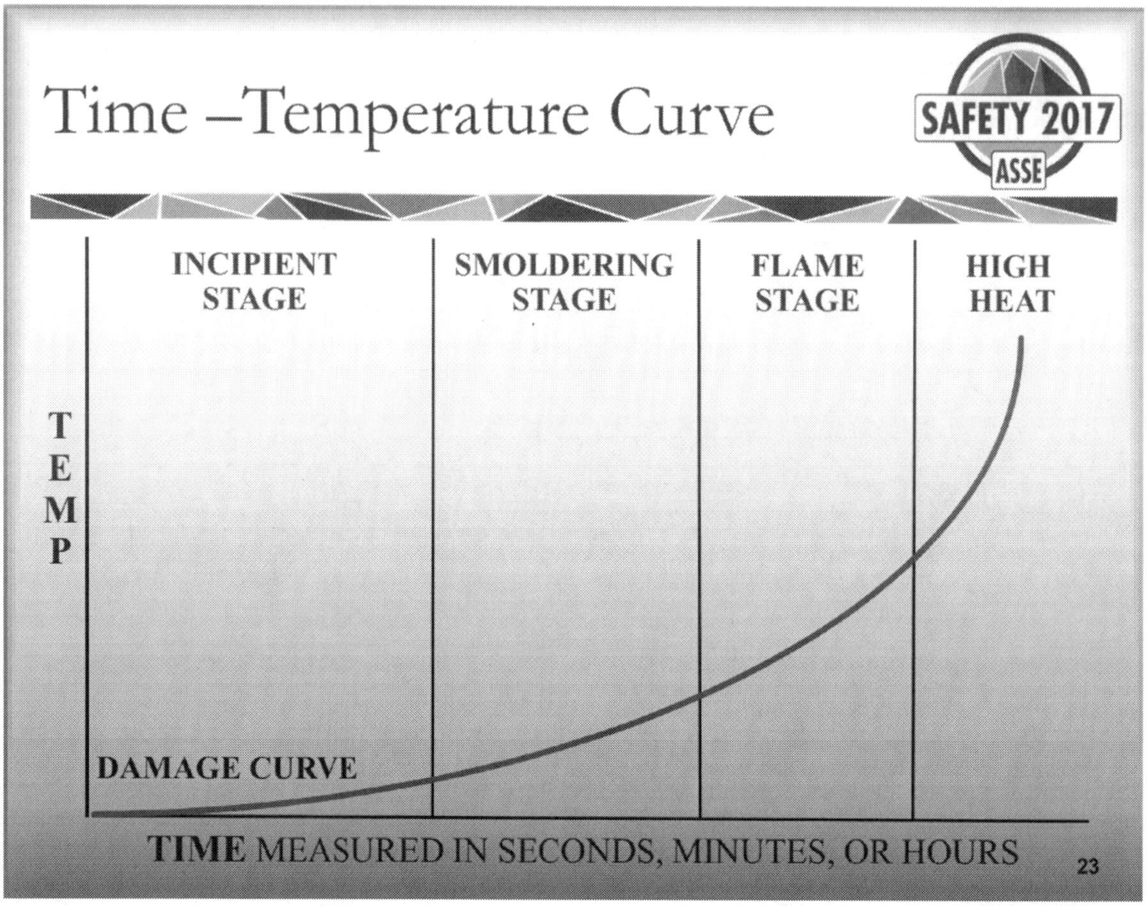
Time –Temperature Curve
SAFETY 2017
ASSE
INCIPIENT STAGE
SMOLDERING STAGE
FLAME STAGE
HIGH HEAT
TEMP
DAMAGE CURVE
TIME MEASURED IN SECONDS, MINUTES, OR HOURS
23

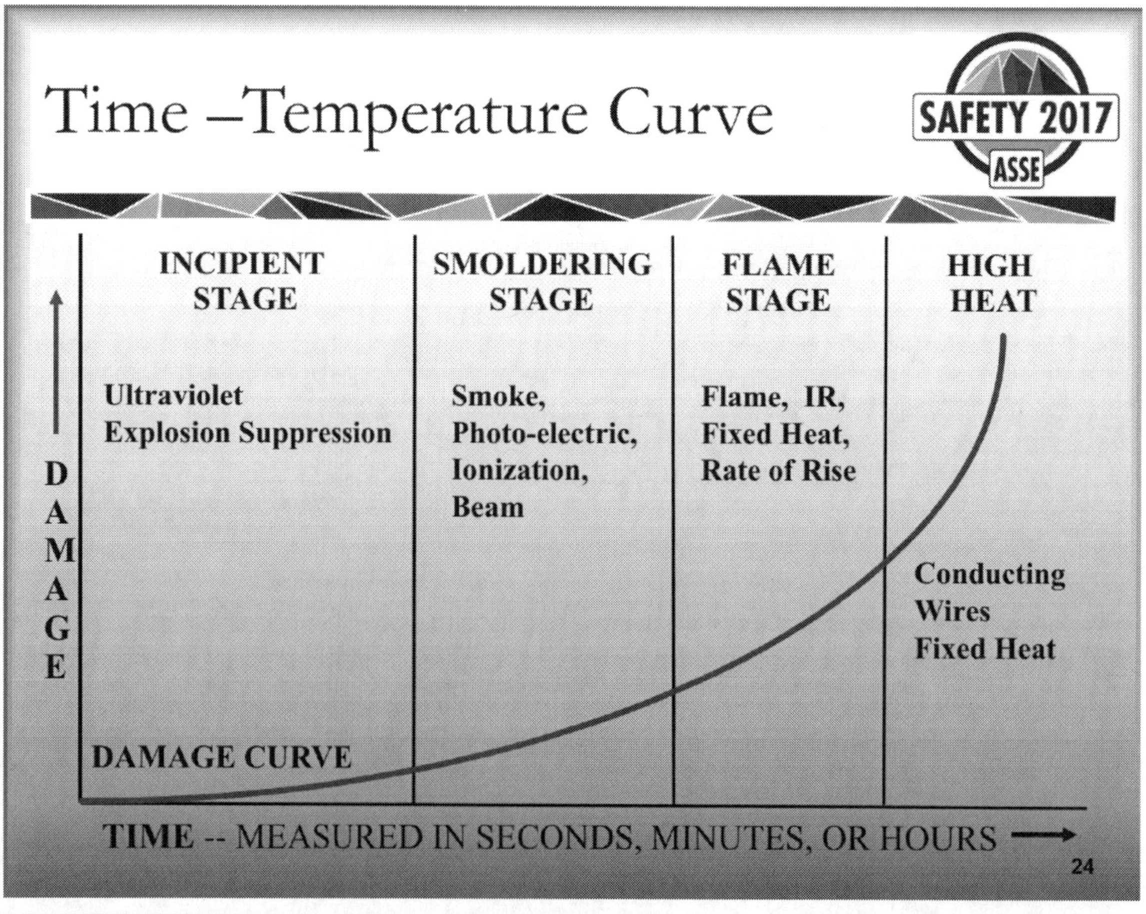
Time –Temperature Curve
SAFETY 2017
ASSE
INCIPIENT STAGE
SMOLDERING STAGE
FLAME STAGE
HIGH HEAT
Ultraviolet
Explosion Suppression
Smoke,
Photo-electric,
Ionization,
Beam
Flame, IR,
Fixed Heat,
Rate of Rise
Conducting
Wires
Fixed Heat
DAMAGE
DAMAGE CURVE
TIME -- MEASURED IN SECONDS, MINUTES, OR HOURS
24

So Far

SAFETY 2017 ASSE

- Principles of Extinguishment
- Know what is burning
- Concept of Flammability Range
- Applied Controls
- Stages of a Fire
- Fire Detection

25

System Controls – Fixed Suppression

SAFETY 2017 ASSE

Sprinkler Systems

- Wet
- Dry
- Anti-freeze
- Deluge
- Pre-action
- Combined Dry/Pre-action
- Cycling On-Off
- Ref: NFPA 13

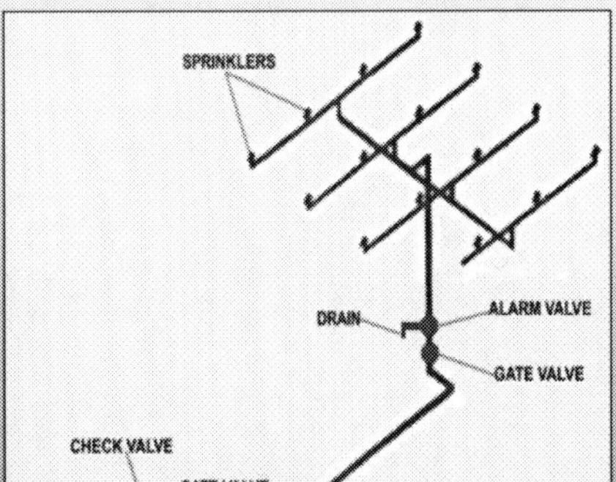

26

System Controls – Fixed Suppression

SAFETY 2017 ASSE

Sprinkler Systems

- Wet
  - Most Common
  - Water in the pipes
  - Very efficient
  - Requires Heat

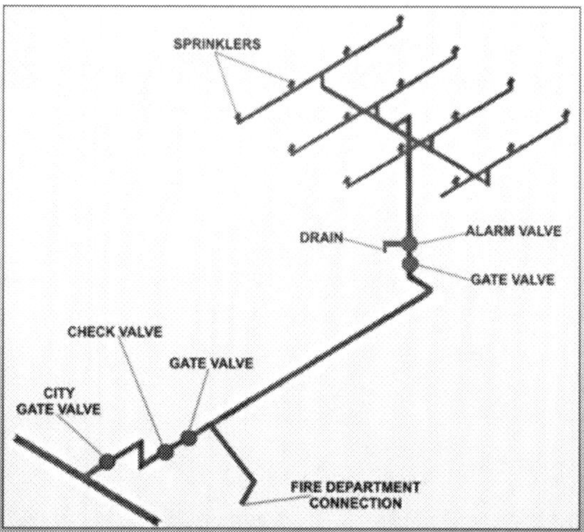

27

System Controls – Fixed Suppression

SAFETY 2017 ASSE

Sprinkler Systems

- Dry
  - Compressed air in pipes
  - Needs more devices
  - Used in areas subject to freezing
  - Limited in size

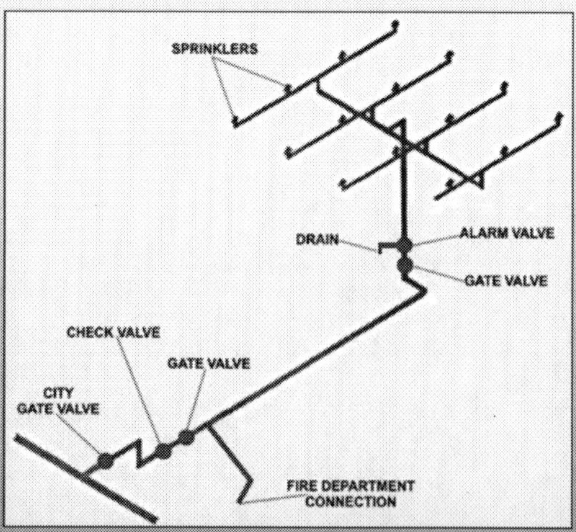

28

System Controls – Fixed Suppression

Sprinkler Systems

- Anti-Freeze
  - Filled with a glycol solution
  - Limited in size
  - Some applications to storage occupancies

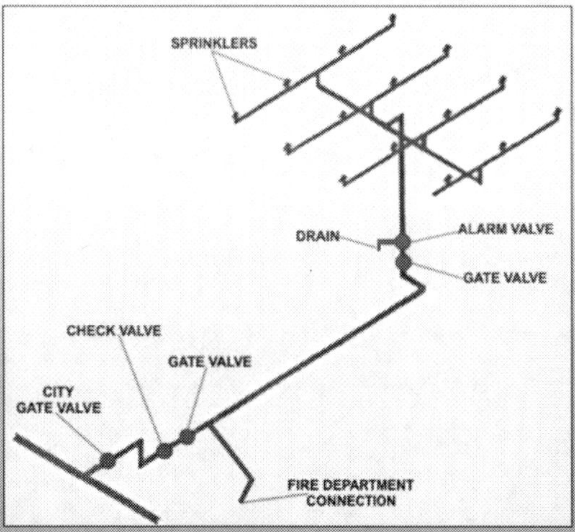

29

System Controls – Fixed Suppression

Sprinkler Systems

- Deluge
  - Nothing in pipes
  - Sprinklers are open
  - Used in High Hazard areas / processes
  - Requires an actuation system

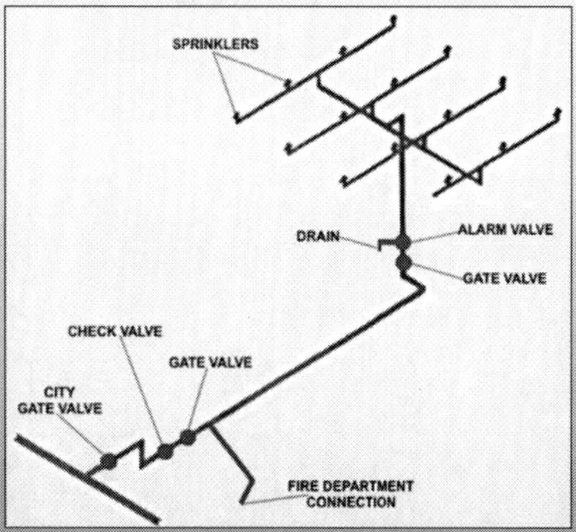

30

System Controls – Fixed Suppression

Sprinkler Systems

- Pre-Action
  - Pipes filled with a compressed air
  - Requires an actuation system
  - Minimizes water damage

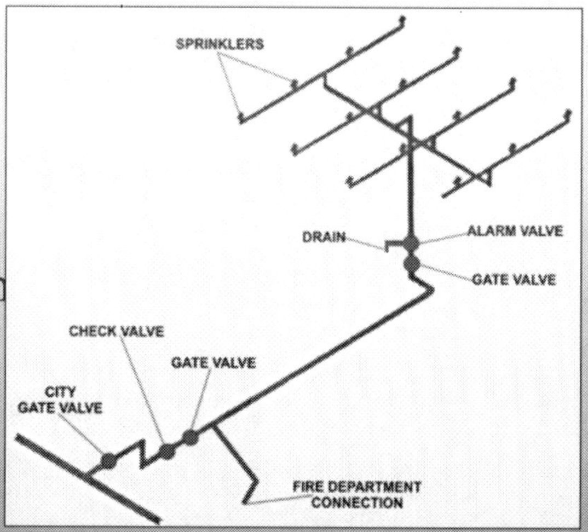

31

System Controls – Fixed Suppression

Terminology

- Pipe Schedule System
  - Pre – 1972
  - Pipes Sized per a schedule
  - Pipes Sized based on Occupancy
    - Light, Ordinary Hazard, Extra Hazard
- Hydraulically Calculated Systems
  - Pipes sized on friction loss
  - Loops and Grids MUST be Calculated
  - Risers Clearly Placarded with Design Info

32

System Controls – Fixed Suppression

Terminology

- Branch lines
- Crossmains
- Feedmains
- Risers
- Sprinklers

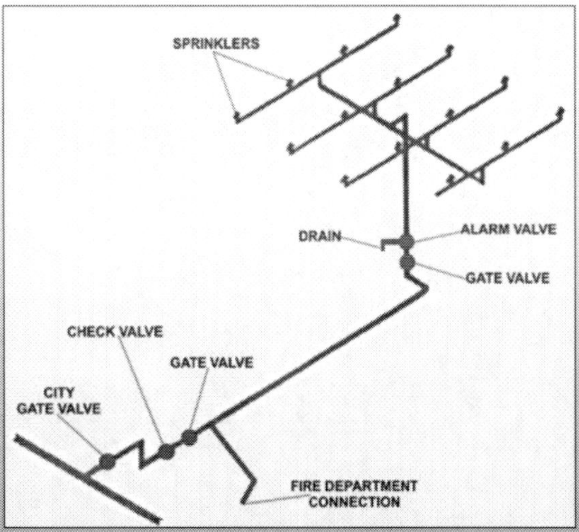

33

System Controls – Fixed Suppression

Sprinkler Systems

- Tree Systems
- Looped Systems
- Gridded Systems

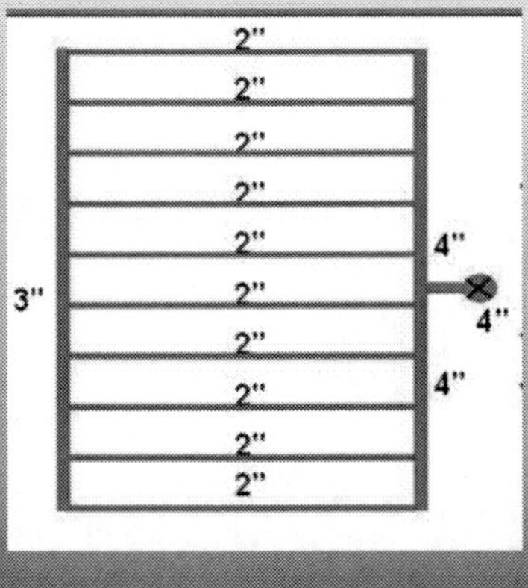

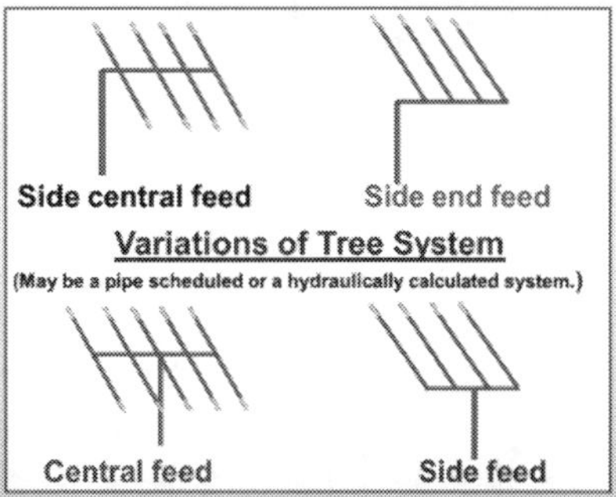

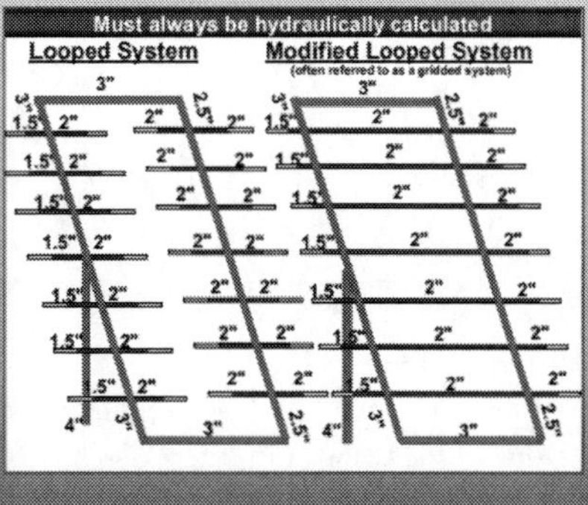

34

So Far

SAFETY 2017
ASSE

- Principles of Extinguishment
- Know what is burning
- Concept of Flammability Range
- Applied Controls
- Stages of a Fire
- Fire Detection
- Sprinkler Systems

35

System Controls – Fixed Suppression

SAFETY 2017
ASSE

Sprinklers

- Upright
- Pendant
- Special Application

36

System Controls – Fixed Suppression

SAFETY 2017 ASSE

Sprinklers

- Upright
- Pendant
- Quick Response
- Fast Response
- Nozzles
- Storage
- Special Application

37

System Controls – Fixed Suppression

SAFETY 2017 ASSE

Sprinklers

- Good for 50 years
  - Must be tested
- Orientation
  - Replace like kind
- Obstructions
  - Adequate Clearance

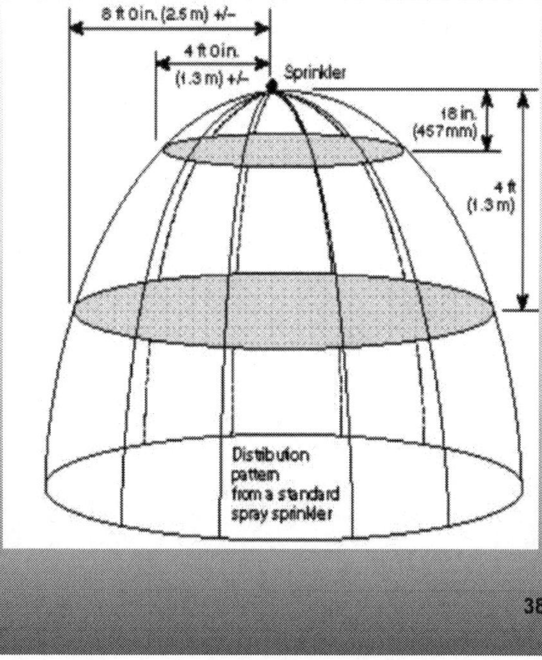

38

So Far

SAFETY 2017 ASSE

- Principles of Extinguishment
- Know what is burning
- Concept of Flammability Range
- Applied Controls
- Stages of a Fire
- Fire Detection
- Sprinkler Systems

39

System Controls – Fixed Suppression

SAFETY 2017 ASSE

Other Systems

- CO2
- Halon 1301 1211
- Clean Agents
  - FM 200
  - Inergen
  - Sapphire
- Dry Chemical
- Liquid Salts
- Explosion Suppression

40

System Controls – Fixed Suppression

SAFETY 2017 ASSE

Other Systems

- Used where water damage is an issue
- Can be used in inhabited areas
- Preferred for Specific Hazards
- Special Maintenance Needs

41

So Far

SAFETY 2017 ASSE

- Principles of Extinguishment
- Know what is burning
- Concept of Flammability Range
- Applied Controls
- Stages of a Fire
- Fire Detection
- Sprinkler Systems
- Special Extinguishing Systems

42

System Controls – First Attack

SAFETY 2017 ASSE

- Hand Held Extinguishers
  - Water
  - Dry Chemical
  - CO2
  - Metal Powders
  - Liquid Salts
- Class Of Fire
  - "A" Paper, Cloth, Wood
  - "B" Oils, Grease
  - "C" Electrical
  - "D" Metal
  - "K" Kitchens

43

System Controls – First Attack

SAFETY 2017 ASSE

- Hand Held Extinguishers
  - Placement
  - Size
  - Correct Extinguisher for Class of Fire
  - Employee Training
  - Maintenance
  - Obsolete Extinguishers

44

Maintenance

SAFETY 2017 ASSE

- Maintenance for Suppression Systems
  - Prescribed by NFPA 25
- Maintenance for Detection Systems
  - Prescribed by NFPA 72
- Fire Extinguishers
  - Prescribed by NFPA 10
- Document the Work / Tests

45

So Far

SAFETY 2017 ASSE

- Principles of Extinguishment
- Know what is burning
- Concept of Flammability Range
- Applied Controls
- Stages of a Fire
- Fire Detection
- Sprinkler Systems
- Special Extinguishing Systems
- Hand Held Extinguishers

WHEW !

46

Questions

SAFETY 2017 ASSE

47

Session No. 602

Essential Leadership Competencies for HSE Professionals

Julius E. Rhodes, SPHR
Founder and Principal
mpr group
Chicago, IL

Introduction

Individuals engaged in the HSE profession have a certain body of knowledge concerning the technical competencies they need in order to execute their roles. However, a critical and often lacking piece is the essential leadership competencies that HSE practitioners need in order to ensure they are seen as effective contributors to the overall business. It is these characteristics and traits that allow HSE professionals to make more positive and definitive contributions to their organizations and be viewed in a manner that lifts all members of the profession. These critical leadership skills must be displayed on a consistent basis so HSE professionals can be seen as business leaders capable of successfully operating throughout the organization in the attainment of its goals.

What Is Leadership

If you were to ask anyone to define leadership you would most likely receive as many definitions about this term as the number of people you asked. Each year there are more than 2000 books written on leadership and while there are many reasons for this, I contend that at its core, like many things in life, there are many ways to solve a problem and no one best way of looking at any concern. Rather to the extent that we understand that different people have different ways of connecting with and executing upon the things that resonate with them, it behooves us to make our approaches at connecting with others flexible to meet differing individual needs. For our purposes, I will simplify leadership as taking action when it is called for that provides the most benefits for the greater good.

Understanding What Makes Leadership Essential

If you work off of the premise that leadership is a function of taking action when it is called for that provides the most benefits for the greater good then the next issue you might encounter is trying to understand what makes leadership essential. When you think about leadership, it is about action and while most of us know what to do (e.g. the responsibilities of our role) we sometimes are unclear about when to execute those roles, especially when it might present a conflict situation with others. This fact is even more complicated when those others are either people we don't have functional authority over (e.g. we do not manage them) or they have functional authority over us such as our managers or managers in another department.

When these issues are encountered it often occurs where delays in executing one's role can mean the difference between missing an opportunity for improvement or outright failure. In business the hardest thing that any of us will have to do is bring truth to power. This means we have to be willing to stand up to those who might be in a position where they manage us or manage others and they have more organizational 'juice' than we do. That being said, it is particularly important for HSE professionals to utilize their voice because our efforts allow, in large measure, for the members of our team to go home in the same safe conditions in which they came to work and no other entity within our organizations can make that claim.

As an HSE professional what you do, when you do it and how you do it is paramount to the effective and efficient operation of your organization and it helps establish the culture and leadership identity of your workplace. If you look at Maslow's Hierarchy of Needs the first thing you see is a need for safety. If our team members do not feel that they are in a safe environment then only a few possibilities exist which are: we will not be able to attract the talent we need, we will not be able to keep the talent we have attracted or we will cease to be an entity as the damage done to individuals and our identity within the area we operate will drive us to extinction. In any of these cases you can clearly see why leadership is essential to what we do, when we do it and how we go about executing our leadership role.

When you look at the essential leadership competencies that I will identify in the remainder of this text, I would like for you to keep in mind that procrastination is often carried out when there are many competing views as to what course of action should be taken. However, leadership does not require the actions of many others but it does require the specific action of a dedicated person with the intent to do the best possible work for the benefit of the greater good. In that case, you become and are that person that if you are committed to leading becomes the conduit through which action takes place that benefits specific situations but also has widespread implications for others within your organization and for you as well. As an HSE practitioner, technical competence is essential but being seen as someone who consistently displays leadership competencies will not only widen your sphere of influence but it will also increase your ability to impact areas outside of the HSE realm.

As you think about leadership in a holistic sense I want you to keep this quote in mind: "Leadership and learning are indispensable to each other". This quote comes from a speech that President John F. Kennedy was to deliver in Dallas on November 22, 1963.

That planned address underscores the need that as HSE practitioners we need to move beyond our comfort zone within our area of expertise and be constantly aware of the fact that as leaders and hopefully lifelong learners we must constantly evolve.

Displaying leadership means that we must always challenge ourselves and that there should never be a time when we are comfortable with the knowledge we have acquired. Each day is an opportunity to get better because if we are not getting better we are getting worse and none of us can afford to get worse. In life we are not measured solely by what we achieve but by what we overcome in the achieving.

Essential Leadership Skills

As I mentioned earlier, if you were to ask anyone for a definition of leadership, you will probably receive as many examples as the number of people you asked. With that in mind, the following list is not intended to be an all-inclusive one. Rather, it is meant to be analogous to a discussion starter or thought provoker around some of the areas you should keep central in your thinking around leadership and for which you can logically extend those thoughts as you delve deeper into your understanding of this area.

So with that in mind I'd like to begin with saying that everything you need to be successful you already have but you have to start at the beginning:

- Understand And Know About Yourself And What You Are Capable Of Doing

By doing a critical rethink of who you are and what you can do you provide the impetus for change and no one can change you unless you are willing to change yourself first which requires knowing yourself. This area also includes your ability to constantly challenge yourself to acquire new knowledge and support others in the acquisition of knowledge so that they develop to their full potential and you're seen as providing the support needed so others can advance.

The next area may seem elementary but it's amazing how often little slippages executed over a period of time have the propensity to change the way people view you and the way you view yourself. Therefore you need to always do the following:

- Hold Yourself To The Highest Standards Of Excellence

We are known bas much for what we do as what we don't do and no one should have a higher expectation of you and your ability to deliver excellence than you do. In that regard it's also important to know that we don't accomplish anything in a vacuum so we must also associate with others that hold themselves accountable and responsible for pursuing the highest standards of excellence.

If you are to accomplish these first two essential leadership competencies then you will most assuredly need the ability to do the following:

- Display A Consistently High Level Of Energy And Work To Energize Others

This energy is not something that I limit to the physical realm but it must also play itself out in the mental realm which quite often is the hardest one to conquer. Many times we fail to complete a task not because our body tells us we cannot but that we allow our minds to do so first. Also, it's true that if you cannot energize yourself, then you will not be able to energize others.

As we move forward I want to re-emphasize the fact that leadership is not a popularity contest and often times it is hard to 'buck the system' but we need to do the following:

- Be Willing To Make The 'Hard' Decisions

The areas that I have identified to this point are impacted by your personal leadership brand which is your ability to market yourself to others and it includes these three qualities:

Standards – those things which denote excellence

Values – closely held beliefs which guide behavior consistently

Unique Selling/Value Proposition – what makes you the person you are and why should anyone care

Your image is what others see but your personal leadership brand is what either attracts others to you or repels them from you and it makes a connection with others on a logical and an emotional basis because when you think about it, a majority of decisions are based upon emotions. Your personal leadership brand therefore conveys ownership and allows you to occupy mental real estate in the minds of others around what they think and feel.

Therefore as we further examine your personal leadership brand, it can be summarized the following way:

The **unfair advantage** every individual has by capturing the value of our Personal Characteristics, Attributes and Life Experiences in a package that we effectively communicate to our audience, leveraging the specific 'use value' that comes from our Unique Value Proposition

Summary

I hope that in reading this composition I have clearly laid out for you why leadership competencies are essential and that you see them as a way of moving forward as part of a journey of lifelong learning. I do want to leave you with a few traits that you'll need to fully execute on your leadership objectives that I believe will provide the traction you need to address this area. Again this is not an exhaustive list but it provides a platform upon which you can build.

To display essential leadership competencies you will need the following:

Initiative – take action that goes beyond what is required of you

Persistence –make repeated efforts or take different actions to overcome obstacle(s)

Systematic Planning – develop and use logical, step-by-step plans and processes to reach goals

Information Seeking – seek information to help achieve objectives or clarify business problems.

Self-Confidence – have a strong belief in self and one's own abilities

Assertiveness – directly confront problems and issues with others

Use of Influence Strategies – leverage relationships with influential people to reach business goals

As you think about the areas I have covered keep in mind that while all of them have their applications, trying to fit them into your role as a set of absolutes would not be beneficial. Rather, like a new article of clothing, you have to identify which parts work best for you first – to thine own self be true. Once you have identified a piece that works for you, make it a strength and then add on additional pieces. I wish you well in your journey.

Session No. 603

Screening and Selecting Safe Contractors and Subcontractors

Mary L. Browne, CSP
HSE Manager

Nathan W. Lutz, CHST
HSE Supervisor

Nordic PCL Construction, Inc.
Honolulu, Hawaii

Introduction

Choosing qualified subcontractors for your project will directly affect your bottom line. Successful projects result in happy owners, future work, lower insurance premiums and better EMR.

Assisting contractors and subcontractors on our projects helps them improve their own programs resulting in retention of good workers, higher employee morale and may present a better opportunity to get competitive bid pricing on future work from these subcontractors.

The success of a construction project is highly dependent on the solid foundation you build on, so starting off with a strong team of competent contractors and subcontractors gives you the best opportunity for success. As a result, we end up with a qualified subcontractor base for repeat business and an owner preferred contactor.

Getting Started

In order to be successful, both the owners and contractors need to have established processes in place. Some of the more common steps in this process may include:

- Solicitation for bid
- Submission of the contract with safety and health requirements to the bidders
- Review statistical data such as incident rates, OSHA citation history, safety and health procedures and EMR history
- Have candid discussion with prospective bidders to learn about their safety and health culture

- Contractors will be selected upon acceptance of their bid
- Every contractor and subcontractor must acknowledge they have received the owner's site specific safety plan and will comply with all safety procedures
- Hold HSE preconstruction meetings with the contractor or subcontractor to review safety submittals and clarify expectations

Solicitation for Bid

In Hawaii, due to the tight knit community, many of the contractors and subcontractors know each other quite well and frequently exchange information. This is how most companies hear about upcoming work.

Meeting prospective contractors and subcontractors at the very beginning of the bid process will help establish a cohesive unit and lead to more open communication as the project progresses. However, at this stage of the bidding process, there may be very limited safety and health specifications and greater emphasis on scope of work and timeline schedules.

The Contract

Some safety contracts state, "The contractor must comply with all legislative jurisdictional requirements". It's important to know that OSHA standards are minimum requirements and to be successful, many companies have procedural requirements that are more stringent than OSHA and may require additional resources on the part of contractors and subcontractors. Some of these requirements may include:

- Mandatory attendance and participation in weekly safety meetings
- Completion of pre-task safety checklists
- Participation in project safety committee meetings
- Fall Protection Task Plans
- Inspections of fall protection equipment; harnesses, lanyards
- Assured equipment grounding program
- Quarterly inspections of ladders
- Auditing Pre Job Instruction plans
- Participation in site inspections

It's important to be very specific and identify safety expectations in this contract in order to hold contractors and subcontractors accountable.

Your contract should include specific actions that will be taken should a contractor or subcontractor fail to meet safety and health expectations. One example might be to have the contractor or subcontractor improve unsafe practices within a specified number of days or face dismissal.

Screening Contractors and Subcontractors

This is where businesses must take great care in their selection to ensure chosen contractors are most able to complete the work safely. Incorrect selection could jeopardize your company's reputation. Many times initial screening is done during the bid stage where the focus is on scope of work and schedule, not so much on safety expectations.

Determine what threshold limits you want to use to pre-qualify contractors and subcontractors. EMR thresholds for most companies is 1.0 or less and TRIR values are most often compared to BLS industry average. Typical screening information usually includes:

- TRIR and LTIR
- EMR
- OSHA citations over the last 5 years (verified on the OSHA website)
- Self-perform work hours
- Company's safety plan
- References from previous projects

For companies who lack the time or resources to process pre-qualification submissions, there are several third party companies who provide this service for a fee.

- ISNetworld
- Browz
- Avetta

There are other companies who offer services such as facility/workplace inspections and audits, compliance assessments, safety manuals, protocols and standard operating procedures. Although available, these services could be cost prohibitive for some contractors.

Some options for contractors and subcontractors who exceed threshold limits could start with getting a safety commitment from company owners. Their commitment could include increased inspections and audits of work, more task specific training for workers and additional safety requirements (safety committee participation, leading toolbox talks, etc.).

Contractor Selection

Provide Estimating with your safety Exhibit specifying Sub Trade safety requirements. This will give prospective bidders their first view of safety expectations and the opportunity to ask questions and clarify requirements. It also allows you to determine whether the Sub Trade will be a good candidate for consideration.

NORDIC PCL CONSTRUCTION

SUB CONTRACTOR/ TRADE CONTRACTOR SAFETY HISTORY QUALIFICATION FORM

Project:			
Sub/ Trade Contractor:			
Point of Contact:		Phone #:	
Scope of Work:			

Safety History

If a "Yes" response entered, please attach additional information.

Subject	Yes	No	Answer Below		
Lost Time Incident Rate for the past three years.	N/A	N/A	Current	Last Year	Year Before Last
Total Recordable Incident Rate for the past three years.	N/A	N/A	Current	Last Year	Year Before Last
List your EMR for the past three years.	N/A	N/A	Current	Last Year	Year Before Last
If your EMR is exactly 1.0 for any policy year, was or is your firm new or too small to have an EMR calculated?			N/A		
Is your firm self-insured for Workers Compensation claims?			N/A		
To whom are incidents reported to in your company?	N/A	N/A			
Do you have an Orientation Program?			If yes, how often?		
Do you hold on-site Safety Meetings with workers?			If yes, how often?		
Do you conduct site safety inspections?			If yes, how often?		
Do you hold (Tailgate/Toolbox) Weekly Safety Meetings?			If yes, how often?		
Do you have a written Injury Illness Prevention Plan?			**If yes, please submit a copy when requested**		
Do you have a written Fall Protection Plan?			**If yes, please submit a copy when requested**		
Do you have a written Hazardous Communication plan?			**If yes, please submit a copy when requested**		
Do you have a written Respiratory Protection Plan?			**If yes, please submit a copy when requested**		
Do you have a written Environmental Plan?			**If yes, please submit a copy when requested**		
Have you been cited by any environmental agency within the past 5 years?			**If yes, please explain when requested**		
Does your company have an individual dedicated specifically to safety? If yes, please identify by name and title.			**Name:** **Title:**		
If no, identify the person (Title) within your company directly responsible for the mgmt. of your company's Safety Program:			**Name:** **Title:**		
Have you been cited by Federal or State OSHA within the past 5 years.			**If yes, please explain why and what you have done to address the citation(s).**		

Please use the previous three (3) years OSHA 300 records to complete the following; please submit a copy when requested			
Number of lost workday cases:	Current	Last Year	Year Before Last
Number of restricted workday cases:	Current	Last Year	Year Before Last
Number of injuries / illnesses:	Current	Last Year	Year Before Last
Number of employee hours worked:	Current	Last Year	Year Before Last

Note: In order to qualify for PCL's bid list, the subcontractor must have a current EMR of 1.0 or less. If the current EMR exceeds 1.0, the subcontractor must provide documentation that it has (or is in the process of) initiated policies and programs that will result in improved safety performance in order to be placed on PCL's bid list. It is at PCL's sole discretion to approve or disapprove a subcontractor who has a current EMR of greater than 1.0.

Exhibit 1. This is an example of a subcontractor safety history request form which provides a quick overview of the subcontractor's safety culture.

Develop an internal risk matrix to rate sub trades you work with. Use their incident rates, OSHA citations, EMR and feedback from your job teams to determine who performed well on previous jobs and who didn't. Update your Data Base following job completion. Get information on Sub Trade personnel to find out how well they worked with the job team and managed their own workers. Supervisors can have a huge impact on job performance and safety. Keep in mind, this risk matrix may not be applicable in certain remote areas where your Sub Trade pool is very limited.

Having a current data base will help identify where you may need to focus your safety efforts to mentor higher risk Sub Trades. You need to make sure your expectations are communicated to your Sub Trades and assist them to clearly understand what they need to do. This could be done through Safe Start meetings with the owners or leaders of each company at the beginning of the project or through Pre-Construction meetings when your project first starts up. This provides an opportunity for the Sub Trades to meet the team, see what other Sub Trades they will be working with, and to get clarification if something is not clear.

Communication with your Sub Trades should continue through the life of your project to help them meet your company requirements.

Post Contractor Selection – Time to Work

Now that estimating is complete and contractors and subcontractors have been selected, it's time to start mobilizing for the project. Make sure your expectations are communicated to your Sub Trades and help clarify what they need to do. This can be accomplished by holding preconstruction meetings so Sub Trades can meet the team, hear what safety expectations are required, get answers to any questions they may have and also network with other Sub Trades they will be working with.

Provide Safety Submittal Exhibits and establish a method to maintain and store these documents. Access to these documents needs to be available to the job team so they can be reviewed and Sub Trades can be held accountable for submitting all requirements prior to starting work.

Some examples of these requirements may include:

- HSE Acknowledgement
- Project Specific HSE Plans
- Trade Contractors Qualified / Competent Person List
- Job Hazard Analysis
- Safety Data Sheet Listing
- Operator Certifications (i.e. Forklift, Articulating, Scissor, etc.)
- Worker Training Records (i.e. Fall Protection, Scaffold, Powder Actuated Tool, etc.)

Clearly convey company safety requirements which exceed OSHA and make sure all Sub Trades understand that full compliance is expected.

Exhibit D
Subcontractor Safety,
Health and Environmental Requirements

Each subcontractor is responsible for the safety and health of its employees. Full compliance with OSHA, MSHA, and ANSI Standards (that are incorporated by reference in OSHA and MSHA) and all other Federal, State, and Local regulations is mandatory. Nordic PCL requires that subcontractors and their employees comply with the Project Specific Health, Safety and Environment (HSE) Plan including the following requirements on all Nordic PCL jobsites, in addition to client requirements.

1) **Safety and Health Orientation**
Each subcontractor employee must attend Nordic PCL HSE Orientation. An additional HSE orientation may be required by the client as specified in the Nordic PCL Project Specific HSE Plan.

2) **Project Specific HSE Plan**
Each subcontractor must provide their Project Specific HSE Plan or written safety and health program, and written Hazard Communication Program to the Nordic PCL Project Manager before mobilizing on the jobsite. Subcontractors whose programs do not meet Nordic PCL's minimum standard will be requested to make improvements or agree in writing to adopt Nordic PCL's Project Specific HSE Plan prior to mobilization on the jobsite.

3) **Personal Protective Equipment and Clothing**
Each subcontractor must verify that its employees have, or are provided with, the required PPE and clothing and are trained in its use.
   a) Safety glasses that meet ANSI Z-87.1 requirements must be worn at all times when on Nordic PCL jobsites, except in administrative areas. Safety glasses must have rigid side shields.
   b) Safety glasses or goggles and a face shield or other protection approved by Nordic PCL must be worn when cutting or grinding, or in designated areas.
   c) Hard soled leather or substantial synthetic material safety boots with steel or composite toes that comply with the ASTM F2413 – 11 Standards are required to be worn on the jobsite, but not in administrative areas or parking lots. Boots must have at least a six-inch upper that vertically extends to cover the protruding ankle bones.
   d) Hard hats are required to be worn and must conform to ANSI Z-89.1, Class B specifications. Bump caps and "cowboy" style hard hats are not permitted. Hard hats must be worn with the bill forward in accordance with the manufacturer's instructions. The hard hat must not be painted and must have the subcontractor's identification logo or name and the worker's name on the front of the hat.
   e) Hearing protection must be worn whenever noise is measured above 85dB, when normal conversation cannot be conducted, or when the area is posted as noise-hazardous.
   f) Abrasion resistant gloves must be worn when working on the jobsite. When using knives or handling sheet metal or glass, cut resistant gloves will be worn. Gloves must not be worn around rotating equipment.

4) **Fall Protection**
   a) One hundred percent fall protection is required when working without guard railings or netting more than six (6) feet above the next lowest level, including operations on scaffolding and structural steel.
   b) Fall protection harnesses and lanyards must be inspected by a competent person quarterly. Colored tape, zip ties or tags will indicate the quarter the inspection was conducted.
   c) Roof leading edge protection must be provided when working within six feet of the edge when that edge is more than six (6) feet above the next lowest level.

   Failure to comply with the fall protection criteria listed above may result in immediate removal of the violating worker.

5) **Ladders**
   a) Only fiberglass ladders are allowed on Nordic PCL jobsites, however, wooden job-made ladders built in compliance with ANSI Standards are allowed.
   b) Extension ladders will be held at the base by another worker until securely tied-off at the top.
   c) Ladders must be inspected before use and quarterly by a competent person and tagged and/or taped according to the Project Specific HSE Plan.

6) **Vehicles and Mobile Equipment**
   a) Each subcontractor employee must have a valid driver's license with proper endorsements to operate any vehicles on the jobsite.
   b) Mobile equipment operators must have documented proof that they are trained and qualified to operate the specific brand and type of equipment.

Rev. 03 January 2015

Exhibit 2. This is an example of contract safety requirements once the contractor or subcontractor has been awarded the bid.

c) Daily pre-use inspections of mobile equipment will be performed and documented. The documentation must be retained for review.

d) All incoming vehicles and mobile equipment must be inspected and safety-checked before initial use and operator training must be documented and available for review.

e) Installed seat belts will be used by vehicle and equipment operators.

f) All mobile equipment and other construction vehicles (not cars or pickups with unobstructed rear views) must have operable back-up alarms.

g) Any mobile equipment used in an enclosed area such as a basement will be propane or electric powered.

7) Environmental Requirements

a) Subcontractors must provide current Safety Data Sheets (SDS) to Nordic PCL before any materials/chemicals are brought to the jobsite.

b) Equipment fluid leaks will be contained, cleaned up and disposed of properly. Leaking equipment must be repaired before continued use.

c) Nordic PCL supervisors must be notified immediately when any chemical or oil spill occurs, no matter how small. **Spill control and clean-up** are the sole responsibility of the subcontractor when the subcontractor causes the spill. All spill control and clean-up must be accomplished by trained personnel.

8) Confined Space Entries

a) When confined spaces will be entered by the subcontractor, and Nordic PCL employees are not in the space, the subcontractor will provide all rescue and emergency equipment, continuous atmospheric monitoring instruments, respirators, fire extinguishers, and any other equipment required by OSHA or MSHA as appropriate.

b) All jobsite confined spaces are to be initially considered "permit-required" and will not be entered unless a permit is signed by Nordic PCL or as specified in the Project Specific HSE Plan.

c) If a subcontractor employee enters a permit-required confined space, the subcontractor must have an on-site trained rescue force available as specified by OSHA or MSHA as appropriate, or arrange for that service from the client/owner or an outside service company.

d) If Nordic PCL employees are also in the confined space, Nordic PCL will conduct continuous atmospheric monitoring. When there are no Nordic PCL employees in the confined space, the subcontractor is required to perform initial and continuous atmospheric monitoring.

9) Lifting and Hoisting

a) Lift plans are required to be prepared for all lifts in excess of 75% of capacity.

b) For all lifts in excess of 90%, the following requirements apply: an analysis must be completed by a Professional Engineer, who is qualified and experienced in heavy crane lift evaluation, and the crane must be equipped with a load indicator, and either a wind anemometer or access to the site wind indicator.

c) Engineering reviews and sign-off of the *Permit For Hoisting Personnel Utilizing A Personnel Basket*, HSEOP-03-07, are required prior to using a personnel basket.

10) Excavations

a) An *Excavation-Ground Disturbance Permit*, HSEOP-05-01, will be obtained from Nordic PCL before digging six inches or more under the surface.

b) Protective systems such as trench boxes or shoring must be in place in all trenches five (5) feet or more in depth unless benching or sloping, in compliance with OSHA Standards, is maintained.

c) Spoil piles must be kept at least three (3) feet back from the edges of a trench.

11) Electrical Safety

a) Assured grounding and Ground Fault Circuit Interrupters (GFCI) are required on <u>all</u> electrical equipment and electrical tools and must be furnished by the subcontractor.

b) All electrical tools, equipment and cords will be inspected prior to use and quarterly for safe ground condition. Colored tape, tie-wrap or tags will denote the currency of the inspection.

c) Electrical cords, hoses, and welding leads must be routed overhead when feasible, to prevent tripping hazards and damage. Hoses, cords, and leads routed on the ground will be protected from damage caused by vehicles.

Exhibit 2 (cont'd).

12) **Hazard Identification and Control**

a) Subcontractors must complete a Job Hazard Analysis (JHA) for each high risk task to verify that the hazards and risks for the activity are identified and that appropriate controls are implemented. The JHA shall be submitted to Nordic PCL for review before the task is begun. All workers involved in the task will review and sign off on the JHA.

b) Each subcontractor crew must complete a Nordic PCL Pre-job Safety Instruction (PSI), attached as Exhibit X-1 or one that is similar in content and purpose. A PSI must be completed at the site of the work prior to each task on every shift and reviewed by each worker after breaks and lunch.

c) Subcontractors must conduct daily informal HSE inspections of their work areas and monthly formal inspections, as required in the Project Specific HSE Plan. Copies of formal inspections will be provided to Nordic PCL.

13) **Miscellaneous Requirements**

a) Scaffolding must be inspected by qualified persons and tagged with Nordic PCL or similar tags approved by the Nordic PCL safety coordinator. If the subcontractor erects scaffolds, it must assign a scaffold competent person.

b) Compressed air hoses and couplings must be inspected daily before use. All hose couplings must have positive locking devices. Chicago-type fittings must be secured with wire or clips.

c) Prior to conducting demolition operations, the *Nordic PCL HSE Checklist For Demolition Operations*, HSEOP-20-01, must be filled out and all requirements must be completed to Nordic PCL's satisfaction.

d) When erected, barricades will have a tag attached explaining the purpose of the barricade, the hazards in the barricaded area and the supervisor's name and phone number who is in charge of the barricade.

e) Subcontractors must ensure that all job-related injuries and illnesses, no matter how minor, are reported to Nordic PCL immediately. A preliminary incident report must be submitted to Nordic PCL within 24 hours of the occurrence.

f) Each subcontractor shall submit weekly:

- Number of workers on the jobsite;
- Total man-hours worked;
- Number of near misses;
- Number of first aid cases;
- Number of recordable incidents;
- Number of work days lost;
- Equipment and property damage incidents;
- Sub-Tier contractors' number of employees and total man hours, as well as incidents listed above.

g) When thirty (30) or more subcontractor employees are on the job, or if safety becomes an issue as determined by Nordic PCL, the subcontractor will be required to provide a safety coordinator approved by Nordic PCL as competent to perform safety responsibilities for the scope and size of the project and, unless waived by the Nordic PCL Project Manager, be present on the jobsite during regular working hours.

h) The Nordic PCL hot work permit, or equivalent, when required due to the nature of the task, must be prepared by the contractor performing the work. The subcontractor must provide a trained fire watch as required.

i) HSE field safety meetings must be conducted weekly and must be fully documented with content and attendance. Copies of handouts provided to subcontractor's workers must be provided to Nordic PCL.

j) When directed by the Nordic PCL Project Manager and/or the Project Specific HSE Plan, the subcontractor will participate in the jobsite behavioral-based observation system. One subcontractor employee from each subcontractor will be trained and tasked to complete four observations per week. Assigned observers will attend the behavioral observers' meeting once each week.

k) Upon request, a subcontractor representative must attend project safety committee meetings.

l) Subcontractors must maintain training records for each of their employees on a Nordic PCL jobsite and will immediately provide those records to Nordic PCL upon request.

m) When conducted by Nordic PCL, subcontractor workers will attend Craft Safety Workshops and participate in safety perception surveys.

14) **Evacuation Map and Alarm Signals**

It is the responsibility of the subcontractor to obtain and communicate to their employees the evacuation routes that are designated to egress their work areas and to identify jobsite alarm signals.

Exhibit 2 (cont'd).

Consider holding a Safe Start meeting with owners and/or principals from high and low risk companies to openly speak and agree about supporting the efforts to keep workers safe as the project goes along and "buy-in" to the safety program. The ultimate expectation is to send every worker home safely at the end of each day.

Project Specific HSE Plan Acknowledgement Form

Project Name:____

After reviewing the policies and practices as outlined in this plan, the company owner, site superintendent, on-site foreman, lead hands, and all trade contractors are to sign off this sheet. The sign – off sheet must be returned to the NPCL Project Manager, before commencement of work-related activities on the jobsite.

I have read and understand this Project Specific Health, Safety and Environmental Plan and will carry out my work within these guidelines.

Company Name:____

Company Owner

Name: ____________________________ Date: __________________

Signature: __________________________ Title: __________________

Company Superintendent

Name: ____________________________ Date: __________________

Signature: __________________________ Title: __________________

On Site Foreman

Name: ____________________________ Date: __________________

Signature: __________________________ Title: __________________

Exhibit 3. This sample form is used to get confirmation that the winning Sub Trade has received, reviewed and will follow the Owner's or General Contractor's site specific safety and health plan.

Watch for signals which may indicate potential unsafe performance by subcontractors on the project. Some "red flag" indicators may include:

- Failure to submit required documents; i.e. training records, incident or work hour reports
- Poor attendance or failure to participate in safety meetings, committees or other programs
- Allowing work to proceed in unsafe conditions or not abating hazards or addressing unsafe behaviors
- Incidents resulting in worker injury, property damage and/or near misses

Corrective Action

Non-compliance needs to be documented and specifically address disciplinary measures. Disciplinary records should be kept for each Sub Trade and tracked. You may consider posting sub deficiencies in a conspicuous place on the project site without the individual's name.

Some other methods to help poor performing contractors and subcontractors may include:

- Meeting with the company owner or principal
- Requesting replacement of contractor or subcontractor supervision on the project
- Mandating more involvement by the contractor's insurance safety department to help monitor operations regularly
- Require additional safety training
- Require the contractor or subcontractor provide a full time safety professional on the project

Summary

Project success is achievable through careful screening and selection of contractors and subcontractors. Companies that spend the time planning and preparing prior to mobilization ultimately will encounter less disruption from poor performance and unsafe contractors.

Have clear procedures in place so you can obtain safety information to help you select the safest contractors from prospective bidders. Make sure contractors provide you with written assurance that they will comply with your safety and health program and understand they will be held accountable for full compliance.

Lastly, have a plan in place to address unsafe behaviors and/or non-compliant contractors.

Bibliography

Baldwin, Rich and Albert Valenzuela. "Best Practices for Screening and Selecting Safe Contractors and Subcontractors". Governor's Pacific Rim conference presented in Honolulu, HI by ASSE and HIOSH, May 2014.

Dolama, Maryam Ghaffari and Farnaz Sadeghpour. 2015. "Factors impacting selection of construction subcontractors". The University of British Columbia. (https://open.library.ubc.ca/cIRcle/collections/52660/items/1.0076446)

Gillen, Matt, Scott Schneider, Steve Hecker, and Linda Goldenhar. 2014. "Safety Culture and Climate in Construction: Bridging the Gap Between Research and Practice". Report from a joint workshop conducted by The Center for Construction Research and Training and NIOSH, Washington, D.C., June 11-12, 2013. (http://elcosh.org/document/3782/d001295/safety-culture-and-climate-in-construction%3A-bridging-the-gap-between-research-and-practice.html)

Safety Science. "Leading indicators of construction safety performance". July 2012 (https://www.researchgate.net/profile/Andrew_Wehle/publication/257356475_Leading_indicators_of_construction_safety_performance/links/558aa48008ae50d602c9592b.pdf

Session No 604

New ANSI & OSHA Confined Space Safety Standards–Host vs. Contractor Employers

Timothy C. Healey
The Hartford Steam Boiler Inspection & Insurance Co.
Hartford, CT

Terry W. Krug, MS, CIH, CSP
EXOSHA, Inc.
Bartlett, IL.

Introduction

This session is designed with enough detail to reach the intermediate level of attendee's knowledge of confined spaces activities, however, comprehensive enough to offer technical information to an advanced audience. The attendee will go away with an appreciation of the new ANSI/ASSE Z117.1 Standard, "Safety Requirements for Entering Confined Spaces," and the new OSHA Confined Space Standard for Construction.

The Host Employer's Perspective

Case Study Scenario: A Mock Business Meeting

A facility owner realizes they need to hire a contractor to complete needed work inside three locations in the plant that have been determined to be 'confined spaces'. The facility simply does not have the expertise to attempt this work. Within these spaces, it is anticipated the work could include the following:

A. Painting,
B. Equipment or machine repair
C. Welding/grinding or related Hot Work, or
D. A thorough inspection and evaluation of the space such as: structural integrity, air quality including oxygen levels, carbon monoxide, flammables and radon levels.

The host employer, after reviewing company needs, budgets and time restraints, decided to do the inspection and evaluation task for two of the spaces and paint the third and largest of the three. He researches for weeks and hires a company that can accomplish the task.

The contractor is given the following information:

- this activity must occur within the month of July;
- the contractor will be entering 3 separate confined spaces; and
- have a total of five days to accomplish the task with results and a written report required within two weeks after the last permit space has been exited.

These tasks are the same for two of the spaces (items C and D above), as these both involve a thorough inspection and evaluation including structural integrity, air quality, radon levels and WGBT measurements. Clad welding may be needed to restore proper thickness in the steel walls or to add bracing to the existing spaces. The third space is to be spray painted with a water base coating the inside of a refrigeration unit. The contractor asks how much lead time he will have to secure people and equipment. The Host says a weeks' notice is usually what they give. Will that be enough time? The contractor says he can make that work. A contract was developed and will be signed by both parties when they meet next week. Payment will be in three installments 1/3 at signing, 1/3 when exiting the last space and the last 1/3 when the final results and report is received.

The two employers agree to meet at the host's facility next Tuesday 9:00 am and the conversation goes like this:

<u>Host Employer–Tim Healey, Plant Manager, No Bull Beef Company</u>

Host and Contractor Conversations

Good morning, Mr. Krug (Owner of Your Space is Our Space, Inc. – hired Contractor Employer)

Good morning, Mr. Healey (Plant Manager of No Bull Beef – Host Employer/Facility Owner Representative)

T. Krug (TK) – It looks like a nice day to see the 3 spaces you want my company to enter.

T. Healey (TH) – Yes! We have lucked out on the weather, it should be in the 60s and sunny.

TH – Before we go to look at the spaces I have a few questions in addition to all that I have asked you last week and my company decided to hire you. You have said that you have been in business for five years and have 8 employees including office & shop employees, is that correct?

TK – Yes.

TH – How long has your company been doing confined space work?

TK – About five years.

TH – How many employees do confined space work and do you participate in the entry team?

TK – Yes! I usually act as the entry supervisor, however if I can't I have George supervise.

TH – Who does your training in safety and health topics?

TK – We hire an ex-OSHA person locally.

TH – Are you familiar with OSHA regulations? Tell me what you know about the regulations that apply to Permit-Required Confined Space entry operations, documentation and recordkeeping.

TK – Proceeds to describe the general industry standards that are applicable; i.e. 29 CFR 1910.146, plus respirator standards, fall protection, rescue, etc.

TH – Are you familiar with ANSI standards?

TK – Yes.

TH – Did you know that ANSI just revised their safety requirements for entering confined spaces, ANSI Z117.1? I use ANSI standards in my facility because I can pick up useful information on topics that OSHA doesn't go into with as much detail. The OSHA regulations are designed for enforcement, where the ANSI standards are developed with industries' best practices in mind and provides references, examples & state of the art procedures.

TK – I am generally aware of that, yes.

TH – Do you own your own test equipment or do you rent them?

TK – Both depending on the job task. I will need to rent a heat stress monitor, but I have a multi-gas monitor and spray painting equipment and a compressor.

TH – I would like you to go beyond the minimum required by OSHA and follow ANSI Z117.1[4] . The chapters that are applicable to your work can be found in this copy I am providing to you. Take a few minutes to familiarize yourself before you agree to comply so we are clear on your intent to follow the recommendations in Z117.1 R2016[4] prior to signing the contract. I believe that by following the ANSI Standard, you will meet or exceed the OSHA 1910.146 regulations, which are over 20 years old. Do you agree?

TK – Yes. I will need to adjust my bid to reflect any extra effort my employees and I will be doing to comply.

TH – Not a problem. I will talk to my boss since we are changing the parameters of the work of our original contract. In the meantime, I have provided an outline of the new ANSI Z117.1[4] revised in 2016. You will want to review this and we can discuss any questions you may have as a result.

Worthy of Note

In construction work, as in general industry as we have just discussed, there are recently updated regulations and consensus standards.

The Occupational Safety and Health Administration (OSHA) has had regulations governing activities on construction sites for over 45 years (29 CFR 1926)[1.] OSHA released their regulations two years ago titled, "Confined Spaces in Construction" (29 CFR 1926.1201-1213)[2]. (The Subparts are listed below.) Since some of the planned activities are considered 'construction work', they will be following those minimum regulations. The Host Employer will be looking for the Contractor to also be guided by ANSI A10.43.

The ANSI A10.43[3] standard was approved in 2016 to specifically address confined space entry on construction sites.

1926.1201-1213, Confined Spaces in Construction[2]

1926.1200 [Reserved]

- ❑ 1926.1201 Scope.
- ❑ 1926.1202 Definitions.
- ❑ 1926.1203 General requirements.
- ❑ 1926.1204 Permit-required confined space program.
- ❑ 1926.1205 Permitting process.
- ❑ 1926.1206 Entry permit.
- ❑ 1926.1207 Training.
- ❑ 1926.1208 Duties of authorized entrants.
- ❑ 1926.1209 Duties of attendants.
- ❑ 1926.1210 Duties of entry supervisors.
- ❑ 1926.1211 Rescue and emergency services.
- ❑ 1926.1212 Employee participation.
- ❑ 1926.1213 Provision of documents to Secretary.

Additional Observation

With multiple 'sets' of regulations, and consensus standards developed for industry-specific needs, it comes as no surprise that are some differences between them. For example, the definition in Z117.1 2016[4] states the following:

2.5 Confined Space. Enclosed area large enough and configured to allow a person to bodily enter and has the following characteristics:

- Its primary function is other than human occupancy.
- Has restricted entry and exit. (Restricted entry and exit is a physical configuration, which requires the use of the hands for support or contortion of the body to enter into or exit from a confined space.)

The 1926.1202 Confined spaces in construction standard[2] defines confined space as the following:

Confined space means a space that:
(1) Is large enough and so configured that an employee can bodily enter it;
(2) Has limited or restricted means for entry and exit; and
(3) Is not designed for continuous employee occupancy.

Because of differences such as these, it is **vital** that all parties agree **prior** to the commencement of any work which regulations and standards will be applied and followed.

Endnotes

[1] Occupational Safety and Health Administration (OSHA). 2017. 29 CFR 1926, http://www.osha.gov

[2] Occupational Safety and Health Administration (OSHA). 2016. 29 CFR 1926.1201-1213, http://www.osha.gov

[3] American National Standards Institute (ANSI)/ASSE. 2016. A10.43 Standard, "Confined Spaces in Construction and Demolition Operations". Park Ridge, IL: American Society of Safety Engineers (ASSE).

[4] American National Standards Institute (ANSI)/ASSE. 2016. Z 117.1 (R2016) Standard, "Safety Requirements for Entering Confined Spaces". Park Ridge, IL: American Society of Safety Engineers (ASSE).

Appendix

ANSI/ASSE Z117.1

Table of Contents

Appendices:

Assessing Physical and Mental Fitness for Work

Peter P. Greaney, M.D.
President, CEO and Medical Director
WorkCare, Inc.
Anaheim, California

Introduction

The workforce is a complex organism that presents multiple challenges for environment, health and safety (EH&S) professionals. The need to determine whether an applicant or employee is physically and mentally fit for work is a common confounder. While this responsibility often falls outside their comfort zone, it is possible for EH&S professionals who lack medical training to confidently engage the fitness-for-work assessment process. This paper provides guidance based on best clinical practices and more than three decades of experience in injury prevention and management. It is divided into four sections:

I. Quantifying the value of fitness-for-work assessments.
II. Implementing the assessment process.
III. Complying with applicable rules and regulations.
IV. Adopting a comprehensive policy.

Quantifying the Value of Fitness-for-Work Assessments

A growing number of employers consider fitness-for-work (FFW) assessments an essential contributor to their organization's comprehensive environment, health and safety (EH&S), sustainability and risk management efforts. The inherent value of an objective assessment becomes even more apparent when one considers workforce characteristics that increase injury risk and are largely out of the control of managers. Such factors include aging, comorbid conditions, poor sleep habits and fatigue, smoking, depression and substance abuse.

<u>Definition</u>
A FFW assessment is a medical examination designed to assess whether an employee or job applicant has the physical, mental and emotional capacity to perform assigned tasks in a competent manner – and in a way that does not unreasonably threaten safety, health or property.

FFW assessments are used to:

- Protect and promote employee and public health.
- Reduce accidents, injuries and property damage.
- Facilitate safe job placement and return to work.
- Manage medical, insurance and legal liability costs.

For many companies, FFW evaluations are part of the onboarding process. They are also viewed as a logical extension of medical surveillance examinations that are performed to establish baselines and measure physical changes in response to exposures such as noise or toxic substances over time. They go hand-in-hand with drug-free workplace policies, and behavioral health and employee assistance programs (EAPs). They also give employers the opportunity to ensure they are in compliance with federal laws and jurisdictional regulations.

Return on Investment

A cost-benefit analysis of FFW assessments largely depends on how an employer defines the value of a work day in their enterprise. Some of this value is associated with *cost avoidance*. An undetected physical or mental impairment may affect an employee's ability to work safely or diminish their productivity (presenteeism). It may be a contributing factor in a work-related incident that results in a workers' compensation claim, medical treatment, recordable injury, lost work time and even long-term disability.

Awareness of physical and mental functional capacity facilitates the use of targeted preventive interventions and a prompt response in the event of an injury or incident. In addition, companies that demonstrate they care about worker fitness are more likely to avoid costly consequences such as damage to brand and image, poor employee morale and the inability to recruit well-qualified candidates.

Injury Trends

The use of FFW assessments as a means to reduce accident and injury rates applies to virtually all workplaces. Despite steady annual declines, certain industries in the U.S. still have high rates of exposure, injury and property damage. Disparities in injury rates among certain industry types suggest work-fitness assessments add value in these workplaces. Figure 1 illustrates disparities in non-fatal injury and illness rates among private industries.

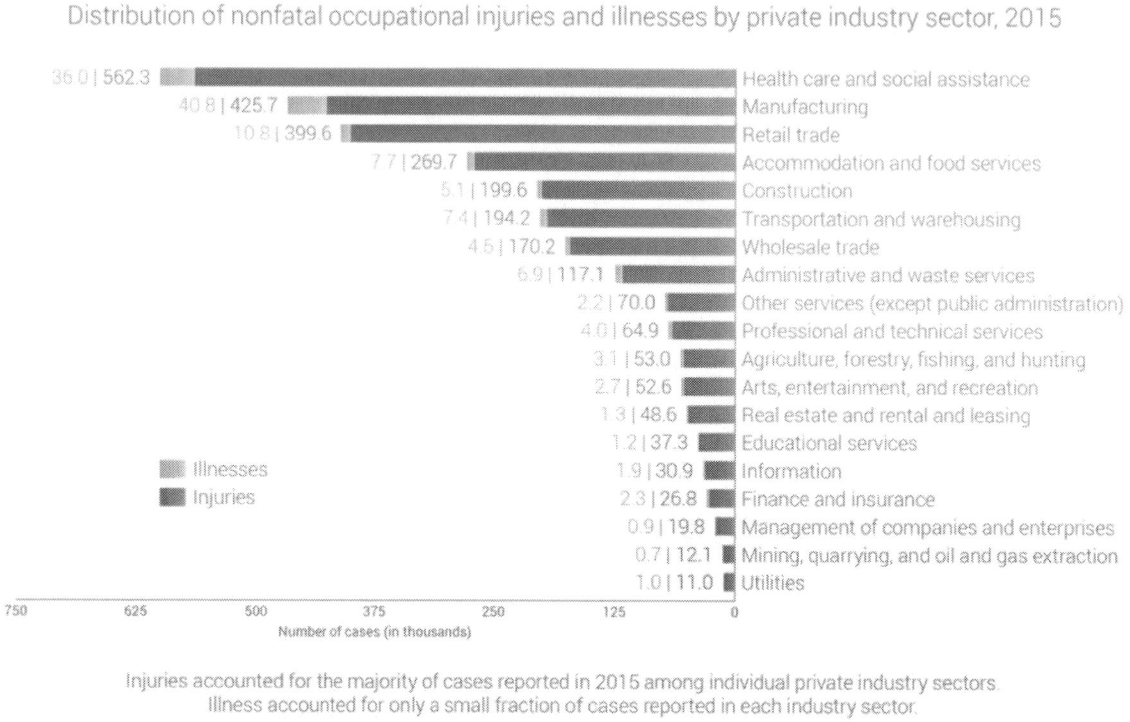

Figure 1. (Source: U.S. Bureau of Labor Statistics.)

Claim trends offer useful insights on the need for prevention and detection of physical and/or mental impairments:

1. In 2012, the National Council on Compensation Insurance (NCCI) studied the impact of comorbidities on workers' compensation claims: the percentage of claims with a comorbid diagnosis increased from 2.4 percent to 6.6 percent between 2000 and 2009, an indicator of the need for return-to-work evaluations.[1]
2. In a Travelers insurance *Injury Impact Report* featuring findings from an analysis of more than 1.5 million claims, material handling was the most common cause of injury.[2] Lifting, lowering, filling, emptying or carrying an item caused 32 percent of injuries, with manufacturing and retail sectors taking the biggest hit. The average cost for the most frequently occurring injuries ranged from $8,000-$42,400 per claim. Strains and sprains were associated with an average of 57 days away from work.
3. The Liberty Mutual Research Institute for Safety's annual *Workplace Safety Index* found that the most disabling work-related injuries (in the claim study year 2013) accounted for nearly $62 billion in direct U.S. workers' compensation costs.[3] Liberty Mutual's findings have been notably consistent for the past 12 years, suggesting there are still critical lessons to be gleaned from the findings.

Research also shows that the presence of comorbid medical conditions such as high blood pressure, heart disease, obesity, arthritis and diabetes affect health outcomes. For instance, an overweight, diabetic employee will likely take longer to recover from a work-related injury or illness than an employee in the same job, with the same type of injury but who does not have any comorbidities. A FFW assessment may be used to evaluate the diabetic individual's physical ability to safely return to work. Sleep disorders provide another illustration. Sleep disruption and fatigue have been linked to a number of generalized health and behavioral disorders, including reduced efficiency while operating a motor vehicle.[4]

Substance Use

Data from Quest Diagnostics' Drug Testing Index™ (DTI) and the National Survey on Drug Use and Health (NSDUH) reinforce the need for FFW assessments. The DTI has been used since 1988 to analyze millions of drug test results by drug category, testing reason and specimen type. The NSUDH is a comprehensive survey that includes metrics such as alcohol and illicit drug use by state and employment status.[5]

DTI data show steady increases in overall drug positivity rates in the U.S. workforce since the study's inception; marijuana positivity rates have increased notably over the past five years. The NSDUH reports corresponding higher self-reported use of marijuana, with 22 million Americans identifying as current users. The Substance Abuse and Mental Health Services Administration (SAMHSA) reports approximately 3 percent of workers are under the influence of an illicit drug at any given time. Between 2009 and 2013, the pharmacy benefits management company Express Scripts found nearly 60 percent of patients in the general population – many of whom are employed – were legally prescribed potentially dangerous mixtures of opioids and other medications.[6]

The human tragedy that plays out in association with opioid misuse and abuse simultaneously creates liability and increases risk for public and private-sector employers. In lawsuits, state courts have found employers and workers' compensation insurers financially responsible when an injured worker fatally overdoses on prescribed painkillers. In the workplace, prescription and illicit drug abuse is associated with impaired performance and judgment as well as:

- Higher work-related injury and illness rates and workers' compensation costs.
- Costly emergency room visits and hospitalizations.
- Long-term disability and lost productivity.
- Sleep disruption leading to fatigue and inattention.
- Damage to company brands and business interests.
- Violent incidents.
- Motor-vehicle accidents.

Mental Health

Undetected and undertreated mental health disorders in the workplace can create a cascade of negative effects. The federal Occupational Safety and Health Administration (OSHA) reports nearly 2 million employees are victims of reported workplace violence a year, with a significant percentage of incidents attributed to working with volatile, unstable people.[7]

In the National Comorbidity Survey, the first large-scale field survey of mental health in the U.S., 18 percent of respondents who were employed said they experienced symptoms of a mental health disorder in the previous month; less than half (40 percent) received professional treatment. In a study, SAMHSA found that among full-time workers aged 18 to 64 years old, the highest rates of workers experiencing a major depressive episode in the past year were found in personal care and service (10.8 percent) and food preparation and serving-related occupations (10.3 percent).[8,9]

According to the U.S. Centers for Disease Control and Prevention, in a national study approximately 80 percent of people with depression reported some level of functional impairment because of their depression; 27 percent reported serious difficulties in work and home life. However, only 29 percent of depressed adults reported contacting a mental health professional in the past year. In a three-month period, patients with depression missed an average of 4.8 workdays and suffered 11.5 days of reduced productivity. Depression is estimated to cause 200 million lost workdays each year at a cost to employers of $17 to $44 billion.[10]

Despite these statistics, mental health issues continue to be ignored by all sectors of society. Lack of awareness and acknowledgment of mental health issues contributes to absenteeism and presenteeism, and it is frequently a comorbid condition found in workers' compensation claims. EH&S professionals who follow a well-orchestrated and robust approach to mental health issues will more easily navigate tricky legal waters and avoid injury management pitfalls.

Implementing the Fitness-for-Work Assessment Process

When individuals are deemed fit for work, it means they are able to perform assigned tasks completely and in a manner that does not threaten the safety or health or themselves or others.

Process

A number of scenarios apply to the assessment process. An assessment may be performed:

- As part of hiring and onboarding.
- To address safety-sensitive duties.
- Prior to a change in work assignment.
- At the request of the worker.
- In response to observed behavior.
- Upon return to work after an illness, injury or prolonged absence.
- When there is reasonable cause to believe fitness may be compromised.

A collaborative, multi-disciplinary approach is recommended. While clinical findings may be clear-cut, a certain degree of interpretation is often needed in order to make the best job-placement decision. The FFW process may involve individuals and departments including employees and applicants, EH&S professionals, frontline supervisors, occupational health/onsite medical, human resources, legal, risk management and senior executives. Other parties such as labor organizations, insurers, benefits administrators, wellness program administrators and behavioral health professionals may also be involved.

Optimally, a physician with training and experience in occupational medicine performs the assessment. The process may be driven by a company's corporate health department with the protocol recommended by occupational health personnel. Whether internal or external, the evaluating clinician should have both medical and behavioral health knowledge. He or she must know how to avoid potential inconsistencies, deploy legally defensible tests, carefully record findings and maintain confidentiality. It is the employer's responsibility to provide functional job descriptions and a safe work environment.

Working During Recovery

Experience and studies show that work during recovery is therapeutic; not working (de-conditioning) delays recovery and return to function. While an injured employee may need to temporarily modify some activities, in most cases he or she will fully recover. An employee who experiences a work-related injury or illness may be in pain and fearful that working will make it worse. A FFW evaluation provides reassurance that the employee is capable of safely returning to full duty or a temporary modified job with minimal risk of re-injury. Returning to work as soon as it is medically safe to do so has numerous benefits:

- Working and staying active contributes to overall health and well-being.
- Work provides a sense of purpose and social stimulation. Not working can lead to feelings of isolation, depression and unnecessary chronic disability.
- Data repeatedly show that the longer someone is off work, the greater the likelihood they will never return to the workplace. This has significant impacts on personal lives, employers and society as a whole.
- Being at work helps sustain productivity and reduces the pressure absence may place on co-workers, the employer and family members.

Post-Offer or Pre-Assignment Clearance

Post-offer or pre-assignment fitness-for-work evaluations typically are used to determine if there is a match between an individual's functional capabilities and the physical requirements of the job. It may include a combination of a physical exam and a functional capacity assessment using testing methods that are considered proven and objective. It is critically important for the evaluator to have access to an accurate, up-to-date and valid functional job description.

Reasonable Cause

The "reasonable person" standard applies to an observed tapering off in performance. This phrase is frequently used in legal settings to denote a hypothetical person in society who exercises average care, skill and judgment in conduct, and who serves as a comparative standard for determining liability. A reasonable-cause assessment is likely warranted when an employee:

- Exhibits pain behaviors (e.g., limping, grimacing).
- Acts in an odd or inappropriate manner.
- Seems to be accident-prone or distracted.
- Has frequent or unexplained absences.
- Shows a steady performance decline.
- Is caught lying or stealing.
- Shows physical or behavioral signs of substance use.

Impairment at Work

Impairment at work often manifests itself as presenteeism. The American College of Occupational and Environmental Medicine's (ACOEM's) Health and Productivity Management Center defines presenteeism as health-related productivity loss while at work. The concept incorporates:

- Time on task: In the workplace but not working, e.g. "sleeping it off."
- Quality of work: Incidence and magnitude of mistakes, e.g., injury rates, product defects.
- Executive functions, e.g., creativity, initiative.
- Quantity of work output.
- Capacity for peak performance.

- Interpersonal factors:
    - Social function (positive morale impact of optimal health)
    - Addictions (manipulative behaviors)
    - Personality disorders (criminal behaviors)
    - Mood disorders (irritable with customers)
- Motivational: Observing a sick employee's low output lowers the expectations of fully healthy workers for their own performance.
- Work culture: The collective health or impairment level creates a culture that has a positive or negative effect on the health of the group.

Presenteeism directly impacts business operations, but it can be difficult to quantify related losses. All employees' experience a less-than-fully-productive day now and then. In one study, a Global Corporate Challenge survey of 2,000 employees using the World Health Organization's Workplace Health and Productivity Questionnaire for validation found the average rate of absence per employee was four days a year. However, respondents self-reported being unproductive on the job an average of 57.5 days a year, or about 75 percent of their productivity level at a cost 10 times higher than the cost of absence.[11]

Impairment may be caused by stress, family tension, ailing elders, medication, physical symptoms, and medical or mental diseases. In a multi-employer analysis of more than 300,000 pharmacy claims, 120,000 medical claims and 15,000 employees' health-related productivity costs, ACOEM and the Integrated Benefits Institute applied an employee survey instrument (the Kessler health and work performance questionnaire) to these data. Ten health conditions most often associated with lost productivity were identified: musculoskeletal complaints, depression, fatigue, chronic pain, sleeping problems, high cholesterol, arthritis, hypertension, obesity and anxiety.[12]

The health and safety risks associated with these types of conditions may only reveal themselves through an assessment. Additionally, an assessment of work fitness may be conducted in response to a suspected impairment, such as when:

- An individual has concerns about his or her own health and requests it.
- Co-workers or family members question an individual's physical ability or mental capacity to work safely.
- An employer observes behaviors that suggest an employee poses a risk of harm to self and others or a pattern such as interpersonal conflicts or failure to meet production quotas.

Protocols

Following a request from an employer, an occupational physician or other qualified medical professional should follow established protocols and best clinical practices to determine whether restrictions are medically indicated, and if so, for how long. All findings must be documented, including any specialist medical opinions or lab test results ordered as part of the evaluation process. The evaluator is expected to:

1. Integrate information obtained from various sources, such as occupational health, human resources, safety, legal, supervisors and confidential medical records.
2. Perform the assessment, either in person or through a surrogate.
3. Issue an opinion, e.g.: Is the individual capable of performing the job with an acceptable degree of risk? Are job accommodations needed?

4. Keep information confidential.

The evaluator is required to advise the applicant or employee, but not the employer, in the event an assessment identifies previously undetected health conditions that have no impact on the job, or if an existing health issue could potentially be aggravated by physical demands or environmental hazards. An employer can have liability if an assessment and follow-up notice to an applicant or employee is found to be neglectful (*Coffee v. McDonnell Douglas Corp.*).

Complying With Applicable Laws and Regulations

Certain laws and workplace regulations enacted to protect employees' medical privacy and right to work in a safe and heathy environment apply to the FFW assessment process. It is incumbent on EH&E professionals to be familiar with the laws and regulations that apply in their jurisdiction and to their specific industry.

The following briefly summarizes applicable federal statutes:

<u>Americans with Disabilities Act (ADA)</u>

This landmark legislation originally enacted in 1990 and later amended in 2008 allows applicants and employee medical evaluations to be performed on a post-offer and return-to-work basis as long as they are job-related, consistent with business necessity and equally applied. All applicants in the same job category must receive the same assessment. A separate psychological exam may be used to identify a mental health disorder or impairment; it cannot be used to single out people with "undesirable" personality traits.

An employer may use work-fitness assessment results to create suitable job accommodations for a worker with physical disabilities or deny placement based evidence of an inability to provide reasonable accommodations. Under the ADA, a reasonable accommodation may consist of job modifications, workstation or schedule adjustments, or a limited leave of absence. An accommodation is not considered reasonable if it poses a danger to an employee, co-workers or third parties, or if it will place an undue hardship on the business.

A FFW assessment helps identify protected disabilities. The discovery of a covered disability triggers a required reasonable accommodation analysis. An ADA-protected disability is defined as:

- A physical or mental impairment that substantially limits one or more major life activities.
- A record of physical or mental impairment.
- Being regarded as having an impairment.

In general, the ADA does not apply to transitory conditions such as a broken leg, conditions controlled by medication or devices (such as eye glasses), and certain physical characteristics. According to the Equal Employment in Opportunity Commission (EEOC), the federal agency that enforces certain provisions of the ADA, not every physical characteristic constitutes an impairment. For example, normal deviations in height, weight or strength are not impairments. However, morbid obesity (weight that is 100 percent or more over what is considered normal) is an impairment and will be considered a disability if the applicant or employee can show that it substantially limits a major life activity.

Family and Medical Leave Act (FMLA)

Under the FMLA, an employer may require an employee returning from health leave to present medical certification verifying fitness to resume work. The certification applies exclusively to the health condition associated with the leave. An employer may contact the medical evaluator to clarify and authenticate certification, but not obtain specific medical information. An employer may require an updated fitness-for-duty certification every 30 days as long as reasonable health and safety concerns exist.

Genetic Information Non-Discrimination Act (GINA)

GINA prohibits the use of family medical history and other genetic information as part of an assessment intended to determine an applicant's or employee's ability to perform a job. It protects applicants who may have a genetic predisposition for a specific injury or illness from being denied placement in a job that could potentially exacerbate the condition.

Health Insurance Portability and Accountability Act (HIPAA)

HIPAA is a far-reaching statute intended to protect patient privacy and ensure data integrity. It applies to the coordination of health care services – including medical surveillance, case management, disability review and onsite medical evaluations. It requires covered entities to make reasonable efforts to limit the amount of protected health information (PHI) that a medical professional uses or discloses to an employer to the *minimum amount necessary*. HIPAA does not apply to workers' compensation cases. However, many employers consider compliance with at least minimal privacy and security standards to be best practice – regardless of the nature of the complaint, delivery setting or circumstances. Typically, only legal and medical departments within companies have truly secure firewalls. At the workplace, access by other departments to medical information should be restricted. Documentation containing PHI must be stored in accordance with OSHA regulations (29 CFR, Part 1910 – Access to Employee Exposure and Medical Records). Medical records cannot be released to an unauthorized party without an employee's express written consent and authorization, except under specific circumstances that must be explained.

Wrongful Discharge

In general, an employer who discharges a person with a disability may do so only if the employer can demonstrate that:

- The person is unable to perform essential job functions.
- No essential reasonable accommodation exists that would enable the person to perform essential job functions.
- The person would create an imminent and substantial danger to himself/herself or a substantial danger to others by performing the job.
- No reasonable accommodation can be made to remove or reduce danger.

Adopting a Comprehensive Policy

To maximize their value, fitness-for-work evaluations should be part of a comprehensive, written workplace policy designed to promote effective management of complex workplace health and safety issues. The document should include supportable job demands for each position. Related components may include:

- Coaching staff on physical, behavioral and emotional signs of impairment and how to intervene.
- Pathways for safe return to work.
- Drug-free workplace policy and drug-and-alcohol testing parameters.
- Exercise, nutrition and disease management programs.
- Employee assistance, counseling and other behavioral health interventions.

A company may also adopt procedures, practices and design considerations that apply to ergonomic solutions in the workplace, including assessment of the fit between the worker and the workstation. While the U.S. does not have a mandated ergonomics standard, OSHA has issued guidance, and other organizations such as the International Standards Organization and the American National Standards Institute have published related standards.

Many employers look to government agencies, including the U.S. Nuclear Regulatory Commission and those operating within the Department of Transportation, for guidance on fitness-for-duty requirements designed to protect employees and the public. Industries with safety-sensitive jobs and organizations representing fire, police and other public safety professions are another reliable source.

Training

Education is key: Senior leaders, EH&S professionals and frontline supervisors should be coached on the management of FFW programs. Employees should have access to resources to increase their awareness of healthy behaviors.

Return-to-Work Pathways

An ACOEM guidance document on *Preventing Needless Work Disability by Helping People Stay Employed*[13] outlines steps employers can take to overcome return-to-work barriers such as not providing temporary modified work; allowing "light duty" to be perceived as a cushy assignment or meaningless work; and failing to intervene early. Employers are encouraged to develop transitional work programs and work with evaluating clinicians who use practical, evidence-based methods to evaluate and document workers' functional abilities. In another paper, ACOEM advises personal treating physicians to make sure their patient's current capabilities match or exceed the physical, psychological, cognitive and social demands of proposed work tasks.[14]

Drug-and-Alcohol Programs

The Drug-Free Workplace Act was enacted in 1988. Since then, Quest Diagnostics' analysis shows that drug use among American workers has declined 74 percent, while the rate of positive test results for certain drugs – including amphetamine and opiates – has continued to climb. In workplaces *without* a drug testing program, Quest Diagnostics' analysis of combined DTI and NSDUH data found:

- The rate of employee illicit drug is 13.9 percent compared to 9.5 in workplaces with a drug testing program.
- There is typically a 30 to 45 percent higher occurrence of self-reported drug use. Workplace drug testing programs have been proven to be valuable tools for deterrence and the promotion of safe and healthy work environments. For the purposes of assessing fitness for duty, alcohol and drug testing may be performed under the following circumstances:
- Pre-placement

- Government-mandated
- Safety or security-sensitive
- Post-damage or injury

Testing should be performed by the least intrusive method possible. Random testing is not typically used in FFW assessment situations.

Exercise, Nutrition and Disease Management Programs

Wellness generally refers to health promotion, disease prevention programs and activities offered to employees as part of an employer-sponsored group health plan or separately as a benefit of employment. Many of these programs ask employees to answer questions as part of a health risk assessment and/or undergo biometric screening for risk factors such as high blood pressure or cholesterol levels. Wellness programs typically provide educational health-related information and classes on nutrition, weight loss and smoking cessation; onsite exercise facilities; and/or coaching to help employees meet health goals. Employers may provide limited financial and other incentives in exchange for an employee answering disability-related questions or taking medical examinations as part of a workplace wellness program.

The EEOC states that an employee health program must be "reasonably designed to promote health or prevent disease." A wellness program that asks employees to answer questions about their health conditions or have a biometric screening or other medical examination for the purpose of alerting them to health risks is considered reasonable. Collection and use of aggregate health risk assessment (HRA) findings to design and offer targeted programs also meet this standard. Asking employees to provide medical information on a HRA without providing any feedback about risk factors or using aggregate information to design programs or treat any specific conditions is not considered to be reasonably designed. An employer cannot require an overly burdensome amount of time for participation, involve unreasonably intrusive procedures or require employees to incur significant costs for medical examinations.

Employee Assistance Programs

A FFW assessment may be used to identify the need for an EAP referral for further evaluation, counseling and other interventions related to mental health or substance use issues. An EAP may be used as a resource for return-to-work monitoring and unannounced follow-up testing in the event an employee has a positive drug test. In many cases, family members may access EAP counseling services.

The Employee Assistance Professionals Association's *Workplace Outcome Suite 2016 Annual Report* features findings from a survey of 13,400 EAP participants.[14] Among results reported:

- Survey participants missed 12.2 of work over a 30-day period before EAP services were offered. After 90 days of EAP services, participants missed 6.5 hours of work over 30 days, a 46.4 percent improvement.
- Participants experienced a 26.7 percent improvement in concentration at work after 90 days of EAP services.

In Summary

There is not a single, stand-alone solution that can be introduced to reduce risks associated with behaviors or conditions that may impair fitness for work. EH&S professionals and their colleagues must apply a combination of strategies in order to provide adequate protections for employees, employers and the public.

Endnotes

1. Laws, C. and D. Colon. "Comorbidities in Workers Compensation." National Council on Compensation Research Brief. October 2012. (https://www.ncci.com/Articles/Documents/II_research-brief-comorbidities-in-workers-compensation-2012.pdf)
2. "Travelers Injury Impact Report." ©2016. (Retrieved March 7, 2017) (http://investor.travelers.com/Cache/1500085614.PDF?O=PDF&T=&Y=&D=&FID=1500085614&iid=4055530)
3. "Liberty Mutual Workplace Safety Index." 2016. Liberty Mutual Research Institute for Safety. (https://www.libertymutualgroup.com/about-liberty-mutual-site/research-institute-site/Documents/2016%20WSI.pdf)
4. Tulio de Mello, M., et al. "Sleep Disorders as a Cause of Motor Vehicle Collisions." International Journal of Preventive Medicine. March 2013, 4(3): 246–257.
5. Jupe N. "Data Shows Escalating Drug Use in the U.S. Workforce." Quest Diagnostics blog post. Jan. 24, 2017. (http://blog.employersolutions.com/data-shows-escalating-drug-use-in-the-u-s-workforce/)
6. "A Nation in Pain." Express Scripts. Dec. 9, 2014. (http://lab.express-scripts.com/lab/insights/drug-safety-and-abuse/americas-pain-points.)
7. "Workplace Violence." Occupational Safety and Health Administration web page. (Retrieved March 7, 2017) (https://www.osha.gov/SLTC/workplaceviolence/)
8. "The NSDUH Report: Depression Among Adults Employed Full-Time, by Occupational Category." Substance Abuse and Mental Health Services Administration, Office of Applied Studies, Rockville, MD. Oct. 11, 2007.
9. National Comorbidity Survey. Harvard Medical School. 2005.
10. "Workplace Health Promotion." Centers for Disease Control and Prevention web page. (Retrieved March 6, 2017) (https://www.cdc.gov/workplacehealthpromotion/health-strategies/depression/index.html)
11. "Presenteeism costs employers 11 weeks per worker." HR Daily. March 15, 2016. (Retrieved March 8, 2017) (https://www.hrdaily.com.au/nl06_news_selected.php?selkey=3893)
12. Loeppke, R. "The value of health and the power of prevention." International Journal of Workplace Health Management, Vol. 1, No. 2, 2008: pp. 95-108.
13. "Preventing Needless Work Disability by Helping People Stay Employed." Journal of Occupational and Environmental Medicine. 48(9) September 2006: pp. 972-87.

14. "The Personal Physician's Role in Helping Patients with Medical Conditions Stay at Work or Return to Work. American College of Occupational and Environmental Medicine Position Paper. Dec. 11, 2008.

15. "EAPs Can and Do Achieve Positive Workplace Outcomes: Workplace Outcome Suite Annual Report 2016." International EAP Association and Chestnut Global Partners. 2016.

Session No. 610b

Contractor Safety Management in Oil & Gas KNPC Refinery

Ali A. Kshawe
Manager-HSE KNPC Kuwait

Narayanan Vasudevan, CSP
Safety Engineer KNPC-MAA Refinery Kuwait

Kuwait National Petroleum Company
Kuwait

Introduction

Ali A. Kshawe and Narayanan Vasudevan have worked in field of HS&E for past 25 years. During their course of work in petroleum refining the authors have been actively involved in developing HSSE requirements for contractors in refinery and implementing the same.

Kuwait National Petroleum Company (KNPC) is a major oil refining company in the world. Like of most of oil and gas production facilities KNPC outsources their maintenance and project construction activities. Oil refining industries have inherent hazards and risks which need to be mitigated during maintenance and project execution. Certain activities of contractors have potential to place either contract personnel or operating personnel or facility at risk.

A sound contractor safety management will mitigate potential risk by having a proper system of contractor selection and monitoring their activities on day to day basis.

KNPC has developed a sound management system for all KPC subsidiaries. It begins with planning which include communication of risks of contract to potential bidders and the company HSSE management system which part of scope of contract. Each potential contractor has to undergo an HSSE prequalification. The objective of the HSE Pre-qualification is to screen potential contractors to establish that they have the necessary experience and capability to undertake the scope of work in an HSE responsible manner, and to effectively deal with the associated risks. The contractors who pass the prequalification are included in list of bidders.

Potential bidders of contractors are invited for a Pre-tender meeting at client facility to brief/clarify on the HSSE expectation during the contract. Contractors HSSE guidelines is part of tender documents which lists are HSSE expectation of KNPC to contractor.

Contractors are expected to present their HSSE plan for all contracts which will include Pre-mobilization and post mobilization requirements. These requirements are audited during mobilization and contractor execution stage by KNPC custodian of contract along with KNPC HSSE department.

Contractors HSSE plan shall include demonstration of commitment of leadership, HSSE policy of the contracting company, HSSE strategic objectives, HSSE organizational structure, HSSE communication, HSSE & Competency training assurances, Risk assessment and critical task procedures, Management of environmental, health & security hazards, HSE-MS implementation plan, HSSE performance KPI & performance monitoring, HSSE audit plan and Management review.

Post contract HSSE Performance/Closeout Report Prepared, Submitted and recorded for Future Reference.

The main objective this presentation is to highlights KNPC contract management system in brief as listed above. Presentation will also address the KNPC method of continuous monitoring of contractors. It will also address KNPC continual improvement plan on contractor HSSE management system.

KNPC Contractors HSSE Management System

KNPC HSSE management system for contractors have various elements and stage gates from the tendering to completion of project which will is being described in the following paragraph

Pre-Qualification:

KNPC has a structured pre-qualification criteria for qualifying a contractor for a project. Prequalification of contractor is judged based on the past work experience, Financial strength, Local support and more importantly the HSE capability & track record of the contractor. Contractor is evaluated by a set criteria which include the HSE management of the company & Past incident records. Only Prequalified contractors are eligible for bidding for a contact.

Pre-Tender Meeting

Pre-tender meeting (PTM) is called by commercial department with all prospective bidders who are pre-qualified for KNPC contract. All the contractors who have bided for tender are invited for the PTM. KNPC is represented in the meeting by commercial department, user department of the contract and HSE department. Contractors are briefed on the scope of contract along with commercial procedure. HSE department representative will brief the HSSE expectation from the contractors and provide all necessary clarifications on the contractor HSSE guidelines which is part of contract document. All clarifications sought and explanation given by KNPC are documented and is part of the tendering process. The main purpose of the PTM is to ensure that all contractors clearly understand the scope of work along with client expectation. This will ensure that contractors will quote appropriately and cost will consider all requirements. An underquote will result in substandard quality of work and HSSE requirement during execution and subsequent operation will be at risk.

Contractor HSE Guidelines

Contractor HSE guidelines is intended to make prospective contractors aware of basic HSE requirements of company. This guideline will supplement contract terms and conditions with respective to HSE issues. Contractor and contractor employees shall make themselves familiar with HSE guidelines for contractors prior to commencing work at company premises.

KNPC contractor guidelines stipulates that all contractors will have a company representative identified who is referred as contract superintendent. All questions with respect to contractor performance is referred to contractor superintendent who will be focal point of contract at KNPC during course of contract. Contractor shall maintain a representative in the field at all times who shall be fully in charge of contractor employees and maintain liaison between KNPC and the contractor.

The company representative (contract superintendent) shall be accountable to ensure compliance of contractors HSE guidelines requirements and shall continuously monitor the contractor HSE performance. Contractor is responsible for strict compliance of this guidelines by informing his employees and sub-contractors employees if any of the requirements and all specific instructions with respect to work execution.

Contract HSE Plan

Contractor as a part of tendering process shall demonstrate compliance to HSE requirements by preparing and submitting its HSE plan within the framework of tender. This shall be appropriate to size, complexity and risk of the contract and shall be meet all requirements stipulated and in contractors HSE guidelines.

The contractor HSE plan shall demonstrate how the contractor and subcontractors if any has an effective HSEMS applicable to the specific work which is contracted. Hazards and effects of the contractor work to the environment, people assets and reputation of the company have been identified assessed and controlled will be documented in HSE plan. The responsibilities of the execution of the job as per HSSE requirements shall be assigned to specific designated persons throughout the phase of the contract. Risks evaluation and measures to reduce the risk to ALARP and acceptable to company shall be addressed.

The contractor HSE plan need to be finalized post contract award in terms of contract award. This HSE plan shall be made available at least one month prior to mobilization of manpower and equipment to site. Progress of the HSE plan shall be presented every six months to KNPC in a workshop conducted in coordination with contract custodian.

Pre-& Post Mobilization HSE Audits

Post award HSE meeting are held immediately after contract award and before execution of work. This meeting covers the review of major risk controls for the contract, confirmation of activities/deliverables described in HSE plan for completion and pre-mobilization requirements. Further competency of workforce, HSE manpower availability as per contract requirements, confirmation of KPI, confirmation of HSE activities like communication meetings, audits & reviews, Review of company & contractor's emergency response plan, Incident reporting & investigation procedures, Incentive & penalty scheme etc. After above meeting a contract HSSE Readiness for commencement will be issued by company.

At the end of contract demobilization is conducted according to HSE plan. The hazards associated with demobilization is reassessed, any new hazards are identified and control measures are implemented to minimize the risk. HSE plan shall be modified accordingly. An HSE clearance certificate for site restoration will be issued by company.

Final HSE Evaluation and Close Out Report

Contractors HSE performance will be jointly evaluated by contractor and company representative. Contract is closed out with a report of HSE performance and lesson learned, providing feedback for future knowledge and improvement. The contractor HSE data is documented and will be considered during bidding for future projects.

Health Safety Environment Management system (HSEMS)

Contractor must have a written HSE management system approved by KNPC HSE department prior to commencement of work at company worksite. The HSEMS prepared for the contract will depend on the variable such as size of firm, size of project, nature of activities and location. Contractors HSEMS will be in line with KNPC system which include following elements

- Leadership & Staff commitment
- Process Safety Management
- Operations procedure and Safe work practices
- Process quality assurance project
- Pre-Startup safety Review
- Mechanical integrity
- Training & performance
- Contractors/Vendor performance
- Incident investigation and communication
- Emergency planning & response
- Occupational Health
- Environment
- Security (Cyber & Physical)
- Auditing & performance management

Site HSE Organization and Responsibilities

The contractor HSE program shall include site HSE organization with respect to overall site organogram. Dedicated HSE personnel requirements based on the manpower of the contract is stipulated in KNPC. Contractor HSE personnel shall be mobilized before commencing of activity at KNPC sites. Nonconformity on this requirement, a penalty of KD 100 per person per working day delay are imposed on the contractor. Maximum number of contractor employees estimated to be engaged in site on any day will be calculation basis if HSE manpower requirements of the contract. In case of major shutdown activities contractors are required to mobilize HSE manpower based on the expected manpower as per requirements stipulated in guideline.

HSE Orientation

KNPC HSE department conducts HSE orientation (induction training) for contractor key personnel. The orientation program covers basics KNPC HSE program, HSE rules within the company, Company incident reporting and investigation procedures, PPE requirements company emergency response plan, Company HSE policy, Company waste management system and roles & responsibilities of employees. Company access badges are given to employees after HSE induction training. Subsequently contractor Key personnel are supposed to provide training to all contractor personnel.

HSE Meetings

Communication is key to HSE management. Contractor shall hold at least monthly HSE meeting chaired by contractor site manager and attended by representative from each level of contractor. The meeting agenda include HSE records and activities, statistics, incidents, PPE, new/revision in KNPC HSE procedures and other employee HSE concerns at site. These meetings are documented and action required are followed up for implementation.

Tool Box meeting is mandatory to be conducted prior to carrying out an activity at site. These are conducted by job supervisor who communicates the hazards of the activity to workers and mitigation plan in place to carry out the job safely.

HSE & Competency Trainings

Contractor shall establish a structured HSE & job competency training program for contract employees. Contract shall conduct a training need assessment and establish a HSE training matrix based on the scope of contract. These HSE training matrix will be in line with KNPC HSE training matrix and approved by KNPC HSE department. Contractor must maintain a training facility.

The contractor shall provide Technical training as required for each craft of the employee as per scope of contract. Employees will be trained on all the technical aspects of performing task/operating the equipment, associated job hazards and methods by which these hazards can be controlled.

In addition, contractor shall identify and develop certified training for each craft. Training may include certified trainers, crane operators, rigging, gas testing, Scaffolding etc. Certified training shall be provided to employees by KNPC approved third party agencies. Contractor corporate manager shall carry out six monthly audit on their training program and ensure that training program meet quality requirements of KNPC

Contractor HSE KPI

KNPC top management sets Key Performance indicators (KPI) of the company at their Safety Health Environment Executive committee meeting(SHEEC) which is chaired by CEO during April every year. This KPI are set considering the past year HSE performance and the identified years of improvement in coming fiscal year. Contractors KPI are derived from the KNPC approved KPI as they are business partners for running refinery operations smoothly.

Common KPI used to monitor contractor performance include HSE training intensity, HSE audits planned against conducted, periodic medical examination status, % of HSE recommendation completed, Lost work case Frequency & severity rates, Trend of non-compliances noted from defined work practices, Cumulative good days etc. These KPI re monitored in periodic HSE communication meetings. Contractor corporate manager shall carry out an annual assessment of their site performance and provide feedback to company representative.

HSE Inspection & Audits

HSE inspections & audits are visible signs to employees that HSE is a top priority and issues are handled appropriately. KNPC expects every contractor to develop audit plan for each level of contractor management. Contractor top management shall conduct periodic HSE visibility visits. Contractor Job managers & Engineers shall carry out safety audit of the work area at least once a week. Contractor Safety personnel shall carry out site inspection daily. The results of the site visits/audits should be used for setting goals for future.

Medical Examinations

Contractor employees should undergo pre-employment and periodic medical examinations as per company requirements. In addition, special respiratory users and confined space entrants undergo fitness examinations. The contents and frequency of such medical examinations are specified in company procedure which are part of tender document.

Waste Management System

Contractor shall ensure to full compliance to company solid waste management system. Waste tracking system is effective system of records, handling and disposal of waste substance arising out of contractor activities. Appropriate waste manifest need to be filled for transporting and disposal of waste. The contractor shall ensure that KNPC approved Material Safety Data Sheet (MSDS) is always available at work site for hazardous materials & chemicals handled by contractors.

HSE Performance Incentive Rewards & Penalties

A periodic quarterly evaluation on contractors HSE performance within the duration of the contract is carried out by Company representative and HSE department jointly. These evaluations are used for providing incentives to contractors on excellent HSE performances. Contractors consistent substandard performance can lead to termination of contract and future hiring of contractor is subject to acceptable past HSE performance.

All contractors and contractor employees shall comply with company HSE procedure, guidelines and safe work practice. Violations of company requirements will lead to penalty which can be monetary or more severe as suspension/dismissal depending on the severity.

Conclusion

KNPC HSSE policy clearly states that HSSE is integral part of business. Contractors are considered as business partner in running business. KNPC management believes and demonstrate that HSSE is of prime importance and protecting people property and environment is a top priority. Company provides all necessary support to contractor in implementation of HSSE system to have a safe, healthy & working environment.

Bibliography

Kuwait National Petroleum Company. Pre-Qualification for Engineering, Procurement and Construction (EPC) Contractors, KNPC-EPCPQ- 001 Rev. 00. (http://www.knpc.com.kw/Documents/Gen%20PQ%20Doc.-%20Rev.%2000%20final.pdf)

Reyes, Jaime M. Health, Safety & Environment Guidelines for Contractors. Kuwait National Petroleum Company, HSE-TSSA-05-1118, rev September 2016.

The New Silica Standard – Assessing Exposure in the Oilfield

Timothy Hicks, MSPH, CIH, CSP
HHS Consulting, Inc.
Charleston, South Carolina

Introduction

Last year, OSHA promulgated a chemical-specific standard for Respirable Crystalline Silica (RCS) in the forms of quartz, cristobalite, and tridymite. Among other impacts of the standard, it lowered the Permissible Exposure Limit (PEL) significantly and established an Action Level (AL) at one-half the PEL that triggers various aspects of the standard. Medical surveillance is required if workers are exposed to greater than the AL for 30 days or more per year. This chemical-specific standard is different from the others in that it requires the use of objective data by employers to demonstrate compliance in industries or work environments known to carry the risk for exposure to RCS. This document focuses on the general industry standard, 29 CFR 1910.1053.

The standard incorporates a phased compliance approach within five years of the effective date, June 23, 2016. Most aspects of the standard become enforceable within next year in 2018. Medical surveillance is required for workers found to be exposed to RCS levels at or exceeding the PEL on that date; however, employers are allowed to delay surveillance activities for workers exposed to less than the PEL but at or more than the AL until 2020. Operations involving hydraulic fracturing have one more year, in 2021, to implement feasible engineering controls. This is likely due to ongoing partnership activities occurring between that industry and OSHA.

The previous PEL was based on epidemiological and toxicological research available prior to the publication of the 1968 ACGIH Threshold Limit Values (TLV's) that were incorporated into Subpart Z at the time of promulgation in the early 1970's. Much more has been learned about the effects of exposure to RCS in the 50 years since. The primary disease associated with exposure is silicosis, or a type of pneumoconiosis that results in chronic progressive obstructive pulmonary disease. Besides the obvious reduction in the quality of life among affected workers, the disease may result in lung cancer. Other diseases, such as those affecting the kidney or an increased susceptibility to tuberculosis are also known to be associated with exposure to RCS. OSHA estimates the exposure limit reduction along with medical surveillance will save 642 workers annually from these occupational diseases.

Standard Applicability

The language of the RCS standard contains several definitions that have a significant impact on how an employer may demonstrate compliance - chiefly, the term "Objective Data". In part, the term is defined as air monitoring data, such as those gathered via area sampling of the work environment or those gathered from the personal breathing zone (PBZ) of workers that are similar in nature such that the data is comparable among several workers. This is the concept of "Similar Exposure Groups" (SEG's). Objective data of this nature may be used to estimate exposures across similar work activities and environments. Another part of the definition includes calculations based on the composition of the substance. Presumably, this may include data on safety data sheets (SDS) for substances that do not contain RCS as a means to demonstrate negative exposure. Objective data is a key concept in exposure assessment.

Another important term found in the standard is "exposure". Exposure is defined as the concentration of RCS in the personal breathing zone (PBZ) of a worker without the benefit of personal protective equipment, such as a respirator. While exposure does not always translate into dose (or worker uptake) when adequate protection is available and used, it may trigger various aspects of the standard such as medical surveillance.

Though very similar, the general industry standard does not apply to employees engaged in construction work. A few differences between this standard and the construction standard, 1926.1153, exist – most notably the responsibility for oversight by a "competent person". Curiously, the standard contains an exemption for unusual (non-regular) tasks occurring in varying work environments. Another exemption occurs when the employer can demonstrate, using "objective data", that no foreseeable exposure to RCS in excess of the AL exists. To demonstrate this, employers would likely need to conduct and document a negative exposure assessment for the work task, material, and work environment. This standard does not contain applicability triggers for the composition of the RCS-containing material. As such, this standard appears to apply to all general industry work not using absorptive clays unless or until employers demonstrate a negative potential for worker exposure.

Exposure Assessment

The purpose of exposure assessment is to determine which employees or groups of employees are exposed to RCS equal to or more than the AL (or PEL) and, consequently, which might be at risk for silica-related disease. Remember, the concept of exposure included in the standard does not take into account any benefit of reduction offered through the use of respirators. Exposure assessment involves the collection and/or evaluation of sample data. As such, this activity should be conducted under the advice or guidance of an industrial hygienist. The standard offers two strategies for employers: A performance option or a scheduled monitoring option. The performance option involves the use of sampling data related to the exposure of each employee individually. This data may include objective data such as area sampling results (as opposed to PBZ sampling), equipment manufacturer data relating to the effectiveness of controls related to that equipment, and may possibly include elements of Table 1 in 1926.1153. It is the author's opinion that this option may be best used defending negative exposure assessment(s).

Alternatively, the scheduled monitoring option allows for the use of SEG's as opposed to data related to individual employees. The advantage to this approach minimizes sampling load such that workers having similar exposures can be grouped in a manner that data from one may be used to represent another. This approach requires employers to sample worst case exposures among members within an SEG. Absent empirical data indicating which are, in fact worst case, it is the author's opinion that employers should adopt guidance offered by NIOSH in Publication 77-133 related to minimum sample size as related to sample group variability. This guidance assists employers in defining an approach that ensures worst case exposures will be captured during a sampling regimen for a particular SEG.

The standard requires that employers reassess exposures following a change in materials, work task(s), or work environment that may potentially increase employee exposures such that they equal or exceed the AL.

If an employer determines employee SEG's equal or exceed the AL, they must repeat the process semi-annually. If exposures are found to equal or exceed the PEL, they must repeat the process every quarter year. This establishes an ongoing routine for workplace exposure reassessment unless or until reduced to below the AL through the use of feasible engineering or administrative controls. If/when the exposure of a SEG is reduced to below the AL, the employer may discontinue the reassessment routine if the reduction is confirmed using the repeat sampling direction contained in the standard. Importantly, if the initial assessment reveals exposure at a level below the AL, the employer may use this for documentation of a negative exposure assessment.

Example Exposure Assessment

One SEG found at hydraulic fracturing operations are employees who operate/control the flow of sand, called "proppant," from storage bins to a blender where where it is mixed with other chemicals prior to being injected downhole. Proppant from the bins are gravity fed onto a horizontal belt. Job duties include monitoring the rate of proppant transfer along with the the remaining level in the bin compartments. The conveyor system may be operated from the ground (Figure 1) or from an elevated operator station at the transferring end of the bin. This SEG may be exposed to dust from handling proppant containing RCS in concentrations much higher than the AL in an uncontrolled environment.

Figure 1. Sand bin operator transferring proppant.

In order the conduct an exposure assessment for Sand Bin Operators, a SEG, the standard requires employers to use an approved analytical method. One method listed in Appendix A of the standard is NIOSH NMAM 7500, respirable silica by x-ray diffraction. This method requires the use of a size-selective particle sampler, a cyclone, used a specific flow rate to deposit only particles averaging 0.4 microns on a filter media. The media is then sent to an accredited laboratory for analysis. The standard requires that the laboratory meet ISO 17025 standards in terms of quality control and assurance. The laboratory will then develop a report with the mass of silica found by x-ray diffraction along with the airborne concentration given the total volume of air passed across the filter during the sampling period. To ensure an accurate sample result, it is critical that the individual sampling the activity takes note of the flow rate and the time spent sampling. Flow rate should be calibrated with a primary standard before and after the sampling event to ensure the rate remained consistent throughout the sampling period. This is one means to reduce sampling error. The results are only as good as the accuracy and precision of the sampling and analytical methods. Together, they have a limit of detection below which the measure is no longer detectable to the method. This limit can sometimes be reduced by increasing sample volume. It imperative this lower detection limit is above the AL. Additionally, one should expect variability in sample results gathered from workers within a SEG. Results will rarely, if ever, be the same.

Sampling workers engaged in hydraulic fracturing operations presents two challenges: Extended work shifts of more than 8 hours, and batch processing. The first relates to the definition of the PEL and AL in that it is an 8 hour time-weighted average, not 12. This means that the sampling period must occur during an 8 hour period comprising the highest (worst case) exposure period of a shift. This strategy would be simple, if that period were well defined during a particular well stimulation (frac-ing) operation. Frac-ing a well occurs in stages defined by periods when proppant and chemicals are blended and injected downhole followed by periods of inactivity. This cycle, or batch, may occur multiple times per shift and is determined largely by the volume of sand, water, and chemicals to be injected during a particular stage. In the example noted in Figure 2 below, six PBZ samples were gathered from Sand Bin Operators during a shift. The uncontrolled worst case (highest) exposure among the six samples was found to be 1220ug/m3 as a 12h TWA. When reduced to an 8 hour TWA using the example data, the exposure increases to 1830ug/m3. This is a 33% increase when the sampling data includes 480m of the both stages pictured in the graph and reduced inactivity time. This is an example of the challenges inherent to sampling workers in the industry.

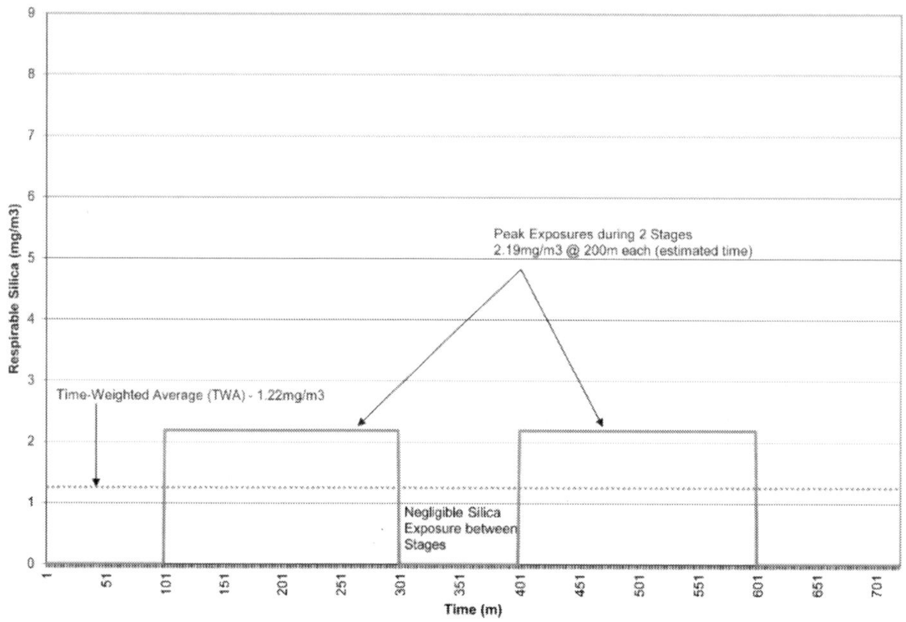

Figure 2. Example worst case TWA exposure during a 12 hour shift.

Exposure Control

Exposure control is a primary disease prevention activity required of employers. The strategy should follow established control hierarchy with emphasis on substitution or elimination of the hazard followed by engineering solutions. One example of substitution for the control of RCS dust generated from proppant transfer activities is the use of ceramic particles in place of sand for this application. Another might be the use of various coating compounds currently available on the market to treat the sand at the mine or during transloading prior to shipment to the well site. In the last 3 years, industry groups have been focused on the development and deployment of dust capture devices as engineering controls. They may be passive control devices that utilize the energy of pneumatic transfer to filter respirable dust or active control devices that suspend dust and deposit it in a central collection device where the exhaust air is filtered. Administrative control must include access control of areas measuring RCS concentrations at the PEL or greater, according to the standard. Personal protective equipment would include the use of respirators in those areas, should they exist at a work site.

Most employers are required to deploy "feasible" engineering controls by next year, 2018. The hydraulic fracturing industry has until 2020 to do so. Feasibility determination will likely be a point of contention for many employers as with occupational noise. OSHA's enforcement policy for occupational noise allows for a cost benefit approach to feasibility; however, RCS standard requires the use of engineering controls even when they are not completely effective at reducing workplace exposures below the PEL. It appears that feasibility determinations will be driven primarily from a technical standpoint. Employers in the hydraulic fracturing industry will find "infeasibility" difficult to demonstrate with the various engineering controls available and currently in use throughout the industry.

For the hydraulic fracturing industry, NIOSH has developed and patented passive dust control devices, or mini-baghouses, that can be retrofitted onto sand bins. Similarly, Sierra Dust Control and Airis

Wellsite Services also have systems available on the market that employ active dust control and collection strategies to reduce workplace exposures.

Regulated Areas

The RCS standard requires employers to identify, demarcate, and control access to areas where the airborne respirable silica concentration exceeds the PEL. In order to determine these areas, employers must measure the workplace concentrations if foreseeable exposure to RCS exists. Employees who enter these areas are required to utilize respiratory protection in accordance with a Respiratory Protection Program.

Various direct reading devices are available on the market that have the capability of measuring respirable dust. Unfortunately, few of these are capable of delineating how much of that dust contains silica. An employer might elect to assume that all respirable dust indicated by a device is silica as a conservative estimate. Otherwise, area sampling should include lab analysis of the respirable dust fraction for silica or other objective data. If a regulated area exists at a work site, then exposure to RCS among employees at the site is certainly foreseeable such that a formal exposure assessment is also required. Conversely, if area sampling does not indicate airborne concentrations at or above the AL, employers may be able to use the information as objective data toward a negative exposure assessment.

Exposure Control Plans

The standard requires employers to develop a written Exposure Control Plan (ECP) if foreseeable exposure to RCS exists in the workplace. This plan incorporates various elements required by the standard related to defining tasks involving exposure to RCS, and describing control strategies associated with each task. Housekeeping measures must also be included in the plan and may include such things as worker decontamination upon leaving a regulated area, the handling of fines, or captured dust particles. The ECP must be reviewed annually by the employer and essentially revalidated.

The scope of an ECP is not well defined in the standard. For example, is one ECP sufficient for a company, a facility, a site? Must there be only one ECP per site or could multiple ECP's at a particular site? It is the author's assumption that the scope of the document will be defined per regulated area on a particular site.

In the hydraulic fracturing industry, one Exposure Control Plan may not fit all sites. The example in Figure 3 illustrates variability among uncontrolled exposures at various sites on the basis of proppant size such that exposure potential increases with decreasing proppant size. Exposure variability among sites may also be driven by differences in engineering controls utilized, or differences in equipment configuration. Other drivers may include the arrangement of the site in terms of the number of sand bins deployed, the volume of proppant used per stage, and the use of gravity feed silos or boxes in place of traditional sand bins.

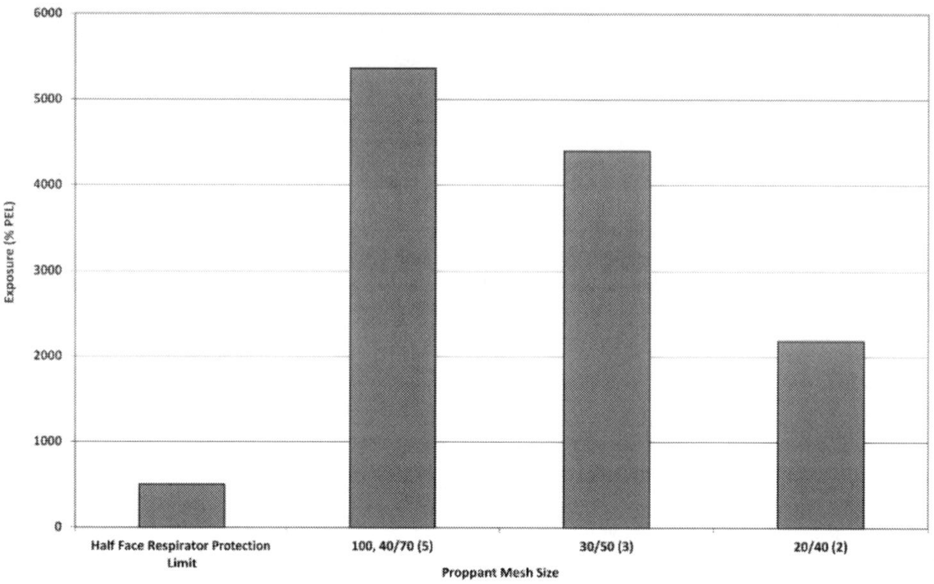

Figure 3. Example exposure assessment by proppant size absent engineering controls.

Figure 3 also illustrates inadequate exposure control with the use of a half-facepiece respirator alone, as the maximum use concentration is limited to 500ug/m3 or ten times the PEL.

Respiratory Protection

If workplace exposures to RCS remain above the PEL following deployment of feasible engineering or administrative controls, employers must offer respiratory protection to employees when engaged in tasks involving such exposures – including those associated with equipment installation, breakdown, and maintenance. Such protection is also a requirement for entry into a Regulated Area, should any exist at a work site. The issuance and use of respiratory protection for this and other purposes is covered under the Respiratory Protection standard, 1910.134. Additional requirements such as medical evaluation of wearers, fit testing, and hazard assessment apply. Importantly, workplace concentrations must not exceed the Maximum Use Concentration (MUC) for the respirator issued. The MUC is a function of the PEL. Lowering the PEL to 50ug/m3 technically decreased the effectiveness of respirators used for protection against RCS. For the hydraulic fracturing industry, meeting this requirement will be challenging absent the use of engineering controls.

Training

The new standard requires employers to integrate RCS into their existing Hazard Communication Program. As such, training must include must include carcinogenicity, effects on the lungs, and effects on the immune system and kidneys associated with exposure to RCS. Employers must also incorporate elements of the Exposure Control Plan into the training specifically related to tasks involving exposure to RCS along with control methods employed. If a medical surveillance program is required, the employer must also describe it. A copy of the RCS standard must also be made available to employees. These elements require an employer to conduct an exposure assessment and develop an ECP before conducting training that meets this standard.

Medical Surveillance

Medical surveillance is a secondary disease prevention strategy required of employers who employ workers exposed to workplace concentrations of RCS equal to or exceeding the AL for 30 days or more per year. The purpose of this program is to identify early signs of silica-related disease among exposed workers, and increase effectiveness of controls if necessary. The program consists of various medical tests conducted by a Professional Licensed Health Care Provider (PLHCP), requires employers to repeat the testing protocol every 3 years, and requires management of associated records. For employers experiencing high employee turnover or those having a temporary workforce, this requirement will likely be difficult to manage. This will make exposure control methods a much more attractive option to most employers in the hydraulic fracturing industry.

Recordkeeping

Air monitoring data and objective data are employee exposure records subject to 1910.1020. Both data sets require specific elements defined in the standard including employee social security numbers. Objective data records must also include the data source. A data source available to members of the hydraulic fracturing community was made available to them by the American Petroleum Institute last year. This source is one example of objective data that could be used to supplement the exposure assessment process. Medical surveillance records are employee medical records subject to the same standard listed above.

Bibliography

American Industrial Hygiene Association (AIHA). 2015. *A Strategy for Assessing and Managing Occupational Exposures, 4th edition* (ISBN 978-1-935082). Jahn, S, et.al.

National Institute for Occupational Safety and Health (NIOSH). 1977. *Occupational Exposure Sampling Strategy Manual* (Publication 77-173). Leidel, N, et.al: Department of Health and Human Services

______. 1990. *Silica, Crystalline, by XRD (filter redeposition):7500.* NIOSH Manual of Analytical Methods. Centers for Disease Control.

Occupational Safety and Health Administration (OSHA). 2011. 29 CFR 1910.1020, *Access to Employee Exposure and Medical Records* (retrieved March 7, 2017) (https://www.osha.gov/pls/oshaweb/owadisp.show_document?p_table=STANDARDS&p_id=10027)

______. 2016. 29 CFR 1910.1053, *Respirable Crystalline Silica* (retrieved March 7, 2017) (https://www.osha.gov/pls/oshaweb/owadisp.show_document?p_table=STANDARDS&p_id=1282)

______. 2011. 29 CFR 1910.134, *Respiratory Protection.* (retrieved March 7, 2017) (https://www.osha.gov/pls/oshaweb/owadisp.show_document?p_table=STANDARDS&p_id=12716)

______. 2010. Federal Register 75:64216-64221. *Interpretation of OSHA's Provisions for Feasible Administrative or Engineering Controls of Occupational Noise* (retrieved March 7, 2017) (https://www.osha.gov/pls/oshaweb/owadisp.show_document?p_table=FEDERAL_REGISTER&p_id=21773)

______. 2016. Federal Register 81:16285-16890, *Occupational Exposure to Respirable Crystalline Silica; Final Rule* (retrieved March 7, 2017)
(https://www.osha.gov/pls/oshaweb/owadisp.show_document?p_table=FEDERAL_REGISTER&p_id=26072)

Fatality Prevention Auditing: Effective Design and Deployment

Peter T. Susca, MS
OpX Safety
Wethersfield, CT

Introduction

The author was introduced to occupational fatalities and serious injuries (FSIs) in the mid 90's while working as a safety professional in the elevator/escalator industry. During that era fatalities and serious injuries were common in the industry. Despite experiencing 8-10 fatalities per year, our company was respected for having one of the best safety records in the business. We maintained a very robust safety management system and design/implementation standards which were created through industry standards, best practices and lessons learned from our FSI events.

Recognizing that our safety approach, albeit successful at reducing OSHA Recordables, wasn't designed to effectively predict and prevent fatalities, we needed some way of focusing on operational risks with the potential to kill or seriously injure our employees – a field FSI risk reality check. To fulfil this need we developed a Fatality Prevention Audit (FPA) process. Our FPA was a globally deployed assessment of actual field operations to known FSI hazards and our required controls. It resulted in an excellent reality check of field processes, behaviors, conditions and precursors of FSI events which ultimately led to a significant reduction in FSIs.

Since those days I have developed and deployed numerous FPA programs for a wide variety of industries such and manufacturing, mining and steelmaking. Even in such a wide variety of operations, the process of designing and deploying an effective FPA process is largely the same.

This paper focuses on the key elements essential to the initial development and implementation of an effective FPA process – independent of your industry or operational environment.

Why Do You Need An FPA?

Is your organization surprised when you have a serious or fatal accident, because you never saw it coming? Even if your organization has never experienced a fatality, asking "Why do we need a Fatality Prevention Audit" is a fair and very appropriate question. If your organization doesn't fully understand and recognize the value of this process, it will become more of an organizational burden than a value. And, if not deployed properly, it will offer a false sense of FSI risk security to management.

Here are some of the attributes of this process that should help distinguish it from safety processes you may already have in place:

1. The FPA focuses only on FSI situations that exhibit "Clear & Present Danger" (C&PD). *Clear & Present Danger* is defined as *"an act, situation or condition foreseen to be a Direct Cause of a fatality or life threatening serious injury"*.
2. The process requires auditors to assess situations where employees and/or contractors are actively engaged in work activities. Though conditions (static work areas, equipment, tools, etc.) are valuable evidence of fatality risk, they are not the prime focus of this process.
3. The process of gathering "what people really do" requires auditors to not only understand the reality of work practices (hence the value of Subject Matter Experts), they must be competent in the art of asking the right questions in an appropriate manner.
4. The FPA tools (e.g., protocols and guidance) and skills are readily integrated into the work practices of employees, supervisors and contactors to assure that fatality prevention is an ongoing priority that is an integral part of daily work activities.
5. The findings from this process must flow immediately to senior management and be respected the same way they would respect a fatality. FPA results must supersede all other safety concerns.
6. Even companies with a robust safety management system (SMS) may have significant FSI risk gaps in their risk gathering, ranking and decision-making processes. A visual representation of an organizations' risk gathering, ranking and decision-making process is depicted in Figure 1. The FPA validates the health of these processes and how effectively they "talk to each other".

Figure 1. Are there FSI gaps in your safety risk data?

Given all of the reasons why a FPA process may add value to your organization, there are some limitations that must be recognized. First, the approach described herein focuses on acute FSI hazard exposures and acute outcomes and does not address long term exposures and chronic impacts such as potentially fatal occupational illnesses (e.g., asbestosis). The process scope focuses on "outcomes" such as knowledge/skills, actions, processes and situations and does not directly evaluate the health of risk control programs (e.g., Energy Control) or management systems. Audited entities are typically required to assess the "systemic" reasons and actions for the findings after the audit. In addition, it is assumed that these "limitations" will be assessed in other SMS validation mechanisms (e.g., programmatic audits, risk assessments and system audits).

Developing an Effective FPA Process

Once you have buy-in and approval from your management team that the FPA approach will be supported and its output will be respected and acted upon the same way they would respond after a fatality, then it's time to develop the process. Typically, the FPA process evolves in phases as depicted in Figure 2.

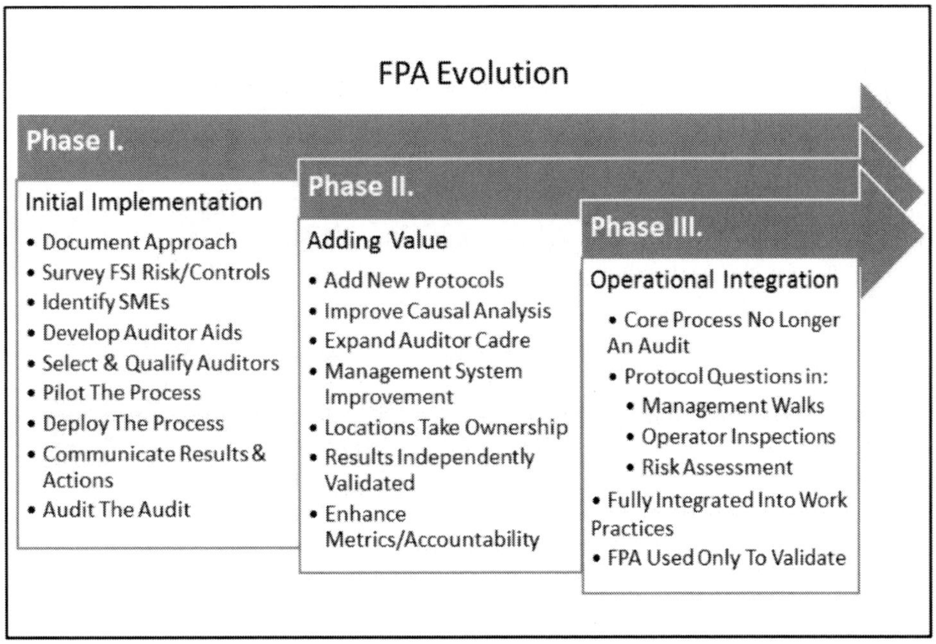

Figure 2. The goal is to evolve the FPA principles into your operational processes.

Though it may take years to evolve through the FPA Phases (and some organizations may not make it all the way) it is important not to slide back or lose ground. The ultimate goal of the evolution is to imbed the FPA principles into preventive safety measures and standard operational processes such as engineering and process design, acquisition due diligence, work procedures, job responsibilities, training, hazard/risk assessments and management walks. The full-fledged FPA then evolves into a validation mechanism that periodically checks the health of robust fatality controls that are integrated into your daily operational approach and SMS.

The initial development and implementation (Phase I.) is where most organizations stumble. Here are the elements that need to be considered as an organization enters into Phase I.

Define the Approach

Process development starts with the creation of an overall program defining the FPA approach for your organization. The program should follow a Plan Do Check Act (PDCA) format and include elements such as:

- Purpose and scope
- Expectations
- Deployment approach
- Ownership/leadership of the process
- Rules of engagement
- Team make-up and selection
- Auditor qualifications,
- Audit scheduling and staffing requirements
- Responsibilities of audited entities
- Reporting and presentation of audit results
- Data tracking/analysis
- Periodic program auditing
- Continual improvement mechanisms

Many strategic decisions need to be made at this stage, such as who will oversee/manage the overall FPA process and whether onsite contractors will be included in the assessment. The program must clearly define the threshold criteria for "FSI Hazards" and situations that constitute a finding such as C&PD.

Survey FSI Risk and Controls

Once the program is completed the next step is to assess the organizations knowledge of its FSI risk. Collect and analyze data from all safety risk gathering processes (Figure 1) and separate out those that meet the defined FSI threshold criteria. Identify and evaluate the existing FSI controls (i.e., many companies have Cardinal Rules or standards for recognized high risk activities/exposures). In addition, industry-standard and best practice FSI prevention controls should be factored into the process. Group the FSI risks and associated control expectations into categories such as fall hazards, mobile equipment, etc., to facilitate the selection of Subject Matter Experts (SMEs).

Identify SME's

Involvement of the right SMEs is critical to effective and sustainable FPA development and deployment. SMEs are "working experts" in the operational aspects (e.g., production, service, repair and maintenance) of your organization. They know the right way to do the job, what it takes to be proficient and what really happens when people are out of management's line of sight. These are the people that you want to create the questions for your field protocols and as a key part of your auditor cadre. Pair them with safety professionals and technical staff in teams aligned to their area of expertise. Bought-in SMEs can add tremendous value outside of the audit process by driving improved work processes, safety measures, education and worker selection/qualification criteria.

Develop Auditor Aids

"Auditor Aids" are elements of the process that ensure auditors focus on the key issues, are consistent in their approach and understand what is expected. They typically consist of protocols (standard field approach, questioning and decision-making aids), supporting information or auditor guidance, report/presentation formats and scoring mechanisms. At the heart of the process are the protocols which are developed using SMEs for each discipline/risk area. The SMEs use their practical knowledge along with the existing and newly developed FSI controls to create protocol questions to field-assess C&PD.

To support these questions field guidance (especially helpful for auditors who do not have technical expertise in each FPA discipline) is created to assist auditors in the evaluation and determination of C&PD findings. Field data (findings, location, number of employees, etc.) are entered into the protocols and ultimately tallied and accumulated into the FPA Report and Presentation. Scoring/rating systems (Pass/Fail, numeric grading, etc.) can also be included in the tools to facilitate the progress tracking of locations, drive accountability and enhance predictive FSI risk measures. It should be understood that in the measurement of fatality risk, 100% Pass is the only acceptable score. Caution should be used with the creation and use of FPA metrics because they can "depersonalize" the data and drive the wrong management behavior.

Select, Train and Qualify Auditors

No amount of protocols, guidance and training can replace a high quality auditor. It is imperative that you select, train and validate your auditors' skills to a high and consistent standard of care. When selecting/developing your initial auditor cadre consider the following traits/abilities:

- Technical Expertise: Their knowledge of how the work (including people, process, equipment and facilities) must be performed to result in a safe and operationally effective outcome. A SMEs operational legitimacy carries a lot of weight when speaking with those at risk.
- Investigative Nature: Having a value, zeal and ability to dig into what is "behind the curtain". Capacity to learn the art of how to effectively ask questions and listen/watch.
- Good Communicator and Relationship Builder: Can converse well with those at risk and first line supervision. Capable of building a level of trust that facilitates the discovery of hidden FSI risk.
- Process Buy-in: Capable of calibrating oneself to the unique focus of the FPA process, staying aligned to the purpose and scope and not getting distracted.
- Savvy: Ability to recognize (with their senses) when something doesn't look, sound, feel and/or smell right and where to go to find the reason(s). For example, a savvy auditor knows that a loaded scrap bin is an indication of operation risk and, in some cases FSI risk as depicted in Figure 3.
- Independence: Ability to separate themselves from the influence of their co-workers, management, etc., to provide a factual and unbiased assessment of the operation. Ability to

step back from and question existing standards of care (e.g., customary FSI controls and poorly respected FSI hazards).

- Intestinal Fortitude: A dedication to the value of finding and reporting risk in order to prevent a FSI and the ability to appropriately report/present factual information that may be difficult for the site (site management) to accept.

Figure 3. Will your auditors recognize this as an indicator of FSI risk?

Once potential auditors are selected a "qualification" process needs to be designed and deployed in a manner that assures these individuals are taught and can apply the requisite skills/behaviors. The process involves well-defined learning objectives and an instructional approach (classroom and field) that provides the defined knowledge and skills and validates competency. Typically, a trainee would participate in a series of actual audits where they are mentored and ultimately approved as a qualified auditor. In addition to auditor qualification, Team Leaders would be required to participate in a similar process focused on FPA management and leadership.

Pilot the Process

Prior to the roll-out of the process it is useful to run a pilot with a select group of auditors. The purpose of the pilot is to validate that the auditor aids are clear, appropriate and reasonable for the conduct of the audit considering the time and resources allotted. The pilot should reveal issues with protocols (content, clarity, applicability, etc.), audit scoping challenges (i.e., what is required to provide an adequate sample), logistical needs (access to the site and operations, space availability for audit team meetings meeting, etc.) and the other factors influencing the effectiveness of the audit. The learnings from the pilot are used to enhance the FPA process prior to deployment.

Deploying the FPA

Now that all of the key process elements are in place it is time to plan for and arrange the deployment of the FPA. The following is a list of factors that if not already built into your program, should be decided upon prior to rolling out the audit process:

- How will you define "Audited Entities"? Audited entities must be defined by the organization to ensure that all areas with potential FSI risk are assessed. The size and complexity of these operations will impact your audit planning.
- How will you make the FPA roles and responsibilities clear? For example, what is the responsibility of the Team Leader vs. the Site Manager in the creation of the action plan?
- What amount of auditing is reasonable considering the constraints of the team (quantity of team members and their capacity), the size/complexity of the site and the time allotted?
- How will the schedule of audits be determined and approved?
- How will you determine audit frequency (e.g., based on risk or time)?
- Will a numeric rating be used – what is the acceptable threshold?
- How and when will the findings be communicated – daily de-briefs, report and/or presentation?
- What is the "Stop the Audit" threshold? For example, imminent danger situations and/or the preponderance/systemic nature of FSI risk.
- What are the action planning and action closure responsibilities and processes?
- What accountabilities are in place to support continual risk and process improvement?

Communicating Results and Corrective Actions

Now that the on-site portion of the audit is complete, what needs to be communicated and acted upon? The audit will result in a series of observations and findings that must be communicated to the audited entity in a manner that facilitates their taking the right action to prevent recurrence of the risk. Along with the presentation of the findings it is helpful to show photographs, enlist support from team members, local SMEs and supervisors to support the significance of the findings.

Following the issuance of the report/presentation of findings, some amount of "coaching" may be necessary to assure that the audited entity understands the significance and the value of correcting the systemic reason(s) for the situations identified. Data from each audited entity should be entered into an overarching database that tracks performance, common issues, corrective action closure to schedule, etc. This information will be used to improve the FPA, provide for improved controls and drive risk reduction and accountability.

Auditing the Audit

As the audit process progresses it is important that the effectiveness of the approach be assessed and validated. To assure that the audit and auditors remain true to the process, a periodic assessment and calibration is required. It is recommended that a calibration audit be performed at least annually to assure that all auditors are applying the process in a similar manner. Consistency of approach assures consistently of outcomes. The calibration process can consist of an audit process owner taking a group of Team Leaders to audit a site together or observing Team Leaders while conducting actual audits.

As a flow down from Team Leader calibration, each Team Leader should evaluate and calibrate Team Members as part of the audit process. All of this qualification data should be maintained by the process owner and used to maintain the validity of the overall FPA approach. In addition to auditing the auditors a program audit should be conducted of the entire FPA process on an annual basis. This audit can be included as part of an SMS audit or a vertical safety audit.

In Summary

While the development of an FPA process may not be for everyone, there is clearly a need to field reality-check FSI risks and controls within organizations that operate in high hazard environments. If you fear that these FSI risks may not be captured or effectively elevated/appreciated by senior management, then a robust FPA process may be in order.

Session No. 614

OSHA, Don't Tell Me What To Do

Jeremy Bethancourt

Mark Cannon

Augspurger Komm Engineering

Introduction

Federal law tells us that, as employers, we are to provide a workplace free from hazards which are likely to cause serious injury or death and to comply with standards, rules and regulations issued under the OSH Act (Williams-Steiger, 1970). The reason for such laws is, because Congress found that "*personal injuries and illnesses arising out of work situations impose a substantial burden upon, and are a hindrance to, interstate commerce in terms of lost production, wage loss, medical expenses, and disability compensation payments.*"[1, 2]

Accordingly, Congress passed a law, signed by President Richard Nixon, which dictates that as far as possible, every working person in the United States should have safe and healthful work and places of work. In a sense, this is an attempt by the government to legislate morality because it implies that, if left to our own devices, our tendency would be to expose workers to hazards, so their protection must come from a source bigger and more powerful than their employer. Sadly, from a historical perspective, such conclusions may have been true then as much as they are today. Without a doubt, in many situations employers have become strictly focused on the legal compliance of the OSHA regulations rather than moral implications of failure to protect workers. Compliance with "the law" rather than moral obligation has resulted in a business strategy where employers avoid punishment for violations of the OSH Act by focusing on the "letter of the law."[3] The paper trail of complying with the regulations benefits employers in an inspection or after an incident, whether or not a worker was protected from an incident, recognized or not.

Workers and their employers would be better served if the employer and workers focused on the intent of the law: to ensure workers can return home at the end of every shift without injury. Rather than

---

[1] Subsequently created the Occupational Safety & Health Administration under the Department of Labor and the National Institute for Occupational Safety & Health under the Department of Health & Human Services.

[2] Federal law applies except where such jurisdiction is retained by states which have an OSH Plan approved by the U.S. Secretary of Labor.

[3] The **letter of the law** versus the spirit of the **law** is an idiomatic antithesis. When one obeys the **letter of the law** but not the spirit, one is obeying the literal interpretation of the words (the "**letter**") of the **law**, but not necessarily the intent of those who wrote the **law**.

just concentrating on satisfying the interminable list of regulations handed down by OSHA, employers should also look at the big picture of protecting their employees. Management can combine the morale decision to personally safeguard workers with risk identification and management tools. This allows employers to successfully and proactively mitigate hazards and exposures and create their own solutions for worker safety without OSHA needing to tell them what to do via the prescriptive guidance found in many regulations. Rather, the organization (which includes the entire safety management system, supervisors, management, and employees, and approach to hazard mitigation) tells OSHA how and why the company makes the workplace safer and then just does it without reservation.

Historical Perspective

The industrial revolution pulled workers from agrarian society into factories increasingly exposing people to crude and unsafe working conditions. It has been said that workplaces in the United States were among the most dangerous (Aldrich, 1). Due to humanitarian and social responsibility, states began passing the Employer's Liability Law in the late 1800's. However, common law defenses made it difficult and expensive to file claims against employers. As part of any defense, employers would demonstrate that workers had assumed the risk, been injured by the actions of another employee, or been partially responsible for the incident. The court system would usually reject accountability on the part of the employer (Aldrich, 1). Based on data compiled during the early part of the last century, it was determined that roughly half of all workers fatally injured were successful in litigation and even then the amount recovered was only about half a year's pay (Fishback and Kantor 2000, 58-66). Due to the relatively low cost of injuring workers, industry had little motivation to focus on anything other than production. Costs associated with worker injuries and fatalities were relatively low (Op.cit.).

Insurance companies that provided monetary protection to employers against the costs of accidents began sending safety engineers to manufacturing plants. They assessed hazards and made recommendations to management to improve safety, reduce injuries and control costs (Blake 1943, 13). The advent of World War II forced society and companies to consider the importance of conserving and using resources wisely and efficiently. This included the realization that human resources (workers) needed to be protected from lost time injuries in order to meet production demands for materiel used in the war effort. Employers realized the value of having internal, full-time safety personnel. Demand for qualified safety engineers trickled down to universities, which developed curricula to train and provide qualified safety people.

Technological advances fueled by the war effort caused standards organizations and committees to create guides and formalize voluntary best safety practices for the design and manufacture of machinery. Indeed, when OSHA was created, many of the regulations were based on safety standards already created by industry. The advantage to this was centralizing, codifying and translating safety knowledge across a broader spectrum of industries. Additionally codifying the practices caused disparate groups to realize there was a cross-functional overlap between safety practices. For example, lockout/tagout started with electrical applications to protect workers from being electrocuted. However, the methodology had applications to other hazardous energy sources and spread across to pneumatic, hydraulic and stored energy of any sort. For example, lockout/tagout started with boilers as shown in Exhibit 1, then migrated to electrical and hydraulic applications.

Exhibit 1. This is an example of a boiler lockout/tagout from 1929.

Motivation for Protecting Workers Should Not Come from OSHA

A safety person who grows up in the age of OSHA may think the reason we protect workers is because it's the law. They would be wrong. We learned long ago that workers are humans and protecting them is the humane thing to do. *The real basis for accident prevention is the avoidance of human suffering* (Berman and McCrone 1943, 3). We tend to assess workers at face value and it would help if we humanized them to management and ourselves. Are they a single parent? Are they supporting a multi-generation family? What kind of challenges do they face at home? Why do we care about this? Because, depending on their circumstances, the worker is thinking about their home life, not about how to avoid getting themselves injured.

Maquiladoras are an example of workers thinking about things besides work. [4] A common scenario at these companies is an assembly line of employees almost robotically putting together parts while visiting with each other to pass the time. They are the most vulnerable people within the safety professional's universe because they, meaning the workers, assume they are always safe. Additionally, workers' skills sets, makes them human capital necessary for the function and operation of the organization. To think of them as expendable is not only abhorrent; it ignores the big picture of the state of the industry. Stratification of skills is narrowing the bands of workers that fit into particular skill populations. There are only so many available.

The recent upturn of construction after the recession created a demand for workers that cannot be easily satisfied (Noguchi 2013). Many of workers left the industry during the recession and they are no longer available (Semuels 2015). Those who are working have an advantage because they realize they are a scarce commodity and employers are competing for them. With this dearth of skilled workers, management cannot afford to lose workers due to injuries. Employers cannot risk missing production schedules due to lost time injuries.

---

[4] A maquiladora in Mexico is a factory that operates under preferential tariff programs established and administered by the United States and Mexico. Materials, assembly components, and production equipment used in maquiladoras are allowed to enter Mexico duty-free. Products made can be exported into the U.S. at lower tariffs than those from other countries.

Besides pay, one attribute of their employer pertinent to workers is whether the employer is safety conscious and cares about their safety. Every worker wants to go home at night and companies must communicate and show employees that is the company's desire as well.

Safety Professionals Should be Afraid

"Wheels of Tragedy", "Safety or Slaughter", and "Blood on the Asphalt" are examples of driver safety movies used during the 60's to scare student drivers into being careful when behind the wheel. Similarly, gruesome workplace safety videos pantomiming gouged eyes and lost scalps are intended to scare workers into being safe. In contrast, the curriculum for safety professionals is rather antiseptic and learning is based on hypotheticals: if the machine is unguarded, the operator could amputate their hand. How many safety professionals have seen what a hand amputation looks like, or seen what it looks like after it's reattached, or read the testimony of an amputee to get their perspective on losing the use of their hand? It is not trivial – more like a boulder dropped in the pond of their life. It affects everything they do: self-esteem, motivation, relationships, attitude, how they view their future. As John Donne said, "no man is an island entire of itself". In this case meaning that when workers are injured on the job it affects everyone else whether they know it or not, by way of increased costs for products, insurance premiums, impact to society, etc.

Losing your hand is a problem. Having OSHA show up at the safety professional's place of work afterwards is an inconvenience. At the most, the injury will cost the safety professional and employer time and money. The amputee will never be made whole. Further, it is unlikely that worker will ever want to operate that machine again, no matter how safe it now is, because they are afraid. Reason makes the point that a common cause of bad outcomes is a lack of early detection of defensive weaknesses i.e. people involved forget to be afraid (Reason 1997, 37). The first goal of a safety professional is to protect workers from human suffering, not to protect their employer from OSHA citations. If the safety professional and thereby the employer is afraid for workers' safety and effectively meets the first goal, secondary goals related to OSHA will almost be a byproduct.

It has been said, that workers can tell if you actually care when you are speaking to them about safety or if you are just going through the motions. The reality is that if workers know you actually care about them, you can have genuine connection with those workers (or at least some of them). People want to be respected so when workers realize that they are more than simply bodies on the job, they tend to be more motivated. The motivation can be channeled into safer productivity.

Safety Professionals Should Also Think in Terms of "Doing Good" to Others

Safety professionals underestimate themselves if they think going to work every day is just a job. Their sphere of influence is more powerful and larger than they think. Every worker that the safety professional passes has the opportunity to benefit from the safety professional's knowledge, insight and ability to assess if something is amiss. "Doing Good" to others dates back to the Quran and the Bible. Doing good, by default, avoids the "don'ts" such as those listed in the Ten Commandments. Google applied the same intent in the early 2000's by using the slogan, "Don't be evil." Later, the parent company, Alphabet, revised the phrase and incorporated it into the company's code of conduct as "Do the right thing." Doing good means following the spirit and intent of OSHA regulations versus just following the plain text of the regulation.[5] OSH professionals must be able to understand and interpret the text correctly toward an action which protects workers.

Just Focusing on OSHA Won't Necessarily Protect Workers

If a company concentrates on following certain regulations, it tends to cause safety professionals and workers alike to be myopic and spend their time ensuring the company is obeying the letter of the rules. Large exposures and the specter of potential environmental incidents and fines tend to overshadow other concerns that may be in the safety professional's immediate area without their realization. The following examples of incidents occurred at facilities where an OSH representative was working full time, yet employer demands created competing interests that did not allow time for a thoughtful risk assessment.

A company OSH representative at a refinery walked right past a contractor's crew for a week while they used a jackhammer in a trench. The jackhammer was used almost continuously for several days. Had that OSH supervisor bothered to look where the noise was coming from, they would have seen an antiquated jackhammer being used by the same worker who was not wearing gloves. That worker eventually became totally disabled from whole-body vibration experienced during that one job. (Wasserman and Wasserman 1999, 56)

A company OSH representative at a sugar plant was aware workers were cleaning inside a lime kiln which would occasionally drop "clinkers" of various sizes.[6] The OSH professional was fairly new and although this task was performed annually without a history of serious incidents, the OSH representative assumed hardhats in the kiln were sufficient protection for falling objects. Unfortunately, on this day, the first clinker knocked the employee's hardhat off, the second clinker disabled him permanently from a brain injury.

A light weight curtain used for safety stopped working on a fairly new robotic assembly machine. After consulting with the manufacturer, the employer was told to bypass the curtain and use E-stops when accessing the interior of the machine so they could continue to use it to meet production demands.[7] The company OSH representative observed the operation of the machine for some time and figured using the E-stop removed the source of energy and was appropriate for regular work. They did not consider what would occur if parts were dropped inside the machine and a worker went in to clean them up, forgetting to hit the E-stop. They found him lifeless, pinned between a robotic arm and another structure inside the machine.

OSHA cited a refinery for not providing fall protection on several of their loading stations for tanker truck drivers while the drivers were atop the tanker platform loading the truck. One station did have a suspended fixed guardrail that surrounded the platform while the truck was parked. To lessen the fine, the refinery created duplicates of that fixed guardrail on the rest of the loading stations. The CSP at the refinery interfaced with OSHA and showed then what had been installed.[8] However, the CSP never evaluated the various size platforms that the guardrail was intended to surround. As a result, a driver fell through the two-foot gap between his platform and the suspended guardrail resulting in quadriplegia.

---

[5] Ibid

[6] Clinkers are limestone that gets hot enough to adhere to the walls of the kiln, cool, then form a solid mass.

[7] An E-Stop (emergency stop) is a readily accessible device intended to be used by workers and/or end users to avert harm or to reduce existing hazards to persons, machinery, or work in progress.

[8] CSP (Certified Safety Professional) is a professional designation provided by the Board of Certified Safety Professions.

A surprise OSHA inspection at a wood mill resulted in a citation due to a lack of a point of operation guard on a jump saw. The employer chose to remedy the citation by building a two-hand control, as shown in Exhibit 2, and sent a picture of the modification to the OSH compliance officer who accepted it as adequate abatement of the hazard.

Exhibit 2. This is a picture of a two-hand control approved by a state compliance officer.

The mill installed the two-hand control. Not long after, the employer became aware that the buttons were taped down yet kept the machine in production, and did not consult OSHA for guidance upon making that discovery. A worker had his hand amputated as a result of the tape being used to bypass the buttons. The mill thought the box followed OSHA regulations, but gaps in the knowledge by both the employer and the local compliance officer allowed the improperly design two-hand control to be put on the factory floor.

Yet another example of how employers strictly following OSHA regulations does not necessarily compel the protection of workers is looking at the high frequency of workers injured by falling off ladders. In one case study, a worker was using a ladder to install plywood sheathing on the vertical end of a modified pre-manufactured truss. While standing on the ladder a worker (who had previously received eight hours of both classroom and on the job safety training) explained that he felt the ladder begin to wobble and had the ability to stop but instead continued with his task. Eventually the ladder collapsed and he fell, causing a compound fracture to his arm. As an employer, our first reaction might be to evaluate the ladder for defects and, when none are found, we focus on the way the ladder was used, how it was selected, and/or if it was the correct tool for the job. This way we could determine if there was a failure by the employer to follow OSHA regulations or if the worker violated the standard by not was using the ladder correctly. As was mentioned, the worker admitted that he continued using a ladder which he recognized was not behaving appropriately. After determining that the ladder was the most feasible tool to use for access, in compliance with OSHA regulations, the employer provided additional training to the worker, and the crew, which in part reminded the workers not to use ladders which appear inadequate and to engage the management team if the use of the ladder does not seem feasible or a safe and viable option. To be clear, this might be seen by some as a scenario dictated by the adversarial dynamic of OSHA enforcement; either the employer violated the standard or the employee misbehaved, but what if it is neither.

As OSH professionals, our duty is to ensure workers are not injured and resources are protected regardless of any dynamic the employer faces through regulator compulsion. The examples mentioned are just a few which can provide an example of where organizations and professionals could have done a better job protecting workers.

Non-Routine Work and ALARP

Manuele recognized that the unusual or non-routine tasks were the sources of a large proportion of workplace incidents (Manuele 2008, 51). He also noted that these types of incidents had "*multiple and complex causal factors*" often related to "*organizational, technical, operational systems, or cultural origins*". Mr. Manuele is describing the dynamics of the construction industry which is a briar patch for the safety professional because its' supposedly routine work has the characteristics of non-routine work. Potential causal factors for an incident include: a transient population where English is most likely their second language, some have never received safety training, a changing landscape of hazards as the structure is erected or project continues, a constant high energy source and a confluence of different safety cultures where machismo may be admired and using PPE is an annoyance and/or considered a sign of weakness. Construction OSH professionals don't always have the luxury of a human resources department to weed out incorrigible employees, sterile classroom training, a common culture, and/or language. Instead, on one hand, they have a general contractor with an "unyielding" production schedule.[9] On the other hand, they have a crowd of unknown standing the in middle of a construction site filled with drying concrete slabs, who want to work but may not know how to, or even want to work safely.

The answer is, we must look beyond prescriptive compliance with OSHA regulations and use what we know as safety professionals to reduce the risk to a level that is as low as reasonably practicable (ALARP) given the available resources (Manuele 2008, 103).[10] If we cannot remove the risk, in this case gravity, then we have to find a way to mitigate its effects given the circumstances of the worker culture, changing landscape, and technological feasibility of available systems.

An example, once again, is ladder usage. By focusing on the training requirements for ladder use, the employer is able to demonstrate compliance with the written standards even though such training does not actually provide protection for workers if there is an unexpected incident of "misuse"[11]. Of course there will be those who would argue "affirmative defense" when an employee "misuses" equipment they have been trained to use, but that does not prevent the injury, financial losses, or suffering of workers.[12] The real issue is, how do we as employers keep workers safe knowing that, statistically speaking, workers are going to "misuse" ladders for a myriad of reasons, including that workers and supervisors may not have foreseen the potential of such "misuse"? The answer is we look beyond mere compliance and focus on eliminating or mitigating the hazard, in this case gravity. If we know workers have a tendency to "misuse" ladders then we must acknowledge that propensity and figure out a way to eliminate the hazard (in this case gravity) or defy it.

The following is an example where a worker used a ladder correctly and neither he, nor his supervisor saw the hidden hazard which caused gravity to engage. A worker is on a ladder using a worm-drive drill. Prior to the task, the foreman and worker discussed the activity. The worker climbed up the ladder using three points of contact, had a coworker hand him the drill, and set himself in position to drill holes into a large glue laminated beam; total compliance with the ladder regulations and tool use regulation. Yet, the worker falls to the ground and breaks his arm. So what went wrong, what did the team fail to see? Well,

---

[9] The term unyielding is meant to convey the understanding that the general contractor may have contract stipulations which impose substantial financial penalties to any sub-contractor for delays regardless of the reason and even if the reason is directly related to increasing worker safety.
[10] ALARP stands for "as low as reasonably practicable", and is a term often used in the regulation and management of safety-critical and safety-involved systems. The ALARP principle is that the residual risk shall be reduced as far as reasonably practicable.
[11] The term "misuse" (for the purposed of this white paper only) is not meant to imply intent (i.e. employee misconduct) of the worker in as much as any deviation from manufacturer instruction which could be by nature classified as a violation by reference from the manufacturer.
[12] Under the OSH Act, affirmative defenses include, but are not limited to, "infeasibility," "unpreventable employee misconduct," and "greater hazard." CFR1910 2200.34(b)(3)

they failed to see what they could not see. The drill bit hit a knot while spinning at 1500 RPM then got jammed. The kinetic energy had to go somewhere and it did - right into the worker who was twisted and thrown off the ladder. In this incident, all regulations relating to ladder use and tool use were followed and no hazard was recognized. So how do we keep this sort of thing from happening again? Do we tell the workers to be careful when using the drill? The issue is that we have to stop focusing on the nuance of each individual OSHA requirement and look at how combining activities changes the situation and potentially introduces hazards which neither regulation may address.

If all we do is focus on OSHA requirements, then we are destined to fail. OSHA sets the minimums required. OSHA does not prescribe guarantees. We can do better. As we have discussed, in previous articles, OSH professionals and managers must think beyond regulatory requirements and use what they know to keep workers safe (Bethancourt 2017, 63). That said, in addressing the incident described above, there is no prohibition against having a worker "tie off" while working on a ladder. In fact, there are situations where regulations support that idea. So if we know workers might fall off a ladder while working, or that, from our own experience, "misuse" on ladders does occur, and we know that we have the ability to provide them fall protection systems which can protect them "misuse" of a ladders occurs, then we should find a way to protect workers irrespective of what the regulations state.

To be blunt, OSH professionals are insane to continue and simply follow the prescriptive pattern of train and retrain where the situation facilitates compulsions of misuse. As a follow-up, the company who provided this example, decided the best way to protect from what they cannot see is to have workers use fall arrest once they get to the working level of a ladder before they begin the actual task.[13] If workers lose their balance, overreach beyond capacity, gets thrown off, or inadvertently makes a mistake, so long as they follow the policy of "tying off" then gravity has been mitigated as a hazard. Coincidentally, regardless of what occurs on that ladder, there should not be a "serious" OSHA violation where the workers are not exposed.[14]

Conclusion

Although this has been a simplistic discussion, OSH professionals and supervisors should ask themselves why they became safety professionals in the first place if they don't think their job is to facilitate the avoidance of human suffering. The panoply of OSHA regulations are merely a formalized means to regulate and enforce processes to protect workers that were in existence long before OSHA was created. Workers go to their jobs to earn a living while balancing other concerns in their lives. They should not be fixated about whether they will be able to go home at the end of their shift. Being fixated, even perhaps fearful, is the safety professional's job and it is our jobs to provide guidance and training to those workers so they can avoid injury and losses. Continual mistrust, vigilance and ongoing risk assessments are attributes that allow the safety professional to proactively and successfully administrate safety practices at their workplace. As long as workers are at risk, OSH professionals must continue to be the experts in protecting workers. To be truly effective, we cannot rely on OSHA to force innovative methods that will guarantee a place of employment free from recognized hazards. That job belongs to OSH professionals and employers.

---

[13] Fall arrest in this instance is a self-retracting lifeline attached to an adequate anchor point which stops the worker within two feet of falling off the ladder.

[14] Technical violation verses employee exposure.

As OSH safety professionals we must remember what actually enabled the creation of OSHA's regulations which so many people refer to as a bible of compulsion. "*Each employer shall furnish to each of his employees employment and a place of employment which are free from recognized hazards that are causing or are likely to cause death or serious physical harm to his employees*" (Williams-Steiger 1). As mentioned in a recent article in Professional Safety:

> "…those are not just words but rather a mandate of moral and legal responsibility. If we truly understand the implication and power of that statement, then we will think beyond the nuance of all the different OSHA standards. We will look at a situation and know what it will take to keep the workers safe, and then we would just do it without reservation" (Bethancourt 64).

For OSH professionals and industry managers, the goal should not be to avoid OSHA fines and penalties. If we engage in our moral responsibility, then compliance with OSHA regulations will simply be the secondary benefit and not the primary reason of our activities. The goal of remaining profitable while simultaneously ensuring workers are safe is not a zero sum game. Since it takes two criteria for a serious OSHA citation to be issued then ensuring workers are not exposed will always ensure compliance with OSHA regulations regardless of the technicality of the regulations. If employers want OSHA to "leave them alone," and not tell them what to do, industry must think beyond regulatory compulsions and do what it takes to make the work place safe for those who are the base of the economy, the American worker.

Bibliography

Aldrich, Mark. *History of Workplace Safety in the United States, 1880-1970*. EH.Net Encyclopedia (retrieved February 15, 2017) (http://eh.net/encyclopedia/history-of-workplace-safety-in-the-united-states-1880-1970)

Berman, H. H. and H.W. McCrone. 1943. *Applied Safety Engineering*. New York: McGraw-Hill Book Company, Inc., 1943.

Bethancourt, Jeremy, "Fall Protection: Overcoming Misconceptions in Residential Construction." *Professional Safety*, American Society of Safety Engineers, March 2017.

Blake, Roland P. 1943. *Industrial Safety*. Upper Saddle River, NJ: Prentice-Hall Inc., 1943.

Cooper, Sheldon. *The Big Bran Hypothesis*. Columbia Broadcasting System. October 1, 2007.

Fishback, Price and Shawn Kantor. 2000. *A Prelude to the Welfare State: The Origins of Workers' Compensation*. Chicago: University Chicago Press.

MacLaury, Judson, "The Job Safety Law of 1970: Its Passage Was Perilous", *The Monthly Labor Review*. March 1981.

Manuele, Fred. 2008. *Advanced Safety Management*. Hoboken, NJ: John Wiley & Sons, Inc.

Noguchi, Yuki, *Homebuilding Is Booming, But Skilled Workers Are Scarce*, All Things Considered, National Public Radio, January 17, 2013. (http://www.npr.org/2013/01/17/169611619/homebuilding-is-booming-but-skilled-workers-are-scarce)

OSHA, Safety and health regulations for the construction industry (29 CFR Part 1926), 2014

Reason, James.1997. *Managing the Risks of Organizational Accidents*. Farnham, Surrey, UK: Ashgate Publishing.

"Reasonably Practicable," WorkSafe New Zealand Fact Sheet, March 2016 (worksafe.govt.nz).

Semuels, Alana, *Where Have All the Construction Workers Gone*, The Atlantic, February 13, 2015. (retrieved March 12, 2017) (https://www.theatlantic.com/business/archive/2015/02/where-have-all-the-construction-workers-gone/385417)

Wasserman, Donald E. and Jack F. Wasserman. *Occupational Vibration: A Brief Overview*. The Institute for the Study of Human Vibration. August 1999.

Wikipedia. *Don't be evil.* (retrieved February 14, 2017) (https://en.wikipedia.org/wiki/Don't_be_evil)

Williams-Steiger Occupational Safety and Health Act. Public Law 91-596, 84 STAT. 1590, 91st Congress, S.2193, December 29, 1970.

Session No. 615

The Lone Worker is Changing, Are You?

Nicole J. Washam, BS, CSP
Risk Control Specialist

Nicole A. Coughlin, BS, MBA, ASP
Risk Control Specialist

IMA Inc.
Denver, CO

Introduction

With a combined experience of approximately 30 years the authors have worked in 10 different industries where the challenge of lone worker safety has existed. In the course of their careers they have helped companies in the identification of lone workers and how to control the hazards presented. In their current role, they consult with a variety of industries, including manufacturing, municipalities, energy, and construction, on various safety topics including the management of lone workers. Through their work experience the authors have identified the need for adapting hazard identification, traditional safety definitions and programs relating to lone workers to our ever-changing world.

The traditional definition of a "lone worker" is changing as a direct result of our changing world and culture. The traditional definition identifies a situation in which personnel work with hazardous materials or in a position with a high degree of risk, otherwise known as a safety-sensitive position. The traditional definition takes into account personnel with high risk; high exposure while the new definition along with an effective hazard assessment will show how low risk personnel can now have high implications on a company. This paper will cover how companies may want to change the definition of a lone worker, the steps to conduct a hazard assessment to quantify the level of exposure in their workplace, and ways to control the risk exposure through safety policies and procedures or emergency management plans. Whether a seasoned safety and health professional, executive or top management, or someone just assigned to safety in their organization, this paper will present various tools for the reader to take back to their workplace.

Redefining the Lone Worker

What is your definition of a lone worker? Several organizations define a lone worker as someone who is alone on the job, or anyone who works alone in a fixed facility, or someone that cannot be seen or heard by another person. Oftentimes a lone worker is categorized as an employee with limited or no access to other employees if an emergency were to occur while working alone and:

- performs work in a safety-sensitive function,
- is the only worker in the work area at any time, OR
- is not directly supervised by his or her employer, or another person designated as a supervisor, at any time

This traditional definition of a lone worker identifies workers with high risk, high impact for an organization. his definition must change as the security concerns in the world are changing, such as workplace violence and/or terrorist threats and attacks. As the internal and external threats to employees change, the identification and classification of a lone worker is also changing. The lone worker is no longer limited to those working in hazardous situations or with hazardous materials, but can affect personnel in a wide range of industries such as municipalities, manufacturing, oil and gas/pipeline, utilities, agriculture or schools, and it must encompass the low risk worker that now has high implications to an organization.

The Changing World

Security Threats

A business may be considered a soft target, which is a location that is relatively unprotected or vulnerable, especially to a terrorist attack. A soft target offers a strategic objective to those wishing to exploit the vulnerabilities of the business. According to Global Risk Insights these strategic objectives include:

- An atmosphere that is inviting to visitors and offers an area that is open, such as a large reception area
- Lack of "visible security guards and detailed screenings prior to entry"
- A lack of training for security guards that are present to handle a terrorist attack

When a business does not have a security infrastructure to act as a deterrent and houses a large amount of personnel, this can then make the business a strategic location for anyone planning to do harm. There is much consultation regarding the selection of a target, and businesses with high population and little security must be aware of employees they have facing the public and their ability to communicate to others in the building in the event of a security threat. Often, this includes a phone call, an action that in most cases can not be completed when confronted with a security threat. Other communication protocols include silent alarms, audible alarms, or silent strobes. Even with those protocols, this is only reactionary to the threat. The best defense is prevention and training to potentially identify a threat before it becomes a tragedy. Prevention includes identifying those that are at risk, such as lone workers, and giving them the training necessary to take note of suspicious activity that may be taking place during a terrorists selection process. This training and awareness can be identified through a proper lone worker hazard assessment.

Workplace Violence

Recent events in the news show that the threat of workplace violence can come from both external (domestic violence) and internal (co-workers) threats. According to the Center for Disease Control and Prevention (CDC) "homicide accounts for almost a third (27%) of work-relate deaths in women – it is the second leading cause of injury death for women in the workplace" behind motor vehicle accidents. On a much larger scale, Workplace Fairness estimates that workplace violence affects around 2 million Americans each year with around 800 of them losing their lives due to workplace violence. Clearly workplace violence is a current issue, so how can we control it by including it in our new definition of a lone worker and ensuring it's captured on a hazard assessment?

The Revised Lone Worker

Given this information, the revised definition of a lone worker may meet any of the criteria listed below. A Lone Worker is someone who:

- is the only worker in the work area at any time,
- is not directly supervised by his or her employer , or another person designated as a supervisor, at any time,
- limited means of communication including receiving help in an emergency,
- is exposed to the public, or
- works in secure or unsecured areas

In our changing world, whether a worker handles hazardous or non-hazardous materials, or works in a safety sensitive function or not, they may still meet the definition of a lone worker. For example, the receptionist at the front desk of an office is exposed to the public at all times. They greet visitors and redirect those that are lost. While there are many other people in the office building, this person is a lone worker in the sense that there is no one around to provide assistance in the event of an emergency. Although having two receptionists is not feasible, a company should assess this position to identify how the lone worker can communicate without a phone. For this case, a panic button they can push that lights up silent strobes in work areas that are not visible from the front desk. This way, when the button is pushed a potential security threat is not aware that the alarm has been sent.

Another example is that of a construction worker assigned to fire watch at the end of the day. As the other workers are leaving the site for the day, the fire watch must remain behind for an hour to ensure no fires occur from the hot work that was conducted. Construction sites are highly visible and often times broken into after dark. If this worker is alone, have we accounted for the threat of thieves coming on site. Or perhaps there is a fire that starts from the hot work, does this lone worker know what to do if an emergency occurs? Does the fire watch try to contain the fire? Exit? Call 911? Who does the fire watch notify within the company?

How about an agricultural employee that goes to the field for note taking? Taking notes in a field seems, on the surface, to pose no hazard. But, what if they twist their ankle on the uneven ground? Has the employer completed a hazard assessment to know if the note taker is allergic to insect bites or bee stings? How far are they from emergency personnel? How far are they from other fields or the home office? If they are hurt and never leave the field, will anyone know?

There are many situations that may seem to pose a reduced risk to employees, but has the company assessed the risk for the worker that is left on fire watch, that drives to field for harvest, or the one that greets the public?

Hazard Assessment

According to the Occupational Safety & Health Administration (OSHA) just about every safety and health program that is to be established in a workplace should be the result of hazard identification. OSHA emphasizes hazard identification as being the first step towards the organization quantifying their risk and how to control the hazard exposure to employees. Hazard assessments allow organizations to break down tasks to identify hazards and risks associated with the task. However, lone workers present unique challenges and in order to address these challenges an organization will need to change how it conducts a hazard assessment.

The hazard assessment for a lone worker is not the typical Job Hazard Analysis (JHA) or Job Safety Analysis (JSA). Where a JHA only takes into account a specific task and identifies the hazards and controls associated with that task, for *anyone* who may perform the task; whereas a lone worker risk assessment must take into account the *individual* worker, the task they're completing and the environment they are working in.

For example, at a work location there are approximately 30 employees on site during the day. There are 4 buildings on location that employees work in. The main office houses nine (9) employees, the processing plants have 20 employees, and the lab has one (1) employee. Although there are 30 employees on site during the work day, the lab employee in this scenario is a lone worker as they work isolated in a building away from other employees. It is not often that others are in the lab during the course of the day. With this we can look at the revised definition of the lone worker at this location to determine if they meet the criteria of the definition:

- Are they the only worker in the work area at any time?
  - Yes, although surrounded by other buildings and employees on site, this worker is alone in a building and not visible to other employees.
- Is the employee directly supervised by his or her employer, or another person designated as a supervisor, at any time?
  - No, supervision does not reside in this building and does not frequent the area or perhaps supervision for this position does not office at this location.
- Is there limited means of communication including receiving help in an emergency?
  - Yes, if anything were to happen, there is no intercom, no audible alarms, or other means of communication to other buildings.
- Is this worker exposed to the public?
  - Yes, the building is located on a main road and has a driveway leading to its door. The property does not have a security fence that prevents unauthorized access.
- Does the employee work in secure or unsecured areas?
  - Depends, the building can be locked making it secure. However other employees have access to the building and may leave a door unlocked or propped open.

This particular location may also work with products or operations that are not viewed favorably by the public, such as medical or GMO research or oil and gas operations in a protected area that brings out activist groups to the location.

Using the example provided the company must conduct a lone worker hazard assessment that takes into account the:

- Environment
- Worker
- Hazards

Lone Worker Hazard Assessment				
Assessment Performed by: Safety Manager and Employee	Date: June 21, 2017			
Job Title: Lab Employee	Location: Denver			
1.0 The Lone Worker	Yes	No	N/A
1.1 Does the worker have direct supervision by his or her employer, or another person designated as a supervisor, at any time?		X	
1.2 Is the lone worker sufficiently experienced for the task?	X		
1.3 Is the lone worker provided with training that allows the equipment, substances, or personal protective equipment to be used safely?	X		
1.4 Is the lone worker provided with information about the risks involved with the task and the precautions to be taken?	X		
1.5 Does the lone worker have any known medical conditions that make them unsuitable for working alone?		X	
1.6 What shift is the worker assigned to?	1st Shift		
2.0 Work Area	Yes	No	N/A
2.1 Will the worker be in a secured or unsecured work area? (For example, is a badge required for access? Is a door left unlocked or propped open?	Secure		
2.2 Is there safe access and egress?	X		
2.3 Can a lone worker safely operate the equipment / machinery required to be used?	X		
2.4 Can all substances and hazardous materials be safely handled by a lone worker?	X		
2.5 Is the worker exposed to the public?	X		

2.6	Can manual material handling, where required, be carried out safely by a lone worker? (For example, if company policy states you cannot manually lift more than 50 lbs and the job requires a manual lift of 75 lbs, a lone worker is not permitted.)	X		
2.7	Has any risk of violence been addressed? (Consider threats that have been made such as, a terror threat level, domestic, protestors, co-workers, animals or insects, and the public.)		X	
2.8	Will the lone worker be exposed to extreme temperatures?		X	
3.0	Communication and Supervision	Yes	No	N/A
3.1	Are there arrangements for regular contact between the lone worker and supervisor/employer?		X	
3.2	Are supervisors able to periodically visit and observe the lone worker?	X		
3.3	Is there access at all times to communicate with the "outside world" and emergency services?		X	
3.4	Are automatic warning devices appropriate for the situation?	X		
3.5	Is there a JSA / JHA for the task?	X		
4.0	Emergency Management	Yes	No	N/A
4.1	Is the lone worker trained in first aid?	X		
4.2	Are adequate first aid facilities available?	X		
4.3	In an emergency can Emergency Management Services locate and reach the lone worker?	X		
4.4	Are alarm systems in working order (such as fire alarms, security system alarms, carbon monoxide alarms or weather related notification)?	X		
4.5	Is there an emergency action plan (EAP) in place for this site?	X		
4.6	If yes to question 4.5, does the EAP address lone workers?		X	

<u>Action Plan:</u>

The last step to complete a hazard assessment is to look at the varying degrees of controls in place or that need to be added. Once these controls, or lack of controls, are identified, the organization should create an action plan to reduce the risk to an acceptable level for the safety and security of the lone worker. If controls are not available to bring the task to an acceptable level, an organization would need to consider multiple employees in the work area. Some examples of controls are listed below:

- Adding more workers to a shift
- Consider removing the human element entirely
- Check in times
- Adding communication and/or tracking devices /man-down alarms
- Panic button/device that is a silent alarm
- Policies & procedures
- Training workers to understand if they are a lone worker or not and how to abide by rules

Using the lab employee scenario of a lone worker, the following is an example of a completed action plan.

<table>
<tr><td colspan="4">Action Plan</td></tr>
<tr><td colspan="4">Activity/Task

Complete the relevant details of the activity being assessed

The lab employee works alone in the lab. This building has entry points that are accessible by the public due to no security fencing. The employee works with research material and documentation that is considered secrety to the company.</td></tr>
<tr><td colspan="4">Significant Hazards

All hazards associated to the activity should be entered here: task, individual, load and environment. Refer to the Hazard Assessment and enter hazards for any question answered "No"

A. There is no direct supervision of the employee.
B. The lone worker is exposed to the public.
C. Potential risk of violence.
D. There are no arrangements for regular communication.
E. Ability to communicate to the "outside world" is not available at all times.
F. The EAP does not address lone workers.</td></tr>
<tr><td colspan="4">Current Control Measures

List current control measures

A. No current control measures are in place as the position does not require direct supervision.
B. Doors can be locked manually.
C. No fencing around the perimteter of the building to prevent unauthorized access.
D. None identified.
E. None identified.
F. There is an EAP on site but it does not currently address what a lone worker should do.</td></tr>
<tr><td>With these controls</td><td>Unacceptable</td><td>Further Controls</td><td>Adequately Controlled</td></tr>
</table>

the risk is (circle)		Required	

Further Control Measures Required
List further action needed to adequately control risks Automatic locking mechanism or fencing around the property. Better means of communication. Revise the EAP.

Action Required	Date of Completion	By Whom
1. Work with a vendor to change out locks on the door to ones that lock automatically or that are key card accessable.		Office Manager
2. Work with a vendor on costs for installing fencing around the property to prohibit unauthorized access to the building.		Office Manager
3. Identify regular communication intervals with office manager or direct supervisor to ensure that the lab employee is okay.		Lab employee/ supervisor
4. Identify whether or not a “man-down” alarm is needed for this position.		Safety Manager
5. Identify communication requiremetns in the event of an intrusion or the threat of personal safety.		Safety Manager
6. Revise the location Emergency Action Plan (EAP) to identify how lone workers should respond to an emergency situation, to include intruders on the property.		Safety Manager

Assessor Name:	
Signature:	
Date:	

Policy, Procedure, and Emergency Management Plan

Policies are written rules that help an organization establish guidelines for how employees act. Procedures offer detailed instructions for employees on how to respond to certain situations or work on a specific piece of equipment. Ultimately, the purpose of policies and procedures are to capture or establish instructions for how decisions are made and what actions take place by outlining the boundaries or restrictions around certain tasks or situations. Whether it's operating a piece of equipment or control of hazards the "how" to both are outlined in a company's policies and procedures.

Another commonly forgotton purpose of policies and procedures is they are frequently referred to in the event of a safety incident such as an employee injury, to see if there was a violation of a policy or deviation from a procedure. It is important to have accurate, conscise, specific and frequently updated policies and procedures as a way to show the organization's controls over given hazards. Safety manuals offer just that. After a hazard assessment is conducted policies and procedures should then be established to identify ways to mitigate hazards. Lone worker exposure is something that, if it exists as a potential hazard in an organization, should be addressed in a company's safety manual.

Emergency management plans are similar to policies and procedures except they outline details on how an employee and/or employer plans for and responds to workplace specific emergencies. Many times there are policies and procedures referred to or included within an emergency management plan. Common sections in emergency management plans that coincide with lone worker safety are alarms, emergency evacuation protocols, and incident notification and response.

To adequately account for lone worker hazard controls it is recommended to include all three aspects: a policy, procedure and emergency management plan. For example, a policy might state that no employee should work alone in a specific part of the facility. The procedure may be that if an employee must work in that area they use a buddy system and wear a radio or communication device. The emergency action plan would capture the steps to respond in the event a lone worker "went down" or was found unresponsive.

In Summary

A lone worker presents unique hazards to an organization whether the lone worker is high risk, high frequency or low risk, high frequency. As security concerns increase, the lone worker may not be those traditionally identified as working alone. As this paper has discussed, even with a staff on site, an employee may still be considered a lone worker due to the lack of interaction between other personnel at the work location. Specifically, as organizations analyze lone workers, the security concerns for those potentially at risk to violence from the public should be considered. Just because an employee may see various personnel or visitors throughout the day does not exclude them from the definition of a lone worker and the threats that they may face. Through preventative actions, such as conducting a Hazard Assessment, organizations can prepare themselves and their employees to identifying threatening situations before they turn into tragic consequences.

Bibliography

Centers for Disease Control and Prevention (CDC). 2013. *Women's Safety & Health Issues at Work* (retrieved January 17, 2017) (https://www.cdc.gov/niosh/topics/women/serious-injury.html).

Denverite. 2016. "Homicide is a Leading Cause of Work-place Death for Women" (retrieved November 21, 2016) (http://www.denverite.com/homicide-leading-cause-workplace-death-women-6193/)

Global Risk Insights. (2016). *Why Soft Target Terrorist Attacks Will Remain a Threat* (retrieved February 13, 2017) (http://globalriskinsights.com/2016/07/soft-target-terrorist-attacks/)

Occupational Safety and Health Administration (OSHA). 2008. E-CFR 1915.152, *Personal Protectcive Equipment* (retrieved February 13, 2017) (https://www.osha.gov/pls/oshaweb/owadisp.show_document?p_table=STANDARDS&p_id=10268&p_text_version=FALSE#1915.152(b))

Workplace Fairness. *Workplace Violence* (retrieved January 17, 2017) (http://www.workplacefairness.org/workplace-violence).

Being Awake is Not Enough: Why You Need a Fatigue Risk Management System

Mike Harnett, BPE, KIN, ESS
VP Human Factors
SIX Safety Systems Inc.
Calgary, Alberta
Canada

Introduction

Our world today operates at a much different pace from a hundred years ago. Through the advent of automation and technology, we are able to reach out to the world in real time. We can work from home, our car, the office, or the coffee shop at any time, day or night. We may work flexible shifts, extended shifts, night shifts, rotating shifts, split shifts, swing shifts or simply 9-5. The reality is, today's world operates in 24 hours. What we do with those 24 hours predicts what we'll be able to do in the next 24 hours and beyond.

From an organizational perspective, it is imperative to understand how they utilize their workforce to meet the 24 hour demands of society and a global economy. Unfortunately many businesses believe that simply having a body present with (hopefully) their eyes open is enough to fill the roster and get the work done. What they fail to understand is that being awake does not necessarily equate to alertness and subsequently, optimal performance.

Furthermore, management often assumes that fatigue is controllable by the individual and out of the hands of the employer. This is a short-sighted and a misguided belief.

There is nothing simple about fatigue. It is very much a complex issue thus requiring a multi-level approach that is driven by a formal management system.

Introducing a Fatigue Risk Management System (FRMS)

A management system is a set of guiding principles that form the strategic framework for your fatigue management program. Incorporating a systems approach is a switch from the more traditional "person" approach. The person approach concentrates on human error and tends to lay blame on worker behaviours such as forgetfulness, inattention, carelessness, and neglection of procedures.[1]

Alternatively, the system approach accepts that humans are not perfect and will indeed make errors, but in this case, the errors are regarded as consequences rather than causes, and largely due to systemic failures.[2] Control, therefore, shifts to changing the conditions under which individuals work rather than solely trying to change the individual.

This is especially true of fatigue, where poorly designed work schedules, extra work assignments, ignorance of commuting risks, excessive light exposures and demanding job tasks prevent the worker from being able to get the amount of restorative sleep necessary for optimal recovery. Simply providing employee training on how to manage fatigue, which is often the first and sometimes only tactic used by organizations, is putting the cart before the horse.

By digging deeper, we often find flaws in the system. It is through the identification of these flaws that we can now implement a stronger plan.

There are a number of reasons why organizations have not adopted a systems approach when trying to solve their fatigue related concerns. Many are due to misguided beliefs and myths.

- Sleep and tiredness is a state of mind and controllable by workers
- Equating sleep with laziness or not caring about the job
- Confusing sleeplessness with vitality and high performance (this is evident with the executives who are first in and last out of the office, skipping lunch breaks, taking work home, etc. and being rewarded for such behavior)
- Worrying about perceptions (we can't pay people to sleep at work)

In many cases, management doesn't realize just how big the fatigue problem is or how it impacts workers and their performance. This commonly occurs when,

- Employees are afraid to admit or talk about fatigue related issues because of potential disciplinary action.
- Employees (and their families) have never been asked about fatigue and how it impacts them.
- Managers and shift schedulers who are in control of shift work activities, workloads, and schedule designs having never experienced shift work themselves.
- No metrics have been gathered to link fatigue to errors, incidents, or other performance factors.

This not to say that employees do not have a role to play in managing their own fatigue levels, but until an organization addresses many of the myths surrounding fatigue and actively involves the stakeholders in the process, management is guessing at what the problem is and how to correct it.

---

[1] Reason J. "Human Error: Models and Management." *BMJ.* (2000): 320.
[2] Ibid.

Paradigm Shifting

If we want to measure fatigue against safety and performance, than it requires a management system that is driven by objective analysis of data, has established targets, integrates with existing OHS management and business operating systems, and collaborates with and respects the stakeholders and their families. [3, 4] This often requires a fundamental paradigm shift.

For this to occur, an organization must first acknowledge that fatigue is a "fit for duty" concept that must be managed just like the policies for drugs, alcohol, physical safety (e.g. ergonomics) and psychological health (e.g. bullying, sexual harassment, etc.). This can be hastened by demonstrating the link between performance and fatigue and how, if managed correctly, improvements in the system can occur.

Finally, management must accept accountability as owners of the system and their role in promoting alertness and mitigating fatigue. This is the difference between laying blame and stopping the investigation at the worker level and finding the flaws at the system level. It takes a mature organization to recognize the role they play in both the problem of fatigue and the solutions towards eliminating their contributions to worker fatigue.

At the end of the day, the goal is to have harmony between fit for duty and fit for worker, as depicted in Figure 1.

Figure 1. Fit for duty/fit for worker continuum (copyright SIX Safety Systems Inc.)

Review of an FRMS Model

It is not unusual for companies to struggle with maintaining safety programs that they introduce. Based on personal experience and being an external consultant for most of her career, this author has noted that much of the downfall is based on the fact that they do not have the strategic

---

[3] Brown D. "Managing Fatigue Risk: Are Duty Hours the Key to Optimising Crew Performance and Alertness?" *Proceedings of the Flight International Crew Management Conference.* Brussels, Belgium, 2006.

[4] Air Line Pilots Association, International (ALPA) 2008. *Fatigue Risk Management Systems: Addressing Fatigue Within a Just Safety Culture* (http:www.alpa.org/portals/alpa/ pressroom/inthecockpit/FatigueRiskMSWP_ 6-2008.pdf.).

framework in place (a management system) to ensure success of the initiative. As a result, employees are often reluctant to buy in to the "flavor of the month" safety program that's being introduced, because they know, based on history, that it will fade away in time until the next flavor is introduced.

It is with this knowledge that the author is presenting an FRMS model that has proven effective in providing a path towards a successful fatigue risk mitigation and alertness plan. It is designed to clearly identify the parameters under which your fatigue program operates, such as the purpose of the program, what it will look like, how it will be run, and how it will remain viable.

To assist in internal acceptance, efforts were made to develop a model that would streamline with existing management system standards so as to make any transition in an organization that much easier. The following international standards were reviewed for guidance in the development of the model.

- CSA Z1000: Occupational Health and Safety Management System (Canada)
- ISO 45001: OHS Management Systems
- CSA Standard Z1002: Hazard Identification and Elimination; Risk Assessment and Control
- ANSI/API RP755: Fatigue Risk Management Systems for Personnel in the Refining and Petrochemical Industries
- ISO 31000: Risk management - Principles and Guidelines
- ISO 14001: Environmental Management Standard

During the review process, common themes emerged and are summarized in the following six key elements.

1. Organizational Culture, Commitment and Engagement

   This section includes the level of commitment required by Senior Management and defines the roles and responsibilities of everyone in the organization as it relates to fatigue risk management. It clearly highlights the participation of the workers in the process.

2. Strategic Plan

   The strategic plan emphasizes the establishment of objectives and targets to measure against. It includes development of policies and scientifically supported procedures that support legal and regulatory requirements as well as the path towards established targets.

3. Hazard Identification and Risk Assessment Process

   This is the tactical component of the plan to reduce fatigue related risk. It begins with identifying all core contributing factors (both individual and organizational) and assessing their impact. From there, selection of appropriate controls can be introduced.

4. Implementation Plan

   Based on the results of the risk assessment process, implementation of control strategies must be thoroughly defined. This includes preventative and proactive measures, resource allocation, competency and training, communications and awareness, design and procurement, and change management strategies.

5. Evaluation and Corrective Actions

   The only way to determine the success of the FRMS is to establish ongoing monitoring and measurement of the process and program initiatives. Subsequent to that, corrective actions will need to be established where necessary.

6. Management Review and Continuous Improvement

   Based on previous experience, this last step is often the most overlooked step. A process needs to be in place whereby management reviews all findings and metrics that arise through program audits and reviews. It is through such reviews that continuous improvement can be achieved.

Most organizations have some of these FRMS elements already in existence, although they may not be well defined, or they exist as part of a larger umbrella category such as their Occupational Health and Safety Management System (OHSMS) with no specific link to how fatigue fits in.

Since it is imperative that your (eventual) FRMS ties in with your OHS Management System, the goal was to develop a model that would allow for metrics to be captured. To that end, the author developed a needs assessment and gap analysis tool (NAGA©) that would clearly identify where the gaps are in their system. A scoring system was applied to collect metrics. Figure 2 illustrates how this metrics model provides insight and direction for change.

Category	Points Available	Company X 2015	Company X 2017
Organizational Culture, Commitment & Engagement (26 questions)	52	15	34
Strategic Plan (19 questions)	38	10	30
Hazard ID and Risk Assessment (13 questions)	26	15	21
Implementation of Risk Controls (46 questions)	92	6	45
Evaluation & Corrective Actions (15 questions)	30	0	14
Management Review & Continuous Improvement (11 questions)	22	0	6
Overall Score	**260**	**46**	**150**

Figure 2. Comparison of Potential NAGA Scores vs Actual NAGA Scores

The company used the scoring system to identify their gaps and flaws and make the necessary improvements to their FRMS. By collecting such metrics, it helps to drive informed decision making and continuous improvement.

The Tactical Side of an FRMS

An FRMS is the strategic framework that supports your tactical "boots on the ground" approach to identifying and mitigating fatigue, referred to as a Fatigue Risk Assessment process, or FRA.

Our approach to assessing fatigue related risk follows a Fatigue Related Incident Trajectory.[5,6] In this model, there are five sequential interception points where controls can be introduced to mitigate risk and thus, prevent fatigue related incidents from occurring. Figure 3 demonstrates this model in action by providing examples of control strategies. Since every company is different, controls need to be based on the identified needs within their own organization.

The first two levels are critical to address with level 1 representing the most significant risk. If a company has a poorly designed shift schedule that doesn't allow the employees to obtain enough sleep in their off hours, then any subsequent controls will fall short. A well rested crew is the best defense against fatigue.

Level 2 recognizes that simply having a good shift schedule in place doesn't guarantee individual compliance. Employees still have to ensure they utilize their off hours effectively to ensure they obtain enough sleep. Second jobs, family commitments, or lifestyle choices can all influence their behaviours and prevent them from getting the recovery sleep they need.

Risk Assessment		**Examples of Control Strategies**
Latent Errors	Level 1: Sleep Opportunity Making sure schedules provide employees with enough time off to get the sleep they need	• Follow relevant and biocompatible standards for safe rostering • Institute formal tracking of all assigned regular and extra work hours • Evaluate commute practices and their impacts • Predict sleep opportunity using a bio-mathematical model on shift roster
	Level 2: Sleep Obtained Determining whether employees actually got the sleep they needed	• Review prior sleep/wake data through self assessments, physiological monitoring (Actigraphy) • Employee surveys • Subsidies for personal assistive devices such as modifications to private homes to improve sleeping conditions; improved commuting practices

[5] Dawson, Chapman & Thomas. "Fatigue Proofing: A New Approach to Reducing Fatigue Related Risk Using the Principles of Error Management." *Sleep Med Rev* 16, no. 2 (2012), 167. doi: 10.1016/j.smrv.2011.05.004.

[6] Transport Canada (TP). 2010. *Fatigue Risk Management System for the Canadian Aviation Industry - Developing and Implementing a Fatigue Risk Management System* (TP 14575).

	Level 3: Fatigue Related Behaviour Keeping an eye out for anyone who displays symptoms of fatigue on the job	• Symptom checklists • Provide training to Supervisors and Team Leads with regard to detecting and intervening with fatigue related behaviours • Provide a mechanism for workers to report fatigue (NO disciplinary action!) • Allow for naps during night shifts • Provide adequate facilities for rest, breaks, meals, and other essential requirements • Consider drowsiness detection technology (e.g. eye gaze/perclos)
Active Errors	Level 4: Fatigue Related Error Strategies to make sure that when fatigue is detected in the workplace, it does not result in errors or incidents	• Human factors assessment of proposed facilities and equipment at the pre-design phase • Implement checklists/double-checking process during identified low points • Develop recovery room in a quiet and comfortable environment • Institute an error analysis system
	Level 5: Fatigue Related Incident An accident or incident investigation procedure that allows the organization to determine whether fatigue was a factor, and if so, how to avoid a repeat incident	• Investigation Checklists (post incident) • Performance Impairment Checklists (for errors and incidents)

Figure 3. Overview of a tactical fatigue risk assessment process based on a fatigue related incident trajectory.[7]

Tools and Technologies

Since we have established that causes of fatigue are multi-faceted, it's logical that we will require multiple, overlapping tools, technologies, and control mechanisms as a defense against fatigue-related incidents.[8]

Perhaps the biggest myth in fatigue management is that organizations will succeed if they invest all of their resources on the latest technology that measures fatigue or levels of drowsiness. It can be easy to succumb to the lure of technology as the answer, and while there is some value in many (not all) of these new and emerging devices such as eye gaze technology, scanning brainwaves, face tracking, cameras and actigraphs, they are each a tool, not a solution.

A 2013 review of fatigue detection technologies available on the market at that time revealed that the majority of 'validation' data that was produced were not found within the scientific peer-

[7] Transport Canada (TP). 2010. *Fatigue Risk Management System for the Canadian Aviation Industry - Developing and Implementing a Fatigue Risk Management System* (TP 14575).
[8] Ibid.

reviewed literature, but within the quasi-scientific, grey or marketing literature. The authors went further, stating that none were sufficiently validated to provide a comprehensive solution to managing fatigue-related risk at the individual level in real time.[9]

The reality is, many people can perform their duties, and in fact do perform their duties, in a fatigued state. In this regard, we have to again remember that it's not about measuring fatigue as much as it's about measuring safety and performance.

While we need to acknowledge the usefulness of technology, particularly in capturing data and driving change in systems and behaviours, it should not act in isolation. It may require a composite of technologies and practical tools to fully take charge of the fatigue risk. There is also a risk that the technology becomes used in ways unintended, whether deliberate or inadvertently, such as a disciplinary tool. If the organizational culture is not mature enough for the technology, this may not be the proper tool to utilize as it may cause more issues than solve.[10]

Again, a well established culture will see the value in incorporating tools for the workers, such as lifestyle training for them and their family, self-reporting of fatigue (without threat of disciplinary action), fatigue self-assessments, rest and recovery rooms, exercise equipment, etc.

Summary

Fatigue affects everyone. Safety professionals deal with the consequences of fatigue-related incidents on a regular basis without even knowing it. There is a need to educate organizations on this hidden hazard and provide specific tools to mitigate risk but none of this can be done until an organization knows how big the problem is within their own doors.

Safety professionals need to start acquiring the necessary data to drive change from within. It requires a fundamental change in the way we currently evaluate errors, record incidents, perform investigations, conduct audits and collect worker input. By incorporating an evaluation of fatigue related risks into each of these documented processes, we can choose the right controls to put in place and foster continuous improvement. But none of this is possible without a strong FRMS to support the safety professional's efforts.

---

[9] Dawson D. et al. "Look before you (s)leep: Evaluating the use of fatigue detection technologies within a fatigue risk management system for the road transport industry." *Sleep Med Rev,* 18 no.2 (2013):141. doi: 10.1016/j.smrv.2013.03.003.
[10] "Managing Fatigue In The Extractive Industries" *Thematic Analysis Of The Icmm/Ipieca/Ogp Workshop* Johannesburg, SA 2011.

Bibliography

Air Line Pilots Association, International (ALPA) 2008. *Fatigue Risk Management Systems: Addressing Fatigue Within a Just Safety Culture* (http:www.alpa.org/portals/alpa/pressroom/inthecockpit/FatigueRiskMSWP_ 6-2008.pdf.)

Brown D. "Managing Fatigue Risk: Are Duty Hours the Key to Optimizing Crew Performance and Alertness?" *Proceedings of the Flight International Crew Management Conference.* Brussels, Belgium, 2006.

Dawson D. et al. "Look before you (s)leep: Evaluating the use of fatigue detection technologies within a fatigue risk management system for the road transport industry." *Sleep Med Rev,* 18 no.2 (2013):141. doi: 10.1016/j.smrv.2013.03.003.

Dawson, Drew, Janine Chapman, and Mathew Thomas. "Fatigue Proofing: A New Approach to Reducing Fatigue Related Risk Using the Principles of Error Management." *Sleep Med Rev* 16, no. 2 (2012), 167. doi: 10.1016/j.smrv.2011.05.004.

"Managing Fatigue In The Extractive Industries" *Thematic Analysis Of The Icmm/Ipieca/Ogp Workshop* Johannesburg, SA 2011.

Reason, J. "Human Error: Models and Management." *BMJ.* 2000: 320;768–770.

Transport Canada (TP). 2010. *Fatigue Risk Management System for the Canadian Aviation Industry - Developing and Implementing a Fatigue Risk Management System* (TP 14575).

Session No. 618

Leadership for Today's Problems

Don Wilson
V.P. of SafeStart
Belleville, Ontario
Canada

Introduction

Companies increasingly view the accumulated skills and intellectual capital of their workforce as a major resource and, in many cases, as a competitive advantage. If you ask a group of CEOs what their business's biggest asset is, one of the most common answers you'll hear is "It's our people that make this company a world leader in our field."

To safeguard that leadership position, companies make every effort to ensure their employees are kept satisfied, healthy and safe at work because their ability to succeed in the marketplace is heavily reliant on the continued health and job satisfaction of their workforce.

To keep employees happy and satisfied companies offer competitive wages and benefits. To help keep them healthy, employers offer world-class health insurance and additional paid programs like smoking cessation programs or subsidized gym memberships to help employees maintain a healthy lifestyle. But what role does traditional workplace safety management play in protecting workers from unintentional injury and death, and how is that role changing in the modern world?

Many safety systems aren't particularly efficient at protecting workers. Most corporate safety programs focus exclusively on workplace safety, which means that for the eight hours a day someone is at work they have the benefit of a safety manager, PPE and other safety measures.

Unfortunately, the vast majority of injuries and accidental deaths occur outside the workplace, and a key issue that many employers face today is how to use their current safety management system to protect employees 24/7.

This paper will demonstrate that businesses will increase productivity and reduce costs by including off-the-job safety in their safety program. It will also outline how corporate leaders can take proactive measures to protect their workers at work, at home and on the road.

The Changing Nature of Workplaces and Population Demographics in the Western World

It would be a considerable understatement to say that workplaces and the types of work conducted by employees in countries like Canada and the USA have undergone a dramatic revolution in the past few decades. One factor influencing this change has been the rise of cheap labor markets in emerging economies. Change has also been driven by the availability of cheaper and more reliable global transportation networks to move goods easily between markets. A third factor is the ability to seamlessly connect engineering and management systems located in high-wage nations like Canada or the USA with manufacturing and service locations in low-wage countries such as China or India via the Internet and telephone. These three factors have led to many low-wage and low-skill jobs that were once the staple of western economies being "outsourced" to low-wage regions. As a result, more of the jobs available in western economies require much higher degrees of skill, education and training.

Trends: Skilled VS Unskilled Labor

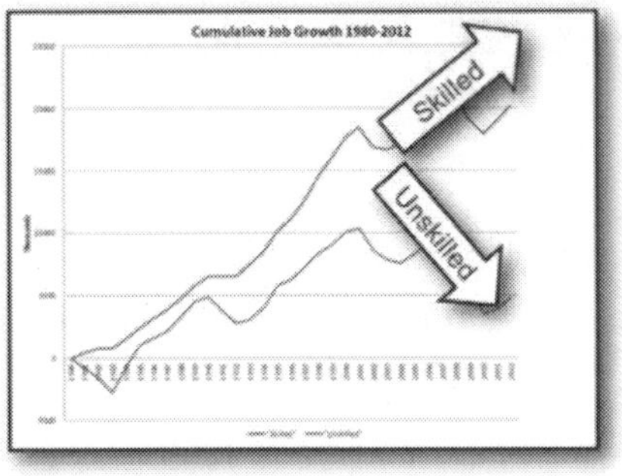

Trends: Looming Skilled Trades Shortages

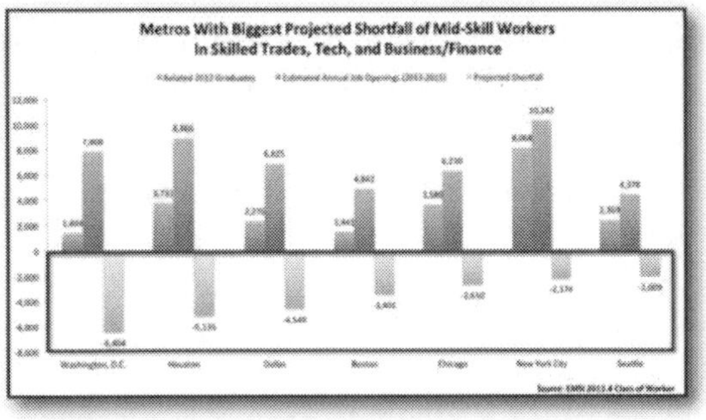

Exhibit 1. Trends in labor skills and availability of skilled workers in 7 metro areas.

Throughout the past 50 years countries like Canada and the United States have also been undergoing a second revolution: a dramatic decrease in average family size. This birth rate decrease is now at the point where most western countries can no longer maintain their present population levels without increasing immigration from countries that still have growing populations.

Unfortunately, while many citizens who live in emerging economies are eager to move to North America, their education, skills and work experience often do not match up with the skills and education that are in high demand in Canada and the United States. Given this, as well as the length of time it takes our education system to produce graduates with the skills and education required for many jobs, employees truly are a company's most valuable asset—and the hardest asset to replace.

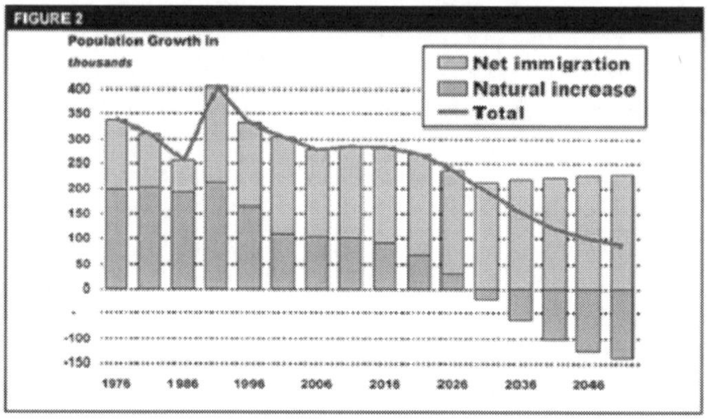

Exhibit 2. Projected population growth of the United States.

The Changing Nature of Work and Workplace Safety

Automation and regulation during the past century have reduced the level of workplace risk. Vast improvements have been made in workplace hazard reduction and in removing humans from harm's way with robotic or remotely controlled production systems. At the same time, regulatory agencies (such as OSHA in the USA) and new due diligence laws (such as Bill C-45 in Canada) have helped to bring attention to workplace safety.

Most ASSE members have their workplace safety process efficiently operating well above industry standards with a well-rounded set of procedures, including JSAs, engineering, guards, and PPE. As a result of this focus on workplace safety, injuries in the North American workplace have slowly but steadily decreased over the last decade.

Statistically speaking, workers are now much more likely to get injured at home or driving to work than on the job. In fact, nearly three times as many workers suffer nonfatal injuries off the job as on the job.

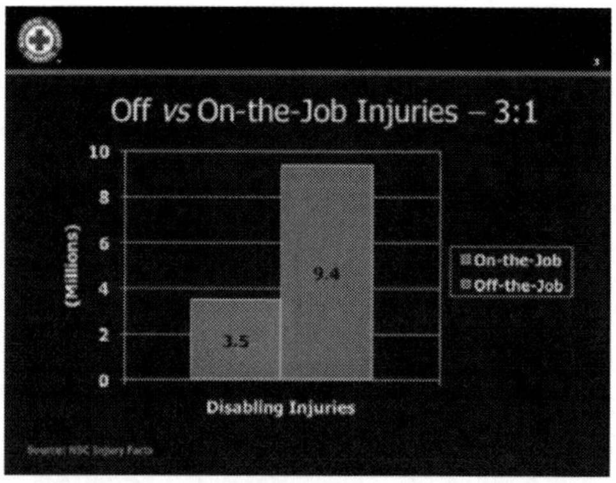

Exhibit 3. Off-the-job injuries vs. on-the-job injuries.

For fatalities the numbers are even starker, with 11 times more people dying from unintentional injuries off the job than on the job.

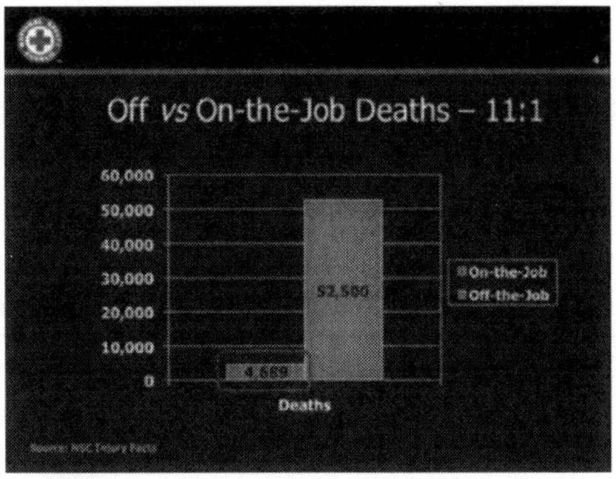

Exhibit 4. Off-the-job deaths vs. on-the-job deaths.

The nature of work and what constitutes a workplace has also changed. Over the past century jobs in the western economy have migrated from being mostly manual in nature to becoming primarily service-based. Often, these new, more cerebral jobs can and are being done anywhere and at any time. As employee teams become less centralized in a single workplace, organizations' ability to enforce safety rules, procedures and regulations has also naturally declined. Gone are the days when an ASSE member's world was the inside of a factory and every employee worked their daily shifts in that workspace.

Companies fail to prevent off-the-job injuries because the methods they use to keep employees safe are effective at work, but they are non-transferrable. Once employees leave work at the end of the day or work remotely, they no longer have their safety manager looking over their shoulder and are less able to rely on the measures designed to keep on-site workers safe. Many corporate leaders also fail to recognize that injured employees present a serious financial cost regardless of whether the injuries occur during or outside working hours.

The Real Cost of Employee Injuries

Most people are aware off the direct cost of injuries, and it's a relatively straightforward task to tally up medical and workers' compensation costs to determine how much a company has to pay out annually due to workplace injuries.

However, these costs are only the tip of the iceberg. To truly account for the cost of employee injury, it is necessary to also consider the financial impact of the following factors:

- time lost from work
- loss in earning power
- lost time by fellow workers
- loss of efficiency due to break-up of crew
- lost time by supervision
- cost of breaking in new person
- damage to tools and equipment
- time damaged equipment is out of service
- spoiled work
- loss of production

- failure to fill orders
- overhead costs
- company/worker morale

These hidden costs all have one thing in common—they affect a company's bottom line regardless of whether an employee was injured at work or at home. If someone gets hurt on the weekend and can't come into work on Monday, their employer is negatively impacted by the worker's failure to fill orders, loss of team efficiency and more. And if a replacement is hired to make up for the loss in production, the new worker's inexperience may lead to damaged tools, additional material waste and a general reduction in efficiency.

The real cost of employee injury isn't medical claims but the loss of productive workers and the sizeable financial investment required to hire and train a replacement. It's in a business's best interest to avoid these costs by protecting employees where they get hurt the most—outside the workplace.

The Importance of Universal Safety Skills

Even though hazards vary from one workplace to the next, and from work to the road and home, the basic causes and patterns of injuries remain constant, and lack of knowledge about hazards is rarely a factor. In fact, most people are injured *despite* knowing about a hazard. So what causes someone to come into contact with a hazard they're aware of? The answer is human error.

There is a universal pattern of mental states that cause people to make one of four errors. These four states and four errors combine to cause over 95% of all unintentional injuries.

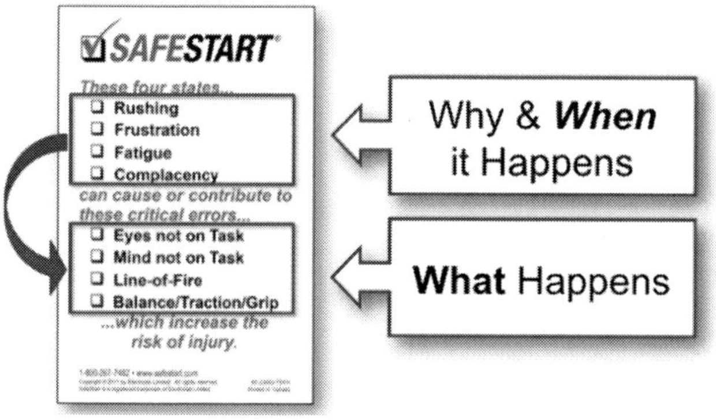

Exhibit 5. The SafeStart state-to-error pattern.

This state-to-error pattern is universal. Not only do rushing, frustration, fatigue and complacency increase the risk of injury but they also increase the potential severity of potential injuries. One of the reasons traditional safety programs are unable to effectively combat off-the-job injuries is because they are only relevant in a specific workplace context. Therefore, solutions to limit injury-causing errors must be as universal as the errors.

Fortunately, it is possible for people to improve their safety awareness, skills and behavior in order to make fewer errors and reduce their risk of injury. In particular, there are four skills that have been demonstrated to provide lasting improvements to employee safety:

- self-trigger on the state
- analyze close calls and small errors
- look at others for risk patterns
- work on habits

Briefly, these four techniques help workers to:

- become more aware of when they're in one of the four states in order to reduce the chances that it will lead to an error
- examine the small errors they make to learn lessons that will help them avoid making larger errors that might lead to more serious injuries
- recognize the dangerous behavior of others and use it as a reminder to avoid similar risk patterns in their own behavior
- develop these and other safety skills through constant practice and habit formation.

These skills are transferrable to any situation and they have a much greater reach than the traditional hierarchy of controls. While these methods of strengthening safety awareness can be extremely effective, simply being aware of them isn't enough to prevent injuries. To truly benefit from these new skills takes practice—at work, at home and on the road—which also takes motivation.

Teaching 24/7 Safety

Experts agree that when it comes to influencing change you need to help people answer two questions positively for themselves before any change is likely to occur. First, people need to decide if they think they can actually do what you want them to do. Second, people must decide if they think it will benefit them personally. Is it easy enough for them to do and is it worth their effort? A great deal of safety leadership consists of finding ways to get employees to answer "yes" to both these questions. From a 24/7 safety perspective this means the personal skills, habits and behaviors need to be easy to use in their daily life and that people need to believe that adopting them for use will actually reduce their injuries.

One of the most effective ways to engage and motivate employees is to focus on their safety agenda rather than the company's safety goals. For example, most people are far more motivated to protect their families than themselves, and given the necessary tools almost every employee will teach their family stronger safety skills.

From a company's perspective, turning each employee into a trainer (even if they're only training their own family) is an ideal outcome. People retain far more information when they actively participate in learning. As shown in Exhibit 6, as a learner's degree of participation in learning increases the amount of information they retain also goes up. When it comes to learning the practical safety skills outlined in the previous section, encouraging employees to teach the skills to their families will help them stay motivated, retain more information, be more likely to practice the safety skills themselves and feel more invested in a safety program.

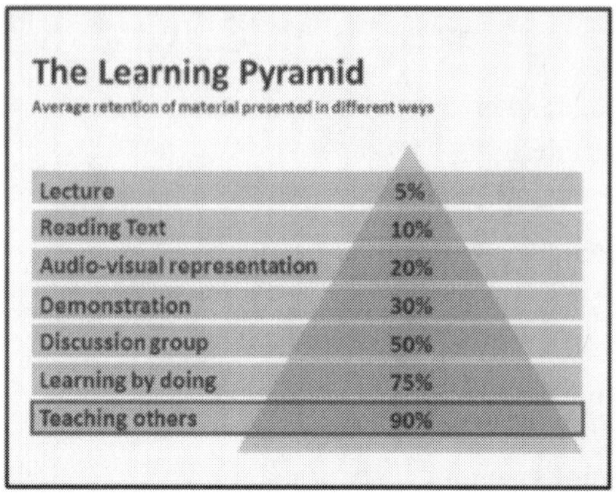

Exhibit 6. Knowledge retention based on teaching method.

Conclusion

Traditional approaches to safety management fail to protect employees by exclusively taking measures to reduce injuries in the workplace. This ignores the fact that people are statistically safest at work and much more likely to get hurt off the job. Additionally, there is an extremely high cost to replacing employees regardless of whether they are on the clock when an injury occurs.

Command and control strategies have little positive effect in improving off-the-job injuries, and it behooves employers to adapt their current safety management system to protect employees 24/7. They can do so by providing workers with skills to increase their awareness and reduce the errors that lead to injuries no matter where they are. If businesses motivate employees by addressing their safety agenda they can effectively influence them to learn skills and adopt habits to reduce injuries. In the process, they will also lower costs and avoid productivity loss.

Session No. 619

Understanding the Proper Fall Protection Methods for Aerial Lifts

Terry Keenan, CHST
Corporate Safety Director
McCormick Construction Inc.
Greenfield, MN

Paper Scope and Focus

This paper will attempt to address and focus on four areas:

- The fall hazards present when using an aerial lift.
- A discussion of the improper methods of fall protection used in industry today.
- A review of the OSHA standards in force that guide us in selecting the proper fall protection methods for the different hazards that we encounter.
- Identify the proper method to use for fall protection when operating an aerial lift.

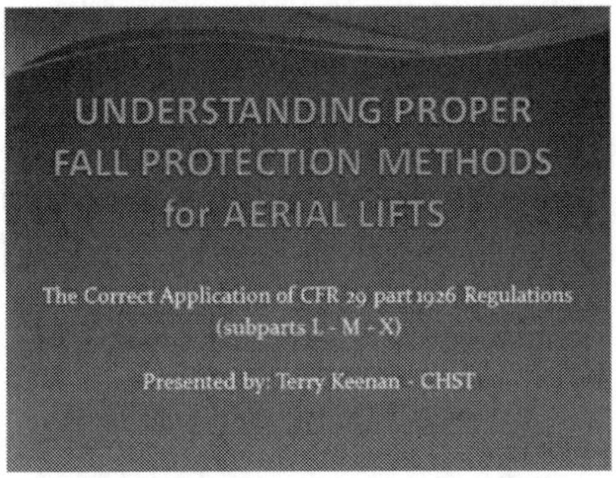

Basic Questions that Need to Be Answered before We Get Started

- What is "Fall Protection" by definition?
- Answer: "Fall" is the act of falling, or to drop. "Protect" is to shield from injury.

QUESTION #1

What is "Fall Protection by Definition?

ANSWER:
"FALL" is the act of falling, or to drop.
"PROTECT" to shield from injury.

- How do we "protect from falls" in Construction Operations?
- Answer: There are many options that need to be applied…the key is selecting the correct method for each situation that places us in an exposure to a fall.

QUESTION #2

➢ How do we "Protect From Falls" in Construction Operations?

ANSWER: There are many options that need to be applied...the key is selecting the correct method for each situation that place us in an exposure to a fall!

When a construction worker is wearing a body harness and a 6 foot rip-stop deceleration lanyard that is anchored to the lift basket, is he protected from the fall hazard?

A standard practice in the construction industry is to have workers "tie-off" for fall protection when working in an aerial lift. Are they truly and properly protected?

- At this point in the presentation there will be a video that shows the catapult effect that is often a hidden hazard during the operation of an aerial lift. The video shows a worker without any fall protection other than the basket guardrails getting ejected out of the basket when the lift drops off of a curb. The worker flies through the air and hits the ground due to being catapulted out of the basket. The video is on You Tube and is titled "Boom Lift Catapult Europe 2014".

If an aerial lift is boomed out and the boom is about 45degrees from horizontal, and if workers are wearing a body harness and are anchored with a standard 6 foot lanyard were to fall, what would be the result?

If a fall protection anchor point is modified to give the worker more mobility, or if the workers were to anchor to the top rail, are they going to be protected from a fall?

- At this point we will be viewing a video that illustrates what happens when workers are catapulted out of an aerial lift with the standard harness and lanyard attached to the basket. The video shows that an operator will be thrown over the side of the lift basket and smash into the boom and/or the ground resulting in serious injury. The video is on You Tube and is titled "Catapult on Boom MEWP – Katapult Effect bei TeleskophubarbeitsbA1/4hnen".

It is important to know that the maximum capacity of the lift unit must not be exceeded or the lift may turn over or the boom may fail. Fall arresting forces are great and a minimum requirement for anchor points is 2 times the intended load of the fall. A 6 foot arresting lanyard is designed to limit forces to 1800 pounds. Most aerial lifts are not rated for more than about 500 – 700 pounds. Using a fall arresting system in an aerial lift will exceed the rated capacity during a fall.

The workers in the construction industry today have been told to "Tie Off" in an aerial lift. The problem is, we have not been teaching them the correct way to accomplish this. Many workers are using the wrong lanyards and an inferior harness that does not allow them to obtain proper fall protection in the lift.

The proper method of fall protection in an aerial lift to best protect the operators and workers is to utilize a "Restrain System". A shorter lanyard (some are adjustable) with a body belt or a harness with side D rings works best to keep the occupants in the basket when the boom bounces.

Is a scissor lift an aerial lift? Do workers need to be tied off when in the basket of a scissor lift? The answer is no to both questions. A scissor lift is classified by OSHA as a mobile adjustable scaffold. As long as the guardrails are in place, no additional fall protection is required. If additional fall protection is required by the client or site specific rules, a fall restraint system must be utilized. These devices are not rated with enough capacity to support a fall arresting system.

What OSHA regulation is applicable to a scissor lift? What fall protection is required? Specifically the CFR 29 part 1926.450 - .452 would be that correct reference that explains what we have discussed above. There is also an OSHA letter of interpretation that clearly states that a scissor lift IS NOT an aerial lift, so the rules in 1926.453 would not apply.

There are many sections in the construction regulations that address fall hazards and fall protection methods. Once again it is important to select the proper method for the specific hazard type that the workers are facing.

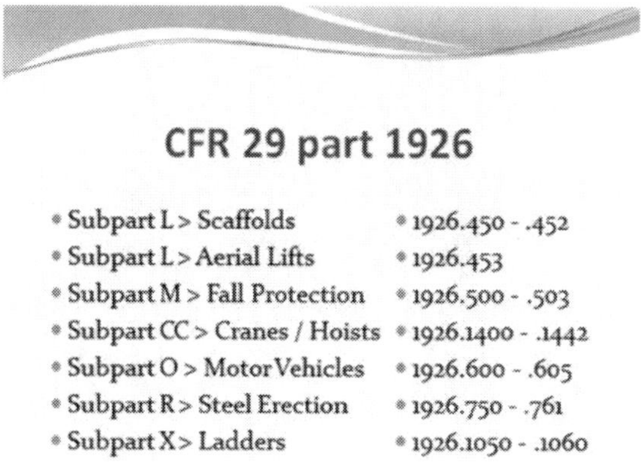

In CFR 29 part 1926.500 we find the fall protection standard. Too often we try to apply the rules there to all fall protection hazards. This is incorrect. In the "Scope & Application" section of this standard it clearly states that these rules are to be applied with Exception. If you are applying the fall protection rules from other "vertical" standards, 1926.500 would not be applied unless it is referenced by the other standards in play. Only if the provisions of the other vertical standards cannot be implemented would we then go to the fall protection standard to select a method of protection.

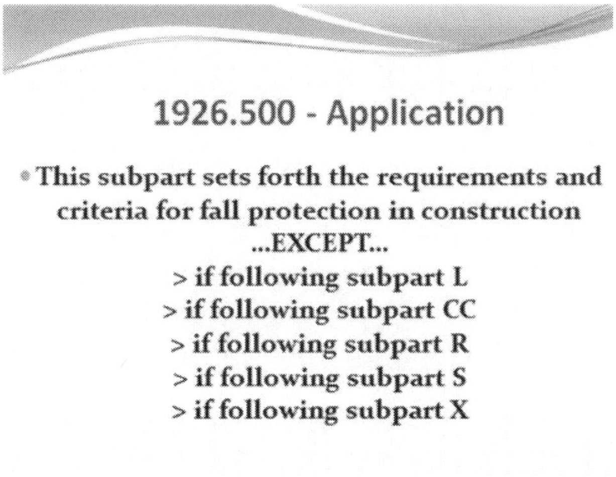

If the provisions set forth in the various specific vertical standards cannot be met or implemented, then we select from the following methods found in 1926.500 – 503.

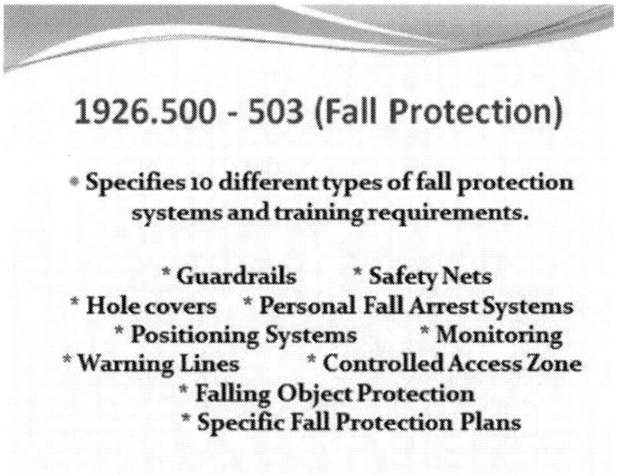

Let's take a look at some of the vertical standards and their provisions for fall protection. First, 1926.450 – 452 subpart L, Scaffolds. Here we see that primary fall protection is the guardrail system and 1926.502(b) is referenced. A proper access is required as specified in ladders & stairs found in 1926.1053 in subpart X. Solid work platforms and other performance criteria is specified in subpart L, 1926.451 – 452.

1926.450 - 452 (Scaffolds)

- **Fall protection provided by guardrail system.**
  ***Specified in subpart M - 1926.502(b)**
- **Access (legal) provided by a ladder / stair.**
  ***Specified in subpart X - 1926.1053**
- **Solid working surface provided by platform.**
  ***Specified in subpart L - 1926.451 - .452**

When you put it all together it looks like this. I think that these guys really know what they are doing!

In CFR 29 part 1926.1431 we find the provisions for hoisting personnel in a crane man basket. These are different from aerial lifts and scissor lifts in that they are supported on a crane with a capacity to support a fall arresting system. The cage / platform must be designed by a Registered Professional Engineer, the guardrail system and a fall arrest system is used together. Either fall arrest or fall restraint may be used due to the crane capacity and the ability to handle the intended loads.

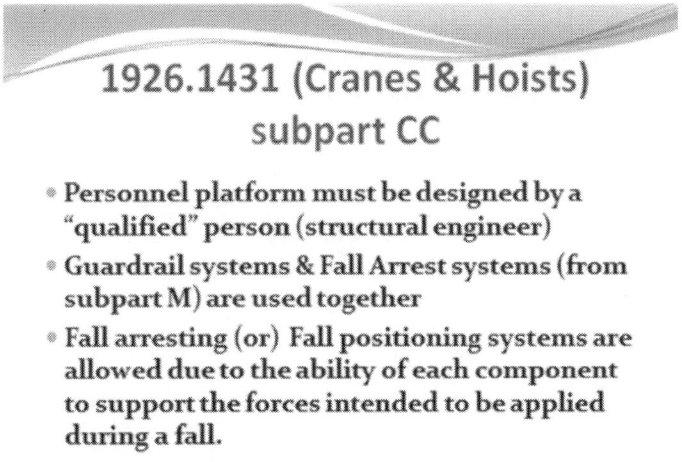

Obviously this man lift platform is not in compliance. There are numerous violations of the standards that would keep the workers safe in this device.

These man-lift baskets look a lot better than the homemade version. Notice that they have a proper enclosure, designated rigging, overhead canopy protection, and the test weights for pre-operational evaluation.

Now let's take a look at the vertical standard that applies to an aerial lift. A good place to start would be to define what an aerial lift actually is. CFR 29 part 1926.453 makes this pretty clear. Aerial lifts are: Extensible Boom Platforms – Articulating Ladders – Articulating Boom Platforms – Vertical Towers – or some combination of these devices. A Scissor Lift is not an aerial lift device!

Aerial Lift - DEFINED

• 1926.453

> Extensible Boom Platforms
> Aerial Ladders
> Articulating Boom Platforms
> Vertical Towers
> or Combinations of these Devices

• A Scissors Lift is NOT an Aerial lift!

Fall protection in / from an aerial lift is also clearly specified in the standard. Some fall protection is afforded by the guardrail & gate system. The Primary fall protection should be a restraint system as specified by 1926.502(e) in subpart M for fall protection methods. This system is designed to prevent falls and keep the workers in the basket.

1926.453 (Aerial Lifts)

• Some fall protection provided by Basket.
*Enclosure & Gate prevent falls

• Primary fall protection should be a Restraint (positioning) System as specified by 1926.502(e)

[safety belt or harness with a short lanyard]
Systems are designed to "prevent" falls

Other specific work safety rules are provided in 1926.453(b)(2) that aid in good fall prevention as well. Do not tie off to adjacent structures when in the basket. Stand firmly on the floor of the platform. Use a body belt & lanyard. Do not exceed boom & basket load limits specified by the manufacturer. If these rules cannot be followed, THEN refer to subpart M section 1926.500 for alternative fall protection methods.

Fall Protection in Aerial Lifts

- 1926.453(b)(2)(iii) - tie-off to adjacent pole, structure or equipment is not permitted
- 1926.453(b)(2)(iv) - stand firmly on the floor / climbing up -or- out is not permitted
- 1926.453(b)(2)(v) - a body belt shall be worn (see note to this requirement)
- 1926.453(b)(2)(vi) - boom & basket load limits specified by manufacturer shall not be exceeded

Again, there are different methods of fall protection for different types of fall hazards. Depending on what you are trying to accomplish you must understand the forces involved with each method. The chart below will contrast some of the criteria for fall arrest versus fall restraint.

Fall Arrest -vs- Positioning

Fall Arrest	Positioning
Anchor - 5000 lbs. -or- (safety factor at least x2)	Anchor - 3000 lbs. -or- (safety factor at least x2)
Lanyard - 6 ft. arresting	Lanyard - adjustable (NA)
Body Harness (only)	Body Harness -or- Belt
Distance till stopped - 9.5'	Free Fall Distance - < 2ft.
Prompt rescue measures are required	Self Rescue - in most cases there is no rescue needed
Arresting Force = 1800 lbs.	Force = Body Weight

In support of my stance that in most cases a fall arresting system would be inappropriate for a worker in an aerial lift, OSHA has issued a letter of interpretation (02-18-99) that explains the reasons for this position. The letter clearly states that either system would be acceptable, but the aerial lift system MUST be capable of supporting the intended load of the fall. If this cannot be achieved, then a fall restraint system must be used.

OSHA Letter of Interpretation (02-18-99)

- **Paragraph #2 - Fall protection is required! (Arresting or Positioning is Allowed)**
- **Paragraph #4 - The total system must be capable of withstanding the intended load imposed by arresting a fall!**

> **If this cannot be achieved, then a restraint system must be used.**

What is your opinion on this picture? How many OSHA violations do you see in this operation?

- At this point of the presentation we will be viewing a video that simulates a catastrophic accident in an aerial lift. The television "reality" show {Myth Busters} placed a manikin in an aerial lift and tried to catapult the dummy over a building. The video is on You Tube and is titled "Catapult Demonstration".

OSHA has issued another letter of interpretation concerning the use of a body belt in a fall arresting system. Body belts are not to be used for fall arresting because they cause too my damage to the victim of a fall, and the wearer may fall out of the belt during this traumatic event. A body belt is still available and may be used for a restraint system as required in an aerial lift.

The best way to avoid confusion that I have found is to issue a body harness with a belt & "D" rings integral with the harness. This way the user only needs to switch lanyards when changing from fall arrest to fall restraint.

Here are some key points to remember:

KEY POINTS!

- Do Not Mix & Match OSHA Standards
- Insure that the proper Standard is being applied to the work situation
- Remember that more than one standard may apply. Make sure that you use the proper interaction / overlap as designed.
- READ the ENTIRE STANDARD!

Put It All Together

- **Inspection Procedures & Criteria!**
- **Donning & Doffing Procedures!**
- **Which Lanyard To Use!**
- **How To Establish A Sound Anchor!**
- **Where To Get The "Right Stuff"!**

Bibliography / Documentation

Occupational Safety and Health Administration 29 CFR Part 1926, Safety and Health Regulations for Construction. Subpart Cc, Cranes and Derricks in Construction; Subpart L, Scaffolds; Subpart M, Fall Protection; Subpart R, Steel Erection; Subpart X, Stairways and Ladders.

Fuller, Thomas P. 2014. *Basics of Occupational Safety and Health.* Itasca, IL: National Safety Council.

OSHA Letter of Interpretation (02-18-99) Fall arrest versus Fall restraint systems

OSHA Letter of Interpretation (03-27-95) Fall protection requirements in various lift types

Youtube Video / Boom Lift Catapult Europe 2014

Youtube Video / Catapult on Boom MEWP – Katapult Effect bei TeleskohubarbeitsbA1/4henen

Youtube Video / Catapult Demonstration (from Myth Busters)

All photographs were original by the author taken on project locations
All power point slides were original creations by the author

Session No. 620

Establishing an Effective Electrical Safety Program: Requirements and Approaches

Paul A Zoubek, CSP, CIH, CESCP
President
Zoubek Consulting, LLC
San Diego, CA

Introduction

Worker injuries resulting from inadequate electrical administrative controls can be debilitating and even fatal. It is the safety professional's responsibility to guide employers on their obligations to maintain a safe work environment through administrative hazard controls.

NFPA 70E electrical safety code provides guidance for workplace electrical safety program establishment and implementation. These programs are intended to reduce arc flash and shock risk to qualified persons. Outlined are practical ideas for developing and implementing programs that are organization and operation specific. Key elements from NFPA 70E (Standard for Electrical Safety in the Workplace®) as it relates electrical safety program requirements are outlined forthcoming.

NFPA electrical safety codes including NFPA 70 (National Electrical Code®), NFPA 79 (Electrical Standard for Industrial Machinery), and NFPA 70B (Recommended Practice for Electrical Equipment Maintenance) provide guidance for engineering controls to electrical hazards. NFPA 70E (Standard for Electrical Safety in the Workplace) requires the employer to implement and document an overall electrical safety program that is appropriate to the risk associated with electrical hazards and as part of EOHS Management System. The program requirements apply to situations where the worker will be exposed to energized electrical systems at some point during servicing and maintenance activities.

Seven Steps to a Proactive Program

All too often health and safety programs are established and are filed in a binder or electronic folder and are rarely accessed for safety program guidance as intended. NFPA 70E (2015 Edition) Article 110.1 outline the required components of the electrical safety program. Outside the usual writing and filing of a safety program, the following steps outline recommendations for a proactive electrical safety program in accordance with 70E Article 110.1

Step 1: List Equipment Maintenance Requirements

The electrical safety program shall include elements that consider condition of maintenance of electrical equipment and systems. It is prudent for a facility to first list equipment which energized work will be performed. This list includes equipment where energy controls procedures would be need to be generated and zero energy state verification would need to be performed. The list should also include equipment and electrical systems where electrical testing and troubleshooting may occur. Once an organization knows that electrical work will need to be performed it can proceed to the next step and make decisions on which performance standards will apply.

Step 2: Establish Performance Standards

Depending on equipment surveyed and listed the following non-inclusive performance standards should be considered as they apply to respective electrical equipment.

a. NFPA 70: National Electrical Code
b. NFPA 79 (Electrical Standard for Industrial Machinery)
c. NFPA 70B (Recommended Practice for Electrical Equipment Maintenance)
d. NFPA 70E (Standard for Electrical Safety in the Workplace)
e. Various ANSI Standards
f. Recommendations from equipment manufacturers
g. Etc.

Once applicable performance standards are researched and applied to equipment, the organization can move to Step 3 and assess risk to electrical hazards as it specifically pertains to equipment.

Step 3: Perform Risk Assessment

An electrical safety program shall include a risk assessment procedure that addresses employee exposure to electrical hazards. The procedure must identify hazards (shock, arc flash), assess the risk (likelihood X consequences), and implement controls according to the hierarch of controls. The following are considerations for risk reduction as they apply to electrical safety program implementation.

a. *Elimination*. Work on a system de-energized. Getting to the elimination stage however may necessitate starting at the bottom of the controls hierarchy with administrative and PPE controls.

b. *Substitution.* It is difficult to substitute electrical energy post-design. Substitution may include lower amperage and overcurrent protection engineering controls with faster clearing times. Using substitution should decrease risk, but most likely will need to be used in combination with other controls.
c. *Engineering Controls.* The most common controls for electrical hazards include engineering controls. Ensuring equipment is maintained in accordance with Step 1 and Step 2 will make engineering controls more effective.
d. *Administrative Controls.* Administrative controls in electrical safety include the electrical safety program itself and approach boundaries listed in NFPA 70E for shock protection plus arc flash boundaries resulting from arc flash risk assessment.
e. *Personal Protective Equipment.* PPE will no doubt need to be utilized in any potentially energized situation either from arc flash hazards, shock hazards, or both. The extent of PPE necessary will result from the engineering analysis performed on the equipment (the arc flash risk assessment).

Annex F of NFPA 70E (2015 edition) provides non-mandatory guidance on performance of the risk assessment. Other guidance/options for risk assessment methods may be obtained in various ANSI Standards (i.e.: ANSI Z10, ANSI Z244.1, etc.), OHSAS, ISO, or applicable EOHS management system methods.

Step 4: Establish Controls

Once risk assessments are completed and decisions are made based on hierarchy of controls, the controls should be implemented. When implementing administrative controls and PPE for energized work the following sequence is suggested.

a. *Identify tasks.* If voltage testing, for example will be performed the task should be identified and corresponded to the incident energy and voltage range identified from risk assessment. Tasks and corresponding protective measures are then transferred to procedures in step 5.
b. *Identify PPE.* Identify arc flash and shock PPE from task. Then transfer data to procedures.
c. *Identify boundaries.* Identify arc flash and shock boundaries from risk assessment. Then transfer data to procedures.
d. *Establish training.* Training requirements and agenda for qualified persons are outlined in NFPA 70E (2015 Edition) Article 110.2. The length and extent of training are subject to interpretation from the employer and in accordance with the risk assessment.

Step 5: Set-up Procedures

All too often PPE-Category Tables are cut and pasted into an electrical safety program which rarely correspond to site conditions, tasks, equipment or voltage ranges. Electrical safety procedures should address actual site conditions and specifically for equipment in use at site. Electrical safety procedures should list of steps executed during energized work. A common example of procedural controls which are site specific are lockout/tagout energy controls procedures. Energy Control Procedures that involve interaction with energized circuit parts should outline voltage testing procedures & related protective means.

Step 6: Institute Job Briefing

Job briefing is a requirement before starting energized work and should be conducted in accordance with procedures established. Job briefing can transpire via equipment specific procedures, energy control procedures, or if required electrical work permit.

Step 7: Ensure Auditing Protocols are Established

Recent edition of NFPA 70E have required that the electrical safety program itself be audited and that field work be audited. The electrical safety program shall be audited to ensure principles and procedures are being followed and may take the form of energy control procedure audits or quality audits if safety programs are encompassed. NFPA 70E requires a triennial audit of the electrical safety program.

Field work shall be audited to ensure principles and procedures are being followed and may be part of aforementioned electrical safety program audit. NFPA 70E requires an annual audit of field work and can be a documented observation of work practices by the qualified person.

NFPA 70E does not require a specific format for auditing protocols but does require that audits be documented.

Conclusions

In recent years, OSHA has been collecting additional data from employers on high risk workplace hazards. In September 2015 OSHA announced that it will move away from tallying each inspection equally and instead will use a weighted system based on how complicated the inspection may be. Electrical hazards in the workplace pose both high risk and complication. Safety professionals are encouraged to focus on the establishment of a proactive electrical safety program outlined in NFPA 70E as well as the guidance standards from other NFPA to mitigate risk to workers from potentially energized electrical hazards via administrative controls.

References

______. NFPA 70, *National Electrical Code*, 2017 Edition. Quincy, MA: NFPA.

______. NFPA 70B, *Recommended Practice for Electrical Equipment Maintenance*, 2016 Edition. Quincy, MA: NFPA.

______. NFPA 70E, *Standard for Electrical Safety in the Workplace*, 2015 Edition. Quincy, MA: NFPA.

______. NFPA 79, *Electrical Standard for Industrial Machinery*, 2015 Edition. Quincy, MA: NFPA.

National Fire Protection Association. 2009. *The Electrical Safety Program Guide.* Quincy, MA: NFPA.

Session No. 621

Strategic Training Design–With a Magic Wand, a Bar Napkin and Speed Dating

Kery Mortenson, MS, CPT
Manager, Training Effectiveness

Sharon Kemerer, MSN, COHN-S/CM, FAAOHN
Corporate Director, Occupational Health & Safety

Baxter International

Introduction

The opportunity to achieve top quartile performance in health and human safety requires new and innovative approaches. We must leverage predictable analysis and evaluation methods combined with innovative blended solutions to enhance the safety in the workplace and workforce. Isolating key behaviors was a critical success factor for the Safety Connection initiative.

Health and safety professionals are conditioned to use training as a primary tool to advance objectives in achieving safety excellence. With limited time and resources, it is imperative to analyze the systematic causal factors that may be entrenched in the workplace and the workforce. Assuming that only a training solution will produce the desired outcome is a risky decision and could derail or impede the desired state. This paper describes a process to analyze, design and measure the impact of blended solutions, cultural indicators and leading safety indicators.

We devised and applied the following principles, methods and tools throughout the initiative:

- ✓ Applied a holistic approach for achieving safety goals
- ✓ Leverage rapid tools to identify desired goal
- ✓ Utilized a bar napkin exercise to discuss solutions and perceptions with stakeholders
- ✓ Created an Impact Map that aligns key safety behaviors to measurable results

✓ Apply the Success Case evaluation methodology to quickly measure the pilots effectiveness

Utilizing this approach required experts within our company who are skilled in Human Performance Technology and who can effectively partner with leaders and experts in safety to discern key barriers that potentially impede success by uncovering answers to the following questions:

- How is your safety strategy perceived within your company?
- What is the desired state for your safety initiative? Who else knows this and cares?
- Are your safety indicators really leading or lagging?
- What are the critical safety behaviors for your workforce and management?
- How do you currently measure those key behaviors?
- What internal and external motivators exist to support the goal?
- What consequences are in place for supervisors or managers who exhibit poor safety?

Determine the Solutions and Perception of Your Plan

To start the conversation around the Safety Connection initiative, we used a simple table top method call the "Bar Napkin" exercise to understand how the current safety strategy perceived within our organization.

This exercise is used to get leaders started down the right thinking path about their initiative. It explores the range of other initiatives they are responsible for, how they are identifying solutions, and what they believe are the perceptions by the end user. Reflection questions are typically added after completion of the exercise to deepen the dialog and enable more innovative thinking about the issues and how to solve them. Below is an example of how the napkin is setup.

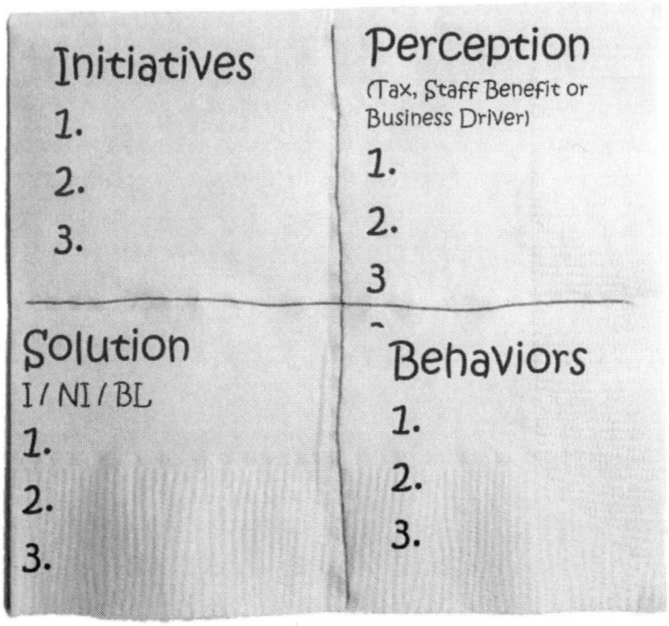

The following are the phases of completing the bar-napkin exercise:

Phase	Action
1	Handout a napkin and ask them label each quadrant with the following - **Initiative** - **Solutions** - **Perception** - **Behaviors**
2	In the **Initiative quadrant**, ask participants to write down the top 3 initiatives they are responsible for.
3	Under the **Solutions** quadrant, we ask them to identify the solution they are pursuing for each initiative using the following labels - **IS**= Instructional solution (Communication, training, assessment) - **NI**=Non instructional solution (software change, process simplification) - **Blended**= combination on Instructional and non-instructional solutions
4	In the **Perception** quadrant, we ask them to identify how the initiative is currently being received by the organization as one of three types in the upper right quadrant - **Tax** (Participate in a "check the box" activity seen a not value added) - **Staff Benefit** (Solution may help employees – Onboarding/mentoring) - **Business driver** (Participants see a true business impact –tied to bottom line)

5	In the **Behaviors** quadrant, we ask them list the top 3 critical behaviors for an initiative perceived as a business driver.
6	Reflect on their responses – then ask them if they have at least one initiative that is a Business Driver?
7	Optional question: Why are you working on projects that are not perceived as a business driver?

Chart Your Course–Begin with the end in mind using the power of a "Magic Wand"

Starting with the end in mind for any initiative makes a lot of sense. The team may have an idea of what that looks like, but is it written down anywhere? Has it been socialized within your organization? For the Safety Connection initiative, this was quickly accomplished using the 'Magic Wand' technique.

Using the magic wand helped the Safety Connection tram step out of the current environment and focus and on the future state. Creating the future goal statement is a facilitated activity that works best with two facilitators (One facilitating the activity and the second person decoding the output).

The desired state is typically created in 15 minutes. It is the first step toward establishing an effective strategy. Lastly, it tends to be a very fun activity – so make sure you stock up on magic wands!

Phase	Action
1	Invite a group of involved stakeholders & participants
2	Introduce the purpose of creating the desired state. Ask a volunteer who in most interested in holding and waving a magic wand
3	Ask the holder of the magic wand to wave it over the room and the participants.
4	Ask the audience the following questions: Tell them to record each response to the questions on a post-it 1. If you were to wave a magic wand over this issue and the next morning you came into work and desired state has occurred for the Safety Connection program– what would it look like? 2. Who is the target audience and what would they be doing differently? 3. What would people be saying about it?

	4. What leading indicators would be in place?
6	One facilitator collects the post it notes from the participants, while a second facilitator categorizes the feedback and begins creating the desired state on a flip chart.
7	Facilitator review s the desired state with the participants at the end of the exercise to ensure agreement.

The safety team created the following goal:

Establish a zero harm workplace where leaders actively and positively engage team members to own safety and team members help each other to see and correct safety issues

Determine the Causal Factors Using a Rapid Gap Analysis (RGA)

After we clarified the desired state with the safety leadership team, we then set out to identify causal factors would prevent their desired state from being realized.

The Rapid Gap Analysis (RGA) is a systematic method to pinpoint causal factors in the environment and individual worker that prevents the desired state from occurring. We completed this activity with the same team who provided the desired state.

We first describe the RGA analysis table below which includes potential gaps rooted in the environment (Data, Instruments, and Incentives). The lower part of the table includes potential gaps rooted in the workforce which include (Knowledge, Capacity and Motives).

We then provided a list of open ended questions that helps the team target potential gaps in each of the six cells of the table. After identifying the potential gaps that exist in one or more cells, we then have the group color code the gaps for each cell using the following criteria:

- Red= Major gaps that prevent desired state,
- Yellow= Minor gaps that prevent desired state
- Green= No gaps and or positive opportunities for desired state

Completing the Rapid Gap Analysis (RGA) was completed in one hour. This activity provides the context for the team to then identify solutions to mitigate the gaps identified in the RGA exercise.

Information	Instrumentation	Motivation	
DATA	**INSTRUMENTS**	**INCENTIVES**	
▪ Data and Information Metrics ▪ Communication and Feedback	▪ Tools ▪ Resources ▪ Work environment	▪ Incentives ▪ Consequences ▪ Rewards	Rooted in the Environment/ Organization
KNOWLEDGE	**CAPACITY**	**MOTIVES**	
▪ Skills ▪ Knowledge ▪ Behavior	▪ Individual capacity	▪ Recognition ▪ Motivation ▪ Expectations	Rooted in the Individual Worker

Thomas Gilbert
Fundamentals of Human Performance Technology
Published by ISPI 2010

Speed Dating for Solutions Using the "21 Activity"

After completing the Rapid Gap Analysis (RGA), we then brainstormed a list of solutions that would mitigate or reduce the major gaps identified in the RGA exercise. and rate the effort required and the corresponding benefit.

To accomplish this, we use a simple exercise called "21 activity". It is an easy process that eliminates bias toward a specific solution or idea. The 21 activity is completed in 45 minutes.

- 15 minutes to generate the solutions and
- 15 minutes to rate the effort and benefit of each solution.
- 15 minutes to select the solution(s) with assigned responsibility

Phase	Action
1	Invite a small or large group of involved stakeholders & participants
2	Provide the instructions for the exercise. Review the major gaps identified in the RPA Handout the 21 Exercise to each participant
3	Ask each participant to write down their best solution that will reduce the gap and

	or move to the desired state.
4	Then ask each participant to provide the solution to someone else in the room Provide instructions on how to rate their solution with two other participants
6	Play seven rounds – each person has to connect with two new people If possible, use a noise maker to indicate
7	The facilitator captures all the solutions with the corresponding point value The ask participants to rate the effort and benefit of each solution using a (Low/Moderate/High rating). Identify the solutions to be implement and who will be responsible

Summary example of using the "21 activity"

Idea/Solution	Activity Score	Effort/Benefit
Identify and communicate Safety Measures for Success	21	H / H
Holds managers and supervisors accountable for including safety conversations	19	L / H
Leadership follows up with managers and supervisors to ensure that safety behaviors and culture are being implemented	18	L / H
EHS communicates safety investigations at Tier Meetings	16	M / H
Leadership includes safety action plans in Manager and Supervisors goals.	15	M / H
Review SOPs and make safety changes	11	L / M

Create a Clear Line of Site Using Impact Maps

Impact maps are an essential tool to establish a clear line of site that aligns team behaviors to business results and ultimately to the initiatives goal.

Creating an Impact map is a facilitated activity that works best with two performance consultants (One facilitating the activity and the second person capturing the output from the participants.

Attached is the Impact Map completed for the safety initiative. The critical behaviors were validated to ensure the key results would be realized as part of the process.

Knowledge and Skills	Critical Behaviors	Key Results	Goal
Recognize hazards and risks for your work area Interpersonal communication skills Coaching and Feedback skills Root cause analysis/methods skills	Conduct daily employee discussions Actively listen to safety concerns. Conduct weekly safety walks. Correct hazards and risky behaviors before they cause injuries. Model safe behavior and use required safety equipment.	Near misses increase initially and then decrease Decrease in reported injuries Safety improvements are visible All significant or major incidents have a root cause identified Safety "culture" metrics improve	Establish a zero harm workplace where leaders actively and positively engage team members to own safety and team members help each other to see and correct safety issues

In the example above, the instructional solutions were identified as part of the overall program.

Evaluate Results and Business Outcomes Using Success Case Methodology

To evaluate the impact of the Safety Connection program, we applied the Success Case Evaluation Method (Brinkerhoff, R., and Advantage Performance Group, 2006) to evaluate the Safety pilot. The purpose of using this evaluation strategy was to discover those instances where the application of the critical safety behaviors led to the achievement of one or more business outcomes.

The goal of a Success Case evaluation is not solely to quantitatively assess and analyze the full impact of these initiatives, nor is it to draw inferences about the average impact. Instead, the Success Case methodology is more focused, efficient, and strategic. It determines first the extent to which program participants leveraged the identified key behaviors along with their experience into important business value. The methodology then explores the most significant of these reported applications to analyze, describe and document them, and to determine those factors that seem to have facilitated or impeded worthwhile results.

Conversely, when participants are found who report little or no impact from their participation in program, the Success Case method uncovers why that is so and what could be done to remedy that negative outcome.

Evaluation Process

Phase 1: Survey the Participants
In order to create an impact survey that would assess the most positive impact with the safety experience produces, the evaluators had a conversation with the project leader and stakeholders. During those conversations, leaders and stakeholders defined the participants. Specific behaviors and key results identified in the Impact Map made up a critical part of the survey.

The impact survey was drafted and validated. The electronic survey was then sent via email to all participants. There were a total of 79 participants who received the survey. 45 participants (57% response rate) completed the Success Case Survey.

Phase Two: Conduct Interviews
Interviews were conducted with selected survey respondents to further explore their responses. After the survey responses were received, we randomly selected respondents from among those who indicated that safety program helped them produce valuable business impact. We also selected survey respondents on the low end of reported impact and those who indicated on their survey that they applied the safety behaviors, but were unsure of the results. And finally, we randomly selected survey respondents who indicated on their survey that they found the behaviors in the safety program a reminder of what they already knew and were doing. A total of 10 participants were interviewed.

Pilot Outcomes Using Success Case methodology

- Participant feedback met or exceeded expectations for value, quality of content, and instructor effectiveness
- A success case evaluation of the critical safety behaviors was performed one month after the Safety Connection Manager/Supervisor Training
- 45 participants (57% response) completed the Success Case Survey and 10 participants were interviewed
- There was a **measureable increase in application with positive results** for all 7 behaviors

Programs met or exceeded participant expectations.

Criteria	Commit to Safety	High Impact Safety Conversations	Hazard Identification	Root Cause Analysis
VALUE	98%	97%	97%	96%
CONTENT	97%	94%	95%	94%
INSTRUCTOR	98%	94%	95%	94%

The survey also revealed 40% of the population were already demonstrating these behaviors

Increase in application for key safety behaviors

Key Behavior	Survey Respondent data	Implications
Conduct daily employee safety discussions	• **33.33% implemented with positive results** • **20% implemented but not sure of results** • 8.89% have not implemented it, but plan to • **37.78% was already doing this** • 0% will not be doing this	The training has encouraged and enhanced daily employee safety discussions, however, in some areas the culture is still in progress of changing.

Measurable Outcomes Being Achieved

Using the behaviors and skill from the Training, what are the measureable outcomes you are achieving?

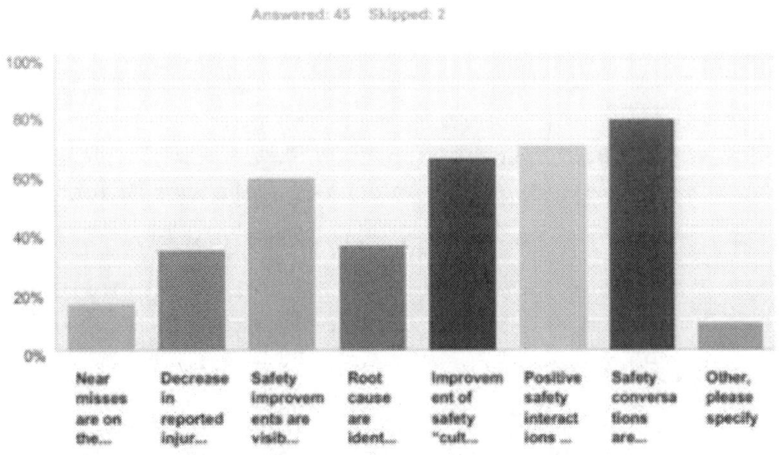

Color	Category
	Near misses are on the increase
	Decrease in reported injuries to team members
	Safety improvements are visible to employees
	Root cause are identified for EHS incidents and near misses
	Improvement of safety 'culture' measurements
	Positive safety interactions are reported at Tier Meetings
	Safety conversations are taking place daily
	Other, please specify

Key Lessons Learned

- The opportunity for health and safety professionals to achieve top quartile performance required a new approach using predictable and rapid analysis and evaluation methods combined with innovative solutions was a critical success factor.
- Deployment of blended solutions and approaches unleashed targeted performance improvement while demonstrating business value.

- Leveraging leaders as teachers had a profound impact on the learning audience and made the supporting actions believable and actionable. This was realized through gathering learner feedback on the quality of instruction.
- Using innovative research-proven techniques for program evaluation provided early indications of changes in the current state and implications for sustained performance.

References

Brinkerhoff, R. and Advantage Performance Group. 2006. *Building skills and knowledge in sales, leadership and business acumen.* San Rafael, CA: Advantage Performance Group.

Gillum, T., and K. Mortenson. 2012. *Ledge walking technology* (PowerPoint presentation).

Goldsmith, M. (2007). *What got you here won't get you there.* New York: Hyperion.

Merriam, S. B., & Associates.2002. *Qualitative research in practice: Examples for discussion and analysis.* San Francisco: Jossey-Bass.

Mezirow, J. (1991). *Transformative dimensions of adult learning.* San Francisco: Jossey-Bass.

Mezirow, J., & Associates. 2000. *Learning as transformation: Critical perspectives on a theory in progress.* New York: John Wiley and Sons.

Scott, S. 2002. *Fierce conversations: Achieving success at work & in life, one conversation at a time.* New York: Viking.

Slatter, J. 2015. A common sense. What's normal? Better Practice blog. http://blog.betterpractice.com/a-common-sense/ (retrieved, April 13, 2015).

______. 2013. *Better Practice™ Guidebook.* Lakewood, CO: Better Practice.

Snowden, D.J. and M.E. Boone. 2007. A leader's framework for decision making. *Harvard Business Review, 85*(11), 68-76.

Van Tiem, D.M., J.M.Moseley, and J.C. Dessinger. 2000. *Fundamentals of performance technology: A guide to improving people, process, and performance.* San Francisco: ISPI.

Zero Gravity: Safety Solutions for the Communications Tower Workforce

Todd Schlekeway, M.Ed
Executive Director
National Association of Tower Erectors
Watertown, South Dakota

Kevin Dougherty
President
MILLENNIA CONTRACTING, INC
New Castle, Delaware

Introduction

Explosive growth in wireless communications and data consumption over the course of the last several years has sparked a significant spike in communication tower industry work. Communication tower contractors and the tower technicians they employ continue to be the "central nervous system" of the wireless and broadcast infrastructure industries. The network installations and broadband deployments they perform on a daily basis are having a transformational impact on industries as diverse as health care, financial services, higher education, agriculture, energy, aviation, utilities, public safety, government and the military.

This enhanced technology is vital to meet both our present and future broadband needs and society's insatiable thirst for data. According to annual survey results released by CTIA – The Wireless Association, Americans used 9.6 trillion megabytes (MB) of data in 2015. This data usage is three times the 3.2 trillion MB of data utilized in 2013. The CTIA data compared this growth to the equivalent of consumers streaming 59,219 videos every minute or roughly 18 million MB!

This exponential growth, in turn, has produced safety and quality issues resulting in accidents and fatalities during the construction and maintenance of wireless sites. Most commonly, these accidents and fatalities are associated with fall hazards. According to the industry news website *Wireless Estimator*, the industry experienced ten fatalities in 2014, four fatalities in 2015 and seven fatalities in 2016. Additionally, fall protection safety violations also consistently land in the top 10 most frequent citations issued by Occupational Safety & Health Administration (OSHA) enforcement officials on an annual basis.

The Federal Communications Commission (FCC) and OSHA have recognized this problem and collaborated to work with the National Association of Tower Erectors (NATE) and stakeholders from throughout the industry to help provide a safer working environment for the communication tower workforce.

In February of 2016, Roger Sherman (FCC Wireless Telecommunications Bureau Chief) and Dr. David Michaels (Assistant Secretary of Labor for OSHA) wrote an article discussing this problem entitled *The Cell Phone in Your Pocket Shouldn't Cost a Workers Life.*

Additionally, the FCC and OSHA have teamed up to host several workshops in Washington, D.C. specifically devoted to communication tower worker safety.

In October of 2014, the FCC and DOL/OSHA hosted a Tower Climber Safety and Injury Prevention workshop to explore issues surrounding tower climber safety, injury prevention, and fatalities involving work on communication towers. The workshop included panel discussions on factors contributing to the high rate of injuries and fatalities and best practices to ensure tower climber safety and prevention of these injuries and fatalities. The participating panelists at the workshop emphasized the importance of adequate fall protection during this event.

In February of 2016, the FCC and DOL/OSHA hosted the 2nd Tower Climber Safety Workshop which included panel sessions devoted to industry best practices, analyzing comments received from OSHA's 2014 Request for Information (RFI), and current telecommunications industry workforce development initiatives.

While training curriculum regarding fall protection in the construction industry is abundant, communication tower technicians face a unique set of hazards not commonly addressed. These hazards include, but are not limited to: remote location worksites; variability in worksites, rural vs. urban differences; worker fatigue; variable weather conditions; climbing/navigating unusual structures; and variable climbing paths. Investing in the quantity and quality of fall protection training in the communication tower industry will make a significant impact by increasing the number of qualified instructors, increasing the number of trained workers, and ultimately, decreasing the number of worker injuries and fatalities.

NATE is currently a leader in developing and promoting safer working conditions for workers in the wireless industry through its development of best practices safety resources and standards. Additionally, NATE is the founder and sponsor of the Wireless Industry Safety Task Force. The impetus behind the task force occurred in October of 2013 when NATE organized the Telecommunications Industry Safety Summit in Dallas, Texas. Top wireless industry safety and operations executives representing the carriers, tower owners, Original Equipment Manufacturers (OEM), and turnkey/construction management firms met with NATE officials to discuss major areas of concern, examine leading indicators and challenge each other to achieve sustainable safety improvements.

Fall protection emerged as a major hazard at the Safety Summit and has continued to be a focal point of the various ongoing initiatives of the wireless industry. Task Force participants have identified tower technicians not being properly tied-off to the structure as being the cause of many tower-site accidents and compromised safety. A consensus has emerged on the Task Force that everyone involved in the industry has a role in emphasizing and enforcing 100% tie-off in order to ensure a safer work environment and prevent future accidents.

Fall Prevention Safety Resources for the Communication Tower Industry

As stated in the introduction, for companies and workers directly involved in working at heights on communication structures, falls from heights continue to represent one of the frequently encountered hazards the workforce is confronted with on a daily basis. Some examples of other hazards associated with communication tower work include inclement weather, falling object hazards, equipment failure, structural collapse and electrical hazards.

Fortunately, the wireless infrastructure industry is in a strong position to mitigate and eliminate communication tower falls. There is an abundance of fall protection standards, training programs and safety resources available to the industry's workforce to ensure that all elevated communication tower work is done safely. Additionally, the industry has come together on some exciting new initiatives that will also help enhance tower technician safety when working at heights.

Industry Standards

It is critical that tower construction and maintenance companies and workers adhere to all of the fall protection standards governing the industry. Simply put, if the fall protection standards are followed, working at heights is a safe endeavor. The following is a listing of the fall protection standards that should be adhered to in order to ensure a safe working environment for the industry's workforce.

- *NATE Tower Climber Training Standard – 4th Edition*
  The National Association of Tower Erectors (NATE) Tower Climber Fall Protection Training Standard (CTS) 4th edition is available to the industry in order to better facilitate training and improve safety in the industry.

  The NATE CTS outlines individual standards for varying levels of tower climber expertise: Authorized Climber, Authorized Rescuer, Competent Climber and Competent Rescuer. The document also contains a Course Training Plan to demonstrate how to implement and utilize the NATE CTS. The CTS is the Association's signature safety resource and provides one more tool in the toolbox to ensure that climber training is consistent regardless of who is conducting the training.
- *OSHA Safety and Health Regulations for Construction – 1926 Subpart M*
  1926 Subpart M sets forth the requirements and criteria for fall protection in construction workplaces. Subpart M covers information including the duty to have fall protection, fall protection systems criteria and practices and fall protection training requirements. Additionally, 1926 Subpart M includes appendix resources devoted to sample fall protection plans, positioning device systems, personal fall arrest systems, guardrail systems and determining roof widths.
- *ANSI/ASSE Z359 Fall Protection Code – Version 3.0*
  The American National Standards Institute (ANSI)/American Society of Safety Engineers (ASSE) Z359 Fall Protection Code package should be adhered to by all elevated workers and is a comprehensive fall protection standard that includes the following information:
  - Definitions and Nomenclature Used for Fall Protection and Fall Arrest
  - Safety Requirements for Personal Fall Arrest Systems, Subsystems and Components
  - Minimum Requirements for a Comprehensive Managed Fall Protection Program

    - Safety Requirements for Positioning and Travel Restraint Systems
    - Safety Requirements for Assisted-Rescue and Self-Rescue Systems, Subsystems and Components
    - Specifications and Design Requirements for Active Fall Protection Systems
    - Qualification and Verification Testing of Fall Protection Products
    - Safety Requirements for Full Body Harnesses
    - Connecting Components for Personal Fall Arrest System
    - Personal Energy Absorbers and Energy Absorbing Lanyards
    - Safety Requirements for Self-Retracting Devices for Personal Fall Arrest and Rescue Systems
    - Safety Requirements for Single Anchor Lifelines and Fall Arresters for Personal Fall Arrest and Rescue Systems
- *A10.48 Standard - Criteria for Safety Practices with the Construction, Demolition, Modification, and Maintenance of Telecommunications Structures*
  The ANSI/ASSE A10.48 Standard is a game changer for the industry as it has the distinction of being the first comprehensive standard encompassing the entire tower construction, service and maintenance industry.

  The standard establishes minimum criteria for safe work practices and training for personnel performing work on communication structures including antenna and antenna supporting structures, broadcast and other similar structures supporting communication related equipment. The complete construction and use standard thoroughly addresses the following categories:

    - General
    - References
    - Definitions
    - Pre-Job Planning
    - Job Site Conditions
    - Fall Protection
    - Radio Frequency/Electromagnetic Energy
    - Base Mounted Hoists Used for Overhead Material Lifting and Personnel Lifting
    - Personnel Lifting Accessories and Processes
    - Rigging
    - Gin Poles and Other Lifting Devices
    - Climbing Facilities and Access
    - Structural Construction Loading Considerations
    - Training Program
    - Capstan Hoist
    - Demolition
    - Helicopter Used for Lifting Loads

The final version of the A10.48 Standard is a by-product of over six years of hard work and diligence by a group of dedicated organizations and subject matter experts. The A10.48 Standard provides the ultimate road map for companies and workers to adhere to in order to raise the bar on safety and quality in the industry.

NATE recently released a *Climber Connection* video highlighting the A10.48 Standard. The video includes practical information on gin poles and rigging class guidelines for the industry to follow in accordance with the standard. This video is an outstanding resource for companies and workers to utilize in order to become familiar with the key provisions outlined in the A10.48 Standard. Embracing and implementing the A10.48 Standard into daily tower activities will enhance the culture of safety in the industry. The A10.48 Standard video can be viewed on NATE's YouTube channel at the following link: https://www.youtube.com/watch?v=33DZTeu8b58&feature=youtu.be.

Training Resources for the Communication Tower Industry

In addition to adherence to fall protection and other relevant industry standards, there are many outstanding training options available to today's elevated workforce that help ensure fall protection safety. Today's workers can receive training in fall protection and rescue from private training companies, employer training programs, apprenticeship programs and community college offerings. Proper training in fall protection and rescue is absolutely imperative prior to work commencing at communication tower-sites.

To facilitate industry training, the Association has developed a website portal called the NATE EXCHANGE. The NATE EXCHANGE is a convenient, consumer-driven, "one-stop-shop" platform for tower construction and maintenance companies and individual tower technicians to gain access to the most sophisticated and up-to-date training courses in the tower industry. The Association is proud to promote the courses listed on the EXCHANGE as valuable resources and benefits to the industry. We encourage you to utilize this website tool to select the courses and training companies that best fit your specific workforce needs. The NATE EXCHANGE can be accessed on the Association's website at the following link: https://natehome.com/nate-exchange.

Personal Protective Equipment (PPE) and Personal Fall Arrest Systems (PFAS) are also vital components of fall protection safety. This equipment should be inspected by tower technicians prior to each use to ensure that it is in satisfactory condition and therefore deemed safe to utilize when working on a communication tower. All tower industry personnel and stakeholders should utilize an equipment checklist before each use to ensure that a culture of fall protection safety permeates throughout each worker.

NATE has developed an Equipment Basics Checklist that allows for tower industry personnel to document the status of their field equipment, take notes and rate them under the categories of satisfactory condition or remove from service. The checklist contains the most comprehensive list of tower industry equipment governing the following areas:

- Daily Vehicle Safety Inspection
- Daily Basic Job Inspection Forms
- Daily Basic Employee PPE Inspection
- Daily Use Rigging Inspection
- Daily Use Tools and Equipment
- Instrumentation
- Hoists
- Man Baskets
- Gin Poles
- Other Miscellaneous Categories

NATE encourages all tower industry personnel and stakeholders to utilize the NATE Equipment Basics Checklist on a daily basis to ensure that a culture of safety permeates throughout the industry. It should be noted that the checklist is not a legally binding document, but rather a list of recommendations based on safety principles supported by the National Association of Tower Erectors. The NATE Equipment Basics Checklist is available on the Association's website at the following link: https://natehome.com/wp-content/uploads/2013/01/Equipment-Basics-Checklist.pdf.

Manufacturing & Engineering Initiatives

Manufacturers and industry experts have met previous challenges facing our industry and there are several current industry led initiatives taking place in this arena to help create a safer environment for tower technicians who work at elevation.

In December of 2016, NATE, in conjunction with the University of Dayton Research Institute in Dayton, Ohio, conducted a Step Bolt Testing Event with structural manufacturers to determine the strength of 5/8" step bolts on monopole structures. This was an important initiative as monopole structures are considered the most difficult communication towers to climb. It is imperative that the step bolts manufactured on monopole towers are of adequate strength to ensure the safety of the elevated workforce. The data and key findings collected from this testing event are currently being utilized to assist structural manufacturers and potentially be placed as a requirement in the Telecommunications Industry Association (TIA) -222 Rev. H Structural Standard that is currently under development.

Representatives consisting of industry associations, wireless carriers, tower companies, contractor companies, training companies and equipment manufacturers have recently established a working group to study the critical role safety climb equipment plays in protecting tower climbers. The goal of the working group is to help ensure safety climb systems are installed, maintained and inspected properly and in accordance with existing standards.

The formal working group has established a plan to create specific industry standards as well as a common vocabulary for safety climb systems used in the installation, maintenance and inspection of safety climb systems. The working group will also work to clarify the requirements necessary for competent contractors to properly inspect these safety climb systems prior to considering their use as part of the fall protection plan for the Scope of Work (SOW) on a telecommunications site, as well as the inspection of this equipment as an appurtenance on the structures.

Unmanned Aerial Systems (UAS) Technology

Unmanned Aerial Systems (UAS) technology is currently in the early stages of integration into the wireless infrastructure industry. UAS technology will play a valuable supplemental role in enhancing safety and quality in the industry as the regulatory environment surrounding the commercial use of UAS technology gains clarity. UAS technology will never serve as a replacement for the tower technician workforce, but there is an obvious supplemental benefit related to UAS utilization when working on communication structures. The safe and appropriate use of UAS in the industry can assist with such functions as pre-site inspections and close-out photos on the tower. These practical functions can serve to reduce the number of times a worker would have to scale a tower during a particular job, therefore helping serve to mitigate the potential of falls from heights.

The National Association of Tower Erectors (NATE) recently developed a 2nd Edition of the *NATE Unmanned Aerial Systems Operations Around Vertical Communications Infrastructure* resource document.

The 2nd Edition of this valuable safety resource is a by-product of collaboration between the NATE UAS Committee and other prominent representatives from the commercial UAS industry. The intended focus of the document is on UAS operations around wireless infrastructure, cellular towers, broadcast towers and utility structures. The 2nd Edition also incorporates updates from the new guidelines and provisions associated with the Federal Aviation Administration's (FAA) Rule 107 for the commercial utilization of UAS technologies.

NATE is the industry leader in the integration and utilization of UAS around communications infrastructure and this safety document will serve as a go-to resource for all workers and stakeholders in the industry to use as a guide in order to expand the safe and efficient commercial operation of UAS. The 2nd Edition of the *NATE Unmanned Aerial Systems Operations Around Vertical Communications Infrastructure* document is available to the industry as a free resource and accessible to be downloaded on NATE's website at the following link: https://natehome.com/wp-content/uploads/2016/05/UAS-Operations-Document-2nd-Edition-Jan-2017-E-FILE.pdf.

Safety is Everything

In conclusion, the abundance of industry standards and training resources combined with innovations in engineering and technology will play a vital role in workforce safety as the wireless networks of tomorrow are deployed. This important work is essential and the industry continues to remain vigilant and committed to ensure the fall protection and safety of our most precious resource...the men and women who make up our workforce.

Bibliography

American National Standards Institute (ANSI)/American Society of Safety Engineers (ASSE) 2016. *Criteria for Safety Practices with the Construction, Demolition, Modification and Maintenance of Telecommunications Structures* (retrieved February 6, 2017) (http://www.asse.org/ansi/asse-a10-48-2016-criteria-for-safety-practices-with-the-construction-demolition-modification-and-maintenance-of-communication-structures/).

ANSI/ASSE Z359 Fall Protection Code Package – Version 3.0 (retrieved February 6, 2017) (https://natehome.com/regulations-and-standards/standards/ansiasse-z359-fall-protection-code-package/).

CTIA Everything Wireless. 2016. *America's Data Usage More Than Doubled in 2015* (retrieved February 6, 2017) (http://sitefinity.dmz.ctia.org/resource-library/press-releases/archive/americans-data-usage-more-than-doubled-in-2015).

Federal Communications Commission (FCC). 2016. *The Cell Phone in Your Pocket Shouldn't Cost a Worker's Life* (retrieved February 6, 2017) (https://www.fcc.gov/news-events/blog/2016/02/11/cell-phone-your-pocket-shouldn%E2%80%99t-cost-worker%E2%80%99s-life).

National Association of Tower Erectors (NATE). 2015. *Tower Climber Training Standard – 4th Edition* (retrieved February 6, 2017) (https://natehome.com/wp-content/uploads/2011/10/CTS-Inside-Pages-March-2015-FINAL-E-FILE.pdf).

National Association of Tower Erectors (NATE) [ClimberConnection Video]. (2017, January 23). *ANSI/ASSE A10.48 Standard* [Video file]. Retrieved from https://www.youtube.com/watch?v=33DZTeu8b58&feature=youtu.be

______. 2017. *Unmanned Aerial Systems Operations Around Vertical Communications Infrastructure- 2nd Edition* (retrieved February 6, 2017) (https://natehome.com/wp-content/uploads/2016/05/UAS-Operations-Document-2nd-Edition-Jan-2017-E-FILE.pdf).

NATE Equipment Basics Checklist.(2012, December). Retrieved from https://natehome.com/wp-content/uploads/2013/01/Equipment-Basics-Checklist.pdf

NATE EXCHANGE. Retrieved from https://natehome.com/nate-exchange

Occupational Safety and Health Administration (OSHA). 29 CFR 1926.500, *Safety and Health Regulations for Construction – 1926 Subpart M* (retrieved February 6, 2017) (https://www.osha.gov/pls/oshaweb/owadisp.show_document?p_table=STANDARDS&p_id=10756).

Wireless Estimator. 2016. *Fatality Tracker: US Tower Structure Related Fatalities* (retrieved February 6, 2017) (http://wirelessestimator.com/content/fatalities).

Actively Caring for People's Safety
How to Cultivate a Brother/Sister Keeper's Work Culture

E. Scott Geller, Ph.D.
Alumni Distinguished Professor
Virginia Tech

The title of this paper is the same as a new book/manual published by ASSE and coauthored by the author and his daughter, Krista S. Geller, Ph.D., who for a decade has worked as a safety professional for three organizations. Her real-world challenges at teaching and leading people-based safety informed the preparation of this book/manual. The label "book/manual" reflects the fact that this scholarship was designed to serve as a training manual for addressing the human dynamics of occupational safety. Seven critical lessons are explained and illustrated, accompanied by discussion questions and behavioral exercises with feedback opportunities. Here I review the seven life lessons on which our new book/manual is founded.

Suppose you were asked to define and explain the top seven lessons you have learned about the psychology of safety, whether from a course in introductory psychology or anything you've heard or read regarding the dynamics of human experience, including a behavior-based or people-based safety workshop. Don't choose the most memorable, important, or most researched lessons, rather choose those evidence-based lessons you believe should be taught and disseminated worldwide to benefit the human dynamics of safety, security, and quality of life. Which would you choose? Before reading further, it would be beneficial to ponder this question and derive your own list of seven life lessons. As you read, compare your list with those discussed here, noting similarities and discrepancies.

Obviously there's no universal answer to the question: "What are seven crucial life lessons from psychology that could benefit safety and human welfare on a large scale?" Answers will certainly be biased by personal experience, including idiosyncratic reading of diverse literature (including "pop psychology") and varied educational and/or research experiences at an educational or research institution and beyond. The seven life lessons presented here are derived from the author's intensive and extensive study of human dynamics–five years in graduate school and 48 years as a teacher and researcher of psychology at Virginia Tech (VT).

The first four lessons connect directly to applied behavioral science (ABS) and the remaining three life lessons reflect humanism, a realm of psychology considered by many to be opposite to behavioral science. In fact, some readers will consider the life lessons derived from humanism to be an overly "radical" departure from behaviorism. Yet, a primary aim of this presentation is to convince you that these life-lessons should be accepted and deemed important by everyone in a work culture, from safety professionals and CEO's to the line wage workers, regardless of bias toward a particular domain of psychological science.

First, let's appreciate the term "actively caring" —the theme of this presentation and six books published this year (Geller, 2017; Geller & Geller, 2017; Geller & Kipper, 2017; Geller & Veazie, 2017a,b). Note how "actively caring" connects to "humanistic behaviorism" with "actively" reflecting *behaviorism* and "caring" reflecting *humanism*.

Actively Caring for People (AC4P)

It's imperative to understand the difference between caring and acting. From the author's conversations with thousands of employees in the corporate world, it's clear no one wants to see anyone injured on the job. This is *caring*. Yet, when asked to muster the courage to say something to a peer who is working at risk and could be hurt, the majority admit they do not *act* on their caring by giving their co-workers corrective feedback. Nor do many feel comfortable providing personal appreciation when observing a coworker's safe behavior. Thus, these employees do not *actively care*. They don't go beyond their intrinsic or natural caring to proactive action.

At a power plant in Maryland, 31 year-old Benjamin Heron fell to his death in August of 2007 after wearing his safety harness 150 feet above ground, but not tying off when he moved on the platform from one tool to another. Two co-workers on the platform noticed he had unhooked and then did not tie back on. But these workers said nothing and Benjamin fell to his death moments later, slipping between the platform and the first rail.

There are too many stories like this, where chances to help are passed up and the ultimate price is paid. It's not that Benjamin's co-workers didn't care; it's that they didn't *act* on their caring. They said nothing to remind Benjamin to tie off. His wife and three children likely wonder every day why someone didn't go beyond caring to acting by giving Benjamin corrective feedback. Of course, they had their excuses (e.g., "We're only contractors and aren't supposed to tell full-time employees what to do"), just as we all rationalize our lack of proactive AC4P behavior. It's likely, however, these men do not experience many days without thinking, "If only I had..."

From Behavior-Based Safety (BBS) to AC4P

Many companies have implemented an AC4P process for safety called BBS (Geller,1996a; Geller, 1998; McSween, 1995) or People-Based Safety (Geller, 2005; Geller, 2008b). Its purpose is to help people develop the confidence and skills to take care of each other every day (i.e., to be brother/sister keepers) before injuries occur—to be continuously proactive. Typically, employees use a critical behavior checklist (CBC) to pinpoint safety-related risks to look for when systematically observing each other (Geller, 1996a; Geller, 1998; Geller, 2001). Then they observe their co-workers and check "at-risk" or "safe" on the CBC, based on the behavioral definitions they previously determined in a group meeting with their work team.

After completing the CBC, the observer provides supportive feedback to the observed worker for the safe behaviors recorded and corrective feedback for any at-risk behavior noted. This behavioral feedback is not given to direct behavior change, but to empower personal acceptance and self-accountability. A nondirective or humanistic approach to feedback delivery is most likely to be appreciated and considered (Geller, 2015; Geller, Perdue, & French, 2004).

The CBCs are given to a data processor who tallies the information and distributes a summary to all employees. At-risk behaviors occurring at the highest frequencies are identified, specifying targets for further improvement. This information is shared with all employees so together they can learn where more mindful focus is needed for injury prevention. The purpose: to cultivate a brother/sister keeper's culture in which 100% of the time, every day, any worker seeing another worker at risk (including breaking a safety rule) speaks up to prevent the possibility of an injury.

The real value of using a CBC is *to help people develop in both confidence and competency, so they provide feedback every time they see someone at risk of being injured.* It's not about acquiring tick marks for a computer program and summary output. The formal observation-and-feedback process helps participants become more skilled and comfortable at providing and receiving interpersonal feedback regarding safe and at-risk behaviors. Such formal observations should be viewed as *practice* to develop people's AC4P competence and courage to provide BBS feedback any time they see someone in potential danger of an injury (Geller & Veazie, 2014).

From Formal to Informal Observations

Consider a winning athletic team. Rigorous long-term practice with behavioral feedback occurs before the first game. With BBS, the practice is formally using a CBC to improve behavior. As with athletics, the practice is important but the purpose of practice is preparation for the game.

The essential intent of the formal BBS process is to enable employees to "play their best game". Do not assume your number of formal observations completed will lead to fewer on-the-job injuries. Only when the observations occur voluntarily, beyond the formal practice sessions, will the connection between the BBS observation-and-feedback process and injury rate most likely occur. The ultimate leading indicator: How often do workers give informal feedback to sincerely commend ongoing safe behavior or warn a coworker of a hazard or an at-risk behavior they observe?

Such interpersonal discussions are routine in a brother/sister keeper's culture, but my SPS (Safety Performance Solutions) Partners and I have found few work cultures with this level of interdependent actively caring. But routine AC4P behavior for safety is essential to sustain an injury-free workplace. The seven life lessons for safety defined and illustrated next are fundamental to achieving a brother/sister keeper's work culture. In other words, when employees at all levels of a corporation understand, appreciate, and practice the following evidence-based seven lessons on a regular basis, an AC4P work culture will be achieved and result in a sustained injury-free workplace.

1. Employ the Power of Positive Consequences.

"The problem is to free men, not from control, but from certain kinds of control" (Skinner, 1971, p. 5). This quotation reflects B. F. Skinner's concern for people's attitudes and person-states, and his antipathy for the use of punitive consequences to influence behavior. In his seminal research-based book, *Beyond Freedom and Dignity*, Skinner explains that control by negative consequences needs to be reduced in order to increase perceptions of personal freedom. Biglan (2015) and Geller (2016) reverberate this critical lesson in their recent books on applying behavioral science to nurture a positive prosocial culture and improve human lives worldwide.

Usually, the same situation can be perceived as: a) control by punishing unwanted (e.g., at-risk) behavior; or b) control by rewarding desired (e.g., safe) behavior. In the author's introductory-psychology class at Virginia Tech (VT), for example, some students are motivated to avoid failure (e.g., a poor grade); other students are motivated to achieve success (e.g., increased knowledge or a good grade).

Which students feel more self-directed and in control of their class grade? Which students have more positive expectations about the class? Isn't the answer obvious? Readers need only reflect on their own attitude or mindset in similar situations where their behavior was perceived to be influenced by positive versus negative consequences.

Consider the four distinct achievement typologies shown in Figure 1. They were derived from the seminal research of Atkinson (1957, 1964). These four mindsets have been researched to explain differences in how people approach success and/or avoid failure (Covington, 1992; Covington & Roberts, 1994).

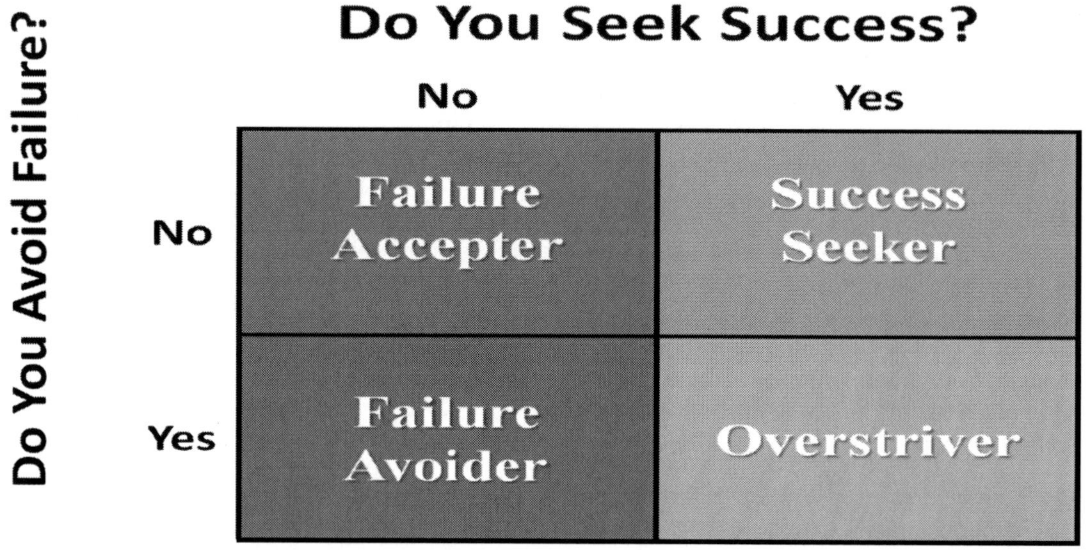

Figure 1. The four motivational typologies defined by achieving success vs. avoiding failure (adapted from Covington & Roberts, 1994).

It's best to be a *success seeker*. These are the optimists who respond to setbacks (e.g., to corrective feedback) in a positive and adaptive manner. They are self-confident and willing to go beyond requirements, instead of circumventing challenges to avoid failure. They wake up each morning to an *opportunity* clock rather than an *alarm* clock. This is a mindset or attitude people can influence in themselves and in others through intrapersonal and interpersonal conversation, respectively.

In contrast, *failure avoiders* have relatively low expectancy for success and a high fear of failure. Much of their behavior is motivated by a desire to protect themselves from appearing incompetent. These individuals are motivated, but they are not "happy campers." These are the students who say, "I've *got* to go to class; it's a *requirement*," rather than, "I *get* to go to class; it's an *opportunity*."

The lesson: Applying soon, certain and positive consequences is the most efficient and effective way to improve behavior and attitude at the same time. But historically, so much safety-related behavior has been supported by negative rather than positive consequences. How do most organizations keep score for their safety performance? Are the conversations more about avoiding failure than achieving success?

Ask workers what they've done for safety this week, or what they intend to do for safety today. How readily will they list desirable safety-related behaviors? Discount replies such as, "I didn't get hurt," or "I won't get hurt." You want to hear behaviors worthy of positive recognition.

It's critical yet challenging to employ this life lesson every day. We live in a "click-it-or-ticket" culture that relies on negative consequences to manage behavior, from the classroom and workplace to our homes, and throughout our travel in between. The political, bureaucratic approach to behavior management is to pass a law and enforce it. For more than four decades now, since the birth of OSHA, more safety-related behavior has been other-directed rather than self-directed and self-motivated. It's not enough to understand and believe this lesson—we need to act on it. Thus, our next lesson.

2. Apply Observational Learning.

A large body of psychological research indicates that observational learning is involved to some degree in almost all human behavior (Bandura, 1969). People's actions influence others to a greater extent than they consciously realize. Children learn by watching their parents at home, and coworkers are influenced by the actions of other coworkers. Usually, people are not mindful of this modeling influence. Consider what children learn about safety by watching the driving behavior of their parents, including their parent's verbal behavior.

Observational learning is key to the dramatic success of BBS and PBS at preventing injuries. Consider a basic BBS and PBS process: 1) Coworkers develop a checklist of critical safe and at-risk behaviors on their job—a CBC; 2) Subsequently, they observe each other while working, and systematically check the safe and at-risk behaviors they observe; and 3) The observer shows the completed CBC to the worker observed, and the results are discussed.

The role of observational learning is clear. The observers note safe work practices, and in the process they might learn new ways to work safely. They might also observe at-risk behavior to avoid, perhaps behavior they sometimes perform themselves. This suggests behavioral observers will subsequently work more safely on similar work tasks. This logical supposition from observation learning has actually been demonstrated empirically (e.g., Alvero & Austin, 2004; Alvero, Rost, & Austin, 2008). Of course, a most important component of the peer-to-peer observation process of BBS and PBS is behavioral feedback—the third leadership lesson.

3. Become a Behavior-Based Feedback Coach.

"Practice makes perfect" is not true; practice makes permanence. Only when practice includes behavior-based feedback can behavior improve. Often behavioral feedback occurs as a natural consequence: the golfer and tennis player see where their ball lands after swinging a golf club or tennis racket. Yet even when the outcome of behavior is observed, behavior-based feedback from a coach is necessary for optimal improvement. This leadership lesson is represented by the letters of COACH: "C" for Care; "O" for Observe; "A" for Analyze, "C" for Communicate, and "H" for Help (Geller, 2001).

Start with Caring. "Know I *Care* and you'll care what I know. Because I care about your safety, I'm willing to *Observe* you and note occurrences of safe and at-risk behavior." The observer records occurrences of certain behaviors on a CBC and jots down environmental factors that could be influencing the observed behavior, from environmental conditions to behavioral consequences. Noting external factors that might influence particular safe or at-risk behavior reflects the *Analysis* phase of coaching.

The *Communication* phase is next. The observer delivers the information derived from *Observation* and *Analysis*. Although most people want to improve, many are hesitant to give and receive the level of communication needed for beneficial change. Some view corrective behavioral feedback as an indictment of their current work style, job skills, or diligence. This reaction is most likely when the recipient of feedback is asked to change dramatically, and when the current risky protocol has been followed for years (i.e., it's habituated).

To thwart resistance to change, behavior-improvement coaches avoid disruptive and dramatic communication, unless the individual is in eminent danger. Instead, they emphasize incremental "fine tuning" of a task. The effective coach facilitates beneficial change to both behavior and attitude by highlighting the positives—occurrences of safe behavior.

Similar to client-centered or humanistic therapy (Rogers, 1942), the perceptions and person-states of the individual being coached are considered. The behaviors and environmental factors are discerned from the perspective of the person observed, and feedback communication is supportive and nondirective (Geller et al, 2004). Typically, feedback is not delivered to direct behavior change, but rather to enable personal acceptance and self-motivation for beneficial behavioral change. Supportive or rewarding feedback is used to acknowledge employees' safe behavior, thereby contributing to a positive safety culture (Williams, 2010).

If the interpersonal communication is effective, the last letter of COACH —*Help*—is realized. The behavior-based feedback was accepted and will be practiced to enhance occupational safety. The four letters of HELP—Humor, Esteem, Listen, and Praise—reflect humanistic strategies to increase the probability a coach's behavioral feedback will be accepted and appreciated.

4. Give More Supportive than Corrective Feedback

"We can't learn unless we make mistakes." How many times have you heard this? While this might make us feel better about the errors of our ways, and provide an excuse for focusing more on other people's failures than on their successes, nothing could be further from the truth. Behavioral scientists have shown convincingly that success—not failure—produces the most effective learning (Chance, 2008; Reed, 2016).

Edward Lee Thorndike, for example, studied intelligence at the start of the last century by putting chickens, cats, dogs, fish, monkeys, and humans in situations that called for problem-solving behavior. Then he systematically observed how these organisms learned. He coined the *Law of Effect* to refer to the fact that learning depends upon behavioral consequences (Thorndike, 1931). And, markedly more learning occurs following positive consequences (success) than negative consequences (failure).

Does an error have to occur in order to solve a problem? You can reflect on your own experiences to answer this question. A pleasant consequence gives you direction and motivation to continue the behavior. You know what you did to receive the reward, and are thus motivated to earn another.

In contrast, a negative consequence following a mistake only tells you what not to do. It provides no specific direction for problem solving. An overemphasis on a mistake can be frustrating and discouraging, and actually de-motivate you to continue the learning process, right?

Errors are not necessary for learning to occur. In fact, when training results in no errors, made possible with certain presentation techniques, learning occurs most smoothly and is most enjoyable (Chance, 2008; Reed, 2016). Errors disrupt the teaching/learning process and can lead to a negative attitude, especially if negative social consequences accentuate the mistake. Even subtle reactions to an error—a disappointed face or verbal tone—can increase feelings of helplessness or despair and turn a person off to the entire learning process.

Effective Supportive Feedback

The most powerful positive consequence to support a learning process is supportive feedback—the theme of this discussion. Five basic guidelines are given next for giving quality supportive feedback.

Be Timely

In order for supportive feedback to provide optimal direction and support, it needs to be associated directly with the desired behavior, as *not* demonstrated in the illustration. When people know what they did to earn the appreciation, they might be motivated to continue that behavior.

If it's necessary to delay the supportive feedback, the conversation should relive the activity deserving recognition. Talk specifically about the behavior that warrants special acknowledgement. Don't hesitate to ask the recipient to recall aspects of the situation and the commendable behavior. This enables direction and motivation to continue the desired behavior.

Make it Personal

Supportive feedback is most meaningful when it's perceived as personal. This can be done by providing an example of your own learning, your own experiences, your own understanding—all to show you can relate to the topic at hand.

Verbal support should not be generic, fit for any situation, as in "Nice job!" Rather, it needs to be customized to fit a particular individual and circumstance. This happens naturally when the supportive feedback is linked to designated behavior. When you recognize someone, you're expressing personal thanks. This shows people you truly care, that you are genuine in your conversation. By making it personal, you can identify with the circumstance.

Sometimes it's tempting to say, "*We* appreciate" rather than "*I* appreciate," and to refer to departmental gratitude rather than *personal appreciation*. But, speaking for an organization or company can come across as impersonal and insincere.

Of course, it's appropriate to reflect value to the organization when giving supportive feedback, but the focus should be personal: "I saw what you did to support the safe behavior of that worker and I really appreciate it. *Your example demonstrates the kind of leadership we need around here to nurture an AC4P safety culture."* This second statement illustrates the next guideline for giving quality supportive feedback.

Take It to a Higher Level

Supportive AC4P feedback is most memorable and inspirational when it reflects a higher-order quality. Adding a universal attitude like leadership, integrity, trustworthiness, or AC4P to your recognition statement makes the feedback more meaningful and thus rewarding. It's important to state the specific behavior first, and then make an obvious linkage between the behavior and the positive character trait it reflects.

Later, we discuss the use of an AC4P wristband to show appreciation for safety-related behavior. In these cases, we offer the wristband as a symbol of AC4P leadership, and worn to show membership in an elite group of individuals dedicated to cultivating an AC4P safety culture. Thus, the AC4P wristband is given to not only reward AC4P behavior, but also to signify participation in a Movement to activate and reward AC4P behavior, from your workplace to your home and throughout the community at large.

The conversation that follows the wristband is critical. For the wristband to carry the level of appreciation needed to affect behavior, the person passing out the wristband must deliver an effective one-on-one conversation. Show genuine compassion when delivering the AC4P message and wristband; take your time, and show the other person that this wristband is a symbol of an AC4P Movement throughout the jobsite, the entire organization, and beyond.

This connection brings the interpersonal supportive feedback to a higher level, enabling positive impact on the recipient's self-esteem, competence, and sense of interdependence and belongingness. Earlier we explained how these person-states enhance the propensity to perform more AC4P behavior. Later we discuss the connection between these person-states and self-motivation.

Deliver it Privately

Because supportive feedback is personal and indicative of higher-order attributes, it needs to be delivered privately and one-on-one. This requires a certain degree of courage for those not comfortable in private, one-on-one conversations, especially with people they don't know well. But consider this: Such verbal support is special and only relevant to one person. So, it will mean more and seem more genuine if it's given personally—from one individual to another.

It seems conventional to recognize individuals in front of a group. This approach is typified in athletic contests and reflected by the pop-psychology slogan, "Praise publicly and reprimand privately". Many supervisors take the lead from this common-sense statement and give workers recognition in group settings.

Indeed, isn't it maximally rewarding to be held up as an exemplar in front of one's peers? Not necessarily, because many people feel embarrassed when singled out in front of a group. Part of this embarrassment could be due to fear of subsequent harassment by peers. Some peers might call the recognized individual an "apple-polisher" or "brown-noser," or accuse him or her of "sucking up to the supervisor".

In athletic events the participants' performance is measured objectively and therefore winners are determined fairly. However, in work settings it's usually impossible to assess everyone's safety-relevant behavior objectively and obtain a fair ranking for individual recognition. Hence, praising one individual in public may lead to perceptions of inequality or favoritism from individuals who feel they did equally well, but did not get praised.

Plus, such ranking sets up a win-lose atmosphere—perhaps appropriate for some sporting events—but not in settings where teamwork is needed to achieve group goals and promote an interdependent injury-free culture.

It's beneficial, of course, to recognize *teams* of workers for their accomplishments, and this can be done in a group setting. Since individual responsibility is diffused or dispersed across the group, there is minimal risk of individual embarrassment or later peer harassment.

However, it's important to realize that group achievement is rarely the result of equal input from all team members. Some take the lead and work harder, while others *loaf* and count on the group effort to make them look good. Thus, it's important to deliver personal and private recognition to those individuals who go beyond the call of duty for the sake of their team.

Let it Sink in

In this fast-paced age of trying to do more with less, we try to communicate as much as possible when we finally get in touch with a busy person. After recognizing an individual's special AC4P effort, we are tempted to tag on a bunch of unrelated statements, even a request for additional behavior. This comes across as, "I appreciate what you've done, but I need more".

Resist the temptation to do more than praise the safety-related behavior you saw. If you have additional points to discuss, it's best to reconnect later, after the supportive feedback has had a chance to sink in and become part of the individual's self-talk for self-recognition and self-motivation.

By giving supportive recognition, we give people a script they can use to reward their own behavior. In other words, supportive feedback strengthens the other person's self-reward system. And, positive self-talk (or self-recognition) is critical for long-term maintenance of desired behavior. Thus, by allowing supportive feedback to stand alone and sink in, we enable the internalization of rewarding words that can be used later for self-motivation to perform the recognized behavior again.

Use Tangibles for Symbolic Value

Tangible rewards can detract from the self-motivation aspect of quality recognition. If the focus of a safety-recognition process is placed on a material reward, the words of appreciation can seem less significant. In turn, the beneficial impact on one's self-motivation is lessened.

On the other hand, tangible rewards can add to the quality of interpersonal recognition if they are delivered as tokens of appreciation. Rewards that include a relevant AC4P safety slogan can help to promote the desired behavior. But how you deliver a tangible reward will determine whether it adds to or subtracts from the long-term benefit of your praise.

The benefit of interpersonal recognition is weakened if the tangible consequence is viewed as a payoff for the safety-related behavior. However, if the reward is seen as symbolic of going beyond the call of duty for another person's safety or well-being, it strengthens the praise. Have the courage to tell it like it is: The AC4P wristband or another tangible reward is a token of appreciation and/or a symbol of going beyond the call.

Consider Secondhand Recognition

Up to this point, we've been discussing one-on-one verbal communication in which one person recognizes another for a particular AC4P behavior. It's also possible to recognize a person's outstanding efforts indirectly, and such an approach can have special advantages.

Suppose, you overhear a colleague talk to another person about your outstanding presentation regarding a certain safe operating procedures. How will this secondhand recognition affect you? Will you believe the words of praise you overheard were genuine?

Sometimes people are suspicious of the genuineness of praise when it's delivered face-to-face. Is there an ulterior motive? Perhaps a favor is expected in return. Or maybe the recognition is seen merely as an extension of a communication exercise and thus devalued as sincere appreciation. Secondhand recognition, however, is not as easily tainted with these potential biases.

Suppose a colleague tells you that someone else told him about the superb job you did leading a certain safety meeting. What will be the impact of this type of secondhand recognition? Chances are you'll consider the recognition authentic because your colleague was only reporting what someone else said. Because that person reported your success to your colleague rather than you, there was no ulterior motive for the indirect praise.

Such secondhand recognition can also build a sense of belongingness or group cohesion among individuals. When you learn someone was bragging about your behavior, your sense of friendship with that person will likely increase.

When we talk about the achievement of others in behavior-specific terms, we begin a cycle of positive communication that can support desired behavior, as well as activate self-talk for self-recognition and self-motivation.

Have the courage to initiate this cycle of positivism. This also sets an example for the kind of interpersonal communication that enhances self-esteem, self-efficacy, personal control, optimism, and group cohesion. As explained earlier, these are the very person-states that increase the potential for the occurrence of AC4P behavior and the achievement of an interdependent culture of compassion (Geller, 2001; Geller, 2016).

Accept Supportive Feedback

Most of us get so little supportive feedback from others that we are caught completely off guard when acknowledged for our commendable actions. We don't know how to accept recognition when it finally comes. Don't shy away when it does come; have the courage to embrace it. Remember the basic behavioral-science principle: Consequences influence the behaviors they follow.

As the first lesson explained, positive consequences increase the probability the behavior recognized will continue. Plus, one's reaction to the supportive feedback influences whether the observer will attempt to provide supportive feedback again. Thus, it's crucial to react appropriately when receiving recognition from others. Let's consider seven basic guidelines for receiving supportive behavioral feedback.

Don't deny or disclaim. Often when we attempt to give supportive recognition, we get a reaction that implies we're wasting our time. We get disclaimer statements like, "It really was nothing special"; "Just doing my job". The most common reply: "No problem." This implies the behavior recognized was not special and did not warrant supportive feedback.

Supportive feedback needs to be accepted without denial or disclaimer statements, and without deflecting the credit to others. It's okay to show pride in your small-win accomplishments, even if others contributed to the successful outcome. After all, in an injury-free culture everyone goes beyond the call of duty on behalf of the well-being of others. Therefore, in this context, numerous people deserve recognition daily.

Accept the fact that supportive feedback will be intermittent at best for everyone. So when your turn comes, accept the positive feedback for your recognized behavior and for the many prior safety-supportive behaviors you have performed that went unnoticed.

Keep in mind your genuine appreciation of the supportive feedback will increase the likelihood the person who gave you feedback will give more behavioral recognition to others. Plus, you might be inspired to do the same, given that you realize personal benefit from the feedback and want others to experience such behavioral support.

Listen actively. Listen actively to the person recognizing you with supportive feedback. Certainly, you want to learn what you did right. Plus, you can evaluate whether the supportive feedback is given well. If the feedback does not pinpoint a particular behavior, you might ask the person, "What did I do to deserve this?" This will help to improve that person's method of giving supportive feedback in the future.

Of course, it's important to show genuine appreciation for the special attention. Consider how difficult it is for many people to go out of their way to recognize others with supportive feedback. So, revel in the fact that you're receiving some behavioral recognition, even if its quality could be improved.

Use it later for self-motivation. Most of your safety-related behaviors go unnoticed and unrecognized. You perform many of these behaviors when no one else is around to observe you. Even when other people are available, they are usually so preoccupied with their own routines they don't notice your extra effort for safety. So when you finally do receive supportive feedback for safety-related behavior, take it in as well-deserved.

Don't hesitate to relive this moment later by talking to yourself. Such self-recognition can motivate you to continue going beyond the call of duty on behalf of the safety of others. As mentioned earlier, self-talk can help you muster the self-motivation to perform more of the recognized behavior. It also contributes to holding safety as a value that remains pre-eminent, even when priorities change.

Show sincere appreciation. You need to show sincere gratitude with a smile, a "Thank You," and perhaps special words like, "You've made my day." Your reaction to receiving supportive feedback can determine whether that person will deliver similar recognition again. So be prepared to offer a sincere "Thank You," as well as words that reflect your pleasure in the memorable positive interaction. Plus, consider the courage the other person might have needed to give you that personal recognition.

We find it natural to add, "You've made my day" to the "Thank-You" because it's the truth. When people go out of their way to offer us quality recognition, they *have* made our day. In fact, we typically relive such situations to improve a later day.

Reward the recognizer. When you accept supportive feedback well, you reward the person for his or her appreciating behavior. This can motivate that individual to deliver more positive feedback, especially if the person is more of an introvert and requires courage to step out and speak up to give interpersonal recognition.

Sometimes, you can even do more to assure the occurrence of more supportive feedback. Specifically, you can recognize the person for recognizing you. You might say, for example, "I really appreciate you noticing my safety-relevant behavior and calling me a leader of the AC4P process to sustain an injury-free workplace." Such supportive and rewarding feedback provides direction and motivation for those aspects of occupational safety that are especially worthwhile and need to become routine.

Ask for recognition. If you feel you deserve supportive feedback, why not ask for it? You will likely view this recognition as less genuine than if it had been spontaneous, but the outcome from such a request can be quite beneficial. You might even receive some words worth reliving later for self-motivation. Most importantly, you will remind the other individual in a nice way that s/he missed a prime opportunity to offer supportive feedback. This could be a valuable learning experience for that person.

Consider the possible beneficial impact from your statement to another person that you are pleased with a certain outcome of your extra effort. With the right tone and affect, such verbal behavior will not seem like bragging but rather a declaration of personal pride in a small-win accomplishment—something more people should feel and relive for self-motivation. The other person will surely support your personal praise with supportive testimony, and this will bolster your self-motivation. Plus, you will teach the other person how to support the commendable behavior of others.

5. Embrace and Practice Empathy

The difference in the feeling states activated by rewards versus penalties is the rationale for giving more supportive than corrective feedback—Lesson 4. Similarly, how an intervention process is implemented can increase or decrease feelings of empowerment, enhance or destroy interpersonal trust, and facilitate or inhibit a sense of interdependent collaboration or belongingness (Geller, 2005; 2015; Geller & Geller, 2017). It's useful to assess the perceptions or person-states that occur concomitantly with an intervention process. This can be accomplished informally through one-on-one interviews and group discussions, or formally with a perception survey (O'Brien, 2000; Peterson, 2001).

Decisions regarding which safety-improvement intervention to implement and how to refine existing safety processes should be based on both objective observations of behaviors and subjective evaluations of feelings or person-states. It's usually possible to apply a personal assessment of *empathy* to evaluate the internal impact of an intervention process. Just imagine yourself going through the same intervention procedures and ask yourself, "How would I feel?"

Empathic listening, diagnosing, and action planning take patience. We're not talking about the high-tech communication like emailing, texting, and tweeting. Conversations at this level are often not efficient, but they are consistently most effective. Through probing and listening, the objective is to first empathize and regard what it's like to be in the other person's shoes. The objective then shifts to developing a corrective intervention that fits the circumstances as mutually understood by each person involved in the conversation. If a personal commitment to follow through with a specific action plan is stated, the coach was empathic and most effective.

6. Manage Behavior and Lead People

Management is not the same as leadership. Managers hold people accountable to perform desirable (e.g., safe) behavior and avoid undesirable (e.g., at-risk) behavior. Leaders inspire people to hold themselves accountable to do the right thing—to follow the safest protocol. Managers affect behavior with an external (or extrinsic) accountability system. Leaders facilitate self-motivation by influencing person-states (e.g., perceptions, attitudes, and/or emotions) that facilitate self-motivation. Self-motivation (or self-directed behavior) often leads to discretionary behavior (Daniels & Daniels, 2005)—more desirable behavior than that required by one's job description.

Self-Motivation
Geller and Veazie (2017) use the C-words of *Choice, Competence,* and *Community* in their realistic narrative to illustrate the three evidence-based perceptions or person-states that actuate self-motivation. Internal, interpersonal, and environmental conditions that heighten these states, presumed to be innate human needs by some psychologists (Deci, 1975; Deci & Flaste, 1996; Deci & Ryan, 1995), enhance personal perceptions of self-motivation.

Guidelines for enhancing perceptions of choice, competence, and community are detailed elsewhere (Geller, 2014a,b), and many of these are consistent with the life lessons discussed here. Consider, for example, how proper application of Life Lessons 1, 2, 3, and 5 can increase a person's perception of competence and fuel self-motivation.

Empowerment
The management literature typically refers to empowerment as delegating authority or responsibility, or sharing decision-making (Conger & Kanungo, 1988). When a manager says, "I empower you," s/he usually means, "Get 'er done." In contrast, the empathic safety leader assesses whether the "empowered" individual *feels* empowered. Three questions, derived from social learning theory (Bandura, 1997), should be asked to assess one's perception of empowerment.

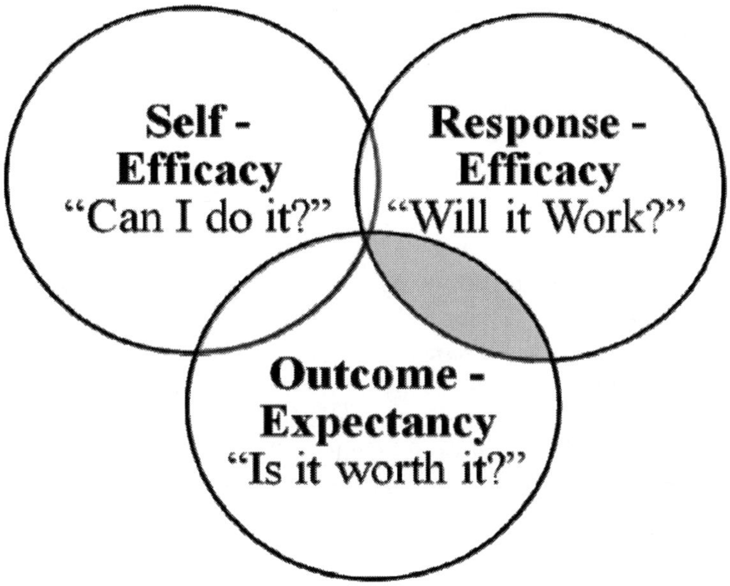

Figure 2. The three beliefs that determine empowerment (adapted from Bandura, 1997).

As depicted in Figure 2, the "Can you do it?" question asks whether the empowered individual or group has the resources, time, knowledge, and skills to handle the assignment. This knowledge and skill component refers to training, and the term "self-efficacy" places the focus on personal "belief." A manager might conclude a worker has the competence to complete a certain task, but the "empowered" individual might feel differently. A "yes" answer implies self-belief that, "I have the competence to handle the assignment and meet the performance-improvement goal."

Knowing what to do is not enough. The "empowered" person must also believe the assignment connects to a meaningful purpose. This is a response-efficacy question. Will attaining the process goal (i.e., performing the desired behaviors) contribute to a valued mission of the organization, work team, or the individual? For safety, this translates to believing a particular injury-prevention process (e.g., an AC4P coaching process) will contribute to achieving an injury free workplace.

More training might be necessary to answer "yes" to the "self-efficacy" question; and education might be needed to obtain an affirmative reply to the response-efficacy question. People might believe they can accomplish a particular process or task, but not believe such an accomplishment will make a difference in achieving an ultimate outcome. In this case, education is needed, including an explanation of an evidence-based principle or theory, and perhaps the presentation of convincing data.

The third question to assess empowerment addresses motivation, meaning, "Is the expected outcome worth the effort?" Performing relevant behavior is motivated by anticipating a positive consequence to achieve or a negative consequence to avoid. Referring back to our first Life Lesson, people feel they have more choice and are more likely to be self-motivated when working to achieve a positive consequence than when working to avoid a negative consequence.

Empowering goals. Safety professionals are well aware of the benefits of behavior-focused goal-setting to activate process activities aimed at achieving a particular outcome. They are also likely aware of a popular acronym used to define the characteristics of an effective goal: SMART. Trainers and consultants present variations of the words represented by this acronym, with "M" representing "Measurable" or "Motivational" and "T" referring to "Timely" or "Trackable," for example.

The author has proposed the following acronym words: "S" for Specific, "M" for Motivational, "A" for Attainable, "R" for Relevant, and "T" for Trackable (Geller, 2005, 2008a), later adding "S" (i.e., SMARTS) for Shared. Because social support can increase one's commitment to work toward a goal, it's useful to share your process goal with family, friends, and/or work colleagues (Geller, 2014c). A rationale and procedural details for each component of effective goal-setting are provided elsewhere (Geller, 2005, 2014c). Here it's instructive to note the connection between SMARTS goals and the empowerment model depicted in Figure 2.

Specifically, SMARTS goals are empowering because they are *Attainable* ("I can do it."), *Motivational* ("It's worth it."), and *Relevant* ("It will work."). This connection makes it clear that both empowerment and goal-setting are similar behavioral antecedents, setting the stage (or activating) certain behavior(s). Figure 3 illustrates the connection between a vision, a goal, empowerment, and consequences. An extremely important point is shown: A vision, a goal, and empowerment are not sufficient to sustain desirable behavior. Relevant consequences are necessary.

People are motivated to work for goal attainment when they expect to acquire desirable consequences and/or avoid undesirable consequences. People need to believe in the vision (e.g., an injury-free workplace) and feel encouragement from leaders to attain process goals that support the vision. Thus, safety leaders provide supportive and corrective feedback to increase the quantity and improve the quality of behaviors consistent with vision-relevant process goals (Life Lesson 3).

Self-motivation vs. Empowerment. Consider a critical difference between these two concepts. Empowerment is a behavioral antecedent or activator; self-motivation reflects the impact of behavioral consequences. When feeling empowered, an individual is ready (or activated) to work toward achieving a designated goal. On the other hand, a self-motivated individual anticipates or has received a consequence that supports self-directed rather than other-directed behavior (Watson & Tharp, 1997).

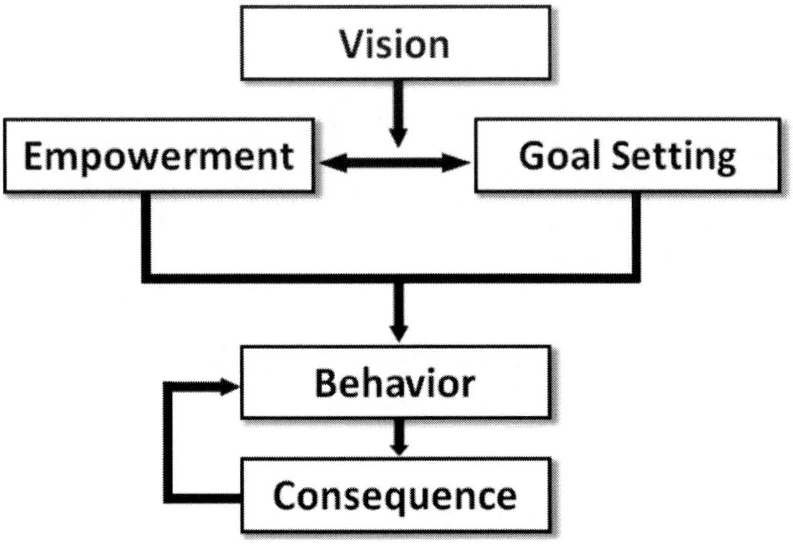

Figure 3. The connection between goal setting, empowerment, and behavioral consequences.

The initial proponents of self-motivation (Deci, 1975, Deci & Ryan, 1995) did not connect self-motivation with behavioral consequences. However, a behavioral scientist associates consequences with motivation, as implied by Skinner's "selection by consequences" legacy (Skinner, 1981). With this behavioral science perspective, consequences that reflect personal choice, competence, and/or a sense of community or belongingness should enhance self-motivation and increase the sustainability of a behavior-change intervention.

Bottom line: An intervention that applies positive consequences to increase the occurrence of a safe behavior is sustainable to the extent that it inspires self-motivation by linking behavioral consequences with perceptions of choice, competence, and/or community. This profound principle of self-motivation is reflected in our final life lesson.

7. Progress from Self-Actualization to Self-Transcendence

Humanism is implicated by this seventh life lesson. However, without behavioral science, it's just a theory with limited practical value. The author proposes that a large-scale transition from self-actualization to self-transcendence is key to cultivating an injury-free workplace.

A Hierarchy of Needs

The most popular theory of human motivation is the "Hierarchy of Needs" formulated by humanist Abraham Maslow (Maslow, 1943). Maslow arranged categories of human needs hierarchically, and presumed people will not attempt to satisfy needs at one level until fulfilling the needs at the lower stages.

People are first motivated to fulfill their physiological needs, including the survival requirements of food, water, shelter, and sleep. After meeting these needs, people are motivated by a desire to feel safe and secure from potential dangers. When workers are engaged for safety, they are working proactively to satisfy this need.

Our social-acceptance needs are next, or the desire to have friends and feel a sense of belongingness. After these needs are achieved, we shift our concern to self-esteem -- the desire to feel worthwhile, respected, and generally successful. Then we climb to "self-actualization." Is this the highest level of Maslow's Hierarchy of Needs? This is how the need hierarchy has been taught for decades in most psychology and leadership classes, but it's not true.

Maslow's Hierarchy of Needs is depicted in Figure 4, and self-actualization is not at the top. Shortly before his death in 1970, Maslow revised his renowned hierarchy and positioned self-transcendence at the top (Maslow, 1971). At this level people go beyond self-interest and perform AC4P behavior, such as reporting a safety hazard or giving coworkers behavior-based feedback about their safe or at-risk behavior.

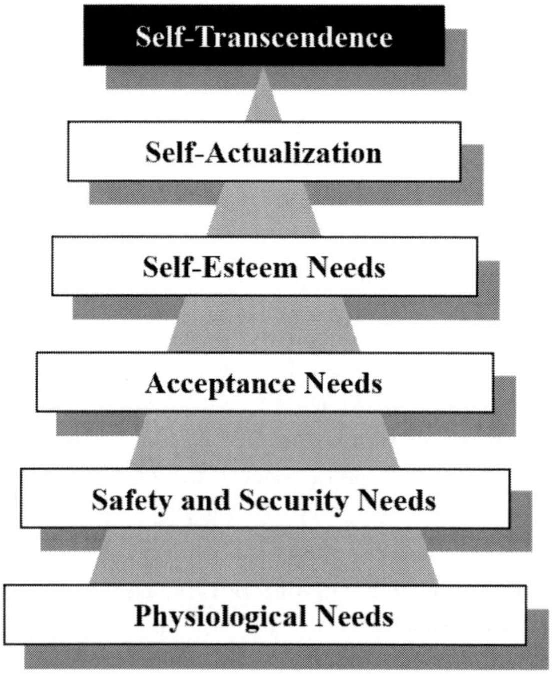

Figure 4: Maslow's revised hierarchy with self-transcendence at the top.

Common sense suggests that various self-needs require fulfillment before self-transcendent or AC4P behavior is likely to occur. But it's possible to think of examples of people performing various AC4P behaviors before satisfying all of their personal needs. For example, Mahatma Gandhi put the concerns of his entire nascent country of India before his own. In his 50-year struggle to help his poor and downtrodden compatriots, Gandhi was imprisoned, suffered extensive fasts, and eventually assassinated.

The connection between Maslow's Hierarchy of Needs and various positive behavioral consequences provides critical insight for fueling self-motivation and sustaining the impact of a behavior-improvement intervention. First, an individual's position in the hierarchy determines what types of consequences are likely to be most reinforcing at a particular time. Without food, shelter or sleep, for example, most people focus their behavior on satisfying these biological needs. But after this need level is satiated, human behavior is motivated by consequences linked to higher-level needs.

Indeed, higher need levels implicate consequences related to self-motivation. Consequences that advance one's sense of connection with others, community for example, satisfy one's need for acceptance or social support. Those consequences that support a person's belief in personal competence to perform worthwhile work connect to the needs for self-esteem and self-actualization. And, reaching beyond one's self-needs to help others through AC4P behavior contributes to satisfying the need for social acceptance and self-esteem, and even self-actualization.

Question: When does the need for social acceptance, self-esteem, and self-actualization get fully satisfied? At what point does a person become satiated on consequences linked to these higher need levels? These rhetorical questions are asked to reiterate the value of delivering rewarding consequences that reflect the three C-words of self-motivation (i.e., Competence, Choice, and Community) and thereby enhance need states in Maslow's Hierarchy that are not readily satiated.

Bottom line: Behavioral consequences that foster perceptions of personal competence, self-worth, belongingness, and/or autonomy also facilitate self-motivation and self-directed behavior, and are likely to be more sustainable than consequences unrelated to these person-states.

The AC4P Movement

The author coined the term "actively caring" in 1990 when working with a team of safety leaders at the Exxon Chemical facility in Baytown, Texas (Geller, 1991). The vision was to cultivate a brother/sister keepers' culture in which everyone monitors each other's safety—a workplace where people "have each other's backs." They routinely go above and beyond the norm to benefit the health, safety, and/or welfare of their coworkers.

"Actively caring for people" was an ideal description for this site-wide paradigm shift. As discussed earlier, most people care about the well-being of others, but relatively few "act" on such feelings of caring. The challenge was to get everyone to *actively care*—to take effective action based on their caring. The vision: a company culture with more interpersonal empathy, compassion, and AC4P behavior.

To promote this AC4P vision, the author began distributing green silicon wristbands embossed with the words, "Actively Caring for People," to recognize individuals for their AC4P behavior. This recognition approach has been used successfully to reduce bullying by promoting and rewarding AC4P behavior in various educational settings (Geller, 2017; McCarty, Teie, McCutchen, & Geller, 2016). Plus, organizations have adopted the AC4P approach for occupational safety to help achieve an injury-free workplace (e.g., Bolduc et al., 2014; Geller et al., 2012; Geller & Veazie, 2014; Ludwig et al., 2014; Roberts, 2014; Williams & Geller, 2016)

For the more recent applications, the AC4P wristbands were redesigned to include a different identification number per each wristband, and a website (www.ac4p.org) enables people to: a) share their AC4P stories (with the number of the wristband they gave or received); b) track worldwide where a particular AC4P wristband has been; and c) order more AC4P wristbands to reward others for their AC4P behavior. The AC4P wristband is depicted in Geller and Geller (2017).

Delivery of an AC4P wristband should be accompanied with words that serve higher-level needs. Never imply the wristband is a "pay-off" for AC4P behavior. Rather, the wristband is only a token of appreciation for the "special servant leadership exemplified by the observed act of kindness." Plus, the wristband recipient is told s/he is now one of many who have joined the AC4P Movement—a flourishing international effort to nourish cultures of interpersonal compassion and interdependent AC4P behavior.

To date, more than 5,000 AC4P stories have been shared on the website, and more than 150,000 AC4P wristbands have been purchased, with proceeds going to the Actively Caring for People Foundation. The mission of this accountability system is to activate and reward AC4P behavior globally and to inspire the nurturing of AC4P cultures in various settings. Our vision: brother/sister keepers' safety cultures of compassion and interpersonal AC4P behavior worldwide, leading to organizations without injuries, families without abuse, communities without violence, and nations without wars.

In Conclusion

This ASSE Proceedings paper first specified seven life lessons derived from applied behavioral science (ABS) and humanism (i.e., humanistic behaviorism) were explained to show how everyone, including managers, can inspire self-motivation for injury prevention in the workplace and beyond. The first four lessons were derived from ABS and are employed in almost every successful intervention developed and implemented to increase the frequency of safety-related behavior and decrease occurrences of at-risk behavior. The next three life lessons connect directly to principles of humanism, which are beyond the empirical, ABS domain of BBS.

Operational definitions were included with each of these humanistic lessons, making it possible to bring them to life with a BBS or PBS process. For example, the concept of self-transcendence was described in terms of interpersonal AC4P behavior, and a practical application of Life Lesson 1 was illustrated (i.e., the AC4P wristband) to increase the frequency of AC4P behavior and to help nurture an AC4P culture.

The life lessons reviewed here have successfully improved the human dynamics of safety in many organizations worldwide. This includes numerous applications of BBS with less-than-optimal quality engagement by all employees. Some organizations have avoided BBS or PBS because of select case studies of BBS failures promulgated by consultants "selling" another approach. Still other organizations are unaware of the evidence-based ABS approach to preventing injuries, and how humanistic principles can enhance the effectiveness of a behavior-change process (Geller, 2015).

While this presentation justifies a behavior-focused coaching approach to injury prevention on a macro scale and explains ways to improve a less-than-optimal BBS process, the author hopes readers will see broader applications of these seven life lessons. It is hoped readers will translate these lessons into practical procedures to optimize quality AC4P engagement throughout their organizations and beyond. An injury-free workplace is contingent on the number of employees performing AC4P behavior for occupational safety, and inspiring others to do the same.

Bibliography

Allesandra, T. and M.S. O'Connor (1996). *The platinum rule: Discover the four basic business personalities and how they can lead you to success.* New York: Warren Books, Inc.

Alvero, A. M. and J.Austin (2004). The observer effect: The effects of conducting behavioral observations on the behavior of the observer. *Journal of Applied Behavior Analysis, 37,* 457-468.

Alvero, A. M., K. Rost, and J. Austin (2008). The safety observer effect. *Journal of Safety Research, 39*(4), 365-373.

American Humanist Association. (2008). Retrieved June 16, 2014 from http://www.americanhumanist.org/

Atkinson, J. W. (1957). Motivational determinants of risk-taking behavior. *Psychological Review, 64,* 359-372.

______ (1964). *An introduction to motivation.* Princeton, NJ: Van Nostrand.

Bandura, A. (1969). *Principles of behavior modification.* New York: Holt, Rinehart & Winston.

______ (1997). *Self-efficacy: The exercise of control.* New York: W. H. Freeman and Company.

Biglan, A. (2015). *The nurturing effect: How the science of human behavior can improve our lives and our world.* Oakland, CA: New Harbinger Publications, Inc.

Bolduc, J. E., T. Plunkett, M. Foy, and J. Dean (2014). Actively caring to keep people safe: Evolving from BBS to AC4P. In E.S. Geller (Ed.). *Actively caring for people: Cultivating a culture of compassion.* (pp. 219-225). Newport, VA: Make-A-Difference, LLC.

Brehm, J. (1966). *A theory of psychological reactance.* New York: Academic Press.

Chance, P. (2008). *The teacher's craft: The 10 essential skills of effective teaching.* Long Grove, IL: Waveland Press, Inc.

_______ (2008). *The teacher's craft: The 10 essential skills of effective teaching.* Long Grove, IL: Waveland Press, Inc.; Reed, D. et al. (2016). Actively caring for higher education. In E.S. Geller (Ed.). *Applied psychology: Actively caring for people* (pp.563-593). New York: Cambridge University Press.

Conger, J. A. and R.N. Kanungo (1988). The empowerment process: Integrating theory and practice. *Academy of management review, 13*(3), 471-482.

Covington, M. V. (1992). *Making the grade: A self-worth perspective on motivation and school reform.* New York, NY: Cambridge University Press.

Covington, M. V., and B.W. Roberts (1994). Self-worth and college achievement: Motivational and personality correlates. In P. R. Pintrich, D. R. Brown, & C. E. Weinstein (Eds.). *Student motivation, cognition, and learning: Essays in honor of Wilbert J. McKeachie.* Hillsdale, NJ: Earlbaum.

Daniels, A. C. and J.E. Daniels (2005). *Measure of a leader.* Atlanta, GA: Performance Management Publications.

Deci, E. L. (1975). *Intrinsic motivation.* New York: Plenum

Deci, E. L. and R. Flaste (1995). *Why we do what we do: Understanding self-motivation.* London, England: Penguin books.

Deci, E. L. and R.M. Ryan (1995). *Intrinsic motivation and self-determinism in human behavior.* New York: Plenum

Geller, E. S. (1991). If only more would actively care. *Journal of Applied Behavior Analysis, 24,* 763-764.

______ (1996). *The psychology of safety: How to improve behaviors and attitudes on the job.* New York: CRC Press.

______ (2001). *The psychology of safety handbook.* Boca Raton, FL: CRC Press.

______ (2005). *People-based safety: The source.* Virginia Beach, VA: Coastal Training Technologies Corporation.

_______ (2008a). *Leading people-based safety: Enriching your culture.* Virginia Beach, VA: Coastal Training Technologies Corporation.

______ (2008b). The tragic shooting at Virginia Tech: Personal perspectives, prospects, and preventive potentials. *Traumatology*, 14(1), 8-20.

_______ (2014a) (Ed.). *Actively caring at your school: How to make it happen.* Newport, VA: Make-A-Difference, LLC.

_______ (2017). *Actively caring for people in schools: How to cultivate a culture of compassion.* New York: Morgan James.

_______ (2014b) (Ed.). *Actively caring for people: Cultivating a culture of compassion.* Newport, VA: Make-A-Difference, LLC.

_______ (2014c). The psychology of self-motivation. In E.S. Geller (Ed.). *Actively caring for people: Cultivating a culture of compassion* (pp. 73-75). Newport, VA: Make-A-Difference, LLC.

_______ (2015). Humanistic behaviorism: Enhancing the positive impact of behavior-based safety (BBS). *Industrial Safety and Hygiène News*, October.

_______ (2001). *The psychology of safety handbook* (Chapter 15). Boca Raton, FL: CRC Press; Geller E.S. (2016). The psychology of AC4P behavior. In Geller, E.S. (Ed). *Applied psychology: Actively caring for people* (pp. 45-82). New York: Cambridge University Press.

Geller, E. S., J.E. Bolduc, M.J. Foy, and J. Dean (2012). In pursuit of an actively-caring safety culture: Practical methods, empirical results, and provocative implications. *Professional Safety. 57*(1), 44-50.

Geller, E. S. and Krista S. Geller (2017). *Actively caring for people's safety: How to cultivate a brother/sister keeper's work culture.* Park Ridge, IL: American Society of Safety Engineers.

Geller, E. S., S. Perdue, and A. French (2004). Behavior-based safety coaching: Ten guidelines for successful application. *Professional Safety, 49*(7), 42-49.

Geller, E. S. and R.A. Veazie (2014). Behavior-based safety versus actively caring: From other-directed compliance to self-directed commitment. *Professional Safety, 59*(10), 44-50.

_______ (2017). *The courage to actively care: Cultivating a culture of compassion.* New York: Morgan James.

_______ (2017). *The motivation to actively care: How you can make it happen.* New York: Morgan James

Ludwig, T., W. Carvelas, C. Engelbrecht, and M. McClintock (2014). Actively caring for safety in South Africa: The power of the AC4P wristband (pp.209-216). In E.S. Geller (Ed.). *Actively caring for people: Cultivating a culture of compassion.* Newport, VA: Make-A-Difference, LLC.

Maslow, A. H. (1943). A theory of human motivation. *Psychological Review, 50,* 370-396.

______ (1971). The farther reaches of human nature. New York, NY: Viking.
McCarty, S., S. Teie, J. McCutchen, and E.S. Geller (2016). Actively caring to prevent bullying in an elementary school: Prompting and rewarding prosocial behavior. *Journal of Prevention and Intervention in the Community, 44*(3), 164-176.
O'Brien, D. P. (2000). *Business measurements for safety performance.* New York: Lewis Publishers.
Peterson, D. (2001). *Authentic involvement.* Itasca, IL: National Safety Council.

Random acts of kindness. NA (1993). Emeryville, CA: Conari Press.

Roberts, S. (2014). Actively caring for occupational safety: Preventing injuries with people-based safety. In E.S. Geller (Ed.). *Actively caring for people: Cultivating a culture of compassion* (pp.180-197). Newport, VA: Make-A-Difference, LLC.
Rogers, C. (1942). *Counseling and psychotherapy.* New York, NY: Houghton Mifflin.
Skinner, B. F. (1938). *The behavior of organisms: An experimental analysis.* Acton, MA: Copley Publishing Group.
_______ (1953). *Science and human behavior.* New York: Macmillan.
_______ (1971). *Beyond freedom and dignity.* New York: Alfred A. Knopf
_______ (1981). Selection by consequences. *Science, 213*, 502-504.
Thorndike, E.L. (1931). *Human learning.* Cambridge, MA: MIT Press.
Watson, D. C. and R.G. Tharp (1997). Self-directed behavior: Self-motivation for personal adjustment (7th Edition). Pacific Grove, CA: Brooks/Cole Publishing.
Williams, J. H. (2010). *Keeping people safe: The human dynamics of injury prevention.* Toronto, Canada: Government Institute.
Williams, J. H. and E.S. Geller (2016). Actively caring for occupational safety. In E. S. Geller (Ed.). *Applied psychology: Actively caring for people* (pp.301-338). New York: Cambridge University Press.

Positive Influence–The Key to Effective Leadership

M. E. "Eddie" Greer, CSP
President and CEO
Eddie Greer & Associates
Horseshoe Bay, Texas

Introduction

I am often asked if I feel there is a difference between a leader and a manager. My answer is always an unequivocal yes. I am a firm believer in the statement that you ***manage things and you lead people***. This is one of the missing pieces when it comes to many safety professionals …leadership training and the ability to influence. I have seen evidence of companies that do all types of company and regulatory training and completely disregard one the most important pieces of training for their leadership. And they sit back and wonder why their safety endeavors are not successful. It can be boiled down to lack of effective leadership and not being able to positively influence people.

Let's cut to the chase. We need leaders that can influence people because we had 4,836 work place fatalities in the United States in 2015. This is the highest number since 2008. Why do we need effective and influential leaders? I think these numbers speak for themselves.

Joseph Grenny, Kerry Patterson, David Maxfield, Ron McMillan and Al Switzer in their book, *Influencer, The New Science of Leading Change* state that what qualifies people to be called leaders is their capacity to influence others to change behavior in order to achieve results.

Look at that one more time. At the end of the day, what qualifies people to be called leaders is their capacity to influence others to *change* their behavior in order to achieve important results.

Definitions

John Maxwell in his book, *Developing the Leader Within You,* simply states that "*leadership is influence*".

Byrd Baggett in his book, *Dare to Lead* states that "the heroes are the leaders who realize the worth of people; those who understand the power of compassion; those who understand and value the dignity of employees. These men and women understand the wisdom of serving others and the *positive influence* that genuine concern for others have on the balance sheet."

Kouzes & Posner in their book, *The Leadership Challenge* states that leadership is the "art of mobilizing others to want to struggle for shared aspirations."

Definition most used by the author simply states that "leadership is the ability to inspire and influence others to exceed their perceived talents through guided empowerment and being enabled." It also helps if you have followers!

Here is a list of what the author feels are the qualities of an effective leader:

- Honesty – Worthy of our trust
- Forward Looking – See the big picture for themselves, workers & the company
- Inspiring – Communicate the vision
- Competent – capable of effective decision making
- Character – What you are when no one else is around.
- Trust – Doing what you say you are going to do
- Servant Leader – Willingness to serve others
- Strong People Skills – Fully understands people
- Motivator – Giving people the opportunity to grow and expand their knowledge
- Integrity – Doing what's right but maybe not popular
- Positive Influence – ability to attract and retain the right people
- Responsible – Doesn't pass the buck
- Positive Attitude – ability to pass on this attitude to the right people
- Problem Solver- Solver not creator
- Desire to continue to learn
- Time Manager
- Unselfish – Gives credit to others
- Self-Discipline – Follow tasks to completion
- Change Maker
- Good Listener
- Legacy Builder – Growing other leaders
- Executor – Gets things done

It is interesting that surveys conducted by Kouzes and Posner eight years apart produced the same top four results on the characteristics of admired leaders. People want leaders that are honest, forward-looking, inspiring and competent.

Expectations

What is it that we expect of an effective leader?

Problem solving is the essence of what leaders do. As leaders, the goal is to minimize the occurrence of problems – which means they must be courageous enough to tackle them head-on long before circumstances force their hand. Leaders must be resilient in their quest to create and sustain momentum for the organization and their people. The reality of the workplace finds leaders dealing with people that complicate matters with their self-promotion, office politicking, power-plays and ploys along with envy. Other issues are lack of resources and budgets or other circumstances that make it harder for people to be productive.

A leader must **never** view a problem as a distraction but rather as a strategic enabler for continuous improvement and opportunities previously unseen.

There are some effective solutions to solving problems.

Open Communication

Problem solving requires open, uninhibited and expected communication to allow all parties to freely express concerns and points of view. Open and effective communication toward problem solving happens because of a leader's ability to facilitate an open dialogue between people who trust their intentions. They also feel that are in a safe environment to share why they believe the problem occurred as well as specific solutions. Open communication allows all voices to be heard and all points of view accounted to allow the leaders, with their team, to map out a path toward a viable and sustainable solution. Open communication fosters a true open-door policy.

Remove Silos

Silos that are present in the workplace invite hidden agendas rather than welcoming efficient cross-function collaboration and problem solving. These silos make problem solving more difficult because leaders are more likely dealing with self-promoters and hidden agendas rather than team members fostered by a cross functional environment.

Breaking down silos allows the leader to more easily engage their employees to get down and dirty and solve problems together. It becomes more about finding resolutions and making the organization stronger and less about office politics.

The Right People in the Right Place

In the end, problem solving is about people working together to make the organization and the people it serves better. If you have the wrong people in critical areas, effective problem solving becomes a long and winding road of misery and failure is almost a surety.

Having the right people in place allow the organization to tackle problems head on and get on with the business at hand. The wrong people in place turn things around to make it more about them and less about what is required to transform a problem into a new opportunity.

This is one area where effective leaders have complete control of the situation; the selection and promotion of their most valuable asset, human capital. To get ahead leaders need to understand the power triangle of communication, recognition, and influence. Effective communication leads to recognition of key team members and this in turn fosters influence.

John Maxwell in his book, *Developing the Leader within You,* speaks of the best investment in the future is a proper influence today. Influence comes in many different coats. A child can and is usually influenced by their Mom in determining what clothes to wear and what they eat. A Dad may influence a son in the direction of liking/playing sports and maybe even a career choice.

We are influenced daily by advertisements pushing the finest and fastest cars, what to take for a headache, how to invest our money and even what deodorant to buy. Influence is all around us.

Mr. Maxwell spells out how leaders can be a major influence with the level of leadership they hold. There are five key levels that are listed below.

"Position" leaders are entry level and their influence comes with a title. This person may be "in control" because they were appointed to a position and lead by intimidation. People who stay at this level get into territorial rights, protocol, tradition, and organizational charts. These become very negative when they are the basis for authority and influence and are poor substitutes for leadership skills. Their job security is based on title, not talent and is often gained by appointment. People working for this appointed leader WILL NOT follow him or her beyond his or her stated authority.

"Permission" leaders get people to work for them when they are not obligated. Leadership here begins with the heart, not the head. People don't care how much you know until they know how much you care. A leader on the "permission" level will lead by interrelationships. On this level, time, energy, and focus are placed on the individual's needs and desires. People who are unable to build solid, long lasting relationships will soon discover they are unable to sustain long, effective leadership. This level allows you to have fun but staying too long on this level without rising will cause highly motivated people to become restless.

"Production" level leaders find good things begin to happen. Profit increases. Morale is high. Turnover is low. Needs are being met and goals are being realized. Good results are happening and this is the reason for the activity. On this level, people come together to accomplish a purpose. The leader used the permission level because people like to get together but they love to get together to problem solve and accomplish something. People sense success and they like what the leader is doing. Problems are fixed with very little effort because of momentum.

"People Development" is where the leader is able to duplicate their effort in growing additional leaders where people consistently demonstrate superior performance. A leader becomes great, not because of his/her power, but because of the ability to empower and enable others. Loyalty is key here when the follower had personally grown (developed) through the mentorship of the leader. This level is where long range growth occurs. The commitment to developing other leaders will insure ongoing growth to the organization and to the people.

The final level is "Personhood". Few people ever achieve this level. It comes from a lifetime of proven leadership and growing people and organizations. These leaders are bigger than life.

Keep in mind with these levels that:

1. The higher you go, the longer it takes.
2. The higher you go, the higher the commitment.
3. The higher you go, the easier it is to lead.
4. The higher you go, the greater the growth.

Having a Long Term Workable Strategy

Effective leaders connect the dots and map out a realistic plan of action in advance. These leaders are comfortable with problem solving because they know how to gather the right people, resources, develop a budget and utilize knowledge from past experience. They fully understand that a solid strategy must be implemented to head off potential problems, create solutions and work toward a common vision.

Leaders take enough time to step back and assess the situation and the opportunities that each problem presents. They have a strategy that serves as the foundation for how the problem will be approached and managed. They are able to anticipate the unexpected and utilize the strengths of their people to assure the strategy leads to a sustainable solution.

Effective leaders understand that they cannot shoot from the hip.

Kouzes and Posner in their book found through extensive research that there are five practices common to personal best leadership experiences.

Model the Way

Leaders understand if they want to encourage their employees to achieve high standards they must lead by example. They know that what they do on a daily basis wins the respect of their employees and not the fact that they have a title. Leaders must be clear about their beliefs and what they hold true. They must also communicate to back up these beliefs with action.

Challenge the Process

Leaders serve as pioneers and are willing to step out into the unknown. They understand that there will be a change from the status quo. Leaders must be good listeners because ideas come from customers, clients, vendors, and their employees on the front line. Leaders must then acknowledge good ideas and act on them. These well thought out actions are the key to the opportunity door.

Inspire a Shared Vision

They look across the horizon of time, imagining the attractive opportunities that are in store when they and their employees arrive at a distant destination. Leaders have the desire to make things happen, to change the way things are, to create something that no one else ever created before. Leaders must also know and understand themselves, their people and their industry.

Enable Others to Act

Effective leaders understand that they don't have all the answers and must rely on their team. Research shows that effective leaders success is determined by how often they use the word we. Leaders involve, in some way, all those who have a stake in the vision. From this, leaders discussed teamwork, trust, empowerment and enabling employees as essential elements of success.

Encouraging the Heart

While the climb to success can be difficult, leaders encourage others to keep going. People become frustrated, tired and disenchanted and the effective leaders encourage the heart of their employees to carry on. Encouragement is critical to a successful business or endeavor. Great leaders know that awards given with the right spirit can build a strong team.

Motivator

The author feels that motivation is something that comes from within but can be influenced by many factors. There are basically two types of motivation: *intrinsic* or internally driven and *extrinsic* or outside driven.

The leader playing on the positive side of *intrinsic motivation* will get employees motivated by giving opportunities to grow or learn new tasks, gain knowledge by putting in more effort and reaching desired goals, and actually mastering specific goals or skills. Great leaders instill a belief and recognition that employees can grow in their career and gain self-confidence.

The extrinsic or outside driven motivation comes from wanting more money, competition among peers, or the fear of punishment. This is definitely the wrong type of motivation and usually does not produce a long term employee.

The "Wrong" Influence

The wrong type of leader can be a major thorn in the side of safety, the welfare of employees and the organization. These type of leaders do not display any of characteristics that have been discussed above. They are interested in only promoting themselves (narcissism) and their influence is the mighty dollar.

Case in point is the CEO of Massey Energy, owner of the Upper Big Branch Mine in Montcoal, West Virginia. On April 3, 2010 an explosion in the mine owned by Massey Energy killed 29 miners. The CEO of the company made it very clear that his motivation was money. He micromanaged all aspects of his company and did not care about the safety of the workers in the mine. He managed to manipulate MSHA inspectors with special codes that would direct employees to clean up or perform differently when inspectors were present.

The CEO's disregard for safety was quite evident in the 515 safety citations issued by MSHA in 2009 and was fined $382,000. These fines were just a minor inconvenience since the mine was producing over $600,000 per day in coal revenues. Revenue that came from disregarding basic mine safety procedures including removing coal dust and maintaining an adequate ventilation system to remove methane gas. Both of which contributed to the fatal explosion.

The subsequent investigation found the physical conditions that led to the fatal explosion were the result of a series of basic safety violations. These safety violations along with other factors including intimidation of miners, advance notice of inspections, and two sets of books with hazards recorded in an internal production and maintenance book but not in the official examination book. ***There was also very clear evidence that the CEO promoted and enforced a workplace culture that valued production over safety, including practices calculated to allow it to conduct mining operations in violation of the law.*** All negative influence.

The really sad part is that the CEO was found guilty in December 2015 of one misdemeanor charge of willfully conspiring to violate mine safety and health standards and faces one year in jail and a $250,000 fine. A very small and unfair result for a man that contributed greatly to 29 miners dying.

Conclusion

The role of the effective leaders in today's workplace is more important than ever. There is a need for safety leadership at all levels to help make a difference and to ensure that all employees get to go home at the end of the workday.

Leaders have to make the extra effort to pick the right people for the job and then do everything in their power to help make them successful.

The Bureau of Labor Statistics (BLS) reported that there were 4, 836 workplace fatalities in the United State in 2015. And this is an increase from 2014. 4,836 people just like you and me that left home to go to work and did not return. And this numbers goes up exponentially when you consider the spouses, children, parents, friends and co-workers that are affected when there is a fatality. Need there be anything else said as to why we need proactive effective leaders that can influence safety and workers.

Bottom line is that we need people to step up and become effective safety leaders to help influence the "C" suite within the organization. This strategy creates a benefit from having a safe work environment and most of all allowing employees to go home each and every day. Now, isn't that what we are all about? Making a difference in the lives and well-being of our employees.

Bibliography

"Upper Big Branch Mine disaster." Wikipedia: The Free Encyclopedia. Wikipedia Foundation, Inc. 5 December 2015.

Bureau of Labor Statistics, "Census of Fatal Occupational Injuries Summary, 2015". U.S. Department of Labor, Economic News Release, December 16, 2016. (https://www.bls.gov/news.release/cfoi.nr0.htm)

Grenny, Joseph et al. 2013. *Influencer The New Science of Leading Change.* New York: McGraw Hill.

Kouzes, James M. and Barry Z. Posner. 1995. *The Leadership Challenge.* San Francisco, CA: Jossey-Bass.

Lopis, Glenn., *The Most Effective Ways Leaders Solve Problems.* Forbes, 2014

Maxwell, John C. 1993.*Developing the Leader Within You.* Nashville, TN: Thomas Nelson, Inc

______. 1995. *Developing the Leaders Around You.* Nashville, TN: Thomas Nelson, Inc.

Onion, Meredith L. and Michael F. O'Toole. 2003.*You've Just Been Made the Supervisor. Now What?,* Itasca, IL: National Safety Council.

Session No. 627

CDOT ROCIP Heavy Civil Construction Insurance Meeting the Challenges

Mark Semonisck, BS/MS Safety Engineering, CSP
Denver, Colorado

Julie Mileham, AIC, ARM-P, CPCU, MBA
CDOT Risk Manager
Littleton, Colorado

Doug Martin, CSP, ARM, CPCU, MBA
Liberty Mutual Risk Control Consultant
Denver, Colorado

CDOT ROCIP Introduction--Relevant Background and History

The Rolling Owner Controlled Insurance Program (ROCIP) described in this paper has been in effect from December 2012 to the present time. It involves the owner Colorado Department of Transportation (CDOT), the insurance broker Lockton Companies, and the insurance carrier Liberty Mutual.

Background

Generally speaking, a ROCIP is where the owner designs and purchases the insurance for basic and critical aspects of the work, enrolls the contractors as policyholders, and then accepts and manages risks and/or claims. What makes it 'rolling' is that not all construction projects bound by coverage will be started simultaneously. Projects are started and completed within the policy timeframe. The Program insurance broker helps with design, implementation and oversight. The Program insurance carrier helps with claims management and assists with risk tracking and reduction. Where terms and philosophies may be the same, each owner and -program's design and implementation may differ. Lastly, the contractor policyholders, both general and sub-contractors are directly engaged and are the key element to a successful Program.

History

ROCIPs are not unique to other state Department of Transportation (DOT) construction projects, or Colorado highways. The Transportation Expansion Project conducted from 2001 thru 2007 was a $1BB plus project that performed under a Partner Controlled Insurance Program (PCIP). This project was the largest of its kind in the United States at the time as it addressed the 14th busiest highway and involved light rail construction along side. This project was notable not only for size and scope, but for being finished under budget, ahead of schedule and with minimal worker injuries or other insurance claims.

Primary Goals & Objectives

The primary goals for any construction project owner's commitment are costs, performance, schedule, quality, environment and safety. The key areas amplified by a transportation project and shared by CDOT's own exceptional needs, is to perform work that pays close attention to the traveling public, property protection, worker safety, public relations, and care for the taxpayer's investment. CDOT's ROCIP program does that in the past, present and future.

Program and Work Conduct Challenges

Program Challenges

The primary concerns for ROCIP Program and road building challenges to be addressed are contractual, roles and responsibilities, inherent construction hazards, unique Colorado working conditions, and protecting the traveling public. Also, these concerns inter-connect, are dynamic and must be addressed continually as work phase's progress, project parameters and scope change, and participants continually evolve. Of note is on one particular project, peak worker count was approximately 300 where nearly 1800 workers went thru the required new worker orientation over the life of the project.

All successful construction projects begin at the contractual stage. CDOT performs approximately $600MM worth of highway projects annually and has been in existence in several forms since 1919. It maintains a world-class level of performance in engineering, maintenance, and construction management. There are well-established construction specifications, contracts, and qualified direct on-site management. All work meets and exceeds all national highway standards and federal, state and local requirements for contracting.

ROCIP roles and responsibilities beyond CDOT engineering and general contractor obligations primarily fall on the CDOT Risk Manager/Program Owner's Representative, Insurance Broker, and Insurance Carrier. The CDOT Risk Manager oversees the program's obligations and assures that program results are on track and communicated to executive levels. The CDOT Risk Manager also assists with communications and helps with CDOT on-site personnel or others if needed. The Program insurance broker has a number of persons directly involved including an assigned executive, manager, claims representative, loss control consultant, injury counselor, and program administrator as the primary daily program performance gatekeeper. The insurance carrier assists with managing claims, monitoring loss experience and provides additional dedicated loss control resources.

Work Conduct Risks and Challenges

The work related challenges are immense. There are millions of vehicle miles traveled each year on every highway and that is growing exponentially for a popular destination state such as Colorado. Like many states, the highway infrastructure is in various life-span conditions with much of it reaching beyond its originally intended life. This leads to potential roadway accidents and insurance claims beyond the construction cycle. Many of the projects in this ROCIP had traffic counts of 90-350k per day. Conducting work while minimizing inconvenience to the traveling public is a challenge. When those motorists are speeding, impaired or distracted, as can be common, this becomes a bigger concern; keeping workers and the public safe means working in close proximity to traffic, year-round, at night, and sometimes during inclement weather. Unique Colorado conditions include high mountain environments, narrow work areas, wildlife, dense urban interchanges, co-existing utility corridors, railroads, pedestrians and cyclists, and environmental protection.

Program Design and Management

ROCIPs are only cost effective and risk beneficial where certain projects, individually and collectively, are considered based on work scope, schedule, and cost of construction. Executive management and risk management determine participants then discuss this shorter list to determine final projects included. It is very important that these projects are clearly identified as ROCIP projects in the contractual, request for proposal (RFP), and bidding stages, as the bids are significantly impacted. Typically the financial results "success math" includes cost of risk minus cost of losses, of which the Owner keeps the difference. What is important here is that in publicly funded work the financial benefit is returned to the taxpayer and/or re-invested for additional work performed.

ROCIP Design

CDOT has vast experience relative to learning from past and other's loss events/, including understanding and assessing risk of highway construction. Using this knowledge from a design standpoint, projects that are longer term and larger scope benefit from the typical advantages of a ROCIP: broader coverage, higher limits, limited litigation, required insurance and industry best management practices for safety and claims performance, and less expensive policies leveraged by an owner's volume pricing being the primary economic benefits. Key components are comprehensive contract and insurance policies, diligent enrolling of contractors and routine payroll reporting, quarterly program financial/performance status reports, and scheduled payments for both premiums and claims payouts.

ROCIP Stakeholder Engagement

Another critical element is working closely with contractor partners and industry groups. In most state DOT construction programs including Colorado, the partner base is well-known entities of local companies or elements of national organizations. They are typically high performing, largely experienced, well known from most likely having worked for CDOT in the past. Highway road building is a highly specialized industry and line of work. Except for some of the service providers such as utility work, concrete placement, or cranes for example, very few if any of the primary contractors work both highway and vertical construction. They understand the risk management and insurance concerns, as well as how to prevent and manage unique concerns. It is important to note that not all contractors are in favor initially for giving up control of their risk management or financial returns from successful programs and this is addressed with outreach and additional partnering opportunities on program design whenever possible. While discussing contractor partners it is also key to discuss normal desires and requirements by federal funding to include a percentage of the work value to include Disadvantaged Business Enterprises (DBE), and small, woman and minority owned businesses. CDOT is fully committed to both their inclusion and success. Why it is important to identify in insurance program design and implementation is that these organizations may have less sophistication, maturity or experience in the higher levels of performance with best management practices. CDOT ROCIP Program management teams work closely with primary or first-tier contractors to offer assistance as needed before they arrive, during work, and ultimately making them a better potential partner in the future. Lastly, a seemingly unique component of Colorado's road-building industry is the Colorado Contractor's Association (CCA). This highly engaged and active group is part of the Associated General Contractors (AGC) but is entirely focused on highway construction related issues. They are a direct and respected representative organization dedicated to CDOT and contractor's mutual interest in protecting the public, workers, and delivering highest quality roads. The CCA participates in state and local legislation, contractual specifications/guidance, best management practices, and risk management concerns of its members.

Current ROCIP Program Financial Expectations and Results

ROCIPs by philosophy and design are expected to perform better than basic insurance programs for the reasons described previously. The third leg of this "milk stool" is constant monitoring and reporting on performance. Using CDOT, broker, carrier and contractor resources to monitor and report keeps the program on track and successful throughout its life. This is done typically through daily contact as needed, weekly ROCIP program status meetings with all participants, monthly and as needed jobsite visits, carrier claims review meetings, and finally through highly detailed and useful quarterly reports published by the broker. Key components of this monitoring and reporting are enrolled contractors, insurance costs, losses, field contacts and feedback, positive and negative events, and results projections.

Program Insurance Broker and Carrier Resources

There are a number of Program insurance broker and carrier resources provided through the ROCIP for underwriting, program management, claims and loss control. For the CDOT ROCIP the broker provided the dedicated Loss Control Manager who provided assistance, oversight, and routine reporting. This person was the team's boots on the ground and eyes on the scene to keep the stakeholders apprised of work and progress, both good news and unfortunate. The underwriting component helps develop the risk management through analysis, program design, and policies. This paper is intended to focus on specific aspects that loss control and safety can address. A good philosophy is that the best way to manage claims is to not have them. First and foremost is having a highly qualified loss control consultant assigned from both the broker and carrier. These individuals have experience with contractor organizations, project performance, regulatory compliance, and best industry practices. In many cases contractors are already clients of the broker and/or carrier, which is an added plus. Secondly, having these individuals directly involved in risk assessments before the work starts, during work, and during any incident is beneficial. The Loss control Consultants can also observe loss trends . Lastly, the carrier loss control consultant is key in helping the project by participating and looking forward in contractor progress meetings, on-site compliance and performance assistance, incident review, and loss trend analysis.

Broker Carrier Loss Control Value Added Propositions

As discussed previously, these individuals are already well connected to the participants and work. Key areas of focus and opportunities to make a positive impact are in the following areas: protecting the traveling public and ensuring CDOT's integrity, public protection is a key element for this owner for the obvious reasons. Protecting the traveling public and the organization's integrity for safety and quality's end product. The biggest risk areas of challenge are with pedestrian, cyclists, and business disruption and project neighbors. Primary impacts can be travel disruptions, noise, dust, vibration, road damage, utility strikes, and construction equipment. The other key area is worker safety in one of the widely recognized most hazardous activities. Even when there are no accidents or losses the hazard is 24-7. This concern is addressed by focusing attention on known risks such as OSHA Focus Four leading causes of fatalities, mandatory Job Hazard Analysis (JHA) at all contract levels, direct participation in weekly site progress meetings to anticipate concerns, 100% fall protection with no unapproved exceptions normally allowed by OSHA, exceptional utility locating, identification and protection, and close monitoring of traffic controls both internal and external.

The primary platforms for monitoring and communicating efforts is thru the monthly jobsite loss control surveys via the carrier risk control representative conducted with CDOT engineer, Contractor and subcontractor representatives, with a detailed report provided to all stakeholders. This effort looks at future work, high hazard work, worker and small contractor coaching, and JHA review and comment. There are both scheduled site-wide reviews, as well as, critical and off-hour work reviews. It is not uncommon for 40% of all work, in particular high hazard or disruptive such as demolition, girder erection or traffic switches, to be performed at night and weekends to provide added safety for both the public and workers. The other means of documentation and communication is provided by the broker LC and involves detailed and meaningful quarterly status reports, annual stewardship report, and claims reviews, weekly or as-requested program status meetings with stakeholders.

Conclusions and Take-Aways

CDOT is embarking on ROCIP II, a similarly structured program from ROCIP I discussed here. The following are a number of recommendations and observations from the participants valuable in both the present program and future program in no particular order or priority.

- Assure that Program objectives, design and requirements are well communicated in the owner's organization; in particular with any persons directly interacting with the Program in the field.
- Assure that Program objectives, design and requirements are well communicated to the contractor's / policyholders organizations. It is particularly best practice if there are opportunities for input in the design or implementation by the stakeholders; i.e., choosing medical providers, deductible limits, or incentives.
- Assure that all desired actions are clear and concise in the contractual requirements. This is important when requirements exceed current laws or regulations, and which are normally considered to be minimal standards. Examples are drug and alcohol testing, fall protection, utility locates and protections, and traffic controls.
- Fully engage any public informational organizations with the owner or contractors.
- Fully engage any local industry groups that may be dedicated to your stakeholders.
- In the case of publicly funded work there may be requirements for disadvantaged or small-business entities that may need additional attention or resources prior to and during work. This should be a joint effort with the owner and general contractor.

- Fully engage all related safety and loss control resources attached to the project. Each contractor normally has its own professional safety, insurance carrier, or consultant partners who may help with the program's success.
- Begin outreach and direct involvement by Program participants with construction work as soon as possible.
- Look for what makes each project site different. For example, traffic counts and type, geographical, meteorological, environmental, and adjoining properties.
- Use a highly experienced and qualified insurance broker and carrier provider.

Session No. 628

Preparing and Practicing for a Successful Failure Using NFPA 1600

Leo DeBobes CSP
Public Sector Practice Specialty Administrator
Assistant Administrator, Emergency Management and Regulatory Compliance
Stony Brook University Hospital

Edward Zimmerman
Fire Protection Practice Specialty Assistant Administrator

Gabriel Miehl. CSP, CFPS
Past Fire Protection Practice Specialty Administrator
Safety Programs Manager
GE Transportation-Erie
Erie, PA

Emergency planning is a systematic process that incorporating practice to ensure success. Employers are required to develop emergency plans in anticipation of a future incident. Plan exercise is a regulatory expectation and risk management obligation to protect people and the environment. In order to be successful and sustainable, the planning and exercise process must be set in a management

NFPA 1600 can be act as a model for planning and practice activities using an all hazards PDCA approach. NFPA 1600 is a readily available framework that provides the basis for an emergency management system. This system can be linked to tools and templates available from DHS and FEMA. Using the PDCA model allows an organization to learn how to effectively develop policies and procedures that can then be practiced and practically applied successfully by the employees tasked to respond to an emergency.

PDCA for Emergency Planning, Execution, and Improvement

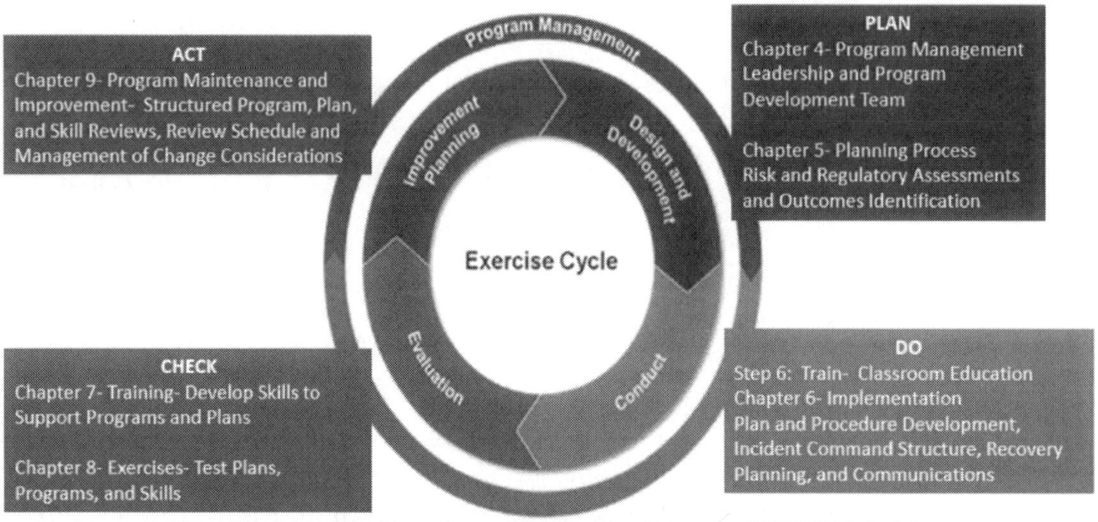

Homeland Security Exercise and Evaluation Program (HSEEP) Model
Overlay with NFPA 1600 PDCA Model

HSEEP1, NFPA 1600

A key to success with program development and execution is to first gain leadership support and involvement. Identify a leader in the organization that has the ability to actively support through resources and active engagement. This leader will act as a sponsor that will develop a sustainable team to work through the PDCA process. Resourcing is key as it will enable a team of individuals that can both actively engage with the PDCA process and can internalize ownership of the program. Program success is most often demonstrated by owners that are passionate about their work.

Ownership of this process should rest with the operations function of the organization. Operations has the staffing, resources, and metrics to drive sustainable change that will occur through the emergency planning process. Ownership by operations will also let the EHS professional act as a participant and technical expert. EHS as participant will enable the organization to learn. EHS as sole leadership during the emergency planning process will hinder the sustainable success of the emergency plan by not allowing operationalization outside of the EHS function.

Once the team is set, NFPA 1600, along with various other tools, can be incorporated into PDCA to generate a functional emergency management system.

Tools: Risk Assessment, Planning, and Training

FEMA comprehensive preparedness guide (CPG) 201, Threat and Hazard Identification and Risk Assessment Guide, is a useful companion to NFPA 1600 as it also incorporates a four step model for emergency planning.

This document references the threat and hazard identification risk assessment as THIRA:

"1. Identify the Threats and Hazards of Concern. Based on a combination of experience, forecasting, subject matter expertise, and other available resources, identify a list of the threats and hazards of primary concern to the community.
2. Give the Threats and Hazards Context. Describe the threats and hazards of concern, showing how they may affect the community.
3. Establish Capability Targets. Assess each threat and hazard in context to develop a specific capability target for each core capability identified in the National Preparedness Goal. The capability target defines success for the capability.
4. Apply the Results. For each core capability, estimate the resources required to achieve the capability targets through the use of community assets and mutual aid, while also considering preparedness activities, including mitigation opportunities."

FEMA has also published the Emergency Management Guide for Business and Industry and describes it as "a step-by-step approach to emergency planning, response, and recovery for companies of all sizes". This guide follows the NFPA 1600 PDCA model closely and provides guidance on forming a planning team, identifying regulatory requirements and customer deliverables, analyzing risk, and then developing plans, procedures, training, and exercises.

Figure 1: The THIRA Process

FEMA1

Table 1: Core Capabilities by Mission Area[3]

Prevention	Protection	Mitigation	Response	Recovery
		Planning		
		Public Information and Warning		
		Operational Coordination		
Forensics and Attribution Intelligence and Information Sharing Interdiction and Disruption Screening, Search, and Detection	Access Control and Identity Verification Cybersecurity Intelligence and Information Sharing Interdiction and Disruption Physical Protective Measures Risk Management for Protection Programs and Activities Screening, Search, and Detection Supply Chain Integrity and Security	Community Resilience Long-term Vulnerability Reduction Risk and Disaster Resilience Assessment Threats and Hazard Identification	Critical Transportation Environmental Response/ Health and Safety Fatality Management Services Infrastructure Systems Mass Care Services Mass Search and Rescue Operations On-scene Security and Protection Operational Communications Public and Private Services and Resources Public Health and Medical Services Situational Assessment	Economic Recovery Health and Social Services Housing Infrastructure Systems Natural and Cultural Resources

FEMA1

Working with Emergency Services

The worst time to meet your emergency services provider is during an emergency. Emergent meeting have a high risk of leading to misunderstanding, misinterpretation, and unneeded and unnecessary stress. The best case scenario is to use NFPA 1600 and the various planning tools available to incorporate emergency services as part of the PDCA process. Inclusion of emergency services will bring a new set of eyes on old problems that the internal planning team may have become organizationally blind to.

Many different organizations can help bring emergency services to the table:

- Local Emergency Planning Committees- Chemical hazard planning and mitigation expertise provided by a diverse group ranging from industry, government, to the media.

- Emergency Services Organizations- County or state firefighter associations. These associations provide a single point of contact for multiple provider agencies.
- Local and Township Officials Organizations- Council of governments that provide a single point of contact for area local officials.
- Trade Organizations- Chambers of commerce or other business associations that often provide planning or training resources to a large customer base.
- Professional Organizations- Local chapters of ASSE, SHRM, or other safety and security organizations that speak the language of emergency planning.
- Insurance Providers- Able to act as a third party to help build contacts with emergency services providers.

Many of these organizations have emergency planning as part of their charter functions or have tools and relationships developed with emergency services providers that can be leveraged.

Table B-2. Notional Fire Department Functions

Notional Fire Department Functions
Mission Essential Function: Fighting fires is the mission of the fire department and cannot be deferred.
Essential Supporting Activity: Keeping the fire trucks operating is an essential supporting activity. This must be done; it cannot be deferred. However, just keeping the fire trucks working does not provide the service that the community expects from the fire department. This **supports the mission.**
Deferrable Mission: Providing community fire prevention education and training is important and may be considered a fire department mission; however, during a disaster, this can be **deferred** until a later time.
Deferrable Supporting Activity: Providing retirement guidance to fire fighters is a good thing; but this is neither the fire department's mission nor something it should spend time and resources on during a crisis.

Table B-3. Functions Categorization Model Worksheet

	Essential	Non-Essential (during an emergency)
Mission	**Mission Essential Function** *Example: Respond to emergency Fire Department calls—fight fires*	**Deferrable Mission** *Example: Provide community fire prevention education*
Non-Mission	**Essential Supporting Activity** *Example: Keep fire trucks operational and ready to respond*	**Deferrable Supporting Activity** *Example: Provide retirement guidance to firefighters*

FEMA 1

Vulnerability Analysis Chart

TYPE OF EMERGENCY	Probability	Human Impact	Property Impact	Business Impact	Internal Resources	External Resources	Total
	High Low 5 ←→ 1	High Impact 5 ←→ 1 Low Impact			Weak Resources 5 ←→ 1 Strong Resources		

The lower the score the better

FEMA 3

Practice

"Culture eats strategy for breakfast."

--Peter Drucker

Why practice? Do you need to check a box for an audit, or is your organization committed to improving the chances of a success during an emergency?

Practice is a necessary regulatory business investment. Many regulations and standards require the leadership of an organization to make their employees rehearse scripted responses to emergencies. The most basic of these plans is the fire drill. You can check the box as required, or embrace the drill as an investment in protecting your people and operation to ensure the continuity of your organization. This investment is best exemplified by the actions of Rick Rescorla during the attack on the world trade centers on September 11th, 2001. Thanks to the leadership of Rick Rescorla, all but six of the 2,700 Morgan Stanely employees escaped the South tower.

"After the truck bombing that year, Rescorla had warned Hill: Next time by air. He expected a cargo plane, possibly loaded with chemical or biological weapons. In any case, he insisted on marching his troops through evacuation drills every few months. The investment bankers and brokers would gripe, but Rescorla would respond with his Seven P's: Proper prior planning and preparation prevents poor performance. He wanted to develop an automatic flight response at Morgan Stanley, to burn it into the company's DNA." ("Tower- A Tower of Courage," Michael Grunwald, Washington Post, October 28, 2001; http://rickrescorla.com/articles/a-tower-of-courage)

Culture will enable your organization to cash in on the practice investment. The key to culture is to find the basic few touch points that will drive the desired outcomes being reinforced by the practice. At the end of the day, as an organization, from the senior leadership at the top to the wrench turner on the front line, you have to have a shared vision of outcome. This shared vision needs to be developed as part of the planning and programming process discussed in the earlier sections of this narrative.

The basic goal of any practice is to develop capacity for people to successfully respond to emergency situation. You define success in the planning and programming phases of emergency management. For a fire drill, you have defined what emergencies initiate an evacuation, how the emergency will be reported, what alarm systems will be used to notify employees, and the actions employees are expected to take on observing the alarm systems. Practice gives employees the muscle memory and confidence to jump to action.

Practice can be developed using the Plan-Do-Check-Act model:

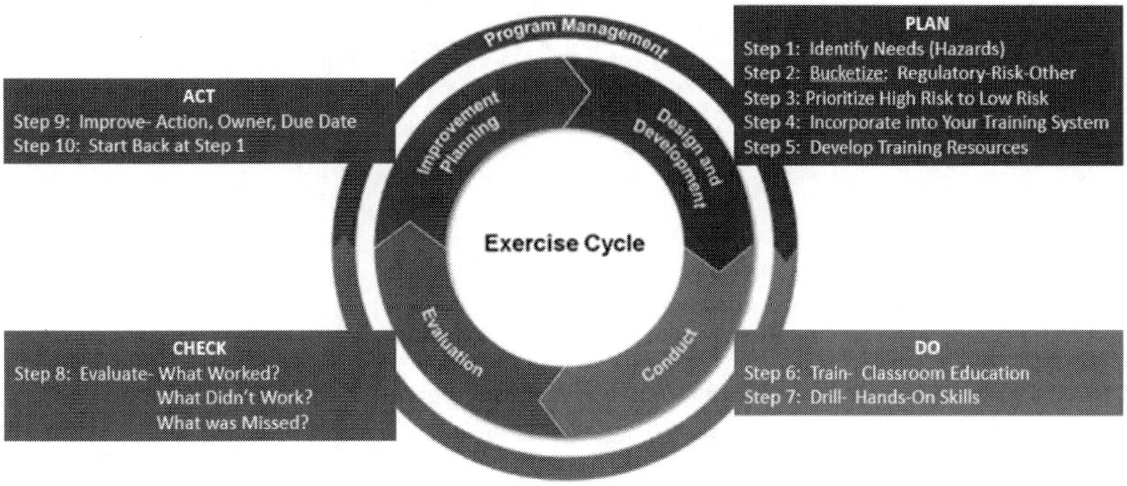

HSEEP1

Plan—Develop Practice Resources

Practice needs will be based on the risk assessment conducting using NFPA 1600. The practice content can then be developed to include goals, objectives, schedules, and resources.

- Goals: Defines what must be done and should be approached with the end product in mind.
- Objectives: Defines the steps, or the "how", to achieve the goals.
- Schedule: The schedule when the objectives will be met and can be both strategic and tactical. The strategic plan will lay out how the objectives will be met over time to reach the goals, exemplified by an annual training plan. The tactical schedule lays out the development of the drill or exercise at a step-by-step or hour-by-hour level.
- Resources: Resource identification is critical to a successful practice session to ensure that planning can be fully developed and implemented. The right knowledge base, subject matter experts to plan and participate in the practice session, is also critical.

This step can benefit from team planning and execution:

FEMA3

The Homeland Security Exercise and Evaluation Program (HSEEP) has several tools that can be used to assist with practice planning and development. Examples of these tools include a "master task list" that lays out an exercise framework and an exercise schedule that facilitates strategic annual exercise planning:

Master Task List [Exercise Name]

Exercise Planning Tasks	Responsible Party	Contact Information	Suggested Timeline	Date Completed	Remarks
Identify exercise design and development elements and begin development of exercise documentation • Scope, objectives, and core capabilities • Evaluation requirements (capability targets and critical tasks) • Scenario threat/hazard • Participants and extent of play • Exercise staffing requirements • Exercise logistics (date, location, including breakout locations or specific exercise play sites, if needed)			During IPM		
Assign responsibilities and due dates for tasks and determine date for next planning meeting			During IPM		
Develop and distribute meeting minutes			[NLT 1 week after IPM]		

DHS1

Do—Getting the Practice Done

Practice execution is a tactical task; however, practice participants must possess a basic working knowledge of the skills that will be demonstrated. Classroom training time and resource will need to be allocated prior to practicing or a process developed to evaluate participants to ensure an adequate skill level to allow for practice participation.

Practice can be in the form of a drill or an exercise and can be a tabletop, walk-through, or full scale:

Complexity ↓ Resource and ↓ Time Needs Increase↓	Drill: Small scale, able to be completed during a time frame of a single meeting slot. Examples: • PPE drills such as timed donning or doffing or successful removal of contaminated gloves. • Extinguishing an incipient fire in a burn pan using a portable fire extinguisher. • Completion of an emergency vehicle operator course. • Evacuation or life boat drill.
	Exercise: Larger scale practice of multiple drill elements, needing an extended time frame to complete. Examples: • Control of a simulated leak on a chlorine rail car in full PPE using a midland kit.

	• Multi-stage evacuation, protect-in-place, or shelter-in-place drill at a large multi-building facility.
	Tabletop: A simulation that primarily practices implementation of the incident command system using scale mock-ups, maps, photos, etc. and focuses on a question and answer interaction between the facilitator and participants. • Transcaer derailment response. • Evacuation and shelter plan activation after a large scale natural disaster.
	Walk-Through: Practice of a skill set that focuses on successful completion without full implementation of a plan or procedure. • Control of a simulated leak on a chlorine rail with identification of key action milestones, designation of ICS staffing, but no PPE use. • Layout of a full technical decontamination line without deployment of PPE or patient contact.
	Full Scale: Complete implementation of a plan or procedure, including use of PPE, ICS, and all necessary equipment to mitigate the emergency situation.

FEMA provides several resources that can be used to build practice sessions. A good example is the question and answer format used during a tabletop exercise is the FEMA Disaster Scenario Exercise for Organizational Planning Tabletop Exercise.

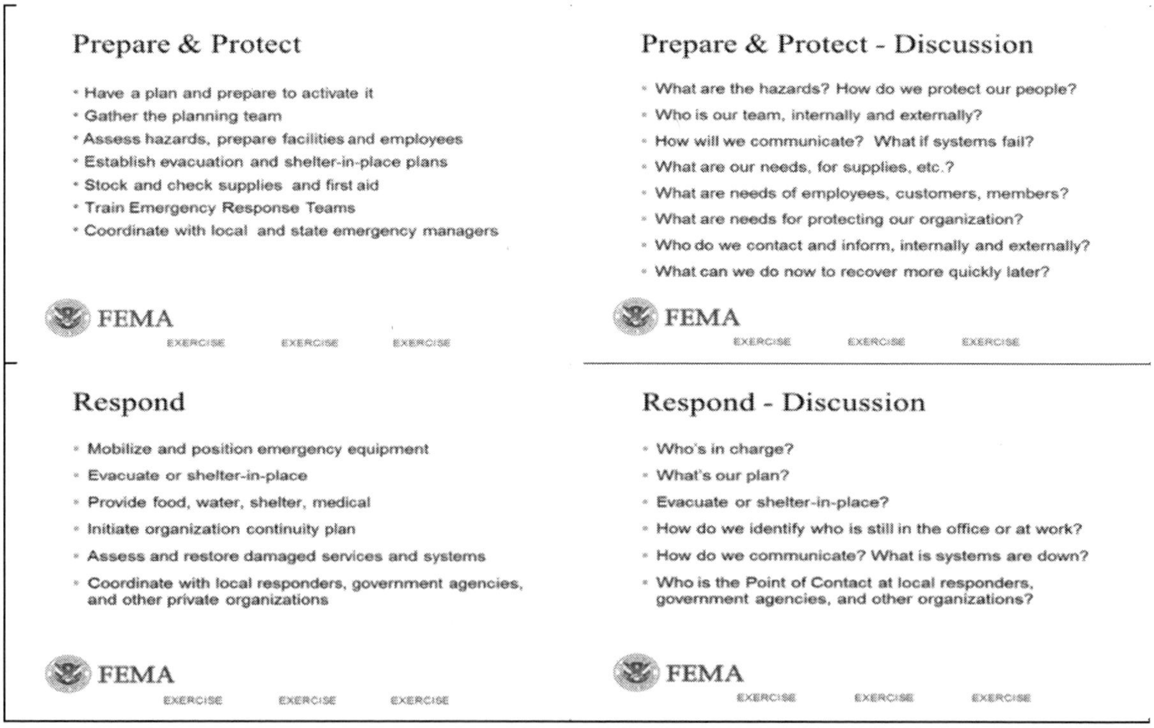

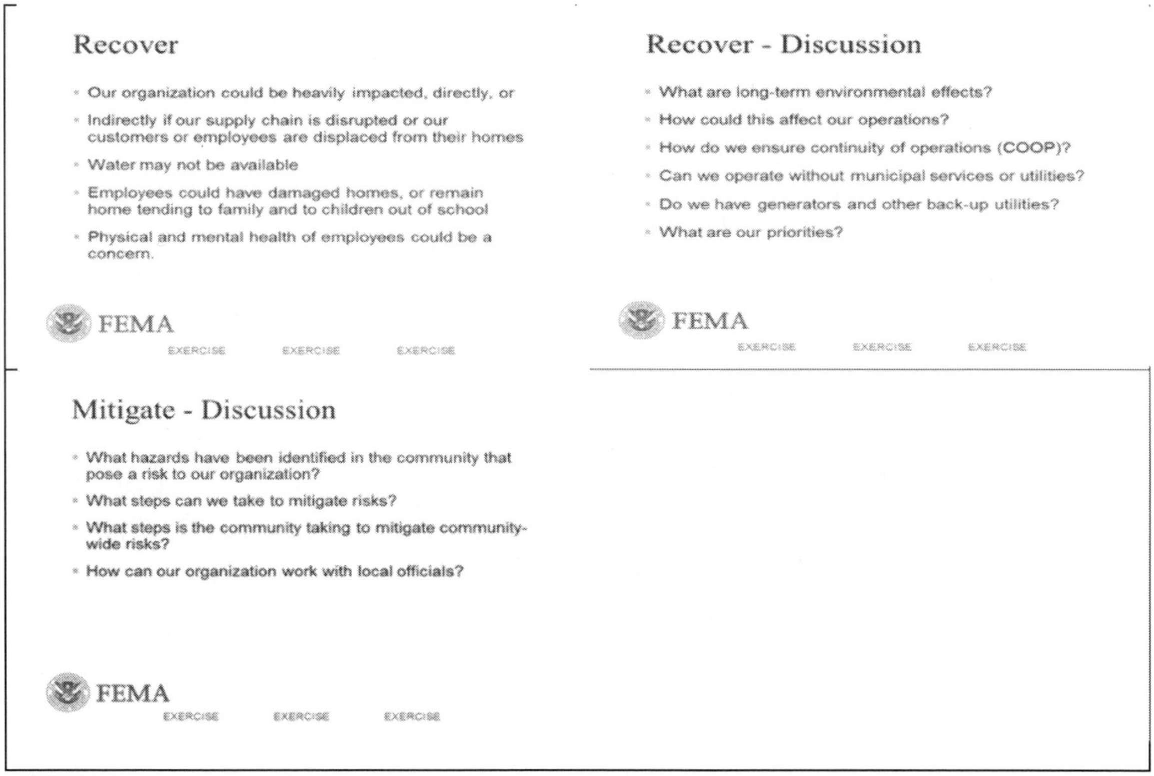

FEMA4

To ensure success, the practice session needs to be scripted out to define specific actions, action owners, and the sequence that each action needs to occur. The same tools used to plan an actual emergency response can also be used during the planning process to provide this script. Examples of these tools include the Incident Command System forms available through FEMA. Incorporating ICS into practice sessions will build capacity to work with responding emergency services providers to successfully implement mitigation strategies and to transfer command to the senior emergency response official.

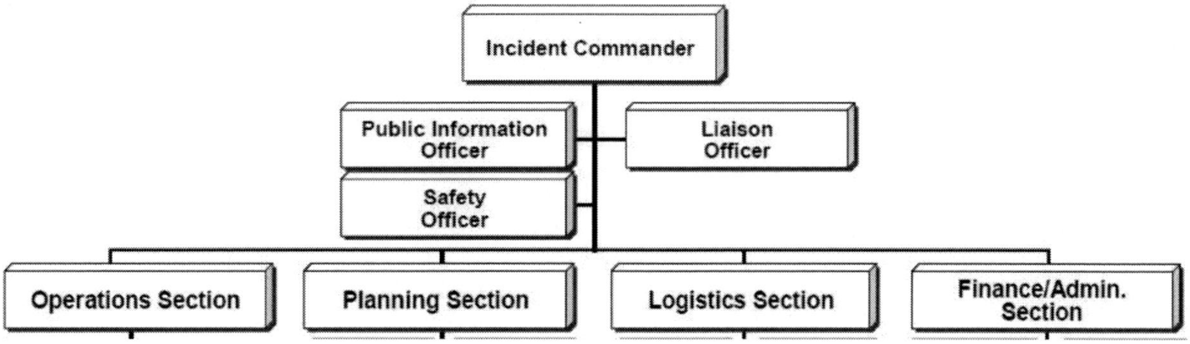

FEMA5

Act Three--Check

Practice cannot be concluded unless an analysis is conducted on how well participants met goals and objectives. There are many different types of analyses, or critiques, that can be used to define how responses can be improved.

- Hot-Wash--A briefing style review conducted directly after the conclusion of a drill. The briefing is conducted by a facilitator with a scribe taking notes. Notes are grouped: what worked well, what did not work, what was missed during the drill, and parking lot items that relate to the drill but were not contained in the main focus of the drill.

FEMA4

- Formal--A step-by-step review conducted soon after a drill. These reviews often make use of a checklist format and may be designed to meet regulatory requirements.

- Questionnaires--A paper or electronic survey designed to quickly gather basic information from exercise participants. These surveys are scripted to obtain an understanding of how the participants performed basic functions.

Evaluations need to be carefully scripted to keep the sessions on track and on time. A good rule of thumb is the **KIS**S principle. Keep the evaluation simple to stay on task to identify gaps and best practices. Keeping it simple will also keep the participants interested and engaged. The overall goal of the evaluation is to improve the process.

Act Four--Act

The evaluation step is only meaningful if follow-up actions are evaluated for implementation and communicated. Communication is key to validate participation in an evaluation. Participants need feedback on how ideas were or were not implemented. The improvement step will be integrated with the standard auditing follow-up for the business and can be as simple as assigning an owner and due date for each action. Actions need to be specific and tied to a regulatory requirement or hazard condition that was identified in step two.

<u>HSEEP Example:</u>

APPENDIX A: IMPROVEMENT PLAN

This IP has been developed specifically for [identify the State, county, jurisdiction, etc., as applicable] as a result of [full exercise name] conducted on [date of exercise]. These recommendations draw on both the After Action Report and the After Action Conference. [The IP should include the key recommendations and corrective actions identified in *Chapter 3: Analysis of Capabilities*, the After Action Conference, and the EEGs. The IP has been formatted to align with the *Corrective Action Program System.*]

Table A.1: Improvement Plan Matrix

Capability	Observation Title	Recommendation	Corrective Action Description	Capability Element	Primary Responsible Agency	Agency POC	Start Date	Completion Date
[Capability 1: Capability Name]	1. Observation 1	1.1 Insert Recommendation 1	1.1.1 Insert Corrective Action 1	Planning	State X EMA	EMA Director	Dec 1, 2006	Sep 1, 2007
			1.1.2 Insert Corrective Action 2	Planning	State X EMS System	EMS System Director	Dec 1, 2006	Feb 1, 2007
		1.2 Insert Recommendation 2	1.2.1 Insert Corrective Action 1	Training	State X EMA	EMA Director	Dec 1, 2006	Jan 1, 2007
			1.2.2 Insert Corrective Action 2	Systems/ Equipment	State X EMA	EMA Director	Dec 1, 2006	Mar 15, 2007
	2. Observation 2	2.1 Insert Recommendation 1	2.1.1 Insert Corrective Action 1	Planning	State X EMS System	EMS System Director	Dec 1, 2006	Jan 15, 2007
			2.1.2 Insert Corrective Action 2	Systems/ Equipment	State X EMA	EMA Director	Dec 1, 2006	Jan 1, 2007

FEMA6

References

DHS1- https://www.llis.dhs.gov/HSEEP/Documents/master-task-list

DHS2- https://www.llis.dhs.gov/HSEEP/Documents/training-and-exercise-plan-template

FEMA- http://www.fema.gov/blog/2014-10-15/why-your-community-should-race-wave

FEMA1- http://www.fema.gov/media-library-data/8ca0a9e54dc8b037a55b402b2a269e94/CPG201_htirag_2nd_edition.pdf

FEMA2- http://www.fema.gov/media-library-data/1386609058826-b084a7230663249ab1d6da4b6472e691/Continuity-Guidance-Circular2.pdf

FEMA3- http://www.fema.gov/pdf/library/bizindst.pdf

FEMA4- http://www.fema.gov/ppt/privatesector/fema_ttx_chemicalAccident.zip

FEMA5- http://training.fema.gov/EMIWeb/is/ICSResource/icsforms.htm

FEMA6- https://emilms.fema.gov/IS921/921_Toolkit/downloads/AAR-IP_Template2007.doc

HSEEP1- https://www.llis.dhs.gov/hseep

National Fire Protection Association. 2016. *NFPA 1600 Standard on Disaster/Emergency Management and Business Continuity/Continuity of Operations Programs*. Quincy, MA.: NFPA.

OSHA (Training)- https://www.osha.gov/Publications/osha2254.pdf

RMP- http://www.ecfr.gov/cgi-bin/text-idx?tpl=/ecfrbrowse/Title40/40cfr68_main_02.tpl

Strategies to Address Environmental Issues During Repurposing and Redevelopment

Maureen T. Hoke, P.E.
Vice President
OBG
Savannah, GA

Introduction

Managers at manufacturing facilities are typically, and appropriately, focused on active production and associated support activities. When economics necessitate repurposing or closure of a facility, this focus must change, and it can be a challenge for many managers to adjust to this change. Management of environmental and safety issues related to repurposing, in particular, can be a challenge. Failure to consider, or plan for these issues can trigger a cascade of events that can complicate the process and cost both time and money. Careful assessment of environmental conditions and regulatory requirements, and development of a detailed management plan early in the process can help facility managers identify, avoid, or control potential problems, resulting in more efficient and cost-effective implementation.

This presentation will outline a three phase approach that will lead to a comprehensive and coherent strategy for environmental management during repurposing and redevelopment. The key decision points at each phase of the process, and criteria that should be considered for these decisions will be identified. Ideally, the process begins, almost immediately after the decision to repurpose a facility is made, with a feasibility study that includes identification of the planned endpoint, evaluation of an approach or approaches to reach this endpoint, and preparation of a budgetary cost estimate and schedule for each approach. It is important to inventory environmental liabilities and assets affected, as well as relevant permits and regulatory requirements, and risk management parameters to help set the scope of the future facility assessment. It is also critical to understand how repurposing, or site closure could affect existing permits, require new permits, or trigger previously irrelevant regulatory requirements. During this phase, it is also important to identify critical environmental data that are incomplete or unavailable, and identify assessments required to obtain these data. Critical safety issues, particularly hazardous chemicals or facility spaces that will require special attention or greater levels of safety documentation and oversight, should also be identified. Armed with the information obtained during this phase, a preliminary approach and schedule for planning and implementation can be developed that will identify key milestones and decision points for subsequent phases. It is important to complete this initial phase as early as possible, as many environmental assessments or permitting efforts can require a long lead time.

Where needed, the feasibility study is followed by the facility environmental assessment. This includes design and implementation of assessments to characterize those areas (and only those) that will be affected by repurposing, demolition, and closure. This assessment will typically include building material assessments, asbestos surveys, and, possibly, soil and groundwater investigations in support of due diligence, closure of permitted environmental units, or known environmental liabilities. The facility environmental assessment should be detailed and comprehensive to avoid costly surprises and delays during later phases.

Permit closure, modification or application will typically begin in parallel with the facility environmental assessment, except when data from the environmental assessment are necessary for the permitting effort. This is a good point to engage the regulatory agencies and identify areas of flexibility, and areas where the agencies may be bound by statutory requirements.

Following the environmental assessment is the planning and mobilization phase. Detailed plans are developed for relevant areas of work, including: waste management, demolition, security, traffic, safety, site restoration, stormwater management, *etc.* Contractor bids for decommissioning, demolition and related construction contracts are then solicited using the plans and information generated to this point, evaluated, and awarded depending upon the owner's contracting strategy. If previous phases have been performed adequately, this will be straightforward, and it will be easy to roll into the implementation phase, which will also be straightforward, as long as plans are followed. While changes in approach can be made at any stage, beyond this phase they become disruptive and costly in terms of both schedule and budget. Finally, it is important to thoroughly document environmental activities (waste disposal, sampling, removal of impacted materials, etc.) during implementation, in order to avoid additional, after the fact, requirements and potential cost impacts.

The following sections will discuss the phases introduced above in greater detail.

Feasibility Study/Budget Development Phase

A feasibility study and overall repurposing/redevelopment budget should be established as early as possible. By thoroughly understanding internal goals, external opportunities, and overall objectives, facilities can start to develop a path forward. The desired end point should dictate planning and actions. Will the facility continue to be maintained or sold? Will infrastructure be repurposed or demolished for future use? Can materials be recycled or sold, or must they be disposed? The answers to these questions, and others like them, can help the owner identify whether elements of the repurposing/redevelopment can be passed on to a future owner, or be ignored entirely, given the anticipated end use. ***Even with uncertainty regarding the ultimate end point, or inevitable changes that will occur during the course of the process, early planning will lead to the development of a realistic timeline and identification of key decision points that will keep the process in control.*** Having a defined end point also provides a meaningful constraint on budget forecasting.

A third-party consultant can help develop these early plans, budgets and schedules, working with the site owner or manager and their in-house environmental staff to identify and incorporate the owner's internal goals and objectives. Early planning is the most important phase. Engagement of an experienced consultant can identify opportunities for repurposing/reuse of materials that will provide savings on the overall project and assist the owner in avoiding pitfalls and liabilities common to these types of projects. Property and assets may be more or less valuable than anticipated, or they might have value in ways not initially considered. Internal costs that would normally be ignored may become budget drivers. The budget needs to be fit to a realistic time frame for repurposing/ redevelopment. It may be that idling a site is costlier than demolition, if the budget is considered over an appropriate time frame.

Even high-level assessments prepared for the purpose of identifying high-spot cost estimates can be helpful in decision making relative to whether to proceed with facility restructuring/ repurposing. Such evaluations would include, in addition to evaluations of plant and equipment value, property value, engineering and support services costs, and environmental liabilities. In such an evaluation, assumptions or decisions regarding the end state of the facility must be made early on:

- Will slabs and foundations be left? Would this be allowable? Is environmental contamination present within these materials and what will the liability be if they are left in place or removed?
- What will the zoning or property use at the end of the process be?
- Will the evaluation cover the entire facility or only portions thereof?
- What are the known environmental liabilities? How should unknown environmental liabilities be assessed?
- Is there value in maintaining existing permits? Is there value in maintaining existing features of the facility, such as on-site rail, power substations, pipelines, port/waterfront facilities, etc., or other features that must be maintained/not damaged during demolition activities?

Assessment and Implementation Planning/Permitting Phase

Assessments necessary to characterize site conditions relevant to repurposing/redevelopment should be performed as soon as possible, so that data gathered can be used to refine the scope, schedule, and budget. The assessments should be designed to match the proposed endpoint. Before performing assessments that will generate new data, detailed reviews of historic data should be conducted. This can significantly reduce expenses and streamline the development of new data to fill remaining gaps.

Assessments should be designed to address regulatory requirements and support key decision-making. The generation of data should be done with purpose and legal counsel may also need to be involved when deciding on the extent of subsurface investigations. Investigations can be designed in a phased approach, where the results of initial phases can be inputs into decisions regarding whether additional phases are required or whether final decisions can be made on the results of the current phase.

Development of detailed, clear, and well-organized plans both for assessments and the repurposing/redevelopment implementation to follow will help to avoid missteps and maintain an efficient schedule; however, it is important to be flexible, as conditions may change. Data gathered should be reviewed in real time, and the impacts to the scope, schedule, and budget should be continuously reevaluated to verify that the program is on track. Communication between the owner and the third-party consultant contracted to conduct the assessments is critical at this stage.

Plans should also include consideration of unknown conditions. Planning for the fact that unknowns will be discovered during the project, and providing a process for systematically dealing with such situations can prevent confusion, significant schedule delays and unnecessary cost increases.

Typical plans developed for large, campus-wide repurposing/redevelopment projects may include:

- Health & Safety
- Facility Assessment
- Hazardous Materials Abatement
- Demolition
- Soil Management
- Waste Removal/Management
- Traffic Control
- Site Security
- Regulated Area (may use other terminology depending on regulatory programs) Closure

Depending on the type of facility, assessment planning can include a variety of items or subjects. Production equipment can often be the source of environmental contamination, or, more commonly, waste that needs to be disposed as part of the repurposing implementation. This would include:

- Oil-filled units, which can potentially include polychlorinated biphenyls (PCBs), such as transformers, capacitors, and hydraulic units
- Equipment containing chlorofluorocarbons (CFCs), such as coolers, air conditioning units, drinking fountains
- Universal wastes and electronics, such as light bulbs; batteries; mercury-containing switches, thermostats and other items; capacitors; televisions and computer equipment
- Asbestos-containing materials, including, but not limited to, thermal systems insulation (TSI), gaskets, caulks, floor tiles, and ceiling tiles
- Lead-based paint and other lead-containing items
- PCB building materials (caulks)
- Radioactive devices
- Concrete (floors, walls, trenches, basins, *etc.*)
- Fluid filled pipes, vessels and tanks
- Hazardous and non-hazardous wastes.

Various agencies likely have jurisdiction over different aspects of the facility. A clear and concise understanding of applicable permits and regulatory programs is critical so that a plan for addressing them can be developed and communicated to the agencies. It is best to begin engaging the regulatory agencies early in the process. ***Collaboration at an early stage can help to avoid problems later.*** While high-level involvement is key, it is also important to include the rank and file agency personnel that will implement programs and conduct site inspections/audits to avoid surprises and delays.

Examples of typical environmental permits and regulatory programs that may be applicable to a facility include:

- Air (Title V or other)
- Storm water discharge
- Waste water discharge
- Ground water withdrawal/use
- Surface water withdrawal
- Cooling water discharge
- Resource Conservation and Recovery Act (RCRA) / Hazardous and Solid Waste Amendments (HSWA)
- RCRA hazardous waste generator (large/small quantity)
- Universal waste generator (large/small quantity)
- Oil Spill Prevention Control and Countermeasure (SPCC)
- Asbestos abatement
- Erosion and Sedimentation Control (ESC)
- Surface disposal/landfilling
- Wetlands

Repurposing / Redevelopment Implementation Phase

With proper planning, implementation should be straightforward. ***The most important aspect of implementation to control cost and schedule is to stick to the approach and sequencing established during the planning phase.*** Delays cost money. Work done out of sequence may affect the performance of subsequent tasks, and can also increase costs. Flexibility can be accommodated when conditions warrant, but it is important to remember that changes may affect the schedule and budget.

Especially for larger repurposing programs, having resources on-site during the implementation phase that are dedicated solely to oversight and documentation of environmental programs is crucial. From the first day of mobilization, making sure that contractors understand the environmental and safety constraints of that particular facility and the sequencing with which the project must be implemented to orchestrate the plan is critical to success. The environmental management/oversight is necessary to:

- coordinate between construction trades (demolition contractor, abatement contractor, grading contractor, disposal contractor, *etc.*) with respect to environmental and health and safety items/issues
- track and document waste removal and disposal
- maintain files to document appropriate work training and required certifications, based on the potential exposure for each trade/craft per task
- inspect waste storage areas, per regulatory requirements, and implement corrective actions as necessary

- coordinate/communicate with regulatory agencies
- oversee demolition contractors to verify wastes are handled safely, stored in appropriate containers, and that wastes are removed prior to demolition activities.
- observe demolition activities and respond to identification of previously unknown environmental conditions (*i.e.*, buried materials/contamination identified during excavation of foundations)
- manage response to unforeseen releases to limit impacts to surrounding environment.
- provide information/documentation for compliance with permits and programs

Case Studies

The following provide two case studies to illustrate the process above, or how not following the process above can create additional problems.

Case Study # 1 - Pharmaceutical Manufacturing Facility, Georgia

The project involved a 1,000,000 square foot (ft^2) former pharmaceutical manufacturing facility, located on 105 acres within a 1,000-acre tract of property along a major river. The facility was originally constructed and began operation circa 1915. The initial facility assessment and plan development for the repurposing program were conducted by a third-party environmental consultant. Decommissioning began in 2007; abatement and demolition were initiated by a demolition contractor in 2010.

Due to multiple contract and safety concerns, the original demolition contract was terminated and a design-build contract was procured for further development of plans and implementation of the repurposing. The contract was awarded as a Construction Management At-Risk (CM At-Risk) contract to an engineering/environmental and CM firm with subcontracted demolition and abatement partners in July 2012. The facility owner was looking to make up the lost time and money spent since the original contract was terminated, and complete the project as quickly and cost-effectively as possible.

The new team started at the beginning, taking work completed by the original demolition contractor into account. Existing documentation was reviewed. Detailed plans were developed: health & safety, facility assessment, facility abatement, demolition, soil management, waste removal and traffic control; and notifications and permit applications were submitted to the appropriate regulatory agencies.

Since previous assessment activities had been limited in scope, a comprehensive assessment was performed, which included over four hundred samples of concrete floors, walls and trenches for waste characterization, as well as verification of previous asbestos surveys and supplemental asbestos sampling where needed. Over two hundred additional samples of TSI were collected to supplement existing asbestos survey data. The facility was surveyed and items of environmental interest were noted and included in the abatement and demolition planning processes.

The scope of abatement and demolition activities also included environmental compliance throughout the project: asbestos abatement, universal waste abatement, vertical demolition, waste management/removal, horizontal demolition, site restoration and closeout. Environmental compliance required providing full time environmental oversight; maintaining spill prevention and control; supervising line breaking and cleaning of pipes, equipment and tanks; tracking environmental items of inventory to final disposition; and responding to discovery of unknown environmental issues throughout the vertical and horizontal demolition progress.

The project was completed in the fourth quarter of 2013, ahead of schedule and under the owner's original budget. Detailed planning, including planning for unknown conditions provided a framework to keep the project on track and an ultimate success-story, even with the initial set-backs due to termination of the first demolition contractor.

Case Study #2 - Pharmaceutical Research & Development Facility, New York

The project involved an operating pharmaceutical manufacturing campus. The campus was comprised of two portions: an active manufacturing facility and a research and development (R&D) facility, both of which included numerous buildings and ancillary structures. The entire campus was to eventually be repurposed, but it was to be done in two phases, with the R&D facility being demolished first.

For the repurposing of the R&D facility an environmental consultant that had experience working at the facility was retained to complete hazardous building material surveys prior to decommissioning and demolition activities; however, PCBs were not delineated prior to the demolition activities. The demolition contractor removed the concrete foundations from the structures and began stockpiling them onsite pending landfill approval for disposal.

Characterization of waste materials required by the receiving landfill prior to disposal identified PCB concentrations in stockpiled concrete from floors and foundations at levels above regulatory criteria for Toxic Substances Control Act (TSCA) characterization as hazardous waste. By this time, a significant volume of stockpiled material had been accumulated, and the designation of the material as TSCA hazardous for PCBs necessitated very specific remedial activities and requirements, adding cost and schedule implications to the project. A second third-party environmental consultant was brought in to provide support in further characterizing and designing a remedial action plan for the TSCA hazardous material.

When the owner was in the planning stages for repurposing of the second phase – for the active manufacturing facility, the second third-party environmental consultant was retained and a significant planning effort was expended to characterize and delineate hazardous building materials, including asbestos, lead and PCBs, prior to the development of abatement and demolition plans, cost estimates and schedules. This strategy resulted in minimization of abatement and disposal costs resulting from TSCA hazardous and other wastes, as the extent of those materials was delineated prior to removal to allow for sufficient segregation during abatement and demolition activities.

Conclusion

Repurposing/redevelopment is likely to occur at any given facility at some point, whether it be a full campus repurposing or merely a single building or process area. Management of environmental considerations inherent to repurposing can be challenging for facility personnel, especially those whose main focuses are the business aspects of the project. However, failure to adequately plan for these considerations can trigger a cascade of events that can complicate the process and cost both time and money. Careful assessment of environmental conditions and regulatory requirements, and development of detailed plans that consider and include steps for identifying the inevitable unknown conditions early in the project can help facility managers identify, avoid, or control potential problems; resulting in more efficient and cost-effective implementation.

Multifactorial Approach to Reduce Workplace Musculoskeletal Disorders

Lori L Peacock, MSPT, CPT, EPN
Physical Performance Solutions
Aiken, SC

Introduction

Musculoskeletal disorders, more commonly known as MSDs, are of a complex etiology, for the origin is as different as every individual. This salient ingredient in safety and health of the employee and company continues to evade typical programs. There are other arenas of business – sports in particular – where there is a strong focus on preventing and maintaining the health of the employee - the athlete. Other industries are no different: oil production and exploration, assembly manufacturing, food processing, farming and construction. All industries and businesses have the same denominator as sports businesses: the employee is an athlete. Therefore using proven preventative sports medicine ideas and techniques is the next wave to subdue musculoskeletal injuries in the workplace. This talk delves into why individuals are vastly different, why this is now more important than ever to incorporate proven, approved methods to a company's existing safety program and to utilize ways for a team approach in a worker's physical safety.

We will also discuss how this type of preventative program changes emotions in workers and look at the current studies on how emotions really affect and effect the worker in the work place.

Musculoskeletal Disorders, MSDs

MSDs continue to be a leading cause of pain, suffering, and disability in the American workplace In a recent publication from the Occupational Health and Safety Administration, OSHA, it is reported that approximately 40% of injured workers are not receiving any type of financial compensation from work related injuries; thus burdening the family, our social system, other subsidized programs and private insurances.

As a clinician, I wonder how many "accidents", not classified as a MSD, are really the result of a MSD that is not reported due to the lack of early intervention programs in the workplace, or the notion of the worker and the company, that the soft tissue complaint is from an "old football injury in high school".

There are business organizations in the US that do realize the significance that a prior injury can affect physical performance and increase the risk for a future injury. Which ones? Professional sports teams!

Although many of us do not think of professional sports as a business, it is. The professional athletes we are accustomed to viewing on the TV screen and the franchise owners also must use workers compensation regulations when a player becomes injured on the field. This becomes complicated as teams travel to different states to play. As an example, a player becoming injured in a state other than where they live and the franchise home office follows various state rulings.

The Reality of Costs

Total costs from one soft tissue injury, without surgical intervention, will vary depending on the following, but not limited to: size of the company, the surrounding events that lead to the reported injury, the job type, or the number of soft tissue injuries that have already occurred in that year. The worker, supervisors, medical department and most environmental health and safety and human resource departments typically do not think of all the costs generated, or how much more product will need to be sold in order to make up the difference for that one soft tissue injury. OSHA and the Bureau of Labor Statistics, BLS, gather information and post on the web for our use. One such site, https://www.osha.gov/dcsp/smallbusiness/safetypays/estimator.html, provides estimates of possible costs from work related injuries. Two MSD type of injuries are located in Table 1. Although your particular job in your company may not be in the financial department, knowing your company's profit margin will certainly help you gain a better understanding of the true costs of one soft tissue worker injury.

Type of Injury	Direct Costs	Indirect Cost	Total Costs in Sales for Company to Recover the One Loss for both Direct and Indirect Costs at 8% Profit Margin
Carpal Tunnel Syndrome	$30,509.00	$33,559.00	$800,850.00
Strain	$33,140.00	$36,454.00	$869,925.00

Table 1. Worker injury calculator.

(From: https://www.osha.gov/dcsp/smallbusiness/safetypays/estimator.html)

The information reported above are estimates; however, the information yields the grim reality of what our companies face in monetary terms of just one soft tissue worker injury. This information is becoming all-important as our workforce is aging.

What Other Business Arenas are doing to Reduce Injuries

Years of research among high school and college athletes tells us much: that the likelihood of an injured area will be injured again. As a health professional in the realm of physical therapy, the information of previous injury, whether there was surgical intervention or not, is a large piece of information: it will help guide all movement interventions in order for the client to obtain the best physical outcome and education about themselves. Business owners of professional sports teams are well aware of this and are now utilizing a multifactorial approach with their players.

Utilizing the Information Safety Professionals have already

Some current knowledge on MSD injuries are:

- Work Risk Factors
  - Forceful Movements
  - Awkward Postures
  - Static Postures
  - Tedious Repetitions

Combined with:

- Individual Factors
  - Eating Habits
  - Physical Fitness
  - Work Habits
  - Previous Injury or Surgery
  - Age

All of the above increase the risk for development of adhesions, that we don't feel, yet will eventually cause in anyone of us fatigue, stiffness, discomfort, movement compensation, pain and loss of function. The question is how do adhesions develop, how many or how much does one need to have before an injury, where are they in the human body, how easy are adhesions resolved, and the questions go on.... The best place to begin is with a general knowledge of the human body in terms of soft tissue.

The Human Body: Same yet different

The growth rate of safety professionals in industry is long overdue and greatly appreciated; large companies may also hire ergonomists and athletic trainers as well to assist in preventing injury and increasing physical abilities of workers. In order to truly begin to lower the rate and occurrence of MSDs, safety and medical professionals in industries must use a Multifactorial approach, which will include either hiring a new facet of professionals or using outside professionals to assist in utilizing components not currently considered mainstream in many industries. This component has to do with the human body, the fascial system and how unseen alterations of this system create the potential for physical injury and metabolic changes.

By adulthood, most of us will have 206 skeletal bones, somewhere between 640 and 850 skeletal muscles, and approximately 100,000 miles of arteries, veins, and capillaries![1] While our nerve impulses travel approximately 324 mph, we are constantly finding that the micro-adhesions, scars, compression and other forces acting somewhere on the nerve decrease this speed significantly. Between the nerves of our nervous system, digestive tract and fascial connections, and their variations and characteristics, it should be no surprise that the area of MSDs continue to be difficult and challenging to prevent.

The fascial system is an area of more recent study and research, and clinicians and health professionals are increasing their understanding with what fascia really is and how strongly it influences what we typically know as muscles. Many researchers regard this tissue as an organ of the body as well: that there are many ways fascia internal structures function, influence and

---

[1]https://www.fi.edu/heart/blood-vessels

move. As safety professionals, the urgency of learning basic biological information of the human body and learning the best resources to hire for the company is now here.

The Salience of learning and using a general knowledge of the human body

Through years of dedicated research and study, there are multiple reasons for our own unique physical makeup and the changes we experience in our lifetime. Table 2 is a very generalized summary of changes of soft tissue the human body can experience and possible physical affects for both men and women. (Chaudry, H. et al., 2016)Whether someone works in administration or in a more physical job such as manufacturing, food processing, material handling or other, the physical abilities of that person are important. By learning generalized currently-known and accepted information about the human body, followed by how to apply this information in a work situation, true early intervention and prevention of a worker can be accomplished.

General Human Characteristics	After Age 30	Generalized Reasons
Soft tissue healing	Increased time for soft tissue to heal	• Increased time for soft tissue healing from altered blood flow, • Changes in hormone levels, • Diet rich in processed foods, • Multiple fascial adhesions • Increased age
Physical Grip Strength	Decreases	• Fascial adhesions • Altered nerve conductivity • Altered blood flow • Decreased oxygen to cells • Increased age
Stamina	Decreases	• Emotional stressors • Nutrition rich in processed foods • Altered hormones from aging • Fascial adhesions • Increased age
Soft tissue flexibility	Decreases	• Decreases due to hormonal changes, • Fascial adhesions, • Decreased blood flow • Increased age
Nerve conduction speed	Decreases	• Slows with aging due to thinning of the lining around the nerve, • Fascial adhesions, • Decreased oxygen, • Decreased blood flow • Increased age

Table 2: General characteristics of some reasons for soft tissue physical changes over age 30.

There is a common denominator in Table 2: fascial adhesions and increased age. We know and understand a lot about the aging body, however, research is now understanding more about what fascia is and does in the human body. Some interesting points we now know about fascia:

- Fascia is connective tissue that ultimately gives us our shape
- The several forms of fascia ultimately cover or wrap around every structure
- Fascia has the ability to contract, similar to that of muscle
- Fascia has more nerve endings than does muscle

The implications for opportunity for enhanced prevention in the workplace now opens up in light of this information to the medical, safety, ergonomist and engineering design professional. No longer can this information be kept solely in the medical field reserved for physical therapists, physicians, chiropractors, bodyworkers or the like. As previously mentioned, our workforce is aging and still working; therefore, the likelihood of MSDs increasing and the general costs increasing due to what is now known about the tissue changes that begin to take place after age 30.

Learning and Applying Injury Prevention Techniques from other Industries

Many of us enjoy our sports! It may be US football, basketball, baseball, or other activities such as skiing, bull riding or gymnastics. What many of us don't think about regarding "our" sport, is that "sports" is an international business competing for your business. When you think of your favorite college or professional athlete, the organization, or business administrators, need to view that athlete, at times, as an operating expense. The athlete costs money weekly; and, the athlete is making a lot of money in exchange for assisting the team to win the championship and increase sales for that organization. The business owners and administrators understand that this athlete must be kept injury free in order to compete at the highest level possible week in and week out. What are these franchises doing to prevent an injury of their athlete? How does this treatment comply with OSHA standards?

A New Paradigm in Biometrics

What many professional and college sports are engaging in to prevent their athletes from injury is multifactorial. They train and issue proper safety equipment for all players, use computer applications for continuing education for each player and utilize a third component: their biometric data.

Biometric screening is the measurement and collection of certain physical characteristics of the human body. Although many industries have adopted this wellness approach, the sports industry has now expanded their biometrics to include much more detail of the athlete such as, but not limited to:

- Height
- Weight
- Sleep patterns

- Movement assessment
- Strength with movement assessment
- Nutrition

The ultimate goal is to prevent an injury and to prevent further trauma from an old injury. Professional sports teams are now partnering with companies that have the technology and background of biomechanical movement and movement retraining that use a methodological approach as one aspect of injury prevention. Combined with other provisions of all types and forms of safety equipment, professional sports organizations are equipping themselves to detect the player "at risk" and to therefore make the necessary corrections prior to an injury.

Benefits of Using a New Type of Biometrics in the Work Place

Many companies use post-offer physical ability testing, especially in work areas that will have a physical component to the job. What we are seeing in other business arenas, such as professional sports is extensive physical biometrics performed on the player as soon as they are hired. This information is more valuable as it will take into consideration all one's strengths and weaknesses. Even without previous injury, there will be some movement or task that is more difficult to perform. The perceived weaknesses can therefore be resolved, thereby reducing the risk for soft tissue injury. Many industries can benefit from this approach. The return on the investment, in addition to the savings of a workers' compensation injury is large and broad. A few benefits are listed below:

- Decrease number of worker injuries
- Decrease experience modification rates
- Better productivity
- Better quality of product made
- Increased sales of product
- Better company culture
- Increased quality communication between administration and workers

OSHA: First Aid Massage

As a company embarks on their mission to further reduce MSDs in their workplace, a review of OSHA is first on the list. Currently, most companies rely on the worker to report physical discomforts, soreness, and pain to their employer. At this point, a medical professional, typically an occupational health nurse who has expertise in the work field, begins a triage with the employee.

At the writing of this paper, OSHA regulation, 1904.7 (b)(5)(ii)(M) allows the use of massage as an acceptable form of first aid treatment for the worker. Following on the first two segments of this paper, reducing fascial adhesions and using preventative sports injury techniques, the use of first aid massage becomes a powerful tool for the company and their employees to help reduce soft tissue injury. Clinical health facilities and individual companies that use first aid massage continue to indicate that using first aid massage is an effective means to reducing soft tissue discomforts. Companies often utilize outside sources for this type of early intervention. These specialized companies show that the return on investment is high in addition to objectively showing increases in worker physical abilities and the workers' reported wellbeing.

As with any health and wellness prevention program, the success depends on:

- The employer's ability to engage employees
- The employee's perception of value
- The employee's perception regarding positive health changes
- The Return on Investment

There is a strong additional positive component when adding in first aid massage with an early intervention program and that is the improved psychological outlook of the worker regarding themselves and the company they work for. This is partly due to the lasting physical benefit of increased range of motion, strength and decreased fatigued at work and at home. More research is currently being conducted across the globe on how emotions affect job performance. Many results are stating that when workers feel that their employers are doing their best to keep them healthy, the worker is happier and more productive.

What We are Learning about Movement

We have an opportunity to learn from our young adult athletes. Their tenacity to get back into the sport that caused an injury is to be admired. Others to be admired are, again, the researchers who tirelessly look at components of human systems to find out the "why". One of the "whys" always being looked at is why was normal movement not fully restored after the injury? Today's technology allows us to peer into the body to look at how muscles are functioning in real time. We now have the ability to view how muscle firing patterns are taking place during movement. The interesting information is now we can now see if muscles are not firing at a normal rate, or if they are firing in a pattern that is reported to be the norm. An even more interesting fact to this is that the person performing the movements are typically not aware that the muscles are behaving abnormally. This is where preventative sports medicine is going: to locate dysfunctional movement patterns through a system of established movement tests combined with using the latest in technology to find muscle tissues that show a problem, and resolve the problem before an injury occurs.

Using Surface Electromyograph in the Work Place

Surface electromyography (SEMG) is a technique in which electrodes are placed on (not into) the skin overlying a muscle to detect the electrical activity of the muscle. Using the equipment does not hurt and it can be used in most work situations, as many of the products are portable and do need the use of cables to attach to the computer. This equipment is used in large institutions and have worked alongside many large companies, such as Toyota and other developers, to assist in design and research of new products.

An astonishing aspect to Surface electromyography all of this information is that it can now be brought to the worker, and that a majority of the information gathered and utilized and problem solving is really a life event. In other words, all that is accomplished for the worker and company, actually goes far beyond the confines of the company. All workers will reap the benefits outside of work into their lives; bringing about a true enhancement of physical wellness and wellbeing.

Developing Teams and Strategies in the Work Place

Using teams in the workplace is invaluable. Input and ideas from across the company can propel ideas into highly sought after reality. In the realm of MSDs, developing a cohesive team is a necessity, as individuals from the actual worker, to EHS, medical, engineering and management will be on the same page as this last frontier is explored. A salient component to the team will be an individual that has an excellent working/professional knowledge of movement and human tissues. As most companies do not yet employ such a person, locating companies that excel in the biomechanics, ergonomics and the resolution of soft tissue dysfunction is a key element.

Once your team is in place, begin to use what you have and already know, share the information and begin to build your strategy to locate missed opportunities at work stations, morning stretch routines, and how current worker discomforts are being communicated to team. The information you are gathering will be well utilized when it is placed into your company's existing metrics. We don't like to view ourselves as a number, however using metrics is a good and valid method to show changes in a company.

One example how teams can collaborate information and use it to make changes regarding soft tissue risks is the following. At Physical Performance Solutions, LLC, we travel to companies to assist in reducing MSDs through using the preventative techniques used in sports medicine. Collecting information is important. Some data that is viewed regarding company A.M. stretches is: what types of morning stretches are used, how often, how many workers participate and are the stretches truly beneficial for those work stations? Adding this information with technology such as SEMG, small, but important changes in a morning stretch program is very valuable in a company. Sometimes deleting a stretch, or adding one simple movement can create a large difference in the workers' physical wellness.

Conclusion

Utilizing the advancements of technology and what other business arenas are accomplishing to reduce workplace MSD injuries will certainly advance the company, it's culture, and the employees' wellness. Reducing the MSD is achievable; and quickly too. Much of the information and tools are in place in every company. They need to be utilized in a fashion that the best possible results are noticed by everyone. What a company may lack, resources of outside companies to assist them are close by and affordable.

Bibliography

Bahr, R. H. 2003. Risk factors for sports injuries—a methodological Approach. *Britsh Journal of Sports Medicine, 37*, 384-392. Retrieved from www.bjsportmed.com

Burn, B. "What is a Biometric Screening- The Process Explained," Retrieved February 19, 2017.from What is a Biometric Screening- The Process Explained - Houston Personal Trainers - 1st Class Training: http://www.1stclasstraining.com/what-is-a-biometric-screening-the-process-explained/

Carney Law Firm. *Carney Law Firm: Worker's Compensation for NFL Players,* 2017, March. Retrieved March 3, 2017 from Carney Law Firm: https://www.carnlaw.com/workers-compensation-for-nfl-players.html

Chaudry, H. et al. Muscle Strength and Stiffness in Resistance Exercise: Force Transmission in Tissues. *Journal of Body Work and Movement Therapies*, 2016.

Congress, K. S. 2012. Changes in Fascia Releated to Repetitive Motion Disorders - Questions and Answers [Recorded by M. F. Barbe]. Vancouver, BC.

Congress, L. A. 2012. Link Between Manual Therapy, Movement, Fascia and Cancer [Recorded by T. M. Findley]. Vancouver, BC, Canada.

Elhert, G. 2001. *Length of the Human Intestine.* Retrieved February 18, 2017 from The Physics Factbook: http://hypertextbook.com/facts/2001/AnneMarieThomasino.shtml

Emamdjomeh, Armand and Ken Bensinger. *Business,* February 1, 2014. Retrieved March 3, 2017 from *LA Times*: http://www.latimes.com/business/la-fi-nfl-claims-20140201-dto-htmlstory.html

Ferrario, B. "The Latest Sports Technology Available to Prevent Sports Injuries." Retrieved February 19, 2017. from Alliance Technoloty Partners: Blog: http://alliancetechpartners.com/the-latest-sports-technology-available-to-prevent-sports-injuries/

Findley, T. E. Transmission of muscle force to fascia during. *Journal of Bodywork and Movement Therapies*(19), (August 2014). 119-123.

Hackbarth, N. et al. March 2017. *Work Place Well Being.* Retrieved from Limeade: http://www.quantamworkplace.com

Improving Workforce Health Productivity. 2016. Retrieved from Willis Towers Watson: http://www.willistowerswatson.com

Langtree, I. "500+ Amazing Human Body Facts and Trivia," Retrieved February 18, 2017. from 500+ Amazing Human Body Facts and Trivia - Disabled World: https://www.disabled-world.com/medical/human-body-facts.php[2/18/2017 2:42:41 PM]

Meeuwisse, W. M. 1994. Assessing Causation in Sport Injury: A Multifactorial Model. *Clinical Journal of Sports Medicine*(4), 166-170.

Michaels, D. *Adding Inequality to Injury: The Costs of Failing to Protect Workers on the Job,* (June 2015).Washington DC: Occupational Health and Safety Administration.

Moore, K. L. 1995. *Essential Clinical Anatomy.* Baltimore: Williams & Wilkins.

National Research Council and Institute of Medicine. 2001. *Musculoskeletal Disorders and the Work Place: Upper and Lower Extremities.* Washington DC: National Academy Press.

OSHA. *USDOL Regulations and Standards*. Retrieved February 19, 2017 from OSHA: https://www.osha.gov/pls/oshaweb/owadisp.show_document?p_table=STANDARDS&p_id=9638

OSHA.gov. 2014. Retrieved from OSHA: https://www.osha.gov/pls/oshaweb/owadisp.show_document?p_table=UNIFIED_AGENDA&p_id=4481

Oswald A. J. et al. 2014. *Happiness and Productivity*. Warwick: University of Warwick.

Peacock, L. "Soft Tissue Injury Preventions," (September 9, 2016) Aiken, SC, U.S.

Preidt, R. "For Many College Athletes, the Payoff is Lifelong Disabilities: Study," *U.S. News*, (2014).

Pronk, N. Placing Workplace Wellness in the Proper Context: Value Beyond Money. *Preventing Chronic Disease, 11*, (July 10, 2014). p. 4.

Reaston, M. (2015, August 11). "Better Approach to Soft Tissue Injury," (August 11, 2015). Retrieved May 11, 2016 from Insurance Thought Leadership: http://insurancethoughtleadership.com/tag/electrodiagnostic-functional-assessment/

Reaston, M. et al. 2011. Musculoskeletal disorders early diagnosis: A retrospective study in the occupational in a work setting. *Journal of Occupational Medicine and Toxicology*, 6-11.

Schleip, R. E. 2011. Strain Hardening Fascia: Static Stretch of Dense Fibrous Connective Tissue can induce a temporary stiffness increase accompanied by enhanced matrix hydration. *Journal of Bodywork and Movement Therapies*, 94-100.

Science Kids. *Human Body Facts*. Retrieved February 18, 2017 from Science Kids: Fun Science and Technology for Kids: http://www.sciencekids.co.nz/sciencefacts/humanbody/nervoussystem.html

The Franklin Institute. *Blood Vessels*. Retrieved Retrieved February 17, 2017 from The Franklin Institue: https://www.fi.edu/heart/blood-vessels

Thomasino, M. 2001. "Length of a Human Intestine,"*The Physics Factbook*. Retrieved February 18, 2017 from http://hypertextbook.com/facts/2001/AnneMarieThomasino.shtml

Wiatrowski, W. *Monthly Labor Review* (June 2014). Retrieved from Beauru of Labor Statistics: https://www.bls.gov/opub/mlr/2014/article/examining-the-completeness-of-occupational-injury-and-illness-data-an-update-on-current-research-1.htm

Session No. 631

Workforce Health Status and Risk: An Emerging Issue in OHS Risk Assessment

Jeanne Puglisi Sherwood, RN
WellAdvantage
Sykesville, MD

E. Andrew Kapp, Ph.D., CSP, CHMM
Underwriters Laboratory Inc.*
Northbrook, IL

Introduction

The declining health status of today's workforce has decreased the predictive value of risk assessment practices in Occupational Health and Safety (OHS), and has contributed to an emerging issue threatening workforce safety. Common risk assessment practices used across industries do not consider today's escalation in debilitating medical conditions among workers and the resulting increase in likelihood and severity of injury for the general workforce population. This paper will serve to expand the OHS professional's knowledge base regarding workforce health and raise awareness of this issue.

Risk assessment is an essential component of any comprehensive safety and health program. This practice has been effective in reducing the likelihood of injury, illness or accident before such an adverse event is experienced. This fundamental practice has contributed to the standardization of safety practices through attempting to address the multitude of factors contributing to potential injury, illness or accident.

Although necessary and utilized widely, the process of identification of potential harm to workers through risk assessment practices is losing effectiveness. This is a bold statement; however, we must acknowledge there is a glaring omission across the most commonly used risk assessment practices today: the underlying health status of the worker. A critical contributor,

*Disclaimer: The views and opinions expressed in this paper are solely those of the author and do not necessarily represent those of Underwriters Laboratories Inc.

underlying health, is not considered in any of the commonly used risk assessment practices and therefore the physical and mental condition of the worker is assumed a constant in risk prediction. Yet this single unaccounted for factor is quite varied across individuals and one of the largest contributors to injury, illness and accident in the workplace.

Underlying Health as a Deterrent to Workforce Safety and Productivity

Health risk factors are physiological conditions or lifestyle habits, such as hypertension or smoking, which can negatively impact physical or cognitive functioning. Underlying health risk factors are associated with the increased incidence of declining health and the onset of diseases such as diabetes, heart disease or mental health issues. These health risk factors and diseases have an impact on basic human function such as potential decreased cognitive function or declining physical capacity. As a result underlying health and the presence of health risk factors are major determinants of both performance at work, and the likelihood and severity of workplace injury and illness.

Research regarding the frequency and severity of injuries substantiates the increased risk of employees with health risk factors. A national study by Lin, Verma, and Courtney (2013) found that obesity (BMI $\geq$ 30 kg/m^2) is associated with a 29% higher odds of sustaining a non-fatal occupational injury. This finding is in line with a study by Østbye, Dement, and Krause on Duke University employees that associates obesity with increased incidence of Workers Compensation (WC) claim activity including likelihood of filing a claim, duration of disability and costs associated with that claim. Among their findings: employees with BMIs $\geq$30 (obese) and employed in high risk occupations had over 7 times the rate of filing WC claims when compared with employees with BMIs below 30 and low-risk jobs. (Østbye, Dement, and Krause, 2013).

In addition the number of health risk factors per employee had a significant impact on outcomes. Looking at both WC and short–term disability (STD) claims at a large single employer (> 20,000 employees), Kuhen, et.al. found that the odds of filing a WC claim for those employees who had 5 or more health risk factors was nearly 3 times that of low risk (0-2 health risk factors) employees. Likewise medium-risk participants (3-4 health risk factor) had 1.5 times higher odds, when compared with low-risk participants to file for STD (Kuhen, et.al., 2009). An earlier study by Musich, Napier, and Eddington (2001) looking at a multitude of health risk factors and their association with WC claims at a Fortune 100 company found smoking and lack of physical activity among the most highly related risk factors with high WC claims.

The research with working populations also supports the unfortunate connection between health risk factors and diminished work performance. Burton et al., (2005) found that 10 health risk factors were significantly associated with a loss of productivity by employees over the previous two weeks including smoking, lack of physical activity, high blood pressure, high cholesterol, and obesity (BMI $\geq$ 30), with those individuals experiencing 5 or more risk factors losing upwards of 12 % productivity over those with two or less.

Health Risk Factors and Disease Conditions

Health Risks and Disease Trends

According to the National Center for Health Statics more than 72 million people – are obese. (Ogden, 2007). The prevalence of disease and health risk factors in the workforce has increased significantly. For example obesity (a health risk factor) has increased from 15% of the adult US population in 1995 to nearly 1/3 in 2015 (Table 1), and diabetes, a disease, has increased from close to 4% to over 9% (Table 2). (American Health Rankings, 2015) This trend shows no sign of stopping and employers must make adjustments.

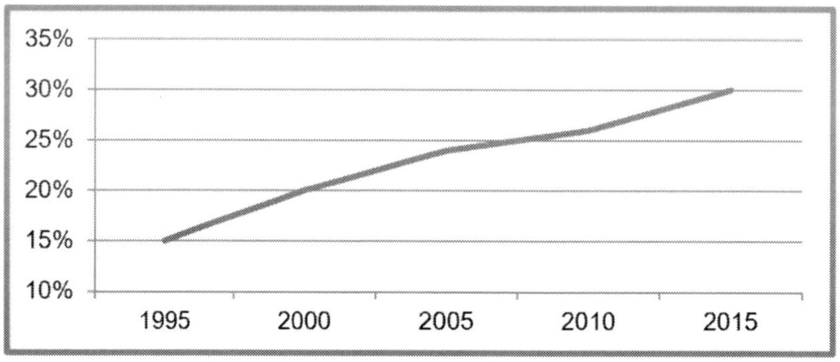

Table 1: Prevalence of Obesity in Adult Population in the US
Obesity (BMI≥30)
Source: America's Health Ranking, 2015

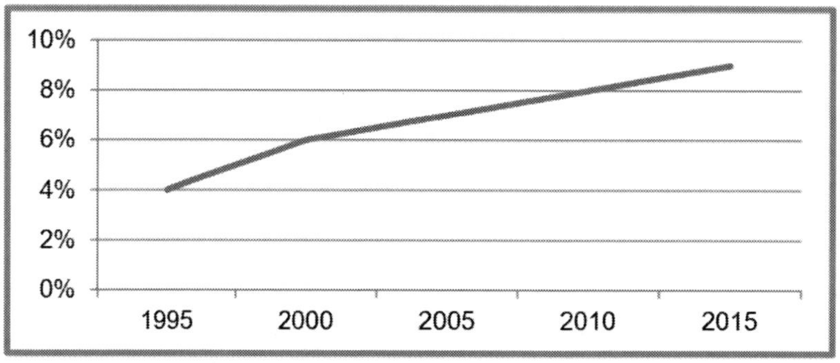

Table 2: Prevalence of Type Diabetes in the Adult Population of the US
(<150 minutes of physical activity Weekly)
Source: America's Health Ranking 2015

Predictive Vulnerabilities Specific to Health Conditions

Certain risk factors and diseases often present with additional health conditions. As an example, obesity is a risk factor that can substantially increase the impact of impaired function due to the likelihood of co-morbid conditions (co-morbidities). Obesity is a complex condition involving the regulation of energy metabolism and is often associated with other health conditions (comorbidities) such as diabetes, sleep disorders, hypertension and osteoarthritis. Each of these risk factors or conditions can result in specific reduced physical and/or mental function of workers.

Due to the frequent onset of certain comorbidities in conditions such as obesity and diabetes, we can anticipate potential vulnerabilities which may be present in these workforce populations. As an example, 31% of obese US adults are diagnosed with arthritis. (Barbour, 2017) Adults with arthritis are 2.5 times more likely to have two or more falls and suffer a fall injury compared with adults without arthritis. (Barbour, 2014) More than 80% of cases of type 2 diabetes can be attributed to obesity. (Brancati, 2000) Diabetic workers are at higher risk for health conditions such as neuropathy and hypoglycemia. Neuropathy causes reduced sensation in the extremities. In the case of a diabetic working in a high risk environment, there may be a delay or reduction in the ability to sense injury such as a burn or cut to the leg or foot. As a result a wound could be more extensive. In addition, the healing time for this population may be delayed, extending the duration of disability. Hypoglycemia, or transient low blood sugar levels, can impact cognition, and cause dizziness/fainting and visual disturbances. Although these occurrences can be minimized by proper management of diabetes, the risk of occurrence can be elevated during times of increased stress and illness.

Addressing Risk in the Workplace

Fitness for Duty

Current safety practices include "Fitness for Duty" when performing safety critical jobs such as public transportation operators, or when workers are returning to a job after a period of disability. However underlying health issues and the impact on human performance are concerns that affect a wider range of workplace activities.

Fitness-for-duty is defined as a physical, mental, and emotional condition which enables the employee to perform the essential tasks of his or her work assignment in a manner which does not threaten the safety or health of oneself, co-workers, property, or the public at large (Kohanna, 2012). There are a number of common triggers for FFD assessments including regulatory requirements, post job offer evaluation, and return-to-work situations. Even observable employee behavior with objective evidence of a problem and employee self-report of a medical sign or symptom can initiate a FFD assessment, but are these the only times when an employee's baseline health should be factored into the risk assessment process?

Common Risk Assessment Practices

There are a large number of risk assessment practices available to the OHS professional today. While each is unique, with its own strengths and weaknesses, none that the authors are aware of (other than FFD) take into consideration the underlying health status of the employee. Two commonly used risk assessment practices are now discussed to illustrate this notable absence.

The Job Safety Analysis (JSA) is a widely used practice for predicting the injury risks of specific tasks and identifying methods to reduce these risks. At its most basic a JSA is a simple three phase process of delineating the steps in a particular task, enumerating the hazards associated with each step, and indicating corrective or preventive measures for each hazard identified. The hazards are determined from observing the job being done and noting the tools, materials, and activities of a worker completing that job. The health status of the individual employee performing the job being analyzed never enters the process.

Hazard and Operability Analysis (HAZOP) is another widely used practice for reexamining the safety-related consequences of process deviations used extensively, but not exclusively, in the process industries. HAZOP is a systematic analysis of a process to assess the safety-related consequences of the malfunctioning/deviations of individual components or functions on the system as a whole. The HAZOP process uses a team approach and a relies on applying a number of guidewords (e.g., no/not/none, more, less, part of, reverse, other than, as well as) to parameters of the process (e.g., pressure, temperature, flow, composition, level) and applies them to each step in the process. The resulting records of a HAZOP include:

- the design intent of the component or function,
- a description of the possible deviations from the intent,
- the possible causes of each deviation,
- the consequences of each deviation, and
- corrective or preventive actions to be taken to minimize the hazard associated with the deviation

While possible causes for each possible deviation are considered, there is no specific reference to consider underlying operator health condition provided.

There are some common reasons why the health status of the employee is not factored into these, and many other risk assessment techniques. These assessments are typically performed by the EHS professional, a Manager, or Supervisor. It may be that none of these individuals are naturally versed in medical science; they are not knowledgeable on these health conditions and their effects on the employee performing the work. When performing the respective risk assessment the assessor's attention is drawn largely to those elements external to the worker or operator performing the task; there is nothing to even trigger the analyst attention in on underlying health condition of worker. Neither technique discussed, nor any other that we are aware of, has the explicit ability to include such underlying health condition information in the assessment if it were even known to the analyst. Finally there are Health Insurance Portability and Accountability Act (HIPAA) concerns with giving analysts access to this personal health information.

To make risk assessment a more accurate process by factoring in employee health condition we must address the issues of 1) Lack of knowledge among risk assessors of employee health conditions and their effects/contribution to injury, illness and accidents; 2) Absence of cues and guidance for the inclusion of underlying health condition of workers in most risk assessment practices in use; 3) HIPAA concerns with giving access of such information to risk assessors. Furthermore we must address the above mentioned issues in a way that leads to a risk assessment process that is practical and scalable for use in an actual employer situation.

Influencing Workforce Health

It has become a best practice from a business standpoint to address workforce health to reduce injury/illness, protect employees, reduce workers' compensation and disability leave, increase productivity and engagement and contribute to the bottom line of the organization (Edington, 2009). An example practice includes "health and well-being initiatives," which may include onsite biometric screenings, healthy workplace features and rewards for healthy behaviors. Programs such as this have become increasing common as the health for workforce populations has declined. These types of initiatives have evolved in effectiveness and complexity over time to include strategies such as outcome based reward structures to encourage high risk participants to actively address underlying health risk factors. Additional strategies such as the CDC NIOSH Total Worker Health model integrates practices for multiple departments such as Safety and Occupational Health as a step towards addressing workforce health and safety from a unified standpoint.

With the evidence strongly supporting the association of underlying health risk factors and disease with the incidence of injury, illness and accidents, the safety profession has the unique opportunity to increase the protection of the workforce populations they serve. The integration of workforce health considerations into safety practices has the potential to make a significant impact on the preservation of health and quality-of-life for workforce populations of today and well into the future.

Acknowledgement

Special thanks to Bonnie Monroe for her contributions to this work.

Bibliography

American Council of Learned Societies. 1994. "Teaching the Humanities: Essays for the ACLS Elementary and Secondary Schools Teacher Curriculum Development Project." New York: ACLS.

America's Health Rankings Annual Report 2015, December 2015 at http://www.americas healthrankings.org/ Accessed on March 10, 2017

Barbour, K.E., C.G.Helmick, M.A. Boring, and T.J. Brady. (2017). "Vital Signs: Prevalence of Doctor-Diagnosed Arthritis and Arthritis-Attributable Activity Limitation --- United States, 2013—2015." Morb Mortal Wkly Rep. March 7. [Epub ahead of print].

Barbour K.E. et. (2014). "Falls and Fall Injuries Among Adults with Arthritis—United States, 2012." Morb Mortal Wkly Rep.;63(17):379-383.

Brancati, F. L. et al. (2000). "Incident Type 2 Diabetes Mellitus in African American and White Adults." Jama, 283(17), 2253. doi:10.1001/jama.283.17.2253

Bureau of Labor Statistics (BLS). 2013. "Fatal Occupational injuries by industry and event or exposure, All U.S." (www.bls.gov/iif/cfoi/cflb0268.pdf).

Burton, W. N. et al. (2005). The association of health risks with on-the-job productivity. Journal of Occupational and Environmental Medicine, 47(8), 769-777.

Deming, W. E. (1991). "Quality, Productivity, and Competitive Position." Four-day workshop presented in Cincinnati, Ohio by Quality Enhancement Seminars, Inc., Los Angeles, CA, May 1991.

Dunjó, J., et al. (2010). Hazard and operability (HAZOP) analysis. A literature review. *Journal of hazardous materials, 173(1),* 19-32.

Edington, D. W. (2009). *Zero trends: Health as a serious economic strategy.* UM-HMRC.

Gould, J., M. Glossop, and A. Ioannides. (2000*). Review of hazard identification techniques.* Health & Safety Laboratory, Sheffield.

Kohanna, F. (2012). Practical Steps for Addressing Fitness-for-Duty (FFD). Presented at the 2012 OSHA Oil & Gas Safety Conference, Dallas, TX , Dec. 4-5, 2012.

Kuhnen, A.E., et al. (2009). Employee health and frequency of workers' compensation and disability claims. *Journal of Occupational and Environmental Medicine, 51(9),* 1041-1048.

Lin T, S.K. Verma and T.K. Courtney. (2013). "Does obesity contribute to non-fatal occupational injury? Evidence from the National Longitudinal Survey of Youth." *Scand J Work Environ Health;39(3):268–275.*

Musich, S., D. Napier, and D.W. Edington. (2001). The association of health risks with workers' compensation costs. *Journal of Occupational and Environmental Medicine, 43(6),* 534-541.

National Institute of Occupational Safety and Health (NIOSH). 2012. "Total Worker Health." (www.cdc.gov/niosh/TWH/totalhealth.html/).

Norris, W. 1981. The Unsafe Sky: The Unvarnished Truth About Air Safety. New York: W.W. Norton & Co.

Ogden C.L. et al. Obesity among adults in the United States— no change since 2003–2004. NCHS data brief no 1. Hyattsville, MD: National Center for Health Statistics. 2007.

Østbye T, J.M. Dement, and K.M. Krause (2017). Obesity and Workers' Compensation: Results From the Duke Health and Safety Surveillance System. *Arch Intern Med;167(8)*:766–73.

Prevention of type 2 diabetes mellitus. (n.d.). Retrieved March 9, 2017, from http://www.uptodate.com/contents/prevention-of-type-2-diabetes-mellitus

Richards, Ellen. 1988. "Four Exercises Women Do Too Many Of." *Self.* 42: 174-177.

Lockout Tagout Best Practices, Real Life Consequences

Kina Repp
Workplace Safety Advocate
Master Lock Company LLC
Oak Creek, WI

Introduction

A comprehensive and effective Lockout Tagout program is paramount for safety within any workplace where employees can be exposed to hazardous energy. The equipment and workforce within each facility is unique, therefore, lockout tagout programs and their implementation can vary significantly between companies. Unfortunately, many businesses often fail to implement an adequate lockout tagout program within their facility, or in the worst case, don't practice the use of lockout tagout at all. This is highlighted by the fact that lockout tagout continues to be a top 10 OSHA violation each year. In FY2014, lockout tagout was the 6th most cited OSHA standard.

Every year, thousands of injuries and hundreds of deaths are attributable to the unexpected release of hazardous energy, and these accidents could have been avoided with the proper utilization of lockout tagout products and procedures. Kina Repp, a nationally recognized motivational and safety speaker, stands as proof that a failure to enforce lock out/ tag out procedures can have a devastating impact.

As a young college student, I was employed at an Alaskan cannery and tasked with cleaning conveyor belts. My first day on the job, I was assigned to clean a large conveyor without safety training, or the supervision of individuals properly trained to operate the equipment, and with no lock out/tag out procedures in place.

What followed was a horrifying chain of events that forced me to fight for survival and changed my life and that of my family. Today, I dedicate myself to sharing the details of that life-shattering experience in hope of preventing injuries and personal devastation for others. My presentation emphasizes why it is critical to:

- Have and strictly enforce lock out/tag out procedures;
- Ensure all personnel associated with energy sources are well-trained in and follow control procedures;
- Look for and be prepared to mitigate hidden hazards;
- To ensure individuals protect the well-being of their co-workers; and

- Ensure new employees understand and follow all safety processes and rules.

My story is an emotional reminder as well about the ripple effect an incident can have others.

This paper outlines the key requirements of OSHA's 1910.147 Control of Hazardous Energy standard and offers best practice recommendations for developing an effective energy control program.

Lockout Tagout Overview

OSHA's 29CFR 1910.147 Control of Hazardous Energy standard was implemented in 1989 in order to reduce the number of injuries and deaths in the workplace related to the unwanted or unexpected release of hazardous energy. In 2002, OSHA estimated that approximately three million workers in the US serviced machinery and equipment and were therefore at the greatest risk of being exposed to hazardous energy. Today, this number is likely to be closer to two million workers due to the downturn in domestic manufacturing and related decrease in manufacturing employment. In 2012, 248 workplace deaths were either caused by electrocution, or from being caught in running equipment or machinery, according to the Bureau of Labor Statistics. In addition, OSHA estimates lockout tagout related injuries lead to an average of 24 days of lost work time for recuperation. Therefore, the loss in workforce productivity and associated medical costs can be staggering.

The 1910.147 standard is applicable to situations where the unexpected start-up of equipment or release of stored energy could harm employees during servicing or maintenance operations. The OSHA standard outlines the minimum requirements for controlling hazardous energy, and includes several key requirements that employers must implement in order to be compliant with the standard. These include:

1. Provide written energy control (lockout) procedures for each piece of machinery or equipment where the control of hazardous energy is necessary during servicing or maintenance
2. Provide lockout tagout devices that meet all of the following criteria
   a. Durable: Capable of withstanding their usage environment for the full duration that they will be in use
   b. Substantial: Capable of preventing removal without the use of excessive force such as bolt cutters
   c. Standardized: Must be standardized within a facility by one or more of the following; color, size, or shape
   d. Identifiable: Must indicate the identity of the employee that applied the lockout device
   e. Used exclusively for safety: Must not be used for any other purpose within a facility
3. Annual inspection of lockout tagout procedures
4. Ongoing training for employees

Compliance with the 1910.147 standard requires lockout tagout programs to include all of the aforementioned elements. Each of these program elements are explained in detail throughout the rest of this paper, and best practice recommendations are provided in order to maximize the effectiveness of lockout tagout programs.

Lockout Tagout Program Elements

Written Lockout Procedures

Written lockout procedures are required for all machines where hazardous energy must be controlled during maintenance and repair operations. At a minimum, each written lockout procedure must include the following:

1. A description of the intended use of procedure
2. Specific steps for shutting down machinery, isolating hazardous energy and securing machinery for maintenance
3. Specific steps for the application and removal of lockout devices
4. Testing protocol to verify the effectiveness of the energy control procedure

A written lockout procedure serves as a guide for the authorized employee to effectively isolate hazardous energy sources and lockout equipment prior to performing maintenance and repair operations.

Best Practice:

Mount visual lockout procedures at each machine. A visual lockout procedure includes the information noted above, but also includes machinery images to aide the authorized employee in identifying the type and location of all hazardous energy isolation devices supplying energy to a machine. These visual aides help ensure that lockout procedures are completed accurately and in the correct order.

Lockout Tagout Devices

Multiple criteria are provided for lockout tagout devices to ensure they are not only appropriate for their usage environment, but can also be easily identified by employees.

1. Durable: Capable of withstanding their usage environment for the full duration that they will be in use
    a. Example: If exposed to chemicals or extreme temperatures during lockout tagout, the devices must be capable of withstanding these harsh environmental factors.
2. Substantial: Capable of preventing removal without the use of excessive force such as bolt cutters
    a. Example: Tags must be capable of withstanding at least 50 lbs. of pull force when used for lockout tagout.
3. Standardized: Lockout devices must be standardized within a facility by one or more of the following: color, size, or shape
    a. Example: Establishing red colored locks solely for lockout tagout purposes. Or, allocating specific colors for different purposes such as red for personal lockout locks and blue for procedural locks.
    b. Best Practice: Utilize a specific padlock model such as a Master Lock 410 Zenex™ solely for lockout tagout purposes, and assign various colors to differentiate between various worker types. For example, electrical employees carry a green lock and contractors carry an orange lock. This practice makes it

very easy for employees to visually identify what workers are performing maintenance while equipment is locked out.

4. Identifiable – Must indicate the identity of the authorized employee that applied the lockout device
    a. Example: Adding an identification tag to a lockout device, or adding labels and/or laser engraving to a safety padlock to identify the employee by name.
    b. Best Practice: Utilize photo ID tags and labels in order to enhance employee identification. Often employees cannot identify employees by name, but they can identify them via a visual image.
5. Used exclusively for safety – Must not be used for any other purpose within a facility
    a. Example: A cable lockout device that is used for lockout tagout should not also be used to secure items within a facility such as a fence.
    b. This ensures that workers can easily differentiate between devices that are being used for lockout tagout, and devices that are utilized for security or other purposes.

Annual Inspection of Lockout Procedures

Facilities must conduct periodic inspections of all lockout procedures at least annually to ensure they accurately reflect the equipment and energy isolation requirements, and to confirm employees are correctly following the procedures. A different authorized employee than the one who is assigned to perform the lockout procedure must complete inspections. The inspector must review the responsibilities with each of the authorized employees who are assigned to complete the lockout procedure, and the employer must certify that the inspections have been completed for each applicable piece of equipment or machinery within a facility.

Employee Training

Employees must be trained on the energy control program within a facility including how to use and apply/remove lockout devices in order to control hazardous energy. Training requirements vary between authorized and affected employees:

1. Authorized employee training: Includes how to identify hazardous energy sources, an overview of the various types and magnitude of hazardous energy sources present within a facility, and the method for locking out each hazardous energy source.
2. Affected employee training: Instruction on the purpose/objectives for utilizing an energy control program.

Summary

In summary, energy control programs are designed to protect employees from injury or death due to the exposure to hazardous energy sources during maintenance operations. The ultimate goal is to ensure employees go home safely each day, and avoidable injuries are eliminated within the workplace. Any successful lockout tagout program requires support from the management team to instill lockout tagout as an essential component of the company culture. Sharing my life changing story helps to bring home personal ownership in safety.

Bibliography

Bureau of Labor Statistics (BLS). 2012. "Fatal occupational injuries by event or exposure, 2012." (http://www.bls.gov/news.release/cfoi.t01.htm).

Occupational Safety and Health Administration (OSHA). 1989. "Standard 29CFR 1910.147, The control of hazardous energy (lockout/tagout)" (https://www.osha.gov/pls/oshaweb/owadisp.show_document?p_id=9804&p_table=STANDARDS)

______. 2002. "OSHA Fact Sheet" (https://www.osha.gov/OshDoc/data_General_Facts/factsheet-lockout-tagout.pdf)

The author is an employee at Master Lock Company LLC responsible for safety awareness for the Safety Solutions portfolio. Master Lock is a manufacturer of Lockout Tagout products, software, and services, and has provided OSHA compliant products for the market since 1990.

Noise and Vibration Measurement with Smartphone Applications

Helmut W. Paschold, Ph.D., CSP, CIH, PE
Assistant Professor
Indiana University of Pennsylvania
Indiana, PA

Introduction

Vibration in the occupational and general environmental settings can be manifested in many ways according to frequency, magnitude, or medium and subsequent effects on the human body. Frequency ranges with potential risk to humans can range from less than one cycle per second (Hertz – Hz) to hundreds of thousands. The vibration ranges to be considered are Whole Body Vibration ranging from 0.5 to 80 Hz and audible sound from 20 to 20000 Hz. Both WBV and noise are acknowledged as having potential to causing harm to the human body.

Safety and health (SH) professionals address hazards through anticipation, recognition, evaluation, and control. The anticipation and recognition of noise hazards can be considered a basic skill set among SH professionals. With the human sense of hearing, we are much attuned to noise in our environments. What is more difficult for the human sense of hearing is the ability to quantify the magnitude of sound. People can categorize sound or its unwanted companion noise as low, moderate, loud, or very loud, but generally fail in attempting to accurately estimate exposures in decibels (dB). Most SH professionals should know the guide to assess if a noise environment is exceeding 85 dB – do you have to raise your voice to be heard at arm's length? This approach is not an accurate means of noise measurement and surely would not stand the scrutiny of an OSHA compliance officer responding to a noise complaint in the workplace. However, this basic technique can satisfy to hazard assessment levels of anticipation and recognition for noise.

WBV is even more difficult for people to assign quantified magnitudes than with noise. While seated in a vehicle seat, such as a forklift traversing a construction site, the human operator can qualitatively describe a smooth, bumpy, or rough ride. However, it would be extremely unlikely that the operator could closely approximate the intensity in terms of actual acceleration in meters per second squared ($m\text{-}s^{-2}$) within the frequency range of 0.5 to 80 Hz.

Why are we concerned with human noise exposure? A very recent CDC study published in the February 7, 2017 issue of Morbidity and Mortality Weekly Report (MMWR) suggests that noise-induced hearing loss (NIHL) may affect 25% of the adults in the United States or a total of 39.4 million Americans. Workplace exposure to loud noise doubles the probability of NIHL. In 1981, OSHA put forth its Hearing Conservation Amendment 1910.95. In section (c)(1), OSHA states "The employer shall administer a continuing, effective hearing conservation program, as described in paragraphs (c) through (o) of this section, whenever employee noise exposures equal or exceed an 8-hour time-weighted average sound level (TWA) of 85 decibels measured on the A scale…". Noise damages hearing.

WBV is an issue of growing concern. Adverse health effects of chronic exposure to WBV include lower back pain, abdominal or digestive problems, cardiovascular disorders, or simple discomfort and annoyance. The European health and safety community has addressed the issue of WBV for decades, with regulatory action levels and exposure limits in place for about two decades. SH professionals in the US have relatively little knowledge about WBV, most likely due to WBV not being included in OSHA standards and thereby falling off the regulatory "radar screen". At the present time, the best-known standard for WBV is the International Organization of Standardization's ISO 2631-1 (1997), which also appears in an ANSI standard and closely modeled in ACGIH's TLVs.

Accurate measurement of noise and WBV exposures requires calibrated instruments capable of reliable results with precision and accuracy. Sound is measured with sound level meters or dosimeters compliant with ANSI standards S1.4-1971 (R1976). Lacking US standards for WBV, WBV instruments are required to comply with ISO standards for measurement. These instruments are colloquially referred to as the "gold standard" of measurement when properly calibrated, maintained, and operated.

Somewhere in the middle of the spectrum between human sensory evaluations and the gold standard instruments is the relatively recent introduction and use of smartphone applications. Smartphones have the core equipment found in noise and WBV instruments – a microphone or accelerometer, a computer processor, and a display. The question begged to be asked: How reliable are these smartphone applications for field measurement of noise and WBV vibration? As safety professionals are called to perform a wider variety of tasks in their jobs, they must learn to use tools readily available. With the inclusion of a relatively accurate microphone and triaxial accelerometer in contemporary smartphone devices, the tools to conduct vibration measurements are at the SH professional's fingertips.

Vibration

Vibration is mechanical wave motion travelling through a medium that can cause a transfer of energy from one object to another. A simple sine wave representing the apparent motion of a particle affected by vibration is characterized by frequency and amplitude. Wave frequency is the distance between the waves and the amplitude the height of the waves. Frequency is expressed in *hertz* (Hz), the number of cycles per second. The magnitude or strength of vibration is its acceleration, measured in $m\text{-}s^{-2}$ for WBV and dB for sound. In older literature, WBV acceleration might be reported in "g" forces, representing the 9.8 $m\text{-}s\text{-}^{2}$ acceleration due to gravity. WBV and

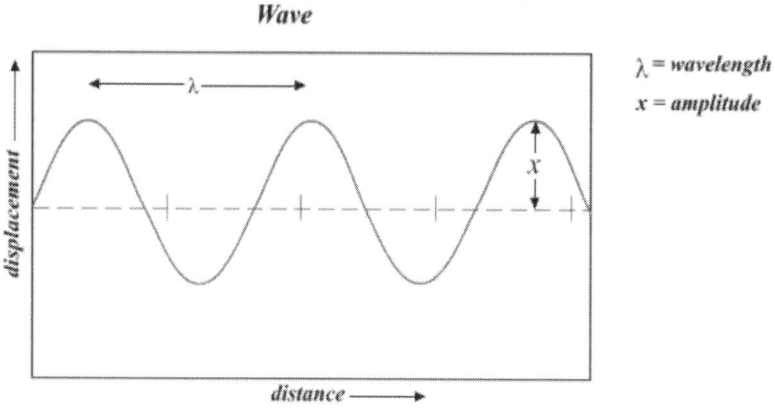

Figure 1. Vibration as a simple sine wave.

noise values are expressed in terms of the root-mean-square (r.m.s.), a mathematical method analogous to the statistical standard deviation function. In the work environment, humans are exposed to multiple waves of differing frequencies, amplitudes, periods, and directions with graphical results seemingly random in appearance. The vibration waves can travel through air, solids, and liquids and transfer energy from the originating source to the human body upon contact with a vibrating medium.

The transmittal of noise vibrations to humans occurs mostly through air to the ear; however, there can be conduction vibration through flesh and bones. In ambient conditions at sea level, sound travels at approximately 1,130 feet per second. The range of hearing for a healthy young person can be between 20 and 20000 Hz. The frequencies of concern for human hearing are in the 100 to 8000 Hz range. The pressure fluctuations caused by sound are compared to a base reference of 0.00002 N/m^2 and converted into a dimensionless logarithmic ratio called the decibel (dB).

Unlike sound, WBV is reliant on contact with the human body with transmission maximized when seated or prone on a vibrating surface. Vibrations transmitted through the legs or arms tend to be attenuated and not cause impact on the torso. WBV is traditionally considered to cover a frequency range of 0.5 to 80 Hz. The frequencies of greatest concern for WBV tend to be between 4 and 20 Hz. WBV is evaluated in three axes – X, Y and Z as shown in Figure 2.

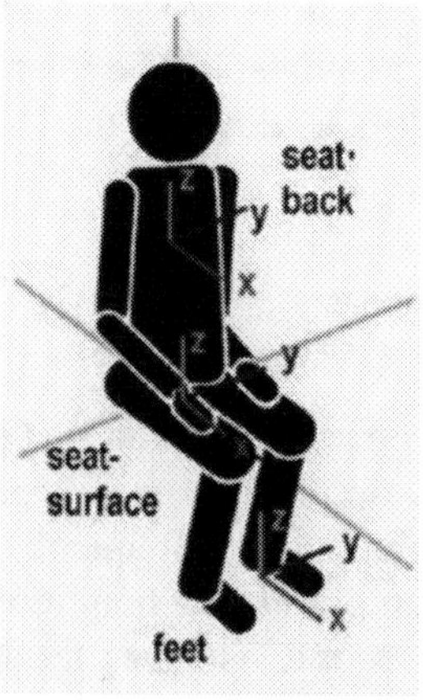

Figure 2. X, Y, and Z axis orientation for WBV.

Traditional Monitoring

Noise monitoring in accordance with OSHA methods requires the use of a Type II instrument (such as a sound level meter), which essentially has accuracy of +/- 2 dB. Noise measurements are recorded with a slow-response setting and use of an A-weighted scale to yield results in dBA. The A-weighted scale places highest emphasis on the frequency bands most closely associated with damage to human hearing. For occupational settings with highly variable noise levels and high employee mobility, the use of noise dosimetry equipment is required, where the instrument integrate noise and time measurements continuously during the monitoring time to calculate a resultant equivalent time-weight average of percentage dose. Accuracy and precision is assured by scheduled factory calibrations and pre and post-measurement bench calibrations. Most SH professional are familiar with noise monitoring instruments as a common industrial hygiene practice. Noise monitoring equipment can range from several hundred dollars to thousands in purchase costs.

Most SH professional well not well-acquainted with WBV measurement. In a previous survey research project, I discovered the self-reported ability to monitor WBV within a group of 2,600 SH professionals was ranked as somewhere between not at all to very little knowledge. Most field WBV instruments consist of an instrument meter connected to an accelerometer via cable. With most WBV measurements conducted on seated workers, the accelerometers are typically seated within a flexible neoprene disc as shown in Figure 3 to provide easy alignment and negligible discomfort to the driver. Meters must comply with ISO 8041:1990(E) (including Amendment 1:1999(E)) and ISO 2631-1:1997 requirements to assure valid readings. Specified intervals are given for factory calibration; however, these units are not field calibrated pre and post-testing as with SLMs. The ISO standard also prescribes a weighting system analogous to noise dBAs. The weighting places emphasis on the lower frequencies most associated with lower back injury causation, using W_d for the X and Y axis, and W_k for the vertical Z axis. The weighting curves for WBV monitoring are depicted in Figure 4. The costs of WBV field monitoring units start at approximately $5,000 at the lower end and can run significantly higher.

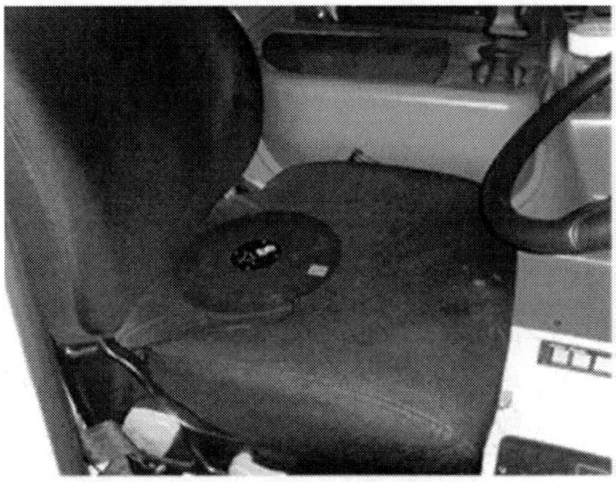

Figure 3. Neoprene disc with embedded accelerometer for WBV measurement.

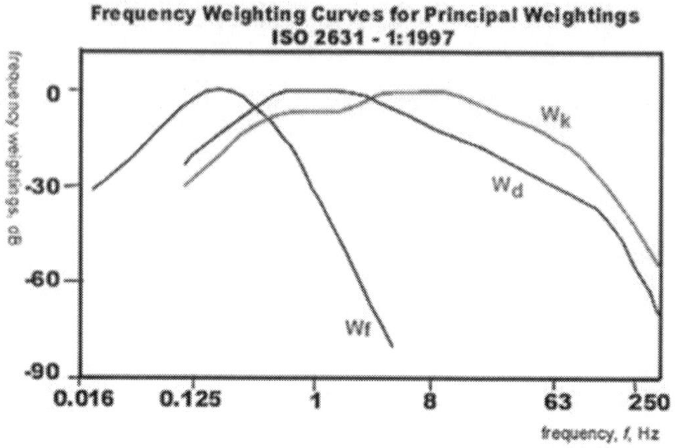

Figure 4. Weighting curves for WBV monitoring.

Smartphone Monitoring

In 2014, NIOSH scientists conducted an exhaustive evaluation of 192 different sound measurement applications (apps) for both iOS and Android platforms to determine potential reliable use for SH professionals. NIOSH researchers narrowed the field to four apps meeting a higher level of testing criteria. From here, NIOSH partnered with an app developer to create a reliable SML app, released in January 2017. The no-cost NIOSH-SLM app was laboratory tested and found to meet the Type 2 criteria for a SLM with the iPhone microphone. With a high quality external microphone capable of calibration, the NIOSH SLM produced accuracy results of a Type 1 SLM. An iPhone with an external MicW i436 professional microphone (used in NIOSH testing) is depicted in Figure 5 and shows the main screen.

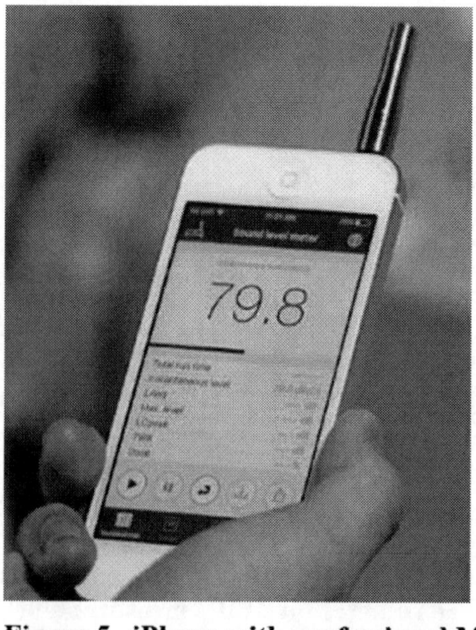

Figure 5. iPhone with professional MicW i436 microphone.

A 23-page instruction manual for the NIOSH SLM app can be downloaded in pdf format from https://www.cdc.gov/niosh/topics/noise/pdfs/niosh-sound-level-meter-application-app.pdf The document contains detailed operations for the use of the app.

A newly released smartphone application, *WBV* by Byte Works, Inc. ©2014, measures WBV using methods prescribed in ISO2631.1. The WBV app was designed by ergonomic scientists from the University of Queensland, Australia. The app can be downloaded free of charge; detailed information can be found at http://www.byteworks.us/Byte_Works/WBV.html regarding WBV app measurements. The key ability which allows an iPhone (in this case a 5th generation) to conduct relatively reliable WBV measurements is the LIS33iDLH MEMs type accelerometer (STMicroelectronics, Geneva, Switzerland). A screen shot of the WBV is shown in Figure 6.

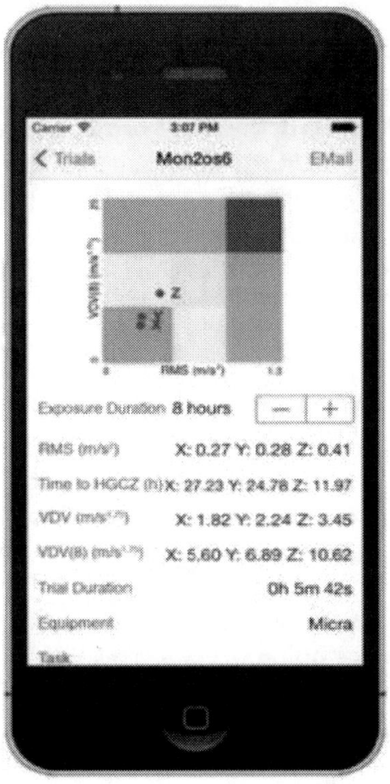

Figure 6. Screen shot of WBV app on iPhone

Discussion and Summary

The use of smartphone apps for vibration monitoring is still a relatively new field. NIOSH research appears to have done a fairly extensive job of corroborating the reliability of results for its SLM in a laboratory setting. Likewise, the University of Queensland researchers have provided detailed justification for the relative reliability of the WBV app. With the apps ready for use, the next step is to gather extensive measurement information to determine field reliability.

The most obvious benefit to the SH professional is the cost of monitoring. Even considering the cost of an iPhone, the flexibility and accuracy of use results in a far more economical method than the "gold-standard" instruments, especially for WBV. The i436 external microphone, highly sensitive and calibrated, can be purchased for under $125. Another microphone tested with the NIOSH app, the Dayton Audio iMM-6) can be found for less than $25. These additions to an iPhone can create extremely useful SLMs. With the high cost of WBV field measurement instrument (and their associated complicity of operation), the app offers a relative easy to use format that is readily available to all. This accessibility is hoped to allow more SH professionals to explore potential WBV risks in the workplace. Additionally, the use of mobile devices for "crowdsourcing" data is increasing in domains such as environmental stressors (noise and vibration). This method provides greatly increased data sets for research analysis. The crowdsourcing techniques can be developed for use with an employed population for enhanced data collection in a workplace and better define total worker exposures, especially with WBV.

Some limitations can be noted for the use of sound and WBV apps. Without an external high-quality microphone, accuracy is expected to be lower and may not be sufficiently reliable for OSHA compliance monitoring. With the small orifice size of an iPhone, the microphone can easily be compromised by dust and debris. Maintenance of the smartphone is crucial in assuring relatively accurate results. The WBV app may not yield the higher accuracy of the "gold-standard" instruments. However, in the absence of an OSHA standard for WBV in the U.S., any assessment is better than none to help SH professional identify and mitigate WBV exposures in the workplace.

The benefits of these apps far outweigh the shortcomings. In any case, even if the apps are not sufficiently reliable for compliance purposes, they are extremely useful to the SH professional in the anticipation and recognition phases of hazard risk management.

Bibliography

Kardous, A. and B. Shaw. "Evaluation of smartphone sound measurement applications (apps) using external microphones–A follow-up study." *The Journal of the Acoustical Society of America*. 2016. 140, No 4

Paschold, H. and A. Seergeev. "Whole-body vibration knowledge survey of U.S. occupational safety and health professionals." *Journal of Safety Research. 2009: 40(3):171-176.*

Song, C.; J. Wu, H. Gong, and B. Gou. "RESen Sensing and evaluating the riding experience based on crowdsourcing by smart phones." *The Eighth International Conference on Mobile Ad-hoc and Sensor Networks*. December 14-16, 2012. Chengdu, Sichuan, China

Session No. 634-a

Preventing Accidents and Incidents: Safety Excellence Journey at Construction Sites

Krishna Nirmalya Sen, F.I.E
Head EHS
Larsen & Toubro MMH
Kolkata, India

Introduction

Frequency rates of severe injury, including fatality, as well as occupational diseases, are notably higher in construction than other industries. Studies also indicate that the safety performance of the construction industry is always under challenge due to various factors, including the inherent nature of high-risk activities, deployment of large number of unskilled people, and use of heavy machinery, as well as various demographic factors.

Construction workers keep moving from one place to the other to pursue their profession. Prior to joining the trade, a miniscule number of them receive some formal skill training and safety training, which are essential for safe working. Consequently, a large portion of them get set to pick up the trade skills on-the-job, which compounds the challenges of ensuring safe working at site.

In this study, spanning 18 months' duration, involving over 20,000 workers from multiple project sites, factors associated with "risk perception," emerge as one of the most common causative contributors for various recordable injury cases.[2] The study highlights that misplaced risk perception by the individuals associated with the activities at various levels have acted as the "trigger," directly or indirectly, leading the event chain to injury outcome.

Subsequently, focused efforts are planned for mitigation by improving risk awareness among all concerned team members through various methods, including on-the-job, classroom training, and experiential learning associated with activity-related hazards and consequences.

This paper will describe facets of risk perception, and its role in risk mitigation, and demonstrate the effectiveness of awareness programs, as well as other initiatives in optimizing risk perception for incident prevention

Risk can be quantified and is a product of probability and consequence of the event. Any activity beyond an acceptable risk level needs to be treated with suitable mitigation measures to bring the risk to within as low as reasonably practicable limits.

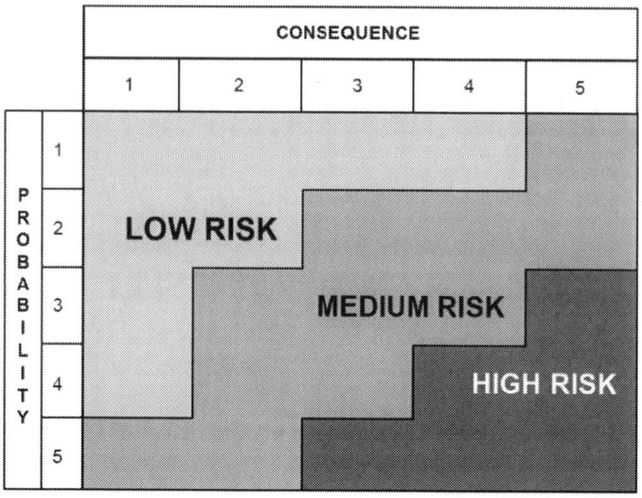

Figure1. Risk Matrix

Site-based Activities

Common site-based activities are selected, and risk perceptions of workers, staff and EHS staff, are analyzed in line with the risk matrix. Some of those activities considered for this survey and the safe practices to be followed during the work are mentioned below.

Manual Material Handling

Manual material handling is related to moving of items by various manual ways, such as lifting, lowering, carrying, pushing, pulling etc. Manual handling injuries are part of a wider group of musculoskeletal disorders (MSDs). Poor man-machine interface and workplace layout are the key factors in many hazardous manual handling tasks. An ergonomic approach is recommended to look at how the task can be fitted to the individual.

Lifting by Crane

The crane should be in order, and crane operators need to have suitable skill and adequate experience in operating the particular crane they are using. They should follow the manufacturer's instructions contained in the rating sheet, manuals, and operating procedures provided with the crane. Operators should also follow their employer's policies regarding regular and pre-start checks. The operator must never operate a crane until they are sure that the conditions are safe. For safe operation of the crane, it is important that each member of the team have a mature understanding of the associated risks for each micro-level task, as it plays a key and continuous role in determining sustained adoption of safe practice in a dynamic, ever-changing situation.

Cladding

Cladding is the application of one material over another to provide skin or layer intended to control the infiltration of weather elements, or for aesthetic purposes. Cladding does not necessarily have to provide a waterproof condition but is instead a control element. The EHS person should ensure the establishment of safety zones around operations at height. Also the provision of safe means of access and safe work platforms should be ensured. Similarly, the working crew should identify the hazards that can occur in the site related to their work, so that it can be assessed, and those risks can be mitigated before work.

Grinding

Grinding is an abrasive material removal and surface generation process used to shape and finish components made of metals and other materials. Grinding employs an abrasive product, usually a rotating wheel brought into controlled contact with a work surface. Damaged wheels, or even wheels suspected of being damaged, should not be used. Safety steps for proper use of grinding wheels include, but are not limited to:

- Always having machine guards in place before turning on a grinding wheel
- Running wheels for at least one minute before actual work begins
- Always using eye protection
- Properly balancing and dressing wheels before use

Risks and hazards associated with grinding work have to be captured, and proper mitigation can be done, to ensure safe and healthy work.

Welding

Welding poses a range of both well-known and subtle hazards to health and safety. These can act quickly or may show up only in the long term. They can be rapidly fatal (electric shock or exposure to cadmium fumes) or have delayed effects (lung changes over time). Only workers with appropriate levels of competence and skill should do welding. Some problems associated with welding are:

- poor quality structural welds
- insufficient or lack of necessary welding procedures for welders

The problems that can happen during welding have to be identified before starting work, so that effective risk management can be implemented.

Crane Marching

There are multiple hazards that can arise regarding cranes in general. Many accidents involve large lift systems like tower cranes and mobile cranes. But hazards do exist with all types of cranes, including overhead cranes, and in all facets of crane operation. The three most common hazards involving overhead cranes include electrical hazards, overloading, and materials falling/slipping from overhead hoists. Taking precautions for all these hazards well before execution of work will improve the quality of safety, thereby making the workplace safer, for which risk perception plays a key preliminary role.

Findings of Risk Perception Survey

Many conclusions can be drawn from the risk perception chart. Risk perception of workers, staff, and EHS officers for various selected site-based activities is represented in the form of a chart. Risk perception is analysed on the basis of risk index, which is the product of probability and severity of a particular work-related hazard. Risk perception of EHS officers is higher than workers and staffs in the case of MMH, lifting by crane, loading by crane, grinding, and transportation. One of the interesting results is that the risk perception of workers in cladding and crane marching is greater than staff and EHS officers. The low risk perception of staff in most of the activities, except gas cutting and welding, hints the lack of alertness among them. Awareness programs or campaigns can be carried out in all sites to reduce the gap of risk perception among various levels. This chart also gives an impression that the root cause of most incidents in a workplace can be linked with the risk perception of

EHS staff, as it is their prime responsibility to identify the risks and avoid/manage them properly.

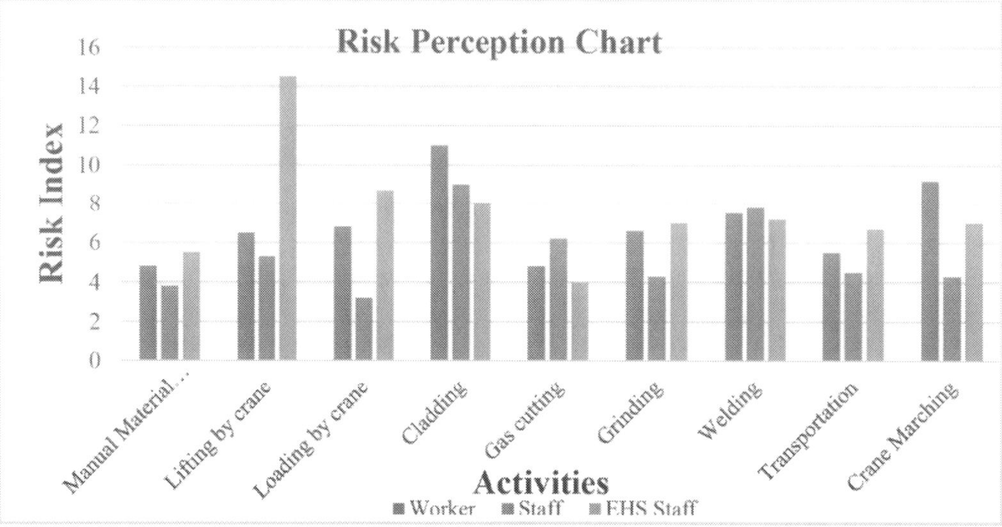

Figure 2. Risk Perception Chart

Risk Control in Construction Workplace

Risk control in construction sites plays an important role in the overall management of the safety and health of the operation. This requires an in-depth analysis of the activities and associated risks in the dynamic situation on a regular basis. To achieve this, factors that could harm people should be clearly identified and checked whether suitable steps are already integrated to prevent adverse consequences. This is known as risk assessment, and should be regularly carried out for best results. Identifying the appropriate mitigation measure and implementing the same to control risks is also a significant part of risk assessment. This could be based on past experience and emerging safety observations, followed by risk rating and risk categorization:

- Steps to assess risks
- Identify potential hazards
- Decide how it can harm people and who will be harmed.
- Evaluate the identified risks and decide on mitigation measures
- Note the major findings.
- Review the assessment and update if necessary

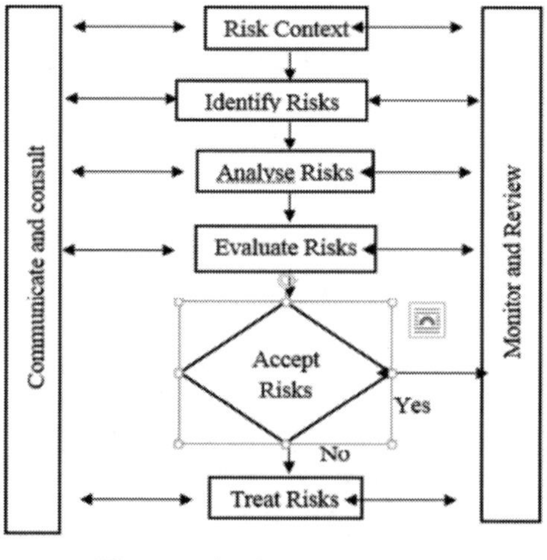

Figure 3. Risk Management Process

References

Campbell Institute. 2014. *Risk Perception: Theories, Strategies, and Next Steps* (retrieved April 1,2017) (http://www.nsc.org/CambpellInstituteandAwardDocuments/WP-Risk%20Perception.pdf).

Hopkin, Paul. 2014. *Fundamentals of Risk Management: Understanding, Evaluating and Implementing Effective Risk Management*. 3rd ed. Philadelphia: Kogan Page.

Kozlovská, Mária, and Zuzana Struková. 2012. "Overview of Safety Risk Perception in Construction." *Journal of Civil Engineering and Architecture* 6, no. 2.

Session No. 634-B

Hiding Incidents, Its Consequences and Analysis within the Company

Muhammad Yasir, Field HSE Advisor,
B. Sc (Petroleum and Gas Engineering), TechIOSH, NEBOSH IGC, Rospa
British Safety Council Level 5 Diploma in Health and Safety
Abraaj Energy Services SAOC
Sultanate of Oman

Umair Ahmed Abbasi, Project HSE Manager,
Postgraduate Diploma (Health, Safety and Environmental Engineering), NEBOSH IGC
British Safety Council Level 6 Diploma in Health, Safety and Environmental Management
Almabani General Contractors
Jeddah, Kingdom of Saudi Arabia

Introduction

In the oil and gas industry while working at the drilling rigs, dangers at workplaces are very high. As they say, in the battle of iron and flesh, Iron always win. Needless to say, working with steel is not easy and we have to prevent ourselves from being its victims. People should always be in defensive mode when it comes to working with steel and should stay ahead of the plan knowing the behavior of it. People working on the off-shore and on-shore drilling rigs are familiar with the dangers at work place and several tools are developed to enhance the understanding of individuals about identifying such hazards and hunting them down to prevent themselves and their fellow workers from being injured.

Recently, one of our drilling units in the fleet found to have issues of not reporting incidents officially. This became serious when one such incident somehow got reported through a different channel (unofficially) where an involved crew member had fractured his ankle but no such thing was reported from the drilling unit. This raised a concern to find out why incidents are not being reported. It became a question of utter importance because many rigs state that they have worked 3 years without lost time injury, five years, some say ten years without any lost time injury. Now, to us, we think that this number may have been achieved by hiding small injuries, however, this could not be suddenly concluded unless/otherwise proved. So, conducted sessions on different drilling units one by one and tested the level of conformance of the implemented incident management process. Several sessions were held with the crews where they were asked about the barriers they see in reporting incidents.

Below is a summary recorded from discussions with almost 115 employees of the drilling crew in eight different sessions on hiding and reporting injuries at work place.

The later part covers another approach of highlighting incidents which were reported during 2014 and 2015 within the company. All those incidents reported in 2014 had several recommendations which were carried over to the next year and implemented slowly. This brought the introduction of new tools, modified procedures, changes in layouts etc. were implemented. The comparison of the incidents in 2014 with 2015 shows reduction in the incidents, which clearly tells that incident reporting has prevented many people from being injured, property from being damaged, environment from being harmed and reputation from being destroyed.

Causes of Not Reporting:

The discussion sessions were planned in advance and the crew from the drilling section was invited to participate in the meeting. They were explained the purpose and background of the discussion on why the incidents are not reported and later on conclusion was shared.

- During the discussion, it was found that the crew has a fear of being disciplined. There are written rules known to everyone on board and violating every single rule will result in taking disciplinary action against the individual. These include violating the twelve life saving rules (12 LSRs), procedural violations, house rules violations, permit to work violations etc. which all fall under the category of disciplining the violators. They understand that the injury could have been prevented if a different approach was adopted. But after being injured, the damage has already taken place and now the time is to face the consequence. Since, they understand that the safety track record is of utter importance to them and all their good performance will be forgotten on one mistake. They prefer to keep quite and not to report to anyone to avoid the entire cross-questions, interviews in the follow up investigation.

- The ratio of promotions in this region is very high and people get hired and promoted quickly. It takes almost a year to jump from a green hat employee to be promoted to a regular employee and then keep on promoting to another level with a difference of 1-3 years, maximum. A floor hand could have an experience of 2-3 years on an average, which means that he barely stayed for about half of his total experience on rig, due to equal on/off working schedule. Now, if such an employee gets injured who has barely completed one or two years on the rig should be worried about this job or being demoted again. This fear comes from the follow up investigations, which cover the employee profile highlighting his experience on his job, current position with the company and total experience. Therefore, some believed that they might get demoted or fired if they report injuries. To them, it was more important to keep their job and then to disseminate the learning from the injury. While some believed if they report injuries, their immediate supervisor and site supervisors may think it had happened due poor involvement in the job and not taking it seriously. So the confidence their supervisors have been put in them to do one job is shattered.

- Another critical factor of hiding injuries was employee's reputation. Since everyone knows about the quick promotions and that only those got promoted who worked hard and contributed to the best of their capabilities. Their immediate supervisors recommended them for promotions based on their outstanding performance. However, when such employees got injured, they preferred not to report, as it would destroy the good image they had built by their hard work and devotion. Investigation from the injury reports may highlight that the employee was at fault; therefore, they consider keeping it between them and not reporting them would be their face saving.

- Every employee has an appraisal submitted by the unit manager on annual basis. This appraisal is important and describes the performance of employee with the company and recommends him raise in salary and allows him to be short listed for next level promotions and gain more benefits. Some had fear if they report injury; it may show their lack of understanding of the company rules and procedures or otherwise, violating them may result in losing points in their appraisals. Bad appraisals may stop them being promoted to the next level. In other words, report an injury and bear the loss financially.

- In drilling operation, completing a well without any incident before the stipulated time makes all the employees eligible for well bonus and incentive program. These well bonuses and incentives are delivered on monthly basis and gets stopped if an incident is occurred or if the well is completed after the target duration. So, in case of an injury, all the employees will lose the well completion bonus, therefore, they hide the injuries and do not report them just to protect their incentives.

- The most important of all is peer pressure and it links to everything that happens at the work. The fellow employees may taunt him all the time for scar he is bringing to the whole shift/crew. The fellow workers may link the loss of incentives because of his unsafe attitude.

Consequences of Not Reporting:

- The obvious consequence of not reporting an injury means no the opportunity to learn. The objective of the incident reporting is to find out reasons that led an incident to occur and adopting changes in the system, process, design and procedures to prevent the same from happening again. If an injury is occurred and not reported, it can happen again to the people exposed to similar conditions next time when all the unsafe sequence of events converge as in previous incident.

- Another consequence of non-reporting is the about many safety concerns which could have helped in preventing the incident from happening may never get addressed. The process that resulted in an injury may require to be changed would never be changed and people would follow the same obsolete and unsafe practices without realizing the exposure of dangers in the specific task.

- Non-reporting would never address the changes that may have required in the layout and design. Incidents have revealed that relocation of equipment have resulted in high potential incidents ignoring the safety aspects of it. Therefore, changes in the equipment or they layout are crucial and not permitted without the management of change procedure. Non-reporting would obviously put more people at risk working in similar surroundings.

- Non-reporting would never bring any improvement in the existing processes whether inadequate or not being enforced. Permit to work system existing, as a paper exercise would never guarantee well-being of the workforce at work. Consequences in terms of failure in the implementation of existing process are result of poor enforcement from the worksite leadership. Without a doubt, the existing processes could be inadequate and may require attention but would never get done unless they are reported.

- Incident findings do not focus on blaming individuals, however, people hiding incidents would never let the company find out how certain equipment could be made safer to use. Findings from the incidents state that the equipment involved in a particular incident was the main cause and it was not a human error. In such cases, changes that may be required in the placement, operation and maintenance of the equipment would never get addressed. Certain new equipment require special tools to operate or use them and some people try to adopt short cuts by making such required tools in the workshops, which are commonly called home made tools. Now, the incidents that involve the use of home-made tools resulting in injuries would definitely put people at risk who have still not been hurt. If such incidents are not reported, the companies who are bound to provide safe working environment would not be able to provide them the right tools. Learning from the use of home made tools will definitely result in procuring/purchasing the right tools for the job or equipment.

- Work instructions, job safety analysis, standard operating practices, risk assessment, management of change, are few important areas that deal with the safety and well being of the individuals at a worksite. These areas need continual improvement on regular basis and if they are not updated or reviewed, then the safety of the entire team is at stakes.

- High potential incidents have occurred when an object, which is not supposed to be placed at one place, is unknowingly kept there without analyzing the risk associated that may result in catastrophic incidents at the workplace. The installation of new equipment, removal of any obsolete devices, relocation of equipment even by few feet shall carefully be recorded and analyzed. Changes made out in the facility should properly be approved and deployed after all the risk is assessed and controls are put in place. Failure to execute the process will obviously result in putting more lives at risk, and ignoring the fact of learning from others' mistakes.

- Certain methods adopted in combating the work place a job differently in a safer manner may never get updated/modified. Similar injuries/events may repeatedly occur. Exposure of risk may never get eliminated or reduced and can even increase with time. Rig employees and third party crew member or bystanders would remain exposed to similar hazards without putting any controls in place. Work areas would remain unsafe to work. Companies may get penalized for not finding solution for the problem; unless they don't get reported, nothing would ever get done.

- Behavior of the people has been one of the major cause of the incidents. The incident statistics show that the equipment has not hurt anyone unless it is improperly rigged up, secured or managed by people. Incomplete rig up, poor inspections, use of faulty devices and taking short cuts to do one job has been a cause of the incidents too. Now such people have not only put their lives at risk, but also teaching the young employees with wrong message too. Reporting has to enhance and such employees should not be allowed to be part of the team who instead of making a workplace safe try to corrode the system from inside.

Impact on Injured Person (IP):

Injuries can become sufferings for a life time. Workplace hazards released can cause minor injuries and but if the potential is high, the injured person (IP) can suffer major impacts leading to permanent total disability (PTD). There are many consequences associated with such injuries on IP's life physically, socially, mentally, psychologically, and financially.

- IP may never get any adequate medical assistance. Trying to save his image, he may not report the injury and try to cover it by putting ordinary bandages. Such injuries may worsen the situation and a small contusion may result in amputation.

- Exposure will stay the same and if not the IP again, one of his fellow employee would surely get suffered. If the circumstances that led the incident to occur are not corrected, it may get repeated again. Nothing would be done to eliminate or at least, reduce the exposure. Some unsafe conditions can become catastrophic unless they are addressed. Those employees trying to hide such high potential or even low potential conditions may become partner in crime with those who saw an unsafe condition and didn't do anything about it.

- IP may need professional assistance and guidance which would never get done. Everyone has a different pace to work and some are quick learners while some take time. The site managers should see the behavior of such employees at work and compare their performance so plans could be laid out in order to fix the problems. Some people with the passage of time, gaining some experience and getting familiarized with operation and equipment start to develop bad habits of bypassing systems. People with such habits shall immediately be reported so they could be given adequate assistance and guidance. An in-time damage control will help the IP from further getting injured.

- IP may need more training and counselling which would never get done. Mid-level experienced employees sometime think they know everything. Workers in the company get promoted quickly do not get the chance of adequate learning on a work place and quick promotion also reduce the time to learn and impart their learnings. Training is very important part in the professional development of the employees and if they lack training and get promoted, they would become hazard for themselves and those they work with. Findings from the incidents could state that the employee may had poor judgement, slow reaction time, poor coordination, low aptitude, under the influence of medication, under depression or any fears or phobias which may express the poor mental state of him. Such employees should not be allowed to work be taken to medical facilities or be given psychological assistance from a psychiatrist. The company is willing to do everything by which an employee could be helped with to understand the gist of having all those procedures in place and their implementation.

- IP may require more focus while doing jobs may never get done. Coaching and Supervision can not only help one understand his job fully, but improve his performance at work too. There could be many reasons why an incident could occur if someone's focus is diverted like people having problems at home, lack of experience, not knowing the next step, not performing while working

- Persons may need open-ness to talk and discuss issues. Pressurizing them during tool box talks will make employees quite. There has to be an air of open-ness where employees do not get afraid to talk about their problems. This starts from the tool box talks where supervisors must ensure two-way communication with them. Supervisors must ask open end questions to see their understandings and get an input from them if they had any issues by doing the job in that specific manner. Let them speak up on how the same job could be done differently. Supervisors should take care of their teams at worksite by avoiding frustration, confusing instructions, conflicting directions, degrading instructions so the employees may not work under any kind of stress.

- Worst case scenario, employees may lose their jobs. Nobody would want to work with such employees who do not give 100% of efforts at work. Why would one want to do two person's job if the other is not willing to do.

- IPs may get a compensation in cash if he has been permanently disabled due to any work related injury, and it may not become possible for him to continue working on the same job role. Hands & fingers are the basic tools of life and without them, many challenges could be faced for the rest of the life.

- No personal professional development for employees with poor working record.

- Compromising personal health and safety over monetary benefits/losses in case if someone wants to save safety bonus of the year or monthly incentives.

Impact on Injured Person's Family:

1. There is a huge emotional impact attached to the injury.
2. Some injuries have long-lasting impacts on the person and his family. There was an incident in Oman when an assistant driller, a crew supervisor, was working with rigging down stabbing board. He was holding the rope and standing underneath the load. The person on the top was disengaging the stabbing board plateform. Since the stabbing board was under tension due to its weight, one of the pins could not be knocked off. The person on top finally released by pushing the stabbing board on one side allowing the stabbing board to free fall from height of about 5 meters on the head of the supervisor holding the guide rope. It resulted in breaking is his back bone at several points leaving his legs and lower part of the body below waist permanently paralyzed. The IP could not longer do things that a normal person can do. He played football regularly in his time off and use to jog on the coast regularly. His life has completely been changed and injury he had left long lasting impacts.
3. Financial losses in supporting families are very common. Keeping in view the example stated above, the person can no longer work on the drilling rigs. He may not be able to support his family as he used to. He may not be able to work as long as he used to do while he was physically fit.
4. Injuries may result in people with no jobs. No company wants to take careless employees or people with bad safety track records. A person who is not a team player cannot be accepted to work any longer with the team.
5. Injuries may restrict the IP enjoy the beauty of life

Impact on the Team IP Works With:

- Peer pressure
- Crew to blame the IP for destroying safety records by his unsafe attitude
- Fears of being side lined by other crew members

Impact on the Safety Culture on the Rig/Unit:

- People ignoring others bad habits and unsafe attitude
- People not implementing HSE tools in its true spirit
- People not motivated enough to participate in safety campaigns
- Supervisors not concerned about their teams' safety
- Learnings not being shared.
- Learnings not being adopted and practiced.
- Others mistakes/incidents/injuries being repeated but narrowly being saved from getting injured
- People not realizing dangerous occurrences
- People not performing hazard hunts and risk assessments
- People adopting conventional and obsolete practices
- People taking short cuts
- People ignoring procedures and standard operating practices, work instruction and thus not carrying out job safety analysis.

Positive Aspects of Reporting:

- Reporting is to fix issues so they don't get repeated
- Reporting is not a blame game. Management commitment is required to be known to all so they understand the objectives of reporting.
- Reporting is to ensure that the recommendations are being addressed
- Reporting is to ensure the modifications in equipment, procedures, processes and layouts get addressed.
- Reporting is to address risks by 'Prevention through Design'.
- Reporting is to share a learning with everyone so the same don't get repeated.

Management Support:

1. Provide resources to address the issues highlighted through stop cards/hazard hunts/inspections/audits
2. Provide adequate training to concerned
3. Analyze crew members behaviors and provide counselling (Onsite/on duty counselling and off duty counselling in town if felt needed)
4. Demonstrate leadership commitment
5. Implement incentive programs. It was found that the incentives are not being paid on time. This is where management needs to look at more closely.

Conclusion

Through the discussions, the EHS team suggested introduction of an anonymous reporting system in which the names of the individuals would be kept secret if they report the incidents on a specific phone line. I see this as a negative sign and it gives an impression that management is supporting those who hide incidents and then adopting a different channel to deal with them. It should not be taken that far and be addressed at the very base level.

We, being EHS professionals along with the maintenance and drilling team supervisors should teach the people to do their jobs correctly. They should learn to use the EHS tools in their daily routine and non-routine activities. Restoring reporting culture organisation is difficult where interventions are not considered important. Site management has to enforce basic EHS Tools such as stop cards. Workers should always be encourage to intervene and stop unsafe acts and report them through the stop cards system. This starts by making more and more interventions. To do more interventions, we need our daily tool box talks to be more effective and remind crew members about their empowerment to stop. They should know that the change in job steps would require a TBT to be repeated. Pre tour TBTs should be focused on the day to day activities, while pre job TBTs should cover the job specific hazards. Visual tools such as HAZ-ID books should be showed to the crew members to see the hazards they are exposed to and how to implement the controls in their activity. Time to time, EHS professionals should run procedural discussions with the crews so they understand their job steps and take necessary precautions. Motivating and acknowledging their efforts to intervene and stopping unpleasant events from occurring would definitely develop a positive reporting culture in an organisation.

OSHA Compliance: Emerging Issues and Inspection Preparedness

Brendan Moriarty, CSP, CIH, MSPH, AOEE
Vice President
Casualty Executive Field Specialist
Chubb North America
Chicago, Illinois

Introduction

Companies invest thousands of dollars in OSHA based safety programs in terms of development time, production downtime for training, software and engineering changes. These same companies spend very little time on actually planning for an OSHA inspection. Many companies today have business contingency plans in the event of a fire or natural disaster but neglect to plan for an OSHA Inspection. Having an OSHA preparedness plan in 2017 will be vital for many organizations due to increased OSHA fines, new OSHA rules and requirements, additional focus on National Emphasis Programs and the use of the General Duty Clause.

OSHA is changing its inspection metric to add weight to complex inspections such as those involving ergonomic hazards, chemical exposures, workplace violence and inspections, which typically involve the use of the General Duty Clause. OSHA deems this new system as its resource-intensive enforcement activity and is focused on some of the most severe work place hazards. (1)

OSHA fines in 2017 are increasing as well; OSHA has decided to raise penalties by about 78 percent, resulting in a maximum fine of $124,709 for repeat or willful violations compared to the previous fine of $70,000. Fines for serious and other-than-serious violations will now have a maximum fine of $12,471, an increase from the prior fine of $7,000 for serious and other-than-serious violations. (2)

Occupational Health Hazards have always been an OSHA priority through the development of National Emphasis Programs and the requirements associated with hexavalent chromium, lead and other chemicals (8). The most recent rule change pertains to silica and includes provisions, which will significantly lower the permissible exposure limits and require employers to use engineering controls or other means to limit worker exposure. (3)

Under a new rule, employers will be required to electronically submit certain records of workplace injuries. This rule would require larger companies and what OSHA considers "high hazard employers" to submit electronic versions of their injury and illness logs. OSHA feels that the analysis of this data will allow them to use its enforcement more efficiently and some of the data will also be posted to the OSHA website. (4)

Companies today need to be prepared for these changes to ensure that all of their safety efforts and programs are up to date and reflect these new regulations. Companies can audit their existing programs by conducting an internal mock OSHA inspection to ensure that all aspects of their written programs are complete and that all employees are fully trained on each applicable OSHA standard. If companies do not conduct a mock OSHA inspection, their first inspection may be completed by an OSHA inspector instead of an internal safety committee or an internal management group.

OSHA Inspections and Severe Injury Reports (SIR)

Employers that report a severe injury may be subject to an OSHA inspection if the injury meets certain criteria. A severe injury is an injury such as a fatality, in-patient hospitalization, amputation, or loss of an eye as the result of a work-related incident. These types of injuries require the employer to report the injury to OSHA within 8 hours. These reports are referred to as Severe Injury Reports (SIR).

Once reported, OSHA will examine the Severe Injury Report and determine if the injury associated with the Employer Report falls into a specific category. OSHA will categorize the injury report as a Category 1, 2 or 3. An onsite inspection of an employer's business conducted by an OSHA inspector will occur if the injury or circumstances surrounding the report meets certain criteria. If OSHA deems that the injury does not warrant an inspection, an employer is expected to conduct its own investigation into the work-related incident and share its findings with OSHA.

These findings must involve written documentation called "abatement certification", which needs to be signed by an official of the company and should detail the abatement measures that have been implemented to correct the condition that was the basis of the employer report.

After OSHA receives a severe injury report from an organization, the OSHA Area Director or his/her designee must determine whether to conduct an inspection or a Rapid Response Investigation (non-onsite). The Severe Injury Reports provided by the employer will fall into one of three categories: Category 1 reports will require an OSHA inspection; Category 2 - reports may result in either an inspection or a Rapid Response Investigation.

A Category 1 inspection will be conducted if any one of the conditions below is reported:

- All fatalities and reports of 2 or more in-patient hospitalizations;
- Any injury involving a worker under 18.
- The employer has a history of the same or similar hazards or incidents within the past 12 months

- Repeat offenders
- History of egregious, willful, failure-to-abate, or repeated citations
- Report of a hazard covered by a local, regional, or national emphasis program.
- Any imminent danger.

OSHA inspections will begin within five working days (except for fatalities and catastrophes) of receipt of the employer report. A Category 2 employer report does not involve any of the conditions contained in Category 1. Category 2 reports are reviewed by the OSHA Area Director to determine if an OSHA inspection is warranted or if a Rapid Response Investigation shall be conducted. This determination is based on the circumstances of the event and consideration of a number of different factors. Some of these factors are listed below:

- Have the work conditions that resulted in the employer report of injury or illness been corrected? What were the corrective measures taken and how quickly were they implemented?
- Can complete abatement of the reported workplace condition be implemented before an inspection is conducted?
- Are other employees still exposed to the hazard that resulted in the reported injury, illness or fatality? If so, how many employees?
- Was the work related incident the result of a safety program failure such as Confined Space, Lockout Tagout or Process Safety Management?
- Did the work related incident involve a serious hazard such as explosive materials, combustible dust or falls?
- Does the employer have prior OSHA inspection history?
- Is the employer an OSHA, VPP, OSHA Strategic partnerships, SHARP or an active Alliance member?
- Did the work related incident involve a health hazard such as chemical or heat stress exposures? (5)

A Rapid Response Investigation (RRI) will begin within one day after receipt of the employer report. The RRI is intended to identify any hazards, provide abatement assistance and confirm abatement. OSHA is looking for the employer to conduct an internal investigation of the incident, provided documentation confirming the investigation and corrective action and post OSHA RRI documentation in the area where the injury occurred. An employer has five days after they have initiated the severe injury report to inform the Area Director in writing of the results of their internal investigation by letter or email.

The written document should confirm that actions have been taken to correct the conditions that resulted in the employer report. If the employer cannot complete its internal investigation and/or completely correct the condition within those five working days, it should notify OSHA in writing, before expiration of the five days, of the reasons why the investigation and/or corrective action cannot be completed. The employer also needs to send an Abatement Verification signed by a company official which describes the completed corrective action.

The employer should post a copy of the RRI letter from OSHA in a conspicuous place where all affected employees will have notice or near the location where the incident occurred.

If OSHA has not received a response from an employer (or a response deemed inadequate by OSHA) within the five working days, the Area Director will decide whether to conduct an onsite inspection or to make further attempts to contact the employer. (5)

OSHA Inspections

OSHA is changing the inspection metric for OSHA Inspectors from the number of inspections conducted by the inspector to the complexity of the inspection conducted by the inspector. Inspectors will get more "credit" for inspections involving ergonomic hazards, chemical exposures, workplace violence and Process Safety Management violations.

The prior inspection metric that gave equal weight to all inspections, which may have discouraged some OSHA offices from committing the resources to complete more complex investigations involving serious hazards.

As a result, OSHA felt that some employers may have thought that OSHA was unlikely to conduct an inspection looking at ergonomic hazards and certain chemical hazards due to the lack of a specific OSHA standard. This new system emphasizes the importance of inspection and enforcement activity that is focused on some of the most hazardous exposures such as ergonomics, heat, chemical exposures, workplace violence and process safety management.

The weighting system assigns a value to specific categories of inspections. The value is known as an Enforcement Unit (EU) and all inspections receive an EU value of at least one. However, these specific categories receive the following EU values: (1)

- Process Safety Management Inspections - 7 EUs
- Ergonomic Hazard Inspections - 5 EUs
- Heat Hazard Inspections - 4 EUs
- Workplace Violence Hazard Inspections - 3 EUs
- Fatality / Catastrophe Inspections - 3 EUs
- Federal Agency Inspections - 2 EUs
- Rapid Response Investigations - $^1/_9$EU

Slips, Trips and Falls

Slips, trips, and falls comprise the majority of general industry accidents. They cause 15% of all accidental deaths, and are second only to motor vehicles as a cause of fatalities. The OSHA standards for walking/working surfaces applies to all permanent places of employment, except where only domestic, mining, or agricultural work is performed. To reduce the potential for these hazards OSHA created OSHA Standard 1910 Subpart D in 1971. OSHA has made modifications to Subpart D, which will impact more than 100 million employees at almost 7 million general industry establishments. The modifications made to Subpart D includes new requirements and policies intended to identify and reduce slip, trip and fall exposures within general industry.

The new requirements address common slip, trip and fall exposure areas such as stairs, ladders, roof, floors and aisleways. The new rule for general industry requires employers to keep walking-working surfaces with the workplace in a clean and orderly sanitary condition free from protruding objects, liquids, spills, snow and ice.

Employers are also required to develop programs to identify and evaluate slip, trip, and fall hazards and provide appropriate personal protective equipment such as fall arrest systems if needed. One of these programs involves an inspection protocol, which includes regular and periodic inspections and maintenance of all walking-working surfaces in an employer's workplace. Any deficiencies or hazardous conditions noted during an inspection must be corrected before an employee uses the walking-working surface again.

Slip, Trip and Fall training must be provided to employees by a qualified person and include: fall hazard awareness, procedures to minimize falls hazards, fall hazards and procedures associated with the use of personal fall protection systems. Retraining must be provided when the employer determines that the employee does not understand the training or when workplace changes occur which render prior training obsolete or inadequate. Employers must ensure employees are trained in these requirements by May 17, 2017. (6)

Dermal Exposure

Skin exposure to chemicals in the workplace is a significant problem in the US. Both the number of cases and the rate of skin disease in the US exceed recordable respiratory illnesses. In 2010, 34,400 recordable skin diseases were reported by the Bureau of Labor Statistics (BLS) at a rate of 3.4 injuries per 10,000 employees, compared to 19,300 respiratory illnesses with a rate of 1.9 illnesses per 10,000 employees.

Most chemicals are readily absorbed through the skin and can cause other health effects and/or contribute to the dose absorbed by inhalation of the chemical from the air. Many studies indicate that absorption of chemicals through the skin can occur without being noticed by the worker. In many cases, skin is a more significant route of exposure than the lung.

This is particularly true for non-volatile chemicals which are relatively toxic and which remain on work surfaces for long periods of time. The number of occupational illnesses caused by skin absorption of chemicals is not known. However, it is argued that an estimated 60,000 deaths and 860,000 occupational illnesses per year in the US attributed to occupational exposure, a relatively small percentage caused by skin exposure would represent a significant health risk. (7)

Dermal exposure can be controlled and prevented. Many times it is as simple as changing the chemicals being used. When that is not an option, there are many types of personal protective equipment (PPE) that are available. Being familiar with the chemicals that are used in a facility and the specific information on routes of entry and employee protective measures contained in the Safety Data Sheet will help reduce dermal exposures to employees.

Employers should follow the hierarchy of controls to prevent occupational skin disease in employees. This would include, elimination (preventing skin contact), substitution (less hazardous compound), engineering controls (isolation booths), administrative controls (training) and personal protective equipment (gloves, coveralls, etc.).

Hand contact is a significant route of exposure. Therefore, proper glove selection is a major means of controlling dermal exposure. Factors that affect glove selection include:

- Type of chemical handled

- Frequency and duration of chemical contact
- Nature of contact, splash, mist immersion, contaminated surfaces
- Concentration of the chemical
- Temperature of the chemical
- Size and comfort requirements
- Glove exposure to tear of abrasive conditions

Common dermal exposure activities include the following:

- Wet or dry cleaning of work tools, equipment and work areas
- Disinfection of work tools, equipment, and work areas
- Contact with solvents
- Contact with monomers of epoxy resins and tacky surfaces or hardening agents (such as glue or epoxy resins)
- Use of preparations containing soaps, detergents, and disinfectants (7)

OSHA National Emphasis Programs

OSHA's National Emphasis Programs (NEPs) focus on significant health and safety hazards OSHA has developed specific guidance and protocols to the OSHA field offices regarding these inspections. National Emphasis programs common to general industry include Isocyanates, hexavalent chromium and crystalline silica. (8)

Local Emphasis Programs are enforcement strategies designed and implemented at the regional office and/or area office levels. These programs are intended to address hazards or industries that pose a particular risk to workers within the specific region. The emphasis programs may be implemented by a single area office or at the regional level and applied to all of the area offices within the region. (9)

These LEPs will be accompanied by outreach intended to make employers in the area aware of the program as well as the hazards that the programs are designed to reduce or eliminate. This outreach may be in the form of documents sent by US mail, training at local tradeshows, or speeches at meetings of industry groups or labor organizations.(9)

Hexavalent Chromium

Welding and metal cutting can be a full time occupation within a facility or can be done sporadically by maintenance employees. Welding can bring forth significant occupational health hazards and the exposure needs to be evaluated regardless of how often welding operations take place within a facility.

Two of the principal occupational health concerns associated with welding involve exposure to hexavalent chromium and manganese. Industrial Hygiene air sampling for welding fumes (manganese) and hexavalent chromium is often required by OSHA and air sampling results will typically be reviewed during OSHA inspections.

Hexavalent chromium is generated during hot work such as welding on stainless steel or other steel material containing chromium metal. Hexavalent Chromium is an A1 confirmed carcinogen. It is estimated that 558,000 workers are potentially exposed to Hexavalent Chromium in the United States. (10)

The OSHA Permissible Exposure Limit is for Hexavalent Chromium and Action Level are very low. OSHA also has targeted inspections associated with employee exposure to hexavalent chromium. High levels of hexavalent chromium can also be found in chromate pigments (paints). If sampled levels exceed the OSHA Action Level or Permissible Exposure Limit, the employer must conduct periodic sampling and meet other requirements in order to comply with OSHA Standards. (10)

Isocyanates

Isocyanates are extremely hazardous and present significant occupational health exposures to employees using these materials. Isocyanates can sensitize workers, making them subject to severe asthma attacks after repeated exposure. There is evidence that both respiratory and dermal exposures can lead to sensitization. Death from severe asthma in some sensitized subjects has been reported. These powerful irritants are used in the manufacture of flexible and rigid foams, fibers, coatings such as paints and varnishes, and elastomers. Isocyanates are common within the automobile industry, auto body repair, and in protective coatings for truck beds, trailers, boats and decks. (11)

The most widely used compounds are diisocyanates, including methylenebis (phenyl isocyanate) (MDI), toluene diisocyanate (TDI), and hexamethylene diisocyanate (HDI). Safety Data Sheets to determine. (11)

Safety Administrators from industries like those mentioned above need to review Safety Data Sheets to determine if isocyanates are present within the materials that they use and give the exposure top priority in terms of developing controls to reduce employee exposure.

Controls such as closed systems and ventilation should be the principal method for minimizing isocyanate exposure in the workplace. Other controls, such as worker isolation and use of personal protective equipment such as respirators and personal protective clothing to prevent dermal exposures may also be necessary. (11)

Crystalline Silica

Occupational exposures to crystalline silica are associated with the development of silicosis, lung cancer, pulmonary and other airways diseases. Close to 1.7 million U.S. workers are exposed to respirable crystalline silica in a variety of industries and occupations, including construction, sandblasting, and mining. (12)

The Occupational Safety and Health Administration (OSHA) has issued a final rule to curb lung cancer, silicosis, chronic obstructive pulmonary disease and kidney disease in America's workers by limiting their exposure to respirable crystalline silica. The rule is comprised of two standards, one for Construction and one for General Industry and Maritime.

OSHA estimates that the rule will save over 600 lives and prevent more than 900 new cases of silicosis each year, once its effects are fully realized. Key provisions of the final rule include:

- Reduces the permissible exposure limit (PEL) for respirable crystalline silica to 50 micrograms per cubic meter of air, averaged over an 8-hour shift.
- Use engineering controls (such as water or ventilation) to limit worker exposure to the PEL;
- Provide respirators when engineering controls cannot adequately limit exposure; limit worker access to high exposure areas
- Develop a written exposure control plan
- Offer medical exams to highly exposed workers
- Train workers on silica risks and how to limit exposures.
- Provides medical exams to monitor highly exposed workers and give them information about their lung health. (3)

Safety Administrators from industries that have a crystalline silica exposure need to review the materials being used or fabricated to determine if they contain even small amounts of crystalline silica and give the exposure top priority in terms of developing controls to reduce employee exposure. The first step would involve conducting industrial hygiene air sampling to quantify the exposure.

If levels exceed established exposure limits, the implementation of controls needs to be immediate to reduce further employee exposure. Typical controls pertaining to crystalline silica exposure includes using wet methods, dust suppression systems, ventilation, enclosure and respiratory protection. (12)

OSHA has set forth the following compliance schedule for Crystalline Silica:
Both standards contained in the final rule take effect on June 23, 2016, after which industries have one to five years to comply with most requirements, based on the following schedule:

- Construction - June 23, 2017, one year after the effective date.
- General Industry and Maritime - June 23, 2018, two years after the effective date.
- Hydraulic Fracturing - June 23, 2018, two years after the effective date for all provisions except Engineering Controls, which have a compliance date of June 23, 2021.(3)

OSHA Recordkeeping, Electronic Submission of Records and Anti-Retaliation

OSHA has developed new provisions which call for employers to electronically submit injury and illness data that they already record on the OSHA 300, 301 and 301A. This new requirement does not add to or change an employer's obligation to complete, retain, and certify injury and illness records. The new requirement mandates that certain employers electronically submit some of the information from these records to OSHA. Some of this data will also be posted to the OSHA website.

The amount of data submitted will vary depending on the size of company and type of industry. OSHA will provide a secure website for the electronic submission of information. The website will include web forms for direct data entry and instructions for other means of submission (e.g. file uploads). Establishments must submit the information electronically and may not submit the information on paper.

OSHA has effective safeguards in place to prevent the disclosure of personal or confidential information contained in the recordkeeping forms submitted to OSHA. OSHA will not collect employee name, employee address, name of physician or other health care professional or healthcare facility name if treatment was given away from the worksite.

OSHA believes that public disclosure of injury and illness information will encourage employers to improve workplace safety and provide valuable information to employees, job seekers, customers and the general public. (4)

The electronic reporting requirements are based on the size of the establishment, not the firm. The OSHA injury and illness records are maintained at the establishment level. Employers need to determine if they need to provide OSHA with the required data for an establishment, by determining the establishment's peak employment during the last calendar year. Each individual employed in the establishment at any time during the calendar year counts as one employee, including full-time, part-time, seasonal, and temporary workers.

OSHA has defined an establishment as a single physical location where business is conducted or where services or industrial operations are performed. A firm may be comprised of one or more establishments. (4)

- The final rule takes effect Jan. 1, 2017, and reporting requirements will be phased in over two years;
- Establishments with 250 or more employees must begin submitting information from Form 300A by July 1, 2017, and must submit information from all forms (300A, 300, and 301) by July 1, 2018.
    - In 2019 and every year thereafter, the information must be submitted by March 2, 2019.
- Establishments with 20-249 employees in certain high-risk industries must begin submitting information from Form 300A by July 1, 2017, and again by July 1, 2018.
    - In 2019 and every year thereafter, the information must be submitted by March 2, 2019.

OSHA State Plan states must adopt requirements that are substantially identical to the requirements in this final rule within 6 months after publication of this final rule. Some states may choose to allow employers in their state to use the federal OSHA data collection website to meet the new reporting obligations. Other states may provide their own data collection sites. (4)

The rule also has provisions prohibiting employers from discouraging workers from reporting an injury or illness. OSHA is requiring employers to inform employees of their right to report work-related injuries and illnesses free from retaliation. This requirement can be satisfied simply by posting the OSHA workplace poster, which is a long-standing OSHA requirement.

The OSHA Act (Section 11(c)) already prohibits any person from discharging or otherwise discriminating against an employee who reports a fatality, injury, or illness. However, OSHA may not act under that section unless an employee files a complaint with OSHA within 30 days of the retaliation.

In contrast, under the new rule, OSHA will be able to cite an employer for retaliation even if the employee did not file a complaint, or if the employer has a program that deters or discourages reporting through the threat of retaliation.

When workers are discouraged from reporting occupational injuries and illnesses, the information gathered and reported is incomplete and inaccurate.

The rule includes *three* provisions that are intended to address this issue:

- An employer's procedure for reporting work-related injuries and illnesses must be reasonable and must not deter or discourage employees from reporting
- Employers must inform employees of their right to report work-related injuries and illnesses free from retaliation
- An employer may not retaliate against employees for reporting work-related injuries or illnesses

This rule does not prescribe specific procedures that employers must establish. Rather, employers are free to establish their own procedures. Organizations should review their reporting procedures for elements that might deter or discourage a reasonable employee from accurately reporting a workplace injury or illness. (13)

Employers must inform employees that they have a right to report work-related injuries and illnesses free from retaliation by their employer. Employers can meet this requirement by posting the current version of the OSHA poster or other by other means such as providing employees a written notice or email of their right to report work-related injuries and illnesses free from retaliation.

The rule prohibits employers from taking adverse action against employees for reporting work-related injuries or illnesses. Adverse action is action taken by the employer that would discourage a reasonable employee from reporting a work-related illness or injury accurately. Examples of adverse action include:

- Discharge, demotion or denying a significant benefit to the employee.
- Demeaning or embarrassing the employee.
- Threatening to penalize or discipline an employee for reporting
- Requiring employees to take a drug test for reporting without a legitimate business reason for doing so

The rule does not prohibit drug testing of employees. The rule prohibits employers from using drug testing, or the threat of drug testing, as a form of retaliation against employees who report injuries or illnesses. If employee drug use could not have contributed to the injury or illness, post-incident drug testing would likely only discourage reporting without contributing to the employer's understanding of why the injury occurred. Drug testing under these conditions could constitute prohibited retaliation.

Employers may conduct post-incident drug testing if there is a reasonable possibility that employee drug use could have contributed to the reported injury or illness. The rule does not prohibit drug testing of employees, including drug testing pursuant to the Department of

Transportation rules or any other federal or state law. Employers may conduct post-incident drug testing pursuant to a state or federal law, including Workers' Compensation Drug Free Workplace policies, because the law does not apply to drug testing under state workers' compensation law or other state or federal law. Random drug testing and pre-employment drug testing are also not subject_to the new rule. (13)

This rule does not prohibit incentive programs. However, employers must not use incentive programs in a way that penalizes workers for reporting work-related injuries or illnesses. Incentive programs should encourage safe work practices and promote worker participation in safety-related activities.

Employers should consider programs that reward:

- Worker participation in safety program activities and evaluations;
- Worker completion of safety and health training;
- Reporting and responding to hazards and close calls/near misses;
- Safety walkthroughs and identification of hazards during safety walkthroughs/inspections;
- Conformance to planned preventive maintenance schedules;
- Compliance with legitimate workplace safety rules (13)

The rule does not prohibit disciplinary programs. However, employers must not use disciplinary action, or the threat of disciplinary action, to retaliate against an employee for reporting an injury or illness. Employers must ensure disciplinary action is based on the violation of specific work rules and is applied consistently with all employees regardless of employee injury.

Employers should review their reporting procedures, programs, and policies for elements that may result in retaliatory actions against an employee for reporting an injury or illness. (13)

References

1. U. S. Department of Labor Occupational Safety and Health Administration (OSHA). *New Enforcement Weighting System.* https://www.osha.gov/dep/enforcement/ews_memo_09302015.html
2. ______. *US Department of Labor announces new rules to adjust civil penalty amounts* https://www.osha.gov/pls/oshaweb/owadisp.show_document?p_table=NEWS_RELEASES&p_id=32745
3. ______. *OSHA's Final Rule to Protect Workers from Exposure to Respirable Crystalline Silica* https://www.osha.gov/silica/
4. ______. *OSHA Injury and Illness Recordkeeping and Reporting Requirements* https://www.osha.gov/recordkeeping/index.html

5. ______. *OSHA Severe Injury Reporting* https://www.osha.gov/dep/enforcement/Interm_Enforcement_Procedures.html

6. ______. *Final Rule to Update General Industry Walking-Working Surfaces and Fall Protection Standards* https://www.osha.gov/walking-working-surfaces/index.html

7. ______. *Safety and Health Topics – Dermal Exposure* https://www.osha.gov/SLTC/dermalexposure/hazard.html

8. ______.*National Emphasis Programs* https://www.osha.gov/dep/neps/nep-programs.html

9. ______. *Local Emphasis Programs* https://www.osha.gov/dep/leps/leps.html

10. ______. *Safety and Health Topics; Hexavalent Chromium;* (https://www.osha.gov/SLTC/hexavalentchromium/index.html)

11. Centers for Disease Control and Prevention (CDC); *Isocyanates* http://www.cdc.gov/niosh/topics/isocyanates/default.html

12. ______. *Silica* http://www.cdc.gov/niosh/topics/silica/default.html

13. U. S. Department of Labor Occupational Safety and Health Administration (OSHA); *Improve Tracking of Workplace Injuries and Illnesses – Employee's right to report injuries and illnesses free from retaliation* https://www.osha.gov/recordkeeping/modernization_guidance.html

Rethinking Risk Assessment: Adding the 3rd Dimension

Gary A. Higbee EMBA
Higbee & Associates, Inc.
Johnston, Iowa

Risk Assessment Process

Risk assessment processes have been around for a number of years. The first time I can remember using one was in the early 1960's. Yes, I'm aware this is almost ancient history but it has a place in this discussion. We were looking at all the ways one could be hurt while performing welding and assembly operations while building an agricultural tool called a roller Harrow. You would have a hard time finding a roller harrow today because technological advances have made it all but obsolete. Back then there was a big demand for the product and it was a difficult product to produce. Lots of heavy lifting, difficult weld positions and so many pinch points you couldn't count them all. We had few safety regulations but not many and OSHA was still years away. We had many injuries and we knew we had a problem. We had to do something. So, we sat down as a team and listed each step of the job and tried to determine any exposure that could cause an injury. Exposures like sharp edges, pinch points, hot material, heavy lifts, etc. Much like an early version of a Job Hazard Analysis or a job safety analysis. The basic risk matrix might look like Exhibit #1 below.

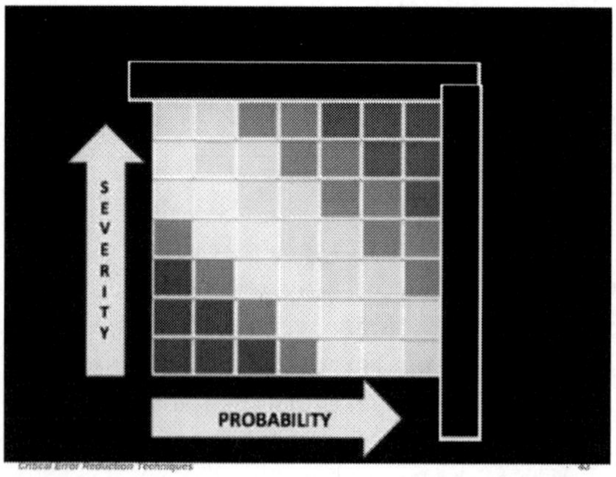

Exhibit 1.

I've taken the liberty to have five levels of risk represented, signified by the colors red, orange, yellow, beige, lime green and green. Most of the risk assessments I've seen have only three colors and of course 3 risk levels.

Essentially, we looked at the likelihood of something happening (Very Likely, Likely and Not Likely). Then we looked at the severity possibility if that something happened. We wanted to know how serious the injury might be (Very Serious, Serious or Not Serious). While this isn't rocket science, it is a system intended to create better awareness, and eliminate a situation we termed unacceptable risk. We use this system even today and for good reason. This system is simple and it gets us to increase our awareness and correct obvious issues.

The logical conclusion is that high probability and high severity required significant action to lower both severity and probability, while low probability and low severity would likely not require any or very little corrective action. Well, this all sounds good but there is an effectiveness issue because it is based only on two dimensions – likelihood and severity. Even back then I knew something was missing in the way the risks were assessed. What about miss-calculation, errors in judgement, issues with skill deficiencies or even simple human error. The risk assessment ignores any of these variables or others for that matter.

Discussion Example

As a long time baseball player and coach we tried to limit errors on defensive. Skill development, Process Positioning, Functional Dexterity, Muscle Memory, you name it we worked on it. We kept all types of data on errors and why they might have happened. The simple fact was the likelihood of an error was determined more by the speed of the runner than it was by anything else. Fast runners had the effect of making the fielder rush to the ball and even rush their throw. The rushing of the throw often caused the throw to be off line and the runner would be safe. Sometimes the throw would be so far off line that the runner was not just safe at first but they might reach second if the ball went into the dugout or the stands. There were even cases when the ball didn't go in the dugout but rolled into the outfield or into foul territory. And then the runner would reach third or maybe even home. The result level of the error was variable. So, when the fielder was stressed the likelihood of an error increased as well as the seriousness of the result.

Another Issue to Consider

Now there is another issue with this situation and it involves the variability of the result. Just because the fielder rushes the throw doesn't mean the result is always a negative result. What if the rushed throw turns out to be a great throw? Now the fielder gets on ESPN because he made the greatest defensive play of the week. Now we have added the issue of a positive result from a negative act. In other words, "Positive Reinforcement for a Negative Act". This can only encourage the rushing of a throw in the future and that rushed throw has all the result variables discussed previously.

Variability and Process Improvement

I'm an engineer. I hate variability. Give me a consistent repeatable process and it is relatively easy to improve. Give me a variable process and I now must make it repeatable and consistent before I have any chance to improve it. So, if you look at risk and include this state to error analysis, the risk might be understated using the simple two-dimensional method.

There are a number of states that can stress the situation, states like Rushing, Frustration, Anger, Fatigue, Fear, Panic and Over Confidence often called complacency. Any one or a combination can cause errors and not just safety errors but also quality and productivity errors.

Complacency

While there are many states as we noted above, none seem to have breadth and depth of complacency. If this is true, and I certainly believe it is, this state will turn our Risk Matrix redder than any other state. Not only that, it will also change the perceived risk dramatically. Let me use Exhibit #2 to explain.

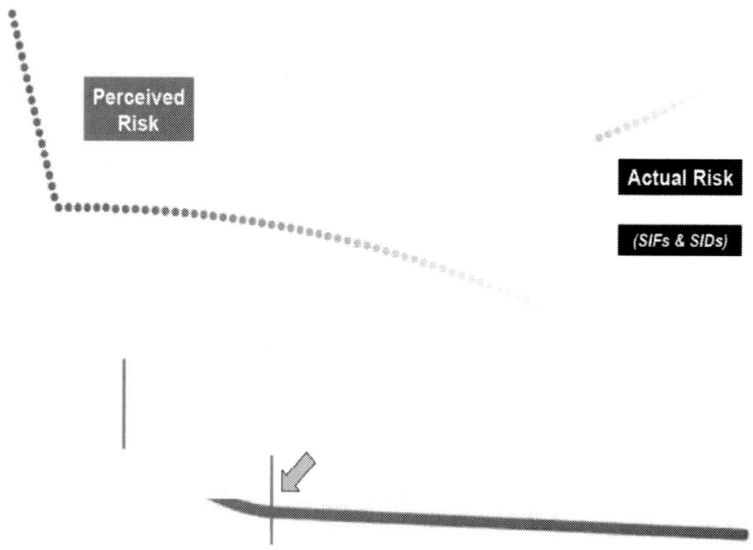

Exhibit 2.

The blue line is for skills starting in the lower left corner and as they gain skill and knowledge over time, the blue line raises. The continual raise of the blue line represents not only continual skill development but can account for newer technologies as they may be introduced into the task. When people have a task to learn like learning to ride a bicycle or a new job they start with low job skills but very high awareness. Awareness is shown by the multicolor line starting in the upper left corner (Huge awareness) and moving down toward the lower right corner (Low Awareness).

When their skill level reaches a point where they are competent that is the first stage of complacency point on the exhibit. They are good enough at the activity they no longer must be thinking about it and they can perform the task thinking about other things. At this point, there are few injuries or property damage events. The progression will likely continue and they will reach the second stage of complacency where their mind is more than wandering. They are at the point something external to them must happen to get their mind back on the task. At this point, the injuries will begin to increase, as well as property damage.

Often, we will calculate the actual risk of a task. This is a numerical evaluation that happens more in Europe than the United States. The method is very involved and while we will not discuss it here, the process is very valuable even though the accuracy is debatable. The actual risk is denoted as the line of solid black dots in the upper part of the exhibit.

The perceived risk is the risk level perceived by the employee. This is noted by the gray dots starting in the upper left corner and moving down over time. At first the perceived risk is higher than the actual risk but as they gain more job skills, the perceived risk will become closer to the actual risk. This is shown when the gray dots are horizontal. So, at this point they have a realistic perception of the risk based on a comparison of actual risk and perceived risk. This requires research to verify so what we are looking at today is a combination of research data and assumptions.

Between stage one and stage two, all the calculations start to separate. The actual risk is often thought to never change but when complacency starts to take over, the actual risk calculation changes with the addition of the Human Factor. At the same time the actual risk is increasing, the perceived risk continues to go down. At this point, the gap between perceived risk and actual risk is exponentially larger than most think possible. This gap can explain some of the increase in Significant Injuries and Fatalities (SIDs).

Stages of Complacency

We show two stages to the phenomenon called complacency on the exhibit but you can think of complacence as having three stages.

Complacency Stage 1 – Confident in one's ability to perform a function to the point they do not have to think about it to have repeated positive results. They can still respond to stimulus that brings them back to the moment. Example: You are driving on a trip and realize you are getting tired and you pull over for coffee. Just simple complacency but managed.

Complacency Stage 2 - Confident in one's ability to perform a function to the point they do not have to think about it to have repeated positive results. They can no longer respond to stimulus that should bring them back into the moment. Example: You are driving on a trip and realize you are getting tired but you did nothing about it. Stage 2 is over confidence not managed.

Complacency Stage 3 - Confident in one's ability to perform a function to the point they do not have to think about it to have repeated positive results. They refuse to respond to stimulus that should bring them back into the moment. Example: You are driving on a trip and realize you are getting tired but you did nothing about it. Then you hit rumble strips repeatedly but refuse to pull over. Excess confidence dangerously unmanaged.

Conclusion

So, what does all this talk about errors and states have to do with the topic of risk assessment? Well, a lot! Returning to the risk assessment tools mentioned above, we are missing something very important if we stick with just two dimensions. When we add a third dimension for Human Factors the matrix changes and might look like Exhibit #3 below.

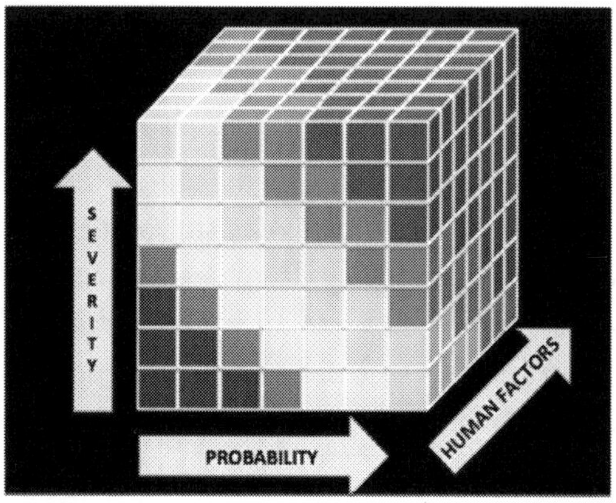

Exhibit 3.

Think of the third dimension as an exponential curve moving from less risk to more risk over time as complacency, frustration, fatigue or rushing increases. Notice how the risks indicated by the red and orange cubes have increased. The work is the same. In fact everything is the same except we have added the Human Factors. The result is more at risk situations identified than we originally thought. Now we have more to work on but we should also see better results.

There is a human factor to all errors, a third dimension if you will. Now this is not intended to blame the player for the error or the worker for the injury, poor quality or lower productivity. What we need to do is change our management practices to limit or remove stress that causes the errors. We can't make that change until we can systematically recognize them. Using a three-dimensional risk assessment tool will do that for us. With three dimensions process you can see the stress as well as the probability and severity. You see, we can cause these stresses even when we have no intention to do so. Just saying we need to get this shipment out by 4 p.m. can cause rushing for example. Work schedules can create fatigue. Work process can cause frustration as well as relationships with supervision and coworkers.

In addition, the employee can bring all kinds of issues with them to work each day that can alter their focus as well. Off the job activities can also cause fatigue and these are usually things the employee has control over. Frustrations with a significant other, child, financial issue are just a few things that can create frustration and sometimes even anger. We may not know those issues the employee has or we may not understand fully how our management practice can cause performance issues but they do affect the risk of performing a task.

Risk probability and severity potential fluctuates based on our interaction with the risk. The human factors can easily increase risk and Human Factors can also decrease risk. While some believe that human factors are invisible, they are very visible to the trained eye. You can determine rushing (going faster than normal), frustrations and fatigue easily. Complacence is more difficult but still possible with a little training.

The risk matrix needs to account for these fluctuations. Using a third dimension is one way we can accomplish the accountability.

We do not always know what state our employee is in but better communication skills, trust and teamwork will help discover the Human Factors. Without the open dialog we must anticipate a certain level of risk in the third dimension – a calculated guess if you will. I know it isn't an exact science but if we leave it out, the existing system isn't an exact science either.

Revolutionizing Risk Assessments--Oh My, Oh My, Bowties!

Tabitha A. Laser, CSP
CH2M
Houston, TX

Introduction

Many companies have embraced the importance of conducting risk assessments as a good safety practice; while others still rely on the findings of incident investigations and regulations to approach risk reduction. Unfortunately, doing the latter can have numerous 'blind spots', which can prevent comprehensive understanding of hazards, risks, and necessary actions to prevent and/or mitigate risk events from occurring.

There are different techniques available for conducting risk assessments to identify hazards, risks and barriers. These techniques range from brainstorming sessions, utilizing the Swiss cheese method, and conducting bowtie assessments. All of these methods add value to the risk assessment process; but the bowtie method is perceived to be the most thorough.

The first 'real' bowtie diagrams appeared in 1979, but how and when the method found its exact origin is not completely clear. After the catastrophic incident on the Piper Alpha platform in 1988, the oil and gas industry was urged to gain more insight in the causality of seemingly independent events and conditions and to develop a systematic/systemic way of assuring control over these Hazards. In the early nineties the Royal Dutch / Shell Group began utilizing bowties, followed quickly by the other oil and gas industry leaders. Recently other industries have started to use bowties as well including aviation, mining, maritime, chemical and health care. (CGE Risk Management Solutions B.V., 2015)

Where Does a Bowtie Fit into the Risk Management Process?

The risk management process is generally comprised of four main phases, which repeat on a regular basis to support continuous improvement efforts. The four main phases of the risk assessment process are outlined below:

1. Identify Risks
2. Assess Risks
3. Respond
4. Monitor and Review

When these phases are understood, thoroughly applied, and supported throughout an organization they result in the following:

- Increased awareness of risks and their barriers,
- Clearly defined action plans to help address gaps in barriers,
- Assurance protocols to help ensure barriers remain strong, and
- Continuous improvement incorporating learnings and emerging
- Prevention and/or mitigation of risk events.

To better understand where the bowtie methodology applies in the risk management process requires understanding the intent for each phase.

Identifying risk is the first phase in the risk management process. This phase entails clearly identifying risks associated throughout the organization. Risks identified can include, but are not limited to, the following:

- Business operational / financial risks
- Safety risks
- Health risks
- Environmental risks
- Quality risks
- Security risks
- Reputational risks

The action of identifying risks is not something that should be done once and considered complete. Instead, risk identification should be perceived as an ever-changing process that is constantly attempting to identify new, changing and emerging risks that could potentially impact the organization.

Having clearly defined processes to identify risks is very important as it feeds directly into the risk management process. In the event risks are not identified, then they will not be able to proceed through the next three phases, which could prevent necessary barriers from being in place, and ultimately lead to incidents.

Assessment is the second phase of a risk management process. This phase is where many organization spend a great deal of time and resources attempting to quantify the risks that have been identified. Generally this is done by utilizing a risk matrix where severity and likelihood are charted to determine a risk ranking or score, which is then utilized to help determine what levels of management need to endorse the risks, and in some cases, the action plans. Generally, the higher the ranking, the higher the level of management necessary to endorse performing work correlated with the risk.

In many cases this is where the assessment phase stops and moves onto the third phase. Some organizations do attempt to assess the risks past this point by utilizing other methods to identify barriers, and understand gaps. These methods include the Swiss cheese method, fault tree analysis, event tree analysis, causal factors charting, and control thinking.

The "Swiss cheese" method, which involves identifying the barriers (e.g.: process, person, and plant) necessary to prevent an incident, has been widely used and accepted for assessing risks. The Swiss cheese model of accident causation illustrates that, although many layers of defense lie between hazards and accidents, there are flaws in each layer that, if aligned, can allow the accident to occur (Figure 1). The model was originally formally propounded by Dante Orlandella and James T. Reason of the University of Manchester, and has since gained widespread acceptance. It is sometimes called the cumulative act effect (James Reason, 1990).

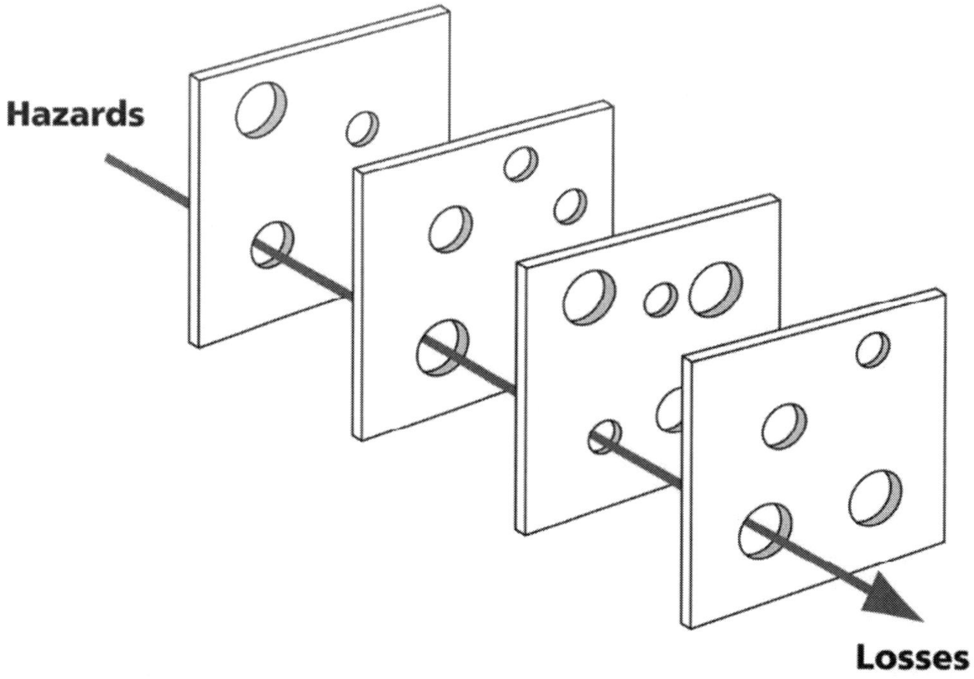

Figure 1. Swiss Cheese Model.

One concern with the Swiss cheese method is that the holes don't always have to align for an incident to occur. Another is that its focus is narrowed by looking at a linear progression from the hazard to the loss, which doesn't always clearly identify both prevention and mitigation barriers.

Other risk assessment methods have also received scrutiny regarding lack of comprehensiveness, over complication, lack of data to support results, and communication difficulties. For example, the fault tree analysis primarily focuses on the causes up to the risk event, and does not acknowledge the response needed after the event to reduce severity.

The bowtie method is actually similar to many of the methods listed above, but more thorough in areas that are weak, and simplified in the areas that can be overcomplicated. The assessment phase in the risk management process is where the bowtie method fits very well. The bowtie method is a risk evaluation method that can be used to analyze and demonstrate causal relationships in risk scenarios. The method takes its name from the shape of the diagram that you create, which looks like a men's bowtie (Figure 2).

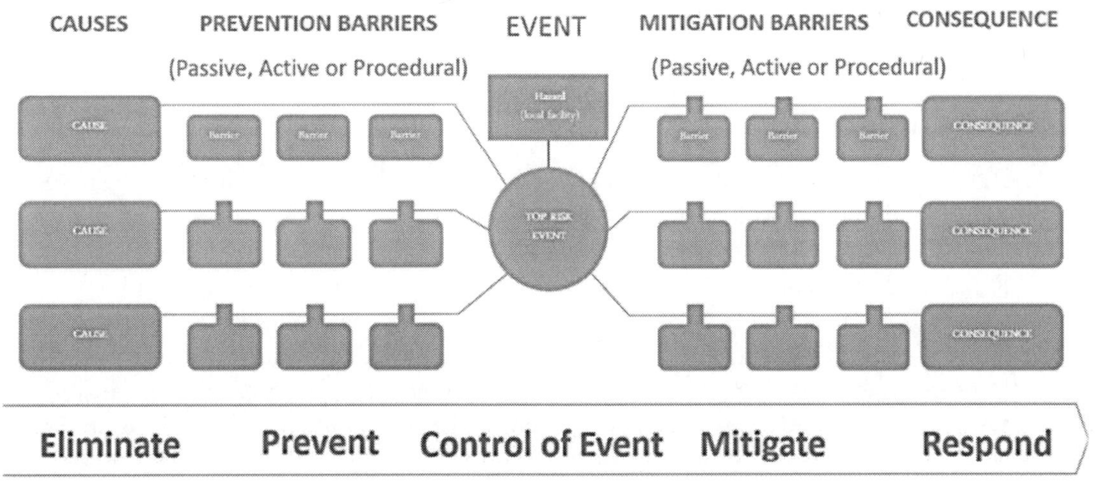

Figure 2. Bowtie Model.

A bowtie diagram does many things. First of all, it provides a visual summary of plausible accident scenarios that could exist around a certain Hazard. Second, by identifying barriers the bowtie displays what an organization does to control those scenarios. Once the barriers are identified, the bowtie method takes it one step further and identifies the ways in which barriers fail, or are weak. By visualizing the interaction between barriers and their escalation factors one can see how the overall system weakens when barriers have escalation factors (CGE Risk Management Solutions B.V., 2015).

Traditionally, the bowtie method has been primarily used to assess high risk scenarios only due the complexity associated with completing and updating the model. Recently, however, this method has been successfully applied to many types of risks as organizations become more familiar with how it works, and increase competencies around using and applying the methodology.

Response is the third phase of the risk management process, which is the phase where action plans are developed to address issued identified in the assessment phase. This is where the bowtie method adds the most value. If using an assessment method that does not thoroughly identify the barriers necessary to prevent or mitigate an event, and does not clarify gaps in barriers, then the action plans cannot be comprehensive enough to be adequate. The comprehensive nature of utilizing the bowtie method helps better understand barrier weaknesses, which:

- Fosters more effective and comprehensive action plan development.
- Promotes standardization of action plans.
- Highlighting where too many/few resources have been exerted.
- Assists with prioritization to reduce risks and support the action plans.

- Allows for communication of risks and barriers to everyone consistently.

The fourth phase of the risk management process is to monitor and review, which includes ongoing monitoring and review of the overall risk management process, but also of the effectiveness of the defined barriers, confirmation of barrier strengths, and of adequate closure of actions defined on the risk action plan.

Bowtie Terminology

Organizations may refer to the sections in the bowtie utilizing different terminology. For the sake of this report the following terminology and definitions apply:

- Hazard - Has potential to cause harm to human, finance, reputation, and/or environment
- Risk Event - Loss of control of the hazard, and what actually causes the harm.
- Cause (Threat) – The situation, or condition that results, or could result, directly or indirectly in an accident.
- Barriers - Should stand alone on the same cause or consequence line, have no shared failures, and not include the cause/consequence they are being used to prevent/mitigate, and be able to completely stop a risk event from occurring without help from other barriers.
    - Prevention Barrier - Barriers to the left of the risk event, which are focused on preventing the likelihood of the risk event from occurring.
    - Mitigation Barrier - Barriers to the right of the risk event, which are focused on reducing the severity of in incident after it has happened.
- Barrier Activities – A barrier can include many activities that enable the barrier to function effectively (e.g.: actions, processes, tools and equipment, competencies, etc.) Over time activities can be expanded to help continuously improve the understanding and effectiveness of the barrier.
- Consequence – The direct, undesirable result of an accident sequence (i.e. death, property damage, reputational damage, etc.)
- Escalation Factor – Gaps or weaknesses in barriers.
- Action Items – Actions identified to address escalation factors.

Conclusion

A bowtie is a comprehensive risk assessment tool that should be utilized in the risk management process to assess risks more thoroughly. Even though the method can be perceived as complicated, confusing, difficult, and even intimidating initially, education, awareness and familiarization with the process will lead to comfort applying the method, and significant reductions in organization's risk portfolios. Taking the step to begin utilizing the bowtie method to assess various types of risks, other than just high level risks, will greatly revolutionize how we understand our risks, resource our action plans, and continuously strive to reduce the likelihood and severity of incidents.

Bibliography

CGE Risk Management Solutions, The History of Bowtie (http://www.cgerisk.com/knowledge-base/risk-assessment/the-bowtie-methodology)

Reason, James. 1990. *The Contribution of Latent Human Failures to the Breakdown of Complex Systems.* (http://www.jstor.org/stable/55319)

Session No. 640

Developing a Communication Strategy to Support a Journey to Zero

Justin Ganschow, MS, CSP, CHMM
Caterpillar Inc.
Peoria, IL

Don Rosenbarger, PE
Delta Companies Inc.
Cape Girardeau, MO

Introduction

At every level of an organization, the way leaders talk about safety has a direct impact on attitudes, beliefs, and behaviors – which drive safety performance. A safety improvement journey fueled by accountability and positive recognition requires a communication strategy that leverages a variety of written and active messaging opportunities to inform, involve and inspire. Through words and deeds, leaders must set a sincere and credible "tone from the top," managers must ensure accountability for safety activities and front-line supervisors must champion a culture of zero-incident performance.

Delta Companies Inc., a heavy highway contractor specializing in asphalt paving and road construction materials, including aggregates, ready-mix concrete, hot-mix asphalt, liquid asphalt, and emulsions headquartered in Cape Girardeau, Missouri, committed to an engagement-based approach to safety improvement in 2015. They learned quickly that effective communications at each level deliver the trust, responsibility and openness required for measurable culture change across state lines and organizational strata. This case study presentation will focus on a strategy within a strategy – Delta's communication tactics to ensure the success of its safety culture initiatives.

Learning Objectives:

- Understand the building blocks of a safety journey communication strategy that goes beyond announcements and into accountability building.
- Distinguish the hallmarks of a quality safety conversation from the characteristics of a traditional approach.
- Prepare to give your executive leaders specific recommendations for demonstrating their commitment to safety.

A Need for Change

COLAS USA, a national leader in infrastructure construction and maintenance, has more than 5,000 employees in 50 offices throughout 25 states. Using a traditional approach to safety management, the organization made year-over-year improvements, but in 2015 leaders decided to reach for world-class performance. The CEO of COLAS USA led eight subsidiary companies to embark on a journey toward "Goal Zero" – using the process Caterpillar developed to achieve a safety culture transformation and world-renown results. This presentation will focus on the experiences of one COLAS USA subsidiary, Delta Companies Inc. (Delta), through the first two years of an expedition to safety excellence.

When the journey began, Delta – with operations in Arkansas, Illinois and Missouri – was experiencing a plateau in injury rate improvement and struggling to get all members of its organization to take ownership and accountability for safety. Working in rock quarries, ready-mix concrete, hot-mix asphalt, over-the-road hauling and road construction, Delta employees were exposed to a variety of hazards on a daily basis in union and non-union environments. Any safety processes or communication tools had to be flexible enough to be effective in any of these variable shift, non-static worksites and consider seasonal layoffs. The union-represented unit faced additional challenges of a changing labor force and geographical boundaries imposed by different union local jurisdictions, making it harder to build trust in and commitment to company values and processes.

Front-line employees, supervisors, middle and senior managers were interviewed at the beginning of the journey to gain their perspective of the current safety culture. Many agreed the culture was shifting from a reactive to a proactive approach to managing safety. The lagging indicators of TRIR and LTR had improved, which led some in management to believe the culture was strong. However, the pervasive belief that lagging indicator data provides the best measure of a company's safety performance has the potential to drive risky attitudes and behaviors. There was a misconception among front-line employees, and at some levels of leadership, that the absence of injuries equals the presence of a strong safety culture. To several interviewees, active involvement in safety meant not getting hurt, not falling off a truck, or otherwise avoiding a recordable incident – thus, luck, over taking specific actions to create a presence of safety. Delta's leadership believed they would benefit from education in how to sharpen their focus and efforts on desirable outcomes (activities that promote safety), rather than undesirable ones (incidents). Hourly personnel believed their leaders were committed to safety, though it did not appear safety had been elevated to an unwavering core value throughout the organization. Follow-up questions revealed that employees' perception of leadership's commitment was fragile because they did not see front-line, middle or upper management actively involved in safety on a regular basis. Nevertheless, the organization was in a good position to build on positive perceptions by getting front-line and middle management more actively involved in safety and by implementing a communication system through which top leadership could convey its commitment in a visible way.

Change Management – Resistance, Behavior Change & the Role of Communication

For every change journey, and every initiative within the journey, there are "Current" (where you start), "Delta" (what you have to do – where the change takes place), and "Desired" (where you are going) states. Through the journey the organization is working to change unwanted behaviors, beliefs, and norms to different behaviors, beliefs, and norms that support the "Desired" state.

Delta began its journey in 2015 and was in its own "Delta" state, using employee data to build a strategy to achieve the "Desired" state. Working with a Caterpillar Safety Services consultant, the cross-functional Delta Safety Steering Team (DSST) launched three employee-driven Continuous Improvement (CI) Teams to build specific processes around safety meetings, new hire orientation and near-miss incidents. These activities were designed to transform unwanted behaviors, beliefs and norms through new and improved processes, and to gain engagement through the ranks for desired behaviors, beliefs and norms. The DSST is comprised of salaried and hourly employees from Arkansas, Illinois and Missouri; and includes individuals representing aggregates, construction, ready-mix concrete, liquid asphalt and administration. The CI Teams are cross-functional in the same manner as the DSST, except there are no salaried employees on the team in a voting role.

Change is difficult for almost everyone. When we are comfortable with specific processes, we've done it one way for our whole career, or we just don't want to take the time to learn a new method, we resist shifts in the requirements imposed on us. It's human nature. However, in order to shift a culture through proactive safety processes, everyone in the organization is required to change their behaviors or work practices according to specific activities defined for their role. A critical component to ensure that happens effectively is a communication strategy that identifies opportunities to communicate, potential roadblocks and their solutions and flexible vehicles to reach the intended audience. As part of the strategy development, the team learned eight typical reasons why people resist change, with the ultimate goal of taking everyone in the organization with them on the cultural improvement journey:

- Don't see the need to change
- Don't see the change as a solution to their problems
- Have too much going on
- Don't believe the leaders really support it
- Don't trust the change agents
- Don't like change
- Current behaviors get in the way
- History of badly-managed change

Beyond the written or electronic communication methods, consistent day-to-day expectations must be demonstrated through behaviors that drive accountability for desired safety expectations. For example, a question on the Safety Perception Survey asked Delta employees if risks were sometimes overlooked in order to get the job done. Over half of the front-line workers indicated risks were, indeed, being taken for the sake of production. Delta did not direct employees to take risks; however, due to the attention management paid to production, the employees perceived that production was of the utmost importance. However, in another question, 100% of all Delta survey takers stated they believed the organization encouraged employees to work safely. Somewhere between their intentions and their actions, messages were getting mixed. Employees will do, first and foremost, whatever it is they believe their direct supervisor holds most important. If a supervisor behaves in a manner that conveys production is more critical than safe production, employees may be tempted to take shortcuts or undue risks to deliver production targets at all cost. Delta elected to establish effective safety norms for all employees to commit and demonstrate at all times.

Working with a Caterpillar Safety Services communication consultant, the cross-functional DSST first developed key messages they believed would build trust, define goals, explain the process, involve people, drive accountability, and celebrate achievements in each safety activity recommended by a CI Team. The DSST first developed twelve "key messages." Then, over a three-month time frame, cut the count to six, and finally honed the messages down to five. The essence of the safety journey at Delta can now be held in one hand – five fingers, five key messages. Each message supports Delta's slogan, "Safety/Our Way of Life," which you see on their hardhats, clothing, machines, vehicles and buildings. The key messages are printed on posters hung throughout every Delta site and on wallet cards distributed to all employees. They are:

1. Go Home Safe. Everyone. Everyday.
2. Be Heard. Communication lets us address concerns and celebrate successes.
3. Work Together. Reach Goal Zero for a better culture.
4. You Drive Change.
5. Trust is a Must. From the senior leader to the newest hire.

Accountability & Recognition – Achieving Desired Behaviors by Measuring and Acknowledging What's Right

Communication is essential to ensure desired behaviors are occurring at each level of the organization and are being reinforced with positive recognition. An indication that safety accountability is lacking may be in the form of these commonly heard statements:

- It's not my job!
- I'm too busy.
- I forgot.
- Nobody told me what to do.
- That's the way we've always done it.
- That's the way I was taught to do it.
- I thought someone else was supposed to do that.
- I was just about to get to that.
- Don't blame me, it was her idea.
- I didn't get the email.

Incident investigation data reveals that more than 90% of injuries are caused by at-risk behaviors, rather than hazardous conditions on the worksite. The job of the cross-functional DSST and CI Teams was to define specific activities within leading indicator processes (safety meetings, inspections, observation programs, training, etc.) for all members of the organization that drive behaviors which create a continual presence of safety. After they are defined and formally communicated to the company leadership, training is developed by the CI Team so everyone knows exactly how to perform their role. The employees that created the safety activities become the trainers for their peers. This ensures they know the process inside out and can convey the reason why the activity was created, how it works, and the intended outcomes. Equally important is an efficient measurement system to monitor not only the quantity, but also the quality of those defined activities. When everyone knows what the goal is, how they can directly impact it, and how to tell if they are winning, they are much more likely to be engaged in the process.

Finally, methods of acknowledgement for those performing the tasks is determined to reinforce those desired behaviors. In order to be effective, recognition for the completion of safety activities must be timely, relevant to the work completed, sincere, confirmed by the employee and frequent. Further, recognition must be significant enough to motivate someone to change his/her behavior. The most effective way to determine what approach will work is to ask employees what will be meaningful for them. During the CI Team events, employees who built the processes determined how they'd like to be recognized. Specifics of these communication methods are discussed in the next section.

Tactical Communication Strategy – The Who, The What, The How

Safety Steering Team Communications Manager

Assigning a communication role to a member of the DSST was a part of Delta's success. While several team members contributed to the variety of formal communication modes, having a single point of contact helped keep the Action Item Matrix (AIM) up to date with clear expectations. Responsibilities of a Safety Steering Team Communications Manager are as follows:

- Participates in all DSST meetings. Leads discussions about communication needs. Anticipates communication opportunities aligned with each DSST action / discussion.
- Builds communication strategy that includes opportunities for active (conversational) and passive (written / media) communications and methods to reach all employees with important messaging.
- Directs development and distribution of communications before / after Goal Zero Process events / workshops.
- Assists leaders in integrating DSST key messages, journey updates, and appropriate recognition into their communication opportunities, both active and passive.
- Serves as a point person for employee questions / concerns about the Goal Zero Process.

Communication Action Item Matrix

The DSST meets on a monthly basis and follows a formal agenda. In order to keep track of open issues, an AIM is maintained. It lists the specific activity, the person responsible, the due date, the completion date and any additional comments. Tied to the AIM is a Communication Plan, maintained by the Communications Manager (see below). It documents the event or milestone, target audience, anticipated resistance, key messages, methods, responsibilities and feedback loops. When properly utilized, it leaves little room for missed opportunities.

Communications Training

When it comes to demonstrating positive safety behaviors to communicate our intentions, not all supervisors and managers are inherently skilled. Delta's Safety Steering Team decided to train all supervisors and managers in Caterpillar Safety Services' "Speak Up!/Listen Up!" and "Recognize It!" workshops. "Speak Up!" instructs participants in easy to use techniques to help employees overcome emotional barriers to speaking up when something doesn't seem right. On the other side of the conversation, "Listen Up!" provides strategies for focusing on the message, not the messenger, in order to receive safety-related feedback that creates a stronger, safer culture. To focus leaders on what their employees are doing right, "Recognize It!" teaches that correcting unsafe conditions and behaviors is a necessary part of protecting employees, but recognizing safe actions is just as critical to achieving safety excellence. Traditionally, individuals in an organization are recognized and even promoted for finding problems and problem employees – catching people doing wrong, and then solving the problem, correcting or terminating the problem employee. The industry does not typically catch people doing right. The "Recognize It!" phase continues to be a struggle for most supervisors.

Communication Vehicles

Active and passive communication modes developed by the DSST come in a variety of forms, with the goal of reaching all employees regardless of their role or location. A few of their passive methods included personalized letters from the president to invite employees to join the Steering Team, postcards sent to employees' homes with Goal Zero messages, announcements included in employee paychecks, safety spotlights in the company newsletter mailed to each employee's home, hard hat stickers, safety slogan decals, key message posters and wallet cards, thank you and recognition notes to employees from senior management, creation of videos featuring employees and company-wide emails after DSST and CI Team events.

Active methods of communication tend to be more impactful, as our actions speak louder than our words. In addition to the spontaneous daily safety conversations fostered through the training mentioned above, formal safety interactions between co-workers, supervisors and managers occur on a regular basis (see next section).

Formal Recognition

Delta firmly grasped the concept of recognition in acknowledging the contributions its employees made to improve the safety culture. Studies show the most impactful form of recognition is one-on-one between an employee and a person he/she respects. It's best when it happens in the moment. Formal recognition, on the other hand, can be done in a group setting after the fact. It may or may not involve a tangible award/reward. Delta leaders have recognized the members of its Steering Team and CI Teams with monogramed shirts, plaques, dinners and celebrations. They've put their pictures in company-wide emails and newsletters for peer recognition. Members of the COLAS USA. management team have also joined in several meetings and recognition events. These efforts to formally recognize early drivers of the cultural transformation made others want to get involved, and reinforced the company really values employee engagement.

Practical Results – Roles in Communication through Safety Meetings

Accountability is the key to driving change. Measuring the quality of proactive leading indicator safety processes that front-line employees, supervisors and managers are completing gives a prediction of the desired outcomes (better injury statistics). For example, below are specific activities supervisors in Delta complete in their bi-weekly safety start-up meetings.

Safety Meetings -ACTIVITIES		Frequency
Who	Activity	Weekly
Supervisor of Front-Line Employees (Foremen, Leadmen, Plant/Paving Supervisors, Project Managers, Estimators)	Initiate safety meetings with employees to go over daily work plan and concerns raised by employees (either volunteers or supervisor assigns to ensure all employees participate).	2
	Provide employees safety meeting form, ensure it is completed.	2
	Ask open ended questions about concerns about previous day's work or potential hazards of the work to be completed.	2
	Recognize employee that have brought up issues or taken action to correct an issue.	2
	Feed issues raised by employees to the appropriate personnel for follow-up. Document action on form.	2
	Turn in Safety Meeting forms in to Manager within 1 work week.	1
	Ensure quality of Safety Meetings: -Employee involvement, Discussion amongst participants, Site-specific (or non-work related) topics, Relevant topics (heat stress in the summer, ice in the winter, etc.), Variety of topics, Issues being escalated	2

Initiating safety meetings on a bi-weekly basis communicates to the employees that no matter what the production schedule says, the Safety Meeting takes precedence. Meetings are led by front-line employees, which shows they are leading the change. The leader asks open-ended questions, conveying they are engaged and not just "checking the box." The supervisor provides feedback and recognition to those that participate, and follow-up with action on employee concerns when necessary. After the meeting adjourns, the supervisor documents what took place to allow measurement and evaluation by their own supervisor. The meeting forms below were created by CI Team #1 in order to measure the quality of the Safety Meetings, track any issues raised, and provide upper management visibility to field activity – so recognition can be given when they visit the sites.

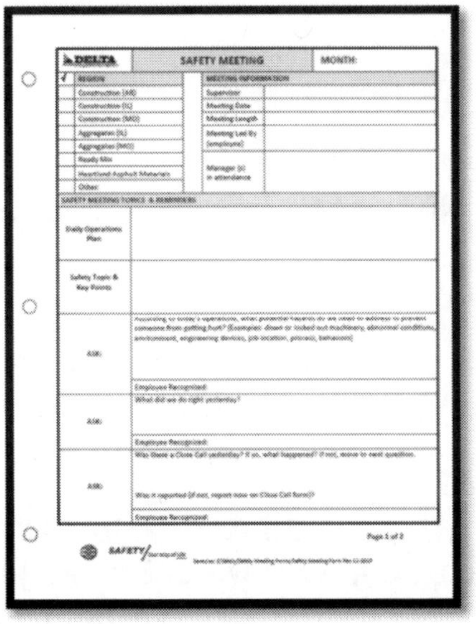

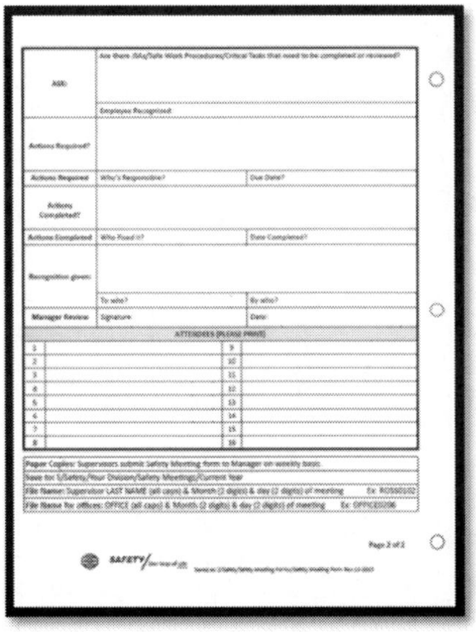

Lasting Results

The efforts Delta invested in communication paid dividends over the past two years. In 2016 alone, Delta experienced over 3,500 employee-led safety meetings, 65 new employees were onboarded, and 27 near-miss "SHIELDS" events were reported and addressed. As a result of these communication-centric leading indicator activities, Delta's safety results reflect a journey to safety culture excellence. Industry safety awards earned for 2016 performance were from the AGC of Missouri, Illinois Association of Aggregate Producers, Missouri Limestone Producers Association, AGC of Illinois, and Southern Illinois Builders Association. Employees across the company were recognized by leadership every month for their specific contributions to the safety culture. And most importantly, Delta employees have gone home safely at the end of their shifts. The Total Recordable Incident Rate (TRIR) has reduced from 4.9 in 2014 to 3.3 at the end of 2016. At Delta Companies Inc., leaders now consistently demonstrate their intentions with actions that positively influence the attitudes, beliefs, and behaviors of their employees. The slogan at the Delta Companies Inc. is "Safety / Our Way of Life," and it is truly becoming so.

Coca-Cola Operational Excellence: A Safety Diagnostic Event

Daniel Callahan, CSP
Coca-Cola Refreshments
East Hartford, CT

Steven Jablon, CSP
Coca-Cola Southwest Beverages
Austin, TX

Introduction

For decades, Fortune 500 Companies have leveraged Operational Excellence (OE) methodologies to drive productivity, employee engagement and culture. By definition, Operational Excellence is a workplace philosophy and methodology focused on identifying and eliminating defects or inefficiencies in a process, most often, a manufacturing process. OE involves focusing on the 'customers' needs, keeping employees positive, empowered, and engaged and continual improvement. It's clear, concise, practical, but most importantly; actionable and teachable.

OE problem solving methodologies and techniques have not historically been applied to a facility, market or region's 'Safety Process' or 'Safety Defects'. It has similarly not been applied to address Safety Maturity and Culture. It is important to remember that Culture drives employee behavior and employee behavior sets the norms, both positive and negative.

Examples of norms resulting from a negative culture include excessive clutter, employees not wearing seatbelts, tires not choked, finger pointing, supervisors not properly training new employees, etc. Norms resulting from a positive culture include the opposite; excellent housekeeping, all employees wear their seatbelts and when someone is not, they are coached by a peer based on the need to protect themselves and others around them, employees are heard, engaged and part of hazard identification and solutions!

If you seek to change culture, pick a norm you want to change, activate a plan to change that norm. Change the negative norm, you change behaviors, you change behavior you begin changing culture.

The OE problem solving tool; D M A I C (Define, Measure, Analyze, Improve and Control) allows your plant or geography to get very specific to the negative norms that drive performance using available data and employees' operational expertise.

The 'Customer' in this example is our employees and the "defects" are injuries, near misses and unsafe behaviors. The goal is to identify the norms and root causes leading to the defects or injuries and change the norm(s) driving unfavorable performance.

This approach uses Root Cause Analysis (RCA) and DMAIC (Define, Measure, Analyze, Improve and Control).

- **D**efine: What is the project's purpose, scope and major milestones.
- **M**easure: Measure the current process performance; "*narrow the problem area*" using process mapping and focusing on injury clusters within the process
- **A**nalyze: Identify the effect (defect) and narrow the variables to pursue and analyze potential root causes "*confirm with data*" using Fish Bone Process and the "5 Whys" – Cause and Effect. These tools will help "Pareto" (or prioritize) the norm to focus on and improve in the next phase of the process, while also considering risk of each cause.
- **I**mprove: Improve the process "develop & pilot solutions". Prioritize solutions to root causes using an OE 2x2 matrix, based on the highest value, lowest-resource-requirement solutions.

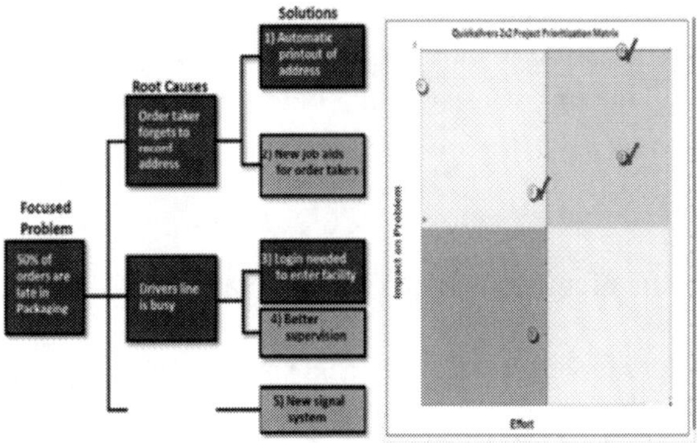

- **C**ontrol: Control the process "maintain the gains & transition to full implementation". Standard Work procedures created. Tracking of process data completed over time, allows the process owner to proactively see when a process, metric, etc., starts to trend out of specification. The Control Plan is a structured process with countermeasures, actions, and owners to quickly identify and correct trends before they are out of specifications. Project financial impact calculated. Project Completed, improvement sustained!

By the end of this presentation, attendees should have a basic understanding of Operational Excellence Principles and how to apply a structured problem solving methodology (DMIAC) to begin changing culture and elevate the safety maturity of their facility or geography.

Operational Excellence – Driving Safety Maturity and Culture

This process uses data and OE techniques to understand where/how losses occur and to identify true root causes or negative norms leading to low safety maturity, negative culture and ultimately poor safety performance.

The objective is to understand the Safety Maturity and Culture at a Facility/Regional level and set in place actions to elevate "Safety Maturity" and "Culture" to a place where safety is embedded into everything our employees do, and they are the driving force for identifying, implementing, and sustaining proactive safety excellence.

This first step in the journey is gaining alignment and commitment to host the workshop event from leadership. A 'Safety Diagnostic Event' can last 4 hours or 4 days (or longer) and will occupy resources for a similar timeframe.

<u>Safety Maturity and Culture</u>
The event begins with a review of 'Safety Maturity'. This is important for a number of reasons but first it allows leadership to truly understand what 'good' looks like. One of the first questions we ask participants is how would they rate the 'Safety Maturity' of their facility or geography (from 1 to 5, 1 being very low and reactive and 5 being truly embedded into everything they do). *Not surprisingly, participants usually rate their maturity fairly high and typically 1 to 2 levels higher than it actually is.*

A review follows of specific elements of each level's maturity

- Level 1 Element – Culture is very reactive at this stage, minimal safety awareness, processes and routines are in place, however still exposes the site to risk. Limited reaction even when/ if issues are identified. Safety- related roles and responsibilities are not well defined, the "Safety Guy" owns Safety.
- <u>Level 2 Element</u> – Documented accountability taken for unsafe actions and behaviors not resulting in accident or injury is greater than or equal to the number of claims found to be a result of unsafe actions during the investigation.
- <u>Level 3 Elements</u> – Near miss reporting process in place and recognition that all safety defects are preventable. Safety Committee is chaired by a front line associate.
- <u>Level 4 Commitment - Elements</u> - Annual Personal Safety Action plans. Each associate identifies actions and commits to improving his/her own safety.
- <u>Level 5 Embedded - Elements</u> - Implementation of a Behavior Based Safety program where peers consistently give feedback on observations. Management and associates have committed

to zero defects as an achievable goal for Health & Safety. Associates actively lead the identification, implementation and sustainability for proactive safety excellence.

As you can imagine, the participant's eyes are opened to what 'good' looks like and what they are capable of achieving. The second and more important piece is they begin to use this as a road map to direct their Safety plan and culture, they can see the destination.

Define - DMAIC

The next phase of the event is to understand/define the problem or 'bound the problem'. It requires a dive into data to understand what and who are driving the metrics or accidents without preconceived ideas. It is important to let the data drive the focus. Too many times, safety professional's state what the 'problem' is based on their experience, tribal knowledge or feedback that can and often is biased without being backed up with data. One sure way to lose commitment from leadership is to not be specific and fact-based.

	Total Claims		Total Incident (OSHA)		Lost Time Claims	
	2015	Prior Year	2015	Prior Year	2015	Prior Year
Drivers	120	93	82	71	38	28
Merchandisers	92	100	63	70	21	25
Warehouse	57	62	40	43	14	16
Account Managers	35	42	28	34	10	11
Cooler Service	15	13	13	11	4	3
ANYWHERE Region Grand Total	319	310	226	229	87	83

**Figures shown are fictitious for demonstration purposes.*

Measure - DMAIC

In this example, drivers are the issue, driving overall quantity of injuries or defects and unfavorable performance to prior year. The focus will be on the Delivery Driver process. The next step is to measure the current delivery driver process performance to "narrow the problem area" using process mapping and focus on injury clusters within the process.

The exercise is to literally map out the driver process from punching a clock, walking in the parking lot, performing pre-check inspection to driving to customer stops, unloading products and returning to the plant. The next step with the group is to map injuries from past years, one by one, and match where they occurred on the process map. This provides an excellent visual understanding (cluster) of where the injuries are occurring most frequently within the delivery driver process. In this case, one would have thought the largest cluster would have been with drivers lifting product, however, it turns out the largest cluster is drivers are getting injured when moving product with equipment.

Step 1: Map the Process Steps
Step 2: Align the safety defects / incidents with each step

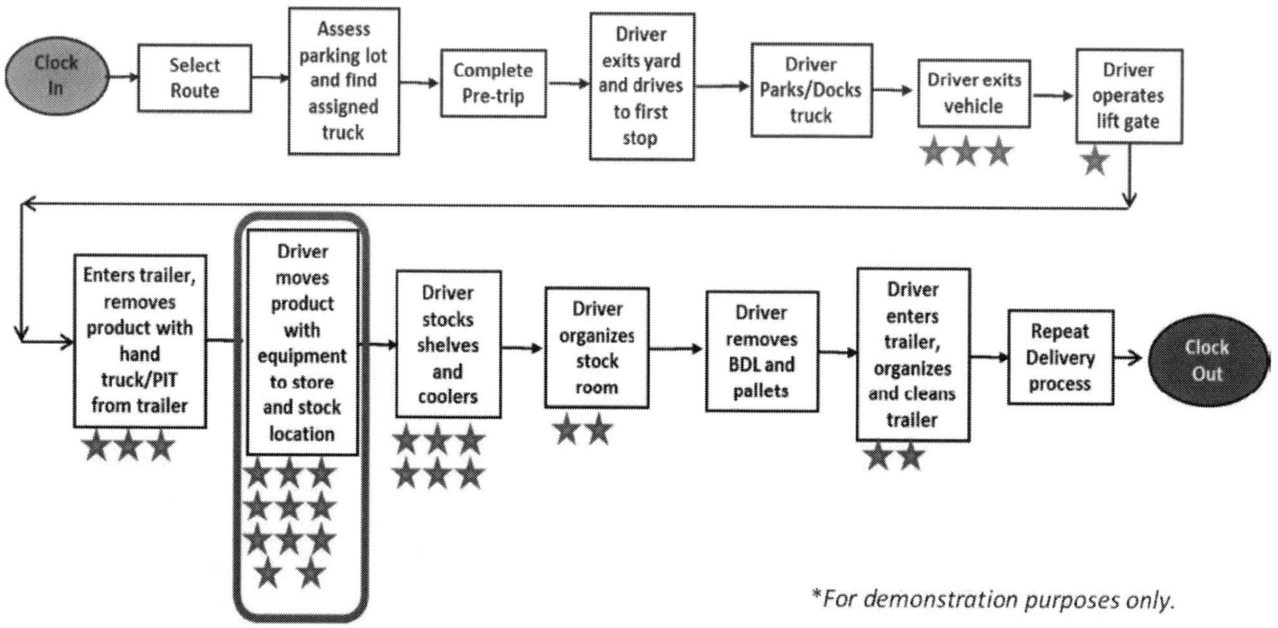

**For demonstration purposes only.*

<u>Analyze – DMAIC</u>

Now that the scope is narrowed to drivers getting injured moving product with equipment, the next step is to analyze potential root causes to 'confirm with data' using Fish Bone Process and the '5 Whys' – Cause and Effect. These tools will help 'Pareto' (or prioritize) what norm(s) to focus and improve in the next phase of the process, while also considering the inherent risk of each cause.

The effect is <u>drivers getting injured when moving product with equipment</u>, the team brainstorms possible root causes using the fishbone diagram and '5 Whys' process. In this part of the process is important to mine your frontline associates knowledge of their job! They should be part an integral part of the event. They know the causes leading to the effect or clusters.

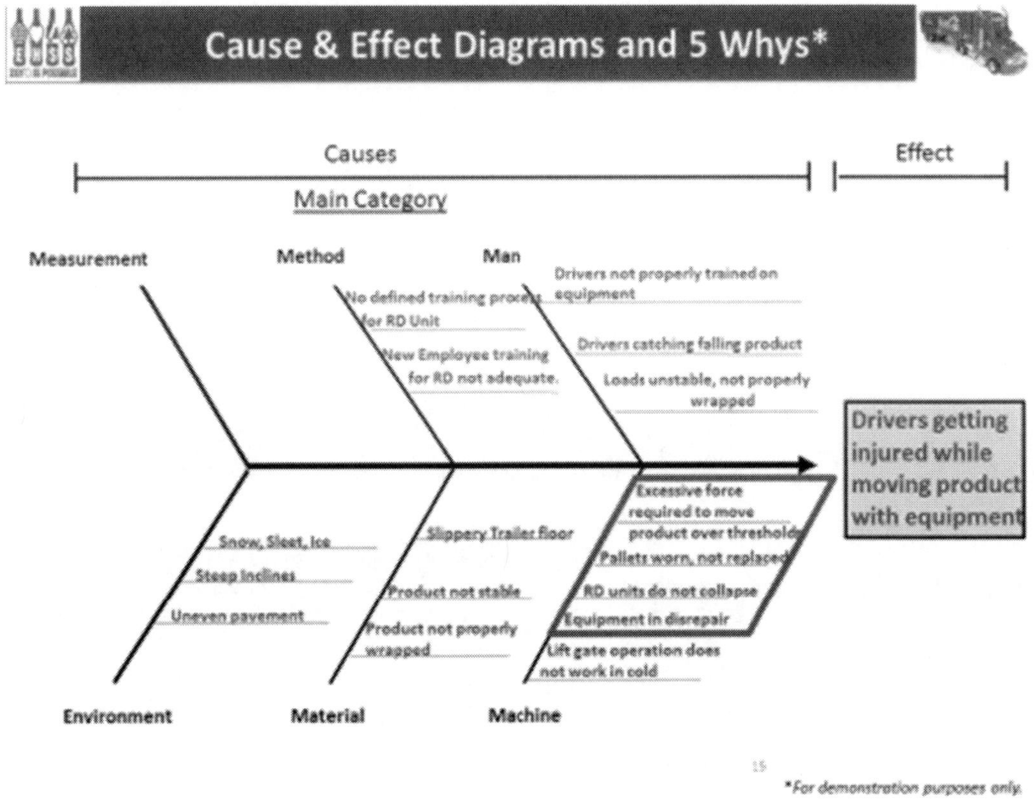

Improve – DMAIC

The next step in the process is to identify common themes of the root causes. Are issues (root causes) around equipment condition? …training on how to operate equipment? …new employees? …training execution? Prioritize the themes and brainstorm solutions focused around the norms/themes leading to the effect. These solutions can now be prioritize using an OE 2x2 matrix, based on the highest value, lowest-resource-requirement solutions.

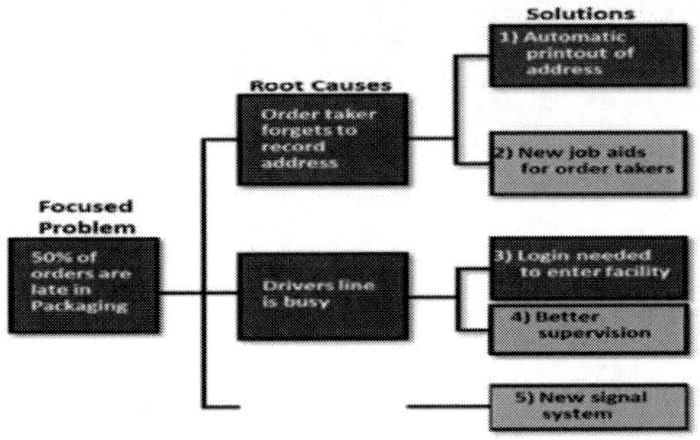

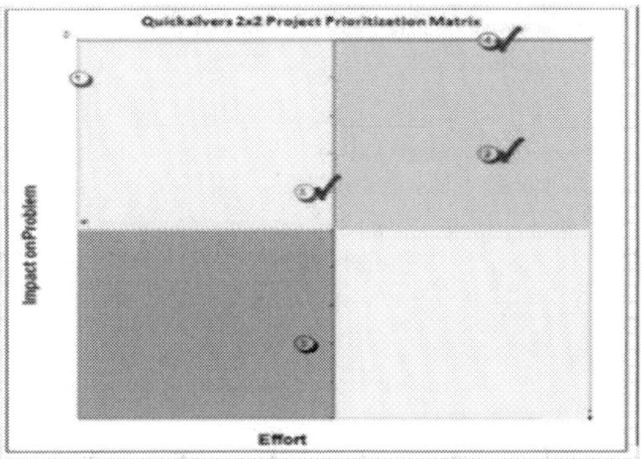

Once a possible solution is chosen using the matrix, the team will work through a current state and future state of the norm they want to change and the solution or future state they want to see.

One major theme that evolved from the fishbone for "drivers getting injured moving product with equipment" is around the Rapid Delivery Cool-lift Units and Pallet condition product is moved on.

- Excessive force required to move product over thresholds
- Pallets worn, not replaced
- RD units do not collapse
- Equipment in disrepair

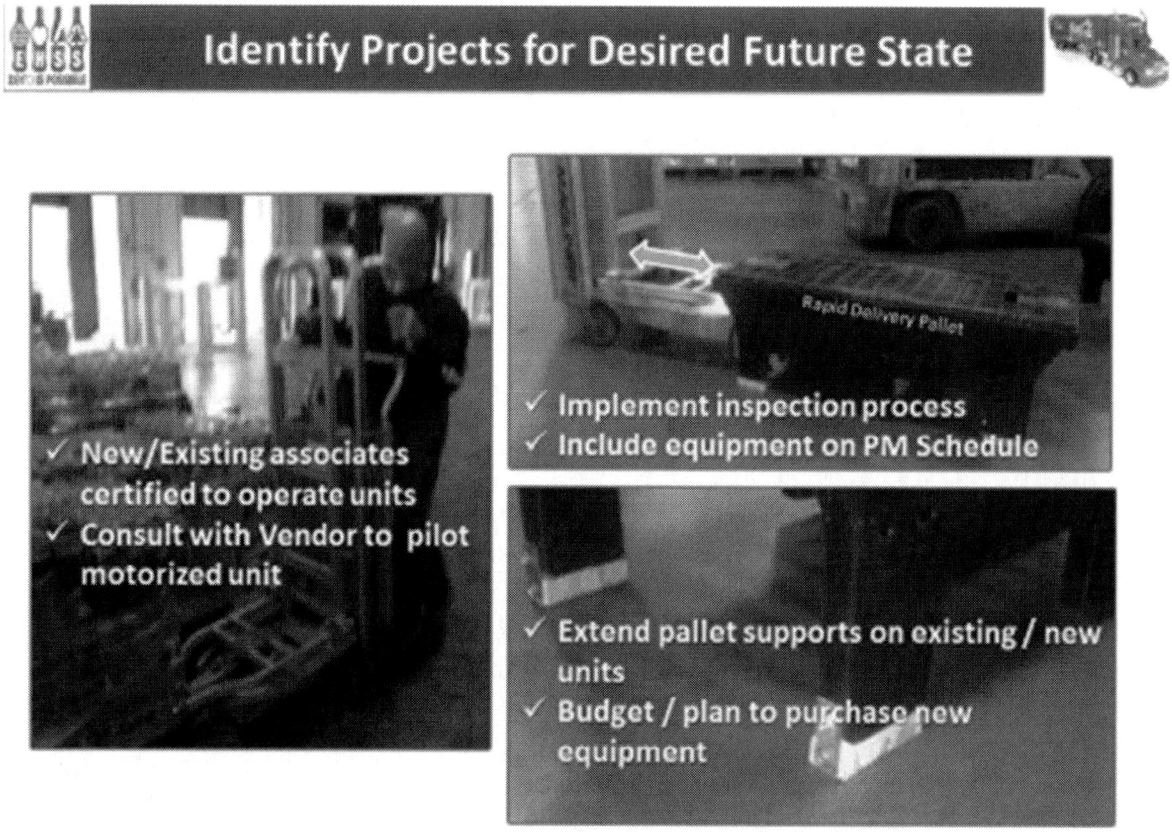

Just like in the analyze phase it is equally as important to mine your frontline associates knowledge of their job and solutions to the causes. They will have the solutions to the causes of the "effect".

Control – DMAIC

The final and the most important piece is to Control (or sustain) the process 'maintain the gains & transition to full implementation'. The Control Plan is a structured process with countermeasures, actions, and owners to quickly identify and correct defects identified.

This is done through a Steering Committee. Steering committee should consist of leaders and associates and be allocated appropriate time and resources for weekly meetings and project work. It starts with choosing the team, meeting frequency, reporting cadence. Team will need to be trained to the process. Steering Committee will revisit the work / process map clusters / root causes / Themes / Actions to address culture. They prioritize actions based on severity/frequency/likelihood of a 'defect' occurring. Projects are prioritized by cost to implement and effectiveness.

The team will select the projects for quick wins ("just do its") and projects based on this methodology. Team(s) focus and report on progress during each meeting and re-visit process maps and themes as needed.

All leaders monitor team progress and stay engaged. Projects may not be handed off.

Closing Thoughts/Take-Always

1) Leverage OE Tools and data to direct you to improvement opportunities at your facilities.
2) Let the data tell you the story.
3) Mine your frontline employees' job experience to identify root causes to defects and solutions.
4) Select quick-win projects to gain early buy-in, momentum.
5) Commit resources and time to sustain the process with robust steering committee.

Remember, change the norm(s) driving the "effects" leading to your facilities defects or injury cluster, you will change your employees' behavior. Change employee behavior and begin changing culture!

The Front Lines of Safety Culture Management: Dynamic Learning for Supervisors

C. Christopher Patton, CSP
UT-Battelle, LLC
Managing Contractor at DOE's Oak Ridge National Laboratory
Oak Ridge, TN

Introduction

Much of the effort to build and maintain a robust safety management system is focused on written documents which outline for employees the step-by-step, rule-by-rule expectations of how they should behave. These include procedures, training, job hazard analysis, and many other forms of instruction. Why is it, then, that even companies with very mature safety management systems in place still continue to experience unwanted events and injuries? Why do the best safety improvement strategies and initiatives sometimes meet resistance and failure? The answer is that often times we fail to recognize and account for the underlying organizational culture. It is the culture that causes organizations to accept or reject change. It is the culture that influences employees' interpretations of the meaning of instructions and select behaviors they think meet the intent and are appropriate or acceptable in that environment.

Managing organizational culture requires constant and consistent reinforcement. The front-line supervisor is ideally positioned to greatly influence the culture. They set expectations, demonstrate behaviors and values, and reinforce employee behaviors and attitudes. However, many supervisors lack an understanding of this important responsibility. Supervisors often become so focused on ensuring the work is getting done, they fail to provide the leadership necessary to nurture a strong, positive culture.

In this paper, the author will review sources of information for the definition of safety culture and share thoughts on the development and implementation of a safety culture management strategy, including the importance of training and engaging front-line leadership. It will also explore the use of situational learning to create dynamic learning opportunities and a model the author has used to develop and engage front-line leaders.

Defining Your Safety Culture

Elements of Safety Culture

It seems we often hear the term "safety culture" used, but rarely with any clarity as to the real meaning. We use it positively to indicate a low number of unexpected events or a high awareness of safety in the workplace. We use it negatively as a vague causal factor in injuries and unexpected events. It is the author's experience that the use of the term "safety culture" is rarely paired with an understanding of the desired traits or principles which make up the safety culture and contribute to the positive or negative perceptions and outcomes.

Several models exist for the traits and defining principles that make for a strong positive safety culture. The nuclear industry has extensive experience in managing safety culture. The Institute of Nuclear Power Operations (INPO) has published a document which is important to the understanding of safety culture titled *Traits of a Healthy Nuclear Safety Culture* (INPO 12-012, April 2013) in which they identify traits and attributes of a strong positive safety culture. While focused on the nuclear power industry, it has broad application across any industry or organization.

The safety culture traits are divided into three categories to align with the categories in the International Nuclear Safety Advisory Group (INSAG-4, *Safety Culture*). These categories and traits are as follows:

- Individual Commitment to Safety
  - Personal Accountability
  - Questioning Attitude
  - Safety Communication
- Management Commitment to Safety
  - Leadership Accountability
  - Decision-Making
  - Respectful Work Environment
- Management Systems
  - Continuous Learning
  - Problem Identification and Resolution
  - Environment for Raising Concerns
  - Work Processes

Similarly, the Department of Energy has published their safety culture focus areas and associated attributes (DOE G 450.4-1C, 9-29-2011, Integrated Safety Management Guide, Attachment 10) in which they identify three traits and several associated attributes they felt offered the greatest potential for achieving a strong safety culture. They are:

- Leadership
  - Demonstrated safety leadership
  - Risk-informed, conservative decision making
  - Management engagement and time in field
  - Staff recruitment, selection, retention, and development
  - Open communication and fostering an environment free from retribution
  - Clear expectations and accountability

- Employee/Worker Engagement
    - Personal commitment to everyone's safety
    - Teamwork and mutual respect
    - Participation in work planning and improvement
    - Mindful of hazards and controls
- Organizational Learning
    - Credibility, trust and reporting errors and problems
    - Effective resolution of reported problems
    - Performance monitoring through multiple means
    - Use of operational experience
    - Questioning attitude

Creating an Envisioned Corporate Safety Culture

The author's experience with developing a vision for an organizational safety culture is within the context of a Department of Energy national laboratory. In the organizations with a nuclear mission, the INPO principles for a strong safety culture were already well adopted. However, the remainder of the organizations exhibited resistance to the rigor of the nuclear industry for fear that it would inhibit the freedom to innovate which is required in the research environment. So, the author's organization undertook an effort to define a set of safety culture principles based on the INPO principles but without the flavor of the nuclear industry.

The author believes that every organization must make a thoughtful effort to customize a set of foundational safety culture principles which fit with the organization's desired values, mission, and organizational culture. These must come from the leadership of the organization, because it is the leadership that will be responsible for institutionalizing these principles.

Developing a Safety Culture Management Strategy

Establish a Foundation

A strategy for implementing a change in the safety culture involves some key steps. The first step is to establish a foundation upon which to begin the process of integrating the desired cultural traits. In a world of policies and procedures, it is important to integrate the safety culture principles into the controlling documents of the organization. This helps provide direction, as well as avoiding conflict and confusion. The safety culture principles should be engrained in the Environmental Safety and Health Policy of the organization. Procedural documents on work control and safety should be aligned with and reinforce the principles. Roles and responsibilities should be defined to incorporate the role each level of the organization has in implementing the principles, and the responsibilities to which those persons will be held accountable.

Communicate and Educate

Staff in the organization will likely not associate with the organization's safety principles if they are not made aware of the value the principles bring. Staff will need help understanding how these principles translate to behaviors in the workplace. One successful tool the author has utilized is publishing positive stories about actions employees and management have taken which were in line with one or more of the organization's safety principles. This helped staff understand how the principles contributed to a successful outcome, and what behaviors are necessary to demonstrate those principles.

Mentor and Coach

All levels of the organization will likely need help maintaining focus and consistency with regard to implementation of the safety culture principles. While a traditional approach is to manage from the top down through the layers of the organization, this must begin with a clear understanding of the principles at the top of the organization. Without the proper education and reinforcement at each layer of the organization, it is likely the message will become distorted and diluted as it moves closer to the worker. This is a natural yet challenging aspect of making change in an organization. Persistence and a high level of communication are necessary to overcome this challenge.

Monitor and Adjust

Similar to any continuous improvement effort, we must close the loop. Organizations will need to monitor implementation and use that feedback to further develop and refine the actions being taken to support the safety culture management strategy. Various methods can be utilized for monitoring implementation. The most traditional method is the safety culture perception survey. The author has also found much success in focus groups, employee feedback, and observation data. Existing data collection systems can be adjusted to collect relevant information. For example, an existing program which involves observing workers and collecting data on safe behaviors could be altered so the observations look for behaviors associated with the desired safety principles.

The Power of Front-Line Leadership

Communications in an Organization

One major challenge for implementing a culture management strategy is created by the disconnects between layers of management in an organization. How often have we heard our senior executives get excited about a big initiative to make a change in the organization, yet within 6 to 9 months the effort has fallen apart? In the author's own organization, there was tremendous "buzz" at the leadership level about improving the safety culture. Yet, in speaking with employees, they had heard nothing. As a result of competing priorities, the message was lost somewhere as it trickled-down through the organization. Penetrating this "clay layer" in an organization can be a real challenge.

Why Supervisors?

There are many definitions of safety culture being used. One key element found in many of them is that the culture includes behaviors modeled by the organization's leaders. For our front-line workers, the majority of our employees, the most frequent, visible representation of leadership is their first-line supervisor. No matter what message is spouted at the senior leadership level of an organization, the supervisor can make or break that message for their employees. They do that by how they deliver the message, how they react to the message, and by their actions to either reinforce or not reinforce it. Employees look to their supervisor to understand acceptable behaviors. If the supervisor consistently behaves in accordance with the safety principles and rewards employee behaviors that exemplify the principles, then employees will generally demonstrate the expected behaviors. They will internalize those principles and act accordingly.

If front-line supervisors are the face of leadership, and if they have the ability to make or break a message, we have to ask ourselves the challenging question, "what have we done to prepare our supervisors for this important role?" Much too often, a supervisor is selected not for their leadership skills, but for their skill in their profession or craft. They rise to the top as a worker and we assume then that they will make a great supervisor. An important piece of the puzzle for successful implementation of a safety culture management strategy is to prepare our first-line supervisors for the task and garner their support.

Using Simulations to Develop and Engage Front-Line Leaders

An Example: The Laboratory Operations Supervisor Academy

The author's organization created an educational opportunity for front-line supervisors with the objective of developing their understanding of the value of the organization's safety principles, their meaning and observable behaviors in their workplaces, and techniques for communicating with and managing employees by utilizing the principles as a means to influence behaviors. The course involves very little traditional instruction, but rather uses situational learning to immerse the supervisor in challenging real-life situations in which they can practice these skills. Each simulation is followed by a facilitated conversation in which they receive feedback from their peers and discuss how the safety principles applied in that situation. This method of learning has proven extremely effective.

One benefit of using simulations for experiential learning is the opportunity for dynamic learning. In other words, each person who participates is able to take away lessons unique to themselves and their own skill-gaps. By participating with a diverse set of peers, the students benefit and learn from each other.

Developing Situational Learning Experiences

There is no shortage of challenging workplace situations to draw from for the creation of scenarios. The difficult part is crafting the scenario in a way in which one or more of the principles can be highlighted through behaviors (or lack of behaviors) to create an opportunity for the student to be able to solve the challenge by reinforcing that principle.

For example, if a scenario is created where an employee is disengaged in a work planning meeting, you have created an opportunity for the front-line leader to have a discussion about why participation is important relative to one or more of the organization's principles. The supervisor may take a more direct approach and admonish the employee, but then misses the opportunity to help the employee understand the fundamental reason why they should participate. Either way, the feedback and dialogue after the simulation will create an opportunity to discuss whether the approach was effective and the impact it may have had on the culture of the group. Through sharing of ideas in a facilitated discussion, each participant learns from the thoughts and experiences of their peers. Their own individual strengths and weaknesses are identified and discussed. In this way, the learning is specific to each person, going well beyond a traditional list of learning objectives.

Summary

An organization seeking to improve safety culture should understand the traits and attributes which make a strong safety culture. Then, review those traits in the context of their own mission and organizational culture and define for themselves a set of foundational principles which underscore the envisioned safety culture.

Implementation is a challenge. While top-down coaching and communication is critical, there is no layer of management more influential to the workers than the first-line supervisor. Without their clear understanding and support, change will not likely occur. Situational learning has been an effective means of educating and engaging front-line supervisors in a way that allows them to develop their own natural leadership style while practicing new skills that promote the organization's desired cultural principles.

Bibliography

Department of Energy. 2011. *DOE G 450.4-1C, 9-29-2011, Integrated Safety Management Guide, Attachment 10.* (https://www.directives.doe.gov/directives-documents/400-series/0450.4-EGuide-1c)

Institute of Nuclear Power Operations. 2013. *INPO 12-012 Traits of a Healthy Nuclear Safety Culture.* (http://pbadupws.nrc.gov/docs/ML1303/ML13031A707.pdf)

International Nuclear Safety Advisory Group. 1991. *INSAG-4, Safety Culture.* (http://www-pub.iaea.org/books/IAEABooks/3753/Safety-Culture)

AUTHOR INDEX

AUTHOR INDEX

AUTHOR INDEX

ASSE Professional Development Conference (Safety 2017)

Denver, Colorado, USA
19 - 22 June 2017

Volume 2 of 2

ISBN: 978-1-5108-5001-9

Printed from e-media with permission by:

Curran Associates, Inc.
57 Morehouse Lane
Red Hook, NY 12571

Some format issues inherent in the e-media version may also appear in this print version.

Printed by Curran Associates, Inc. (2017)

For permission requests, please contact American Society of Safety Engineers (ASSE)
at the address below.

American Society of Safety Engineers (ASSE)
520 N. Northwest Hwy
Park Ridge, IL 60068
USA

Phone: +1 847 699.2929

www.asse.org

Additional copies of this publication are available from:

Curran Associates, Inc.
57 Morehouse Lane
Red Hook, NY 12571 USA
Phone: 845-758-0400
Fax: 845-758-2633
Email: curran@proceedings.com
Web: www.proceedings.com

TABLE OF CONTENTS

VOLUME 1

VOLUME 2

Prevention through Design Goes Virtual: A Case Study

Marjory E. Anderson, CSP
Architect of the Capitol
Washington, D.C.

Kevin E. Wilcox, PE, CSP
LJB Inc.
Miamisburg, OH

Introduction

While Prevention through Design (PtD) may be considered a "hot topic" within the safety industry, many organizations have been slow to adopt its practices. Reasons for this range from lack of information, limited examples to validate its worth, and the changes required to typical project development. This case study provides background on this emerging topic and offers details of how PtD is being effectively applied on a complex project. The resulting safety improvements and cost savings realized on this project illustrate the benefits that can be achieved by implementing PtD. Armed with this knowledge, safety professionals can help educate architectural and engineering (A/E) consultants—the "designers" in the Prevention through Design system—so PtD can be utilized to its maximum benefit.

This case study shows how the Architect of the Capitol (AOC) used a virtual inspection of a design model to prevent fall hazards from being designed into a building renewal. AOC often undertakes design and building projects as part of fulfilling its mission to serve Congress and the Supreme Court, preserve America's Capitol, and inspire memorable experiences. AOC is responsible for the operations and care of more than 17.4 million square feet of facilities, 580 acres of grounds on Capitol Hill and other areas in the District of Columbia, Maryland, and Virginia, and thousands of works of art. Many of these facilities are historic, with limited space and, in some cases, outdated systems. AOC must make use of available technology and tools to design modern systems and incorporate the latest safety advances into these facilities, while preserving the historic character and aesthetics the American people expect.

Why Use Prevention through Design?

Organizations that have applied PtD programs have benefited from meaningful results: life-threatening work hazards are reduced, productivity is improved, and costs are lowered.

Implementing PtD ensures that safety measures are evaluated and implemented during the programming and design phases of a project. In addition to increased safety for operational personnel, incorporating design safety in the earliest stages of a project saves money because designers don't even need to erase lines on their drawings – the safety aspects are simply incorporated into the design. Although it may be easier to recognize safety hazards in an existing structure, it is safer and more cost effective to eliminate hazards during design and implement protection measures during construction or renovation.

Save Lives and Money

When safety professionals—and front line employees who know the work best—are brought in at the beginning of a project, safety concerns can be addressed within the project's original timeline and budget. Evaluating and implementing safety measures early in the project ensures optimal safety for workers—from construction into operations, and throughout ongoing maintenance.

Besides these safety benefits, applying PtD significantly reduces the cost of addressing fall protection issues. The impact to bottom line cost can be dramatic through addressing life-cycle costs. For example, eliminating the need to erect, use and dismantle scaffolding to solve fall protection issues after project completion—even in just one area—can result in significant savings and reduced risk over the life of a building. When fall hazards are eliminated or managed through engineering control solutions, there are fewer fall protection issues to contend with long term. This translates to fewer equipment purchases, less training and a reduced amount of elements to manage, such as system installation, PPE purchases, and system inspection and certification. Beyond this cost savings, the organization can realize substantial improvements to worker safety and productivity.

The Importance of Buy-In

Garnering support from the facility owners or end users is critical to effectively applying PtD. When managers hold design teams accountable for minimizing risk, everyone is more motivated to ensure that safety is considered early and often. Ideally, this becomes part of an organization's culture, and everyone embraces the idea of addressing safety throughout a project's continuum. When PtD is used, safety issues can be minimized throughout the project's lifecycle.

For PtD principles to work—whether a formal PtD program is in place or not—safety personnel need to be engaged early in the design process. Simply because of their background and focus on safety issues, safety personnel can recognize potential hazards while they're still lines on paper or models on a computer—before funds are spent on construction. These potential hazards will become even more apparent when the safety professional has been involved in previous projects, and has experience identifying safety issues in facilities, as well as addressing post-construction safety issues.

Case Study

In 2011, the Architect of the Capitol (AOC) Safety Division began work on two parallel initiatives:

1. <u>PtD Program Development: Working towards routinely implementing PtD on all AOC projects</u>
   For many years, the AOC Safety Division desired to implement a process where safety was designed into projects—what is known today as PtD. This effort was not immediately accepted agency wide, so several initiatives were undertaken to gain buy-in. First, detailed research into PtD and its use in industry was performed. Next, AOC-wide management support and buy-in was obtained and a safety policy based on the research findings was incorporated into the AOC Safety Manual. The Safety Engineer personally engaged with project managers and engineers at AOC to explain the process and how it could benefit AOC projects. Over time, buy-in began to occur and increased interest was shown by various engineers, architects, and project managers. Customized training for AOC personnel involved with project design and construction was developed. This training is currently underway and is being well received by those who will ultimately put this policy into action.

2. <u>Interior Fall Hazard Risk Assessment: Assessing interior building fall hazards in all facilities</u>
   The effort to assess interior fall hazards was an off-shoot of a previous project to address roof fall hazards. AOC identified that areas within buildings could pose fall hazards to employees. Where these hazards existed, some type of control would be required. When AOC began assessing the House Office Buildings, it became obvious that the same type of assessment was not practical for the Cannon House Office Building (CHOB) because a major renewal project was being planned (see Exhibit 1). The team acknowledged that everything in this building could change as part of the renewal, making a physical assessment invalid.

Exhibit 1. The Cannon House Office Building is undergoing a complete renewal, which is expected to be complete by 2024.

<u>The CHOB Project</u>

With the recent development of its PtD program, AOC decided to engage a consultant to perform a third party review of the CHOB design—with a goal of eliminating or mitigating fall hazards through the construction process. Built in 1908, CHOB is the oldest Congressional office building outside of the Capitol Building. The century-old building has developed safety, health, environmental and operational issues that continue to worsen over time. The renewal process is scheduled to take approximately 10 years—ending in 2024—with each of the five phases aligned to fall between Congressional move cycles.

Why Use a BIM Model?

The AOC had already determined that Building Information Modeling (BIM) would be used to develop the renovation plans. Typically PtD is implemented primarily through design team meetings, where safety and operational personnel engage with the larger design team to review drawings and develop solutions to eliminate or mitigate safety hazards, some of which was done on this project. However, to reap the full benefits of PtD, the safety consultant used the Cannon Renewal Project BIM to help identify safety hazards and specifically to identify and minimize fall hazards inside the building. For a project of this magnitude, using a BIM model instead of 2D drawings allowed safety personnel to view all the integrated drawings, making it easier to evaluate interferences and devise abatement solutions acceptable to all stakeholders.
Cannon Renewal Project Cannon Renewal Project

Advantages

The process of implementing PtD using BIM was well received by the A/E and AOC teams. The project realizes the following advantages:

- **AOC expects to see significantly fewer hazards in the completed CHOB due to PtD.** With the ongoing fall hazard risk assessment for the interiors of its buildings, AOC was able to learn about fall hazards in other similar buildings. This information was shared with the CHOB design team and allowed the team to "design out" common hazards, avoiding the introduction of these same issues.
- **Engaging multiple stakeholders in developing solutions and making changes during design requires less money and effort and always adds value during construction/operations.** Based on previous experience, operations and maintenance safety issues created as a result of construction may go unidentified and/or unaddressed, due to budget and schedule constraints which often prevent it from being improved. In addition, if a challenging issue was identified within the model, the appropriate stakeholders could engage with the issues and develop the most beneficial solutions from a variety of perspectives. Once construction is complete, the options are typically limited and are likely less optimal than if the issue is addressed during design. When these items are acknowledged and corrected in the design phase, the effort is reduced and the benefits are increased.
- **Using BIM has tangible advantages for mechanical and plumbing design as well as safety.** The AOC Cannon Renewal team uses this modeling to work out conflicts before construction begins and the contractor cannot begin construction until a successful BIM model is produced. This proved to be beneficial not only for improving safety, but also for minimizing conflicts with mechanical and plumbing disciplines.

Disadvantages

While the Cannon Renewal Project Team was pleased with the process and outcome of using BIM to eliminate or mitigate safety hazards, there are some disadvantages that should be acknowledged:

- **Even with a collective mindset geared toward improving short- and long-term safety, all adjustments can be negatively viewed as a change.** The process is iterative in nature and frequently requires multiple revisits of a work product that may have been perceived as completed.
- **The fall hazard assessment project had not reached the point for assessment of the CHOB until the architectural and mechanical designs had already begun development.** PtD and subsequent evaluation work for this project began later in the design process than the Cannon Renewal Project team would have preferred. There were certain elements that could not be changed and workarounds had to be created. To be optimally effective, PtD should be incorporated and focused on eliminating and mitigating hazards at the outset of the project. Still, it is always better to implement PtD before construction/operations, since it's very difficult and expensive to make changes at that stage.
- **A virtual model is never quite as accurate as real life.** While the modeling on the Cannon Renewal Project is very detailed, conducting a safety assessment in a virtual world is never as straightforward as doing so in real life. Software tools allow the user to move around the model in a way that approximates real life, but capturing the nuances of the environment, such as lighting, constricted spaces, or the position of people doing the work should not be expected to replicate real life.

Other Considerations

The following items also need to be considered when contemplating use of BIM as a tool for virtual implementation of PtD.

- **Using this approach during the design process requires direction from owner level.** The traditional design approach is that each system is designed and coordinated from a construction interference standpoint. If something can't be built, it is generally accepted that at least one of the designers must change something (mechanical engineer, electrical engineer, etc.). Safety is not always considered as an equivalent design practice. So, if it is determined that something cannot be easily and safely maintained, there is resistance to changing the design based strictly on that single criteria. The owner must make it clear to the design team that safety is equally important to the other traditional design disciplines.
- **Performing a virtual assessment requires training and experience.** While the use of BIM is no longer innovative and safety assessments certainly aren't new, combining the two isn't commonplace. To use BIM, a powerful computer is needed, and may require the use of software packages that are new to safety professionals. If the safety representatives aren't accustomed to seeing a 3-D world on a 2-D screen, they may need to practice visualization skills. It can be challenging to obtain an appropriate sense of scale. Since the model can be zoomed in and out, it is easy to lose sight of what the screen is displaying, which requires training and experience on the user's part.
- **Designing renovations into an historic building requires unique and creative insight.** Designing renovations for older buildings is not the same as designing for newer, modern buildings. The design team must be knowledgeable not only of modern standards but also of the historic aesthetic and past construction methods of such buildings. If safety representatives aren't aware of both, then designing out hazards will be inherently difficult and potentially ruin the visitor experience or inappropriately mitigate the hazards.

Example Outcomes

Exhibits 2-6 show examples of safety issues that were identified in the model.

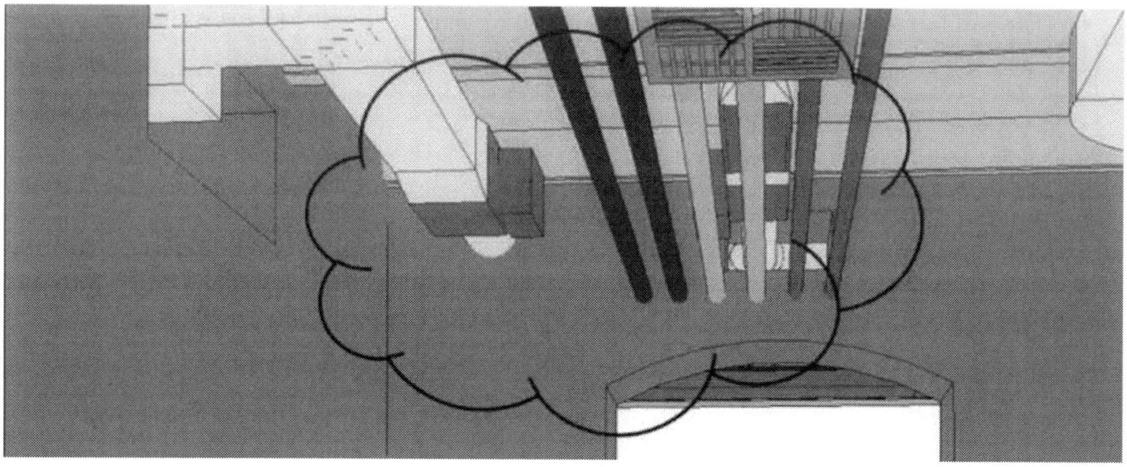

Exhibit 2. This image shows HVAC equipment that is obstructed by piping, making access difficult. The team determined that the piping or HVAC equipment could be relocated so that both are accessible for maintenance.

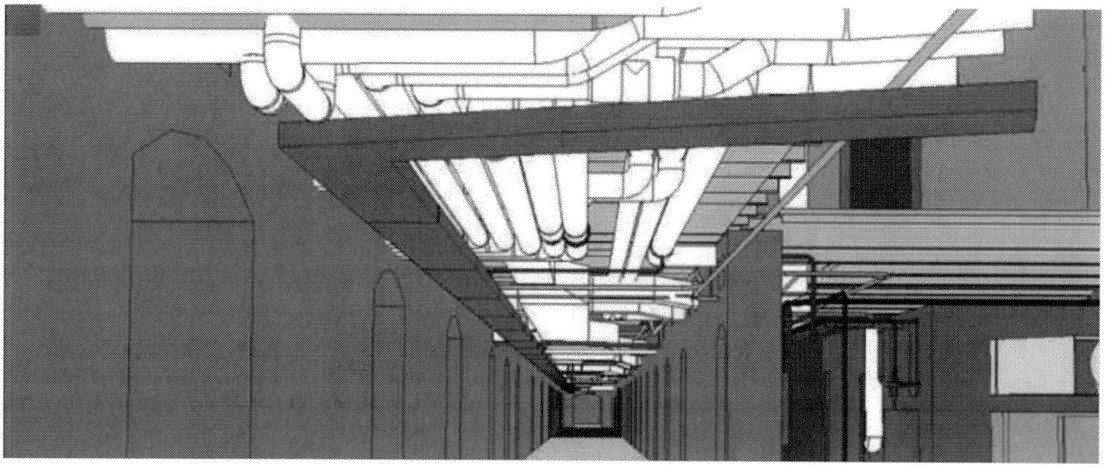

Exhibit 3. This graphic shows no valves or other items needing access are blocked by the cable tray, just as prevention through design would suggest

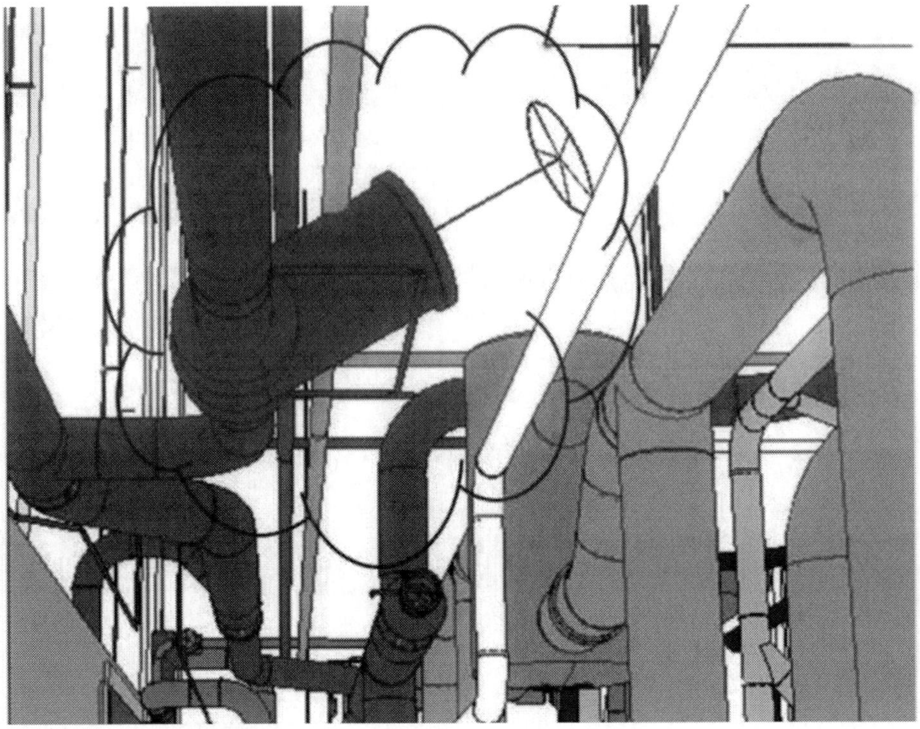

Exhibit 4. While this valve is expected to be accessed rarely, the piping below it would present a challenge in accessing it when necessary.

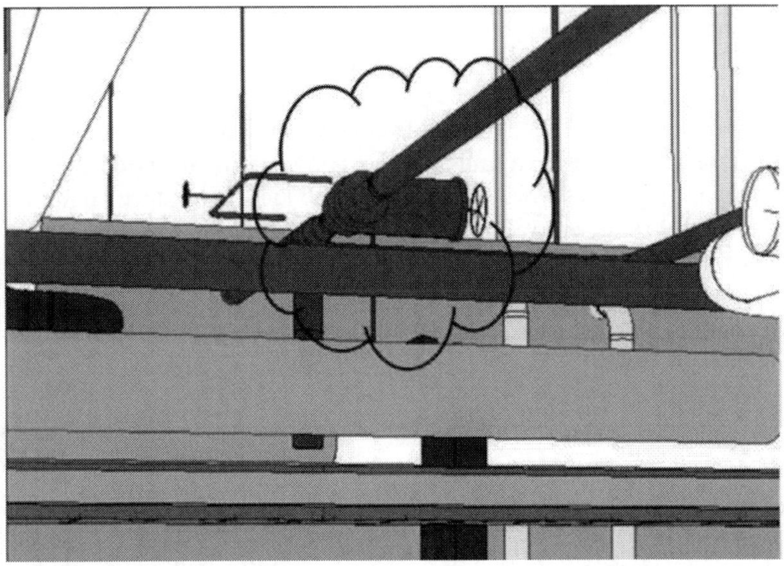

Exhibit 5. This valve is located such that access can be achieved with a portable ladder or rolling stairs. Relocating the valve so that it could be accessed while standing on the floor was considered but ruled out.

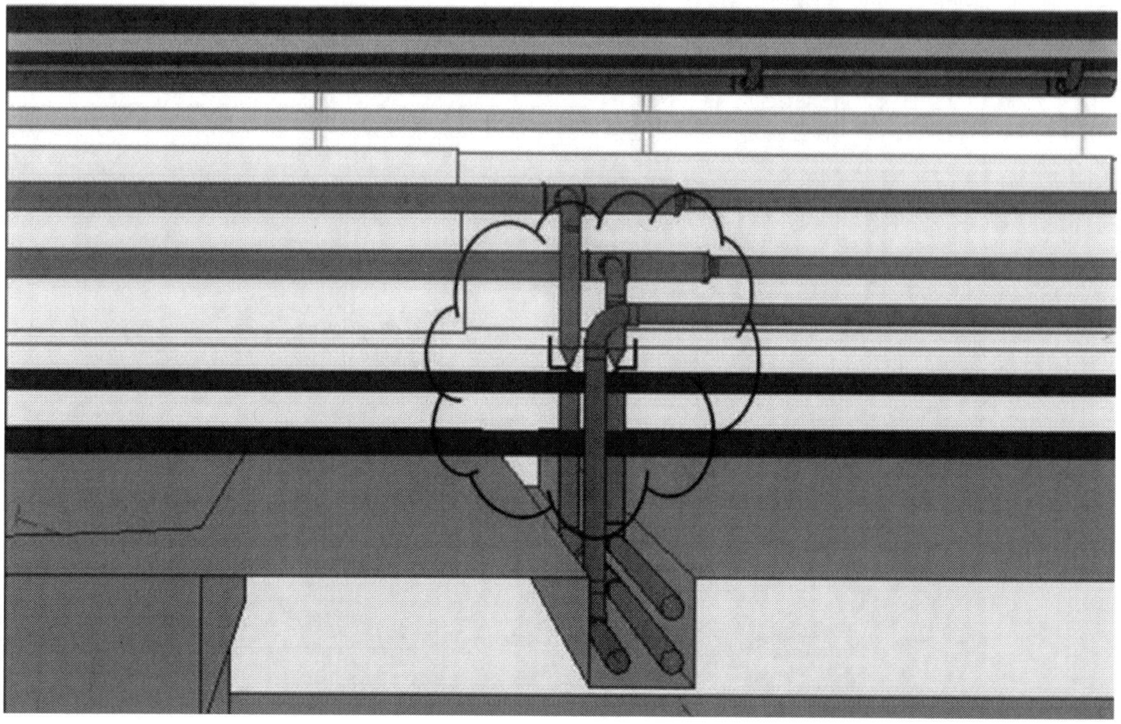

Exhibit 6. Valves located between wall and ductwork allow for access from a step ladder. In the existing building, valves like these are often located directly above ductwork that is added after the piping is in place.

The above examples show that by changing configurations of piping and/or ductwork, it is possible to eliminate the need for purchase of specialized equipment to achieve access to obstructed structures and prevent employees from climbing on or over items not meant to hold human weight. When valves, filters, and other equipment components are obstructed, maintenance personnel may need to find alternate and unsafe routes of access, placing themselves in increased danger. To protect workers from falls, the ideal is to place components at floor level or in areas where hand-railed catwalks can be provided. When the design cannot incorporate the ideal, it becomes necessary to purchase scaffold or mechanical lifts. If the design further makes the use of these infeasible, then ladders are the typical fallback. However, as safety professionals, we know that conducting work from a ladder--especially when tools must be carried or personnel must reach beyond the ladder to conduct a task—makes it more likely to encourage violation of safety standards and adversely affect employee safety.

Conclusion

As illustrated by this case study, Prevention through Design—applied in this instance using a virtual risk assessment through a BIM model—can result in significant risk reduction. When the building owner, design team and safety personnel all work together to improve safety, the benefits are realized during construction and make a significant impact on operations as well as throughout ongoing maintenance. And, the adjustments required for incorporating PtD pale in comparison to the benefits of increased safety and reduced cost.

For AOC, the project will continue for a number of years, and PtD will continue to be used to ensure that safety is designed into the renewed CHOB. This will in turn save on maintenance costs while providing a safe working environment for AOC and Congressional personnel. AOC recognizes the importance of safety and PtD and strives for a culture of people first, safety always. Where hazards are proactively designed out of projects, workplaces become safer, reducing the potential risk for incidents, injuries and property damage.

CSB Investigation: Williams Geismar Olefins Plant Reboiler Rupture and Fire

Lauren Grim, P.E.
Chemical Incident Investigator
U.S. Chemical Safety and Hazard Investigation Board
Denver, Colorado

Abstract

This paper presents findings and lessons learned from the U.S. Chemical Safety Board's investigation of the 2013 Williams Olefins Plant reboiler rupture and fire. The investigation found that serious deficiencies in the site's process safety management program, including Management of Change, Pre-Startup Safety Review, and Process Hazard Analysis, were causal to the incident. This article presents lessons learned from the incident that apply broadly to the chemical process industry.

Introduction

In June 2013, a reboiler that was isolated from its pressure relief valve catastrophically ruptured when heat was applied to it during an operations activity. Two operations personnel were killed. An investigation into the incident found that a series of process safety management program deficiencies over the 12 years leading to the incident caused the reboiler to be unprotected from overpressure. Weaknesses in the site's Management of Change (MOC), Pre-Startup Safety Review (PSSR), and Process Hazard Analysis (PHA) programs contributed to the incident. This article discusses lessons learned from the 2013 reboiler overpressure incident at the Williams Geismar Olefins Plant, which was investigated by the U.S. Chemical Safety Board.

Incident Background

The propylene fractionator (a distillation column) in the Williams Geismar Olefins Plant originally operated with two reboilers (Reboiler A and Reboiler B). In 2001, the site installed block valves on the shell (process) side and tube (hot water) side to allow for operation of only one reboiler at a time, with the other reboiler on standby. The tube side of the reboilers periodically fouled and required cleaning, and installing the valves allowed for single reboiler operation and continuous operation of the distillation column when the reboilers required cleaning.

The pressure relief valve for both reboilers was located on top of the distillation column. The block valves installed in 2001 could isolate the reboilers from their pressure relief valve, introducing an overpressure hazard to the reboilers (**Exhibit 1**). At the time of block valve installation, the site did not identify the reboiler overpressure hazard. In the subsequent 12 years leading to the incident, overpressure protection was not effectively applied to the reboilers.

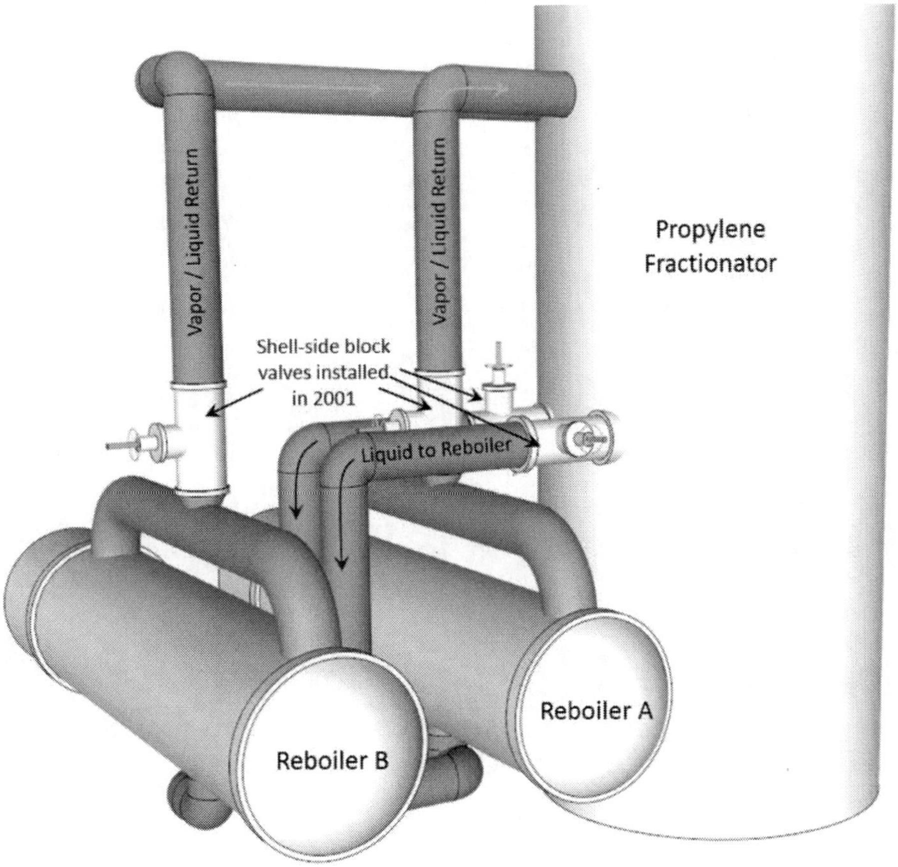

Exhibit 1. Four block valves were installed on the reboilers in 2001. The block valves isolated the reboilers from their pressure relief valve, which was on top of the propylene fractionator.

The Incident

On June 13, 2013, while working to identify the cause of a process flow problem, an operations supervisor opened the hot side (tube side) of the standby distillation column reboiler. The cold side (shell side) remained blocked in. Three minutes later, the reboiler exploded (**Exhibit 2**).

Propane and propylene process fluid erupted from the ruptured reboiler and from the propylene fractionator due to failed piping. The process vapor ignited, creating a massive fireball. The force of the explosion launched a portion of the propylene fractionator reboiler piping into a pipe rack approximately 30 feet overhead. A Williams operator working near the reboiler at the time of the explosion died at the scene. The operations supervisor succumbed to severe burn injuries the next day. The plant remained down for 18 months and restarted in January 2015.

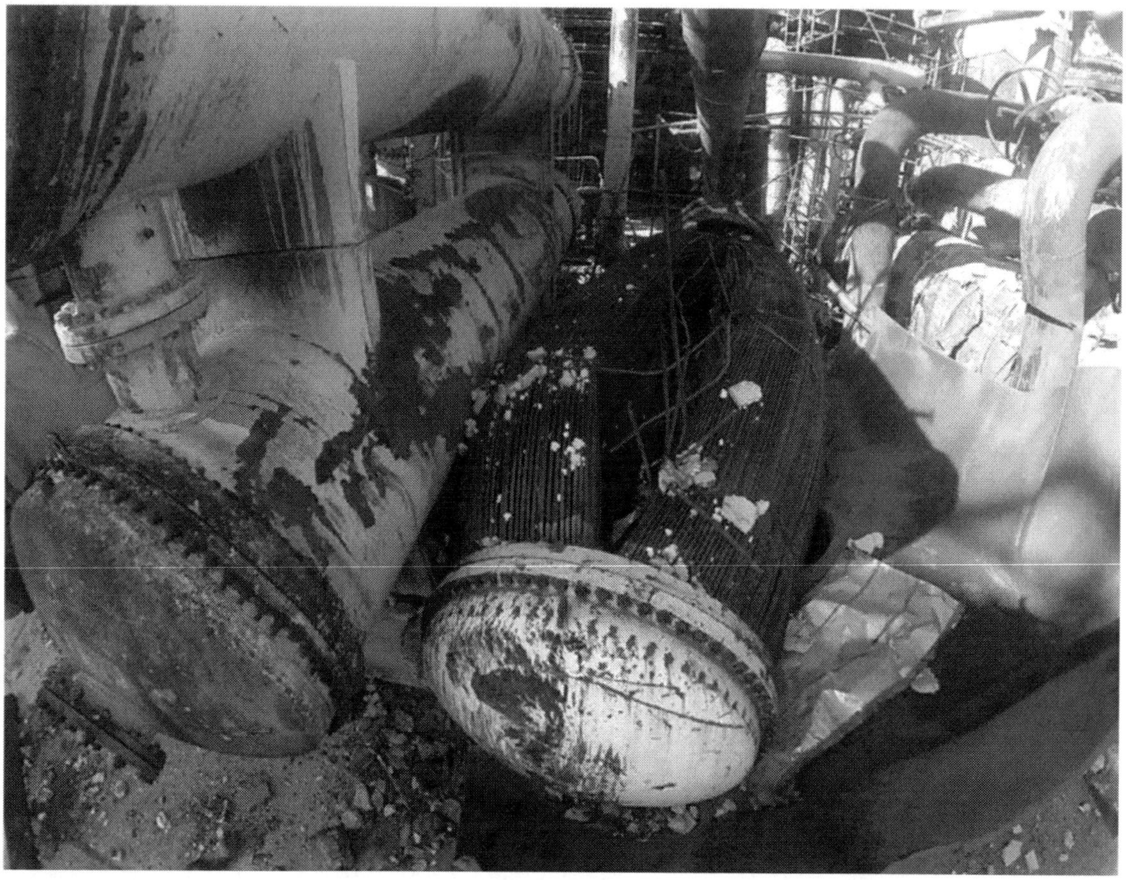

Exhibit 2. This is a post-incident photo of the ruptured Reboiler B.

Technical Analysis

Metallurgical testing and analysis found that the reboiler failed, resulting in the formation of a crack, at a high internal pressure estimated to be between 674 and 1,212 psig. The CSB concluded that a pressure of this magnitude was likely the result of liquid thermal expansion of liquid propane confined within the reboiler shell.

Williams performed maintenance on Reboiler B in February 2012. Following this maintenance activity, workers left Reboiler B on standby, reportedly filled with nitrogen and isolated from the process by a single closed block valve on the inlet piping and a single closed block valve on the outlet piping. The CSB determined that between the 2012 maintenance activity and the day of the incident—a period of 16 months—flammable liquid propane from the propylene fractionator accumulated on the shell side of the standby Reboiler B (**Exhibit 3**).[1] The propane could have entered the standby reboiler via a mistakenly opened valve, leaking block valve(s), or another unknown mechanism.[2] (Depending on the scenario that allowed propane to enter the reboiler, the nitrogen could have compressed and/or been pushed from the reboiler into the process.) Williams had not installed instrumentation to detect process fluid within the reboiler. As a result, Williams personnel did not know that the standby Reboiler B contained liquid propane.

---

[1] Discussed in the CSB report on the incident, the reboiler was at least 65.5 vol% full of liquid propane.

[2] Large gate valves such as the ones installed on the Williams reboilers are known to leak. The American Petroleum Institute (API) specifies allowable leakage rates through closed valves. For 16-inch and 18-inch valves such as the inlet valve and outlet valve on the propylene fractionator reboilers, API specifies an allowable leakage rate of 64 and 72 bubbles of gas per minute, respectively, during leak testing of the valves. (See API Standard 598, 9th ed. *Valve Inspection and Testing*, September 2009, p 10.) The reboiler block valves were leak tested following the incident. Their leakage rate was within that allowed by API Standard 598. While valve leakage likely allowed some process fluid to enter Reboiler B while it was on standby, a different mechanism could have introduced the bulk of the process fluid to the standby reboiler.

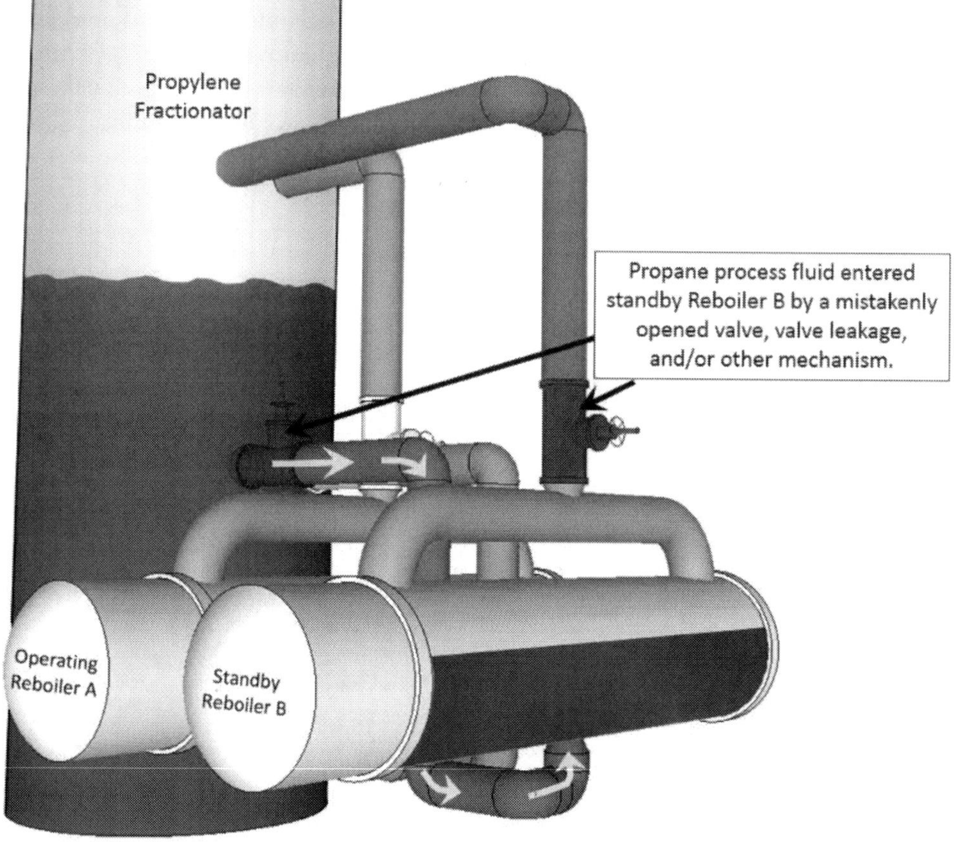

Exhibit 3. Propane process fluid entered standby Reboiler B by a mistakenly opened valve, valve leakage, and/or another mechanism.

Post-incident field observations identified that the Reboiler B tube-side hot quench water valves were in the open position. The shell-side process valves were closed, which isolated the shell of Reboiler B from its protective pressure relief valve on the top of the propylene fractionator.

When the Reboiler B hot quench water valves were opened, the liquid propane within the standby Reboiler B shell began to heat up. This caused the liquid propane to increase in volume due to liquid thermal expansion, filling any remaining occupiable vapor space within the shell. When the liquid could no longer expand due to confinement within the blocked-in Reboiler B shell, the pressure rapidly increased[3] until the internal pressure exceeded the shell's mechanical pressure limit (**Exhibit 4**), and the reboiler shell failed. The sudden and catastrophic failure of the reboiler caused a boiling liquid expanding vapor explosion (BLEVE). The process vapor ignited, creating a massive fireball.

[3] The liquid propane expanded and pressurized the reboiler faster than the vessel contents could escape through the leaking block valves.

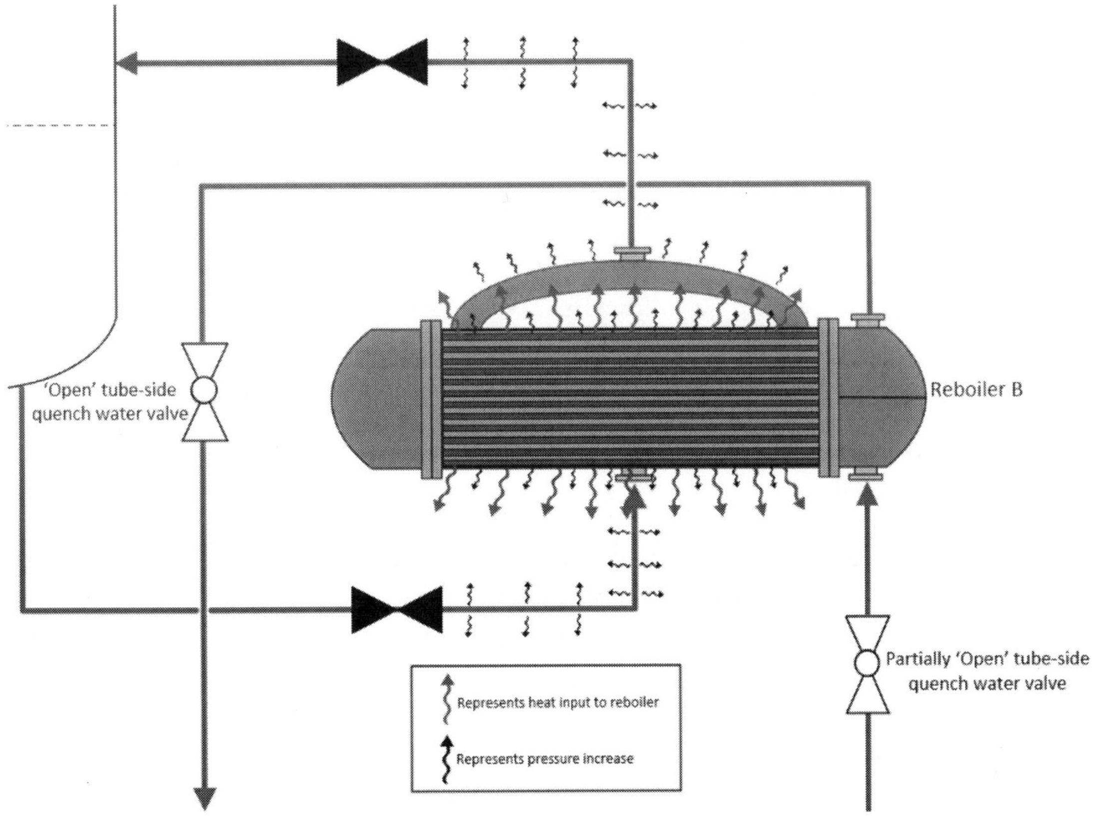

Exhibit 4. Expanding shell-side liquid propane could not sufficiently increase in volume due to the lack of overpressure protection and the closed shell-side process valves. As a result, shell-side pressure increased until the reboiler shell failed.

Process Safety Management Program Weaknesses

Weaknesses in the site's Management of Change (MOC), Pre-Startup Safety Review (PSSR), and Process Hazard Analysis (PHA) programs contributed to the incident. The following sections discuss those weaknesses and lessons learned, which can apply broadly to the chemical process industry.

Weak MOC Analysis

Performing an MOC is a key opportunity to identify and control new hazards introduced by a change to a process. The MOC performed for the 2001 block valve installation did not identify the overpressure hazard introduced to the process:

1. The MOC for the valve installation did not identify or control the overpressure hazard introduced to the reboilers. The MOC reviewers indicated that the valves did not have to be car sealed open, which had they been locked open would have provided overpressure protection for the reboilers. The option of using a car seal was the only specified overpressure

protection method on the MOC checklist, even though in this case installing pressure relief valves could be a better option. Nevertheless, the MOC reviewers did not identify that the reboilers required overpressure protection—through either an open path to a pressure relief device using a car sealed open valve, or by installing a pressure relief device on each reboiler;

2. The MOC reviewers indicated that the change did not require a Process Hazard Analysis (PHA), a more robust hazard evaluation option performed at the discretion of the MOC reviewers. The installation of the valves introduced a serious overpressure hazard to the reboilers, and a formal PHA would have been the best opportunity to identify and control that hazard;

3. MOC reviewers indicated either the valves met all codes and standards, or that the question was not applicable. The addition of the valves without ensuring overpressure protection for the reboilers, however, did not meet requirements within industry codes and standards by the American Petroleum Institute (API) and the American Society of Mechanical Engineers (ASME); and

4. The 2001 MOC was performed after the block valves were installed and the unit was operating with the new equipment. This limited the MOC analysis to an after-the-fact activity to address regulatory requirements rather than an effective tool used to identify and control new process hazards.

   The selected responses in the MOC checklist indicate that the reviewers focused largely on managing documentation and maintenance requirements for the new valves, such as needed process safety information updates and inspection requirements, and not on how the addition of the valves could affect the operability and safety of the overall process.

Not only does this "focus-on-the-new-equipment-only" approach to conducting Management of Change not meet the intent of regulatory requirements,[4] it can be dangerous. Williams introduced hazards that it did not fully understand or control.

PSSR was Incomplete

The site's PSSR was performed by filling out a 21-question form. The completed and management-approved PSSR documentation, however, had several of the questions left unanswered (**Exhibit 5**). One of the questions on the PSSR form prompted, "PRV's [pressure relief valves] lined up and block valves car sealed open? Pressure release systems in place and operational and traced where appropriate?" No response was given on the PSSR form, and overpressure protection was not effectively applied to the reboilers.

[4] The OSHA PSM and EPA RMP regulations require that Management of Change procedures shall ensure that the "impact of change on safety and health" is considered and addressed prior to the change. See 29 C.F.R. § 1910.119(l)(2)(ii) and 40 C.F.R. § 68.75(b)(2).

	Circle the appropriate response		
1. Has a process hazard analysis been completed, recommendations resolved, and incorporated in design as deemed appropriate?	Yes	No	Not Applic.
4. Is operations properly trained to startup and operate the facilities?	Yes	No	Not Applic.
6. Are all necessary operating procedures in place and current for safety, environmental, operating, emergencies, maintenance and technical?	Yes	No	Not Applic.
7. Is operations familiar with any operating changes for new or modified equipment?	Yes	No	Not Applic.
17. PRV's lined up and block valves car sealed open? Pressure release systems in place and operational and traced where appropriate?	Yes	No	Not Applic.
Are procedures available for new and modified equipment?	Yes	No	Not Applic.

Exhibit 5. This is a selection of responses on the checklist filled out during the PSSR performed following the 2001 reboiler valve installation. Several key process safety questions were not answered or were answered incorrectly.

Lessons Learned

- MOCs reviews should analyze hazards in the entire process *affected* by the change. Similar to PHAs, conducting MOC reviews as a multidisciplinary group—composed of individuals with different experiences and different areas of expertise—can assist in identifying hazards introduced by a process change.
- MOCs should be conducted before implementing a change in the field. Otherwise, they can turn into a paperwork or check-the-box exercise rather than an effective tool to identify and control hazards.
- PSSRs are key opportunities to verify effective implementation of design intent, accuracy of process safety information, and proper installation and configuration of field equipment. PSSRs should be effectively conducted before placing equipment in service.
- Management should ensure that MOCs and PSSRs are performed on-time and are complete. Process safety culture is influenced by management's actions. When incomplete hazard analyses and incomplete documentation are seen as acceptable, overall process safety at the site can deteriorate.
- Maintain a high level of vigilance when implementing PSM programs. Partially or ineffectively conducting MOCs, PSSRs, and PHAs can cause significant hazards to be overlooked, and this can lead to catastrophic incidents, sometimes years later.

Process Hazard Analysis

Three Process Hazard Analyses (PHAs) were conducted in the twelve years following the 2001 valve installation and prior to the 2013 incident. None of the PHAs were effective in controlling the reboilers' overpressure hazard.

A PHA was conducted in 2001. That PHA team did not identify that there was an overpressure hazard if the hot side (tube side) of the reboilers operated while the cold side (shell side) was blocked in. No recommendations were made to provide overpressure protection to the reboilers.

PHAs conducted in 2006 and 2011 addressed the overpressure hazard, but recommended actions that could have controlled the hazard were ineffectively implemented due to system errors, lack of communication, and misinterpretation of recommendation intent.

2006 PHA

The 2006 PHA, like other PHAs, was led by a contracted PHA facilitator, and technical, operations, and safety personnel were members of the PHA team. The team did not discuss that in practice only one propylene fractionator reboiler operated at a time; the PHA facilitator was under the incorrect impression that both reboilers operated simultaneously. This incorrect impression led the team to issue the following recommendation:

> Consider locking open at least one of the manual valves associated with each of the propylene fractionator reboilers [...] so that the relief valves on top of the propylene fractionator can provide thermal relief protection for these reboilers.

Locking open a process valve on the standby reboiler to provide overpressure protection is not a typical overpressure protection strategy; the reboiler would not be operating, but would still be open to the process and filled with process liquid.[5] Personnel charged with implementing the recommendation likely misunderstood the intent of the recommendation and only car sealed open the *operating* reboiler. As a result, the high-risk overpressure hazard remained in the standby reboiler. Even though the 2006 PHA recommendation intent to provide overpressure protection to both reboilers was only partially implemented, it was communicated to the Process Safety Management (PSM) coordinator that the action item was complete, and the coordinator marked the PHA recommendation action item as "complete" in the site's action item tracking system. No verification of action item implementation—through documentation or field verification—was required, and so the error in implementation was not identified.

In addition, the installation of a car seal to lock open the reboiler process valves—as recommended by the 2006 PHA—was a significant process change that required an MOC and a PSSR. But Williams did not perform an MOC or a PSSR for the installation of the car seal. The field verification portion of the PSSR should have provided an opportunity to identify that the PHA action item to car seal open a process valve on *both* reboilers was not complete. Yet, the PSSR was never performed.

The site did not perform an MOC for the car seal installation likely because key operations personnel did not understand that an MOC was required. Also, before the June 13, 2013 incident, at times the site began fieldwork on a process change without a completed and approved MOC.

---

[5] This was a low corrosive and minimally fouling environment, and this configuration likely would not harm equipment. This configuration, however, would have left an unnecessary inventory of hazardous chemicals in the process, and an inherent safety review should identify the opportunity to minimize the hazardous chemical inventory by blinding the standby reboiler from the process.

Due to these errors, overpressure protection was never provided for the reboiler that was in standby mode, even though the overpressure hazard was identified and corrective actions were recommended in the 2006 PHA.

2011 PHA

Relying on the action item tracking system from the 2006 PHA, the PHA facilitator documented in the 2011 PHA tracking spreadsheet that process valves on both reboilers were car sealed open. The PHA team did not discuss that only the operating reboiler was car sealed open in the field, and no field verification of safeguards was performed prior to the PHA. As a result, the PHA team and PHA facilitator did not identify that the car seal safeguards that were documented in the PHA spreadsheet did not actually exist in the field.

The PHA team identified, however, that the applicable P&ID did not show the car seals, so the team recommended updating the P&ID to indicate car seals on both reboilers. The individual who updates the P&IDs caught the discrepancy. He updated the P&ID to show only the active reboiler was car sealed open, and he informed the PSM coordinator that he implemented the PHA recommendation this way. However, management approved this implementation and the action item was marked as "Complete" with no qualifying information that it was implemented differently than worded by the PHA team. As a result, the 2011 PHA did not effectively control the significant overpressure hazard in the standby reboiler.

Lessons Learned

- Develop a method to confirm whether safeguards listed in PHA documentation and that are shown on P&IDs actually exist in the field. One way to do this is by field-verifying P&IDs prior to conducting PHAs.
- Ensure that PHA action items are implemented to address the original safety concerns of the PHA team.
- Ensure action items have been effectively implemented before closing them out. Field verify and/or provide supporting documentation that the action item was effectively implemented.

Pressure Relief Valve to Protect Equipment from Overpressure

After the overpressure incident, the site installed pressure relief valves onto each reboiler (**Exhibit 6**). Pressure relief valves are active safeguards, and are higher on the hierarchy of controls than car seals, which are administrative controls that can be more prone to failure

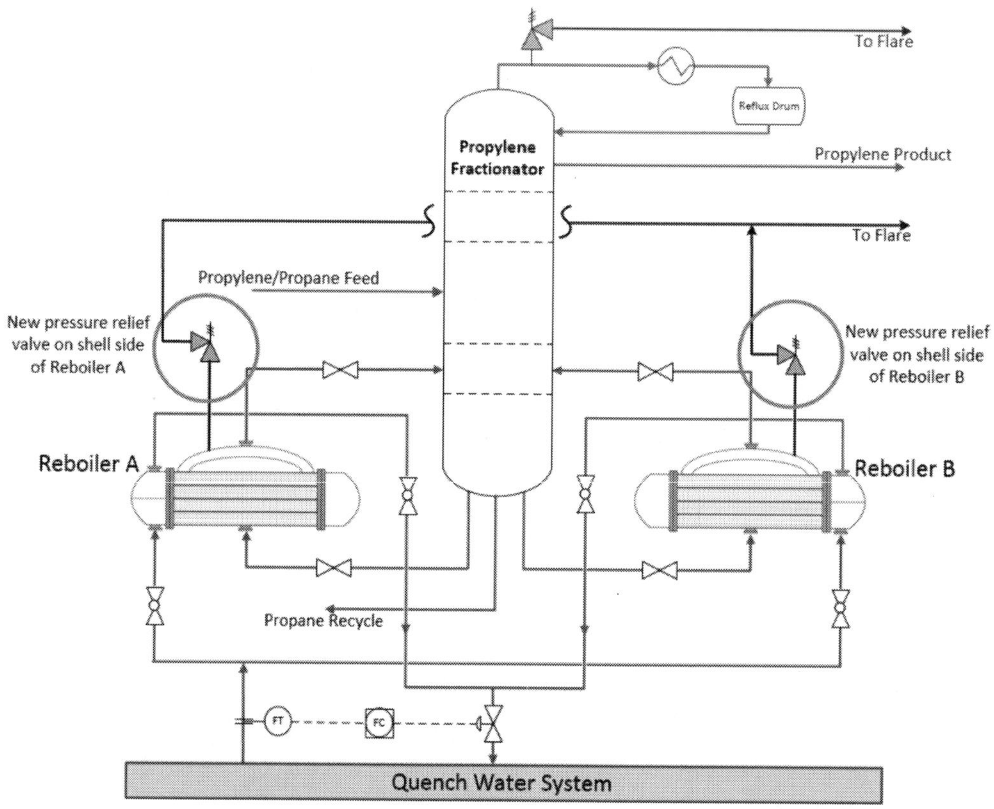

Exhibit 6. Post-incident, the Williams Geismar facility added pressure relief valves to the shell side of Reboiler A and Reboiler B.

Lessons Learned

- PHA teams should ensure that all pressure vessels have effective overpressure protection. At a minimum, a pressure relief device should protect process equipment from overpressure scenarios where internal vessel pressure can exceed design code limits.
- PHA and MOC teams should effectively use the hierarchy of controls whenever feasible when evaluating and recommending safeguards (**Exhibit 7**).

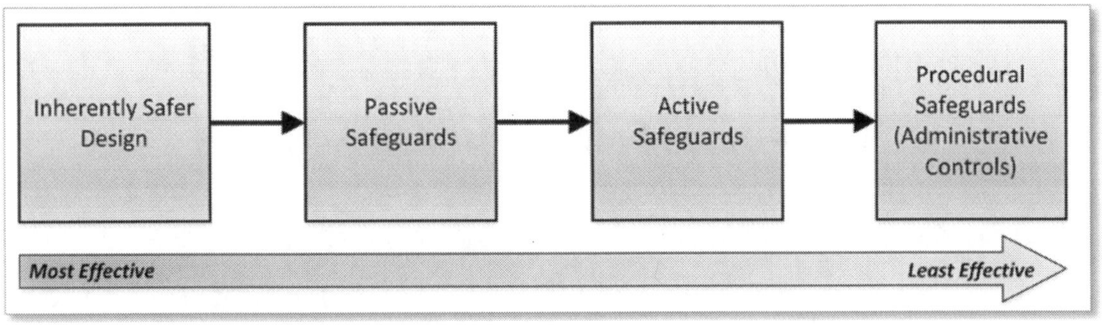

Exhibit 7. This is the hierarchy of controls. The higher in the hierarchy (further to the left), the more effective the risk reduction achieved.

Conclusion

In the 12 years leading up to the June 2013 incident, a series of deficiencies in the site's process safety management program caused a reboiler to be unprotected from overpressure, leading to catastrophic rupture of the reboiler and two fatalities.

- The applicable MOC for the installation of block valves that could isolate the reboiler from its pressure relief valve was performed after the valves were already installed, and the analysis of the valves' effect on the safety of the process was superficial;
- The PSSR performed for the installation of the block valves was incomplete and important safety questions were not answered;
- There was a lack of communication by the PHA team to discuss that in practice only one reboiler operated at a time. This caused the PHA team to issue a recommendation to car seal open both the active and inactive reboiler. Confusion by individuals who implemented the recommendation likely caused only the active reboiler to be car sealed open, and the overpressure hazard remained in the standby reboiler;
- When PHA action items were not implemented as intended by the PHA team, the site did not identify the implementation error and the significant overpressure hazard remained in the standby reboiler.

This incident reinforces that there are often many opportunities throughout the life of a process to identify process hazards and implement safer design or mitigative safeguards. It takes diligence and attention to detail to identify hazards and opportunities for improved safeguards.

To read the full report on this incident, see U.S. Chemical Safety and Hazard Investigation Board, *Williams Geismar Olefins Plant—Reboiler Rupture and Fire*, Geismar, Louisiana; June 13, 2013. Report 2013-03-I-LA, October 2016. The report is at http://www.csb.gov/williams-olefins-plant-explosion-and-fire-/.

Team-Based Safety Training Activities

Linda M. Tapp, ALCM, CSP
SafetyFUNdamentals
Madison, NJ

Introduction

Accelerated Learning principles, applied to safety training, can be very powerful tools - especially when you consider the topics that safety trainers often try to teach. These are often life and death issues and the absorption and retention of this information is extremely important. Traditional safety training usually involves showing a video and then discussing the video as it relates to the company's program and policies or showing a PowerPoint presentation, reading the slides to the audience and then handing out a quiz. Accelerated learning is much different. Accelerated learning is focused on the results – not the methods. Whatever learning tools work to increase and enhance learning can be called accelerated learning methods. Many safety professionals spend a good amount of their work week training others. Accelerated learning principles can help you spend less time creating training and can help trainees learn faster (and remember more) so accelerated learning applied to safety training can be a win-win situation for everyone.

Everything in an accelerated learning class is focused on the results and not the materials or activities themselves. For example, a safety trainer announces that everyone is going to get up play twister with the goal of getting everyone relaxed and "ready to learn". While this might seem like a fun icebreaker, the twister game is fun only for the sake of being fun. None of the information to be covered in this training is being reinforced by this activity. In an accelerated learning class, we might also play games but the games have a different focus – on the results, instead of the activity. For example, a BINGO game could be played where instead of a number and letter being called out, a clue is called. Instead of numbers being on the BINGO card, answers directly related to the training content are listed. This game is still fun, and gets everyone involved, but includes accelerated learning principles, especially when you have small teams work on each BINGO card instead of individually. Learning from peers is a key concept of accelerated learning and is the focus of this paper. Safety training that relies on the group members' participation in team based activities can not only enhance the quality of the training but can increase the amount of training material retained.

A sample BINGO game (Exhibit 1) used an Ergonomics Training Class as well as the accompanying clue sheet (Exhibit 2) follows.

safetyFUNdamentals
SafetyBINGO

Office Ergonomics

Glare	Wrist Rest	Eye Strain	Twisting	Monitor
Right Angles	Arm's Reach	Carpal Tunnel Syndrome	Knee Height	Head-aches
Awkward Posture	Mouse	Repetitive Motion	Tendonitis	Flat On The Floor
Close To The Body	Shoulder Height	Lower Back	Contact Stress	Neck Pain
Task Lighting	Eye Level	Force	Breaks	Elbow Height

 safetyFUNdamentals

Exhibit 1. Sample BINGO Game for an Ergonomics Training Class

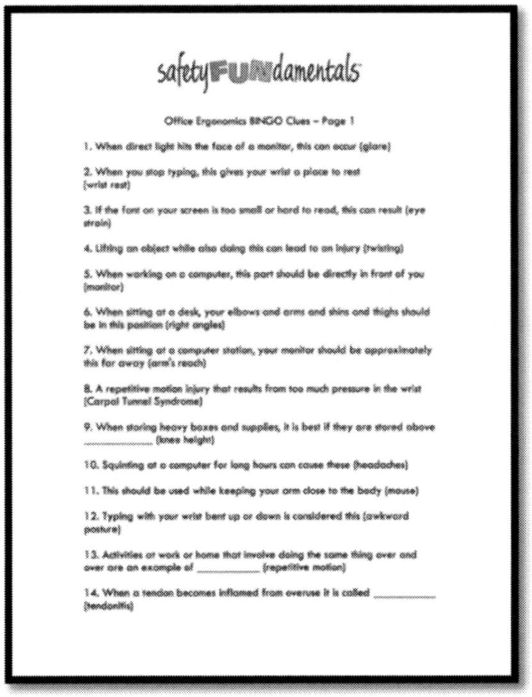

safetyFUNdamentals

Office Ergonomics BINGO Clues – Page 1

1. When direct light hits the face of a monitor, this can occur (glare)

2. When you stop typing, this gives your wrist a place to rest (wrist rest)

3. If the font on your screen is too small or hard to read, this can result (eye strain)

4. Lifting an object while also doing this can lead to an injury (twisting)

5. When working on a computer, this part should be directly in front of you (monitor)

6. When sitting at a desk, your elbows and arms and shins and thighs should be in this position (right angles)

7. When sitting at a computer station, your monitor should be approximately this far away (arm's reach)

8. A repetitive motion injury that results from too much pressure in the wrist (Carpal Tunnel Syndrome)

9. When storing heavy boxes and supplies, it is best if they are stored above ____________ (knee height)

10. Squinting at a computer for long hours can cause these (headaches)

11. This should be used while keeping your arm close to the body (mouse)

12. Typing with your wrist bent up or down is considered this (awkward posture)

13. Activities at work or home that involve doing the same thing over and over are an example of ____________ (repetitive motion)

14. When a tendon becomes inflamed from overuse it is called ____________ (tendonitis)

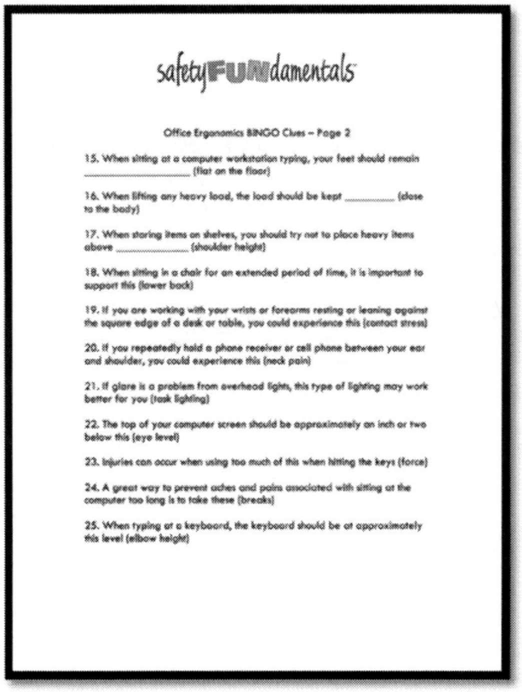

safetyFUNdamentals

Office Ergonomics BINGO Clues – Page 2

15. When sitting at a computer workstation typing, your feet should remain ____________ (flat on the floor)

16. When lifting any heavy load, the load should be kept ____________ (close to the body)

17. When storing items on shelves, you should try not to place heavy items above ____________ (shoulder height)

18. When sitting in a chair for an extended period of time, it is important to support this (lower back)

19. If you are working with your wrists or forearms resting or leaning against the square edge of a desk or table, you could experience this (contact stress)

20. If you repeatedly hold a phone receiver or cell phone between your ear and shoulder, you could experience this (neck pain)

21. If glare is a problem from overhead lights, this type of lighting may work better for you (task lighting)

22. The top of your computer screen should be approximately an inch or two below this (eye level)

23. Injuries can occur when using too much of this when hitting the keys (force)

24. A great way to prevent aches and pains associated with sitting at the computer too long is to take these (breaks)

25. When typing at a keyboard, the keyboard should be at approximately this level (elbow height)

Exhibit 2. Sample Clue Sheet for an Office Ergonomics Training Class

Principles of Accelerated Learning

Team-based safety training activities rely on the principles of accelerated learning to be successful. The major principles of accelerated learning are:

Total Learner Involvement Enhances Learning

Learning should involve the whole mind and body. Learning does not just take place in the head but involves the whole body with all its senses, emotions and receptors. Training should be "SAVI" which means it applies to the somatic, auditory, visual and intellectual parts of a learner.

Learning Is Not the Passive Storage of Information but the Active Creation of Knowledge

Training needs to be created by the trainee and not necessarily given to them. When learners learn, they are creating new meanings, associations and neural networks within their existing self. One thing that good safety training always tries to accomplish is to get the trainee to understand how the training applies to him or her in a realistic way.

Activity-Centered Learning Events Are Often Superior to Presentation-Centered Ones.

Learning takes place by absorbing many things at once, not one fact at a time. The brain works as a parallel processor and is challenged when it is asked to do many things at the same time. If you ask someone to sit still and just look at a bunch of slides or listen to a speaker, learning will not be as great as if the same material was presented along with an activity that would relate to the material at hand. Activity, or doing the work itself, enhances learning. We have all heard the famous Confucius quote "I hear and I forget. I see and I remember. I do and I understand." In almost every area of life, we learn by doing. For example, you can have someone read endless safety policies and manuals and watch videos about safely driving a forklift but until the person actually gets on and drives it, real learning will not take place.

In a traditional training class, the trainer is the dispenser of information, solely responsible that learning occurs The student is defined as a passive receiver of information and few if any of the resources they bring to the training are incorporated into the learning process (Michaelsen). By using team-based activities in safety training, there is less pressure on the Trainer to "perform" and more reliance on the trainees to use the knowledge and skills they previously learned. Additionally, by working in teams, trainees learn from each other since everyone's background, education and training is different.

It has also been found that team-based training provides empowerment (Michaelsen). Empowerment can be defined as the "means, ability, or opportunity to do." By providing opportunities for trainees to learn from one another, the Trainer is empowering them to take more control and responsibility for their own, and their co-workers, safety education.

Also related to activity-centered learning is the use of the "image brain". Our nervous system is more of an image processor than a word processor so any time we can take concepts and make them visual, the brain is more likely to learn. Activities are a great way to turn the safety concepts we must teach into actual images that will help the trainee learn more and faster.

Finally, activity-centered learning events can be designed in a fraction of the time it takes to design presentation-centered ones. When we put together a training class, we often spend a great deal of time on the PowerPoint presentation, the hand-outs and making copies of related materials like sections of OSHA standards. When you realize that people learn much more, and much faster from experiences with feedback, you will begin to see how accelerated training methods, such as the use of team-based training activities, can save you time.

How Do People Learn?

There is a great deal of research into how the brain works and how we learn. When this research is applied to learning, many of the ways we were taught to believe about learning are now no longer correct. We were most likely taught to learn by sitting still and straight in our chairs all day, memorizing information and working on our own. Learning was often stressful as we feared being punished and was very often based on competition. Considering what we now know the physical, mental and emotional needs that must be met in trainees in order for them to learn, the following are basic guidelines to keep in mind.

Create a learning environment that reduces stress and creates positive feelings. You would not want an environment will you single people out and possibly embarrass them. If your trainee is sitting is a dingy dirty training room where they can hear the company intercom endlessly paging them, they are not going to be relaxed enough to learn well.

Positive emotions have also been found to be very important in enhancing learning. If someone is sitting endlessly in a lecture, they probably won't have positive emotions for very long. Activity can help with these positive emotions. If someone is stressed or bored or angry, their learning will be inhibited. If learning is positive, relaxed and engaging, learning will be increased. Team-based training activities can help to keep the trainee engaged. Peer pressure from others on the team will keep them involved and doing their share of the work.

Give your class attendees problem-posing and information accessing exercises that get them to think, make connections and make meaning for themselves. As mentioned earlier, it is very important with safety training that we help trainees learn to take the information that is presented to them and then have them customize it for their specific needs. For example, in a hazard communication training class, you would not just review a sample Material Safety Data Sheet/Safety Data Sheet (MSDS/SDS) but you would have the training teams 1) locate the MSDS/SDS for a product he or she actually uses on their daily job and 2) ask them to determine which types of personal protective equipment are required when using that chemical and 3) where they would find that personal protective equipment. Strive for total learner involvement, i.e., having the trainees totally and actively involved with full responsibility for their learning.

Design in opportunities for the attendees to collaborate with one another and be social. Team-based safety training activities are a perfect way to do this. When your trainees work together, they will use more of their total brain and learning will increase. If you present the class with a problem, an exercise or even a game, have them complete it in teams so they can work together on finding the solution.

Have plenty of opportunities for people to get up and out of their seats and move around. Activity can increase the learning process, keep the attendees more involved and help chase away boredom. Getting people physically active and using as many senses as possible will help awaken the body and improve circulation. This will have a positive impact on learning. The part of the human brain involved with movement is the motor cortex and it is located right next to the area used for problem solving. If you restrict bodily movement, you restrict the total brain from functioning at its best. Learning is hampered when we separate the body from the mind. Many people find it difficult to concentrate when their bodies are not doing something physical. Think about the people you sit next to in training classes that are constantly doodling, twirling their pen or bouncing their leg up and down. These are probably people who learn better when they are active.

Team-Based Safety Training follows all of these principles. When team members participate in an activity, they are involved and creating their own learning around that activity. When these activities are fun, the trainees are also relaxed and enjoying themselves which then leads to increased learning.

The Advantages of Working with a Team

Most of the safety training activities available can be completed by trainees working on their own but a great deal of the benefits from accelerated learning would be lost in this case. Accelerated learning research has found that trainees learn more when they learn from each other so opportunities to have trainees work together whenever possible should be pursued.

A few of the benefits of having your trainees work in teams include:

- The opportunity to learn from each other and gain unique perspectives from each team member because what can be learned by one team member's experience is entirely different from what can be learned by another's experience.
- Opportunities for sharing ideas including the possibility of identifying new approaches to problems or ideas for hazard elimination
- Increase in creativity when people work together in a team
- Building trust and establishing strong relationships with co-workers plus development of conflict resolution skills with co-workers which can transfer back to their day to day job responsibilities.
- Encouragement of team members to feel proud of their contributions in tackling problems together and creating a work product
- Encouragement of healthy risk-taking. Safety training activities should never run the risk of embarrassing individuals. Trainees less confident in their abilities to participate in activities because of lack of knowledge, experience or even command of the language, will benefit from team-based activities. With team-based training activities, if the team fails to complete the activity correctly, the lack of success will fall on the team and not an individual.
- Opportunity d=for senior or more experienced employees to mentor less experienced workers
- Opportunity to cover more material. Researchers at Stanford found that team-based learning allowed professors to cover more material than in a typical lecture class.
- Involvement of everyone. A significant advantage of team-based training is that a trainee cannot stay "hidden" as they could in a typical training class. It is very difficult to be anonymous in a small team
- Confidence building. As the popular saying goes, "there is no "I" in team" but that doesn't mean that team-based training doesn't benefit the individual. Working with others to complete a difficult challenge builds confidence in everyone involved.

Team-Based Activities Disadvantages

Although the advantages outweigh the disadvantages, there are a few potential issues to keep in mind when using team-based safety training activities.

- Some activities may take longer when using teams. Often when a group of trainees that know each other very well get together, they may get sidetracked and talk about other issues either related or not related to the training topic.
- When the assigned team-based safety training activity leads to passionate discussion, it can also make it difficult for the Trainer to get trainees to wind down the activity and finalize their work. Although, anytime trainees want to keep discussing safety issues, it isn't necessarily a bad thing.
- In larger teams, a member or two may attempt to "free-ride" by not fully participating in the training activity since they know others will work on the activity and/or finish the project.

Things to Consider Before Creating or Using a Team-Based Safety Training Activity

Consider what you want the trainees to be able to do after completing the activity and ensure that the activity will help meet those objectives. There are many team-based safety training activities to consider. Make sure you are selecting the right one for the class objectives. You should also determine what information the team will need to have before attempting to complete this activity. Do they already have the background knowledge or do you need to deliver additional information in a lecture or other format? Can you assign pre-class work that will have them ready to participate in the activity when class starts?

The nature of the activity that teams participate in have an enormous impact on the quality of the learning experience the activity can provide. To be most effective the team-based training activity should have a tangible output. Otherwise it is going to be difficult to ensure that learning took place. Even something as simple as a list of items produced by the team after completing the activity will show that the task was completed successfully.

Ideally, a team-based safety training activity would be impossible to complete by only one team member. Much of benefit of team-based activities is collaborating with peers so a training activity is stronger if the only way to successfully complete it is by working with others and relying on teammates.

The team-based safety training activity that the teams are assigned to complete should be designed so that most of the time the team spends working on the activity is spent doing what teams do best - identifying problems (such as safety hazards), thinking of solutions, discussing alternative views and possible solutions and making decisions. The activity, whenever possible, should provide practice dealing with real-life situations and conditions that the team members could face. By working with teammates to apply what they have learned to the activity, especially if it is personalized to their workplace, trainees will see great benefits. By reinforcing concepts, issues, ideas, rules and procedures through the training activity, trainees will have a higher level of retention (Reis).

To increase the value of a team-based safety training activity, mechanisms should be put in place to measure each team's performance. You can do this by publically comparing teams or asking for peer evaluations for longer more complex team-based training. For example, for a safety training activity that involves hazard identification, tell the training class before the activity begins that the number of hazards identified by each team will be posted and the team with the highest number of identified hazards will be declared the winner.

Specific Safety Training Activities Involving Teams

Hidden Pictures/Hazard Hunt - Hidden Pictures or Hazard Hunt is a game that includes a small team working together to identify hazards in a supplied photo or drawing. A photo or drawing should have at least 6 hazards included and the hazards should not all be obvious. If possible, the hazards should be related to class content although this is a great game for general safety training tips such as training on hazard recognition. With most safety training games and activities, it is recommended the trainees be divided into small groups. When using Hidden Hazards, each smaller group of trainees can be given a different photo or drawing or they can all be given the game drawing for comparison purposes. If the same drawing is used, the groups should be spaced far enough away from each other so that they cannot hear each other discussing and identifying the hazards. After 10 minutes or so, stop the group discussions and ask each group to report how many hazards they identified. It is highly likely that the various groups will have identified a different number of hazards. Start with the groups with the biggest list and ask them to identify their hazards. If they name something that another team did not include it will likely lead to lively discussion and that is the purpose of this exercise.

PPE Match Up - PPE Match Up is a good activity to use when providing chemical hazard awareness training or any training that involves the proper use of personal protective equipment. Before class, several pairs or pieces of personal protective equipment that is mentioned in the class should be collected. If possible, samples of various hazardous materials, particularly those mentioned in the class, should be collected. If a material is too hazardous to bring into the classroom, then a empty bottle or other container can be used but with a label on it saying "Pretend Sample for Training Only – (name of material). After you have presented the applicable training information, divide the training class into teams. Give each team a set of the PPE you have collected. Assign each team a different color and then place a large sheet of paper in each of these colors in front of the various hazards you have brought into class. When you say go, each class will need to put the PPE they have been given in front of the hazard for which it is bet suited. Each team's PPE should be placed on the corresponding color paper so it is obvious which team's equipment is in front of each of the various hazards. After all teams have finished, go to each of the different hazards and see what the teams thought was the most important PPE for providing protection against that hazard. Review the right and wrong selections, and why.

Safety Sequence - Safety Sequence is a good activity for training topics that include a series of steps or a safety procedure, such as lock out tag out or confined space entry. Safety Sequence (with pictures) can be a real learning tool. To organize the game, you will need to find a simple illustration or photo (one you take yourself is fine if the picture is clear and clearly shows the step) of each step of the procedure. For a simple lock out tag out exercise, the photos you might want to include would be 1) Determine the Hazards, 2) Notify Others, 3) Isolate Energy Sources, 4) Remove Potential Energy, 5) Verify Lock Out, 6) Perform Work, 7) Prepare to Start up again, 8) Remove locks and tags, 9) Test Equipment, and 10) Notify others when work is complete. Make copies of these photos or drawings so you can put one step on a small card and make several sets of cards (one set for each group of trainees). After the material has been presented, shuffle the cards and give a set to each group. Tell them they have 1 minute to get the cards in the proper order, with the first step of lock out first in the pile and the last step in the lock out tag out procedure should be last. After one minute has passed, tell the groups to stop and review the steps in the proper order. Ask each group if they have the same order or if they disagree. If the groups complain about having only one minute to do the exercise, begin a discussion about how difficult it can be to make important decisions when rushed and why it is very important to thoroughly understand critical safety procedures. Note: Instead of having the team simply put the cards into the correct sequence, a great way to add involvement to this activity is to give each team member one of the steps in the sequence (you need the same number of team members as steps in this case) and have the team members line up in the correct order and then announce when they believe they have the sequence correct.

Pin the Pain – Pin the Pin is another exercise that is great with any topic related to chemical hazards or any topic that covers injury and illness prevention for specific body parts. To play "Pin the Pain", you can make a simple line drawing or find a simple body outline that is not copyright protected. After presenting the class information, give each team a copy of the body outline and a pack of push pins or small colored stickers and ask them to "Pin the Pain." Based on the information you just presented, tell the trainees they are to identify where injury or illness could occur as a result of improperly working with the chemical or activity you are describing in the training class. For example, if you were presenting a class on the safe handling of hydroflouric acid you would expect each group to "Pin the Pain" on the eyes, skin, heart, nose and lungs for example. After each team has had the chance to "Pin the Pain", ask them to present their body outline and explain why they pinned the areas they did.

A-Z Race - A game called "A-Z Race" is a popular way to summarize and wrap up a training class while involving all trainees. All that is needed is a piece of paper and pen for each trainee team. Again, small groups of 3-5 work best. It is also helpful but not necessary to have a large piece of flipchart paper or a whiteboard available to display the results at the end of the exercise. To begin, ask each team to make two columns on their paper. Down the left side of the page, the first half of the alphabet (A-M) should be

written and down the side of the second column, the second half of the alphabet should be written (N-Z). Tell the class that this is a race. Each team must think of a word related to the training topic that begins with each letter of the alphabet and the first team to finish will be declared the winner. It helps if you can offer a prize to the winning team members, even if the prize is only a piece of candy, but it is not necessary. When one team finishes, stop the game and ask the team to shout out what they wrote next to each letter. Write these words on the flipchart or whiteboard for all to see.
Teams participating in the exercise mentally go through all of the key points of the class and conduct their own review of the material. When the winning team shares its answers with the class and the words are written on the flipchart or whiteboard, a second review takes place. Learning occurs without participants really thinking about it. (Note: to see samples of these training materials, visit www.freesafetygames.com.)

Million Dollar Madness - Million Dollar Madness is a great way to open a longer safety training class. To play, you first divide the class into teams. If there are different departments represented in the class it might be a good exercise if people from the same departments or work areas were put on the same team. Tell the class that an anonymous donor dropped off a bag with 100 million dollars on the company doorstep last night with the stipulation that the money be used only for safety improvements. With money not an issue, each team has 5 minutes to develop a list of improvements they would make or new equipment they would buy using that money that would directly improve plant safety. Remind the class that money is no object. After all teams are finished, ask the teams to share their wish list with the class. It is a good idea to have someone write all the ideas down (or you can do it on a flip chart) so that all serious ideas can be saved and shared with management for possible consideration. Encourage discussion by asking "How did the fact that you have unlimited funds change your thinking"? "How can you make some of these ideas come true"? and "Which ideas surprised you as being more or less expensive than you expected?"

Safety Lotería - Safety Lotería is a version of a popular Mexican game called Lotería. If you have played Safety Bingo, you know that the game boards have words or short phrases related to the class topic. Safety Lotería is like Safety Bingo but instead of words on the game boards, there are small pictures. The game can be played very simply by the Trainer randomly selecting a picture and then showing the card with the picture to the trainees. If a trainee has that picture on his or her card, they mark it off. Just like in regular Bingo, the first person to get all of the pictures in a line (across or down) marked off, wins. While this is an amusing game, it does not do a good job of teaching new skills related to safety. To make the activity more educational, the picture selected should not simply be shown but a hint or even a small riddle should be said that describes the picture. (If your audience is not English speaking, this description or riddle will need to be said in the language of the trainees). When the game takes place in this way, trainees need to think about the class content before selecting the appropriate square on the board. An effective game in safety training is one that is an equal mix of chance and skill. Without the short description of the picture, Safety Lotería is all chance. By asking trainees to figure out what safety picture you are talking about by a short description or question, skill is involved and the game is more effective as a safety training aid.

Summary

In addition to increasing interaction and retention for the trainees, team-based safety training games and activities can be designed in a fraction of the time it takes to design typical classes. When you realize that people learn much more and much faster from experiences combined with feedback, you will begin to see how accelerated learning principles applied to safety can be a winning formula. Team-based safety training activities are one way to encourage collaboration and learning from peers which is a core principle of accelerated learning.

Bibliography

Brame, Cynthia J. "Team-based Learning." *Center for Teaching*. Vanderbilt University, n.d. Web. 09 Mar. 2017. <https://cft.vanderbilt.edu/guides-sub-pages/team-based-learning/>.

Duggan, Tara. "Training Programs for Effective Teamwork." *Chron.com*. Chron.com, 03 Nov. 2012. Web. 09 Mar. 2017. <http://work.chron.com/training-programs-effective-teamwork-12359.html>.

Furjanic, S. and L. Trotman. 2000. *Turning Training Into Learning*, New York: Amacom.

Gratton, Lynda, and Tamara J. Erickson. "Eight Ways to Build Collaborative Teams." *Harvard Business Review* (2007): n. pag. *Harvard Business Review*. HBR, 01 Nov. 2007. Web. 09 Mar. 2017. <https://hbr.org/2007/11/eight-ways-to-build-collaborative-teams>.

Mattson, Dave. "6 Benefits of Teamwork in the Workplace." *Sandler Training*. Sandler Training, 19 Feb. 2015. Web. 09 Mar. 2017. https://www.sandler.com/blog/6-benefits-of-teamwork-in-the-workplace.

Medsker, G. J., and M. A. Campion. 1997. "Job and Team Design." *Handbook of Human Factors and Ergonomics*. Hoboken, NJ: Interscience, 450-89. Print.

Meier, D. 2000. *The Accelerated Learning Handbook*, New York: McGraw Hill.

Michaelsen, Larry K., Arletta Bauman Knight, and L. Dee Fink. 2002. "15/Team-Based Learning: A Strategy for Transforming the Quality of Teaching and Learning." *Team-Based Learning: A Transformative Use of Small Groups*. N.p. Westport, CT: Praeger, N. pag. Print.

Michaelsen, Larry K. "Building Learning Teams: The Key to Harnessing the Power of Small Groups in Higher Education." *The Center for Teaching and Learning*. UNC Charlotte, n.d. Web. 09 Mar. 2017. <http://teaching.uncc.edu/learning-resources/articles-books/best-practice/large-classes/building-learning-teams>.

Pike, R. 1994. *Creative Training Techniques Handbook*. Minneapolis, Lakewood Books.

Reis, Rick. "Benefits of Learning Teams." *Tomorrow's Professor Postings*. Stanford, n.d. Web. 09 Mar. 2017. <https://tomprof.stanford.edu/posting/750>.

Sibley, Jim. "39/Readiness Assurance Process." *Getting Started with Team-based Learning*. Place of Publication Not Identified: Stylus, 2016. N. pag. Print.

Stolovitch, H. and E. Keeps. 2002. *Telling Ain't Training*, Alexandria, VA: ASTD.

Tapp, L. *Accelerated Learning Methods for Safety,* ASSE PDC Proceedings, June 2005.

______. "It's All Fun and Games: Twenty Five Activities to Make Training Great", ASSE PDC Proceedings, June 2006.

______. 2009. *SafetyFUNdamentals: HazardHunt* Cherry Hill, NJ: Ennismore Publishing.

Thiagarajan, S. 2003. *Design Your Own Games and Activities*, San Francisco, John Wiley & Sons.

American Society of Safety Engineers

Professional Development Conference June 19-22, 2017

SAFETY 2017 ASSE

Session: 646

Fundamentals of SH&E: Workplace Health

C. David Langlois, CSP

Langlois, Weigand & Associates, Inc.

Workplace Health

SAFETY 2017 ASSE

Safety – usually associated with acute injuries, short-term, traumatic exposures

Health – usually directed at chronic exposures, persistent, prolonged, repeated

2

Definition

SAFETY 2017 ASSE

Industrial Hygiene – "that science and art devoted to the anticipation, recognition, evaluation, and control of those environmental factors or stresses, arising in or from the workplace, which may cause sickness, impaired health and well-being, or significant discomfort, and inefficiency among workers or among the citizens of the community"

3

Environmental Stresses

SAFETY 2017 ASSE

Chemical – environmental concentrations

Physical – noise, heat, ionizing, etc.

Ergonomic – man-machine interface

Biological – micro living organisms

4

States of Matter

Solids – definite shape and volume

Liquids – definite volume but no definite shape

Gases – neither definite volume nor shape

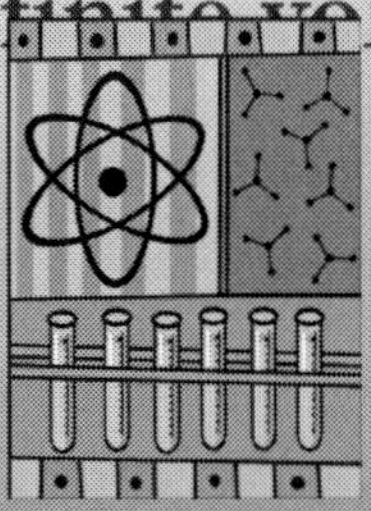

5

Chemical Stresses Chart

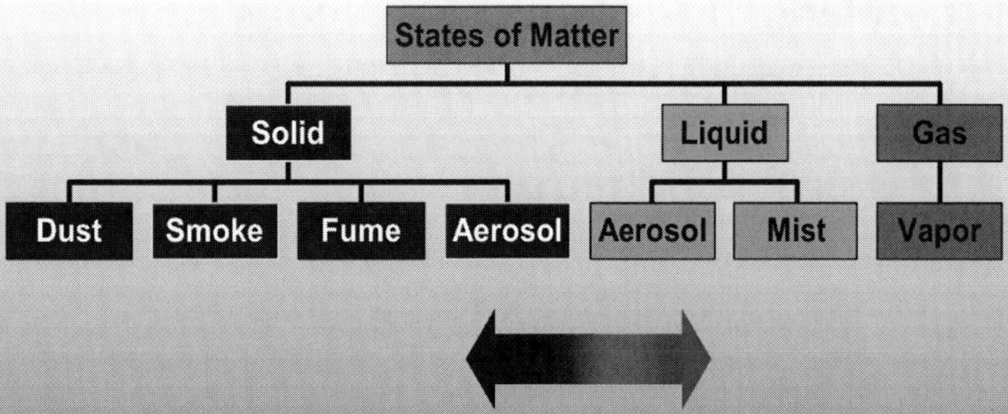

6

Chemical Stresses

Dusts – solid particles from handling, grinding, crushing, impact (1-25 microns) (cotton, grain, grinding wheels)

Smoke – carbon or soot particles less than .1 microns in size and are the products of incomplete combustion (fire, gas engines)

7

Chemical Stresses

Fumes – solid particles generated by condensation from the gaseous state (welding, soldering, brazing)

Aerosols – solid particles or liquid droplets of fine enough size to remain dispersed in air for a prolonged period of time (powder sprays, paint sprays)

8

Chemical Stresses

Mists – suspended liquid droplets generated by condensation of liquids, or by breaking up a liquid into a dispersed state (mixing vats, maintenance degreasers)

Vapors – volatile form of a substance normally a liquid or solid at STP (paint thinners, nail polish remover)

9

Terms to Express Concentrations

TLV – threshold limit value - airborne concentration under which it is believed that nearly all workers may be repeatedly exposed day after day without adverse effect (ACGIH)

PEL – permissible exposure limit (OSHA)

10

Terms to Express Concentrations

TWA – time weighted average – average exposure over a workday

C – ceiling – level not to be exceeded at any time

STEL – short term exposure limit (15 minute exposure not producing harm)

11

Terms to Express Concentrations

REL – recommended exposure limit, NIOSH term to designate agency's maximum concentration

BEI – biological exposure index; advisory levels adopted for some substances by ACGIH based on blood, urine, or expired air – TLV-TWA for eight hours

12

Terms to Express Concentrations

ppm – parts per million

mg/m^3 – milligrams per cubic meter

f/cm^3 – fibers per cubic centimeter

13

Routes of Entry

Inhalation – breathing, most common

Ingestion – swallowing

Absorption – penetration through the skin

Injection – forcing by mechanical means

14

Effects of Exposure

SAFETY 2017 ASSE

Concentration of substance

Probability of substance to produce injury

Rate of generation of material

Control measures

15

Toxicity

SAFETY 2017 ASSE

Capacity of a material to produce injury or harm; Depends on dose, rate, method, site of entry, general health of individual, diet, temperature

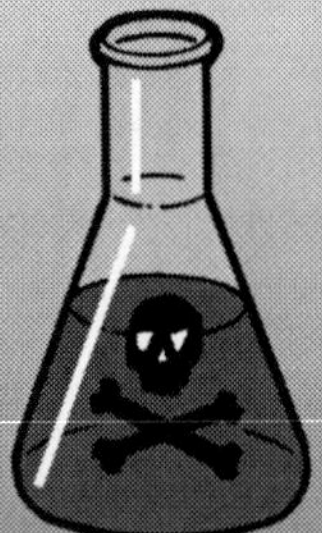

16

Toxic Effects

SAFETY 2017 ASSE

Irritants – inflame surfaces of the body

Systemic poisons – attack organs or system

Depressants – affect the central nervous system

17

Toxic Effects

SAFETY 2017 ASSE

Asphyxiants – prevent oxygen from reaching body cells (simple – nitrogen; chemical – carbon monoxide)

Carcinogens – cancer causing (benzene)

Teratogens – affect the fetus (lead)

Mutagens – affect the species (radiation)

18

Physical Stresses

Noise 

Temperature extremes 

Ionizing radiation 

Non-ionizing radiation 

19

Noise

Noise – Unwanted sound

Sound – pressure variation in air or other medium, that humans can detect

20

Noise

Decibels – pressure related to 20 micronewtons per square meter ($20\mu N/m^2$)

Frequency – one cycle per second (hertz)

PEL – OSHA enforcement 90 dBA/ 8 hrs

NIOSH – Recommends 85 dBA/ 8 hrs

21

Temperature Extremes

Climatic conditions

Work demands

Clothing

22

Radiation

Ionizing – electromagnetic or particulate radiation capable of producing ions, directly or indirectly, by interaction with matter

Non-ionizing – magnetic fields, radio-frequency/microwave, optical radiation, lasers

23

Ergonomics

The study of human characteristics for the appropriate design of the living and work environment

Matching the person to the task, as opposed to matching the task to the person

24

Ergonomic Stresses

- Mental workload
- Physical workload
- Energy cost
- Work classification
- Work/rest cycles

25

Biological Stresses

- Microorganisms – bacteria, viruses, fungi, and their products
- Arthropods – crustaceans, arachnids, insects; bites and stings
- Allergens and toxins from higher plants – poison ivy, poison oak

26

Biological Stresses

Bacteria – single-cell plants living in soil, water, organic matter, plants, or animals

Virus – microscopic agents able to live only in living cells

27

Biological Stresses

Fungus – parasitic lower plants including molds, mildews, mushrooms

Parasite – organism living in/on another living organism

28

Biological Stresses

SAFETY 2017 ASSE

Protein allergens from vertebrate animals; urine, feces, hair, saliva

Current topics - OSHA bloodborne pathogens, sick-building syndrome, weapons of mass destruction

29

Hierarchy of Controls

SAFETY 2017 ASSE

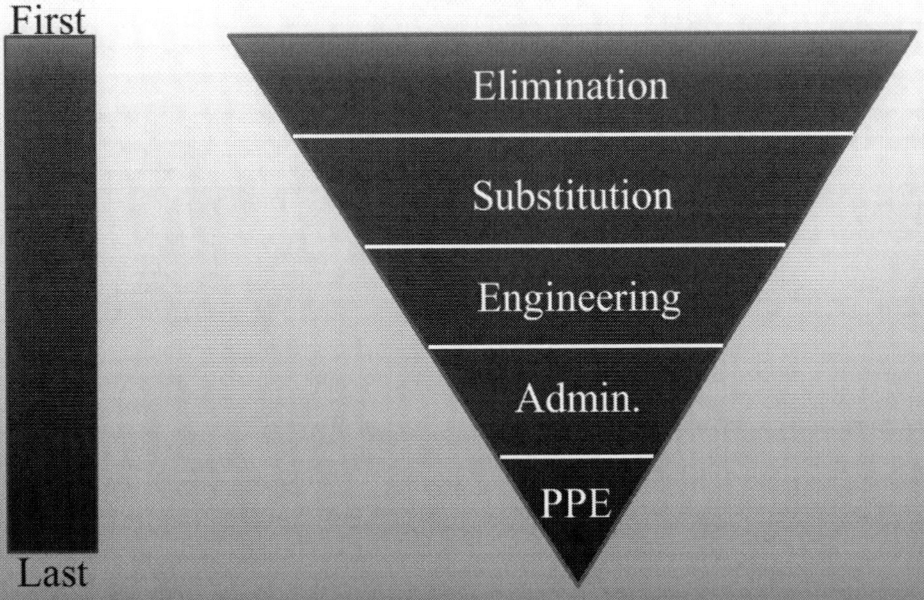

30

Industrial Hygiene Controls

SAFETY 2017 ASSE

- Substitution
- Process change
- Isolation
- Wet methods
- Local exhaust
- General ventilation

31

Industrial Hygiene Controls

SAFETY 2017 ASSE

- Personal hygiene
- Housekeeping
- Waste disposal
- Special controls
- Medical controls
- PPE

32

Conclusion

Questions

Comments

Evaluation form

References

33

Session No. 650

Leading the Leaders

James "Skipper" Kendrick, CSP
Kendrick Global Enterprises, LLC
Hurst, TX

Introduction

Much has been written and said on the topic of leadership. It is widely accepted that one of the key tenants of a productive safety process is management leadership and commitment. A prime example of this tenant were comments made by executives during the 2016 ASSE PDC Executive Summit. These executives expressed their views on leadership and how we professionals need to be viewed as leaders within our own organizations. One nugget within the Summit was a comment from one of the participants;".....talk to me in my language, the language of business."

Years of anecdotal research have shown that there are few formal educational venues or programs that prepare organizational leaders for their role in leading OSH within their respective organizations. They typically receive little to no training or preparation as to how their leadership skills and talents may be utilized regarding the organization's vision, values, beliefs, roles and responsibilities relative to OSH. Many "don't get it" until faced with a significant emotional event (citation, fine, serious injury, death, etc.) And if impacted in this manner, many actions are performed in a "knee jerk" manner and may not have lasting impacts.

An additional challenge is that in many organizations, the OSH function is a staff or advisory function with no direct operational authority. In these situations, a different leadership set is required for OSH professionals. The ability to "influence" is most probably the skill needed most. This skill need be applied both upward and downward throughout the management hierarchy. To provide maximum influence, it is critical that the OSH professional know, understand and identify leadership traits and skills of those within the organization.

This session will explore the styles and qualities of leadership and facilitate a discussion as to tools and techniques that can be used to guide organizational leaders as they strive to visibly demonstrate leadership within the OSH arena.

A Discussion on Leadership

Various Leadership Styles and Traits

Much has been written and published over time relative to the subject of leadership. The following is an admittedly short and brief synopsis of the field known as leadership. It is not meant nor intended to be all encompassing. Participants wishing more detail on any of the following may research and gain additional volumes relative to each style or trait.

Autocratic
- Authoritarian
- Commands and Controls
- Makes Decisions Without Outside Input
- Is the Total Authority
- Frequently Imposes Their Will on Others

Democratic
- Participative
- Engages Others
- Values Others Input
  - Still Retains Ultimate Decision Making

Laisssez-Faire
- Delegative
- Provides Vision Then Removes Themselves
- No Direct Supervision of the Work
- Feedback is Rare

Charismatic
- Engaging
  - Creates Energy
  - Liked & Respected
  - Inspirational
  - Emotional Appeal

Servant
- Values Others & Their Opinion
- Assists Others to Attain Goals/Objectives
- Seen as a Tool to Enable Success
- Humble and Evokes Humility

Transactional
- Goal Setting
- Clear Expectations
- Followers Agree to Follow
- Rewards & Consequences to Drive Performance
- Closely Monitors Progress

Transformational
- Motivator
- Highly Visible

Transformational – continued
- Sees the Big Picture
- Excellent Communicator

Situational
- Tell Others What to Do
- Great Sales Person
- Actively Engages Others
- Delegates to Groups

Authentic
- Solid Sense of Self
- Solid in Their Beliefs
- Openly Acts on Values & Beliefs
- Trustworthy
- High Ethics

Social-Charismatic
- Values of Others vs Values of Self
- Ethical Model
- Not Fully Defined in Current Terms – Still Studying

Thoughts on Leading Leaders

Once again, volumes have been written on the qualities of the leaders represented in the above styles. One prominent author and researcher, John Maxwell, has offered suggestions as to what characteristics are incumbent within modern leaders. The following are brief samples from Maxwell. Again, for further detail, more information is available through the studies and writing of Maxwell. Due to time constraints, many other prominent authors/researchers present similar and additional thoughts and materials which are not covered or utilized in this presentation

Lessons from John Maxwell – Critical Qualities/Abilities

Courage

Focus

Listen

Problem Solver

Relation Builder

Take Responsibility

Self-Discipline

Teachable

What Leadership Traits Would You Need to Deal with a Particular Leadership Style?

From your earlier definition and selection of a leader within your organization, what traits should you use the address the selected leader/style? You may choose to use the form provided at the end of this paper.

Now is the time to commit to action. What three specific activities could you use to influence your chosen leader?

Activities

1.

2.

3.

LEADING THE LEADERS WORKSHEET

NAME A LEADER WITHIN YOUR ORGANIZATION WHOSE EFFORTS, IF APPLIED, WOULD FURTHER THE PROGRESS OF YOUR SAFETY PROCESS:

OF THE LEADERSHIP STYLES/TRAITS PRESENTED, WHICH WOULD BEST DESCRIBE YOUR LEADER. LIST ONE OR TWO RECENT EXAMPLES OF HOW THIS STYLE/TRAIT WAS MANIFESTED WITHIN THE ORGANIZATION.

STYLE/TRAIT	EXAMPLES:

DRAW ARROWS FROM THE CHARACTERISTICS DISCUSSED TO THE STYLE/TRAIT TO BE INFLUENCED.

Courage	Autocratic
Focus	Democratic
Listen	Laissez-Faire
Problem Solver	Charismatic
Relation Building	Servant
Take Responsibility	Transactional
Self-Discipline	Transformational
Teachable	Situational
	Authentic
	Social-Charismatic

LIST THREE SPECIFIC ACTIVITIES THAT MAY BE TRIED TO INFLUENCE THIS LEADER:

1.

2

3.

Session No. 651

Persuasion Power--
How to get it, use it, and keep it

Michael E Rega, CSP
MSM Managing Partner
The Ecliptic Consulting Group, Inc.
Tampa, FL

Introduction

Beginning in 1978, The Ecliptic Consulting Group (ECG) engaged the professional services world to bring leading strategy and people management concepts to practice in technical fields.

For many professional people in EH&S, we are given tremendous responsibility but very little positional power in our organizations. Most times the end result is unnecessary delay, lack of efficiency, or a message that is not acted on which may lead to dire consequences for us and our organizations.

Based on years of organizational and behavioral research by titans of the industry like Dr. Robert Cialdini, Dr. Sally Fuller, Dr. Jeffrey Pfeffer, and Daniel Pink, *Power* is a discussion of what actually is happening in organizational behavior as opposed of what we think is happening. How to get things done through others, achieve business results, and how to rise to leadership positions will be explored. It is common to believe that good work and dedication are the main drivers in the success of a professional career. A mountain of recent data strongly suggest otherwise.

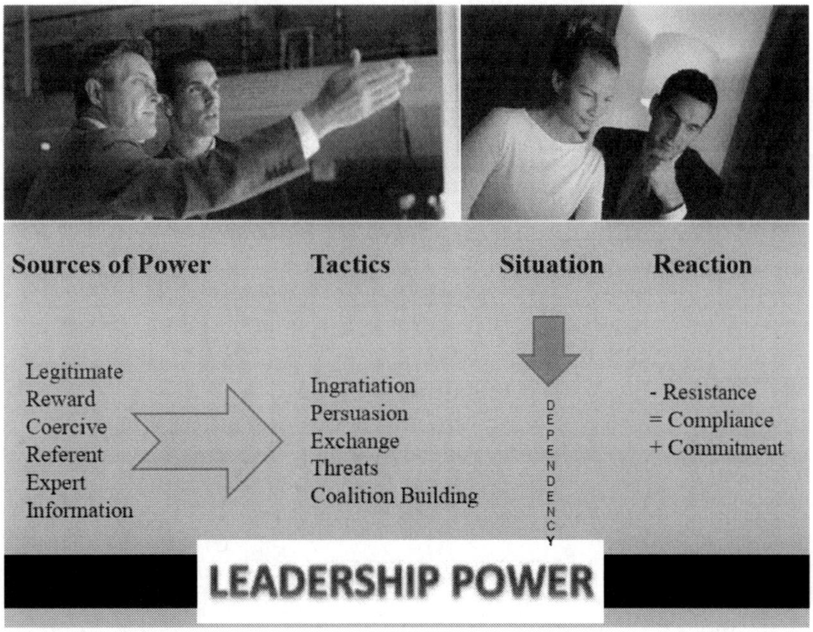

The author is a 25 year veteran of consulting with technical professionals, has a master's degree in Leadership/Organizational Sustainability, and is a strong supporter of using business topics as leverage to open up and expand critical business relationships.

This paper will uncover modern organizational power and suggest that instead of the power structure being a mystical free-for-all, it is actually quite orderly and predictable. A distinction will be made between our historic understanding of organizational power and the future of this fast moving and often confusing topic.

Gaining Persuasion Power

Organizational leaders understand the power levers that create change, enhance careers, and drive results. Successful personal power creation is action, not position. EH&S professionals can learn strategies how to enhance their organizations by uncovering and demonstrating principles of personal power. There are five main sources of organizational power that if properly deployed, anyone in the organization can and will begin to gain influence with the people around you. If you have ever wondered how some people in businesses achieve incredible results, while other equally qualified and bright individuals wallow, the power tools discussed here will provides the tools to maximize your effectiveness and grow your personal organizational value.

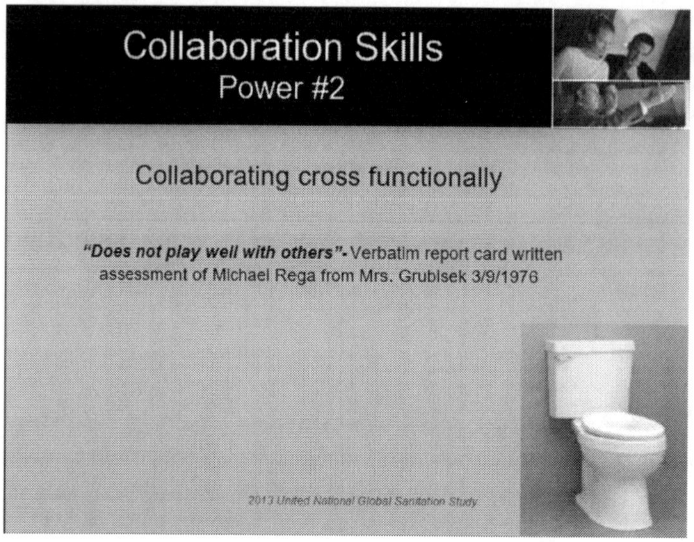

Many things influence human beings but perhaps the strongest instinctual drive is the need to be associated with a group. It almost does not matter what the group stands for, how it behaves, or who its leaders are. The overarching most important thing is that we *belong* to something, anything. There are some strong indications our group need is traced ancestrally to when humans banded together to form tribes to better hunt, mate, and survive. It would seem difficult for prehistoric man to live more than a generation as a hermit. The harnessing of that simple human requirement is what all great groups do.

In modern business circles it is a proud description to say you work for Apple, Google, or any of the high-flying success stories of the past decade in business. Outside the world of business, the US military spends tremendous effort branding their services. So much so that most retired service people describe themselves, even late in life, as "former Marine, or ex Navy" etc.

Our understanding of group mentality as described in the graphic below is key to capturing the power of the group and implementing that power into whatever group you belong to, civic, academic, religious, family, or business.

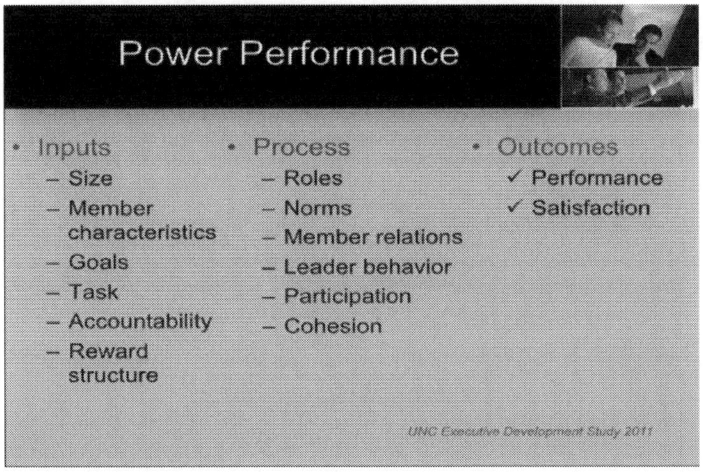

The deep understanding of what makes a group really perform is a power enhancing skill. As we go through life we may have all been members of groups or teams that for whatever reason were miserable. They may have even delivered the work, won plenty of games, or successfully implemented some new concept, yet still the dissatisfaction of the members was palpable. In contrast, opposite situations have likely happened to all of us. Your team or group delivered horrible results, lost almost every game, but your bonded anyway and most members remember a positive group experience.

The important component in grasping and building group power is first understanding the task to be completed, than build the group based on the qualities (inputs) that use and interdependency model. Most business graduate programs are based on this model and the end results speak for themselves. An MBA is the most popular graduate degree in the world and beyond degree completion, more than 83% of B-school graduates report a positive or very positive experience. There is power in understanding how group power is implemented and whether it is a college basketball coach or a project manager, wielding that power effectively can change careers.

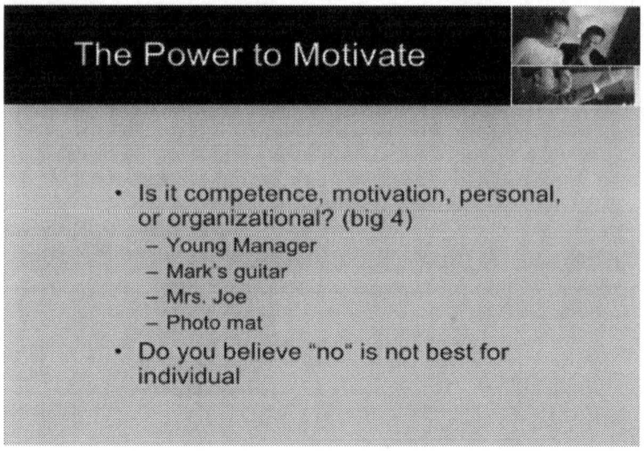

After you have power, keeping power becomes an interesting task. Frank Underwood of the acclaimed Netflix program, House of Cards, in season 1 said, "Money is the Mcmanansion in Sarasota that starts falling apart after ten years. Power is the old stone building that stands for centuries, and I cannot respect a person who does not understand the difference." A pessimistic quotes from a calculating, and at times, evil character. Yet consider what he is actually saying. Building and keeping power can be a career enhancing decision while chasing a paycheck may be a good job and a poor career move.

Keeping organizational power requires persuasion mastery. I define that as *allowing people to do what you need them to do—willingly*. It is the willingness of others in an organization that is an essential key to long-term power keeping.

Jeff Pfeffer in his groundbreaking research discovered that one the most important things individuals can do in organizations is to be spoken about (in a meeting, phone call, etc.) positively when they are *not present.* This is after all how legends are made. Daniel Boone, Wild Bill Hickok, and Jesse James just to name a few were larger-than-life characters because of the power of enhanced story telling. Their legends grew so much that in the end it is likely they didn't even remember the line between reality and fantasy. People spoke about these legends when and where they were not present, and it elevated their status literally beyond belief.

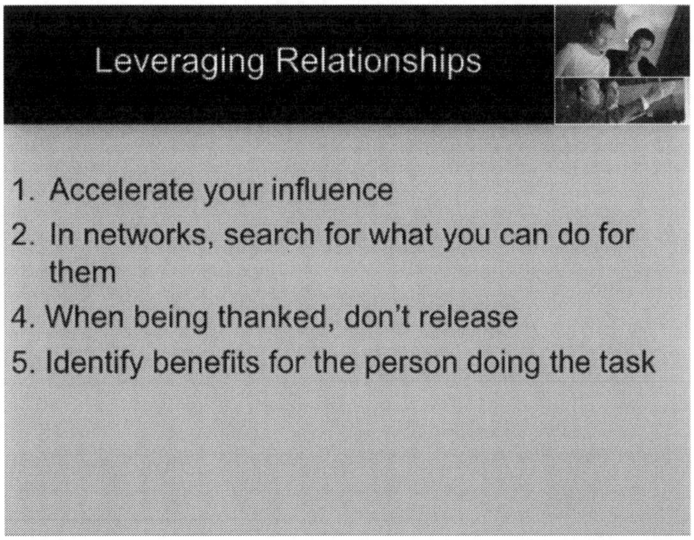

Finally, there is the massive power of persuasion. The oldest and perhaps still best example of a persuasion model was crafted originally by the ancient Greek Socrates and later put to text by Plato, his student. In his day, orators were very powerful people that governments feared. People like Socrates could change the course of history, alter the masses thinking, and change the very fabric of society. Socrates of course paid the ultimate price for his persuasion, yet we can learn a great deal from his model. It is a six-step model that has stood the test of time, and although the names of the steps change occasionally, the core substance has not altered for centuries. The stages of the model are not necessarily ever in a neat process and can take weeks or months depending on the complexity of the persuasion. However, if you trace major North American social change, you'll begin to recognize the stages in societal progression: anti smoking, use of seatbelts in automobiles, low fat diets, even the passing of the Affordable Care Act followed Socrates's brilliant model.

The Persuasion Model

1. Preliminary Rapport-the power of "liking"
2. Comparison Illustration-the power of normative, group behavior
3. Suggestion-simply put, the power of suggestion
4. Explanation-the analytical cost/benefit of any kind of change
5. Confirmation-proving that the concept actually works and normative
6. Commitment-this is called confirmation bias and is a powerful self persuasion method

Socrates and most every very persuasive person have mastered forms of this model and know how and when to deploy it. Generally they intertwine the power of storytelling to deliver the model as is the case with the famous "Dream Speech" Martin Luther King delivered in 1963. That speech resonates through the generations due to his use of the entangled storyline, which gives the speech an amazing punch.

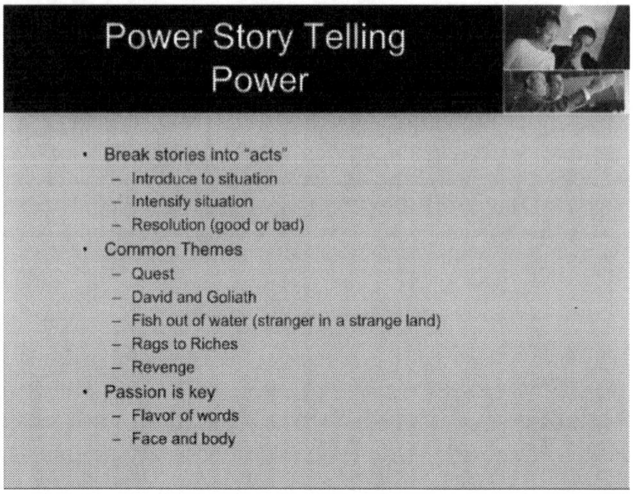

Libraries have been dedicated to the art and science of power. This brief paper is meant to capture the finest and most modern components of power and make them useful for the organizations of 2016. I believe that anyone can learn, master, and deploy power for the benefit of the organization, themselves, their teams, and society.

Bibliography

Ardry, R. 1970. *The Social Contract.* New York: Antheneum.

Gallo, Carmine. 2005. *Ten Simple Secrets of the World's Greatest Business Communicators.* (Naperville, IL: Sourcebook, 13.

Lubin, Joann. "Managers and Managing—Boss Talk" *Wall Street Journal,* December 27, 2000.

Nosanchuk, T.A. and J. Lightstone. "Canned Laughter and Public and Private Conformity." Journal of Personality and Social Psychology Vol 29 (1) Jan 1974, 153-156.

Opportunities and Challenges when Working with a Latino Workforce

Jerry E. Rivera
Director of Safety
Washington, DC Chapter
National Electrical Contractors Association

Jorge A. Otalora
Division Safety Director
Vienna, VA
Hoar Construction, LLC

Introduction

According to Dodge Data & Analytics, the construction industry is starting to see some strong growth in 2016. Dodge forecasts that the total value of construction starts could reach $612 billion, 9% from 2014. While this is great news for the construction industry in general, within the shadows of this economic prosperity lies a harsh reality, Hispanic/Latino/Latino workers continue to experience higher than average injury and illness in the workplace. Troubling, while the data demonstrates higher incidents, the reality is much likely worst as most Hispanic/Latino/Latino workers fail to report accidents in the workplace for fear of reprisal. This paper will review some of the challenges and opportunities that are present when working with a Hispanic/Latino/Latino worker.

Objective

The overall objective of this Proceedings paper is to identify training opportunities and challenges when working with a Hispanic/Latino/Latino workforce. The proceeding will review some of the statistics and findings that shall serves as a call to action for safety and health professionals. The proceeding will help identify some practical solutions to consider when training Hispanic/Latino/Latino construction workers.

Bureau of Labor Statistics Data

The Bureau of Labor Statistics (BLS) published in 2014 shows fatal job injuries involving Hispanic/Latino/Latino workers divided in two categories, Foreign Born and Native Born. Statistics for the last 10 years has consistently showed some significant disparities, foreign born Hispanic/Latino/Latinos bearing the highest burden. If we review the data on Table 1 foreign born Hispanic/Latinos carry the largest fatal injuries than the native born. This is quite interesting because data reveals that native and foreign born have differences in fatal injury trends.

A series of factor could influence this outcome between both groups such as the literacy rate, knowledge of workers' rights, willingness to assume risk or cultural difference between native and foreign born workers. A common factor that unites both groups is the increase in fatal injuries.

The data reveals that there are distinct cultural challenges when training foreign born Hispanic/Latino workers. For the most part, the United States has seen increased number of immigrants from Central and South America. Central American immigrants are more likely to come from war torn countries.

Immigrants from war torn countries in general have consistently demonstrated more tolerance to assuming greater levels of risk. Studies show that these risk assumptions are deeply rooted in their infancy and presumptuously carried into adulthood. In summary statistics demonstrate that even with safety related training foreign born Hispanic/Latino/Latino worker are more likely to assume greater levels of risk in the workplace.

Summary of Statistics

Fatal work injuries among Hispanic/Latino workers were 7.5 percent higher – 804 in 2014 compared to 748 in 2012. Of the 804 fatal work injuries incurred by Hispanic/Latino workers, 513(or 64 percent) involved foreign-born workers. Since 1992, work-related fatalities among Hispanic/Latino Workers in the United States have increased by more than 50%.

Accuracy of Injury Data

While the data available to date shows that Hispanic/Latino workers have experienced higher injuries, illnesses and fatalities these numbers don't account for unreported events. This is of significance because underreporting of injuries, illness and fatalities amongst Hispanic/Latino workers can be driven by immigration status. Hispanic/Latino workers fear that not only will they face job loss but potentially deportation.

There is an urgent need to use community based organization as a venue for educating Hispanic/Latino workers of their safety and health rights in the United States. Community based organizations are of interest because they often have a relation of trust with the Hispanic/Latino community.

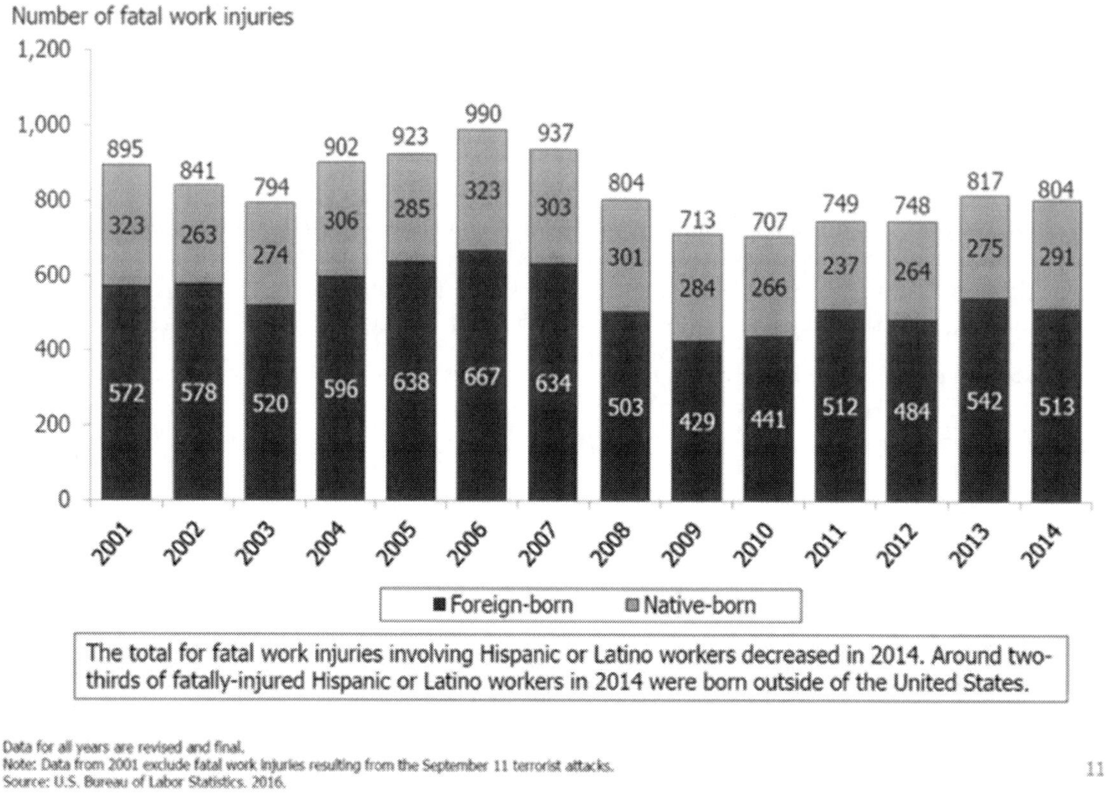

Table 1. Number of Fatal Work Injuries Involving Hispanic or Latino Workers, 2001-2014.

Economic Burden on the United States Economy

The economic impact for fatal and nonfatal injuries in 2007 is estimated at more than $6 billion. The cost workplace injuries are even greater acting as an anchor on the United States Gross Domestic Product (GDP). Hispanic/Latino workers who suffer injuries, illness and fatalities in the United States share a big chunk of the economic impact.

Therefore, it is in the best interest of the construction industry to invest in effective safety and health efforts that help reduce injury, illness and fatality of Hispanic/Latino workers. Without a doubt, the cost of inaction makes the United States economy less competitive.

Historical Training Development in the United States

Safety and health training in circulation today focuses on the literal translation of material from English to Spanish. This poses a unique challenge as the intended safety and health message is often sacrificed to accommodate the translation. The focus of translating material to Spanish should consider the intent of the original message and while the translation might not be literal, the core message has a better chance of surviving the translation. Major hurdles in the translation of Spanish safety and health resources have suffered the lack of consideration for both languages in the translation process.

Further research should be conducted on the effectiveness of translated safety and health training material. This could be best achieved by integrating academia, community based organization, Spanish linguist, engineers, tradespersons and safety practitioners. The main point is that the solution need to come from collaboration of the most parties affected.

Dissemination of Training

If the availability of safety and health resources today suffers from inadequate translations, dissemination efforts in safety training will pose its own unique set of challenges. While hundreds of safety and health training have been translated to Spanish many have not yet reached its intended audience. The dissemination of safety related training for Hispanic/Latino workers should seek lasting partnerships with community based organizations. These organizations have gained the respect of the Hispanic/Latino community where they eradicate.

Safety professionals should seek to align their efforts beyond their organizational boundaries to the communities in which they operate. This will establish employers as valuable members of the community and result in broader outreach efforts for current and future employees.

Getting the Job Done

While Hispanic/Latino workers value training in general, many reports that they are less likely to dedicate significant time to safety training because they hold work to be at a higher value to their livelihood. The most valuable asset for Hispanic/Latino workers is their ability to earn a living wage. For Hispanic/Latino workers making a living wage is a symbol of the American Dream, for this dream they are willing to assume greater risk.

Best Practices in Safety Training Today

When developing safety and health training consider the intended message, the accuracy of translation and the cultural differences. A combination of text, images (preferably pictures) and scenario based training might be best suited for Hispanic/Latino workers. This approach helps account for the language barriers that are often present in learning environments for Hispanic/Latino worker.

Those who have been most successful in the development of safety and health training in Spanish utilize a decentralized approach of translation that involves the consideration of the content from both the languages.

Scenario based training is the most effective form amongst adults and Hispanic/Latinos in the construction industry. Many Hispanic/Latino workers have learned their construction trade via on the job training which makes the scenario based approach a highly attractive tool.

Most importantly safety training must be aligned to professional development in the workplace. If the safety training is strategically aligned with career progression it will have more of a sustainable impact.

NIOSH Studies on Overlapping Vulnerabilities

The term "**occupational health disparities**" refers to increased rates of work-related illness and injuries in particularly vulnerable populations [Centers for Disease Control and Prevention (CDC), 2011].

Vulnerable populations are often described using a single characteristic (age, race, income, employment, etc.) [Cutter et al. 2005].

A growing body of research explores how a particular characteristic, such as being an immigrant, a racial minority, or a temporary worker, can increase an individual's risk for occupational injury or illness and suggests effective ways to improve their safety and health.

However, as occupational health disparities research demonstrates, workers frequently belong to more than one of these vulnerable groups that affect their overall risk of injury or fatality [Krieger 2010].

Young Hispanic/Latino immigrant workers are never singularly Hispanic/Latino one day, immigrants the next day, and young persons the following day; they are all of these every day.

A worker with overlapping vulnerabilities is simultaneously a member of two or more at-risk groups such as immigrants, temporary workers, new workers, older workers, small business employees, and non-union workers.

Each vulnerability has characteristics that add unique barriers to the worker's OSH (for example, an immigrant worker's fear of deportation for reporting unsafe conditions) [Flynn 2010] or that intensify existing barriers to safety that are common for all workers (such as lack of training in small businesses due to financial constraints) [Cunningham et al. 2014].

Numerous efforts have shown promise in addressing the unique needs of vulnerable workers. The following examples of intervention certainly do not cover the entire range of positive efforts, but they do provide several elements of interventions that OSH professionals may consider for more effectively assisting vulnerable worker and employer groups with OSH:

- Development of binational partnership with the Mexican government to identify and address occupational health inequities among immigrant workers [Flynn et al. 2013].
- Implementation of programs to teach immigrant day laborers to serve as peer safety leaders. The program trained the laborers to recognize safety and health hazards, to communicate effectively with coworkers and supervisors, and to facilitate tailored OSHA 10-hour training. By 2014, more than 500 workers in New Jersey and New York received OSHA 10-hour cards through this program [Scruggs and Arroyo 2014].
- Integrating foundational workplace safety and health skills as missing life skills into curricula by integrating "Talking Safety" in school-and community-based programs, youth development organizations, professional organizations, and temporary employment agencies, may be one way to ensure that every individual, before entering the workforce, has the basic workplace knowledge and skills necessary to engage in safe and healthy work.

Available Training Resources that are Useful when Working with the Hispanic/Latino Workforce

The following are some of the resourcs available for the training of the Hispanic/Latino workforce that we have successfully used in the past. These are not all inclusive. There several other resources, but for the purpose of this presentation, we will refer to those we have used and obtained good results.

ASSE

The **Safety Professionals & the Latino Workforce** (SPALW) common interest group addresses the increase in incidents among the Hispanic/Latino workforce due to misunderstanding of cultural gaps, communication barriers and inappropriate safety training. SPALW provides one-on-one support through mentoring in both English and Spanish and embraces our nation's richness and ethnic diversity and to address issues of safety and cultural diversity, language, customs and traditions among different ethnic groups

OSHA

The **National Institute for Occupational Safety and Health** (NIOSH) provide national and world leadership to prevent workplace illnesses and injuries. In its website, OSHA has a section called "Spanish-Language Compliance Assistance Resources." This page was developed to assists primarily English and bilingual users (Hispanic/Latino employers and workers) to be in compliance with OSHA regulations.

The Center for Construction Research and Training (CPWR)

CPWR not only provides training resources for Hispanic/Latino construction workers, but also it fosters and conducts researches on critical safety issues like "Falls among Hispanic/Latino Construction workers" or "How to better protect day laborers".

The electronic Library of Construction Occupational Safety and Health (eLCOSH)

This electronic library was developed and maintained by CPWR. it contains nearly 1,000 photos of bad and good safety practices as well as videos, PowerPoint presentations and some others printed material. Each one of them offers a great opportunity to address any safety concern using visual aids which is an enhancement.

Practical Solution for Closing the Knowledge Gap when Working with the Hispanic/Latino Workforce

Instructor Selection

First and fourth most it is important to note that instructor should be qualified safety professionals as defined in the "ASSE Guide to Hiring the Right Occupational Safety & Health Professional." The criterion included below is intended to be supplemental skills required when training a Hispanic/Latino workforce.

Cultural – Instructors should have a broad understanding of the audience's cultural differences.

Language – Instructor should be proficient in the language of understanding of participants, preferably in reading, writing and speaking ability.

Qualified – Instructor should be proficient in the subject matter being covered. While an instructor might be qualified to teach trenching and excavation the instructor might be ill equipped to teach electrical safety. See OSHA definition for Qualified Person.

Curriculum Development – Instructor should be familiar with ANSI/ASSE Z490.1-2016 Criteria for Accepted Practices in Safety, Health and Environmental Training and ANSI/ASSE A10.2 Training Management, Development, Delivery and Evaluation.

Scenario Based Training

This approach integrates predefined safety critical task to be avoided based on risk exposure. For example, when conducting safety related training about fall protection, attendees should be shown a scenario where the use of fall protection application is warranted. This gives workers a visual and practical approach for understating fall protection requirements.

Allow ample time for participants to demonstrate knowledge gained in application. The instructor should take this opportunity to assess participants understanding and making on the spot corrections. Small group participation, with a preference on groups no larger than 8 participants is encouraged.

Peer to Peer Safety Training in Small Groups

Many construction safety professionals have adopted an expanded version of the safety toolbox topic that integrates small groups, a targeted objectives and hands on evaluation. The approach works best when the safety training is directly related to their daily task. This expanded version of the toolbox topic training integrates a meaningful approach because of peer to peer learning is most likely to occur. When training is conducted in groups smaller than 8 who remain together, peer to peer coaching is likely to occur.

The small groups approach to toolbox for Hispanic/Latino workers serves as a setting that is less formal and more engaging with a more lasting effect of knowledge gained.

In closing this paper is not intended to be all encompassing of all the challenges faced by Hispanic/Latino workers. The intended outcome is for the information to serve as a call to action for all safety practitioners that there is further research needed to find sustainable solutions to enhancing safety related conditions of all workers.

There will be no silver bullet, instead a long road ahead that will lead us to advance safety and health for all, especially those who carry the greatest toll during economic prosperity.

On the Job Training

An old yet effective mean of training is on the job training. When using on the job training for safety related subjects it is best to pair up an inexperienced worker with a seasoned worker that is proficient in the subject matter. Before using this approach, safety professionals should identify what are the learning objectives that we want to accomplish and the skills that the seasoned employee should pose to ensure successful peer to peer learning. Emphasis has been placed on instructor lead training historically but on the job training has proven to be of great value.

1. Identify safety tasks to be observed
2. Develop a knowledge skills assessment list
3. Defined acceptable parameters for demonstrating competence level

Coaching

The goal of job coaching is to improve an employee's performance. Coaching focuses on the individual needs of an employee and is generally less formal than other kinds of training. There are usually no set training sessions. A manager, supervisor, or veteran employee serves as the coach. When he or she has the opportunity to get together with the employee being coached, the focus should be on:

- Answer questions
- Suggest more effective strategies
- Correct errors
- Guide toward goals
- Give support and encouragement
- Provide knowledgeable feedback

Translated Material to be used for Training

When developing, or selecting training material for Hispanic/Latino workers caution should be exercised as not to use material that has gone through literal translations. Curriculum should be developed with the targeted audience in mind as to ensure that the learning objectives are met.

Training material that has been translated by translation software often fails to capture the learning objectives. When using translation software safety professionals should ensure that the training material is reviewed by a designated qualified instructor (see above instructor selection criteria).

Conclusion

While the authors of this Proceeding paper have provided some insight on Opportunities and Challenges when Working with Latino Workforce, further research is needed on the subject. Readers can appreciate that historical attempts have been focused on a single solution to a multi factorial problem. Moving forward this proceeding paper should serve as a call to action for community based organizations, academics, industry and safety professionals to combined efforts for identifying practical solutions that will help reduce the Hispanic/Latino rising injury trend.

Flash Fire and Explosions Caused by Electrostatic Discharges: Analysis, Prevention, and Control

Vahid Ebadat, Ph.D.
DEKRA Insight (Chilworth Technology)
Princeton, New Jersey

Abstract

Many processes and operations in a chemical plant involve handling and processing of liquids and powders, most of which could under certain conditions be flammable. Flammable gas, vapor, and dust cloud atmospheres can be ignited if a sufficiently energetic ignition source is present. It is noteworthy that most flammable gas/vapor atmospheres and some dust clouds can ignite by very small energy ignition sources. One potential ignition source that, in many cases could be an inherent part of the process/operation is an electrostatic discharge. Electrostatic discharges result from the generation and accumulation of electrostatic charges. Electrostatic charge generation most commonly occurs whenever two materials, liquids and/or solids, make and then break contact with each other. The accumulation of the electrostatic charge can result in electrostatic discharges. This suggests that there are many occasions during the transfer, handling, processing, and packaging of flammable liquids and powders in a chemical plant flash fire and explosion hazards exist.

This paper shall discuss the practical measures that can be considered to prevent/control electrostatic ignition hazards that are associated with some common liquid and powder handling/processing operations in the chemical industry.

Introduction

What is Static Electricity?

Static electricity is the charge that is generated whenever any two contacting surfaces move relative to each other. Static electricity is generated during many common chemical processing operations, such as when liquids flow relative to pipe walls or when powder particles come into contact with the surfaces of the processing and conveying equipment, among other mechanisms.

Contrary to some common belief, static electricity is predictable, quantifiable, and controllable and in fact the predictable and controllable nature of electrostatic charge is used to our benefit in a multitude of applications, including photocopying, air pollution control, and spray coating. There are also many negative instances of static electricity, including electrostatic nuisances, problems, and hazards. The shock that we receive from a doorknob after walking across a carpeted floor is certainly a nuisance. Static electricity can also cause materials handling and product quality problems by causing bulk solids and powders to agglomerate and adhere to the walls of vessels and conveyances. Actions taken to control electrostatic hazards and nuisances may have the collateral benefit of resolving such electrostatic problems. Measures for correcting quality and materials handling problems associated with static electricity are beyond the scope of this chapter, however.

Of course, the greatest concern posed by static electricity in an industrial setting is the risk of fire and explosion due to the ignition of flammable atmospheres by electrostatic discharges. Flammable gases, liquids, and powders are used in many chemical processing operations and precautions are required to prevent their inadvertent ignition. One area of precautions involves the exclusion or elimination of potential electrostatic ignition sources from locations where flammable atmospheres may reasonably be expected to exist. It is this area that is the focus of this paper and it is the controllable nature of static electricity that makes such precautions possible and effective.

Electrostatic Charge Generation

There are several mechanisms by which electrostatic charge is generated, including contact electrification, double layer charging, and induction charging.

The most common method of electrostatic charging in practice is "contact electrification" - also known as triboelectrification. Contact charging occurs when any two materials come into contact and then (quickly) separate. There is a rearrangement of electrons at the contacting surfaces where one surface is left with a surplus of electrons and is said to be negatively charged, while the other surface is left with a deficit of electrons and is said to be positively charged. Contact electrification occurs when for example powder particles collide/rub against the conveying chutes during pouring and when the soles of footwear make contact with insulating flooring. The level of charge that is generated by contact electrification is influenced by factors such as the (chemical) nature of the two contacting surfaces, the true contact area, and surface conductivity. The handling and processing of bulk solids and powders typically result in the development of relatively high levels of static charge because many powders are insulating in electrostatic terms and because of the relatively large surface area available for contact electrification.

Electrostatic charge is generated during the handling and processing of liquids by a mechanism known as "double layer charging". Double layer charging can occur at liquid/solid and liquid/liquid interfaces. Ions of uniform (like) polarity are adsorbed at the interface. The adsorbed ions attract ions of opposite polarity within the liquid forming a double layer within the liquid near the interface. This layer is thin for conductive liquids and more diffused in relatively less conductive liquids due to reduced ion mobility. As the liquid flows relative to the interface, the oppositely charged ions repel and move apart increasing the potential within the liquid. This effect -- combined with the competing effect of ion recombination at the vessel wall -- results in an unbalancing of charge within the liquid. This is the type of static charging that occurs as liquids flow through pipes. The factors that can aggravate charge generation in liquids include:

- Presence of immiscible solids and liquids;
- Presence of strainers and filters;
- Splashing and spraying; and

- Stirring and mixing.

"Induction charging" occurs when a conductive object that is electrostatically isolated from ground enters an electric field, such as that emanating from an adjacent charged object. The charged object causes a polarization of the otherwise balanced charges (equal numbers of positive and negative charges) in the conductor. If an isolated conductor sufficiently charged by induction is subsequently exposed to ground, an electrostatic discharge may result. It is generally possible to control such induction charging through the grounding of conductive objects.

Charge Accumulation

Generation of electrostatic charge is not in and of itself hazardous. However, if static charge is allowed to accumulate, it can give rise to hazardous electrostatic discharges. Electrostatic discharges occur between objects or surfaces at different potentials (voltages). As an object or surface accumulates charge, the electric field strength above it intensifies. If the field strength between the two objects or surfaces exceeds the breakdown strength of the atmosphere between them, an electrostatic discharge will occur. Electrostatic charge can accumulate in liquids, on insulating solids, and on conductive objects, and personnel isolated from ground.

Electrostatic Discharges

When the difference in the electrical potential between two objects exceeds the breakdown strength of the atmosphere between the two objects, an electrostatic discharge could occur, provided that at least one of the objects is conductive in electrostatic terms. Such discharges may be sufficiently energetic to ignite proximate flammable atmospheres, including gases, vapors, and dust clouds.

Flammable Atmospheres

Perhaps the greatest hazard posed by an electrostatic discharge is as a potential ignition source for flammable atmospheres. Three elements are required for a fire or deflagration to occur: (1) a fuel; (2) an oxidant, and (3) a sufficiently energetic ignition source. These three elements are commonly referred to as the fire triangle. The first two of these elements -- the fuel and oxidant -- when in an appropriate ratio are referred to as a flammable atmosphere.

For an electrostatic discharge to ignite a flammable atmosphere, the discharge energy must be greater than the Minimum Ignition Energy (MIE) of the flammable atmosphere. Minimum Ignition Energy (MIE) is defined as the smallest electrostatic discharge energy that is capable of igniting the flammable atmosphere at its most easily ignitable concentration. The MIE of a flammable atmosphere can be compared with the effective energy of potential electrostatic discharge in order to determine the extent of required precautions by identifying which type of discharge may pose a threat. The MIE of most hydrocarbon gases and organic vapors ranges from 0.1 - 1.0 millijoules (mJ), while the MIE of most dispersed dusts ranges from less than 3 mJ to over 1000 mJ. While some literature data for MIEs of gases, vapors, and dusts are available, it is generally recommended that such data be developed on a process specific basis since slight variations in composition or physical and chemical characteristics can have significant effects. For example, the MIEs of most powders and dusts decrease markedly as particle size and moisture content decrease.

Electrostatic Characteristics of Materials and Plant

The first step in assessing electrostatic problems and hazards is to establish the electrostatic properties of the materials processed, handled, and/or used in the plant, including:

- Minimum Ignition Energy of flammable atmospheres
- Resistance-to-Ground of conductive (metal) objects and items of plant
- Electrical Resistance of operators' body to ground (footwear and flooring)
- Electrostatic Chargeability of powders and liquids (alternatively, measure surface voltage or electric field during processing)
- Volume Resistivity of powders
- Conductivity of liquids
- Surface Resistivity of solid objects such as plastic containers and liners

National Fire Protection Association (NFPA) Standard 77, "Recommended Practice on Static Electricity" provides valuable assistance in controlling the hazards associated with the generation, accumulation, and discharge of static electricity. NFPA 77 provides a basic understanding of the nature of static electricity, provides guidelines for identifying and assessing the hazards of static electricity, as well as the techniques for controlling the hazards of static electricity.

Controlling Electrostatic Hazards

This section discusses some of the measures that could be considered in order to reduce the risk of flash fires and explosions due to electrostatic discharges from some common operations in chemical industry.

Conductive (Metal) Plant

The electrical resistance-to-ground of all conductive items of plant including metal pipes, equipment, vessels, and containers should be checked. If resistance is greater than 10 ohm, direct ground connection will be required. It is important that the ground connections are checked regularly and their purpose is known to the operators and maintenance personnel.

Personnel

Personnel wearing insulating footwear and/or working on insulating flooring (e.g. epoxy coated or painted surfaces) can typically attain potentials of 10 to 15kV by just doing various manual operations such as walking around, putting on or taking off overalls, or pouring powders and liquids out of containers. Sparks form electrostatically charged personnel could have maximum discharge energies of 20mJ to 30mJ. The following measures should eliminate the possibility of electrostatic sparks from operators.

- Personnel should be grounded so that their resistance-to-ground is less than $1x10^8$ ohm. This may be achieved by a conductive or static-dissipative flooring in combination with wearing static dissipative footwear with a resistance-to-ground between $5x10^4$ ohm and $1x10^8$ ohm.
- Resistance of the floor/surface on which the operator, wearing static dissipative footwear, is standing should also be $<1x10^8$ ohm.

Insulating (Plastic) Materials

Examples of insulating materials include plastic bags, liners, drums, pipes, hoses, and plastic or glass lined pipes and vessels. Non-Conductive (Insulating) Materials typically have a *Surface Resistivity > 10^{10} Ohm/Square.* Non-conductive materials such as plastic containers and liners can become electrostatically charged by manual rubbing causing "Brush" discharges. Brush discharges can in practice only ignite flammable vapor / gas atmospheres with Minimum Ignition Energy less than about 4mJ. More significantly, pneumatic conveying of powders through non-conductive (plastic) hoses and pipes can highly charge the inside surfaces of the hose/pipe giving rise to "Propagating Brush" discharges. Additionally highly charged liquids and powders entering non-conductive containers and liners can charge the inside surfaces of the container and/liner and giving rise to "Propagating Brush" discharges. "Propagating Brush" discharges can ignite all flammable vapor and gas atmospheres and dust clouds with Minimum Ignition Energy up to about 1,000mJ.

It is important to note that grounding of non-conductive materials would not facilitate the relaxation of electrostatic charges to ground. To avoid "Brush" and "Propagating Brush" discharges one must consider using conductive or static dissipative materials with a Surface Resistivity less than 10^{10} Ohm/Square.

Liquids

Static electricity hazards can arise in various liquid handling operations including filling, sampling, mixing, and filtration. During liquid handing and processing operations electrostatic charge can accumulate on; 1) bulk liquids in non-conductive containers and vessel; 2) on low conductivity liquids, and; 3) on liquid mist regardless of the liquid conductivity. The following suggestions can reduce the electrostatic ignition hazards:

- **Use Electrically Grounded Conductive Plant** - All items of plant including pipes, vessels, containers etc. should be electrically conductive and/or static dissipative and grounded
- **Increase Liquid Conductivity** - An antistatic additive may be used in very small concentrations in order to raise the liquid conductivity
- **Filters and Valves** - Use valves with maximum diameter possible. Locate filters and valves as far as possible from the entrance to the receiving vessel
- **Control of Liquid Entry to the Vessel** - Liquid should enter a vessel through a dip leg with submerged ends or bottom inlet point
- Control of Flow Velocity
  - Liquids with conductivity > 100pS/m, no flow velocity restrictions
  - Liquids with conductivity < 100pS/m and no immiscible components, flow velocity should be less than 7m/s
  - Liquids with conductivity < 100pS/m and containing immiscible components, flow velocity should be less than 1m/s

Powders

Highly charged powder particles entering a vessel can give rise to discharges on the surface of the bulking powder during filling of vessels called "Bulk" or "Cone" discharges. The energy of these discharges depends on the powder "Volume Resistivity", "Electrostatic Chargeability", particle size, and vessel dimensions with a maximum energy of about 25mJ.

"Bulk" discharges will not occur if the powder has a Volume Resistivity $< 10^9$ Ohm.m and is handled in grounded conductive plant. However, if the powder is insulating (Volume Resistivity $> 10^9$ Ohm.m), has a Minimum Ignition Energy less than 25mJ, and if the Electrostatic Chargeability test results show that the quantity of electrostatic charge on the particles is sufficient to cause discharges from the surface of the bulking powder then one of the following measures should be considered:

- Installation of inert gas blanketing, or
- Installation of explosion protection and isolation, or
- Measures to reduce electrostatic charge generation and accumulation

Effect of Humidity on Charge Accumulation

If water vapor is present in air, it absorbs onto surfaces and forms a slightly conducting surface layer. The extent to which the water absorbs and the increase in conductivity depends on the nature of the surface and the humidity of the atmosphere. Although there is not a definitive relative humidity where the conductivity of all materials changes, holding the relative humidity at above about 65% at about 20°C may significantly reduce electrostatic effects.

Explosion Protection

If the formation of a flammable atmosphere cannot be prevented or all ignition sources cannot be reasonably excluded or eliminated, then the risk of a fire or explosion will persist and measures must be taken to protect against their consequences. One method of preventing a flammable atmosphere from forming is inert gas blanketing or purging. This involves the introduction of an inert gas to a vessel or system to reduce the relative oxygen concentration to a level below which a fire or explosion cannot occur, i.e., the limiting oxygen concentration (LOC).

Explosion protection measures include containment, relief venting, and suppression. Containment is the design of vessels and equipment so that they are sufficiently robust to be capable of withstanding and containing the maximum explosion pressure. Relief venting is the installation of special panel(s) in vessels and equipment that are designed to open at a pressure lower than the design pressure of the balance of the vessel and equipment. This protects the vessel and equipment from overpressure and relieves explosion pressure and products to a safe area. Suppression is the rapid introduction of a suppressant to the incipient explosion. This prevents the course of the explosion from reaching its maximum pressure and thereby protects the vessel and equipment from overpressure. These measures are often coupled with explosion solation, which is the use of equipment or methods to prevent the propagation or communication of fires and explosions from one unit to another.

Concluding Remarks

Electrostatic charge generation most commonly occurs whenever two materials, liquids and/or solids, make and then break contact with each other. The accumulation of the electrostatic charge can result in electrostatic discharges. This suggests that there are many occasions during the transfer, handling, processing, and packaging of flammable liquids and powders in a chemical plant when flash fire and explosion hazards exist. The assessment of electrostatic ignition hazards normally involves:

- Determining the magnitude and polarity of the charges that are generated;
- Investigating the ability of the processing equipment and materials to retain and accumulate the static charge;
- Determining the type and igniting power of the discharges that could result from electrostatic charge generation and accumulation; and
- Determining the sensitivity of the flammable atmosphere to ignition by the resulting discharges (Minimum Ignition Energy)

If the energy of the electrostatic discharge is greater than the Minimum Ignition Energy of the flammable atmosphere a fire or explosion hazard will exist and measures should be taken to avoid the conditions under which such discharge could occur. If the exclusion or elimination of electrostatic ignition sources is not practical, other measures such as avoidance of flammable atmospheres or explosion protection must be considered.

Finally, whatever preventative and protective measures are considered (hardware or procedural), they have to be effectively implemented, installed, and maintained. Clearly, training of staff to understand electrostatic hazards is also important.

References

Britton, L. G., *Avoiding Static Ignition Hazards in Chemical Operations*, New York: American Institute of Chemical Engineers (1999).

Ebadat, Vahid et al., "Reducing Electrostatic Hazards Associated with Chemical Processing Operations," Chapter 36, Handbook of Chemical Health and Safety, Oxford, UK: Oxford University Press, 2001.

NFPA 77, "Recommended Practice on Static Electricity," Quincy, MA: National Fire Protection Association, 2014.

Session No. 656

New Tools for an Emerging Issue: How ANSI/ASSE Z-10 can help with Indoor Air Quality (IAQ) problems

Peter Shelling, MS, CSP, CIEC
Progressive Insurance
Mayfield, Ohio

Indoor Air Quality–A Modern Occupational Health Issue

Modern building construction and operations have come a long way since the first energy crisis of the 1970's. Both occupant comfort and cost-effective management practices have improved compared to ages past. Yet while building construction has proven the ability keep warm or cool air sealed in and thereby reduce operating costs of Heating, Ventilation and Air Conditioning (HVAC) systems, the same energy-efficient design has had a paradoxical effect on select human health concerns. Additionally, people spend more and more time than ever indoors according to the American Lung Association.1 These two facts of modern life have brought on a growing phenomenon of 'Sick Building Syndrome' and 'Building –Related Illness.' (BRIs)

One of the downsides of energy management in the nature of building construction is to keep contaminants that originate in building operations or are brought in by occupants sealed in along with the warm or cool air. Symptoms that make these contaminants suspect sources include anything from upper-respiratory infections and itchy, watery eyes to asthma. Known causes include mold mycotoxins, bacteria and other organic particulates. These biological etiologies are encouraged to grow, using the building materials and environments for food, moisture and required temperature. Other sources of certain BRIs can be found in the building materials themselves. These include asbestos, silica dust or chemicals and other materials used in the operations taking place in the building.

The increase in BRIs is even more pronounced when natural or man-made disasters strike and create conditions for accelerated growth or airborne dispersal of contaminants. Mold follows flooding, wildfires beget distributions of any manner of particulates, and any event which causes structural damage can increase the concentrations of asbestos or particulates from other building materials in the air being breathed by building occupants. Even a simple electrical power outage can render HVAC systems all but useless for several days, creating perfect environments for growth of biological contaminants. With the requirements of modern business, protecting the

[1] American Lung Association http://www.lung.org/our-initiatives/healthy-air/indoor/at-work/

health of workers and the public is only the beginning of recovery- but sometimes actually gets overlooked as symptoms and even sources aren't always imminent.

This fact- combined with modern architecture and construction techniques- has led to an increase in occupational illnesses based upon building occupancy. Putting aside the common complaints of discomfort and colds and flu, aliments such as asthma, mold sensitivities and allergies

Shouldn't this be something of concern for EHS Professionals?

Outside of a few specific concerns in Industrial Hygiene or asbestos, our own 'body of knowledge' has been of minimal help on the subject. Why do we think that is the case?

Three main reasons:

1. There isn't much to assist us as far as OSHA or ANSI standards
2. Many of the issues are perceived as building design issues.
3. Injuries and illnesses are frequent but rarely severe enough to warrant our attention over other more pressing matters. If we can't see blood, we trend to not worry much about it, or it just doesn't get the press that falls and trench collapses get.

Regarding #3: Both employee health and productivity dollars add up with small things, and bad IAQ can be like 'death by a thousand little cuts' for a workplace.

For Issue #2: If we understand the building sciences and certain principles of architecture, we can assist with 'Prevention through Design', or at least be better prepared when inevitable IAQ problems happen.

In case #1: We will find some assistance in other industry standards, as we can combine them together to utilize one of our favorite new tools – the **ANSI/ASSE Z 10 Occupational Safety and Health Management System.** Such discussion will be the basis of this presentation.

What are the common symptoms of Building-Related Illnesses?

Common symptoms include:

- Cough, sneezing or watery eyes that go away once the person leaves the building.
- Fatigue
- Allergic reactions, including dermatitis
- Headaches or nausea

As you can tell, these symptoms are common for many different type of illnesses-- sometimes cause by poor IAQ, sometimes it's just cold and flu season. Thus an investigation often begins with a process of elimination.

Illnesses that often involve poor IAQ as contributing factors include:

- Hyper-sensitivity pneumonitis
- Occupational asthma
- Certain forms of cancer-especially lung cancer.

What are the most common causes of IAQ problems?

The most common causes of IAQ problems in buildings are:

- Not enough ventilation, lack of fresh outdoor air or contaminated air being brought into the building
- Poor upkeep of ventilation, heating and air-conditioning systems,

- Dampness and moisture damage due to leaks, flooding or high humidity
- Occupant activities, such as construction or remodeling
- Improper housekeeping (too little or too much)
- Environmental sources such as radon

What are we as EHS professionals supposed to do about all this?

OSHA doesn't give us much in the way of standards other than some specific chemical PELs and the General Duty clause. There are a few guidance documents such as **OSHA 3430-04 2011** "Indoor Air Quality in Commercial and Institutional Buildings."[2] The US Environmental Protection Agency offers a bit more, but is directed more at IAQ in schools and specific sources such as mold or radon. Private standards such as the American Society of Heating, Refrigeration, and Air Conditioning Engineers (ASHRAE) 62.1-2016 "Ventilation for acceptable Indoor Air Quality" among others can be helpful tool as well.

Managing IAQ is often as much art as science, and this allows us as safety professionals to use one of our favorite tools-- the American National Standards Institute's ANSI Z10 system for risk management and hazard reduction. By using Z-10 as the 'glue' to put all of these other standards and practices together, we can support continuous improvement, anticipate problems more effectively, and help all stakeholders better understand their roles in keeping a healthy indoor environment.

ANSI/ASSE/AIHA Z-10 – What tools can we use to help?

Our first and favorite is the Hierarchy of Controls: (Z-10 E5.1.2):

The hardest part about managing IAQ may be identifying our pollutant sources. After that the higher-ordered controls will be the tools we find ourselves using more frequently.

Elimination:

We want to try and get rid of materials or processes that contribute to hazards. With Indoor Air Quality, that means basically two things: Prevention Though Design and source elimination. With the first, we need to be in the conversations with the architects, builders and building owners (which is admittedly rare), and should we be so lucky it will help if understand the basics of building science and HVAC engineering. More likely, we will be invited into the discussions after occupation. - and we will find that many of our problems will associated with occupancy and occupant activities. Examples include mold and other bio-aerosols and carbon monoxide emissions. These are contaminants sources we can eliminate once identified.

Substitution:

A common source of IAQ problems is chemical vapors and emissions. This can be controlled by substituting less hazardous chemicals for the more hazardous varieties. The classic example is that we no longer use bleach any longer to kill mold--, rather use mild, biodegradable detergents instead. 'Green' cleaning products tend to offer fewer complaints, and low-VOC -emitting paint are also good substitutions.

[2] https://www.osha.gov/Publications/3430indoor-air-quality-sm.pdf

Engineering:

Engineering controls are one of the hallmarks of a strong IAQ program (and of course can be a double-edge sword and contribute to the problem as well if not used properly) HVAC systems are the primary method air is exchanged with the outside, and filtered along the way. Perhaps you have heard the maxim "Dilution is the Solution to Pollution"- which is not always true, but in the case of many air quality issues is a good start. The HVAC system is often the first place we look to control many issues, but can also distribute contaminants and pathogens further if there is a source that is not managed (such as an outside air intake located in close proximity to the trash dumpster or loading dock areas). Auxiliary units such as portable fans can also be utilized when needed.

Warnings

There are many controls that fall within the realm of warnings that impact IAQ: The information on an SDS for example, or even a thermometer or Carbon monoxide (CO) detector.

Administrative Controls

As with all safety issues, administrative controls may be our most frequent method of improving IAQ, and are typically associated with occupant activities. How well is the 'no-smoking' policy enforced? Are employees allowed to bring in their own house plants (which can bring in other biological entities, or cause allergic reactions to other workers.)? Certainly, housekeeping procedures such as effective vacuuming and dusting can help maintain desirable IAQ- or contribute to symptoms when too many short cuts are taken.

Personal Protective Equipment

We should not get this far with the average office worker, but there have been instances such as nearby wildfires or open landfill activities in which tools such as N-95 masks might need to be utilized. We need to ensure that we are training people properly in such instances,

Plan > Do >Check> Act

One of the main driving forces of Z-10 success is the continuous improvement generated by involvement of stakeholders at all levels. A standard protocol for both investigations and of complaint events and preventative measures will include the following:

Task -1- Visual assessment

Review existing ventilation, maintenance and housekeeping procedures to document conditions that indicate possible IAQ challenges or might lead to contaminant accumulation. This task includes an extensive walk-through of the facility and will include visual examination of the HVAC air-handling equipment as well as accessible ductwork and plenum areas. Problems identified during this task should include roof leaks, visible staining, and obvious housekeeping or HVAC maintenance deficiencies.

Task 2- Measurement of general environmental conditions

Complete measurements of environmental conditions in numerous areas including modular offices, hallways, conference rooms and open areas. Obtain measurements of temperature and relative humidity throughout to document thermal variation, humidity variation and air circulation. Measure CO2 (carbon dioxide) to provide indications of the quantity of outside fresh air introduced via the ventilation system, potential contamination from outside the building, and evaluation of the effectiveness of the ventilation system in providing comfortable conditions to the building. Carbon monoxide levels may be obtained to detect possible infiltration of combustion exhaust gases into the facility.

Task 3- Sampling of air in complaint areas

Depending upon the nature of complaints and possible sources that may have already been ruled out, sampling may consist of bio-aerosol cassettes, measurement of airborne particulates using a direct reading aerosol monitor (laser particle counter), or VOC sampling using a vacuum canister or other capture method. Action levels may vary depending upon local ordinances or simply the nature of the occupancy, but there are standards to guide such decisions.

Task 4- Data Evaluation, interpretation and report.

Assemble data and information from sampling results and physical observations into a written report. Data should be compared with existing guidelines and recommended levels published by such groups as ASHRAE, NIOSH, OSHA and EPA, or others as appropriate. Data will also be compared with previous IAQ baselines to assess changes in variables and determine action, if any.

Task 5 – re-evaluate once controls have been applied

In some cases, such as a verified mold event, additional sampling may be required at a later date to verify the efficacy of action taken. Of course, this part takes time and often can only be considered effective when 'no news is good news'. When the majority of complaints have ceased and desisted, people are productive again and we can share 'lessons learned' to help prevent future IAQ issues, the project will be considered complete.

Challenges and Opportunities

The biggest challenge of IAQ management may be getting stakeholders to understand what are- and aren't- real threats. For example, there have been a few high-profile cases of 'black mold', yet the vast majority are rather benign. However, several hundred people die of Carbon monoxide poisoning (and many others suffer serious, long-term illnesses) every year in the USA.[3]. Managing the actual threats and managing perceived threats require different skill sets. As most Safety professionals know, we often wear many hats ourselves. But more importantly is the relationships we have with our business partners, human resource professionals, engineering staffs and other technical experts we may need to call on for assistance.

Given the cost of poor IAQ to employers, we also have the opportunity to contribute to the bottom line. Often that contribution goes unrecognized until the next crisis, but that is all too often the nature of our work. Tools available within our ANSI/ASSE Z-10 kits can help us measure our own contributions. One such is our favorite, the 'Risk assessment matrix'

Example: When a particular building (some might even call this an example of a 'sick' building) may seem to have several possible IAQ problems, we can prioritize with an RA matrix:

---

[3] https://www.cdc.gov/mmwr/preview/mmwrhtml/mm6303a6.htm

Hazard/ Risk level. (1=low, 5=high)	1	2	3	4	5
Mold growth in restroom (*Stachybotyrus chartarum)*				x	
HVAC system not producing adequate Air Changes per Hour (ACH)		x			
Chemicals in use (maintenance, janitorial)			x		
Carbon Monoxide from underground garage			x		
Office dust (many possible sources)		x			

Table 1.1 –Risk Assessment of IAQ issues before Z10.

Hazard/ Risk level. (1=low, 5=high)	1	2	3	4	5
Mold growth in restroom (*Stachybotyrus)*	x				
HVAC system not producing adequate Air Changes per Hour (ACH)	x				
Chemicals in use (maintenance, janitorial)		x			
Carbon Monoxide from underground garage		x			
Office dust (many possible sources)	x				

Table 1.2 Risk Assessment after Z10 controls applied.

Then we can also take a look at how we get these reductions:

Hazard/ control	Elimination	substitution	engineering	warning	administrative	PPE
Mold growth in restroom (*Stachybotyrus)*	X				X	
HVAC system not producing adequate Air Changes per Hour (ACH)			X			
Chemicals in use (maintenance, janitorial)		X			X	
Carbon Monoxide from underground garage			X	X	X	
Office dust (many possible sources)					X	

(And we didn't even need to bring out the N-95 masks!)

References

ANSI/ASSE/AIHA (2012) *Occupational Safety and Health Management Systems* (ANSI/ASSE/AIHA Z-10-2012). Des Plaines, Il: American Society of Safety Engineers.

American Society of Heating, Refrigeration and Air Conditioning Engineers (ASHRAE) *62.1-2016 Ventilation for Acceptable Indoor Air Quality*. Atlanta, GA: ASHRAE.

Bas, Ed. 2004. *Indoor Air Quality: A Guide for Facility Managers,* 2nd Edition. Lilburn, GA, The Fairmont Press.

Burroughs, H.E. and Shirley Hansen. 2011 *Managing Indoor Air Quality,* Fifth Edition. Lilburn GA: The Fairmont Press.

Manuele, Fred. 2008. *Advanced Safety Management: Focusing on Z10 and Serious Injury Prevention*. Hoboken, NJ: John Wiley and Sons.

May, Jeffrey C. 2006 .*My Office is Killing Me: The Sick Building Survival Guide.* Baltimore MD: Johns Hopkins University Press.

NIOSH (Rev. 2015) "Indoor Environmental Quality." Atlanta, GA: Centers for Disease Control. https://www.cdc.gov/niosh/topics/indoorenv/

OSHA (2011) 3430-04 2011 "Indoor Air Quality in Commercial and Institutional Buildings. Washington DC: U.S. Government Printing Office.

USEPA (2002) "Indoor Air Quality Building Education and Assessment Model." Washington, DC. https://www.epa.gov/indoor-air-quality-iaq/indoor-air-quality-building-education-and-assessment-model

Walaski, Pamela. Risk Assessment Certificate Program. (Three-day ASSE workshop presented in Las Vegas, Nevada at ASSE Seminar Fest, February 2015.)

Welch, Laura S. "Severity of Health Effects Associated with building-Related Illnesses." *Environmental Health Perspectives* vol 95 (1991) pp 67-69.

Employee Safety Engagement in the Middle East – Overcoming historical cultural barriers

Nicole Zerbel Ivers, M.S., CSP, CIH
AECOM Middle East Limited
Doha, Qatar

Introduction

Concern for worker safety and working conditions in the Middle East, and especially Qatar, is currently on the world stage (Cronin, 2017). This paper aims to present and discuss many of the challenges to achieving world class safety in this area, as well as the root causes and the successes and improvements that have been achieved. The solutions are not straight forward though. The barriers to achieving world class safety in the Middle East are deeply rooted in culture and are extremely complex.

Working for the past six years in a safety leadership role, as well as being a female, I have been able to provide a unique approach and view to the situation in the Middle East and identify opportunities to overcome many cultural barriers to an injury free workplace. This paper presents information and first-hand experience I have gained working in the construction industry in the Middle East for the past six years, as well as much study and research into the complex cultural roots. As you read, it is important to remember that "culture" is not simply nationality or religion. The "culture" within the workforce and on our projects is a result of knowledge, experiences, values, attitudes, perceptions, social habits, etc.; as well as many years of conflict between countries and people.

Current Economical Drivers in Middle East

Before addressing the safety challenges, it's important to understand what is driving the rise in development and subsequent construction in the Middle East.

The Gulf Region – or the Gulf Cooperation Council (GCC) - of the Middle East consists of Kuwait, Bahrain, Oman, Qatar, Saudi Arabia and the United Arab Emirates (UAE) (GCC n.d.). These are the areas surrounding the Arabian Gulf (Exhibit 1). These countries have undergone rapid economic, demographic and social changes in the past twenty years – mainly due to their oil and gas exports. However, this has also allowed the region to develop in the financial sector, infrastructure, tourism, and major events such as the FIFA 2022 World Cup (FIFA 2017) and Vision 2030 development plan (QNV 2008) in Qatar and the 2020 World Expo in Dubai (Expo

2020). As well as the basic stadiums for events such as the World Cup, Qatar is building everything from new infrastructure such as roads, expressways, and drainage to a new airport and soon an expansion; multiple rail systems; a new port and conversion of the previous one into a cruise ship port; hotels; malls; and multiple "multi-use" developments which will be cities themselves. According to the Finance Minister of Qatar Ali Shareef al-Emadi, the main projects associated with the World Cup are spending $500 million a week (EDESKNEWS 2017). According to him, the same pace of the money will be spent up to 2021, but the overall budget will be about $200 billion (EDESKNEWS 2017). This incredibility grand scale of projects and the timeframe has created a huge demand for migrant labour. The labour and circumstances under which they are being hired and brought to these countries contributes to safety culture in the GCC today.

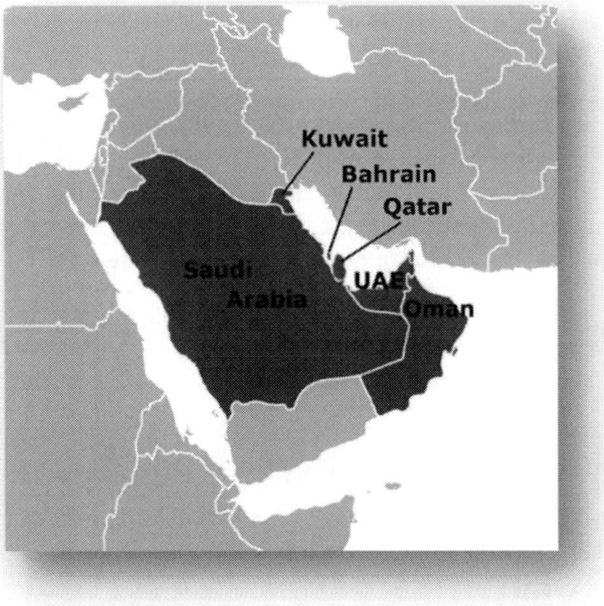

Exhibit 1. The map of the Gulf Cooperation Council (GCC) countries. (GCC n.d.)

Occupational Safety Laws and Regulations

Contrary to popular belief, there are actually many occupational laws and regulations to protect workers in the Middle East. For example, the United Arab Emirates (UAE) has the Labour Law - UAE Federal Law No.8 Year 1980 (as amended) - as well as a number of supporting legislation, regulations, and guidance including, but not limited to, the Ministerial Decision No. 32 Year 1982 (Protection of employees against occupational hazards); UAE Ministerial Decision No. 37/2 Year 1982 (Levels of Medical Care to be provided by employers); UAE Fire and Life Safety Code of Practice; and Abu Dhabi Occupational Safety and Health System Framework (AECOM, 2016). Qatar has its own labour law - Law No 14 of the Year 2004 - as well as the Qatar Construction

Specification (QCS) and numerous Ministerial Decisions (AECOM, 2016). Oman, and even Saudi Arabia have similar labour laws as well regulating safe work practices (AECOM, 2016). The majority of these documents are consistent with United Kingdom regulations, as well as OSHA. The problem with the laws and regulations is that there are different degrees of enforcement of the laws. In my experience, Dubai and Abu Dhabi have made advancements in this area compared to other countries within the GCC. In the UAE, the government will respond to incidents and the Contractor receives "black points" against them which factor in to future contracts. However, in Qatar, the Contractor is rarely, if ever, held accountable for their safety record except by individual clients (who also have varying degrees of regard for safety). As far as preventative enforcement – like random OSHA inspections – they don't exist to a large extent anywhere in the Middle East. In Qatar, we rely mainly on enforcing the contractors' employment contracts which make reference to the laws, regulations, and various specifications. Non-compliance to these laws result in a "Non-conformance report" (NCRs) or a low score on their Key Performance Indicators (KPIs) which both in turn affect the contractors' payments. However, this is a long process which often loses its effect once the contractor implements corrective actions. It does not get a proactive result.

In my opinion, the biggest impact to improving the safety culture, in absence of a strong enforcement agency, is the employers themselves. The large international engineering firms, who often perform supervision or program management roles, are raising the standards in the Middle East. The expectation for safety performance– regardless of the repercussions of not doing it – is driven by these companies' corporate safety programs and visions. If all companies working in the Middle East would adhere to the same standard then we would not have such varying degrees of safety standards and Contractors and labour would understand the expectation from project to project.

Migrant Labour in the Middle East

Before being able to address how to improve safety standards and discuss solutions, one must first understand the characteristics of the workforce.

Workforce Nationality

Because the GCC has a limited workforce – especially when it comes to labour trades - the region has been almost fully dependent on a migrant workforce for everything including their oil and gas industry; construction; hospitality; transportation; medical care; education; etc. For example, as of 2017, Qatar's population was almost 2.5 million, of which Qataris (local nationals) make up only about 300,000 of those (Snoj, 2017). As the demand for migrant workers – specifically labour – has increased, so has the numbers of recruiting agents, which has driven wages down (ILO, 1996).

Table 1. Nationality breakdown of Qatar (Snoj, 2017).

Nationality	Population	Percent of total*	Nationality	Population	Percent of total*
India	650,000	25.00%	Serbia	2,000	<0.10%
Nepal	>350,000	13.50%	Portugal	1,500	<0.10%

Nationality	Population	Percent of total*	Nationality	Population	Percent of total*
Qatar	313,000	12.10%	Brazil	1,500	<0.10%
Bangladesh	280,000	10.80%	Netherlands	1,432	<0.10%
Philippines	260,000	10.00%	Albania	1,200	<0.10%
Egypt	200,000	8.60%	Macedonia	1,000	<0.10%
Sri Lanka	145,256	5.60%	Ukraine	1,000	<0.10%
Pakistan	125,000	4.80%	New Zealand	989	<0.10%
Syria	54,000	2.20%	Japan	944	<0.10%
Sudan	50,000	2.10%	Denmark	900	<0.10%
Indonesia	43,000	1.70%	Poland	700 – 800	<0.10%
Jordan	40,000	1.80%	Belgium	600	<0.10%
Iran	30,000	1.50%	Cuba	600	<0.10%
UK	25,000	1.10%	Austria	500	<0.10%
Lebanon	24,000	0.92%	Bulgaria	500	<0.10%
Ethiopia	22,000	0.91%	Croatia	500	<0.10%
Palestine	20,000	0.85%	Sweden	470	<0.10%
Tunisia	20,000	0.77%	Argentina	400	<0.10%
Kenya	14,000	0.55%	Mexico	400	<0.10%
USA**	>11,000	>0.43%	Venezuela	337	<0.10%
China	>10,000	>0.41%	Kyrgyzstan	330	<0.10%
Eritrea	10,000	>0.41%	Hungary	300	<0.10%
Canada	9,000	0.35%	Singapore	300	<0.10%
Iraq	8,976	0.40%	Switzerland	250	<0.10%
Turkey	8,000	0.31%	Senegal	A few hundred	<0.10%
Nigeria	7,500	0.29%	Belarus	200	<0.10%
South Africa	6,000	0.23%	Finland	200	<0.10%
Australia	5,500	0.21%	Kazakhstan	200	<0.10%
Algeria	5,000	0.20%	Colombia	200	<0.10%
France	5,000	0.20%	Norway	160	<0.10%
Ghana	5,000	0.21%	Moldova	154	<0.10%
Malaysia	4,848	0.19%	Gambia	135	<0.10%
Thailand	4,500	0.18%	Azerbaijan	120	<0.10%
Afghanistan	3,500-4,000	~0.2%	Ecuador	100	<0.10%
Spain	3,500	0.14%	El Salvador	100	<0.10%
Uganda	3,000	0.11%	Slovakia	100	<0.10%
Romania	2,500	0.10%	Czech Republic	100	<0.10%
Greece	2,200	<0.10%	Benin	82	<0.10%
Italy	2,100	<0.10%	Dominican	44	<0.10%

Nationality	Population	Percent of total*	Nationality	Population	Percent of total*
Germany	2,000	<0.10%	Liberia	40	<0.10%
Ireland	2,000	<0.10%	Brunei	20	<0.10%
Russia	2,000	<0.10%	Latvia	10	<0.10%
South Korea	2,000	<0.10%	Liechentstein	1	<0.10%
Vietnam	2,000	<0.10%			

. *Percentages correspond to the total population of Qatar at the time the data is derived from.
**Only Military personnel. Data for overall US population in Qatar not available.

Workforce Skill

The skill of the workforce is highly variable since workers come from more than 80 different countries – all with different levels of training and construction experience. For professionals, there is a comprehensive system of degree and/or diploma attestations required to prove one's education and experience. However, for labour – which is traditionally the carpenters, scaffolders, masons, electricians, etc., there is little to no verification. Therefore it is often up to the contracting company to confirm a worker's competency before being placed in a job role. Unfortunately, this often doesn't happen and we end up come with incidents and injuries because a worker does not know how to properly operate a circular saw (for example) even though he claims to be a carpenter. Of course when it comes to safety, most of these new migrant workers have never been exposed to basic PPE, much less systems and procedures of safe work methods. That said; there are also some of the most experienced people working in the GCC – designing and constructing some of the largest and most complex mega projects in the world. It takes a highly flexible and determined person to be successful in environments with the varying levels of competency, copious different languages and cultures, unpredictable risk perceptions, and, to top it all off, extremely high temperatures.

Competency is therefore a major challenge in the Middle East. The Contractors, for the most part, are not prepared to train thousands of workers, test capability and qualifications, observe, and supervise closely before each worker is deemed competent. The majority of workers have their word taken that they are competent in their particular trade and unfortunately, it's not until there is a near miss or an incident, it is determined that they are not competent.

Working and Living Conditions

While it's easy to look online or in the newspaper and see media reports about the "poor working and living conditions" in the Middle East – and specifically in Qatar - this is definitely not representative for the whole country or the GCC. Just like any country – including the United States, exceptions can be found. While I have seen some substandard levels of safety and deplorable living conditions, it is not what the majority of employers accepts or provide for their workforce. Most of the onsite safety and the worker accommodations of the large global Contractors that I have worked with have provided high standards for their workers. Many workers are provided an immense amount of on-the-job training, medical care, hot food catering onsite and in their accommodations, laundry service, computer labs and training, and recreation.

Exhibit 2. Non-compliant housing in Qatar (Personal Photo, 2016).

Work Conditions

On construction sites, there are requirements for basic "welfare" such as proper shelters for rest breaks, toilets, access to water, and a safe place to pray. Workers typically work from 6am to 6pm with an hour lunch a multiple additional breaks throughout the day for rest or prayer. The labour law limits the typical work day as a maximum of eight hours plus two hours of overtime for a total of ten hours. Transport time in the morning and the evening are not paid however so this can create a long working day. Because not all employers adhere to the labour laws, workers may work longer hours, and some are not paid on time. Further, if the worker is not provided catered food and laundry service, then much of their rest time is spent cooking and washing. This is the reason for the welfare standards described further below.

Extreme temperatures in the Middle East create a harsh working environment. While many of the workforce is somewhat acclimatized to a hot, humid climate, many – for example – those from Nepal, are not. The temperatures in the peak of the summer can reach up to 120 plus degrees Fahrenheit or more (Climate of Middle East, 2017). Because of these temperatures, the labour laws of the Middle East countries provide limits to the times and temperatures under which people can work outside and /or in direct sunlight – usually called the Summer Working Hours or Midday Ban. These labour laws have greatly reduced the number of workers suffering from heat related illnesses and death.

Worker Accommodations

Worker Accommodations, or "labour camps" as they have been historically referred to as, have come a long way in the past five to ten years. Originally, the standard worker housing consisted of four bunkbeds to a room with eight men (Exhibit 2).

Few received catered food and laundry services. Most did not even have the simplest form of fire detection or prevention and due to this; many camps were destroyed by fires (Gulf News, 2016). While we still find some housing like this, it is now the exception.

Today, there are many standards for worker "welfare". In Qatar, clients such as the Supreme Committee for Delivery and Legacy, Ashghal, and Qatar Foundation all have standards that cover everything from the worker's recruitment to the number of toilets in their accommodations. The living conditions are now very good for many of the workers. Most worker accommodations provide hot catered meals based on ethnicity; rooms with just two to four per room (Exhibit 3); laundry service; access to education (Exhibit 4) including computer classes; satellite television for most major nationalities; recreation rooms with games and exercise equipment; organized sports; onsite banking, shopping, and barber; medical clinic and onsite treatment by licensed doctors; and excursions into town or shopping areas on the weekends. The reason for ensuring that catered food and laundry service is provided is so the workers do not have to spend their resting time doing more work – although many of them enjoy cooking as a social aspect of their culture and therefore many have set up facilities for them to cook if they wish to do so.

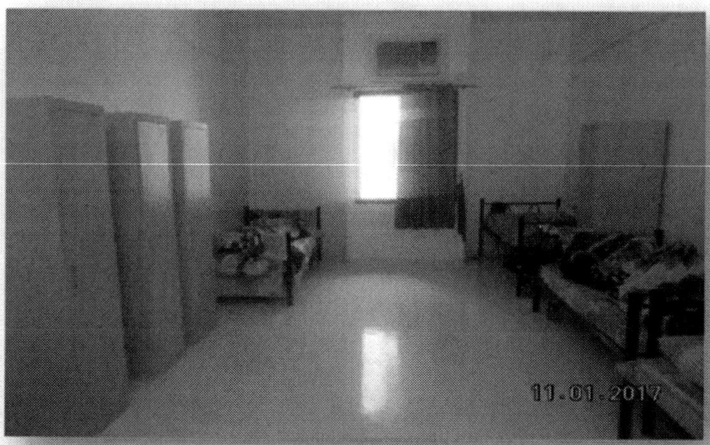

Exhibit 3. Photo of compliant worker housing in Qatar (Personal Photo, 2017).

Exhibit 4. Computer class at a Worker Accommodation in Qatar (Personal Photo, 2017).

Again, not all workers have this level of living conditions but thousands of workers do; and as more companies wish to work for clients with worker welfare standards, they too will be required to provide a high level of housing and welfare.

Engagement and Overcoming Barriers

Even with the many nationalities, lack of law enforcement, and varying working and living conditions discussed thus far, these challenges simply set the stage for the everyday challenges on site. Below are some of the largest obstacles and how we have been able to engage with workers and develop solutions to improve safety performance.

Language and Communication

One of the foremost barriers to effectively communicating with each other in the GCC is language. While English is considered the "business language" in most countries in the GCC, it is not spoken or understood by everyone – especially in the construction trades. Without effective communication, a worker's experience and competency cannot be effectively assessed. Their medical history cannot be obtained. Training and safe work methods cannot be conveyed. Hazards cannot be adequately described. In the event of an injury or an incident, the investigation is difficult to conduct. They cannot convey to us issues or concerns about their safety. Personally, one of my weaknesses as a safety professional is that I am fluent in only one language. The majority of my coworkers often speak two to three languages while many of my safety officers speak up to nine difference languages and dialects.

Unfortunately, language is not often considered when Contractors build their workforce. For example, one of the Chinese contractors who work on one of our major projects (there are a

Exhibit 5. Photo of the Corporate and New Orbital Highway Contract 3 safety team representing twelve languages spoken between them (Personal Photo, 2016).

number of Chinese contractors working in the GCC so this does not depict one company specifically) have a particularly difficult challenge with communication and language. Most of their upper management (managers and engineers) are from China and they speak some English but are not as comfortable speaking it as they are their native language. Therefore a number of details can be easily lost in translation. Next, the labour hired for the project is mostly from Nepal, India, Bangladesh, and Egypt. Most of these labourers and foremen cannot communicate with the Chinese engineer supervising their work – and they definitely cannot get the fine details of performing their work safely. We rely on a few people who can translate for us from Chinese into the languages of the workforce – a workforce of 4,000 plus people.

However, Contractors cannot always choose where their workforce comes from even if they want to. The government issues the Contractor with "block visas" which are for a specific country. For example, one Contractor was given a block visa for workers from Vietnam. The Contractor had no experience with Vietnamese workers, their skill, their language, or even their food. The Contractor was not prepared with adequate translators upon their arrival. This caused a lot of confusion and resulted in a number of injuries as the Contractor trained and accommodated these workers into their workforce.

One of the ways we have been able to improve the language communication on our projects is by employing staff who speak multiple languages and identifying the contractor staff who can translate for us – especially for the scarcer languages such as Chinese and Vietnamese. For example, the team shown below (Exhibit 5), put together strategically for a large road construction project, speaks twelve languages between them: English, Tamil, Malayam, Hindi, Arabic, Urdu, Pashto, Punjabi, Telugo, IBO (Nigerian), Tagalog and Pilipino. This covers the majority of the workforce on this particular project.

Exhibit 6. Translator idendification stickers to enhance communication on the site. (Personal Photo, 2017).

True engagement with workers about safety cannot occur without effective communication. Therefore, another tool we have used to aid in communication is a translator identification system. In order to help identify translators for both the supervision team and the contractor workforce, we use a system of flag stickers which display the languages that someone speaks and therefore can translate (Exhibit 6). This has shown an immediate increase in communication between workers and supervisors on site. HSE Officers have observed being approached much more often by workers to ask about site issues and HSE concerns. The assumption is that workers who do not speak anything other than their native language are hesitant to speak to someone that may not understand them. It also creates an interest for the workers when they see their country's flag on another worker or supervisors hard hat. The door becomes open for them to approach and talk openly.

Risk Perception

Another obstacle to safety in the Middle East is "risk perception". Risk perception is a judgement made about a risk, based on what we believe will be the outcome, the likelihood of the outcome actually happening and how concerned we are if it does happen (Risk Perception, n.d.). Varying risk perception is one of the largest challenges when trying to implement safe methods of work. Many of our workers are coming from countries where risks are taken every day.

For example, in Bangladesh, it might be normal practice to climb and ride on top of a train to ride into town (Exhibit 7). The photo below was taken by a colleague of mine working in Dhaka, Bangladesh which shows a routine activity with a low perceived risk. A Bengali man may have ridden the train like this every day of his adult life, never saw anybody get hurt, and therefore has developed a "judgement" that the chance of falling off and being killed is very low. And further, if this person has no other option for transportation, and is trying to provide for his family, he may not be very concerned about the possible outcome anyway. It may be worth whatever risk he does see in it. Fast forward to this same Bengali man arriving on a construction site to work. This man exchanges his sandals for heavy steel toe boots, coveralls, high visibility vest, hard hat, safety glasses and possibly a fall prevention harness. He is told that it's not safe for him to work from heights, or even ride in the back of a pickup truck. Again, what one person perceives to be perfectly safe and routine, could be seen as extremely dangerous to another. Especially when they have never seen an activity is performed any other way. They don't know what they don't know.

Another factor that appears to affect the risk perception for many workers in this region are beliefs associated with their religion. Firstly, this is not particular to any specific religion. While there is a strong belief in fatalism in religions such as Islam and Hinduism, many of the workers report a sense of destiny when it comes to their own actions. For example, the Qur'an says that before one is born, a person's destiny is decided, and that everything that happens to is the will of Allah (Islam and Karma, 2017). Many people report to me that they do not need to take precautions for their own safety; that if it's God or Allah's will – or their fate - for them to fall off of a building and be killed, so be it. In my experience though, this is easily overcome with the right engagement and education on safety provisions.

There are many theories and studies trying to explain why risk perception varies (Oswald, 2014). A study on risk perception and various theories by David Oswald (2014) has concluded that voluntary risks under one's own personal control are more likely to be accepted (and under-

Exhibit 7. Train transportation in Dhaka, Bangladesh (Personal Photo, 2017).

rated) than in-voluntary risks. For example, a voluntary risk of getting into a car accident when driving your car versus an involuntary risk of being a victim of crime or a natural disaster. If voluntary risks taken by driving are any indication of how someone may take other voluntary risks, it may explain the differences in working behaviors we see in the Middle East. According to the Global Road Safety Partnership's Annual Report (2014), the Middle East and North Africa has the second highest road fatality rate globally. Africa has the highest.

Oswald (2014) also concluded that other influences on the perception of risk included work pressures and time-saving benefit. This is a very relative finding as most of our projects here in the Middle East have begun under unrealistic schedules and the pressure for production is immense. Therefore, overcoming the varying risk perceptions of people from 80 plus different cultures and backgrounds which is also influenced by production pressures is not an easy task. First it's important for us as safety professionals to try to understand the factors influencing the workers' risk perception. Is it because they have never been exposed to the task and don't understand the hazards? If yes, then we use task hazard assessments and awareness training to help them understand the hazards. Often we identify safety champions within their own ethnic groups to engage with them and remind them of their wives and children and the impact on them if they were to be hurt or killed. This is why it is so important to have trust and communication with the workers and not be the "safety police".

Hierarchal Society and Social Structure

The Middle East (and many of the surrounding countries) is a highly hierarchal society – especially in the construction field. "...Social structure is the patterned social arrangements in society that are both emergent from and determinant of the actions of the individuals" (Social Structure, 2017). This includes not only culture or ethnicity but gender as well. Unfortunately, I have observed very often that your ethnicity determines the "class" you fall into and therefore your job and pay.

For example, India, the country where the majority of expat workers originate from, has been known for their strong caste system (Caste System, n.d.). This social stratification system still exists and creates inequality within their own country (Caste System, n.d.). That same system of inequality appears to be applied on our project sites as well in the trades. Many countries also have long histories of conflict between them and their people mostly due to politics and religion. Having so many cultures, nationalities, and religions working together on one site is an inherent safety risk. Fortunately this is one area where I have seen the majority of workers peacefully and respectfully working together and not bringing political or religious differences into the jobsite. However, I believe that when it comes to taking directions or guidance from someone of another religion or nationality (which one may have historical differences with), that worker will be less inclined to follow or trust. I also believe that supervisors often believe they are "better than" those in the lesser skilled trades and may be more willing to put these people into unsafe situations. I feel that this is a challenge that is unlikely to be overcome as we are talking about thousands of years of values and beliefs. However, we are able to "keep the peace" onsite by understanding and respecting the various ethnic groups and their lifestyles and not forcing them to completely integrate. I believe allowing them to form their "cliques" which keep them in their comfort zone, allows tensions to stay low. Using safety champions within their own groups to help us to gain trust with, engage with and train the workers.

Training

As discussed previously, not all migrant workers who come to the Middle East to work are skilled and/or competent - and "competency" tends to be somewhat subjective any way. For example, a new worker arrives in the Middle East to work. Once they arrive to the construction project that they are assigned to, they will likely receive their PPE, site badges, and a basic "safety induction". The safety inductions are typically presented by a safety officer and offered in a couple of the main languages spoken by the workers. Not all workers will always receive an induction in the language they understand. Further, many inductions are taught in a classroom setting with a power point presentation and text which the workers may or may not be able to read – even if it's their native language. For contractors without a strong safety culture or budget for training, this might be the extent of safety training a worker may receive.

Training is improving and where we spend the most of our time. The major contractors working on the sites I have been involved with have quite robust training programs - typically including courses such as "basic safety awareness"; working at height / fall protection (Exhibit 8); confined space entry; fire extinguisher training; heat stress prevention; Permit to Work; PPE awareness; etc. Training is taken very seriously by many contractors trying to provide a higher level of safety skill and is often taken out of the classroom and into the field to perform in practical, hands on way. However, as discussed in previous sections, the challenges are the communication barriers which may not allow for a worker to fully understand and retain the information. Further, their perceived risk about the task may prevent them from taking the training seriously. To overcome this, we provide training in ways and settings which gain the most trust of the workers and keep their attention. Some of our best trainings have been conducted while sitting down on site (in a safe location) and sharing a drink and a snack together.

Therefore, it is very important for "comprehension" to be confirmed at the end of any training. This can consist of a written or verbal test, or a practical exercise. This is a step that is often skipped. The worker signs their name, sits in the course for the allotted time, and they receive a certificate. This in no way confirms that they comprehended the information and can apply it.

Exhibit 8. HSE Manager working directly with a Supervisor, Foreman, and their labour and educating them on fall protection through practical, hands on training.

Supervisor Accountability

While inductions, training, and pre-task specific meetings, or "toolbox talks", clearly have benefit, I have observed that all this becomes ineffective when you don't have the commitment and support of the supervisor (or higher management in general). All training and knowledge about safe procedures often go out the window when a worker is told by his supervisor to perform the task or be fired. As much as we would like to believe that everyone has "stop work authority", few workers trust that they can object to unsafe work without retribution, and as a result, the task often gets performed in an unsafe way. I am yet to work with a contractor, or any company for that matter, who has truly established a safety accountability program for supervisors. Job descriptions often talk all about production and quality control but don't say a word about safety (OHS Insider, 2010). They should be specifically worded and action-oriented (OHS Insider, 2010). These potential supervisors should be assessed at the interview stage. If they don't already have the skills and qualifications necessary to meet the safety expectations of the job, they should not be hired, or expect to provide them the necessary training. The following are some examples of Supervisor safety responsibilities which can be included in a job description.

1. Know and understand safe work procedures, hazards, and risks pertaining to his / her job;
2. Be able to explain all safe work procedures, hazards, and risks to employees;
3. Assess competency of employees and ensure they are trained;
4. Consistently enforce the safety requirements and hold employees accountable for following them;
5. Participate in task planning, risk assessments and conduct pre-start or toolbox talks with employees;
6. Conduct daily inspections of work area and correct unsafe conditions and behaviors;

7. Submit a weekly report addressing safety issues, including incidents and near misses, results of inspections, etc. and submit it to their line manager;
8. Submit monthly recommendations to improve safety on the project and/or change employees' behaviors and/or attitudes to safety.
9. Respond to all safety-related incidents and near misses, participate in the investigation, and write a report that explains what happened and identifies the cause and recommends corrective action within 48 hours of the incident's occurrence. Supervisors shall follow up as necessary to ensure that corrective action is taken (OHS Insider, 2010);
10. Recognize employees for their commitment to safety.
11. Act as an example for team, displaying the correct behaviors and attitude for safety.
12. Maintain records showing evidence that he/she has performed the above responsibilities and for use in performance assessments.

That last line is the key. While getting the expectations set at the start of one's employment as a supervisor, it is just as important to evaluate the supervisor and link their safety performance to pay raises and bonuses – just as they often are to production and quality (OHS Insider, 2010).

Finally, supervisor accountability for safety is a very much cultural in this region. Many of the countries the supervisors come from do not have strong safety cultures – especially not where supervisors are responsible for safety. Safety is most often the "safety guy's" job. Even some of the laws and regulations in the GCC undermine supervisor accountability by requiring a certain number of safety "officers" per number of workers. For example, the Qatar Construction Specification (QCS, 2014), requires one safety officer per 50 workers. This often lets supervisors "off the hook" by always having a safety officer nearby to "do safety". Here in Qatar, I have found that supervisors range from "a blatant disregard for safety" (Exhibit 9) with bullying attitudes to those that embrace it and have a true value for safety (although this is usually rarer). The Middle East has a predominant "blame culture" (Middle East Construction News, 2010). An incident happens so you fire the worker. Problem solved? Not usually. In my experience, the majority of supervisors simply do not have the knowledge and accountability to perform their work – and ensure the work of their team – is performed safely.

Exhibit 9. Example of a supervisor not following the safety requirements and not setting an example for his team (Personal Photo, 2017).

Because of these reasons, we have concentrated much of our efforts on our sites on the supervisor staff. We identify safety champions who are willing to be trained and support safe working practices of their workers and colleagues. We routinely give rewards and recognition for workers who display the right behaviors and work safely (Exhibit 10).

Exhibit 10. Reward and recognition ceremony on a Qatar project (Personal Photo, 2016).

We also use incidents or near misses as an opportunity to re-train supervisors, clearly set expectations, and provide guidance and coaching. It is a large cultural barrier to overcome though – especially when the supervisor's own supervisor or manager doesn't hold him accountable for anything but production. Of course that's why safety is led from the top down. But we try to tackle it from top, bottom and the middle. It takes a while but we have found that many supervisors will improve and be more accountable for safety. However, if a supervisor is found to repeatedly put himself and/or his teams' safety at risk, then disciplinary actions or removal is warranted. In my experience, the largest improvements in the safety of a workforce occur only once the management and supervisors commit to it.

Exhibit 11. Supervisors giving a pre-task safety briefing to their employees prior to working on night shift in Qatar (Personal Photo, 2017).

Conclusion

The numerous cultures working together in the Middle East, all with their own languages or dialects, experiences, skill levels and training, values, attitudes, perceptions, and social habits present a huge challenge to obtaining consistently safe behaviors and working environment. As I hope you can see from the discussion presented here, it's not as easy as simply telling workers to follow safety rules; or merely telling supervisors that they should take responsibility for safety and their workers. It takes continuous hard work and dedication to engage with all levels of workers in the Middle East in order to change the safety culture. Being a safety professional is often a thankless job but I am constantly reminded by the improvements I have witnessed in the region that we are making progress. We have projects exceeding 50 million manhours since its last lost time injury; projects with better supervisor leadership and commitment than I have seen in the United States; and most importantly, thousands of workers going home safely at the end of each day. As with countries like the United States or United Kingdom, change doesn't happy overnight but progress is being made in great strides in the Middle East.

References

2022 FIFA World Cup Qatar (FIFA). 2017. (Retrieved March 7, 2017) (http://www.fifa.com/worldcup/qatar2022/index.html).

AECOM. July 28, 2016. Register of SHE Legal and Other Requirements, Revision 8.

Climate of, Climate of Middle East. 2017. (Retrieved March 8, 2017) (http://www.climateof.com/middleeast/index.asp).

Cronin, Sean. 2017. "Qatar happy to be in the spotlight on world stage." *The National*. March 7, 2017. (http://www.thenational.ae/business/industry-insights/economics/qatar-happy-to-be-in-the-spotlight-on-world-stage).

EDESKNEWS. February 8, 2017. *Qatar has said it is spending $500 million a day at the world Cup* (Retrieved March 1, 2017) (http://edesknews.com/qatar-has-said-it-is-spending-500-million-a-day-at-the-world-cup/).

Expo 2020 (Expo). 2017. Expo 2020 Dubai United Arab Emirates. (Retrieved March 7, 2017) (http://www.expo2020dubai.ae/).

Global Road Safety Partnership. 2014. *2014 Annual Report*. (https://www.grsproadsafety.org/).

Gulf Cooperation Council (GCC), Wikipedia, n.d. (Retrieved March 1, 2017) (https://en.wikipedia.org/wiki/Gulf_Cooperation_Council).

Gulf News. June 3, 2016. 11 dead in Qatar labour camp fire. (Retrieved March 8, 2017) (http://gulfnews.com/news/gulf/qatar/11-dead-in-qatar-labour-camp-fire-1.1839703).

Gulf Talent. 2007. *Federal Law No 8, For 1980, On Regulation of Labour Relations*. UAE Labour Law. (Retrieved March 8, 2017) (https://www.gulftalent.com/).

International Labour Organization (ILO). March 6, 1996. *Migrant Construction Workers Facing Pressure on Wages and Working Conditions* (Retrieved March 2, 2017) (http://www.ilo.org/global/about-the-ilo/newsroom/news/WCMS_008082/lang--en/index.htm).

Islam and Karma, n.d. *Concepts of Fate in the Karma Belief System and in The Qur'an.* (Retrieved March 8, 2017) (http://www.islamandkarma.com/fate.html).

Krause, Thomas and Kristen Bell. 2015.*7 Insights into Safety Leadership, First Edition.* The Safety Leadership Institute.

Middle East Construction News, 2010. *Construction industry calls for no-blame culture.* (Retrieved March 10, 2017) (http://meconstructionnews.com/957/construction-industry-calls-for-no-blame-culture).

OHS Insider, 2010. *2 Ways to Hold Supervisors Accountable for Safety*. (Retrieved March 10, 2017) (https://ohsinsider.com/do-diligence/2-ways-to-hold-supervisors-accountable-for-safety).

Oswald, D., F. Sheratt and S. Smith. June 2014. *RISK PERCEPTION AND SAFETY BEHAVIOUR: AN ETHNOGRAPHIC STUDY.* Proceeding of CIB W099, June 2014, Achieving Sustainable Construction Health and Safety, Lund, Sweden.

(https://www.researchgate.net/publication/279767040_RISK_PERCEPTION_AND_SAFETY_BEHAVIOUR_AN_ETHNOGRAPHIC_STUDY)

Qatar Construction Specification (QCS), 2014. Qatar.

Qatar National Vision (QNV). 2008. Qatar National Vision Plan. General Secretariat For Development Planning. (http://www.qdb.qa/English/Documents/QNV2030_English.pdf)

Snoj, Jure. 2017. *Population of Qatar by nationality - 2017 Report* (retrieved March 3, 2017) (http://priyadsouza.com/population-of-qatar-by-nationality-in-2017/).

The Economist. 2009. *The GCC in 2020, Outlook for the Gulf and the Global Economy, A report from the Economist Intelligence Unit, Sponsored by the Qatar Financial Centre Authority* , Economist Itelligence Unit, London http://graphics.eiu.com/marketing/pdf/Gulf2020.pdf .

Wikipedia. n.d. Caste System in India. (Retrieved March 8, 2017) (https://en.wikipedia.org/wiki/Caste_system_in_India).

______. n.d. Risk Perception. (Retrieved March 8, 2017) (https://en.wikipedia.org/wiki/Risk_perception)

______. n.d. Social Structure. (Retrieved March 5, 2017) https://en.wikipedia.org/wiki/Social_structure

Have We Really Learned From Major Accidents?

Rajender Dahiya, CSP, PPSE-MIChemE
Sr. Technical Services Manager – Loss Control
AIG – American Insurance General
Houston, Texas

Abstract

Considerable reductions in process safety incidents have not been realized over the past 40 years despite the extensive developments in technology. Root causes for these incidents remain the same but the likelihood and severity of consequences have been slightly reduced. The debate continues – Why do we continue to experience similar incidents even with establishment of new tools, technologies, management systems and lessons learned?

Is it the lack of process safety competency, corporate culture, or complexity of current management systems and associated tools contributing to the continuation of the learning failures?

This paper will review the mutual causes of learning failures from past major process safety incidents and examine what has not changed over several decades. Improvement opportunities are provided for the most common learning gaps.

Introduction

Learning from incidents is not limited to just incident investigation, but much beyond that, learning from incidents is also ingrained into corporate policies, engineering and design standards, and operating management systems. So, it is important to identify and eliminate the gaps in these processes and systems to prevent the recurrences of major accidents.

With enormous overall technological developments in the last several decades, processes, tools and management systems are getting more complex leading to inconsistency in operating management systems and hamper the effective implementing.

The design, construction, and operation bring several opportunities for errors and mistakes during the life cycle of a project. That's why every project, when handed over to operations, has a long list of design and installation deficiencies.

During the initial startup, operating set points are changed and a number of safety systems and alarms are bypassed. Some of these bypassed critical systems are never restored due to a lack of effective management programs.

Consistent gaps are discovered in the operating management systems and process safety knowledge of the workforce. The most common causes of learning failures and improvement opportunities are clarified in order for each of the identified gaps to advance the learning process and prevent the recurrences of process safety incidents.

Have We Learned From Accidents?

Failure to identify the hazards and manage the risks of highly hazardous materials and processes is one of the primary root-causes of an incident, and failure to learn from the past incidents is the leading cause of recurrences of incidents. Identifying the causes of learning failures and improving to prevent the recurrences are consistent challenges for industries. Figure 1,

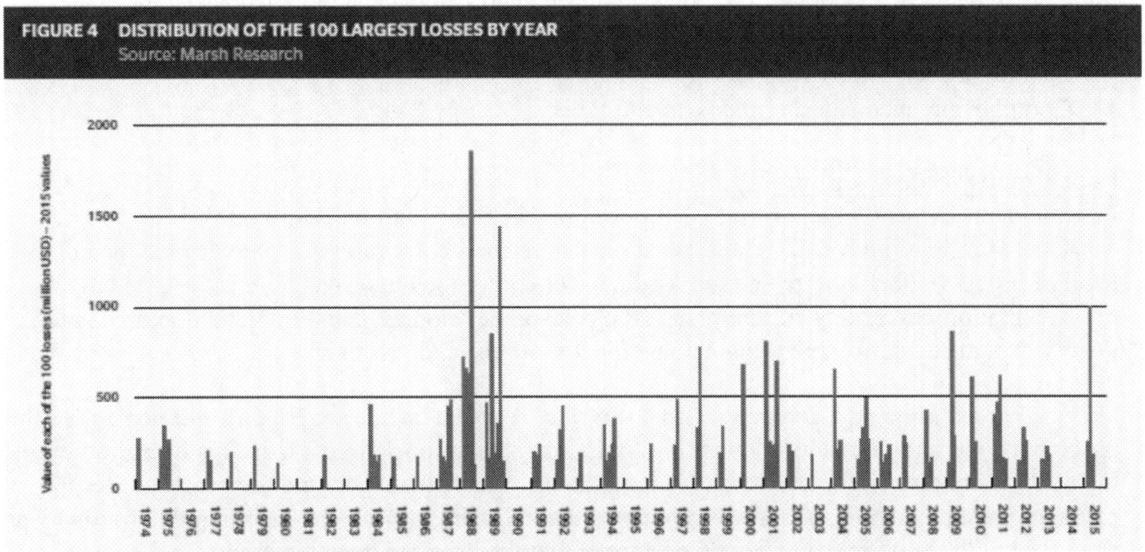

Figure 1. Distribution of the 100 largest losses
Source: Marsh Report

"Distribution of the 100 largest losses," demonstrates clearly that the losses from the incidents have decreased since the last 40 years [1].

As per the Marsh report [1], the top 20 largest losses from 1974-2015, are distributed as follows: 15 explosions, 2 blowouts, 1 fire, 1 storm, and 1 collision. Notice that explosions are the number one cause of the losses.

Investigation outcomes of major accidents are very consistent, and validate that the causes and consequences are very identical in spite of enormous overall development in technology, tools, techniques, and learning. The following were the cohesions in the past accidents:

- All were process safety accidents
- Root causes and consequences were alike
- Several precursors – warning signs / weak signals
- Good occupational safety records
- Learning's were shared.

Most importantly, all accidents were preventable.

Commonalities in past accidents clearly indicate that there are steady gaps in the learning process and companies are facing persistent challenges in identifying and eliminating those learning gaps.

Most Common Causes of Learning Failure

There have been several common and consistent challenges for many decades including the last 20 years, which had stricter codes and regulations. Some of the challenges and causes of learning failures are listed below and are explained in more details in the following sub sections:

- Poor process safety culture
- Lack of process safety competency
- Complex and inconsistent processes and tools

Poor Process Safety Culture

Past incidents evidently demonstrate that poor process safety culture is still one of the leading causes of learning failures [2]. The biggest challenge for a corporation is to determine the current level of process safety culture and the roadmap for continuous improvement. Industries have realized that process safety culture plays vital roles in development and implementation of high quality design standards, operating management systems, and processes.

For example, in a company with a runaway reaction hazard, board operator controlled the process by operating outside its operating limits with the auto trip bypassed. This operation upset incident was not reported. Operation manager, maintenance manager, and the process safety manager were not aware about the trip bypass status. During the closing meeting, this finding was discussed with the plant manager and he agreed that this trip had been bypassed for the last 6 years without any documentation. The company has experienced similar accidents in the past causing serious losses. This scenario demonstrates that due to the poor process safety culture, safety bypass procedure is not implemented effectively, and no inquiry was made in the last 6 years about this highly hazardous process and the condition of the safeguards. These types of incidents are warning signs and must be investigated, and action must be taken immediately to address the root cause in maintaining the integrity of safety critical systems.

In a good process safety culture, this finding could have alerted the management team and an investigation could have initiated immediately to find the root causes and resolve the situation. Investigation would then be extended to review other high hazard scenarios and the status of the availability of safeguards. Other examples are: overdue high priority action items including code violation, no written operating procedures, no testing and maintenance of safety critical systems, overdue inspections of piping, tanks, vessels, etc. Occupational safety and production is still given priority over process safety.

Root causes of many learning gaps presented in this paper are due to poor corporate process safety culture. Refer to CCPS vision 2020 [3] examples of good process safety culture, how an executive should behave, and what type of questions to ask while visiting an operating plant. Process safety training is equally important for management to learn the importance and role of process safety in incident prevention.

Lack of Process Safety Competency

Process safety competency plays a vital role in effective implementation of process safety management and improving the company's process safety culture. A competency gap is realized from top management to the shop floor workforce. Inadequate and ineffective training by incompetent trainers is the major cause of this gap. For example – a company provides a 45 minutes web based training to all safety critical employees but it is inadequate. In another company, 3 day training covers only hazard identification and risk assessment. Incompetent management and employees will not be able to develop and implement the management programs and tools effectively and consequently leading to the poor maintenance of safety systems.

Process safety training matrices should be developed for each employee based on the roles and the level of training required. An initial face to face detailed training should be provided to each safety critical employee, covering all the 14 PSM elements. Annual refresher trainings should be need-based, instead of a required repetition of already known facts.

Complex and Inconsistent Processes and Tools

Management programs and processes are getting complex and inconsistent day by day due to the availability of unlimited information from numerous sources. The challenge is to make these processes fit for purpose and user friendly. Good corporate process safety culture combined with a competent management and workforce greatly improves these processes. The most frequent discoveries which contributed to learning failures are explained in the following activities:

Hazard Identification and Risk Assessment

The hazard identification and risk assessment starts early in the design stages and continues throughout the lifecycle of the project. After investing considerable amount of time and resources for designing a new facility, a long list of design deficiencies is generated for every project. This clearly indicates that there are certainly gaps in the learning process during engineering and design. So, a newly installed facility always comes with the design deficiencies and possibly unidentified and/or unmitigated hazards/risks. The causes of learning failures and improvement tips are provided below.

Poor quality safety studies – HAZID, PHA: At the design stage, safety studies are conducted by incompetent leaders and/or teams without the availability of correct and updated documents and drawings. For example, consequence scenarios and risk matrix numbers are adjusted to lower the risk to an acceptable level. Nonspecific, generic safeguards are used to fill the empty space in the worksheet with content such as operating procedures, operator training, inspection, operator rounds, etc.

A competent leader and team cancel the study when the appropriate documents are unavailable. They never allow the team to go back and forth to adjust the risk numbers and eventually make it "favorable". Recommendations are written specifically to address the causes and meet the company procedure requirements.

At the operation stage, some companies revalidate their PHA without integrating the correct expertise, without incorporating the changes (MOC), and failing to apply the lessons learned from past incidents. Example: A plant manager and a contractor revalidated the PHA and missed several existing safeguards (SG) for a runaway reaction scenario and a very high risk was reduced to an acceptable level by adding only one safeguard.

This practice is more common in small size operations covered by OSHA-PSM. If SGs are not identified and listed in the PHA report, the chance of SGs being maintained properly diminishes. A competent team must revalidate the PHA and ensure that all SGs are listed and changes are incorporated.

Complexity in design: Design engineers focus more on the safeguards rather than challenging the design and applying inherently safer techniques. Previous designs are accepted, more safeguards are added, and projects are repeated without the implementation of the latest technologies and innovations.

Conducting a high quality hazard identification (HAZID) study at the early stages of a project is very critical in order to identify inherent hazards. The safer and simpler designs can be achieved by applying a hierarchy of design controls at every stage of the project. Inherent safety in design (ISD) techniques helps to make the design safer, simpler, and cheaper thus making it easier to build, operate, and maintain [4].

Figure 2, "Simplified design," depicts how an inherently safer design can make the process safer and simpler by either eliminating the needs for an independent protection layer (IPL) altogether or by minimizing the number of IPL's which decreases the design and maintenance costs while encouraging a safer and simpler process.

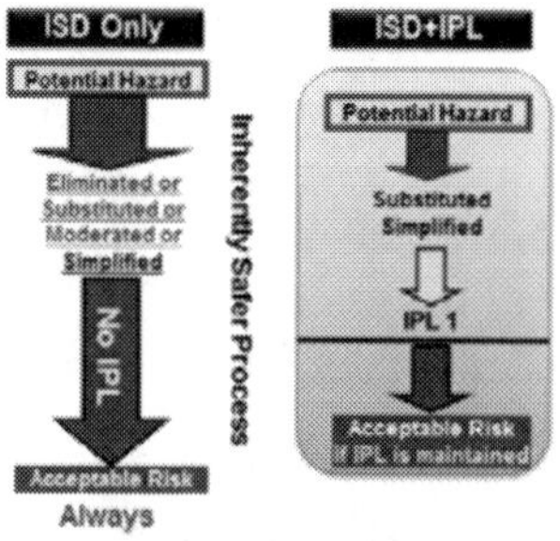

Figure 2. Simplified design

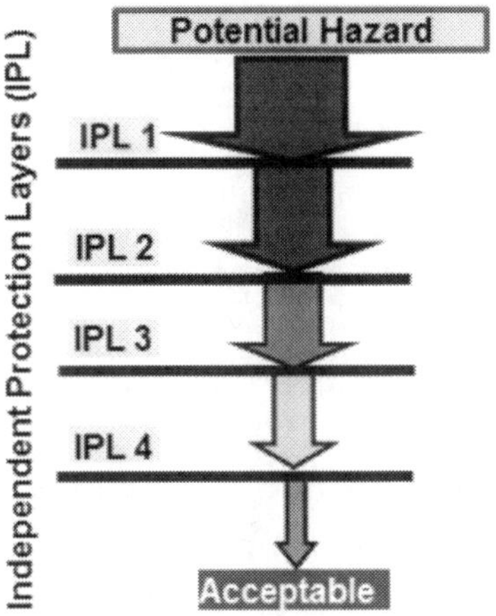

Figure 3. Risk status at design stage

Mechanical Integrity of Safety Critical Equipment

Management failure to maintain the integrity of safety critical systems is one of the most common root causes of past major accidents while also being the most repeated recommendation in insurance loss control survey reports across all types of industries. There were several accidents in which all the preventive as well as mitigation safeguards were either bypassed or not maintained. In many cases, the plant management was not aware about the risk status due to the poorly implemented management of change, or safety bypass programs and the plant is operated with higher risk. Figure 3, "Risk status at design stage," demonstrates that each identified risk is mitigated to an acceptable level using an appropriate number of independent protection layers (IPL) during the design stage. Causes of safety systems failures and tips for improvement are given below:

Lack of quality assurance: Defects during procurement, construction, and installation are caused by a lack of quality assurance and positive materials verification (PMI) process. Many companies depend on the vendors and suppliers and don't have any PMI process. A lack of time and rush to meet the projects deadlines ends up sacrificing the quality of the overall design.

These errors and mistakes can be captured before procurement and installation if the quality assurance and PMI process are implemented without pressure of meeting the project schedule and compromising the quality design and construction.

Process set point changed: Operating outside the "integrated operating windows" (IOW) [5] is considered normal during initial startup and startup after shut down of any facility. Some of these changes are never reinstated. Some companies may not even have their design documents to confirm the IOW and operate the plant at unknown operating risks.

A well implemented safety bypass/set point change/MOC program eliminates the possibility of operating the plant with an unacceptable risk level. All design documents including IOW must be readily available for use and reference.

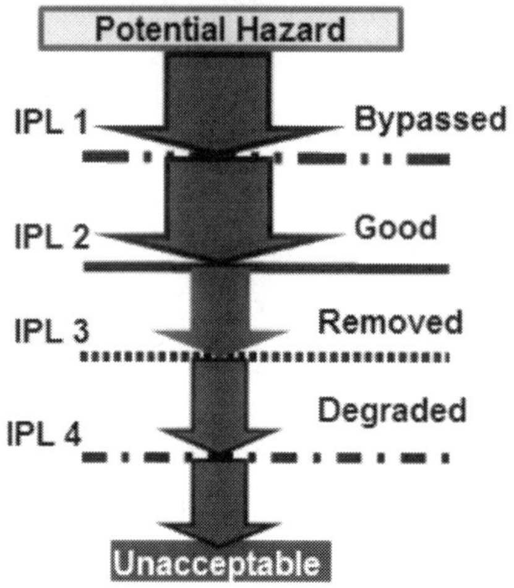

Figure 4. Risk status at operation stage

Safety systems bypassed/disabled: A general statement, "my plant is built to codes" keeps the management and its employees in a false sense of security. Compare "built to codes" and "as operated" facility after 1, 2, 3 …10 years. Figure 4, "Risk status at operation stage," shows how the safeguards are bypassed, removed, or degraded, and a plant is operated with high/unacceptable risks.

A complex design with excessive alarms and safeguards is the reason for the bypassing and disabling of safety systems. Furthermore, the poor implementation of safety bypass programs and a lack of process safety culture and competency is the second leading cause of bypassing or disabling safety systems.

Many companies are facing the challenges to maintain their safety systems and implementing repeated recommendations from insurance surveys and auditors. Incomplete records are often observed in the control room and employees are unable to explain the current status of the safety systems.

Example: Safety bypass procedures, forms, logs, shift logs, and the actual status of the safety system are inconsistent more than 70% of the time. In some cases, MOCs are not available for a bypassed system.

An effective safety bypass program helps in addressing this gap by assessing the risk before bypassing any safeguard and ensures that the safeguard is brought back to operation. A strong process safety culture with competent personnel eliminates the risk of operating the plant with higher risks without appropriate safeguards.

Lack of testing and maintenance: Poorly implemented management programs, a lack of process safety culture, and low competency are other leading causes of lack of inspection, testing and maintenance of safety systems. Many companies have no list of the safety systems and an undeveloped PM program. Others have several systems that are overdue for inspections and testing.

A written safety requirement specification and a list of all safety systems must be developed in the design stage of a project and maintained for the life of the plant. Preventive maintenance and inspection frequency should be developed based on at least safety requirement specifications and vendor guidelines, while all safety systems must be tested and maintained to meet the integrity level as designed.

Incident Investigation and Learning

Learning from incidents is not a simple standalone process and cannot work in isolation. It is embedded into corporate policies, operating management systems, and engineering standards. Having a consistency and alignment among all these systems and processes is vital in order to apply the learning. Companies are struggling to design and implement a simpler and fit for purpose process and tool which can deliver useful data for learning. Incident investigation and learning process comprises of mainly five steps. The continuing gaps and tips for improvement are provided below for each step:

Incident reporting: Many companies do not report process incidents such as operations upsets, safety system challenges, operating outside the "Integrated Operating Windows"[5], and failure of safety systems tests. These incidents are the warning signs and weak signals for a high potential process incident and must be reported. The procedures, incident reporting forms, or tools must be updated to include these incidents.

Incident investigation: The majority of process safety incidents have low actual, but high potential consequences. A root cause analysis is only conducted for actual high consequence incidents. Incidents with low actual consequences must be investigated for actual as well as potential consequences using root cause analysis techniques. Potential consequences for the same incident may be a disaster.

Root cause analysis: Many companies lack competent people and tools to conduct root-cause analysis while others have the tools and competent people but investigation is closed by listing the immediate causes. Competent teams should conduct a root cause analysis instead of stopping at the immediate or contributing causes and then closing the investigation. Without finding the root causes, recurrence cannot be prevented. This omission is one of the major causes of learning failure.

Recommendations: Generic recommendations do not address the root causes and are not measurable while specific recommendations are not tracked to closure and not verified for effective implementation. Example: a hot work was performed on an operating gas platform with several violations including: using a contractor's permit, no gas testing, flammable materials not removed, drains not covered, etc. The root cause of this violation was management failure to implement the work permit system. The actions should address this cause but management fired the contractor and a disciplinary action was taken against the operator to correct this serious violation. The root cause still exists and this same violation will continue until a serious accident occurs. Recommendations must address the root cause which, if implemented effectively, should prevent the recurrence of a similar incident. The quality of action implementation is another cause of learning failure. A checkbox does not help.

Learning sharing: Lessons learned are not shared within or outside the company due to a lack of effective learning process. This is a total failure and defeats the overall incident investigation objective. The best industry practice is to produce a one pager incident summary with causes and actions to prevent the recurrences. A monthly process safety beacon issued by CCPS is the best example [6].

Employees prefer simpler and more user friendly changes which end up making their work more effective. Companies should develop or update their procedures and then design the tools to meet the company requirements. End user involvement is very critical in the entire process. The tools should be need based, and not what the vendor wants to sell. Complex tools fall in the data dumps and employees keep using their paper copies for reporting and investigating in isolation and ultimately the whole learning process fails.

Actions Tracking and Closure

Action tracking from various sources such as PHA, audits, inspections, incident investigations, loss control surveys, etc. is another challenge industries are facing. Some tools in the market are very efficient and user friendly with auto tracking features, while others are impractical and fail to deliver useful results. The continuing problems with action tracking are given below:

- Central tracking system: complex systems that are not designed and implemented effectively are data dump and don't deliver beneficial outcomes such as current status, priority, and overdue actions. Companies continue using individual spreadsheets and actions are not auto tracked to closure.
- Individually tracking systems: lacking auto tracking feature and unable to produce status of all actions from various sources for management review. Actions are not filtered based on priority and the source.
- Action implementation verification: A poorly implemented action is another leading cause of learning failure. Lack of competency and economic reasons force an action to close without proper verification. Checkbox still prevails.

A user friendly tool which should be able to at least import actions from various sources, prioritize, auto track, provide status for management review, and input in performance measurement.

Management of Change (MOC)

A poorly developed and implemented MOC program is the root cause of inconsistency in various management programs and processes. Like other programs as stated above, the MOC process, which includes procedure, form, risk assessment, timings, approval, and the status are not aligned with each other. Most importantly, a complex eMOC tool hardly complies with the procedure and is poorly implemented due to complexity and user unfriendliness.

There are several faults in the MOC process including procedure, training, implementation and eMOC tool. The following gaps were noted from audits, inspections and investigation:

- Inadequate procedure – safety critical systems not covered

- Lack of training and ineffective implementation
- Complex and inconsistent procedure and eMOC tool
- Risk assessment missing or poorly done
- Generic recommendations, not prioritized, and not tracked to closure

Backlog of MOC and recommendations tracking is one of the major challenges in the industries. Example: a company bought and implemented an eMOC tool off the shelf which does not conform to the company procedure. After more than a year, management decided to update the procedure to comply with the eMOC tool which was poorly implemented. On many occasions, management failed to provide a copy of MOC for bypassed safety critical systems such as disabled pressurization of a critical MCC, pump trip bypass on a reactor, plant fire alarm system out of service for upgrades etc.

It has been observed that even the eMOC tool subject matter experts are unable to browse through it and provide the required information.

MOC procedure must be written or updated using learning from the previous processes and from the best industry practices before designing an eMOC tool to ensure that it is fit for purpose, consistent, and deliver the useful results such as status of current MOCs and recommendations tracking. An effective training should be provided to all involved with a continuous technical support.

Prestart Up Safety Review (PSSR)

A well written and implemented PSSR process could prevent the incidents not only during startup but also later in the lifecycle of the facility. PSSR is a cold eye verification process to ensure that the facility is constructed as per the design, all operations and maintenance procedures are in place, and employees are trained [7]. One of the major challenges for an effective PSSR is pressure from the project and operations management teams for startup urgency, which leads to unidentified or manipulated deficiencies for a timely startup. The following are some of the key gaps:

- Inadequate procedure or checklist
- Incorrect timing and incompetent team
- Business pressure – confrontation with project and operations management teams
- Checking the box is the most common cause – just a paper formality
- Adjusting the actions priority to start the facility in time.

Example: Commissioning a 1000 miles gas pipelines in a potentially unstable country with a recommendation to post warning signs along the pipeline. This recommendation was changed to low priority due to economic reasons. There are several examples of serious accidents which could have been prevented if PSSR was done correctly and actions implemented before starting a hazardous process.

Generally, PSSR is conducted for initial plant startup and only after the MOC. The best industry practice is to conduct the PSSR after each major maintenance or turnaround, and startup after an emergency shutdown and startup of a mothballed process/equipment. PSSR should be conducted by a team independent from the commissioning or maintenance teams to diminish bias.

Performance Measurement
Lack of process safety performance measurement is another root cause of recurring incidents. However, the struggle of identifying and measuring the leading and lagging indicators is continued. Key learnings are [8]:

- Tier 1 & 2 PSE are lagging indicators and very a few in numbers which do not provide enough data for trending, learning and improvement.
- Tier 3 & 4 PSE are company defined and leading indicators. Companies are struggling to identify, report, and analyze. These PSE provide advance warning signs of a potential incident and weaknesses of the safeguards or management programs. Tier 3 incidents are challenges to safety systems and end up not being reported or investigated. Tier 4 PSE are very informal, not recorded, or not taken seriously and are sometimes invisible.
- For most of the past major accidents, warning signs were not captured, investigated or actions were not implemented. This is another major gap in the incident prevention learning process.

Industries have learned that leading indicators provide warning signs and weak signals for a high potential incident and should be investigated for the potential consequences to prevent a serious accident.

Conclusion

Conducting gap assessment in process safety management systems and developing strategic action plans will help in improving the overall learning process for accident prevention. Recommendations from regulators, third party/insurance reports, findings from corporate audits and inspections, and incident investigation can provide a good start.

Tips for improving the learning gaps;

- Improve process safety culture through management commitment and employee involvement
- Improve process safety competency by formal training at all levels
- Develop effective communication: top-bottom/bottom-top and across the organization
- Develop and implement simple, fit for purpose, and consistent processes and tools
- Maintain the integrity of safety critical equipment and systems
- Collect and analyze quality data ONLY – for continuous improvement
- No Checkbox at all

Consistency, fit-for-purpose operating management systems, simpler processes, tools and techniques, and a good process safety culture with competent workforce all come hand in hand to prevent the failures to learn from past accidents. Overall learning to prevent the recurrences of incidents can undergo an improvement when the above stated factors receive more attention and reform.

Endnotes

[1] The 100 Largest Losses, 1974-2015, Large property damage losses in the hydrocarbon industry, 24th edition, March 2016. www.marsh.co.uk

[2] U.S. Chemical Safety Board. Accessed at www.csb.gov

[3] Vision 20/20 – American Institute of Chemical Engineers, Center for Chemical Process Safety. https://www.aiche.org/ccps/resources/vision-2020

[4] Center for Chemical Process Safety, *Inherently Safer Chemical Processes: A life Cycle Approach,* John Wiley & Sons, 2nd Edition, 2010

[5] API Recommended Practice 584, Integrity Operating Windows, 1st edition, May 2014.

[6] "Process Safety Beacon," American Institute of Chemical Engineers, Center for Chemical Process. https://www.aiche.org/ccps/search/beacon

[7] Center for Chemical Process Safety, Guidelines for Performing Effective Pre-Startup Safety Reviews.

[8] American Petroleum Institute, ANSI/API, Recommended Practice 754, *Process Safety Performance Indicators for the Refining and Petrochemical Industries,* Second Edition, Washington D.C., 2016.

Machine Risk Assessment and Reduction, an Ongoing Journey

Paul J. Prince, CSP
Senior Loss Control Consultant
ESIS® Health, Safety and Environmental Services
Hendersonville, TN

Introduction

In a "perfect" safety management system, in an industrial production environment, we would expect to find all potential workplace hazards identified with effective controls in place to prevent each hazard from resulting in or contributing to a workplace accident. For this to happen, I would envision a company with a solid, centralized safety management system in place, adequately staffed and funded at all levels to carry out the safety function. This group of safety professionals would have pro-actively and meticulously assessed the work environment to identify all hazards. They would have compared the existing hazard controls against the hierarchy of controls; eliminating and/or provided engineering controls for each hazard where possible. For remaining hazards where elimination and engineering control was not possible they would have implemented a perfect behavior management system. In this regard there is a oneness of purpose that separates safety from the factors associated with production and quality. (Haight, Yorio, Rost & Willmer 2014)

For most of us who work in general industry this truly is a vision commensurate with Dorothy in the Land of Oz. In the real world we don't find many companies with strong centralized safety management, well developed and implemented top down instructions and sufficiently funded and staffed safety departments. I do believe that the vision of pro-active risk assessment is now being recognized as the way toward an achievable system of continuous improvement as it relates to hazard identification and risk reduction. One subset of workplace hazards are those hazards associated with machinery. Control of these hazards alone requires a major management initiative. Even today in the year 2017 we find production equipment "home grown", custom made in house, along with other equipment originally made decades earlier. When we examine this equipment we may find some guarding, no guarding or evidence of guarding that was removed. Often, the existing machine hazard control is the simple awareness of the hazard and hopeful avoidance by the operator. One reason why some machine controls remain in this realm is that installation of guarding is not cheap and it may not even be possible for some equipment. It can be a tough sell to convince a plant manager or corporate level finance officer to install $100K guarding on a machine that's been in production for decades and there has never been an accident on the machine. Many industries have relied upon administrative controls in the absence of machine guards for years and decades before this reliance has resulting in an unsafe behavior producing a serious or fatal injury to the operator or other employees in the workplace. The reality is that even when a serious machine hazard is recognized proactively, it remains until an accident manifests and corrective actions finally go beyond procedural control of the hazard.

To prevent serious injury and/or fatalities a pro-active approach is needed to identify hazards and start a management process to compare existing machine safeguards against a solid, written machine guarding standard. Top management must recognize this as the only way to improve and avoid the serious lifelong disabling or fatal accident. It requires top management recognition of the problem and a willingness to establish a system of continuous improvement with allocation of resources to get the job done. This presentation explores a management systems process and what is necessary to develop an effective machine safeguarding program.

What constitutes proper machine guarding?

For our machine guarding to be "effective" as a control we want the guarding to require a very conscious and deliberate action on the part of the employee to access the hazard area. We can ask the question, how conscious and deliberate an action? There is a need for a written defined standard for machine guarding. Once we have this source document, we can assess our equipment to determine where we meet this standard and where changes are necessary to comply. For many companies today there is no program guidance as to "what is" proper machine guarding. What is needed is consolidated information to clearly define what constitutes proper machine guarding with no grey areas or ambiguity. OSHA regulations for the most part establish a performance standard. OSHA 1910.212 establishes a general performance requirement in that, " one or more methods of machine guarding shall be provided to protect the operator and other employees in the machine area from hazards...." The standard goes on to state that the machine point of operation shall be guarded. This is a general guarding requirement. In other parts of 1910 Subpart O we have more prescriptive standards that address specific equipment such as; woodworking machinery, abrasive wheel machinery, mills and calendars (for rubber and plastic), mechanical power presses and other equipment. There are numerous types of equipment that are not specifically addressed in the regulation. Interesting to note that when we see OSHA citations for machine guarding, often the primary reference is the General Duty Clause, from the OSH Act of 1970, which states, "Each employer, shall furnish to each of his employees employment and a place of employment which are free from recognized hazards that are causing or likely to cause death or serious physical harm to his employees;". What responsible managers want to know is clear prescriptive machine guarding information. When is our machine guarding OK? If I install a guardrail fence with warning signs around a turret punch, does this meet proper guarding standards? To establish this we need to go to modern industry "best practice" standards (such as

ANSI and ISO). Best practice industry standards are the "how to" standards that tell us what constitutes proper machine guarding. There are hundreds of industry "best practice" standards that address machine guarding. Standards in some cases address guarding by type of machine. Other standards address requirements for guard opening size, perimeter guard design and height, control circuit design and reliability, design and operation of safety devices, control color and labelling and many, many more. What we find in these standards is that guarding is now very prescriptive. A central problem is that who has the time to find all pertinent industry standards, learn the standards and use them to conduct an assessment. This is where we find a great need for a top down initiative to develop company specific guidelines and guarding specifications based upon these "best practice" standards. Once this is done the company requirements become clear, prescriptive and can be found in a central location for easy reference. We now have a written standard we can use to bench mark where we are with our existing machinery and also to ensure any new equipment is specified, designed and built to meet these standards.

There are fundamental guarding premises that come out of industry "best practice" standards, which are necessary to meet OSHA performance requirements, these are;

- We cannot access the hazard from over, under, around or through the guarding or safety devices.
- Fixed guards are fastened in a manner that requires a special tool for guard removal. Guards cannot be easily removed by hand or use of materials in the work area.
- Devices can only be used on equipment with reliable stopping action. Design standards for safety circuit "control reliability" are spelled out and follow a hierarchy based upon the risk assessment process.
- The risk assessment process is integrated into a process that determines the level of machine guarding and design of the safety control system.

So now when we look to the ANSI or ISO standards we find clear prescription of what constitutes proper machine guarding in written form. This is where we would find that a guardrail violates the first fundamental principle or proper machine guarding in that we can crawl over or through the guardrail fencing. In these standards we would find clear prescription of how high the barrier has to extend above the floor and the permissible size of any openings in the barrier based upon the location of the hazard behind the fencing.

The Pro-Active Approach to Machine Guarding: Identifying hazards and applying risk reduction before the accident.

Hazard Identification and Assessment

At the core of the philosophy of Safety Management Systems is the pro-active search, identification and assessment of hazards in the workplace. Machine hazards are one sub-set of workplace hazards and machine risk assessment must be extended to this area. You will find that risk assessment of machinery is mandated by current ANSI and ISO standards. Company policy must be developed which incorporates hazard assessment methodology and requires the review of existing equipment. This policy should establish standards for machine hazard review, risk assessment and reduction strategy for both existing and new equipment. Policy content should address the following:

- Assessment Methodology-- All existing equipment should be assessed to determine whether existing hazards are properly guarded. This assessment should identify each hazard and subjectively quantify the potential **severity** presented by the hazard, **frequency of exposure** to the hazard and the **probability of an injury** occurrence. Existing machine guarding must be considered for an associate's **ability to deviate from the existing control**, i.e. can the guard be easily removed, does the guard have openings that permit easy access to hazards, are interlocks easily bypassed, etc. This is a "scored" assessment. Each hazard has a risk score. The point of this should be obvious; few industries that have many pieces of equipment have resources to fix everything at one time. They can however implement a process of continuous improvement with focus of their resources first directed to the highest rated machine hazard.

- New Equipment Specification-- The idea is that all new equipment is designed utilizing risk assessment. Each hazard is identified in the design phase and guarding is applied following principles outlined in industry "best practice" standards. If a proper review does not take place, then equipment may hit the production floor with improperly controlled hazards. Management should have procedures in place that clearly specify how they want a new piece of machinery built. A general reference in a purchase contract stating that the equipment must meet all OSHA requirements is not sufficient. Formal risk assessment should be mandated in the specification. For large, custom made production equipment specific construction standards should be identified in the specification. Procedures should call for a "factory acceptance test" (FAT) where the original equipment manufacturer (OEM) demonstrates before equipment delivery that the new machine is designed in accordance with the company specification. Deficiencies that are identified in the FAT are corrected and only then is the equipment delivered to the production facility where a site acceptance test (SAT) is conducted by the OEM. This protects a company's interest and ensures only a safe and well-designed piece of equipment makes it to the production floor. This process of design and acceptance should be spelled out as part of the new equipment contract.

- Management of Change-- This is a vital part to any management system, particularly as it relates to safety. Change can create new risks that, if not properly managed, can create conditions that contribute to accidents resulting in injuries and death. Change in equipment and process design mandates a new risk assessment to ensure new hazards are not created or introduced without proper controls. The management of change is a core process that is part of an effective system of management. A formal process that addresses this change is necessary. (ASSE 2012)

Important to note that once programs like that noted above are developed, it is important to ensure personnel are trained on the program content. I have audited machine guarding programs and found new equipment in production that did not meet standards. The company had a well-developed new equipment specification program, however newly hired engineers were not trained on the program existence and it was not followed.

The above elements will ensure that all new or modified equipment is built to meet modern safety requirements and all existing equipment is in line for review and upgrade. The net effect is a workplace that continuously improves with machine related risks under continuous evaluation and upgrade.

Risk Reduction

An identified hazard for a new piece of equipment should result with either the elimination of the hazard or a proper machine safeguard. Risk reduction for existing equipment is often not as simple. It may not be possible to eliminate or easily safeguard machine hazards associated with the operation of older equipment. If this is true, then usually something can be done to reduce risk even if the risk cannot be eliminated or safeguarded to meet our standards. The objective of the process is not to create the perfect work environment overnight but to establish a system of continuous improvement. Risk reduction should follow the basic hierarchy of controls below. (Machine Safety 101)

1. Eliminate the hazard- design it out.
2. Isolate the hazard with hard guarding.
   a. Add fixed or interlocked guards.
   b. Install devices that sense the presence of a person and stops the machine (hazardous motion.
3. Implement administrative controls- procedural training and awareness.
   a. Signage and audio/visual warnings.
4. Use personal protective equipment – goggles, gloves, shields, etc.

Important to note is that the guarding effectiveness is reduced as we go down this list. Our concept of taking the human factor out of the risk reduction measure would apply. For items 1 and 2 above we can significantly reduce risk by elimination of the hazard or by guarding in a manner that ensure a person must significantly go "out of their way" to gain access to the hazard. Items 3 and 4 are completely administrative and are not desirable, although in the absence of proper guarding they may reduce risk until proper safeguarding is installed.

The ultimate goal is to achieve a level of control for a respective hazard that does not rely on human awareness and avoidance of the hazard. The hazard is eliminated by design or safeguards in the form of fixed guards and/or devices are installed to ensure personnel are separated from the hazard.

Control reliable design is a key feature necessary when utilizing guard interlocks and devices such as; light curtains, pressure sensitive mats, laser scanners and other systems that sense the presence of a person and initiate the stopping of hazardous motion. When devices are in use as part of a new or revised guarding system, modern standards (ANSI and ISO) require the controlling circuit design to fail safe and ensure the stopping action is achieved. Modern standards initially addressed this issue in 1995 when EN 954-1, Safety of Machinery-Safety Related Parts of Control Systems, Part 1: General Principles for Design. These principles are applied under the latest machinery standard, ISO 13849-1. The key point to recognize is that when machine stopping action is critical to safety of the employee, the stopping action is reliable.

A machine guarding program mandates the review of machine guarding. It generates a work list from the scored machine risk assessments. A team reviews the existing hazard controls and seeks the higher level of risk reduction.

Training-- An essential ingredient

Even the most sophisticated machine guarding can be defeated by a motivated employee. A training program is necessary to develop and help sustain the proper culture and behaviors as related to machine guarding purpose and maintenance. The best time to do this training is at the start of a machine guarding initiative and then as necessary to sustain the culture. Training for all personnel is needed to establish a fundamental understanding; machine guards are not to be removed by anyone, unless they are authorized by management and have special training to do so. One common observation in industry are machines missing guards. Guards are removed and not replaced for a variety of reasons. Sometimes the guards lose fasteners and cannot be replaced upon removal for maintenance. The understanding and culture must be established that "we don't run equipment with guards missing"! There are new fastener designs where the fasteners can be loosened to remove a guard and the fasteners will not separate from the guard, consequently they can't be lost.

Higher levels of training must be provided to personnel involved in guarding assessment. These individuals should have a core understanding of the following:

- The philosophy of machine guarding and what constitutes proper machine guarding.
- Machine hazards that must be guarded.
- Machine guarding standards and options; fixed guards, interlocked guards, devices such as light curtains and optical scanners.
- Guard interlock requirements.
- The concept of machine control reliability.
- The machine hazard assessment process and hazard scoring process.
- Machine specific requirements; i.e. mechanical power press, injection mold machines, robots.

I consulted at a production plant that hired an outside specialty company to come in and install machine guards on all their equipment. They spent over $500k on machine guarding and when I returned to the plant several months after the project I saw large, new guard sections in dumpsters outside of the building. The guarding design did not facilitate production and equipment maintenance. This demonstrates that the best people to involve in machine guarding of existing equipment are maintenance and production operators. Once they understand machine guarding intent and principles they will come up with the best ideas on how to reduce associated risks. If they have input into the problem and the solution they are more likely to help sustain the solution.

Establishing An Effective Risk Reduction Program

A corporation with multiple machine laden facilities would best be served through a top down, centralized management process with procedures that establish standards and required activities. In addition to this, prescriptive guidelines are also needed. For a machine guarding program we have already outlined the need for company procedures that contain the following;

- Clear standards as to what proper machine guarding is and looks like.
- A mandate for risk assessment of existing equipment for the purpose of establishing a prioritized action list. A list of machines in order of corrective action, i.e. examination for further risk reduction options including replacement.
- A clear standard for new equipment acquisition that includes risk assessment and proper guarding in accordance with industry "best practice" standards. This new equipment specification is specific and specifically sites safeguarding standards. It should require accountability for review and "sign off" as insurance that new equipment enters a facility with all machine hazards properly safeguarded.
- A Management of Change procedure to ensure all equipment process changes are closely examined for introduction and proper control of any new hazards created from the change.
- A formal ongoing process of machine risk review. As a consultant I reviewed an accident at a facility that manufactured nylon web strapping. We found that an interlocked had failed on the machine involved and corrective action was to change the circuit design and also relocate the control panel where it would not be subject to excessive vibration from the machine operation. The cost of the change was approximately $4000. The location had 120 identical machines. They corrected the issue on all machines over a 5 year period and developed an interim administrative control for the machines that were not immediately fixed. My point in this example is that they were able to address the issue over time and move forward with a program of continuous improvement. Unfortunately, the drive for the issue was reactive to an accident incident rather than pro-active.

Summary

General Industry in America is filled with production equipment that presents serious and even fatal exposures. A gypsum board plant experienced a veteran worker getting caught and killed under the board flippers at a new facility.(OSHA 5/2011) Another experienced maintenance worker was checking temperature settings below one die on a carousel producing car seat foam.

His head was crushed between the protruding upper mold handle and a guardrail support column. (OSHA 11/2009) The hazardous exposures were obvious and visible yet guarding changes were not made until after fatal accidents. Such exposures are prevalent in many industrial settings. To prevent serious injury and/or fatalities a pro-active approach is needed to identify hazards and start a management process to reduce associated risk. Unfortunately the release of funding for equipment change is often based upon "return on investment". If the return cannot be documented and presented as a solid factual return on the invested capital funding is not necessarily appropriated. A successful management system must have top leadership and an established culture to support any risk reduction initiative. (Haight, Yorio, Rost & Willmer 2014) This is very true as it applies to the establishment of an effective machine guarding program. Success of the program cannot be measured on a "return on investment" basis, but rather the continuous improvement of hazard controls (in consideration of the hierarchy of controls).

I have been consulting industry on machine guarding for 30 years. I have only seen a handful of enlightened safety professionals and corporate level managers who have adopted the strategy I have presented above. Successful implementation will result in preventing a fatality that would have happened ten years from now. How do you put a dollar figure on that? For these few corporations the process has been a journey. It's taken years to develop the company policy/procedures, which are living documents under constant revision. For one aggressive company it took two and a half years to complete company- wide risk assessments of all equipment. They now allocate part of the annual safety budget toward equipment upgrade or replacement using a ranking system that was part of the original risk assessment. They recognize that "an elephant is eaten, one bite at a time" and now have a system of continuous improvement. The size and scope of the project for this company (as with many other companies) is huge. The process of continuous improvement will go on and it will be years, even decades before most machines will meet their written safeguarding standard. During this period it is likely they will continue to have some machine related accidents, but they will be on the right path to deal with this problem.

Bibliography

American Society of Safety Engineers (ASSE) 2012. "Managing Change in Manufacturing & Production Facilities" (retrieved February 20, 2017) (http://www.asse.org/assets/1/7/carl_harrison_article.pdf)

Haight, J.M. et al. "Safety Management Systems – Comparing Content & Impact," *Professional Safety,* May 2014.

Machine Safety 101. "Understanding the Hierarchy of Controls" (retrieved February 24, 2017) (http://machinerysafety101.com/2011/02/28/understanding-the-hierarchy-of-controls/)

"OSHA Weekly Fatality/Catastrophe Report," May 2011 (retrieved February 1, 2017) (https://www.osha.gov/dep/fatcat/fatcat_weekly_rpt_05282011.html)

"OSHA Inspection Detail," November 24, 2009 (retrieved February 1, 2017) (https://www.osha.gov/pls/imis/establishment.inspection_detail?id=313928525)

Session No.661

Pedestrian Slip Safety: A Guide for Busy Professionals

Rebecca Findlay-Debeck, BA, LLB, MAPlan, WACH
Slip Assessment Services
Sydney, New South Wales
Australia

"Whether you live in a city or a small town, and whether you drive a car, take the bus or ride a train, at some point in the day, everyone is a pedestrian."[1]

Anthony Renard Foxx

Introduction

While pedestrian safety is an issue which effects everyone, the identification and control of slip, trip and fall ('STF') risks is an often misunderstood and vexed area of safety management. This paper aims to demystify STF risks, providing a plain English introduction to key elements in pedestrian safety — contextualizing slip risk; identifying testing options; assessing sites and surface materials; and using matting as a design control.

Risk Context

Causative Elements

There are a range of environmental, design and human factors that contribute to STF risks. From a design and environmental perspective, elements which impact pedestrian safety include, but are not limited to, surface materials type, treatment and maintenance; lighting/visibility; handrail presence and design; slope; and exposure to climatic conditions. In terms of human factors, STF risks are impacted by factors such as footwear; age; gait; and levels of pedestrian attention.

---

[1] Foxx, A. 2013. Everyone is a Pedestrian August 5, 2013 (retrieved January 1, 2017) (https://www.transportation.gov/fastlane/everyone-pedestrian).

However, it is generally a combination of elements which results in actual STF injuries. So why focus on surface safety in the management of slip risk?

Surface materials are statistically significant in slip and fall causation. The United States of America (USA) National Floor Safety Institute (NFSI) estimates that 55% of all slip and falls are attributable to hazardous surface materials.[2] Similarly, in Spain — after the introduction of mandatory (barefoot) pedestrian slip safety testing for public buildings (such as swimming pool facilities) in 2014 — audit results indicated that more than 60% of surfaces failed to meet legislatively required slip resistance standards.[3] From a risk management perspective, the levels of slip risk associated with flooring materials (and their treatment) are controllable, measurable and capable of enforcement — resulting in surface safety being the central focus of pedestrian STF mitigation strategies.

Incidence of Slip, Trips and Falls

Within the United States of America, it is estimated that 80 million falls (and 9.9 million falls related injuries) occur each year, evenly distributed across adults from 18 years to 65 plus years.[4] The overall economic cost of slip and falls is estimated at $111Billion per annum, with 25,924 deaths; 693,500 hospitalizations and 5,022,536 emergency room admissions each year.[5]

For comparative purposes, the prevalence of slip and fall incidents within Westernized nations is most easily demonstrated through workplace injury statistics. In Australia, slip and fall injuries have consistently represented 20% to 23% of all *serious* workplace injuries. Since 2010, there has been a progressive rise in reported incidents: 2007–2008 being 21%; 2009–2010 being 20.9%; 2010–11 being 21.2%; 2011–2012 being 22.4%; and 2013–2014 being 23%. A comparative breakdown of the 2013–2014 statistics for workplace reported injuries is provided in Figure 1.

---

[2] National Floor Safety Institute (NFSI). 2017. *Untitled video [home page]* (retrieved March 1, 2017) (https://nfsi.org/).

[3] Frías. E, Queipo de Llano. J. 2016. "Spanish Pre-regulatory Research in Slip Resistance". Paper presented at: *Developments in Slip Resistance Part II. Proceedings of the inaugural Slips, Trips and Falls 2016: The Way Forward* conference; London, United Kingdom.

[4] 32.3% of injuries occur between ages 18–44; 35.3 % between ages 45–65; and 32.3% for adult 65 years and older. See: Liberty Mutual Institute for Safety. 2016. *From Research to Reality: Fall Injuries Inevitable or Preventable* (retrieved March 1, 2017) (https://www.libertymutualgroup.com/about-liberty-mutual-site/research-institute-site/Documents/FallInjuries.pdf).

[5] Ibid.

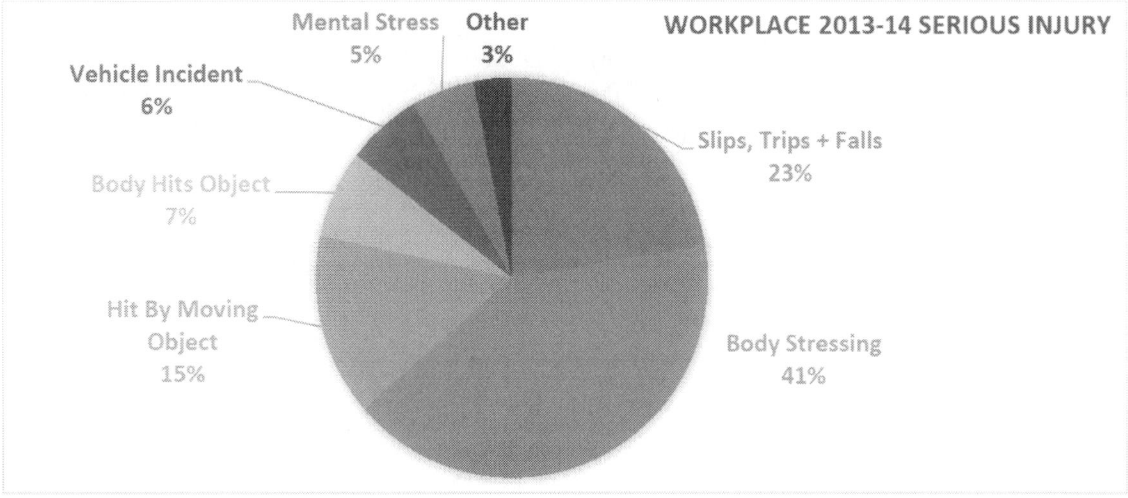

Figure 1. This chart provides a comparative breakdown of all serious Australian workplace injuries reported in FY2013-2014.

Similarly, in the United States of America, slip and falls currently (2015) represent 27.9% of all *serious* workplace injuries and claims, being 309,060 cases. While this demonstrates a reduction in total workplace slip and fall injuries across both government and private sectors since 2014 (29.3%), within Local Government, STF still comprise a staggering 51.1% of all workplace injuries (up from 49.9% in 2014).[6] See Figure 2 for a breakdown of the 2015 USA serious workplace injury statistics by causative event.

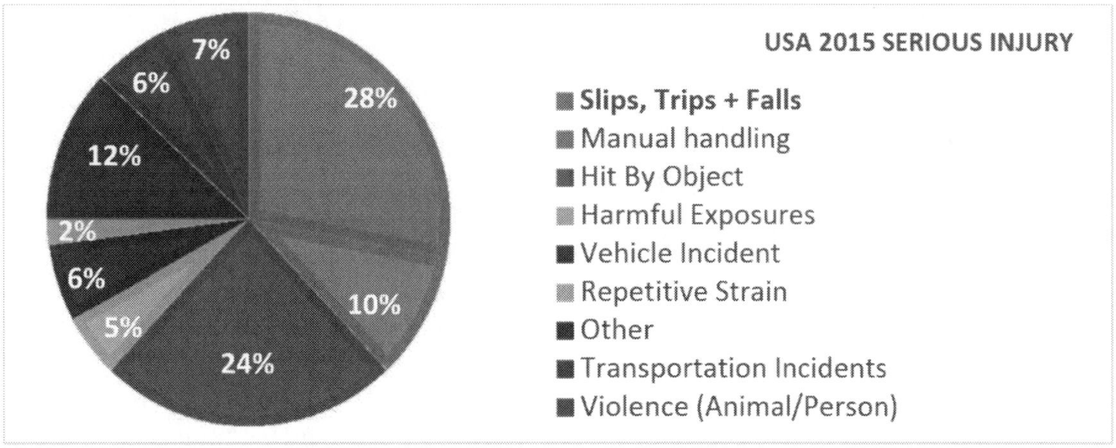

Figure 2. This chart provides a comparative breakdown of all serious USA workplace injuries reported in 2015.

---

[6] Bureau of Labor Statistics (BLS). 2017. *News Release: Nonfatal Occupational Injuries and Illnesses Requiring Days Away From Work, 2015* November 10, 2016 (retrieved January 1, 2017) (https://www.bls.gov/news.release/pdf/osh2.pdf).

In terms of history, there has been relative consistency in the number of (non-fatal) private industry slip and fall incidences in the United States of America: 2011 being 229,630 cases; 2012 being 223,700 cases; 2013 being 247,120 cases; and 2015 being 238,610 cases.[7] In 2013, slips, trips and falls represented 16% of all workplace fatalities (734 persons). In 2014, this rose to 17% with 820 deaths but reduced to 16.5% in 2015, resulting in 800 deaths. Fatal injury distribution rates are reported as being generally equivalent across gender.[8]

Risk Management Drivers

Commercially, the management of pedestrian STF risks is driven by three main motivations:

- Productivity
- Legislation
- Litigation

Productivity Drivers

Productivity based considerations are both internally and externally driven. The cost of STF events on workplace performance is primarily witnessed through staff injury impacts (actual days lost; alternate duties; skilled worker replacement etc.); insurance claims cost and premiums; and workers compensation payments. Secondary productivity based impacts can include corporate reputational damage; negative brand association; product performance or output reductions (market consequences); and staff attrition (with succession planning impacts).

The cost to industry of STF incidents in the United States of America is estimated at $18 Billion per annum. The *2016 Liberty Mutual Workplace Safety Index* (reporting on 2013 statistics) indicates that 16.4% or $10.17 billion of injury costs to business resulted from falls on the same level; 8.7% or $5.40 billion related to falls to lower levels; and 3.8% or $2.35 billion concerned slips and trips without falls.[9]

Legislative Drivers

Legislative motivation for proactive STF risk management largely arises from:

- workplace safety laws (including fines and prosecutions); and
- building code requirements and laws

OHS Laws

Legislation governing workplace health and safety is found within most Westernized legal systems. In Australia, the *Workplace Health and Safety (National Uniform Legislation)* Act 2011 (WHS Act) and its subordinate Regulations was enacted at the Federal level with mandatory

---

[7] Bureau of Labor Statistics (BLS). 2016. *Databases, Tables & Calculators by Subject (Falls, Slips, Trips)* (retrieved March 1, 2017) (https://data.bls.gov/timeseries/CSU00X4XXXXX6E100).

[8] Bureau of Labor Statistics (BLS). 2016. *Distribution of fatal injury events by gender of worker, 2014* (retrieved March 1, 2017) (https://www.bls.gov/iif/oshwc/cfoi/cfch0013.pdf).

[9] Liberty Mutual Institute for Safety. 2016. *2016 Liberty Mutual Workplace Safety Index* (retrieved March 1, 2017) (https://www.libertymutualgroup.com/about-liberty-mutual-site/research-institute-site/Documents/2016%20WSI.pdf).

adoption by all states.[10] The legislation imports a positive obligation to ensure the provision and maintenance of safe work environments, which extends to health and safety education, training and the regular monitoring of workplace conditions and risks (WHS Act s.19).

The responsibility to ensure safe workplaces in Australia is qualified by the 'reasonable practicality' of hazard reduction, determination of which requires the balancing of: the likelihood of risk, the degree of probable harm, and knowledge or expected knowledge of the risk; with the methods, cost and availability of risk elimination or minimization options (WHA Act s.18). Through workplace incident statistics and related worker's compensation claims, the likelihood of occupational slip and falls; the consequences of injury; and expected employer awareness are readily established. Given the range of remedial options available for STF risk reduction — the cost of which is substantially lower than litigation, damages and insurance claims/workers compensation — the defense of 'impracticality' for pedestrian safety management at work cannot be established.

In the United States of America, under the federally enacted *Occupational Safety and Health* Act 1970 (OSH Act) employers are responsible for providing safe workplaces, that is, workplaces that do not pose or present serious hazards to workers.[11] Employers are also obliged to follow Occupational Safety and Health Administration (OSHA) health standards, and to remediate safety problems.[12]

In terms of surface safety, reference is generally made to *Standards 1910.21 Walking-Working Surfaces* which requires employers to ensure clean walking-working surfaces (1910.22(a)(1)); safe means of access and egress (1910.22(c)); regular surface inspection and maintenance (1910.22(d)(1)); and the repair or correction of hazardous conditions, supervised by a qualified person (1910.22(d)(2); (d)(3)).[13] However, the general duties provisions of section 5(1) of the OSH Act would seem equally applicable to managing STF risk:

> "(a) Each employer …
>
> > (1) shall furnish to each of his employee's employment and a place of employment which are free from recognized hazards that are causing or are likely to cause death or serious physical harm to his employees…".[14]

It is established that surface safety is a 'recognized hazard' and, as demonstrated by workplace injury statistics, STF injuries are known to cause both death and serious physical harm

---

[10] This legislation replaced various state-based *Occupational Health and Safety* Acts. Its scope includes all workers (including volunteers) and all workplaces.

[11] Employees who work for state and local governments are not covered by Federal OSHA, but have OSH Act protections if they work in states that have an OSHA-approved state plan. See: Occupational Safety and Health Administration (OSHA). 2016. *Workers Rights* (retrieved March 1, 2017) (https://www.osha.gov/Publications/osha3021.pdf).

[12] Occupational Safety and Health Administration (OSHA). 2006. *All About OSHA* (retrieved March 1, 2017) (https://www.osha.gov/Publications/all_about_OSHA.pdf).

[13] Occupational Safety and Health Administration (OSHA). 2006. *OSH Act of 1970* (retrieved March 1, 2017) (https://www.osha.gov/pls/oshaweb/owasrch.search_form?p_doc_type=OSHACT).

[14] Ibid.

to employees. In this regard, section 5(1) of the OSH Act could be read as importing a strict and pro-active responsibility on employers to effectively monitor and manage pedestrian slip safety in their workplaces.

Prosecution of Australian employers for occupational STF injuries has generally been in response to high profile incidents — for example, hot oil related slips at McDonalds Warragul, Victoria ($65,000.00)[15] and Hungry Jacks Adelaide, South Australia ($90,000.00)[16]. The United Kingdom has been more proactive in regularly prosecuting and fining employers for workplace STF injuries, investigating issues ranging from inadequate matting to cleaning systems.[17] In 2016, the United Kingdom also introduced revised sentencing guidelines which increased workplace safety fines to between £2million and £50million pounds (company size scaled) and removed company director protections (now personally liable).[18] Since the Guidelines' introduction, there has been a three-fold increase in workplace safety fines (with a 102% average increase in penalty levels applied).[19]

Building Codes & Laws

In terms of Building Codes, Australia requires slip compliance certification for all buildings containing stairs, stair landings and ramps as part of the *National Construction Code* 2015 (Cth) (incorporated into the *Building Code of Australia*).[20] This legislative approach is prescriptive and mandatory, with quantifiable safety thresholds required to be met before building occupation is

---

[15] Mickelburough, P. 2016. *Herald Sun: Camera captures moment McDonald's worker falls into bucket of boiling oil* March 4, 2016 (retrieved March 1, 2017) (http://www.heraldsun.com.au/news/camera-captures-moment-mcdonalds-worker-falls-into-bucket-of-boiling-oil/news-story/03aad16b2db72484368ee5db507863c1).

Naegelen, J. 2016. *Nine News: McDonald's owner fined $65k after worker suffers third-degree burns* March, 4, 2016 (retrieved March 1, 2017) (http://www.msn.com/en-au/news/australia/mcdonalds-owner-fined-dollar65k-after-worker-suffers-third-degree-burns/ar-BBql3Rc).

[16] $45,0000.00 fine and $20,000.00 legal costs: Littlely, B., Fewster, S. 2015. *The Advertiser: A Industrial Court fines Hungry Jack's $90,000 for employee's slip and fall into vat of boiling oil* January 30, 2015 (retrieved March 1, 2017) (http://www.adelaidenow.com.au/news/south-australia/sa-industrial-court-fines-hungry-jacks-90000-for-employees-slip-and-fall-into-vat-of-boiling-oil/news-story/a6c4e6aaf33d055e396b3af8aed7a613).

[17] Health and Safety Executive (HSE) (UK). 2017. *Prosecutions* (retrieved March 1, 2017) (http://news.hse.gov.uk/category/about-hse/prosecution).

[18] Warburton, C. 2015. *Health and Safety at Work: New court guidance could see safety penalties top £10 million* November 3, 2015 (retrieved March 1, 2017) (https://www.healthandsafetyatwork.com/regulation/sentencing-guideline-2015).

[19] Knutt, E. 2017. *Health and Safety at Work: Total UK safety fines triple in first year under new sentencing guideline* January 30, 2017 (retrieved March 1, 2017) (https://www.healthandsafetyatwork.com/regulation/sentencing-guideline-one-year-on-iosh-osborne-clarke).

[20] Australian Building Codes Board (ABCB). 2015. *National Construction Code* (retrieved March 1, 2017) (https://www.abcb.gov.au/Resources/Publications/NCC/NCC-2015-Complete-Series_).

permitted. Slip safety assessments of flooring surfaces (beyond stairs, ramps and landings) can also be conditioned under State based planning legislation, such as the *Environmental Planning & Assessment Act* 2000 (NSW), compliance with which is a pre-requisite for occupation certification.[21]

In the Unites States of America, the *International Building Code* 2009 and state based ordinances also provide legislative incentive to manage STF risks. Although 'strict' ('*surfaces shall be slip resistant*')[22], this requirement is generalized and non-prescriptive, and may not be enforceable unless provisions are expressly adopted by state or local statute.[23]

Slip Safety Standards
While there is a range of technical standards that relate to the conduct of slip tests in the United States of America, there is no legislative *requirement* to undertake pedestrian safety testing. Notwithstanding this, as the slip resistance of surface materials cannot be accurately determined without testing, it is difficult to establish legislative compliance in its absence.

Litigation Drivers

Perhaps the strongest incentive of for STF risk minimization arises from litigation. Despite legislation limiting civil claims in New South Wales, Australia,[24] this state has seen increasing slip and fall litigation — with damages typically ranging from $200,000.00 to $400,000.00, although several judgements exceed $1million.[25] In the United States of America, damages for STF injuries are substantially higher, often resulting in publicly reported multi-million dollar settlements.[26]

There have been some significant legal decisions in Australia that have impacted industry STF surface management procedures and cleaning practices. In *Strong v Woolworths*, the Court

[21] New South Wales Government. 2017. *NSW Legislation* (retrieved March 1, 2017) (http://www.legislation.nsw.gov.au/#/view/act/1979/203).

[22] The International Building Code (IBC) is a model building code developed by the International Code Council (ICC), in use or adopted in 50 states, the District of Columbia, Guam, Northern Marianas Islands, NYC, the U.S. Virgin Islands and Puerto Rico. See: International Code Council (ICC), 2009. *International Building Code* (retrieved March 1, 2017) (http://codes.iccsafe.org/app/book/toc/2015/I-Codes/2015%20IBC%20HTML/; http://www.iccsafe.org/international-code-adoptions/).

[23] *Mary Todd v Rent-a-Center and Midwest Inc* (2011) January 25, 2011, State of Michigan Court of Appeals V No. 295370 Kalamazoo Circuit Court (retrieved March 1, 2017) (http://publicdocs.courts.mi.gov:81/opinions/final/coa/20110125_C295370_36_295370.OPN.PDF).

[24] *Civil Liability Act* 2002 (NSW). See: New South Wales Government. 2017. *NSW Legislation* (retrieved March 1, 2017) (http://www.legislation.nsw.gov.au/#/view/act/2002/22).

[25] See, for example: Malouf & Partners Law Firm. 2017. *Case Summaries* (retrieved March 1, 2017) (https://www.gerardmaloufpartners.com.au/CaseSummaries.aspx).

[26] For searchable examples, see: *Jury Verdict Alert* (accessed March 1, 2017) (http://www.juryverdictalert.com/).

determined that "...*reasonable care required the inspection and removal of slipping hazards at intervals not greater than 20 minutes*".[27] Likewise, in *Irena Alat v Franklins Pty Ltd*, the Court found that 16 minute cleaning rotations were insufficient for reducing slip risks in supermarket chilled food aisles.[28] The scope of responsibility for managing STF risks has also been broadened. In *Indigo Mist Pty Limited v Palmer*, the Court held an architect partly responsible (contributory negligence) in a nightclub slip and fall claim due to their failure to regard pedestrian slip safety in designing the premises.[29]

Similarly, in the United States of America, litigation has helped define acceptable industry practice. For example, in relation to a 2008 slip and fall in Starbucks, San Diego (Melrose Drive, Vista), the jury found that floor signage displayed by the coffee house after wet floor mopping was inadequate.[30] This case confirmed that 'bookending' wet floor areas with 'warning cones' or signage was standard industry practice.

When to Test?

Testing Approaches

There are four key stages within building/project development where pedestrian slip safety testing is recommended:

1. Design or flooring section stage
2. Pre-installation (sample or prototype testing)
3. Installation
4. Occupation

Design

Slip safety testing at the design stage assists in the selection of appropriate flooring materials to achieve statutory compliance, or desired safety standards, at a site. In simplest terms, testing determines the 'fitness for purpose' of the flooring materials/product. Established suppliers will often provide slip certification as part of marketing or supply agreements, so the availability of slip testing documentation should be investigated at the design phase to inform decision-making.

---

[27] *Strong v Woolworths Limited* [2012] HCA 5, March 7, 2012, Case Number: S172/2011 at 38 (retrieved March 1, 2017) (http://eresources.hcourt.gov.au/showCase/2012/HCA/5).

[28] *Irena Alat v Franklins Pty Ltd* [2012] NSWDC 104 (retrieved March 1, 2017) (http://www.austlii.edu.au/au/cases/nsw/NSWDC/2012/104.html).

[29] *Indigo Mist Pty Limited v Palmer* [2012] NSWCA 239, August 9, 2012 (retrieved March 1, 2017) (https://www.caselaw.nsw.gov.au/decision/54a638583004de94513d9d84).

[30] The plaintiff was awarded $7.5 million in damages for permanent brain injury: *Anthony Zaccaglin, et al v. Starbucks Coffee Company*, San Diego Superior/37-2009-00084460-CU-PO-NC judgement delivered December 22, 2011 (retrieved March 1, 2017) (http://www.juryverdictalert.com/jury-verdicts/item/premises-liability-2/slip-and-fall/zaccaglin-v-starbucks-slip-and-fall-in-starbucks).

Pre-Installation
With increasing globalization in materials supply, it is essential to verify that the product purchased is, in fact, the product delivered. The rise of low-cost, inferior grade building materials (including counterfeit or fake materials), is having serious impact on construction developments internationally.[31] Variations in flooring materials, especially tiling, can be subtle and significant —rarely assessable by touch or vision alone. For large scale developments, pre-installation testing provides an inexpensive means of confirming product quality, allowing issues to be resolved without delays in project completion. Prototype testing also reduces consequent costs, materials wastage and product warranty disputes that can arise with materials or product defect issues.

Installation
Slip safety testing is often required at the installation stage to obtain final building/occupation certification. Surface treatments (such as sealants) are frequently applied to flooring materials upon installation, altering original product safety certifications. Testing at the installation stage (prior to pedestrian use) is important to ensure that materials/products have achieved expected slip resistance. Once flooring is in use, supplier warranties can be voided through subsequent flooring treatment (whether surface sealants or chemical cleaners). Slip testing at the installation stage can off-set later claims regarding surface materials quality, and establish baseline safety compliance for building insurance coverage and public liability.

Occupation
The regular testing of pedestrian surfaces is recommended once flooring is in use. Testing allows for the 'lifespan' of materials, or rate of product degradation, to be more accurately determined, aiding maintenance and capital works/asset management planning. Importantly, annual or bi-annual slip testing creates the flooring conditions history essential for effective claims management and the defense of any litigation. Proactive slip testing also satisfies safety monitoring responsibilities under workplace health and safety legislation, and is fundamental for STF risk control.

Testing Options

When planning slip safety testing, it is important to first establish which testing method(s) and device(s) are favored by industry, insurers and the Courts within your jurisdiction. Once established, qualified and experienced testers should be engaged [32], and equipment calibration or

[31] See, for example: Morris, M. 2015. *ABC News: Imported construction materials putting buildings, lives at risk* 25 June 2015 (retrieved March 1, 2017) (http://www.abc.net.au/news/2015-06-24/imported-construction-materials-put-lives-at-risk:-specialists/6571398).

Wambui, M. 2016. *Kenyan Builders Grapple with Fake Products from Asia* 8 February 2016 (retrieved March 1, 2017) (http://www.kenyahomesguide.com/1347/kenya-fake-building-products/).

[32] ANSI/NFSI B101.0-2012 provides guidance on 'qualifications'. See: American National Standards Institute, Inc (ANSI). 2012. *American National Standard: B101.0 Walkway Surface*

currency information obtained (if relevant). The frequency of testing will depend upon pedestrian use; surface conditions and treatment (including anticipated rate of product degradation); and corporate risk appetite.

Testing Devices

Pedestrian slip safety testing devices ('tribometers') are multiple and varied. They can test surface slip safety 'dynamically' or 'statically'; be self-automated or tester controlled/propelled; be portable or laboratory dependent. However, it is important to recognize that different jurisdictions have established requirements and preferences in terms of testing devices which should be acknowledged or observed.[33] As research has consistently demonstrated that the devices used for measuring surface safety do not produce consistent, compatible or, often, repeatable results — being strongly dependent upon device operators, as well flooring materials, treatment, contaminants and conditions — following established (industry supported) testing methods and guidelines is recommended.

Dry Coefficient of Friction

Dry coefficient of friction ('dry CoF') testing measures the level of surface contaminants — such as dust or chemical treatment (cleaning product) residue — in smooth, dry flooring conditions. This test is often used to establish the efficacy of cleaning and maintenance regimes, and is a useful performance monitoring tool for service contracts. Dry CoF testing is done with portable devices that are generally used only on internal, dry surfaces. This test is not considered suitable for highly profiled surfaces, carpets, wet or outdoor areas. Testing results are registered as being above or below an established dry CoF slip resistance threshold. In Australia, this threshold is: <0.40 CoF or >0.40 CoF.

It is important to highlight that dry CoF does not equate to, or relate to, wet coefficient of friction ('wet CoF'). A flooring surface may have effective slip resistance under wet conditions, but not under dry (and *vice versa*). Wet CoF and dry CoF testing methods are complimentary, and when conducted in tandem, provide a comprehensive slip safety profile of a flooring surface.

Wet Coefficient of Friction Testing

Wet CoF testing measures either the static or dynamic CoF of flooring surfaces under wet conditions (indicating 'slip' and 'skid' risk). It is estimated that more than 90% of falls occur on wet (or food contaminated) flooring surfaces.[34] Wet CoF testing devices are portable; can be used both externally and internally; either on *in-situ* surfaces or on non-fixed materials (sample form).

Auditing Procedure for the Measurement of Walkway Slip Resistance, Southlake, Texas: National Floor Safety Institute, 2012.

[33] In Australia, permissible tribometers are legislatively mandated. In the United States of America, tribometer specifications are regulated, with specific devices subject to industry approval processes: National Floor Safety Institute (NFSI). 2017. *NFSI Approved Tribometers* (retrieved March 1, 2017) (https://nfsi.org/ansinfsi-standards/nfsi-approved-tribometers/).

[34] Health and Safety Executive (HSE) (UK). 2017. *Slips on Wet and Contaminated Floors* (retrieved March 1, 2017) (http://www.hse.gov.uk/food/slips.htm).

The relevant standards for wet CoF testing in the United States of America are ANSI/NFSI B101.1-2009 (static CoF) and ANSI/NFSI B101.3-2012 (dynamic CoF). These standards relate to both laboratory and on-site testing, outlining approved methods for determining the wet CoF of hard flooring surfaces such as tiling, concrete, synthetic and wooden materials.[35] These standards define acceptable tribometers; provide testing procedures; articulate slip resistance bandwidths; and specify result reporting requirements.

One of the most common methods of wet CoF testing is (dynamic) Wet Pendulum Resistance testing.[36] This is the preferred method of testing internationally[37], and in the United States of America, is governed by ASTM E303–93(2013) *Standard Test Method for Measuring Surface Frictional Properties Using the British Pendulum Tester*[38]. These testing devices utilize a pendulum (with a specified rubber footer) swung to replicate pedestrian heel strike on a wetted surface. The resistance achieved correlates to established bandwidth readings, indicating slip potential. Testing results are generally recorded as numerical values followed by either BPN (British Pendulum Number), PTV (Pendulum Test Value) or SRV (Slip Resistance Value). Wet pendulum testing devices can be used to determine both pedestrian and vehicular slip and skid resistance.

Surface Microroughness

Surface microroughness testing is used to indicate the slipperiness of a surface under wet conditions. It measures the roughness of flooring materials, and can be used to monitor surface wear. It is unsuitable for use on carpets or very rough surfaces. The results are recorded within Rz parameters (generally above or below 10-20 µm), being the correlation of various 'peak to valley' surface measurements. Devices used in this testing method are considered inexpensive, quick and highly portable, although microroughness testing is generally recommended in supplement to wet CoF testing. This testing method is primarily used in the United Kingdom, and is not commonly used in Australia.[39]

Oil and Barefoot Wet Ramp Testing

Ramp testing methods were developed in Germany and are laboratory based. There are two main forms of ramp tests: 'oil' and 'barefoot wet' (for swimming pools, showers, change rooms surfaces). Ramp methods are considered most effective in testing gratings, heavily profiled surfaces and surfaces designed to trap contaminants.

The oil ramp test (commonly known as DIN 51130) uses an aerially harnessed operator wearing heavily-cleated safety boots walking on a ramp treated with (motor) oil. As the ramp is

---

[35] ANSI/NFSI B101.1-2009 expressly excludes carpet and highly polished materials.

[36] This device is also known as: pendulum co-efficient of friction tester; portable skid resistance tester; the British pendulum; and the TRRL pendulum.

[37] Wet Pendulum Resistance testing is used in more than 49 countries worldwide.

[38] American Society for Testing and Materials (ASTM) International. 2017. ASTM E303-93 (2013) *Standard Test Method for Measuring Surface Frictional Properties Using the British Pendulum Tester* (retrieved March 1, 2017) (https://www.astm.org/Standards/E303.htm).

[39] Health and Safety Executive (HSE) (UK). 2017. *Surface Roughness* (retrieved March 1, 2017) (http://www.hse.gov.uk/slips/sat/faqsurf.htm).

mechanically inclined, the operator registers a result as they start to slip on the surface. Similarly, in the wet barefoot test (commonly known as DIN 51097), an aerially harnessed operator wearing no shoes walks on a ramp treated with soap solution. As the ramp is mechanically inclined, the operator registers a result as they start to slip on the surface. The results achieved correlate to a pre-established slip result range, generally identified as R9 to R13.

It is worth noting that ramp testing has been treated cautiously in some countries — with claims that the contaminants used are not representative of actual surface conditions, and that results reporting processes are highly tester dependent and non-replicable (subjective).[40]

Slope Adjustment

It is central to remember that *all* slip safety testing results (even for 'ramp' based testing methods) are given at a level surface reading (generally taken as being <1.5° of slope). Where the materials or products tested are to be applied to inclined surfaces or ramps, the testing results achieved will need adjustment. It is the *adjusted* value that becomes the relevant 'result' for safety compliance assessment. The Australian slip safety standards provide conversion equations and tables for slope correction which may be instructive.[41]

Site, Product & Materials Assessment

Flooring Product & Materials Certifications

When assessing product slip certifications, the compliance standards quoted should correspond to those applicable in your jurisdiction. This is readily determined by reference to the stated testing method, or if the testing method is unspecified, to actual testing results. Most testing results have inherent indicators for the method used. For example, a R9 to R13 rating identifies 'ramp testing' (although whether testing was 'oil ramp' or 'barefoot wet ramp' would still need to be established).

To determine product suitability, testing results need to be evaluated in terms of the characterization of the site. It is important to note that slip testing methods are not comparable (nor readily convertible) to each other. If the relevant standards in your jurisdiction requires a SRV rating, then an R rating would be not legally acceptable. Notwithstanding this, some countries have provided equivalence guides to assist in understanding slip safety testing results. For example, the British Health Service Executive (HSE) has provided a (qualified) approximation of the German DIN51097 and DIN51130 ramp testing methods (R rating results)

[40] Health and Safety Executive (HSE) (UK). 2012. *Assessing the slip resistance of flooring: A technical information sheet* (retrieved March 1, 2017) (http://www.hse.gov.uk/pubns/geis2.pdf).

[41] Standards Australia. 2013. *AS4586-2013, Slip Classification of New Pedestrian Surfaces; AS4663–2013 Slip Resistance Measurement of Existing Pedestrian Surfaces*, Sydney: SAI Global Ltd, 2013.

to wet pendulum CoF results (BPN rating results).[42] Whether a converted result would be acceptable in your jurisdiction should be the subject of independent legal advice.

Fundamental to any assessment of supplier provided certification is establishing the independence of the testing (and tester) from the materials/product producer. The use of independent testing/testers reduces any perceived conflict of interest (or bias) in the event of STF claims or litigation, providing bi-partisan confidence in the reliability of results and shortening evidential hearing times. Certified and experienced testers or testing laboratories should always be preferred.

Characterizing Sites

While most slip standards specify *minimum* levels of acceptable safety, current industry practice and/or industry best practice may mandate a higher level of surface safety, particularly when use or user-related risks are assessed. In this context, accurately characterizing a site is central to effectively managing STF risks.

In Australia, the classification of surface safety is dependent upon both the flooring materials and the flooring use. So, for example, tiling that may be acceptable for use in an Australian office (P1 required) may not be acceptable for use in a bathroom (P3 required).

There are two main models in characterizing flooring use:

1. Primary purpose
2. Highest level of risk

The 'primary purpose' approach considers the primary or intended use of the surface. The 'highest level of risk' approach focuses on actual use or subsequent conditions which present a higher risk to pedestrians (not necessarily corresponding with the intended use of the surface). The latter assessment is particularly relevant where a flooring material has been selected for 'primary or intended use', but is unsuitable for 'actual use' conditions.

To demonstrate this, Exhibit 1 shows a pedestrian walkway in the central business district of Sydney, New South Wales, Australia. While the 'primary purpose' of the walkway is as a pedestrian thoroughfare between Martin Place and Angel Place (actively used pedestrian locations), the subsequent parking of motorcycles within the walkway has significantly altered the level of pedestrian risk. The adequacy of the flooring materials is now brought into question with (design) unforeseen impacts from oils, petrochemicals, vehicle weight and movement on the surface.

---

[42] Health and Safety Executive (HSE) (UK). 2012. *Ramp Tests* (retrieved March 1, 2017) (ww.hse.gov.uk/slips/STEP/education/advanced/8E7F777B-3B84-49FE-A3D6-D0324E25A801/HSLCourseTemplate/28531/slidetype1_130599.htm).

Exhibit 1. Pedestrian walkway between Martin Place and Angel Place, Sydney, Australia.

Exhibit 2. Commercial lobby in Parramatta, New South Wales, Australia which addressed new slip risks (associated with the introduction of a lobby café) through matting.

Some hybrid flooring characterizations develop from market changes and design trends. The introduction of coffee bars and café annexes within commercial building lobbies is one such use transition. Flooring materials selected at building development do not always anticipate subsequent impacts from operational changes — for example, food and beverage services introduce new and increased flooring contaminants, as well as an intensification of surface use. Exhibit 2 shows a commercial lobby in Australia which introduced a café with patron seating in its entry area. Slip testing indicated that the surface was now non-compliant with pedestrian safety standards. In this instance, pedestrian STF risk was managed through the introduction of extensive matting.

Similarly, the installation of design features such as vertical gardens and 'green walls' (Exhibit 3) within buildings can also see unanticipated changes in surface safety, particularly though leaf litter and water spillage (if not pro-actively managed). Whether the flooring material installed is suitable for design change impacts requires confirmation through testing.

Exhibit 3. Examples of corporate building 'green walls' in Sydney, New South Wales, Australia.

Exhibit 4. Examples of multiple 'surfaces' (for the purpose of slip safety assessment) within the same flooring areas.

Surface Identification

In addition to site characterization, accurately identifying the number of surfaces at a site is central to pedestrian slip risk management. Variation in flooring materials— including patterning, design or break tiling — can create a different surface for the purpose of safety assessment (Exhibit 4). Similarly, areas of the same flooring which have significant wear or variation in surface finish (e.g. matt or gloss) may also constitute a 'new surface', requiring separate testing to determine overall flooring slip resistance.

Design Controls

There are a range of design controls available to remediate or reduce pedestrian STF risks, varying in cost, permanence and effectiveness. However, floor matting remains one of the most popular, accessible and inexpensive means of addressing STF risks in building management.

Matting

Matting is generally accepted as an effective design control for capturing ingress water and pedestrian traversed surface contaminants within building safety management. It serves the three-fold purpose of:

1. reducing slip and fall risk through the absorption of water and capture of external contaminations/debris;
2. protecting flooring from external impacts, extending the lifespan of surface materials; and
3. lowering operational costs through reduced cleaning and maintenance demands.

In 2012, the United States of America released standards for the selection, installation and maintenance of entrance matting in commercial facilities: *ANSI/NFSI B101.6-2012 Standard*

Guide For Commercial Entrance Matting In Reducing Slips, Trips And Falls.[43] While this standard provides comprehensive guidance on mat selection considerations, the specifications are, primarily, performance based (non-prescriptive). Key considerations such as entrance matting minimum lengths and widths are not specified.

The use of entrance matting is supported under Australian standards for pedestrian safety, however no guidance is provided on the types or size of matting adequate to reduce STF risks.[44] Instead, the British standards, *BS7953:1993 Entrance flooring systems. Selection, installation and maintenance* and *BS8300:2009+ A1:2010: Design of buildings and their approaches to meet the needs of disabled people – Code of Practice* inform Australian matting selection.

The British standards state that *"...the ingress of soil and surface moisture to buildings...should be reduced to the lowest practicable levels, e.g. through the use of appropriate entrance flooring systems" (BS 7953).* BS7953 specifies the minimum width for entrance matting as 2.1 meters in the direction of pedestrian traffic (accommodating 26″ wheelchair wheel dimensions). Industry specialists suggest that matting between 4 to 6 meters in the direction of traffic is most effective, with up to 94% of surface contaminants removed at 6 meters of length.[45]

To appropriately reduce STF risk, matting should accommodate the design, location and use of the premises, as well as adjoining land features (geography based impacts). As contaminant types vary — grease, water, soil, grit, sand, dust, debris — matting should be tailored to address dominant and known risks. The volume and type of pedestrian traffic is a key consideration, including pedestrian configurations such as the ratio of visitors to regular users of a site (effecting familiarity with flooring conditions). Matting materials — natural or synthetic fibers, pile density and thickness — will depend upon site assessment.

Functionally, matting needs to be capable of:

- removing contaminants (such as through fiber-based abrasion or embedded grid scrapers);
- retaining contaminants (limiting spread to internal flooring surfaces);
- releasing contaminants (ease of mat cleaning); and
- recovering (matting fibers need to return to an 'as new' state post cleaning).

Other considerations may include relevant compliance requirements, fire resistance/retardation, electrostatic potential or matting anti-bacterial properties. Amenity based criteria may have regard to sound absorption, visibility and integration with interior design. In terms of slip resistance, mat specifications often factor load and impact resistance, as well as moveability or migration risks (balancing mat stability with ease of maintenance). Necessarily, the use of matting to control

---

[43] American National Standards Institute, Inc (ANSI). 2012. *American National Standard: B101.6 Guide For Commercial Entrance Matting In Reducing Slips, Trips And Falls*, Southlake, Texas: National Floor Safety Institute, 2012.

[44] 'Transitional areas' design controls. See: Standards Australia. 2013. *AS4586-2013, Slip Classification of New Pedestrian Surfaces*, Sydney: SAI Global Ltd, 2013.

[45] Rodgers, P. 2016. "Effective Entrance Flooring Systems". Paper presented at: Developments in Slip Resistance Part I. Proceedings of the inaugural *Slips, Trips and Falls 2016: The Way Forward* conference; London, United Kingdom

STF risks needs to be monitored to determine effectiveness and wear, especially avoiding rippling, buckling, curling and fraying, which can create additional trip risks.

Conclusion

The conference presentation concludes with a case study on two integrated pedestrian/road developments in Adelaide, South Australia. The management of STF risk in these projects serves as a cautionary tale on the consequences and cost of failing to address slip safety in large-scale public works.

Bibliography

American National Standards Institute, Inc (ANSI). 2012. *American National Standard: B101.6 Guide For Commercial Entrance Matting In Reducing Slips, Trips And Falls*, Southlake, Texas: National Floor Safety Institute.

______. 2012. *American National Standard: B101.3 Test Method for Measuring Wet DCOF of Common Hard-Surface Floor Materials*, Southlake, Texas: National Floor Safety Institute..

______.2009. *American National Standard: B101.1 Test Method for Measuring Wet SCOF of Common Hard-Surface Floor Materials*, Southlake, Texas: National Floor Safety Institute.

______.2012. *American National Standard: B101.0 Walkway Surface Auditing Procedure for the Measurement of Walkway Slip Resistance*, Southlake, Texas: National Floor Safety Institute..

American Society for Testing and Materials (ASTM) International. 2017. ASTM E303-93(2013) *Standard Test Method for Measuring Surface Frictional Properties Using the British Pendulum Tester* (retrieved March 1, 2017) (https://www.astm.org/Standards/E303.htm).

Anthony Zaccaglin, et al v. Starbucks Coffee Company, San Diego Superior/37-2009-00084460-CU-PO-NC judgement delivered December 22, 2011 (retrieved March 1, 2017) (http://www.juryverdictalert.com/jury-verdicts/item/premises-liability-2/slip-and-fall/zaccaglin-v-starbucks-slip-and-fall-in-starbucks).

Australian Building Codes Board. 2015. *National Construction Code* (retrieved March 1, 2017) (https://www.abcb.gov.au/Resources/Publications/NCC/NCC-2015-Complete-Series_).

British Standards Publications (BSI), 2010, *BS8300:2009 + A1:2010: Design of buildings and their approaches to meet the needs of disabled people – Code of Practice,* online delivery available from https://www.bsigroup.com/.

Bureau of Labor Statistics (BLS). 2016. *Databases, Tables & Calculators by Subject (Falls, Slips, Trips)* (retrieved March 1, 2017) (https://data.bls.gov/timeseries/CSU00X4XXXXX6E100).

______. 2016. *Distribution of Fatal Injury Events by Gender of Worker, 2014* (retrieved March 1, 2017) (https://www.bls.gov/iif/oshwc/cfoi/cfch0013.pdf).

______.2017. *News Release: Nonfatal Occupational Injuries and Illnesses Requiring Days Away From Work, 2015,* November 10, 2016 (retrieved January 1, 2017) (https://www.bls.gov/news.release/pdf/osh2.pdf).

Foxx, A. 2013. *Everyone is a Pedestrian,* August 5, 2013 (retrieved January 1, 2017) (https://www.transportation.gov/fastlane/everyone-pedestrian).

Frías, E. and J. Queipo de Llano. 2016. "Spanish Pre-regulatory Research in Slip Resistance". Paper presented at: Developments in Slip Resistance Part II. Proceedings of the inaugural *Slips, Trips and Falls 2016: The Way Forward* conference; London, United Kingdom.

Health and Safety Executive (HSE) (UK). 2017. *Prosecutions* (retrieved March 1, 2017) (http://news.hse.gov.uk/category/about-hse/prosecution).

______.2017. *Surface Roughness* (retrieved March 1, 2017) (http://www.hse.gov.uk/slips/sat/faqsurf.htm).

Indigo Mist Pty Limited v Palmer [2012] NSWCA 239, August 9, 2012 (retrieved March 1, 2017) (https://www.caselaw.nsw.gov.au/decision/54a638583004de94513d9d84).

International Code Council, Inc (ICC). 2009. *International Building Code*. Illinois: International Code Council Inc Publications, 2009.

Irena Alat v Franklins Pty Ltd [2012] NSWDC 104 (retrieved March 1, 2017) (http://www.austlii.edu.au/au/cases/nsw/NSWDC/2012/104.html).

Jury Verdict Alert (accessed March 1, 2017) (http://www.juryverdictalert.com/).

Knutt, E. 2017. *Health and Safety at Work: Total UK safety fines triple in first year under new sentencing guideline* January 30, 2017(retrieved March 1, 2017) (https://www.healthandsafetyatwork.com/regulation/sentencing-guideline-one-year-on-iosh-osborne-clarke).

Liberty Mutual Institute for Safety. 2016. *From Research to Reality: Fall Injuries Inevitable or Preventable* (retrieved March 1, 2017) (https://www.libertymutualgroup.com/about-liberty-mutual-site/research-institute-site/Documents/FallInjuries.pdf).

______.. 2016. *2016 Liberty Mutual Workplace Safety Index* (retrieved March 1, 2017) (https://www.libertymutualgroup.com/about-liberty-mutual-site/research-institute-site/Documents/2016%20WSI.pdf).

Littlely, B. and S. Fewster. 2015. *The Advertiser: A Industrial Court fines Hungry Jack's $90,000 for employee's slip and fall into vat of boiling oil* January 30, 2015 (retrieved March 1, 2017) (http://www.adelaidenow.com.au/news/south-australia/sa-industrial-court-fines-hungry-jacks-90000-for-employees-slip-and-fall-into-vat-of-boiling-oil/news-story/a6c4e6aaf33d055e396b3af8aed7a613).

Malouf & Partners Law Firm. 2017. *Case Summaries* (retrieved March 1, 2017) (https://www.gerardmaloufpartners.com.au/CaseSummaries.aspx).

Mary Todd v Rent-a-Center and Midwest Inc (2011) January 25, 2011, State of Michigan Court of Appeals V No. 295370 Kalamazoo Circuit Court (retrieved March 1, 2017) (http://cases.justia.com/michigan/court-of-appeals-unpublished/20110125_C295370_36_295370.OPN.PDF?ts=1396125979).

Mickelburough, P. 2016. *Herald Sun: Camera captures moment McDonald's worker falls into bucket of boiling oil* March 4, 2016 (retrieved March 1, 2017) (http://www.heraldsun.com.au/news/camera-captures-moment-mcdonalds-worker-falls-into-bucket-of-boiling-oil/news-story/03aad16b2db72484368ee5db507863c1).

Morris, M. 2015. *ABC News: Imported construction materials putting buildings, lives at risk* 25 June 2015 (retrieved March 1, 2017) (http://www.abc.net.au/news/2015-06-24/imported-construction-materials-put-lives-at-risk:-specialists/6571398).

Naegelen, J. 2016. *Nine News: McDonald's owner fined $65k after worker suffers third-degree burns* March, 4, 2016 (retrieved March 1, 2017) (http://www.msn.com/en-au/news/australia/mcdonalds-owner-fined-dollar65k-after-worker-suffers-third-degree-burns/ar-BBql3Rc).

National Floor Safety Institute. 2017. *Untitled video [home page]* (retrieved March 1, 2017) (https://nfsi.org/).

New South Wales Government. 2017. *NSW Legislation: Civil Liability Act* 2002 (retrieved March 1, 2017) (http://www.legislation.nsw.gov.au/#/view/act/2002/22).

______.2017. *NSW Legislation: Environmental Planning & Assessment Act* 1979 (retrieved March 1, 2017) (http://www.legislation.nsw.gov.au/#/view/act/1979/203).

Occupational Safety and Health Administration (OSHA). 2006. *All About OSHA* (retrieved March 1, 2017) (https://www.osha.gov/Publications/all_about_OSHA.pdf).

______.2006. *OSH Act of 1970* (retrieved March 1, 2017) (https://www.osha.gov/pls/oshaweb/owasrch.search_form?p_doc_type=OSHACT).

______.). 2016. *Workers Rights* (retrieved March 1, 2017) (https://www.osha.gov/Publications/osha3021.pdf).

Rodgers, P. 2016. "Effective Entrance Flooring Systems". Paper presented at: Developments in Slip Resistance Part I. Proceedings of the inaugural *Slips, Trips and Falls 2016: The Way Forward* conference; London, United Kingdom.

Standards Australia. 2013. *AS4586-2013, Slip Classification of New Pedestrian Surfaces.* Sydney: SAI Global Ltd., 2013.

______.2013. *AS4663–2013 Slip Resistance Measurement of Existing Pedestrian Surfaces.* Sydney: SAI Global Ltd., 2013.

______.2014. *HB198:2014 Guide to the Specification and Testing of Slip resistance of Pedestrian Surfaces.* Sydney: SAI Global Ltd., 2014.

Strong v Woolworths Limited [2012] HCA 5, March 7, 2012, Case Number: S172/2011 at 38 (retrieved March 1, 2017) (http://eresources.hcourt.gov.au/showCase/2012/HCA/5).

Wambui, M. 2016. *Kenyan Builders Grapple with Fake Products from Asia* 8 February 2016 (retrieved March 1, 2017) (http://www.kenyahomesguide.com/1347/kenya-fake-building-products/).

Warburton, C. 2015. *Health and Safety at Work: New court guidance could see safety penalties top £10 million* November 3, 2015 (retrieved March 1, 2017) (https://www.healthandsafetyatwork.com/regulation/sentencing-guideline-2015).

<u>**Session No. 662**</u>

How to Use ISO 45001 to Get H&S Risk Management on Your Business Agenda

John F Stevens, CMIOSH, MASSE, FCIPD
Assistant Administrator – RM/I
Go-Ahead Group H&S Advisor (Bus)
Hatfield, UK

Lawrence Bamber, CFIOSH, FIRM, MASSE
Risk Solutions International
Poynton, UK

Summary

This paper and the presentation will demonstrate - via The Plan, Do, Check, Act cycle (PDCA) - how to use ISO 45001 to get H&S risk management on your business agenda.

We will demonstrate how to drive enhanced business excellence by integrating H&S risk management into the overall business framework in such a way that H&S risk management becomes part of the normal business processes of the organisation and second nature i.e. BUILT IN BUT NOT BOLTED ON!

The process is one of evolution, not revolution, as the Standard provides a template that focuses on the positive aspects of effective business management, such as: Leadership; Planning; Support Mechanisms; Operational Management; Performance Evaluation and Continuous Improvement.

The key parts of the Standard will be examined, and the presentation will highlight the PDCA cycle, and the goodness of fit between H&SRM systems and an overall business excellence model. The implementation of ISO 45001 into organisations will not only improve H&SRM, but will also enhance the way organisations conduct their operations, with increasingly organisations realizing that the management of H&S risks must be at the top of their agenda Accredited businesses will have a leading edge over their competitors which in turn results in business excellence, whilst not-for-profit organisations can be reassured that the management of their H&S risks will improve their operations and reduce the risks to their workers.

Globally figures from the International Labour Organisation show there are annually:

- 2.3 million people killed by work accidents and disease
- 6,300 workplace deaths per day (one every 15 seconds)

- 313 million non-fatal work accidents
- 160 million people with occupational disease

Clearly more needs to be done to reduce worker deaths' injuries and occupational ill-health, and the clear way forward is to adopt and implement a business focussed and integrated health and safety management system. The new ISO 45001 Standard will assist the global awareness of this business focussed and integrated approach, which will support the drive towards business excellence, coupled with other key business management methods.

Increasingly, whether by a formal structured process or an evolutionary process, there is the recognition that to make a step change in the management of H&S risks requires two key actions:

1. Management at all levels must accept that they have the ultimate legal and ethical responsibility and moral imperative to manage H&S risk so they pose no harm to employees and any others who may be affected by the business operations

2. Clear and demonstrable worker engagement & involvement (an important new feature of ISO 45001) is key to establishing policies, procedures and methods of work that are sensible, practicable, safe AND can actually be adhered to.

Without the active commitment and involvement of management and workers the management of H&S risks will not be placed on the 'business agenda' and ISO 45001 at worst will not be considered or at best will only be a paper exercise.

What Does ISO 45001 Look Like?

ISO 45001 is the first worldwide H&SMS Standard, with an aim to help prevent injury & ill health in workplaces across the world, which also fits in with some of the Business Excellence goals.

This is especially relevant as the ILO data mentioned above also states that approximately 4 % World GDP is taken up by the costs of work accidents & disease.

The Standard therefore places greater emphasis on leadership; worker engagement and involvement; context; and continual improvement; in order to reduce the impact of poor OHSMS on the world.

In order to achieve this and to contribute to business excellence, the Standard requires that leaders in organisations need to be visibly & demonstrably involved in the health & safety of their workforces, and in ensuring that the OHSMS works in practice. This is vital, as leaders set the tone for the rest of the organisation.

A good and effective OHSMS is business focused and is designed to be fully integrated with business, commercial and operational processes. Leaders and staff must be fully involved in the design, creation, and implementation and monitoring; without which there will be limited buy-in and compliance.

ISO 45001 Format

As with all ISO management standards the format follows William Deming's PLAN - DO - CHECK - ACT (PDCA) concept. This provides a framework for organisations to **<u>'Plan'</u>** what they need to put in place in order to minimise the risk of harm to employees and others. The control measures should address concerns that can lead to long term health issues & absences from work (e.g. occupational stress), as well as those that give rise to injury accidents.

The Standard then requires that top management (i.e. the leaders) takes ownership, and visibly demonstrates commitment via positive, pragmatic leadership, so as to ensure that all workers have the appropriate skills & competence to use the effective control measures put into place in the **'Do'** phase. This recognises the value that worker engagement / involvement has in the implementation of effective OHSMS.

The **'Check'** phase identifies all the key elements that should be regularly addressed to ensure the system is working as designed, and provides useful feedback to management on how to mind the gaps!

The **'Act'** phase determines opportunities for continual improvement, by ensuring that all recommendations arising out of the **'Check'** phase are implemented within a finite, agreed timescale.

As with its predecessors, PDCA is a closed loop system; the more times you go round, the more your system improves in its effectiveness and reduced losses. Auditing is applied to each step - the logic being that the more times you go round the loop, then the system improves each time e.g. Continuous/ Continual Improvement! (See Figure 1 below.)

Figure1.

Standard Principles

OSH should be seen to be an integral part of the business; thereby contributing to productivity, profitability, competitiveness & sustainability – all business excellence components.

The key success factors for Standards Compliance (& Business Excellence) are:

- Good, positive leadership
- Sound business processes
- Trained & skilled workforce
- Operating in an open & trusting working environment

There are sections covering all of the above factors within the Standard

ISO 45001 v OHSAS 18001

Compared to OHSAS 18001, the new ISO 45001 includes more in respect of the following:

- Context (of the organisation)
- Leadership (management)
- Worker engagement (consultation)
- Documented information (document control)
- Risk & opportunities (identify risks AND opportunities)
- Hierarchy of control (details included as per UK HSG 65)
- Supply chain (importance of managing input and output risks)
- KPIs (use of leading and lagging KPI's)
- Compliance status (internal audit to check compliance)
- Continual improvement (always looking to learn from system performance and new initiatives)

So why is ISO 45001 an improvement on OHSAS 18001?

- Risk based approach – its adopts a risk-based approach rather than a requirements focus

- No conflict with existing standards – any potential conflict with existing standards has been avoided

- An integrated management system – it promotes an integrated management approach

- Provides a streamlined approach for all sizes of organisation – the Standard can be adopted for any size or complexity or organisation
- No transition arrangements from OHSAS18001; this means a fresh start application procedure, and a new auditing regime

- It will be a globally agreed Standard with virtually all the world's countries' National Standards bodies contributing to its development & introduction

It will gradually replace 18001, which will be withdrawn.

Main Requirements of ISO 45001

Top management /Leaders will have to take ownership & show a commitment to a positive culture of good OSH management within their organisation.

There will be a requirement to take into account the context of the organisation, in terms of its geographical, geopolitical, supply chain, & public expectations.

To create internal systems & provisions for worker engagement/ involvement including training & consultation

To identify & take opportunities for continual improvement at all levels within a positive OSH culture

Conformity to the Standard should demonstrate compliance to National Standards requirements for OHSMS, because of the risk based approach.

Detailed Content of 45001

The content & layout of the Standard follows the agreed ISO protocol as outlined in Annex SL.

The Key sections are No's 4 to 10.

4. Context of the Organisation : including organisational understanding; needs of employees & others; and scope of OSHMS

5. Leadership & Worker Participation : including commitment; OSH policy; roles and responsibilities, accountabilities & authorities; participation & consultation

6. Planning, including actions to address risks & opportunities; hazard identification; risk assessment; legal requirements; action planning; OSH objectives & their achievement

7. Support, including resources; competence; awareness; communication; documented information; creating& updating

8. Operation, including planning and control; control hierarchy; change management; procurement; outsourcing; contractors; emergency preparedness & response

9. Performance Evaluation, including monitoring; compliance evaluation; internal audit; management review.

10. Improvement, including incident data; non conformity; & corrective action.

Main Business Agenda and Business Excellence

The introduction of ISO 45001 will encourage more organisations to:

- Place the management of H&S risks firmly on their main business agenda
- Include an improvement in their management of OSH as a key business goal and strategy.

Organisations will appreciate that as ISO 450011 is a globally recognised Standard then it can be used to demonstrate to customers and their supply chain that they are a responsible and socially focused organisation.

Historically, the global focus on organisations has been on the quality/safety of products, environmental best practices and employee pay and general working conditions. The new ISO

45001 will enable the prevention of death, injuries and ill-health to be brought much more into an overall business focus and public awareness. Additionally, the management of supply chains will be further enhanced as stakeholder expectations are raised of the 'parent' organisations need to focus on the health and safety of the workers employed in its supply chain.

Organisations who currently adopt a business excellence model will be able to integrate ISO 45001 with their business processes; with more clarity than previously with OHSAS 18001.

In particular ISO 45001 will have the following benefits:

- Globally
  - Cut human, economic & social toll
  - Address changing world of work
  - Strategic attention, leadership & resources
- Organisations
  - The Big 3 - Leadership; Engagement and Competent Advice
  - Transparency & Accountability
  - Corporate Social Responsibility/ Sustainability & Supply Chain Management
- OSH Professionals
  - Preparing & Auditing Boards/CEOs
  - Understanding Context & Stakeholders
  - Leading, Procurement & Reporting

As stated above, ISO 45001 follows the Deming Plan - Do - Check - Act model. So does Business Excellence! Indeed Deming's 1982 book "Out of the Crisis" argued that in order to achieve business excellence i.e. to be successful, then organisations needed only 2 things:

1. Commitment
2. Ability to open themselves up to new thinking

Just as ISO 45001 requires!

Deming is considered by many (including the authors) to be the Father of Quality who outlined 14 Management Principles to be followed in order to significantly transform both the quality & effectiveness of their operations, thereby achieving Business Excellence. These 14 guiding principles are as follows

1. Adopt a new philosophy management must lead & know their responsibilities

2. Create constancy of purpose towards continual improvement of all products & services to become competitive & stay in business

3. Build quality into the product/service at the design stage; thereby minimising the need for inspections

4. End the practice of awarding business on the basis of price alone

5. Continual improvement of all management systems e.g. better quality; productivity; decrease costs; and [Authors' Note enhanced, business focused and integrated health & safety management]

6. Institute on the job training [Authors' Note : Worker engagement]

7. Institute clear leadership management / supervision and help workers & equipment to do a better job.

8. Drive out fear so that everyone works effectively & efficiently [Authors' Note: a just / blame free culture]

9. Break down barriers between departments / silos

10. Eliminate Zero Defects programmes no targets on productivity etc as these create adversarial relationships [Authors' Note: Zero Defects = Zero Accidents?]

11. Eliminate work quotas on the factory floor: eliminate management by numbers & numerical goals

12. Remove barriers that rob the hourly paid of his / her right to worker pride

13. Institute a vigorous programme of education & self-improvement

14. Put everyone in the organisation working towards accomplishing the transformation of the organisation: it's part of everyone's job!

Arguably Deming's 14 Principles are more relevant today than they were in 1982, especially when aligned with ISO 45001. They provide a route to business excellence, whilst at the same time, if applied via ISO 45001 compliance / accreditation, with a strong leadership and focus for a business focused and integrated approach will help to achieve a positive OSHMS leading to fewer injuries & cases of ill health.

A Win Win situation!!

For Further Reading

"How to create economic incentives in occupational safety and health: A practical guide" Working Environment Information – Literature Review, Dietmar Elsler, Andreas Heyer, Klaus Kuhl, and Lieven Eeckelaert, (eds.), Agency for Safety and Health at Work. (https://osha.europa.eu/en/tools-and-publications/publications/literature_reviews/guide-economic-incentives)

De Weerd, Marjolein, Rory Tierney, Brigit van Duuren-Sturrman, and Evelina Bertranou. "Estimating the cost of accidents and ill-health at work: A review of methodologies." European Agency for Safety and Health at Work. Luxembourg: Publications Office of the European Union, 2014.
(https://osha.europa.eu/en/tools-and-publications/publications/reports/estimating-the-costs-of-accidents-and-ill-health-at-work)

Knight, Alec, et al. "Economic incentives to improve occupational safety and health: a review from the European perspective." European Agency for Safety and Health at Work, Dietmar Elsler (ed.) Working Environment Information. Luxembourg: Publications Office of the European Union, 2010.
(https://osha.europa.eu/en/tools-and-publications/publications/reports/economic_incentives_TE3109255ENC)

Case Study: Driving Safety Culture with a Proven Near Miss/Safety Concern Program

Kelly Bernish, CSP
Rachel Bunn
Safety, Security & Risk Management, City of Fort Collins
Fort Collins, CO

Introduction

Improving Safety Performance and reducing employee injuries, measured by both leading and lagging indicators, is a fact of life for many safety professionals. Everyone is always looking for new and creative ideas to make improvements. This session will give proven, real-life, innovative implementation strategies and ideas that attendees can take back to their organization to implement or revitalize and improve a near miss/safety concerns reporting program that they may already have. The ultimate goal is to prevent injury through awareness, relationship building and communication.

Why are we still having injuries even though we are doing all this safety "stuff"?

Sometimes you are too close to a situation or "you don't know what you don't know" and it takes a fresh look. The City of Fort Collins took the initiative to bring in some experts to help.

Through a series of site visits, interviews and focus groups to better understand the true safety culture of the City, it was evident there were several missing pieces but failure to identify, report and act on near misses and safety concerns, jumped to the forefront and seemed like a likely, fruitful step. No longer was the City going to focus on how many days it had been since the last lost time injury, the focus will be how many near misses were reported, corrected and most importantly, shared.

Piloting, Learning and Back to the Drawing Board

Piloting

Some areas of the City already had a process for recognizing and reporting near misses and safety concerns. Learning from those areas, their processes were gathered, taking the best of all worlds and combining them into one robust, city-wide program. Due to the scale of the City, a decision was made to start with a pilot program. One department in the City was chosen that had a current program to build upon. Information was gathered through focus groups, attending meetings and just simply observing their everyday work environments. Our regular interaction helped build trust and relationships with the employees and ultimately a lot of buy-in. Deploying the actual program with the department helped identify opportunity for improvements.

The pilot program helped identify the need for additional components such as a safety team to champion the changes that were needed and a more robust tracking system. The department decided it would be helpful to create a safety team that would compile, monitor and help correct and communicate the reported concerns with the help of their safety partner. This team took ownership over the reporting and drove change within the safety culture of their department. Collecting the reports ultimately gave way to the need for a more robust tracking system and a way to establish metrics and trends. The most important element in the tracking system was to have a process for monitoring and ultimately closure. The worst thing that could have happened was to have employees submit near misses and safety concern reports and then never hear any follow-up or resolution.

One of the most vital components was setting up systems to make reporting easy for everyone through a variety of different means. As you can imagine, the City has a vast continuum of work scenarios. The key was to appeal to our diverse workforce; all areas, ages and work environments and make reporting simple:

- Online reporting form
- Email message
- Text message
- Hotline (this went to a monitored voicemail, not a live person)
- Drop Box
- Verbal communication

It was important to recognize the pilot department for their three months of work towards the development of the program. This group became our key collaborators and supporters who also became a resource for other areas of the City during the rollout. Relationships are the key to safety success!

Focus Groups

Focus groups were essential in the planning of our program, starting with the pilot department and then city-wide. Ready to start planning for the city-wide rollout, we started by scheduling various focus group sessions throughout the City, incorporating employees from different departments. We kept the groups to about 10 employees to allow for intimate discussion. The purpose was to find out what would work for them and what they knew.

Questions asked

- What is your definition of a near miss?
    - What definition of a Near Miss resonates with you?
- Have you ever reported one?
- Is there a difference between near miss and a safety concern – and is that important?
- What is the easiest way for you to report?
- What barriers did you foresee in reporting?

Representation from frontline employees gave a clear picture of what would work for them and their areas. These groups not only gave us insight into what workgroups really wanted but it also gave them some ownership and buy-in which proved to be helpful in the rollout. Employees wanted:

- Follow-up and closure
- Education on what the program would do for them/their department
- Should they fear retaliation and punishment
- Anonymity?
- How is this sustained?
- Don't let this become just another "safety program of the month".
- Don't get hung up on safety concern vs. near miss – let's just report them all!

Conducting focus groups can also become an opportunity for airing a lot of discontent. It's important to not only have a facilitator who will keep the group on track and focused, but allow for the raw conversation and feedback to develop.

Back to the Drawing Board

- What vulnerabilities could we identify early on?
- There may be an unclear definition of what a near miss is.
- Are the adequate means of reporting or maybe even too many to manage?
- Is there a fear of retaliation or punishment?
- Will these reports create more work when it comes to mitigating and finding a resolution? .
- Do other issues such as unions and language barriers play a role?

Be prepared to revisit and educate over and over again. As employees start to address their own safety concerns, they may have the best of intentions but still not have it quite right or may have a perspective that is not in line with the appropriate philosophy, for example posting number of days since the last lost time incident or celebrating days without a work injury. This is a great opportunity to be leveraged as a safety partner, allowing employees to reach out to their safety professionals as a resource to calibrate. Encourage and acknowledge employees that take ownership to correct issues.

Ready for Rollout

Relationship Building

As well intentioned as your communication may be, buy-in at all levels of the organization may not occur the first time around. Take the time to present to leadership. Meet with supervisors one-on-one. Anticipate rejection and the need for more education and time. It may take getting everyone on the same page individually and then trying again as a group. Is there a perception that you are forcing this? Communication is key. Their idea of safety and yours may be different. This is a great opportunity to listen and consider the norms of the individual, department or organization.

Take every opportunity to educate, discuss and present. Get on agendas, attend meetings, create communications and share what others have reported.

Encourage storytelling versus metrics. All areas of the city start meetings off with a near miss or safety moment. Pictures are the BEST! Encourage employees to utilize photos whenever possible, they really help create the story that resonates with others.

Be patient. You may need to get more feedback and visit with key partners and senior leaders. You can't repeat yourself enough. The change and awareness doesn't occur overnight and repetition will make it happen.

What to Anticipate

The rolling hills effect - some months may have more reports than others. Be patient and keep communicating, a thousand times sometimes isn't enough. When you reach that lull in reporting, it may be time to wage another round of education and outreach. Engage your leaders. Make sure the metrics are visible to encourage healthy competition.

Be ready to talk money. Depending on where your organization is on the safety evolution spectrum, sometimes you will have more projects pending than there are resources for, particularly funding. Consider a prioritization matrix or process to decide what to seek funding for. Understand your businesses unique processes for funding. This is where business acumen is really important as a safety professional.

Controversial, sensitive and unrelated reports – along with all of the great information received, there will be topics and issues that arise that will present some unique challenges. Our philosophy has been to never turn anything away in an effort to keep people coming back. There will always be issues that arise that cause people to be defensive and sometimes outright not very nice. There are very few times when otherwise, but we generally take the high road in an effort to be known as the partners that really get things done and are not afraid to take on some long standing issues. Again, this is where business acumen comes into play, where you want to make business cases for what needs to be done. Sometimes that requires a lot of research, investigation, meetings and solid, well thought out, planning. Know when ethical issues arise, you will have to take those on, as uncomfortable as they may be. Check in with whomever handles your ethics issues in your organization to ensure you can work hand in hand with them should those types of issues arise. Above all, never let issues slip through the cracks and go unaddressed. Nothing will kill the credibility of what you are doing quicker.

Safety 24/7 – Embrace reporting and messaging around safety regardless of where it occurs; work, home, play and just everyday stuff. This encourages a general safety mindset and will change the way people relate to safety. Nothing creates more buy in than truly meaning you care about someone, on and off the job…and their families too!

Safety Alerts – Take some of these stories and create written communication or safety alerts as need be. Not only does this drive others to look at similar situations they may have, but it also creates a strong sense of pride for those reporting and correcting. Encourage widespread posting of the alerts and using them for "safety moments" at meetings and gatherings (attachment 1).

Branding – If you have the resources, consider branding (attachment 1) your safety effort in a way that will cause recognition and common understanding. We adopted the CitySafe brand for everything we do including our near miss initiatives. Lanyard cards (attachment 2) were created to remind employees of all of the different ways to report safety concerns and near misses.

Where We Are Now

Tracking

A master electronic list on the City intranet shares reports that have been submitted and can be viewed by everyone. Editing rights are controlled. This list has the capability to show pictures if there is one associated. Viewers can see the steps that have been taken to resolve the concern. This has been essential for collaboration between departments across the City. We share monthly and annual reports and discuss results with departments that may still need more education on the process.

Sustaining the Momentum

Share results, share experiences, share stories! As informative as metrics can be, at the end of the day you're going to remember someone's story or personal account, not the injury rate of the organization. Take every opportunity and get creative. Highlight a report and share the resolution. Don't be afraid to incentivize reporting. Push a three month reporting challenge within a department and even across the organization.

Metrics

In an effort to move away from focusing on lagging indicators, our first leading indicator was created. It is as simple as raw numbers by service area on the monthly operating report. In the background is the detail information and even greater detail can be found on the intranet site. The idea of course is more is better and over time to maintain a constant level of reporting and quality.

Attachment 1

CitySafe

City of Fort Collins

SAFETY ALERT!

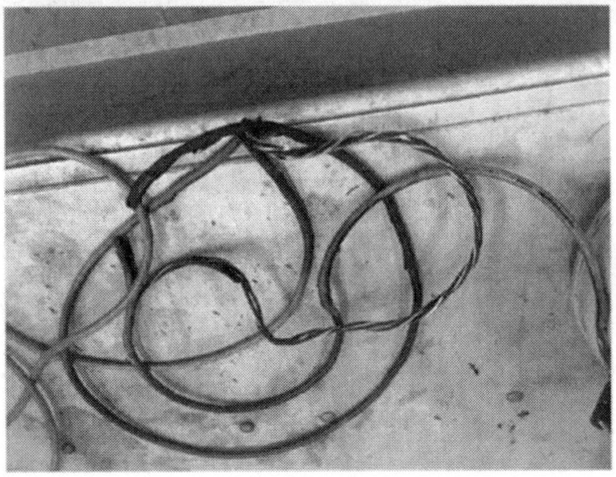

SAFETY CONCERN
Damaged Cords

"In our maintenance shop there are two power cord reels that are hung from the ceiling. These reels were installed in 1986. While doing some work we had the need to pull one out past where we usually need to. We were shocked (not really but close) to find that the rubber outside sheathing was cracked worn and missing. It was very close to becoming a fire risk. It was scary to say the least. Both reels have since been replaced with new cord reels."

ACTION REQUIRED
- Regularly pull retractable cords all the way out for full inspection.
- Regularly inspect cords on power tools
- Take any damaged cords out of service immediately

TO REPORT A NEAR MISS OR SAFETY CONCERN IN YOUR WORKPLACE VISIT http://citynet.fcgov.com/riskmanagement/nearmiss/submit

Attachment 2

CitySafe

SPEAK UP FOR SAFETY

REPORT A NEAR MISS OR SAFETY CONCERN

- **Online** Reporting Form CityNet.fcgov.com/ssrm/
- **Text** or **Email** safe@fcgov.com
- **Hotline** 970- 416-4200 or x4200
- **Talk** to your supervisor or SSRM
- **Access Fort Collins** online or phone app

Session No. 664

5-Why: How This Simple Tool is Confused, Misapplied and Highly Underutilized

Mark Galley, CRE
President
ThinkReliability, Inc.
Houston, Texas

Introduction

5-Why is one of the most recognized problem solving tools. The intent is to ask several Why questions to reveal a more thorough explanation of how and why a particular incident occurred. The concept is simple, but application varies substantially. Different groups within the same company don't necessarily conduct a 5-Why the same way. The tool is widely misinterpreted and misused. There isn't a clear understanding of fundamentally what a 5-Why is and when it's supposed to be used.

Significant opportunity exists for management and front line personnel to improve the way they explain the details of an incident. Miscommunications add to the risk. Some managers don't use 5-Why, because they see it as too simplistic for complex issues. Other managers have a strict interpretation that the fifth Why reveals the root cause. A common criticism of 5-Why is that it's not repeatable because three different people each get a different 5-Why. When people know different aspects of an issue they're not supposed to get the same 5-Why. That flawed premise contributes to the decades-long confusion about the tool.

These simple misunderstandings are symptomatic of other managerial disconnects in a company. What should be a baseline to begin an accurate cause-and-effect analysis has become a starting point for debate. These disagreements undermine a cooperative dialogue between front-line employees and management.

Simple changes in the way an organization uses 5-Why improves the way people communicate detail and ultimately prevent problems. This presentation will explain the basics of 5-Why and provide several examples of how management can affect front line problem solving as well as how the front line can affect management.

5-Why Basics

History

Asking Why five times has a long history within Toyota. Before Toyota entered the automobile industry in 1933, it made looms to produce fabric. Sakichi Toyoda, the founder of Toyota, built his first wooden hand loom in 1890 (Toyota Boshoku). He is credited with developing the 5-Why discipline within the company. I can imagine him tracing the causes of a thread break through the different parts of a loom to solve a problem. Taiichi Ohno is credited with establishing what is known today as the Toyota Production System starting in 1948 through 1975. He described the 5-Why method as "the basis of Toyota's scientific approach. By repeating Why five times, the nature of the problem as well as the solution becomes clear" (Ohno). 5-Why has become part of different problem solving approaches such as Lean and Six Sigma. Toyota formalized 5-Whys as a tool with their company, but Why questions have always been the basis of understanding something better.

4-Year-Olds

Anyone who has spent time with a 4 year-old has witnessed their repeated Why questions firsthand. Kids ask a lot Why questions. They use the answer to their question to ask the next Why question. Without any training in quality tools Why questions come naturally to 4-year-olds. Do you remember asking Why questions of your parents when you were four?

The typically response to a Why question begins with the word *because*. Why can't I ride my bike? You can't ride your bike *because* the rear tire is flat. The effect is the child can't ride their bike. The cause of that effect is the tire is flat. Note the root of the word because is *cause*. The word *because* means 'caused by.' Why questions help identify cause-and-effect relationships. Cause-and-effect is what explains how and why something occurred. 5-Why is a tool to better understand the cause-and-effect relationships in a problem. Several Why questions together identify several different cause-and-effect relationships which provides for a more thorough explanation of an issue.

Welding Robot Stopped

Following is an example 5-Why from the Toyota website. The problem being analyzed is a welding robot. Here are their five Why questions starting with the problem, robot stopped:

1st Why	"Why did the robot stop?"
Answer	The circuit was overloaded, causing a fuse to blow.
2nd Why	"Why is the circuit overloaded?"
Answer	There was insufficient lubrication on the bearings, so they locked up.
3rd	"Why was there insufficient lubrication on the bearings?"
Answer	The oil pump on the robot is not circulating sufficient oil.
4th	"Why is the pump not circulating sufficient oil?"
Answer	The pump intake is clogged with metal shavings.
5th	"Why is the intake clogged with metal shavings?
Answer	Because there is no filter on the pump.

The process begins with a problem. In this case the welding robot stopped. Writing each Why question as a complete sentence is a common format with the answer written beneath the question. That answer then becomes the basis for the next Why question. This continues for 5-Why questions. Some companies teach their employees that the answer the 5th Why question is the root cause. While the concept is simple - start with a problem then ask five Why questions - there is significant opportunity to improve the use of this improvement tool.

Complete sentences are not the ideal way to explain cause-and-effect. The five sentences above have 81 words. There is duplication since each answer is rewritten as a question in the next sentence. It's much simpler to communicate cause-and-effect relationships visually in boxes. Here is the same set of cause-and-effect relationships from the welding robot shown visually.

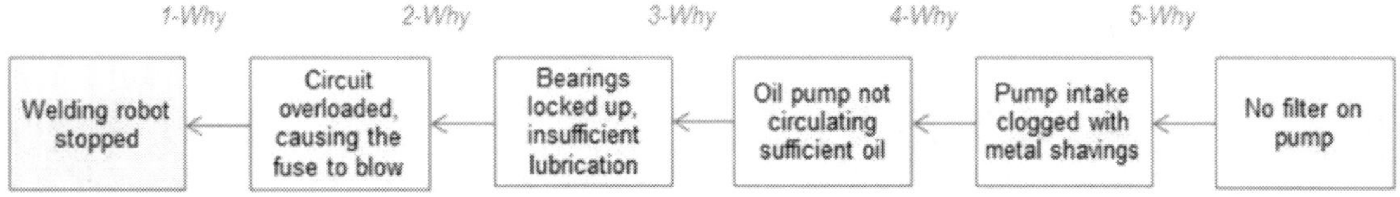

Exhibit 1. 5-Why shown visually.

Showing the cause-and-effect relationships visually uses only 31 words. Because each box represents a cause and an effect the does not have to be repeated. The box *Oil pump not circulating sufficient oil* is a cause of *Bearings locked up* and an effect of *Pump intake clogged with metal shavings*. Organizations strive to make improvements in their operations by finding every opportunity to reduce waste, yet that same organization may choose to use almost three times as many words to communicate the same information. Companies don't always apply a lean approach to their problem solving tools. Simply asking the people responsible for conducting investigations whether they prefer sentences or boxes would provide valuable insight.

There are also issues with the way the sentences are worded. Answers to Why questions should not contain a cause-and-effect relationship. For example, *Circuit overloaded, causing the fuse to blow* needs to be broken into an additional cause-and-effect relationship. The *fuse blew* because the *circuit was overloaded* is clearer. Longer sentences make it easy to mix causes. Showing simple cause-and-effect relationships breaks the problem down into parts. The same mistake was made for *Insufficient lubrication on the bearings, so they locked up.* This sentence should be broken down into one more cause-and-effect relationship. Those two boxes now look like this:

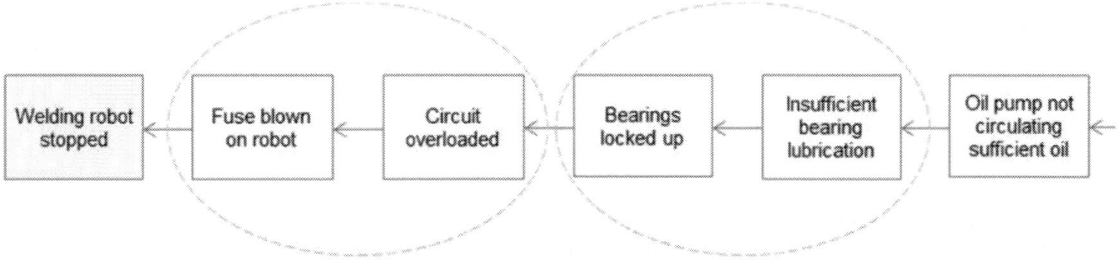

Exhibit 2. Causes broken down into simpler relationships.

This analysis in Exhibit 2 shows one cause-and-effect relationship at a time. Written sentences can mix details together. The visual approach makes it to review and challenge the information to ensure it's accurate. By breaking out those causes the 5-Why expands to a 7-Why that still has 31 words (two of the causes are not shown in Exhibit 2).

Cause-and-Effect Basics

Many organizations operate on the premise that complex issues with multiple factors can't be explained with basic cause-and-effect. They believe that explaining an issue is different that identifying the causes (Hollnagel, 2004). Cause-and-effect is precisely how problems are explained – all problems. It's a consistent approach that doesn't change based on the type or magnitude of an issue. The investigation of a major incident can start with a 5-Why that expands into as much detail as needed. It's built one cause-and-effect relationship at a time in a systematic, structured way. Exhibit 3 shows how the Welding Robot example can be broken into a 17-Why.

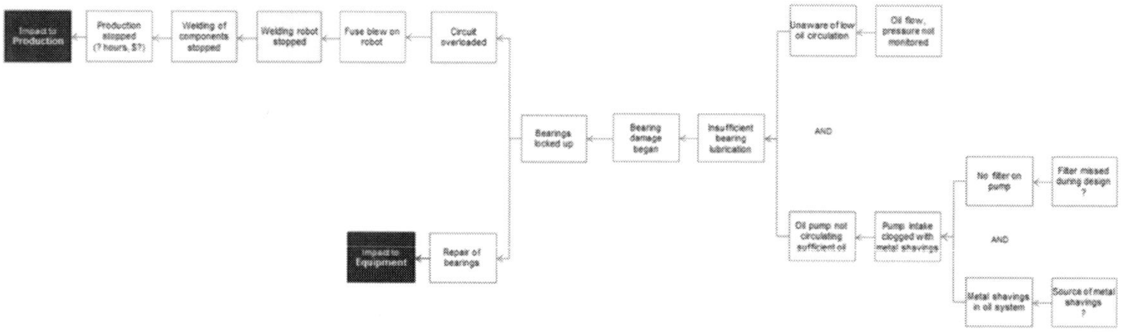

Exhibit 3. Welding Robot as a 17-Why.

Without a fundamental understanding of the cause-and-effect principle 5-Why could inadvertently restrict the analysis by stopping the analysis too early. Five is simply a reminder that there are multiple causes to an issue. People sometimes summarize issues by providing only one Why. Five is a good way to ensure that people dig a little further. Think of a 5-Why as an early phase of an investigation. A thorough explanation will require more causes which is just more Why questions.

Think of a 200 mile drive. Five miles is a great start, but it doesn't get you to your destination. You need to keep adding miles, one at a time. For this trip you'll need to drive a total of 200 miles. And the trip, just like all trips, begins with the first mile. Then you complete the second mile, the third, the fourth, all the way through to the end. This is how 5-Why progresses. It starts with one Why question, then two, then three, all the way through to as many causes as are needed to thoroughly explain the issue. The analysis, like the drive should not be farther than it needs to be. This is why small issues have fewer causes and bigger incidents have more causes – 15, 30 or more. All of the investigations still begin one Why and pass through five along the way as they expand.

The linear nature of 5-Why creates some of the confusion. Cause-and-effect relationships occur in series *and* in parallel. The 5-Why from the welding robot explains that the

Pump intake was clogged with metal shavings because *there was no filter on the pump*. This cause-and-effect relationship implies that a filter should have been on the pump. Was it missed in design or not installed after maintenance? It's also unusual that the source of the metal shavings is not identified. Metal shavings are not supposed to be in the lubrication system. Where did they come from? This question is not in the 5-Why. For this pump intake to be clogged with metal shavings the *pump intake had not filter* and *there were metal shavings in the oil system*. The metal shavings are a cause of the pump intake clogging. Without the metal shavings the pump intake doesn't become clogged.

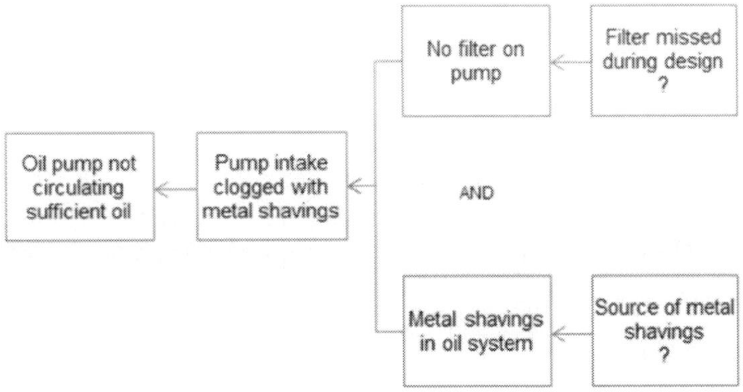

Exhibit 4. Welding robot – Parallel cause paths.

This parallel cause-and-effect relationship shows how two causes are required to produce one effect. Since both causes are required for that effect to occur only one of the causes has to be controlled to prevent that effect. Parallel cause-and-effect relationships are a normal part of a complete analysis. The series and parallel causes are essential for providing a complete explanation and finding different solution options. The 5-Why is a great way to start, but it doesn't provide a thorough explanation of the issue.

Criticisms of 5-Why

The two common criticisms of 5-Why is that it's too simple and not repeatable. The fact that 5-Why is simple is one of the advantages of the method. Start simple and then expand as needed. 5-Why is not a complete method. It's a phase of an investigation. Establishing a 5-Why baseline for problem solving in your company is a great way for people to learn cause-and-effect thinking. Frontline people become an incredibly important component of problem solving when they can easily start a 5-Why on their own.

The premise that 5-Why is not repeatable is a mistake. 5-Why is not supposed to be repeatable. If you ask the plant manager, the electrician, the operator and the maintenance person "Why did the robot stop?" you'll get different answers. People don't answer Why questions the same way. But that doesn't mean there's a contradiction. The incident happened a particular way and your investigation should reflect that. People can provide different answer, yet all their answers can be accurate. If the maintenance person says that the welding robot stopped because *the pump intake was clogged with metal shavings* and the electrician says "No, it was because the fuse blew" there is no contradiction in those two statements - once causes the other. Verbally these cause-and-effect relationships are difficult to understand. Visually it's much simpler. Those two cause-and-effect relationships fit together precisely the way the incident occurred.

The electrician and the mechanic may think they're in an argument, because they provided different answers. Once the cause-and-effect relationships are laid out visually, it's much easier to see how both are telling the truth. This is a subtle, but significant point in changing the way people communicate detail within a company.

There are many pointless debates and unnecessary arguments because the basics of cause-and-effect are not understood. People hear different answers to the same question and think it's a dispute. Just because people don't respond to a Why question in exactly the same way there is not necessary an issue. Electricians, operators and maintenance technicians are experts in different areas. They know different information so they naturally response with different answers. These are simply different causes on the analysis – either in series or in parallel. This is how a complete explanation is developed - with input from those who know.

Benefits of 5-Why

Thinking that the 5th Why provides the root cause is a mistake. But, 5-why is a great tool when used appropriately. These are some of the benefits of establishing a 5-Why baseline in your business.

Simple
A clear understanding of cause-and-effect is fundamental for becoming proficient at explaining and preventing problems. One Why question is an incredibly simple way to start an investigation. That 1-Why can be expanded into more, but keep it simple. Even the most complex issue can be dissected into individual cause-and-effect relationships. This approach is so simple most organizations don't do it. Companies have a tendency to make things more complicated that they need to be.

Accurate
The cause-and-effect relationships for an incident fit together a particular way. Regardless of how simple an analysis begins, it should be accurate. Here are two examples of accurate 1-Whys: The Titanic sank because it hit an iceberg is accurate, it's just not complete. The Challenger Space Shuttle was lost because of an O-ring failure is also accurate, just not complete.

Quick
It doesn't take a lot time to start the analysis. One why question takes just a few seconds to answer and document. Even with only some basic information a 5-Why can be laid out in 60-seconds for any type of issue regardless of industry. If there are unknowns then maybe only two or three Whys can be answered. The Why questions will provide direction on what information needs to be collected.

Engaging
With a 5-Why only a few cause-and-effect relationships are laid out in straight line. The analysis may not necessarily be complete, which is an advantage for expanding it. When others, those who are familiar with the details of an issue, review a 5-Why they can't refrain from commenting on what's missing. Starting the analysis simple is an effective way to get details. Additional information may not contradict the existing analysis, but simply add to it.

Communication & Culture

The front line people within a company are an underutilized problem solving resource. Those closest to the work are well aware of what goes well, what goes poorly and what has become normal. Those on the front line not only have information about problems, they also have insight and ideas for solving them. Accessing this information requires changing the way people communicate the details. Problems are not always explained as clearly or as thoroughly as they should.

5-Why is a way to engage those closest to the work in the problem solving process. It doesn't have to take a lot of time. It's a simple and quick way for people to share what they know. Each of these simple 5-Why analyses can be reviewed and expanded as needed. Changing the problem solving culture form reactive to proactive requires a more engaged front line and more receptive management that asks for information instead of providing prescriptive directions. Any solutions driven by management that don't match what actually happened within an issue undermine management's credibility. Those incidents provide evidence to the front line that management just doesn't get the real world. Changing the way an organization explains and prevents problems – between management and front line - changes culture.

In Summary

5-Why is an excellent way to begin an investigation. It's less of a method and more of a phase of an investigation. There is no need to restrict the number of causes in an analysis to five. Cause-and-effect relationships can be captured in both series and parallel to thoroughly explain an issue. The analysis can even begin with one Why question that expands as needed. Including front line people in a 5-Why problem solving process is an effective way for management to access insight and ideas that is presently unused.

Bibliography

Hollnagel, Erik, 2004. Barriers and Accident Prevention. Surrey, England, Ashgate Publishing ISBN 978 07546 4301 2

Taiichi, Ohno.1988. *Toyota Production System: Beyond Large-Scale Production.* Portland, OR: Productivity Press ISBN 0915299-14-3.

Toyota. 2006. *Toyota Traditions "Ask 'why' five times about every matter."* (retrieved February 2017) (http://www.toyota-global.com/company/toyota_traditions/quality/mar_apr_2006.html)

Toyota Boshoku. 2008. *Our Roots – The History of Toyoda Boshoku.* (retrieved February 2017) (http://www.toyota-boshoku.com/global/special/discover/history01/index.html)

Session No. 665

Leveraging Generational Differences to Enhance Workplace Safety

Robert Jameson
Vice President

Scott Ratliff, P.E.
Senior Engineer

Central Mine Services, Inc.
Danville, KY

Introduction

The landscape of the American workforce is changing in a variety of ways. Based on a 2015 analysis performed by the Pew Research Center (Pew 2015), Millennials have now surpassed both Baby Boomers and Generation X as the largest living and working generation. Furthermore, a recent Gallop Poll (Gallop 2014) indicates that nearly 50% of Baby Boomers plan to continue working beyond the age of 65. Consequently, employers will soon be facing the most generationally diverse workforce in American history, and many companies are already confronting the challenge of managing three groups of workers with completely different perspectives on life, work, and the world in general.

Perhaps the most significant differences can be found when comparing the traits, strengths and weaknesses of the "Baby Boomer" and "Millennial" generations. While the differentiating characteristics of these two groups are often associated with their cultural and societal perspectives, the real opportunity is provided by the different strengths that each group offers. Baby Boomers possess a wealth of knowledge and skills, developed through many years of work and life experiences. Millennials, who are still fairly new to the workforce, are technologically savvy and bring with them an affinity for change that leads to an increase in efficiency and a better sense of community. However, generational strengths such as these are too often being idled due to the current acceptance of stereotypes reported frequently by the media and the structural hierarchy that most managers have grown accustomed to. Once companies begin to leverage these strengths by challenging that structural hierarchy, embracing the cultural shift and encouraging effective mentoring and collaboration, they can take advantage of the generational diversity and the opportunity that it provides.

Generations in America

The term "generation" is commonly used to describe a group of people who were born around the same time. While this is technically true, it does not fully describe the true nature of what forms a generation and what makes that generation unique. In their book, *The Fourth Turning*, the authors, Mr. William Strauss and Mr. Neil Howe, describe a generation as "the aggregate of all people born over a span of roughly twenty years, or about the length of one phase of life: childhood, young adulthood, midlife, and old age." They further explain that "members of a generation … encounter key historical events and social trends while occupying the same phase of life" (Strauss 1998).

According to the Pew Research Center (Pew 2015), generations are analytical constructs and developing a consensus on what marks the boundaries between one generation and the next takes time (Pew 2015). Currently, the following time boundaries are used by Pew Research Center staff to differentiate the generations:

- The Silent Generation – 1928 to 1945
- The Baby Boom Generation – 1946 to 1964
- Generation X – 1965 to 1980
- The Millennial Generation – 1981 – mid 1990s

The Silent Generation

The name "Silent Generation" was coined in the November 5, 1951 cover story of *Time* in reference to the generation that was coming of age during that time (Younger Generation 1951). Strauss and Howe, in their book *Generations,* also used the designation in reference to the generation in America that was born from 1925 to 1945 (Howe 1991). The term "silent" refers to the generation's silence during the McCarthy era and their willingness to assimilate into the social order.

The silent generation brought the strong work ethics of their parents into the factories of industrialized society. They grew up during lean times, including the Great Depression and World War II, and believe in the value of hard work and saving for the future. The majority of this group is now in their seventies and most have retired from the workforce. According to Pew Research, as of 2015, the Silent Generation only represents 2% of today's labor force (Pew 2015).

Baby Boomers

The term "Baby Boom" was first used in Landon Jones' book "*Great Expectations: America and the Baby Boom Generation*" to define the demographic group born after World War II. According to Jones, Baby Boomers were the wealthiest generation up to the era in which they arrived, and were amongst the first to grow up genuinely expecting the world to improve with time (Jones 1980). For several decades now, Baby Boomers have had an outsized presence in the workplace and their strong work ethic has propelled them into the most influential roles within many organizations.

Ernst & Young published a study in 2013 based on the input of 1,200 professionals across various industries and three generations: Baby Boom, Gen X and Millennial (EY2013). During the study, participants were asked about the strengths and weaknesses of workers from different generations, based on the perceptions of their peers. Baby Boomers were recognized for their productivity, as well as for being "hard-working", "team players", and good mentors for others. Conversely, Baby Boomers ranked the lowest in adaptability and their ability to collaborate.

Generation X

As with most generation labels, "Generation X" is a loaded term. The label was first coined, and later disowned, by Douglas Coupland in his 1991 book titled "*Generation X: Tales for an Accelerated Culture*". For Coupland, the letter "X" was meant to signify the generation's random, ambiguous, contradictory ways (Coupland 1991).

While initially identified as the "MTV Generation" and viewed as lazy and cynical, members of Gen X are now perceived as having the greatest ability to generate revenue, and possess traits of "adaptability", "problem-solving" and "collaboration" (EY 2013).

Millennial

When Neil Howe and William Strauss coined the term "Millennial" in their 1991 book "*Generations*", they weren't sure if the title would stick. Colloquial use seems to have come much later, as Google Trends data indicated nearly zero interest in the term in 2005. More recently, however, searches for Millennial/Millennials have grown, picking up speed around 2013 and skyrocketing in 2015 (Sharf 2015).

When compared to other working generations, Millennials are believed to be the most technologically-savvy, use social media to leverage opportunities, and be "enthusiastic" about their jobs. When compared to other generations, however, Millennials are generally not perceived as "hardworking" or "a productive part of the organization" (EY2013).

Generations in the U.S. Labor Force

Today, three of the four working generations make up the vast majority of the U.S. Labor force. The Baby Boomer, Gen X and Millennial generations each represent roughly one third of the overall labor force (Pew 2015). While the population of Gen X in the U.S. could grow slightly, the Millennial population is expected to continue to increase and widely surpass that of other working generations. This, in turn, will result in the Millennial culture having a massive influence on the workplace and society as the whole.

Although Millennials now have the largest presence in the workplace, Baby Boomers are choosing to work longer than generations before them. According to an Employee Benefits Research Institute survey, in 1991, 50% of all American workers planned to retire before they reached the age of 65. Today, that number has declined to 24% (EBRI 2016). Some Baby Boomers are delaying retirement for financial reasons, while others say that their job defines their identity and that they are engaged, enthusiastic employees who are committed to their work.

The level of experience that Baby Boomers bring to the workplace is unparalleled and they tend to hold many managerial, decision-making roles. This is both a strength and a weakness for many companies, as Boomers tend to stick with what has made them successful and are often reluctant to change. Millennials, on the other hand, are generally very comfortable with change, have an affinity for technology, and want to make a difference. With the cultural influence from Millennials that is beginning to shape the workplace, and the Baby Boomers' stronghold on top, decision-making roles, and reluctance to change and adapt, a major generational clash has ensued.

Baby Boomers vs. Millennials

The workplace is one area in which this generational clash is having a profound impact, largely due to the different perceptions, values and interests of each generation. Unfortunately, mainstream media stereotyping often indicates that these differences are so significant that Baby Boomers and Millennials are going to have a very challenging time working together. Baby Boomers are described as "living to work" while Millennials are "working to live". Millennials are viewed as "wanting a trophy for just showing up" and Baby Boomers are seen as "workaholics driven to earn their spot". It is in this space where the generations are portrayed as being so different that they cannot work together and will become disengaged, frustrated and, in the case of the Millennials, inclined to change jobs.

Obstacle or Opportunity?

Based on the mainstream stereotypes, coexisting in the same workplace appears to be a nearly impossible task, and, if not addressed and managed properly, these generational differences can result in devastating consequences for any company. However, once action is taken to better understand each generation and identify how each is influenced, inspired and motivated, companies can begin to recognize the unique opportunity that is available in today's labor force. That is, the opportunity to leverage the experience, managerial skills and work ethic of the Baby Boomers with the Millennials' affinity for change, ability to adapt and apply technology, and overall efficiency. If companies can begin to recognize the true nature of these generations and the value that they add to the workplace, the talents of each can actually be leveraged to enhance a company's performance in areas including safety, productivity, and quality.

Leveraging Generational Strengths

A list of the strengths of each generation would be quite lengthy so, in an effort to describe the generations, the follow predominant points are being highlighted:

Strengths of Baby Boomers

- Experienced
- Management skills
- Strong work ethic
- Resourceful

Strengths of Millennials

- Technologically savvy
- Self-confident
- Efficient
- Team-oriented

How to Leverage Strengths

Perhaps the easiest place to begin leveraging the strengths of each generation is to recognize that a cultural shift is occurring in organizations and, rather than avoid it, actually embrace it. Organizations that fail to recognize the changing demographics and shifting cultural norms of society and their workforce will be left behind when it comes to attracting and retaining the best employees. Employee expectations and societal norms, coupled with rapidly changing technology are bringing significant change to the workplace. Company leaders who recognize, embrace and champion these changes will benefit by overcoming any staff weaknesses within one component of the organization with the strengths of others on staff.

In addition, companies will benefit from examining and challenging existing structural hierarchies. Traditional structure, which typically relies on the most senior or highest-ranking employee making decisions, may not be the most effective means for making the best decision. The knowledge and understanding on a specific topic of younger, lower-seniority staff member may lead to a better decision than the "we always have done it that way" experience of an employee in a managerial position. For example, the best person to advance a company's social media presence is likely the person who regularly uses a variety of social media platforms, rather than the managerial level employee who may not utilize social media applications at all. This is likely best accomplished through a more flat organizational structure and cross-generational mentoring.

Cross generational mentoring is the key to leveraging the strengths of each generation. The weaknesses of one generation can often be overcome by the strengths of the other. Companies that recognize this, and do not allow stereotypes to limit their potential, will be successful. Cross-generational mentoring is different than classical mentoring. Mentoring is often thought to be the process where an entry level employee learns from a member of the company's leadership team. In cross generational mentoring, long term employees are partnered with less seasoned, but still very knowledgeable staff members, and the pair train each other. This requires each to be candid about their weaknesses and areas of needed improvement. This can be difficult if the employees work in "boss/subordinate" roles and an area of needed improvement is perceived to be seen as a weakness. If each participant is honest and open to learning, the exchange of information can make decision making easier and very effective. For a Baby Boomer who has worked hard to earn their spot and be in a decision-making position, it can be difficult at times to have to admit when they do not know or understand a topic, and actually learn from a Millennial. It can be even more difficult for that same Baby Boomer to relinquish some authority and let the younger person make the final decision. For a Millennial, it can be challenging to fully follow the knowledge a Baby Boomer shares as it may often require the Millennial to move out of their comfort zone and rely on developing interpersonal business relationships and a work/life balance that doesn't always meet their expectations. In addition, Millennials have come of age and entered adulthood at a time of rapid change. By not having the experiences of a much-more seasoned Baby Boomer, Millennials have a tendency to become impatient when changes aren't implemented as quickly as they may expect, both within their organization and their own professional career.

Opportunities to Enhance Workplace Safety

The generational differences found in today's labor force can impact an organization on a variety of levels and could potentially inhibit progress within any department or division. The most dangerous and unfortunate of these, however, is employee safety. It is imperative that the decision-makers within these organizations' safety departments recognize that a cultural shift has arrived and prepare to embrace and adapt to those changes. If this is done correctly, safety programming could be enhanced in a way unlike ever seen before.

Companies are now equipped with a workforce that is more experienced and generationally diverse than ever before, and this unique situation provides them with the potential to advance in areas including safety, productivity and overall efficiency. Particularly, employee safety and overall safety programming can be largely influenced in three key areas: Talent, Technology and Training.

Talent

Attracting, developing and retaining talent is paramount to any organization's long-term safety program success. This is as important now as it has ever been, and the different perspectives and values of the generations within the workforce make it as challenging as ever.

Attracting Talent

One of the most significant factors in attracting talent is the culture of an organization, which now must meet the expectations of a wide variety of potential employees. The benefit packages and organizational structure that many Baby Boomers are accustomed to are not likely to attract many Millennials who have a different perspective. An example of the difference in expectations could easily be observed through a conversation between Baby Boomers and Millennials involving a simple benefit such as paternity leave. Paternity leave was mostly unheard of when Baby Boomers were having children, so they often don't see it as a valued, or even necessary, benefit; however, paternity leave is quickly becoming an expected and valued benefit for most Millennial parents. Baby Boomers, especially at this stage in life, tend to be much more interested in retirement benefits than maternity/paternity benefits. A "one size fits all" benefits package for all generations doesn't fit as well as it may have in years past. Benefits such as these often represent the foundation of a company's culture and how potential employees will expect to be treated if they should choose to go to work for that organization.

Companies must recognize what potential employees are seeking when selecting a new employer. Prospective employees will likely want to know how the company will support them in balancing their work with their personal lives, how they will be integrating technology into their work and will their work be meaningful. The difficult answer for most companies is found in meeting the expectations of each generation. Millennials have a different set of expectations as a group than do Baby Boomers. Progressive organizations recognize that neither is right or wrong. The organizations most likely to attract the best applicants, regardless of the generation they come from, are those that embrace the value of developing a culture that recognizes, and makes every effort, to support the goals of their individual employees, both at work and in their personal life. Work is no longer just a place to which one goes to make money. It is an integrated part of the employee's life both on a professional level and a personal level.

Developing Talent

While they are often perceived as a "work to live" generation, Millennials are often extremely self-confident in their abilities and worth; therefore, they value professional development and, due to the fast-paced culture in which they were raised, are often looking to develop and advance their careers in a variety of ways. When Baby Boomers began their careers, they likely saw a path to a position of authority within a company's Safety Department that required hard work and a willingness to patiently wait for their opportunity to advance. For Baby Boomers, that opportunity typically presented itself once their superior retired. Millennials are overall less patient and a pathway to a leadership role is not as clearly defined as it has been in the past. Baby Boomers, who hold many of these senior-level positions, are not retiring as early as previous generations. Therefore, the path to a decision-making position is often less evident to the Millennial who is eager to develop the skills to move into that desired position. This is where mentoring can be imperative to the development of newer employees. Having the opportunity to develop skills via a mentor is extremely helpful in an employee's overall professional development, when advancement opportunities are not yet present. Millennials can learn the value of interpersonal relationships as well as benefit from a Baby Boomer's experiences when they have a focused mentor.

Baby Boomers may believe they have developed fully once they've reached the position of authority and decision making they have sought in their career. This can lead to decision making based on "what has always worked for me". Unfortunately, the many changes that have and are currently taking place within organizations makes a Baby Boomer's previously effective thoughts, beliefs and norms less impactful. This is a place where cross generational mentoring can be very helpful. When Baby Boomers and Millennials mentor one another, each learn insights into what has formed an organization's culture to date and how it needs to adapt to meet the needs of a rapidly changing workforce.

An area many companies need to avoid is giving all safety technology related projects to Millennials. Many managers make the mistake of giving technology work to their Millennial staff members "because they are all about technology." Millennial generation employees may be comfortable with technology, but their interests are far more diverse than that alone. They are interested in development which helps them to become a well-rounded, more promotable employee. Limiting an employee's assignment to technology will make it difficult to retain staff members.

Retaining Talent

The final component of leveraging generational differences regarding talent is employee retention. The presence of social media in the everyday lives of Millennials has led to a concept called "FOMO", or the "Fear of Missing Out". Young workers today have access to more information than ever before. Included in that information, is the current status of their peers, both personally and professional. This information can be a real threat to employee retention and many Baby Boomers are not yet aware of that risk.

Employees who spend time on social media know a tremendous amount of information about their former classmates, friends, family members, etc. The information includes work-related topics such as promotions, travel, benefits, bonuses, and so on. Additionally, employees may also have access to observing other big ticket items such as homes, cars, vacations, etc. being a symbol of someone's "terrific job". Social media users can suffer greatly from a perception that "the grass is greener on the other side". When most Baby Boomers joined the workforce, they were only aware of the other staff members they saw frequently and information from the annual Christmas card they received from colleagues. Millennial staff members are well-informed about the "culture" of other companies and how their employees are progressing in life. It is important for organizations to recognize this high level of information exchange and ensure they are offering employees' ways to be enriched in their current safety department positions and see that opportunities exist in house. This is an area in which mentoring again can be very helpful. An employee who believes they are learning skills and doing meaningful work are much less likely to suffer from FOMO.

Retention of staff from the Baby Boomer generation is different. Baby Boomers are much less likely to be concerned with missing out on what their peers are doing in their careers. They are more likely to be interested in stability that allows them to stay with the company until their retirement. Baby Boomers, as a group, have found a significant part of their identity in their jobs and that will continue in the later stage of their careers. They want to have work that brings them a sense of accomplishment and allows them to flex their "organizational muscles". Millennials are equally interested in work they can feel good about, but it often comes from an interest in work they find personally meaningful and in some way, contributes to their sense of purpose.

Technology

Technology has brought significant change to the workplace and it appears that the pace at which this change is taking place is becoming faster. From our computers to our smartphones, our abilities to communicate with one another or retrieve and manipulate data quickly has brought rapid changes to the workplace. When we purchase a cellular phone today, we expect a newer, better model to be available within a year or two. In the meantime, however, we also expect regular software updates and new applications that improve the device's functionality and application. This rapid change and improvement is a part of the Millennial culture. The technology available today has been transformed, largely by younger minds, into the devices and applications that we enjoy. That same innovation and technological advancement can be applied to the workplace and specifically to safety. Every day, technology is being transformed into a new way in which employees can work more safely. Unfortunately, the decision-makers too often do not see the need and/or application of this technology, and are reluctant to embrace changes with which they are not comfortable.

Daily integration / Champion

For some employees, change will be challenging and for others it will be embraced. When safety technology brings about change, people will react according to their understanding and comfort level. In general terms, Millennials are technologically-savvy and will embrace these advancements. For Baby Boomers, it may be a bit more challenging to accept. As technological advances continue to be developed, companies that seek ways to integrate these changes into their employees' workday will likely be the most successful.

In order to be a leader in technological advancement, companies must allow those employees with an affinity for technology to seek out new technologies that will benefit the company, and utilize these employees to develop the means for helping all employees, regardless of their generation, to use the new technology. This type of cultural development within an organization demonstrates that new technology is a part of the company's growth and everyone will be supported in integrating the change into their work. This, in turn, inspires creativity in applying technology but also takes away the challenge and uncertainty of how the technology will impact an individual.

This type of development can be easily observed by companies that utilize technology in safety or toolbox meetings. A meeting can easily be conducted without any technology, but it can be enhanced by presenting the training on a video screen or tablet. Some people are more visual learners, so offering information with that additional support can help some meeting attendees better understand the message. For a supervisor who is comfortable with technology, this may be an easy addition, but for others, this could lead to frustration and a feeling of incompetence.

System Enhancement

Safety departments have gathered safety data for decades, and many safety professionals honed their skills with pencils, checklists and hours of data analysis and manual corrective action tracking. With advancements in technology, companies can now gather real time data that can be analyzed quickly. Safety professionals are able to utilize tablets, along with software, to monitor a wide variety of critical risk management business processes in an integrated system. The systems are typically cloud based so that multiple users can store and retrieve information easily and include functions such as hazard identification, inspections/audits, accident investigations, employee training, etc. The data from these activities is then stored for analysis and corrective action development. And once this is done, systems track corrective action to completion. Technology allows users to easily gather and analyze data with far less administrative time and the system does the tracking of items with automated "reminders" to ensure important items don't "fall through the cracks".

For a Safety Department employee who embraces technology and change, this type system is a dream come true. For a seasoned veteran who is less familiar with technological advancements and is reluctant to change, this is a nightmare. Setting up a process for helping each of these is critical to the implementation and integration of these systems. Companies who embrace this concept and leverage the technological strengths of their Millennial workers can make this type of transition much more smoothly.

Training

Employee safety training should be required when a new employee is hired, when an employee is undertaking a new task, when new equipment or new processes are put in place and when employees transfer to a different work area. Furthermore, routine training is necessary to ensure employees understand and are refreshed on their initial training. Companies that conduct training this frequently tend to use the same techniques or resources to deliver the message, often leading to a lack of attention or interest. By leveraging the strengths of the generations in the workforce, companies can bring new and effective methods to enhance and improve employee safety training.

Explore New Platforms

Employees are often asked to watch the same training videotapes and DVDs, or sit through the same PowerPoint presentation, year after year. While these can be very effective methods for training employees, training expectations and needs appear to be changing. Millennials are utilizing new platforms such as YouTube for their news and educational needs. According to Google (Google/Ipsos 2015), 91% of smartphone users turn to online resources such as YouTube doing a task and the largest demographic group doing these "how-to" lookups is Millennials. Progressive companies are recognizing these trends and are finding that training and communications must be delivered in a variety of platforms. Having technology in place allows trainers to access resources such to bring informative and entertaining content into their training programs.

Life Skills

While we may never see the qualifications "likes to tweet" or "good at video games" on a job posting, companies should recognize that some life skills may translate well into work skills. Employees that are comfortable with technology and utilize it as part of their leisure time are more likely to be comfortable with operating equipment or directing a company's social media campaign than someone who does not. As we've noted earlier, Millennials are tech-savvy and employers can leverage this affinity to help them integrate new technologies or platforms into their operations.

Baby Boomers also play a vital role in assisting less experienced employees with implementing change within a company. An experienced employee can draw from their array of interpersonal relationships, both at work and in their personal life, and help a tech-savvy trainer recognize that not all learners have the same foundation and how effective different training approaches will work.

Cross-Generational Mentoring

Millennials have skills that Baby Boomers do not have and are needed by companies. The exact same thing can be said for Baby Boomers. Imagine the abilities of the two working together. Companies would have an experienced employee who fully understands how change can be implemented in an operation coupled with the creative energy and a vast understanding of the power of new technology and social media. Baby Boomers will learn that Millennials are not the one dimensional stereotype found in the media and vice versa. Millennials should be able to gain knowledge from the experiences their mentor has had in their career while Baby Boomers will learn how new products or services can be brought into a company's operation.

Cross generational mentoring will help people to better recognize the strengths of others and allow projects or functions to be led by the most capable person and not necessarily the boss. This is a leap of faith for some organizations and could cause issues with a member of the company's leadership team that thinks seniority trumps all. If long held norms of decision making and authority can be set aside, companies can allow less experienced but skilled employees to lead at some times and seasoned veterans to lead at other needed times. There can still be a structured organizational chart that will be needed, but it will need to be less rigid than in times past to allow flexibility in leadership and decision making. This is the starting point of leveraging each generation's strength.

Conclusion

As noted earlier, one of the simplest ways to begin leveraging the strengths of each generation is to recognize that a cultural shift is occurring in organizations and embrace it. Organizations that fail to recognize the changing demographics and shifting cultural norms of society as well as their workforce, will be left behind when it comes to attracting, developing and retaining the best employees. Employee expectations, coupled with rapidly changing technology, are bringing significant change to the workplace. Companies and their leadership teams who recognize and champion these changes will benefit significantly as they will be able to leverage the strengths of their employees in a variety of areas including productivity, product or service quality and employee health and safety programming.

Bibliography

Coupland, D. 1991. Generation X: Tales for an Accelerated Culture. St. Martin's Press.

Employee Benefit Research Institute. 2016 survey by the Employee Benefit Research Institute (EBRI 2016).

Ernst & Young (EY). 2013. *Younger Managers Rise in the Ranks.* http://www.ey.com/us/en/issues/talent-management/talent-survey-the-generational-management-shift

Forbes, What Is A 'Millennial' Anyway? Samantha Sharf, August 2015.

Gallup. 2014. *Many Baby Boomers Reluctant to Retire.* (http://www.gallup.com/poll/166952/Baby-Boomers-reluctant-retire.aspx).

Google/Ipsos. August 2015. *Consumers in the Micro-Moment.* U.S. n=1291. Online smartphone users 18+.

Howe, Neil and William Strauss. 1991. *Generations: The History of America's Future, 1584 to 2069.* New York: William Morrow & Company.

Jones, Landon. 1980. *Great Expectations: America and the Baby Boom Generation.* New York: Coward, McCann & Geoghegan.

Pew Research Center. 2015. Millennials overtake Baby Boomers as America's largest generation. (http://www.pewresearch.org/fact-tank/2016/04/25/millennials-overtake-Baby-Boomers/).

Strauss, William and Neil Howe. 1998. *The Fourth Turning: An American Prophecy.* New York: Broadway, (print).

"The Younger Generation." *Time.* November 5, 1951.

Session No. 667

Harnessing the Power of Safety Data to Improve Compliance and Performance

Steve Ludwig
Rockwell Automation
Cleveland, OH

The State of Safety

Workplace safety is a multifaceted issue for many manufacturers and industrial operators. It includes both machine and process safety, and is vital to protecting workers, avoiding production interruptions and achieving operational excellence. But it also comes with a number of challenges, including:

- Worker behaviors: A safety-system design should consider every human-task interaction within the machinery or equipment. Operators may bypass poorly designed safety systems that don't take these interactions into account. And even in a well-designed safety system, workers may not follow standard operating procedures, which can result in unnecessary downtime.
- An evolving workforce: A major workforce shift is underway worldwide, and the safety implications are significant. Older workers who are nearing retirement but still on the job are at higher risk for certain injuries and can take longer to recover. And the younger, less experienced workers taking their place are more prone to injury and tend to have more serious injuries.
- Machinery stoppages: Stoppages can happen for any number of reasons: jams, misfeeds, adjustments, changeovers, maintenance and more. However, companies often have minimal visibility into why or when these stoppages occur. This prevents them from understanding if stoppages can be attributed to specific workers, machinery, lines or shifts. It also limits their ability to remedy issues.
- Regulatory compliance: Compliance with complex global safety standards is increasingly essential but also challenging. At the same time, these standards allow for the use of more advanced safety technologies that enable companies to address safety and productivity in new ways.
- Data management: Many safety professionals continue to rely on outdated data-collection and reporting methods. Most often, safety data is manually entered for inspections, compliance logs, incident reports, training and other processes. And the systems in which this data is stored are not connected to plant-floor systems.

This is in sharp contrast to the growing use of real-time data and seamless connectivity in the rest of the production environment. Everyone from operators and technicians to corporate executives are using "smart" capabilities to gain new efficiencies, improve product quality and make operations more responsive. Rockwell Automation calls this The Connected Enterprise.

Rise of The Connected Enterprise

The Connected Enterprise begins with the convergence of enterprise-level information technology (IT) and plant-level operations technology (OT) systems. Merging these historically separate systems into a single, secure network architecture provides a seamless foundation for real-time connectivity and information sharing.

With this foundation in place, companies can harness the power of enabling technologies, including:

Industrial Internet of Things (IIoT) technologies include any device connected via the Internet Protocol. They enable companies to access production, quality, safety and other data that, until now, has been trapped in machines, manufacturing processes and supply chains.

Big data and data analytics help manage massive amounts of data and contextualize it into actionable information that is relevant to each worker. This can help improve performance management on the plant floor, such as with predictive analytics, and supports better decision making and frictionless productivity companywide.

Wireless and mobility technologies can access, capture and communicate data in new ways. For example, mobile video communications can put an expert based in Chicago on a plant floor in China at a moment's notice.

Altogether, these technologies are helping manufacturers and industrial operators improve:

- Visibility and data management
- Operational efficiencies
- Quality management
- Asset utilization
- Inventory management
- Workforce utilization and knowledge retention

Smarter Safety

So how can safety professionals use The Connected Enterprise to replicate the benefits already seen on the operations side to drive safety performance improvement?
It begins with the use of contemporary safety technologies that combine machinery and safety control into one platform. These systems are less susceptible to nuisance shutdowns than hardwired safety systems, which can help improve productivity and profitability.

But they also offer another key benefit: access to safety-system data, including:

- Device status
- Operational status
- Error or fault codes
- Event counters or timers

- Event sequences
- Stoppage codes
- Motion monitoring
- Notification of safety logic changes

By harnessing this data and the greater connectivity available within The Connected Enterprise, safety and operations professionals can work together to:

- Improve safety-system visibility
- Better understand safety risks
- Enhance safety
- Reduce safety-related downtime
- Transform operations with safety in mind
- Evaluate safety system use or misuse
- Improve compliance

Put Safety Data to Use

The ability to access safety system data and convert it into meaningful information has enormous potential to transform how safety professionals monitor and manage safety. A key opportunity: incorporating safety information into EHS management systems to identify discrepancies between how policies and procedures are defined and how they're actually followed in day-to-day operations.

For example, e-stop buttons are intended for use only in emergencies. But workers can easily misuse them, such as to clear a jam or stop production at the end of a cycle. Such inappropriate usage can lead to increased scrap and longer machine start-up times.

The data for these actions can't be captured in most plants today. This lack of insight hinders safety professionals from identifying misuse of safety systems and can present vexing challenges for those seeking to drive continuous improvements in their EHS performance. It also limits plant managers from truly understanding and addressing downtime issues.

But safety-system data can be captured from information-enabled devices in The Connected Enterprise. Attainable data can include an e-stop activation's time stamp, downtime duration, and the line and shift associated with each activation. Stoppage reason codes also can be built in to convey why a machine was stopped, such as for jams, misfeeds, cleaning and more.

Safety professionals can then use their existing alarm-and-events and metrics software to analyze the data and identify if e-stops are being used at an abnormally high rate. From there, they can investigate the issue's root cause, such as insufficient standard operating procedures or improper machine design. They also can identify if higher activation rates are associated with specific production lines or shifts, which could indicate the need for additional training for a select group of workers.

Similar visibility can be extended to other individual safety functions, including light curtains, safety mats, guard doors or gates, and lockout/tagout (LOTO) procedures. And it can be extended to examine the sequence timing and relationships between multiple safety functions.

These capabilities ultimately allow safety professionals to identify and resolve areas where safety's execution on the plant floor falls short of the safety strategy defined in the EHS management system.

Better Understand Safety Risks

Risk assessments are vital at the design stage to identify machine safety risks and risk-reduction measures. After the design stage, however, the assessment data is rarely ever used again.

The availability of safety data in a Connected Enterprise brings renewed purpose to risk-assessment data in the form of a risk calculator. This tool can easily be configured as a basic table within the FactoryTalk® VantagePoint EMI (enterprise manufacturing intelligence) software.

Within the tool, safety professionals can enter the anticipated use frequency data from a machine's risk assessment as a baseline for safety performance. They can then compare this against the machine's actual use frequency data. The result is an unprecedented capability to measure anticipated risk against actual risk for each machine access point.
Use frequency that is significantly higher than anticipated could be the result of a product or process change that needs to be addressed. And use frequency that is lower than expected could indicate that a safeguard is being defeated and needs to be re-evaluated.

Enhance Safety

Connecting people, equipment and worksites creates new opportunities to enhance worker and environmental safety, including:

- Remote access: Transportation incidents are a top cause of fatal work injuries in the U.S. Remote monitoring of dispersed or isolated operations can reduce worker travel demands, such as the need for oil and gas workers to travel across sites to check on wellheads, pump stations and storage sites.
- Operations visibility: Visibility into manufacturing process states, environmental conditions and other factors can be critical when working with hazardous materials or in harsh conditions. For example, pharmaceutical or chemical companies can use real-time data to be sure hazardous chemicals are not released into the environment.
- Worker locating: Network connectivity can be a company's only link when its employees are working in isolated locations. For example, wearable sensors can locate workers in underground mines and other hard-to-reach places during an emergency. Video, voice and display technologies also can help monitor and communicate with these employees should a safety incident occur.
- Information delivery: Especially for older workers nearing retirement, improving ergonomics can help reduce the likelihood of injury and help retain workers with valuable experience. Wireless and mobile technology can support this by delivering information in a convenient way and helping reduce the strain put on older workers.

Reduce Safety-Related Downtime

Better visibility into safety-system performance and stoppages can help determine root cause of shutdowns. Safety and production data also can be combined to understand the frequency, duration, time and location of safety-related shutdowns.

Armed with this information, safety and operations professionals can work together to develop mitigation plans to improve productivity. This could be as simple as having discussions with operators on a single production line or shift where downtime issues are most frequent. It also could involve updating procedures if specific downtime issues are common throughout an operation.

Additionally, improved safety-system diagnostics can alert operators and technicians to where safety-related failures are occurring during runtime operations. This can simplify troubleshooting and lead to faster downtime resolutions. The diagnostics even can be used in predictive analytics to identify leading indicators and address machine issues before they become machine failures.

Ease Compliance

The safety data required for compliance and reporting purposes is largely collected through manual audits today. This can be a time-consuming process that requires valuable work hours, results in production downtime and can be subject to human error.

By integrating auditing functions into the HMI and controller, organizations can automate and speed up the auditing process, free up personnel to focus on other priorities, and reduce the likelihood of errors. Also, any abnormalities can be annunciated in the HMI dashboard or reports to help plant personnel quickly spot and address potential issues.

Transform Operations With Safety in Mind

Connected operations present opportunities for companies to create inherently safer operations.

For example, manned topside platforms used in off-shore oil and gas production can be vulnerable to potentially catastrophic events, from explosions to ship collisions. They also often require helicopter transportation for supplies and staff, which can be dangerous. Ethernet-connected subsea platforms that pump gas directly to onshore production facilities can reduce the need for manned topside platforms.

The growing use of connected, autonomous technologies in mining operations also is creating new opportunities to help minimize safety risks. Mining companies already are using autonomous trucks and trains to transport materials. The driverless vehicles can be tracked and controlled from a central location, resulting in true "pit to port" connectivity.

Implementation

While all of these benefits are more easily realized in a fully Connected Enterprise, they also can be achieved in an iterative process as companies become increasingly connected. Regardless of the approach a company takes, some key steps are crucial to making the most of safety in connected, information-enabled operations.

1. Active EHS Representation

Connected operations span people, processes and technology. As a result, bringing people together from across an organization is critical when deploying a Connected Enterprise or simply expanding connectivity in a smaller manner. A cross-functional team should be formed to include not only operations and IT stakeholders but also environmental health and safety (EHS) professionals.

Only by having a seat at the table can EHS professionals define their goals as part of a larger, companywide connected strategy. This should come naturally to any company that is committed to the three Cs of safety – culture, compliance and capital – which are part of a holistic safety approach and can be measured using tools such as the Rockwell Automation Safety Maturity Index.

2. Assess the Current State

A company's roadmap to improving safety in a connected operation begins with understanding where it's at today. Key questions to ask during an assessment include:

- Are we using contemporary, integrated safety systems?
- Is safety data manually collected or is it automated in a data-collection system?
- What safety data is already available? How is it reported?
- Does equipment or machinery have central network connectivity?
- Could safety data be collected using an existing data-collection platform?

3. Determine Meaningful Data and Information

This is where safety, operations and other team members specify the safety-system data they need to support their goals. Important considerations at this stage include identifying:

- What data must be captured
- Where and how it will be collected
- How it will be contextualized and delivered as actionable information
- Who will receive the information
- What actions will result from the information

4. Implement or Upgrade Safety Systems

Safety-system data is already available in an integrated safety controller. Implementing changes merely involves specifying the right data coming out of the controller by creating new tags for each access point. This data can then be contextualized within a plant's existing EMI software and delivered to safety and operations personnel in the form of relevant, actionable information via EMI dashboards.

Safety data also can be incorporated into other existing software, such as FactoryTalk Alarms and Events. Operators can create alarms and events for a wide range of safety scenarios, including light-curtain faults, opened panel doors and triggered e-stops. They can then track and analyze this information right alongside the other machine or fault data they're already monitoring.

Incorporating safety data into FactoryTalk Metrics can provide insights into machine or equipment performance in relation to safety events. For example, a plant manager can pinpoint a machine that has experienced multiple light-curtain faults as the source for excessive downtime and missed production targets during a specific production period. Meanwhile, a safety professional can look at the same data and identify those faults as a safety abnormality that needs to be addressed.

A library of preconfigured faceplates is available for each safety controller and component within the software programs. They provide detailed status information for each device and links to help information. Additionally, the software supports mobile capabilities, giving safety and operations professionals the freedom and flexibility to view safety information while they're roaming the plant floor, at a conference or at home.

5. Analyze and Optimize

Getting information to workers is important, but so is getting workers to act on the information.

That's why incorporating safety-system information into daily operations is critical. For example, analytics should be included in daily production meetings, and standard procedures should be developed for collecting, analyzing and interpreting data.

With these elements in place, safety professionals can monitor and refine all aspects of safety in a Connected Enterprise as part of a continuous-improvement program. This could include expanding the amount of data collected, or it could include setting higher goals in areas such as improved visibility, fewer incident rates and reduced downtime. Safety professionals also can use the safety-system data to better measure and meet their corporate safety goals.

It's important to remember that a Connected Enterprise is an ongoing journey. The technologies and opportunities within it will continue to evolve. Companies should be mindful of this and constantly seek opportunities to improve their future state.

Summary

The ability to access, analyze and act on safety-system data in a Connected Enterprise represents a turning point for safety compliance and productivity. It creates opportunities to better understand risks and safety-system usage, enhance worker and environmental safety, reduce safety-related downtime, and even reimagine operations as innately safer.

Resources

The Connected Enterprise: http://www.rockwellautomation.com/global/innovation/connected-enterprise/overview.page

Rockwell Automation machine safety solutions: http://www.marketing.rockwellautomation.com/safety-solutions/en/MachineSafety

Rockwell Automation Safety Maturity Index: http://www.rockwellautomation.com/global/products-technologies/safety-technology/safety-maturity.page

Session No. 669

Make Your EH&S Training Come to Life–And Change Behaviors!

David A. Casavant
Executive Director
Sustainable Workplace Alliance
Lake Wales, Florida

Introduction

If you're responsible for your organizations training needs, especially those that involve technical subjects, you have likely experienced the difficulty of training adult students. As adults, we have comfort levels, experiences and fears that mold our ability and desire to learn something new. If you're going to train your coworkers, you must reach them in a way that encourages them to change their behavior.

In a recent survey conducted by the Sustainable Workplace Alliance, we found that the overarching problem with technical training was "How to make the training interesting". It's true, teaching regulatory topics may not be the most exciting of topics, but does that really mean that the training must be boring?

Absolutely not! In fact, by focusing on just four critical areas, we can move the needle from boring and forgettable to exciting and memorable.

Using Icebreakers to Open Up Your Audience

Studies show that nearly 75% of the population fears public speaking. This is especially true when teaching difficult EH&S compliance topics.

In fact, there's a name for this fear: Glossophobia. Glossophobia is the fear of public speaking. The word "Glossophobia comes from the Greek root "glosso" (which means 'tongue') and the root "phobia" (which means 'fear')

Wouldn't it be nice to deliver a polished presentation, not be so nervous and get your audience involved and interested?

This can be achieved by starting the presentation with an icebreaker. The icebreaker is a low-stress opening that puts both the audience and the trainer at ease. If the trainer is less stressed out, she can deliver a memorable presentation. If the audience is at ease, they will be able to pay attention, feel free to ask questions and really absorb the main points of the discussion.

What does an icebreaker look like? Perhaps discussion of a current event, a relevant statistic or an emotional story. Be sure to ask relevant questions of the audience during the icebreaker. This puts them at ease and signals that your presentation will be an interactive one.

Participatory Training Experiences Are Critical

To effectively change your audience's behaviors, it's critical that you introduce "participatory training experiences" (PTE) into your training presentation.

Simply put, the PTE encourages your audience to interact, ask questions, share experiences and solve issues faced in their day-to-day jobs. Otherwise, you're simply lecturing and it's very possible that you're not in-tune with the unique needs and concerns of the audience.

Adult learners need to know that the training they receive will be relevant to what they do and will assist them in doing the task more efficiently. Participatory training allows the audience to see relevancy and resolve task oriented issues that they are facing.

The PTE concept is based on a study conducted by Dr. Edgar Dale, commonly known as the "cone of learning" (see Figure 1). The findings of this study are based on the realization that your students are more likely to remember, retain and practice what they were taught, if they were actively involved in the learning experience.

Dr. Dale, found that after two weeks the students would remember only 10% of what they read, 20% of what they heard, 30% of what they saw and 50% of what they saw and heard. These are passive forms of learning and are quickly forgotten by the students.

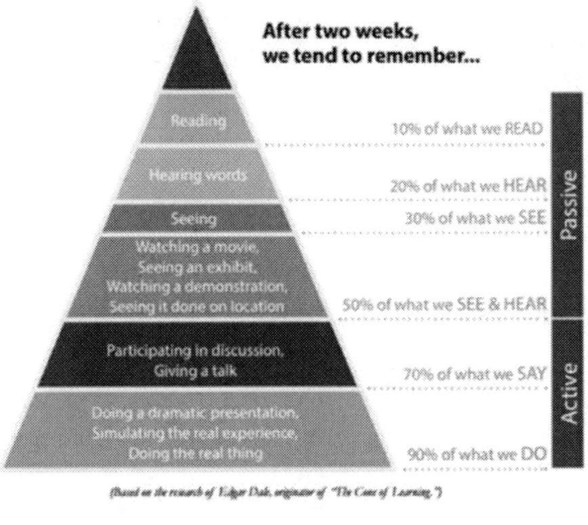

Figure 1.

Conversely, Edgar Dale found that students remembered 70% of what they said in a training class (i.e. participating in a discussion) and 90% of what they did in the training class. For instance, simulating the real experience. These last two forms of learning are active in nature and better support a learning scenario that is both interesting to the participant and is more likely to change behaviors.

Another study that drives home the PTE concept was conducted in 1970 by Carver, Johson & Friedman. This study found that the average person speaks at a rate of about 125-175 words per minute. Your audience has an ability to listen and process (albeit poorly) approximately 450 words per minute (See Figure 2)

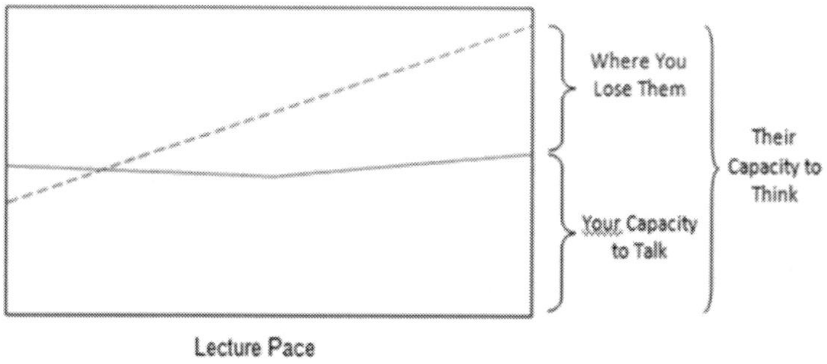

Figure 2.

The 75% time differential between the trainers talk rate and the audiences listen rate is where we lose them. If we don't make the training engaging, interactive and interesting, the audience will begin to wander. They multi-task. And the information we are sharing with them doesn't resonate. The solution to this time differential is to move away from lecture mode and incorporate a participatory training experience.

As we interrupt the lecture mode (which we can admit is important when teaching technical topics) and drop in interactive, participatory segments we regain the audience's attention just as they start to drift (see Figure 3). In other words, we mix up the presentation medium and delivery to accommodate the often-short attention span of our audience.

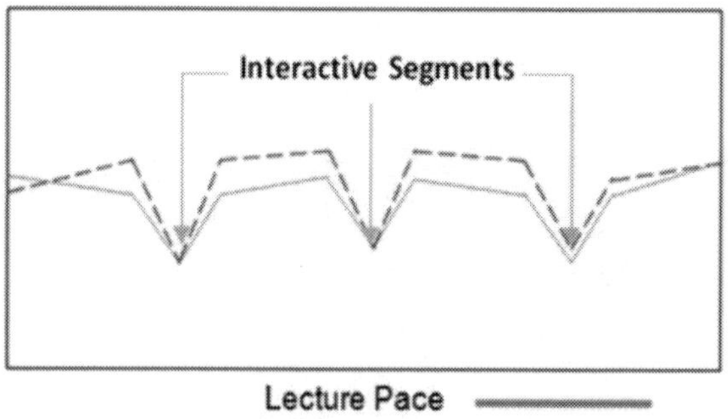

Figure 3.

Using Technology to Get the Audience Involved

Another way to introduce participatory training experiences into your presentation is to include interactive technology.

Technology is great, but the goal is to keep the technology simple to use for both you the trainer and your audience. With that in mind, a very easy to use and engaging presentation tool is known as Poll Everywhere (www.polleverywhere.com).

The premise of this technology is you conduct a live survey of your audience and they participate in this poll using their smart phones or tablets. For instance, you can display a question with multiple choice answers and the audience will use their cell phone to anonymously select the answer they believe to be correct (see figure 4).

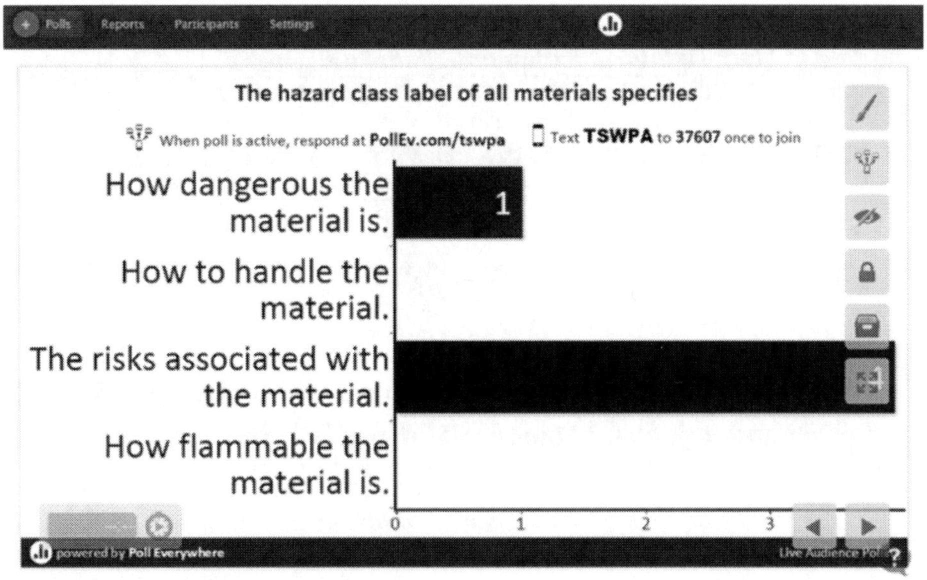

Figure 4.

This level of interactivity is useful to the trainer as it allows you to immediately get the pulse of the audience and determine if the information you are conveying is sticking.

But it's also fun for the audience and allows them to participate anonymously, which for the more introverted person is a real benefit.

Death by PowerPoint

PowerPoint is another form of technology, however its one that is often misused by trainers (especially those who train technical topics). The key to remember is PowerPoint should support your training. It's not the sum total of your training. PowerPoint can add life to your training by adding a visual element. It's not meant to contain every word that you're going to say while on stage.

A basic formatting rule should be followed when developing our PowerPoint slides. This rule is often called the "Rule of 55". This rule states that no sentence should contain more than five words and there will be no more than five lines of text on a single slide.

The concept is to declutter and simplify the screen. Your PowerPoint slide should supplement what you are saying from the front of the room, not merely repeat what you are saying.

Simplifying the screen can be a challenge when presenting certain documents, for instance the OSHA Form 300 (see figure 5)

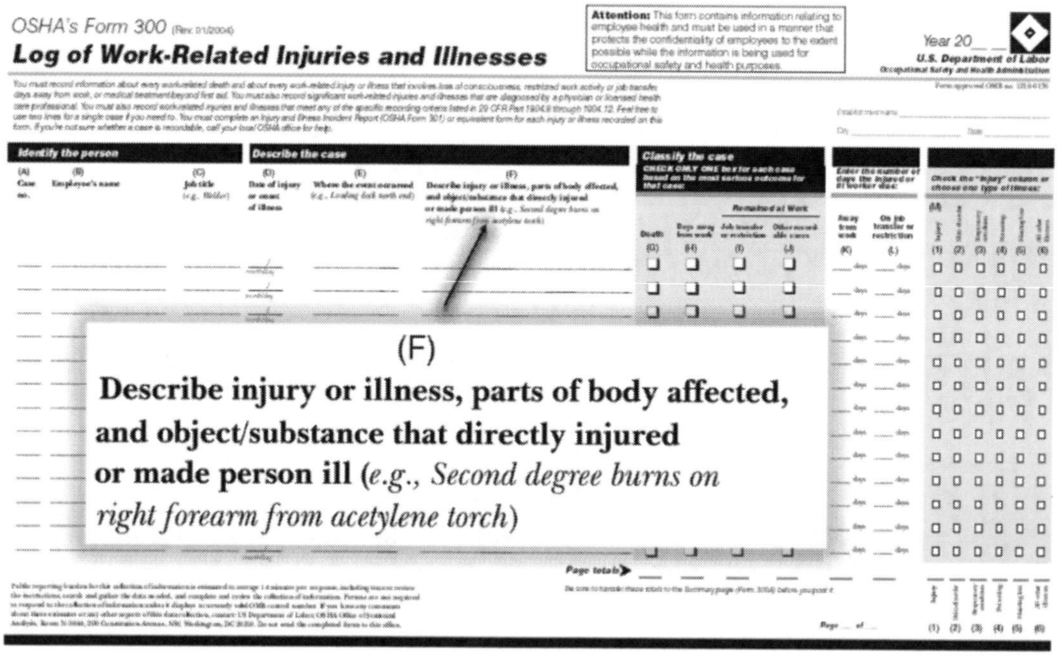

Figure 5.

In the example of the OSHA 300 Form, we have many column and rows that need to be explained. The problem is the text is small and difficult to read. This is true if displayed on a monitor or screen and its equally true if the student is reading from a handout.

The solution if to show the entire 300 Form so the student can get a sense of the layout, but as you are referencing individual columns, create magnifications that are easy to read. These magnifications can also be used in your handouts, thus helping the student to easily follow your presentation.

In Conclusion

Training technical topics need not be boring, tedious or difficult to follow for the audience. Nor does this training need be a stressful event for the trainer. With the implementation of a few best practices, both the audience and the trainer can enjoy an experience that will motivate and change behaviors.

Session No. 670

High Impact Training: Using Life Events and Social Media for Employee Engagement

Joseph Sanna, Jr., CSP, MS, CPSI
JoAnn H. Goshorn, CIH, CSP, MS, ARM

The PMA Companies
Camp Hill, PA

Introduction

Employee training is a common task of safety professionals during their daily work activities. Hours may be spent in development of training materials. Some organizations expend substantial portions of their safety budgets to purchase training materials. Despite the time and funds expended, sessions can be ineffective in influencing worker behaviors. Incorporation of information available via the internet can be used to create more impactful training sessions which result in greater employee engagement in safety training.

In many cases, training involves a routine session that covers regulations, policies, and procedures. Inattention and disengagement of the audience may result from boredom during the session. A component that could be missing is a convincing discussion of why the information is important to the employee. Statistics about workplace accidents and fatalities may be incorporated into the training, but the impact can be minimal if the information is provided without a context and the information does not resonate with the listener.

We can incorporate internet based information into training materials to create customized training sessions that will increase employee interest. When the listeners are interested in the story behind the training, and can identify with the content of the training, the session is more meaningful. This provides a greater opportunity to impact employee behavior.

Adult Learning and Employee Engagement

Much of the safety training conducted today entails a lecture style of meeting. An instructor conducts classroom training, often using presentation software and videos as training aids. Some research shows that lecture and audiovisual styles of training result in limited retention rates for students (National Training Laboratories, Bethel, Maine). These training techniques may fail at engaging employee attention thereby limiting the effectiveness of training sessions. Adult learners are interested in the goal of the training, what they are going to learn, and how the information could impact them personally.

A plethora of information is available via the internet. Internet based resources can be used to incorporate real life stories into training in a manner that creates a more personal impact to workers. Information can be obtained from resources such as the local news, OSHA Fatal Facts, You Tube, and even social media sources. The information can be used to create a training session that engages workers by showing them relatable real life scenarios. These scenarios help workers envision how their actions can impact others in their family, workplace, and social settings. From this prospective, the trainer goals and the employee goals can match. The desired outcome is to instruct employees to work safely, so they can continue to live and work in a manner that meets their personal goals. This corresponds to the employer's goal of maintaining a safe and healthy workplace. Both goals can be met.

Real Life Stories

Sharing real life stories provides a great opportunity to achieve employee engagement. The stories could be examples that have occurred within the organization or situations that employees have experienced in their own lives. A real life story based on the experience of a family member of a seriously injured worker is used to illustrate the impact of a work related injury on the community around the worker. The immediate family of the injured worker has to react when notified of a medical emergency. This includes communication from the employer to the family, communication among family members, traveling to a medical facility, and meeting with medical providers. The process results in an emotionally charged atmosphere that is stressful to everyone involved. A serious injury or a fatality forces family members to make difficult decisions under duress, which may negatively impact everyone involved. The outcomes of these situations could influence the worker and the family for years into the future. There may also be an impact on co-workers, as well as management of the employer.

A family tree graphic is used to illustrate the impact an accident could have on an extended family unit (Figure 1). The example in the real life story presented encompasses four generations. The concept of impact on the family is expanded to illustrate the impact on friends, co-workers, and the community around the injured worker. A generic family/acquaintance tree is provided and used to prompt the listener to think about the number of others who would be impacted in the event he or she was involved in a workplace accident. The family tree can be used as a guide to conduct an exercise for employees during a training session. Employees can use the family tree to list family members and acquaintances that would be affected if they were injured or killed in a workplace accident. This provides a personalized training experience that reinforces the importance of safety training.

The events of a family in a real life scenario can be used as a backdrop to challenge employees to define their ideas of personal success. Employees are asked to think about how their goals for personal success would be impacted as the result of a workplace accident. This provides the context for training that raises training from a routine chore of a safety professional to a meaningful experience for both the instructor and the attendees.

Internet and Social Media

Internet and social media can be mined for information that is applicable to supplement safety training sessions. Search engines provide tools to locate information about types of work related accidents and specific types of incidents. Searches for information such as "construction worker

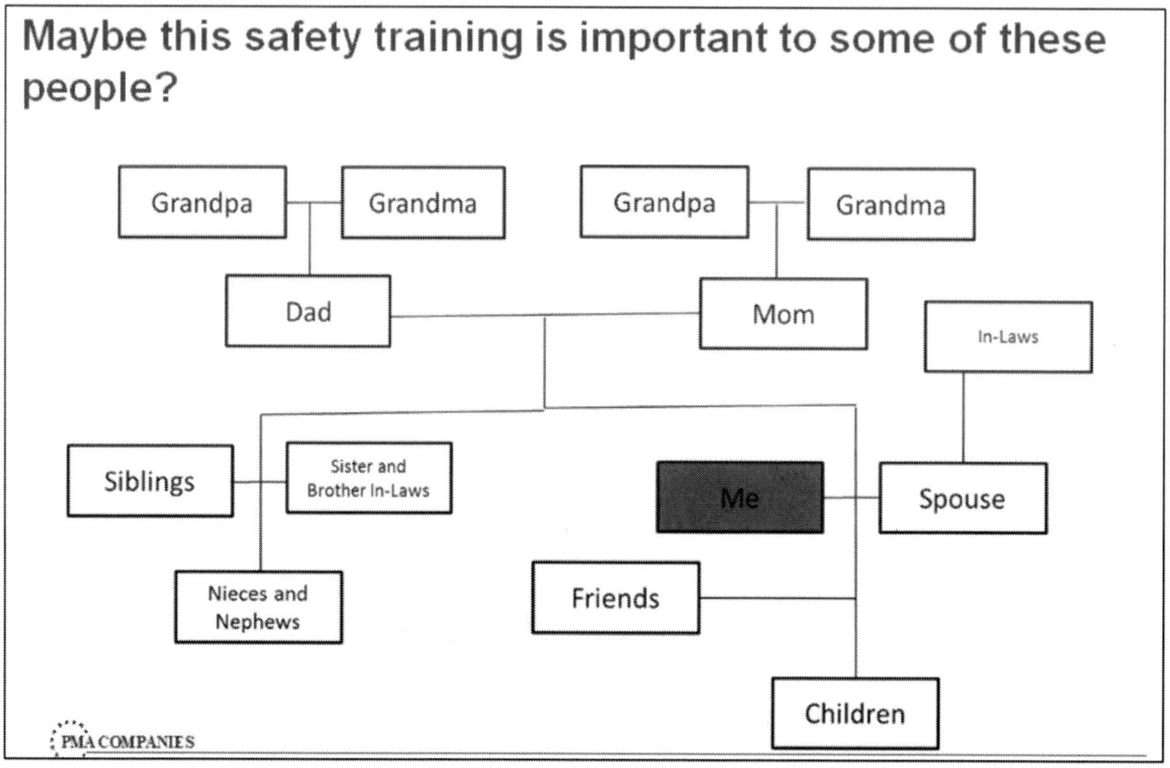

Figure 1. Family Tree Diagram.

fatality" or "industrial accident" can be entered in a search to find information. Searches can be refined by customizing the search to target information for trades such as roofers, masons, or others, or to define a specific geographic area such as a state or city, or a time frame, such as 2016.

Another resource is the OSHA website. This site contains a streaming list of fatal workplace accidents, the state where the accident occurred, the name of the worker, and a brief description of the accident. An internet search using this basic information will provide further details that can be incorporated into a training session. Use of information about specific incidents and the potential impact on a worker plus those around him or her generates a more personalized training session. The context of an accident that impacts multiple lives makes the training more relatable to employees. When employees can see themselves and those around them in the lives of other injured workers, there is a greater chance for the training information to positively impact employee behavior.

Other media resources contain information, descriptions, new articles, photographs, and even video clips of events surrounding workplace accidents. Internet searches can provide information for use during training sessions. Some examples of phrases that can provide material are:

- Worker dies
- Worker electrocuted
- Confined space death
- Ladder death

- Machine crusher worker
- Arc flash accident
- Employee dies

There are numerous phrases that can be entered into a search engine to obtain potential safety related resources. With some practice at developing phrases that target your specific needs, you will narrow results of the search. Use of search filters to define specific time frames or delete unwanted material will provide more targeted results. An example of a more refined phrase for entry into a search is "industrial accident, Denver, Colorado, 2016." This simple search phrase leads to a number of articles about workplace accidents, some resulting in fatalities, in one city of the US, during a specific time period.

Another example of information available via the internet that can be used to supplement safety training is from an industrial accident that occurred at a beverage plant in Jacksonville, Florida. An internet search using the phrase "bottling accident in Jacksonville, Florida," will lead to various news articles about the accident, information about the OSHA citation that resulted from the accident, and a video of the scene, which is available for internet based viewing. This example could be used during training for new employee orientation, use of temporary workers, machine guarding, lock out/tag out, supervisor leadership, or other topics.

The OSHA home page provides a running list of fatal accidents that occur in the US every year. According to OSHA statistics, more than 4,500 workers die on the job each year. This averages to approximately one fatal work related injury every two hours. Entering the worker name and state, along with the phrase workplace fatality, or similar wording will provide information about these specific incidents.

Conclusions

Safety training can be ineffective in impacting worker behaviors if there is a lack of engagement of the audience during training. Training sessions can be made more effective by including real life scenarios that are designed to enhance the safety training topic. Information about work place injuries and fatalities is readily available through internet resources.

A key to developing training sessions that can create a greater impact is to include information that leads employees to see a more personal side to workplace accidents. Incorporation of real life scenarios as an educational tool will lead the training attendee to visualize his or her own family and the impact of an accident on the people most important in their lives. This training method can lead to a higher level of employee engagement and will make your training sessions achieve greater results.

Bibliography

Grabell, Michael; Olga Pieree, and Jeff Larson. "Temporary Work, Lasting Harm." ProPublica, December 18, 2013. www.propublica.com (retrieved 2 February 2, 2017).

Grabell, Michael, Krista Kjellman Schmidt, and Minhee Cho. 2013. *First Day on the Job Was His Last: What Happened to Ray Davis*. (Video file). Retrieved from https://www.youtube.com/watch?v=cNcsTRQNZLE/2017/02/26.

Heard, Dr. Wildav. "What is the learning pyramid?" March 6, 2013. https://drwilda.com/2013/03/06/what-is-the-learning-pyramid/2017/02/26

Occupational Safety and Health Administration. "OSHA Fatal Facts." www.osha.gov (retrieved February 2, 2017).

______."Workplace Fatalities." www.osha.gov (retrieved February 2, 2017).

Reilly, Robyn. "Five Ways to Improve Employee Engagement Now." Gallup, January, 7, 2014, www.gallup.com (retrieved March 1, 2017).

Implementation of the New FDA Sanitary Transportation of Human and Animal Food Rule

Brian Hammer, MPA CDS
Senior Transportation Consultant
Nationwide Insurance
Des Moines, Iowa

Introduction

The new The Sanitary Transportation of Human and Animal Food rule enacted by the Food and Drug Administration's (FDA) Food Safety Modernization Act (FSM) is now final. It was published in the Federal Register on April 6, 2016. The rule further advances the FDA's efforts to protect food from farm to table by helping to keep them safe from contamination during transportation.[1]

The rule builds on safeguards envisioned in the 2005 Sanitary Food Transportation Act (SFTA). Illness outbreaks resulting from human and animal food contamination during transportation, as well as incidents and reports of unsanitary transportation practices, have caused concerns about the need for regulations to help ensure that foods are transported safely. Learn more about the Sanitary Transportation of Human and Animal Food rule on the FDA website.

History

Congress passed the Sanitary Food Transportation Act (SFTA) of 1990 (49 USC 5701) in response to concerns about how food was being transported in trucks that also haul a large variety of products that were consider inconsistent with hauling food products safely.

It is not hard to comprehend that the Federal Highway Administration, the precursor to the Federal Motor Carrier Safety Administration, was charged with developing regulations to keep foods safe during transportation and did not have the expertise to do so.

The next big transportation bill was the National Economic Crossroads Transportation Efficiency Act (NEXTEA). It transferred the responsibility to the FDA and required that they develop regulations to ensure that the transportation and storage of food was done in a safe manner no matter what mode of transportation it was taking. Some of the highlights are:

- Regulations to forbid commingling of food stuffs and hazardous materials
- Required training for inspectors at the FDA and USDA level but did nothing to insure that DOT inspectors, who would actually interact with the carrier and food while in transit.
- Required DOT inspectors to notify FDA or USDA of possible contamination of food while in transit. The USDA has control over meat and poultry.

In 2005, Congress passed the 2005 Sanitary Food Transportation Act

- Requires the Secretary of HHS to issue regulations setting forth sanitary transportation practices to be followed by shippers, carriers by motor vehicle or rail vehicle, receivers, and others engaged in food transport; and
- Requires the Secretary of DOT, in consultation with the Secretaries of HHS and USDA, to establish procedures for transportation safety inspections for the purpose of identifying suspected incidents of contamination or adulteration of a food.

Who does the rule cover

The Sanitary Transportation of Human and Animal Food rule applies to shippers, receivers, loaders and carriers who transport food in the US by motor or rail vehicle, whether or not the food is offered for or enters interstate commerce.[2] It also applies to:

- Persons, such as shippers, in other countries who transport food to the US directly by motor or rail vehicle (from Canada or Mexico), if the food will be consumed or distributed in the US.
- Persons, such as shippers, in other countries who transport food by ship or air and arrange for the transfer of the intact container onto a motor or rail vehicle for transport within the US, if the food will be consumed or distributed in the US.
- Companies involved in the transportation of food intended for export until the shipment reaches a port or US border.[3]

The rule does not apply to exporters who ship food through the US (for example, food being transported from Canada to Mexico) by motor or rail vehicle if the food does not enter US distribution.[2]

The following are also exempt from the rule:

- Shippers, receivers or carriers engaged in food transportation operations with an average annual revenue of less than $500,000
- Transportation activities performed by a farm
- Transportation of food that is transshipped through the US to another country
- Transportation of food that is imported for future export and is neither consumed or distributed in the US
- Transportation of compressed food gases (for example, carbon dioxide, nitrogen or oxygen) authorized for use in food and beverage products or food-contact substances

- Transportation of human food byproducts for use as animal food without further processing, such as those sold directly to farmers to be fed to livestock. (Byproducts transported to facilities to be manufactured into feed or pet food are *not* exempt.)
- Transportation of food that is completely enclosed by a container, except food requiring temperature control for safety
- Transportation of live food animals, except molluscan shellfish[3]

Note: Please be aware of potential contamination risks when transporting to farms as mud, debris and manure can be transferred onto trucks or trailers.

Key requirements

The Sanitary Transportation of Human and Animal Food rule establishes requirements for:

- Vehicles and transportation equipment – The design and maintenance of vehicles and transportation equipment should not cause transported food to become unsafe. For example, vehicles and equipment should be adequately cleanable, suitable for their intended use and capable of maintaining temperatures necessary for safe transport.[4]
- Transportation operations – During transportation, measures should be taken to help ensure food safety, such as maintaining adequate temperature controls, preventing the contamination of ready-to-eat food from touching raw food, protecting food from contamination by non-food items in the previous or same load and protecting food from cross-contact, such as the unintentional incorporation of a food allergen.[5]
- Training – Carrier personnel should have documented training in sanitary transportation practices. It is required when the carrier and shipper agree that the carrier is responsible for sanitary conditions during transport.[6]
- Records – Records of written procedures, agreements and training should be maintained. The required retention time for these records depends upon the type of record and when the covered activity occurred, but does not exceed one year.[7]

Key Requirements for Shippers

Shippers should develop and implement written procedures to help ensure that:

- Equipment and vehicles are in appropriate sanitary condition
- Previous cargo does not make food unsafe when transporting in bulk
- Adequate temperature control is maintained for food safety[4]

Key Requirements for Loaders

Loaders were added after the proposed rule and play a very important part in food transportation. When training loaders, it is incumbent that the important role they play in the transportation chain is explained to them

The loader must determine that:

- The transportation equipment is in appropriate sanitary condition before loading food that is not completely enclosed by a container
- Each mechanically refrigerated cold storage compartment is adequately prepared for refrigerated transportation before loading food requiring temperature control[5]

Loaders should be trained on how to properly inspect the trailer and to look for signs of contamination. It is important that they know the washing requirements for the food they are about to load and can verify that the trailer has received that type of wash.

It is important that loaders understand how to insure that a trailer has been pre-cooled to required temperature and how to operate the system to make sure it is turned off during the loading process.

They need to know how to inspect packages to look for signs of defect, leaking or other deficiencies. Loaders should be instructed on proper methods to pulp check for temperature without contaminating the product or causing damage to the packaging.

Proper load segregation and securement are tantamount for the safety and security of food during transportation, so a loader should be trained to accomplish that task.

Key Requirements for Carriers

Transportation operations should accomplish the following:

- Assign competent supervisory personnel
- Conduct operations to help keep food from becoming unsafe during transportation
- Take measures to help prevent the cross-contamination of raw foods and non-foods by using techniques of load segregating, isolating or packaging
- Isolate, segregate or use other measures to help prevent contamination or cross contact of food not completely enclosed in containers or transported in bulk vehicles
- Take effective measures to help ensure the adequate transportation of food requiring temperature control for safety.[5]

If a shipper transfers responsibility to the carrier, the carrier must have written procedures that:

- Specify practices for cleaning, sanitizing and inspecting vehicles used for food
- Describe how it will maintain and record temperature control as required by the shipper
- Describe how it will make provisions for the use of bulk vehicles, including maintaining information/records on the vehicle's previous load and maintaining cleaning/wash records as required by the shipper[5]

The carrier is required to train personnel in sanitary transportation practices and have a documented certification of training. While there is no mandated length or syllabus for the certification training, suggested topics include:

- How to inspect trailers for signs of contamination and damage that may cause loss of temperature or allow contamination
- Load segregation, isolation and securement
- How to record and maintain temperature control
- Hand washing and general hygiene
- Food safety items that may be unique to the shipper's or carrier's specific operation

If a carrier takes on the responsibility for food safety from the shipper, they should first consult with an attorney who can more properly outline the additional risk that the carrier is taking on. Unfortunately, that risk may have to be assumed as it may simply be a condition of receiving the load and the carrier will have little if any push back if they in fact would like the opportunity to handle the load. Once they do, then the training and recordkeeping requirements kick in.

Carriers who routinely haul food should have a basic food transportation training for drivers as a base training module followed up by any additional training that might be required either by the shipper or by the type of food being transported that may be specialized beyond the basic training. (See: Appendix A, "Delivery Driver Lesson Plans".)

Compliance Dates

Recognizing that businesses, especially small businesses, may need more time to comply with the requirements, the FDA has adjusted compliance dates accordingly.

Small businesses, other than motor carriers who are not also shippers and/or receivers, employing fewer than 500 persons and motor carriers having less than $27.5 million in annual receipts must comply two years after the publication of the final rule, making their compliance date April 6, 2018.

Other businesses that are not small and are not otherwise excluded from coverage must comply one year after the publication of the final rule, making their compliance date April 6, 2017.

References

[1] 21 CFR § 1.900-1.934. Sanitary Transportation of Human and Animal Food. US FDA.
[2] 21 CFR § 1.900. Who is subject to this subpart? US FDA.
[3] FSMA Final Rule on Sanitary Transportation of Human and Animal Food. US FDA. May 9, 2016.
[4] 21 CFR § 1.906. What requirements apply to vehicles and transportation equipment? US FDA.
[5] 21 CFR § 1.908. What requirements apply to transportation operations? US FDA.
[6] 21 CFR § 1.910. What training requirements apply to carriers engaged in transportation operations? US FDA.
[7] 21 CFR § 1.912. What record retention and other record requirements apply to shippers, receivers, loaders and carriers engaged in transportation operations? US FDA.

Bibliography

Food and Drug Administration (FDA), 2016, "Final Rule on Sanitary Transportation of Human and Animal Food".

Keener, Larry ."Transportation: Squeaky Wheel of the Food Safety System" *Food Safety Magazine,* November 2003.

Appendix A

Delivery Driver Lesson Plans

Category:	**Subject:**	**Prepared By:**
Driver Training	Delivery Driver Training Sanitary Transportation of Food	

Overview & Purpose	**Time: 90 minutes**
The purpose of this class is to train drivers on the FSMA Transportation Rule and to educate them on issues of food contamination, safety and security	

Topics	Required Knowledge	Required Demonstrations	Other Resources
FSMA Sanitary Transportation Rule	• Student must know the requirements of the rule • Student must understand what parts are applicable to them	Written examination	• Writing instrument: Pencil or pen

Topics	Required Knowledge	Required Demonstrations	Other Resources

	• Student will be made aware of why rule is important
	  • Regulatory compliance
	  • Maintaining adequate temperature controls (if required)
	  • Preventing ready to eat foods from touching raw foods
	  • Preventing food from being contaminated by non-food items in previous or current load
	  • Protecting food from cross contact, such as unintentional incorporation of a food allergen.		

Employee Hygiene	• Student must know how to properly sanitize and when they should do so • Identify locations within their workplace for washing	Written examination and classroom demonstration	

Topics	Required Knowledge	Required Demonstrations	Other Resources
Inspection of vehicle	• Student shall know the methods to fully inspect vehicles • Student will be given examples of what to look for when inspecting vehicles • Student shall know how to examine wash records to determine if type of wash is acceptable for prospective load. • Before loading a food not completely enclosed by a container, the loader must determine that the transportation equipment is in appropriate sanitary condition and eliminated non-listed allergens. • Before loading a food requiring temperature control, the loader must determine that each mechanically refrigerated cold storage compartment is adequately prepared for refrigerated transportation.		• Fact sheet on transporting food safely • Company Policy • Access to operator's manual of any refrigeration unit or tanker

Topics	Required Knowledge	Required Demonstrations	Other Resources

Temperature Monitor	• Should know where to look on manifest for specific shipper requirements for temperature
	• Should know how to pulp check load prior to loading
	• Should be trained on how to work refrigeration unit as they do differ in makes and models and advances in technology
	  • Driver shall be knowledgeable on how to proper record temperature readings.
	  • Driver shall know how to report temperature deviations from requirements to insure that the food is not introduced into the food chain prior to being inspected by a qualified person
		Examination
		
		Demonstrate use of thermometer to pulp check load	

Topics	Required Knowledge	Required Demonstrations	Other Resources
Loading	• Students should understand the principles of load segregation • Students shall understand the principles of load securement and package integrity • Students shall understand basics of weight distribution and it's effect of vehicle stability	Examination and demonstration of use of a variety of securement devices	Load Segregation Matrix
Transportation Security	• Driver will be instructed on company security procedures 1. Locations to park 2. Door locking requirements 3. Routing • Driver shall be able to look for signs of tampering	Examination	
Summary	At the end of the session, the student shall be instructed as to the exact policies of their work place. This training covers a generic overview of the regulation. The company is required to certify the training and document it for recordkeeping.		**Additional Notes**

Session No. 701

Conflict Management & Resolution for EHS Professionals

Jonathan Klane, M.S.Ed., CIH, CSP, CHMM, CET
Ira A. Fulton Schools of Engineering
Arizona State University
Tempe, Arizona

Safety professionals often have to manage or resolve conflicts. These conflicts may be between workers, management, trainees, etc. Traditional "positional" conflict management often fails to satisfy either party. In the following creative nonfiction stories, we will look at what works, what doesn't, and why more progressive techniques work best. The following case examples occurred; details have been changed for pretty darn obvious reasons. Read on and enjoy!

Facilities Flooring Fiasco

"Why should we believe you?! You're being paid by *them*!"

The accusation was directed at me by one of a small group of very disenfranchised staff at a university many years ago. She said it with obvious distrust, almost spitting out the words with venom. She would suffer no fools, that's for sure.

I nodded my head politely, "You're right, I am being paid by management. Someone has to pay for my services; I'm not doing this as a pro bono case. Of course I can't ask you to pay me, so yes, the university as represented by management – them – has to pay me. But that won't change or affect what I say." I tried to look contrite on my client's behalf and nodded toward them off to the side. "They know they've made mistakes, that's why I'm here. So, it's up to you all. I'm here to resolve it to everyone's satisfaction. After this, I will meet with you all privately if you are willing to meet with me. How does all of this sound to you all?"

Karla studied me, "We'll meet with you, the three of us without *them* but I'm not guaranteeing anything."

"Fair enough, thanks. Where do you want to meet?" I wanted them to choose as much as possible, to have whatever control of it that they could.

"Back here," Karla turned toward a larger drafting and sign-making area in the back. I followed her and the other two women who'd been affected by renovations gone wild. We entered the room and I closed the door. It was a glass wall for more light. There were drafting stations and all sorts of plastic sign-making supplies with signs in various states of completion. I read a few: "Danger," in red of course; "No Parking;" and "Safety is Number 1" – ironic. I looked around at the work materials. Everything was neat and tidy, no open containers of paints or coatings. It looked like a healthy workplace; boy, was I wrong. We all sat down. Karla wasn't exactly still scowling at me, but her face wasn't even close to a smile.

"Okay, thanks. So, please tell me what's happened. And feel free to ask me anything."

"Are you going to make them clean it all up?" Joyce asked me.

"Well, I hope so, but I need to know everything that's happened, so I can figure out the best course of action."

"Best for whom?" Karla demanded.

"Yeah, who?" Linda chimed in.

I looked at them, "You all, hopefully. But I need your help. Let's talk. Tell me about it all. Give me all of the details please." I'm not sure that I was prepared for as bad as it kept getting over the course of the story they told me. Geez, how could such smart people keep making such bad decisions?!

It basically went like this. A couple of weeks ago, they'd come in on a Monday to a flood throughout the large open area of cubicles. All of the carpeting was soaked and any porous materials like boxes, drawings, and files on the floor were also wet. Certainly, some of the baseboards and lower walls must've soaked up some of the water. It turns out that a valve had busted on the toilet and it had poured out over the weekend. No one knew exactly when it started, but it likely lasted between 12 and 60 hours, with a best guess of 24-48. The maintenance crew came in and started tearing up the wet carpeting, with the women present. In this space were all sorts of very old, large, and heavy pieces of furniture. Big metal and wooden desks, file cabinets full of paper, drawing and map holders with those huge thin drawers to hold huge drawings, stuffed full. You name it, that old furniture was there. So they left it all in place and only removed the carpeting they could easily access.

"We told them that the carpeting below each thing was wet. Would they listen to us? No, we were just dumb women. Idiots!" Karla was exasperated. The others agreed, "We did try to tell them!"

"Oh my god!" I laughed, "Yup, sounds like they were."

The women all smiled at my agreeing with them. They went on. The wet carpet and some wallboard that got and stayed wet started to stink; it was mold. The women complained. They each had health-related issues: mold allergies, asthma, or other complications. It put them all at a greater risk. Then it got worse.

They tore up the carpeting from below the furniture, and they ripped off the old wooden baseboards exposing the mold, still with the women working in the space. They decided to apply a bleach solution to kill the mold. Since it obviously had gotten worse by not dealing with it right away, they decided they'd use a stronger than normal concentration. They mixed up a fresh batch of 50% bleach solution and sprayed the space down, *with the women still present.* Needless to say, the bleach acted as a strong pulmonary irritant, causing additional problems for the women. To say they weren't happy is putting it mildly.

"It was awful!" Linda complained, "I was coughing something wicked."

"We all were," Joyce continued. "At times I felt like I was choking it was so bad."

"Wow, I am so sorry," I apologized for my client. "That sounds horrible."

Karla looked at me and cocked her head slightly, "Thank you, that's the first that anyone has apologized to us for this."

I nodded, "Well, that's one of several mistakes. Let's hope we can correct it." Then it got worse yet again.

The carpet had left behind lots of the glue, which didn't provide a good, solid base for the new tile and its mastic. So, they had a contractor come in with a solvent to remove the carpet glue: yes, with the women *still* working in the space. Ugh.

"It made me nauseated," Linda reported.

"And I felt light-headed when I was in the space," Karla gestured with her arms and looked around continuing, "The smell even got into our space back here where we'd been forced into to do our work."

"That's when we said we'd had it and demanded they bring in someone from outside to handle it. That's where you come in," Joyce nodded at me.

Karla looked at me, "So, what are you going to do?"

"Good question. I've got a meter–it's called a photoionization detector or PID for short–and I'll use it to see if there are still detectable levels of any of the solvent remaining. You can watch me as I do it, and I'll tell you my results if you'd like. I'll also evaluate the walls for remaining mold. We'll discuss revising the plan to finish the floor refinishing with mechanical means."

"Won't that raise all sorts of dust though? I don't want to breathe that in, too," Linda said.

"Me neither," Joyce chimed in.

"Yes, you're right," I agreed. "I won't approve a plan with you still working in an area with concrete dust; that'd be unhealthy. They'll need to install plastic barriers and have the work done on the weekend. We can use a dust meter or a particle counter to verify that levels are back to normal before we allow you all to re-enter your work areas. How does that sound?"

The three women looked at each other checking in before answering. There were little head nods. "That actually sounds good," Karla admitted, "I didn't think you'd help us when I heard they were bringing in some consultant of their choosing. But now I see why."

I smiled, "Thanks Karla, I appreciate it. If it's okay with you all, I'll get started. Now comes the not so fun part, telling them how badly they screwed up and what they need to do. But I think they will accept pretty much whatever I come up with as long as it seems reasonable, which all of this is."

"Thanks Jon," Joyce gestured to her co-workers, "we all appreciate what you're doing for us."

"Yes, very much," Linda added, while Karla actually smiled at me.

"You ladies are all quite welcome; it's what I'm here to do. Let's get started."

And we did. I was able to find a pocket of solvent in a floor crack that needed cleaning up. I worked with my client to design an acceptable floor finishing plan with plastic barriers and work done after hours. I emphasized the need to share the plan with the ladies who were, in effect, my new clients and the biggest stakeholders in its success.

I did a lot of education and training for the maintenance guys. I found areas with visible mold needed abatement, and had them remove some more baseboard. I showed them what to look for and asked them to take on looking for more. They got into it and showed me areas that they were able to also find. They became my aides and part of the team trying to solve it. I checked in with the ladies once it was all done.The message I got back was quite rewarding.

> *Hi Jon,*
>
> *Linda, Joyce, and I just want to thank you again for all the help. You actually listened to our complaints and took us seriously – that meant a lot to us. Our work area and our relations with "them" are both better and healthier.*
>
> *Thank you,*
>
> *Karla*

Sometimes, you win a few friends in the process of making things better for everyone. Here are some of my lessons learned and techniques that helped:

- Honesty and sincerity
- De-escalation
- Don't be defensive
- Okay for people to be upset and express anger
- Be genuinely interested in what they want to tell you
- Don't get dragged into an argument but freely admit mistakes
- Solve the inherent problem to everyone's benefit and satisfaction

***

The "Employer Whisperer"

"Um, we've got an employer who's not very happy at all and she is making it plainly obvious," Louise told me. Louise works for me, and we were at our Career Fair, providing safety oversight based on how many students we get for the event with a limited number of spaces. It gets quickly crowded.

"Hmm, okay, what seems to be the issue?" I had an idea but wanted to know for sure before I tried to de-escalate it.

Louise frowned, "It's the usual, we asked her if we could move her table back for more student line space and she blew her top about it. I think Bobbi wants to talk it over with you first."

Bobbi was our Director for our Career Center, and she is so nice, as are all of our Career Center staff. We often talked about situations and context before I'd intercede, so it made sense.

"Okay, I'll find Bobbi first. Thanks for the heads-up!"

I found Bobbi out in the already busy hallway. "Oh Jon, I was looking for you. We need you to have a chat with an employer who's upset about her table."

I smiled, as this happened every so often over the years. "Yeah. So I heard. What happened?"

Bobbi sighed, "Louise asked her really nicely, and she had a snit fit, demanding to speak with Bryan, our financial guy. She said she wanted to examine her contract and what she paid for."

"Okay, I'll get Bryan and we'll go and see her."

"Thanks, and good luck!"

I smiled again, while going to get Bryan, "We may need it."

Bryan and I approached the employer in question. He and I smiled. "Hi there, I'm Jonathan, our bald-headed safety guy," I doffed my high viz ball cap. "And this is Bryan, our financial guy. How can we help you?"

She smiled a bit at my introduction, "Thank you, I want to know why I have to move my table back and I don't have more space? How much did my company pay for? I always had more space in the past. By the way, is that really your title?"

I beamed at her polite question and decided to answer that one first; focus on something pleasant when you can. "Yes, here, check out my card," as I handed her one.

"Bald-headed safety guy, that is funny," she looked at me. "Okay, Mr. bald-headed safety guy, what's up with the space and my table? Why do I have to push it back?"

"Actually you don't have to move it for us."

"Wait, what? But she said that I had to push it back to the wall; I don't want to." She looked puzzled.

"I'm sorry about the confusion. It's really up to you. You can if you want to gain more room in front but you don't have to."

"Well, okay, but why would I want to push it back?" she asked.

"Sure. The Fire Marshal requires us to maintain our fire lanes and exits in here. The yellow tapes mark the lanes we have to keep open. As long as the students aren't standing in the yellow lane, he's happy. The blue tapes are the lines we do for each employer like for you. If you'd like more space in front of your table we can move your table back only as far as you want it. If you're happy with it here, that's fine, too. Some don't want so many students in front of their table; it's really up to you. What would you like us to do for you?"

She looked from me to the blue lines and the yellow line and then back to me. "Um, well, I think I'd like a little more space for the students then. Can you help me move it?"

I smiled and turned to Bryan, "We'd be happy to take care of it for you. Here, let's just lower your display for a minute while we move the table." We moved it about halfway back. "There, how's that?"

"Perfect," she smiled. "Thank you both so much! Would you like some of the swag I have here? Please why don't you take something?"

I smiled, "Aww, that is so kind of you. If you insist, I'd like a USB, please. Bryan?"

"Same please."

"Here you go, and take some candy, too!"

"Okay, thanks. If you need anything else, please just ask for me."

"The bald-headed safety guy," she laughed, "I think I can remember that!"

We left to go back to our designated headquarters and check-in area. Bryan turned to me, "Wow, what the heck was that?"

I laughed, "Yeah, she just needed to feel in control of her space."

"Well, that was pretty cool."

"Thanks. You know what they say; even a blind hog finds a few acorns."

"Hah! I suppose so.

Sometime later, Bobbi caught up with me. "Well, I don't know what you did but you worked your magic again. She's happy and telling me how helpful you and Bryan were. She's like your new best friend now."

"Yeah, when I checked back or walked by, she smiles and waves and wants me to take more swag so I grab a candy; it seems to make her happy to return the favor."

"Or to apologize," Bobbi added. "Well, we appreciate it as always. We're going to call you the 'Employer Whisperer' from now on."

I smiled, "I'll put it on my card right, under bald-headed safety guy."

Sometimes, you can turn things around in ways that no one expects. Some of my lessons learned:

- Cede control to them
- Apologize
- Put the responsibility on a third party they don't know
- Use existing standards
- Help them to find ways to reframe gracefully
- Accept any gratitudes offered

***

“Do We Really Have to Watch Some Movie?”

“Professor Klane, do we really have to watch some old movie for homework?”

It was one of my Conflict Management & Resolution class students. “No, Mary, you don’t have to watch it. It’s for extra credit, so it’s optional.”

Mary looked at me like I was an idiot. “Don’t you know, for students ‘extra credit’ isn’t optional; it’s required. We feel compelled to do whatever our profs offer: even extras, bonus questions, you name it.”

“Oh, I see, I didn’t realize. Sorry.”

“Don’t be sorry, it’s good to get extra credit. So is this movie any good?”

“Hmm, define your version of good.”

“Well, what kind of movie is it?” Mary asked.

“It’s an action movie. Do you like action?”

“Um, sometimes. What’s it about?”

“Well, it’s about an attempt to assassinate our President while he’s in Spain. It covers a short span of time, maybe 15-20 minutes.”

“How does it do that? Isn’t it a full-length movie?”

“Oh yes, it is. It depicts the time and action from a variety of different perspectives.”

“We see the same scene over and over? Sounds boring.”

“Well, it might seem like it would be boring, but it’s anything but boring. It’s actually an edge of the seat action flick. It kept me glued wondering what would happen next.”

“Really? That good?”

“Yes, I think you’ll like it. Plus, it shows that there are many perspectives and stories we need to gather and knit together before we really can understand everything going on.”

“Oh, I see what you’re doing. Very clevah of you.”

I smiled, ‘thanks Mary. Let me know what you think of it in the paper.”

“Oh, I will. How long for the paper?”

“Only 1 page; that’s all I want to grade for an extra credit paper for you all.”

Mary smiled, “Like I said, you’re clevah. Okay, see ya!”

“See you, Mary.”

The following week Mary turned in her paper. "Hey Professor Klane, here's my paper!"

"Thanks Mary. So, I'm sure I'll read it, but what'd you think of the movie?"

Mary smiled, "Oh, I liked it a lot! Like you said, edge of your seat, right? I was captivated and couldn't have predicted how it ended. Wow!"

"I'm glad you enjoyed it. Anything else?"

"Yeah, it did a really great job showing all of those different views and experiences and how they all added up to a complete story. Thanks for the recommendation."

"You're quite welcome, of course. I guess that the film makers were the clevah ones, eh?"

Mary smiled, "Wicked clevah!" A true Mainah if I evah heard one.

Sometimes I learn to reframe as much as they do. Other lessons for me:

- "With great power comes great responsibility"
- Humor helps (but not aimed at them)
- Give info where it helps ease their worried mind and furrowed brow
- Never say "I told you so," even politely and in context
- Shared experiences can help

***

Well, I hope that you've enjoyed these true vignettes; there are plenty more. Each depicted the issues and the solutions through creative nonfiction. Feel free to reach out to me with any comments or questions on the techniques depicted or this preferred style of writing to convey science in a more compelling and easier to read manner.

Remember, conflicts are an opportunity for better relations!

What challenges do you face?

Session No. 702

Vision 20/20: Preventing Opportunity Blindness

J.A. Rodriguez Jr., CSP, SGE

Introduction

Very often, opportunities are abundant for those who can see them. The issues surface when we shade our vision in such a way as to only allow certain information to come through. We then make decisions on that limited data set. We make assumptions and we act on them to the point of creating opportunity blindness.

Think of the last time you listened to part of a conversation between two parties and wondered if everything is what it seems. Was an inaccurate conclusion drawn? The answer is likely to be a resounding, "yes", because context of the message is often missing.

Understanding the impact and power of assumptions, the relationship between vision and blindness, what an opportunity is, how to create them, how process thinking plays a role, how human behavior is influenced, and how to view any situation in a different light are the keys to unlocking the magic behind achieving the intended outcomes.

The Power of Assumptions

Understanding the impact and power of your assumptions is the first step in establishing clarity of vision or blindness. We all make significant assumptions every day, sometimes unconsciously. You are commuting to work in your automobile. The assumption is that there is no need to look side to side when the light turns green at the intersection. The vehicle's accelerator is depressed micro-seconds after the traffic light indicates it is safe to proceed. You go through. Nothing happens. Everything is okay. What were the significant assumptions we all make in similar situations? There are several. That every driver follows the traffic laws 100 percent of the time 24/7/365. That the "other" driver is not texting and driving while approaching the intersection. That the traffic engineers have the traffic lights synchronized properly. That our journeys' are not going to be interrupted, and that personal injury is not imminent.

Are these assumptions safe? It depends. If you had to interview the family whose loved one was killed by a distracted driver at an intersection, the answer is no. If you base this answer on your experience, the answer is likely yes.

Assumptions can cause blindness if they are not kept in check. A worker takes a picture of a colleague at a remote site in the Arctic. The image shows a white winter wonderland with an employee exiting the facility buried in snow. Upon returning back to the home office, both employees look at the picture and noticed a large polar bear glancing at them. Neither worker noticed the polar bear's presence when the picture was taken. Neither worker realized the danger they were in over those few minutes while walking to their vehicle. What were the assumptions that caused blindness to this real hazard? To start, the employee taking the picture assumed all is good because nothing like this has ever happened before, so he did not see the hazard staring right at him. The employee whose picture was taken assumed that he did not need to worry about situational awareness relative to the real polar bear danger. There is no need to look around for polar bears even though they are spotted often in the area. After all, nothing has ever happened before this event. This seemingly innocent assumption could have resulted in life-changing events for both employees.

Relationship Between Vision and Blindness

Vision and blindness have a polar opposite and a complementary relationship. Polar opposite in that where there is no vision, there is total blindness. Where there is total blindness, there is no vision. In the polar opposite relationship case, unconscious assumptions relate to unconscious bias and unmitigated risk. If we think about the polar bear example, total blindness towards a real hazard set in when an unconscious assumption led to an unmitigated risk.

A complementary relationship exists when mitigated vison leads to mitigated blindness, mitigated blindness leads to mitigated vision, and some mitigated vision drives some mitigated blindness. This occurs when conscious assumptions drive conscious bias, which leads to mitigated risk. We experience an example of this every day when we drive our automobiles. Most of us do not jack up our cars to perform a detailed inspection of the braking system every time we drive. These inspections are scheduled for maintenance, and we thereby assume everything will work perfectly in the meantime. This conscious assumption then drives a conscious bias (that braking systems are reliable if properly maintained), which leads to a mitigated risk. We then enter our vehicles without thinking twice about questioning the braking system.

How Most of Us View Opportunities

Most people only see what they allow themselves to see. When presented with a situation, they only see the situation, not as presented, but as they perceive it. Think of a straight line. Draw one across two sheets of paper. Hand one of the sheets to an engineer and the other to an artist. Ask each of them what they see. The likelihood is very high that each will see something different. Each will visualize this opportunity from a varying point of view.

When people see straight lines when presented with straight lines, opportunity blindness is in a strong position. The inherent mindsets of those that see only what is presented include:

- Opportunities just pop up
- Great opportunities are rare
- Just keep going, and opportunities will surface someday
- Someone else is responsible for making it happen

- In a perpetual waiting pattern for opportunities to drop

When people see dips and valleys when presented with straight lines, opportunity blindness is in a weak position. The inherent mindsets of those that see more than what is presented (innovative people) include:

- Opportunities are created
- Opportunities are there… find them
- Moving forward while glancing back at lessons unleashes vision
- I am responsible for making opportunities happen
- In perpetual pursuit of opportunities

Focus on understanding what is not visible so you can identify and create opportunities where others cannot see them.

Seeing Opportunities Where Others Simply Cannot

In a small Massachusetts town, there is a large hill/small mountain. A local businessman decided he would purchase the property from the current owner who felt, for various reasons, that the property was worthless. These reasons included lack of accessibility, no building permits could be issued, and the terrain was rough with no access to utilities. The current owner was glad to "dump" this property over to a well-educated buyer at a bargain price. He felt that, other than the nostalgia of saying he owned a small mountain, there is no value in holding on in consideration of the taxes he was paying to own the property.

The local business man buys this "worthless" piece of land. His friends and colleagues thought he was crazy to take good money and waste it on bad land. He was criticized for making a bad deal, even if it was at a bargain price.

The previous owner saw a worthless mountain. The new owner saw an opportunity. Soon after the purchase, the new owner started a gravel and fill dirt company. Day by day, truck by truck, gravel and fill dirt was sold from his small mountain. Month by month the big hill got smaller and smaller until the day came when it was leveled. The new owner converted his small investment into a multi-million dollar business.

Now that the land was level, he decided to continue selling more soil and gravel. Day by day, truck by truck gravel and fill dirt was sold from his land until the hole got bigger and bigger. His gravel and fill dirt business flourished making the new owner even richer.

At some point, the hole became a crater. There is no more gravel and fill dirt to sell, so he closes down the business and opens up another one. He dove into the landfill business. Day by day, truck by truck, customers paid him to landfill their waste. Month by month the crater creeped back to a large hole and then to level land once more. His landfill business flourished, making the new owner even richer.

When it was all said and done, this individual was paid to take down his small mountain, to dig a crater and to fill it back up. He eventually donated the property to the city for a park. This business man saw a tremendous opportunity where his friends and the previous owner simply saw a useless piece of land.

How Process Thinking Fits in

All work-related outcomes are the result of their processes, planned or unplanned. Everything produced is the result of a series of events and/or actions that created it, regardless of intention.

Looking at the big picture first offers insights into why things happen. Thinking in terms of process facilitates the identification of sustainable solutions. It opens up the possibility of other opportunities. For example, we often arrive at a conclusion that blames employees for injuries, when in fact, we are missing an opportunity to identify a process step or two that needed to be modified, deleted or included. If we believe that every process works exactly as it was designed, then the design is faulty if it produces an injury or other undesirable event. While employee behavior is a very important process element, it is definitely not the only one. The error is in believing that it is.

If the eventuality of human behavior is often guided by assumptions and risk analysis, then the starting point to changing it is by modifying the process that encouraged it in the first place. When you fix the process, you often fix the behavior, not the other way around.

Behavior-based safety focuses on people and on getting people to make the right choices, based on their understanding of the consequences. The difference here is that there is more, much more. By modifying the process, you minimize choices and decision points, thereby allowing for more consistency of performance and preventative action sustainability.

Process thinking allows you to see opportunities normally hidden from view. This tool can expose areas needing improvement and prevents blindness to sustainable solutions. Unidentified risks lead to unmitigated ones, which introduces vulnerability to expected outcomes.

Process thinking is a mindset that is minimalistic. The more you focus on the process the less distractions there will be, the better your vision, the greater your chance of success will be. Systems thinking is the next tool in identifying how all the pieces fit together.

How to Leverage Process Thinking and Human Behavior to Prevent Opportunity Blindness

Looking for what is not there is usually the best way to find it. Let's consider the case of too much work and, apparently, not enough resources.

Carmen and her team have won several pursuits within the same period. The demand for her team's time is high. Several high-profile milestones are coming up in a few weeks.

Phillip, a project engineer, is talking to his boss, Carmen, the group manager: "Hi Carmen. Can we talk? I just don't know how I am going to get all of this work done in the timeframe you need me to get it done. I mean, I have five projects I am working on. They all have tight schedules. Commitments were made that we may not be able to meet."

Carmen: “I know. It is a good problem to have but it is still a problem. What do you need?”

Phillip: “I need more help working the project. It is overwhelming.”

Carmen: “Is it a prioritization issue?”

Phillip: “No, it is a too much work and not enough hours in a day issue.”

Carmen: “As you know, this is an extremely busy time for us all. Stop by later today with what you have and let’s see what I can do.”

Jessy, a fellow project engineer, overhears their conversation. James, a supply chain specialist, and Sammy, a mission assurance lead, are brought into the mix by Jessy.

Jessy: “James, Sammy…I just heard Phillip ask Carmen for help getting his workload reduced. The fact is we are all too busy. I hope his excess workload does not come my way.”

James: “We are all busy Jessy, but I wonder if there is something we can do to help?”

Sammy: “Perhaps James, but who is looking for more work?”

Jessy: “I know, I have just about as many projects as Phillip does.”

James: “I don’t know, perhaps as a team we can all come together, lay it all out for Carmen, and identify the real priorities, perhaps increase our efficiency.”

Jessy: “That all sounds good, but in the meantime we are losing valuable time that we can use to continue working.”

Sammy: “All I know is that something has to give.”

At issue: Phillip perceives he is overloaded with work and is very concerned about meeting his commitments to Carmen. Jessy and Sammy perceive they are very busy as well. James is trying to figure it out.

This scenario plays out all too often. Process thinking leads us to believe that this outcome is the result of a leadership process, one that produced an outcome exactly as it was designed. In this case, Carmen overloaded Phillip, and the team is torn about how to proceed. Alleviating Phillip’s load will drive a different behavior. The key is to identify the opportunity to do so.

The only co-worker who sees an opportunity in this chaos is James. It is true that if they all come together and presented a solution to the boss, the team will increase its productivity and relative efficiency. James has the vision, but does he have the leadership skills to achieve buy in from the team? He must take the next step and encourage collaboration. By looking to fill the gap, James takes advantage of an opportunity to develop additional leadership skills and potentially set himself up for a future promotion.

Conclusion

Seeing what others do not is a learned skill. It takes awareness, diligence, and practice. Opportunities are there for the taking as long as we keep our vision clear.

Be aware and leery of assumptions. They can impact your life or the life of others and your career. Understanding them adds a new dimension to self-awareness and the relative successes we achieve.

Vision and blindness run on a tight rope where one affects the other in ways that impact life's outcomes. Conscious assumptions lead to mitigated risks through conscious bias.

Every process works precisely as it was designed. Outcomes, expected or not, desired or not, are the results of their process. Therefore the process is not faulty for not producing the intended results, the design of it is.

Leverage process thinking and human behavior to prevent opportunity blindness. Find the gem in chaos.

Session No. 703

Dropped Object Prevention: From Afterthought to the Forefront

Abby Ferri, CSP, MEHS
The Ferri Group LLC
Minneapolis, MN

Preventing workers from falling is important, and continues to be a leading cause of death in all industries. Preventing injury or death due to falling and dropped objects is often an afterthought.

Dropped objects caused 247 deaths in 2015, five percent of fatalities. This is an increase from 243 deaths due to dropped objects in 2014. No industry is immune from this hazard. Safety professionals must incorporate all hazards of work at-height into their fall protection, fall prevention, and at-heights safety plans.

There is no specific "dropped objects" or "objects at-height" standard. In the United States, but the Occupational Safety and Health Administration (OSHA) mentions the topics in a few regulations:

Scaffolds, 1926.451(h) : "Falling object protection"
Fall Protection, 1926.501(c) : "Protection from falling objects"
Steel Erection, 1926.759(a) : "Securing loose items aloft"

Though there are few specific standards, it is important to keep in mind that the General Duty Clause, Section 5(a)(1) of the OSHA Act of 1970 states that, "each employer shall furnish to each of his employees employment and a place of employment which are free from recognized hazards that are causing or are likely to cause death or serious physical harm to his employees."

Many state plan states, including California, mention this topic:

Cal/OSHA 3273(e) : "Protection from falling objects"
Cal/OSHA 1511(a) : "Proper precautions have been taken to protect the employee while doing such work"

Another angle for standards related to dropped objects is equipment performance standards. Currently, not all equipment used for preventing dropped objects is created equal. In the past year, several leading safety equipment manufacturers have been working in coordination with the

International Safety Equipment Association (ISEA) to develop a standard for dropped object solutions. The working name for ISEA Standard 121 is "American National Standard for Dropped Object Prevention Solutions," and it is in the draft stage. The goal of the standard is to provide manufacturers guidelines for consistent classification, testing procedures, and performance requirements for dropped object solutions. Once finalized, it is likely that the American National Standards Institute (ANSI) will adopt the standard.

Regardless of applicable regulations, when an employer identifies dropped objects or objects at height to be a hazard to workers, controls must be developed to protect personnel. While there are no specific training standards that related to dropped objects, any training must be relevant to the controls that personnel will use to meet OSHA standards and employer requirements. Per OSHA and ANSI Z490.1-2016, training documentation should answer this question:

"Did the Employee Receive Adequate Training to Do the Job?"

Keep this question in mind for all dropped-object prevention training, as well as any other training conducted for personnel related to hazards identified in the workplace.

Initial training should be conducted before an employee's exposure to dropped-object hazards and should include instruction on hazards, control, and the inspection, storage, and use of any dropped-object prevention equipment that will be used on the site. Personnel should also understand how to report incidents and near misses. A general guideline for refresher training is at least once every two years, or when there are new hazards or changes to the dropped object prevention equipment or the company's policy.

Levels of required competency for personnel should be established by the employer. The employer should determine the level of training that end users, managers, supervisors, and safety personnel should have.

Staying Hands-free on Walking-working Surfaces

As word continues to spread of the solutions available to prevent dropped objects, savvy owners and general contractors are beginning to require their subcontractors and vendors have a "tool tethering program" for work on construction sites under their control. This type of directive is short sighted, though tool tethering is thought to be synonymous with dropped-object prevention. While tool tethering is an important part of a dropped-object prevention plan, it is not the only control available.

Whenever possible, attempt to limit the amount of tools, materials, or equipment carried by or attached to the worker. Use mechanical means to move items to at-height work places, or use hoist containers that have secured tops to prevent items from falling out.

OSHA's revised Walking-Working Surfaces standard addresses hazards that general industry workers face, including climbing technique while using ladders. The requirements of 1910.23 standardize some controls that have been considered best practices for quite some time:

Workers should face the ladder when climbing up or down it.

Workers should use at least one hand to grasp the ladder when climbing up or down it.
No worker should carry an object or load that could cause them to lose balance and fall while climbing up or down the ladder.

Besides providing sturdy, rated, and topped containers for hoisting and lowering, other containers, like tool pouches and tool holsters, should be used when a worker must climb up or down a ladder with tools and supplies. These solutions can be attached to worker's fall protection harnesses, tool belts, or the structure being worked on. There are many solutions that are specifically designed for common tools and equipment, like tape measures, cell phones, and drills.

When choosing containers, look to equipment that is manufactured, tested, and rated for the purpose of safely transporting gear to at-height work areas. Traditional pouches, drywall buckets, burlap sacks, gym bags, and other improvised transport solutions can fail, putting workers at risk of injury or death.

A Roadmap for Dropped Object Prevention

After the initial hazard analysis and identification of controls, a roadmap for deploying dropped objects solutions can include some intermediate steps to ensure the program is thoughtful and takes into account the intricacies of at-height work.

The dropped-object prevention plan should encompass tasks that include the following:

- Work that is four feet (general industry) or six feet (construction industry) above the ground level.
- Any work that is elevated relative to other personnel or sensitive objects, equipment, or processes.
- Raising and lowering equipment, tools, gear, and materials to and from elevated work areas.
- Where otherwise determined or required by a safety manager or other authority.

<u>Pilot First</u>
A great first step with a dropped-object prevention plan is to identify the work areas or tasks with the highest risk for dropped objects. Choose 6-10 tools that might be higher risk, and identify ways to transport them to the elevated work area for safe use. Often, a suitable anchor point can't be found for securing topped containers independent of a worker, so other avenues must be explored. Solutions may include topped containers or retrofit connection points for tools that do not have captive connection points as part of their design.

Common challenges at this phase include tools that do not have an inherent captive connection point; tools that have multiple parts that could separate and fall (i.e., socket wrenches); hammers and other striking tools that do not have a head-locking mechanism; and tools that are not equipped with non-slip handles. Ensure the tools are in sound condition, per their manufacturer requirements, prior to attaching a retrofit connection and tethering solution. It is also critical to choose the correct connections and tethering solutions based on the weight of the tool and other characteristics, as described above. Never use homemade tools and equipment. Always ensure that the tools and equipment are appropriate for the expected elements and conditions of the worksite. For example, a non-sparking or non-marring tool should be paired with a connection and tether that meets the same criteria.

Even though there are no standards governing the testing and marking of dropped-object prevention solutions, manufacturers are deploying these proactive practices. Look for equipment that is tested to a 2 to 1 safety factor and is marked with certified capacity information.

During the pilot phase, monitor workers closely to catch bad habits and correct them before they are ingrained. The chance of misuse of tool tethers is high, and the chance for improvisation is even higher. Training, informative postings, and frequent inspections are key.

Equip and Educate

Workers who will be using dropped-object prevention solutions should be provided training regarding the selection, use, inspection, storage, and maintenance of the items. Information about how dropped object prevention solutions fit in with their existing fall protection equipment is critical.

Many construction sites require workers to provide a small set of their own tools. If tool tethering or containers are required as part of the employer's dropped-object prevention plan, this includes employee-owned tools as well. Currently, a variety of dropped object prevention solutions are available to workers via online order or tool suppliers, and workers have purchased such items and use them by choice at jobsites. An employer may choose to control this risk by prohibiting employees from using their owned dropped-object prevention solutions on the employer's tools.

As discussed earlier in this paper, it is the employer's responsibility to establish the required competencies for different levels of employees. Supervisors and safety management staff may have overall site and equipment inspection authority and responsibility, while end users can receive specialized training on the equipment they will be using.

Plan – Do – Check – Act

Follow the classic Deming Cycle after initial implementation and launch of your dropped object prevention plan. This may mean that you repeat the pilot phase multiple times with various work areas or tasks. As standards are developed for equipment manufacturers to follow, your equipment selection may change, requiring refresher training for workers.

As with your other safety programs, the dropped-object prevention plan should be reviewed at least annually and as new tasks or equipment are introduced to the worksite. In the coming years, dropped-object prevention solutions will evolve, so your program will too. Lean on the expertise of the equipment manufacturers, as well as your industry's best practices, as you build your program.

Bibliography

The Deming Institute. 2016. *PDSA Cycle.* (https://deming.org/management-system/pdsacycle).

Dropped Objects Prevention Scheme (DROPS), 2017. (www.dropsonline.org).

Ergodyne. 2017. Technical Bulletin. *Forthcoming Dropped Objects Standard Being Drafted.* Minneapolis, MN: Ergodyne.

Ergodyne. 2017. White paper. *How To Be Hands-Free: Solutions To Help You Climb Safely To Heights.* Minneapolis, MN: Ergodyne.

Occupational Safety and Health Administration (OSHA), 2017. *Walking-Working Surfaces standard.* (https://www.osha.gov/walking-working-surfaces/).

Emergency Response Plans: How Good Is Your Plan Under Fire

Kristin R.B. White, J.D.
Member, Jackson Kelly PLLC
Denver, Colorado

Ryan Burr, CSP
Delta Air Lines, Inc.
Atlanta, Georgia

Introduction

The many thousands of workplaces subject to Occupational Safety and Health Administration (OSHA) regulation are faced with a variety of challenging safety issues over the course of completing a project or operating a facility. Under OSHA standards, all workplaces are required to develop emergency response plans in order to effectively respond to the many and varied safety crises that businesses face in the modern economy. Considerable effort is expended every year by safety and legal professionals to create and further develop quality emergency response plans. Unfortunately, no matter how much time and money is expended to prevent crises, they occur, and can have devastating consequences. When a crisis does occur, the affected company must implement their emergency response plan and investigate the accident or incident. Taking the time and effort to construct a thorough emergency response plan is invaluable when a real-world crisis scenario occurs. However, even the most elegant emergency response plan requires dedicated effort to implement correctly.

The purpose of this paper is to provide a proactive and cautious approach to investigating accidents and incidents, with particular attention paid to securing and documenting the scene of the accident, communications and the attorney-client privilege, interviews and statements of third parties, documenting the crisis, providing assistance to the families of those affected by the crisis, and producing quality and useful internal investigation report.

Securing and Documenting the Scene

After taking emergency response measures designed to protect the lives of those involved in the crisis incident, the first priority for an employer should be to secure and document the scene of the incident. Securing the site is of vital importance, as it is incumbent upon employers to preserve material evidence that will be necessary to complete a root-cause investigation. Access to the site of the incident should be heavily restricted, and limited only to potential law enforcement, OSHA compliance officers, and the employer's investigative personnel. Preventing access to the site may be accomplished through the use of barriers, such as caution tape or cones identifying the restricted area.

Documenting the conditions present at the scene of the incident is essential for achieving a full understanding of the cause of the incident. At a minimum, employer investigators must document facts, such as the date of the incident, the date of the investigation, the identity of the injured employee, the identity of any witnesses, a description of the injury sustained, whether the injury is permanent or temporary, and the location of the incident. All individuals present at the time of the incident should be pulled aside to be interviewed, as discussed more fully below.

Many different tools now exist to document the site conditions present at the time of the incident. Investigators should take many pictures of the scene of the incident, the site conditions, and any other relevant locations. Site investigators should ascertain whether the incident was captured by video-recording devices, such as security cameras, and, if so, obtain the footage from those devices.

Employers should develop forms that guide employer investigators through the investigation and ensure that all proper steps of documenting the incident are taken. A standardized investigation form or guideline will help produce data in a way that is easily used by investigators when producing their investigation report.

Communication and Privilege

The occurrence of a crisis event will require employers to communicate with several outside parties, particularly OSHA personnel investigating the incident, and members of the media seeking additional information. Employers should be very careful when interacting and communicating with third parties in order to protect company information and maintain control over the investigation process. Although communications with third parties is unavoidable, it should be minimized to the greatest extent possible. All external communication should be routed through approved individuals to ensure that accurate information is relayed to the appropriate parties.

Two privileges exist that protect from disclosure the results of any internal investigation conducted by the employer as a result of the crisis event. It is important to note that it is not always possible to protect internal investigative materials from disclosure. These are the attorney-client privilege and the attorney work-product privilege. Employers should be aware that these privileges may be waived through the actions of the employer and their employees. Considerable care should be devoted to maintaining the protections provided by these privileges.

The attorney-client privilege protects communications between an attorney and their client from disclosure. To gain the protection of the privilege, the communication must be for the purpose of seeking legal advice. This privilege promotes full and frank disclosure by clients to ensure that attorneys have a full understanding of the situation. The attorney-client privilege can be waived through disclosure of the communications between attorney and client to a third party. Even inadvertent disclosures can waive the privilege. It is therefore essential to safeguard information and documents sought to be covered by the attorney-client privilege.

The attorney work-product privilege protects from disclosure all work product prepared by an attorney in anticipation of litigation. To receive the protection of the attorney work-product privilege, the documents must have been prepared at the direction of an attorney and in anticipation of ligation. Documents prepared in the routine course of business will not be protected from disclosure. In a crisis event, it is reasonable to assume that litigation will occur, resulting in many of the documents prepared as a result of the crisis being protected under the attorney work-product privilege. However, it should be noted that investigation reports are generally considered to have been produced in the routine course of business and are not protected from disclosure.

As a result, a key issue for every employer is when and how soon you bring an attorney into the investigation process. Typically, the larger the incident, the sooner it makes sense to seek legal guidance whether from inside your company or outside your company. The protections afforded by the privilege typically will not exist until an attorney is involved in the investigation.

Interviews and Statements from Employees or Third Parties

Interviews and/or witness statements are essential in obtaining enough information to accurately assess the root cause of the incident. An employer must decide if it wants to conduct interviews of relevant witnesses, with notes taken by the interviewer as the record of the conversation, or whether witnesses are each asked to write their own statement detailing the facts they witnessed. In larger, more complicated investigations, it may be more beneficial to have the interviewer asking questions in order to ensure the interviewee answers the necessary questions. In those instances, the only record of the interview is the interviewer's notes, which as discussed above, may be protected from disclosure. Because employers will not always have the luxury of time to conduct one-on-one interviews on every investigation, smaller, more routine cases can be handled through statements written by each witness. The purpose of employee and third-party interviews or statements are to determine the cause of the incident, evaluate inconsistent statements made in the interviews, and gain knowledge to be used to prevent the incident from occurring again.

When interviews are conducted, they should be thoroughly documented to ensure their usefulness in creating the final investigative report. However, witness interviews can be high-stress situations for the individuals who witnessed the incident. Witnessing an emergency, particularly one involving a fatality, can cause significant emotional and psychological distress. Interviewers should be aware of this, and take steps to ensure that they acquire the information they need without unduly disturbing the witnesses.

For witness statements, employers must decide who to request witness statements from, and whether the statements should be delivered in written or oral form. It is important to ensure that these statements are based on fact and not the opinion of the individual making the statement. Statements are best prepared after the person preparing them has an idea of the facts of the situation and the legal position of the employer.

Gathering and Providing Documents to OSHA

During the course of the incident investigation, OSHA will certainly demand extensive documentation covering many aspects of the work performed at the workplace where the incident occurred. Employers should be prepared to respond to these requests in ways that protect the company while being responsive to the requests. Employers should prepare all investigation documents as though they will be reviewed by an adverse party in litigation. Gathering and retaining documents should be addressed in the document control policy of the employer.

In addition to all the documents generated by the investigation itself, attention should be paid to other documents created during the routine course of business. This includes emails and text messages. OSHA will likely request many documents that are produced in the routine course of business, including daily inspection reports, attendance reports, and other documentation produced as part of project advancement and completion. Most significant of these are training records, which will certainly be requested by OSHA. Requesting these records will allow OSHA to obtain a more complete understanding of the conditions present at the site during the incident.

Document disclosure can be extremely complex, requiring close attention from even the most experienced safety practitioners. Disclosure of documents or other materials to enforcement agencies, such as OSHA, may waive privilege as to third parties involved in the incident. Any document disclosure should be handled by individuals with sufficient knowledge and expertise to protect the company from inadvertent disclosure of privileged information.

Providing Family and Employee Assistance

Providing assistance to family members of those involved in a workplace incident should be part of any robust emergency response plan. Notification and support of the family members of those involved in a workplace incident, particularly a fatality, is an important priority in a time of crisis. Whether the incident results in a fatality or not, employers need to be proactive when reaching out to family members and understanding their wishes. Family members may want to have some control over knowledge about the incident. This should be respected to the extent an employer can do so while complying with all reporting obligations. This task should be assigned to a designated individual with sufficient training and experience to ensure that appropriate contact is made. If family members approach the company, their questions or concerns should be directed to the same designated individual.

Employers should take active steps to inform family members of those involved of the benefits they may be entitled to under the law or under company benefit plans. Any pension, 401(k), life insurance, death benefits, or other benefits should be explained to beneficiaries to allow them to understand their options in the wake of the crisis.

Employers also should be aware that other employees may experience feelings of anxiety, grief, or even guilt over another workers' death or injury. Larger employers typically have access to employee assistance programs, wherein employees can reach out for counseling at no charge. There is benefit for all employers in providing onsite grief counseling to employees in the days after an incident occurs. By providing onsite assistance, employees are able to "stop by" while at work, without having to make an appointment during off-work hours. Further, appointments should be scheduled for any witnesses to a trauma, in case employees feel too embarrassed to reach out and ask for help.

The Internal Investigation Report

Employers should decide, prior to an incident, whether they will prepare an internal investigation report for each incident or only for certain incidents. When developing the internal investigation report, employers should maintain focus on the fact that the results of the investigation may end up subject to outside scrutiny. Every action performed by the employer in conducting the internal investigation and drafting the internal investigation report should be taken with the purpose of the report in mind, finding the root cause of the incident, protecting all employees from future harm, and protecting the reputation of the employer. Employers should be aware of several pitfalls they face when conducting its investigation that can have a serious impact on the internal investigation report. Employers should avoid producing premature, inaccurate, speculative, or interim reports that could reflect poorly on operations. These reports often include poorly planned or designed incident reenactments and premature consultation with equipment manufacturers. Assigning fault should be avoided during the development of the initial internal investigation report. The employer's reputation is on the line when producing these reports, necessitating the production and delivery of a complete and finished product.

Employers should take care when conducting the internal investigation. Development of a useful investigation report will require the contributions of many individuals and disciplines at the company. Those involved should include management, technical, and legal staff. Meetings to discuss theories of the incident create witnesses who may be later required to testify at an OSHA tribunal. The notes taken at these meetings may become evidence. Statements made without full information and sufficient deliberation can cause credibility concerns for the employer before both OSHA and the public.

Conclusion

Times of crisis put emergency response plans developed by employers to the test. Even employers possessing the most thorough response plans will be presented with important decisions during a crisis event. Effectively deploying an emergency response plan requires employers to actively plan for and consider many factors, including, as discussed above: securing and documenting the scene of the accident, communications and the attorney-client privilege, interviews and statements of third parties, documenting the crisis, providing assistance to the families of those affected by the crisis, and producing a quality and useful internal investigation report.

Acknowledgement

The authors would like to thank Peter Almaas, Jackson Kelly Associate, for his help in drafting this paper.

Workplace Wellbeing: Bridging Safety and Health

John Dony
Director of EHS&S and Campbell Institute

Katherine Smith
Program Manager
Campbell Institute
National Safety Council
Itasca, IL

Executive Summary

Employee health and wellbeing is not just an emerging topic in the EHS field, but one that is quickly gaining momentum. Research, medical, and academic groups are pioneering the concept of integrated health and safety, which is *the blending of health and safety programs along a continuum of organizational, personal, and occupational activities to enhance overall worker wellbeing and prevent work-related injuries and illnesses*. Underlying this definition is the belief that a true culture of health and safety is dependent on not only a strong safety program, but also a program that focuses on worker wellbeing. Campbell Institute members have already signed on to this belief, which is why many have had wellbeing programs in place for several years.

The Campbell Institute has conducted a research study on worker wellbeing programs at Institute participant organizations. Nine organizations were interviewed about the development, implementation, and lessons learned of workplace wellbeing and safety initiatives. The five key takeaways from the research are:

1. Pilot health and wellbeing programs with stakeholder input
2. Craft good communication
3. Experiment with incentive structures
4. Engage employees through organized activities
5. Connect wellbeing to safety

This paper outlines the major successes and challenges Campbell Institute organizations have experienced when developing and maintaining their wellbeing programs. Additionally, this paper addresses other issues for further discussion, such as the difficulty in calculating a return on investment for wellbeing programs, and some of the barriers to truly integrating health and wellbeing with occupational safety.

Introduction and Background

The business case for workplace safety is well documented and has been argued for by safety professionals and organizations, including the Campbell Institute and its participants. The latest focus for maintaining a sustainable business enterprise has moved beyond just workplace safety to include overall employee health and wellness. With the advent of recent programs like NIOSH Total Worker Health® and academic partnerships between American College of Occupational Medicine (ACOEM) and UL (Loeppke et al., 2015), it is obvious that employee health and wellbeing is not just an emerging topic, but one that is quickly gaining momentum. These groups and others are pioneering the concept of integrated health and safety, which is *the blending of health and safety programs along a continuum of organizational, personal, and occupational activities to enhance overall worker wellbeing and prevent work-related injuries and illnesses* (Loeppke et al., 2015). Underlying this definition is the belief that a true culture of health and safety is dependent on not only a strong safety program, but also a program that focuses on worker wellbeing. Campbell Institute members have already signed on to this belief, which is why many have had wellbeing programs in place for several years.

The connection between worker health and safety seems fairly obvious from an industrial hygiene and ergonomics perspective; exposure to adverse workplace environmental conditions or awkward working postures is bound to negatively affect worker health and safety that impedes the ability to perform work effectively. But worker wellbeing and safety are connected in other ways that may not be as obvious. For instance, age and excess body fat can put a worker at risk for certain musculoskeletal disorders (Schulte et al., 2012). Those who smoke or those with certain diseases or skin ailments can be at greater risk for occupational exposure (Schulte et al., 2012). Psychosocial stress and/or prescription medications can lead to fatigue or distraction, which increases the risk of injury, and overall stress from the workplace environment can result in less organizational trust in the employer, reducing the likelihood that an employee will be compliant with the safety policy (Schulte et al., 2012).

While the above outlines several ways in which worker safety can be compromised by worker health (and vice versa), there is also evidence of the benefits of integrated workplace wellbeing programs championed by NIOSH and ACOEM/UL. For example, if occupational injury is operationalized through workers' compensation claims, then there are numerous studies that can demonstrate the connection between improved safety and good physical condition, good mental health, and absence of chronic illnesses (Bunn et al., 2010; Hymel et al., 2011). Workplace wellbeing programs are also tied to lowered prescription drug and medical costs and reduced hospital admissions. Another study shows that wellbeing programs are far less effective in workplaces where there are unaddressed safety issues (Sorensen et al., 2004). In other words, the workplace must have a strong foundation of safety efforts in order for a wellbeing program to take hold and yield positive results.

A recent study found that workplaces that foster a broader "culture of health" experience increased job satisfaction among employees and reduce turnover (Kwon & Marzec, 2016). Having a positive "culture of health" entails not merely offering weight loss or smoking cessation programs for people to participate in, but actually making structural and policy changes to the organization to foster healthier behavior, such as encouraging stair use, implementing no-smoking policies, and offering healthy foods in cafeterias. The broader "culture of health" approach includes gaining higher-level support and aligning leadership priorities with management-level implementation strategies. Such a strategy is in keeping with previous safety research, showing that cultivating a deeper culture of safety is essential for motivating employees to stay in compliance with safety protocols (Ford & Tetrick, 2011; Hambach et al., 2011).

Healthcare costs have been and continue to be of great concern to employers, especially because one's health insurance is generally connected to an employer. Determining a company's total health care cost should take into account both direct costs, such as the cost of healthcare coverage and workers' compensation, and indirect cost, such as those related to productivity losses (Condon, 2016). When it comes to general workforce health costs, the Centers for Disease Control (CDC) postulates that if "tobacco use, poor diet and physical inactivity were eliminated, 80 percent of heart disease and stroke, 80 percent of Type 2 diabetes and 40 percent of cancer would be prevented" (Mensah, 2006). Achieving this would save over half a trillion dollars a year, with most of that staying in the pockets of employers, who cover almost 62 percent of the population not eligible for Medicare (National Center for Health Statistics, 2010). Regarding indirect costs, general productivity losses that were related to personal or family health problems cost U.S. employers $1685 per employee per year, or about $225.8 billion annually (Stewart et al., 2003).

Overall healthcare costs in the U.S. have been steadily rising since the Centers for Medicare & Medicaid Services have been collecting data on national health expenditure accounts, starting in 1960. In 2014, U.S. healthcare spending increased 5.3 percent from the previous year, totaling $3.0 trillion or $9523 per person. This amount accounted for 17.5 percent of the national's gross domestic product (GDP). For the sake of comparison, the total amount spent on health expenditures in the year 2000 was $1.37 trillion, or $4857 per person. In that year, healthcare costs accounted for 13.3 percent of the U.S. GDP (Centers for Medicare & Medicaid Service, 2015).

Because of increasing direct health expenditures and indirect healthcare costs, Pfeffer (2010) believes that more attention should be paid to employee physical and mental health. Like environmental sustainability, Pfeffer (2010) sees investment in employee wellness as organizational sustainability, and argues that employer concern for worker wellness is part of being a socially responsible organization. Employers can create an organization that is devoted to total worker health by implementing wellness programs, better managing work hours and work stress and providing paid sick days. Numerous workplaces in the past decade have implemented health and wellness programs, with positive outcomes that cannot be ignored.

In 2003, Vanderbilt University initiated an employee wellness program and enlisted the help of researchers to track the program's statistics for the next seven years. In a program that mainly involved completion of a health risk assessment and lifestyle management tool (plus up to $20 per month added to participating employees' paychecks), Byrne et al. (2011) reported a 6.4% increase in employees exercising one or more days per week during the program's first year. Additionally, both the smoking and obesity rates of participating Vanderbilt employees saw sharper declines than the smoking and obesity rates of Tennessee and the U.S. as a whole. A related study of Salt Lake City government employees looked at both the health and cost effects of a workplace health program. Merril et al. (2011) evaluated the Healthy Lifestyle Initiative Program (HLIP) from 2004 to 2008. The program included free annual screenings, financial incentives and educational programs to raise awareness of health issues. In addition to HLIP participants being more likely to exercise and having better self-perceived health, the total cost savings over four years due to lower prescription drug and medical costs was over $3.5 million. In short, that means that every dollar spent on HLIP saved the county $3.85.

Evidence nodding to the benefits of workplace health programs can be found in review articles that analyze multiple studies. Cancelliere et al. (2011) looked specifically at workplace health promotion programs to improve presenteeism, or showing up to work when ill. The authors found not only that workplace health promotions are effective at improving presenteeism overall, but that certain measures appeared to increase the level of success: exercise programs, depression screenings, health-risk assessments, positive work environment and monetary incentives. In a similar review, Pelletier (2011) analyzed 27 studies performed between 2008 and 2010 on the clinical and cost outcomes of workplace wellbeing programs. In this time period, health promotion initiatives tended toward randomized clinical trials and focused on disease-specific programs (e.g., obesity, back pain, lupus). The majority of the 27 studies indicated positive cost and clinical results, although Pelletier (2011) also notes the bias toward published research that focuses on statistically significant, positive results. Still, the number of studies from 2008 to 2010 focusing on wellbeing programs suggests that more employers are implementing and evaluating workplace health initiatives.

The business case for workplace wellbeing programs has not escaped the notice of the *Harvard Business Review*. Berry et al. (2010) conducted a review of several leading companies' wellbeing programs, and came to three overall conclusions: (1) investing money in prevention can have a return on investment of 300% or more; (2) wellbeing programs reduce absenteeism and increase healthy working days; and (3) wellbeing programs increase employee morale and retention. The authors also identified some key characteristics of successful programs, such as having engaged leadership and aligning the program with overall company identity and goals. Successful programs should also be comprehensive in scope and quality, and easily accessible in terms of scheduling and cost.

Due to the scientific community's overall opinion that wellbeing programs are both cost and health effective, the Center for Disease Control and Prevention (CDC) in partnership with the National Institute for Occupational Safety and Health (NIOSH) launched the Total Worker Health™ (TWH) initiative in 2011. The CDC and NIOSH essentially saw a gap in caring for the wellbeing of workers; workplace safety and health programs tend to focus solely on safety and protecting workers from hazards associated with the work environment, while workplace health programs tend to focus only on lifestyle conditions outside of the workplace that may put workers at risk. The Total Worker Health™ program was designed to integrate workplace safety protection with off-the-job health promotion, armed with research evidence to suggest that this integration is the most effective way of protecting workers.

Paying close attention to the "H" in EHS has been shown to be sustainable not only in terms of personnel, but also profit. While there are many types of successful programs and numerous ways for programs to improve, workplace wellbeing programs appear to be another best practice for maintaining an efficient and sustainable business organization. As the lines between work hours and leisure hours become increasingly blurred, it becomes more difficult to draw a distinction between work-related and non-work-related injuries. Cherniak et al. (2011) asserts, "Prevention of chronic disease factors, as well as efforts to maintain high function and effectiveness cannot be confined to a 40-hour work week." As this quote implies, Campbell Institute members and partners have realized that worker health and wellbeing has become a 24/7 proposition to be integrated with worker safety.

Methods

To study the best practices of Campbell Institute members and partners regarding employee health and wellbeing, the Institute conducted a series of in-depth, one-hour interviews with nine Institute participants: BNSF, Cummins, ISN, Mosaic, NASA, Norfolk Southern, Owens Corning, United Rentals, and USG. Interviews started in the winter of 2015, and concluded in the spring of 2016. Those contacted were responsible in some way for the administration, development, and implementation of the organization's employee wellbeing initiatives. The interviewees were frequently human resources managers or persons in charge of employee benefits, but also included a safety and health director or in-house physician.

The interview questionnaire asked how and why Campbell Institute organizations implemented employee wellbeing initiatives, the types of activities and incentives that comprise their programs, and the major challenges and successes they have experienced along the way. Some interviewees also provided supplemental information, such as written documentation outlining their wellbeing initiatives.

Key Takeaways

Regarding the development, implementation, and maintenance of workplace wellbeing and safety initiatives, the interviews from Campbell Institute participants revealed five key pieces of advice:

1. Pilot health and wellbeing programs with stakeholder input
2. Craft good communication
3. Experiment with incentive structures
4. Organize frequent fitness/wellness competitions
5. Connect wellbeing to safety

Piloting Health and Wellbeing Programs

When starting health and wellbeing programs at a company, Campbell Institute participants began by piloting key aspects of their programs at select locations before rolling out the whole program to the entire company. Member company BNSF began by offering general health education, training, and coaching at three pilot cities to see how well they were received before expanding to the rest of the organization. Research participants also recommended gathering the input and support from employees and leaders to get a better understanding of what they would like to see in a company wellbeing program. Both Owens Corning and United Rentals formed focus groups and town halls to ask these stakeholders about their expectations for wellbeing programming and which activities would be most successful. USG carried out one-on-one interviews with employees and conducted surveys to understand the workforce's needs and desires of a workplace health program. Even after launching their health program, United Rentals periodically conducts roundtable discussions and a semi-annual employee survey to stay current with employee feedback and questions.

Crafting Good Communication

Research participants agreed that good communication is key to maintaining effective and engaging employee wellbeing programs. Communication could take multiple forms. Because Institute organizations are known to have excellent cultures of safety and exceptional ways of communicating safety messages, many like Owens Corning and USG have found the best method for relaying health information is to piggyback on the existing company safety communications, adding a tip on health and wellbeing in periodic email newsletters. BNSF holds quarterly campaigns that focus on specific health topics that are timely and/or relevant. Similarly, ISN sends monthly updates to raise awareness of different health topics and encourage employees to be proactive about their health.

Enlisting employees to serve as ambassadors of the company health and wellbeing program was another common method for gaining support for the program and communicating the importance of it. Cummins encourages employees to make individual connections with others in the company to make them aware of wellness offerings and get them connected to the offerings that are relevant to them. One of these employee networks is called "Health Champions." Norfolk Southern also enlists employee ambassadors ("Hot Shots") to communicate the importance and benefits of their program. Owens Corning gathers testimonials from employees to create posters and other messaging to convey how well the health initiatives are working.

The issue of personal health information being kept confidential is a hurdle that nearly every organization with an employee health program must confront. Several research participants underscored the importance of emphasizing the privacy of employee health information. Norfolk Southern even mentioned their efforts to "over-communicate" the confidentiality of worker health data. Their ultimate goal is not to play "Big Brother" with employee health information, but to do right by workers in protecting their total health. All the organizations in this study emphasized that individual-level health information is never known to the organization; any information is viewed and analyzed in the aggregate. Mosaic utilizes the aggregate health information compiled by a third party to identify gaps in care and proactively address risks at an organizational level. At NASA, employee assistance programs (EAP) are co-located with medical clinics and other site functions, so there is no way for anyone to know if a person is obtaining counseling, getting a regular check-up, or just going to work.

Experimenting with Incentive Structures

A third piece of advice from the research participants was to experiment with incentive structures to find one that works best for your employee population. Finding the right balance of "carrots" and "sticks" to maximize worker participation and achieve optimal results is not an easy task, but it is something that Campbell Institute participants have and continue to test. One member, ISN, found that employees were actually more likely to maintain participation in a steps challenge if they paid for their Garmin watch instead of having ISN subsidize the cost, requiring them to put some "skin in the game." USG offers occupational health testing and a smoking cessation program free of charge to encourage employees to take advantage of these services. United Rentals offers a \$1000 incentive for being tobacco-free and a \$600 incentive for taking part in health assessments and health coaching. At Mosaic, employees are frontloaded with a \$60 wellness credit that they keep if they complete a health screening and assessment after a few months. Mosaic employees also have the opportunity to earn up to \$200 in gift cards for participating in a health coaching program. Because it is a government agency, NASA cannot offer incentives in the form of health insurance premium discounts. Regulations also prohibit the agency from awarding "prizes" and trinkets (e.g., swag). However, agency-wide recognition and awarding two trophies and two plaques have shown to be enough incentive and motivation to gain substantial employee participation. For the past three years, during their annual fitness challenge, participation went from 2100 participants in 2014 to more than 3300 in 2016.

Maximizing participation in health programs, while simultaneously driving up the effectiveness of the programs, is a constant challenge for organizations. Experimenting with participation-based and outcome-based incentives structures is one path to achieving a good balance between participation and effectiveness. At Owens Corning, incentives are loaded as a payroll credit to offset the amount of insurance premiums. Employees can earn up to $500 in credits: $100 for submitting their biometric numbers, $200 for completing a health improvement program, and $100 each for meeting certain biometrics targets for Body Mass Index (BMI) and blood pressure. Owens has found that participation-based incentives are successful at getting employees to sign onto the program, but that outcome-based incentives (based on meeting certain targets) is best at improving program effectiveness.

Engaging Employees

To keep employees engaged with health and wellbeing programming, the research participants found that organizing frequent fitness/wellness competitions was an effective way of keeping workers plugged into wellbeing messaging. Owens Corning holds an annual Fat Out Challenge, which was actually started by an individual site and has since expanded to more locations. ISN offers both team-based and individual fitness competitions for their employees at their headquarters in Dallas, TX, and their seven other office locations. Over the course of 2015, NASA offered over 30 sports competitions or run/walk events at their various facilities across the country.

There is a myriad of ways to engage employees in health and wellbeing, either through programs and activities that are "homegrown" within an organization (such as the examples above), or through third-party applications, like Virgin Pulse. Members Cummins and Norfolk Southern utilize Virgin Pulse to help employees set goals in a number of wellbeing areas: nutrition, sleep, stress, financial health, and physical health. As part of a steps-counting challenge, employees in these companies are provided with pedometers or use the wearable technology that they already have to track their individual progress and/or in competition with other teams. A platform like Virgin Pulse rewards employees with "HealthMiles" for completing annual health screenings or for participating in competitions, and these HealthMiles can, in turn, be used to offset medical plan contributions.

Finding the right platform or application for engaging employees, like finding the right balance of incentives, can be challenging. Off-the-shelf programs may not work for an organization, depending on the organization's size and composition, or a company may not have the resources necessary to implement a third-party wellbeing program and must rely on in-house engagement efforts. For comprehensive programs, like Virgin Pulse, that connect yearly health assessments, biometric screenings, and participation in fitness activities with discounts in insurance premiums, it can be easier to keep employees engaged in the long term and throughout the year, instead of only one- or two-month intervals when the company issues a fitness challenge.

Many Institute participants have international locations, and there is a desire to engage these employees in health and wellbeing efforts as well. In these cases, multi-site and multi-country team competitions, like at Owens Corning and ISN, have been most successful in gaining global support for health and wellbeing. In general, Institute participants found that these frequent methods of engaging employees promote team building and improve morale while helping employees get healthier, in all the various ways that "health" implies.

Connecting Wellbeing to Safety

Research participants recognized the importance of better connecting wellbeing to safety. While the departments responsible for managing safety and wellbeing may be separate in an organizational chart, Campbell member organizations have found ways to informally connect them by communicating to employees that being physically and mentally well enables a person to perform work better and safer. An equivalency they draw between safety and wellbeing is that being proactive about your health is similar to performing preventive maintenance on machinery: Both actions allow workers to read trends and keep up with changes, so that there are no surprises or catastrophes down the line. USG has noted that certain programs, like their medical monitoring program, began purely as a safety initiative, but now have very overt ties to wellness. Keeping an eye on one's health numbers and acting quickly when an issue arises has helped USG employees not only be healthier but stay safer on the job, avoiding absenteeism, presenteeism, and days of restricted work. Seeking to knit together safety and health, United Rentals has launched a new metapostures program, which was developed with the help of physical and occupational therapists, and teaches employees to stretch in ways to strengthen muscles and lubricate joints. This program is one way that United Rentals has connected safety with health from an ergonomic perspective.

To connect safety with their wellness strategy, Mosaic partnered with two different organizations; one focused on traditional occupational health, and other focused on wellness and episodic care. These two organizations have combined their efforts to form the Wellness and Occupational Health Clinics at Mosaic sites. This effort has fed into Mosaic's total health management strategy as clinicians from the wellness and episodic care side can communicate with clinicians on the occupational health side to identify potential risks and gaps in care. Mosaic employees have also developed a level of trust with the clinicians on either side, making it easy for the wellness and occupational safety halves to cross-reference each other and contribute to total worker health.

Major Successes in Wellbeing Programming

Participants in this research noted the major successes that they have experienced so far with their health and wellbeing programs, which ranged from increasing participation to building deeper culture change within the organization. In addition to seeing a large uptick in participation in Virgin Pulse, Norfolk Southern includes the sharing personal success stories around weight loss and smoking cessation as some of their program's biggest victories. Similarly, United Rentals has seen an increase in program participation due to employees being advocates of the program and giving testimonials. United Rentals is also proud of the commitment to wellness on the part of their CEO, who pledged to lose 25 pounds or donate $25,000 to the United Rentals compassion fund, a 501(c)(3) organization.

The health and wellbeing team at Cummins has reached out to the medical community to improve their program offerings and help inform them of additional therapies for medical conditions. Cummins has partnered with the American College of Preventive Medicine and the American College of Lifestyle Medicine to find remedies to chronic conditions that consist of lifestyle changes, such as diet, exercise, and meditation. These partnerships have created the foundation for evidence-based health and wellbeing programming at Cummins.

While hard return on investments for wellbeing programs can be difficult to calculate, ISN has seen softer returns for their efforts through improved health outcomes among employees, along with better employee engagement and morale. Even though it is trickier to determine the return on these types of outcomes, they most certainly have a positive effect. Through their efforts, ISN has encouraged the many young people at their company to be more proactive about their health and pay attention to their health metrics, both now and going forward.

BNSF and Owens Corning count among their successes the small organizational changes that have led to cultural change and awareness of wellbeing. Both companies note that providing healthier food options, charging less for healthy food options, and eliminating large soda cups are small ways to persuade employees to make better choices. Encouraging employees to walk, take stretch breaks, and take the stairs are other methods to lay a foundation for a broader "culture of health" across the organization.

Biggest Challenges and Issues for Further Discussion

Biggest Challenges

Participants in the study were asked about the biggest challenges they have or continue to experience in maintaining their wellbeing initiatives. One of the main hurdles, especially in the manufacturing and railroad environments, is the demographic issue of appealing to a mostly male, middle-aged workforce and getting them to care about their physical, mental, and emotional wellbeing. This is a population that does not typically visit a doctor regularly and may or may not have ready access to a computer, which makes online health coaching problematic. To tackle this challenge, companies like Norfolk Southern, BNSF, and Owens Corning have campaigns that are meant to target this population in particular, and are working on finding good health coaching mobile applications for use on a smartphone.

The technology issue is another challenge that a couple participants face. Cummins has had to vet several vendors to find those that offer what Cummins wants in terms of programming, and with the ability to pull claims data with the right amount of privacy. Often, there are no off-the-shelf programs available, and they have to develop programs in house. Technology, in the form of wearables, has been a small challenge for ISN, specifically finding a piece of wearable technology that tracks both steps and heart rate. As these technology offerings evolve and improve, ISN is confident that they will find fitness trackers and programs that will keep employees engaged in the long term, not just in three- or six-month intervals.

At NASA (similar to other federal agencies), many variables may impact health promotion and wellness programs, both locally and agency-wide. The most significant variables are budget and human resources, especially since these programs at NASA are primarily implemented by support contractors. At any given time, the contractors responsible for implementation of their center's health and wellness promotion outreach may change. Reorganizations and reassignments of programs and division that handle health promotion and wellness may also influence program strength. Regardless of the many changes NASA has experienced over the years in their occupational health programs, health and wellness continue to be recognized as important components contributing to employee productivity, effectiveness, and efficiency.

The communication of health and wellbeing messages remains a small hurdle for some participants. Mosaic notes that their employees regularly receive numerous messages from the company on many topics, including safety, career activities, and volunteer/giving opportunities, that it is easy for health and wellbeing messages to get lost in the shuffle. Mosaic also wants to communicate to employees that "wellbeing" is more than just the physical aspect and includes emotional, mental, and even financial aspects.

Truly Integrating Health and Wellbeing with Occupational Safety

While Campbell Institute members and partners recognize the call to fully integrate health and wellbeing with occupational safety, there are sometimes structural or cultural barriers that prevent organizations from completely combining their safety and health efforts. One structural barrier is the data privacy restrictions that companies must abide by to stay in compliance with the Health Insurance Portability and Accountability Act (HIPAA). Cummins, for instance, is performing a study to assess how sleep relates to healthcare costs, incident rate, and severity rate, but otherwise has not found much to explicitly link wellbeing to safety because of health data confidentiality. Norfolk Southern must tread carefully in integrating wellbeing with safety because of their heavily unionized workforce, and the desire to not violate any collective bargaining agreements. Like other Institute member organizations, human resources has ownership of wellness initiatives at Norfolk Southern, but wellness is not tied to workers' compensation.

Other barriers to full integration are more cultural or political in nature. At BNSF, for example, the wellness initiatives have deliberately been kept separate from safety because of the traditionally "punitive" or "mandatory" perceptions of safety. BNSF wants their employees to see health and wellbeing as completely voluntary and positive, without fines or penalties for non-participation. For other companies, it is difficult to make an overt assertion that measures like TRIR, LTIR, or DART have been reduced because of wellbeing programming, if only because it is difficult to perform the necessary correlation analyses when these rates are already close to zero. Also, those in charge of running wellbeing programs at companies are loathe to taking much credit for impacting traditional lagging safety metrics, as this is seen as discounting the efforts of the safety team.

Calculating a Return on Investment

Although there are numerous studies to show that workplace wellbeing programs produce a return on investment (ROI), these numbers are sometimes difficult to calculate accurately and consistently. Data privacy issues can play into this, as it is problematic to compute the cost savings on preventive care when a company is unsure if a person visited the doctor for a routine checkup or for an illness. And, as noted earlier, it is difficult to parse out the impact of wellbeing initiatives from other contributing factors, such as occupational safety and even sustainability. This research discovered that Campbell Institute members and partners in general have not performed hard ROI calculations because this is not their primary motivation for offering wellbeing programs. While cost control is important, it is merely an ancillary goal to the organization's main objective of seeing that employees go home in the same or better condition as when they came to work. All of the Institute organizations in this study expressed that the purpose of their health and wellbeing programs is to do right by employees, not to realize a return on investment.

Instead of calculating an ROI, Owens Corning instead focuses on overall health numbers, like aggregate measures of blood pressure, weight, and body mass index to see if sites are improving. Similarly, BNSF evaluates their program from a population health-management perspective; if the aggregate numbers steadily improve, the program is seen to have a positive effect. USG reminds itself and other organizations that investments in health and wellbeing require a long-term view when evaluating return, which is why health and wellbeing requires a commitment on the part of employers (and employees). The longitudinal studies on health and wellbeing programs bear out this finding.

Discussion and Conclusion

The Campbell Institute organizations included in this research would be the first to say that their health and wellbeing programs are not perfect. Just like their safety management systems, these programs are in a continuous improvement cycle, going through much iteration to make each cycle better than the last. The organizations in the study typically began small or piecemeal with a collection of health-related activities, such as smoking cessation, physical fitness, weight loss, and biometric assessments, that, over time, became parts of an overall health and wellbeing program across the organization. Institute members are also diligent in researching and gathering input from employees, leadership, and other stakeholders to ensure that they are rolling out programs that are relevant and of interest to the organization's people.

Continuously improving health and wellbeing programs is also a matter of finding the right incentives to encourage employee participation. Through experimentation, Institute members have found a good balance of "carrots" and "sticks" to help employees continue with weight loss programs, engage in physical fitness, quit smoking, or seek assistance with any type of health issue. This does not mean, however, that these organizations have stopped trying to improve their incentive structures.

Similarly, communicating about relevant health topics and offering timely health tips is a continuous process. Institute organizations recognize that pairing health information with safety information is a natural fit, particularly when their culture of safety is already strong. Institute members have also realized that there is no such thing as too much reassurance to employees that their health data is kept private and secure. Communicating on this issue and obtaining positive testimonials from employees are ways for organizations to garner employee trust on health and wellbeing and increase participation in programs.

The conclusions of this research and what is not currently included as a "takeaway" point to other aspects of health and wellbeing that should be incorporated in future research and discussions. A recent benchmarking session among Campbell Institute members and partners revealed that many are still taking introductory steps to addressing mental health, work strain, and stress as part of an overall wellbeing program. Many would like to see ways to improve employee assistance programs, counseling, and fitness for duty checks to ensure that employees are undistracted mentally and emotionally before beginning work. This is an area where the Campbell Institute could take pointers from its members with headquarters in European countries, where the issues of stress and mental health have received more attention and gained more traction regarding workplace safety.

Building an overall "culture of wellbeing" appears to be a goal for the participants of this study. Institute members and partners long ago realized that implementing a couple of ad-hoc safety initiatives was not enough to actually instill a culture of safety within the organization, and the same may be said of wellbeing. A true culture of wellbeing cannot be brought to bear simply by installing a smoking cessation program or a weight-loss challenge. As has been realized with safety, offering incentives will only go so far. With OSHA discouraging the use of incentives to meet safety goals, it's not a stretch to think the same will be the case for wellbeing initiatives. Carefully considering how wellbeing can become ingrained within an organization's culture is an area for follow-up when studying this topic in the future.

Finally, it can be concluded that Campbell Institute organizations understand the value in integrating health and wellbeing with occupational safety. The connections between the two have been clearly demonstrated in a host of peer-reviewed, academic research: Healthier employees are at less risk of industrial exposure and perform work more safely with lower rates of absenteeism and presenteeism. There still exist both institutional and cultural barriers, however, that prevent organizations from truly integrating health with safety. The nature of wellbeing programs also makes it difficult to calculate the true return on investment for these programs. Given time for more research and innovative methods, making the business case for a fully integrated safety, health, and wellbeing system will be more straightforward, paving the path for more employers to follow suit.

The organizational practices detailed in this paper are best described as good wellbeing *programs* that have yet to become excellent wellbeing *systems*. Just as workplace safety has undergone development, transformation, and maturation throughout the past few decades, this research points to a similar maturity curve for health and wellbeing, with organizations moving beyond a simple compliance-driven attitude to one that is driven by values and deeply held principles. Given the current trajectory, reaching a point of achieving excellence in health and wellbeing systems is on the horizon.

Bibliography

Berry, L.L., Mirabito, A.M., & Baun, W.B. 2010. "What's the hard return on employee wellness programs?" *Harvard Business Review*, December 2010, 104-112.

Bunn, W.B., Stave, G.M., Allen, H., Naim, A.B. 2010. "Evidence-based benefit design: Toward a sustainable health care future for employers." *Journal of Occupational and Environmental Medicine*, 52, 951-955.

Byrne, D.W., Goetzl, R.Z., McGown, P.W., Holmes, M.C., Beckowski, M.S., Tabrizi, M.J., Kowlessar, N., Yarbrough, M.I. 2011. "Seven-year trends in employee health habits from a comprehensive workplace health promotion program at Vanderbilt University." *Journal of Occupational and Environmental Medicine,* 53(12), 1372-1381.

Cancelliere, C., Cassidy, J.D., Ammendolia, C., Cote, P. 2011. "Are workplace health promotion programs effective at improving presenteeism in workers? A systematic review and best evidence synthesis of the literature." *BMC Public Health,* 11, 1-11.

Centers for Disease Control/National Institute for Occupational Safety and Health (CDC/NIOSH). 2015. *Total Worker Health™.* (http://www.cdc.gov/niosh/twh/totalhealth.html)

Centers for Medicare & Medicaid Services. 2015. National health expenditures; aggregate and per capita amounts, annual percent change an*d percent distribution: Selected calendar years 1960-2014.* (http://www.cms.hhs.gov/NationalHealthExpendData/.)

Cherniak, M., Henning, R., Merchant, J.A., Punnett, L., Sorensen, G., Wagner, G. 2011. "Statement on national worklife priorities." *American Journal of Industrial Medicine*, 54, 10-20.

Condon, M. 2016. "A health workforce: How workers' comp and wellness programs go together." *Proceedings of the Safety 2016 Professional Development Conference,* June 26-29, Atlanta, GA.

Ford, M.T., Tetrick, L.E. 2011. "Relations among occupational hazards, attitudes, and safety performance." *Journal of Occupational Health Psychology,* 16(1), 48-66.

Hambach, R., Mairiaux, P., Francois, G., Braeckman, L., Balsat, A., Van Hal, G., Vandoorne, C., Van Royen, P., van Sprundel., M. 2011. "Workers' perception of chemical risks: A focus group study." *Risk Analysis,* 31(2), 335-342.

Hymel, P.A., Loeppke, R.R., Baase, C.M., Burton, W.N., Hartenbaum, N.P., Hudson, T.W., McLellan, R.K., Mueller, K.L., Roberts, M.A., Yarborough, C.M., Konicki, D.L., Larson, P.W. 2011. "Workplace health protection and promotion: A new pathway for a healthier – and safer – workforce." *Journal of Occupational and Environmental Medicine,* 53(6), 695-702.

Kwon, Y., Marzec, M.L. 2016. "Does worksite culture of health (CoH) matter to employees? Empirical evidence using job-related metrics." *Journal of Occupational and Environmental Medicine,* 58(5), 448-454.

Loeppke, R., Hohn, T., Baase, C., Bunn, W., Burton, W., Eisenberg, B., Ennis, T., Fabius, R., Hawkins, J., Hudson, T.W., Hymel, P., Konicki, D., Larson, P., McLellan, R., Roberts, M., Usrey, C., Wallace, J., Yarborough, C., Siuba, J. 2015. "Integrating health and safety in the workplace: How closely aligning health and safety strategies can yield measurable benefits." *Journal of Occupational and Environmental Medicine,* 57(5), 585-597.

Mensah, G.A. 2006. "Global and domestic health priorities: Spotlight on chronic disease." *National Business Group on Health webinar*, May 23, 2006.

Merrill, R.M., Hyatt, B., Aldana, S.G., Kinnersley, D. 2011. "Lowering employee health care costs through the healthy lifestyle incentive program." *Journal of Public Health Management Practice,* 17(3), 225-232.

National Center for Health Statistics. 2010. "Health, United States, 2009." Hyattsville, MD.

Pelletier, K.R. 2011. "A review and analysis of the clinical and cost-effectiveness studies of comprehensive health promotion and disease management programs at the worksite." *Journal of Occupational and Environmental Medicine,* 53(11), 1310-1331.

Pfeffer, J. 2010. "Building sustainable organizations: The human factor." *Academy of Management Perspectives,* February 2010, 34-45.

Schulte, P., Pandalai, S., Wulsin, V., Chun, H. 2012. "Interaction of occupational and personal risk factors in workforce health and safety." *American Journal of Public Health,* 102(3), 434-448.

Sorensen, G., Barbeau, E., Hunt, M.K., Emmons, K. 2004. "Reducing social disparities in tobacco use: A social contextual model for reducing tobacco use among blue-collar workers." *American Journal of Public Health,* 94, 230-239.

Stewart, W.F., Ricci, J.A., Chee, E., Morganstein, D. 2003. "Lost productive work time costs from health conditions in the United States: Results from the American productivity audit." *Journal of Occupational and Environmental Medicine,* 45(12), 1234-1246.

Session No. 708

The Behaviorial Safety System: Understanding Commitment, Courage, and Leadership

Bob Veazie
President
People Powered Leadership

For decades, the safety professional has led us to believe, through an abundance of consulting services, that the path to reducing injuries is obtaining a culture where people speak up when they notice others at risk. Many organizations have been sold that they need to have a behavior-based safety process. Safety leaders have thought that, if you have an employee-led team, you will get most of your employees involved in a couple observations of others a month with a checklist, and increase the volume of observations in your organization, that the level of risk decreases, and injuries will be reduced.

There is wisdom in the behavior-based safety process in that it does engage more employees, it does teach risk needs to be managed, and it does get managers excited when they see numbers go up (the number of observations each month). The problem is that, in focusing on the behavior safety process, much of the necessity to study and shift behavior across a broader system of behavior – focusing more on mindfulness and leadership – was lost. Sometimes, an over-focus in one area creates an unintentional under-focus across a broader system. And while companies worked hard at engaging employees and increasing the number of observations, in many cases they lost the connection to the purpose of the mechanics of the behavior-based safety process

The overall objective of the behavior-based safety process had little to do with the mechanics of the process, the number of observations turned in. Rather, it was about the development of people. The purpose of behavior-based safety was to develop a culture where, much more frequently, employees would speak up and help a peer choose and act to be safer. Unfortunately, many organizations failed to change the real measure of success. They live with employees speaking up only 30-40% of the time. In other words, the desired culture of "actively caring" for one another was not fully achieved because the focus, over time, disconnected from the purpose (people-based), the entire focus became the process and the number of observations, instead of the real challenge of personal change.

Behavior and mindfulness remain primary concerns of safety, and all organizations that have safety as a value must continue to support and improve increasing employee engagement and ownership of managing risks. This will only be done by working on all three elements of the entire behavior safety system.

The Behavior-based Safety System

In the book, *The 7 Habits of Highly Effective People*, author Stephen Covey highlights the seven behaviors for being effective on anything in life. These seven behaviors, based on research and principles, are organized into two primary categories according to Covey: the private victory and the public victory. In addition to these two areas, Covey defines the seventh habit as that of "continually sharpening the saw."

Covey's first three habits, starting with habit #1, "Be Proactive," make up the "private victory" starting point for all human effectiveness. He would suggest that effectiveness starts with being proactive versus reactive and it starts with yourself (not somebody else). Isn't that where safety excellence begins, with a personal commitment to safety? If so, why has so much of safety training and education the last 30 years been on behavior-based safety, which is about focusing on someone else, versus committing and growing yourself to be more safe? I am not stating that a focus on behavior-based safety methodology and principles is totally. Yet, I am suggesting there is a more powerful starting point to focus on increasing personal safe behavior.

The second set of three habits of the overall seven are called the "public victory." After explaining, through the first half of the book, how effectiveness begins with yourself, Covey is very clear that we live in an dependent world and our effectiveness, while beginning with ourselves, must eventually be connected to how we work with others. Daniel Goleman called this our level of "emotional intelligence." The focus on behavior-based safety was to develop a culture of "actively caring," which Covey's fourth habit called "create win-win agreements." In safety, these win-win agreements are really about how I will watch out for your safety: if I see you doing something that appears unsafe, I will say something, if you will, in turn, watch out for my safety. I believe Covey as well would see behavior-based safety as related to "win-win agreements," and argue that just as the many safety consultants do some form of developing a culture where people speak up and help each other to be safer, which is critically important. Going further, he might have stated actively caring for each other is an important element of safety effectiveness, but out of order, if significant focus is not first given on "the private victory" or the system how I manage my own risk by being more mindful and catching and correcting myself more out of my own commitment to safety.

The third component of the behavior safety system is that of the overall context or support system for people: leadership. Without specific, focused reinforcement of behavior, people don't excel as far. It is researched that what is reinforced grows as a behavior. Therefore, if an employee focuses on two-to-three core risks one day and does their activity extremely mindful of these tasks, producing a safer day, wouldn't it be then more complete to have someone, or others around them, letting them know how well they did? Leaders many times think that leadership is the measuring of results (how many observations?), yet real leadership is that which reinforces others and increases the culture of consistent safe behavior. Leadership is in the conversations and relationships with others, not in the numbers.

The Behavioral Safety System is the integration of the three core behaviors required for maximum or extreme safety: (1) personal commitment to managing risks; (2) observing and acting when someone is doing something very safe or providing feedback to someone who may be having a mindless moment; and (3) the leadership energy required to fuel and sustain the continuous developing of the culture of these three behaviors.

Commitment-based Safety

The first core area of focus for safety excellence must be how each of us identify and manage the key risks we face. If I'm driving a car, what is my key risk(s), and how am I mindfully managing these? If I'm tempted to text while I'm driving, how do I mindfully commit myself to not be distracted while driving, and perhaps even actually feel good at the end of the drive that I chose that commitment and met it? If working on a ladder for the day, do I focus on not reaching to the side too far, to have three points of contact, and not follow my temptation to get "that little bit more" by going up to the top rung?

The first behavior is the personal management of risk that I call commitment-based safety because it takes a personal commitment to really live a life of accelerated risk management.

Basic psychology proves the constant battle we have with safety as consequences to behavior are strong driver of our choices and actions. B. F. Skinner's research demonstrated that people do what they do for the consequence or the outcome they expect to get. Why do people speed while driving? It is mostly because they want to save time. The consequence they expect of saving time causes actions that put people at risk. If a person doesn't choose and mindfully go to work on a personal commitment to manage specific risks, they will more often take the risk to get the consequence of saving time that is such a powerful temptation.

Another reason for ensuring a strong system of engaging employees to personally manage risk comes from the work of Daniel Kahneman who won the Noble Peace Prize. In that, Kahneman discloses that most of the time human beings are functioning, they are doing so in "Mind System 1" or a low-energy, less-mindful state. This could be considered "auto pilot." Living on auto pilot is sure, over time, to have us miss seeing some risks and decrease our probabilities of being safe. One of the reasons I'm convinced that it is essential for sustained zero injuries to have employees actively care for each other is that we, as human beings, all have moments of mindlessness, auto-pilot. A person working next to me might be in "Mind System 1," and not even notice how often they are reaching into those moving machine parts or notice the fact they forgot to lock out the tool they are in because they entered the machine thinking about fixing the production problem, not mindfully taking care of their personal safety first. I don't blame employees when they are injured, as many times they are simply being human with an average amount of "System Mind 1" going on.

Kahneman also described a higher level of alertness. He called this "System Level Minds 2." It is where we put forth the energy and focus, and our attention is raised to a very high level. This is similar to how most of us drive in snow or rainstorms when we tense up, clutch the wheel, and lean forward, watching carefully through the storm. Most fully recall those moments of high alertness in such situations. Yet, this level of alertness or mindfulness is hard to continue because of the energy it requires. And not creating a system around people for a full and frequent "Plan-Do-Check-Act" process leaves people without highest degree of the positive social consequence of daily support. How can we be our most safe if we are not shifting the level of time we are in "Mind System 1" to "Mind System 2?" It was this shift, this potential we all are capable of, that drove the honoring of Dr. Kahneman to receive the Noble Peace Prize for his decades of research and conclusions.

Commitment-based safety is a system of engaging employees every day in the focus on risks they will face, the choice on the critical few with highest risk, based on frequency and severity, and the follow through to reinforce the work every day. It is interesting how often employees are asked to write down their top three-to-four risks on their job that day, and so few after one minute have more than one or two risks. How do you manage what you don't know, what you haven't identified, and what you are not mindful about?

So many employees work without such focus, and they work in average safety programs that still produce numerous near misses, injuries, and serious injuries with a constant chance or probability of the worst case: fatality. Commitment-based safety uses the principles and method of Dr. W. Edwards Deming: simply recognizing the power of Planning, Doing (with committed focus), Checking and then changing based on what we learn. Any organization that does not set in place a system of daily choice and self-accountability with positive social consequences leaves workers more exposed.

Actively Caring Culture

One of the greatest experts in the field of safety is Dr. E. Scott Geller. Geller, a professor in the field of psychology, brought some basic principles from psychology to the field of safety in the 1970s, and the world of safety, for many people, has never been the same. The founder or "grandfather" of what is now known as "behavior- based safety," Geller has worked with thousands, and spoken in front of millions on the greater ability of people to be safe when we all have the courage to speak up when we notice risks.

These principles first introduced by Geller into the work world have, over decades, morphed into the behavior-based safety process. While the process mechanics and counting outputs of the process was never Geller's intention when working with organizations, it seems to be the part of the work that stuck more over time. The behavior-based safety process, when continually connected to the purpose of human development and reinforced for the behavior change versus the managing the metrics of the process, can deliver exceptional results. Geller's interest has always been focused on a "people-based" approach, which is far more about people growing and changing themselves (human development) over time as they work together to be increasingly safe. The behavior-based safety (BBS) process mechanics were to serve the human development intention, not become the purpose of the work itself. In fact today, because so few who do BBS have understood the intention of behavior-based safety, and don't lead the structure to the desired behavior change, Dr. Geller will less often be heard calling the system "behavior-based" any longer: it's "people-based," he says, having people truly change and grow in their hearts, not only filling out forms. It's amazing how we tend to get so fixated on structures, and forget the essence of what the structures were intended to emerge. Doing things is easy; becoming better through changing one's own behavior is much more difficult.

But, it is true, the second behavioral component of the strongest safety systems is the degree people change and are willing to talk with others about their personal safety; reinforcing the good, mindful choices, and speaking up in a moment of someone else's mindlessness or bad choice (choosing a shortcut for the consequences). This by some is called "actively caring," which is intentional, about going beyond just caring for other people to really acting on what you care about. If you see kids running around the side of a pool, do you care they don't get injured and watch, or do you "actively care" and talk with them about the potential slip/fall that could happen and the potential doctor visits (thus ruining their fun day). With over 38 million Americans each year getting injured, and having to visit a doctor's office for care, we could surely all get better in having the courage to act (such as conversing with others) when it comes to their safety.

Leadership: Reinforcing Safe Behavior

What behaviors get reinforced, they get repeated more and more. One of the greatest acts of leadership is to talk with employees about their safety, about the risks they are choosing to manage, and let them know how much that means to the leader. It's amazing how much time is spent by people in management positions in meetings ***away*** from people, when the conversations that are held with people are likely the best advantage to excellent performance in an organization.

Another aspect of excellent reinforcing leadership is those leaders that continue to educate their employees, to remind them of "WHY" we do what we do that is beyond the structure. One of the greatest chances to take a process based in sound principles to become yet another "Flavor of the Month" is to have the mechanics of the process drift away from the knowledge of the principles of the process, or the "why" we are doing this. I frequently ask managers for their personal "why" on a process like commitment-based safety or behavior-based safety, and I am told "to be more safe." Generic overarching reasons don't emotionally connect employees to change themselves. People can't emotionalize "be safe;" they have heard it over and over and over since they were a little kids. Great leaders have "personal whys" for their key organizational processes that are built in sound principles or research for why that works: why we have chosen this method, and they communicate these whys often.

How many companies have worked with behavior-based safety where employees now hate it, or only do it because they are incented to complete some observation forms? This frequently evolves into the process itself is no longer delivering lowering risks; it's only delivering numbers that managers get excited about. Don't think for a minute that the number of observations in an organization is a solid representation of the degree risk is managed. This can cause a false sense of security, and leave the workforce exposed.

Human commitment is driven by three key areas of focus: (1) education, (2) choice (autonomy), and (3) reinforcement (accountability – positive works best). The role of leaders for highest performance in organizations is where they are mindful about continuous education (not training on mechanics or how to do something, but a purposeful connection), provide choices for employees to fully be empowered, and commit themselves to the best performance they can contribute, and then ensuring reinforcing conversations and systems are in place to continually learn, grow, and contribute more.

In Conclusion

Human behavior will always be a major component of all incidents and injuries. That is not to blame people for the outcome, but rather to invite them to be a stronger part of the proactive management of risk necessary to shift the odds so far that the probabilities become extremely low for a major incident.

There are three areas of behavior that need continuous improvement in any organization that values safety and wants to continually battle to decrease the odds of an injury happening in their organization. These three areas of behavior begin with a personal commitment to know and manage risks by everyone every day. At the end of each day, people aren't injured either because they were lucky (took risk and didn't get injured) or they "earned their safety." Safety is a produced outcome and, at the end of the day, when you have focused on key risks, and can be personally accountable to tell how you did that day on managing your risks, you have minimized the mindless time of relying on luck, and optimized increased mindfulness and personal choice to pick the safest path to produce your safety that day.

After each of us is in a system that promotes our personal risk management, it is essential we work within a culture that grows our courage to speak up and actively care for others to be safe. Knowing all of us have off moments of mindfulness, and are tempted sometimes by consequences that drive the thought and action to "just take the risk this time, I'll be OK – this time," it is important that we help each other remain true to making best choices, and being most mindful with risks.

And lastly, leadership is the fuel that sustains personal and collective behavior. The greatest act of leaders is to reinforce the behavior of employees. Organizations that profess they "value safety" must then have leadership behavior expected and performed to have integrity to that professed value. Leaders as well must have daily and weekly actions (conversations with employees) outside the meetings that consume them to energize and reinforce safe employee behavior.

References

Covey, S. 1989. *The 7 Habits of Highly Effective People.* New York: Simon & Schuster.

Darley, J., & Latane, B. 1968. "Bystander interventions in emergencies: Diffusion of responsibility." *Journal of Personality and Social Psychology*. 8:377-383.

Dekker, S.W.A. 2014. *Just Culture: Balancing safety and accountability.* 2nd ed. Boca Raton, FL: CRC Press.

Deming, W.E. & Orsini, J.N. 2013. *Leadership Principles from the father of quality.* New York: McGraw Hill.

Geller, E.S. 2001. *The psychology of safety handbook.* New York: Lewis Publishers.

Geller, E.S. 2011. "The human dynamics of injury prevention: Three new E-words for occupational safety." *EHS Today*. 4(5): 42-44.

Kahneman, D. 2011. *Thinking, fast and slow.* New York: Farrar, Straus and Giroux.

Session No. 709

Pronouncing the H in HSE

Bruce Packard, MD, MPH
Axiom Medical Consulting, LLC
The Woodlands, TX

Introduction

According to the Center for Disease Control (CDC), between 1900 and 2000, the life expectancy in the US increased from 46 years of age to 74 years of age for men (48 to 79 for women).[1] Most people can identify the top three causes of deaths in the US: heart disease, cancer, and stroke. However, most people are surprised to find out that the top three causes of death in 1900 were due to infections: pneumonia, tuberculosis, and gastrointestinal infections. By 2000, deaths due to infection do not even make the top five causes of death.[2] Most people would think that the reason for the drop was due to the introduction of antibiotics; however, deaths from infection dropped significantly before antibiotics were discovered. How could that be?

The reason for the drop is due to prevention, not treatment. Having a clean water system through chlorination, and an effective method of managing waste started the improvement. Education on bacteria, cleaning, better housing, and the introduction of vaccines continued the improvement. While important, antibiotics have had a minor impact on the decrease in deaths from infection. Prevention is the major reason for the drop in deaths from infections.[3]

The Occupational Safety and Health Administration (OSHA) and safety professionals know that prevention is the essential element of any effective safety program. In 1982, OSHA initiated the Voluntary Protection Program (VPP) with the stated goal that:

> Participants develop and implement systems to effectively identify, evaluate, prevent and control occupational hazards so that injuries and illnesses to employees are prevented.[4]

Prevention is key to developing a strong health and safety program.

---

[1] Center for Disease Control (CDC). "Health, United States, 2010." (https://www.cdc.gov/nchs/data/hus/2010/022.pdf).
[2] Center for Disease Control (CDC). "Leading Causes of Death, 1900-1998." (https://www.cdc.gov/nchs/data/dvs/lead1900_98.pdf).
[3] *MMWR Weekly*. "Achievements in Public Health, 1900-1999: Control of Infectious Diseases." 1999:48(2);621.
[4] Occupational Safety and Health Administration (OSHA). 2009. FactSheet, "Voluntary Protection Plan." (https://www.osha.gov/OshDoc/data_General_Facts/factsheet-vpp.pdf).

The Occupational Health Event Timeline

When reviewing an event timeline, there are three phases: prevention, event and response. A common model for occupational health programs focuses on monitoring health through OSHA-mandated surveillance programs, has plans to respond to an emergency and focuses on the treatment of any injured employees.

	Prevention		Event		Response
Common Model		Exam	Emergency Response	Treat	

However, there is much more that can and should be done to prevent an injury or illness. To start, effective programs recognize the connection between industrial hygiene (IH) and occupational medicine (OM). Both programs complement each other to form an optimal occupational health (OH) program, and should work closely together in all three phases.

	Prevention				Event		Response	
Optimal OH	Cutting Edge	ID Exposure	Policy& Procedure	Exams FFD	Emergency Response	Treat	Case Mgmt	RTW

Effective Occupational Health

Policy and Procedures

Many programs start by focusing on required policies and procedures. Of course, policies and procedures are essential. Given your participation in this ASSE conference, it is expected that you understand the importance of and have adequate policies and procedures. So, this paper will only note the importance of having current and appropriate policies and procedures for your worksite. Make sure that your OH team is involved in writing and/or reviewing your health and safety policies.

Identify Exposures

What is this? Of course, the company has identified the hazards with programs for hearing, respiratory or other programs that are typical to the work. Identification of routine or common exposures is easy, and most companies will have the obvious exposures covered. But, while it sounds simple, both workers and management can become blind to routine activities.

It is well known that heavy metals and NORM (naturally occurring radioactive material) can be entrained in oil and gas streams. Since the occurrence is only found under certain conditions or in certain fields, it is not always found until after the exposure occurs, and can lead to a high level of concern as employees try to figure out their level of exposure and what it means to them and their family. Despite being a known risk, exposure events still occur where the NORM is identified well after the work has been completed.

In the last five years, OSHA and NIOSH released two Hazard Alerts related to common practices in the petroleum industry. In June 2012, they released a Hazard Alert titled, "Worker Exposed to Silica during Hydraulic Fracturing." In February 2016, they released an alert on the age-old practice of tank gauging titled, "Health and Safety Risks for Workers Involved in Manual Tank

Gauging and Sampling at Oil and Gas Extraction Sites." While tank gauging is an age-old process, OSHA reported nine deaths between 2010 and 2014 related to this practice. Both of these alerts were related to well-known hazards; however, not all workers were being alerted to the problems. This can be especially problematic when the worker is a contractor that only works there part time or for a few weeks before moving on to another job.

Without a concerted effort to look for unidentified hazards, even office buildings are guaranteed to have blind spots, hazards of which no one is aware of the hazard until and event occurs.

Monitoring and Exams

It often takes an outside perspective to see blind spots to which workers have become blind. Continual monitoring of worksites is essential for prevention. Are all sites being visited and how often?

Worksites are not static locations that never change. Rather, they continually try new innovations to improved productivity. In general, if the most recent data is three-to-five years old, the data will contain at least some inaccuracies. Changes can and do occur over a period of time such that workers do not think to update risks or hazards (whether better or worse) or become immune to the risk as they become comfortable with the new procedure.

Over several years, this author worked with a company that dutifully collected noise data; however, when asked to provide noise data for a specific site, it was determined that instead of visiting all of the approximately 30 sites, the collectors kept returning to the same six sites. In short, there was no data to assist with a hearing loss determination. It can be tempting to assume that a few sites are representative of all sites; however, that is rarely true. Different sites may do the same basic activity, but how they do it can vary widely. Some sites use noisy compressors; some sites do not. And, in sites that use noisy equipment, the workers may or may not spend time in those high noise areas.

Just as important, how is the data kept? Is it kept in a form that can be readily searched for trends, or is it in paper in multiple locations that cannot be searched until it is organized and filed in a central location?

An example of searchable data involves compliance with medical surveillance exams. Are employees completing their exams on time? Are there certain locations or supervisors with better compliance rates than others? When there are challenges, how are they handled? One effective method of notification and compliance involved monitoring medical surveillance exams. The employee was notified to get the exam two months before the exam was due. If the employee failed to get the exam by the due date, the supervisor was notified. After another month, if the employee still failed to complete the exam, the supervisor's supervisor was notified. Compliance with the exams quickly reached 100%. But, this was only possible because the system was in a database, which was searchable and could provide automated notifications.

To have an effective monitoring system, IH and OM must work closely together to identify and resolve bottle necks.

Fitness for Duty

Fitness for Duty (FFD) is an area that can help. How are your medical surveillance programs being handled? Are you tracking their health, or are you focused on the employee's FFD? How about when an employee returns to work (RTW)? The American's with Disabilities Act (ADA) and other regulations impact these decisions, but simply assuming that the employee is FFD is not a safe practice.

Many physicians have little knowledge of the workplace and will rely heavily on what their patient, your employee, tells them. Without a process to determine an employee's FFD, you can return an employee to work before they are ready. While FFD determinations are dependent on many factors, like the essential functions of the jobs, returning an employee to work who is not FFD poses a safety risk to the employee or their co-workers. Of course, determining that an employee is not completely FFD does not necessarily mean that they cannot RTW. Rather, it allows safety and the supervisor to determine if there is work that can be done while they employee gets reconditioned. When handled correctly, evaluating the FFD of an employee is a win-win situation for the employee and the company.

Case Management

Case Management (CM) is an important area that OM can support the employee and the business. There are many factors that can only be handled effectively by OM specialists. Obviously, there are many regulatory issues to consider. For example, there are regulations related to the American with Disabilities Act (ADA), the Family Medical Leave Act (FMLA), and HIPAA. While these regulations can be challenging, they do not prohibit effective CM, and in fact, make the need for CM more important.

Starting CM early is essential to getting the full benefit. The medical system in the U.S. is complex and can be confusing to people who are not familiar with how to access care. Depending on the local availability of specialists, getting an appointment can take time. If the employee doesn't know which specialist to see, they could be referred to another doctor. All of this takes time, sometimes weeks, just to get a diagnosis and start treatment. Case managers should be looking for delays in care and helping to prevent or shorten unnecessary delays.

Most businesses and supervisors are surprised to find out that non-occupational physicians receive little or no training about the workplace. The CM functions as a intermediary between the employee, the employee's physician, and the company. With the employee, the CM can assist with assessing appropriate care and addressing delays in care. For the physician, the CM can help the physician understand the work and assist with the safe RTW. For the company, the CM can assess health-protected information and inform the company of work-related issues, such as appropriateness of the absence and the RTW.

A good CM helps the patient seek the appropriate level of care. Sometimes, that is getting referred to a specialist and sometimes, it is seeing a primary care doctor. In some cases, it can be reassurance and watchful waiting. Companies can get about a 300% return on investment by helping to get employees to the appropriate level of care.[5]

Another area where the CM helps is with the RTW. Did you know that the greatest cost of an employee being absent for a medical reason is not the medical care but the time off? This cost comes from paying the employee's time off and not getting work done or paying someone else to do the work. When possible, evaluating the employee for transitional job duties will get them back to full duty faster than requiring them to be 100% ready to work. It is important to

[5] Internal data from Axiom Medical Consulting, LLC.

understand that transitional duty is temporary, and requires monitoring to ensure that the employee is indeed progressing; otherwise, the transitional duty can morph into permanent restrictions. The CM can monitor the progress, and notify the supervisor and human resources if the employee is not progressing as expected.

Obviously, case managers cannot prevent sickness absence; however, they can shorten absences by helping the employee to get appropriate care, and ensuring the employee's safety when they are ready to RTW.

Cutting Edge

Cutting-edge programs are programs that are dealing with challenging area of occupational medicine and change over time. Some items identified as cutting edge may be surprising: FFD, wellness, fatigue, and emerging work changes, to name a few.

FFD and wellness can work together. Dr. Ostbye reviewed workers' compensation data from Duke Health and determined that, "there was a clear linear relationship between BMI (body mass index) and rate of claims." This held true for indemnity-claims costs, medical-claims cost and lost workdays. Like Dr. Pollack, Dr. Ostbye concludes saying:

> Maintaining healthy weight not only is important to workers but should also be a high priority for their employers given the strong effect of BMI on workers' injuries. Complementing general interventions to make all workplaces safer, work-based programs targeting healthy eating and physical exercise activity should be developed and evaluated.[6]

Wellness is an area that is developing a strong correlation with safety and company performance. Dr. Goetzel analyzed the stock performance of companies that won the C. Everett Koop Award. This award is given to companies that have developed a nationally recognized wellness program. He found that:

> Our analysis of Koop Award winners' stock performance compared with the S&P 500 highlighted one aspect of corporate social responsibility; a company's treatment of its workers and, more specifically, its provision of an exceptional workplace health program. The question posed by this analysis was whether offering award-winning workplace health promotion programs, as recognized by The Health Project in the form of the Koop Award, was correlated with the value placed on the company by Wall Street investors.[7]

The Koop Award winning companies significantly outperformed the S&P 500 Index between 2001 and 2014 (325% versus 105%); see Figure 1.[8] The researchers acknowledge that many factors likely contribute to the improved S&P 500 stock performance, but that this study and other ongoing research demonstrates the value of health promotion and protection programs.[9]

---

[6] *Archives of Internal Medicine*, 2007;167:766.

[7] *Journal of Occupational and Environmental Medicine*, 2016;58(1):13.

[8] *Ibid.*

[9] *Ibid,* p14.

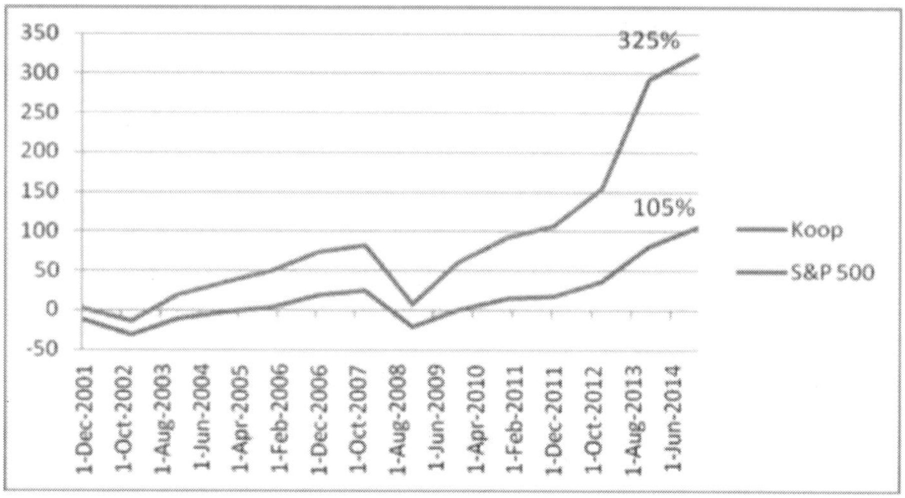

Figure 1. Cumulative stock performance (in %) of Koop Award Winners compared with the S&P 500 Index (2001-2014) *(Copyright © 2015, Wolters Kluwer Health, Inc.)*

Conclusion

Like safety programs, occupational health programs take many forms. Thinking of OH as surveillance exams and work-related injuries is like limiting safety to training and work permits. Yes, those are important, but both perceptions only scratch the surface of a full-service safety or OH program that brings to bear the full range of preventative services that need to be performed if the company wants a truly effective HSE program.

Bibliography

Center for Disease Control (CDC). July 30, 1999. "Achievements in Public Health, 1900-1999: Control of Infectious Diseases." *MMWR Weekly*. 48(29);621-629.

Center for Disease Control (CDC). "Leading Causes of Death, 1900-1998." (https://www.cdc.gov/nchs/data/dvs/lead1900_98.pdf).

Center for Disease Control (CDC). Health, United States, 2010. "Table 22. Life expectancy at birth, at 65 years of age, and at 75 years of age by race and sex: United States, selected years 1900-2007." (https://www.cdc.gov/nchs/data/hus/2010/022.pdf).

Goetzel, R, *et.al.* January 2016. "The Stock Performance of C. Everett Koop Award Winners Compared With the Standard and Poor's 500 Index." *Journal of Occupational and Environmental Medicine*. 58(1):9-15.

Occupational Safety and Health Administration (OSHA). 2009. FactSheet, "Voluntary Protection Plan." (https://www.osha.gov/OshDoc/data_General_Facts/factsheet-vpp.pdf)

Ostbye, T; Dement, JM. April 23, 2007. "Obesity and Workers' Compensation: Results From the Duke Health and Safety Surveillance System." *Archives of Internal Medicine*. 167:766-773.

Session No. 710

Positive Investigation Methodology: Investigating Success

Michael Tooma, BComm, LLB (Hons), LLM, CSP,
Partner and Global Head of Safety
Clyde & Co.
Sydney, NSW
Australia

Introduction

On 4 November 2010, 469 people on a flight from Singapore to Sydney were in the centre of dramatic events that had the potential to go down as one of the world's worst air disasters. Shortly after leaving Changi Airport, an explosion shattered Engine 2 of Qantas Flight QF32 – an Airbus A380. Hundreds of pieces of shrapnel ripped through the wing and fuselage, creating chaos as vital flight systems and back-ups were destroyed or degraded. The crisis showed every sign of ending tragically. Instead, the plane landed safely a few hours later at Changi airport in Singapore with all people on board unharmed.

At the height of the crisis, the A380's electronic centralised aircraft monitoring system (ECAM) was spitting out hundreds of alerts on what was wrong with the plane and instructions on what the pilots were required to do to address those defects. It was at this pivotal moment that Captain Richard de Crespigny made a profound decision. As he recalls in his book on the crisis:

"*I was growing tired of being reactive to the ECAM and I wanted something positive to focus on. There were too many alerts, too many things broken and not much to be achieved by dwelling on them.*"[1]

At that moment, Captain de Crespigny had what he called an epiphany:

"*I inverted the logic. I remembered what Gene Kranz, NASA's Flight Director said during the Apollo 13 mission: 'Hold it, gentlemen, hold it! I don't care what went wrong. I need to know what is still working on that space craft.' We went back to basics and it became easy.*"[2]

Captain de Crespigny's actions in focusing on what was right with the plane on that day rather than what was wrong with it, allowed him and his crew to land the plane safely, saving the 469 lives on board. Are these lessons for safety professionals from this?

---

[1] De Crespigny, R. 2012 , *QF32* (McMillan) 218
[2] Ibid.

Investigating Success

Things go right far more often than they go wrong. Yet we currently restrict all our safety learning to those situations where things go wrong – incidents. What if we invented that logic? What if we started to learn from things getting right? What if we investigated those projects delivered without incidents?

Investigating success will allow us to better understand the conditions that facilitated it. This, in some cases, will be by design – good planning, competent personnel, adequate resource, effective communications and consultations, good processes and appropriate supervisions. In other cases, the success may have been achieved despite the system. That is, workers innovated to achieve success. Those innovations may have been appropriate – competent personnel making risk based decision – or it may have been inappropriate – at risk behaviour where the adverse outcome did not eventuate by chance and not design. In the latter situations, the "success" is illusive. It is not "success" at all. Investigating it allows us to understand it and correct it. In the case of appropriate innovations – red "success" – investigating it allows us to capture it and systematise it. This is work led lessons for improving safety. A real ground up approach. The theory here is that work is planned rarely resembles work as done. By keeping up the pretense that it does, we deny ourselves the opportunity to adjust the theory to reflect reality or at least to correct the reality to a safer outcome.

Near Misses

How do we learn from what went right? Let's take the example of a high potential incident or a serious near miss. The reason such incidents are a concern is because a serious adverse outcome, typically a fatality or serious injury, was narrowly averted. There are typically many absent or failed defences which led to the incident. However, the adverse outcome was averted. That is something(s) intervened to shift the trajectory of the incident from one where the adverse injury causing outcome would occur, to the one where the outcome was averted, albeit narrowly. Those things may be effective defences. They may be unplanned or unintended defences. Understanding which defences, planned or unplanned, work, and harnessing them, may be as instructive as understanding which defences failed or were absent.

The investigation of these effective or unplanned defences begins with firstly identifying what happened. The next line of inquiry in a near miss is understanding what could have happened but did not happen – the adverse outcome associated with the "high potential incident". Next we need to discover what "went right" to allow that adverse outcome to be averted. This is achieved by asking "what went right?" repeatedly until the possibilities have been exhausted. The aim is to do so five times but in certain circumstances it may be appropriate to do this more and in others, the inquiry will be exhausted sooner.

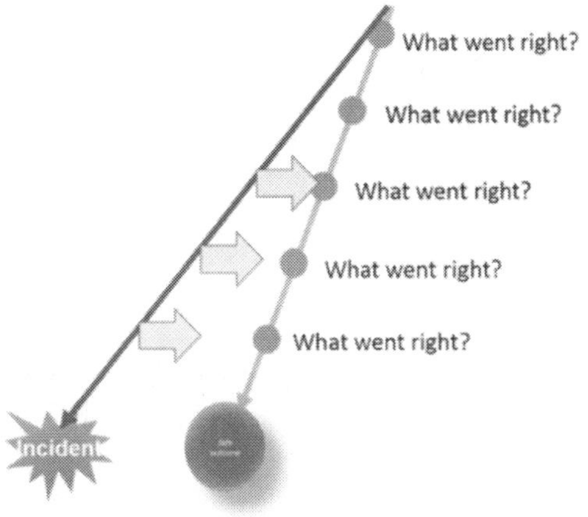

Figure 1. Near Miss.

Systematically adopting that approach however, requires validation that the "accidental control" is indeed universally valid, such that including it in a system will increase its resilience. As such, the control needs to be tested through a series of "what if" scenarios to test its resilience in changed conditions. These changed conditions include:

- change in the natural environment, such as extreme weather
- change in the plant being used
- change in the process being adopted
- change in the consequences, such as the height at which the work is performed, or the speed at which the vehicle is travelling
- change in the feature of the worker, such as weight, height, literacy or language skills.

Such a process will promote greater resilience in the safety management system. It will capture the controls that work and systematically implement them to improve the system. Only then can we guarantee that the systematic learning and improvement are part of the fabric of the organisation. Only then can we claim that we are ensuring the health and safety of workers and other people that may be affected by the conduct of our business or undertaking.

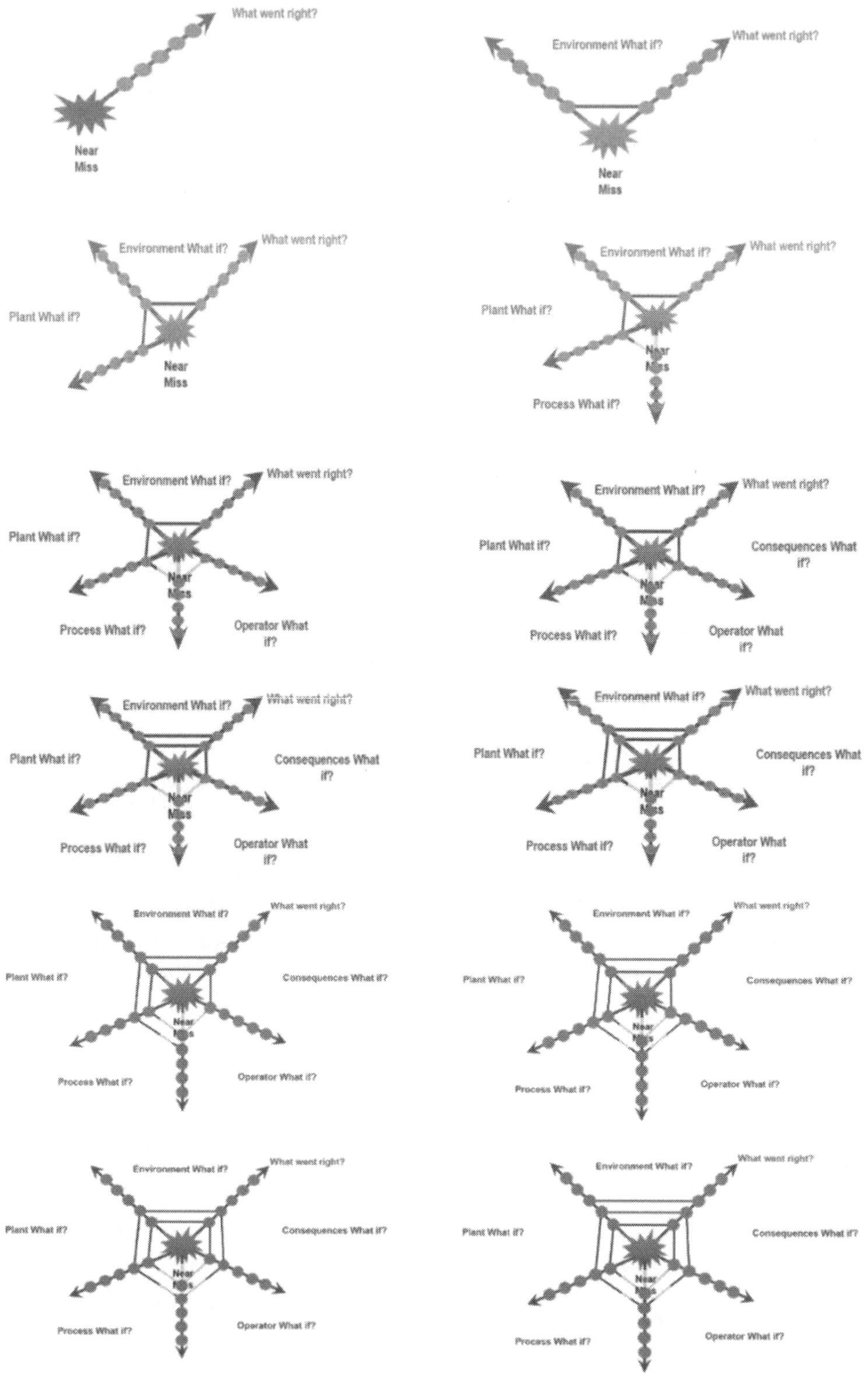

Figure 2. Spider web diagrams – Positive Investigation Methodology.

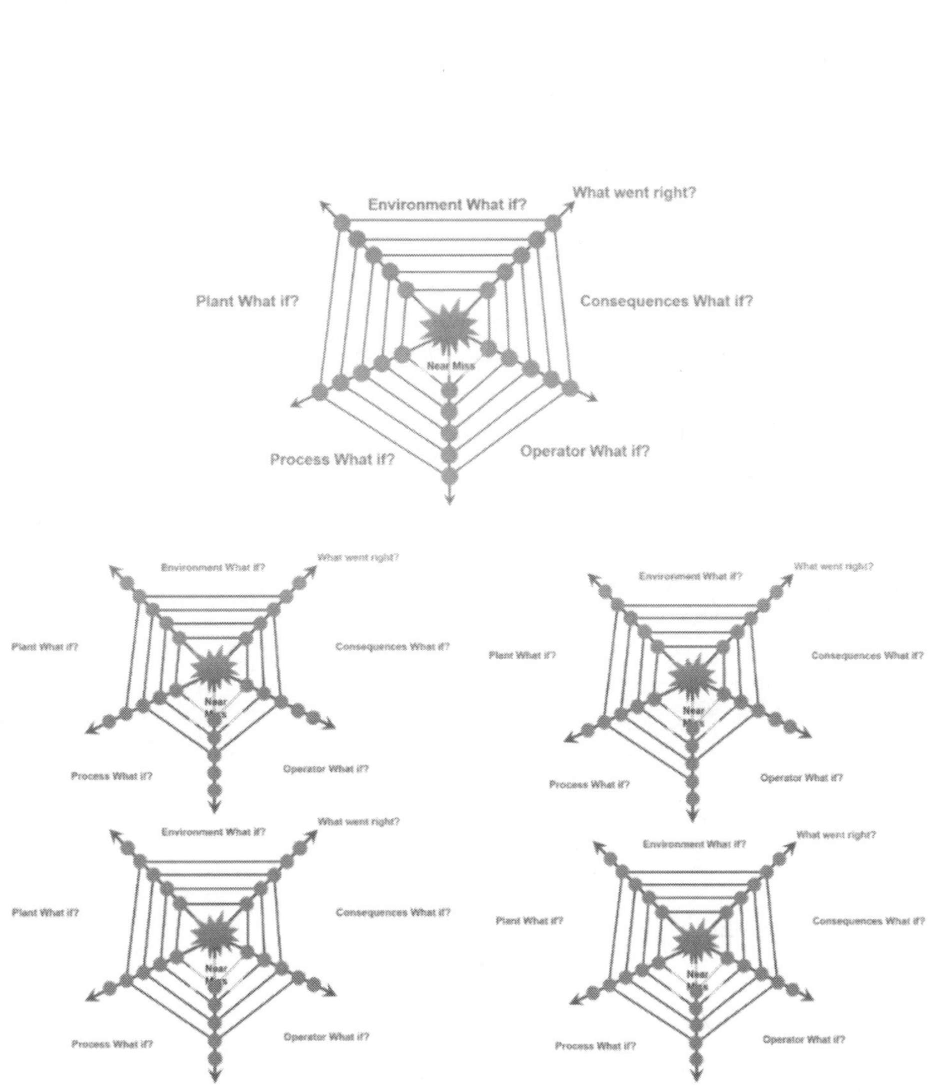

Figure 2 (cont'd).

In adopting this positive methodology, we can capture the practical lessons of what actually works, rather than the theoretical lesson of what should be working, but is not working. This creates more powerful feedback loops. Reinforced through key performance indicators for undertaking such positive investigations, and identifying what went right and converting those lessons into appropriate controls, can create a system that is a very powerful tool in incident prevention.

When identifying controls under that methodology however, innovation is critical. The idea is that you take what accidentally occurs and systematise it. That is typically achieved through technology not procedures.

Serious Incidents

The Positive investigation methodology (PIM) can also be useful in the context of serious incidents. Even in serious incidents, there is often something worse that could have occurred that did not occur because of effective mitigation factors or an effective incident response. It is useful to recognize this in the context of an investigation because it is a recognition of the efforts of the individuals involved.

The focus in this approach is two-fold. Firstly, understanding what could have happened if circumstances were different as well as recognising and learning from what made the incident less serious. Secondly, verifying that whatever controls are being implemented would be effective in different circumstances.

If we build on the approach developed above, an incident event occurred because of a series of primary events leading up to it, each causally related to the incident event. These are the direct causes of that event. By asking why these events happened, we uncover the root causes of that event. In the “5 why’s” model, “why?” is asked 5 times. As we have discussed, for every event sequence, there is a road not travelled because of some control that worked or some other barrier, whether intended or unintended. That alternate path can be uncovered by asking “what if”. This would disclose an alternate event sequence, together with its own root cause factors.

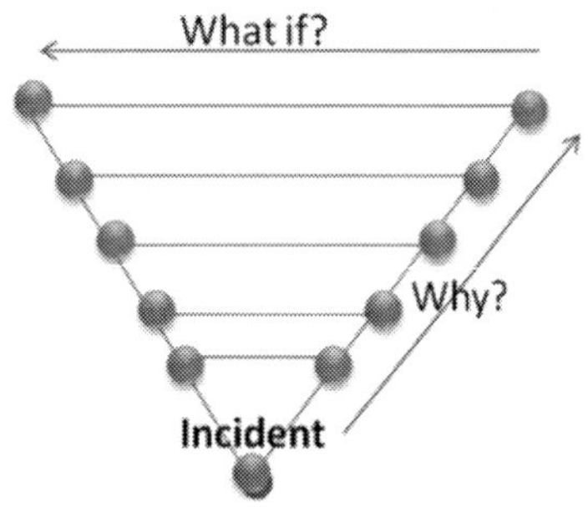

Figure 3. Causal Analysis -- Actual

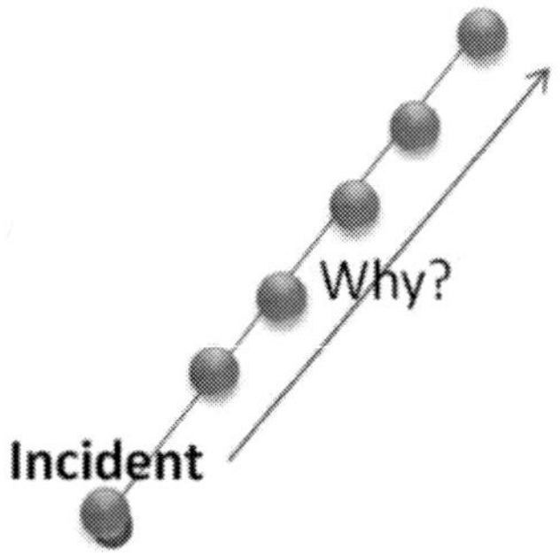

Figure 4(a). Causal Analysis – First What if?

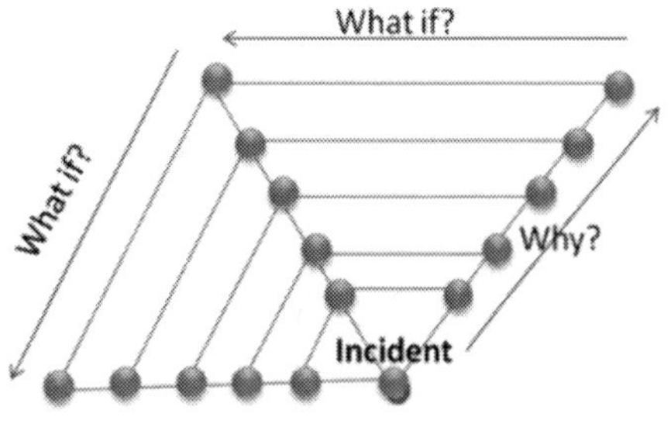

Figure 4(b). Causal Analysis – Second What if?

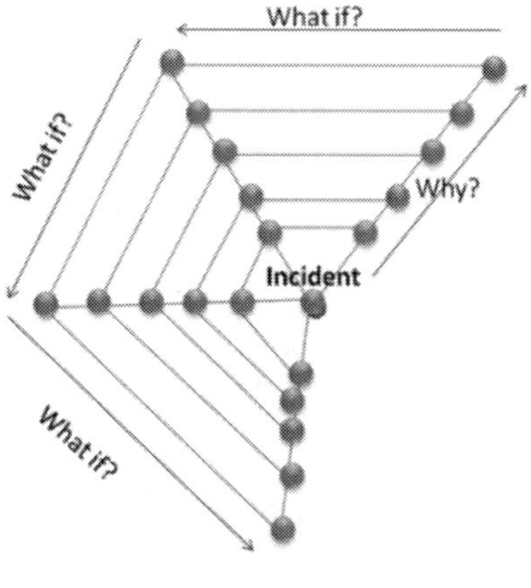

Figure 5. Causal Analysis – Third What if?

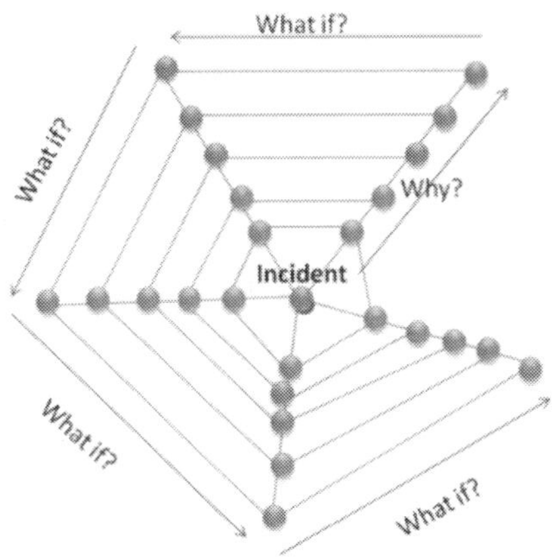

Figure 6. Causal Analysis – Fourth What if?

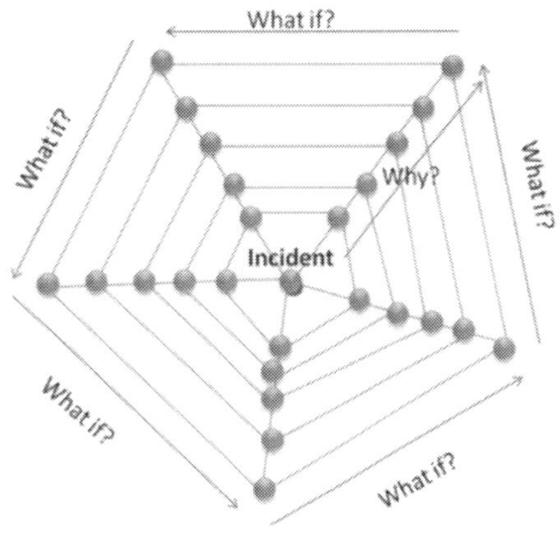

Figure 7. Causal Analysis – Fifth What if?

If we ask the "what if" question four additional times, we get four alternate realities in addition to the actual sequence which occurred. That provides a m9999999uch richer tapestry of system deficiencies and therefore recommendations for system improvements.

Conclusions

In safety we need data points to analyse performance and adjust the system to improve that performance. Ironically, the better we get at safety, the less incidents we have, the fewer the data points. The danger with this is that when things go wrong, they will go wrong in a big way. That appears to be the experience with recent disasters. Adjusting our tense to seek out data points in what goes right and not just what goes wrong allows us to monitor, analyse, learn and respond to deviations from the planned approach in a controlled manner. Over time, that will create the cycle of 'boom and bust' which is safety data in hazardous industries.

Bibliography

De Crespigny, R. 2012 , *QF3*. Fitzroy, Victoria (AU): McMillan.

Tooma, M. 2012, *Due Diligence: Incident Notification, Management and Notification*, North Ryde NSW (AU): CCH Australia Limited.

Session No. 711

Serious Injury and Fatality Prevention in Oil and Gas Exploration

Ian M. Threadgold, BSc, CSP, ASP
Threadgold Safety Management
Katy, Texas

L. John Barrett, CSP
BHP Billiton Petroleum
Houston, Texas

Introduction

Recent studies have shown that, while the incident rates of less serious injuries are still being reduced, serious injuries and fatalities (SIF) in many industries are declining at a much slower rate (Manuele, 2008; Krause, 2012). In some industries, the rate of reduction is non-existent, and has been for some years.

One of the fundamental "truths" in modern safety management was that, by focusing on minor injury reduction, we could equivalently reduce incidents that resulted in serious injuries and fatalities: the famous "Heinrich Triangle" (Heinrich, 1931; Bird & Germain, 1985). Incident occurrence data and research over the last few years disproves this theory (Anderson & Denkl, 2010; Manuele, 2011; Krause, 2012).

This contradiction with traditional safety thinking has identified a necessity to approach serious injury and fatality causes differently than those that result in less serious outcomes. One industry that has made inroads into identifying and applying different techniques in serious injury and fatality prevention is in upstream oil and gas exploration, namely the geophysical exploration industry.

This paper will describe how the contractor and client sides of the geophysical industry have systematically shared incident information for over 60 years, developing solutions based on the shared learnings and the resulting outcomes, which have had an impact on serious injuries and fatalities across the industry.

History of Geophysical Surveys and Safety

Geophysical surveys are used to study the earth's crust to identify structures and traps in the rock where oil and gas might accumulate. The most common geophysical survey technique is the reflection seismic method, often just called "seismic." Seismic uses knowledge of earthquakes and wave theory to estimate the properties of the earth's subsurface by creating a measureable soundwave on the surface, and then detecting and measuring the reflected waves as they reflect back from the subterranean structures. The process is effectively creating and measuring earth echoes (Dragoset, 2005); an energy source creates the soundwave on the surface, and sensitive receivers laid out over a vast area listen for and record the "echo." Over time and with repetition, an image of the earth's subsurface is developed.

Seismic surveys are the epitome of a dynamic worksite, which can be defined as "work settings that experience constant change in people or the working environment" (Cooper, 2009, p 37). On land seismic surveys, workers range across the countryside in all types of terrain and in all types of weather, using local labor (very often inexperienced) and resources in order to position the tens of thousands of sound detectors, and to drill the thousands of holes for the dynamite charges, or operate the specialized equipment, in order to generate the seismic soundwave. Land crews can be anywhere from 150 to 1500 personnel in size, and have operated in most geographical areas of the world.

Historically, marine seismic surveys were just land surveys on water, but today they generally take place from large, dedicated vessels towing the seismic detectors and soundwave generators behind them. Crews of 80 to 100 staff deploy 10 to 20 cables behind the vessel; each one is over five miles long and being towed just 300 yards apart, covering over 2.5 square miles.

The reflection seismic technique first appeared during World War I, as listening devices were developed to listen for underground noises to help detect tunnelling efforts under the front lines, or use triangulation to locate large artillery pieces behind the front lines.

Seismic surveying developed after the war as the search for oil expanded. The first seismic exploration company was incorporated in 1920 in Oklahoma (Sweet, 1978), and the first commercial discovery that was attributed to reflection seismic was also in Oklahoma, just eight years later. In 1930, the Society of Exploration Geophysicists (SEG) was formed as an early professional association for petroleum geophysicists.

In the early days of seismic, dynamite was used exclusively as the energy source. As can be expected, this was the cause of many of the early fatalities in the industry. In 1934, seven men were killed when a surface charge accidently exploded, along with an estimated 200 pounds of dynamite that was stacked nearby (Lawyer et al, 2001). In the 1940s, the Bureau of Mines took action and developed regulations and guidance for the handling, and use of explosives by the seismic industry.

Even with these improvements in place, in the early 1950s, an employee working in geophysical operations, looking for oil and gas deposits, was five times more likely to get seriously injured than someone in the broader petroleum industry (Westby, 1953). Seismic was, and still is, an inherently dangerous business to work in. As late as 1977, a hovercraft used for deploying the energy source in very shallow water was destroyed when the wrong charge was connected, killing all 11 people aboard (IAGC Aide Memoir, 2017). Explosives are still a high-risk hazard in the geophysical industry, and their use, handling, transport, and storage is still a prime concern. The two industry trade associations have worked with explosives' manufacturers to develop strict process and procedures, to develop safer products, and to advocate improved competency standards among personnel responsible for explosive's handling with local regulators. Between 1985 and 2015, seven single fatalities were reported, making death due to explosives now the 12th most frequent cause of death in seismic (IAGC Aide Memoir, 2017).

Sorge (1952) described the formation of the SEG Seismic Safety Committee in 1951, with the goal of trying to improve safety performance in the industry. At this time, seismic crews were operated by both contractors and oil companies. By the late 1970s/early 1980s, there were few to no oil company-owned seismic crews remaining; it was all contractor-owned operations.

The SEG, through this Seismic Safety Committee, were the stewards of safety in the geophysical industry from the early 1950s until the mid-1980s, when the stewardship role was taken over jointly by two trade associations: the International Association of Geophysical Contractors (IAGC), formed in 1971; and the International Association of Oil and Gas Producers (IOGP), formed in 1974. Table 1 below is a timeline of some of the key events over the last century in the seismic industry, from a safety perspective.

1920	First seismic company incorporated (Geological Engineering Company, April, 1920)
1921	First reflection seismograph geologic section (August 9)
1928	First commercial oil discovery due to seismic (4 December 1928)
1930	SEG founded
1951	SEG Seismic Safety Committee established
1971	IAGC formed
1974	IOGP predecessor (E&P Forum) established
1982	IOGP creates safety, health and personnel competence committee
1985	SEG Seismic Safety Committee disbanded
1985	IAGC and IOGP each set up seismic safety committees
1988	Piper Alpha Disaster
1992	IAGC 7th Edition of *Land and Marine Safety Manuals* (first edition for global use)
1994	HSE Management System guidance issued by IOGP (Report 210)
1995	IOGP first HSE model contract schedules for geophysical operations
1999	IOGP renamed, publishes guidelines for working together in a contract environment
2000	IAGC publishes *Geophysical Environmental Guidelines*
2006	IAGC *Aide Memoir for Geophysical Risk Assessors* developed

Table 1: Timeline of Some Key Events

A Culture of Sharing Incident Information

A History of Sharing

In his 1952 paper, Sorge described how the SEG Seismic Safety Committee established a "geophysical accident information exchange" as part of their goal to improve safety performance in the industry. A summary of this shared information was eventually published by the Seismic Safety Committee in the form of an *Annual Geophysical Accident Report.*

The earliest report found in the literature was included in the SEG publication called *Geophysics*, in the January 1976 edition, as the *1974 Geophysical Accident Report.* It also contained historical accident rate information back to 1964 for trending purposes. Additional annual reports were also found in the *Geophysics* publication for 1975, 1977, and 1979. The 1975 report stated that, "the accident frequency and severity rates indicate that geophysical work is still the most hazardous in the oil industry." Proubasta (1987) reported that the SEG Seismic Safety Committee produced its last annual report in 1982.

Dujmovich (1992) talks about quarterly "incident information exchange" meetings in the early 1990s between the IAGC and IOGP (under its earlier name of the E&P Forum). Threadgold et al. (2010) described how this information exchange has continued, with Joint IAGC/IOGP HSSE Forums occurring every six months, a format that continues to the present day. The spring forum is held in North America (frequently Houston), while the fall forum is held in Europe (frequently the UK or possibly another international location). The forum lasts two days, and is a combination of an update of each other's activities and current work program, and speakers on key topics. An example of real success at these joint meetings has been the in-depth sharing and discussion of high-potential (HiPo) incidents and accidents. Krause (2012) describes a precursor to a serious injury or fatality as "an unmitigated high-risk situation that has a high probability of resulting in serious injury" (p. 54). This perfectly defines what HiPos are. HiPos can be considered as a free learning event for the industry, and an alert of a precursor of a serious injury or fatality event.

The forums are a time when geophysical contractors openly and honestly bring out their "dirty laundry," and present it to their potential future clients while also sharing it with competing contractors. The result is open discussion and group learning on ways and methods to reduce incident recurrence, and an understanding by the client oil and gas companies of some of the problems faced by the geophysical contracting community when mobilizing a seismic survey, always with some appreciable level of risk.

These industry-wide sessions have invariably led to joint solutions, such as the development of HSSE management systems, best practice risk mitigation guidance, and industry model contract terms, some of which are discussed in more detail later.

An Aha Moment for Many

An Aha moment, for many in the combined geophysical industry safety community, came at one of these joint HSSE forums in 2005, when they saw a slide very similar to Figure 1 below, presented by Eelco Sixma (2005).

Figure 1 below is after Threadgold et al. (2010), and shows the ratio of lost-time injuries (LTI) to fatalities for different activity categories. The data behind the chart shown below is from 2007, as published by the Bureau of Labor Statistics (BLS), for all industries. The year of the data is not critical because a similar picture is seen for any recent year.

The key message is from the two extremes of the chart. The right side is where there are thousands of LTIs per fatality; therefore incidents are rarely fatal. This includes slips, trips, and falls, etc., and refers to what is commonly known as personal safety.

The left side has only a few LTIs before a fatality occurs, and can therefore be considered frequently fatal. These activity categories can also involve multiple fatalities on many occasions, and include fires, explosions, and transportation or security incidents. These low-probability, high-consequence risks are evident at the operational level, and can be termed operational safety risks.

On the left side, you realize there are so few LTIs, or learning opportunities before a fatality occurs, that everyone in the industry needs to work from the collective knowledge of the industry to prevent similar future occurrences. A single company will not have enough learning opportunities to work on recognizing and therefore preventing these kinds of accidents from re-occurring.

This was the Aha moment for many, which validated the years of sharing incident information by both sides of the industry. It also brought a renewed focus on the high risk activities, which are seen on the left side of Figure 1.

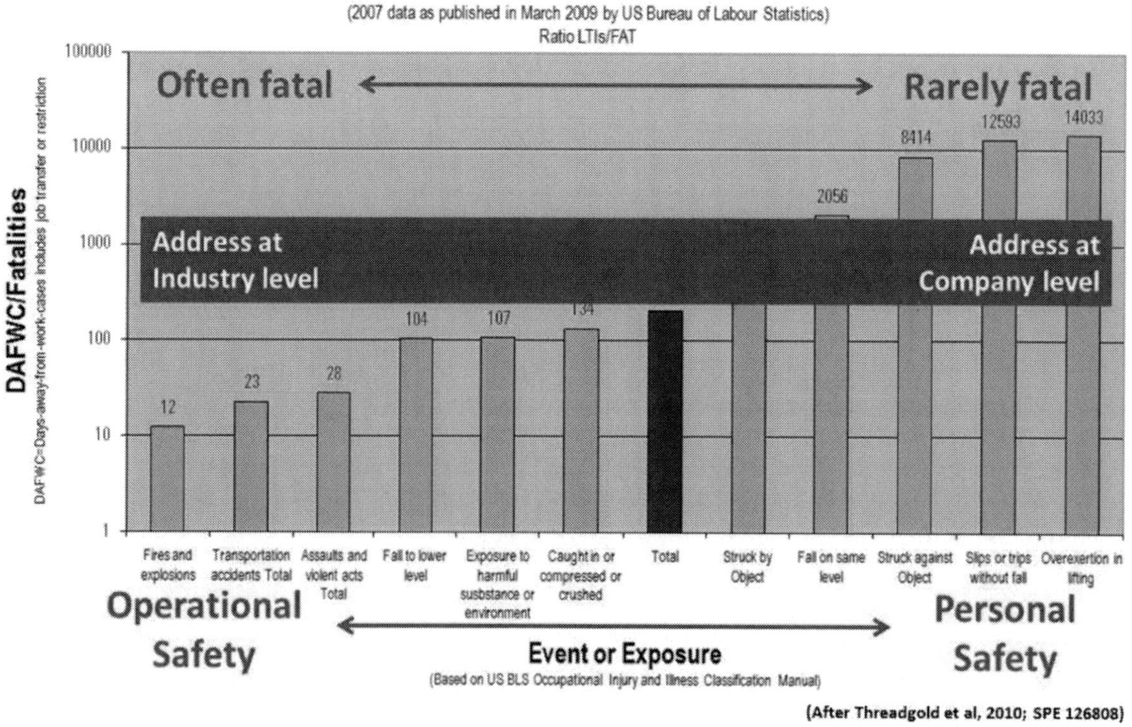

(After Threadgold et al, 2010; SPE 126808)

Figure 1: A Plot of Lost Time Injuries per Fatality Ratio for Different Activity Types,

The Aide Memoir for Geophysical Risk Assessors

An Industry-wide Fatality Database

Following on from the ideas shared earlier regarding which incidents were precursors to a fatal outcome and which were rarely fatal, Sixma (2007) presented an idea at a joint HSSE Forum in 2006 that would lead to a highly influential outcome: the Aide Memoir for Geophysical Risk Assessors. He saw the Aide Memoir as a 'memory' database of fatal events for the geophysical industry. Sixma reasoned that risk assessment requires the use of a "memory function;" a way for risk assessors to refer to "real events" when identifying and analyzing potential risk. Up to this point, fatal accident narratives had been kept by individual companies; there was no industry repository, or "memory."

For high-probability, low-consequence events, most companies had knowledge and "memory" among the risk assessors of these having happened; they had experienced them or knew someone who had. These referred to the slips, trips and falls on the same level, minor scrapes, cuts and contusions: the right side of Figure 1.

For serious injury and fatality events, this was often not the case. These events generally happened so infrequently in any one company that there was often no "memory" among the risk assessors of them having happened, so they were often overlooked or incorrectly risk assessed. What is more, the causes of these serious injury and fatality events are quite different to the causes of the lesser severity events (Manuele, 2008). The venerable Dan Petersen (2001) urged that, "There are different sets of circumstances surrounding severity ... If we want to control serious injuries, we should try and predict where they will happen" (p 11).

Sixma (2007) had arrived at the same conclusion, prompted by a concern he had about the future of risk assessment in a highly cyclical industry, where a significant number of experienced employees would be retiring in the next few years: a phenomenon the industry has called "the Great Crew Change."

Many methods of qualitatively assessing risk in the oil and gas exploration and production (E&P) industry define probability in risk assessment matrices (RAM) in terms of how someone might remember an occurrence in a continuum, such as: "*Has occurred in the E&P industry*" and "*Has occurred in the operating company,*" through "*Occurred several times a year in operating company*" to "*Occurred several times a year in the location*" (ISO 17776:2000). This infers the need for a "memory" among risk assessors (Sixma, 2007). He posited that "risk assessment for low frequency, high consequence incidents (fatalities) needs to be done on the basis of collective industry experience" (p 2) and "[the industry needs] a repository for industry experience: an *Aide Memoir for Geophysical Risk Assessors*" (p 2).

Other benefits that Sixma (2007) foresaw would be using the Aide Memoir as a learning tool for the next generation of risk assessors. It would hopefully prevent reoccurrence of some events, and allow for more consistent risk assessments across the industry. The Aide Memoir also became a way to capture some of the incident information that had been shared for so many years. Ultimately, it helps industry personnel to answer the question from the RAM of, "*Has this happened in the industry?*"

Building the Database

Initially, a request was made to all IAGC and IOGP members to populate the flat-file database with fatal events the members remembered, accompanied with a brief narrative of what happened. Company names, details of the victims, project details, and geographic location were left out to ensure anonymity, encourage reporting (Sixma, 2007), and satisfy the legal experts.

At the beginning, the database was populated with anecdotal reports from reliable sources, and then supplemented with IOGP annual safety performance indicators reports since 1980, IAGC and CAGC data HSE alerts and high-potential reports, STEP change Sadie database alerts from the UK, and specific reports from a number of participating companies. The result, after a year, when Sixma presented the first version at another of the joint meetings, was a database of 379 fatalities in the geophysical industry from 219 events. An interesting point is that from about 1980 onwards, the numbers of fatalities reported in the Aide Memoir matched (or in a few cases exceeded) what had been reported through more formal avenues.

Among the fields recorded were *Year*; *Number of fatalities in event*; a summary narrative and comments including *identified causes,* if available; plus columns for *keywords,* including *type of geophysical operation* (Land / Transition Zone / Marine); *type of accident*; *work activity;* and finer details. Sixma had supplemented the geophysical fatalities with 217 fatalities from 181 events that had occurred in oil field activities similar to geophysical exploration; i.e., ones that "could have happened in geophysical operations" (Sixma, 2007, p. 7). The table was then able to be analyzed using Excel pivot tables. By applying filters, the assessor could "drill down" into the data. The results were dramatic and have impacted risk assessment in the geophysical industry since, showing what can be achieved by sharing accident information and acting upon it as an industry.

An Early Success Case for the Aide Memoir

One of the more remarkable success stories to date was related to **drowning incidents**. While it was known and understood that drowning was a risk for any operation on or near water, it wasn't until the Aide Memoir that it was recognized as being the second highest cause of death after vehicle accidents in geophysical operations, both land and marine.

Drowning deaths had occurred every year in the geophysical industry between 1985 and 2006 (except the year 2000). During that 22-year period, 71 people had died in 66 fatality incidents (an average of 3 fatal incidents per year). Causes of those fatal events were chiefly people trying to swim across rivers, small boats capsizing or sinking, people losing their footing and falling into the water from the decks of vessels or river banks, or people swimming recreationally.

Very often, people were not wearing flotation devices. The problem with immersion risks is that there is often very little precursor evidence to indicate that a fatal incident might occur (Anderson & Denkl, 2010). Accidental immersions in water escalate directly from being a near accident to a fatality: The affected person is either wet or dead!

Once the realization of the scale of drowning deaths became known, the industry reacted swiftly. The guidance from an earlier IOGP publication titled Report 355, *Watercraft & water in geophysical operations–A guideline to operations & management* (IOGP, 2004), was adopted more thoroughly by both contractors in their process and procedures and E&P companies in their contract terms, especially with regards the use of life vests, swim tests, and restrictions on recreational swimming.

Following this, a workgroup was formed by IAGC to develop guidelines for small boats, resulting in publications for members providing guidance on life-saving equipment specifications (life jacket with integrated personal locator beacons), marine, small-boat operating guidance, and competency and training criteria for small-boat operators.

Recently, a combined IAGC/IOGP workgroup was formed to work with another industry group, the International Marine Contractors Association (IMCA), to develop guidance for vessel-to-vessel transfer of personnel while at sea, including the adoption of heave-compensated gangways for some of the larger vessels.

In the nine years between 2007 and 2016, there have been three reported drownings and one boat-involved fatality (the person was struck on the head when two small boats collided); that is less than one fatal drowning incident every two years!

Additional Direction from the Aide Memoir

The Aide Memoir identified which activities were the most frequently fatal in geophysical operations, and led to a renewed focus on developing risk controls and mitigations for these activities and risks on an industry-wide basis. It is the latest phase in the ongoing story of the client and contractor sides of the geophysical industry working together in identifying and discussing risk in order to prevent serious injuries and fatalities.

Besides the activities mentioned above related to the risk of drownings, information from the Aide Memoir database has been used to establish working groups on other specific topics and develop additional joint industry guidance documents between the IAGC and IOGP. Further analysis of the major fatality types in the Aide Memoir showed areas where the industry could focus:

- The main cause of **vehicle-related fatalities** was high-speed, head-on collisions and rollovers, followed by pedestrians being struck, personnel working along roads being struck by highway vehicles, and off-road rollovers. Solution:
  - Threadgold et al. (2010) describe some of the progress made on implementation of geophysical industry recommendations for passenger trucks (the most common way of transporting people in both an on-road and off-road environment).
  - Also, a broader adoption of IOGP Report 365, *Land transportation safety recommended practice*.
- The main cause of **struck-by fatalities** was being struck by a falling tree or branch, mainly to do with chainsaw use and tree felling for line clearance and camp construction. Solution:

    - A web page is currently under development on the IAGC site devoted to chainsaw and tree-felling safety, and the joint community has adopted Canadian tree-felling standards as a minimum standard for geophysical operations, where none existed previously on a global basis.
- **Aircraft fatalities** were largely (82%) helicopter accidents during external sling activities. Solution:
    - Creation of an industry guidance document, IOGP Report 420, *Helicopter guidelines for land seismic & helirig operations,* which is focused exclusively on geophysical operations, where a fatal helicopter accident was five times more likely to occur than in other parts of the oil industry (Scott, 2014).
    - Also, broader adoption of IOGP Report 390, *Aircraft Management Guidelines.*
- **Crushing fatalities** were mainly due to people being caught in rotating equipment or being crushed by a load during lifting operations. Fatalities have also occurred with personnel working around heavy vehicles. Solution:
    - Broader adoption of IOGP Report 376, *Lifting & hoisting safety recommended practice.*

Jointly Developed Industry Guidance Documentation

As new lessons are learned at each joint HSSE Forum from the shared incident information, these are eventually built into industry best practice by inclusion in jointly developed industry guidance documents. Some of the additional guidance documentation is discussed below.

Model Contract Schedules for HSSE

The model contract schedules for HSSE requirements were mentioned earlier. The earliest versions of these were first published in 1994 and 1995 by the IOGP. These have been updated several times jointly between the IOGP and IAGC. The third version was IOGP Report 432, *Managing HSE in a geophysical contract*, published in 2009.

This guidance document includes two tables, which are generic contract schedules for geophysical projects in the areas of both management system requirements and also control of specific risk areas. These contract schedules are used by client oil and gas companies when they tender for geophysical services, as a way of specifying the clients HSSE requirements. With tenders becoming more standardized, the contractor HSSE professionals can more quickly respond to a tender and, therefore, spend more time on managing HSSE than responding to variable tender packages.

Version 4.0 will be published later this year, with the same title, but a different IOGP Report number, and will include much more of the learnings from the Aide Memoir in the minimum contract standards. It will also involve a supplemental report on *Risk Management in Geophysical Operations,* which includes a greater discussion on risk assessment using the Aide Memoir.

Geophysical Industry Auditing Guidelines

Joint work in the mid-1990s led to the publication of IOGP Report 245, *Guidelines for HSE auditing in the geophysical industry*. This has been a widely used document since its publication in 1996. Its wide acceptance and the improvement of auditing guidelines in general have meant that this has not needed to be updated since, unlike some other guidance.

Geophysical Industry Competency Guidelines

Additional joint efforts have led to the publishing of three versions of IOGP Report 292 since 1993. Version 3.0 was titled, *HSE competency management for the geophysical industry*, published in 2014. Over the two decades or more, the content has gone from more discussion of just training to also include competency management systems guidance (Threadgold et al., 2016).

Ongoing Collaboration Efforts

Other ongoing areas that the joint HSSE groups are working on are:

- Addition of all the controls mentioned and lessons learned into the model HSE contract clauses included in IOGP Report 432, *Managing HSE in a geophysical contract* (IOGP, 2009).
- Inclusion of lessons learned from the Aide Memoir in current and future versions of the *IAGC Marine and Land Geophysical Safety Manuals* (IAGC, 2012).
- Improving the quality of the high-potential incidents discussed during the semi-annual joint IAGC/IOGP HSSE Forums by identifying what an actual precursor is in the geophysical industry.

Contractor Management

A final IOGP guidance document worth mentioning was developed by the broader oil and gas industry within IOGP, and then supported by other trade associations like IAGC. This is IOGP Report 423, *HSE management–Guidelines for working together in a contract environment*. It was first published in 1999, and a version 3.0 will be issued in early 2017. This has been widely used to define contractual relationships between clients and contractors in the geophysical sector.

IOGP makes their extensive library of HSE and technical documentation available for free download at: http://www.iogp.org/Our-library

Discussion

Impact of Sharing

In the early 1950s, an employee working in geophysical operations looking for oil and gas deposits was five times more likely to get injured than someone in the broader petroleum industry (Westby, 1953). The geophysical industry is now at par with the injury and fatality rate averages across the global petroleum industry, and has exceeded those in some other sectors (Threadgold, 2015).

Krause (2010) feels that "developing an effective strategy for preventing SIF events requires a new framework that accounts for the variety in type and intensity of exposures in the workplace and that provides organizations the capacity to address them adequately" (p 4). He goes on to say that organizations must understand their specific precursor events, and once they are understood, to put into place more focused efforts in the prevention of serious injuries and fatalities (Krause, 2010). Anderson and Denkl (2010) recognized that there was a need for a method to share incidents across the industry in order to learn from these, because there is not enough information in any individual company's records, and indeed their work stemmed from their geophysical industry experience.

In the geophysical industry, both client and contractor companies have learned how to work together collaboratively in order to identify "the type and intensity of exposures" that have the potential for a serious injury or fatality through shared experiences and standardizing aspects of safety on an industry-wide basis, and in using and analysing the Aide Memoir. The contractor and client parts of the geophysical industry have systematically shared incident information for the last 65 years, on a joint voyage of discovery, and continue to do so with new and important initiatives in order to make the industry safer through relationships based on mutual respect and trust.

Aide Memoir Update

Sixma and Edgecombe (2016) presented the latest version of the Aide Memoir at the IAGC/IOGP Joint HSSE Forum in Paris, France, in September 2016, before Sixma retired from the industry. The Aide Memoir now contains 1078 actual fatalities, with 587 of these from the geophysical industry itself. High-potential events have also been added. This latest version was used in this discussion, particularly the reports between 1985 and 2015, which appear to have better reliability and completeness. Industry reporting became more reliable, and the Aide Memoir more completely populated after 1980.

The latest version of *Aide Memoir for Geophysical Assessors* is available in the public domain at: http://www.iagc.org/hsse.html, along with links to the presentations of high-potential incidents over the last few years, and to IAGC safety documents.

What Other Industries or Sectors Can Learn

Rhinehart (2013) recognized that, "the most basic driver for the trade association is that together we can accomplish what cannot be done single-handedly" (p. 1). He was talking about the specialty coffee industry, but his words resonate for any industry nonetheless. Together, the IOGP and IAGC have accomplished important improvements in safety that was not possible alone. What is more important here is that the members of two trade associations have learned to build relationships based on trust, and become involved in collaborative efforts from different commercial sides of an industry sector for the benefit of both sides. The benefits of developing industry fatality databases, and the added value this can create in terms of improving risk assessment and reducing fatalities across the particular industry, helping develop better industry guidance and building better industry-wide HSSE model contract schedules, are manifest.

Other dynamic work settings, such as construction, mining, agriculture, forestry, fishing and hunting, and infrastructure projects, as well as other sectors of the oil and gas industry (e.g., drilling, productions operations, etc.) have much that they could learn from the geophysical industry's efforts to reduce SIFs. Using the advantages of client and contractor trade associations,

HiPo precursor events can be shared, a simple flat-file database can be used to record serious injuries and fatalities across an industry, key words added, and then these can be analyzed to identify "precursor" events, predict where they might happen again, identify solutions, and set standards and mutually acceptable contract terms for HSSE.

The IAGC Aide Memoir web page (http://www.iagc.org/aide-memoir.html) makes a plea that is relevant for any industry:

"We all owe it to the victims of these tragic accidents that we do remember what happened and prevent this from happening again. Please help your industry do this!"

Bibliography

Anderson, M. and Denkl, M. 2010. "The Heinrich Accident Triangle - Too Simplistic A Model For HSE Management in the 21st Century?" *Paper SPE 126661, presented at the SPE International Conference on Health, Safety and Environment in Oil and Gas Exploration and Production,* Rio de Janeiro, Brazil, 12-14 April.

Bird, F.E. and Germain, G.L. 1985. *Practical Loss Control Leadership*. Loganville, GA: International Loss Control Institute.

Cooper, M.D. 2009. "Behavioral Safety Interventions: A review of process design factors." *Professional Safety,* 54(2): 36-45.

Dragoset, B. 2005. "A historical reflection on reflections." *The Leading Edge* 24, s1 (2005); pp. s46-s70, SEG@75 special issue.

Dujmovich, T.J., 1992. "Proactive Safety Approaches in the Seismic Industry." *Paper SPE 24616 presented at the 67th Annual Technical Conference and Exhibition of the SPE* in Washington, D.C., October 4-7.

Heinrich, H.W. 1931. *Industrial Accident Prevention: A Scientific Approach.* New York: McGraw-Hill.

International Association of Oil & Gas Producers (IOGP). 1994. Report 210, *Guidelines for the Development and Application of Health, Safety and Environmental Management Systems.* London, UK: IOGP.

_____. 1996. Report 245, *Guidelines for HSE auditing in the geophysical industry.* London, UK: E&P Forum (prior name for International Association of Oil and Gas Producers).

_____. 2014. Report 292, *HSE competency management guidelines for the geophysical industry.* Version 3. London, UK: IOGP.

_____. 2004. Report 355, *Watercraft & water in geophysical operations - a guideline to operations & management.* London, UK: IOGP.

______. 2016. Report 365, *Land transportation safety recommended practice.* Version 3.0. London, UK: IOGP.

______. 2006. Report 375, *Lifting & hoisting safety recommended practice.* London, UK: IOGP.

______. 2013. Report 390, *Aircraft Management Guidelines*. Issue 5, July 2008, updated August 2013. London, UK: IOGP.

______. 2013. Report 420, *Helicopter guidelines for land seismic & helirig operations.* Version 1.1, June 2013. London, UK: IOGP.

______. 2010. Report 423, *HSE management – guidelines for working together in a contract environment*, Version 1.0. London, UK: IOGP.

______. 2009. Report 432, *Managing HSE in a geophysical contract.* Version 3.0. London, UK: IOGP.

International Association of Geophysical Contractors (IAGC), 2012. *Land Geophysical Safety Manual.* 10th ed. Houston, TX: IAGC.

______. 2012. *Marine Geophysical Safety Manual.* 10th ed. Houston, TX: IAGC.

______. 2017. *The Aide Memoir for Geophysical Risk Assessors.* http://www.iagc.org/aide-memoir.html (accessed 5 January 2017).

International Organization for Standardization (ISO). 2000. ISO 17776:2000. *Petroleum and natural gas industries —Offshore production installations —Guidelines on tools and techniques for hazard identification and risk assessment.* Geneva, Switzerland: ISO.

Krause, T.R. 2010. "A New Paradigm for Fatality Prevention." *Proceedings of the Professional Development Conference of the American Society of Safety Engineers.* Des Plaines, IL: ASSE.

Krause, T.R. 2012. "Safety Leadership: Serious event prevention. What a new study tells leaders about safety's past - and its future." *Safety and Health Magazine.* January 1, 2012.

Lawyer, L.C., Bates, C.C. and Rice, R.B. 2001. *Geophysics in the Affairs of Mankind: A Personalized History of Exploration Geophysics.* 2nd ed. Tulso, OK: Society of Exploration Geophysicists (SEG).

Manuele, F.A. 2008. "Serious Injuries and Fatalities: A call for a new focus on prevention." *Professional Safety.* 53(12), 32-39.

Manuele, F.A. 2011. "Reviewing Heinrich: Dislodging Two Myths From the Practice of Safety." *Professional Safety.* 56(10), 52-61.

Petersen, D. 2001. *Safety Management: A Human Approach*. Des Plaines, IL: American Society of Safety Engineers.

Proubasta, D. 1987. "Safety versus production." *The Leading Edge*, 6, no. 6, 33–49, http://dx.doi.org/10.1190/1.1439403.

Rhinehart, R. 2013. *The Power of a Trade Association*, Specialty Coffee Association of America (Accessed 12 February, 2016.) (http://www.scaa.org/chronicle/2013/03/10/the-power-of-a-trade-association/.)

Society of Exploration Geologists (SEG), Seismic Safety Committee. 1976. "1974 Geophysical Accident Report." *Geophysics* 41(1), p160. (http://dx.doi.org/10.1190/1.1440602.)

Scott, I. 2014. "Seismic Aviation." *Presented at the September 2014 IAGC/OGP HSE & S Global Forum.* The Hague, Netherlands.

Sixma, E., 2005. "Lagging Indicators, Icebergs and Risk Focus." *Presented at the September 2005 IAGC/OGP HSE & S Global Forum,* Gatwick, UK.

_____. 2007. "Aide Memoir for Geophysical Risk Assessors." *Presented at the September 2007 IAGC/OGP HSE & S Global Forum.* Gatwick, UK.

Sixma, E. and Edgecombe, K., 2016. "Aide Memoir for Geophysical Risk Assessors." *Presented at the September 2016 IAGC/IOGP HSE & S Global Forum* Paris, France.

Sorge, Bart W. 1952. "Safety and Geophysical Exploration." *Geophysics* 17, p 236. (http://dx.doi.org/10.1190/1.1437751.)

Sweet, G.E. 1978. *The History of Geophysical Prospecting*, vols. 1 and 2, 3rd ed: Science Press.

Threadgold, I., Jeffrey, C., Sixma, E., and Garcia, O. 2010. "Working Together – Saving Lives." *Presented at the SPE International Conference on Health, Safety and Environment in Oil and Gas Exploration and Production.* Rio de Janeiro.

Threadgold, I.M. 2015. "Reflections from the past: a brief history of safety in seismic operations over the last century." *The Leading Edge*, Volume 34 No. 9, September 2015; (http://dx.doi.org/10.1190/tle34091080.1.)

Threadgold, I.M., Hallows, S.G. and Ingram, R.P. 2016. "Competency Management Guidance for Geophysical Industry Personnel with HSE Responsibilities." *Presented at the SPE International Conference and Exhibition on Health, Safety, Security, Environment and Social Responsibility in Oil and Gas Exploration and Production.* Stavanger, Norway.

Westby, G.M. 1953. "A Safety Program – Managements Responsibility." *Geophysics* 18, p 532. (http://dx.doi.org/10.1190/1.1437909.)

Session No. 712

Preparing Employees for Interviews by Regulatory and Investigative Agency Representatives

Steven P. Pereira, CSP
Professional Safety Services
Denham Springs, LA

Who Conducts Employee Interviews?

During the course of operating a business, employers may be faced with requests to interview hourly employees and/or members of management, including first-line supervisors. Requests may come from law enforcement and regulatory agency representatives, such as the Occupational Safety and Health Administration (OSHA), the Mine Safety and Health Administration (MSHA), the Environmental Protection Agency (EPA), and the United States Coast Guard (USCG). Investigative bodies, such as the National Transportation Safety Board (NTSB) and the Chemical Safety Board (CSB). may also conduct employee interviews following a major transportation accident or chemical spill.

Requests for interviews may be the result of incidents, complaints, routine inspections or enforcement actions. In any case, employers and employees have rights that should be clearly understood and communicated to all parties involved. This requires preparation. In this paper, we will focus primarily on interviews conducted by OSHA, MSHA, and the CSB.

Disclaimer: The information presented in this paper is based on the author's experience as a safety professional, representing clients and attorneys in dealings with OSHA, MSHA, and the CSB. It should not be considered as legal advice. When developing procedures for managing employee interviews by regulatory agencies and investigative bodies, employers should seek advice from attorneys with experience in these areas.

Why Manage Employee Interviews

Regulatory agencies, such as OSHA, MSHA, and the EPA, have the authority to issue citations, levy fines and, in some rare instances, recommend criminal prosecution. Erroneous information or poor communication can adversely impact the investigation process, possibly resulting in prolonged investigations, costly litigation, and adverse publicity.

It is important to understand that employee interviews are just one aspect in an incident investigation or regulatory agency visit. It is a good idea to have a procedure for managing regulatory agency visits and outside agency investigations.

Rights and Responsibilities

OSHA and MSHA Rights Relative to Employee Interviews

Both agencies have the right to interview hourly employees in private if the employees consent. For details on how OSHA and MSHA handle their interviews and employer/employee rights, the reader may wish to review:

- MSHA Handbook Series, Handbook Number PH05-1-4, dated August 2005[1]
- OSHA, Directive Number CPL-02-00-160, *Field Operations Manual (FOM),* Effective date 8/02/2016[2]
- OSHA, 3021-11R, *Workers' Rights,* 2016[3]
- MSHA, *Miners' Rights and Responsibilities Under the Federal Mine Safety and Health Act of 1977,* OT2, Revised 2015[4]

Both agencies have the right to have truthful responses to their questions.

Employee Rights

Employees have a right to a one-on-one interview with an OSHA or MSHA representative.

Employees are protected from discrimination or retaliation for exercising their rights under the OSH Act.[5]

Employees can decline to be interviewed or have a representative present. In a union environment, the employee representative is usually the union steward, but who represents the employee is up to the individual. He or she may request a personal attorney to be present. In most cases, neither agency will object to a union representative or personal attorney being present, as long as the representatives do not interfere with the interview.

Both OSHA and MSHA representatives do not want a member of management present in the hourly employee interviews. Likewise, they frown on attorneys who also represent management to be present for the interviews. They consider it to be a potential conflict of interest. It is important to understand that the employee has the right to choose his or her representative. If the agency does not want to permit the employee to choose their representative, the employee can decline to be interviewed. **Note:** This can sometimes be contentious and may require the agency

---

1 Mine Safety and Health Administration (MSHA). 2005. *Special Investigations Procedures* (retrieved March 20, 2017) (https://arlweb.msha.gov/READROOM/HANDBOOK/PH05-I-4.pdf).

2 Occupational Safety and Health Administration (OSHA). 2016. *Field Operations Manual* (retrieved March 20, 2017) (https://www.osha.gov/OshDoc/Directive_pdf/CPL_02-00-160.pdf).

3 OSHA. 2016. *Workers' Rights* (retrieved March 20, 2017) (https://www.osha.gov/Publications/osha3021.pdf).

4 MSHA. 2015. *Miners' Rights and Responsibilities Under the Federal Mine Safety and Health Act of 1977* (retrieved March 20, 2017) (https://arlweb.msha.gov/s&hinfo/minersrights/minersrights.pdf).

5 OSHA. 2017. *OSH Act of 1970* (retrieved March 20, 2017) (https://www.osha.gov/pls/oshaweb/owadisp.show_document?p_table=OSHACT&p_id=2743).

representative to check with their superiors or government attorneys. Both agencies can obtain a subpoena to force the interview, but the employee still has legal rights.

The author has represented employees during interviews following several major accidents. In another instance, OSHA denied an employee's request to have him sit in on his interview following a fatal accident. The employee involved had been onsite eight (8) hours prior to OSHA's arrival. Rather than remain onsite for another few hours while logistics were worked out, the employee relented but consented to a post-interview conversation with the author afterwards. In another high-profile incident, one of the author's co-workers was selected by employees to be their representative during interviews, while the author served as management's representative.

Other hourly employee rights include but are not limited to:

- They may decline to provide personal information, such as their home address and phone number.
- They may end the interview at any time without explanation.
- They are not required to sign their statements or summaries thereof.
- They may decline to be audio- or video-recorded or photographed.

Management Interviews
Management employees, including first-line supervisors or foremen, are considered to be agents of the company. A manager's knowledge of, and indifference to, a hazard may likewise be attributed to the company for purposes of classifying the severity of detected violations, conditions or practices. For this reason, a company representative or an attorney should be present for all interviews with regulatory or investigative agency representatives.

Developing a Protocol for Responding to Requests for Employee Interviews

Requests for Interviews
Request that the agency representative(s) provide the company with a list of employees to be interviewed. Coordinate interviews to be as disruptive to the organization as possible.

Hourly Employee Interviews
In advance of the interviews, inform employees to be interviewed of their rights. Hourly employees have a right to have a representative present. In a union environment, this is usually a shop steward. This can typically be anyone of their choosing. They have a right to be interviewed one on one in private if they so choose. They can decline to be interviewed but may be requested to do so under a subpoena. They should be told to listen to the questions carefully and answer them truthfully. They should be informed if they don't know an answer, it's okay to say, "I don't know." They should be informed that they are not obligated to sign any statements or summaries of the interviews. **Under no circumstances should an employee be discouraged, threatened or intimidated concerning their right to talk with a Compliance Safety and Health Officer (CSHO). It is illegal to do so.**

After an agency representative interviews an employee, it's a good practice to ask the individual if he/she is willing to discuss the interview. Many will. This can help you identify potential areas of focus relative to the agency inquiry. However, this must be done with great care. Under **NO** circumstances should you coerce or intimidate an employee into discussing his/her interview. It is illegal to do so.

Management Interviews
For interviews of management personnel:

- All members of management should be informed of their rights concerning employee interviews, and that a management representative will be present. A "management" employee for this purpose is:
  - Anyone in a supervisory position
  - A working foreman who has responsibility for supervising a crew, even if in the bargaining unit

- They should be told to listen to the questions and answer them truthfully. They should not volunteer any information. They should also be informed that if they don't know an answer, they should say so.

- As part of your management training program, consider adding a section on handling regulatory and investigative interviews. Such a program might include mock interviews with legal counsel on how to tactfully and truthfully handle leading or confusing questions such as but not limited to:

  - "We have been told about a number of serious "near miss" events over the past two (2) years. Please tell me about the ones that concerned you most."

  - "With respect to environmental, health and safety issues, which area of the facility concerns you most?"

  - "We have heard a number of anonymous complaints about environmental, health and safety issues at this facility. Have any been reported to you? If so, please elaborate.

Tips for an Effective Management Interview
The following are some tips for effective management interviews:

- Listen carefully to the question.
- Make sure you understand the question.
- If you don't understand the question, say so.
- Answer the question being asked.
- Don't be afraid to say, "I don't know" or "I don't recall."
- Don't volunteer information not requested.
- Don't guess or speculate.
- Take your time and control the pace.
- Silence is okay; don't fill in the gaps.
- Stick to the facts.
- It is okay to disagree, but don't argue.

Summary

Visits from regulatory or investigative agencies are likely to occur out of and in the course of business. In some situations, multiple agencies may respond to an event and want to conduct separate interviews. This can be a very trying situation for both management and employees.

Many managers mistakenly think they should not talk to their employees prior to their being interviewed by a regulatory or investigative agency representative. Keep in mind most personnel will be apprehensive about such an interview because they don't know their rights or what to expect. Those who are about to be interviewed can be put at ease if they know their rights and what to expect. This requires advance planning.

Proactive managers recognize that employee interviews are part of the process and have a plan to manage the interviews to the extent possible. They know and understand the rights of all parties involved:

- Agency representatives
- Hourly employees
- Management employees

Prior to starting the interview process, they inform hourly and management employees of their rights and responsibilities. No attempt is made to coerce or intimidate an employee into providing false or misleading information. Managers and supervisors have been trained on how to effectively and truthfully respond to questions.

In addition, the proactive companies have identified and made prior arrangements with law firms that specialize in environmental, health and safety regulations and investigations.

How well is your company prepared for visits from regulatory and investigative agencies?

Most regulatory and investigative agency representatives are courteous and professional. However, in some rare instances, they have become antagonistic, intimidating, and verbally abusive. Under no circumstances should this type of behavior be tolerated. If you receive a report from a member of management or the hourly ranks relative to a Compliance Officer's intimidating or threatening behavior, you should request a conference with the Compliance Officer's immediate supervisor or Area Director or Supervisor.

Session No. 713

The New Face of HazCom: SDS and Label Compliance in the Post-GHS Deadline Era

Chuck Haling
Vice President of Sales
VelocityEHS
Chicago, IL

Glenn Trout
Chief Executive Officer & President
VelocityEHS
Chicago, IL

Introduction

On June 1 of last year, OSHA expected the millions of employers across the country covered by its HazCom Standard to be in full compliance with the changes arising from alignment with the United Nations' Globally Harmonized System of Classification and Labelling of Hazardous Chemicals (GHS). While the transition process certainly created areas of ambiguity— many of which still remain—one thing is crystal clear. OSHA is still citing workplaces for violations of the HazCom Standard at a vigorous rate. OSHA's own enforcement statistics show that HazCom ranks #2 on its most frequently cited standards list, behind only fall protection. This is part of a trend towards increasing focus on HazCom. HazCom moved up from the #3 spot four years ago, and has continued to be a priority for the agency.[1] Penalties for violations have also become more serious. In August of 2016, OSHA increased fines by 80 percent, and has said it will update fines yearly to account for inflation. This means that the penalties for failing to manage the requirements of the HazCom Standard sting more now than they have in many years.

OSHA issued 5665 HazCom citations in FY 2016, and labels and SDSs figured prominently in those citations. Some of the top HazCom citations were for:

- Failing to train (on SDSs, labels and more);
- Failing to provide barrier-free access to SDSs;
- Failing to develop SDSs for hazardous chemicals produced or imported; and

[1] *Safety + Health* 2016.

- Failures related to labeling, explaining labels to employees and having an appropriate workplace labeling system.[2]

In short, when we are talking about HazCom violations, the bulk of them include issues with SDSs and labels.

As we proceed through 2017, we know that the obligations created by GHS alignment are ongoing. Challenges persist. And missed deadlines upstream still have the potential to impact downstream users of chemicals. For many employers, there are still more questions than answers. Some employers may be confused because of upstream adoption delays; others because of general misunderstandings surrounding what has long been required under the HazCom Standard versus what became newly required as a consequence of GHS alignment. And the constantly changing regulatory landscape, now in flux in a way unrealized in many years, doesn't help either!

Accordingly, employers need to be serious about ensuring they have a compliant HazCom Program in place, and that includes having a written plan, and ensuring workplace labels and safety data sheets are GHS-aligned, and that employees have been trained on all newly identified physical and health hazards of the covered chemicals in their workplaces.

At the same time, special dispensations still exist for chemical manufacturers and distributors, allowing them, in certain situations, to enjoy "good faith" exemptions until December of 2017. Understanding these exemptions is important.

Requirements related to HazCom are likely to continue to change in the months and years ahead. For example, OSHA representatives have indicated the agency's intention to realign with a newer revision of GHS—probably, Revision 7—in the near future.[3] However, by becoming familiar with OSHA's priorities, and how it expects to enforce compliance with HazCom 2012, employers of all types can position themselves for success in a changing regulatory landscape.

Please Note: *OSHA refers to the pre-GHS standard as "HazCom 1994" and to the GHS-aligned standard as "HazCom 2012." These naming conventions will be used throughout the rest of this paper.*[4]

OSHA HazCom Responsibilities

The Hazard Communication Standard calls for chemical hazard information to be shared with downstream users of chemicals.[5] Under GHS alignment, three groups still have responsibilities. **Chemical manufacturers** are required to evaluate and now also to classify the hazards of the chemicals they produce, and they must also provide labels and SDSs to downstream users. **Chemical distributors and importers** must provide labels and SDSs to the employers (e.g., end users of chemicals) that they ship to. And **employers and end users of chemicals** have five key responsibilities, which include:

---

[2] Sven Rundman, *2016 SCHC Fall Meeting Presentation*.
[3] OSHA *HazCom Stakeholder Meeting*, 2016.
[4] OSHA *Small Entity*.
[5] OSHA *Small Entity*.

1. Maintaining a written hazard communication plan, and ensuring it's available to covered employees;
2. Maintaining a written chemical inventory of onsite hazardous chemicals;
3. Ensuring the proper use of labels and warnings;
4. Maintaining safety data sheets and providing employee right-to-know access; and
5. Training employees on the specific hazards of the chemicals to which they are exposed.

Some businesses are still struggling to come into compliance with the above requirements, and that includes the ones related to SDSs and labels. However, there are signs that year over year, businesses are moving closer to compliance. For example, according to *VelocityEHS* obtainment efforts, an average of about 78 percent of the safety data sheets provided by suppliers are GHS-aligned SDSs (as opposed to MSDSs). That number is up from about 35 percent at the start of 2015.

To get a stronger sense of what OSHA expects, let's look at how it arrived at GHS alignment, what it hoped for—and actually achieved—in terms of hitting deadlines, and where it has indicated that it may be willing to meet businesses halfway. When it comes to enforcing compliance for SDSs and labels, OSHA will be taking into account a number of factors.

The Transition Process

OSHA chose to implement alignment with GHS through a phased-in process that had four major deadlines. All of these deadlines brought requirements directly impacting labels and SDSs.[6] At the outset, it's important to note that all of the deadlines have now passed. GHS alignment is here, and by briefly reviewing these deadlines and their requirements, we can gain a greater understanding of how OSHA sees labels and SDSs working together under HazCom 2012.

Recognizing that updated labels and SDSs could begin arriving in workplaces shortly after the publication of the final rule on May 25, 2012, OSHA positioned the **December 1, 2013 employee training deadline**—the first deadline—early in the adoption process to ensure the safety of workers who would be faced with interpreting the new information. Starting on this date, employers needed to train their workers on GHS to make sure they were prepared for the GHS-aligned labels and safety data sheets they would soon begin seeing.

The next deadline was the **June 1, 2015 manufacturer deadline.** By this deadline, manufacturers were required to complete chemical reclassifications and update the corresponding SDSs and labels. Beginning June 1, 2015, manufacturers were supposed to send out only HazCom 2012-compliant SDSs and labels with their chemical shipments.

The next deadline was the **December 1, 2015 distributors and importers deadline**. By this deadline, manufacturers and importers were required to begin shipping products with newly formatted labels and SDSs. OSHA provided these groups with an extra six months beyond the manufacturer deadline to account for the lag time that might exist between when they receive product shipments from manufacturers and when those products were scheduled to be distributed to downstream users.

The final effective date was **June 1, 2016, the deadline for full employer compliance.** By this date, employers were required to update their HazCom written plans as necessary, complete additional employee training for any newly identified physical or health hazards based on manufacturers' reclassification of chemicals, update alternative workplace labeling systems as

---

[6] OSHA *HCS* 2012.

needed, and ensure that SDSs and labels for chemicals in their inventory were up to date. OSHA gave employers a full year after the manufacturer deadline to comply, expecting that this would provide a sufficient amount of time for their chemical inventories to turn over, and for fresh shipments of products with appropriate HazCom 2012-aligned safety data sheets and labels to arrive.

So all three groups covered by HazCom—manufacturers/importers, distributors and employers/end-users—had obligations during the transition, and these obligations were interconnected and ongoing. The actions of one had, and continue to have, important impacts on the others.

Downstream Delivery Issues

During the GHS alignment process, many employers were not completely clear about how the MSDSs in their library would be replaced with SDSs. However, as the June 1, 2016 deadline for complete alignment drew nearer, employers began to have a better idea of how the process would work, and were able to take appropriate actions, when necessary, to help ensure that their workplaces would be compliant. For example, some end-users had expected documents to be sent to them by or on June 1, 2015 from their suppliers, but realized that wasn't necessarily going to be the case, since OSHA's requirements regarding employer distribution of these updated documents didn't change.

OSHA only requires a safety data sheet to be sent with the initial shipment of a chemical, with the first shipment *after* a change has been made, or when explicitly requested by a downstream user. Employers who receive an inadequate or incorrect SDS, or who fail to receive one at all, need to have a process in place to try and obtain one. Employers will want to reach out to suppliers to request an updated, compliant SDS, and should document their request. If the upstream supplier provides a date by which the SDS will be provided, that date should be documented by the employer. If the supplier is unwilling or unable to provide an indication of when an update will be ready, make note of this, and consider switching suppliers or contacting OSHA for assistance.

As previously mentioned, not all documents released in the supply chain are updated GHS-aligned SDSs, so even though the last of the GHS alignment deadlines has passed, many employers still have both GHS-formatted SDSs and pre-GHS formatted MSDSs in their libraries. It's important to keep in mind that, for as long as employers maintain both SDSs and MSDSs, employees must be trained on both formats. This situation may apply even to employers who have taken all necessary steps to become fully GHS-aligned and compliant. For example, if an employer has a manufacturer's chemical in their inventory, and that manufacturer has gone out of business, OSHA permits the employer to continue using the MSDS, while ensuring that the employees are trained on it.[7]

Safety Data Sheets

Under GHS alignment, the name changed but not the overall purpose when it comes to safety data sheets. Though material safety data sheet (MSDS) changed to safety data sheet (SDS), the function remained the same. A safety data sheet still serves as the linchpin to an employer's hazard communication program, providing downstream users with comprehensive safety information about hazardous chemicals in their workplace, aiding in chemical inventory

[7] OSHA *Letter of Interpretation.*

management, and connecting the dots to container labels and other forms of warnings in the workplace.

In addition to being renamed as SDSs, safety data sheets were also given a strictly prescribed 16-section format. Under GHS alignment, these 16 sections must always appear in the same order on every safety data sheet. This is essentially the American National Standards Institute (ANSI) Standard, with which many in the EHS space were already familiar.

The 16 required sections under GHS alignment are:

1. Identification
2. Hazard(s) Identification
3. Composition/Ingredient Information
4. First-Aid Measures
5. Fire-Fighting Measures
6. Accidental Release Measures
7. Handling and Storage
8. Exposure Control/Personal Protection
9. Physical & Chemical Properties
10. Stability & Reactivity
11. Toxicological Information
12. Ecological Information
13. Disposal Considerations
14. Transport Information
15. Regulatory Information
16. Other Information

The advantage of making this order consistent becomes apparent when employees need to look up something quickly, as is often the case during an emergency. For example, if an employee needs to quickly find information about first-aid measures, he or she knows to always look in section 4 of the safety data sheet.

It should be noted that the information required in sections 12-15 is not mandatory under HazCom 2012 because these sections represent areas outside of OSHA's jurisdiction. However, even if these sections are left blank on an SDS, the main headings and numbers must still appear. Also, it's a good idea to train employees on these sections so they know what they are when they see them.

At *VelocityEHS,* we sometimes get the question: "Is NFPA and HMIS information still required on SDSs after GHS alignment?" But this is a bit of a misguided question, because NFPA and HMIS information was never required on MSDSs under HazCom 1994. It was optional. Consequently, it continues to be optional under HazCom 2012. If manufacturers see value in including NFPA and HMIS information, they can continue to include it, and it's possible that their customers may even request it. OSHA has not said in which section it must be placed if it is used. Some companies are electing to add it to Section 2, Hazard Identification. Others are adding NFPA and HMIS information to Section 5, Fire-Fighting Measures. Still others are placing that info in Section 16, Other Information. Though all of these options are acceptable to OSHA, to help mitigate confusion and possible misunderstandings, we recommend placing this information in Section 16 as an industry best practice. It keeps the GHS HazCom 2012 classification numbers a good distance away from these other system numbers, which are based on an opposing hazard-rating scale, where the higher the number, the more severe the hazard.

Some people also ask about the rules for including GHS pictograms on SDSs. Unlike the requirement for shipped labels, OSHA doesn't require the GHS pictograms to be printed in color on the SDS. In fact, OSHA doesn't even require a pictogram's hazard symbol to appear on the SDS at all; the text description of the pictogram can be used in lieu of the symbol.

For example, it could have:

- A pictogram with a red diamond border
- A pictogram with a black diamond border
- The designated words that describe what the pictogram is, but without the symbol (e.g., "Skull & Crossbones")

Any of these variations would be compliant on a HazCom 2012 GHS-aligned SDS.

Labels

In order to ensure workers understand the identity of the chemical(s) they work with, as well as any hazards and related information, OSHA makes labeling a vital component of the HazCom Standard.

With just a few exceptions, such as stationary process containers, and containers used in the transfer of chemicals kept under constant observation, containers of hazardous chemical must be labeled in the workplace at all times. Under HazCom 2012, OSHA sets forth specific requirements for what has to be on shipped container labels, but outside of specifying that there has to be a product identifier, what exactly must appear on a workplace label is largely performance based. Let's look at both of these systems, and get a sense of how they function in tandem as a crucial part of workplace hazard communication.

Shipped Labels

Under GHS alignment, labels on shipped containers have a strictly prescribed set of six label elements that must appear.

The six required HazCom 2012 shipped label elements are:

1. **Product Identifier**
2. **Supplier Information**
3. **Pictograms**
4. **Signal Words**
5. **Hazard Statements**
6. **Precautionary Statements**

Let's take a moment to review each of these.

The **Product Identifier** is the name of the chemical inside the container. It should match the corresponding chemical identifier on the safety data sheet. It should also be included on the employer's written chemical inventory list.

Supplier Information should include the name, address and phone number of the responsible party for the chemical.

Pictograms are black hazard symbols on white backgrounds with red diamond borders. At the UN level, GHS pictograms cover the range of recognized health, environmental, and physical hazards. Under HazCom 2012, all but environmental pictograms are required.

Signal Words are the words used to quickly orient the user to the severity of the hazards of the chemical. There are two signal words in the GHS system: Danger and Warning. Danger is considered to be more severe.

Hazard Statements are short statements that provide information about the hazards associated with a chemical.

Precautionary Statements are statements that provide information on how to prevent an incident, how to respond to an incident involving the chemical, how to safely store the chemical, and how to safely dispose of the chemical.

Note that OSHA *does* allow other information to appear on the shipped label in addition to these six elements, as long as it does not cast doubt or contradict the required information. For example, expiration dates, batch numbers and an indicator of the personal protective equipment (PPE) to wear when handling the hazardous chemicals are all permitted so long as they don't impede or conflict with the required information.

Workplace Labels

Under HazCom 2012, the effectiveness of your labels is performance-based, as it was under HazCom 1994; that is to say, compliance or noncompliance is determined based on if or how well the labels successfully communicate all the hazard information. The specifics for what *must* be present on a workplace label are comparatively vague, which is why this has continued to be such a vexing topic. It boils down to what performs for each unique work environment and workforce. Under HazCom 2012, OSHA says that workplace labels must include *either* the contents of a shipped label, or:

> *Product identifier and words, pictures, symbols, or combination thereof, which provide at least general information regarding the hazards of the chemicals, and which, in conjunction with the other information immediately available to employees under the hazard communication program, will provide employees with the specific information regarding the physical and health hazards of the hazardous chemical.*[8]

The shipped label essentially serves as the benchmark. If an employer chooses not to simply replicate the shipped label, that employer will need to be able to make the case that whatever system has been chosen can perform *just as effectively.* According to OSHA's July 9, 2015 enforcement directive, employers using alternative labeling systems must ensure that their system makes employees aware of all information required under the HCS.[9] During an inspection, an OSHA inspector might pick up a workplace container in front of an employee and ask: "What's this? Tell me about the contents of this container." It will then be up to that employee to draw upon: (1) the information on the label before them, (2) their training, and (3) any other information available to them in their immediate work area (like signs, placards, or process sheets) to provide that information. The quality of the employee's answer will determine whether or not the label is compliant.

---

[8] OSHA *HCS* 2012.

[9] OSHA *Instruction*.

Keeping this in mind, there are three approaches to workplace container labeling that OSHA says are acceptable. We'll look at all three here, but it is our opinion that the first option is the best and safest. Remember: The bar is low for what physically needs to go on the label, but high for what the employee must be able to understand from your label.

Option 1 is to simply replicate the shipping label. OSHA feels these six elements help do the job of communicating the breadth of dangers to which employees may be exposed when working with a chemical. The information the employee needs will be right in front of him or her, affixed to the container. What's more, from the compliance measurement aspect, having all this information in front of them should make it easier for an employee to respond to an inspector if asked whether the hazards are of the chemical.

Option 2 is to use a combination of GHS label elements, plus the product or chemical identifier that matches the product identifier on the safety data sheet, and other information immediately available to the employee, along with training to communicate the hazard information. This combination must convey the same level of information about the hazards of the chemical that an employee would have garnered from the manufacturer-shipped label.

Option 3 is to use an entirely "homegrown" system. OSHA says any system that was compliant under HazCom 1994 is still viable today, including this kind of system, so long as employees still have all necessary health and physical hazard information, and that nothing conflicts with or contradicts GHS information. So when would employers find themselves using or pursuing this sort of labeling option? It might be the best option for companies that have been using a system that includes unique elements fully understood by employees, perhaps one that includes a color-coding system that is connected to processes in their workplace, or is built into existing label capabilities and resources.

It might also be applicable in facilities where employees are experts and have vast knowledge of the chemicals they are working with, and would be competently prepared to protect themselves when using a homegrown system that has less information presented than what would be necessary to protect workers with less chemical expertise. If, on the other hand, an employer's workforce has never worked with chemicals before or doesn't speak English well, the homegrown system would have to account for that as well, or it wouldn't be doing its job of protecting workers and would likely not pass an OSHA inspection.

What about NFPA and HMIS system elements? While they are not, and have never been, required under HazCom, it's common for companies to be using these systems, and OSHA says it's acceptable to keep including elements from these labeling systems on workplace labels as long as:

- The product identifier is provided (it is not automatically included under the NFPA system, for example).
- The information on the label is consistent with, and does not contravene or contradict, required HazCom 2012 information.
- The labeling system communicates all hazard information of a chemical, both short- and long-term exposure risks (NFPA, for example, only addresses short-term, immediate, acute hazards).

Whatever labeling option is chosen, the labels must be in English. This same requirement applies to shipped labels as well. Other languages are permitted in addition, but English has to be present on a label at a minimum. Also applicable to both types of labels is the requirement to immediately replace a label if it becomes damaged or is removed. And, at all times, labels are to be legible and prominently displayed.

To recap, when it comes to workplace labels, they either perform or they don't, and the inspector will see that in a variety of ways. Remember, the shipped label serves as the benchmark in determining how well a workplace labeling system performs. Employers should consider this very carefully when selecting a labeling option for their workplace.

Labeling Small Containers

Some manufacturers run into a challenge when it comes to the requirement to label individual containers, because these containers can be very small. OSHA has always expressed—and it reaffirmed in a 2013 letter of interpretation—that it expects the immediate container of a hazardous chemical to be labeled.[10]

OSHA has recommended using pull-out labels, fold-out labels, tags, or similar methods to ensure the containers travel with their respective hazard information. This is even true in cases in which several smaller containers of hazardous chemicals are inside of larger shipped containers that are labeled. OSHA wants to ensure the immediate container is always labeled at a minimum. To provide some flexibility to manufacturers dealing with this issue, OSHA has offered a practical accommodation. OSHA says that in situations where it is difficult to fit all the information onto a small label, and where that chemical is traveling insider a larger container, the immediate container can be labeled with less information than it normally would have to include, so long as the outer container includes the full label information. Below are the elements that must appear on the small shipped container label in order to satisfy the exception provision:

- Product identifier;
- Appropriate pictograms;
- Manufacturer's name and phone number;
- Signal word; and
- A statement indicating the full label information for the chemical is provided on the outside package.[11]

As the last point indicates, the full label information must then be made available on the outside of the master package containing the smaller individual containers. OSHA knows that this is a concern for manufacturers and has said it is aiming to meet manufacturers halfway with this accommodation.[12] If a manufacturer faces the task of labeling very small containers, a great place to start is to become educated on the options that are allowed by OSHA and to pick the best one. But remember, simply telling an OSHA inspector, "Labeling would have been too expensive," does not qualify as an exemption for the requirement.

Older Chemical Container Labels

As mentioned earlier, when OSHA selected the phase-in timeframe for GHS adoption, it did so with the thinking that most employers would turn over their entire inventory of products and

[10] OSHA *Standard Interpretations*.

[11] OSHA *Letter of Interpretation*.

[12] OSHA HazCom Stakeholder Meeting, 2016.

corresponding SDSs and labels in that time. However, as noted, this has not been the actual case. While OSHA doesn't specifically require employers to alter manufacturer-shipped labels on older shipped containers that have already been labeled under HazCom 1994, there is still the responsibility in place for employers to ensure their employees are aware of all the chemical hazards.[13] So if the employer becomes aware of any newly identified hazards associated with a chemical but hasn't received a new shipment with that new information conveyed on a label, it might be easiest to ensure the information is successfully communicated by updating the shipped label or possibly creating a workplace label. If the manufacturer is still in business, another option is to request an updated shipped label from them. OSHA also requires that all chemical containers be relabeled if the labels are removed, defaced, or become otherwise illegible. Whatever route they choose to take, employers should make it a priority to see that employees are aware of the hazards of chemicals in older containers.

SDS and Label Responsibilities Up and Down the Supply Chain

GHS alignment in the United States gave different groups different responsibilities when it came to SDSs and labels. These responsibilities are ongoing. By looking at these responsibilities in more depth, we can understand how to ensure we meet all requirements. It can also give us insight into how GHS alignment is functioning up and down the chemical supply chain.

Chemical Manufacturers

Chemical manufacturers were required reclassify their chemicals using the new GHS criteria, and to produce GHS-formatted SDSs and labels. But how were they expected to do that? OSHA required the following using prescribed criteria outlined in the appendices of the updated standard:

- They had to gather relevant data pertaining to the chemical.
- They had to review data to ascertain hazards of the chemical.
- They had to determine if the chemical would be classified as hazardous.
- If determined to be hazardous, they had to classify and also categorize health and physical hazards of the chemical by severity where appropriate.

Under GHS alignment, a chemical manufacturer and/or distributor is required to classify a chemical according to prescribed, standardized criteria from the appendices of the rule to determine if a chemical is considered hazardous, and, if so, the degree to which it's hazardous. No longer is it sufficient for a chemical to be classified as hazardous; work must be done to answer the question of *how* hazardous is it. This concept of severity is new with GHS adoption. The degree and nature of the hazard must be indicated on the label and the SDS, using prescribed information provided in the appendices of the rule.

HazCom 2012 also introduced a new definition of "hazardous chemical." Under the standard, a hazardous chemical becomes any chemical that poses a physical or health hazard. Simple asphyxiants, combustible dusts, and pyrophoric gases are also added to the definition of "hazardous chemical."[14] Additionally, a chemical known to be hazardous that does not fit neatly into one of the categories established above is classified as a Hazard Not Otherwise Classified (HNOC).

---

[13] OSHA *Brief: Hazard Communication Standard: Labels and Pictograms.*
[14] OSHA *HCS* 2012.

Note that simple asphyxiants, combustible dust and pyrophoric gases are specifically mentioned. This is because OSHA wanted to ensure that when it moved to align with GHS, it did so in ways that would maintain or enhance protections, not lessen them. When OSHA examined the UN's articulation of physical and health hazards, it identified that these three hazards would be omitted under the UN definition of a hazardous chemical. Therefore, OSHA adopted the UN definition of a hazardous chemical, but made a slight modification to include these three additional hazards to ensure it wasn't losing any protections it had in place under HazCom 1994.

HNOCs are also added to account for chemicals that, through the classification process, simply don't fall into one of the predetermined classification categories. The classification is designed for chemicals that do not meet the specified criteria for the physical and health hazard classes, but which are still hazardous. For example, a chemical manufacturer might produce a chemical that the manufacturer knows is definitely hazardous, but which does not qualify for one of the specified hazard categories. Instead of letting this hazardous chemical get a pass, OSHA says it should be classified as an HNOC. For example, OSHA says a polymerization of monomers should be classified as an HNOC because:

> *Polymerization of monomers is a chemical reaction that is often exothermic (evolves heat). Loss of control of polymerization reactions can result in fires, explosions, or toxic releases. [Therefore] Polymerization is considered an HNOC because of these hazards.*[15]

OSHA has said its intent is that the HNOC classification will be an interim measure, used until harmonized criteria for the hazard being covered can be adopted at the UN subcommittee level, and incorporated into a new version of GHS.

In a change in guidance in late 2016, OSHA said it would allow the use of the exclamation mark pictograms for HNOCs, with the stipulations that "Hazard Not Otherwise Classified" or "HNOC" would appear beneath the pictogram. OSHA has also said it is working to harmonize HNOCs under HCS 2012 with Health Canada's treatment under WHMIS 2015.[16]

Most of the general health and physical hazard classes remain the same under GHS alignment, but new hazards, like Corrosive to Metals and Germ Cell Mutagenicity are added (and marked in this list with an asterisk): [17]

Health:
- Acute Toxicity
- Aspiration Toxicity
- Skin Corrosion/Irritation
- Serious Eye Damage/Eye Irritation
- Respiratory or Skin Sensitization
- *Germ Cell Mutagenicity
- Carcinogenicity
- *Reproductive Toxicity
- Target Organ Systemic Toxicity–Single and Repeated Dose

---

[15] OSHA *Letter of Interpretation*.
[16] SCHC 2016 *OSHA Presentation*.
[17] OSHA *HCS* 2012.

Physical:
- Explosives
- Flammable–Gases, Aerosols Liquids, Solids
- Oxidizers–Liquids, solids, gases
- Self-Reactive Substances
- Self-Heating Substances
- Pyrophoric–Liquids, solids
- Organic Peroxides
- *Corrosive to Metals
- Gases Under Pressure
- Water-Activated Flammable Gases

Added Pressure for Chemical Manufacturers

One consequence of these classification changes is that chemical manufacturers have been required to revisit their existing chemical products and author new, updated GHS-aligned SDSs. Unfortunately, many chemical manufacturers were unfamiliar and underestimated the amount of time authoring compliant safety data sheets would take. Some chemical manufacturers believed authoring new documents would take two or three hours per safety data sheet, only to find that, when they got started, the process took more like two to three days per SDS.

If there is a hidden benefit to this "rude awakening," hopefully it is that manufacturers authoring safety data sheets will be better prepared for the process in the future. OSHA has repeatedly expressed its intention to align with newer versions of GHS, as formulated by the UN. If OSHA does align to a new version of GHS in the near future, hopefully chemical manufacturers will have a better understanding of how much time they will need to devote to SDS authoring in order to be prepared and compliant. OSHA had previously hinted at the possibility of aligning to Revision 6.[18] However, as noted earlier, OSHA representatives mentioned Revision 7 as appearing attractive for realignment at a November 16, 2016 Stakeholder Meeting.

Chemical Blenders

The June 2015 deadline was challenging for chemical manufacturers that blend chemicals because they rely on information passed down from their upstream suppliers. Not all of those upstream suppliers were able to provide the information blenders needed to meet OSHA classification, label, and SDS deadlines. This left many blenders in a lurch, and affected their downstream customers that were on the lookout for updated SDSs and labels. In response to this delay, multiple HazCom guidance and enforcement policies were issued.

On February 9, 2015, OSHA issued *Interim Enforcement Guidance for the Hazard Communication Standard's (HCS) June 1, 2015 Effective Date* (later included in and subsumed by the OSHA Instruction *Inspection Procedures for the Hazard Communication Standard, July 9, 2015 Effective Date*).[19] In this enforcement memo, OSHA gave blenders new options. Specifically, OSHA said blenders might avoid being cited by OSHA inspectors, even if they were out of compliance in respect to SDSs and labels, if they could show that they made a "good faith effort" to do the following:

- Obtain classification information and SDSs from upstream suppliers;
- Find hazard information from alternative sources (e.g., chemical registries); and

[18] *Safety + Health* 2014.
[19] OSHA *Instruction*.

- Classify the data themselves using other resources

To make a case for this effort, blenders were and are encouraged to maintain as much evidence as possible to show their attempts to take the above steps. At a minimum, this evidence should include accounts of communications with suppliers, as well as with downstream customers about when such customers can expect updated information.

Distributors

In this same interim enforcement mentioned above, distributors were given until December 2017 to get their SDSs and labels updated if it was the case that upstream suppliers were the cause for the delay. The final enforcement guidance from July 9, 2015 maintains this and extends the special enforcement guidance to distributors that aren't blenders. Special enforcement was expanded to cover them as well, giving them the same allowances granted to chemical blenders, so long as the delay was rooted upstream and the other criteria applied. Going forward, even employers will want to keep an eye on this December 2017 deadline, given its applicability to more suppliers in the chain.[20]

Employers

Employers—that is to say, end-users of chemicals—were required to be in full compliance with GHS alignment by the June 1, 2016 deadline. To do this, they needed to review the five main elements of a compliant HazCom program and update their own programs where appropriate.

Let's review these five elements in more detail.

- **Written HazCom Plan:** The most important thing to know about this requirement is that the Written HazCom Plan must be specific to a facility. It cannot be something generic that is printed off the internet, with an employer's name written in at the top. OSHA expects to see (among other things) the name of the person responsible for the written plan, not just a job title, as well as the person(s) responsible for SDSs, training, and labels. Further, the plan should reflect how HazCom is actually being managed and by whom, and, as such, should be reviewed regularly and updated as needed. It should, for example, account for any changes taking place with inventory, or the process for managing SDSs, labels, or training, or when the personnel involved in HazCom-related responsibilities have changed.

- **Chemical Inventory List:** This requirement is a list of all the hazardous chemicals used or stored in a facility to which employees may be exposed. Assembling this list is a great opportunity to do a complete walkthrough of a facility, recording and cataloging every hazardous chemical. Remember, while chemicals used in the main line of business certainly need to be recorded, it's also necessary to include things like the hazardous chemical cleaners a maintenance staff uses and the chemicals a grounds crew uses. The chemical inventory list must also include chemicals that are no longer used, but are still stored onsite in the facility, since they still pose an exposure risk.

- **Proper Labels and Warnings:** OSHA requires a label be affixed or attached to every hazardous chemical container, with a few exceptions (such as stationary process containers and containers used in the transfer of chemicals remaining in the constant care of the person making that transfer). When it comes to shipped container labels, OSHA has prescribed six label elements that must be present in order to achieve compliance. When it comes to

[20] OSHA *Letter of Interpretation*.

workplace labels, on the other hand, the format is largely up to the employer. However, the workplace label must successfully communicate the chemical identity and the required health and physical hazard information. (Using workplace labels that have all the information found on shipped labels is an excellent way to do this!)

- **(M)SDSs and Right-to-Know Access:** Employers are required to provide barrier-free access to safety data sheets for every hazardous chemical onsite. This access must function so that safety data sheets are always available to employees in their work areas, during their work shifts. Employers can provide this access with electronic, cloud-based solutions that make safety data sheets available on computers or mobile devices, provided employees are trained on how to retrieve the information on their own.

- **Training:** Employees must be trained on the HazCom Standard and the specific hazards of the chemicals in their workplace. They must be retrained whenever they show signs of needing it.

Reasonable Diligence and Good Faith Efforts

As noted above, when inspectors find situations where a lack of compliance exists in regard to labeling or safety data sheets, they are directed to see if employers that are manufacturing chemicals have made "reasonable diligence and good faith efforts" to obtain updated information when considering a citation. In its July 9, 2015 directive, OSHA specifies that inspectors will look for evidence of an employer's efforts to:

- Obtain classification information and SDSs from upstream suppliers;
- Find hazard information from alternate sources, such as chemical registries; and
- Classify the data themselves.

If shipped container labels don't appear to be updated to the new GHS-aligned format, inspectors will look to see if manufacturers have:

- Documented a process for gathering updated classification information from upstream suppliers.
- Documented efforts to obtain the information from alternate sources.
- Maintained written, dated accounts of communication with upstream suppliers to try to obtain the updated information, as well as to distributors to inform them of why compliance has not been possible.
- Developed a course of action and timeline for making the necessary updates once information becomes available.

OSHA says it will evaluate all of these factors, but may consider one or any combination of them acceptable. It should be noted that OSHA also does not guarantee that these efforts will always mean that a citation is avoided.

Strong Clarifications about Enforcement

Throughout the process of aligning the HazCom Standard with GHS, OSHA has released enforcement directives that have further clarified how GHS alignment will be evaluated when it comes to labels and SDSs.[21] These directives have made clarifications that are still impacting businesses in a number of ways.

Electronic SDS Management

The July 9, 2015 enforcement directive contained explicit clarification that providing required employee access to SDSs, using an electronic provider can be compliant, but that relying on internet searches to obtain SDSs is not compliant. OSHA has always made the assertion that electronic SDS management solutions were compliant. However, some businesses took this to mean that using search engines to find safety data sheets would be compliant. In this clarification, OSHA spelled out in black and white that having employees perform internet searches on engines like Google or Yahoo is never a compliant means of providing SDS access. It puts the onus on the employee to find and evaluate the fitness of an SDS, which is actually the responsibility of the employer. In the same directive, OSHA reiterated that using an electronic SDS management provider continues to be an acceptable option for compliance.

An SDS Unique to Each Chemical Manufacturer

In the July 9, 2015 directive, OSHA also addressed businesses using a single SDS for the same chemical from different manufacturers. OSHA had found that some businesses with containers of the same chemical from different manufacturers were asking themselves if an SDS from one manufacturer might be able to cover or "count" for the others. If it's all the same chemical, they reasoned, wouldn't the SDS information all be the same, too? In its directive, OSHA said that, no, it's not acceptable to do this. SDSs from each chemical manufacturer must be maintained. To that end, OSHA clarified that each manufacturer from the SDS should correlate to the manufacturer on the label. This supports part of the reasoning for having a version in place for each supplier version of a chemical that's used: traceability and the ability to piece each container back to its SDS.

SDSs on Multi-Employer Worksites

The July 9, 2015 directive also addressed how multi-employer worksites should provide SDS information. In short, OSHA said employers must have a system in place to ensure workers have access to SDSs for all chemicals, both those originally on-site and those that may have been brought by visiting employers. For example, if a construction company is building a house, and it brings a paint crew on-site to paint the house, then any chemicals the paint crew brings should be disclosed to the construction company ahead of time so the construction company can ensure that SDSs for the paint crew's chemicals are present at the time the crew arrives. Further, when they arrive, the paint crew should be shown how and where to access SDSs, not just for the chemicals that they have brought, but also for the chemicals already used by the construction business. During an inspection, an OSHA inspector will want to see that a coherent protocol is in place to ensure this happens.

Training Temporary Workers on SDSs, Labels, and More

OSHA continues to focus on situations where temporary workers are present at a host employer facility. In the July 9, 2015 directive, OSHA makes it clear that *both* the host employer and the staffing agency are responsible for keeping temporary workers safe from chemical hazards, and for training them on SDSs and labels. In the directive, OSHA says both entities can be found

---

[21] OSHA *Instruction*.

culpable in the case of an incident. OSHA continues to issue additional directives concerning temporary workers as part of its Temporary Worker Initiative.[22]

North American GHS Alignment by Late 2018

It's also important to note, as we prepare for compliance in the following months and years, that all of North America is on schedule to be aligned with GHS by the end of 2018. This news has the potential to have substantial impact for the SDS and label requirements of companies that do business with companies in Canada and Mexico.

In Canada, alignment was announced in February of 2015. Canada has expressed a desire to coordinate its GHS alignment as closely as possible to that of the United States; this was articulated in the original publication in the *Canada Gazette,* and again in a United States–Canada Regulatory Cooperation Council Joint Forward Plan.[23]

For instance, regarding SDSs, the goal, as described by the Regulatory Cooperation Council (RCC), is to eventually have a single document that can be accepted by regulatory agencies in both countries. However, each country will have its own requirements. For example, in Canada, under WHMIS 2015, safety data sheets need to be in both English and Canadian French. Additionally, classification complications are likely to arise for the immediate future because Canada is basing its GHS alignment on Rev. 6 of the Globally Harmonized System, whereas, the United States is currently aligned with Rev. 3.

Canada is implementing GHS alignment in a three-stage transition that is scheduled to be completed by December 1, 2018; however, individual provinces and territories are able to postpone certain aspects of the new requirements beyond this date, and currently federally regulated workplaces are being given until May 31, 2019.[24]

In Mexico, alignment with GHS was announced in October 2015. Mexico had previously announced "GHS Guidelines" in 2011, but these were non-mandatory suggestions for how to become more GHS-aligned. If things proceed according to schedule, Mexico will implement full GHS alignment by October 2018.

With all of North America scheduled to reach full alignment with some version of GHS by the end of 2018, it's hopeful that the process of importing and exporting chemicals between the countries will become easier. However, if a business ships to or from these countries, they will need to keep an eye on how their implementation goes, looking especially for changes or delays in the months ahead, and ensuring that they're meeting the unique requirements for each country.

Step into Compliance

Whether a business was able to easily meet its deadlines during the GHS transition, or whether it *still* has come catching up to do, taking the following steps can help everyone position themselves for compliance and success for the rest of 2017 and beyond.

---

[22] OSHA, *Protecting Temporary Workers*.
[23] *Joint Forward Plan*, August 2014.
[24] *Canada Gazette*, 2015.

Designate a GHS Leader:

- Assign responsibility for keeping up with GHS to a specific person or persons. Remember, GHS is not a "one-and-done" event. OSHA has indicated its intention to continue reviewing newer versions of GHS, as formulated by the UN, and aligning with them as warranted
- Make sure the GHS leader keeps up with current events and any future directives that might have the potential to impact compliance in the workplace

Identify Person(s) Responsible for HazCom:

- Assign responsibility for overseeing the entire HazCom program, including ensuring new obligations flowing from GHS, are met on an ongoing basis.
- Ensure coverage for additional HazCom responsibilities, as addressed below.
- Confirm workplace systems and equipment are in place to handle current requirements.

Manage the SDS Library:

- Designate someone to be responsible for obtaining and maintain SDSs and related tasks.
- Have a procedure in place for acquiring missing SDSs:
    - Request missing SDSs as soon as possible; and
    - Document all related communications.
- Do not allow workers to use a hazardous chemical for which the SDS is not readily available.

Manage Labels:

- Designate someone to be responsible for workplace and/or shipped labels.
- Enact procedures for ensuring labels are correct, and for replacing labels if they are damaged or defaced.
- Have a process in place for contacting suppliers when a required label on incoming container is inadequate.

Maintain the Written Plan:

- Designate someone to be responsible for the written plan.
- Make sure the location of the written plan is known to all employees:
    - Maintaining the written plan electronically is acceptable, provided all employees have access.
- Keep the written plan updated, and treat it as the playbook for compliance.

Maintain a Chemical Inventory:

- Include a comprehensive chemical inventory list as part of the written plan.
- Build a comprehensive chemical inventory list by doing a physical inventory.
- Update the chemical inventory as necessary to keep it current.
- Ensure the chemical name on the inventory list corresponds with the one on labels and SDSs.
- Keep in mind that chemicals that were not considered hazardous under HazCom 1994 might now be considered hazardous under HazCom 2012; be sure the list accurately reflects post GHS alignment classification.

Train Employees:

- Designate someone to be responsible for training employees on HazCom.

- Train employees on HazCom rules in general, and on the specific hazards of the chemicals to which they are exposed.
- Don't let employees work with chemicals they haven't been trained on.
- Train on both SDSs and MSDSs if both exist in the safety data sheet library.
- Train on chemical hazards associated with non-routine tasks.
- Train on the written plan, the chemical inventory list, labels (shipped labels and workplace labels), and SDSs.
- Training must be delivered in a language the employees understand. It must also be delivered at a "level of understanding" appropriate to employees.
- Retraining must occur when employees when they no longer retain the necessary information.

Conclusion

HazCom excellence and GHS alignment is a marathon, not a sprint. Keeping this in mind is key to label and SDS compliance now and in the future. Yet as we prepare, it's important to remember why we're running this race in the first place: to keep employees and surrounding communities safe. Additionally, the harmonization of chemical hazard communication between countries makes it easier for businesses to ship chemicals between jurisdictions, while at the same time, standardization of information on SDSs and labels helps ensure workers throughout the chemical supply chain find the information they need quickly and efficiently to transport, use, store, and dispose of chemicals in a safe and responsible manner.

As this paper has identified, many challenges still exist related to GHS adoption, including: complications arising from stringent chemical classification, labeling, and SDS creation requirements; complexities related to countries adoption of different versions of GHS; and the incumbency of keeping employees trained on HazCom and its evolving mandates. The good news is that there are more resources than ever before to help businesses stay in compliance and adopt industry best practices. By keeping a focus on *why* we have taken the trouble to align with GHS, and by using the resources that OSHA and others have made available, chemical manufacturers, distributors and employers can work in concert to reduce risk, improve communication and keep one another safe.

Bibliography

Canada Gazette. *Statutory Instruments 2015* (February 11, 2015) (http://gazette.gc.ca/rp-pr/p2/2015/2015-02-11/pdf/g2-14903.pdf)

Occupational Safety and Health Administration. 2012. 29 CFR 1910.1200, *Hazard Communication Standard* (retrieved February 19, 2017) (http://www.osha.gov/pls/oshaweb/owadisp.show_document?p_table=standards&p_id=10099)

Occupational Safety and Health Administration, *HazCom Stakeholder Meeting*, November 16, 2016. (https://www.osha.gov/dsg/hazcom/public_meeting_announcement_20161116.html)

Occupational Safety and Health Administration. 2015. *Inspection Procedures for the Hazard Communication Standard.* (retrieved February 19, 2017) (https://www.osha.gov/OshDoc/Directive_pdf/CPL_02-02-079.pdf)

Occupational Safety and Health Administration. 2016. *Letter of Interpretation: National Association of Chemical Distributors [1910.1200]*. (retrieved March 6, 2017) (https://www.osha.gov/pls/oshaweb/owadisp.show_document?p_table=INTERPRETATIONS&p_id=30305)

Occupational Safety and Health Administration. 2013. *Letter of Interpretation: NIST Labeling of Small Packages [1910.1200]*. (retrieved February 19, 2017) (https://www.osha.gov/pls/-oshaweb/owadisp.show_document?p_table=INTERPRETATIONS&p_id=28805)

Occupational Safety and Health Administration. 2013. *Letter of Interpretation: Pictogram requirements for product labels under the revised Hazard Communication Standard [1910.1200]*. (retrieved February 19, 2017) (https://www.osha.gov/pls/oshaweb/-owadisp.show_document?p_table=INTERPRETATIONS&p_id=28646)

Occupational Safety and Health Administration. 2015. *Letter of Interpretation: Nichole Shoshensky, Authoring Team Lead, MSDSonline*. (retrieved March 8, 2017) (https://www.osha.gov/pls/oshaweb/owadisp.show_document?p_table=INTERPRETATIONS&p_id=31022)

Occupational Safety and Health Administration. 2013. *OSHA Brief, Hazard Communication Standard: Labels and Pictograms* (retrieved February 19, 2017) (http://www.osha.gov/Publications/OSHA3636.pdf).

Occupational Safety and Health Administration. 2012. *OSHA Brief, Hazard Communication Standard: Safety Data Sheets* (retrieved February 19, 2017) (http://www.osha.gov/Publications/OSHA3514.html).

Occupational Safety and Health Administration. 2015. *OSHA Instruction: Inspection Procedures for the Hazard Communication Standard, July 9, 2015 Effective Date* (retrieved February 19, 2017) (https://www.osha.gov/OshDoc/Directive_pdf/CPL_02-02-079.pdf)

Occupational Safety and Health Administration. 2013. *Protecting Temporary Workers*. (retrieved February 19, 2017) (https://www.osha.gov/temp_workers/index.html)

Occupational Safety and Health Administration. 2014. *Small Entity Compliance Guide for Employers That Use Hazardous Chemicals* (retrieved February 17, 2017) (https://www.osha.gov/Publications/OSHA3695.pdf)

Occupational Safety and Health Administration. 2013. *Standard Interpretations*. (retrieved February 17, 2017) (https://www.osha.gov/pls/oshaweb/owadisp.show_document?p_table=INTERPRETATIONS&p_id=28805)

Office of Information and Regulatory Affairs. 2015. *Current Regulatory Plan and the Unified Agenda of Regulatory and Deregulatory Actions* (retrieved March 8, 2016) (http://www.reginfo.gov/public/do/eAgendaMain?operation=OPERATION_GET_AGENCY_RULE_LIST%C2%A4tPub=true&agencyCode=&showStage=active&agencyCd=1200)

Rudman, Sven. *SCHC 2016 OSHA Presentation*, Fall Meeting Presentation, September 27, 2016. (http://www.schc.org/index.php?option=com_jevents&task=icalevent.detail&evid=7)

Safety + Health. 2014. *Latest OSHA agenda hints at update to hazcom standard.* November 24, 2014. (http://www.safetyandhealthmagazine.com/articles/11484-latest-osha-agenda-hints-at-update-to-hazcom-standard)

Safety + Health. 2016. *OSHA's Top 10 Most Cited Violations for 2016* (retrieved February 14, 2017) (http://www.safetyandhealthmagazine.com/articles/14927-2016-oshas-top-10-most-cited-violations)

United States-Canada Regulatory Cooperation Council. *Joint Forward Plan,* August 2014. (http://www.trade.gov/rcc/documents/RCC_Joint_Forward_Plan.pdf)

Session No. 714

Leveraging JHAs Beyond EHS Compliance: The Value of JHAs

Ashley Kunz, CSP
EHS Manager
Boise, Idaho

Mollie Anderson, MSSM
Principal Consultant
BSI EHS Services and Solutions
Hillsboro, Oregon

Introduction

To remain competitive, manufacturers are challenged to be nimble with necessary technology changes, including fast integration of new equipment and processes brought into manufacturing sites on a regular basis. As a result, site management must implement efficient and sustainable change management business processes that continually evaluate, update, and periodically reassess to ensure the safe operations and maintenance of the equipment and facilities. A strong job hazard analysis (JHA) process integrates well, and creates environmental, health, and safety (EHS) value, with necessary business change-management processes in modern manufacturing and research and development (R&D) sites.

JHAs are often used towards compliance to Occupational Safety and Health Administration (OSHA) workplace hazard assessment requirements (Title 29 CFR 1910.132(d)(2). However, as EHS professionals are keenly aware, this can also result in mediocre EHS assessments at best when JHAs are conducted as stand-alone compliance tools, and those which are quickly out of date. JHAs run the risk of being viewed as only a compliance tool, and typically subject to becoming just another passing EHS phase.

This paper is based on a confidential JHA project conducted at a global, process-intensive manufacturing site. This manufacturing site as well as other sites within the company use JHAs as a standard business process to meet regulatory compliance requirements and drive risk reduction. Due to the rapid pace of new process and equipment introductions supporting manufacturing and R&D equipment, the JHAs were quickly becoming obsolete, and a formal process for updating the assessments was not in place. A management decision was made to implement a robust management process to ensure that JHAs not only met regulatory requirements, but also remained current to match the pace of business change.

Cost Burden of Injuries

The escalating cost of injuries is increasingly significant to these industries, too, and to the potential for business interruption. Poor change management practices further contribute towards the growing cost of injuries. Safety management of worker protection and injury/illness prevention has historically been approached from two perspectives: (1) "*Reactive,*" where the focus is to rectify existing problems that lead to significant costs due to injuries; and (2) "*proactive,*" where the focus is on identifying potential concerns in advance of installation.

The 2017 Liberty Mutual Workplace Safety Index cites the top four injury causes as accounting for nearly 50% of the total cost burden in 2014. *Overexertion involving outside sources* was ranked first among the leading causes of disabling injury. This event category, which includes injuries related to lifting, pushing, pulling, holding, carrying, or throwing, cost businesses $14 <u>billion</u> in direct costs and accounted for 23% of the overall national burden.

Increasing in frequency and costs over the previous year, *falls on same level* and *falls to lower level* rank second and third, respectively. With direct costs of $16 billion, these *fall* categories together accounted for the next 27% of the total injury burden. *Struck-by object or equipment* ranked fourth at $4 billion in costs and 7% of the cost burden.

A JHA project is appropriate for assessing these types of injury events.

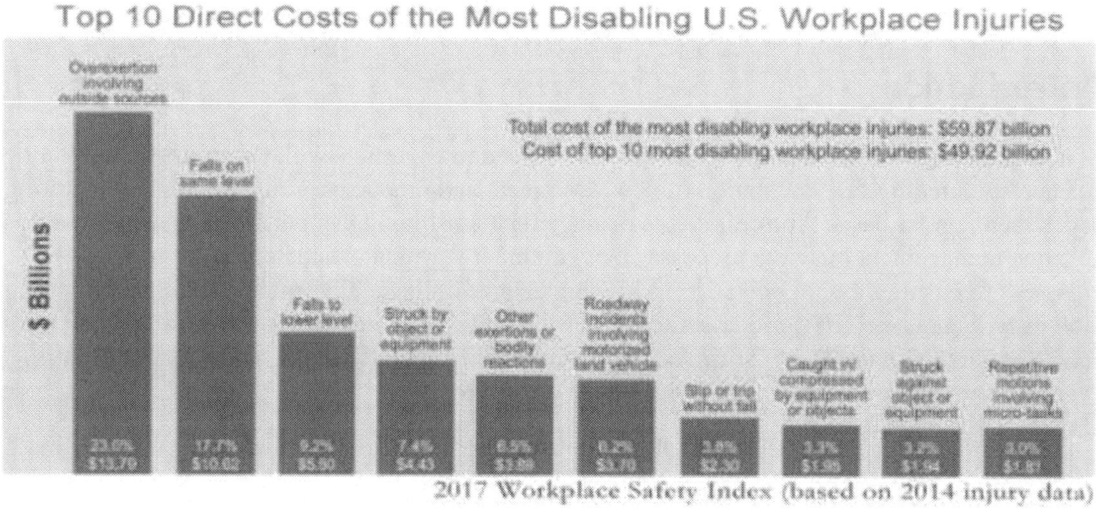

Figure 1. Liberty Mutual 2017 Workplace Safety Index (based on 2014 data).

<u>Understanding JHAs and Their Business Applicability</u>

A JHA project can be narrow in scope or be very broad. It has a wide range of applicability across industries, including high technology/semiconductor, oil & gas, bio-pharma, general industry, construction, government, municipal utilities, and global applications. Common within these differing industries, forward-thinking and socially responsible business is committed to conducting its operations in a manner that protects the health and safety of its workers, sustains the environment, and reduces business interruptions as a key part of its corporate programs.

Some of the primary objectives of a solid EHS program include a systematic and thorough review of potential hazards for operation or maintenance of an activity or procedure, and ensuring credible incident potentials and hazards are not overlooked. A JHA project provides an excellent means by which to complete this thorough review. The results of a JHA project provide benefits and value to the business, and ultimately, to providing for safer, more reliable procedures for operating and maintaining equipment and facilities, lending to safe, environmentally responsible, and efficient manufacturing.

The JHA may be for a single tool, a system, or a complete facility, and provides hazard controls based on frequency, severity, and probability of occurrence. While this may be perceived as resource-intensive, it does provide a means to prioritize hazard and risk control areas and implement corrective actions and requirements. Updating JHAs periodically, as defined by the needs of the company, is not an uncommon practice where there is a desire and a need to improve. Business commitment towards the necessary internal and external resources is important to comprehensively evaluate the business across the organizational lines.

Business commitment also focuses not only on the JHA process itself, but also about how JHAs become relevant for business, and their results viewed as an impactful and beneficial business opportunity. Delving deeper, these benefits also include additional ways in which JHAs have business applicability as being beneficial beyond compliance:

1. JHAs should not only identify procedural steps and hazards, but more effective risk controls for worker protection. Poorly designed protective measures do not adequately reduce risk. As an example, a light curtain may be needed for work efficiency, but improperly designed, has greater worker injury potential.
2. Increased efficiency is also seen with the reduction in time and fewer procedural steps. Increased efficiency is clearly advantageous, but it's important to note that JHAs would also pinpoint the proper execution of these steps, leading to an equal increase in safety.
3. Greater engineering and substitution opportunities are presented with JHAs, and may reduce the consumption of raw materials to support an operational or maintenance practice.
4. Business interruption potential may also be decreased where the results of a JHA project document proper procedures. As an example, an improperly executed maintenance procedure may produce an unsafe condition, such as early warning of an electrical short and smoke detection reinforces the broader view of JHAs, also focusing on the resultant business interruption, where smoke can have an impact to very sensitive product.

JHAs: A Supporting Component of Integrated Hazard Analysis

The integrated hazard analysis process looks at assessing the regulatory and industry codes and standards, including a company's best-known methods (BKMs) and safety guidelines. These results are then melded with the results of the integrated hazard analysis, including specific derived requirements, and ultimately, back to the JHAs. Thus, JHAs are a supporting component of an overall integrated hazard analysis, and resulting in standardization on EHS approach and validation.

This section further examines these areas of EHS regulatory and jurisdictional requirements, and requirements derived from hazard analyses as melded with the integrated hazard analysis.

EHS Regulatory and Jurisdictional Requirements

As part of the integrated hazard analysis, including the JHA project component, applicable regulatory codes and requirements are reviewed, and research completed on hazards and industry best practices.

An integrated analysis will assess equipment, processes, and installation/facilitization specifications against multiple requirements and standards. Included here are federal, state, and local EHS regulatory and jurisdictional requirements, industry standards and consensus guidelines, the company's internal standards and best practices, as well as external requirements specific to customers.

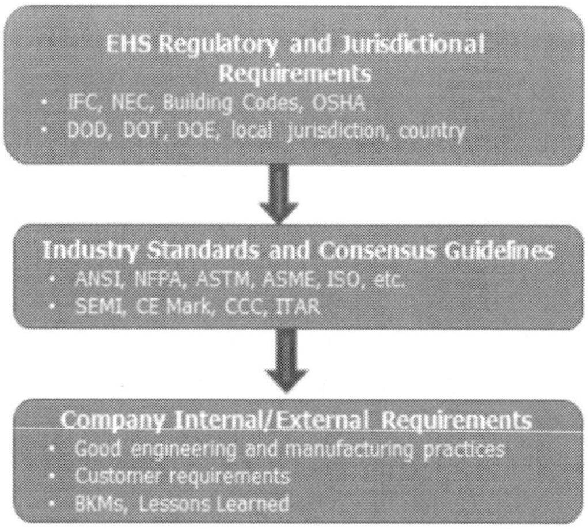

Exhibit 1. EHS Regulatory and Jurisdictional Requirements.

Hazard Analyses and Derived Requirements

These regulatory, industry, and company requirements are then overlaid with the other components within the integrated hazard analysis. This process also looks to recognizing the peripheral influences to the business, including the site, business, and external requirements of the internal and external customer. Managing risks proactively with EHS risk management considerations results in systems and procedures or tasks, which have undergone appropriate hazard analyses during design through all anticipated operational and maintenance activities.

This effort culminates in a formal JHA for each operations or maintenance procedure, with recommendations for mitigation or risk controls for each identified hazard within each maintenance step. A JHA can also be important for pinpointing specific derived requirements for operational or maintenance procedural planning. This also leads to enabling a prioritization of hazard and risk areas, and their associated controls and mitigations. Injury frequency and severity reduction is also evident, and certainly the benefit of these not being repeated or at least the severity significantly reduced.

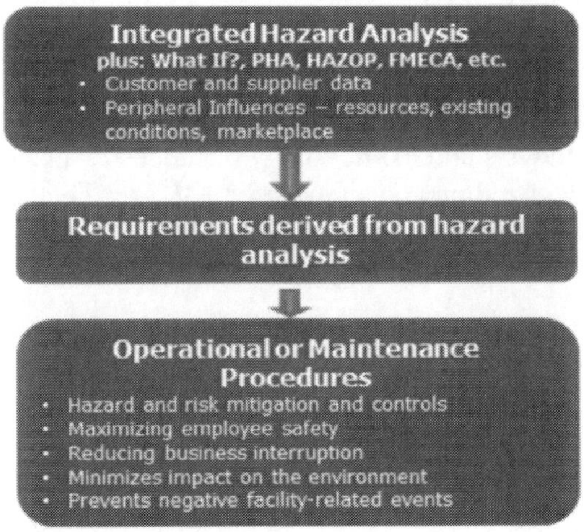

Exhibit 2. Hazard Analyses, Derived Requirements, and Improved Procedures within an Integrated Hazard Analysis.

In summary, the integrated hazard analysis is applicable within the overall business risk process, with JHAs as an integral component. With this process in place, the company's management of change strengthens the proactive EHS practices of business operations, both up and down the responsible supply chain. Companies may also leverage their responsible supply chain design for EHS programs, which is particularly important, as it includes provisions to hold suppliers, subcontractors, and even customers accountable for safe and compliant design and delivery.

This proactive and integrated approach to EHS in the business environment provides for a responsible supply chain, and a *safe* equipment, process tool, or facility integration, without the often costly production startup delays, rework of equipment or facilities, and significant labor cost to remedy the identified risks, and potential for personnel injury.

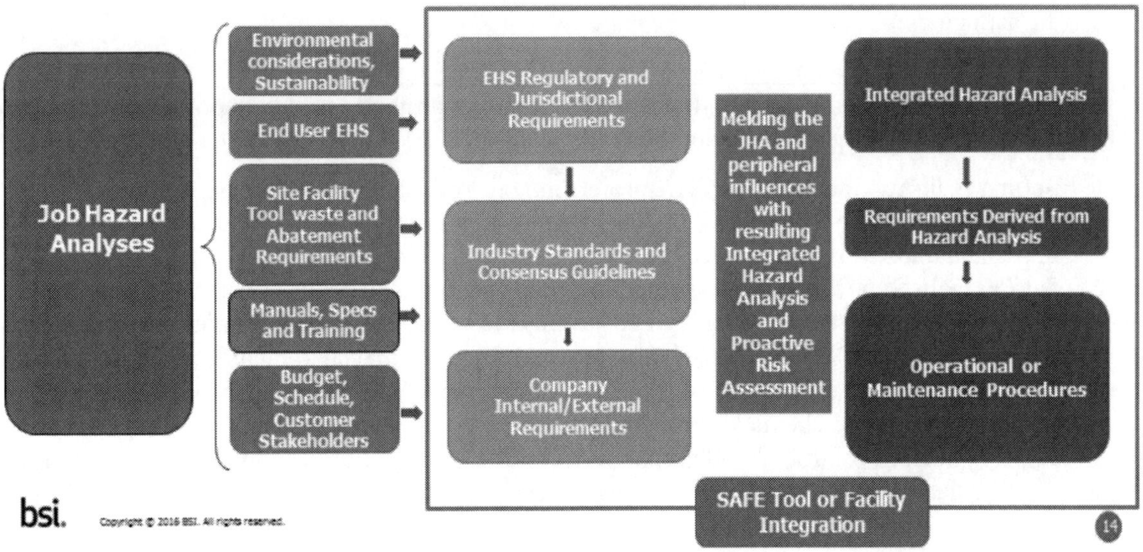

Exhibit 3. Methodology of Process: JHAs as a Component of Integrated Hazard Analysis.

Optimizing the JHA Process: Management Understanding and Support

The principles reviewed in the previous section address the objectives and benefits of a JHA project, plus the utilization of an integrated hazard analysis and methodology of process. This section focuses on the keys to gain management understanding and support for a JHA project, or in other words, 'pitching' to management for support, approval, and funding for a JHA project.

What's the 'Pitch'?
The first step in gaining understanding and support is to provide solid project advantages. These may include:

1. Supporting component to integrated hazard analysis
2. Proactive focus on EHS
3. Good corporate citizen
4. Better-improved EHS protection
5. Minimizes risk to business interruption
6. Prioritized risk control, minimizes downtime
7. Identifies hierarchical business risk from process and facilities risk
8. Production: reduces time and steps

What's the 'Pitch'?
Careful consideration and upfront discussion of several aspects:

1. Careful planning and a path to execution. Prioritization of work practices and procedures typically performed by the company, based on their specific need, such as for facilities.
2. Resource allocation. Development of a project scope within the resources available, be it internally or by contract.
3. Understanding of applicable codes and industry consensus standards. Also, the inclusion of results from the integrated hazard analysis:
    a. Understanding of a higher number of recommendations for each identified hazard. As example, there may not be a single best safeguard, but rather a prioritization of safeguards.

Obstacles to Understanding: Showing Value of Proposed Solutions
Showing the value of proposed solutions and overcoming obstacles is also important by providing realistic timeline and budget considerations:

1. Setting realistic timelines. The JHA project may not be just immediate suggestions of safeguards, but also those which may take time for phase-in. This includes follow-up and implementation of recommended mitigations and control.
2. Setting realistic budget considerations. Included is a potential need for capital expenditures and understanding the potential high cost of injuries. This latter includes not only direct costs, but a business reputational risk. These proposed solutions may also include:
    a. Substitution or elimination of processes or equipment. As example, the elimination of a hand-pouring of chemicals procedure.
    b. Engineering controls vs. administrative controls. As example, the installation of local ventilation vs. the continued wearing of respirators.

Facilitating an Open Forum Towards Culture Change

A key, yet less obvious, benefit from a JHA project, is the facilitation towards an open forum towards culture change. This value is most notably acknowledged in forward-thinking companies, but can also be subtler in benefits.

Different points of view and expertise can generate additional ideas. These, in turn, contribute towards understanding of existing hazards and a brainstorming of recommended safeguards for the system, equipment or tools, or may identify previously missed hazards.

In all, an open forum broadens management and crew understanding towards a positive culture change. It may also bolster employee confidence that hazards, perhaps previously unaddressed, can now be recognized and controlled or eliminated. This will lead to an increased awareness, transparency, participation, management support, and thereby a resulting reduction in risk and the often the costs of incidents.

Mapping the JHA Process to the Business

Understanding the JHA process itself is important, and often vital to the project success; this is termed a mapping of the process to the business, whether it be R&D, manufacturing, facility services, etc. This, too, serves to engage top management and the affected crew leaders. Frequent status meetings and presentations serve to continue this engagement and bring buy-in through consistent technical reviews and discussions. The prioritization of work practices and perceived risk of tools, processes and hazards are also discussed as part of the efficient process work flow.

Also prior to conducting the JHA, tasks and procedures are initially prioritized by perceived risk of tools, processes, and hazards for performing various procedures. JHAs are then conducted for each operation or maintenance procedure by observing work practices, by interviewing workers, supervisors and engineers, and by reviewing maintenance procedures and specifications.

Optimizing the JHA Process: The JHA Process Flow

Every JHA project can be different and tailored to the company's facility, equipment, and processes. Follows are eight general steps from start to finish:

1. Prioritization:
   - Prioritization of work practices and procedures:
     - Based on specific need: facilities, operations, etc.
   - Risk rank procedures or tools and areas:
     - A: Higher risk/hazard exposure
     - B: Medium risk/hazard exposure
     - C: Low risk/hazard exposure
   - Prioritize based on:
     - EHS knowledge
     - Prior incidents or near misses
     - Workers' compensation injuries
     - Perceived risk
     - Project schedule
     - Knowledge and experience of key personnel towards hazard potentials
2. Specification and requirements review:
   - Review of the process and tasks associated with the job tasks or procedures:

- Process
- Operational
- Maintenance
- Review of applicable specification, regulatory, and derived requirements
- Identify existing safety precautions
- Identify existing control methods
- Review tool manufacturer's safety specifications

3. Observe:
   - Schedule in advance with tool owners and supervisors:
     - Consider scheduled down-time and critical processes
   - Observe procedures from beginning to end:
     - Document work practices, all hazards, current controls
     - Take pictures
   - Interview tool owners, technicians, operators, maintainers:
     - Previous and current safety concerns and near misses
     - Ask clarifying questions:
   - Compare work performed to procedures specified
     - Identify gaps, shortcuts, or non-documented work steps
     - Look for inefficiencies and areas for improvement

4. Fill out JHA:
   - Complete JHA documentation
   - Develop recommended controls based on hierarchy of precedence
   - Interview management, engineers, line supervisors:
     - Ask clarifying questions
     - Discuss feasibility of recommendations
     - Discuss missing information
   - Revise steps and recommendations as needed for accuracy
   - Perform risk ranking of findings to prioritize controls and recommendations
     - Use or develop a defined risk matrix to create risk scores

5. Presentation:
   - Present to management, engineering, EHS, line supervision
   - Summarize findings and recommendations:
     - New or alternative control methods for each identified hazard
   - Provides understanding and buy-in of JHA process:
     - Work practices and procedures
     - Present hazards and recommended controls
     - Scope and resources

6. Review and Rank
   - Open forum for discussion:
     - Generates ideas
     - Augments hazards definition, recommendation
     - Implementation strategy
   - Review of recommendations and rank:
     - Feasibility
     - Estimated completion time
     - Example: A process/chemical change may take years

- PPE change can be implemented immediately today

7. Implementation:
   - Schedule for implementation, based on:
     - Hazard severity or final risk ranking
     - Resources available
     - Return on investment
     - Completion time for the recommendation
   - Management of change, and process
   - Follow-up
8. Finalize:
   - Following implementation: verify all recommended actions:
     - Update JHA
     - Recalculate risk
     - Perform periodic reassessment of potential recommended control options
   - Schedule regular interval reviews
   - Update based on procedural change, tool or process change, etc.

Risk Assessment and Solution-based Risk Ranking

A broad yet primary objective of risk management within any given industry, and a key component to an effective EHS program, is to identify and eliminate or control hazards. Best practices have demonstrated that an organization should aspire to design out hazards at the highest feasible order of precedence; then proceed to consider engineering controls or substitution; and lastly, administrative controls, such as procedures and personal protective equipment (PPE). A procedure is an administrative control, whereas a JHA is a formal documented hazard analysis.

An integrated hazard analysis, including the JHA component, includes rationale for mitigation/controls selection, hazard ranking, and other components perhaps not found nor verified in a procedure.

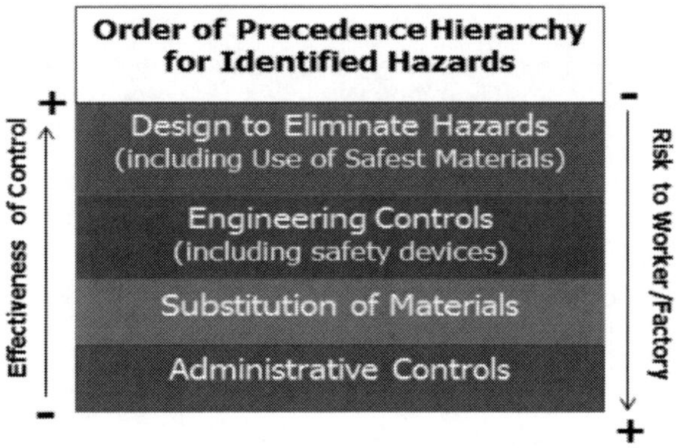

Exhibit 5. Hierarchy of Precedence.

Just as the above hierarchy of controls is molded and applies to differing companies and their systems, so too is the risk assessment and ranking process. There are many risk assessment matrices used throughout industry. The categories and definitions may change slightly, but basically, these matrices compare the severity of the injury or illness to the likelihood or probability of occurrence or exposure.

Below is an example of a traditional risk ranking model from ANSI Z10, *Occupational Health and Safety Management Systems*; however, this is also typical, with some modifications, to other frequently used risk ranking models, such as those based on SEMI S-10, *Safety Guideline for Risk Assessment and Risk Evaluation Process*, and DoD Mil-Std-882, *Standard Practice System Safety*.

			Severity of Injury or Illness			
			Catastrophic (CAT) Death or permanent total disability (Unable to return to work)	Critical (C) Disability in excess of 3 months Hospitalization of at least 3 people per event	Marginal (M) Minor injury, lost workday incident	Negligible (N) First aid or minor medical treatment
			4	3	2	1
Likelihood/Probability of Occurrence or Exposure	Frequent (F)	5	High 20 Operation not permissible	High 15 Operation not permissible	Serious 10 High Priority Remedial Action	Medium 5 Take Remedial action at appropriate time
	Probable (P)	4	High 16 Operation not permissible	High 12 Operation not permissible	Serious 8 High Priority Remedial Action	Medium 4 Take Remedial action at appropriate time
	Occasional (O)	3	High 12 Operation not permissible	Serious 9 High Priority Remedial Action	Medium 6 Take Remedial action at appropriate time	Low 3 Risk Acceptable, Remedial action discretionary
	Remote (R)	2	Serious 8 High Priority Remedial Action	Medium 6 Take Remedial action at appropriate time	Medium 4 Take Remedial action at appropriate time	Low 2 Risk Acceptable, Remedial action discretionary
	Improbable (I)	1	Medium 4 Take Remedial action at appropriate time	Low 3 Risk Acceptable, Remedial action discretionary	Low 2 Risk Acceptable, Remedial action discretionary	Low 1 Risk Acceptable, Remedial action discretionary

Exhibit 6. Example of Traditional Risk-ranking Categories.

	Likelihood	Frequency of Exposure	Frequency of Occurrence
F	Frequent	Task is performed several times an hour, and/or duration may approach at least 4 hours in a day.	Injury or has occurred repeatedly, within the company or at this site.
P	Probable	Will probably occur or task is performed several times a day, duration may approach 1-4 hours a day.	Has happened at this location recently.
O	Occasional	Is performed several times a day, or in typical durations under one hour.	Has happened in the company recently.
R	Remote	Not likely to occur, or task is performed less than one or two times a day, or duration may be under a few hours a month.	Injury occurred in the industry.
I	Improbable	May occur only under exceptional circumstances, or so remote as to be near zero in probability of exposure.	Have not heard of this occurrence.
FSI			Occurrence is NOT calculated for an Fatal and Serious Injury (FSI) potential as the frequency of occurrence is not a valid predictor of likelihood or Probability.

Risk Rating			Remedial Action	Examples
L (Low)	Acceptable	1-3	Remedial Action may not be necessary	These risks may rely more on warnings, administrative, PPE and other devices that may require operator intervention.
M (Medium)	Tolerable	>3-7	Remedial Action at an Appropriate time	Controls such as elimination, substitution, isolation and barriers are still preferable. These risks may rely more multiple Administrative and PPE controls.
S (Serious)	Tolerable	>7-11	Remedial Action on a Priority Basis	Controls such as elimination substitution and engineering controls are preferable. If reliance on warnings and administrative, these should be redundant to additional controls, or additional barriers, guards and other protective devices. (Also, monitor controls based on severity levels)
H (High)	Not Acceptable	>11	Immediate Remedial Action is Expected	Use controls or multiples of controls (defense in depth), such as elimination, substitution or engineering controls, controls with built in redundancies, physical devices that do not require adjustment or operator intervention, or provide positive, ongoing indication of operation. (monitor controls)

Exhibit 7. Example of Traditional Risk-ranking Score Matrix.

Basically, the traditional risk ranking provides an estimation of severity, frequency, probability or likelihood of occurrence. These can be numerical and calculated risk values, or they can be more qualitative. Essentially, though, the severity is multiplied by the frequency or likelihood with the product a number determining risk acceptability. Looking deeper, probability and frequency are synonymous terms within the risk assessment world. Probability or frequency is an estimation of how often the failure event may occur in the absence of controls. Assigning separate values to frequency and probability, and using these as individual values in the risk calculation, greatly diminishes the severity aspect of the failure mode. A prime example is a case where low frequency and low probability values are estimated that, for even a catastrophic hazard, will result in an acceptable risk per this approach.

As example for high scores of minor injury (severity) and high frequency: Light curtain as safeguard–using a hand tool to push a tray with an integrated circuit board into processing tool, several times an hour.

- High scores for perceived "low-risk" tasks
- Frequently occurring & low severity:
  - Severity: 2
  - Frequency: 5
  - **Result: 10**
  - **Hazard: Serious**
  - **High-priority remedial action**

As example for low scores with a severity of disability > 3-mos, but only performed remotely or less than a few hours each month:

- Working at heights, and potentially falling from a ladder
  - Low scores perceived "high-risk" tasks
  - Less frequent & high severity
  - Severity: 3
  - Frequency: 2
  - **Result: 6**
  - **Hazard: Tolerable**

An EHS professionals and, intuitively, addressing the potential fall from a ladder should take precedence in mitigating and controlling the action, then using a hand tool to push a tray with an integrated circuit board into processing tool several times an hour.

An alternate approach, which is developing a customized risk matrix, or what we term a solution-based risk ranking, as shown below, adds a 3rd variable: Probability. This approach has shown to provide:

- Increased level of accuracy
- Better matches common sense
- Easier for management understanding and implementation of prioritized safeguards
- May modify for each company, as needed

Severity Categories

Severity Group	Health Hazard Potential
5-Catastrophic	Fatality or multiple persons severely injured
4-Major	Long-term disabling injury or illness
3-Severe	Offsite Medical Treatment needed with loss work time or restricted work activity
2-Moderate	Minor Medical Treatment may be required
1-Minor	Nuisance or First Aid Injury only

Frequency Categories

Frequency Group	Frequency of Task
5-Frequent	10 or more times per shift
4-Likely	Once per shift
3-Possible	Weekly
2-Rarely	Monthly
1-Unlikely	Quarterly or Yearly

Probability Categories

Probability Group	Probability of Exposure
5-Very Likely	Exposure expected or required for task
4-Likely	Exposure likely
3-Possible	Exposure possible
2-Unlikely	Exposure possible, but unlikely
1-Very Unlikely	Exposure highly unlikely or impossible

Exhibit 8. Three-tier Risk Assessment Categories.

Risk Score Matrix

	5-Frequent	4-Likely	3-Possible	2-Rarely	1-Unlikely
5-Catastrophic	125	80	45	20	5
4- Major	100	64	36	16	4
3-Severe	75	48	27	12	3
2-Moderate	50	32	18	8	2
1-Minor	25	16	9	4	1
	5-Very Likely	4-Likely	3-Possible	2-Unlikely	1-Highly Unlikely

Critical	Major	Severe	Moderate	Low	Minor
125-80	79-48	47-27	26-16	15-5	4-1

*Risk Score = Severity (S) x Frequency (F) x Probability (P)

**Tasks will be reviewed and mitigated based on risk score. Critical Hazard tasks will be first, Major Hazard Tasks will be next, and so on.

***Any task with a Risk Score greater than 27 is considered "High Risk"

Exhibit 9. 3-Tier Risk Assessment Score Matrix.

Without this third variable, many tasks believed to be a lower risk are calculated as high, and vice versa. By adding in the even probability of occurrence to the traditional risk matrix increases the level of accuracy for the hazard and risk identified. Using our same examples, a more reflective risk ranking results:

- High scores for perceived "low-risk" tasks (using tool to push IC tray into tool)
  - Frequently occurring & low severity
  - Severity: 2
  - Frequency: 5
  - Probability: 1
  - **Result: 10**
  - **Hazard: Low**

- Low scores perceived "high-risk" tasks (working from heights on a ladder)
  - Less-frequent & high severity
  - Severity: 3
  - Frequency: 2
  - Probability: 4
  - **Result: 24**
  - **Hazard: Moderate**

Of course, this may also be modified for each company to fit their specific needs, provided it remains consistent across the company.

Summary

Simply put, to conduct a JHA, tasks and procedures are initially prioritized by perceived risk of tools, processes and hazards for performing various procedures. JHAs are then conducted for each operation or maintenance procedure by observing work practices, by interviewing workers, supervisors and engineers, and by reviewing maintenance procedures and specifications.

This proactive and integrated approach to EHS in your business environment provides for a responsible supply chain, and a SAFE tool or facility integration, without the often-costly production startup delays, rework of equipment/facilities and significant labor cost to remedy the identified risks, and potential for personnel injury. Actual results are demonstrated where significant reductions were achieved from the severe/high potential risk at the time of the initial JHAs to moderate/low potential risk for post JHA recommendations.

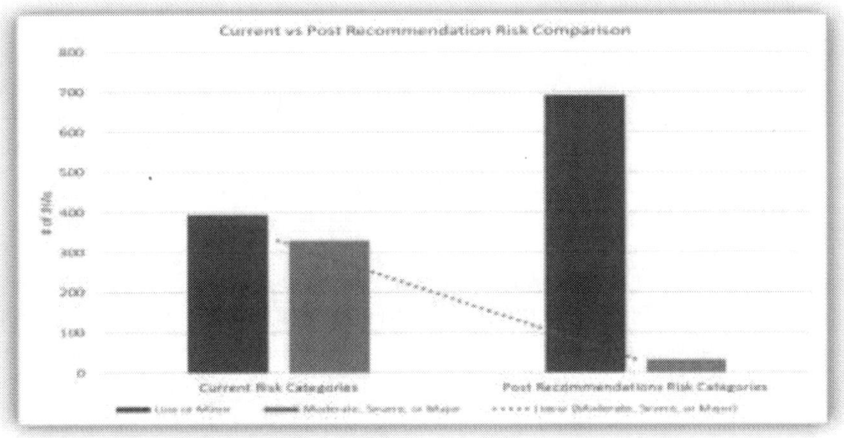

Exhibit 10. Overall Risk Reduction Results.

The JHA process can demonstrate to the employees and management, through the methods described above, that safety is not just a concept, but a fundamental tenant of a company's structure and philosophy. Using the JHA process as outlined in this paper can provide an introduction towards a culture change throughout all tiers of an organization or strengthening an organizational resilience.

Thus, EHS is strategic to the organization. When organizational resilience is viewed from a company-wide perspective, as below, it's clear how EHS supports the three organizational resilience domains:

1. Operational resilience
2. Information resilience
3. Supply chain resilience

Exhibit 11. Optimizing Operational EHS Compliance.

Acknowledgements

Special acknowledgement to Andrew McIntyre, CIH, Executive Vice President and Managing Principal, BSI EHS Services and Solutions, for his wide-ranging inputs on this topic. His support and feedback during the development of this article have been greatly appreciated.

In addition, our thanks to:

- Confidential client and BSI team at confidential client site
- Zevenbergen, Brian and Andrew Dagnan. "Effective Job Hazard Analyses Implementation – Engaging Management and Ranking Risks by Probability." *Presented at the SESHA Symposium 2012.*
- Manuele, Fred. 2013. *On the Practice of Safety*, 4th ed. Hoboken, NJ: Wiley.

Session No. 716

Goal Setting for Continuous Safety Improvement

Michael Fackler, MS, CSP, ARM
Eastern Alliance Insurance
Carmel, Indiana

Introduction

While it is easy to understand ***why*** an organization would want to improve safety performance, it is much more challenging to prescribe ***how*** an improvement will be achieved. Traditionally, organizations have measured the success of their safety improvement efforts by how well they avoid and prevent employees from getting injured, as evident in the various accident-based metrics used to measure and compare and contrast "safe" organizations from "un-safe" ones. Success or failure often hinges upon the absence of adverse events. Efforts to improve safety are typically based on improving some accident-based metric, and often with little consideration as to how these goals will be achieved.

Narrowly focusing on improving accident-based metrics may result in some marginal safety improvement, but more likely will lead to the active concealment and underreporting of hazards and employee injuries (United States Dept. of Justice, 2013) . Accident-based metrics are poor indicators of safety performance and just do not provide the insight necessary for organizational leadership to make informed decisions regarding safety and risk. A more effective approach to achieving sustainable safety improvements is needed.

Through the process of incorporating formal goal-setting mechanisms into existing management processes, management can move beyond simple reductions in accident-based metrics and shift toward the implementation and integration of management processes and systems that bring about a greater clarity and focus to an organization's safety management and improvement efforts.

The Issue with Accident-based Metrics

As early as 1996, influential safety pioneer Dan Petersen wrote how measuring performance through accident based metrics was a "waste of time" and "meaninglessness." (Petersen, 1996, pp. 15-33; Petersen, 2001) Petersen went on to advocate that organizations should use "anything but accident based metrics" to measure the performance of management, and instead focus on the active participation in activities designed to improve safety performance.

Let me be clear; results matter! Preventing injuries and accidents matter! Demonstrating measurable improvement in our ability to prevent accident and injuries, through recognizing and reducing risk, matter! Organizations should never abandon their "zero-injury" mindset, yet be mindful that having a "zero-injury" mindset is not a sustainable safety improvement strategy.

Asking workers to have zero accidents doesn't tell them how to have zero accidents (Conklin, 2012). Telling employees what not to do (make mistakes/have injuries) is not the same as providing the necessary leadership, support, and direction to reduce risk and prevent accidents/injuries.

To sustainably improve safety; it's much more important to understand how results are achieved. Accident-based metrics only provide the end result of our efforts, and provide little insight into specifically how the results were obtained, or how well your current processes are performing to prevent the next accident. In short, accident-based metrics only tell you what has happened in your organization, not what is currently happening. Without available information of how well the organization is performing now, organizations cannot proactively respond to changes in the work environment that may increase risk in some significant way, and can only respond to events as they occur. As a result, accident-based metrics tend to drive reactive, rather than proactive, safety improvement efforts. Managers will spend their time "putting out fires," as opposed to tackling safety issues within an organization with an applied, focused, and coherent strategy.

Further, accident-based metrics do not measure safety: Accident-based metrics measure the *lack* of safety. Measuring safety by counting the number of people injured during their work is a measure of the lack of safety, not its presence. So logically, it just makes sense to stop using accident-based metrics and look for metrics that provide a more accurate measure of safety within an organization. Admittedly, measuring "safety" is difficult. One cannot count what doesn't happen. With safety, one cannot count the infinite amount of accidents that could have occurred but didn't. What we can measure are the things we do on a daily, weekly, annual basis to reduce the level of risk associated with our operations and the likelihood of an employee getting injured. More on this later.

Lastly, narrowly measuring safety performance through accident-based metrics creates a paradoxical situation known as the "Regulator's Paradox." The paradox is that if we measure how safe an organization is by the number of accidents it has, and the organization is then successful in driving the number of injuries to zero, there is then nothing to measure or monitor performance moving forward. If an organization is successful in achieving zero injuries, there will be little to no information available to indicate how well an organization is performing, or how "close to the edge" an organization is from experiencing some adverse event. Without information to gauge performance, management is unable to intervene or make corrections that could improve performance and prevent injury (Hollnagel, 2014, pp. 11-14). Organizations would never have the information available to tell them how close they are from having a serious injury until they have it.

The Case for Organizational Goal Setting

Organizations striving to improve safety must recognize that individual performance is a product of the conditions and environment in which people work. Organizations must recognize this fact and instead, work to improve their capacity to record, measure, and analyze the activities and conditions that affect people's work. The safety professional must look at how our systems and processes unintentionally create opportunities for failure.

The benefit to a holistic approach to safety has enjoyed extensive support from many of the safety field's top practitioners and theorist. In 2001, Petersen wrote

> An unsafe act, an unsafe condition, an accident: all these are symptoms of something wrong in the management system. (Petersen, 2001)

The focus on management systems has not waned, and research into this line of thinking has grown significantly over the years. In 2009, the Department of Energy (DOE) established a series of guidelines for all DOE facilities, titled *Human Performance Improvement Handbook Volume 1: Concepts and Principles* (Department of Energy , 2009). Originally born out of an effort to improve operational safety at nuclear power generation sites, the *Guide* contended the following:

> ...human performance is a system that comprises a network of elements that work together to produce repeatable outcomes. The system encompasses organizational factors, job-site conditions, individual behavior, and results. The system approach puts a new perspective on human error: it is not a cause of failure, alone, but rather the effect or symptom of deeper trouble in the system.

A strategy to improve safety must include a focus on organizational systems and processes. As Dekker so eloquently stated in his *Field Guide to Understanding Human Error;* "a system isn't automatically safe; people have to create safety through practice at all levels of the organization" (Dekker, 2014, pp. 4-6). This paper helps organizations implement a continuous process to improve its safety management systems and processes.

Putting It All Together

Asking people to "be safe" is insufficient. Organizations striving to reduce the risk associated with their operations, and improve safety in a sustainable manner, should seek alternatives to accident-based metrics to measure safety performance. Organizations need more reliable information and insight to measure and gauge safety performance beyond that which accident-based metrics can provide. To provide sustainable safety improvement, organizations need to implement structures and systems that support critical thinking regarding how risk is managed, facilitate communication across the organization, and to instill a continuous improvement mindset. In response to inevitable changes that will occur, organizations will need to monitor the performance of these systems and adapt and adjust them accordingly.

The necessary systems and processes needed to facilitate change will vary depending on size, industry, organizational culture and safety program maturity. Below are a few of the formal systems and processes high-performing organizations can implement to improve safety. This is by no means an exhaustive list, but rather only a small list of the possible solutions available to organizations wishing to improve safety;

- Accident Investigations
- Ergonomic Teams
- Hazard Inspections
- Job Hazard Analysis
- Leadership Training
- Loss Trend Analysis
- New Hire Orientation
- Observation Processes
- Safety Committees
- Safety Management Gap Analysis
- Safety Meetings
- Safety Training
- Toolbox talks

Understandably, knowing intuitively where one can begin the journey to safety improvement can be overwhelming. This is where one can start to see the benefits to the process of formalized goal setting. Goal setting initiates a starting point, and allows for necessary adjustments as the organization matures. In the following, we will review why the process of formal goal setting can be the single most important thing your organization can do to achieve sustainable safety improvement.

Why Formalized Goal Setting Is Critical to Sustainable Safety Success

In the 2012 ANSI/AIHA Z-10 standard for Occupational Health and Safety Management Systems, the alignment and integration of Deming's Plan-Do-Check-Act (PDCA) concept are highly encouraged (see Figure 1). The PDCA concept is a model for continuous improvement, which the committee chair Vic Toy described as an "interconnected, organic way to maintain focus on continual improvement" (Manuelle, 2014, pp. 129-135).

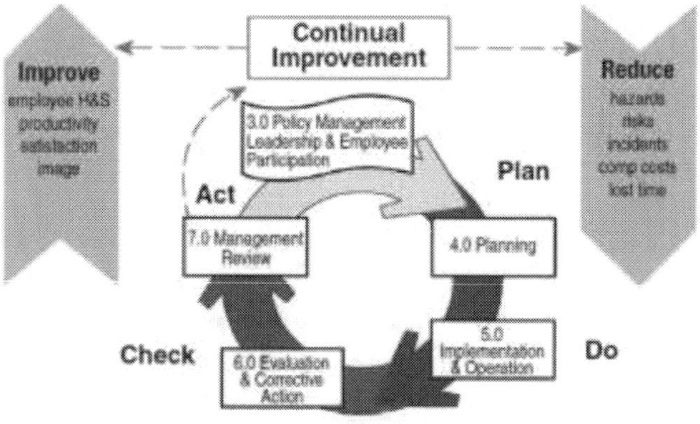

Figure 1. ANSI/AIHI Z10 PDCA Concept Model

Formalized goal setting aligns perfectly with the PDCA concept in that it provides the necessary structure on which the aforementioned PDCA continuous improvement process is built. Any organization, regardless of size, structure, industry, or safety program maturity, can benefit from goal setting. A simplified representation of how the goal-setting process aligns with the PDCA process is as follows:

1. PLAN: Establish goals and objectives targeted at organizational weaknesses and loss leaders.
2. DO: Implement solutions.
3. CHECK: Monitor and measure the implemented processes and systems and report the results.
4. ACT: Take necessary action to improve performance.

Also, the process of goal setting serves as the catalyst for which other safety and organizational performance-management processes are built. Below is a simplified model demonstrating how the integration of the PDCA concept manifests itself through the various management structures built upon the initial goal-setting process. An example of how goals provide the foundational support for other ancillary safety management practices and processes are as follows (Figure 2).

Figure 2. Example of how goals provide the foundational support for other ancillary safety management practices and processes.

1. Annual Goals and Objectives: The organization identifies an area(s) of improvement and identifies processes (or changes to existing processes) it will implement over the next 12 months to improve safety performance.
2. Personal Performance Metrics: Once organizational priorities are set, individual performance metrics would be developed to measure the level of participation and engagement among senior management and front-line supervision. These activity-based performance metrics will serve as a significant component of the individual's performance appraisal.
3. Senior Management Review: Safety managers develop a series of metrics based on individual and organizational performance and provide feedback to management of the efficacy of the interventions taken. The activity-based metrics will allow senior managers the ability to gauge overall support and participation, along with the year-to-date results.
4. Loss Trend Analysis: Some time before the establishment of the following year's goals, the organization will study the efficacy of the intervention(s) and if any modifications, adjustments, or changes are needed.

In short, goal setting provides a formal structure that requires the organization to think critically and strategically regarding its risk reduction and safety improvement efforts.

The Benefits of Formal Goal Setting

Goals, in their rawest form, are simply what an organization, or individuals within an organization, is consciously trying to accomplish. Since 1990, there have been over 1000 studies on the benefits of goal setting to improve individual performance. Time and time again, it has been proven that people and organizations that set goals consistently outperform those that either hasn't set goals or set vague, abstract goals such as a "do your best" goal (Locke & Latham, 2013). An analogy relevant to safety would be the disparity in performance between organizations that set formal safety improvement goals against organizations that either don't set goals or have only "work-safe" or "zero-injuries" goals.

Over the years, researchers have consistently shown that individual performance of participants who set specific and challenging goals was over 250% higher than others who failed to set goals or set easy goals (Locke & Latham, 2013). The higher levels of performance achieved through formalized goal setting are realized through the effect goals have on behavior. Goal setting has the effect of directing attention and action (choice), mobilizing energy expenditure or effort, prolonging effort over time (persistence), and motivating the individual to develop strategies (cognition) for goal attainment (Latham, 2012). These manifestations of these effects are reviewed in the following section.

Goals Provide Enduring Motivation

Through the practice of goal setting, management signals its priorities and values and directs the attention, effort, and resources necessary to accomplishing its stated goals (Locke E. A., 1996). An organization that explicitly states what it wants to achieve is more likely to improve performance in the area for which improvement is needed. It is important to note the overwhelming evidence that goals set must be challenging, yet attainable, to provide sustained enthusiasm and higher levels of performance (Kerr & LePelley, 2013).

Just as an organization's mission statement can drive an organization's culture to adopt certain values, so too can goals and objectives. Goals help individuals decide where to direct their efforts or on which tasks to focus their attention. Setting goals also reduce uncertainty in what constitutes acceptable performance, while also developing ancillary skills, such as problem-solving, critical thinking, public speaking, etc. (Latham, 2012). Employees also experience pride and a sense of accomplishment leading to higher job satisfaction when goals are achieved. (Latham, 2012). Locke and Latham summarized the motivational power of goals in the following passage:

> Nothing breeds success like success. Conversely, nothing causes a feeling of despair like perpetual failure. A primary purpose of goal setting is to increase the motivation level of the individual. But goal setting can have precisely the opposite effect if it produces a yardstick that constantly makes the individual inadequate. Consequently, the supervisor must be on the lookout for unrealistic goals and be prepared to change them when necessary. (Locke & Latham, 1984)

Goals Ignite Effort and Attention

Goals have a positive impact on employee behavior and performance. (Locke & Latham, 2002 cited in Lunenburg, 2011). Explicitly stated goals provide a clear measure of performance. A goal that is both specific and challenging eliminates ambiguity as to what constitutes unacceptable, acceptable, and high levels of performance. (Locke & Latham, 2013, pp. 4-5).

When employees know their performance will be evaluated in terms of how well they attained their goals, the impact of goals increases (Lunenburg, 2011). The most recent evidence of the motivational power of goal setting occurred in the 2016 scandal involving Well Fargo. In 2016, Wells Fargo Bank employees opened up to 2 million bank and credit card accounts without customer authorization in order to meet lofty sales goals. In an interview with CNN, a former employee was quoted as saying, "the sales pressure from management was unbearable." The scandal has since led to the resignation of several high-ranking officers, including CEO John Stumpf, and the firing of four other senior level executives (Sweet, 2017).

In the Wells Fargo example, employee performance was directly tied to meeting/exceeding the unrealistic sales goals placed upon them by management. Managerial compensation, in the form of bonuses, was also directly correlated to how well their teams met/exceeded the sales targets. The Wells Fargo example is an example of an organization that placed unreasonable and unattainable goals and, as a result, led to frustration, anger, and condemnation among its employees. Studies have found that employee performance suffers and deteriorates when employees view goals as unreasonable, or impossible to achieve (Locke & Latham, 1984). It is also an example of the undeniable motivational power goals have on behavior.

Goals Spawn Critical Thinking

The process of goal setting has a natural and immediate effect on individual cognitive processing in that once a goal is communicated, people immediately and instinctively begin formulating strategies as to how such goals can be attained. People recognize goals require plans, and plans will vary from very simple to complex. Establishing goals require employees think critically about how goals can be achieved, and what resources will be needed to succeed.

No one wielded the concept of goal setting to produce significant improvements in multiple areas of an organization better than former General Electric Chairman, Jack Welch. Welch was notorious for his use of goals, including "stretch goals" as the catalyst for creative thinking and to stimulate unconventional solutions (Kerr & LePelley, 2013). With the use of stretch goals, coupled with an aggressive Six Sigma process, GE was able to save $12 billion between 1996 and 2000 (Kerr & LePelley, 2013). Regarding his experience with goals Jack Welch said:

> The people with whom I have been associated have worked hard, enjoyed it more, although not always initially, and in the end, gained increased self-respect from accomplishing more than they previously thought possible (Tichy & Sherman, 2005).

Even Abraham Lincoln recognized the power of setting goals to spawn critical thinking. In 1848, in a speech to the United States House of Representatives regarding how best to meet the challenges of providing improvements to the interior of the United States, Lincoln was quoted as saying;

> Determine that the thing can and shall be done, and then we shall find the way.
>
> (Basler, Pratt, & Dunlap, 2001)

Lincoln knew what scholars since that time came to discover: Goals motivate employees, and will challenge and reveal their critical thinking abilities to find new and often innovative solutions to existing problems, and achieve more than they ever thought possible.

Goal Setting Helps Develop Future Leaders

Lastly, individuals who succeed at leading teams through the goal-setting process are often good at planning overall. Planning involves problem identification, prioritizing solutions, resource allocation, strategy, conflict resolution, team building, etc. These skills have broad application in the multiple areas of the organization and life.

Beginning the Goal-setting Process

Developing a Team

Formalized goal setting is a process of continuous improvement. To better position itself to identify opportunities for improvement and to formulate possible strategies, organizations need a broad range of perspectives from across all levels of the organization. Collaboration generates "buy-in," when employees have an active role in the goal-setting process.

The goal development team can take many shapes. There is no one prescribed number of employees that belong to the goal-setting team. It will be for each organization to gauge an adequate number of people that need to be involved to gather enough information to make informed decisions about where the organization's strengths and weaknesses and vulnerabilities exist in regards to safety and health.

Gathering Information

Once a team has been formed, it will be critical to gather information that will illustrate opportunities for improvement. The information available to the goal-setting team will vary, depending on the organization's size and safety program maturity. There are, however, some basic sources of information available to any organization. Conventional sources of information include, but are not limited to:

- Loss Runs
- Loss-trend Data
- OSHA 300 Logs
- Employee Surveys
- Accident Reports
- Safety Suggestions
- Employee Interviews
- Inspection Results

It is important to have open, candid, and non-judgmental conversations amongst members of the goal-setting team regarding opportunities for improvement. The process of identifying opportunities for improvement should occur in two separate and distinct sessions. The first meeting is the "learning phase," where employees focus only on finding and understanding where the greatest opportunities for improvement lie. Opportunities should be listed and discussed openly. At this stage, the team should concentrate more on identifying opportunities for improvement, and less on any potential solutions. Consider where organizational weaknesses lie and explore the factors that allow these vulnerabilities to exist within your system. Below is a list of questions that may generate initial conversation:

- What does the data tell us?
- Do we see any trends?
- What are our loss leaders?
- What is the largest potential for F&LC (fatal and/or life-changing) injury?
- Where do we struggle the most?
- Where can we improve?
- How can we improve?
- What do the experts say?

Employees often benefit from additional time to process the information and generate ideas about potential solutions. It is best to allow the group to step away from the discussion and allow for deeper reflection and analysis of the issues identified. After the break, the team should come together and prioritize the list of opportunities identified in the earlier session. Only once the group has agreed upon what the priories are, can the team begin formulating a strategy its weaknesses.

Elements of Effective Goals

Once an agreed-upon list of priorities has been established, the group must now begin the task of writing out formal goals and objectives. Goal-setting organizations must take into account a number of considerations before the process of goal setting can begin. Those factors include the type of potential goals, as well as the characteristics of an adequately formulated (SMART) goal.

Goal Type

Goals ensure processes are continuously refined and improved. As stated at the beginning of this paper, organizations should focus more on improving the process, rather than focusing narrowly on improving results. Petersen provides four distinct categories for which goals can be set; they are:

1. ***Routine Goals:*** These are objectives that relate to the regular, repetitive, commonplace activities that are typical, basic duties and requirements of the job. *Example:* Complete at least one (1) safety observation daily.

2. ***Project Goals:*** These goals are for new or special projects that might be undertaken by management. They also could relate to problem-solving or emergency-action activities that are part of a supervisory job. *Example:* Complete at least one (1) ergonomic improvement project quarterly.

3. ***Creative Goals:*** These are goals that relate to innovation or process improvements in the department or in the management system. *Example:* Develop a more comprehensive safety orientation process.

4. ***Personal Goals:*** Goals related to individual efforts that employees might make to supplement their job skills to increase their overall effectiveness. *Example*: Earn the safety leadership certification by year end.

(Petersen, 1996)

SMART goals

Effective goals share five common characteristics. While it is not necessary for goals to contain all five of the SMART characteristics, it is critical that each goal comprises at least one of the SMART characteristics. While one can find many variations to the SMART acronym, for the purpose of goal setting to improve safety performance, SMART goals come to mean goals that are:

- ***Specific:*** Goals should specify what needs to be done and the improvement desired. As goals become more specific, employees are better able to understand performance expectations and formulate strategies to achieve stated goals.
  Non-specific Example: Reduce muscle strains and sprains.
  Specific Example: Complete at least one (1) ergonomic improvement project each quarter.

- ***Measurable:*** When possible, goals should be quantifiable, so that performance can be accurately measured. Performance should be objective, not subjective. As goals are used to measure organizational and individual performance it is important to consider two questions: (1) How will you know whether or not a goal was achieved; and (2) How can/will you measure individual effort?
  Non-measurable Example: Improve safety awareness.
  Measurable Example: Each supervisor shall perform at least one (1) safety observation daily.

- ***Attainable:*** Employees respond to goals that stretch their existing capacity, but are achievable with considerable effort. However, goals are only effective if the employees possess the appropriate knowledge, skill, and abilities necessary to achieve a goal. Without the means to achieve a goal, goals will have a demotivating effect. Even if success is possible, but not probable, employees are more likely to commit to goals if they believe partial success is meaningful (Locke E. A., 1996).

- ***Relevant:*** For employees to commit to a goal, it must be viewed as important. Goals should focus on opportunities for improvement within the organization, including loss leaders, fatal and life-changing (F&LC) exposures, noted areas of concern, growth opportunities, etc. Another distinct, but necessary, expression of a goal's importance is the feedback mechanisms implemented to measure and gauge performance continuously. For people to achieve their goals consistently, they need some means of measuring performance so as to make necessary adjustments required to improve (Locke E. A., 1996).

  Several years ago while performing safety and health gap analysis for a large corrugated (cardboard) manufacturer, I asked the plant manager what his two biggest risks were in the facility. He quickly claimed "hand-placement" and "fork-truck" operations as the two things that concerned him the most (and also accounted for the majority of the firm's previous injuries). Later in the day when reviewing their safety observation data, the supervisor was gushing about the number of safety observations his team had performed over the previous twelve months. A closer look at the data revealed that out of approximately 790 observations, less than five were related to hand-placement or fork-truck operations. What is the lesson to be learned? The organization had not focused on the activities that presented the biggest risk. To provide a sense of purpose and improve safety in a meaningful way, concentrate on the biggest challenges facing your organization. Focus on what matters!

- ***Time Bound:*** Setting deadlines creates a sense of urgency, which helps spur progress. Employees are more likely to invest considerable effort if goals are to be accomplished by a specified date and time.

Four Simple Questions

The process of writing goals can be simplified ***if*** organizations develop each goal by answering the following four questions;

1. What do we want to accomplish?
2. Why do we want to do it?

3. How are we going to do it?
4. How will we measure progress and success?

If you can answer these four questions, your goals will be written in alignment with the SMART concept of goal setting. If you cannot answer these four questions, than any goal you set will be incomplete and less likely to deliver results.

Who Does What, by When?

Once you have determined what it is you want to accomplish, and how you are going to do it, the only thing left is to assign responsibility for the task. It's important to assign responsibility to individuals with the ability and authority to "make things happen,"

Monitoring Performance and Feedback

For goals to work, organizations must continuously track and measure progress towards goal achievement. When organizations provide feedback on how well the organization is performing in relation to achieving its stated goals, leadership and employees can decide if more effort or a different strategy is needed to attain their goal (Locke & Latham, 2013, pp. 6-7).

There is no limit to the number of ways an organization can provide feedback to employees regarding progress toward goal achievement. Many organizations already have mechanisms in place by which other types of goals (i.e., production, sales, quality, customer service, etc.) are tracked and communicated to employees. Communicating safety performance data in conjunction with existing organizational performance metrics signals to employees the value, importance, and significance of the activities designed to improve safety performance.

Goal Failure

Goal setting is a process for achieving positive and sustainable safety improvement. By their nature, goals are intended to stretch our current capacities and improve safety performance beyond current levels of performance. Positive change can result even when organizations fail to achieve their stated goals. Make no mistake; organizations set goals to achieve them. However, in the face of goal failure is important to consider the following two questions:

1. Has meaningful progress been made toward the goal?
2. What effect has the goal had upon performance? That is, how does our performance compare to what would have been had we not established a goal in that area?

(Kerr & LePelley, 2013)

Even when we fail to meet our goals, we can realize positive change in how we manage our work; reduce risk and the likelihood of experiencing serious accident/injury.

Conclusion

Organizations looking to improve the safety performance should adopt formalized goal setting as a central element to their safety management efforts. Goal setting is simple and easy to do but requires time and effort to be effective. Goal setting works because it forces organizations to think more critically about weaknesses facing the organization and how best to address those opportunities for improvement. The process of goal setting facilitates greater communication throughout the organization and helps build a continuous improvement mindset at all levels. Goal setting cannot solve every organization's problem, but without goal setting, efforts to improve will always fall shorter than they would have had goal setting taken place.

Bibliography

Basler, R. P., Pratt, M. D., & Dunlap, L. A. 2001. *Collected Works of Abraham Lincoln. Volume 1.* Ann Arbor, MI, MI, United States: University of Michigan Digital Library Production Services. (Retrieved February 22, 2017) (http://quod.lib.umich.edu/l/lincoln/lincoln1/1:498?rgn=div1;view=fulltext).

Conklin, T. 2012. *Pre-Accident Investigations.* Burlington, VT: Ashgate Publishing.

Dekker, S. 2014. *The Field Guide to Understanding 'Human Error.'* Burlington, VT: Ashgate.

Hollnagel, E. 2014. *Safety I and Safety II: The Past and Future of Safety Management.* Burlington: Ashgate Publishing.

Kerr, S., & LePelley, D. 2013. "Risks, Possibilities, and Best Practices." In E. A. Locke, & G. P. Latham, *New Developments in Goal Setting and Task Performance* (pp. 21-31). New York: Routledge.

Latham, G. P. 2012. *Work Motivation.* London: Sage Publications.

Locke, E. A. 1996. "Motivation through conscious goal setting." *Applied & Preventive Psychology*, 117-124.

Locke, E. A., & Latham, G. P. 1984. *Goal Setting: A Motivational Technique that works!* Englewood Cliffs, NJ: Prentice-Hall.

Locke, E. A., & Latham, G. P. 2013. *New Developments in Goal Setting and Task Performance.* New York: Routledge.

Locke, E. A., & Latham, G. P. n.d.. :Building a Practically Useful Theory of Goal Setting and Task Motivation: A 35-Year Odyssey."

Lunenburg, F. C. 2011. "Goal-Setting Theory of Motivation." *International Journal of Management, Business, and Administration.*

Manuelle, F. A. 2014. *Advanced Safety Management.* Hoboken, NJ: John Wiley & Son's Inc.

Petersen, D. 1996. *Analyzing Safety System Effectiveness.* 3d. Danvers: John Wiley & Sons Inc.

Petersen, D. 2001. *Safety Management: A Human Approach.* Des Plaines, IL : ASSE .

Sweet, K. 2017, February 17. *ABC News*. (Retrieved February 22, 2017) (http://abcnews.go.com/Business/wireStory/wells-fargo-fires-senior-managers-sales-scandal-inquiry-45637854.)

Tichy, N. M., & Sherman, S. 2005. *Control Your Destiny or Someone Else Will: How Jack Welch Is Making General Electric the World's Most Competitive Company*. New York: Collins Business Essentials.

United States Dept. of Justice. 2013, April 12. *Former Shaw Group Safety Manager At TVA Nuclear Sites Sentenced To 78 Months In Prison For Major Fraud Case Against The United States*. (Retrieved 2 20, 2017) (https://www.justice.gov/usao-edtn/pr/former-shaw-group-safety-manager-tva-nuclear-sites-sentenced-78-months-prison-major)

Session No. 717

Chicken, Egg, Cheeg? A Holistic, Integrated Approach To Create Breakthroughs Results

Michael Topf, MA
President
Topf Initiatives
Wayne, PA

Are you confused by academic terms, catchall phrases, and claims by all the different internal and external consultants when it comes to choosing a behavioral improvement process for your company? Do I call or use this or that? Do I address attitudes or behaviors? How do I make sense out of everything I hear and read? How do I sort it all out?

Controversy exists regarding the most effective methods to changing an organization's SH&E culture and preventing accidents, injuries and health/environmental incidents. Just like the age-old question, "Which came first, the chicken or the egg?", in safety today, we ask, "Is behavior-based safety the most effective path?" Many claim yes. We say. by itself, no!

A key question in the minds of most safety professionals, managers/supervisors, and labor leaders in organizations today seems to be, "How do we insure that the men and women that work in our company are integrating safety, health, and environmental excellence into their daily routine?" For safety, health and responsible treatment of the environment to become integral to your organizational culture, each of these elements must be viewed as essential to the productivity process. Safety, health and the environment must be on par with quality, production, cost management, etc. Everyone must recognize their responsibility and accountability for the well being of one another and the world we live in. Critical to achieving this objective is leadership's commitment to choose those actions that will demonstrate a concern for the person and their safety, health as well as the environment we all live in.

Even though people do not want to get hurt, they sometimes, either deliberately or non-deliberately, behave in an unsafe manner. These actions may result in an accident, injury, or environmental incident. Positive or negative circumstances, during times of personal or business change, can be such a distracting influence that incident rates can rise.

Internal and external safety professionals and consultants today are often involved in bantering conversations about what methodologies are most effective to improve a safety culture, as well as influence and change the unsafe attitudes and behaviors of the men and women that work there. Given the various theories and approaches that are being implemented, along with assertions on which are most effective in producing short- and long-term results, many people are confused by the academic terms, phrases, and claims that are used and made by all these people. What should we call it? Do we address the attitudes and thinking that underlies the behaviors first, or do we address behaviors directly first? Do we work with line employees, or do we work with management? Do we set up observation teams to give others feedback, or train our supervisors? And on and on and on…

From Our Experience, We Say Do All of It!!

The root causes of most unsafe behaviors and/or incidents result from either a loss of focus or attention on whatever activity one is doing or from counterproductive attitudes, beliefs and values, which influence the individual to place other priorities first, such as time, comfort, convenience, money, pride, etc. From past experience, individuals can also develop a false sense of security regarding their ability to control or protect themselves as they carry out their work. Incidents or injuries can occur to an individual as a result of their own awareness, attitudes, and behaviors or the awareness, attitudes and behaviors of other people. When corporate and site leadership place primary emphasis on production goals and profits, many employees all at levels will knowingly take shortcuts to achieve these objectives.

There are many influences on all levels of employees, from a variety of sources within and without their work environment, including management, co-workers, upbringing, and families, which shape our personality, values, thinking, and behaviors. These influences can shape both safe and unsafe attitudes and behaviors, on and off the job. Risk taking has always been a part of our culture, where statues are built to those persons who took the greatest risks for personal or social benefit.

Getting caught up in which is the "right" approach or the "right" theory, method, or terms, can end up being a waste of time and energy (as well as creating confusion). Do yourself a favor, forget what you call it! If the process improves your safety, health, and environmental culture; changes the unsafe attitudes and behaviors of all levels of employees, both labor and management, as well as addresses workplace hazards and system changes; and prevents incidents and injuries, it works!

Today's changing world of business and industry, bringing forth times of downsizing, upsizing, right-sizing, re-engineering, restructuring, etc., present challenges for the safety, health and well being of all levels of employees on and off the job. For safety, health and responsible treatment of the environment to become integral to your organizational culture, during these and/or stable times, each of these elements must be viewed as essential to the productivity process. As stated, safety, health and the environment must be on par with quality, the production process, cost management, etc. Everyone must recognize their responsibility and accountability for the well being of one another and the world we live in. Critical to achieving this objective is management and labor leadership's commitment to choose those actions that will demonstrate a concern for the safety and health of each person, employee or private citizen, and the environment. The organization's culture must be enhanced so that safety, health, and the environment are viewed and demonstrated by all as key values and priorities. Consciously or unconsciously, times of change can transfer time, energy, focus, attention, and resources to other priorities. During "lean times," when survival issues come to the forefront, and jobs and livelihoods are at stake, safety, health, and the environment often get "put on the side burner" for those other priorities that mean dollars into the company.

Each industry has it's own particular set of circumstances that can make it challenging to ensure that all employees are behaving in the safest manner possible. Many companies have permanent management, supervisory and line personnel, while others have temporary employee populations that are growing and may be constantly changing. Many of the older, highly trained and experienced employees are being retired, and new employees or those with lesser experience are taking over their jobs and responsibilities. Training people in proper safeguards and practices when their time to be employed is temporary may not be a top priority for some. Many companies don't consider it cost effective when their permanent or temporary employees are limited by time. With production pressures still high on the priority list, having a "body" on the job may be the #1 issue, well trained or not.

Another issue key to the safety, health and well being of employees and the community is whether corporate and site leadership attitudes, beliefs, and thinking are aligned to provide the resources to insure the safeguards necessary to prevent incidents of any type are in place.

Clearly, the cost of not investing in proper training and safety assurance has to be measured against the cost of investing in this area. What measures can be taken to make certain people don't hurt themselves and others, regardless of the length of their employment? Can a dollar amount be placed on safety, health, and the environment, long or short term, when it comes to human suffering?

We are sure the vision for all people and their co-workers, regardless of their position, labor or management, is to be able to come to work and go home in the same or better condition than they arrived. To have all employees, customers, and private citizens safe, in good health, as well as keep the environment clean, is an objective that most people can align behind.

It is important to build a common safety, health and environmental culture where everyone involved can work together towards achieving their vision with objectives and goals that support it. To accomplish this, it is essential to identify any barriers to achieving the vision, and strategize to resolve or overcome any barriers. We encourage you to work with management and line employees to create a vision for safety, health, and environmental excellence, and to identify the barriers that could prevent them from achieving this vision.

It has been estimated by many that 96% or more of all accidents and injuries result from unsafe acts. From our perspective, these acts can be from management or line employees, whether conscious or unconscious, aware or unaware. Moreover, we have found that many safety, health, and environmental incidents resulting from equipment failure could have been prevented if timely maintenance and repairs were initiated, or if proper construction or installation had occurred in the first place.

Two types of events can occur: On one hand, individuals can easily become unconscious and unaware, due to daydreaming, inattention, repetitive tasks, stress, and distractions from a variety of sources. As mentioned, pre-occupation with both personal or positive and negative organizational changes, such as upsizing, increased production priorities, downsizing, re-engineering, job security, the future of the company, changes in product lines, etc., can cause a person to lose focus and get injured. The other factor involves conscious or pre-meditated behaviors, where in these instances, an individual will talk him/herself into shortcuts, such as not following all procedures or avoiding wearing personal protective equipment (PPE). These pre-meditated, unsafe behaviors tend to be supported by reasons, rationalizations, and justifications and are often based on factors such as time, comfort and convenience.

When production schedules need to be met, often where meeting contractual agreements with a customer or gaining a financial bonus or some other reward for achieving quotas are in play, overt and covert messages to get work completed quickly can result in shortcuts and bypassing procedures. These unsafe actions may also be overlooked by foremen and supervisors, especially where their compensation is affected as well. In companies where downsizing or fear of downsizing exists, employees have told us in interviews that they take shortcuts to get their job done faster because they do not want to be perceived as slow. "The slow ones will be the next to be laid off."

Frequently, our safety programs plateau, and systems flounder or fail because the negative attitudes, beliefs and behaviors of those participating in the system sabotage its functioning. Programs and systems are abandoned because people are not contributing constructively or participating at all! Quite simply, **systems don't work unless people do!** Organizational habits and changes can promote attitudes and beliefs such as, "We can't afford to spend the money or time on safety, we need to focus on production and profits." Traditional means of statistical tracking of accidents and injuries do not measure safety behaviors or ensure all levels of employees are behaving safely and are carrying out the activities that will contribute to the prevention of accidents, injuries, and environmental incidents. Low statistics can cause people to become complacent, and direct their attention and energies elsewhere. To ensure continual improvement, constant attention and energy must be applied to safety, health, and environmental performance.

You CAN change behaviors by addressing attitudes, beliefs and the counterproductive circumstances surrounding people. A holistic, integrated approach will address the entire person and entire system and structure that will support your vision for safety, health, and environmental excellence and provide essential strategies to make necessary changes and improvements at all levels. This approach will support safe practice both on and off the job.

A holistic approach will produce lasting change in the behaviors of all levels of employees. Such an approach addresses a person on the physical, mental, emotional, and self levels, and improves attitudes, changes behaviors, increases trust, opens communication, fosters personal responsibility and participation with all employees, and heightens their ability to meet new challenges safely.

The typical behavior-based approach employs techniques that focus on defining behaviors and then provides periodic feedback and reinforcement of those behaviors. While we take no issue with the use of this technique as a valuable method of supporting changes in behaviors on an ongoing basis as part of a larger process, by itself, it has limitations. Given the complexity of human beings, one theory or method will not work with all employees.

What works most effectively is a systematic blend of techniques derived from the schools of cognitive, behavioral-based, affective, and reality-based learning. Skills that will enable a person to better focus and pay attention to whatever task they are doing in the moment will help prevent accidents and injuries that can result from being distracted and resulting inattention.

Lasting behavioral change occurs when a person's values, attitudes, and beliefs are addressed directly. When people believe in the value of safety for themselves and others, they will make appropriate decisions and act safely wherever they are. All levels of employees must be included in the process.

The goals of an improvement process should be the (a) permanence of the change, and (b) generalizability of the change; that is, people will be able to apply their learning and thinking to all situations, on or off their job.

The leadership of the company looking to produce continual improvement in safety, health, and environmental performance must identify and address the negative influences that shape the attitudes and behaviors of all levels of employees. The attitudes and behaviors of leadership must also reflect the awareness and understanding that the focus and attention on safety, health, and the environment needs to be heightened at this time, not diminished.

Essential resources for proper education, training, equipment, and resolution of safety, health, and environmental issues must also be allocated. To diminish resources from this area to "save money" can be far costlier in the long run. This cost can be in dollars or in human suffering, which, as stated above, is quite difficult to put a price tag on.

Defining and tirelessly communicating desired values and commitments can ingrain safety, health and environmental excellence as esteemed values. The attitudes and values of employees at all levels must be examined and measured against the vision for the organization. Expectations must be clarified where there is a lack of agreement. The most core expectation must be that each person exhibits responsibility for his or her own safety and health, as well as the safety and health of co-workers, family, the company, the community, and the environment.

To create breakthroughs in performance requires a holistic, integrated approach, emphasizing the importance of the whole and interdependence of the parts. In other words, all levels of employees, both management and line or labor, must be involved. Unsafe attitudes, beliefs and behaviors from any source can influence people in a way that causes incidents to occur. All employees need to take responsibility for their own and other's behaviors, as well as system improvements, and be willing to intervene (be proactive) and assist any person regardless of their position, employee, contractor or visitor in behaving in a safe, healthy, and environmentally sound manner.

The levels of your organization that need to be addressed in your improvement process are:

a. **The self-management level**: Each person must have skills to observe themselves and take personal responsibility to manage their own attitudes and behaviors on and off the job.
b. **The peer/team support level**: People need to develop the attitude that they care about and have a responsibility to encourage and support their co-workers to be safe. Skills are needed to carry this important interaction in a constructive manner.
c. **The leadership-management, supervision, and labor level**: All levels of leadership need to be able to put aside political and business issues to create and maintain a vision and environment where everyone can come to work and work safely, regardless of external or internal circumstances.
d. **The organizational level:** Insuring the culture, which includes the norms, values, beliefs, attitudes, and commitments of the owners and leaders of the company and its employees,

hold safety, health and the environment as esteemed values, and the attitudes and behaviors that reflect these values are encouraged, supported and acknowledged formally and informally.

Each level is underscored by responsibility and accountability. Total employee involvement is the most effective way to produce and accelerate positive improvement in your safety, health and environmental process. Using the knowledge and experience of your employees to determine the best ways to improve and ensure safe work practices is one of the most effective, low-cost investments of time and money that can pay off in preventing injuries, as well as keeping workers' compensation costs down.

When a person is internally motivated by the value and benefit of safety, health, and environmental excellence for themselves, their willingness to participate and be involved in improvement efforts is greater, and resulting changes in behaviors tend to be long lasting.

External motivational strategies are most effective when in combination with internal motivation. Safety leaders (line and management) must be skilled in coaching and counseling techniques to constructively lead and interact with all employees regarding safety, health, and environmental issues, attitudes and behaviors. This also applies during times of organizational change. Leadership must continually convey the message that all employees must adhere to safety, health, and environmental requirements and procedures for their own well-being and the well-being of the company. Costs from injuries and incidents, both in terms of human suffering, sidelining of valuable team members, and financial liabilities, can push a company over the edge or seriously impair its' viability. Fines and other penalties from regulating agencies can also contribute to making a company "unhealthy."

The process we have found most effective is to:

a. Assess the culture. Discover the prevailing attitudes, beliefs and thinking of all levels of employees that influence and shape safe/healthy or unsafe/unhealthy behaviors. (Includes corporate and division leaders and key influencers).

b. Train all employees in 'self-observation and self-management' skills: Essestial skills include: increasing awareness levels, focusing and paying attention, making safe choices resulting from safe attitudes, beliefs and thinking, working as safety teams, and willingly giving and receiving constructive feedback.

c. Train all leadership, management, supervision and line/labor leadership in skills for effective leadership, coaching and counseling, communication, problem-solving, role modeling, and other skills to lead, manage and support all employees in your safety, health and environmental process.

d. Implement a behavioral observation and feedback process for all employees, up and down the line, to support the awareness, attitudes, thinking, and behaviors that people have learned for safety, health, and environmental excellence. Involve both management and line employees in safety support teams. Show a united front for everyone to be healthy and safe.

e. Implement employee involvement problem-solving processes and cross-functional teams to identify and solve safety, health, and environmental issues and concerns.

f. Provide ongoing support and reinforcement. Every improvement process or program, to be successful and continue, needs ongoing support. Insure that necessary resources, whether human, financial, equipment, etc., are provided and are readily available to prevent breakdowns or failure in any structure or system, human or mechanical. Again, management and labor leadership must be consistent with their support and communication of the commitment to everyone's safety, health, and well-being.

g. Acknowledge and appreciate contribution and success. Humanizing your reporting of progress and/or incidents demonstrates a concern for people, not just the numbers!

Continual safety, health, and environmental improvement is an ongoing process, not a training program or an event! A culture shift represents permanent changes in the kinds of attitudes, values, and thinking that influence desired behaviors from all levels of employees over an extended period of time.Optimum opportunities for breakthroughs in performance occur when corporate and line or site employees, as well as, other stakeholders, reflect these values and beliefs in safety, health, and environmental excellence. Accomplishing this shift requires consistent and sustained effort. A genuine value for safety, health, and the environment will continue, regardless of the internal and external circumstances.

Improving attitudes and behaviors around safety, health, and the environment will result in improvements in attitudes and behaviors in all areas of productivity, including, production, quality, costs, morale, creativity, and innovation. Benefits and savings from an investment in these areas can be tremendous.

Increased awareness of one's own attitudes, beliefs, thinking and behaviors is the starting point to responsible actions that will correct personal and organizational behaviors and issues.

As was mentioned, some of the primary causes of accidents and injuries include inattention, daydreaming, and distractions. These causes cannot be stopped completely, but they can be minimized through increased awareness. Awareness is the starting point. Specific skills and actions are then necessary to help maintain focus and control over one's actions and interactions with their environment on and off their jobs.

Behaviors are often difficult to change in ourselves and others because of the human tendency to resist those things we don't like, disagree with, or find inconvenient. Our thinking process, attitudes, and beliefs largely dictate our positive and negative behaviors.

Addressing behaviors alone, without addressing these underlying attitudes and beliefs, will not generally produce lasting behavior change. By increasing awareness through education, training, and constructive intervention, individuals (both management and line) can discover their counterproductive attitudes and beliefs around safety, health, and the environment, and the possible negative consequences of their behaviors. When safety measures are disregarded by any level of employee and are overlooked by coworkers, unsafe behavior is implicitly condoned. Personnel may also develop a belief that other employees (line or management) don't care about their welfare when unsafe behaviors are overlooked and when resources are taken away or safety departments downsized when times are lean.

Common beliefs are: "It won't happen to me," "I know it's risky, but I don't have time to do it safely," and "We can't afford to spend the resources, both money and people on safety at this time." These and other individual and group attitudes and beliefs will shape safe and unsafe behaviors on and off the job.

As noted, employee involvement is another key feature in safety improvement. With trimmer staffs, everyone must see themselves as a safety leader. All personnel must be empowered in proactively finding ways to identify and solve safety, health, and environmental problems. Sometimes, employees assume the posture of a "victim" around safety and safety improvement. They may blame the company for being too lax or other individuals for being careless or the equipment as inadequate. Some employees are like spectators at a game and passively observe; they may comment on how the game should be played, but we find that they seldom take effective action to help others be safe. This often occurs because they may have had negative interactions in the past with coworkers or with line or management leadership. Many employees need to be encouraged to participate, as well as requiring training in how to be more effective in their interpersonal interactions. Systems and structures must be created to provide an opportunity for people to be involved, identify problems and solutions, and receive feedback on progress. This is essential for continuous improvement. Effective communication between all levels of employees is key.

Lastly, your commitment and support is essential! Safety, health and environmental leaders must be SALES PEOPLE AND CHANGE AGENTS! Be clear on your vision, be educated in all aspects of what you believe is necessary to improve performance and prevent incidents, do your homework, enroll all persons from labor and management necessary for the approval of necessary resources and programs, and provide ongoing support and success of your process! LET YOUR DEDICATION AND COMMITMENT DRIVE YOUR ACTIONS! WISHING YOU CONTINUED SUCCESS!!!

SUMMARY

I. Almost all the problems of modern safety processes can be placed in one of four categories; they are:

- **Counterproductive attitudes, beliefs and behaviors**
- **Interpersonal and team ineffectiveness**
- **Lack of leadership from management and labor**
- **Lack od needed resources abd/or systemic support**

II. The process that we have found works to create a breakthrough in safety performance and meet objectives and goals is:

- **Cultural Assessment: Discover current norms, values, attitudes, beliefs, and commitments that influence or drive safe and unsafe behaviors.**

- **Train all employees to increase their awareness and provide skills to support and manage themselves and others to stay healthy, safe, and protect the environment.**
- **Train leadership, management and labor/line to evaluate, clarify expectations and guide other employees in safe attitudes and behaviors.**
- **Create management and line-safety support teams to interview and support safe practice and correct unsafe conditions.**
- **Provide ongoing reinforcement, resources, systems, and support for continual improvement.**

III. Commitment is essential and ceaseless.

Leader Development and Culture Change in a Depleted Environment

Gary L. Winn, Ph.D.
West Virginia University
Morgantown, WV

Ava C. Dykes, Ph.D.
National Institute for Occupational Safety and Health
Morgantown, WV

Introduction and Literature Review

One of the basic assumptions about establishing a "values-based" safety or engineering culture is that topmost management is supportive and begins the process by pushing these changes downward. But not all organizations have supportive CEOs or Vice Presidents who will foster a supportive leader-development environment; some may be disinterested, some may be unaware, and some may not appreciate internal change. Some upper managers may be happy with simple compliance. Can an individual or small group far downstream develop leaders independent from the organization? Can a motivated individual or small group move the needle toward a supportive values-centered culture, even when upper management doesn't seem to care? We think it's entirely possible to do so by applying some rules and research generated by organizations with successful leader development cultures.

The fields of occupational safety and engineering have embraced the study of leadership with open arms in the past five years or so. The American Society of Safety Engineers (ASSE) and the American Society for Engineering Education (ASEE) have sessions and entire conferences dedicated to the topic of creating leaders and studying what the best leaders do. As most of us know, since 2004, ASSE has offered an annual conference for future safety leaders who are current students studying occupational safety and health. The conference offers the somewhat optimistic opportunity to "learn how to be an effective leader and communicator," and to get important career advice from seasoned professionals about career paths and improving interview skills.

ASSE offers onsite seminars about the principles of leadership; its Foundation offers leadership grants to "help safety professionals expand their knowledge and advance their education." Local chapters include "safety leadership" as PDC topics almost as frequently as they include the word "excellence." There is no end in sight to ASSE's interest in leadership.

Our sister organization, the American Society for Engineering Education (ASEE), has created an entire division named the "Engineering Leadership Development Division" (LEAD) the purposes of which are to address changes at the university level (promoting courses that stress ethics and leadership, for example) or fostering concordance between academics and industry. A fairly steady stream of papers address the observed disparity between academics who say they produce graduates with leadership and ethics training and industry, which says these skills are not strong among new hires in engineering. As with ASSE, there is no end in sight for the discussion of leadership and ethics.

Public institutions have embraced leadership as well. The National Institute for Occupational Safety and Health (NIOSH) follows a similar trend set by the Centers for Disease Control and Prevention (CDC). CDC University's School of Leadership and Management Development began the Initiative for Leadership Enhancement and Development, or ILEAD, in 2008. This program was designed to provide opportunities for CDC management to develop leadership competencies necessary for guiding the agency. Some leadership competencies or key behaviors include accountability, cultural awareness, emotional intelligence, performance management, and personal leadership, among others (CDC, 2017).

But nobody seems to be talking about what to do when the management, from board level to the department boss, simply doesn't care that midstream managers develop leaders. Can a midstream safety professional create a culture where values affect every decision at a fundamental level, particularly for the crafts workers who are on the front lines of safety every day? Can a project engineer who is often charged with safety responsibilities do the same thing away from the corporate office?

Left in the lurch, then, what about looking to academics for guidance of teaching about leadership or ethics? Have academics considered how to create a culture of values-based behavior when upper management is disinterested? In a word, no. As it turns out, academics aren't even sure how to teach ethics, much less leadership. Indeed, as we pointed out a year ago in a paper presented to an engineering group:

> It is fairly recognized in academic circles that ABET (formerly and perhaps more descriptively called the Accreditation Board for Engineering and Technology) and its sister accrediting agency for the related sciences including safety management, ASAC (Applied Science Accreditation Commission) are themselves confused, at least in our opinions, about how much to teach ethics and how to measure it. As Barry and Ohland (2012) suggest, among other authors who have comments on this topic, "Even after multiple cycles of ABET accreditation, many engineering programs are unsure of how much curriculum content is needed to meet the requirements of ABET's Criterion 3.f" [Note: while ABET itself prefers the use of performance-based objectives in subparts a-k which it requires for accredited schools, "an *understanding* [emphasis added] of professional ethical responsibility" (part 3.f) is sufficient.]
>
> Barry and Ohland conclude, in part, that "The primary impact of this study is that it dispels the myth that more courses or course time on professionalism and ethics will necessarily lead to positive engineering education outcomes. Much of the impetus to add more curriculum content results from a lack of conclusive feedback during ABET

accreditation visits." And somehow, we engineering and safety educators must figure how to measure a student's understanding, and also determine how much time-on-task to provide, even though such studies are inconclusive and ABET site visitors are not much help, at least for the ethics part.

We suspect that it isn't going to be any easier to teach about leadership and values than it has been to teach about ethics (Winn and Slagley, 2016).

Leader Development in a Depleted Environment

Absent guidance from professional organizations or academics, we are proposing a model called the "Leader Development in a Depleted Environment" (LDDE) for use downstream in the management structure. While the term "depleted" may connote many things, it is limited here to management that is dismissive, disinterested, hostile or simply unaware of motivated junior leaders' desire or ability to foster leader development and promote values-based behaviors at lower management echelons.

The LDDE model begins with the expression of a set of beliefs about individuals and their duty to each other with a particular reference to safety and health. This means simply that employees at any level must be willing to act ethically even when if nobody is watching. These are "values-in-use" actions, and not "values artifact" actions: They are based on trust and honest caring for subordinates and self, not doing the right thing only when the spotlight is on them; last year, we said:

> Authentic leaders can exist at all levels of an organization; a forklift driver does not need the title "CEO" or "foreman" to manifest his actual values in voluntarily instructing a new employee in the virtues of inspecting the truck's brakes every day whether the rules require it or not. When a safety professional thinks a behavior is manifested because it's the right thing to do, it reflects authentic leadership and not just colorful corporate literature or fancy business cards.
>
> Organizational research conducted over the past twenty years by Schein (1992, 2004) or Ott (1989) suggests that any mismatch between espoused values and values-in-use causes morale problems because employees see that, "what we say we do is not what we actually do." In fact, as Hannah (2007) points out, authentic leaders are pretty much always under some sort of scrutiny and things will quickly "come crashing down should an [authentic] leader lapse or be uncovered as pseudo-authentic." The once-authentic leader has a difficult time recovering.
>
> Does it matter if leaders say one thing and do another thing? In safety, it does.
>
> Project engineers often end up managing all or part of the safety function. Safety professionals coordinate carefully with the project engineer on material selection, equipment purchases, scheduling, pre-engineered designs, discipline and especially modeling safe behavior on the job. Any inconsistency between these coordinating leaders' messages can have devastating consequences. If words and actions are

incongruent, subordinates will see through the inconsistent behavior, and worst of all, people can die (Winn and Slagley, 2016).

Step 1: The Place of an Honor Code

The simplest place to begin the LDDE is for the foreman or department lead to say, "Here is what we believe in." An honor code is a tool for both accountability and feedback. Even without a larger, formalized ecosystem of leader development, small groups can benefit from drawing their own line in the sand for what is, and what is not, acceptable behavior.

One example of a simple but vibrant code of behavior is West Point's, as seen in Figure 1 below:

Figure 1. West Point Cadet Honor Code. (Photo credit: G. Winn (2016))

Every cadet memorizes this thirteen-word message as early as their first day: A Cadet Will Not Lie, Cheat, Steal, or Tolerate Those Who Do." The code represents an institutional aspiration that trust is unassailable among members of our group, and that those who break this trust are not welcome. A small-group leader can surely have more influence and more immediate impact on his or her small work group than upon a large organization; the impact of stating a simple honor code is real and genuine when presented by a motivated junior leader face to face; at the department level, these discussions are received as honest expressions of what we believe and how to act.

A small-team leader can have a ceremony to adopt the honor code, or members can simply nod in agreement; from this point forward, the group will be expecting values-congruent behavior. The person's word is an indelible bond if they say it is so. Infractions of the code can be considered at the same level. People who don't comply are ostracized in ways determined by the peer group, not some corporate policy statement.

Elsewhere we have said:

> Is there a need for an honor code in a company without a complete ecosystem to support values-in-use? We say "yes" because safety leaders and engineers can never, ever, take employee trust and the public safety lightly. When people can die on your watch, and when there are no do-overs, our future leaders must get it right the first time. In turn, employees should be trusted to do the same thing.
>
> Even when nobody is looking.
>
> *Especially* when nobody is looking.
>
> A dishonest worker may say that the electric power is turned off before telling another worker that she can change a breaker in a service panel. A dishonest supervisor can say he has performed a walk-through of morning conditions on the construction site and not bother to inspect the new utility excavation, allowing a highlift to back into it. When people can be injured or even die on your watch, there needs to be unassailable trust at every level, without asking, with nobody looking. Your subordinates must trust that you do what you say, and vice versa.
>
> Having an honor code is a public expression of the desire to try to do the right thing at all times, under all conditions, and not tolerating those who will equivocate. This thirteen-word code can be easily adapted to use in organizational materials, such as into new-hire orientation, and it can easily be knitted into real stories [shared at the water cooler or the tailgate.] (Winn, 2016).

An honor code can be modified to fit the circumstances of place, industry, and certainly culture. To say, "Employees of the welding department do not lie, cheat, steal or tolerate those who do," is enough to start.

Is there a generational conflict possible? At first, we thought that Millennials would think an honor code would be corny, but we were wrong. Although we don't have hard data on this observation, we note that, when discussing the matter over a period of four years now, graduate students seem to think the honor code is a good idea. "Nobody ever talks about this but they should," they will say.

Our data shows Millennial graduate students who were surveyed across three similar academic institutions are altruistic but have increasingly less practical work experience when they enter their chosen career field (Winn, Williams and Heafey, 2013). Similar findings are seen among Howe and Strauss's (2004) seven important characteristics of Millennials that differentiate them from Baby Boomers or Generation Xers:

- They are **confident and optimistic**; they are perhaps over-confident sometimes and sometimes they boast.
- They see themselves as **achievers;** at school, they go to class; they have pretty good moral compasses.
- They offer and hold **conventional values systems** which are similar to, or are more conservative than, their parents, many of whom openly rebelled in the 1960s at their own parents' value systems.

So, in a general way, Millennials seem to appreciate the notion that self-awareness and having a foundation of morality is the first step toward leader development. The opportunity is to take advantage of this generation's motivation, altruism and hope. An honor code is the place to start, even at the department level. It only takes a sincere and values-congruent junior leader to take charge. There is nothing magic or compelling about the code we have proposed here. The idea is more important than the words.

Step 2: *Be–Know–Do*: A Model for Leadership at Any Level

The next part of our model has been used extensively in private industry and the military. At first glance, one of its authors would unlikely to be selected as a speaker for a safety symposium on leadership; Ms. Francis Hesselbein, who turned 100 in 2015. She rose from being a stay-at-home mom in Western Pennsylvania running a Girl Scout troop to ,running the entire Girl Scout organization, to be the woman who 30 years later, was named Fortune Magazine called the "Best Non-Profit Manager In America." She is Chair of the Peter F. Drucker Foundation for Nonprofit Management and Editor in Chief of *Leader to Leader*, the best-known journal about leadership in the U.S., if not the world.

Ms. Hesselbein' s contribution to the study of leader development centers on this single quote for which she is renowned, "Leadership is more about what a leader is than what a leader does." Essentially, leadership is about *character* more than anything else. In 2002, she summarized how to make changes in leader development if the bosses seem uninterested or if we are not at the top [of the management chain]:

> You can begin wherever you are, whatever your job. You can bring new insight, new leadership to your team, your group. It is the leader's job to identify the critical issues in which his or her organization [or department] can make a difference, then to build effective partnerships based on mission, innovation and diversity to address those issues (2002).

The second author of *Be–Know–Do* is Eric Shinseki, who served as Army Chief of Staff and the Secretary of Veterans Affairs under President Barack Obama. Following the Army tradition of not developing leaders in the traditional top-down style of corporate America, but instead creating leaders from the bottom up in a dispersed way, General Shinseki shows that the U.S Army ranks as a world-class corporation for leader development. But how it does it is largely unheralded, except by those with firsthand experience. (For a more complete discussion of how the military creates leaders, see Winn and Banks in *Professional Safety*, January, 2014).

Perhaps surprising to most readers, the U.S. Army trains its junior leaders to take the initiative, rather than wait on orders and lose the initiative. Preparing to do that is a journey that requires self-awareness, a strong sense of selflessness, and personal courage when circumstances call for it. As it turns out, the *Be–Know–Do* model is a perfect model for adoption on the civilian side, and particularly in safety. When minutes count, the subordinate must be right the first time. An employee who is motivated to stop an unsafe act simply because it is values congruent is an authentic leader in his or her own right.

Rather than ask this junior leader to wait on a decision from an upper-management structure th may that may be unaware, disconnected from the day-to-day operation, or, worse, disinterested, the subordinate must be prepared to act. This is precisely why the U.S. Army's middle managers, its non-commissioned officers and lieutenants, are far more critical to mission success that its colonels and generals. Everyone will readily admit as much.

For example, Peter Drucker wholeheartedly endorsed this approach for industry and nonprofits when he reviewed the seminal book by Hesselbein and Shinseki which outlines the *Be–Know–Do* process, "The Army trains and develops more leaders than all other institutions altogether—and with a lower casualty rate. *Be–Know–Do* shows how this is being done, and how it can be adapted by the nonmilitary: businesses, colleges and universities, nonprofits, and churches."
A summary of *Be–Know–Do* is shown in Figure 2 below.

***Be* the ideal and Be proud of it**. Personally represent the company's core values and standards, even if other people don't. Be the person with integrity and a duty to protect less experienced workers. Exemplify the highest ideals and live them at work and at home through your department's honor code. Don't be afraid to do what's right.

***Know* what you are supposed to do** and when you are supposed to do it. Be trained and seek regular competency and skill updates. Be aware of trends in safety compliance and research. Acquire the right technical skills to lead. Know what motivates your direct reports and how your own culture operates. Know where the power centers are, and who are the gatekeepers to information.

***Do* the right thing each time and every time**. Have the interpersonal and communications skills to get your message across accurately and efficiently. Follow the rules and procedures, even when you're by yourself. Don't be that guy who only acts right when his superiors were watching. Junior leaders don't need upper management's permission to show backbone and character.

Figure 2. The *Be–Know–Do* Model as Part of the LDDE.

Step 3: Storytelling, Non-Material Rewards and Personal Courage

Consider this: break rooms, locker rooms, and board rooms are equally useful places to tell stories with a solid learning message. Telling a story about somebody we know who had a close call can be a powerful motivator, but story telling should follow a pattern in order not to be strictly didactic. Stories should be told with these rules:

1. The story must not be directly in response to an injury that just happened but rather a close call.
2. The story must not be condescending or heavy handed.
3. The storyteller must not name names unless the story is a bit funny. No one wants to be the butt of a safety joke.
4. The storyteller can be a craft worker or manager, or even the safety professional. Credibility is most important.
5. The story is most often about a person who *discovered* the right thing, even if he didn't actually *do it* at the time.
6. Stories must be true, plain and simple.

Dr. Winn has used this particular story in class for years because it is about college age-workers and values-not-in-use. Even after years of telling it, the effect is the same.

> Between my freshman and sophomore year at college, my crane-crew buddy was supposed to walk out on these 22-foot-tall concrete sewer pipes standing on end–sometimes wet with rain and always coated with grease to prevent concrete from sticking to the steel end-ring. My buddy made it to the end of summer before he fell backward into a pipe and broke a dozen bones, including his spine. We couldn't even find him for over an hour. Nobody had done any training, nobody did a hazard analysis, and nobody provided a walk board. The company had lots of safety posters and had a safety coordinator, but my buddy nevertheless graduated from College in a wheelchair because there were no values-in-use that day.

Steve Denning writes for *Forbes* online about radical management and innovation. He believes just as much as we about the utility of storytelling to pass along values-in-use of the organization in ways that even a brilliantly written mission statement about core values just can't put across. These are "dead artifacts" of the organization, as he says:

- Storytelling is more than an essential set of tools to get things done: It's a way for leader–wherever they may sit–to embody the change they seek. Rather than merely advocating and counter-advocating propositional arguments, which lead to more arguments, leaders establish credibility and authenticity through telling the stories that they are living.

- Storytelling translates dry and abstract numbers into compelling pictures of a leader's goals.

- Storytelling is often the best way for leaders to communicate with people they are leading–communicating who you are, enhancing the brand, transmitting values.

- Charts leave listeners bemused. Prose remains unread. Dialogue is just too laborious and slow.

- By contrast, conventional management focuses on lifeless elements–mission statements, formal strategies, programs, procedures, processes, systems, budgets, assets–the dead artifacts of the organization (Denning, 2011).

What we are trying to do through storytelling is to teach about personal responsibility or a duty to intervene; that is, acting on the very values we have agreed upon when we adopted an honor code. These kinds of stories are remembered much longer than preaching, corporate newsletters or some silly, plastic values card. Plus, stories are easy to retell. The idea is to make values-in-use automatic. Can storytelling by a mid-level junior leader serve to change the culture of his local organization? Absolutely.

Next, we come to rewards for appropriate behavior and the paradoxical value of *non-material rewards*. What can a midstream leader do to reinforce values-in-use without support from above and perhaps without any real budget anyway? The answer is: probably a lot.

The research literature shows that there are two kinds of reinforcements for appropriate behavior. One is *material rewards,* such as money or a tool picked out of a Grainger catalog, for example. Despite some serious research to the contrary, material rewards continue to be popular with safety leaders because it looks like they are doing something positive. But among drawbacks to material rewards are: hiding injuries; the disappearance of appropriate behavior once the incentive is used up; and factionalism when a single reward is spread far too thin among potential qualified recipients.

Research suggests that the second type of reward, *non-material rewards* lead to longer lasting behavior change and require no budget beyond creative thinking. In a paper published in 2004 about the use of non-material rewards in the construction industry, it was noted that "natural incentives, which we have defined as non-durable behavior reinforcers, are used to modify employee safety behaviors. Natural reinforcers include the use of:

- performance feedback (coaching, for example) to reinforce appropriate behavior;
- employee-determined work schedules; and
- independent work, resulting in less frequent visits by the safety manager, so long as safe behavior continues (Winn, Seaman and Baldwin, 2004).

The work was summarized as saying, "Performance feedback about individual employee safety performance indicates an effective alternative to the use of disciplinary action or material rewards to encourage compliance with safety rules," and we concluded that, "The crafts people in the study sample actually improved their behavior in response to natural incentives," even though in conversation, they said they preferred material incentives such as money. In real life, craft workers preferred responsibility and involvement over money.

This study also looked at worker satisfaction with management leadership in offering natural incentives. "In embedded question 3 (QE3), workers were asked if they wanted to be even more involved in safety inspections than they had been in the past. Ninety-one percent responded that they wanted more involvement. In the final embedded question (QE4), fully 88% of respondents said that they felt management was fully committed to reducing fall hazards. Each of these questions significantly exceed ($p < .010$) the hypothesized level of 50% and were judged to be accurate indicators of worker satisfaction with the safety program." This suggests that the use of reinforcements for safe behavior is not only preferred under these test conditions by craft workers, but that the same workers felt its leaders were committed, and that they wanted to be more involved in their own safety in the future.

These findings are not unique at all. Once the junior safety leader realizes that material rewards (the traditional incentive for safe behavior) can be effectively replaced with what amounts to leadership itself as a natural incentive; that is, responsibility, involvement and work direction, workers are eager to be more involved and will trust their leaders' motives.

Finally, we come to uses of **personal courage** to demonstrate, when all else fails, that the junior leader must act values-congruent at all times, even when his own safety may be at stake. The junior leader may jeopardize his own career or safety but he takes action based first on selfless service to subordinates. The board room may never know about the junior leader's actions, but the break room and the locker room will know. We are not talking about heroics or superhuman courage—only the courage to act according to those values we established earlier.

> The board room will never know about the junior leader's actions that day, but the ***break room*** and the ***locker room*** will know.
>
> ...and that's what matters.

Almost everyone has a story of a "regular Joe" who took extraordinary risks to his career or his own safety. We all know about fictitious Atticus Finch who jeopardized his law career to defend the African-American Tom Robinson in 1930s Alabama. This metaphor is apt here because Atticus Finch represents the challenges of morality when faced with challenge of his career. Finch displayed the kind of personal courage that is easily understood by crafts people, but then, maybe a lawyer is somewhat removed from everyday line work. So we dig deeper, as anyone can.

Mark McGeehan, a local football player from Wheeling, West Virginia, was a B-52 pilot in the early 1990s, when huge air shows were all the rage in the U.S. McGeehan refused to allow his subordinates to fly with another pilot, McGeehan's superior, Bud Holland. Holland was a pilot had shown reckless disregard for standard flight procedures during low flights and risked his crew repeatedly after using these heroics in Viet Nam. On the day of an air show, with dozens of dignitaries in attendance, Holland made a few exceedingly low passes over the base. A chilling video was taken as Holland's plane passed over the air field close the asphalt, and circled around at an angle that exceeded the recommended design limits of the aircraft. His B-52 crashed in full view of the congressmen in attendance, the Secretary of the Air Force, and sadly, McGeehan's own family. Still, McGeehan saved the lives of each of the crew members he pointedly forbade to fly with Holland. McGeehan displayed an incredible kind of values-congruent personal courage that cost him his life.Maybe even Air Force pilots and huge aircraft don't resonate with the crafts people with whom the junior leader is trying to make a point about the need for personal courage when things go really bad.

Consider the case of our own friend close to home, a coal miner named George Maxwell. George had responded in 1994 to a mine's roof fall as part of his duties as a mine rescue expert:

> Another hour passed and everyone knew that time was running out for the trapped miners. George decided he should act and so he entered the mine on his own accord and out of sight of the state officials who had gathered there.
>
> Working near a recent roof fall is a terrifying condition that only a few miners ever experience. Tons of rock up to five feet thick and a hundred feet long come crashing down unexpectedly, flattening heavy equipment and people in its path. The number one rule in all of mining has always been: don't work under unsupported roof. And that's just what George had to do as he crawled prone, inching his way over the top of rocks and debris on the way to his fellow miners, all within a 14 to 16 inch head space. He had to work for an hour under unsupported top if he was going to rescue his fellow employees.
>
> And as you'd expect, George got to the miners with enough water, a few air packs and now, with a radio to confirm their location, and so outside rescuer crews could pick up the pace of the rescue and use heavy equipment until they approached within a few feet. All of the rescued miners, plus George, were standing outside in the sun in about 24 hours (Winn, 2016).

George didn't have to attempt this unconventional rescue, but his near-three decades in the mines had made him the real leader among his peers, without a fancy title and displaying calming, low-key leadership that was almost unknown to his superiors. He displayed extraordinary personal courage at the risk of his own life.

"Almost unknown to his superiors?" That's just the way George liked it.

> What a coal miner like George ***does*** about safety speaks louder than what he ***says***.
>
> … and that's authentic leadership

Summary

The literature on leader development is so large that it generates millions of Google search hits. The topics of leadership and leader development dominate conferences of key professional societies and even some public institutions. Despite the interest, culture change and leader development universally seems to be a top-down affair, with CEOS in corner offices deciding on mission and culture, then printing colorful advertisements heralding their company's move to core values.

In this paper, we propose a simple model we call the LDDE; it is open to improvisation by the very people who practice it; it doesn't need fancy brochures to foster leader development. We have scoured the available literature and determined that, under favorable circumstances, leader development and culture change can also happen at lower levels that are removed from the overt power structure. Even when upper management is dismissive, disinterested or merely unaware, the motivated junior leader can change culture on his or her own, using this model as a guide and modifying it as local needs warrant.

While having a local honor code might seem out of fashion for Millennials, these have been in place and changing lives for decades at our nation's military institutions. So why not have them industry, too? After all, an honor code may be as simple as the thirteen words at West Point or the Virginia Military Institute, and is essentially the same as the code of ethics for ASSE or the National Society for Professional Engineers. And it is encouraging to note that the Millennial generation seems sufficiently altruistic and motivated to accept such a code of conduct, particularly when lives may be at stake.

The structure provided by Hesselbein and Shinseki known as *Be–Know–Do* is the glue that underpins the U.S Army's entire leader development program because it anticipates initiative and leadership from mid-level soldiers who, if sufficiently preferred, can act in time to prevent losses. Remember the whole point is that this kind of preparation and training allows, even encourages, mid-level soldiers to act on their own to save time and human capital. How different is using this structure for training mid-stream soldiers to be leaders in small groups from training mid-stream crafts workers who often work independently toward similar objectives for their respective organizations?

Storytelling clarifies important messages in informal ways, and the best stories are those generated by the workers themselves. There is an authenticity to a story told in the break room that has credibility with crafts people, and the ability to alter the local safety culture far beyond the leadership charts and mission graphs in some corporate annual report. When a junior leader chooses natural rewards over material rewards, that leader is more likely to further change a local culture with long-lasting effects.

Finally, those craftspeople who display personal courage out of the view of upper management sometimes want it that way. Their strength of character—what they do speaks louder than what they say—makes them a local icon, a local hero, even if it's just among departmental regulars. But then, isn't that the entire point: to decentralize safety responsibilities?

Of course, while the model we present here for leader development and culture change under less than ideal conditions is based on only limited empirical research, the model itself needs a field test to clarify these suggestions.

Bibliography

Barry, B. and Ohland, M. 2012. "ABET Criterion 3.f: How Much Curriculum Content Is Enough?" *Science and Engineering Ethics* 18 (2):369-392 (2012)

Center for Disease Control (CDC). 2017. 01/18/201. "School of Leadership and Management Development." *CDC Connects*. (Retrieved 1/18/2017) (http://intranet.cdc.gov/od/hcrmo/CDCU/schools/ilead.shtml.)

Denning, Steve. 2011. *The Leader's Guide to Storytelling: Mastering the Art & Discipline of Business Narrative*, San Francisco: Jossey-Bass.

Hesselbein, F., and Shinseki, E. K. 2004. *Be, Know, Do: Leadership the Army Way: Adapted from the Official Army Leadership Manual.* San Francisco, CA: Jossey-Bass

Hannah, S. T 2010. "The Authentic High Impact Leader." In *Leadership Lessons **from** West Point*. New York:Jossey-Bass.

Hesselbein, F. 2002. *Leader to Leader*. San Francisco: Jossey-Bass.

Howe, N., & Strauss, W. 2004. *Millennials Rising: The Next Great Generation.* 3rd ed. New York: Random House

Ott, J. S. 1989. *The Organizational Culture Perspective.* Chicago: Dorsey Press.

Schein, E. H. 2004. *Organizational Culture and Leadership.* San Francisco: Jossey-Bass.

Winn, G.L. 2016. Practical *Leadership Skills for Safety Professionals and Project Engineers*. New York: CRC Press.

Winn, G.L. and Banks, B.B. "Safety Leadership: Insights from Military Research." *Professional Safety*, January, 2014, pp. 33-38.

Winn, G.L., Williams, A. and Heafey, M. 2013. "A Research-Based Curriculum in Leadership, Ethics and Protocol for Safety Management and Engineering Students." *Safety 2013: Proceedings of the 2013 Professional Development Conference, American Society for Safety Engineers,* June 24-27, Las Vegas, NV

Winn, G.L., Seaman, B. And Baldwin, J.C. 2004. "Fall Protection Incentives in the Construction Industry: Literature review and Field Study." *International Journal of Occupational Safety and Ergonomics.* No. 1, Vol. 10, p. 5-11.

Winn, G.L. and Slagley, J. R. 2016. "Values-Congruent vs. Values-Artifact Leadership: How

Are They Different?" *Proceedings of the American Society for Engineering Education Annual Conference.* New Orleans, LA, 2016.

Acknowledgement

The authors would like to express their appreciation to CRC/Taylor and Francis Group, and in particular to Ms. Cindy Carelli, for extending to us permission to cite key passages in this manuscript from the book, *Practical Leadership for Safety Professionals and Project Engineers,* 2016.

Session No. 720

Process Safety Management Best Practice: Safe Work Permit Management System

Jonathan Zimmerman, M.S., CSP, CHMM
EHS Manager
Kellogg's
Cincinnati, Ohio

Bryan Haywood, M.S.
Founder and President
Safteng.net
Milford, Ohio

Introduction

Many job tasks that expose workers to serious hazards in all industries are non-routine, maintenance-type activities that must be managed to control risks. These hazards are controlled in industry through written safework permits that are completed prior to initiation of these non-routine activities. Safe work permits are administrative controls and must be developed, implemented, and managed properly in order to be effective at managing risk. Examples include hot-work permits, contractor-work permits, confined-space-entry permits, and line-break permits. Having a safe work permit management system to manage and review the effectiveness of these safe work permits is a proactive best practice. This management system must follow the continuous improvement–plan, do, check, act–process. Safety professionals and process safety practitioners can use this paper to guide them in the development of their own safe work permit management system and program.

The Need for a Safe Work Permit System

A survey conducted by a safety agency in the UK showed that a third of all accidents in the chemical industry were maintenance-related, the largest single cause being a lack of, or deficiency in, safe work permit systems. In a study of small- and medium-sized chemical factories, the HSE found:

- Two-thirds of companies were not checking systems adequately;
- Two-thirds of permits did not adequately identify potential hazards;
- Nearly half dealt poorly with isolation of plant, electrical equipment, etc.;
- One-third of permits were unclear on what personal protective clothing (PPE) was needed;

- One-quarter of permits did not deal adequately with formal hand-back of plant once maintenance work had finished; and
- In many cases, little thought had been given to permit form design.

Safe Work Permit System Defined

A safe work permit system (SWPS) is a formal written system used to control certain types of work that are considered non-routine and present potential hazards or take place in potentially hazardous work locations. A safe work permit (SWP) is a *written* document that specifically defines the work to be done *and* the specific precautions to be taken. However, we *must* recognize that SWPs are merely an administrative control and will *only* bring the level of safety desired/needed when the permit-to-work system is *developed, implemented*, and *managed* properly. Safe work permits provide instruction and direction on the task at hand and ensure that steps are not forgotten or misunderstood. These permits serve as checklists that provide the means to review hazards and protective measures to all personnel involved in the task.

A SWP is needed when hazardous work activity can only be carried out if normal safeguards (i.e., written and approved procedures) are not in place to manage the risks associated with the work or when new hazards are introduced/created by the work. Examples are entry into vessels, hot work, elevated work, live electrical work, excavating/trenching and line breaking. In some industries, "contractor work permits" have become a popular means to control work performed by contractors within PSM/RMP process boundaries.

Examples of tasks that require a safe work permit:

- Major and minor maintenance work
- Contractor Work
- Inspections
- Hot Work
- Cleaning
- Confined Space Entry
- Excavation
- Working at Heights
- Scaffolding
- Line Break
- Live Work Permits–Electrical

Safe Work Permit System Development

When our safe work permit system is *developed* with operators, maintenance, engineering and EHS personnel we should have a solid, well-rounded permit to use. When our permit to work system is *implemented* properly, which includes addition to the "new employee orientation," the "contractor orientation" program, and *all* management personnel are trained in its use and function, we should have laid the groundwork for the program to be successful. However, no management system is successful if it is *not* managed properly. This means that there *must* be an auditing element of the system, as well as measurable goals of the system.

One place to start when developing a safe work permit system is a review of all legal and regulatory requirements and Recognized and Generally Accepted Good Engineering Practices. A safe work permit system must be established in writing, and the scope of written program must specify the high-risk tasks requiring work permits and that it applies to all personnel completing the high-risk task, including contractors. Regulatory requirements should also be listed in the written program. Responsibilities and accountabilities for the safe work permit system must also be clearly defined. A risk assessment of critical tasks (especially maintenance and cleaning tasks) must be done early in the safe work permit system development process with consideration given to probability and hazard(s).

Qualifications of person(s) that can issue safe work permits must be clearly defined in the written program. Details on how the safe work permit system works must be written into the program including the following:

- When to start the permit process
- Where to get permits
- How to fill the permit out
- Where to hand in completed permits
- Who must be informed of permit use

General information common to safe work permits includes:

- Name of employees and contractors performing work
- Exact details of work to be performed and location of work
- Date/time
- Hazards identified
- Preparatory requirements (air monitoring, lockout, ventilation, etc.)
- Steps to follow to complete work safely
- Personal protective equipment (PPE) required
- Emergency response equipment needed
- Means to summon emergency help
- Signature of safe work permit issuer and all workers working under the work permit
- Closure of permit with date and signature

After the safe work permit system is finalized and approved, it must be communicated to all personnel and contractors. Initial and refresher training must also be provided. Remember that permit issuers must have more detailed training and testing to ensure they qualified to issue safe work permits. Finally, to close the plan–do–check–act loop, the safe work permit system must be tested, measured, and evaluated. The results of this review must then be addressed in order to continuously improve the system. Feedback should be sought from all personnel and contractors involved in the safe work permit process for ideas on how it might be improved. Each facility must determine and define what the permit review process should cover at a determined frequency (i.e., monthly or quarterly). Examples of evaluation processes to be employed include:

- Spot checks of the work permits while they are being utilized
- Review of completed safe work permits
- Investigations of any incident involving a safe work permit or lack of utilization of safe work permit.

Safe Work Permit System Assessment

If you have an existing safe work permit system or are trying to develop a safe work permit system, a best practice system and written program must address the following questions:

- Is there clear identification of who may authorize particular jobs (and any limits to their authority)?
- Is there clear identification of who is responsible for specifying the necessary precautions (e.g., isolation, emergency actions, etc.)?
- Is the permit clearly laid out?
- Does it avoid statements or questions that could be ambiguous or misleading?
- Is it designed to allow for use in unusual situations/working conditions?
- Does it cover contractors?
- Are those who issue permits sufficiently knowledgeable concerning the hazards and precautions associated with the plant and proposed work?
- Do they have the training and experience to ask enough "what if" questions to enable them to identify all potential hazards?
- Do staff and contractors fully understand the importance of the SWP system, and are they trained in its use?
- Does the permit clearly identify the work to be done and the associated hazards?
- Can plans and diagrams be used to assist in the description of the work to be done, its location, and limitations?
- Is the plant adequately identified, e.g., by discrete number or tag to assist issuers and users in correctly taking out and following permits?
- Is a detailed work method statement given for more complicated work?
- Does the permit to work system require the removal of hazards and, where this is not reasonably practicable, effective control?
- Are the requirements of the applicable OSHA standards (1910.146, .252, etc.) and other relevant codes/standards (NFPA 70E) known and followed by those who issue *and* work under the permits?
- Does the permit state the precautions that have been taken and those that are needed while work is in progress? For instance, are energy isolations specified, and is it clear what PPE should be used?
- Do the precautions cover residual hazards and those that might be introduced by the work (e.g., welding fume(s) and vapor(s) from cleaning solvents)?
- Does the permit contain clear instructions about how the job should be controlled or abandoned in the case of an emergency?
- Does the permit have a return-to-operations procedure incorporating statements that the work has finished, and that the plant has been returned to production staff in a safe state?
- Are time limitations included, and is shift changeover managed?
- Are there clear procedures to be followed if work has to be suspended for any reason?
- Is there a system of cross-referencing when two or more jobs subject to permits may affect each other?
- Is the permit displayed at the job?
- Are jobs checked regularly to make sure that the relevant permit to work system is still relevant and working properly?

Conclusion

A safe work permit system is a critical cog in the process safety wheel ensuring that covered processes are safely operated and maintained. As with any other administrative control, a safe work permit system must be constantly reviewed and improved following a formalized plan-do-check-act cycle. Safety professionals and process safety practitioners have been provided tips and best practices to ensure that their own safe work permit programs are appropriate or, if a program does not exist, a new safe work permit program can be created and implemented.

Bibliography

Industrial Accident Prevention Association (IAPA). 2007. *A Health and Safety Guideline for Your Workplace – Work Permits.* (www.iapa.ca).

UK Health and Safety Executive. *Permit to Work Systems.* (www.hse.gov.uk/comah/sragtech/techmeaspermit.htm).

Does Participatory Safety Education and Training Increase Learning?

Menekse Salar, Doctoral Candidate
M. Fehmi Capanoglu, Doctoral Candidate
Ashley Garner, Undergraduate Research Assistant
Xuanxuan "Avery" Zhang, Doctoral Candidate
Richard Sesek, PhD
Tom Devall, MAS
Robert Thomas, PhD

Auburn University
Auburn, Alabama

Introduction

The way college classes are taught influences the way the material is learned, retained and, one day, used in the real world to impact real people. In order to make a class not only interesting and entertaining but impactful and lasting, a professor must fully engage the students. In all levels of education, students are expected to attend classes. It is believed that getting knowledge from the class and being successful in the class is more likely to happen with attending classes. Several research studies have agreed that students benefit from attending classes (Clark et al., 2011; Cohn et al., 2006). The idea behind this practice is that the additional exposure to the material in a more active, hands-on environment will increase a student's understanding of what is being taught. This should, in theory, show overall grade improvement in the course. Another researcher claimed that attending classes was one of several factors that positively contribute to a student's academic success (Dollinger et al., 2008). In the literature, a negative correlation between absences and course grade has been shown, and this relationship strengthens as the number of absences increases (Silvestri, 2003).

Experiential learning is learning through experience and hands-on activities both in the classroom as well as outside the classroom. These activities provide opportunities to fully apply the material in practice. By doing so, students should understand the concepts taught in class more wholly and, thus, perform better in the course. Thorne (2000) examined whether providing extra-credit opportunities in the class can encourage student participation as well as motivating them to learn class materials well. We are interested in studying the impact of participation in these voluntary, class-related extracurricular and extra-credit activities on classroom performance. We wish to study how such participation impacts performance in terms of final grades in the course. The purpose of this study is to see how participation in voluntary out-of-the-classroom, but classroom-related, extra-credit assignments influences overall performance in the class.

Methodology

This study examines student success as it relates to both attendance and participation in extra-credit opportunities. It has been shown that students who miss class for circumstances beyond their control might be impacted by absence to a lesser or greater extent than those that simply choose not to attend (Durfee et al., 2012). For the present study, excused absences were counted as attendance, since students with excused absences were provided video of the lectures they had missed. Only the total percentage of attendance was recorded for each student. Data were collected from an undergraduate occupational safety and health course ("Occupational Safety and Ergonomics," INSY 3020) at Auburn University. Data were aggregated from two separate offerings of this course (Spring 2015 and Spring 2016) with 279 total students. Absence data were collected using the iClicker system, which is a classroom response tool that can be used for quizzes, student polls, and, in our case, was used for both quizzes and recording attendance. This system facilitates rapid collection of student quiz and attendance information. Classroom polling devices, such as the iClicker system, can help minimize cheating related to copying another's quiz. In addition to attendance, the impact of extra-credit participation on overall success was also measured. For extra-credit opportunities, students had the chance to submit extra credit until the end of semester. This Occupational Safety and Ergonomics class is offered to a variety of students, with disciplines ranging from business to engineering, and can have as many 150 students. There are five possible extra-credit opportunities that are presented to students beginning on the first day of class.

Data were analyzed at the completion of the semester after all student-identifying information was removed from the gradebook (e.g., names, identification numbers, etc.). The anonymous database was analyzed by comparing final grades, stratified by participation in various extra-credit opportunities and by attendance. To determine the impact of each activity, final grades were computed both with and without the points related to the extra-credit activity. Classes met twice per week for 75-minute lectures, delivered in a large seminar/conference room. For this analysis, absolute grading standards (A ≥ 90, B ≥ 80, and so on) were used. Attendance accounted for 5% of each student's final grade. However, all analyses were run excluding these attendance points. A total of up to 5 % extra credit was available to students, making it possible for students to score above 100 % in the course. Several outputs were considered, including "raw" grades with no extra-credit or attendance points considered, and a "total grade" that included all extra points and attendance points that the student had earned. The extra credit consisted of five different activities, each worth up to 1%, that would be added to compute a total overall grade. It was possible for students to earn greater than a 100% total final grade. The five activities were:

1. Joining a professional safety- and health-related society, such as ASSE, HFES, AIHA, etc.;
2. Submitting questions for exams or quizzes;

3. Bringing, sharing, presenting relevant safety material in class;
4. Participating in online safety and health discussion boards (Canvas Learning Management System)' prompted by faculty and teaching assistants (TAs) and contributed to by the class; and
5. Attending a safety-and-health-related seminar and uploading a summary of lessons learned to an online submission system (Canvas Learning Management System).

Results

Comparisons were made to the student's unadjusted, raw scores (without points for extra credit or for attendance) to determine if active engagement activities resulted in better academic performance, as measured objectively by quizzes and exams. The results were clear: Students who participated in extra-credit activities performed better than students who did not. Furthermore, the greater the level of participation, the better the students performed. This preliminary study demonstrated a clear association between participation level and higher grades. However, the direction of causality cannot be definitively established. For example, while increased participation may lead to better grades, it is also possible that more motivated, higher performing students simply participate more. Ongoing research is considering incoming student grades so that this factor can be controlled.

Table 1 shows the mean benefit of participation in extra credit beyond the extra-credit points themselves (e.g., raw score differences). The mean raw scores of students participating in every extra-credit activity were statistically significantly higher than the mean scores of students who did not participate in each of those activities. Attending a safety-related seminar was associated with the largest difference between participators and nonparticipators, with a difference in overall raw grade of ~4.3 percentage points. Even something as simple as joining one or more of the student safety society chapters (ASSE, HFES, AIHA, etc.) was associated with better overall raw grades (~3.4 percentage points higher).

Table 1. Comparison of Participators and Nonparticipators in Extra-credit Activities

	In-Class Discussion Material	Submit exam questions	Online Forum Participation	Safety Society Membership	Attendance of a Safety-related Seminar
P-value	0.0219*	<.0001*	0.001*	0.0011*	0.0007*
Difference Raw Mean Score of Those Submitting Minus Raw Mean Score of Those NOT Submitting	2.79	4.28	3.48	3.44	4.37

Figure 1 illustrates the increasing trend of overall raw score with participation in multiple extra-credit activities. The greater the percentage of extra credit earned, the higher the raw score. Figure 1 also includes total score (includes extra credit and attendance credit) to illustrate the actual benefit enjoyed by those participating in extra-credit activities.

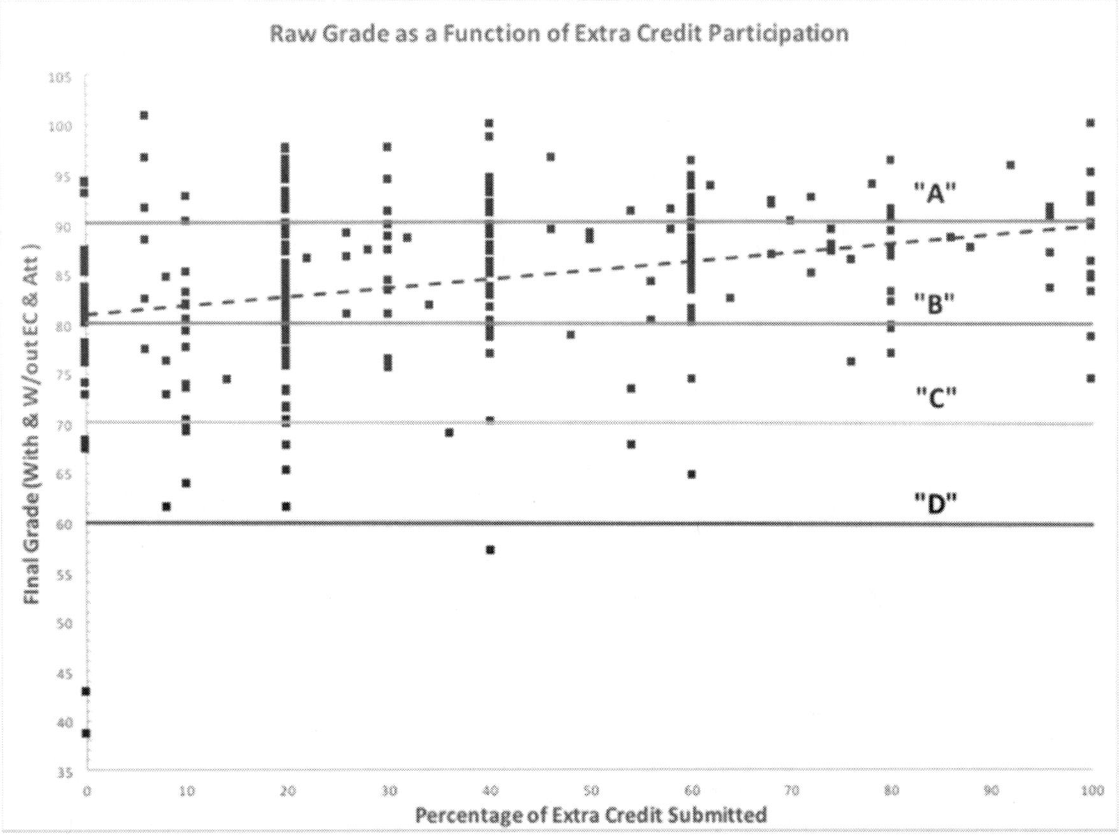

Figure 1. Grade as Function of Extra Credit Participation Percentage

Figure 2 shows the boxplot of mean raw scores, graphed as a function of number of activities submitted. The plot clearly illustrates the upward trend of scores with increasing participation. It should also be noted that less participation was not only associated with lower mean scores, but also a wider variance in scores and decreases in the observed lower bounds.

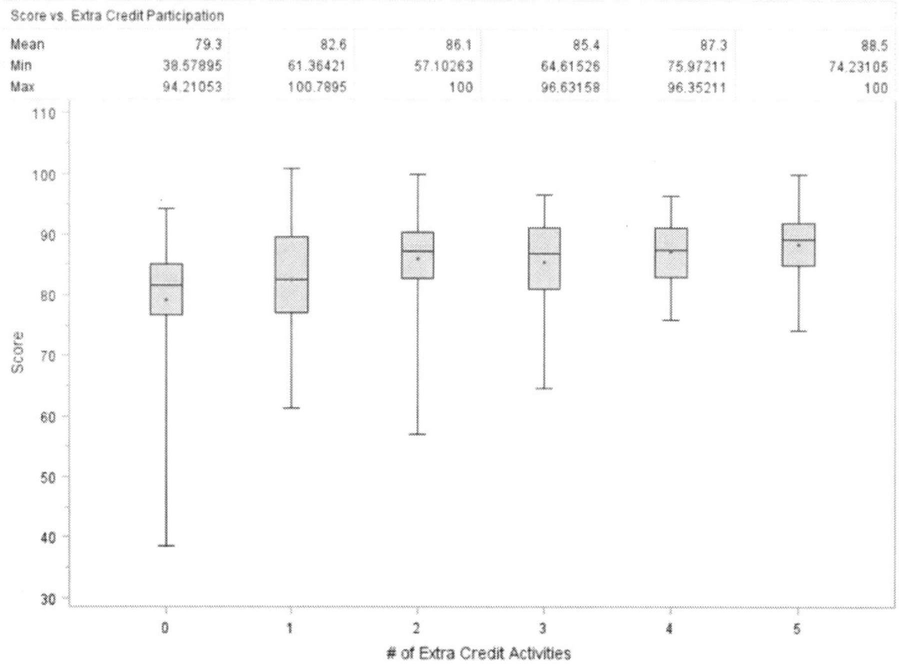

Figure 2. Boxplot of Mean Raw Scores as a Function of Number of Extra-credit Activities

As with previous studies, attendance was clearly and positively associated with both raw and total grades. Figure 3 shows the trend. The more that students attended class, the higher both their raw and total grades. Similar to the extra-credit trend, lower attendance was related to lower grades, greater variance, and smaller lower bound (e.g., the lowest grades occurred when participation was lowest).

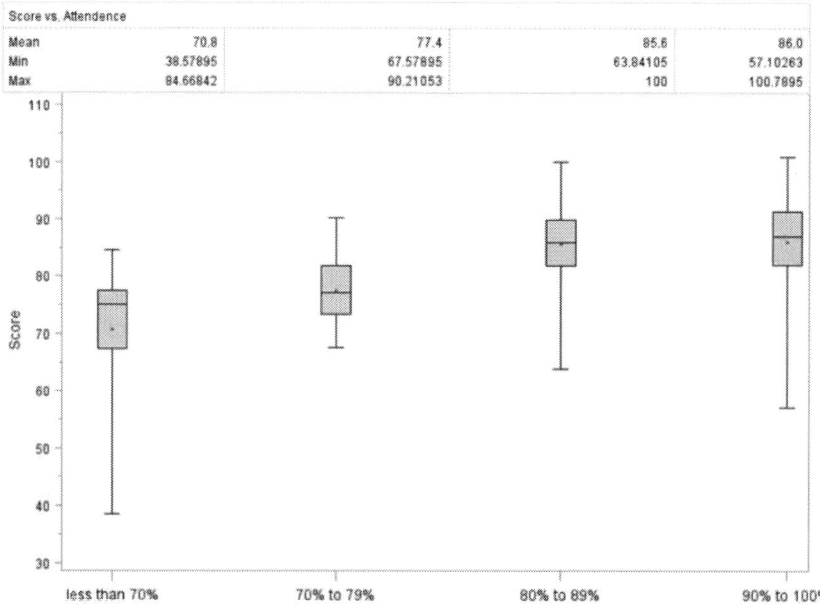

Figure 3. Boxplot of Mean Raw Scores as a Function of Attendance

An interesting observation is that students who participated in 4-5% extra credit actually averaged a raw score of "B+" (88.5%). Those not participating in any extra credit at all averaged a "C+" (79.3%).

Discussion

There are clear, positive and statistically significant associations between overall academic performance and both attendance and participation in extra-credit activities. The greater the attendance or participation level in extra credit, the greater the overall grade. Of course, differences were both greater and more significant when the points for extra credit and attendance were factored into the grade. Another interesting association is that the activities that involved the most effort by students were also associated with the largest differences from non-participating peers. Attending a safety-related seminar and submitting a summary report, as well as providing example test and quiz questions, required the most effort and involvement on the part of students. It is not surprising, therefore, that these activities demonstrated the largest and most statistically significant differences (e.g., potential benefits beyond the extra credit itself).

This study did not control for the incoming grades of students and, therefore, may be measuring differences between "good" and "poor" students, with the higher achieving students more likely to both participate in extra-credit activities and attend class. Another study is underway that considers these grades to provide a more complete picture of the potential benefits of extra-curricular, safety-related content.

It is also hypothesized that the apparent benefits observed here would be similarly realized in occupational training environments. Engaging employees in activities outside of the actual training class might significantly increase their safety learning. For example, students who simply participated in online discussion forums averaged 84%, while those that did not averaged 81%. Providing workers with similar online discussion opportunities could similarly improve retention of safety-related information. Likewise, asking them to photograph examples of both good and poor safety to bring to the next safety training meeting might also improve engagement and, therefore, learning and retention. Having employees "practice" safety outside of work might actually motivate them to develop safety habits that translate to a safer home environment, as well as reinforcing their commitment to safety in the workplace.

Bibliography

Clark G., Gill N., Walker M., Whittle R. 2011. "Attendance and Performance: Correlations and Motives in Lecture-Based Modules." *Journal of Geography in Higher Education*. 35(2): 199-215.

Cohn E, & Johnson E. 2006. "Class Attendance and Performance in Principles of Economics." *Education Economics*. 14(2):211-233.

Dollinger S., Matyja A., & Huber J. 2008. "Which factors best account for academic success: Those which college students can control or those they cannot?" *Journal of Research in Personality*. 42(4):872-885.

Durfee, J. K., Loendorf, W. R., Richter, D. C., Eastern, T. L. D., & Munson, D. M. 2012. "A Formal Research Study on Correlating Student Attendance to Student Success." *American Society for Engineering Education*.

Silvestri L. 2003. "The Effect of Attendance on Undergraduate Methods Course Grades." *Education*. 123(3): 483.

Thorne, B. M. 2000. "Extra credit exercise: A painless pop quiz." *Teaching of Psychology*. 27 (3): 204-219.

American Society of Safety Engineers
Professional Development Conference June 19-22, 2017

SAFETY 2017
ASSE

Session: 722

Fundamentals of SH&E: Basic Construction

Carl W. Heinlein, CSP, ARM, CRIS, CSHM
Senior Safety Consultant
American Contractors Insurance Group

Safety 2017

Fundamentals of Basic Construction OSH

Culture – Values – Zero Injury

Carl W. Heinlein, CSP, CRIS, ARM
Senior Safety Consultant

2

SAFETY 2017 ASSE

3

Learning Objectives

SAFETY 2017 ASSE

- Share specific construction OHS best practices
- Discuss emerging trends in construction
- Exchange ideas on how to manage continuous improvement

4

5

The Olympic Story

6

Olympic Pride
2012 London Olympics

SAFETY 2017 ASSE

Top Olympic medal winners

Final tally – Sunday, August 12th

COUNTRY	G	S	B	TOTAL
United States	46	29	29	104
China	38	27	22	87
Russia	24	25	33	82
Britain	29	17	19	65
Germany	11	19	14	44

7

2012 London Olympics
Enormous Construction Challenge

SAFETY 2017 ASSE

- 5 major venues
- 10 new railway lines
- 100 hectares of parklands and waterways
- New infrastructure
- Residential high-rises

8

2012 London Olympics Manpower

46,000 people worked at the sites

14,500 peak workforce

77 million manhours

9

Olympic Safety Record

Barcelona'92 — 2 Deaths

Atlanta 1996 — 1 Death

Sydney 2000 — 1 Death

ATHENS 2004 — 14 Deaths

Beijing 2008 — 10 Deaths

Rio2016 — 11 Deaths

10

Olympic Delivery Authority (ODA) Under the Direction of Former Olympian Sebastian Coe

To deliver venues, facilities and infrastructure:

- On time
- Fit for purpose
- Protecting the health and safety for the project team

11

Olympic Delivery Authority (ODA)

Challenges and Solutions

- **Twelve (12) different languages on most sites**
- **Putting things in pictures**
- **Communication**
- **Pre-planning**
- **Recognizing success**
- **Daily stretch and flex**

12

2012 London Olympics

SAFETY 2017 ASSE

Difference Makers

- Respect
- Trust
- Empowerment
- Motivating the labor force to excel
- Clarity
- Transparency
- Shared commitment

13

Construction Industry Facts

14

The Construction Industry – OHS

SAFETY 2017 ASSE

- It saves lives
- Morally, the right thing to do
- Adds to company profitability
- Enhances company image
- Etc.

15

SAFETY 2017 ASSE

16

Common Safety "THOUGHTS"

SAFETY 2017 ASSE

- Safety is about preventing accidents!
- Accidents are "caused" by careless employees!
- Zero Accidents = SAFE!
- Accidents are unplanned, unexpected, fortuitous events!
- Safety is the job of the Safety Function!
- Discipline improves performance!
- Safety programs minimize loss costs!

17

SAFETY 2017 ASSE

18

US Construction Industry Facts

SAFETY 2017 ASSE

636,000 construction companies *

>7,000,000 persons employed in construction *

- General Builders
- Street and Road
- Industrial
- Homebuilding
- Specialty – Subcontractors

Construction vs. All Industries (risk)

- 17% work-related deaths
- 80,000 injuries

19

Construction Industry Facts

SAFETY 2017 ASSE

Out of 3,929 worker fatalities in private industry in calendar year 2013, 796 or 20.3% were in construction-that is, one in five worker deaths last year were in construction.

Eliminating the Fatal Four would save 468 workers' lives in America every year.

- Falls — 294 out of 796 total deaths in construction in CY 2013 (36.9%)
- Struck by Object — 82 (10.3%)
- Electrocutions — 71 (8.9%)
- Caught-in/between — 21 (2.6%)

20

21

OHS's Return on Investment

22

SAFETY 2017
ASSE

23

SH&E's Return on Investment

SAFETY 2017
ASSE

A recent survey shows:

- 61% of executives say $3 or more for each $1 invested in safety
- 95% of executives say workplace safety has a positive impact on a company's financial performance
- The survey also shed additional light on the two types of costs:
  - Direct & Indirect
  - 93% of the executives surveyed said they clearly see the relationship reporting that $1 of direct costs generates at least $3 - $5 of indirect costs

24

SAFETY 2017
ASSE

25

Safety Program vs. Safety Culture

26

Safety Management (R)evolution
Changing Organizational Culture
Quiet Transparent Integrated Equal
WORLD CLASS
The "Stealth" Organization
Progressive
Safety Responsibility:
Management Driven/Employee Owned
Perceived:
Good Business Investment
Management Characteristics
Accidents: Intolerable, no excuses
Safety is a measure of management effectiveness
Decisions: time consuming and difficult
Planning: long-term; 3-5 years
Responsibilities/Expectations: clearly defined
No glitz or hype
Employee Involvement: win/win
Communications: informal, open, encouraged
Accountability established, measured, recognized
Safety Is:
A CORE VALUE
Safe is how we do business
Critical Thinking Shift
Systemic
Process Design
Leader-Driven
Collaborative
Excellence
Safety Is:
A PROCESS
A system we manage
Radical Organizational Change
Line/Staff Conflict
Blood Cycles
Symptoms (Behaviors/ Conditions)
Committees
Quick Fix Programs
The NORM
Naturally Occurring Reactive Management
Traditional
Safety Responsibility:
Not Understood (a staff function)
Perceived:
A Cost (part of doing business)
Management Characteristics
Incidents: excused
Compliance is the goal
Likes "cookbook" approach
Little accountability
Lack of employee involvement
High visibility and glitz
Safety Is:
A PROGRAM
"Activities" we do
Significant Financial Crisis
High Insurance Costs
Excessive Losses
Adversarial Employee Relations
Litigation Statutory Ignorance
SWAMP
Safety Without Any Management Process
Safety Is:
IGNORED
That's why we buy insurance
27

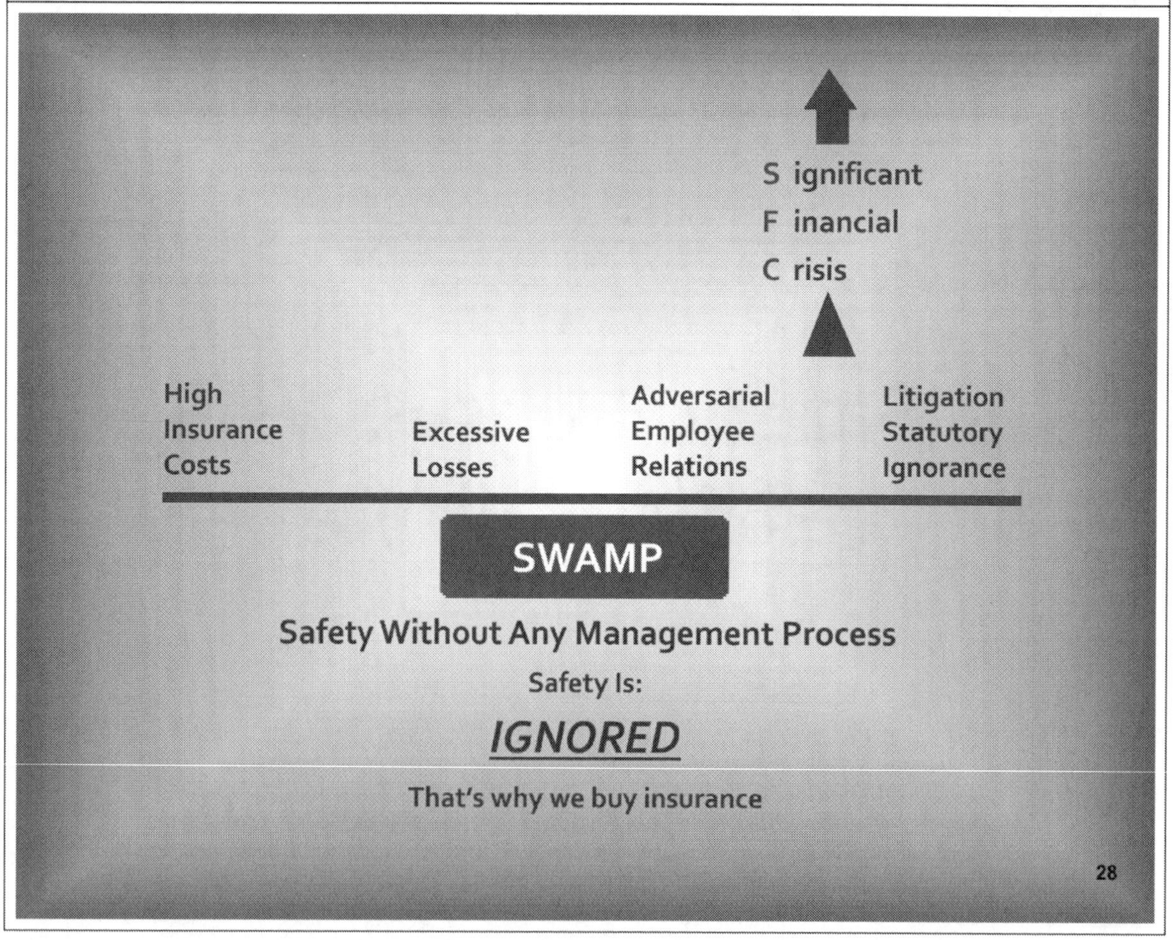
S ignificant
F inancial
C risis
High Insurance Costs
Excessive Losses
Adversarial Employee Relations
Litigation Statutory Ignorance
SWAMP
Safety Without Any Management Process
Safety Is:
IGNORED
That's why we buy insurance
28

Line/Staff Conflict | Blood Cycles | Symptoms (Behaviors/ Conditions) | Committees | Quick Fix Programs

The NORM

Naturally Occurring Reactive Management

Traditional

Safety Responsibility: Not Understood (a staff function)

Perceived: A Cost (part of doing business)

Management Characteristics

- Incidents: excused
- Compliance is the goal
- Likes "cookbook" approach
- Little accountability
- Lack of employee involvement
- High visibility and glitz

Safety Is:

A PROGRAM

"Activities" we do

29

Industry Leadership – "Best Practice"

Critical Thinking Shift

Systemic | Process Design | Leader-Driven | Collaborative

Excellence

Safety Is:

A PROCESS

A system we manage

30

Quiet | Transparent | Integrated | Equal

WORLD CLASS

The "Stealth" Organization

Progressive

Safety Responsibility: Management Driven/Employee Owned

Perceived: Good Business Investment

Management Characteristics

- Accidents: intolerable, no excuses
- Safety is a measure of management effectiveness
- Decisions: time consuming and difficult
- Planning: long-term; 3-5 years
- Responsibilities/Expectations: clearly defined
- No glitz or hype
- Employee Involvement: win/win
- Communications: informal, open, encouraged
- Accountability established, measured, recognized

Safety Is:

A CORE VALUE

Safe is how we do business

31

Your Construction Company

SAFETY 2017 ASSE

So, where does your company fit in....

SWAMP __________

NORM __________

EXCELLENCE __________

WORLD CLASS __________

32

SAFETY 2017 ASSE

33

Proven Actions and Focus Areas

34

Construction Industry Institute (CII) Zero Accidents Study Findings – 1993

SAFETY 2017 ASSE

High-impact, zero accident techniques

- Pre-project/Pre-task planning for safety
- Safety orientation and training
- Written safety incentive programs
- Alcohol and substance abuse programs
- Accident/Incident investigations

35

Best Practice Results

SAFETY 2017 ASSE

Over 30 key findings revealed projects utilizing these best practices have significantly lower recordable injury rates.

36

Construction Industry Institute (CII) Zero Accidents Study Findings – 2001

SAFETY 2017 ASSE

Getting to Zero

- Demonstrated management commitment
- Staffing for safety
- Safety planning – pre-project/pre-task
- Safety training and education
- Worker involvement and participation
- Recognition and rewards
- Subcontractor (General) management
- Accident/Incident reporting and investigation
- Drug and alcohol testing

37

Other OHS Exposures

38

Other Construction OHS Exposures

SAFETY 2017
ASSE

- *Fleet – MOST OVERLOOKED ASPECT OF AN OHS PROGRAM*
- Workforce – Hire Smart/Fire Smart
- Health Wellness – Aging Workforce
- Emerging Issues – Technology
- Subcontractors/Vendors/Suppliers – Valued Partner?

39

SAFETY 2017
ASSE

40

Path to ZERO!

41

Your Company Zero Injury Safety Values

SAFETY 2017
ASSE

- We conduct every aspect of our business in a manner that avoids getting people hurt.
- Performance of our work follows proper planning to ensure safe completion.
- Every accident is preventable. Until injury of our people is eliminated, we have not done enough.
- We do not compromise our safety values based on perceived schedule, cost or production needs.
- Each employee shares the responsibility for the safety of their co-workers and will fulfill that responsibility by their own work practices and by bringing problems to the attention of others.

42

Your Construction Company

SAFETY 2017 ASSE

Your Safety Program is…..

43

SAFETY 2017 ASSE

44

The Personal Challenge

SAFETY 2017 ASSE

- Safety is for YOU and your Company family
- Safety has to become one of your core VALUES
- YOU have to "want to"
- Accountability:
  - The little things ARE important
  - Set a good example
  - Tell your people what you expect from them
  - Correct bad behavior
  - Discipline
- BE UNREASONABLE!

45

Robert W. Campbell, 1914

SAFETY 2017 ASSE

"Safety work is today recognized as an economic necessity, and one of the most constructive movements that has ever come into our National life. It is the study of the right way to do things and is… one of the greatest aids to shop efficiency and economy."

Robert W. Campbell, 1914

46

Resources

SAFETY 2017 ASSE

- Occupational Safety and Health Administration (OSHA)
  - https://www.osha.gov/SLTC/etools/construction/index.html
- National Occupational Safety and Health Institute (NIOSH)
  - http://www.cdc.gov/niosh/construction/
- Electronic Library of Construction Occupational Safety & Health (elcosh)
  - http://www.elcosh.org/index.php
- The Center for Construction Research and Training (CPWR)
  - http://www.cpwr.com

47

Resources

SAFETY 2017 ASSE

- Associated Builders and Contractors (ABC)
  - https://www.abc.rog
- Associated General Contractors of America (AGC)
  - https://www.agc.org
- National Association of Homebuilders (NAHB)
  - https://www.nahb.org

48

SAFETY 2017
ASSE

Carl Heinlein

Senior Safety Consultant

American Contractors Insurance Group

Direct: 724-935-1650

Cell: 412-818-1161

carl.heinlein@acig.com

Building better builders through Safety and Quality excellence.

49

Thank You

SAFETY 2017
ASSE

Questions?

50

Session No. 725

No Weak Links in Your Supply Chain

Ann Minzner Conley, ScM, CSP
Chubb Limited
Whitehouse Station, New Jersey

Lesli Johnson, CSP, ARM
Chubb Limited
Portland, Oregon

Introduction

A basic definition of supply chain is the "series of business relationships specific to the production of a product, responsible for bringing a product to market." The supply chain can be made up of a few entities in a tight regional area, or it can be a broad international network, influenced by a variety of regulatory schemes, customs, and contract specifications.

Supply-chain disruptions can have a variety of adverse impacts on an organization, including loss of income, market share, reputation and even decreased shareholder value. They can impact product integrity, weaken infrastructures, and even put workers at risk. Costs to respond to supply-chain disruptions and continue operations can incur significant extra expense over standard operating expenses. Many companies carry business income insurance, but even with immediate funds available, without adequate planning and action, a company may close following a disaster or significant supply-chain disruption. Industry studies repeatedly show the struggles companies have in recovering from natural and man-made disasters, data breaches, and other-business interruption events.

While many are aware of the highly publicized, large-scale events with immediate global supply-chain impact, companies face supply-chain challenges of varying degrees on a continual basis. In the 2015 Global Cleantech Risk Survey conducted by the Chubb Group of Insurance Companies and Cleantech Group, 61% of the 300 Cleantech executives surveyed had experienced a supply-chain disruption in the past three years, with the majority (84%) of these events resulting in an impact on the companies' bottom lines. Yet 70% of companies surveyed did not have a written supply chain disruption plan, which could have mitigated some of the affects[1] (Exhibit 1).

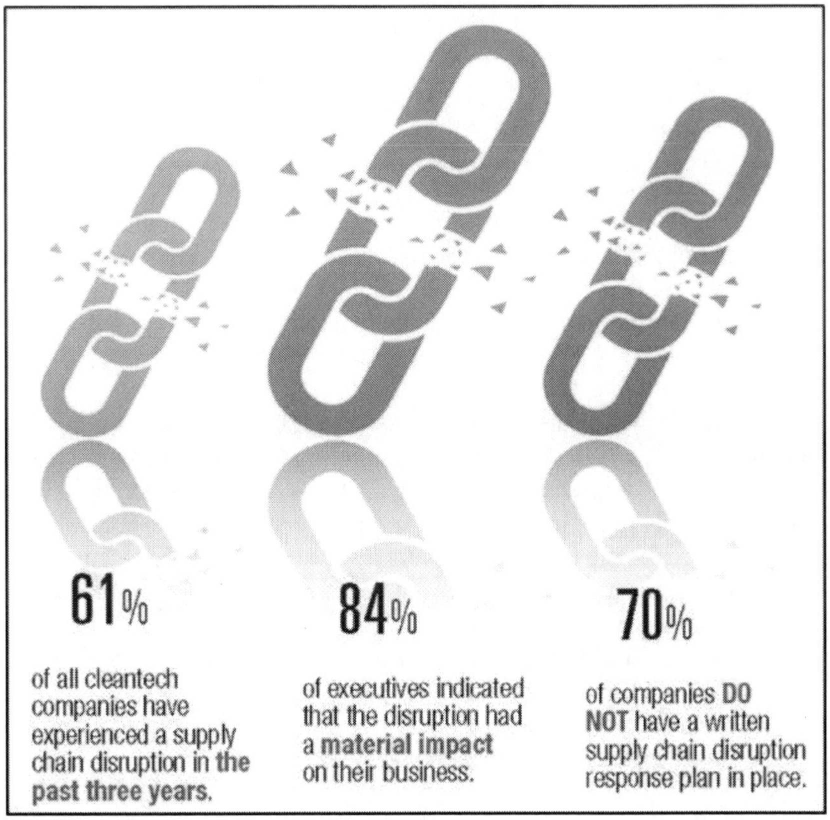

Exhibit 1. 2015 Global Cleantech Risk Survey

The purpose of this paper is to address risks to product and process integrity that can be addressed through supply-chain resiliency measures, as well as to highlight the role the safety professional can have in identifying supply-chain risks and contributing to the development and implementation of risk-reduction measures within their organization.

Examples of Incidents

Events in the last 20 years that have had significant supply-chain impacts include Tropical Storm Katrina in Louisiana in 1999; Europe's Winter Storm Lothar in 1999; the Eyjafjallajökull volcanic ash plume in 2010 that disrupted air travel in northern Europe; the 2011 Tōhoku earthquake, which triggered a tsunami that damaged the Fukushima nuclear plant in Japan; and most recently, the Tianjin port explosion. In 2013, cyber issues contributed to over 50 percent of

[1] Chubb Group of Insurance Companies and Cleantech Group, 2015. "The Hidden Challenge: What Lies Beneath Cleantech's Global Plans."

supply chain disruptions.[2] Companies can have first, second, third and even fourth tier suppliers that are also vulnerable to such disruptions. Sub-tier events are often out of the company's control, which can make it difficult to trace and control these risks.

In March of 2012, an explosion at a German plastics manufacturer, Evonik, killed two workers and destroyed most of the facility. It was thc company's sole production unit. The company had highly specialized manufacturing processes that produced engineered plastic resins that were used to make a certain type of nylon, valued for its high thermal and corrosion resistance. Due to these unique manufacturing and product features, the company was the primary global source for raw materials that were used to manufacture critical components in a wide variety of industries, such as the automotive industry, which relied on the nylons made from Evonik's resins for the production of automotive fuel lines and brake hoses. In the immediate aftermath of the explosion, two of Evonik's customers that supplied the automotive and medical device fields invoked, *force majeure* contractual provisions, due to their inability to fill customer contracts.[3]

In another example, a chemical exporter that purchased chemicals in bulk, repackaged and distributed them to the U.S. and elsewhere, was struggling to produce comprehensive Safety Data Sheets (SDSs). They sought consensus from their insurer to revise expectations regarding completeness of their safety data sheets. By engaging an insurance loss-control engineer, the insurer was able to: (1) make the foreign company aware of U.S. requirements for safety data sheets; and (2) provide resources to assist them in developing purchasing criteria focused on only sourcing chemicals from suppliers that could provide all of the information needed to create an adequate safety data sheet. This collaboration between the safety professional, the multi-national's risk management department, and their purchasing department resulted in improved processes, as well as enabling to company to continue to meet its stated corporate goals of global leadership in the areas of environmental health and safety.

Evaluate Suppliers

From a product integrity standpoint, factors that can lead to increased supply-chain risk include:

- Significant differences in regulatory climate between the supplier and the OEM
- Sourced materials and products that are vulnerable to tampering and counterfeiting
- Sourced materials are commodity products that would be difficult to trace back to the original supplier

Suppliers of raw materials, component parts, services, and finished products should be mapped out to determine how critical they are to profitable company sales. An important part of this assessment is to work back through the supply chain and determine which second- and third-tier suppliers could have a significant impact on business and should therefore also be classified as critical suppliers. A single piece of critical machinery or use of a third-party testing lab could create a bottleneck and, if either one is unavailable, could stop production or sales.

[2] Johnson, Denise. "The Changing Landscape of Business Interruption Claims." *Claims Journal,* June 3, 2015.

[3] Smock, Doug. 2012."Lessons from the Nylon 12 Supply Disruption." (http://www.mypurchasingcenter.com).

From a process integrity standpoint, assessments of companies should include evaluation of business contingency plans, geography (natural hazards and transportation), and labor practices track records. Supply-chain risk-management efforts might need to be tailored to the relatively critical nature of each component of the supply chain. Companies will need to decide what aspects of supply-chain risk management they retain in house, and when they will rely on third-party services, such as third-party testing labs for overseas suppliers.

Strategies

From most effective to least effective, continuation and recovery strategies include:

1. **Duplication and redundancy**. This is the most expensive strategy and therefore used primarily by companies that need duplicate operations for business reasons and companies that cannot afford even one minute of downtime.
2. **Alternate sites can be located in multiple regions**. When put into use, there may be some delay in operations, and staff may need to be relocated.
3. **Mutual aid agreements may be an option for some industries.** This option is less expensive but may include extra expenses. Capacity could be an issue at the other site. Confidentiality should be addressed before pursuing this option.
4. **Manual procedures.** Many companies have older equipment that can be used as backup or revert to more manual operations. This option can be started fairly quickly, but production may be much slower.
5. **No Plan**. Doing nothing is always an option. It has no ongoing cost or testing, but in the event of a disruption, the lack of a plan can significantly increase the impact, time to recover, financial loss, and loss of customers.

Exhibit 2 displays the continuation and recovery strategies reported by executives that participated in the 2015 Global Cleantech Risk Survey.[4]

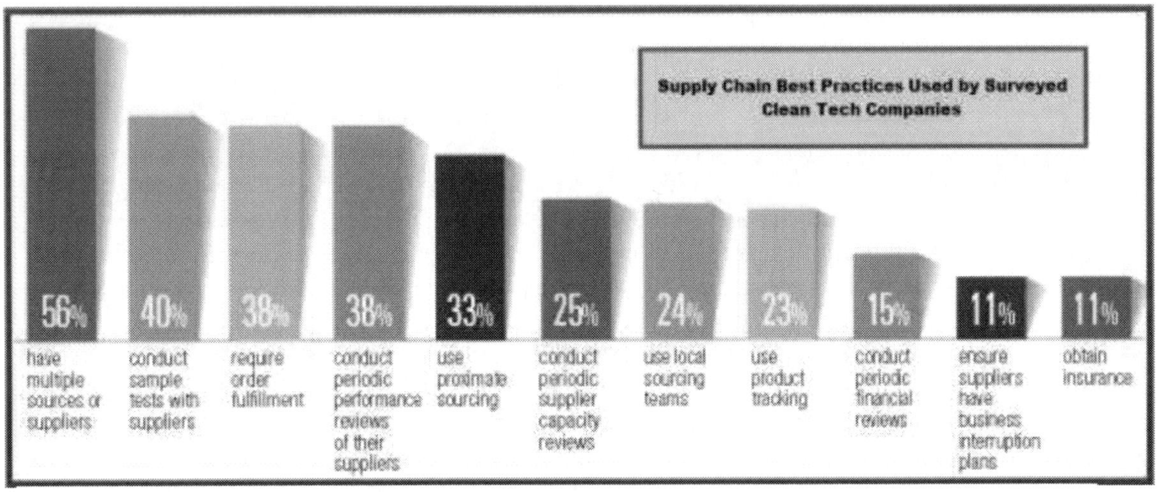

Exhibit 2. 2015 Global Cleantech Risk Survey

[4] Chubb Group of Insurance Companies and Cleantech Group. 2015. "The Hidden Challenge: What Lies Beneath Cleantech's Global Plans."

Risk Management Programs

The four components of a business contingency plan are: avoidance, emergency plan, continuity plan, and recovery.

Avoidance

Avoidance requires proactive planning and activities that will reduce the likelihood of an interruption. Examples include selecting locations that aren't prone to disasters, and housing operations in protected buildings built to current structural codes. For supply-chain exposure, this means evaluating suppliers' exposure to catastrophes. Where it is critical to company operations, supply-chain resiliency should be a part of supplier qualification. This can include requiring that the supplier provide a business continuation plan, a product recall plan, as well as clear communication of performance and quality metrics, along with establishing consequences for failing to meet these metrics.

As data becomes more transparent, it is becoming easier for companies to monitor and validate their supply chain. For example, Underwriters Laboratories, Intertek and others offer online portals to look up a company's product testing certificates. These online portals provide a cost-effective method to begin qualifying alternative suppliers.

Emergency Plan

The emergency plan addresses actions to take during an event to reduce potential damage. The most obvious examples are emergency response and evacuation procedures and product recall plans. In today's connected environment, also consider the need to manage the news of a supply-chain disruption via social media and other news outlets. Remember to notify regulators as appropriate.

Business Continuity Plan

The business continuity plan outlines actions to take to ensure a company stays in business immediately after a disaster occurs. The actions taken at this point are critical, because the longer it takes a company to restore operations, the more likely it becomes that the company won't recover. Having pre-qualified secondary suppliers is critical to this part of the plan.

Restoration Plan

A restoration plan typically has several phases, based on timeline and/or operations. It starts with restoration of the most critical activities, and adds more operations over time until the business is back to full capacity. This phase may take months or even years, depending on how significant the event was. Some key items to consider are adequacy of funding for extra expense, time needed to replace machinery and raw materials, and building permits.

Emergency, continuity, and recovery plans should be evaluated for all critical suppliers (including those in the second- and third-tier positions) since actions taken before, during, and after a disaster can mitigate the amount of damage and time required to function again. Evaluation should include analysis of vulnerabilities, priorities, adequate planning (solutions, costs, and resources), recovery-time objectives, and regular testing with updates may to reflect the test results. Plans can be evaluated against NFPA 1600, *Standard on Disaster/Emergency Management and Business Continuity Programs*. Some industries might have their own standards, such as the *SEMI Business Continuity Guideline for the Semiconductor Industry and its Supply Chain*.

The extent of the plans should be based on the overall exposure. For example, real-time data operations or laboratories using specialized research animals should have well-developed plans. Plans should address local, regional, and widespread disasters. Companies, such as those in the automotive industry that operate on just-in-time principles, need to be especially rigorous in assessing and managing supply-chain risks.

Plans can be developed in house using templates or using a consultant that specializes in creating custom plans. In either case, it is critical that production managers, engineering, quality control, regulatory compliance, and legal have input. The team determines the threats and potential impact of business-specific, local, regional, and business disasters. Once the risks have been identified, critical business functions and specific recovery strategies are determined. The emergency, continuity, and recovery steps must be documented. Testing the plan and updating the plan when the business, market, and suppliers change are critical components of the plan.

The response by the automotive industry to the explosion at Evonik's resin production facility, established a resiliency response model that many could benefit from. In the weeks following the explosion, two hundred companies convened in Detroit, from automotive manufacturers, to component part suppliers and raw material providers. They established validation guidelines for qualifying new materials, such as identifying the critical performance characteristics unique to the parts that were dependent on the plastics produced by Evonik. They collaborated on research and development to qualify new materials. The result was the automotive industry reporting no disruption to production as a result of the loss in Germany.

From a safety professional's perspective, the model implemented by the auto industry in response to the explosion in Germany is one that is common to the profession, where information on risks, as well as risk reduction, management and response measures, is regularly shared for the sake of avoiding worker injuries and other losses. Perhaps this provides an insight into the types of influence the safety professional can bring to supply-chain risk-management measures within their organization?

Another supply-chain resiliency success was demonstrated by Nissan in the wake of the 2011 tsunami in Japan. For Nissan, six of its production facilities, and fifty of the firm's critical suppliers, suffered severe damage. Yet, six months after the disaster, Nissan reported 3.8% production decrease, compared to industry-wide production decreases of nearly 25%. And Nissan ended 2011 with an increase in production of 9.3% vs. an industry reduction of 9%.[5] Key measures that were attributed to Nissan's success include comprehensive risk assessment of various scenarios, including earthquakes, and transparency throughout their supply chain. In other words, the same supply-chain resiliency measures Nissan applied to their own operations, they worked to establish within the operations of their suppliers. This approach of robust risk assessment at the front end, and transparency of solutions for the sake of risk reduction, is intrinsic to how many in the safety profession operate, establishing a competency that is readily transferable to any entity's supply-chain resiliency process.

---

[5] Schmidt, William and Simchi-Levi, David. "Nissan Motor Company Ltd: Building Operational Resiliency." *MIT Sloan Management*, Case Number 13-150.

Establishing safe operating conditions is one of the most controllable factors in a supply chain resiliency strategy. As such, the safety professional can play a key role in qualifying critical suppliers. Perhaps a third party safety survey has been performed for insurance or other purposes that can be reviewed by the safety professional, enabling them to identify process hazards, construction vulnerabilities or other features that could impact continued operations. For critical sole source suppliers, the safety professional could even join the team that visits and qualifies that supplier, adding a new element to the scorecard used to determine the supplier's adequacy to meet a company's needs.

Resources for Creating a Supply-chain Resiliency Plan

National Fire Protection Association (NFPA). 2013. NFPA 1600, *Standard on Disaster / Emergency Management and Business Continuity Programs* (widely used in North America).

International Organization for Standardization (ISO). 2012. ISO 22301: 2012 and ISO 22313, *2012 Business Continuity Management Systems* (widely used in Europe).

American Society of Industrial Security. *Business Continuity Guideline.*

Business Continuity Institute (BCI). *Good Practices Guidelines.*

Conclusion

Traditionally, the supply-chain risk management functions have operated distinct from the safety function in most organizations. Yet, there are many competencies inherent to a safety professional that are transferrable to process of establishing supply-chain resiliency.

Bibliography

American Society of Industrial Security (ASIS). 2008. *Business Continuity Guideline: A Practical Approach for Emergency Preparedness, Crisis Management, and Disaster Recovery.* Alexandria, Virginia: ASIS.

Business Continuity Institute (BCI). 2013. *Good Practice Guidelines.* Reading, UK: BCI.

Chubb Group of Insurance Companies and Cleantech Group. 2015. "The Hidden Challenge: What Lies Beneath Cleantech's Global Plans."

Federal Emergency Management Agency (FEMA). 2013. *Small Business Continuity Planning Integrated with Fire Department Pre-Plans.* Washington, DC: FEMA.

International Organization for Standardization (ISO). 2012. *Business Continuity Management Systems.* (ISO 22301). Geneva, Switzerland: ISO.

International Organization for Standardization (ISO). 2012. *Business Continuity Management Systems – Guidance.* (ISO 22313). Geneva, Switzerland: ISO.

Johnson, Denise. 2015. "The Changing Landscape of Business Interruption Claims." *Claims Journal*.

National Fire Protection Association (NFPA). 2013. *Standard on Disaster / Emergency Management and Business Continuity Programs* (NFPA 1600). Quincy, MA: NFPA.

PWC and the MIT Forum for Supply Chain Innovation. 2013. "Making the Right Risk Decisions to Strengthen Operations Performance."

SEMI Business Continuity Council. 2003. *The SEMI Business Continuity Guideline for the Semiconductor Industry and its Supply Chain*. San Jose, CA

It's All in the Presentation – Techniques to Enhance Your Presentation Skills

Patrick J. Karol, ARM, CSP
EEC Environmental
Philadelphia, PA

Camille P. Oakes, CSP, MEng
MSC Industrial Supply
Atlanta, GA

Introduction

Engaging stakeholders, getting buy-in from employees and management, and becoming a trusted advisor are frequent goals of safety professionals. To accomplish these goals, the place to start is by evaluating and honing your presentation skills. Everyone has attended a meeting, a conference session, or a presentation in which the presenter, presentation, or delivery of the message was lacking. This can be, at best, distracting and at worst, career damaging. A successful safety professional is one that stakeholders can trust and believe in. To be that, he or she must also be someone that people want to listen to, who communicates well, and who delivers the right message at the right level.

In this paper, the authors will discuss methods that a safety professional can use in order to enhance their presentation skills, thereby helping them along the path to achieve career success. Presentations could be anything from reporting audit results to an executive, talking about safety on the floor with a front-line supervisor or employee, giving a training class, or presenting at a large conference. Even a skilled presenter will be able to take away new and valuable information.

The three main steps to enhance presentation skills are to (1) Know Your Audience; (2) Craft Your Message; and (3) Deliver Your Message. This paper will provide an in-depth discussion of these steps and provide valuable information as to how to design, develop and deliver a powerful, engaging, and memorable presentation.

Know Your Audience

Before developing a presentation, the first step is to understand your audience, which will allow you to tailor your message so that it is received well and buy-in is accomplished. Though there will be similarities, presenting to an executive leader will generally be different than presenting to a front-line supervisor, a maintenance team member, or a front-line team member. A presenter must also understand the culture of not only the organization which you are presenting to, but also the culture across that specific team, unit, facility or department. The more information you have about who your audience is, the more successful your presentation to them is going to be.

In order to connect with your audience, you need to understand why the topic you are discussing is important to them. What does the audience expect to learn from your presentation? A presentation to an audience of other safety professionals will likely be more technical and at a higher-level than to employees on the floor. It is important to understand the level of knowledge that your audience has about the particular topic. It can certainly be insulting to present basic information to a highly knowledgeable audience, and conversely will be unsuccessful to speak at too high a level for a novice audience (Ricci 2012).

Speaking with Executives

As a presenter, very little can be as nerve-wracking as having to present to senior leadership. As a rule, they are busy individuals who are required to quickly make decisions that may have tremendous impact on the organization as a whole. If you have been in a board room or on a call with executives, you have seen that they will not, "won't sit still for a long presentation with a big reveal at the end. They'll just interrupt you before you finish your shtick" (Duarte 2012). If you do not expect this, you may become flustered, lose your place, and walk away feeling like you botched a big moment. Consider these recommendations from the Harvard Business Review:

Summarize Up Front

When you begin designing the presentation, lead with the information the audience will care about. Essentially, cut to the chase. Show high-level findings, conclusions, recommendations, and any calls to action. State the information, "clearly and succinctly right at the start, and then move on to supporting data…" (Duarte 2012).

Set Expectations

Know that at this level, you will more than likely be dealing with an impatient audience. Let them know, succinctly, that you will first cover your summary, then recommendations and use the additional time for discussion (Duarte 2012).

Give Them What They Asked for

If you were invited to give a short update to a group of executives on the results of the company's most recent ergonomic initiative, do this before covering anything else. Remember the group is time-pressed and you were invited to provide *specific* information (Duarte 2012). You may want to use this rare opportunity to bring up another concern or idea, but be sure that you first address and answer their request.

Rehearse
Presenting at this level means you need to be quick on your feet and confident in what you are presenting and proposing. The importance of rehearsing for this cannot be overstated. Rehearse with a trusted coworker, manager, or even a small team to get feedback. Anticipate any response from the executives; what questions will they ask? Have the team pelt you with unexpected and/or difficult questions. Have the team interrupt you while you are presenting. Doing this may be a lot of work, however successfully presenting to executives can result in, "people with a lot of influence [becoming] strong advocates for your ideas" (Duarte 2012).

You may also tailor your message to a specific executive. For example, if you are speaking to a group of Human Resources executives or team you may emphasize how safety can be used to improve employee engagement and reduce turnover. If the Chief Financial Officer is your audience, you may stress the relationship between a strong safety program and the cost of workers comp or insurance premiums. It's important to remember that an executive's job is to create value or grow the business. Playing the "moral" card or "it's just the right thing to do" doesn't always resonate with an executive leader.

Speaking with Front Line Management

Tailoring our message is important whether we are speaking to leadership or a front line employee. Everyone must understand the benefits of a safe operation, yet no one in the organization is more critical to the success of our safety program than the front line supervisor.

Supervisors are in a unique position. They have firsthand knowledge of the operation and the employees. They know which procedures are followed and which are not, and why. They know how effective the training is or is not. They know what works and what does not. They are a critical link in the safety chain. The problem is they are a critical link in EVERYBODY'S chain: Human Resources, Quality, Training, Employee Relations, Unions, Customer Service, etc. We need their support, but they are bombarded with requests and issues from other functional areas.

Catering our Message to the Front Line
Safety is an ambiguous term. An effective safety message for supervisors defines safety in operational terms and minimizes or even eliminates the use of the word "safety." Why? As you get closer you get to the front line, two things happen. First, safety gets more ambiguous. It means something different to everyone. Second, the word "safety" tends to have a negative connotation. We can be our own worst enemy when we use terms like "accident," "investigation," "corrective action," and "disciplinary action." Each of these words have negative connotations.

Rather than using the word "safety" or saying "be safe", define "safety" in operational terms. For example; if you work in a warehouse and have a big shipment coming in, you may craft your message to a front line manager by saying, "be sure to remind all employees that forklifts will be active today, so stay within the pedestrian only areas and look both ways before crossing an aisle. Also, remind the forklift operators to stop and look before crossing a pedestrian pathway."

Another option is to leverage the tools in a supervisor's tool box. Supervisors have the familiar tools, like recognition programs, and tools they may have never used, like story-telling.

These tools can make their job as the front line of defense against incidents much easier. By showing supervisors how to leverage these tools to achieve a high level of safety performance, safety becomes an easy sell without using the word "safety." A big challenge for a front line manager or supervisor is to get employees to do what they want them to do, whether it is safety related or not. Showing how to reinforce the behavior they want by using recognition to reinforce that behavior gets a clear message across.

Presenting to a front line manager in this manner reduces any anxiety or preconceived negative notions about safety they may have and helps them solve a problem.

The front line supervisor is the key to a safe and efficient operation, but they are often pulled in different directions by different departments, or they simply get caught up in the day-to-day demands of the operation. Safety can quickly fall down the supervisor's list of priorities. Successfully presenting to the supervisor will position safety as a value and not a priority.

Speaking with Employees

The employees are the heart of the organization and have a tremendous impact on the safety culture, incident and injury rates, and all other safety metrics of a company. They can be a great asset or a great hindrance and most of this depends on how you communicate with them.

Recognize Their Worth

When communicating or presenting to employees, like with all other levels of an organization, the most important tenant is to respect them as human beings. These are the people who are on the floor, doing the work on a day-to-day basis. No matter what, a safety, health, and risk professional will, at least initially, be seen as an outsider, due to you not being the person doing the work. In order for you to be listened to, or for you to get real, true information from employees, you have to be someone who can be trusted. In order to be trusted, you must engage with employees, hear their thoughts, understand their motivations, and present solutions that will work in practice, not just in theory. This increases your credibility.

Understand and Address Their Motivation

With executives, you will want to cut to the chase. The questions you hear in a presentation will typically be related to costs and benefits to the company, timeline, justification, return on investment, industry standards, etc. With employees, it is likely they will want different information. What drives them is different than what drives an executive or supervisor. Let's go through an example. You are a Safety Manager at a distribution center and you wish to implement a behavior based safety program wherein the employees are now expected to complete safety observations in their departments several times a week. You have senior and front line leadership support and you are now ready to present this to employees. You pull all employees together during the quarterly meeting and present the new requirement. One employee asks how their productivity will be affected with this new requirement. Your response is, "we do not believe that will be an issue." Your new idea is received poorly. The room devolves into chaos. Why? It is because you did not deliver the message correctly? Poor eye contact? Poor vocal projection? It is received poorly because you did not anticipate the concerns of the employees and address these concerns with your presentation. As the local safety manager, you should know that each employee receives additional pay for meeting their productivity metrics. Now that you have assigned them a safety task, their worry would be that they would not meet their production goals and thus their take home pay would be reduced. Much like with executives, if you do not address the root of the problem with your presentation, good visuals, graphics, or verbal communication will not make up for this. A better presenter would know their audience, the safety culture and climate, and address this concern on the front end. Instead of saying, "we do not believe that will be an issue," the presenter should communicate that over the past two months these observations were tested in a department by other employees, additional time was given so this did not affect productivity metrics, all leadership is aware and on board with this, and all employees testing this process in that department met their metrics and received their full pay. It is also very helpful to communicate the purpose of these observations. The presenter should communicate the problem they were trying to solve. Even better, include what the benefits are to the employee by solving the problem using this approach. This is a different type of information required than to front line leadership and senior executives, however, the inherent idea is the same: Know your audience and tailor your message to that audience.

Craft Your Message

Before beginning a presentation, you must understand what you are trying to convey. You will need to define your learning objectives and what takeaways you wish your audience to have. You must consider your role, audience, format and topic. From there, you can craft your message to be effective, clear and engaging.

When considering your message, use the RAFT technique (Allen Simon 2017) . The RAFT technique is used for students but certainly can apply to adults in the workplace. It allows you to effectively communicate your ideas and mission so that the receiver can understand. It allows you to write for different audiences and additional will assist you to clearly convey the message and stay on topic.

- **R**ole of the Author/Speaker: Who are you as the author or speaker? Are you an outside consultant? Facilitator? Trainer? Local technical expert?
- **A**udience: Who is your audience, what is their motivation and level of knowledge of the subject?
- **F**ormat: In what format are you communicating? How much time will you be given? Will you be doing a formal presentation to a large group or an informal one-on-one session?
- **T**opic: What is the topic of your presentation? What information are you trying to get across?

As a safety, health and risk professional, it is likely your role at some point will be to deliver technical information and/or bad news to employees, clients, senior leaders or front line management. Consider the following suggestions on how to succeed at any of those tasks.

Delivering Technical Information

You will at some point have to deliver technical information related to safety, health, environmental or risk. A technical briefing is a speech that conveys technical information to a specific audience. Technical briefings should be presented in a way that allows an audience to understand and apply critical information. In the safety world, technical briefings can range from audit results to industrial hygiene survey results to a new policy roll-out. Follow the steps below to ensure your technical briefings are as effective as they can be ("Delivering Technical Briefings" 2017).

- Avoid using too much industry jargon or material that is too technical for the audience to easily understand.
- Avoid words like "obviously" or "clearly" as this can alienate your audience if they do not understand the topic.
- State the purpose of the technical briefing in one or two sentences and use this summary as the focal point for the entire presentation.
- Arrange the material into an outline containing an introduction, main points and a conclusion.

Delivering Bad News

It is inevitable that a safety, health, and risk professional will at some point have to deliver bad news. This can range from poor performance on an inspection or audit, poor results from an employee safety survey, poor safety metrics in general or informing affected employees of a serious incident. If you are a consultant, this may be even more delicate to present as your client may not be open to receiving or sharing negative information with their peers or superiors. Forbes.com provides some beneficial information on how to deliver bad news:

- **Never surprise your audience.** Bad news should never come as a surprise. If the first time your clients, senior leaders, or employees are hearing about poor safety performance is in a major presentation, you have missed your responsibility. Metrics should be communicated often so that the opportunity to improve is given. If you are consulting, spend some time during your audit or assessment with your client directly discussing the results rather than keeping your findings to yourself until a closing briefing in front of the client's peers or superiors.
- **Communicate the silver lining.** You may have heard of the sandwich technique for delivering criticism. This is where you sandwich the criticism in between two compliments. Similar to this, when communicating bad news, also include positive attributes you found. "Focusing on positives will help leadership keep employees productive, upbeat, and future-oriented" (Bies 2012). However, when you present this information, make sure, "that hope [is] grounded in reality" (Bies 2012).
- **Bring solutions.** Forbes recommends that when presenting bad news, you also present solutions or an action plan to solve the problems that led to the bad news. This will help the audience receive the bad news more openly.

Preparation is key in presenting information to your audience. First and foremost, understand your audience and their motivations. Secondly, design and craft your message using the RAFT method. In the likely event you have to deliver highly technical information or bad news, follow this advice to ensure your message is understood and trusted.

Deliver Your Message

Presentation Design

We have all certainly gone to a formal presentation and critiqued the order, layout, graphics and design. So how can you design your presentation to instead be engaging and have them peak and keep your audiences' interest?

Toastmasters International is a world leader in communication and leadership development. The Mission of Toastmasters is to empower individuals to become more effective communicators and leaders. To that end, Toastmasters is an excellent resource in how to better deliver your message, starting with how to design you presentation.

Toastmasters recommends this following when it comes to presentation design ("Visual Aids And Props" 2017):

- Visual aids and props should be colorful and unique, but not so dazzling that they detract from the speaker's presence.
- Diagrams, graphs and charts should always coincide with what is being said in the speech.
- Maps should be simple and easy to understand, with key places or points clearly plotted or marked.
- PowerPoint slides should present main points as short sentences and bullet points and should never be read verbatim by the speaker or presenter.
- Lists should be kept to a minimum. Four or five listed items are usually enough.

Below are some examples of good versus bad slides made with Microsoft PowerPoint.

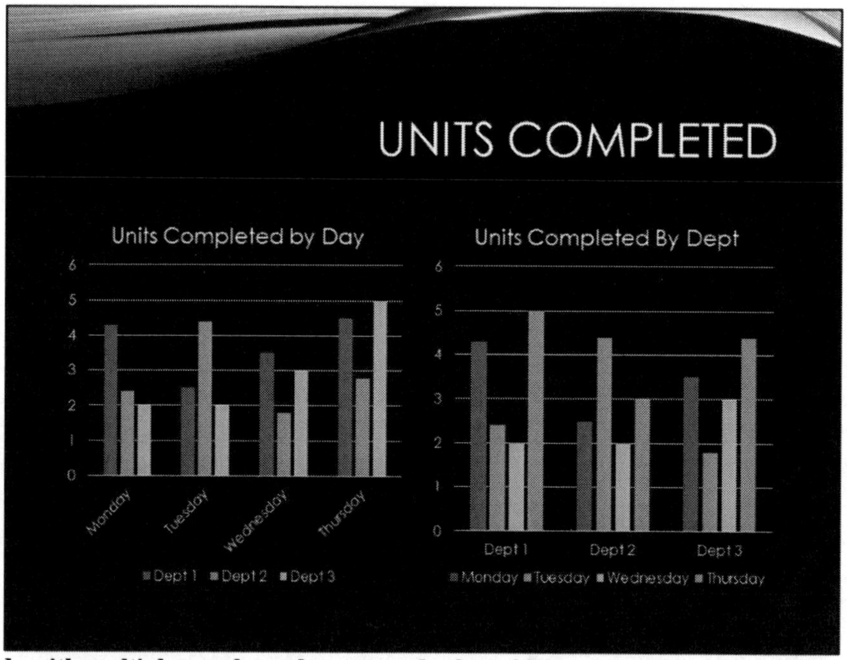

Figure 1. Slide with multiple graphs and overuse of color which may confuse the audience and disrupt readability

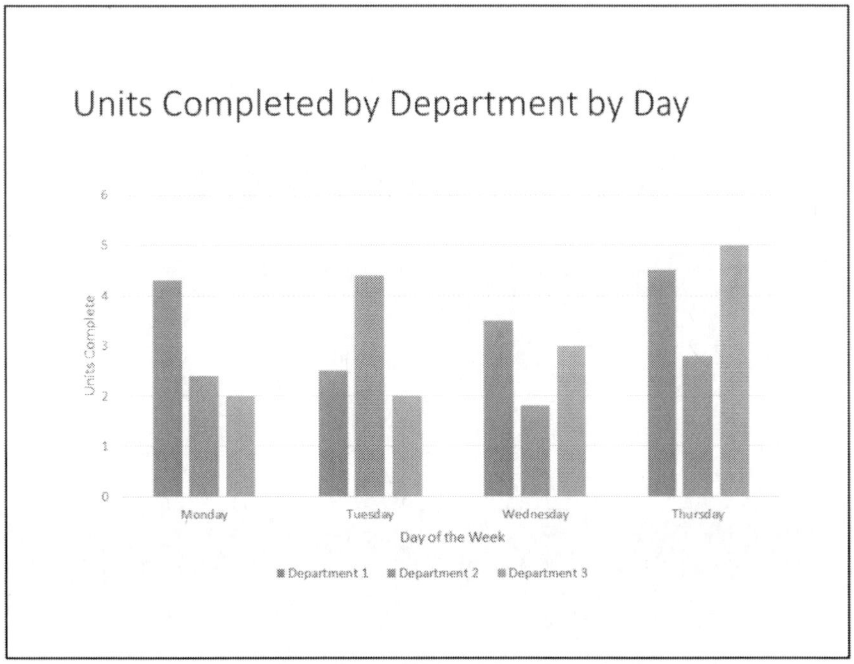

Figure 2. A More Clear Slide with only a single graph

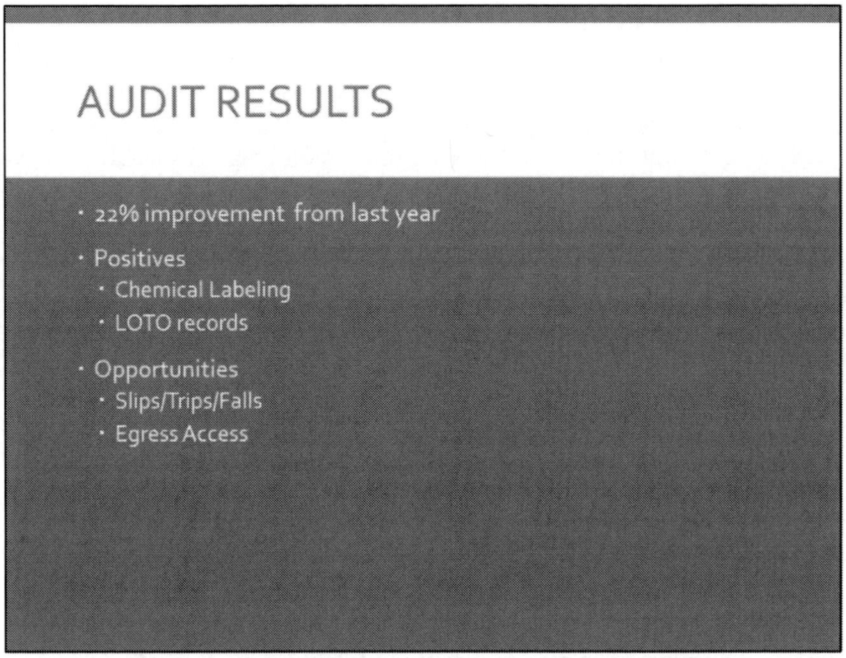

Figure 3. Slide with Limited Text- clear and readable

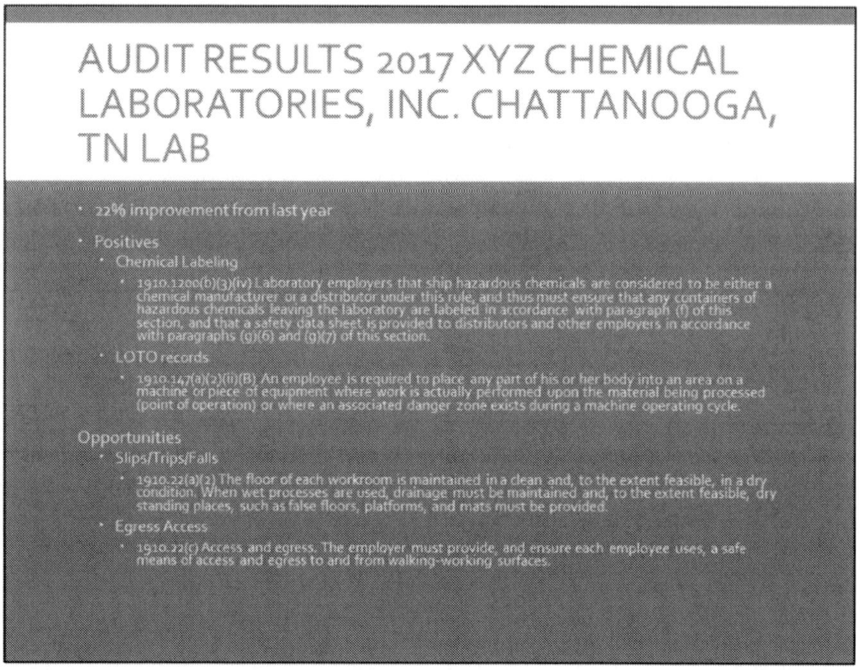

Figure 4. Same slide with significant amount of text limited readability

Visual, Vocal and Verbal Communication

In order to be successful presenters, we must also understand what makes an audience walk away from a presentation and consider it a success. A study by Alber Mehrabian, a UCLA professor emeritus of psychology, found that there were three elements of communication that influenced an audience.

1. **Visual**: This does not mean presentation design or good graphs as detailed above, but rather your own body language, eye contact, and gestures.
2. **Vocal**: This is about your voice, including your rate, volume and inflection.
3. **Verbal**: This is about the content and the actual words you are saying.

Very surprisingly, Mehrabian found that verbal, or literally what you say, accounted for only 7 percent of what influenced the audience. Vocal, or how you say it, accounted for 38 percent of audience influence. In total, 55 percent of audience influence was based on visual, or how you act and look as a speaker. This shows that while content, the words you use, and sounding confident matters, how you present yourself as a speaker is going to have the greatest effect on your presentation (Sudakow 2017).

Improving Visual Communication

Speakers generate a great amount of emotion and interest through the use of non-verbal communication, often called gestures or body language. A speaker's body can be an effective tool for emphasizing and clarifying the words they use, while reinforcing their sincerity and enthusiasm. Toastmasters recommends the following in order to use gestures effectively ("Gestures And Body Language" 2017).

- Eye contact establishes an immediate bond with an audience, especially when a speaker focuses in on individual listeners rather than just gazing over the audience as a whole.
- Control mannerisms. Mannerisms are the nervous expressions a speaker might not be aware of such as putting their hands in their pockets, nodding their head excessively, or using filler words like "um" and "ah" too often. Use good body posture.
- Put verbs in to action when speaking to an audience by physically acting them out with the hands, face or entire body.
- Move around the stage as topics change and move toward the audience when asking questions, making critical connections, or offering a revelation.

The Very Real Fear of Public Speaking

According to Nick Morgan of Forbes.com, about 10 percent of the population loves public speaking and about 10 percent are genuinely terrified. Those that are truly terrified are considered Glossophobic (having an extreme fear of public speaking) and, "will go to great lengths to avoid speaking in a group situation, and will experience nausea, panic attacks and extreme anxiety" (Morgan 2011). This paper and its included advice is unlikely to help extreme glossophobes. However, for the rest of us, the 80 percent in the middle who get anxious and perhaps do not sleep much the night before a big presentation, we can use the fear, or adrenaline, to become better public speakers.

How to Harness Your Adrenaline and Be a Better Public Speaker

- Be prepared. Rehearse enough to be confident, not only in your abilities, but also in the content you are presenting.
- Practice out loud. You can do this in front of a small audience or a mirror. This will allow you to understand how you are coming across.
- If possible, visit the location you will be speaking. If you are doing a presentation at a conference, practice in your session room. If you will be speaking in front of a webcam, turn your webcam on and give your speech.
- Know your strengths and weaknesses as a speaker, and focus on your strengths. Susan Cain recommends:

  > *If you have a great sense of humor, use it. If you're not a natural cut-up, do not try to be. Instead, focus on what you do best. Thoughtful and thought-provoking is every bit as powerful as dynamic and entertaining (Cain 2017).*
- As soon as you enter the room, consider yourself "on stage" and work the room. Engaging in conversations with your audience before your presentation will help loosen you up.
- Start strong. Begin your speech with a powerful opening that will grab your audience's attention, such as a startling fact or statistic, an interesting story or a funny joke.
- Be conversational. Do not read your speech or slides word for word, even if you are afraid. Refer again to tip number one.
- Speak with passion. If you're truly invested in what you're saying, you'll be better able to keep your audience's attention.
- Be patient. Expect the unexpected. You may be interrupted with hard questions; the audience may become hostile; you may run into technical difficulties. If you are prepared and confident in your material and content, you will be better able to roll with the punches.

Make Your Story Stick

According to Chip and Dan Heath in their book *Made to Stick,* messages that most stick with us have common characteristics. Those messages are:

- Simple – It is a bandwidth issue. The more we reduce the amount of information in our message or idea, the stickier it will be. Define your core message.
- Unexpected – the first challenge of communication is getting your audience's attention and the most basic way is to break a pattern. Common sense is the enemy of sticky messages. Expose the parts of your message that are uncommon sense otherwise your audience will dismiss it by saying, "I already know that! It is common sense."
- Concrete – "Safety" is an abstract word, as noted earlier. Define your message in a transparent, universal language that everyone understands.
- Credible – Telling stories using real people is highly credible. Consider using an influential person to deliver your message. Use someone who has benefited from your program, methods or tool. The honesty and trustworthiness of the person, not the status, will determine credibility.
- Emotional – The goal of making a message emotional is to make people care because feelings inspire people to take action. We make people care by appealing to the things that matter to them. As a safety professional, it means taking off our analytical hat and create empathy for specific individuals. We show them how our message or idea is associated with what they care most about. For instance, we discuss going home safe and healthy so they can enjoy their favorite hobbies, pets, time with their family, etc.
- Story – A good story provides inspiration and inspiration drives action. Safety is full of emotional stories. A story about an individual's path to recovery from an injury and how it changed them is a better and more emotional story than talking about the growing injury trend.

You do not need to be a creative genius or the world's best speaker to deliver a successful message. Following some simple steps, designing a good presentation, practicing your speaking skills, and making your story stick will indeed help your presentation be received well.

Conclusion

It is imperative and absolutely necessary in today's world for the safety, health and risk professional to be able to deliver an excellent presentation. If you focus on the three main steps to enhance your presentation skills, (1) Know Your Audience, (2) Craft Your Message and (3) Deliver Your Message, you will have increased success and be a more capable professional.

And one more thing…don't use the word "safety."

Bibliography

Allen Simon, Cathy. 2017. "Using The RAFT Writing Strategy - Readwritethink". *Readwritethink.Org*. http://www.readwritethink.org/professional-development/strategy-guides/using-raft-writing-strategy-30625.html.

Bies, Robert. 2012. "The 10 Commandments For Delivering Bad News". *Forbes.Com*. https://www.forbes.com/sites/forbesleadershipforum/2012/05/30/10-commandments-for-delivering-bad-news/#10b90ef72169.

Cain, Susan. 2017. "10 Public Speaking Tips From My Year Of Speaking Dangerously". *Quiet Revolution*. http://www.quietrev.com/10-public-speaking-tips-from-my-year-speaking-dangerously/.

"Delivering Technical Briefings". 2017. *Toastmasters.Org*. https://www.toastmasters.org/Resources/Public-Speaking-Tips/Delivering-Technical-Briefings.

Duarte, Nancy. 2012. "How To Present To Senior Executives". *Harvard Business Review*. https://hbr.org/2012/10/how-to-present-to-senior-execu.

"Gestures And Body Language". 2017. *Toastmasters.Org*. https://www.toastmasters.org/Resources/Public-Speaking-Tips/Gestures-and-Body-Language.

Heath, Chip and Dan Heath. 2010. *Made To Stick*. 1st ed. New York: Random House Books.

Morgan, Nick. 2011. "Why We Fear Public Speaking And How To Overcome It". *Forbes.Com*. https://www.forbes.com/sites/nickmorgan/2011/03/30/why-we-fear-public-speaking-and-how-to-overcome-it/#41e9f0d2460b.

Ricci, Tom. 2012. "Public Speaking: Know Your Audience". *Asme.Org*. https://www.asme.org/career-education/articles/public-speaking/public-speaking-know-your-audience.

Sudakow, James. 2017. "Great Presenters Do 1 Thing That Most Of You Don't, Science Says". *Inc.Com*. http://www.inc.com/james-sudakow/great-presenters-do-one-thing-that-most-of-you-dont-science-says.html?cid=email.

"Visual Aids And Props". 2017. *Toastmasters.Org*. https://www.toastmasters.org/Resources/Public-Speaking-Tips/Visual-Aids-and-Props.

Session No. 727

Culture Always Trumps Your Risk System–So How Do You Make It Work for You?

Ewan Alexander (B.E.Sc.)
RISKmap International Inc.
Vancouver, BC
Canada

Abstract

HSE professionals and caring leaders, we all work hard on the safety and risk systems to do the best for our employees and companies. However, no matter how strong the system is, it gets trumped when the culture overwhelms it. Even companies renowned for the strength of their risk and safety management systems continue to experience large and significant incidents, and in some cases catastrophes. It is only when leadership, culture and systems work in complete harmony that the resilience of an organization to manage its significant risk is at an optimal level. And even then, it is a precarious balance to maintain that optimal mindset in the organization.

In this presentation, a case study is presented that shows where the maturity of the company's risk systems and safety culture have been benchmarked against a global data base of companies. The risk maturity model and the scientific basis of measurement of culture are presented. The case study will demonstrate the practical application of the approach, and how the outcomes of the study provide input to leadership to make the necessary changes to the culture and the risk systems. And this can be done before significant events strike.

It will also highlight how the process provides input to build systems that are less vulnerable. Within the presentation, Ewan will discuss a case study of a large South American miner, and highlight their efforts to perform a deep dive into their leadership, culture and risk management systems.

The presentation will appeal to those who want to use culture and risk system maturity as a benchmarked key leading indicator to maximize the performance of their risk and safety management systems.

Building and Executing an Effective Risk System for the Mining Industry

As human beings, we have managed risk since the dawn of the ages, and have been adept at doing so. In order to survive, we hunted or gathered. If we did not take enough risk, we died. If we took much risk, we died. As individuals and tribes, risk had to be optimized to maximize our chances to thrive, as individuals and as tribes. (Figure 1)

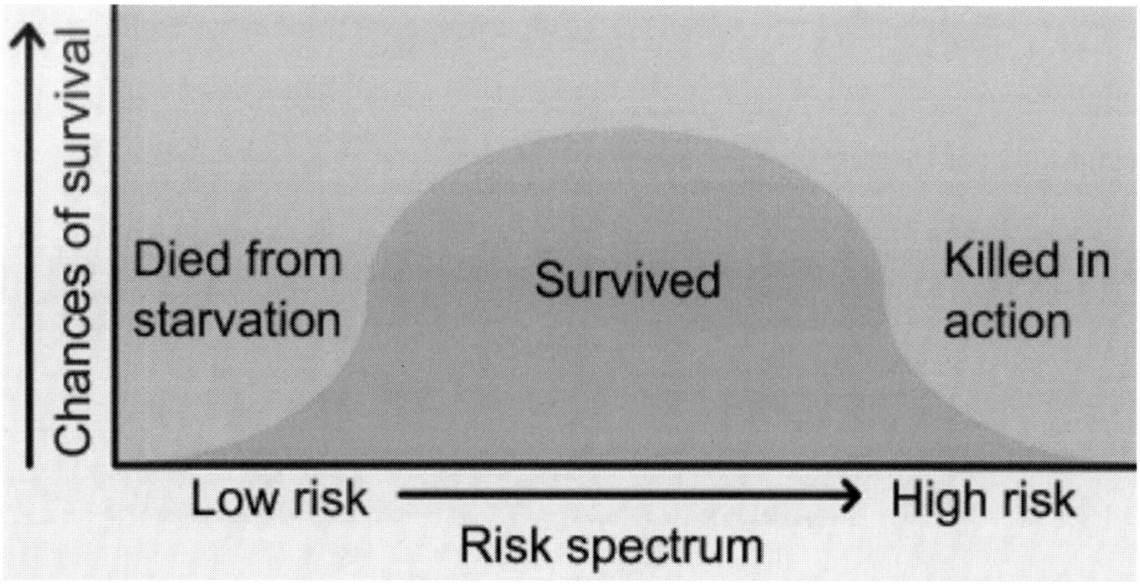

Figure 1: Source: Human Behaviors in Hazardous Situations. (Jan M. T. Dallmans)

In today's world, the risks are vastly different, but to thrive, these must be managed. And no industry knows this as much as mining. In fact, the International Council of Mining & Metals (ICMM) has proposed the Critical Risk Model as a means of identifying critical risk, critical controls and associated performance standards. The ICMM focuses on sustainable development performance in the mining and metals industry. It has 23 members, representing some of the largest mining companies in the world. (Figure 2)

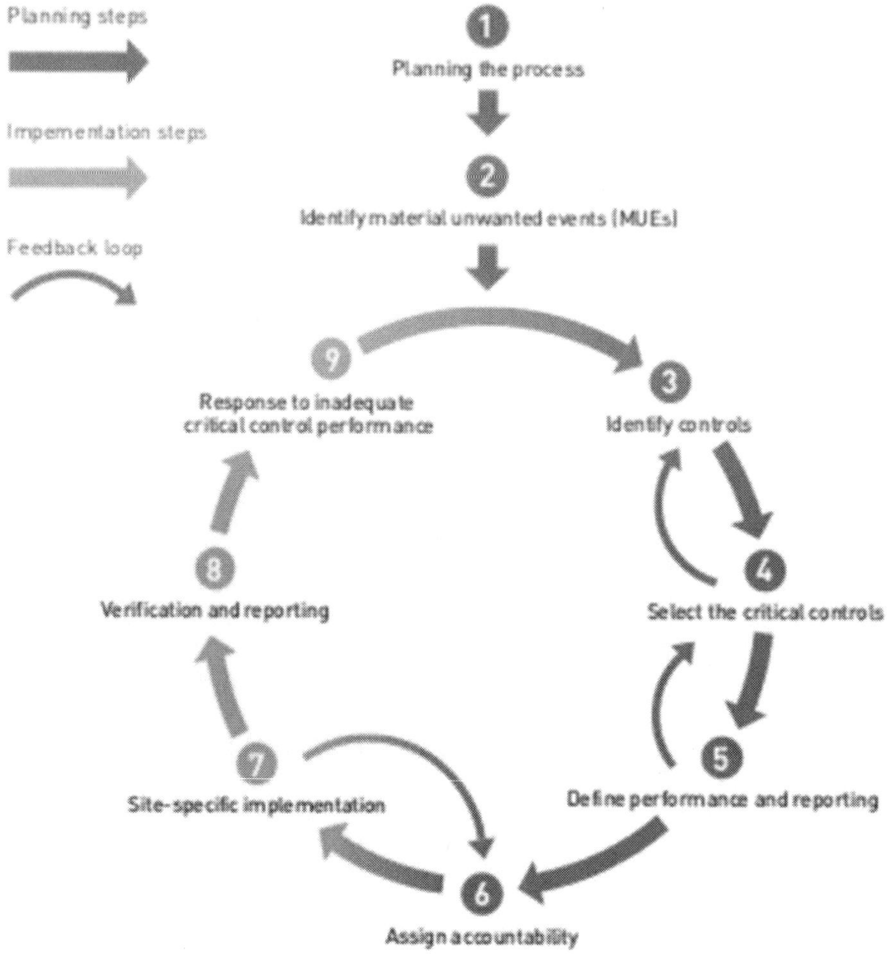

Figure 2: ICMM model for Critical Risk Management.

This approach is helping the mining industry to reduce its fatality rate. But it must be applied sustainably, with the acknowledgement that the system alone cannot be effective. We can say this confidently as even companies known for the strength of their risk and safety management systems continue to experience significant incidents, and in some cases catastrophes. Why does that happen when all safety metrics including injury rate are pointing in the right direction?

It is only when leadership, culture and systems work in complete harmony that the resilience of an organization to manage its significant risk is at an optimal level. It remains an ongoing leadership challenge to maintain that balance.

Let's start with a basic premise. Leadership, supported by simple effective systems, creates culture which creates outcomes. Everyone going home safely is an outcome, as is cost effective operations, and to achieve those outcomes efficiently, the organization must focus on the inputs. And those inputs include effective leadership, effective systems (risk and safety systems), good planning, stable operations and competent workers. (Figure 3)

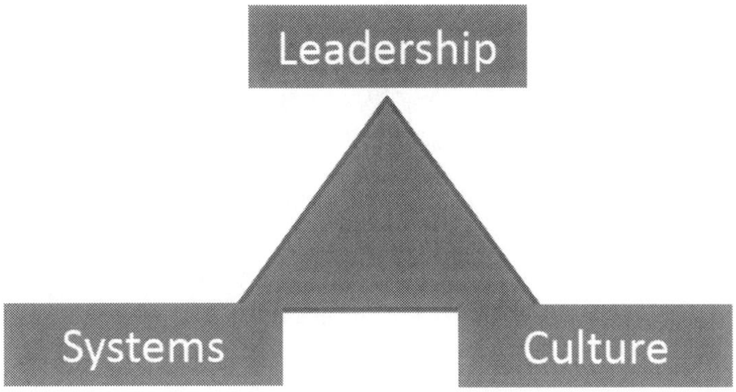

Figure 3.

In order to become risk competent, a company must have a risk management system. These systems range from great complexity, to elegant simplicity. The level of detail is determined by the hazards being managed, and there is tremendous skill in determining the optimum level of detail to manage the associated risks. Too much detail and the system becomes unmanageable and caves in under its own weight. Too little and its controls cannot mitigate or prevent all possible causes. It is a delicate and precarious balance.

An effective system is one that creates critical controls, measured against objective performance standards and this will help the mining industry in the following ways:

- It will focus leadership on managing those risks that are material to the business and its employees. "Material" in this usage, means "critical to business health, and employee wellbeing."
- Supervisors will focus on critical controls. These controls are the barriers between a hazard and the risk that the hazard becomes a reality (incident). As an example, the first thing that a supervisor will focus on when going out in the field is whether the controls protecting their employees are effective.
- Employees will know the controls that are protecting them and what it takes for them to be effective. They will be risk competent. They will know what conditions must exist for the controls to work, and will challenge control effectiveness as part of a continuous improvement cycle. They will follow procedure and know when it is not safe to follow procedure.
- These controls will be tested regularly against objective performance standards. Objective in this case specifically means that it requires a yes or no to answer. There either **is, or is not**, an engineered anchor point for employees to tie off when working at heights. There is no opinion involved, it either **does or does not** exist. The harness either **has or has not been** inspected by a competent individual within the prescribed review period.
- There will be a management of change process that will allow thinking individuals to effectively manage the risk associated with the changed hazards.

A risk competent organization also knows the limits of its system in terms of what it can and cannot do. It will be aware of the "blind spots" created by the system. These blind spots emerge for a number of reasons, the most significant being as follows:

- Overly simple or overly complex. The organization must be clear of how it aggregates its risks. When managing hazardous atmosphere underground, the cause of a hazard could range from noxious vapors, equipment exhaust to fire and explosion. The critical controls are similar between all of these hazards, but not the same. Ventilation is common, vehicle maintenance impacts noxious vapors and possibly vehicle fire. It is a delicate balance learned from experience.
- Inadequate risk identification. Hazards exist in terms of how work is done. So hazard identification must evolve from work planning and standard operation. Too often, the hazard is generally identified, but too broadly defined, resulting in controls that are overly conceptual, and don't relate to the way that workers need to complete their work. As an example, hydrocarbons are identified as a potential for explosion, but the risks of storage, versus loading or unloading are completely different. The hazardous analysis/risk assessment needs to focus on the individual steps of loading, storing and unloading. The controls are likely to be different in each case.
- Operation instability. Stable, well-planned operations and construction, that are consistent are far more effective, and, easier to control. Sources of instability in the process should be the focal points for management. Make the process stable, and controls are more likely to be effective. If employees are required to manage around instability then this means that they will start to improvise, which will either defeat, or undermine the controls that management thinks are in place. Workers can be counted on to manage risk, but make it easier for them, so that they can focus on the low-probability, high-consequence events.
- Everything in the green. Critical controls are all working, and a desire to report to management that all controls are effective results in complacency. An organization must continually be able to challenge how well the controls are working for them, and how they might be vulnerable to random events. Inclusion of randomness is a key skill. The manager of the Piper Alpha rig, which blew up in the North Sea in 1988, testified that "I knew everything was okay, because I never got a report that anything was wrong."
- Risk migration. Because one risk is controlled effectively, a new risk can be created. For example, an openpit mine banned passing of haul trucks when a fatality occurred. This caused light vehicles to race ahead of the haul trucks, eventually resulting in a collision and fatality.
- Risk homeostasis. This is a known human behavior. Once a person feels safe, they will take more risk. Further, as risks are successfully managed, over time this consistency results in overconfidence. It can also result in people paying less attention to the critical controls, no continuous improvement cycle, and redundancy-induced overconfidence, as occurred within NASA in the fateful Challenger launch.
- Failure to take advantage of learning from significant incidents. This can occur when:
    - The mine is focused primarily on U.S. Mine Safety and Health Administration (MSHA) recordable injuries and injury rate, and does not adequately pay attention to actual or potential events where someone might be killed.
    - The company fails to adequately investigate and learn from significant events.
    - The organization simply fails to update and test critical controls based on the experience gained from an incident. Or does not apply the learning to sister sites
- A strong inclination to blame the worker for events that occur. When investigations blame the worker more than 10 percent of the time, then systemic causes and remedies are missed.

To enable companies to benchmark themselves, RISKmap, a division of SAFEmap, has

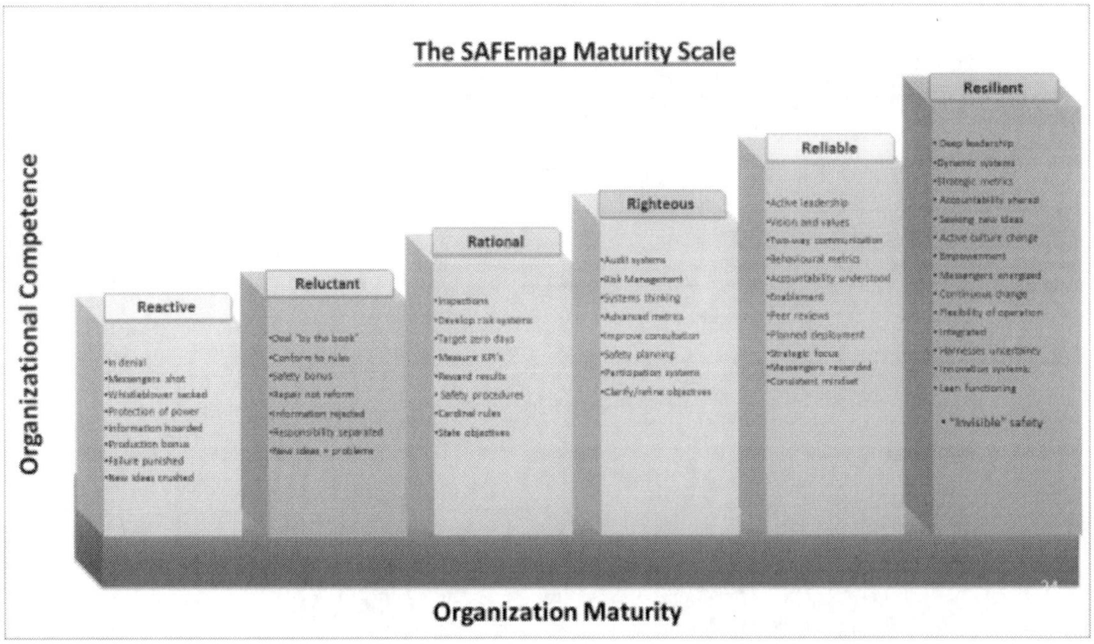

Figure 4.

developed a progression system that measures the organization's maturity in terms of, leadership, systems and safety culture.

The safety culture model was developed 16 years ago for an Australian mining industry study and has since been applied at more than 250 sites world-wide with participation of approximately 300,000 employees (Figure 4).

Organizations without risk systems start at a reactive level of maturity and its people rely totally on their own instinct to survive. Organizations can progress to become resilient in their capability to execute their business effectively in a way that cares for their people.

Organizations benchmarked to date live in the reluctant to reliable continuum. And the most important aspect is those that have completed the benchmark gain feedback which can drive improvements in each of their leadership, systems and culture, in a way that assures that all three pillars can work together (Figure 5).

The benchmarking process is fairly simple. Data is gathered from electronic surveys, focus groups involving less than 10 percent of the population, and a risk deep dive involving two to three-day site visits. Companies then get feedback based on benchmarked performance similar to the three charts below, a strengths, weaknesses, opportunities and threats (SWOT) analysis and specific recommendations.

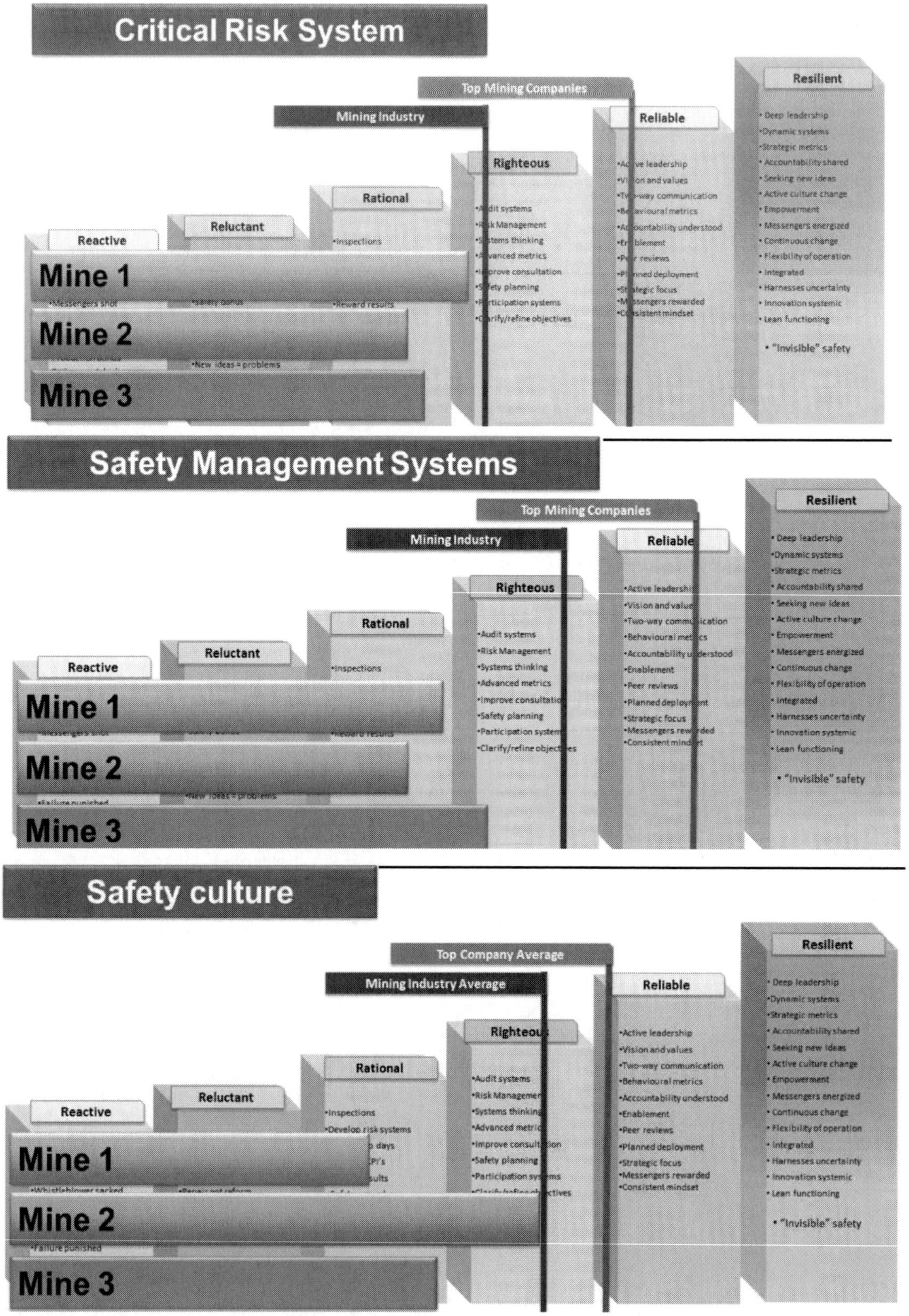

Critical Risk System
Top Mining Companies
Mining Industry
Resilient
Reliable
Righteous
Rational
Reluctant
Reactive
Mine 1
Mine 2
Mine 3
•Inspections
• Deep leadership
•Dynamic systems
•Strategic metrics
• Accountability shared
• Seeking new ideas
• Active culture change
• Empowerment
• Messengers energized
• Continuous change
• Flexibility of operation
• Integrated
• Harnesses uncertainty
• Innovation systemic
• Lean functioning
• "Invisible" safety
Safety Management Systems
Top Mining Companies
Mining Industry
Resilient
Reliable
Righteous
Rational
Reluctant
Reactive
Mine 1
Mine 2
Mine 3
•Inspections
•Audit systems
•Risk Management
•Systems thinking
•Advanced metrics
•Safety planning
•Active leadership
•Two-way communication
•Enablement
•Peer reviews
•Strategic focus
• Deep leadership
•Dynamic systems
•Strategic metrics
• Accountability shared
• Seeking new ideas
• Active culture change
• Empowerment
• Messengers energized
• Continuous change
• Flexibility of operation
• Integrated
• Harnesses uncertainty
• Innovation systemic
• Lean functioning
• "Invisible" safety
Safety culture
Top Company Average
Mining Industry Average
Resilient
Reliable
Rational
Reluctant
Reactive
Mine 1
Mine 2
Mine 3
•Inspections
•Develop risk systems
•Audit systems
•Safety planning
•Active leadership
•Vision and values
•Two-way communication
•Behavioural metrics
•Accountability understood
•Enablement
•Peer reviews
•Planned deployment
•Strategic focus
•Messengers rewarded
•Consistent mindset
• Deep leadership
•Dynamic systems
•Strategic metrics
• Accountability shared
• Seeking new ideas
• Active culture change
• Empowerment
• Messengers energized
• Continuous change
• Flexibility of operation
• Integrated
• Harnesses uncertainty
• Innovation systemic
• Lean functioning
• "Invisible" safety

Figure 5.

This approach was used most recently with a large South American mining company. The feedback, provided to the board, was considered extremely effective. The SWOT approach, broadly provided the following feedback. *Note that these are generalized for the purposes of this article.

Strengths	Weaknesses
• The business had effectively and consistently executed their risk system (in this case it was the ICMM Critical Risk Management (CRM) model).	• Risk system as applied was overly complex creating a significant and costly operational burden to the company.
Opportunities	**Threats**
• Significantly simplify the controls and measure against objective performance standards. • Harness the tremendous enthusiasm in the organization to progress to the next level of capability.	• Some elements of leadership were not aligned with the program. • Narrow and prescribed risk focus left material risks outside of scope.

For example, the organization had done an outstanding job of identifying controls where incidents had occurred but limited their effectiveness through:

- Not delegating control ownership far enough in to the organization. This resulted in superintendents owning more than 20 controls. The organization was eager to be more involved, and the superintendent was overloaded, creating the possibility of inadequate control effectiveness reviews).
- A focus solely on past events meant that other 'new' risks (for example, probable natural disasters, or loss of control of vehicles, were not included in the analysis and subsequent focus of the risk system).
- Effective integration of planning with risk controls was done extremely effectively in the pit. The leadership assured and convinced their employees that speeding in the pit was not necessary to achieve the production plan, and until doing so, employees continually defeated the critical control of haul road speed. Until convinced, employees believed that speeding was necessary for them to achieve their bonus, and that desire would always trump the critical control process. Integration of planning, with speed control, with production bonus, resolved the issue.

In Conclusion

There are no silver bullets. But, an organization that focuses on its leadership, system and culture, measures each element, then acts to assure that each element works in harmony, is more likely to be successful. And the fact that each of these elements can be measured and benchmarked provides leadership with the tools needed to improve their overall risk competence.

Using Lean Thinking to Reduce Manual Material Handling Injuries

Jessica K. Ellison, M.S., CSP, CPE
Principal Consultant
BSI EHS Services and Solutions
San Jose, CA

Introduction

Injuries in the healthcare field continue to be a challenge. According to the Bureau of Labor Statistics, while injuries in this field have decreased from 2012 to 2015, the injury rate for hospitals is 1.49 times higher than the average for private industry. Nursing Assistants have the fourth-highest incidence rate (348.5 injuries per 10,000 full-time workers) in the United States. Hospitals also employ Food Preparation Workers, which have the tenth highest injury rate (290.8 injuries per 10,000 full-time workers), Janitors and Cleaners (excludes maids and housekeeping) are thirteenth on the list (277.4 injures per 10,000 full-time workers) and Maids and Housekeeping Cleaners are the sixteenth-highest injury rate (255.6 injuries per 10,000 full-time workers). In fact, out of the top forty jobs with the highest injury rates, hospitals employ workers in sixteen job classifications on the list. (BLS 2016) Musculoskeletal disorders are the most common injury type in the healthcare industry. In 2015, the incidence rate of musculoskeletal disorders for Health care and social assistance in state government was 89.4 injuries per 10,000 full-time workers compared with the national average of only 32.2 injuries per 10,000 full-time workers. (BLS 2016)

To combat this injury problem, many healthcare organizations have refocused their efforts to decrease worker injuries and they have started to use Lean concepts to help improve processes across the organization. This paper will discuss how to utilize Lean concepts, specifically the A3, to identify the root causes of injuries caused by poor ergonomics and how to move implementation of the identified solutions forward.

Lean Management

Lean thinking and management was formally developed by the Toyota team of Kiichiro Toyoda, Taiichi Ohno, and others in the 1930s. They developed a system called the "Toyota Production System" which changed the focus of production from the individual parts of the production line (people and machines) to the flow of the product across the production line. The ideas of the Lean System are to make sure that the value of the customer is well understood and defined, that all of the steps in the process are necessary (there is no waste), the product flows continuously, the system pulls what it needs, and a system of continuous improvement is implemented to drive towards perfection, continuing to work toward increasing quality and decreasing time. (Lean Enterprise Institute 2017)

While many see Lean Management systems as just a way to reduce the time it takes to complete a task, it can also be used as a way to systematically solve challenging issues. It's also often thought of as a system that can only be applied to manufacturing, but this is not the case. "Lean can be applied in every business and every process. It is not a tactic or a cost reduction program, but a way of thinking and acting for an entire organization." (Lean Enterprise Institute 2017-2) Lean is a problem-solving system that helps to identify where there are challenges, the root cause of those issues, and ways to help teams work together to solve those challenges. It is also a system that encourages documentation of standard work practices and training to those work practices to achieve results across all the employees in the work space, department, or process.

There are many tools that have been developed as part of the Lean Management System. Most of these tools help to create a visual of the process and the challenges associated with the current methods. The next section will help to outline some tools which can be utilized to tackle challenging ergonomic or safety issues in the healthcare environment.

Using Lean Tools to Solve Challenging Ergonomic or Safety Issues

There are many different tools within Lean. This paper will describe how to use the A3, Process Flow Diagram, Fishbone Diagram, and Value Stream Map to visualize the process and help to drive to the root cause and then influence change.

The A3 is a Lean tool used to develop a "picture" of the entire program and plan for action. This tool is a one page document on an A3 size paper (a standard international unit of measure of paper size defined by the ISO 216 standard) which describes the problem statement, current state, analysis, recommended countermeasures, action plan, and follow-up plan. Figure 1 below shows an example of an A3 that was developed to review the tray handling in the Sterile Processing Department of a hospital.

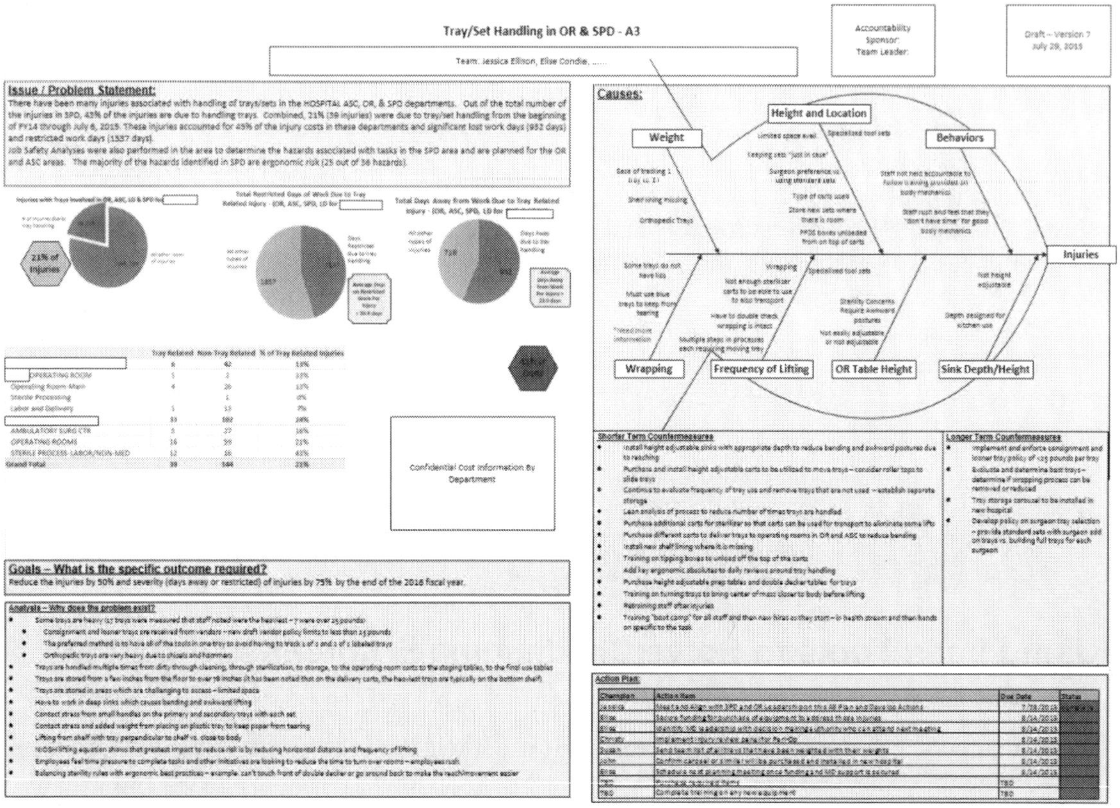

Figure 1. Example A3.

The problem statement needs to be a factual statement outlining the primary issue that needs to be addressed. It should be something that is clear, specific, and tells what is really happening. It is important to keep the problem manageable. For example, tackling injuries due to moving dirty linen by the environmental services team in the primary care units would be a better problem to tackle than injuries due to all lifting activities within the hospital. It is very important to remember not to write a solution into the problem statement. "We need to reduce the tray weight to reduce the number of injuries" is an example where the author of the statement has already assumed that the tray weight is the root cause of the injuries, which may not be the case. An example of a good problem statement could be, "There have been many injuries associated with handling of trays/sets in the HOSPITAL ASC, OR, & SPD departments. Out of the total number of the injuries in SPD, 43% of the injuries are due to handling trays. Combined, 21% (39 injuries) were due to tray/set handling from the beginning of FY14 through July 6, 2015. These injuries accounted for 45% of the injury costs in these departments and significant lost work days (932 days) and restricted work days (1537 days). Job Safety Analyses were also performed in the area to determine the hazards associated with tasks in the SPD area and are planned for the OR and ASC areas. The majority of the hazards identified in SPD are ergonomic risk (25 out of 36 hazards)."

The next piece of the A3 is the description of the current state. For this portion of the A3 it is important to go and see the work being done in action or "Gemba" and to gather facts about the task. (iSixSigma 2017) These facts may include the organizational structure around the process, injury data, production data, the process flow, or any other piece of information you learn from speaking with the staff during the "Gemba." When meeting with staff about their work, make sure to treat them with respect. They know their process better than anyone else and will also be invested in trying to make the process better. Ask them why things are the way that they are to start to uncover underlying issues. This is typically a good place to use a Value Stream Map or a Process Flow map to outline the steps followed and potential hazards. Figure 2 below shows an example of a value stream map with potential hazards. This map shows each of the steps, who completes them, which department is responsible for these steps, what tools or equipment are used to perform this task and what hazards may be present. This value stream map should be developed through a "Gemba" to find out the actual steps utilized and confirm with those who complete the task on a regular basis. Often when reviewing an outline of the process with the end users you will find out important information. For example, someone might point out that this is how the process works on the day shift but that it operates differently on the night shift. Or they may say that this is how it is supposed to work, but in reality it happens this other way which they can then describe.

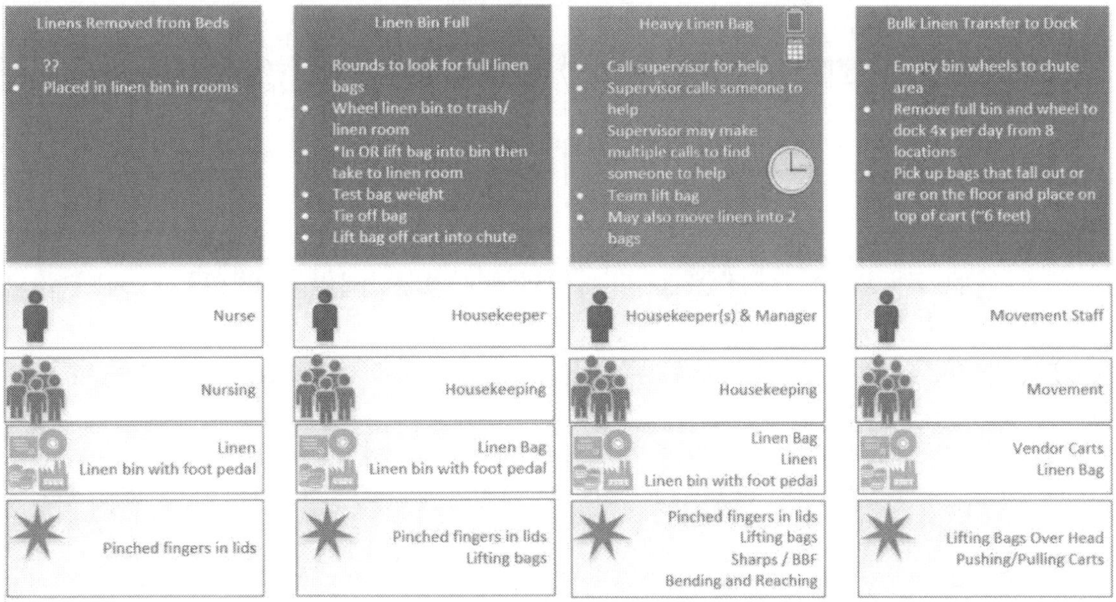

Figure 2. Value Stream Map.

Once the current state is outlined, the next step is to complete an analysis of the current situation. Here is where it can be helpful to use an ergonomic assessment tool to measure the ergonomic risk or a fishbone diagram to outline the root causes of the problem being reviewed. There are several different ways to use the fishbone diagram which could include the primary inputs of Man, Material, Method or Machine or People, Process, Plan, Product. For an ergonomic assessment it may be appropriate to use the primary risk factors of Force, Repetition, Awkward Postures, Static Postures, Psychosocial Factors, Macro Ergonomic Factors. The fishbone is primarily used to organize the items into a more visual picture to help identify the root causes of the issue being reviewed. An example could look like Figure 3 below.

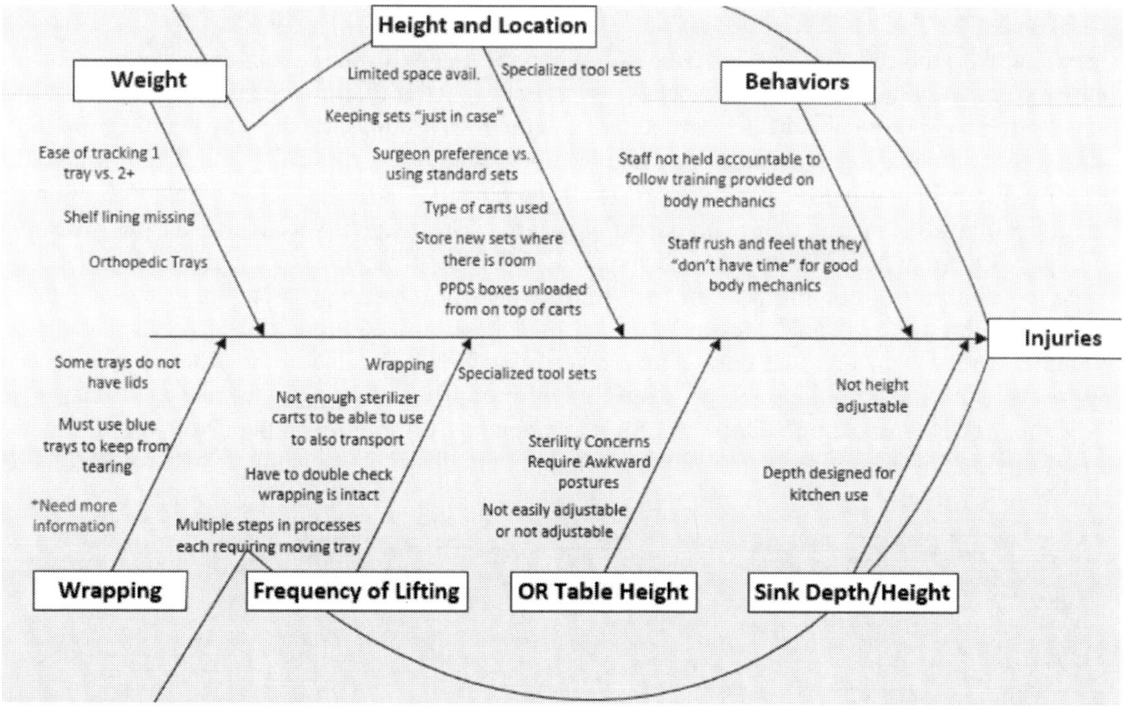

Figure 3. Fish Bone Diagram

The next step of the A3 is to develop an action plan to try some of the possible solutions identified during this process. These possible solutions in Lean are frequently called "Countermeasures" because Lean is a system of continuous improvement. This action plan should describe who is going to complete each task and by when, and have a method for noting the status.

Lastly, the A3 should include a plan for follow up to check on the project once the countermeasures are in place. Ideally a measurable metric should be included so the team can agree when a project has met its ideal goal.

It should be noted that A3s are working documents designed to allow the user to review the problem with key stake holders, staff, colleagues, and other interested parties to allow for a dialog and input to the problem as it is being worked. Once new information is received, the A3 should be updated. Some even recommend writing the A3 in pencil so that it can be easily updated and not seen as a permanent document.

It is also very important to make sure that the A3 has a management or executive sponsor of the project to help make decisions around any funding or organizational change that may be above the authority of the working group. This person should be identified at the beginning of the process, before or just after the issue/problem statement is identified to ensure the team's success.

Summary

In summary, Lean tools can be used to help solve ergonomic issues. These tools help to add in psychosocial factors and organization aspects that may be key to the root cause of the issues being reviewed and can also be important in implementing the change needed to reduce injuries. Traditional ergonomic tools simply look at one task and the ergonomic risk associated with that one particular task. This tends to lead to the recommendations provided in these reports just sitting on the shelf and not being implemented or if they are implemented, they may be tried for a short period and then abandoned with no follow up. An A3 helps to get everyone, "on the same page" with all of the information about the problem, the possible solutions and the action plan all in the same place. This helps the workers involved in the process be part of developing the solution as well as helps the team to dive deeper into the organization issues that may be preventing changes from being implemented. Using an A3 and lean tools helps to drive lasting improvement and employee engagement to tackle challenging ergonomic issues.

Bibliography

Bureau of Labor Statistics. 2016. *Economic News Release – Nonfatal Occupational Injuries and Illnesses Requiring Days Away from Work, 2015* (retrieved February 24, 2017) (https://www.bls.gov/news.release/osh2.nr0.htm)

iSixSigma. 2017. *The Many Sides of a Gemba Walk* (retrieved March 7, 2017) (https://www.isixsigma.com/methodology/lean-methodology/many-sides-gemba-walk/)

Lean Enterprise Institute. 2017. *A Brief History of Lean* (retrieved March 2, 2017) (https://www.lean.org/WhatsLean/History.cfm)

______. 2017-2. *What is Lean* (retrieved March 2, 2017) (https://www.lean.org/WhatsLean/)

You Can't Fix Stupid: Stop the Blame Game and Understand Human Behavior

Timothy D. Ludwig, Ph.D.
Appalachian State University
Boone, North Carolina

Labeling is Easy: Dig Deeper to Change Behavior

It's quite easy to give ourselves a label, isn't it? We look at our behavior, and we look at how it affects others, and we give ourselves a label. I live up in the mountains of North Carolina and drive back roads all the time. Last summer I was driving in a city, was looking around at the unfamiliar surroundings, and drove right under the traffic light into an intersection. Just in case I wasn't aware of my error, a guy in a big Bronco SUV blasted his horn and pulled beside me staring angrily. I looked at him and pointed to my head and mouthed "Stupid". He seemed to agree and the confrontation was over. I had interpreted my own behavior with a label, "Stupid", and that simple adjective seemed appropriate.

In fact, labeling is quite popular in modern business where management training often involves some personality test like the MBTI where we learn everyone's label in hopes of better collaboration. "I'm an Introvert which explains my discomfort working in big teams." "I'm a Judger which explains why I'm so critical." Somehow these labels seem to be the magic elixir that make business work better. But they don't. Everyone goes back to the same environment and acts the same way, nothing changes.

We overuse labels when dealing with the safety of our work crews and managers. The implication is: if workers can't follow rules and procedures that are clearly in the manuals and training, and then they get hurt, they're "Stupid", "Noncompliant", or "Lazy" or "__________" (you can fill in the blank – please keep it rated "PG").

The problem is that you can't fix a label. All the exhortations in the world emphasizing "Don't BE this" won't work. But we do that in our training, in our incident investigation summaries shared with workers, and in our personal conversations. But nothing changes. And you get frustrated. You can't fix it. You're left with nothing, except getting more and more upset.

Instead, consider what behavioral science tells us. Instead of asking a person to BE something, focus on how you can help them DO what is required to be safe. Don't pretend and try to change someone. Leave that arrogance behind. Instead, be a servant. Recognize that

EVERYONE wants to be safe and act safe. It is your job to remove the barriers that put them in the position to, knowingly or unknowingly, take that risk. When you get away from the label, you'll be much more likely to see what those barriers are.

A Lesson Learned

I was invited to go to an automobile parts plant that had a lot of safety issues in a neighboring state into. Its injury rate was too high and a new general manager was brought in, young guy. He had been there for about a half a year and, from all accounts, he had been doing a really good job of just picking the low hanging fruit. You know, such things as putting guards on hazardous equipment, upgrading P.P.E., writing some S.O.P.s, and cleaning up stuff.

I was there to do a safety culture assessment. I had assembled a focus group of line employees and some maintenance folks in a front office board room seeking to understand why their team's safety culture scored the most negative in the whole corporation. The general manager and safety pro was also there. One of the problems that they talked about was they had a lot of aging machinery that would mist and drip a lot of oil. The conversation started quite diplomatic when they acknowledged that things had gotten better. It used to be they couldn't even see the clock on the wall because of the oil mist in the air.

The manager had put in filtration systems and some "band aids" on the machines. He had been working hard at it and the employees, who definitely had some unspoken grievances, did say they appreciated his work. But the manager didn't reciprocate. In fact, when I got them talking more honestly (I am a psychologist after all) the manager had gotten upset. His hands were animated when he said, "I've done all I can. The workers just don't care. I'm spending tens of thousands of dollars fixing equipment and our facilities, but the workers just don't care. They just don't get it. They keep doing idiotic things and get hurt. There's something wrong with them. There's something in the head."

I remember this redhead woman sitting over to the side. I remember her vividly, stewing, and shifting in her seat until she bravely blurted out: "You calling us stupid?" Not a moment passed before he looked at her and said: "Stupid is as stupid does." I was aghast. My mouth was open. Did he just say that? So I knew something was about to go down. The room blew up and the negative safety culture was emerging right in front of me.

The implication was, if workers can't follow the rules... policies, manuals, procedures, training... and then they get hurt... They're stupid. That was the implication the manager threw down in the board room in front of everybody. But you know what? *You can't fix stupid*.

That's why the manager was so frustrated. He used the label: "You employees are stupid." What can you do about Stupid? What can you do? Do you coach the person taking risks? That rarely works. Fire the person? Then your going to hire someone just as... stupid. You can't fix stupid. So you're left with no solution, except getting more and more upset.

The problem is that you can't fix a label. The manager was telling them "Don't be stupid". The famous management guru (and one of my heroes – look him up!) W. Edwards Deming called these exhortations. The definition of exhortation according to Google is *a communication emphatically urging someone to do something*; it's Latin origin is the word for "pleading". But we use exhortations in our training, in our incident investigation summaries shared with workers, and in our personal conversations. But nothing changes. Exhortations, telling folks "Don't BE this" simply does not work. Instead of asking a person to BE or NOT-TO-BE something, focus on how you can help them DO what is required to be safe.

I had had enough of the arguing so it had come time for me to have confidential discussions with the group of workers which is part of my process. This gave me an opportunity to kick the manager out. The manager left the room for about that hour-and-a-half. When he came back around lunchtime, he was ashen. He had no color in his face. He was visibly shaken, definitely embarrassed. He had his head down. The redheaded woman noticed and asked, "What's going on, sir?" To his credit, he was honest. He said, "You know, I decided, I blocked the whole day for this focus group. I didn't anticipate this extra time, so I decided no to go back to my office and get caught up in email. I wanted to keep my eye on the ball and figure out what you folks are talking about when you complain about the plant and your work. You were challenging me to put myself in your shoes so I took that opportunity to go out and work in the plant… see what you guys are talking about."

"I'm no laborer and I'm not a machinist, so I don't want to go out and mess things up by working on or around with the machines. So I thought I'd do maintenance." He joked, "even a manager can change light bulbs." That's what he did. There was one area of the plant that was dark so he got some bulbs and went to work. The manager paused and took a deep breath, avoiding eye contact, and continued, "Well, before I knew it, I found myself with one foot on the top of the step ladder, on the tip of my toe reaching for the light bulb on the ceiling. My other knee was up on piping trying to get more lift." The step ladder kicked off, he slid on the oily pipe but was somehow able to grab the pipe and stop the fall. He was then able to climb down. It had just happened before he returned to the board room.

The redhead jumps up, right out of her seat, "*What are you… stupid*?" It was funny, everyone laughed. They were showing him. "You label us and we label you. What are you going to do now?" He simply agreed, "Yeah, I was stupid."

I took the opportunity to insert some behavioral science, do a smarter analysis, and teach them how to change behavior. Trying to gain control of the situation, I pulled the flip chart over and did an ABC analysis. We first described the behavior that put him at-risk: "standing on top of a step ladder". We then asked the manager: "Why did you do it?" He predictably replied, "Well, it was stupid." I asked if we can fix stupid. The group looked perplexed, they couldn't fix stupid… this was the manager. I finally said, "Instead of trying to fix your manager here let's actually focus on something we can fix. Let's figure out what it is."

The ABC analysis is the core of behavioral science. Its practical. You do it in your head already. When faced with a decision to do something the safe way or risky way you ask yourself:

> "*In this situation…*
> *If I act this way …*
> *Then this will happen*".

When our focus group at the auto parts manufacturing plant started talking about it, we noted that the proper ladder, the six-foot ladder, was located at the shipping dock. To get to the shipping dock the manager would have had to walk all the way through the factory, about a 7 minute walk (times 2 for the round trip). Imagine that. He'd have to all the way across the factory and then walk back through all that hazardous oily machinery carrying a 6-foot ladder… just to change that light bulb. Alternatively, there was a step ladder in a housekeeping closet about 20 steps away.

What is the consequence of getting the correct 6-foot ladder? You would have to walk for about 15 minutes through the plant and be exposed to more hazards. Do you think this consequence reinforces or punishes the desirable behavior? Let's play it out:

In this situation- the correct ladder is in the loading dock;
If I act this way- go retrieve it;
Then, this will happen- it will take 15 minutes, I'll be tired, and I may slip or bang the ladder on equipment.

You'd only have to walk through the plant a couple times with heavy objects to begin to avoid doing so in the future. The behaviors involved in getting the correct ladder is punishing due to this consequence. Behavior avoids punishment. You don't go get that ladder.

There are always alternative behaviors waiting in the wings. What alternative do you have if you do not go retrieve the 6-foot ladder? Well, you could not fix the light bulb. Not an option. You could walk right over to the available step ladder and use it. Humm, lets play that out:

In this situation- the step ladder is near by in the cleaning closet;
If I act this way- go retrieve it;
Then, this will happen- it will take a couple seconds, its not that heavy, and I can get the job done quicker.

You'd only have to execute the easier, more convenient option a couple times to begin to choose doing so in the future. The behaviors involved in getting the incorrect ladder is reinforced due to this consequence. Behavior seeks out reinforcement. You use the step ladder.

We knew the behavior choices facing the manager: he could go get the correct 6-foot ladder in the loading dock or grab the step ladder close by. We know there were a couple different possible consequences for the at-risk behavior of using the step ladder to reach a ceiling fixture. On one hand, it was quicker and easier. It would save 15 minutes and he wouldn't have to lug a 6-foot ladder through the plant. But, on the other hand, he could fall and get injured. So we completed the consequence chart evaluating these consequences.

We took the time to do this simple analysis. We found that we did not have to call each other stupid. We did not have to label. We can't fix stupid, but we can fix the situation! Our analysis allowed us to arrive at a solution: make the safe behavior result in a quicker and easier consequence.

It was one of those "easy button" moments. Managers love the easy button. Our manager hit the table, "We'll procure ladders and put them in the housekeeping closets around the plant. Heck, yeah." The solution was made even easier because he had created these housekeeping closets because of the oil problem. Part of his solution to reduce oil hazards was to get folks to mop the floors when there are oil spills. He outfitted each housekeeping closet with a mop and water and a pail. So, he would just need to procure the six-foot ladders and put them in the closets. This pleased the maintenance guys around the table who had been asking for these for years. Easy button.

We had reached a solution and this fact cheered up the focus group and they started talking about other things to put in these housekeeping closets so their tools would be in closer proximity to where they need them. But when we started talking about the oily floors you see the manager starting to sulk. He stood up and interrupted the brainstorming, "Those housekeeping closets, where we put the mop and the pails, workers don't use them. I'm trying to clean up the oil problem and all I'm asking you to do is to clean up the floor when you see spills. But your NOT, I still see oil on the floor all over the plant. YOU PEOPLE ARE LAZY."

Another label. You're lazy. It's a label. The redhead woman, much calmer now, stopped him, "Sir, we have mops and pails, and we mopped. We tried it, sir. But after we finish mopping the water in the pail is all oily. The next person that uses the mop gets the oily water. After a while, we're just taking the oil mopped up from before and putting it back on the floor."

The manager was dumbfounded, "Change the water. If you changed the water, you wouldn't be putting oil back on the floor." You could almost hear the manager say "Duh!" under his breath. The redhead said calmly, "Sir, where's the plumbing? Where's the hose for the clean water?" Guess where it was, all the way back at the loading dock. To get clean water, a worker would have to push a pail full of oily water through the machinery to the loading dock to change out the water, and then come all the way back. I would suspect more than 15 minutes would be wasted, so they didn't do it.

We didn't need to waste our time labeling. I could have let them argue about stupid and lazy the whole time. It is counterproductive. Instead, by the end of our focus group, the manager was talking about how to retrofit plumbing throughout the factory.

The lesson learned here: instead of labeling make the proper tools more available at the point where the work is done. Change the environment, and by changing the environment, we change behavior. Workers no longer have to find themselves find myself in a situation that encourages them to take that risk any longer.

A Lesson Not Learned

I recall a recent discussion I had with a new friend while we were having dinner with his fellow corporate HSE directors overseas. He was telling me about a supervisor who had a close call with a high-pressure hose that nearly took off his head. Two years later, this same individual was teaching a new worker to do the very same procedure. This time physics was not so forgiving and the worker lost an eye.

The question to me during dinner was, from a behavioral perspective, how can you account for people like this supervisor who intentionally took short-cuts during a procedure that put him at risk? And, to make it worse, teaching others to do so?

This company was heavily engaged in fracking where there's a lot of piping holding enormous amounts of pressure. At the end of the fracking services the well is now producing and you're disassembling the whole complex. Part of process when you're breaking down, is removing the hoses from the huge pumps on the back of flatbed trucks. When you're fracking, you produce lot of pressure to send underground, so when you're breaking the hose and pump assemblies down you have to make sure there's no more pressure in these hoses. It is important that the field operator to do a double check on each hose before they decouple it based on the chance that their may still be latent pressure hiding. There's a 2-minute process, where field operators check the local pressure of the hose pressure and, if neccessary bleed it out fully before opening the coupling and removing the hose.

The story I was told was about a supervisor who was decoupling hoses on his own. He had been with the company a long time so he had done this task many times. He was in a hurry to finish the job. He's kind of fatigued being in the field doing 12-hour shifts for weeks. He was taking short cuts by skipping over the 2-minute pressure test. How many times had he done this in the past and found no pressure, again and again? Unfortunately, this time, he opens the coupling and there was latent pressure inside. The casing explodes off and sends a projectile through the cab of the flatbed holding the pump. Fortunately, no one was in the cab.

Shocked, I asked, "Was he okay? Was everybody okay?" Everybody was okay. I was curious about the incident so I started asking questions. Yes, the pressure check was in the Standard Operating Procedures, yes, it was on the break-down checklist and reviewed in the J.S.A. I kept asking questions but my host stops me, "No, Tim, no. That's not the point."

Three years later, this same supervisor was teaching a new worker, a green hat, to do the very same procedure. In the interest of time, he had taught the green hat to skip the pressure check. This time Physics and chemistry were no so forgiving. This time the one-of-ten-thousand happened and there was still energy in the hose. The casing blows out but this time the projectile ricochets off the pump and strikes the young man in the head. He suffers brain damage. Horrible incident.

"How can you tell me we do not blame that supervisor?" My new friend was clearly personally upset with this supervisor. He labeled the supervisor with such words like "stupid", "reckless", and "*word-that-shouldn't-be-uttered-in-mixed-company*". I'd be upset too. This is a person whose intentional actions hurt others.

Consider the label "stupid". If that's your conclusion for an incident investigation what are you left with? You can't fix stupid. Your left with no solutions. Worse, if we get in the habit of labeling blame we tend to use it as an easy excuse not to do the harder work of a behavioral assessment of the incident. We go no further… we solve nothing … the incident will likely happen again, and this time take out an eye.

As my friend and I started drilling down into the supervisor's behavior we started getting an audience of his fellow HSE leaders. We discussed equipment and tool changes that could help make the risk avoidable. We discussed how the older generation had grew up in a safety culture where supervisors taught short cuts and encouraged them. They now saw a need for targeting supervisor refresher training and safety selection methods of promoting new supervisors.

But then I took them back to the original event, when the supervisor was all by himself, blew the casing off and damaged some equipment. I asked, "What happened after that?" "He got coached." I imagine the coaching was pretty forceful and threatening. "Yeah, okay, he got coached." I reflected, "Did that change his behavior?" No. He was still doing it five years later.

What else happened after the first incident? "We filled out the incident reports. We wrote him up and made it a Hi-Po. A Hi-Po is a close call that gets sent all the way to our top executives because of the potential severity of the incident. His first incident went all the way to the top; it got everybody's attention." "OK," I asked, "Then what?" "What do you mean?" "You said you did this Hi-Po, right? You filled out this form detailing the investigation into the Hi-Po. I'm sure that somewhere on the form, hopefully, there's a big square with the big conclusion as to why the Hi-Po happened. It's in a big square because it highlights what we've got to fix so it doesn't happen again."

"So," I asked, "what was the root cause?" His reply "Human error." Human error was found to be the root cause of this Hi-Po. Translation: "*he was stupid*".

It was with this statement we got to what I considered a root cause. The Close Call report concluded "Human Error" as the cause of the high potential incident. The "Close Call" was filed, the supervisor was talked to, and the work went on, only to have the risk repeated ... how many times by how many people? We then concluded that the Close Call reporting system is what needed to be fixed. It was too easy to blame the worker and not uncover truly actionable changes to the work and the work system that could have stopped the risk from happening, not only with this one individual but throughout the workforce. The Close Call system had to be actionable, it wasn't and the incident was forgotten…until two years later.

If you have "Human Error", "Human Factor", or "Stupid" as an option in your incident investigation forms it is all too easy to just stop there, turn out the lights, and keep on promoting the environment that caused the risk in the first place.

Instead, approach the incident with a clear understanding of the cause and effect relationships between the behaviors that were related to the risk and the reasons why that person, knowingly or unknowingly, was put in the position to take that risk.

Understand the cause of behavior and you'll begin to see a pathway for solutions: A) Ask critical questions around training, tool availability, equipment flaws, perceived time pressure, missed reporting, supervision, and work flow. B) Ask yourself about the person's experience doing the task in the past. Was the safe way to do the task cumbersome or ineffective? Did they adopt the risk in an effort to save time or trouble? What can be changed in the task to alleviate the "costs" of the safe alternatives. C) Then take a close look at your Safety Culture. Was this person told how to do the risky short-cut by others? Alternatively, would a peer be likely to stop and coach the person if they saw the risky behavior occurring?

My Job as a Labeler

Earlier in my career I worked with a severely mentally disabled individual named Violet. I was working my way through school learning behavioral science. Our reach in behavioral science goes far beyond its impact on the safety world making a difference in schools, health care, public policy, sustainability, and with human services such as our work with autistic and mentally disabled individuals. (Get on Behavior.org and learn all these areas that may touch your life and see what we're trying to do sharing behavioral science.)

I was working with a group of about 15 folks with severe mental disabilities and other challenges, so much so they couldn't live life on their own. As such, they were wards of the state. They lived together in their residence where the professional staff fed them, cleaned them, groomed them and provided what training and social life they could.

They had me come in and create programs to try to shape basic life and social behaviors. They had a psychiatrist who administered medications for the mental disorders experienced by this population. They had a social worker as a director, a new guy named Bob. He was kind of a funny, goofy guy. But Bob was one of the wisest people I've ever met.

Back then we used the label "Mentally Retarded" and Violet fell squarely into that category. I conducted a special intelligence test on her and her IQ was below 15 (the average IQ is 100). Her verbal abilities were strikingly limited, typically reduced to sounds resembling simple words.

Violet was also a scary woman. We called her "Violent" because of her volatile behavior. It came to a point where she slammed her fist through a sheet rock wall. We knew this incident was coming because she would slap her right arm against her left shoulder when she'd start getting more and more agitated. We knew something had to be done.

So we had a team meeting. We had the psychiatrist there, me as a psychologist, the staff and Bob. The psychiatrist was treating her for hallucinations. The staff said she often hallucinated that there was an evil pony on her left shoulder. She was hitting herself on the shoulder in an attempt to get rid of the pony. Her left shoulder had deep bruises from the abuse.

For these hallucinations, Violet was on Haldol, a drug given to schizophrenics. It's a strong tranquilizer. If you take it, believe me, you'll reduce everything, not just your daydreams. Violet was on a pretty good dose of it. In the wake of the incident the psychiatrist was recommending to triple the dose during our team meeting. That's a lot of chemicals impacting the brain. The psychiatrist had labeled her "Schizophrenic".

For my part of the meeting, I had all these behavioral programs worked out trying to shape behaviors to counter her violent behavior. I had labeled her "Self Injurious". We had a long discussion about the precursors to her behavior and the behavioral steps the staff should take to reduce the likelihood of violent behaviors. The staff had even more suggestions in what I thought was a fruitful team discussion.

Bob was quiet and after listening to all our input he said: "Hold on... Before we do chemical or behavioral interventions, let's consider physical solutions first". He wanted to find out what's going with Violet as a person... not some label.

Bob took her to get a physical, the first in many years, and found that she had a metal screw in her ankle. Violet was probably 40 at the time. Her body was deformed with a humpback and other contortions. So the big bulge around her ankle must have been considered just another deformity.

In reality, Violet had broken her ankle pretty badly in her youth and surgeons put in this metal screw for healing. Her caregivers forgot to have it taken out. Violet wasn't verbal, she couldn't tell anybody. In fact, she probably had no idea of it being there. All she experienced was pain in her left side due to the calcification and swelling around the screw remaining in the ankle.

Violet was in pain. She couldn't understand what was happening, she couldn't tell anyone about how she felt, and her self-injurious behavior was in reaction to it. Not because she was "Schizophrenic" or "Angry" or "Violent"…. all labels. It was because she was in pain and she couldn't tell us.

Bob sent Violet to surgery where the screw was removed and the calcium deposits shaved away. She then spent a couple months in physical therapy.

When Violet came back she was a different woman. She no longer was hitting the left side of her body. She no longer fit the label "Self- Injurious". She wasn't in pain anymore. Soon thereafter, Violet went off the Haldol completely. She no longer fit the label "Schizophrenic".

Much to my amazement Violet started using words! My previous intelligence test assigned her a clinical label of "Severely Mentally Retarded." Now, with the pain gone, she started using the words she had always been capable of. Moreover, she started grooming herself and interacting with other residents and staff persons. I did a new intelligence test and her IQ (a measure that is supposed to stay stable over your lifetime) jumped more than 15 points with this seemingly new verbal and self-management behavior. *All those skills were always there, we had failed to remove the barriers to her performance.*

Bob (the-Wise) wasn't done yet. He found out that Violet had cataracts. She couldn't see. After cataract surgery she came back smiling, an interactive, shall I say "affectionate" person. I left the job soon thereafter but I still remember my last day vividly. Violet, yes "Violent," came up and hugged me. She used a full sentence with my name at the end, something that I thought she was too "Retarded" to do. It was evident that even my new IQ Label was inaccurate and too low. I was in tears.

I was there to help her. I was being paid to help here. I'd been trained to help her. I had a PhD. The doctor had an MD. The social worker, Bob, didn't. Who among the "Smart" people was the one that really viewed Violet as a person, not a label? It was the person who was with her every day. The person who said, we're not going to label. Instead, we're going to look at her world and understand why she exhibits this behavior.

The lesson learned here for us is that labeling is counterproductive for our safe work environments. Instead, we need to be wise like Bob and understand what is going on in our workers' world and do something about that... instead of trying to do something about them. We then change the environment, and by changing the environment, we change behavior.

Do workers need different, more available tools to work safe? Better, more behavioral training? Procedures that lessen fatigue? Supervision that doesn't encourage short cuts? Etcetera... Etcetera.... It's your job, "Bob," to do the analysis to find the environmental causes of the behaviors so your workers don't have to find themselves in a position to take those risks any longer. We don't need to label.

Folks, dig deeper. Labels are artificial. Labels are an illusion. There's always a reason for the behavior that we get. We can all overcome our labels!

Postscript: A Celebration is in Order

When I was teaching the junior youth Sunday School class at my church I had in my class a young lady who was born with Downs Syndrome. An IQ test would have labeled her as mentally disabled because of this. Miranda was also one of the sweetest kids I got to interact with but I knew she had a limited future, probably performing at a 6th grade level the rest of her life.

About 5 years ago I got word that my University was extending its disability services to include mental disabilities including mental retardation. Most Universities require professors to make accommodations for students with disabilities in much the same way as they provide ramps and the like for students and staff with physical disabilities. Typically, I have students with learning disabilities like dyslexia or attention deficits (ADD) who get extra time on exams and other accommodations so they can perform up to their intellectual abilities less encumbered by their disability.

But I was skeptical about allowing mentally retarded individuals into the University system because their intellectual capabilities would not allow for the learning to take place to begin with. We would be setting them up to fail to live up to the academic standards of our courses regardless of any accommodations we provide.

Last month Miranda graduated with a Bachelors of Arts from Appalachian State University. I was at Miranda's graduation, in my learned doctoral robes, once again humbled by the potential of human performance.

Timothy Ludwig's website is Safety-Doc.com.

Contractors' HSE Performance: Pragmatic Evaluation, Benchmarking and Way for Improvement

Ahmed M. Qabazard
Team Leader HSE (WK)

Ajay Kumar Mohanty
Safety Engineer HSE (WK)

Bala Siva Srikanth Adivi CSP, CMIOSH
TPL Specialist I (HSE) HSE (WK)

Kuwait Oil Company
Kuwait

Overview

Improvements to contractors' HSE standards and performance can be ensured by continuous monitoring and review of their HSE performance. To achieve this, an objective HSE Performance Evaluation (HPE) framework is necessary & essential. Although there are various conventional methods of HSE performance evaluation are available in the industry; however, a more comprehensive HPE framework with pertinent HSE parameters will aid to systematic monitoring approach. In line with this, a comprehensive HPE framework, which takes into account all HSE parameters pertinent to a contracting organization (Business Partner) and its project, has been developed & followed at West Kuwait Directorate (one of three main important assets of Kuwait Oil Company (KOC). Such kind of approach and analysis give better insight and will give an opportunity to Subject Matter Experts (SME's) of the organization to deliver appropriate & effective professional outcome out of it. In this HSE Performance Evaluation, the pertinent factors are critically analyzed, recorded the compliance & deviations w.r.t. the applicable procedures / standards, prepared "SWOT analysis" and suggested the appropriate recommendations appropriate to the risk-based gaps besides benchmarking. This analytical framework of HSE Performance Evaluation (HPE) model can be applied for all business partners, which would facilitate the benchmarking process.

Introduction

Role of Contractors (Business Partners) in industrial development especially in oil & gas sector is very vital. Industrial sector cannot develop with the current growing demand without the assistance of proper contracting agencies. The contractors are key players in the sustainable oil & gas sector projects from predesign and design, through construction and commissioning & project closeout. They also provide immense support in operation & maintenance of some production facilities also.

In this context a question arise in our mind that, how do the performance (especially on HSE compliance) of a contractor is monitored effectively? Our Company's procedure prescribes some of the monitoring processes to ensure HSE compliance / performance of the contractors through Management Site Verification Visits (SVV's), Leadership HSE Visits, audits, etc. In order to bring the effectiveness of these monitoring tools, HSE West Kuwait (WK) team of KOC had done brainstorming exercise critically and developed a pragmatic method of HSE Performance Evaluation (HPE) framework for all contractors working under WK Directorate. It is not only an analysis, but also goes beyond doing the analytical study of the strength, weakness, with prevailing threats & available opportunities for each of the contractor and suggests remedial measures for improvement.

Figure 1. HSE Performance Scale

The comprehensive analytical framework is an effective tool that utilizes critical inspection databases (SVV, Work Site, Site Office & MyHSSE data) to measure the overall Contractor's compliance on KOC HSE Management System (HSEMS) Procedures / Standards implementation at contractors' work sites and their site offices. This is invariably required to have a safe & successful operating journey between KOC and Contractor.

Necessity

Globally it is acknowledged that an improved HSE implementation and compliance through risk based gap analysis has direct and proactive bearings on functional and operating safety standards of the facilities in general and productivity in particular. This will also enhance the employee morale. Hence ensuring positive HSE performance through systematic approach is essential and need of hour in an industry especially in Oil & Gas.

Objective

The purpose of this analytical & benchmarking tool is to evaluate the HSE performance of WK-based contractors by reviewing their level of HSE management systems, compliance, policies, statistical performance, adherence to standards, etc. The positive observations and performance gaps identified through this exercise were used as a basis for the analysis and resultant recommendations, in order to improve upon HSEMS procedural compliance, thereby HSE performance of WK contractors' workforce and of the company as a whole.

The long-term intent is to build a project findings to document best practices, lessons learned, and examples of various documents to benefit KOC and the Contractors in terms of benchmarking, evaluation & enhancing HSEMS performances.

Methodology Adopted

The above evaluation process is unique and different from various conventional methods, which focuses not only on level of implementation but also focusing on the causes of deviations. The following steps are involved in the above process:

- To develop a comprehensive analytical tool / framework utilizing various inspection databases such as
  - ➢ HSE Inspections @ worksites
  - ➢ Observations from Site Verification Visits (SVV's)
  - ➢ HSE Inspections @ Site Offices
  - ➢ Other HSE Leading and Lagging Indicator performance from 'MyHSSE' data

  The above parameters are selected to measure the overall Contractor HSE Performance on KOC HSEMS procedures/Standard compliance.
- Evaluate HSEMS procedural/ Standards implementation & compliances based on the risk level at Work Sites & Site Offices of all KOC West Kuwait Contractors.
- Identification & Analysis of Positive observations/Compliance on KOC HSEMS Procedures/Standards.
- Identification & Analysis of critical risk based gaps/ non-compliance of HSEMS procedures/Standards.
- Benchmarking the HSEMS Procedures/Standards performance compliance for all WK Contractors.
- Sharing Results/ Contractor Performance status with respective Controlling Teams to improve Contractor Performance.
- Closely follow-up with the respective Contractor through the Controlling Team for tracking and closure of the opportunities of improvement (OFI's).

Evaluation Process

The evaluation of the above parameters was done systematically for a period of four years (in the phased manner) in the following way:

❖ **Establishing and Maintaining a Contractor HSEMS Performance Compliance Analysis & Benchmarking Program**:

This has been made with defined strategic HSEMS Performance Evaluation Criteria based on established KOC Risk Assessment Procedures.

❖ **Collecting & Storing Observation Data to Assess & Analyze Performance**

Input has been taken from the following:

- Determining Data Needs (MyHSSE, SVV, Worksite Inspection & HSE Inspection data)

- Data storing requirements(segregation of the findings per clause(with respect of the KOC HSEMS Procedure) & risk based observations
- Risk rating/ranking on individual observations as per KOC risk matrices

❖ **Analyzing, Reviewing, and Finalizing Performance Data**

Three-step process has been established as follows:

- Step 1: Critical review and analysis of observations
- Step 2: Develop individual charts for different HSEMS procedures/standards.
- Step 3: Recommendations based on the individual chart analysis

❖ **Performing SWOT Analysis**

A SWOT analysis (alternatively SWOT matrix) is a structured planning method used to evaluate the **S**trengths, **W**eaknesses, **O**pportunities and **T**hreats involved in a project or in a business venture. We have made an effort to utilize **SWOT model** for identification of effective contractors' HSE Management System Strategies.

The study has been made methodically to identify and analyze internal factors (strengths and weaknesses) and external factors (opportunities and threats) of the contractors. Subsequently, by means of the SWOT model, the identification of all internal and external factors are made for evaluation of proper strategies in order to enhance opportunities and strengths and analyze the weaknesses and threats.

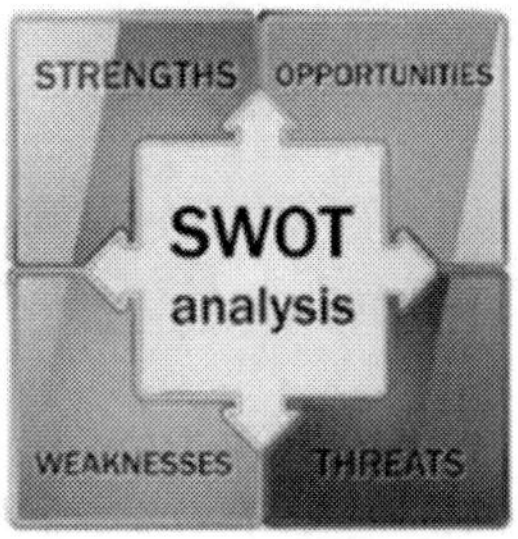

Figure 2. SWOT Model

❖ **Using Performance Information to Drive Improvement**

Three-step process has been established as follows, including critical component i.e. communicating the results to the Controlling Team(s).

- Step 1: Communicate "Contractor Performance Report" to the respective Controlling Teams.
- Step 2: Organize meeting with Controlling Team to explain the evaluating process and the expected improvement action plan on the recommendations.
- Step 3: Follow up on the recommendations & action plans to close the gaps during Monthly Contractor Meetings.

The Process of Annual Analysis:

<u>Phase I & Phase II:</u>

Each the above phase involves the following steps:

- Systematic building of databases for site-checked SVV's, Leadership HSE Visits, construction/project worksites, site offices inspections and 'MyHSSE' reports.
- Selection of HSEMS performance parameters and criteria (i.e. SVV elements, leading/lagging indicators from MyHSSE, HSEMS application standards as per Scope of Work, Permit to Work procedures and other applicable KOC HSEMS procedures and standards
- Thorough review and analysis of databases, including SWOT analysis (Strength – Weakness – Opportunities – Threat) of performance levels, results from the generated site reports.
- Risk ranking of resultant performance gaps, in accordance with KOC HSEMS Risk assessment procedures
- Identify the performance – gaps and positive observations, and develop individual and dedicated analytical charts of the same.
- Brainstorm to list – thorough analysis, specific remedial recommendations to close/improve non-compliance and performance gaps.
- Communicate the Contractor's performance reports and recommendations to the respective KOC Controlling Teams, to initiate and arrange for the necessary corrective and improvement actions.
- Share the recommendations and positive observations with the respective contractors during one-to-one meetings or during Monthly HSE Contractors meetings.
- Track the improvement recommendation(s) compliance action plans to close the identified gaps.

    ✓ Phase I:

    HSEMS Performance Compliance Analysis for the duration (Dec 2013 to June 2014)

    ✓ Phase II:

    HSEMS Performance Compliance Analysis for the duration (July 2014 to Jan 2015)

    ✓ Phase III:

    The comparison process-evaluation between both Phase I & Phase II and HSEMS Performance Compliance Analysis for the duration (April, 2015 to Dec 2015)

Phase III involves the following steps/process:

- Comparison of HSEMS procedures performance parameters compliance levels of Phase II & I.
- Identify the common/repeated site-observations in both phases, in addition to the improving and declining performance compliance levels and parameters.
- Analyze related performance charts for gaps and positive observations' variations and trends.
- Evaluate related risk levels in declining performance gaps.
- Brainstorm to list conclusion and remedial recommendations, in accordance with each chart and its analysis separately.
- Communicate the Contractor's performance reports and recommendations to the respective KOC Controlling Teams, to initiate and arrange for the necessary corrective and improvement actions.
- Share the recommendations and positive observations with the respective contractors during one-to-one meetings or during Monthly HSE Contractors meetings.

✓ Phase IV:

HSEMS Performance Compliance Analysis for the duration (April 2016 to Dec 2016)

All the above phases and no. of contractors that were evaluated in each phase are listed below

Phases	Period	No. of Contractors evaluated	Details
Phase I	December 2013 - June 2014	17	Contractor HSE Performance Compliance Analysis
Phase II	July 2014 - January 2015	13	Contractor HSE Performance Compliance Analysis
Comparison I	December 2013- January 2015	12	Comparison Analysis
Phase III	April 2015 - December 2015	15	Contractor HSE Performance Compliance Analysis
Comparison II	July 2014- December 2015	11	Comparison Analysis
Phase IV	April 2016 to December 2016	19	Contractor HSE Performance Compliance Analysis

Table 1. Phase-Wise Analysis Data.

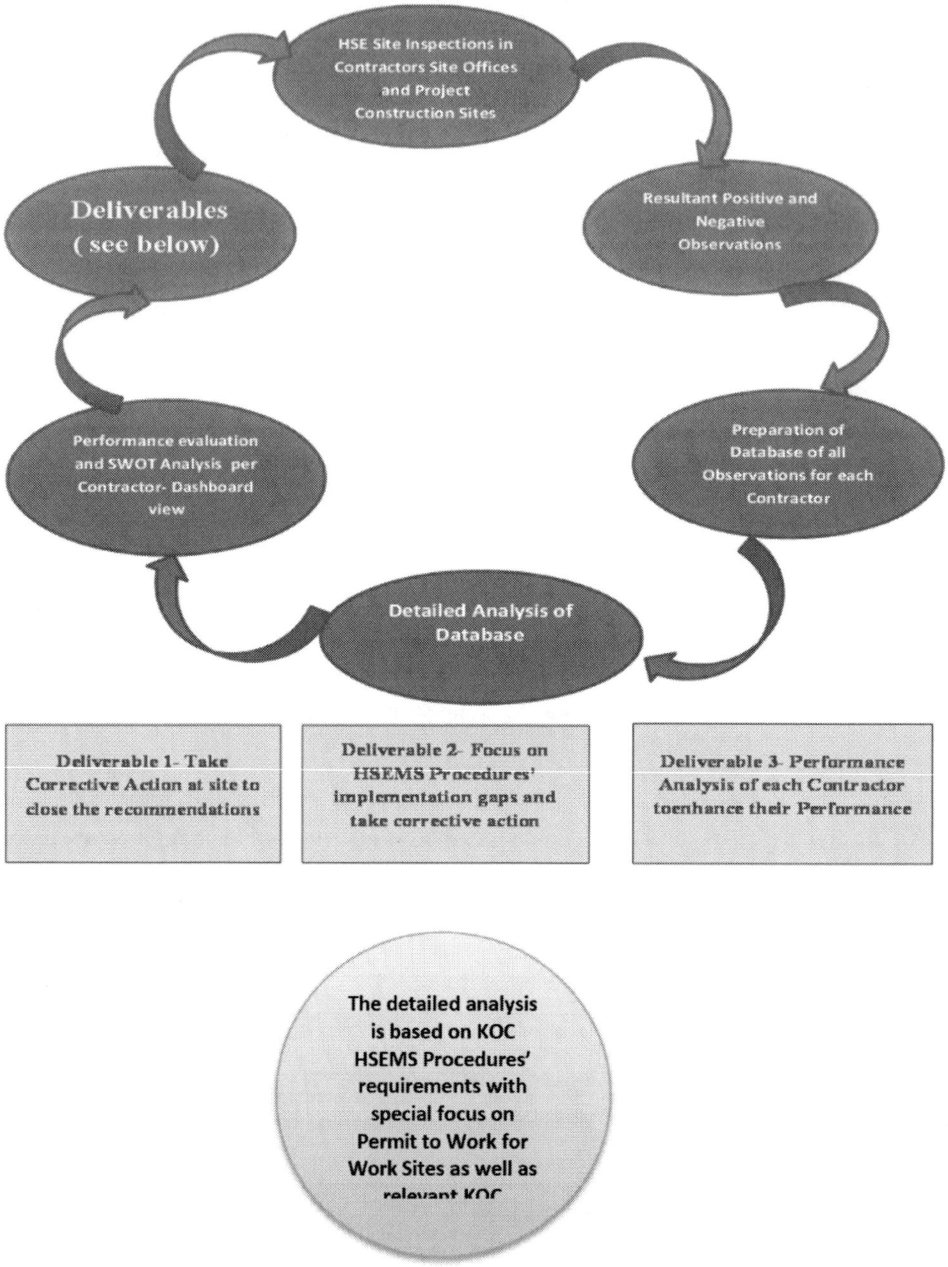

Figure 3. Performance Evaluation Flowchart.

Evaluation Process Structure

Data obtained from 'MyHSSE'

MyHSSE Reporting System is a Web Based Performance Monitoring Tool for Health Safety, Security and Environment at Kuwait Oil Company (KOC). This software being used by KOC to report, record and track HSE measures such as Incidents, Near Miss reports, Hazardous conditions, SOC (Safety Observations and Conversations), SVVs (Site Verification Visits) of CAEs/ NCAEs etc.

The following data of various contractors is taken from 'MyHSSE'

Lagging Indicators	Lost Work Day Case, Industrial Non Disabling Injury Incidents, Property (Asset) Damage Incidents , Fire Incidents, Motor Vehicle Accidents (MVA's)
Leading Indicators	Near Miss Incidents, No of Speed Violations,

Table 2: Input taken from 'MyHSSE'.

Results Gained and Advantages of this Analysis

- Zero Lost Workday Cases (of Contractors) in WK Fields since May 2013
- Provided a reporting mechanism on risk based performance on Contractors HSE compliance to the Controlling Teams and Contractors Management.
- Performance of WK Contractors is benchmarked, which is utilized to enhance their performance compliances.
- Benchmarking encourages healthy competition amongst WK Contractors to strive for better HSE Performance.
- Positive HSEMS procedural compliances have increased significantly, which resulted in positive HSE statistics for WK fields and for the organization as a whole.
- Provides a platform for the Controlling Team to effectively monitor Contractor HSE performance compliance level.
- Effective tool for competent workforce hiring, promotions, work assignments, rewards/penalties of Contractors.
- Provides a structured approach focusing on contractors' strategic performance objectives.
- Transfer best practices across other KOC Directorates for improving Contractor HSEMS compliance performance.

Conclusion

"Contractor's HSE Performance: Pragmatic Evaluation, Benchmarking & Path to Improvement" is a practical approach to assess, evaluate the HSE performance of contractors by reviewing their compliance-level to KOC HSEMS Implementation Procedures and standards. The performance gaps and positive observations identified through this analysis are used as a basis for the analysis and resultant recommendations, in order to improve upon

HSEMS procedural compliance and by extension of HSE performance of West Kuwait contractors' workforce. This analysis not only provides a practical approach for measuring HSE performance to a high accuracy level but also offers guidance on developing a benchmark of HSE performance measures relevant to the organization.

The long-term intent is to build a project findings to document best practices, lessons learned, and examples of various documents to benefit KOC and the Contractors in terms of benchmarking, evaluation & enhancing HSEMS performance. This kind of initiative clearly signify organization's commitment "We Care".

References

HSE Management System (HSEMS) Procedures of Kuwait Oil Company.

Kuwait Oil Company. "HSE Management System Guide".

______. HSEMS Performance Reports of Contractors.

______. MyHSSE Data.

Session No. 734

5S + Safety Finds a New Home at an Old Refinery

Damon Shodrock, MS, CSP
Calumet San Antonio Refinery
San Antonio, Texas

Introduction

The old petroleum refinery is located on the south side of San Antonio. This refinery was built in the 1950s and expanded by various owners over the years. When Calumet Specialty Products purchased the San Antonio refinery in 2013, many operational and organizational changes were desperately needed. At that time, several efforts to improve housekeeping at the refinery were made; however, short-term gains were quickly lost. Within months of any facility-wide cleaning and organizing campaign, the refinery reverted back to disarray.

A New Approach Was Needed

The author practiced and promoted 5S + Safety at previous employers specializing in biotechnology research & development and semi-conductor manufacturing. Those employers demanded high productivity and efficient workplaces. The author was directly involved implementing 5S programs where work areas and processes were analyzed, waste identified and removed and a new order established for improved efficiency.

"5S" is the name for the workplace organization method based on five Japanese and English words: *seiri, seiton, seiso, seiketsu,* and *shitsuke.* These five Japanese words and their English translations begin with the letter "S".[1] 5S methodology describes how to organize a work space for efficiency by identifying and storing the items used, maintaining the area and items, and sustaining the new order. A workforce and management dialogue about standardization usually occurs which drives decision making about how work should be effectively and consistently performed (See Figure 1).

2016: 5S + Safety Launch

In January 2016, Calumet San Antonio Refinery fully launched a 5S + Safety program. All supervisors and managers were trained on the fundamentals of 5S + Safety at the January 2016 Refinery Leadership Team meeting. The author lead this training effort with a MS-PowerPoint presentation defining each 5S category: sort, straighten, shine, standardize and sustain. Several before and after pictures were shown at other facilities that implemented 5S methodology (See Figures 2 & 3).

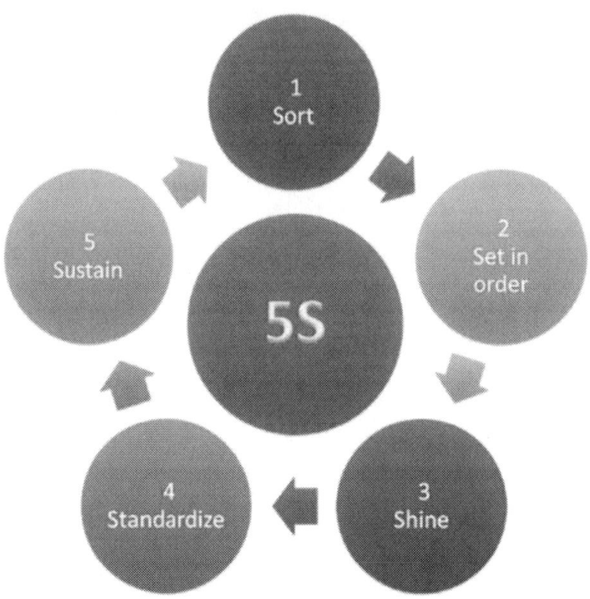

Figure 1. 5S Circular Flow.

Figures 2. Semi-Conductor Fabrication Plant-Hazardous Waste Storage Room: Before 5S.

Figures 3. Semi-Conductor Fabrication Plant-Hazardous Waste Storage Room: After 5S

2016: 5S + Safety Procedure and Checklist Established

The fifth "S", sustain, is accomplished by frequent auditing of workplace conditions to measure the progress of the 5S + Safety Program. The written procedure (Figure 4) and accompanying 5S Checklist (Figure 5) were reviewed and approved through the Calumet San Antonio Refinery (CSAR) Management of Change (MOC) process.

CALUMET

San Antonio Refinery　Standard Operating Procedure

Procedure Name: General Refinery – 5S + Safety Procedure	**Document Number:** SAHSSE-Audit001

Figure 4. 5S + Safety Procedure.

Scope

Many manufacturing facilities adopted a "5S" workplace organizational and housekeeping methodology as part of continuous improvement or lean manufacturing processes. The term refers to five steps – sort, set in order, shine, standardize and sustain – that are also sometimes known as the five pillars of a visual workplace.

The 5S methodology is a simple and universal approach that works in companies all over the world. It supports other manufacturing improvements as just-in-time (JIT) production, total quality management (TQM) or six sigma initiatives. 5S is also a great contributor to making the workplace a safer and better place to spend time.

5S methodology was introduced to Calumet San Antonio Refinery (CSAR) early in 2016. A 5S Audit Checklist is used to evaluate each "S" across all areas.

Instructions

1. Evaluate the audited area for each 5S Category.
2. For each statement, assign 1 point if you observe evidence confirming the statement. Assign 0 points if you see no evidence for the statement and write corrective actions to the right of the point score.
3. Each audit is entered into RegAction Incident Management System (IMS) under the "Internal Audit" classification and preventative actions are tracked within the "Quality" section.

Comprehensive Annual Review

This procedure and the 5S Audit Checklist will be reviewed annually.

Area:		Date:	
Auditor:		Category	Score
5S Audit Checklist		Sort	
		Straighten	
INSTRUCTIONS: 1. Evaluate the audited area for each 5S Category. 2. For each statement, assign 1 point if you observe evidence confirming the statement. Assign 0 points if you see no evidence for the statement and write corrective actions to the right of the point score.		Shine	
		Standardize	
		Sustain	
		5S SCORE	

Category		
Sort	Point	Observations
1. Area is free of unnecessary clutter		Actions
2. Only the required tools are in the work area.		
3. Only the required materials are in the work area		
4. All required Personal Protective Equipment in work area.		
5. Area is free of slip/trip/fall hazards		
Straighten		
1. Locations for tools and equipment are clearly marked and labeled		Actions
2. Frequently used items are within easy reach of work area.		
3. All dormant or excess items are stored properly and out of the way.		
4. Aisles are straight, clear, and well-marked?		
5. Spare parts are stored in their proper single designated location.		
Shine		
1. Everything in workspace is "like new"		Actions
2. No dust or dirt anywhere.		
3. Trash and recycle bins are emptied daily (more frequent if needed)		
4. No painted items are chipped or worn.		
5. All cleaning equipment is stored properly and readily available.		
Standardize		
1. Checklists for 5S activities are available and followed.		Actions
2. Results of previous audit is posted.		
3. All areas for improvement from previous audits are completed.		
4. All charts and metrics in the area are current.		
5. Area signage is clear and well posted.		
Sustain		
1. Time and resources are given to 5S activities in the area		Actions
2. Recognition is given to teams that are involved and excel in 5S activities.		
3. Was the Area Owner present for this audit?		
4. Was at least one hourly employee for the area present for this audit?		
5. Your section of the 5S board is up-to-date.		

Figure 5. 5S Audit Checklist.

2016: 5S + Safety Results

One week after the 5S + Safety training session, auditing in the refinery began. The refinery was divided into 10 zones, and then employees audited one zone per week. Auditors evaluated each zone using an audit checklist by 5S category: sort, straighten, shine, standardize and sustain. Five inspection questions were asked per category and overall score was given with a maximum/perfect score of 25. At the start of the program the average baseline score across the refinery was 9.5 ± 3.1 (~ 2S). By the end of 2016, the average 5S score was 15.2 ± 5.2 (~ 3S) which represents a significant improvement only 10 months later.

2017: 5S + Safety Program Evaluation and Refinement

During the annual review of the 5S + Safety Procedure feedback from 5S auditors was solicited. After one year of practice slight program adjustment were needed. Areas where an hourly employee and the area owner were present during the audit were significantly improved and consistently maintained compared to areas with absentee owners. As a result, two questions were substituted in the Sustain section of the 5S + Safety Checklist:

- Was the Area Owner present for this audit?
- Was at least one hourly employee for the area present for this audit?

In 2017, the audit scope of the program was expanded to include non-refinery areas: Brooks analytical labs, Brooks administration building and Brooks warehouse. The revised checklist was used to audit these areas. Also in 2017, an "Adopt a Refinery" tactic (similar to "Adopt a Highway") was deployed during weekly audits in the refinery. Each auditor was given a trash bag to pick up debris while they were auditing the assigned area.

Conclusion

After reflecting on 2016, 5S methodology has allowed CSAR to make significant and sustained advances in refinery unit effectiveness by identifying stored items, maintaining areas, and sustaining the new order. As a result of 5S implementation, organizational safety culture has improved by eliminating unsafe conditions and behaviors. Over the summer of 2016, this 5S + Safety program was recognized by Calumet corporate leaders for demonstrating Operational Excellence and was promoted to all Calumet facilities. All employees played a role in cleaning, organizing and auditing the refinery which created a safer and more efficient workplace. While comprehensive programs like 5S + Safety take several years to fully establish, this program has a new home at an old refinery.

Endnotes

[1] "What Is 5S? - Sort, Set In Order, Shine, Standardize, Sustain". 2015. Matt Wastradowski. (https://www.graphicproducts.com/articles/what-is-5s/)

Bibliography

Grover, John. 2012. "5S Workplaces: When Safety and Lean Meet." (http://ehstoday.com/safety/5s-workplaces-when-safety-and-lean-meet-0).

Kaizen Institute USA. 2017. "What is 5S?" (https://us.kaizen.com/knowledge-center/what-is-5s.html).

Semiklose, Tom. 2014. "Safety: Is It the Sixth 'S' in a 5S System?" (http://www.plantengineering.com/single-article/safety-is-it-the-sixth-s-in-a-5s-system/c620b2a6c91aa3777203dbf859e024f5.html).

United States Environmental Protection Agency. 2006. "Lean and Environmental Training Module 5 - 6S (5S + Safety). (https://www.epa.gov/sites/production/files/2015-06/documents/module_5_6s.pdf).

Risk Communication: A Critical Component in Achieving ISO 45001 Objectives

Geoffrey Peckham
Chair, ANSI Z535 Committee for Safety Signs and Colors
Chair, U.S. TAG to ISO/TC 145 – Graphical Symbols
Chair-elect, ISO/TC 145 – Graphical Symbols
Member, U.S. TAG to ISO/PC 283 – ISO 45001

Founder, Clarion Safety Systems, LLC
Milford, PA

Introduction

Accurately communicating residual risks to workers, subcontractors and guests is a critical component of a management systems approach to improving workplace safety. As a safety professional, if your organization sets a high priority on safety, actively managing the communication of residual risks in your facilities is part of that task. This session will explore how organizations can use recent advances in standards-based approaches to safety signage to more accurately communicate and reduce risk – all for the purpose of creating safer workplaces and public environments.

The global acceptance of using a management systems approach to OH&S is set to change when ISO 45001 is published, in either late 2017 or early 2018. The importance of ISO 45001 to the safety profession cannot be overstated; it is the first global standard for OH&S management systems and it will replace OHSAS 18001. ISO 45001 is one of the three major ISO management systems standards, which include:

- ISO 9001 (quality)
- ISO 14001 (environment)
- ISO 45001 (OH&S)

Like ISO 9001, the expectation is that the marketplace will be the driver for companies to seek ISO 45001 certification. This means that companies and countries will be telling organizations they need to be ISO 45001-certified to be able to do business together. It's anticipated that organizations of all sizes will adopt ISO 45001 and use the latest standards-based best practices to improve workplace safety.

As of the writing of this paper, the final draft of ISO 45001 has not been finished. This is important to note because the exact words the standard will use will be critical to how safety professionals and certification bodies interpret compliance with ISO 45001 management objectives.[1] That said, as a member of the U.S. TAG to the ISO project group writing ISO 45001, ISO/PC 283, I can tell you that the ISO committee's intent is to give organizations a standard that sets the framework organizations of all sizes can use to effectively manage OH&S-related risk.

ISO 45001 and Risk Communication

To understand ISO 45001, you should know that its structure is identical to ISO 9001 and ISO 14001. All three standards have the same ten sections, as follows:

1. Scope
2. Normative references
3. Terms and definitions
4. Context of the organization
5. Leadership
6. Planning
7. Support
8. Operation
9. Performance evaluation
10. Improvement

A very clear case can be made that communication, in one form or another, is included in every section of ISO 45001. The degree to which communication takes place will be part of the compliance requirement auditors will look at to verify that organizations are actively implementing their OH&S management system. Throughout the ISO 45001 standard, statements are made concerning the need to inform, communicate, and increase awareness of hazards and risks. The following are excerpts from the standard's latest draft[2]:

0.2 Aim of an OH&S management system
f) increasing awareness of its OH&S hazards and risks, and associated operational controls, through information, communication and training;

0.3 Success factors
d) processes for communication and consultation;

7.3 Awareness
Persons doing work under the organization's control shall be made aware of:
e) OH&S hazards and risks that are relevant for them.

---

[1] When ISO 45001 is published, it will be available directly from ASSE since ASSE is the Secretary of the ANSI TAG to this ISO committee. If you're not familiar with ISO 45001, now is the time to take advantage of the learning opportunities ASSE will host on the topic.
[2] This list of quotations will be updated when the second DIS is released in spring 2017.

7.4 Information and communication
The organization shall determine the need for internal and external information and communications relevant to the OH&S management system including:
a) on what it will inform about and communicate;
b) when to inform and communicate;
c) who to inform and with whom to communicate:
1) internally among the various levels and functions of the organization;
2) with contractors and visitors to the workplace;
3) with other external or interested parties;
d) how to inform and communicate;....

The organization shall take into account diversity aspects (for example language, culture, literacy, disability), where they exist, when considering its information and communication needs.

8.6 Emergency preparedness and response
d) the communication and provision of relevant information to all workers and at all levels of the organization on their duties and responsibilities;
f) the communication of relevant information to contractors, visitors, emergency response services, government authorities, and, as appropriate, the local community.

A.5.4 Participation and consultation
Means of encouraging cooperation and the participation of workers, and where they exist, workers' representatives, in order to assist them in exercising their functions in the OH&S management system and avoiding barriers, can include:
c) creating awareness of hazards and OH&S risks;

A.7.3 Awareness
To ensure they work or act in safe and healthy conditions, the organization should make workers sufficiently knowledgeable of:
a) actions to take in emergency situations;
b) the consequences of individual actions and behaviour in relation to OH&S risks;
c) the benefits of improved OH&S performance;
d) the need to conform to OH&S policies and established working practices and the potential consequences of departing from requirements of the OH&S management system.
Awareness programmes should be provided for contractors, temporary workers and visitors, etc., according to the OH&S risks to which they are exposed.

Clearly ISO 45001 will make communicating residual risk a key part of an organization's OH&S management system. How to best communicate those risks will be left to each organization to decide. Examples of risk communication measures include worker training, bulletin board notices, company meetings, and/or the posting of permanent or temporary safety signs. What's important to understand is that in the near future, the ISO 45001-compliant organization will have risk communication that is planned and measured by ISO 45001's continuous cycle of "Plan, Do, Check, Act" methodology.

When you look at the use of safety signs in facilities today, it is clear they are not the result of a continually updated, planned, executed, checked and revised-when-necessary process of risk communication. With the advent of ISO 45001, this will change; attention will be given to the best practice standards for safety signage and to implementing processes that include this standards-based visual communication technology into an organization's risk reduction measures.

Part of the task of managing OH&S risk is to alert people to the residual risks that exist in the workplace. According to ISO 45001, the hierarchy of controls will be used by organizations to eliminate or reduce risks to workers. Because not all risks in the workplace can be eliminated, communication of residual risk takes on special importance. The hierarchy of controls gives companies a range of risk-mitigating steps to implement to help reduce the possibility of interaction with potential hazards. The safety profession has embraced the concept of "hierarchy of controls" for many years – it's nothing new. But what is interesting is that many of the safety professionals who use the hierarchy of controls believe the steps described in a typical hierarchy of controls chart are choices to be used in an "either/or" kind of way, not in a layered "both/and" approach. It's the latter approach that defines how the hierarchy of controls is meant to be evaluated and used as a means to control risk.

For example, a machine has moving parts that create an entanglement hazard. The hazard cannot be eliminated so the equipment manufacturer installed a guard on the machine (let's assume the guard is removable to clear jams and maintain the equipment). The manufacturer has also placed a warning label on the guard that says, "Entanglement hazard. Do not operate with guard removed. Lockout/tagout before servicing." The equipment manufacturer has done three things related to the hierarchy of controls to mitigate this risk associated with their machine:

- Installed a guard.
- Warned about the hazard.
- Instructed how to avoid the hazard.

Now let's take this same example and explore it from the perspective of the safety professional looking to mitigate the risk of entanglement in the machine in their workplace. They can do the following:

- Put in place a procedure that inspects equipment on each shift to make sure the machine's guards are in place.
- Put in place a procedure that inspects the machine monthly to make sure its warning labels are intact and legible.
- Implement safety training for all machine operators and maintenance personnel who could possibly interact with this machine's entanglement hazard to instruct them on how to safely clear jams and service the equipment.
- Install a facility safety sign that instructs people on how to safely lockout/tagout the equipment as a means to reinforce the safety training that has taken place. Additional signage could be installed that communicates the need to wear proper PPE when operating or in the vicinity of the machine (e.g. eye protection, hearing protection, wearing or not wearing gloves, no lose clothing/jewelry/long hair, etc.).

As you can see, using the hierarchy of controls does not limit the person whose job it is to mitigate risks to implementing a single step of control. Instead, more often than not, the more effective way to reduce risk is to layer multiple controls. This is not to say every hazard needs every measure. But it does mean steps in the hierarchy of controls should be considered and that more than one can be used at a time.

A Brief History of Safety Sign Standards

I would venture to say that most people in the safety industry are uninformed about recent advances in U.S. and international safety sign standards. Few people understand how these standards are used to design and install systems of safety signs that effectively communicate and mitigate risk. When most safety professionals think of the signage used in their facilities, they see their signs as something that's put up on the wall as a means to satisfy a compliance requirement – not as a true means of risk reduction.

Over the past 100 years, what's considered "state-of-the-art" for safety signs has undergone significant change. In the beginning, in 1914, crude arrow-style signs with simply the word 'DANGER' were used in the workplace to point to hazards. In 1941, the American Standards Institute (which later became ANSI) published the first national standard for safety signs, with sign designs that contained a short text message to identify the hazard. See Figure 1. This standard governed U.S. safety sign design for the next fifty years.

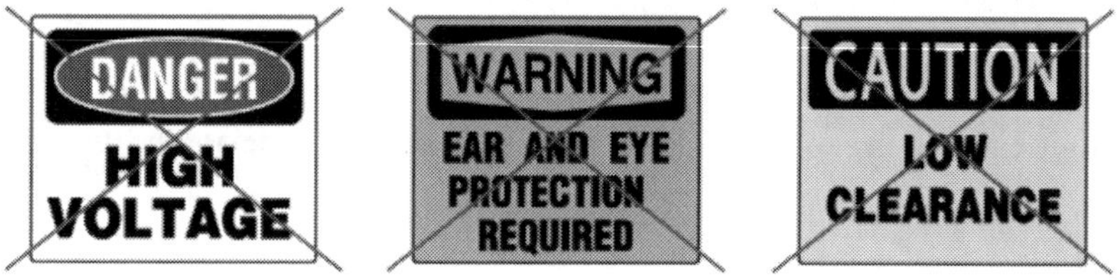

Figure 1: Old-style, 1941-era safety signs

While current knowledge of safety sign design is usually limited to these outdated signs, they only inform people who can read English that a hazard exists. These signs do nothing to inform the viewer about how to avoid the hazard. And without symbols, these signs have no context for the person who cannot read English.

In the 1980s, new safety sign technology started to be developed. Graphical symbols were becoming recognized for their ability to communicate quickly and across language barriers. Risk and hazard analysis methodologies were developed to reduce risk in both public areas and workplaces. Global trade was increasing. And U.S. court precedents set a new legal framework for liability that hinged on the "duty to warn" and "inadequate warnings."

In 1991, the ANSI Z535 standards were approved for publication, ushering in the modern age of U.S. safety sign design. These standards established a three-tier, color-coded system of hazard severity recognition. A new symbol standard used ISO methodologies for comprehension testing and encouraged the use of symbols and pictograms to more effectively communicate safety.

In 2002, the ANSI standards added the option to include symbols and content on the consequence of interaction with the hazard and how to avoid the hazard. The 2007 revision of these standards then realized a decade's worth of international harmonization efforts by incorporating ISO safety sign design principles. Another sign of harmonization occurred in 2006 when NFPA replaced many of its fire and emergency safety symbols with the official ISO symbols having the same meanings.

In sharp contrast to the old-style signage, the latest standards and building codes related to safety signage give safety professionals a whole new level of risk communication technology. See Figure 2. This technology is globally-accepted, graphically-based and more effective at accomplishing the goal of risk communication.

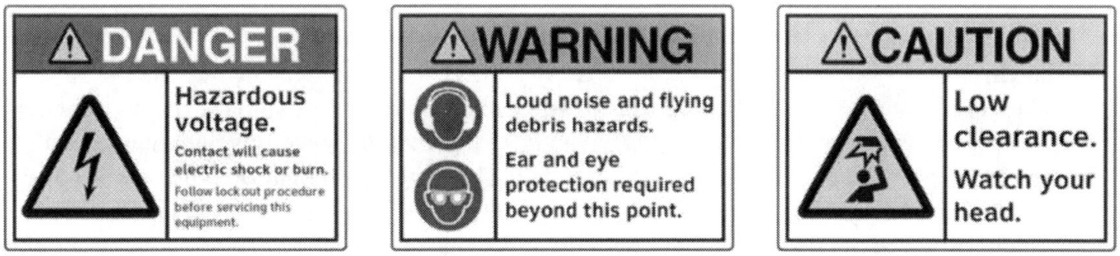

Figure 2: Best practice ANSI Z535/ISO safety signs[3]

Using ANSI and ISO Standards to Create Facility Safety Sign "Systems"

Companies implementing ISO 45001 will benefit from using the latest ANSI and ISO best practice standards to build out their safety sign systems.[4] Well-designed safety sign systems compliant with these standards give safety professionals a means to communicate risk in line with the latest risk assessment methodologies. As part of their foundation, these sign systems use consistent formatting and messaging to give viewers the visual signals, the visual cues, they need to distinguish safety signs from all of the hundreds – if not thousands – of visual messages they see in a typical day. That, in itself, is the first step in successful visual communication: making sure it is noticed and seen.

In the U.S. harmonization of the domestic and international standards in many areas has occurred over the past two decades and it's taken place with the family of ANSI and ISO standards relating to safety signage. Though there are varying tactics taken by different organizations, a harmonized approach can be implemented that incorporates the best ANSI and ISO standards-based design rules. Harmonized systems of signs are built on three key principles:

[3] Illustrations provided courtesy of Clarion Safety Systems, LLC. Best practice sign designs ©Clarion Safety Systems. All rights reserved.

[4] Note that both OSHA and the NFPA have adopted the ANSI Z535 standards into their standards whenever safety signs are needed. In addition, NFPA has adopted the ISO symbols for fire equipment location and evacuation path marking.

1. **Use a common "system" of design elements.** Consistent use of sign design elements will allow signs to be recognized as safety messages. Sign design elements include:
   a. Layout
   b. Color-coding
   c. Indication of risk severity
   d. Graphical symbols
   e. Size of sign and size of text to meet visibility and legibility requirements

The theory behind designing systems of safety signs that use standardized design components is that through the common use of uniform principles for layout, color and content, people will more easily notice, recognize and understand the sign system's safety messages. The overall desire of every organization should be to join the effort to communicate safety in public and private environments using globally-accepted design elements.

2. **Understand and use the five categories of signs to build a system.** The ANSI and ISO standards clearly define "categories" of safety signs, each with their own distinct purpose. When used together, the five categories of signs create a *system* of signage meant to better protect people from harm. Once the safety sign categorization system is understood and implemented, your organization's safety signs will work in combination with one another to convey safety in ways that match up well with the rest of your OH&S management system objectives; the signs will coordinate with the other means you select from the hierarchy of controls to reduce risk. The five categories of signs are:

- Accident prevention signage (that clearly identifies hazards and how to avoid them)
- Fire equipment location signage/safety equipment location signage
- Security and access policy notices
- Egress pathmarking signs and markings, including evacuation escape plan signs
- Hazardous area/substance/chemical identification markings

3. **Use a process of risk assessment.** To a large degree, mitigating risk is everyone's job. It will be very interesting to see how ISO 45001 makes implementing OH&S improvements a daily habit for everyone in a workplace. What distinguishes the safety professional in the task of risk reduction is that they apply a methodology that enables creative solutions to come to the surface and be used over a wide variety of situations. These solutions are informed by past decisions and by the knowledge and experience other people bring to the situation, including the knowledge and experience of the people writing industry standards.

The process of risk assessment enables organizations to identify hazards and opportunities to reduce the likelihood people will interact with those hazards. Because safety sign design, when done well, is a *systematic* process dealing with the same information that goes in and comes out of a risk assessment, the system of signs that uses risk assessment as its foundation will be logical and relatively simple to implement.

Taken together, use of these three principles as the foundation for the design of your organization's *system* of safety signs will result in more risk communication that is in line with ISO 45001 objectives.

Conclusion

ISO 45001 will usher in a new era for the safety profession. It will be an opportunity to see OH&S elevated to a top priority. In my opinion, the success of an organization's OH&S management system will, to a large degree, depend on the willingness of safety professionals to incorporate the latest standards-based best practices into their facilities. When it comes to bringing about awareness of residual hazards in the workplace, using the latest standards-based best practices in signage will be key to effective risk communication. With better risk communication comes the possibility of significantly reducing injuries and deaths in the workplace. And that is our common, shared goal.

Bibliography

American National Standards Institute. *ANSI Z535 standards, parts 1-6* (2011). New York: ANSI.

International Organization for Standardization (ISO). *ISO 3864 standards, parts 1-4* (2011-2016). Geneva: ISO.

______. *ISO 45001 draft standard.* Geneva: ISO.

Session No. 738

Collaborating Between Safety and Risk Management to Improve Safety Program Results

Tom Heebner, MBA, CSP, ARM, CLCS, ABCP
HUB Risk Services
Chicago, IL

Guy Gioino, CSP, ARM, CHMM, ABCP
HUB Risk Services
Summit, NJ

Introduction

It is important for the safety professional to understand the role of the insurance/risk manager within their organization. Although the disciplines overlap they are often operating in silos and don't do a good job working with each other to optimize results and effectiveness. The session will cover the following items.

Understanding Risk Managers Goals and Mindset

- Perspectives of those outside the profession on what a risk manager does and worries about.
- The risk management function, items risk managers get involved in and what they influence
    - Enterprise Risk Management
    - Strategic Risks
    - Financial Risks
    - Operational Risks
    - Hazard Risks

- The comparison of the risk manager and safety roles
    - Risk Manager
        - Risk Avoidance
        - Risk Reduction
        - Risk Transfer
        - Risk Retention
    - Safety Manager
        - Navigation, interpretation, and application of regulations
        - Risk assessment/audit
        - Design of engineering, work practice, and administrative controls
        - Training and communication
- Goals of the risk manager
    - Ideally be integral component of all business operations
    - Minimize premium and insurance costs while maximizing insurance coverage
    - Manage claims and related expenditures
    - Regulatory compliance (SOX, OSHA, DOT, EPA, ACA, EEOC, FMLA, etc.)
    - Favorable Contract Terms (Risk Transfer)
    - Minimize impact of lawsuits
    - Crisis Management

Leveraging Risk Management

- Building the relationship with risk management
    - Get to know risk management staff within your company
    - Share goals, metrics and action plans
    - Get involved in major insurance line renewals (workers compensation, general liability, auto liability, property, etc.)
    - Help coordinate onsite activities required by insurance carriers
    - Show the risk management team that you can be an asset
    - Be open with things that you need help with
- Utilizing risk management
    - Build business case for safety improvements
        - Quantify ROI
        - Impact premium reduction
        - Risk reduction

        - Claims case studies
    - Embedding safety needs in risk management service providers services plans
    - Budget $$$
    - Responding to serious events and complaints
    - Loss analysis and trending
    - Using risk information and risk management expertise for decision making, prioritization and action planning

Understanding Service Provider Capabilities

- Risk Management Service Providers
    - Brokers
    - Carriers
    - Claims Professionals/TPA
    - Auditors
    - Actuaries / Data Analysts
    - Nurse Case Managers
    - Health and Wellness Providers
    - Attorneys
    - Engineers
    - Background Check Providers
    - Online/In person training
    - Security/Crisis Management Experts
- Services Available
    - Inspections
    - Audits
    - Training
    - Security
    - Crisis Management
    - Business Continuity
    - Compliance Guidance
    - Property Engineering
    - Facility Planning
    - Natural Hazard Evaluations
    - Industrial Hygiene
    - Claims Management

- Alternative RTW Placement
- Ergonomics Evaluations
- Loss Analysis and Trending
- Safety Committee Development
- Fleet Safety Management

Data, Information and Metrics

- Insurance Loss Ratios
  - Percentage of the dollar spent
  - Combined Loss Ratios
  - Incurred Losses
  - Expenses
  - Earned Premiums
- Risk Management Information Systems
  - Availability
  - Capabilities
  - Functionality
  - Resources for Safety Professionals
  - Incident Reporting and Tracking
  - Document sharing
  - Recordkeeping
- Total Cost of Risk
  - Premiums
  - Retained Losses
  - Expense and Fees
    - Premiums
    - Collateral
    - Broker
    - Claims admin
    - Losses
    - Internal Risk Management Cost
- Experience Modification Factors
  - Expected Losses
  - Incurred Losses

    - Impact on the Business
    - Long Term Impact of Claims on Costs
    - Using as a Sales Tool
- Catastrophe Modeling
    - Looking at potential events
        - Natural
        - Man-Made
        - Probability
        - Financial Impact
- Exposure Data
    - Used for Risk Assessments on Locations and Ranking Risk Levels
        - Flood
            - Elevation/Slope
            - Flood Zones
            - Flash Flood Potential
        - Hurricane
        - Straight Line Winds
        - Tornado
        - Wildfire
        - Earthquake
        - Landslide
        - Crime
        - Local Fire Service Abilities
        - Maps of Hazards

Summary / Takeaways

- **Build partnerships outside of your safety and facility team**
- **Find ways to help risk management and in turn your safety program will benefit**
- **Understand and utilize the financial metrics of the risk management group in making a case for your investments and action plans**
- **Maximize the resources available to you through risk management**

References

"All About Careers Ltd." *Graduate Jobs, Internships & School Leaver Jobs*, www.AllAboutCareers.com (retrieved March 2, 2017).

"Risk Manager." *Risk Manager - Insurance Glossary,* www.IRMI.com., n.d. (retrieved March 2, 2017).

"A Checklist for Hiring the Perfect Safety Manager." *Hiring the Perfect Safety Manager,* www.IRMI.com, n.d. (retrieved March 2, 2017).

Session No. 739

To See is to Understand--
Exploring a 3-Dimensional Safety Model

Kristina M. Bowden
President
West Coast Resource Services, Inc.
Los Alamitos, California

Introduction

Take a moment to imagine the following; you are a safety manager responsible for multiple worksites along the West Coast of the United States, ranging from Southern California to the Pacific Northwest. These worksites, known as marine terminals, vary in size, ranging from fifty to four hundred acres. The stevedoring industry, at times, has been defined as high-hazard with a heavy man/machine interface. This is a twenty-four-hour operation composed of three working shifts. As the sun breaks the Eastern horizon, today is just like every other day on a marine terminal; you have no idea who's coming to work to move cargo; IWLU membership will fill the jobs. However, who precisely will be determined as that shifts' workforce arrives through the terminal turnstiles. Welcome to the world of stevedoring and marine terminal operations.

This document provides a brief overview of West Coast containerized cargo operations and the unique safety challenges presented with a transient workforce. I will share my prior experience in the role as a West Coast safety manager and the solutions generated by reducing not only the traditional incident frequency rate, but also, the quantity of severe injuries sustained within the workplace. The most meaningful solution developed was a visual three-dimensional model relating two traditionally non-relatable spheres of safety: regulatory requirements and critical safety "to-do's" with the concept of workplace culture. The design is meaningful as all individuals within the work environment can visualize the components required of an effective safety management system. To see is to understand, and with understanding comes forward momentum.

This presentation will examine the fundamental building blocks of the model, the structure they support, and the resulting culture. Exploration of the model, in its current state, can be applied to most work environments. I will share a few particular "wins" impacting both the company's resulting culture as well as the industry.

The ILWU

The International Longshore and Warehouse Union (ILWU) is a labor union which primarily represents dock workers on the west coast of the United States and in Hawaii, Alaska, and British Columbia, Canada.[1] The core of the union, historically, has been the Longshore Division, which established the Union through its victory in the 1934 West Coast maritime strike. The Longshore Division is made up of approximately 30 locals, divided among longshore workers, clerks, and foremen. The larger locals are in Los Angeles/Long Beach, San Francisco/Oakland, Seattle, Tacoma, and Portland.[2] More than 14,200 ILWU workers are employed at West Coast ports.[3]

These men and women, known as Longshoremen, are dispatched daily to marine terminals across the West Coast to move cargo. Longshore workers primarily handle the loading and unloading of cargo and containers, but also, control vessel lines, and maintain stevedoring gear and equipment. Cargo on the West Coast is a combination of both containerization, bulk, and breakbulk. In addition to the longshoremen, another set of workers within the ILWU, known as Clerks, process cargo information for delivery and shipment. The Walking Bosses, also known as Foremen, are direct supervisory representatives in charge of loading and unloading operations and report directly to the stevedoring management.

Work on the waterfront, both loading and unloading of ships and barges in marine terminals, has historically been performed by a workforce employed on a daily basis.[4] Longshoremen are hired for single work shifts and may be asked to return the following shift to complete work. This daily work assignment can be described as transient in nature as the work at one particular marine terminal lasts for a short time. On a daily basis, employers order a specific number of longshoremen, clerks, and foremen to fill jobs and perform all the activities related to cargo handling. Be aware that not every position is reopened daily as there are specific steady jobs, but the majority of the longshore work is completed shift by shift.

The waterfront and everything within it are significant. Vessels are massive in size; container handling equipment is immense; container terminals are hundreds of acres in size. In Southern California alone, ILWU membership numbers are just over nine thousand. Those men and women are dispatched to work in the ports of Los Angeles and Long Beach at over a dozen employers. In 2016, the Port of Los Angeles handled more than 8.8 million units of cargo, setting an all-time record for all ports in the Western Hemisphere.[5] The magnitude of the stevedoring and maritime environment is impressive. However, with this greatness comes unique safety challenges. The skill sets and understanding of risk among these workers are not consistent and

[1] "International Longshore and Warehouse Union." *Wikipedia*. Wikimedia Foundation, 22. Feb. 2017. Web. 28 Feb. 2017 (https://en.wikipedia.org/wiki/International_Longshore_and_Warehouse_Union)
[2] *How the Union Works*. Atlanta: Textile Workers of America, CIO, n.d. Web. 28 Feb. 2017. (https://www.ilwu.org/about/how-our-union-works/. Web.)
[3] "The ILWU Workforce." *PMA Pacific Maritime Association*. N.p.,n.d. Web. 28 Feb. 2017. (http://www.pmanet.org/the-ilwu-workforce)
[4] "Pacific Maritime Association – PMA Annual Report 2015." Pacific Maritime Association – PMA Annual Report 2015. n.d. Web. 28 Feb. 2017. (http://pma.uberflip.com/t/44932-pacific-maritime-association)
[5]Port of Los Angeles,. "Port of Los Angeles: America's Port®." Port of Los Angeles: America's Port®. N.p., n.d. Web. 01 Mar. 2017.

fluctuate with on-the-job experience. The Pacific Maritime Association, a representative of employers on the West Coast, works on behalf of member companies to administer contracts and provide general safety training to the ILWU membership. However, daily management, implementation, and ultimate responsibility for safety fall to each of the individual employer companies. Every day, based on bargaining agreements, longshoremen, clerks, and foreman move cargo.

Within a marine terminal, the working relationship between the ILWU and waterfront employers can be contentious. Historically, the union fought hard and sacrificed much to hold the jurisdiction of work on the West Coast. At that time, the longshore industry was notorious for its corrupt hiring practices, brutally long shifts with no overtime pay, speed-ups or dangerously fast-paced work sessions, and callous ship owner disregard for job safety.[6] Cargo was expected to be moved fast, described as "not letting the hook hang." The ILWU's official slogan, adopted and used from initial organization efforts is "an injury to one is an injury to all."

Within today's work environment, the pressures of high-paced work remain and can be wearying on both union and management. Tensions and strife are often experienced, particularly around contract negotiations. Two master contracts govern the industry: the Pacific Coast Longshore Contract Document and the Pacific Coast Marine Safety Code. These contracts are negotiated at agreed-to intervals, the most recent negotiation occurring in 2014. The Pacific Coast Marine Safety Code is a well-developed, evolved document which has been a part of the industry since 1929. For many years it was considered a voluntary code, until it was included in the November 1946 return-to-work agreement in the Longshore Contract between the Waterfront Employers Association of the Pacific Coast and the ILWU.[7] In additional to Federal and State regulations, it is the primary safety standard, contractually agreed to for the industry. However, as so many safety professionals can attest to, just because there is a set of well-defined regulations or standards, it does not ensure workers will abide by those standards. Moreover, if they do work according to the standards, it is not guaranteed it will be one hundred percent right, one hundred percent of the time.

Therefore, the safety challenge is set: have a well-functioning safety management system aimed at preventing injuries with heavy man/machine interfaces applied to a daily revolving workforce of about fourteen thousand people. Simple, right? Not exactly. Grand resolutions demand nonconventional thought.

Safety and my Experience on the Waterfront

I am a graduate of the California Maritime Academy. In 2005, I started work as a superintendent at a container facility in Long Beach, California. I cut my teeth within the industry at that terminal. New-hire safety training consisted of hanging out with the safety guy, shadowing him at an industry safety meeting, going to the local shoe store for new boots, and having a copy of the Pacific Coast Marine Safety Code tossed your way with the instructions of "read this." After

---

[6] Schwartz, Harvey. *Solidarity Stories: An Oral History of the ILWU.* Seattle: U of Washington, 2009. 10. Print.

[7] *Pacific Coast Marine Safety Code*: 2014 Revision. Place of Publication Not Identified: International Longshoremen's and and Warehousemen's Union, 2014. Print.

working as a superintendent for two years, a safety position was posted for hire within the organization. Having zero formal training in safety, yet having encouragement and support from coworkers, I applied for the job. My colleagues said, "You write excellent injury reports... go for it." Soon after applying for the job, my company issued phone rang. "Congratulations and welcome to the safety team."

It was the most exciting and terrifying day. Excitement because I just signed my offer letter and was given a new title, assistant safety manager. I was terrified because I had no idea how to be a safety manager, nor was I given the "How to Be a Safety Manager" manual. Let me show you a snapshot of what safety looked like for the organization at that time. The program was entirely ISO (International Organization for Standardization) driven. On my new desk lay a large three-ring binder containing all the safety documents for the enterprise. That binder was the safety program. Fresh off the front, I clearly remember never seeing or knowing any of those documents contained within that three-ring binder, the point being that safety people did safety and operational staff moved cargo.

Communication about safety was exclusively related to injury prevention. The goal consisted of making safety our number-one priority, but there was never an explanation given on how to achieve that goal. There was also the safety versus production debate, and the management was terrified regarding the loss of production. I was sent to a general industry OSHA class and returned more confused than when I left. Within the industry, there's a contractually mandated joint accident prevention committee that meets once a month. Both representations from labor and management attend, but after the first few months, I was living in the movie, *Groundhog's Day*. Each sides' position and response to various concerns became predictable. It was a stalemate event I soon dreaded attending. The only safety measurement tracked and shared with senior management was the (LTIFR) Lost Time Incident Frequency Rate and if a worker's compensation claim from the previous week had changed status from non-lost time to lost time.

As I was brand new to the safety game, I kept low, worked hard, and tried to make sense of it all. The primary focus and directive from senior management was to prevent injuries and keep costs down. From a geographical position, the safety staff was divided regionally. Two folks in SoCal, two in NorCal, and two in the Pacific Northwest; one of which was the West Coast director. Most of those managers were seasoned in their roles and felt content with the program. The two folks in SoCal were the boss and me, and our responsibilities included breakbulk in Port Hueneme, in San Diego, three container terminals in Los Angeles/Long Beach, an auto dock, and a cement and paper dock. In June 2009, an announcement was made that my organization was merging with a much larger stevedoring outfit. When the merger was completed, that remaining structure would be the nation's largest stevedore and terminal operator.

Our hearts stood still in fear as this merger brought forward uncertainty. The unanswered question remained: what did this mean for the safety department and the current safety program? My company had a defined ISO-based program. In hindsight, the program was basic and a far cry from a management system. It was, however, what we knew and thought best at the time. The other company, too, had a well-defined safety program, deeply entrenched in DuPont belief. The two were complete opposites. The merger identified the vast differences between East and West Coast programs. An internal safety committee was developed to review both existing programs and make recommendations on what to keep and what to scrap. After several months of discussions and meetings, it was decided the DuPont methods would remain, and everything else was to go. Goodbye ISO safety program. Now, let me introduce you to your new safety goal. Zero harm.

The West Coast was angered and lost. Not only had we lost a program which we believed in, but the West Coast safety director was also unexpectedly released from service. Oh, captain, my captain, but our safety captain was gone. An east coast safety manager then became the director of the organization, but the safety restructuring did not stop there. I vividly remember my manager walking into my office, "We need to go see HR." I gulped and felt sick to my stomach. As we sat in HR's office, I was informed, "Your current position has been eliminated." I thought I might throw up right then and there. Sliding a letter across the table, HR then said, "We have an opportunity we would like to offer you." With the restructure, additional safety headcount had been added to the organization. For me, this meant I would now have the responsibility to manage one large container facility and would report directly to the vice president for that line of business.

Accepting the offer before me, my shaking hand signed the dotted line. Turning to my left, I shook hands with the vice president sitting at the table. "Welcome to the team." All of this monumental change created heaps of operational staff emotion, particularly around the safety program. I quickly transitioned into my new role on terminal accepting the title of Operations Manager Health, Safety, and Environment. At the time, the management headcount was around 62 on a 300-acre facility. Business was good, with high weekly volumes. The primary challenge set forth by my new senior management was: reduce injuries and create an environment of zero harm for all who work at the facility. Through the written word, supporting the safety program had always been outlined as part of the front-line superintendent's job responsibility. However, the attitude towards zero harm was anger. It was a drastic change in concept for the front-line staff, one they resented. Superintendents and managers would come into my office frustrated beyond belief and ask, "How can I be responsible for zero harm?" My initial response was, "I do not know, but we will figure this out together."

These superintendents were cast in fighting a battle, one they would surely lose. Additional anger stemmed from zero harm being tied into their annual personal goals. DuPont made the claim that all could be made right in the world through behavior-based safety. Now, I clearly want to state my position on the matter. I believe that in the correct setting, with the right leadership, behavior-based safety can be effective. It should be thought of as one tool contributing to an overall safety system. For this industry, with the transient workforce, behavior-based safety was a poor misstep. The program was currently successful on the East Coast, but the labor structure was significantly different with alternative senior leadership, creating a positive outcome for success. The culture and labor between East and West Coast operating environments should have never been brought under one program.

The tool to be used in this behavior-based, zero-harm endeavor was the walkabout. The concept was simple; a conversation was to be had when a union member was observed performing an unsafe act. The interaction would start with something along the lines of, "thanks for wearing your safety vest" or "I appreciate what you are trying to accomplish," followed by a conversation about the observed unsafe act. Questions were asked in a non-threating way to understand why the person was choosing to work unsafely. The person who did the unsafe act would then identify any issues they were experiencing in the workplace, followed by a commitment of not repeating the dangerous act. Once the walkabout was complete, and before the end of the shift, the manager would then go back to the office and record the walkabout within the system. Each terminal within the organization was given a required weekly walkabout target to meet. The notion of the walkabout was genuine; creating a conversation between management and membership to understand why unsafe choices were made.

For many front-line managers, the role as a superintendent, is an entry-level management position obtained post-college. Walking into the role, most have little to no real-world experience. Those that do come to this industry from others are thrown for a loop as to the actual culture of the waterfront. In the eyes of a superintendent, safety is often considered a chore instead of an opportunity for change. At times, operational communication is challenging enough. During a walkabout, and for a superintendent, anxiety levels are in a constant state of elevation. With the mandate of the walkabout, safety is now despised, as this program has forced superintendents to interact with a confrontational workforce. It was soon discovered that a portion of superintendents was creating false walkabouts or recycling previously entered walkabouts. The power of copy and paste was active. The response to those who honestly attempted the walkabout were often negative. The union was suspicious as to why these safety conversations were happening, and some foreman felt their jurisdiction was being threatened. At times, the walkabout resulted in a small win or feeling of success, but those successes were not happening enough. The result was most superintendents and front-line managers were turned off from the program altogether.

Adding to the confusion, superintendents were not given a direction or an area where to conduct the walkabout. The scope was too broad and unspecified. Superintendents needed guidance, but instead, it was a scattered free-for-all. However, as long as the site met its weekly target, all was well in the eyes of senior management. Lastly, the system which collected the walkabout information was a total and complete black hole. Once the information was put in, data could not be taken out. As data could not be extracted, there was no way to analyze or share any of these findings in a meaningful way. The safety struggle was real. There's a silver lining in most situations. Mine was my boss. He was energetic, motivated, and eager to do the right thing. The DuPont method hung its hat on the walkabout method, but I was living the madness and was not a believer. From its inception, the concept of zero harm for our workplace was nothing more than a fairytale. It is a beautiful story, but Prince Charming is never going to ride into town on his horse named zero harm.

Knowing alternative concepts, ideas, and programs that exist, I searched to find the right mixture for our workplace. I took a deep dive into other safety theories and philosophies, I read Geller and Manuele, I joined the American Society of Safety Engineers, and I searched for solutions within similar high-hazard industries with transient workforces. Unfortunately, the Longshore and Stevedoring industry are the only industries built as such. In my continued state of safety discovery, I gave the entire program a silent middle finger. Looking back, it is not DuPont's method I found upsetting, but rather, the poor decisions made by senior management to introduce a program that should have never been accepted in this workplace.

Within several months, I abandoned DuPont's principles and created our own. After much thought, I drafted a plan and called a meeting with the Vice President and General Manager. I said, "we are going to do things our way," and the revisions began. Replacement started with clearly-defined expectations for those who chose to work at the facility. I wanted to abandon walkabouts altogether, but we could not because they were corporate mandated weekly targets. Instead, based on injury history and predesignated categories of work, the staff audited basic safety compliance and days away without injury. Knowing what I know now, I would have chosen to express a different metric other than days away, but the positive part was that the site was talking. The program continued to thrive when a commitment was made to labor in the form of an all hands on terminal safety meeting. The first several sessions were raw, honest, and emotional because the labor had a lot to say. We stood in front of them as if we were on the receiving end of a firing squad. We held to our commitment of quarterly meetings, and the tone evolved. Action item lists were created to address concerns between meetings. The site would work to resolve open issues and outcomes, and the topics were shared during each session. The issues that could not be addressed were discussed and explained. The site senior leadership committed to attending these meetings, and the known contention between management and membership decreased drastically. In a five-year period, the LTIFR rate dropped from over 4 to 1.89, the lowest in the company's history. I am fully aware the LTIFR is a lagging indicator, but at the time, that was how success was measured.

After much hard work by all parties involved, the concept of safety started to evolve. The culture of the terminal progressed because of genuine commitments to safety and each other. A moment I will always remember occurred early one morning at check. I was at the equipment corral saying morning hellos when I saw something that made my world stand still because, at that moment, I knew we were doing something right. A senior-ranked UTR longshore operator was having a conversation with a casual UTR longshore operator about her lack of proper footwear. I overheard the senior UTR operator explain to the casual operator that safety at this terminal is real. You follow the rules, and here we wear our safety shoes. At that moment, the glorious safety gates of heaven opened. Angels sang a safety song and unicorns danced in my heart. The significance of this time was the self-monitoring between union members. A small victory had just been experienced. However, as the program progressed, anxiety resided within me as I knew much work still needed to be done. PPE compliance was low-hanging fruit. I needed to develop a plan to pick the fruit at the top of the tree.

However, for continued sustainment of the current culture, this could not just be any plan. This had to be a plan that superintendents who were tasked with doing the work could believe in. They needed to understand their exact role in the program and how they could accomplish it day in and day out. In all reality, the plan was as much for them as for me. As we were making great strides with our program and the result in culture, I needed a system of organization to keep on top of everything: training, communication, messaging, regulatory review, compliance, auditing, the dreaded walkabouts, and incident investigation. Then there was this word of "culture" appearing in safety articles and books describing its need to be excellent.

I needed to sort and organize by priority everything that is a safety program. Most importantly, this plan required to be readily understood by all who were accountable to it. I am grateful for my time as a superintendent because it kept me grounded in understanding their daily capacities. The question had to be asked. Was the organization supporting the superintendents for safety success? Were correct tools for the desired outcome provided? What I needed but lacked was a visual image; one in which all the moving pieces of the program could be visually seen, communicated, and understood. I needed to integrate two traditionally non-relatable spheres of safety; regulatory requirements and critical safety "to-do's" with the concept of safety culture. Little that I know; inspiration will be found in the most unlikely of places.

Safety in the City of Lights

"A walk about Paris will provide lessons in history, beauty and in the point of life."
– Thomas Jefferson

In March of 2013, I had the great pleasure of traveling to Paris, France with my mom. It was our first visit to the City of Lights. Our itinerary consisted of the must-sees: the Eiffel Tower, the Louvre, a dinner cruise on the Seine, and; of course, Notre Dame. The weather was fresh and crisp. Enchanting city streets beckoned walking, for it is a city best enjoyed on foot. Our hotel was comfortably seated on the right bank; minutes from so much of the city's great offerings. When traveling in a new city, I am a big believer in tours. Yes, they can be called touristy, but I always develop a better understanding or new perspective of the destination. The tour guides at Discovery Walks are fabulous. Alec, a guide for several of our tours and a local Parisian, was a phenomenal guide. We booked several tours with Discovery Walks, including a walking tour of Notre Dame.

Here's a quick history lesson. Notre Dame de Paris, meaning Our Lady of Paris, also known as Notre Dame Cathedral, or simply Notre Dame, is a medieval Catholic cathedral located on one of two remaining natural islands on the Seine.[8] Built in 1163, it's located in the center of Paris, and it is widely considered to be one of the finest examples of French Gothic architecture. Most pictures of Notre Dame depict the western façade, showing the two towers. On this tour, our guide, Alec, walked the entire grounds of Notre Dame, describing how the architecture creates the magic surrounding the cathedral. We observed the spire and the east side of the cathedral. Notre Dame was among the first buildings in the world to use flying buttresses.[9] A buttress is an architectural structure built to support or reinforce a wall. The cathedral was not originally designed to include the flying buttresses, but they were added as a necessity to support the walls of the cathedral. Notre Dame is a sanctuary of masterful architecture and a sacred destination.

Forty-five minutes later, we thanked Alec for the grand tour and parted ways with au revoir. Our tour ended where we started, at the West end of the cathedral, in front of Notre Dame's two distinct towers. As we had already visited the exterior of the cathedral, the interior walk was a must. We crossed the threshold of the cathedral and entered a sacred space. The weight of the air was heavy, and respect was demanded for this place of distinct spirituality. I was humbled to be in a domicile of enormous history and master craftsmanship; to put it plainly, you knew you were somewhere special. The Western rose window at Notre Dame is 10 meters in diameter and exceptionally beautiful. The central theme of the West rose is human life and depicts the laborers of the months. While absorbing the greatness of Notre Dame, a thought entered my mind, hitting me like a ton of bricks. I internally cried, "Eureka!" and sighed in relief. On the other side of the world, in a cathedral built eight centuries prior, I solved my safety dilemma. This idea, with such clarity, filled my mind and it has not left since.

That evening, back at the hotel, I sketched out my eureka moment, a small drawing that, in layman's terms, made sense of it all, meaning everything that is safety. Notre Dame, as it stood,

---

[8] "Notre Dame De Paris." *Wikipedia*. Wikimedia Foundation, 19 Feb. 2017. Web. 02 Mar. 2017.
[9] "Notre Dame De Paris." *Wikipedia*. Wikimedia Foundation, 19 Feb. 2017. Web. 02 Mar. 2017.

provided the visual inspiration I needed to structure my thoughts. The flying buttresses provided a visible support structure to the feelings contained and fostered within the cathedral. My required and necessary components of safety no longer floated randomly in space. They now created my structure and were the foundational building blocks critical to safety success.

Strategically placed, the foundational safety building blocks serve the same purpose for me as the flying buttresses serve Our Lady of Paris. They provide a visible support. The void space beneath the visible support is the resulting culture experienced in that environment. For Notre Dame, it is a holy place of sanctity. For me, it is the site's resulting safety culture. Small unicorns once again danced in my heart.

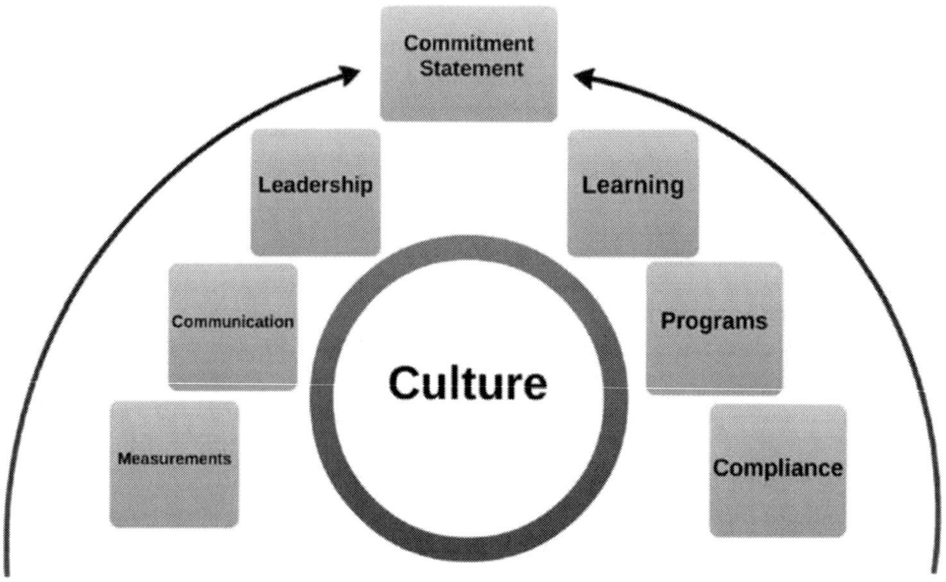

Exhibit 1. Image of three-dimensional safety model.

A 3-Dimensional Safety Model

This is my eureka moment, a three-dimensional safety image. It is referred to as a three-dimensional model because when used in the workplace, it is a three-dimensional environment. For this text, it is represented as a two-dimensional image.

The model overlaid on top of a workplace picture is representative of the same powerful feelings inspired by Notre Dame. One can view the external support functions, and in this case, the fundamental building blocks are shaped like an arch. The components are visible and well defined. The concept is that when a person enters the workplace, or walks into Notre Dame, a distinct feeling arises. On a marine terminal, the culture is known by observing those working in it. Notre Dame was a place of holy sanctity, and visitors acted accordingly. The culture within a container terminal or work environment can be understood by observing those performing the

Exhibit. 2. Image of three-dimensional model in a workplace.

work and the behaviors exhibited. In exploring the model, each building block is strategically placed. Blocks represent a significant portion or component of a safety management system. This model correlates critical program functions with the worksite's culture, creating one visible image. When viewed, the model shows the worksite's safety status, and, if necessary, the work to be completed for the advancement of the system.

The building blocks are constructed based on my experience and needs of a safety management system within a marine terminal. Other safety professionals, when viewing the model, might feel rearrangement of the blocks necessary to meet the demands or needs of their workplace. I have also recognized that between components, one idea might seem appropriate or fit well into multiple blocks. The concept with the blocks is to categorize necessary elements of a safety management system so all who see can understand what must be accomplished in the name of safety and how what is done or not done directly affects the culture. Within safety, at times, redundancy can be positive, and the same holds true for this model. If an idea, topic or program needs repeating or morphing between blocks, the concept and strength behind the model remain.

The two functional blocks at the base of the model are regulations and measurements. Regulations include all protective workplace safety and health standards for a working environment. For a marine terminal located in the state of California, those regulations would be both federal and state OSHA regulations and well as the Pacific Coast Marine Safety Code.

Safety activities and their results must be measured. Safety performance management programs consist of determining what to measure, identifying data collection methods, and collecting the data.[10] Measurements within this building block include both leading and lagging indicators.

Moving up one level within the model identifies programs and communication. The building blocks of programs are where plans of preparation are acted upon, and where the rubber meets the road for worksite staff. The contents of the programs' building block can be vast. In the lower level of the model, the regulations' block identifies minimum necessary standards for a workplace. The programs' block spells out how those standards are acted upon within the workplace. The identification and assignment of what needs to be done, and to whom, is executed within the said program. In addition to the meeting of standards, programs are developed to industry needs or through the identification of hazards. Within a marine terminal, examples of programs would be:

- Auditing, observations and risk assessments
- Incident investigations
- Committees, meeting commitment and frequencies
- Reasonable suspicion assessment™ program
- Hazard communication program
- Fall prevention program
- Lockout/Tagout program

(Note: The above list is not all-inclusive, and was provided merely as an example.)

Communication is what is given to receive understanding. The communication building block identifies who delivers what message, how it is delivered, and the frequency of delivery. As communication is a two-way process, this block also defines how communication is received within the organization.

Moving to the third level of the model, learning and leadership blocks are identified. Learning is the acquisition of knowledge or skills through experience, study, or being taught.[11] Training is a significant component of the learning building block. Most regulatory requirements and programs identify a training need to develop competencies and demonstrate effectiveness. As technology now surrounds us, training may be completed through traditional class sessions or eLearning scenarios.

Senior managerial leadership sets the tone and direction for the organization's safety beliefs as well as providing resources for accomplishing the said program. However, the act of leadership can be demonstrated by any individual within the organization. Based on experience, senior leaders of the organization have the ultimate power to make or break a management system. Although these senior ranking officials within the organization usually understand business, they do not understand the complexities of safety. This model can be used as a tool to demonstrate those complexities and interworkings. The safety professional also provides

---

[10] Janicak, Christopher A. *Safety Metrics: Tools and Techniques for Measuring Safety Performance.* Lanham, MD: Bernan, 2015. Print.

[11] "Google." *Google.* N.p., n.d. Web. 03 Mar. 2017.

leadership within the organization. The safety professional often serves as a pivot piece, guiding and educating senior leadership in the ways of safety, as well as lower-level management and front-line workers. The safety professional is a resource to all.

At the top of the arch is the organization's commitment statement. This top building block could also be the company's' mission, vision, and value statements or their core working beliefs. It is important that these written words be clear and easily understood by all, as this block represents the keystone of the arch. In architectural terms, a keystone is a central stone at the summit of an arch, locking the whole together.[12] It also can be thought of as the central principle or part of a policy or system which all else depends.[13] This keystone is critical, as it holds the blocks of the structure together; without it, the blocks would collapse on themselves and the system would fail. The ILWU's slogan is, "an injury to one is an injury to all." The union has lived by those words for many years, residing in the memberships' belief system. The power and effectiveness of those nine words are in the simplicity in which they are written. When development or revision is needed within an organization's commitment statement, look to organizations or groups experiencing long-term success. They believe in the work, and the words merely capture their beliefs.

Residing under this visual arch is the organization's resulting culture. Culture is defined as the values, practices, and beliefs shared by the members of the group.[14] As a young safety person, I contemplated my organization's culture and racked my brain with how to improve it. Looking back to my days of site safety management, I recognized when I convinced senior leadership to move away from the exclusive use of behavior-based safety and focus on building strong fundamental concepts; the culture demonstrated we had made a right pivot. In a relatively short time, word had spread throughout the waterfront that at our terminal, safety was real. It was not a gimmick or a quick fix. It was a genuine commitment to the idea that whoever worked at that facility for the day morphed into our way of operations. It was never about "Follow the rules or you are fired." It was about moments of coaching which came with the explanation "I understand that other facilities do not operate in this fashion, but we do here." Those who chose not to work safe eventually migrated to other facilities. Those who continue to work at that facility expressed their gratitude in leadership's commitment to providing a safe work environment.

Conclusion

This paper was written for several reasons. The first serves as an introduction to the stevedoring industry and the unique safety challenges presented with a transient workforce. The second and primary reason is to share the three-dimensional safety model. The concept of the model was explained, detailing the reasoning of the arched blocks and the ideas within each of them. Safety, at times, can be a complicated endeavor. The model can serve as a visual guide regardless of what stage of safety you or your organization are currently in. Contents of the blocks can remain simple, providing a general high-level reference. Alternatively, the arch can be filled with

[12] "Google." Google. N.p., n.d. Web. 03.2007.
[13] "Google." Google. N.p., n.d. Web. 03.2007.
[14] Reh, F. John. "7 Ideas to Help Drive Change by Leveraging a Company's Culture." *The Balance*. N.p., n.d. Web. 03. 2017.

detailed interworkings, metric results, and documents. The standard of detail remains with the users.

After working as a safety person for almost ten years, I believe strongly in the power of human connection. Throughout the years, some of my most meaningful and enlightening moments came through simple conversation. In all of that talking and listening, I have learned and experienced that people want to be heard. They want to know that what they are saying matters. Human connection gives much meaning to life. If we, as safety people, can help facilitate that connection, we are fulfilling part of a greater purpose. My hope is this model can be a part of that.

Session No. 740

A Data-Driven Approach to Identifying Physical Fatigue

Lora Cavuoto, Ph.D., AEP
University at Buffalo
Buffalo, NY

Fadel Megahed, Ph.D.
Miami University
Oxford, OH

Introduction

Ergonomics practice involves the identification, and control, of risk factors resulting from physically demanding work, including repetitive motions, forceful exertions, and awkward postures. However, most traditional tools available for identifying these risk factors are limited to characterizing the mean or peak postures, forces, or repetitions, and do not account for interactions between these factors. In addition, the tools often do not account for the development of fatigue. Research has shown that the stressors associated with physically demanding work can result in physical fatigue, which can subsequently lead to decreased productivity, changes in work performance, and increased injury risk (Kumar 2001, Côté et al. 2002, Selen, Beek, and van Dieën 2007).

Reduction of injury risk is dependent on identifying fatigue so that interventions can be put in place. Since the traditional ergonomics assessment tools fail to capture risk factor interactions or fatigue, alternative methods are needed. In a recent survey of manufacturing workers, it was identified that workplace fatigue is typically accompanied by changes in work postures and pace (Lu et al. Under Review). Therefore, methods that capture any changes in worker behavior may provide reliable indicators for monitoring. How to capture this information and what should be done with the data, must first be considered. Wearable sensors provide a means for data collection of any necessary work parameter. Data can be collected continuously from a number of workers for relatively low cost across a range of conditions. However, to date there is a lack of consensus on how the collected data can be integrated for monitoring and individualized risk detection. For effective technological approaches to fatigue measurements, it is essential that the system can predict fatigue (before there is a detrimental productivity and safety impact), measure and monitor fatigue in the operational environment, and intervene when deficits are identified. In addition, considerations of individualized baseline conditions are necessary, but often ignored in a population-based approach to safety. This presentation will include an introduction to the development and examination of the use of sensors for data-driven occupational exposure assessments in work environments. In addition, it will include a description of predictive models based on analytics of the wearable sensor data.

Information on Available Wearable Sensors

The first step to data-driven detection and prediction is ensuring high quality data collection for input to any model. Questions to answer in this case include: Which combination of sensors will be most sensitive and reliable? What is the optimal number of sensors to use? Which measures should be collected? Where should the sensors be placed on the body for accurate assessment? Answering each of these questions involves starting with a detailed description and understanding of the work to be monitored. This can be captured by a combination of in-person and video observation, employee survey, and evaluation of safety and performance records.

The number of available wearable sensors has increased over the past few years and there is a wide selection of commercially available devices that potentially have the desired features for workplace assessment. Table 1 below includes a listing of some available wearable sensors that have been included in field-based measurement studies or have received recent media attention. Characteristics extracted about each device include a description of specifications, available outcomes, and wear locations. While many of the systems incorporate a similar set of sensors, the algorithms for how this information is integrated and the results that are obtainable vary. Overall, a main limitation of most systems is the lack of control and flexibility in the data extraction and analysis. In addition, due to their development for athletics implementation, most commercially available sensors rely on measurement on the chest, and while torso movement may be indicative of overall movement and of trunk posture for low back injury development, it is important to consider whether variability in the task being monitored will be observable at the torso. For many ergonomics and safety applications, the following locations for sensing may be of interest: wrist (for hand movement and upper extremity variability); chest (posture monitoring and task identification); hip (whole body dynamic monitoring, e.g., walking); and ankle (walking pattern variability).

Table 1. Characteristics of commercially available wearable sensors including fatigue-relevant measures

Sensor System	Website for Information	Sensor Wear Location	Measures Captured	Available Outcomes from Software
Polar Heart Rate Monitor	www.polarusa.com	Chest strap and wrist watch	Heart rate; R-R interval on certain models	Heart rate; Heart rate variability; Data visualization
Shimmer3 IMU	www.shimmersensing.com	Any location	Acceleration and gyroscopic data; Galvanic skin response (GSR); Electrocardiography (ECG) for heart rate; Electromyography for muscle activity	Multi-sensor integration; Data visualization; Event annotation

Table 1 (cont'd).

Sensor System	Website for Information	Sensor Wear Location	Measures Captured	Available Outcomes from Software
Catapult	www.catapultsports.com	Upper back	Acceleration and gyroscopic data; GPS	Position, speed, and distance-based metrics; "Athlete Analytics"; Activity profiles; Data visualization; Team dashboard
BioHarness	www.zephyranywhere.com	Chest strap	Heart rate; Breathing rate; Aceeleration data	Heart rate; Heart rate variability; Heart rate recovery; Activity intensity; Trunk posture; Data visualization
Hexoskin	www.hexoskin.com	Smart shirt	Heart rate; ECG; Breathing Rate; Breathing volume; Acceleration data	Heart rate; Heart rate variability; Heart rate recovery; Minute ventilation; Activity intensity; Steps; Cadence; Sleep positions; Data visualization
PUSH Strength	www.trainwithpush.com/	Forearm, just below the elbow	Acceleration and gyroscopic data	Velocity, force and power of strength exercise
WHOOP	www.whoop.com	Wrist	Heart rate; Heart rate variability; Ambient temperature; Acceleration data; Sleep	Team dashboard; Data visualization; Recovery; Strain
Equivital	www.equivital.co.uk	Chest vest	Heart rate; ECG; Breathing rate; Skin temperature; Acceleration and gyroscopic data; GPS; Galvanic skin response; SpO_2 and PPG	Heart rate variability; Body position, motion, and fall detection; Data visualization

Additional Practical Considerations for Analysis and Modeling

Once an appropriate sensor is selected, several considerations remain for data collection and analysis. In terms of data collection, conditions should be fully documented and characteristics about the monitored individuals recorded. Another consideration is the appropriate storage protocols in terms of database structure and privacy protections. In many cases the information collected contains health information about the worker and should be treated as confidential. After data collection, data preprocessing includes: 1) data cleaning, where missing/erroneous data must be detected, synchronized for timing, and sampled to an appropriate frequency; 2) filtering to remove noise caused by outside sources, including surrounding machinery; 3) calculation of any derived metrics, such as heart rate variability indicators from the measured electrocardiogram inter-beat intervals; 4) application of dimension reduction techniques to reduce the size of the data without losing significant information; and 5) feature extraction of potential predictors from the multiple sensors incorporated into each wearable system. This final component is dependent on the question of interest for the analysis, e.g., whether you are interested in the peak trunk flexion of a worker or whether you want to evaluate the average number of steps they take in a day. Then the appropriate data analysis approach should be selected based on the available data and the question of interest. It is recommended that both feature extraction and model selection begin with a pilot test and visualization of the data to gain an understanding of what the sensor data will or will not be able to show for the desired task. While wearable sensor systems provide an opportunity to gather a large amount of data, successful implementation depends on an appropriate sensor to capture the needed data and a well-planned approach for both systematic collection and analysis.

Acknowledgements

This research was made possible by a grant from the American Society of Safety Engineers Foundation entitled “Advancing Safety Surveillance using Individualized Sensor Technology”.

Bibliography

Côté, Julie, Pierre Mathieu, Mindy Levin, and Anatol Feldman. 2002. "Movement reorganization to compensate for fatigue during sawing." *Experimental Brain Research* 146 (3):394-398. doi: 10.1007/s00221-002-1186-6.

Kumar, Shrawan. 2001. "Theories of musculoskeletal injury causation." *Ergonomics* 44 (1):17-47. doi: 10.1080/00140130120716.

Selen, L., P. Beek, and J. van Dieën. 2007. "Fatigue-induced changes of impedance and performance in target tracking." *Experimental Brain Research* 181 (1):99-108. doi: 10.1007/s00221-007-0909-0.

Session No. 741

UCOR: Creating a Strong Safety Culture

Libby Gilley, STS
UCOR
Oak Ridge, Tennessee

Introduction

Maintaining a safety-trained and qualified workforce is essential to success in today's business environment. Creating and sustaining a strong safety culture is the foundation for worker safety, investment worthiness, and company profitability. This paper discusses ways in which URS | CH2M Oak Ridge LLC (UCOR), the U.S. Department of Energy's (DOE) lead cleanup contractor in Oak Ridge, Tennessee, overcame challenges in a high-hazard work environment to meet and exceed client expectations while accomplishing a historical feat—the first ever complete demolition of a gaseous diffusion complex. UCOR's safety culture encompasses a highly engaged workforce that is committed to:

- The implementation of the Core Functions and Guiding Principles of a well-established Integrated Safety Management System (ISMS)
- Continuous improvements associated with Voluntary Protection Program (VPP) tenets
- A solid safety culture built on mutual respect among the client, contractor, employees and the community

UCOR's workforce was awarded DOE's VPP Star status, a designation given to only the safest organizations in the DOE complex. The company has an impressive safety record of more than 8.5 million hours extending over the past three years without a lost workday away case. UCOR received the 2016 Medgate Environmental Health and Safety (EHS) Impact Award and has been recognized as a Tennessee Healthier Work Place.

At the heart of UCOR's safety culture are three focus areas:

- Leadership
- Worker/employee engagement
- Organizational learning

Background

UCOR combines the talents of two global program management companies–AECOM and CH2M–and small business partner Restoration Services, Inc. (RSI). UCOR manages the cleanup of the 2,200-acre East Tennessee Technology Park (ETTP) for its DOE client and partner in Oak Ridge.

The site was contaminated with radioactive, hazardous and industrial wastes generated by more than 40 years of national defense and energy missions.

The ETTP, once called the Oak Ridge Gaseous Diffusion Plant, was built as part of the secret Manhattan Project in the 1940s to enrich uranium for the atomic bombs that would end World War II. The site later produced enriched uranium for commercial and defense purposes. Operations ceased in 1985, and the site was permanently shut down in 1987. DOE then began cleanup operations, which include demolition of many of the site's buildings.

This massive, multi-decade cleanup mission is executed through close partnerships among the DOE, UCOR, community stakeholders and the represented workforce. Cleanup is paving the way to achieve DOE's goals of creating a safer environment and making land available for future economic development. These partnerships work to advance environmental cleanup, enable new and enduring missions, and support the shared vision to reindustrialize the ETTP, turning it into a vibrant private sector industrial park and the nation's 409th national park.

UCOR's safety culture, reinforced by strong partnerships, has been integral to the project's progress and success.

The results have been impressive and, in some ways, unprecedented. In August 2016, the partnership completed DOE's Vision 2016 for ETTP, achieving the first-ever successful demolition of a former uranium enrichment complex's processing buildings. This historic achievement, which was safely accomplished well ahead of schedule and under budget, was made possible by UCOR's committed workforce, which keeps safety at the forefront.

In the past three years, the severity of occupational injuries and illnesses at UCOR was reduced. The company's Day Away, Restricted, Transferred (DART) injury rate was reduced by over two-thirds, steadily declining from 0.72 to 0.20.

At the same time, the frequency of all recordable injuries declined dramatically, as UCOR's total recordable case (TRC) rate dropped from 0.96 to 0.34.

Leadership

When it comes to safety, UCOR strives for the following leadership attributes:

- Demonstrate leading-edge safety
- Make risk-informed, conservative decisions
- Management engagement and time in the field
- Recruit, select, retain, and develop personnel dedicated to safe operations

- Foster open communication and an environment free from retribution
- Provide clear expectations and accountability

UCOR partners with the three unions that represent the bargaining employees at ETTP at both the local and national levels. These unions are the Atomic Trades & Labor Council (ATLC), the Knoxville Building and Construction Trades Council (KBCTC) and the United Steel Workers (USW).

Seven craft safety advocates who represent workers from each of the bargaining units serve to enhance employee involvement, improve lines of communication and strengthen employee commitment to safety. These experienced individuals play a key role in identifying and resolving safety and health concerns while representing fellow workers in decisions affecting safety and health. As with all employees, these advocates/representatives are empowered and have authority to pause or stop work to address concerns with any level of supervision or management.

Locally, UCOR's President and Project Manager, along with senior leadership, promote and attend a monthly Safety Advocate and Stewards meeting where all the unions work together to resolve safety issues. In addition, during regular meetings between the presidents of UCOR, ATLC, KBCTC and USW, union leaders provide insight into upcoming work, budget and challenges and are encouraged to work jointly to improve processes, create efficiencies and resolve challenges.

On the national level, UCOR's President and Project Manager meets quarterly with senior leadership from the National Building Trades Union, AFL-CIO, and the Metal Trades Department, AFL-CIO, to provide similar updates. These partnerships have enabled UCOR and the unions to develop effective safety programs with management and union leadership commitment and encouragement.

In 2017, UCOR President and Project Manager, Ken Rueter, was elected to the Center to Protect Workers' Rights (CPWR) Board of Directors for the Center for Construction Research and Training. This is the non-profit safety and health research and training arm of the AFL-CIO Building-Construction Trades Department (BCTD) and is the largest safety and health operation in the labor movement. CPWR oversees medical surveillance programs for former workers at DOE nuclear weapons sites and provides claims assistance under the Energy Employees Occupational Illness Compensation Program Act for employees with illnesses resulting from work at these sites. In addition, there is the strong collaboration of CPWR, DOE and DOE contractors in the development of safety culture training for both front line leaders and workers across the DOE complex.

As the single board position representing the cleanup industry, opportunities to educate about DOE safety culture and learn about other best practices are created. For example, in November 2016, UCOR safety professionals reviewed CPWR's newly developed safety culture climate assessment tool (S-CAT) and provided feedback to local unions, DOE and CPWR with findings that indicate its alignment with DOE ISMS guide. UCOR facilitated a presentation of the (S-CAT) to the DOE Safety Culture Improvement Panel (SCIP) where it was well received with dialog centering around opportunities for DOE and CPWR to further network and share tools and resources. CPWR, DOE and DOE contractors collaborate in the development of safety culture training for both front line leaders and workers across the DOE complex.

As a result of the trust that has developed between the partners during this contract and the collaborative safety initiatives that ensued, the safety performance record of UCOR has improved dramatically. These relationships are based in UCOR's Zero Injury philosophy, which promotes the concept that even in high-hazard work environments, injuries are avoidable with proper work planning and execution.

Establishing and clearly communicating the company vision, expectations and goals through every level of the workforce are essential to success in every type of industry. At UCOR, each individual is encouraged and expected to be a safety leader. All employees are responsible for their own safety and for that of their coworkers.

The UCOR management team vigorously takes actions to create a strong safety culture by funding, developing, and implementing ongoing, cost-effective processes and programs for projects. These strategies are firmly anchored in a written corporate charter that sets the expectations for accomplishing work safely while fostering a safety conscious work environment (SCWE) where employees are not just encouraged but expected to have a questioning attitude. The corporate ISMS Policy, Program Description, the SCWE Policy statement and the UCOR Safety Culture Sustainment Plan further delineate those expectations.

UCOR's approach for safety culture sustainment was conceived in the context of a key DOE document - DOE G 450.4-1C, *Integrated Safety Management System Guide*, Attachment 10 Safety Focus Areas and Associated Attributes. This is a foundational tool for UCOR's culture because it provides a view of desired behaviors for each individual and team.

Communication within large organizations that operate multiple work locations with a diverse workforce can be challenging. While face-to-face interaction is preferred, it is not always feasible. As a result, UCOR communicates safety and health information through a tool known as the "media blitz." This initiative provides clear, consistent safety messages, updates, and information through a variety of avenues and ensures that all personnel have access to accurate and current safety information.

Acknowledging that individuals absorb information differently, safety messages are communicated verbally, in written form, and visually. Verbal messaging is performed by supervisors at Plan-of-the-Day meetings, walk-downs, and pre- and post-job briefs, and staff meetings. Written communication includes company-wide emails and texts through the Wide Area Rapid Notification (WARN) system and announcements posted on the UCOR homepage. Other written communications include Tool Box Tuesdays, Safety Flashes, monthly newsletters, and on-line availability of written reports and lessons-learned. Visual communication includes monthly Safety Focus presentations, videos, information monitors, bulletin boards, posters, and flyers. Every meeting begins with a safety share that raises worker awareness of the importance of safety in operations.

UCOR implements a Safety Culture Sustainment Plan that uses the three DOE safety focus areas—leadership, employee/worker engagement and organizational learning—as its architecture. The plan is considered a living document that not only drives change but also reflects changes within the organization as work scope evolves. The plan matrix presents activities taken and those planned to improve and sustain UCOR's safety culture and SCWE. The matrix is reviewed and revised as needed by the UCOR Safety Culture Monitoring Panel (SCMP) and senior management for continuous improvement.

At UCOR, all senior managers are expected to complete the DOE TLP-200, *Safety Culture for DOE & DOE Contractor Senior Leaders* course. More than 20 Level I and Level II managers have completed the coursework. This coursework provides the opportunity to synchronize the safety culture thinking and behaviors of UCOR and DOE senior leaders.

UCOR management provides a positive example and sets expectations for safety performance as evidenced through actions, communications, field visits and interfaces with employees. *Safety First - Every Task, Every Activity, Every Time is* the mantra of the UCOR senior management team communicated throughout the organization. UCOR's Monthly Senior Management Safety Checklist makes it easy for managers to visit work areas in that it provides a time commitment, meaningful reason, and pre-determined locations with project contacts for ease of access. New focus areas provided each month sharpen safety observation skills while the rotation between operations imparts knowledge and understanding of the diversity of scopes of work, project personnel, and facilities.

The tool also provides routine opportunities for workers to engage with senior managers in their own environment, affording them the ability to showcase their work, accomplishments, and challenges. Workers freely engage, provide feedback and raise issues during these interactions. As a learning organization, the observations provide a fresh and timely perspective on conditions and behaviors across the project. Communication of issues ensures that appropriate personnel are engaged to facilitate issue closure in a timely and satisfactory manner. Results of observations are shared throughout the project via the media blitz for awareness and lessons learned. In addition, every checklist includes questions regarding SCWE. Feedback from the manager is provided while in the field and their completed checklists are provided to the ESH&QA manager for trending and further distribution.

Worker/Employee Engagement

The ultimate credit for UCOR's strong safety culture goes to our dedicated workforce that implements innovative ways to incorporate ISMS into its daily activities, both at work and at home. Through their involvement, employees have improved work processes to reduce potential hazards and have assisted in identifying and resolving safety issues and concerns.

These tools help UCOR encourage:

- Personal commitments to everyone's safety
- Teamwork and mutual respect
- Participation in work planning and improvement
- Mindfulness of hazards and controls

The commitment to safety is a personal one. This commitment can be seen through the creation of a Personal Safety Action Plan and a Family Safety Action Plan. Each January, UCOR employees purposefully think through the safety commitments they choose to make for the new year and then capture those commitments in writing. Employees are encouraged to post their plans where they can be reminded and others can view them all year long.

Mission Possible: Zero was the VPP safety continuous improvement initiative for 2016. Activities associated with this employee involvement and recognition program focused on worksite analysis, hazard prevention and control, and safety and health training. The goal of Mission Possible: Zero was to encourage each individual to take ownership of safety and become personally involved in safety and health during daily activities. Participants earned safety wings (lapel pins) by completing a series of quarterly safety activities and were then invited to attend a "Wingfest" celebration at the end of each quarter. Participants who completed activities for all four quarters became eligible to receive a Top Gun Safety Leader sweatshirt. The initiative garnered the positive support of the workforce and an average of more than 730 participants each quarter.

Target: ZERO is a 2017 initiative that is highlighting UCOR's pursuit of safety excellence. It is designed to reinforce the Core Functions and Guiding Principles of ISMS. Individuals complete one criterion from each of the five ISMS Core Functions and one criterion from each of the three provided ISMS Guiding Principles for a total of eight criteria during each quarter. After completion of the eight criteria at the end of each quarter, participants attend a lunch where they are recognized for participation. Each quarter begins with a blank slate where all employees (regardless of previous participation) can start to work on completing the current quarter criteria. Those who participate for four consecutive quarters will receive special recognition at the end of the year.

In accordance with UCOR's SCWE Policy, employees are encouraged and expected to express safety concerns and provide feedback. Avenues for feedback include:

- Supervisory/management chain
- Employee Concerns Program (both UCOR and DOE)
- UCOR Ethics Helpline
- Differing Professional Opinion Process (both UCOR and DOE)
- DOE's Office of Inspector General

In a healthy safety culture, people feel free to question when something does not appear or feel right, or if a process seems like it could be improved. In 2012, an event caused the company to recognize that if a questioning attitude had prevailed, the event clearly would have been avoided. In response, UCOR implemented the Questioning Attitude Recognition Program to recognize those who exhibit a questioning attitude and to reinforce that way of thinking and positive behavior in others.

All personnel, including UCOR employees and subcontractors, are eligible to be nominated. Anyone can make a nomination and employees are encouraged to discuss good examples of a questioning attitude during Plan-of-the-Day meetings and staff meetings or with their supervisor or functional manager. Examples can be brought to any Environment, Safety and Health (ES&H) staff member, craft safety advocate, or member of management. Nominations can also be made online through the UCOR intranet. A selection committee meets monthly to review the nominations and select monthly winners and quarterly "best-of-the-best" examples. Monthly winners receive a $50 gift card and quarterly winners get $100. No one is a loser though – all individuals nominated receive a token of appreciation and recognition by their respective teams.

Since its inception, UCOR has recognized more than 690 individuals for having a questioning attitude. UCOR realizes the importance that supervisors play in this aspect of the culture; more than 274 supervisors have also been recognized for encouraging their staff for exhibiting a questioning attitude. More than 272 examples have been collected for sharing.

UCOR established seven Local Safety Improvement Teams (LSITs) to provide another opportunity for direct hire, staff augmentation, and subcontractor personnel to be actively involved in safety. These LSITs cover virtually every project and office location. LSIT members are selected from various work groups and disciplines within a project. The teams are maturing through the efforts of natural leaders and a high degree of management support.

For the second year, management has set aside funding for each of the LSITs to use at their discretion for promotional items and safety recognition events. These cross-functional teams promote employee involvement to create a safety culture that will minimize at-risk behaviors and correct at-risk conditions using the safety observation process. Teams review and analyze safety observation data, provide feedback to their coworkers, identify safety-related concerns, and work with management to recommend improvement actions. LSITs were responsible for facilitating the closure of more than 1,000 safety-related issues/concerns across UCOR worksites and administrative areas. In addition to active team members, each LSIT has a management champion and an elected chairperson. Memberships generally rotate on a staggered basis annually.

Safety Observation Cards are another employee engagement tool used to capture and share observations of both safe and at-risk behaviors and conditions. There are two versions of the cards – one for the field and one for the office environment. Individuals who submit cards can provide their contact information or may submit cards anonymously. Cards are collected at multiple easy-access locations across the site or may be submitted to any supervisor, ES&H professional, member of management, or LSIT member. Results from the observations are reviewed for trends and shared via the media blitz.

New employee orientation includes a one-hour interactive classroom session on UCOR's commitment to safety providing all new hires a clear sense of the company's unique safety culture and the many ways they can become engaged on "day one" of their employment. This training helps new employees quickly adapt to the UCOR safety culture and ensures that, prior to their deployment, employees know who to contact regarding safety questions or concerns. In 2016, UCOR provided New Employee Orientation Safety Culture Training to more than 390 new hires.

UCOR has an active employee-driven Wellness Committee that focuses on education and awareness. Some of the creative activities offered include:

- Biggest Loser contest
- Walk Across Tennessee and Walk Across the Reservation contests
- Wellness Rooms
- Wellness Lunch & Learn sessions
- Wellness Instant Messaging and company-wide educational communications

Organizational Learning

UCOR promotes individual and organizational performance improvement by learning from our own experiences and the experiences of others. These tools help establish:

- Credibility, trust and a confidence in reporting errors and problems
- Effective resolution of reported problems
- Performance monitoring through multiple means
- Use of operating experience and lessons learned
- The use of a questioning attitude

Every 18-24 months, independent safety culture assessments identify strengths and weaknesses of the UCOR safety culture, help focus future safety culture program activities, and track changes from previous assessments. In accordance with DOE and the Energy Facility Contractor Group guidance, UCOR uses an independent expert (Oak Ridge Associated Universities) to perform these assessments. The assessments provide an independent and consistently strong means to gain feedback from all personnel in the organization. Each assessment includes (1) a written employee survey (administered on-line or hardcopy), (2) a series of senior management interviews and (3) a series of focus groups with key demographics including workers, front-line supervisors, mid-level managers, and new employees. The SCMP is then tasked with developing resulting improvement initiatives. A comprehensive set of improvement actions are developed under the sponsorship of the SCMP, which monitors progress in completing actions and the effect on the safety culture.

The UCOR SCMP has become the heartbeat of the safety culture. It is tasked with positively influencing the culture by continually evaluating the overall health of the safety culture and making recommendations concerning improvement initiatives. The panel reviews and communicates information in the context of the safety culture focus areas, the Core Functions and Guiding Principles of ISMS, and the tenets of VPP. The panel includes a cross-sectional and diverse mix of employees, subcontractors, and staff augmentation personnel. Representatives from the bargaining units are active panel members.

Within the broader range of ISMS monthly performance objectives, measures and commitments, UCOR uses a suite of metrics to monitor and sustain its safety culture. The measures, re-evaluated annually, include the following indicators:

- Management field presence
- Employee concerns and differing professional opinions
- Senior management attendance at the President's Accident Prevention Council meetings
- Timeliness of emergency response drills and exercises
- Number of safety observation
- Attendance at monthly safety focus meetings
- Utilization of external operating experiences/lessons learned

Monthly Safety Focus meetings provide current information to all UCOR and subcontractor personnel. The meetings include a review of all injuries, including first aid cases, an overview of company safety performance, and discussion of current safety topics. Tool Box Tuesdays are another mechanism to integrate ISMS into the project weekly.

The Board of Certified Safety Professionals has recognized UCOR for its Safety Trained Supervisors (STS) certification program. The STS certification, an internationally recognized accreditation, provides a means for employers to verify safety and health knowledge of first-line supervisors and managers. The program requires applicants to meet minimum education and experience requirements and demonstrate knowledge of basic safety and health standards and practices.

More than 65 percent of all STSs in the State of Tennessee are sponsored by UCOR. To support this strong safety culture tool, UCOR provides an on-site, 8-hour examination preparatory classes for individuals seeking the STS-Construction certification and pay for all registration, examination, and annual renewal fees. UCOR has sponsored the attainment of more than 240 UCOR Safety STS-C certifications since 2012. Approximately 70 percent of the company's line managers hold STS certifications. Certification recipients are recognized at a lunch and receive a watch bearing the STS and UCOR logos.

The UCOR Human Performance Improvement Implementation Plan provides a method for evaluating processes and procedures to identify and eliminate latent organizational weaknesses and improve effectiveness of existing UCOR programs with the goal of reducing accidents/injuries and improving work control and conduct of operations. A total of 66 individuals have completed the HPI 32-hour Practitioner Course. Practitioners are strategically placed across the project and actively engaged both in the field and in functional support groups.

To maximize positive human performance and support this community of practice, an HPI webpage is available to all UCOR personnel. Practitioner working group meetings and project-level meetings take place regularly where practitioners can continue to grow in their expertise and hone their skills. In addition, practitioners participate in review of processes after an unwanted event to advocate the effective use of HPI tools.

UCOR provides a 2.5-hour HPI Overview and Tools class to all project personnel. This training helps employees identify error precursors and the tools that can be applied to help prevent errors. Since its inception, more than 715 members of the workforce have attended. This class is scheduled to be delivered to 600 additional personnel in 2017; UCOR's goal is to provide this training to the entire workforce.

Looking Forward

Continuous improvement in the area of performing work safely can only be achieved through the willingness of each individual to maintain situational awareness, express a questioning attitude, and execute work in accordance with the functions and principles of ISMS. Our current improvement initiatives were developed with direct input from our workforce provided through the safety culture assessments and the work of our SCMP. We're expanding safety culture training to our first line supervisors, focusing on coaching and mentoring and streamlining our safety programs and communications. These tools and strategies also support our preparations for a formal re-verification of our ISMS, which is scheduled to take place in September 2017. As

DOE and UCOR advance toward final cleanup of ETTP, this robust safety culture will remain the foundation for cleanup success.

Session No. 743

Prevention through Planning: Work at Height

Paul S. Adams, Ph.D., PE, CSP, CPE
3M Company
St. Paul, MN

Jackson Dalton, M.S., CSP

Introduction

Workers falling from elevation is a major concern for safety professionals. Falls to lower level (648) accounted for 13.4 % of workplace fatalities in 2015, and were the second leading cause of workplace deaths (BLS 2016). In a 2010 Water Research Foundation study (Adams and Grieser 16), participating water utilities identified falls from elevation as one of four hazards most likely to cause severe injury or death in water plants. This study developed decision models for preventing falls and other high hazard incidents through facility design that are broadly applicable to General Industry. Some of that research is incorporated into this paper.

A common question facing facility designers is, "How should workers be elevated to perform various tasks?" That is, "Should we install fixed platforms, use boom-lifts or scissor lifts, or simply rely on ladders." Experienced safety professionals know that how this question gets answered will affect both personnel safety and operating costs for the life of the facility.

This paper will examine the problem of falls from elevation in industrial settings and how they can be prevented through planning. After presenting an overview of why workers fall, a set of strategies for preventing fatal falls will be presented. This will be followed by a discussion of the variables that need to be considered when selecting the means for elevating workers and factors to consider in this decision-making. Finally, an overview of facility features for fall protection and fall arrest will be identified to aid facility and process planners.

Why Workers Fall from Elevation

The authors contend that fall arrest systems should be considered secondary protection, and preventing the fall from ever occurring should be the primary objective in fall prevention planning. To meet that objective, it is important to understand why or when workers fall from elevation. Falls from elevation in occupational settings occur when one or more of the following four situations occur:

1. Loss of balance near an unprotected edge
2. Lack of awareness of an opening or unprotected edge
3. Task performance places a worker in a fall permitting situation
4. Mechanical failure of a supporting structure.

Loss of balance near an unprotected edge

Edges of roofs, holes, platforms, and other standing surfaces without guardrails or similar barriers are considered to be unprotected edges. Loss of balance can occur as the result of trips and stumbles, slips, external forces pressing against the body, certain health conditions, or simple lapses in coordination. All ambulatory humans have experienced a momentary loss of balance during their lifetime, and standing or walking near an unprotected edge when this occurs presents a risk of falling off the edge.

Fall Prevention Rule #1: **Anyone could lose their balance at any time.**

Lack of awareness of an opening or unprotected edge

When workers are unaware of an unprotected edge, they can inadvertently step over that edge and fall. One of two situations create the conditions for this type of fall. The edge and opening may be sufficiently hidden that the worker simply does not see the hazard. Examples include skylight panels and floor openings in darkened rooms. Second, a worker may be attending to work activities or other stimuli and momentarily lose situational awareness. A common example is someone stepping backward into an unprotected opening. It is unreasonable to expect humans to maintain awareness of unprotected edges and openings at all times.

Fall Prevention Rule #2: Humans will not be situationally aware of their surroundings at all times.

Task performance places a worker in a fall-permitting situation

Fall-permitting situations are present when a fall from elevation is a realistic possibility if a work activity does not go as planned or intended. Oftentimes, these situations arise due to a mismatch between worker capabilities and the task requirements; i.e., poor ergonomics. They also occur when something goes wrong while elevated, such as an injury that predisposes a person to a fall. Examples include:

- Forceful exertions or sudden changes in forces on the body resulting in instability
- Long reaches that cause the center of mass to extend beyond the point of stability
- Prolonged standing in a constrained posture or location, resulting in fatigue and the need to shift postures to reduce discomfort
- Constrained foot positions that do not accommodate movement to maintain balance
- Ice or other contaminants that compromise footing
- Unstable movements to enhance lines of sight or visual acuity
- Any of the above that encourage a worker to leave a stable walking/standing surface and move onto an inappropriate surface such as a guardrail, or to assume an unstable posture, in an attempt to perform a task.

- Injury or the threat of an injury associated with the task that causes a sudden, unstable movement in response; e.g., an electrician falling off a ladder after an arc flash incident.

Fall Prevention Rule #3: Work activities will not always go as planned; avoid fall-permitting situations that can be problematic when work does not go as planned.

Mechanical failure of a supporting structure

Falls occur when surfaces fail to support the worker. This is an obvious cause of falls, and can occur for a number of reasons. Common sources of this class of causes include poor maintenance, inadequate structural design, and a worker's reliance on a surface that was not intended to provide support or protection against a fall. In short, these falls occur when the supporting surface fails or when the worker misjudges the supportive capacity of a surface or brace. Therefore, a sound strategy is to design and maintain structures in the work environment that are capable of supporting workers, or take measures to prevent/discourage them from being used that purpose.

Fall Prevention Rule #4: If a surface or object looks like it will support a worker, it may eventually be used to support a worker.

Planning to Prevent Fatal Falls

Just as ANSI Z10 directs readers to use a Hierarchy of Controls as a guide for systematically considering safety controls, the authors propose six strategies that should be systematically considered when planning work at height.

1. To the extent feasible perform the work at ground level
2. Minimize the time spent working at elevation
3. Use ergonomic principles to design the task to be as easy and stable as possible
4. Select an appropriate means for elevating workers based on the task requirements and other considerations
5. Provide protection to prevent falls that might otherwise occur
6. Provide fall arrest systems for those tasks with residual fall risk.

Note that these strategies are not exclusive and it may be best to implement multiple or even all six strategies for a given task.

Strategy #1 – Bring the work down, rather than raising workers up.

Eliminating the need to perform the task while elevated eliminates the hazard of a fall from elevation, at least for the portion of the work performed at ground level and away from an unprotected edge. Working on the ground has the added benefit of being more efficient in most situations. Common examples of this strategy include:

- Building subassemblies on the ground
- Installing meters and other displays so they can be read from the ground
- Route piping systems so that pumps, valves, and filters can be serviced from the ground
- Provide remote level sensing on tanks
- Enable the use of cameras, drones, or even smart phones on long selfie sticks for inspection

When it is not feasible to bring work to ground level, it may be feasible to group tasks at an elevation where a fixed platform provides a stable and safe walking/working surface.

Strategy #2 – Minimize the time required at elevation

In addition to minimizing the potential fall exposure time (Strategy #1), there is a sound ergonomic reason for minimizing the time spent working at elevation. It is postulated that work at elevation involves greater energy expenditure than if that work is performed on the ground. It has been the author's observation and experience that elevating the worker reduces their efficiency and increases their fatigue rate compared to the same task performed on the ground. To the extent feasible, it is recommended that designers incorporate quick release fasteners, minimize the use of bolts with fine threads, and add self-centering features for parts or assemblies that must be installed from any surface other than a fixed platform.

Strategy #3 – Make elevated work easy

Ergonomics is even more important to consider when working at elevation than it is for ground level tasks. As the perceived level of stability decreases, muscle recruitment to maintain stability automatically increases, resulting in reduced capacity for performing useful work. For example, if positioning a replacement valve requires 50% of the worker's strength while on the ground, that same task performed from a scissor lift might require 70% or more of the worker's capacity. The result is a much higher risk of a muscle strain or a dropped object. The more unstable the base, the more pronounced this effect is. This phenomenon underlies much of the decision-making for selecting means of elevation discussed later.

Some key ergonomic principles to consider when planning elevated tasks include:

- Design for neutral postures and facilitate worker positioning to avoid long reaches, twisting, and other awkward postures.
- Minimize the required force levels of exertions.
- Provide for positive coupling to objects and secure footing.
- Avoid the need for precise or stable movements. For example, to the extent feasible, design objects so that they self-center when moved into position. Provide hand supports or other surfaces for stabilizing the hands and arms when precise movements are required.
- Avoid task elements that involve prolonged holding of an object
- Except for work performed on fixed platforms, assume the feet must remain stationary and cannot be readily repositioned to maintain stability. Forceful exertions should be expected to result in instability /loss of balance and the need for supplemental support.
- Avoid activities that involve sudden movements, or changes in applied force levels and directions. Keep work smooth and predictable.
- To the extent feasible, provide both visual and tactile feedback when securing and releasing parts.

If elevated tasks are easily performed, there is less temptation for workers to use unsafe strategies that could increase the risk of falls. For example, presenting work within easy reach prevents workers from climbing guardrails to reach more comfortably, or using guardrails as braces for forceful exertions. Reducing task demands also reduces fatigue and preserves coordination necessary to maintain balance and stability.

Strategy #4 – Select the best means of elevation for the task

As suggested in Strategies 1 – 3, fixed platforms provide many of the same ergonomic benefits as working from the ground. However, fixed platforms may not be feasible or practical, and it is important to consider alternative means of elevation when designing facilities. Selecting and planning for various means of elevation is addressed in more detail in the subsequent section.

Strategies #5 and #6 – Protect, then arrest

Fall arrest systems should be viewed in much the same context as seatbelts in cars. They are critically important for reducing injuries caused by an accident, but it is always preferable to prevent the accident altogether. Strategy #5 is to prevent the fall that would occur if not for the protection system. Guardrails, covers, and other barrier systems are the primary means of preventing falls for those workers who lose their balance, forget they are working near an edge, or suddenly become unstable due to external forces. Although guardrails must be designed to withstand 200 lbs. of force, they should not be used to enhance leverage for forceful exertions or as the primary means of achieving stability. Guardrail failures can and do occur as a result of improper design, construction, maintenance, or misuse (see Cause #4 above).

Fall arrest systems are the last line of defense against a fatality. That is, except when used for positioning, their primary function is not to prevent a fall, but rather to prevent a hard landing. Sudden, unexpected falls that are arrested by a harness and lanyard still have the potential to result in injuries, and if not followed by prompt rescue, can end in tragedy due to suspension trauma. As will be discussed later, recent improvements in fall arrest systems reduces the risk of suspension trauma. If a task must be performed that relies on a fall arrest system for fatality avoidance, then it is incumbent on the task planner and facility / process designer to facilitate the use of that system through careful positioning of anchorage points and other design features.

Selecting the Means of Elevation

Common means of elevating workers include:

- Fixed platforms
- Temporary stationary platforms or scaffolds
- Vehicle mounted scissor lifts
- Boom lifts
- Forklift mounted platforms
- Manually propelled mobile platforms
- Portable rolling steps
- Portable ladders.

Selecting the appropriate means for elevating workers requires consideration of performance demand, situational factors and management variables. Performance or task demand variables are functions of the type of work being performed, such as task duration, difficulty, and exertion requirements. Situational variables address environmental factors, external hazards, elevation, and accessibility. Management variables include resource issues like time and cost, as well as risk acceptance and the task frequency. The frequency that a task is performed may be determined by management through preventive maintenance schedules, or by the process itself.

Although there are numerous variables that may be considered, it is not practical to establish decision guidelines that involve dozens of variables. However, by consolidating the list of potential variables, one can structure a solution that is somewhat manageable. Criteria can be developed to assist safety professionals in auditing existing means of elevating workers. Engineers can also use such criteria to specify appropriate systems for new and remodeled facilities.

The following variables are suggested as a starting point in the development of workable guidelines.

Task Demand Variables

- Duration – How long the worker maintains his stance or position at a single elevation (other than the floor or ground level).
- Frequency – How frequently a worker performs a task at the elevated location.
- Number of Hands – How many hands are required to perform the task; i.e., whether one hand is available to provide the third point of contact on a ladder?
- Difficulty – How difficult a task is based on the level of exertion and stability required to perform the elevated work. Exertion is a biomechanical function of both strength/force and posture. Stability requirements are a function of the accuracy of required movements, the security of the base, and the presence of contaminants.

Equipment/Environmental Variables

- Elevation – The maximum distance a person could fall.
- Work Envelope – The maximum distances that a person would need to reach to perform the tasks while elevated.
- External Hazards – A relative ranking of the external hazards, including the presence of ground or overhead (crane) traffic, the presence of adjacent energy sources (physical, electrical, chemical, thermal, etc.), and the floor condition.
- Access – Ground level access requirements; i.e., access for powered equipment and space for stairs.
- Levels – The number of levels at which work is performed along a vertical line.

In addition to the variables given above, the following should also be considered as applicable, and may affect selection decisions. *(Note: This list is a suggestion and is not exhaustive or complete.)*

- Hazards to people and property below the work, such as from dropped tools, parts and welding sparks.
- Size, weight and configuration of objects being lifted, carried, or positioned.
- Personal protective equipment being worn.
- Presence of oil, ice or other floor contaminants and the potential for compromised footing.
- Available visual field.
- Availability of tie-off points for ladders and personal fall arrest systems.
- Training required for employees using the equipment.
- Exceptional environmental conditions, such as floor hazards, traffic, sudden loud noises, etc.
- Lighting, including the need for providing temporary spot lighting.

- Cost. (Only for comparing acceptable alternatives, and never for justifying an unsafe approach.)
- Ease of rescue.
- Ability to communicate with other personnel.
- Individual factors, such as coordination, mental capabilities, skill and experience, sense of balance, comfort with working at heights, fatigue, medications, etc.

Designing to Accommodate Elevated Work

In the preceding sections, strategies for designing elevated work tasks were discussed. Facility design can have a major effect on how elevated work can be performed, and often constrains how workers can be elevated. It is important for both the safety and design professionals to recognize facility features that facilitate elevated work and those that make it more difficult. In this section, some of the facility characteristics that affect how elevated work can be performed are presented.

Accessibility

Stairs are the preferred way to provide access to fixed platforms, but they require significantly more floor area than fixed ladders. If mobile equipment such as scissor lifts are to be used, open floor space such as an aisleway must be available. This means that pipes, electric lines, and other obstructions cannot block the path to where the elevated work will be done, or prevent the worker from standing directly under or alongside the work. That is, enabling the basket to be positioned close to the work prevents the need for long reaches from a scissor or bucket lift. Even ladders require sufficient floor space to assure proper footing and angles of inclination.

Ground/floor surfaces

The condition of the ground or supporting floor for mobile equipment must be level and stable. If portable ladders are used, it is important for the floor or ground area to also be free of contaminants or weak spots that could cause the ladder to tip or slip. If the elevated work is to be performed outside, the surface should be designed to account for variable weather conditions.

Traffic protection

Traffic patterns need to be considered when locating valves, busways, conveyors, and other equipment overhead that may need to be serviced from mobile equipment. When scissor and boom lifts or portable ladders are used, it is necessary to protect them from contact with other vehicles and protect pedestrians who might walk through the drop zone below. Barricades and warning tape are traditional methods for isolating overhead workers from activities below. Closing off roadways or aisles can restrict normal traffic patterns, and if these passageways are critical to operations or for emergency access, special accommodations may be required. These needs may warrant location of the overhead equipment toward one side of the passageway, or even over a cleared shoulder or pull-over area.

Outage planning

Overhead work on pipelines, busways, conveyors and other equipment often requires outages for serviced processes and plant areas. The duration of these outages can be minimized or reduced through careful planning for accessibility, subassembly design to accommodate easy and fast exchanges of components, and segregation of facility services so that interferences are avoided. Where service interferences cannot be avoided, it may be necessary to install permanent protective barriers such as guarding around busways. In addition, planning for material handling and other service needs and installing load beams or hoist connectors can expedite overhead tasks and further reduce outage durations.

Roof service locations

By strategically locating penthouse entrances, HVAC units, and other rooftop equipment away from the edges of buildings, the need for barrier systems and personal fall arrest equipment might be reduced or eliminated. When feasible, powerlines should be routed away from HVAC and other equipment that must be periodically serviced or inspected.

Roof guardrails and edge protection systems

When rooftop work must be performed near a roof edge, fall protection must be provided. Installing guardrails, tie-off cables, and other features to facilitate worker protection creates a safer work environment and expedites task performance.

Anchor points

Personal fall arrest systems are only effective if engineered anchorage points are provided and used. This may require the installation of load beams, tie-off rings, cables, or other supportive structures and devices. To encourage use, anchorage points need to be located so that they are easily reached, and preferably designed for lanyard attachment with one hand.

Cranes overhead and below

Cranes pose a variety of elevated work challenges. First, crane maintenance is frequently performed at elevation using personal fall arrest systems. When work such as lighting, sprinkler system, or other ceiling level maintenance is performed, cranes may be operating below the worker and steps must be taken to prevent collisions between mobile lifts and crane booms or bridges. If elevated work is being performed on top of equipment, cranes working overhead can strike the worker with loads or blocks, or snag tie-off lanyards. Facility or crane features should be provided to accommodate tie-off for crane maintenance and also prevent hazardous crane travel.

Electrical, gas, chemical, steam and water lines

Busways and piping systems are typically either buried or built at elevation. In many facilities, maintenance on overhead valves or switchgear involves the use of personal fall arrest systems. In the absence of proper anchorage points, it is tempting for workers to use a choker lanyard around a pipe or busway and rely upon it as an anchor point. (See Fall Prevention Rule #4 above.) This unsafe practice can be avoided by strategic placement of anchor points.

When valves and switchgear are brought down to facilitate ground-level servicing, the pipes and busways can introduce barriers that restrict mobile vehicle movement. Careful layout planning is again the key to accommodating scissor and boom lift access and negating the need for ladders.

Designing for Fall Arrest and Recovery

Elimination of risk through fall prevention is always preferable to fall restraint or fall arrest. For those situations where fall restraint or fall arrest is necessary, suitable anchor points, connectors, and descent and rescue systems need to be specified by a qualified, competent person. When planning facilities and operations that will require personal fall protection systems, it is helpful to know whether a qualified and competent person will be readily available to identify anchorage points when needed. If such a resource will not be available to make judgments in the field, the identification of acceptable anchorage points may need to be more comprehensive in the design/planning phase.

A number of anchorage point designs are available for a variety of surfaces and structures and some are non-penetrating by design. Temporary anchorage points may be used for jobs of short duration and frequency. Permanent engineered anchorage points are ideal for tasks of indefinite or prolonged duration and high or routine frequency.

A variety of anchorage connectors are available including self-retracting lifelines, vertical lifelines, and shock-absorbing lanyards. The selection of an anchorage connector should be determined based on the nature of the task and proximity to other hazards such as dangerous processes or limited fall clearance. The potential for swing falls or pendulum effects should be evaluated.

Each worker's body should be connected to the anchorage point and anchorage connector through the use of a full-body harness. Full-body harnesses are available in individual sizes and can accommodate a working load limit of 110-420 pounds, depending on the model and design. Task specific harnesses are also available for high-voltage work, work with potential for arc-flash or arc-blast, painting, suspension work such as window washing or confined space entry, etc. The task, duration, frequency, and size of worker should be considered when selecting a full-body harness.

If the task performed requires a high degree of horizontal mobility, then the goal of 100% tie-off must be accomplished by system design. Mobility may be provided by horizontal lifelines, vertical lifelines, twin-leg shock absorbing lanyards, and/or a strategic placement of anchorage points along the working area.

Finally, descent and rescue must be considered when planning for tasks that will require a fall restraint or fall arrest system. There are a number of rescue solutions available. Facilities and tasks need to be planned to accommodate an assisted rescue, and this accommodation is necessary even though modern fall arrest systems may be equipped with self-rescue or suspension trauma prevention features. Both assisted rescue and self-rescue require extensive planning, specialized equipment, and hands-on training.

Summary

Workers fall from elevation when one or more of the following situations occur:

- Loss of balance near an unprotected edge
- Lack of awareness of an opening or unprotected edge
- Task performance places a worker in a fall permitting situation
- Mechanical failure of a supporting structure.

Six strategies identified for preventing falls through planning include:

1. Bring the work down, rather than raising workers up.

2. Minimize the time required at elevation
3. Make elevated work easy by applying ergonomic principles
4. Select the best means of elevation for the task
5. Protect workers with guardrails or other barriers that prevent falls
6. Provide for fall arrest, but fall arrest is not fall prevention.

Selecting the appropriate means for elevating workers requires consideration of performance demand, situational and management variables. Performance or task demand variables are functions of the type of work being performed, such as task duration, difficulty, and exertion requirements. Situational variables address environmental factors, external hazards, elevation, and accessibility. Management variables include resource issues like time and cost, as well as risk acceptance and the task frequency.

Facility design can have a major effect on how elevated work can be performed, and often constrains how workers can be elevated. Facility characteristics that affect how elevated work can be performed include:

- Accessibility
- Ground/floor conditions and stability affordance
- Traffic protection and accommodation
- Outage planning and minimization
- Roof service locations
- Guardrails and other edge protection systems
- Anchor point design and location
- Cranes operating above and below workers at elevation
- Placement and characteristics of busways and piping systems.

If elevated work requiring fall arrest equipment is anticipated, planning is necessary to enable this work to be performed safely and efficiently. Among the features affecting the design of facilities and tasks are:

- Availability of qualified/competent persons to evaluate anchorage points and other fall arrest devices
- Task duration and frequency and the choice or temporary or permanent anchorage points
- Potential for swing or pendulum effects following a fall
- Task specific hazards that affect harness specification
- Number and placement of anchorage points, especially if the work involves horizontal or vertical movement
- Accommodations for assisted rescue.

Bibliography

Adams, P. and B. Grieser. *Workforce Health and Safety: Prevention through Design. Water Research Foundation Report #4236.* Denver: Water Research Foundation, 2014.

Beyer, R.; R. McCarthy, and C. Wood. "Ergonomic Analysis of Extension Ladders." *Proceedings of the Human Factors and Ergonomics Society 38th Annual Meeting*. 1994. 816-820.

Bloswick, D. and D. Chaffin. "An Ergonomic Analysis of the Ladder Climbing Activity." *International Journal of Industrial Ergonomics*. 1990. Vol. 6: 17-27.

Bureau of Labor Statistics. (https://www.bls.gov/iif/oshwc/cfoi/cfch0014.pdf).

Fulmer, S.et al. "Ergonomic Exposures in Apple Harvesting: Preliminary Observations." *American Journal of Industrial Medicine Supplement*. 2002. 3-9.

James, M. "High Stakes and Ladders." *The Safety and Health Practitioner*. July 2000: 32-34.

Kinel, L. et al. "A Walkthrough Ergonomic Checklist for Evaluating Slip/Fall Potential in Lead Abatement Tasks." *Proceedings of the IEA 2000/HFES 2000 Congress*. 2000. 659-662.

McGregor, E. "Will My Safety Harness Save My Life?" *Occupational Safety & Health*. Dec. 2003: 30-34.

Rundstrom, I. "A National Campaign for Increasing Safety and Health in the Use of Ladders and Scaffolds." *Proceedings of the IEA 2000/HFES 2000 Congress*. 2000. 477-478.

Young, S. and M. Wogalter. "On Improving Set-Up Angle Accuracy for Extension Ladders." *Proceedings of the IEA 2000/HFES 2000 Congress*. 2000. 111-114.

Session No.744

NFPA 350 Guide for Safe Confined Space Entry and Work

Nancy A. Pearce CIH
Sr. Fire Protection Engineer, NFPA
Quincy, MA

Robert A. Rusczek, CIH, CSP
Certified Safety & Health, LLC
Wilbraham, MA

Introduction

OSHA's *Permit Required Confined Spaces* general industry standard was issued in 1993, and in 2015 the *Confined Spaces in Construction* standard became law. However, despite these regulations and employers' obligation to provide a safe and healthy workplace, on the average 100 employees in the United States a year die in confined spaces, and many more are injured or barely escape. Some of incidents are from a simple lack of compliance with existing standards. But others appear to result from gaps in existing standards and confusion about what is required for confined space entry. NFPA 350, *Guide for Confined Space Entry and Work*, with the first edition published in late 2015, addresses these gaps and provides practical guidance on how to apply requirements in existing standards to improve confined space safety.

NFPA 350 - Guide for Confined Space Entry and Work

Confined Spaces Are Dangerous!

The Bureau of Labor Statistics (BLS) reported a total of 136 workers were killed in incidents associated with confined spaces in 2016. Often times one incident results in multiple fatalities. The BLS reported that during the 1995 to 1999 time period there were 13 confined space incidents that resulted in 24 fatalities. In addition to fatalities, serious disabling injuries and illnesses occur in confined spaces. The majority of the fatalities are due to atmospheric hazards, such as oxygen deficiency. One of the root causes of many of these incidents is lack of knowledge of the applicable regulations and confined space hazards.

Background

Three of the key standards used in confined space entry in the U.S.—OSHA 1910.146, *Permit Required Confined Spaces*; OSHA's relatively recent 1926 Subpart AA *Confined Spaces in Construction*, and ASSE Z117.1, *Safety Requirements for Confined Spaces*—provide *minimum* performance-based requirements. Performance-based standards focus primarily on the desired outcome—in this case, preventing injuries and fatalities in confined spaces. However, performance-based standards sometimes lack sufficient guidance on how to reach the desired outcome. This is where NFPA 350 fills the gap, providing many of the "how-to's" for confined space safety that will help employers comply with one of the performance-based confined space standards. A good analogy is how NFPA 70E, *Standard for Electrical Safety in the Workplace*, provides detailed guidance on *how* to protect employees from electrical hazards, such as arc flash, whereas OSHA Subpart S electrical regulations specify only that electrical safety must be provided. NFPA 350 provides guidance beyond the minimums, fills in some gaps in safety, and simplifies terminology that can create confusion when applying confined space standards.

Designed for Continuous Program Improvement

NFPA 350 is organized so that an employer can utilize the document as a whole, or use particular chapters to bolster an existing confined space entry program. For example, employers who wish to improve their confined space entry program can implement NFPA 350 in stages, perhaps improving their gas monitoring program, then moving toward improving hazard controls or competencies as the program evolves.

Confined Space Terminology

Perhaps one of the most confusing parts of confined space safety is the terminology used by safety professionals and regulators to describe the various classifications for confined spaces. The terms "confined space" and "non-permit confined space" can be used interchangeably to describe a confined space where no hazards are present. However, this same space becomes a "permit-required confined space" the very next day when an employee enters it to weld or paint. The type of space is different from day to day and depends on what is being done inside the space.

Furthermore, there are several different procedures that can be used for entering permit-required confined spaces. "Reclassification procedures" are allowed in spaces where all hazards have been eliminated. "Alternate procedures" are allowed where the only hazard is atmospheric and is controlled by mechanical ventilation. A full permit is needed for entry where other hazards exist. Even confined space professionals have expressed difficulty and confusion over how to best explain the differences in requirements for entry to employees and employers.

To minimize confusion, NFPA 350 uses one term— confined space—throughout the document, instead of renaming a space based on the presence or absence of a hazard. All spaces that meet the OSHA definition of a confined space are referred to simply as "confined spaces" in NFPA 350. The guide also establishes one procedure for all of those spaces, starting with pre-entry evaluation (a pre-entry evaluation form is included in 350). NFPA 350 does not distinguish between a reclassified space and an alternate procedures entry. All spaces are evaluated on their own merit based on hazards that are inherent, introduced, and/or adjacent to the space. Whether a permit is required for entry depends on the hazards or potential hazards identified in the pre-entry evaluation. This approach of using a checklist to evaluate all spaces prior to issuing a permit is already in practice in some industries. The change in terminology does not conflict with OSHA or ASSE regulations or standards. Rather it promotes safety and simplifies the application of requirements: look at the space, determine whether it's a confined space, and if so, perform a hazard assessment and document it as part of the pre-entry evaluation.

OSHA requires a cognitive assessment of a confined space to determine whether hazards exist that would make it a permit-required confined space. If no hazards exist or potentially exist, and/or hazards eliminated before entry, the space can be downgraded to a non-permit confined space, with certification from the employer, and then no further action is required. Currently, the OSHA general industry standard does not require documentation of a hazard assessment, and the evaluation of whether a hazardous atmosphere potentially exists can be performed without monitoring. The OSHA construction standard requires that before an employer begins work at a construction worksite, each employer must ensure that a competent person identifies all confined spaces in which one or more of the employees it directs may work, and identifies each space that is a permit space, through consideration and evaluation of the elements of that space, including testing as necessary. For some worksites, or even projects on the same worksite, it may be confusing which OSHA standard applies.

In NFPA 350, all confined spaces are evaluated using a pre-entry evaluation form, a signed checklist used to identify and document hazards that are inherent, introduced, and/or adjacent to the space. NFPA 350 takes one additional precaution at this juncture and recommends atmospheric monitoring prior to any confined space entry. Those results are also recorded on the pre-entry evaluation form. The default 350 position is to perform atmospheric monitoring prior to entry into any confined space. The availability, reasonable cost, and ease of use of gas monitors facilitates this premise. Because atmospheric hazards continue to be the predominant source of confined space fatalities, the NFPA 350 Technical Committee decided that the few minutes it takes to perform atmospheric monitoring is time well spent to verify that there is no hazardous atmosphere (or if found, prevent entry until corrective actions are taken and appropriate engineering controls instituted. If no physical or atmospheric hazards are found, and the tasks to be performed in the confined space will not introduce new hazards, the pre-evaluation form is signed and no permit is needed. The pre-entry evaluation form can also be used as the written certification required for entry using reclassification ("downgrading") or alternate procedures required by OSHA's general industry standard, and for construction site confined space evaluations for the construction industry. If hazards are identified, then a permit is needed and the procedures required for safe confined space entry apply.

Development of NFPA 350

NFPA is a non-profit organization that develops codes and standards through a consensus process approved by the American National Standards Institute (ANSI). This process brings together volunteers representing varied viewpoints and interests to achieve consensus on fire, and other safety issues. In addition to input from Technical Committee volunteers, the NFPA process encourages interested parties outside the committee to provide input on the document. NFPA 350 and all other NFPA standards and best practices can be viewed online free of charge by going to NFPA's website at www.nfpa.org/[document number]. Individual copies can be purchased in a variety of formats.

The Technical Committee for NFPA 350 includes experts in confined space safety from industry, labor, enforcement, government, and rescue. The preliminary draft of NFPA 350 was posted for public input in September 2013 and was revised based on that input. The Committee developed a second draft that was posted for public comment in September 2014. The final version of NFPA 350 was revised after considering those comments. NFPA 350, Guide for Safe Confined Space Entry and Work, was first published in 2015. NFPA 350 was open for public input again in 2016. In April 2017 the Committee met to review the comments received since NFPA 350 was first published. The Committee is now preparing a draft of 350 revisions, which will be available on the NFPA website for public comment in September 2017. Any new comments received will then be reviewed, with additional revisions of 350 made for the 2018 edition.

Like all NFPA standards, NFPA 350 is an ever-evolving document that will be revised every three to five years, making the information current and relevant and allowing for public review several times during each revision cycle. To be notified about any activity related to NFPA 350, such as where and when meetings will occur or when the document is open for public input and comment, sign up for e-mail alerts at **www.nfpa.org/350**.

350 references hazards that are *inherent, introduced, and/or adjacent* to the confined space. While most confined space entry programs incorporate the evaluation of inherent and introduced hazards, few address adjacent hazards. Examples of adjacent hazards could be chemical vapors being emitted by a nearby process, exhaust from vehicles, and powered industrial trucks operating near the confined space. The incorporation of adjacent hazards is an important safety consideration for all confined space entry since hazards adjacent to the space can significantly affect the hazards in the confined space.

Some of the confined space incidents that have occurred since the implementation of OSHA's Permit Required Con fined Spaces standard include fatalities related to hazards such as hot work in and adjacent to confined spaces, fatalities related to inert atmospheres, and fatalities due to lack of knowledge and understanding for those involved in confined space entry.

There have been several incidents where fatalities were documented or suspected when an employee was exposed to a hazardous atmosphere adjacent to a confined space that caused the employee to fall into a confined space. NFPA 350 recommends signage for inerted tanks to alert employees of the hazard of such atmospheres adjacent to a confined space.

NFPA 350 Provides "How To's"

OSHA confined space standards require hazards to be evaluated and controlled but do not provide much guidance on how to approach the process of hazard evaluation and control. OSHA tells you what to do; NFPA 350 provides you how to do it by providing supporting information on how to identify hazards, perform gas monitoring, control hazards, and ventilate. For example, Chapter 7, Atmospheric Monitoring, explains how to select an appropriate gas monitor, calibration procedures (including zero and bump test), and interpretation of results, including information on interferences and limits of detection. The chapter also includes best practices for gas monitoring such as initial air monitoring before entry and mechanical ventilation, testing to determine the amount of time for a sample to travel through a length of tubing and be detected at the sensor, and establishing safe levels for entry. Chapter 9, Ventilation, explains the limits of natural ventilation in a confined space and explains how to select and configure ventilation equipment in a confined space. Information is provided on ventilation of inert atmospheres and bonding and grounding of flammable or combustible atmospheres. Drawings of typical ventilation configurations are provided in an annex.

Competencies

NFPA 350 recognizes that demonstrated competencies are critical for workers involved in confined space entry and work. In addition to the entrant, attendant, and entry supervisor roles defined by OSHA, NFPA 350 provides recommended competencies for those who perform tasks such as gas monitoring, ventilation, and rescue. This doesn't mean that NFPA 350 recommends additional workers be assigned to the confined space entry. However, an attendant who is also responsible for atmospheric monitoring and ventilating should have demonstrated abilities and knowledge to perform those tasks. For example, the "Gas Tester" as defined in NFPA 350 should be qualified beyond simply knowing how to turn on the gas monitor and record results. This individual should also understand how to calibrate and interpret atmospheric monitoring results. Likewise, an attendant who is also responsible for ventilating should understand the equipment and method required to maintain safe atmospheric conditions in the space.

Confined Space Rescue

A number of NFPA 350 Technical Committee members are rescue experts who recognized gaps in rescue provisions in existing regulations. NFPA 350 now provides the organizational elements of emergency preparedness that are normally in place in a fire department but not necessarily in a facility rescue program, and provides a tiered approach to response. In addition, the rescue chapter includes information on pre-incident planning and evaluation, rescue gear, rescue configurations, and rescue competencies. The document works in conjunction with NFPA 1670, *Standard on Operations and Training for Technical Search and Rescue Incidents,* for technical aspects of confined space rescue.

Management of Change

Since many confined space incidents relate to change, a chapter on management of change (MOC) has been included in NFPA 350. The MOC system identifies and evaluates potential effects of modifications to confined space configurations, equipment, materials, content, and work tasks. A sample MOC form, provided in an annex, serves to document that the effects of change have been considered.

Prevention Through Design

NFPA 350 provides a chapter on prevention through design (PtD) specifically for confined spaces. The PtD concept seeks to initiate a design process to reduce or eliminate inherent risks and hazards associated with the design of facilities, equipment, and products. PtD can minimize retrofitting control costs and the use of labor-intensive administrative hazard control measures. PtD targets two types of interactions: the construction and/or installation of new confined spaces; and the redesign, retrofit, and/or renovation of confined spaces to eliminate, control, or minimize hazards. While the implementation of methods provided throughout NFPA 350 will improve confined space safety, the only way to fully prevent confined space incidents is to eliminate the con- fined space completely through design or redesign.

Conclusions

Fatalities and other incidents that occur during confined space entry and work can all be prevented. The OSHA regulations establish performance criteria that must be met to keep employees and contractors safe, and to be in compliance. NFPA 350 is a best practice guide that provides the needed guidance to meet the regulatory requirements, and most importantly, maintain a high level of safety for all involved with confined space entry and work.

Session No. 745

Safety, Health and Environment (SH&E) Training that Builds Safety Culture

Peter Gregory, MPH, CSP, STS
NY Metro Area Safety Manager
AECOM
Clifton, New Jersey

Camelia Pacurar
Corporate SH&E Advisor, Middle East
AECOM
Abu Dhabi, United Arab Emirates

Marc Wetter, CSP, CHST, STS
EIC CC&M Safety Manager
AECOM
Englewood, Colorado

(In association with AECOM SH&E Innovation & Collaboration Excellence (ICE) Committee.)

Executive Summary

The purpose of this project was for a team of Safety, Health & Environment (SH&E) professionals from across AECOM divisions to conduct research on industry best practices in training, to develop company-wide training matrices and to describe methods of tracking compliance that will optimize SH&E training and employee development across global AECOM.

To benchmark industry best practices for SH&E training, a survey was administered to two separate consortiums of high-performing companies. The survey was developed to understand both the typical levels of effort expended in SH&E training and the best practices in training for high performing organizations. The survey assessed how many SH&E training hours per year an average worker receives (office vs. field); the duration of new hire SH&E orientation; the duration and frequency of driver training; training delivery methods (live vs. computer-based); and the metrics that drive SH&E training practices. Results of this survey are contained below.

The following is a summary of recommendations from the team's research. These recommendations are detailed further below.

- Tier II SH&E Training Matrices were developed through researching industry best practices and identified training courses that would have the greatest impact on SH&E outcomes across global AECOM.
  - Every new AECOM employee, including interns, should receive a general safety orientation training to understand AECOM's Safety for Life program. Employees should also take an in-house safety refresher training every five years, or less for employees who return to AECOM after an absence of six months or greater, to facilitate employee feedback on training effectiveness, the overall SH&E program, and suggestions for improvement.
  - Defensive driving should be a mandatory course for all employees driving on company business, and a strongly encouraged course for all other employees.
  - All executives should undergo supervisor training in responsibility and recognition training and an executive-based course on basic safety and hazard recognition.
  - All project managers should receive formalized, AECOM-specific training on safety supervision and leadership as part of the project manager certification process. To validate the safety supervision and leadership training, all project managers directly overseeing revenue-generating projects with any aspect of work conducted outside of an AECOM office should obtain certification as a safety qualified supervisor.
  - All staff performing work in the field, regardless of frequency, should undergo basic hazard recognition and analysis training prior to performing work in the field. All AECOM staff should be required to choose from Hazardous Waste Operations and Emergency Response (HAZWOPER), mine safety, or field safety, based on the nature of their projects and worksites, as well as more extensive hazard recognition and analysis training for worksites possessing high hazard activities.
- A robust tracking system should be developed for SH&E training across AECOM. AECOM group leadership should develop Tier III SH&E training matrices specific to their workforce, business lines, and geography.
- As part of new employee orientation, every employee should undergo a Training Essentials Assessment process with their direct supervisor/manager and an SH&E representative.
- A companywide survey assessing SH&E training needs should be administered. The survey should be developed to assess organizational structure, determine performance and competency gaps, measure perceptions and beliefs related to safety, and identify trends within the various layers of global AECOM.

As a professional consulting/construction services firm, we need to offer our clients talent and expertise. It is critical that we invest in the development of our professionals to integrate safety in every aspect of our work. The key to a solid return in SH&E training investment is to focus on skills, behaviors, and desired competencies that will give AECOM professionals the ability to perform any and all client work injury free. A heavy emphasis should be placed on nurturing future SH&E leaders within operations and strategically focusing SH&E training on appropriate personnel.

Based upon the team's experience, research, and interviews, AECOM will be able to positively influence the overall safety culture of the company through the recommendations summarized above and contained in the body of the paper below. The desired state of the AECOM safety culture is one where leaders effectively promote their commitment to the well-being of their people. In return, the people put forth maximum efforts to protect and advance the well-being of one another and the organization.

Purpose & Scope

The purpose of this project was for a team of Safety Health & Environment (SH&E) professionals from across AECOM divisions to conduct research on industry best practices in training; develop company-wide training matrices; describe methods of tracking compliance; and optimize SH&E training and development across global AECOM. The scope of the project included Tier II (Global) and Tier III (Group and Geography) levels within AECOM.

Benchmarking SH&E Training

To benchmark industry best practices for SH&E training, identical surveys were administered within two organizations.

The first was a consortium of high-performing companies of all sizes and sectors, seeking to help one another achieve and sustain safety excellence through the sharing of proven knowledge, actionable solutions, and best practices. The consortium provides the platform for member organizations to disseminate SH&E content through partnerships, events, and research.

The second consists of safety executives from the top US construction firms. The organization exists to communicate best practices in construction safety, share lessons learned, and evaluate emerging processes and technologies.

The 10-question, multiple-choice survey was developed to better understand how member organizations within these two groups implement SH&E training across their firms. The goal was to understand both the typical levels of effort expended in SH&E training and the best practices in training for high-performing organizations. The survey assessed how many SH&E training hours per year an average worker receives (office vs. field); the duration of new hire SH&E orientation; the duration and frequency of driver training; training delivery methods (live vs. computer-based); and the metrics that drive SH&E training practices.

Survey Results

Thirty-six surveys were completed between the two organizations. The results are summarized in Tables 1 through 4 below.

Q1. What industry do you work in?						
Respondents represented a wide range of industries, including construction and engineering (18); consulting services (4); technology and services (2); transportation and logistics (2); electricity (1); aerospace (1); chemical and petrochemical (1); government (1); heavy manufacturing (1); institutional safety (1); light manufacturing (1); marine construction/dredging services (1); mining and agriculture (1); conglomerate and diversified (1); and oil and gas (1).						
Q2. For a new employee at your organization, how much time is spent on safety-specific training during the new hire orientation?	**Hours**	**1–2**	**2–4**	**4–8**	**8–16**	**>16**
	Response Rate (No. of Responses)	5.56% (2)	27.78% (10)	24.00% (9)	13.89% (5)	27.78% (10)
Q3. How much classroom or web-based safety-specific training does an average employee in your organization receive each year?	**Hours**	**<4**	**4–8**	**8–24**	**24–40**	**<40**
	Response Rate (No. of Responses)	11.11% (4)	16.67% (6)	44.44% (16)	11.11 % (4)	16.67% (6)
Q4. For an employee who is an office worker, how much safety-specific training is given in a typical year? (Includes classroom, web-based, and informal training, like a lunch and learn).	**Hours**	**<1**	**1–2**	**3–4**	**4–8**	**>8**
	Response Rate (No. of Responses)	8.33% (3)	27.78% (10)	8.33 (3)	36.11% (13)	19.44% (7)
Q5. For an employee who performs work outside of the office setting (i.e., construction site, manufacturing plant, etc.), how much safety-specific training is given in a typical year? (Includes classroom, web-based, and informal training, like a lunch and learn).	**Hours**	**<4**	**4–8**	**8–24**	**24–40**	**>40**
	Response Rate (No. of Responses)	0.00% (0)	11.11% (4)	30.56% (11)	25.00% (9)	33.33% (12)

Table 1. Annual SH&E Training Hours for New Hires, Employees, Office Workers, and Field Workers.

Findings

- All of the respondents indicated some level of SH&E training for new hires.
- Nearly half of the respondents indicated that new hires complete two to eight hours of safety orientation training, with an additional 27 percent (%) of the respondents indicating greater than 16 hours for new hire safety orientation.
- Of the respondents, 44% indicated that their employees receive an overall average of eight to 24 hours of safety training per year.
- Of the respondents, 36% indicated that their office workers receive four to eight hours of annual safety training, with the next highest response rate of 27% receiving one to two hours of annual safety training.

- Annual SH&E training hours for field employees were distributed relatively evenly across eight to 24 hours (30%), 24 to 40 hours (25%), and greater than 40 hours (33%).
- Respondents reported that all field employees receive at least four hours of SH&E training annually.

Q6. Does your organization require driver safety training for personnel who drive on company business?	**Criteria**	**No Response**	***Yes***	***No***			
	Response Rate (No. of Responses)	2.77% (1)	63.89% (23)	33.33% (12)			
Q7. How often is driver safety training required?	**Criteria**	**No Response**	**Once**	**Annually**	**Every 2 Years**	**Every 3 Years**	
	Response Rate (No. of Responses)	30.56% (11)	19.44% (7)	16.67% (6)	19.44% (7)	13.89% (5)	
Q8. How many hours is your organization's standard driver safety training?	**Hours**	**No Response**	**1**	**2**	**3**	**4**	**>4**
	Response Rate (No. of Responses)	30.56% (11)	13.89% (5)	27.78% (10)	5.56% (2)	16.67% (6)	5.56% (2)

Table 2. Driver Training.

Findings

- Of all the respondents, 64% reported that they require employees to complete driver training.
- Six of the 12, despite selecting "No" responses on driver training (Question 6), responded with specific frequencies and/or lengths of driver training in the subsequent Questions 7 and 8.
- No significant trend was established for frequency of driver training, as there were similar percentages of companies for each answer category.
- Driver trainings were reported to be predominantly either two hours (28%) or four hours (17%) in length.

Q9. On average, what percentage of the SH&E training is live, in-person training as opposed to computer-based training?	**Criteria (%)**	**0%**	**25%**	**50%**	**75%**	**100%**
	Response Rate (No. of Responses)	2.78% (1)	30.56% (11)	19.44% (7)	41.67% (15)	5.56% (2)

Table 3. Training Delivery.

Findings

- Of the respondents, 92% indicated that they utilize a blend of live and computer-based SH&E training delivery methods.
- Of the respondents, 66% indicated that they provide 50% or more of SH&E training through a live delivery method.

Q10. What metrics are used within your organization to evaluate the overall effectiveness of SH&E training (select all that apply)?	Criteria	No Response	RI Rate	LTI Rate	WC Claim Cost	EE Turnover	None
	Response Rate (No. of Responses)	5.56% (2)	77.78% (28)	61.11% (22)	50.00% (18)	19.44% (7)	11.11% (4)

KEY: RI (Recordable Incident); LTI (Lost Time Incident); WC (Workers' Compensation); EE (Employee)

Table 4. Metrics that Drive SH&E Training.

Findings

- Of the respondents, 78% indicated that they use one or several lagging indicators (recordable incident rates, lost time incident rates, and workers' compensation costs) to drive their SH&E training program.
- Of the respondents, 20% indicated that they consider SH&E training as a potential factor in employee turnover.

SH&E Training Matrices

The project team set out to identify training courses that would have the greatest impact on SH&E outcomes across global AECOM. Training matrices were developed by the team through researching industry best practices and also in consideration of the training survey results. Courses were chosen to achieve desired competencies and improve the following:[1]

- visible upper management commitment,
- active middle manager involvement,
- focused supervisor performance,
- active participation,
- flexible systems to accommodate site cultures, and
- positive perceptions by the workforce.

Global SH&E Training Matrices (Tier II)

The Tier II SH&E Training Matrices below broadly define employee categories and identify recommended training at each level of employee. Brief summaries of integral training courses identified in the matrices below are provided in a later section of this paper.

[1]Petersen, Dan. "Safety Excellence--Six Critical Components".

Exhibit 1 – Strategic Training for Safety Excellence for All Employees.

Exhibit 2 – Strategic Training for Safety Excellence by Position.

Geography and Group SH&E Matrices (Tier III)
AECOM group and geography leadership should develop a training matrix for workers in their group or geography and implement the requirements of AECOM SH&E procedures for all employees. Geographic and group SH&E matrices should be continually monitored, measured, and evaluated. To facilitate this, a more robust tracking system needs to be developed for SH&E training. The system must be flexible enough to allow for simultaneous usage at the global, divisional, regional, and project/office levels. The database must be user friendly to facilitate tracking courses taken internally, as well as documenting certificates obtained from external vendors. A comprehensive SH&E training database would allow SH&E to better assess safety competencies; provide guidance on what SH&E courses are needed and where; and help improve Tier III matrices across the company. Ideally, the selected database program should be capable of generating Tier III matrices for all levels of the organization where required.

Global SH&E Training Courses

A brief summary of each of the training requirements prescribed in the Tier II Training Matrices is presented below.

All Employees

Safety Orientation
Every new AECOM employee, including interns, will receive a general safety orientation training on the AECOM Safety for Life program, including incident management, behavior-based safety, defensive driving requirements, "Just Drive" distracted driving awareness, stop work authority, short service employee mentoring, accessing SH&E resources/tools, and a clear expectation of zero injuries.

For employees who will work primarily from an AECOM office, the following additional general safety issues should be addressed in this orientation: office emergency plans, reporting unsafe or unhealthful working conditions in Lifeguard, general office safety, and office ergonomics.

In addition to office safety, employees who work at external project sites should also undergo orientation that addresses obtaining and reviewing Health and Safety Plans and Safe Work Plans, including Job Safety Analyses (JSAs); implementing the daily task hazard assessment process; obtaining personal protective clothing and equipment; reviewing field emergency action plans; undergoing a field site orientation; discussing hazards unique to the project site; reviewing task-specific hazards (e.g., falls, confined spaces, excavations, hot work, electrical hazards, working near water); and covering any other project-specific requirements.

The safety orientation training is intended to be a brief overview and is not intended to take the place of task-, program-, or project-specific health and safety training. The safety orientation should be **five to eight hours per employee** and should be a combination of computer-based (or online) modules and live training conducted by a project manager and/or representative from the SH&E department. Managers should participate in this orientation at some level. The computer-based (online) modules should be standardized globally across AECOM and available in various languages.

In addition to initial safety orientation training, all employees should be required to take a general safety in-house refresher training every five years. The refresher training reinforce the management systems described above, to update them on any changes since their last orientation training was conducted, and to allow them to ask questions or provide feedback on how safety management systems could be improved or enhanced.

Defensive Driving

According to the *Global Status Report on Road Safety 2015,* issued by World Health Organization, about 1.25 million people die each year on the world's roads and between 20 and 50 million sustain non-fatal injuries. The report reflects information from 180 countries. In the last three years, 17 countries have aligned at least one of their laws with best practice on seat-belts, drunk driving, speed, motorcycle helmets or child restraints. Taking into consideration the difference in legislation across the globe, AECOM should have consistent best practices that all AECOM employees must follow.

Research obtained from another AECOM team revealed that, of 927 total high-consequence events within AECOM and its legacy companies over the past five years, 196 were vehicle-related (Munuera, 2015). In other words, approximately <u>21% of all high-consequence events</u> were vehicle related. The report defined a high-consequence incident as follows:

- Lost-Time Injuries
- Serious Vehicle Accidents
- Critical Near Misses

Behavior Based Safety Program

A behavior based safety program emphasizes that, in safety, silence is not golden and it is the responsibility of every employee to speak up whenever they see something that's not right. The training should demonstrate how each employee can contribute to workplace safety by learning simple, easy-to-use feedback techniques since effective feedback doesn't usually happen automatically. A behavior based safety program could help employees learn how to give safety-related feedback to coworkers, demonstrate the internal challenges we face when receiving feedback, and present ways to overcome those challenges.

<u>Executives</u>

Industry data from major US companies has revealed leadership support as being the most effective approach to decreasing the number of incidents, injuries, and environmental accidents. In addition, safety is closely tied to other aspects of corporations, including job satisfaction; quality of life; client confidence; financial stability (i.e., direct costs, insurance, and productivity); and overall execution of projects. These case studies reveal that exceptional organizations all have safety cultures in which leaders provide commitment from above and the people look out for each other on the ground. Ultimately, safety is more likely to become engrained as a companywide value if it is promoted from the top operational executives down to all other levels. When leaders effectively promote their commitment for the well-being of their people, the people in return give maximum effort to protect and advance the well-being of one another and the organization.

AECOM's Safety for Life program embraces this concept, expressed through Life Preserving Principle #1 – Demonstrated Management Commitment. AECOM executives are required to demonstrate their personal commitment to safety, lead the SH&E improvement process, and continuously exhibit support by discussing opportunities for weekly SH&E improvement; actively participating in project safety training and orientation; actively participating in incident investigations; analyzing monthly safety reports to develop safety actions plans; and championing safety meetings. AECOM also includes leading indicators in its Core Value Metrics that track Regional and Area operational manager Senior Management Observations conducted at project sites.

It is critical that AECOM executive and senior leaders are able to systematically conduct and communicate the safety-related activities described above with insight and authenticity. The following trainings were selected for AECOM executives to maximize their influence and impact on the corporation's overall SH&E outcomes.

Supervisor Training in Responsibility and Recognition
All AECOM executives should complete training on responsibility and recognition. This training will establish the concept of how executive leaders can set the expectations that will shape the organization's safety culture. The responsibility is expressed in supervisor accountability for worker's safety, reinforced through a culture of caring. The desired expectations to be communicated are that injuries are avoidable through planning and that workers need to look out for each other. Also, the training will help shift the focus from blaming people when things go wrong to holding managers and supervisors responsible for monitoring and recognizing (rewarding) expected behaviors.

The anticipated time investment to complete this training is **two hours per executive**.

Safety for Executives
Participating in senior management observations, having hazards pointed out to them, seeing examples of work being done safely and at risk, and seeing how safety control measures and equipment operate can give executives a "real world" safety education. In theory, this provides AECOM executives and senior leaders a chance to understand SH&E from front-line managers. However, front-line workers will not always be candid or frank about safety issues, and many senior executives do not have the safety education in their background to be able to ask the questions to facilitate this process.

AECOM executives and senior leaders should participate in basic safety training to be used on their site visits. This training should cover, at a minimum, hazard recognition, behavior-based safety, risk analysis techniques, and incident/injury investigation. A greater impression will be made on the front lines if executives and senior leaders are making well-informed statements and effective SH&E inquiries while conducting senior management observations. This will also reinforce the credibility of the AECOM safety culture to the front lines. The anticipated time investment to complete this training is **two hours per executive**.

Managers

Project managers are crucial in the SH&E process because they are responsible for translating management's policies into action. They are responsible for promoting safe and healthful work practices directly to the employees since they have the greatest control over the work and worker. Managers and supervisors not only know the work and its hazards; they also know the limitations of their staff. Their overall attitudes toward health and safety will be reflected by the employees in their department.

Five principal SH&E activities of a project manager are:

- **Making Safety Contacts** – Communication with direct reports focused on safety.
- **Project Safety Briefings** – Meetings that focus on identifying ways to mitigate hazards, plan the safest way to approach a task, and set a practical standard for safe work.
- **Lifesaving Procedures: Application and Verification** – Field visits to directly verify the real-life application of SH&E procedures while ensuring that hazards are not originating from unsafe acts.
- **Physical Hazard Identification** – Visually scanning beyond workers' actions to look for potential exposures in surrounding conditions and equipment.
- **Incident Response and Root Cause Analysis** – Immediately responding to care for an injured worker and taking mitigating steps to address exposures.

SH&E is closely tied to many other important duties that project managers normally have, such as maintaining order, communication, work planning, schedule, and equipment; establishing work methods; instructing workers and keeping them productive; addressing complaints while maintaining morale; controlling quality and costs; and assigning jobs.

It is for the reasons listed above that AECOM project managers must be able to effectively conduct and communicate safety-related activities. The trainings selected for all managers and supervisors are summarized below.

Safety Supervision and Leadership Training

A key function related to managers is to influence employees to make right decisions by removing variables that drive them to make wrong decisions. Developing and maintaining a high level of safety knowledge and skills and/or business savvy are critical. As such, AECOM managers should receive formalized, AECOM-specific training on safety supervision and leadership. Specific training topics can include such things as the following:

- roles of the manager;
- safety program documents (i.e., health and safety plans, job safety analyses);
- safety leadership;
- safety meetings;
- stop work authority;
- safety observations;
- incident investigation;
- injury management; and
- behavior-based safety (i.e., "4sight", dealing with unsafe behavior, discipline).

To validate this training, each project manager directly overseeing revenue-generating projects with any aspect of work conducted outside of an AECOM office shall complete certification as a safety qualified supervisor. The certifications below should be required as an integral part of project manager training.

SQS Training

The SQS program is intended to encourage and recognize AECOM managers in safety performance at levels consistent with AECOM's intent to be best-in-class globally. Recognizable and quantifiable benefits of managers and supervisors obtaining the SQS accreditation include:

- improved safety culture;
- increased safety awareness among employees;
- validation of employees' fundamental safety knowledge;
- demonstrated competency of its employees by examination;
- reduced workers' compensation claims and reduced insurance premiums;
- reduced need for safety professionals on smaller projects or assignments;
- improved productivity from better communication;
- higher confidence within work groups;
- higher profits from safe work;
- recognition by having employees who hold a nationally accredited credential;
- demonstrated knowledge of fundamental safety practices;
- opportunities for increased job responsibilities or employment;
- increased value to an employer;
- recognition for safety leadership from an employer or other employees, and hold a nationally recognized and accredited credential; and
- increased confidence when dealing with safety and health matters.

An AECOM supervisor who is certified as an AECOM SQS shall be held to the same professional standards as a certified safety manager in accordance with the AECOM Code of Ethics.

The SQS has three prerequisites: 30 hours of documented safety training, good standing with AECOM at the time of the application, and 24 months of supervisory or project/task management experience. The SQS candidate must complete **eight hours of AECOM-sponsored SQS training** before testing. The training curriculum shall be specified and standardized across AECOM but shall have variations in form depending on the business unit. The SQS certification will be valid for five years requiring **at least 20 hours of safety-related training within the previous five years** as an SQS.

Rewards and Recognition Training

Rewards and Recognition Training is a manager training program that shall present effective recognition strategies. By recognizing and acknowledging strong performance and safe behaviors, managers can expect to benefit from increased productivity; proactive values; clearer accountabilities and expectations; better morale; and less inhibited environment where people are more willing to speak up. This training, at face value, aligns with Life Preserving Principle #6 – Recognition and Rewards of the Safety for Life program and would serve to reinforce this important concept throughout the organization.

Staff

All AECOM staff performing fieldwork, regardless of frequency, will undergo some basic hazard recognition and analysis training prior to performing work in the field. All AECOM staff should be required to choose from Hazardous Waste Operations and Emergency Response

(HAZWOPER), mine safety, or field safety. Two of the training options are further described below.

HAZWOPER
AECOM personnel engaged in hazardous waste operations should take 40-hour Occupational Safety and Health Administration (OSHA) HAZWOPER training, or 24 hours of an equivalent course for territories not regulated by OSHA. Subsequently, eight-hour annual refresher courses should be taken through the AECOM Learning Management System. The refresher courses are designed to provide an annual review of topics pertinent to hazardous waste operations, including, but not limited to, hazard recognition, site control, health and safety plans, air monitoring, and personal protective equipment, as well as AECOM company-specific health and safety issues. This course also assists employees in maintaining compliance with regulatory training, where mandated, for hazardous waste operations.

Field Safety
The Field Safety course is designed to provide non-HAZWOPER employees with a review of common health and safety topics pertaining to field operations at AECOM, including, but not limited to, hazard recognition, safe work plans, common engineering and administrative controls, personal protective equipment, incident reporting, and AECOM company-specific health and safety standards. Field safety consists of two hours of on-line modules combined with two hours of live instruction and should be required twice a year.

In addition to the HAZWOPER/Field Safety training requirement, AECOM staff who work in the field in some capacity should also initially participate in basic hazard recognition training that covers a variety of general industry safety and health hazards, as a best practice. This training should emphasize hazard identification, avoidance, control, and prevention, not just compliance standards. The anticipated time investment to complete this training is **10 hours per staff member**. Topics should include:

- walking and working surfaces, including fall protection;
- exit routes, emergency action plans, fire prevention plans, and fire protection;
- electrical safety;
- personal protective equipment;
- hazard communication;
- confined space entry;
- working over water;
- hazardous materials;
- materials handling;
- machine guarding;
- introduction to industrial hygiene;
- bloodborne pathogens; and
- ergonomics (office and/or field).

Training Essentials Assessment (TEA)

Overview

All persons working under the control of AECOM need to be competent to account for hazards and SH&E risks in their work. The competence requirements are not limited to those doing the work or to those who have or can be exposed to SH&E risks but also those who manage a function or undertake a leadership role that is critical to achieving the intended outcomes of the AECOM's Safety for Life program. In determining the criteria of competence for each role within AECOM, managers and supervisors should account for:

- the education, training, qualification, and experience necessary to undertake the role;
- the work environment;
- the hazards identified and associated SH&E risks;
- the preventative and control measures resulting from the risk assessment process;
- requirements applicable to AECOM's Safety for Life program;
- the rights and responsibilities of persons based upon applicable legal and other requirements;
- the importance of compliance with the SH&E policy, applicable procedures, and applicable legal and other requirements;
- the potential consequences of compliance and noncompliance, including the impact on SH&E performance; and
- the value of participation in the SH&E management system.

All AECOM personnel should be evaluated to ensure that they possess the necessary training for their roles. When appropriate, training gaps should be filled by providing additional education, training, and experience. Workers should be re-evaluated periodically to ensure that they have the necessary training for their roles. An evaluation should also be carried out whenever changes have occurred that can impact the activities undertaken in worker roles.

Process

Employee training requirements will be dictated by the work each employee performs (or is expected to perform) and the geographic area(s) where the employee performs these activities. As part of new employee orientation, the employee's direct supervisor/manager and an SH&E representative will utilize training evaluation forms to determine which additional training courses the new employee must complete prior to being assigned to specific job tasks. The training evaluation forms should contain the most commonly required courses, the frequency they need to be taken, and the expected participants. Forms should be developed that are applicable to all Business Groups of AECOM, as well as to individual Business Group-specific requirements. This entire process will be identified as the TEA. The TEA approach should be implemented at the local level.

Future of SH&E Training

A broad-based approach to assessing the SH&E training needs of AECOM as a whole involves administering a companywide survey. The survey should be developed to assess organizational structure; determine performance and competency gaps; measure perceptions and beliefs related to safety; and identify trends within the various layers of global AECOM. Results of this survey, as they relate to this paper, would be to help:

- SH&E determine trainings needs companywide;
- SH&E focus training resources where gaps exist; and
- Develop training material for AECOM personnel varying in culture, socioeconomic status, language, gender, motivation, ability/disability, personal interests, etc.

Cost-Benefit Analysis

Designing and operating AECOM University in the aforementioned optimal state, developing SH&E course materials, providing Information Technology (IT) support, and rolling out required SH&E training to the entire company does have a substantial related cost. However, the expected returns on this investment are even more substantial.

Workplace fatalities, injuries, and illnesses cost the United States billions of dollars every year. In its 2012 Workplace Safety Index, Liberty Mutual estimated that employers paid almost $1 billion per week for direct workers' compensation costs for the most disabling workplace injuries and illnesses in 2010. Employers that implement effective safety and health management, including a strong training system, may expect to significantly reduce injuries and illnesses and reduce associated costs, including workers' compensation payments, medical expenses, and lost productivity. In addition, employers often find that process and other changes made to improve workplace safety and health may result in significant improvements to their organization's productivity, profitability, and morale.

As with all industries, indirect costs (i.e., insurance costs) can undermine our financial performance. As professional consultants, meeting SH&E prequalification criteria established by our clients is a discriminator and wins us work. Noncompliance with global safety standards due to lack of training can create a poor work environment and damage our reputation and brand. Our business's most valuable asset is its good name, i.e., its brand and reputation. Strong brand reputational value equals great profits. For instance, a company with a reputation for quality and safety can charge more for the same product than their competitors because many customers put a price premium on quality products and services that give them positive experiences. Companies with strong brands also retain employees. A recent study suggests that 80% of employees between the ages of 18 and 30 will leave a company if they believe it has a weak brand or no association with ethics.

SH&E training is essential for business sustainability not only from the direct and indirect costs from injuries and accidents but also from the costs of turnover. Industry data has shown a large percentage of employees who receive poor job training leave their positions within the first year, citing the lack of skills training and development as the principal reason for moving on. With one fewer worker, the company's productivity slips. Current staff members are required to work more hours. Morale can suffer. To find a replacement, time is spent screening and interviewing applicants. Once someone is hired, that person needs to be trained. The costs of staff turnover add up. Learning and upgrading employee skills makes business sense. The short-term expense of a training program helps to keep qualified and productive workers who will help the company succeed in the long term.

The project team recognizes that funding SH&E programs in today's economy is challenging for the organization, but, as a professional consulting services firm, we require the ability to offer our clients talent and expertise. It is critical that we invest in the development of our professionals to perform their work safely. The key to a solid return in investment is to focus on SH&E skills and desired competencies that will give AECOM professionals the ability to perform any and all client work injury free. Furthermore, heavy emphasis should be included on nurturing future SH&E leaders.

Recommendations

The following recommendations are based on the team's research previously described in this paper.

- A robust tracking system should be developed for SH&E training across AECOM.
    - The system must be flexible enough to allow for simultaneous usage at the global, group, regional, and project/office levels.
    - The database must be user friendly to facilitate tracking courses taken internally, as well as documenting certificates obtained from external vendors.
    - The database should be capable of generating Tier III matrices for all levels of the organization.
- AECOM group leadership should develop Tier III SH&E training matrices specific to their workforce, business lines, and geography.
    - Group SH&E leaders should monitor measure and evaluate Tier III matrices through trend analysis of incidents and reported safety culture non-conformities (i.e., management commitment, employee participation, subcontractor management, etc.).
    - SH&E leaders, with support from corporate SH&E training, should improve upon Tier III matrices, based upon these trend analyses.
    - Trend analyses should be performed at least annually to help determine training needs and programs for the following year.
    - Approaches should be focused on prevention and not blame or punishment.
- Every new AECOM employee, including interns, should receive a general safety orientation training so that they understand the Safety for Life program.
    - The safety orientation training should be a brief overview and would not take the place of task-, program-, or project-specific health and safety training.
    - The safety orientation should be a combination of computer-based modules and live training conducted by a project manager, supervisor, and/or representative from the SH&E department.
    - All employees should also be required to take an in-house safety refresher training every five years, or less for employees who return to AECOM after an absence of six months or greater.
    - Subsequent safety refresher training should:
        - Reinforce AECOM's Safety for Life program;
        - Provide updates on changes since the last orientation training; and
        - Allow the opportunity to provide feedback on how SH&E could be improved or enhanced.
- As part of new employee orientation, every employee should undergo a TEA process with their direct supervisor/manager and an SH&E representative.
    - The TEA process will utilize training evaluation forms to determine which additional training courses the new employee must complete prior to being assigned to specific job tasks.
    - The TEA evaluation forms should contain the most commonly required courses, the frequency they need to be taken, and the expected participants.

    - This process should be implemented at the local level.
- Defensive driving should be a mandatory course for all employees driving on company business, and a strongly encouraged course for all other employees.
- All executives should undergo responsibility and recognition training and an executive-based course on basic safety and hazard recognition.
- All project managers should receive formalized, AECOM-specific training on safety supervision and leadership as part of the project manager certification process. To validate the safety supervision and leadership training, all managers directly overseeing revenue-generating projects with any aspect of work conducted outside of an AECOM office should obtain certification as SQS.
- All staff performing work in the field, regardless of frequency, should undergo:
    - Basic hazard recognition and analysis training prior to performing work in the field;
    - HAZWOPER, mine safety, or field safety, based on the nature of their projects and worksites; and
    - More extensive hazard recognition and analysis training for worksites possessing high hazard activities.
- A companywide survey assessing SH&E training needs should be administered. The survey should be developed to assess organizational structure; determine performance and competency gaps; measure perceptions and beliefs related to safety; and identify trends within the various layers of global AECOM.

References

Conference call with Scott Harris, Ph.D., UL Director of EHS Advisory Services on November 30,2015.

Emirates Driving Institute, <http://www.edi-uae.com/images/news%20pdf/September2009.pdf>, accessed on December 6, 2015.

"Excellent Experience" by David Crouch, Discussed with David Crouch of Caterpillar Safety Services on November 16, 2015.

First Break All the Rules" by Marcus Buckingham, Discussed with David Crouch of Caterpillar Safety Services on November 16, 2015.

Forbes Magazine, "Your Brand Reputational Value is Irreplaceable. Protect It" Brigham & Linssen, February 1, 2010.

Galford, Robert. "The Trusted Advisor", Discussed with David Crouch of Caterpillar Safety Services on November 16, 2015.

Global status report on road safety 2015, < http://www.who.int/features/factfiles/roadsafety/en/>, accessed on December 6, 2016.

"Good to Great: Why Some Companies Make the Leap…Others Don't" by James C. Collins, Discussed with David Crouch of Caterpillar Safety Services on November 16, 2015.

Go2HR, Employee training is worth the investment, <https://www.go2hr.ca/articles/employee-training-worth-investment> accessed on October 5, 2015.

HSSE in Shell <http://s08.static-shell.com/content/dam/shell-new/local/corporate/corporate/downloads/pdf/HSSE-in-shell.pdf> accessed on September 22, 2015.

ICE COMM Team Project #2, "AECOM High Consequence Events Analysis and Reduction Program" by Alberto Munuera dated December 2015.

Industry Week, Best Practices – DuPont: Safety Starts at The Top, <http://www.industryweek.com/articles/best_practices__dupont_safety_starts_at_the_top_1108.aspx>, accessed September 2015.

ISO 45001 *Occupational Health and Safety Management Systems – Requirements* (Draft version.), March 2015. Geneva: ISO.

Learning_and_Development_Procedure_H3ME_900_PR1, Revision 3, April 25, 2013.

Liberty Mutual Research Institute for Safety, 2012 Liberty Mutual Workplace Safety Index, <http://www.asse.org/assets/1/7/WSI_2012.pdf>, accessed September 2015.

Petersen, Dan. "Safety Excellence-Six Critical Components, < http://www.ishn.com/blogs/16-thought-leadership/post/98384-safety-excellence-the-six-critical-component>, accessed on November 16, 2016.

Queen Mary University of London." Mandatory Health and Safety Training – Guidance Notes" <http://www.arcs.qmul.ac.uk/policy_zone/health_safety/mandatory_health_and_safety_training_guidance_notes.pdf> accessed on September 22, 2015.

URS & DBU Health, Safety and Environment (HSE) Program Requirements, March 4, 2013.

URS Safety Management Standard 055INT, Health, Safety and Environment Training, Revision 1, April 2013. San Francisco: URS Corporation.

Session No. 746

Crash Frequency Benchmarking and behavioral Approaches to Fleet Safety

Peter Van Dyne, MA, CSP, CFPS
Technical Director
Liberty Mutual Insurance

Don Tolbert, CSP
Technical Director
Liberty Mutual Insurance

Session Objectives

Examination of the root causes of motor vehicle crashes provide evidence that driver behaviors contribute more significantly than driver knowledge or lack of skill. Drivers make behavioral choices that increase the risk of a crashes that include following other vehicles with less than the recommended following distance, driving while impaired, driving distracted, driving while fatigued and driving aggressively.

Training, in general, has shown to have less impact on behavioral choices than point-in-time systemic factors. People make decisions on the basis of perceived need weighed against perceived risk. This indicates that driver training, in and of itself, should not be expected to have significant effect on the behaviors that contribute to crashes when system influences point to needs met by those behaviors. This paper and our presentation will address comparing liability crash frequency to similar operations and incorporating behavioral or organizational practices concepts into fleet safety programs. This will allow a company to compare their operations to others and identify ways to strengthen their fleet safety programs.

Crash Benchmarking

Crash benchmarking is an approach to comparative assessment of the impact and effectiveness of a fleet safety program. Most crash benchmarking compares the number of crashes to the number of vehicles or the miles driven. This is more effective when used to compare companies or locations that use similar vehicles, operate them in similar ways, and within the same industry. Most benchmarking studies have historically not taken the state of operation into account. This can create distorted comparisons when more vehicles are used in areas with higher crash rates.

Data comparisons by the authors suggest that expected crash rate is influenced by three main variables: industry, type or size of vehicle, and the state of garaging. Considering these three variables allowed calculation of expected crash frequency. Including company size, based on total premium, in the analysis revealed differences between larger and smaller operations.

The expected crash rate is for property damage (PD) and or bodily injury (BI) events. Crashes with physical damage were excluded unless PD or BI claims were incurred as a result of the same crash. Physical damage was excluded due to the wide range of deductibles. Claims with no incurred costs were excluded from the frequency calculations. The most effective way to explain this is with examples changing one of the variables at a time. The first example uses a contracting fleet with 100 medium trucks (10,001 to 20,000 GVW). The expected range varied by over 100% between the lower and higher expected frequencies.

State	Expected Crash Frequency (rounded to the nearest 10th)
Florida	4
Kansas	2.6
Massachusetts	6
Oregon	2.8
Rhode Island	4.3
All State Average	4

This example shows the impact of various vehicle types using a 100 vehicle contracting fleet in the state of Florida

Vehicle Size (GVW	Expected Crash Frequency (rounded to the nearest 10th)
Light Truck 0 – 10,000	3.4
Medium Truck 10,001 – 20,000	4
Heavy Truck 20,001 – 45,000	4
Extra Heavy Truck Over 45,000	5

The next example shows difference based on industry groupings. The sample was 100 medium trucks in Colorado.

Industry	Expected Crash Frequency (rounded to the nearest 10th)
Contracting	2.9
Mercantile	2.9
Service	3.0
Energy and Utilities	3.7
Wholesale	4.2

The example below examined a 100 vehicle contracting fleet in Missouri of medium trucks and compares large vs. medium sized companies.

Company Size	Expected Crash Frequency (rounded to the nearest 10th)
Medium	3.3
Large	3.4

When comparing the crash frequency of a company to a larger group it is important to understand the benchmark parameters used in calculating the expected or average crash frequency. Most operations will have unique exposures and benchmarks should be viewed as a comparative estimation.

Elements of an Effective Fleet Safety Program

There are many sample and/or model programs that can be found with an internet search and consensus standards such as ANSI Z 15.1 describe fleet safety program elements that should be considered. While topical phrasing of element category headings may vary between these programs, similar exposures or controls are addressed. Author studies of fleet safety programs have identified program elements and practices shared by companies that have lower crash and vehicle related injury rates. These high-performing companies:

- Select drivers based on their driving history and ability to perform the job
- Establish and communicate expectations on how jobs should be performed and have supportive management practices
- Monitor task completion and performance to the expectations using metrics that measure results and risk
- Provide feedback on performance for both workers and management
- Adapt process to influence behavior that meets expectations
- Document the policies and adaptations

Select drivers based on their driving history and ability to perform the job

Selecting drivers based on their driving history and ability to demonstrate compliance with expectations should also apply to anyone that operates a motor vehicle on company business. For this to be effective, criteria must be established and followed. If exceptions are made they should follow an exception policy that limits the exceptions and how exceptions can be made. Exceptions should be made for some criteria and not for others. For example, an exception could be made for similar experience operating vehicles but should not be made for someone with a major violation that would disqualify them from operating motor vehicles on company business.

The process of selection and on-going reviews should be documented and periodically audited to verify it is followed and that the documentation supports the selection criteria. Management should have objectives to comply with the process which are outlined in expectations and mentioned as part of the program responsibilities. Regular performance reviews should include the safe operation of vehicles.

Establish and communicate expectations on how jobs should be performed and management practices

One of the more commonly cited organizational practices relative to safety management systems is comparing performance and behaviors to policies and expectations. For this to be effective, an organization should identify & objectively state the performance they expect, effectively communicate it, and have documentation to support the process. Some fleet safety programs have relied on commercial driving training products to communicate expectations. Commercial driver training products can support company expectations but may fall short in communicating them for driver performance if the expectations are not documented in policies as such.

Objective statements of expectation are essential to performance management in organizations. They should be phrased to be:

- Within the *Performer's Control* – the person can complete the expectation without action by someone else
- Specific – the situation in which the expectation is met is described in precise terms
- Measureable – Observable under normal circumstances (Not requiring unique circumstances for observations
- Active – requiring action for the expectation to be met (Versus no action)
- Reliable – meaning is clear; two or more people observing the same action agree on whether it meets the expectation
- Tangible – performance can be appraised with a "Yes" or "No"

For example, safe driving theory on following distance can be incorporated into an objective performance expectation. Since the space between vehicles obviously increases with speed, a widely accepted separation distance measure that allows enough time for the driver of the following vehicle to react to speed changes in the followed vehicle is 4 seconds. A company's documented expectation that meets the six criteria above could be: *"Drivers maintain four seconds of distance between their vehicle and vehicles they are following"*.

The context of the message is important and should cover a wide range of topics that include driving vehicles, distractions, fatigue, vehicle inspections, vehicle maintenance, trip planning/journey management, and crash or incident reporting. Careful crafting of expectation statements not only allows performance to be measured but sets the stage for incremental performance improvement.

Monitor task completion and performance to the expectations using metrics that measure results and risk

Most driving performance data are related to aggressive events. Expectations for driving to avoid aggressive events should take into account that some aggressive events help a driver avoid a collision but a high frequency of events may indicate a higher risk of becoming involved in a crash or a disregard for the company fleet safety policies. Using data to identify trends and aggressive event outliers can have more productive results if the range of performance is understood and event rates are calculated.

The statistics shown below are for aggressive events from telematics data collected over a short period of time from a fleet of 70 vehicles. They illustrate the range that can occur with similar vehicles used in a variety of locations.

Maximum	73
Median	3
Mean	7.2
Minimum	1

The wide spread between the mean and median is not unusual for telematics events and event rates. There were 7 drivers with event counts 3 times the median and 7 times the mean. This group would be the drivers initially reviewed for root causes of the aggressive driving. The root causes could include one or more of the following contributors:

- Effectiveness of past coaching discussions and in vehicle observations
- Motor Vehicle Record (MVR) violations prior to and after hire
- Driving expectations for vehicle operation
- Driver Knowledge of the expectations
- Vehicle and work experience or lack of that impacts performance
- Fatigue
- Scheduling or changes during the day or inadequate travel times
- Routing of drivers for work to be completed

- Compensation systems that may unknowingly encourage aggressive driving
- Data integrity/telematics device performance
- Consequences for performance that do not discourage aggressive driving
- Driver outside work responsibilities/situations
- Multiple jobs requiring late starts to on time completion
- Commuting times
- On Time departure at Start Of Work Day
- Distractions
- Vehicle condition
- Breaks and lost time during the work day
- Work flow or scheduling exceptions
- Health and wellness

Additional performance monitoring could involve use of technology, seat belt use and vehicle entry and exit protocols.

Provide feedback on performance for both workers and management

Once performance that does not meet expectations has been identified, drivers and their managers need feedback on the performance. For drivers on a focused improvement plan, more frequent feedback, increased interactions to help them understand the driving expectations, and coaching in techniques that can decrease the frequency of aggressive events may be needed. The goal does not need to be elimination of events but reducing situations where aggressive events are needed to react to traffic situations.

Coaching, feedback and ongoing discussions should be directed at root causes identified as part of the root cause investigation process. Identifying opportunities for systematic change which can decrease reliance on aggressive events for crash avoidance should be the focus.

Adopt process to change behavior that does not meet expectations

When examining performance at the individual or organizational level, it is important to understand the basics of antecedents, behavior and consequences. If the consequence of performance that does not meet expectations is "training", the undesired behaviors may not change. Some performance or behavior requires significant and immediate consequences such as removing a driver from jobs that require the operation of motor vehicles. Having and communicating the range of consequences for performance should be part of process that goes beyond just driving.

Document the policies and actions

Fleet safety programs should have two primary objectives: 1) Reduce the risks for motor vehicle crashes. 2) Protect the company assets threatened by loss from crashes. The program policies and practices are designed to reduce the potential for losses. The program documentation, implementation, and ongoing audit/review allow a company to demonstrate that they have & follow a formal program if the company's efforts are questioned after a large loss. Documenting programs/protocols and verifying that they are followed can be very helpful to claims handlers when developing a settlement strategy.

Theory Behind Well-Performing Organizations

Safety does not operate, nor has it evolved, in a vacuum. The safety profession has produced & proven a wealth of tools for refining the effectiveness of people doing things together (organizations) that draw on these essential components. The ways in which a company's performance can be measured is also expanding as technology to collect real time data on how processes (activities) are working continues to improve opportunities for gaining insights on how to systematically improve them.

Complexity - increasing numbers of system components interacting in more variable ways - can challenge approaches to understand them, much less influence their effects. Safety theory & practices that were foundational to risk reduction in simpler times/circumstances, and implemented with increasing effectiveness by progressive companies, are being taxed by compounding complexity.

There are few activities carried out by more people, more often and in the face of more compounding complexity that are better examples of the phenomena described above than "Operating Motor Vehicles". In little more than a generation past, the technologies accepted as common in vehicles today were conceptual or experimental if not outright science fiction. Distraction/inattentiveness/driver error were significantly associated with crashes when vehicles were simpler, with only the most basic protection from harm to occupants when crashes occurred. New technologies converged to produce exponentially safer vehicles & roadways while significantly increasing the variety and flow of distraction sources for the driver. Progress can be described as trading one set of risks for another. In the presenters' opinion, residual risks associated with operation of modern vehicles that have sophisticated capabilities for crash avoidance, occupant protection and even data gathering are good trades. This is particularly true when considering the parallel evolution of increased motivation for companies to change how risk in complex systems is gauged, understood, and reduced.

As change is influenced and improvement made, horizons expand and more effective control of harmful events is affected. Reaching plateaus of achievement is normal and reveals new opportunities. Some become more recognizable through higher proficiencies – or attentiveness - earned through the improvement process and some through shifting priorities. Other opportunities are revealed as problems that do not respond to the same methods used to achieve the higher operational/organizational performance that is emblematic of well-performing organizations.

"Continuous Improvement is better than Delayed Perfection." (Mark Twain)

High functioning companies have integrated risk reduction that sustains safety performance well beyond industry norms. These best-in-class companies build capabilities to produce certain management system features that influence what can be described as essential organizational practices.

Mandate - a tangible *commitment*, top down, to act with specific purpose

Accountability – assurance that meeting expectations associated with the mandate is *systematically* supported

Process – Repeatable sequences of actions that produce a *predictable* output

Execution – Measured process performance that supports increasingly *effective* output

Improvement – Continuous adjustment of processes that verifies & demonstrates the mandate

One of these features, in particular, can be a centerpiece of improvement opportunity for commercial fleet operators. For many if not most organizations, the concept of accountability manifests in a culpability framework for addressing/correcting human error as a cause of incidents or process failures. This is evidenced by the mindset mentioned previously that aggressive driving events can be decreased solely through more or better driver training. While there are situations in which interventions for decreasing aggressive driving events should act on patterns of behavior observed in individual drivers, this traditional approach to accountability is akin to using the rearview mirror when driving. Accountability frameworks should not be limited to examining the path just traveled and correcting as needed. They should be expanded to consider & tap into the knowledge of people involved on systemic opportunities for improvement. These opportunities do not favor people or systems in improvement strategies but embrace the notion that it is people *in* complex systems that make the systems resilient and highly effective (Dekker).

Organizational Practices and Fleet Safety

In 1980, Dov Zohar introduced the concept of Safety Climate (Zohar, 1980) as workers' shared perception of an organization's policies, procedures and practices as they relate to the true/relative value and importance of safety. Since that time, Safety Climate research – including that from the Liberty Mutual Research Institute for Safety Center for Behavioral Sciences - continues expand and mount evidence supporting its utility in predicting future safety outcomes (Christian, Bradley, Wallace, et al., 2009). Practical use of the conceptual underpinnings, supported by ongoing science, reveals opportunities for companies to leverage interactions between functional levels to strengthen shared perception of safety as a priority (Huang, et.al, 2017).

If not implicit in this description of Safety Climate, it is measureable with unique perception surveys based on research supported themes pertinent to the Organizational ("Top Management") and Group ("My Direct Supervisor....") levels. Survey findings inform strategies for strengthening the cascading interactions that influence situational decision-making in safety critical activities. It should be pointed out that these strategies rely on participatory approaches for implementation

The organizational practices listed above, in the context of Safety Climates for commercial fleets, provide a framework for focus areas:

1. Mandate - a tangible commitment, top down, that no need met by operation of the vehicle fleet supersedes prevention of harm to the driver or the traveling public in its operation
    a. Established in writing over the signature of the CEO
    b. Expressed as a value to which other essential values are related (E.g. Productivity, Customer Satisfaction, Sustainability, Community Relations, etc.)
    c. Supported by improvement initiatives at all levels on par, and integrated with other improvement initiatives (E.g. Quality, Productivity, etc.)
    d. Enacted through cited relationships between achievements in other improvement initiatives

2. Accountability – assurance that meeting expectations associated with the mandate is *systematically* supported
    a. Develop Fleet Management Systems that:
        i. Define specific Fleet operating parameters that are unwavering
        ii. Recognize & embrace the reality and need for discretionary decision making by drivers
        iii. Define where the unwavering parameters end and discretionary decision making begins
    b. Develop expectation-driven, cross functional engagement that:
        i. Reinforces blame-free discussion of situational circumstances that threaten the mandate
        ii. Generates ideas for improvement/adjustment for circumstances
        iii. Motivates ongoing involvement in identifying & transferring Best Practice driver decision-making

3. Process – Repeatable sequences of actions that produce a *predictable* output
    a. Supervisory engagement with drivers that enable & reinforce the Fleet Operating Parameters
        i. Expectation-driven Daily Huddles
            1. Reiteration of mandate
            2. Review of Fleet Operating Parameters
            3. Surfacing of situational changes
        ii. Performance reviews that reinforce achievement in meeting expectations
    b. Cross-functional team meetings to continually examine the Fleet Operating Parameters relative to situational circumstances that:

    i. Are scheduled and agenda driven
    ii. Are sponsored by a senior member of management
    iii. Produce Meeting minutes/summary, with improvement recommendations, to the Sponsor
    iv. Begin with discussion of submitted recommendation status
  c. Management stakeholders observe improvement process, respond to recommendations, and celebrate contributions of the cross functional team

4. Execution – Measured process performance that supports increasingly *effective* output
  a. Establishment of Performance Measurement that:
    i. Identifies outcomes (downstream/lagging/dependent variables) to which fleet operating parameter performance may be related
    ii. Identifies process (upstream/leading/independent variables) measures for the fleet operating parameters identified as related to outcomes
    iii. Tracks process measures
    iv. Engages cross functional team in identification of trends to inform adjustment recommendations

5. Improvement – Continuous adjustment of processes that verifies & demonstrates the mandate
  a. Adoption of progress measurement that gauges gap closure (Percent improvement from Baseline) in process metrics identified
  b. Comparison of deliberate change made to process with observed changes to outcomes
  c. Celebration & Reaffirmation
    i. Top down celebration of contributions that achieve milestone improvement
    ii. Statement of value added by the mandate
    iii. Call to action to reaffirm the Mandate

There are proven best-practice elements that should be considered as foundational to Fleet Operating Parameters

Best Practices

Reviewing your program for inclusion of these ten best practices is a way to identify improvement opportunities.

- **Identify all jobs that require the operation of motor vehicles** and include operation of motor vehicles in the job description of listing of job functions. Employees or contractors operating under your authority should be included. Operation of vehicles from full time driving to incidental use on an occasional basis should be included.

- **Adopt a policy that only employees on the list of "approved drivers" may operate motor vehicles on company business**. Criteria to be placed on the list of approved drivers should include meeting the company driving record criteria prior to and during employment, gaining and acknowledging an understanding of the fleet safety policies, completion of a road test to verify the skills and behaviors for operating vehicles as outlined in the fleet safety program are present, acknowledging and agreeing to comply with distractions polices and agreeing to comply with other aspects of the fleet safety program. Managers and supervisors should agree to enforce the policies and comply with them.
- **Establish, follow and audit a motor vehicle driving record verification program**. The program should outline what defines an acceptable driving record, the minimum frequently driving records will be checked and consequences for not meeting the driving record criteria. This should include a policy on reporting of all citations and convictions. Where state driver notification programs exist they should be used.
- **Establish, document, communicate and reinforce distractions policies**. The policy should prohibit a wide range of activities including use of technology and tasks that take a driver's eyes off the road, hands off the controls and mind off of driving. Management must agree to comply to set an example. Operational policies should be reviewed to identify ways drivers are distracted by routine business and practices modified to reduce the potential for drivers engaging in work tasks while driving.
- **Identify use of non-owned vehicles and establish controls** to reduce the risk from this exposure. Drivers of non-owned vehicles should be included in the fleet safety program and drivers must be on the list of approved drivers to operate non-owned vehicles on company business. The controls should include driver qualifications, completion of a non-owned vehicle agreement, providing documented insurance limits meeting those found for a typical vehicle and that drivers are expected to provide maintenance that meets the manufacturer's minimum recommended maintenance schedules.
- **Establish, document and reinforce driving behaviors for the operation of motor vehicles on company business.** Concepts covered in commercial training materials should be explained in the context of being expectations and should address how vehicles are operated and maintained. Topics should include: speed limits (company and posted), following distance, use of mirrors, vehicle maintenance and inspections, keeping windows clean, proper tire inflation, operation in bad weather, locking vehicles, planning routes, driving in parking lots, prohibition on U turns, securing equipment within the vehicle, not operating while impaired or fatigued, use of turn signals, scanning while driving, use of restraint systems (seat belts) and use of technology for directions. Management should agree to the expectations and set an example by complying with policies.
- **Documenting crash reporting protocols and crash scene documentation.** What is expected of drivers, supervisors and other management should be documented and communicated. The procedures should address crash scenes, reporting crashes, crash root cause investigation, crash reviews by management and consequences for crashes. Other types of loss such as theft, fire or storm damage loss reporting should be included.

- **Establish vehicle maintain standards and vehicle inspection process.** This should include maintenance to be performed based on mileage and time. The responsible party should be identified and facilities to be used. Records should be kept to show vehicle maintenance occurred and was performed per the schedule. Vehicle inspections schedules for damage and interior condition should be established and documentation maintained to show inspections were performed. Audits should be performed to verify the process and schedules were followed.
- **Identify fatigue triggers and establish a process to keep fatigued drivers from operating motor vehicles on company business.** Fatigue can have many sources including medical, work scheduling, commuting times, family issues and personal choices. A process to allow work to be done by others or delayed should be established to reduce the potential for a fatigued driver operating a vehicle on company business. Management should look for signs of fatigue and have a process to keep fatigued drivers out of vehicles used on company business.
- **Establish a process to monitor driving performance.** Driving performance should be measured using a combination of observations, data and program audits. Observations could include random observations and scheduled inspections for policy compliance and vehicle condition. Data can come from a variety of sources and include fuel usage reports, telematics data and in vehicle video event recorders. Data should be reviewed to identify outliers and individual performance plans created for drivers needing improvement. The action plans should be based on root causes identified as part of a analysis to identify root causes. Crashes are more commonly associated with driver behavior than lack of knowledge and action plans have a higher potential for changing behaviors when they do not rely just on training.

References

Christian, M.S.et al. 2009. "Workplace safety: A meta-analysis of the roles of person and situation factors." *Journal of Applied Psychology*, 94(5), 1103-1127.

Dekker, S. 2012. *Just Culture: Balancing Safety & Accountability.* Burlington, VT: Ashgate Publishing Company, 80-84.

Huang, Y. et al. 2017. "Safety Climate: How You Can Measure It & Why Does It Matter." *Professional Safety,* 28-35

Zohar, D. 1980. "Safety climate in industrial organizations: Theoretical and applied implications." *Journal of Applied Psychology,* 65, 96-102.

Session No. 750

Safety Discipline, Accountability, Punishment, Blame, Coaching and Feedback

Christopher Goulart, CSP
Risk Control Consultant
Aon
Denver, CO

What are they, how do they work or not work, and when should they be used? This paper will attempt to define each concept and provide a framework for their application. It will also provide some context about the challenges faced by Safety Professionals as they work through using each one of these management practices to optimize performance of employees and to help them reduce risky behavior.

Any Safety Professional who has worked to improve safety and reduce risk within an organization has invariably been faced with the challenge of having to give feedback, provide accountability or discipline, coach, or even punish employees. However, the use of each of these different approaches to interacting with employees comes with possible perils and benefits. This paper will provide some perspective into each of these concepts and will outline the basics of 1) what are these different actions, 2) when is their use appropriate and when should it be avoided, and 3) what model is most effective at motivating employees to work safely.

One aspect of this discussion that this paper will not attempt to evaluate is whether or not Safety Professionals should be the agents of administering discipline, punishment, or any other outcomes. Based on a number of years of experience and observation of hundreds of organizations, it is true that there are many options when considering if Safety Professionals should have the ability or responsibility to hold people to account if they are at risk. In some instances having this ability reside with the Safety Group makes sense as they can immediately intervene in at risk actions and create immediate consequences when they are necessary. In other instances, the Safety Professional has a more collaborative role as a coach, mentor and facilitator rather than as a disciplinarian. In either case, this part of the discussion is beyond the scope of this article and will not be covered here.

The Source of the Conflict

Whether or not to punish, discipline, hold accountable, coach, or give feedback has been a question that has challenged managers, leaders, and safety professionals for decades. The source of the quandary lies at the junction of providing appropriate and meaningful consequences while being fair and ensuring the interactions are productive and provide the most desirable outcomes. This paper delves into each concept and attempts to provide some guidance and insight into how each of these concepts can be used appropriately.

The true source of the conflict appears to reside with the fundamental human motivation to react to adverse events. When an accident or injury happens, there is a deep-seeded desire to "do something". Humans are strongly driven to try to find solutions to problems and solve them as quickly as possible. This often creates a need for people to react quickly and decisively when an accident occurs. Among the options that can be used as potential responses to accidents, a range exists from Punishment to Coaching. The conflict arises when a stronger form of response is used than is necessary and results in the withdrawal of the employee.

Using punishment and blame for mistakes and errors, instead of employing coaching and feedback is one of the most common challenges and creates a host of problems for an organization. The possible issues can range from a loss of morale among most of the employees to an expression of overt hostility from employees toward management. Further, organizations that do not sort out the optimal means of providing negative consequences risk the loss of ethics and integrity. This can occur if the company punishes employees for actions that are not their fault or for conditions that are beyond their ability to control. It is generally considered unethical for punishment and discipline to be applied to persons who have had no intention of causing harm and have made mistakes or engaged in actions that were previously not subject to punitive measures. When this kind of event occurs, the reputation of the organization can be severely impacted with the employees. It is not uncommon for employees could develop a deep-seeded distrust of management and the overall safety process. Avoiding the misapplication of punishment or blame must be a cultural normative value that all business entities strive to achieve on an ongoing basis.

Ensuring that an organization is ethical and practices integrity is not only fundamentally the right thing to do, it also makes for good business. According to 80 different studies that examine the correlation between corporate social responsibility and ethics, 55% point to a positive relationship, 24% find no relationship, and 4% find a negative relationship. The remaining 19% yielded mixed results.[1] Further, a recent article in the Denver Business Journal clearly draws the link between ethical behavior by companies and corporate success.[2] According to one case study covered in the article the advantages that arise when being an ethical corporate citizen are that ethics:

- It builds trust and loyalty (and referrals) from customers.
- It motivates sales staffers because they really believe in the products and services they're selling.
- It creates a culture of honesty, especially important in a business where employees work in customers' homes and can't be closely supervised all the time.
- It attracts top employees whom want to work for a company with standards they share and respect.

What are the Goals of Using these consequences to Motivate Future Behavior?

In the most direct sense, the **proximal** goals of providing negative consequences are to correct error, change behavior, or remove employees who have violated acceptable practices. However, the **absolute** goal is to help create an organization where performance is optimized, risk is reduced, and injuries and accidents can be made less likely to occur. The conflict that we have been discussing arises when the objectives of the ultimate goal are compromised by the actions attempting to attain the proximal goal. In other words, if people are held accountable or punished for actions where a learning opportunity would otherwise exist, they will not change their behavior in any way other than avoiding the activity where they received a penalization. This will compromise the ultimate goal by reducing the likelihood of employee participation in organizational processes (such as the Safety Process), eliminating the chance of dialoging with the employee about the true source of the at risk act, and lose the opportunity to evaluate the systemic causes of risk.

In attempting to reach the ultimate goal of increased safety and reduced risk, organizations should consider that each of the actions for providing negative consequences mentioned (Blame, Punishment, Discipline, Accountability, Feedback, and Coaching) can be used beneficially, but like any tool must be used in the right way at the right time. When used inappropriately, these various forms of consequences can be seriously deleterious to an organization. Consider the following situation… A forklift operator enlists the help of a spotter to place items in overhead racking. The areas where the items are to be placed is immediately behind a hanging HVAC unit. The forklift operator, who was instructed to move those materials to that awkward location by his supervisors, proceeds to the area and begins to move the materials into the overhead storage location. The forklift operator is unable to observe anything in the overhead storage area as the hanging HVAC unit completely obstructs his view and he is utterly reliant on instructions from the spotter. The spotter begins giving verbal commands to the forklift operator, but makes a mistake and tells the forklift operator to move the load to the right rather than to the left. The spotter immediately corrects himself, but not before the operators has struck the HVAC unit with the load and done over $1,000.00 in damage.

This is a classic example of human error. The instructions of the supervisor were being followed and the directions of the spotter were also being followed. However, despite the best efforts an accident still occurred. This is a real world example, and what followed illustrates the introduction of losing sight of the ultimate goal to achieve the proximate objective. Both the forklift operator and the spotter were given written warnings for a safety violation. The spotter received the written notice of a safety violation for making the mistake when giving directions and the forklift operator received the written safety violation because he was operating the forklift and caused the damage. At this organization, documented safety violations never come off a person's record and employees are automatically terminated upon the receipt of three safety write-ups over the course of a career.

In this real-world example, the proximate objective of preventing either employee of ever experiencing the same error caused a reduction in the ultimate objective of reducing risk. Since the company never removed the hanging HVAC unit, nor did it stop storing materials in the overhead area behind the unit, the driver and spotter who had the mishap would likely be some of the best operators to continue to place materials in that spot since they had an appreciation for how difficult it was to complete the task. However, because both the spotter and the forklift operator were disciplined, punished and blamed, they will avoid that task, and any other safety sensitive work, in the future. The result is the proximate goal (delivering a consequence) being achieved at the expense of the ultimate goal (reducing exposures and increasing employee participation in the safety efforts).

Consider the activity of conducting a root-cause investigation. If an event creates a loss and is investigated, it is not uncommon for some element of employee behavior to be involved in the accident. However, when an investigation concludes that Human Error is the root cause of the event, the investigator will have no reason to examine the systemic components of the workplace that may have facilitated the at risk employee behavior. Indeed, any good accident investigation methodology or process will always recommend that the investigator remove the filter of their own experience to ensure the objectivity of the investigation. It is not uncommon for those conducting incident investigations to believe that most or all accidents and injuries are the fault of the injured person. Removing the bias created from personal experience may be needed if the investigator hopes to achieve the aim of ascertaining the true, and often complex, causes of the accident. In order for the investigator to develop effective and systemic corrective actions to keep a similar accident from occurring again in the future, they must arrive at the controllable root causes that go far beyond blaming the injured employee for their actions.

Finally, as eloquently stated in his book, *The Field Guide to Understanding Human Errors,* Sidney Dekker talks about the need that some organizations have to make an example. When an employee experiences an accident, there is often a rush to judgment and a need to prove to the rest of the organization that the company acts swiftly and decisively. This act of making an example of an employee for a failure removes any opportunity for growth and learning.[3]

Ensuring that the consequences provided do not impede the ultimate outcome is the pivotal concern. To quote a common saying, we need to be careful to *"see the forest through the trees"*. It is imperative that actions relating to giving negative consequences are done in a manner that appropriately corrects behavior, engages employees, provides them with instruction for future actions, and always helps the organization achieve the ultimate goal… reducing the likelihood of future accidents and injuries.

Punishment

Punishment is the imposition of a punitive consequence administered by an authority figure. It is intended to create a significant negative effect on the party receiving it. Punishment sends the message that the offender has intentionally violated normative values and must be dealt with punitively. This is used when the authorities perceive a loss of control and illustrates their domination over another. The use and over-use of punishment is common in authoritarian based organizations. Indeed, because the use of punishment is reinforcing, it is not unusual for an organization to develop a "culture of punishment". This type of normative value is a common element in traditional approaches to workplace safety, where the organization looks for at risk acts to punish and reprimand.

The application of punishment results in behavior change by the person or entity being punished. Usually, this behavior change is one where the employee will avoid whatever activity they were engaged in that led to the punishment. This avoidance behavior can lead to the perception that the punishment worked, as the employee will likely replace the at risk behavior with an unrelated activity, thus avoiding the chance of additional punishment. It is tempting to view this behavior change positively as the desire to "do something" when an accident or injury occurs is very strong. Unfortunately, just because punishment causes an at risk behavior to go away, it does not mean that a safe behavior will be the automatic replacement. Managing the use of punishment is difficult and getting it right is even more difficult.

One of the larger challenges with the use of punishment is that it is not uncommon for rules and regulations to be unevenly enforced and for punishment to be administered after some negative event, not always used for poor performance. For example, an employee might work unsafely for many years, but not have an accident and avoid punishment. However, when an employee does have an incident and gets hurt, the employee may receive a written notification of his or her infraction. This documentation often includes the language *"up to and including termination"*. This likely occurs because, according to those in charge, the at-risk behaviors did not warrant attention at the time. But once the threshold of having an injury occur is passed, then punishment is used.

There are several additional challenges with punishment. First, if punishment is used only when an accident or injury occurs, then the employee probably received positive reinforcement for the at risk behavior in the form of tacit approval. Tacit approval occurs when unacceptable actions are observed by others, but do not receive comment. This is often due to the wish to avoid conflict or a desire not to get involved. In either case, the use of punishment without a full understanding regarding all the nuances of the situation can have serious deleterious effects. Another challenge relating to punishment is administering the appropriate level of punishment commensurate with the offense. Discharging an employee for a breach of policy that does not have clearly defined consequences can create a sense of uncertainty for all members of the organization. If the use of punishment is viewed as random at best or as a form of favoritism at worst, then it is likely morale will be negatively impacted in the entire organization.

Punishment teaches employees that they are not in control of their situation. It shows them that their grip on their employment is precarious and when it is used too frequently, it creates a toxic environment where no one will put forth any serious discretionary effort to improve his or her safety or the safety of others. Ample punishment can be devastating to an organization's culture as it will hamper teamwork, collaboration, and ensures that employees will do just enough to get by. Employees in an environment that prioritizes punishment will never seek success. They will only strive to avoid failure. Organizations that focus on avoiding failure will never achieve greatness.

Punishment is best used when an employee has intentionally violated a safety rule that they know and have chosen not to follow. In these situations punishment and blame (described in the next section) are appropriate and the employee is in need of a dramatic intervention such that removal from the organization may be the only viable option. This generally occurs when the employee knows the acceptable course of action and yet makes a decision not to follow procedures. This concept can be illustrated through the use of an example. If an employee steals from their employer, then punishment is absolutely appropriate. The act of theft is unacceptable in all circumstances, and the employee has the intention of committing a crime. In safety, the parallel can also exists when employees make at risk decisions. If an employee makes a decision to intentionally violate a known safety rule, and so puts themselves or others in harm's way, then punishment is equally appropriate.

The key is intentionality and the underlying intent of the employee. If an employee has INTENTIONALLY refused to follow established practices, has documented knowledge of the safe work practices and chooses to ignore them, and has created a situation of excessive risk or harm, then the practice of applying punishment is absolutely appropriate and is likely needed. It is only under these narrow set of circumstances that an organization can be confident that punishment is used appropriately.

In the real world, all negative events require that alternatives to punishment must first be considered. Because using punishment for an oversight or an honest mistake is never helpful, it is crucial that the organization be meticulous in determining if punishment is appropriate.

Human beings often learn best when mistakes are made. Learning from mistakes and rising past adversity forms the cornerstone of many management and leadership philosophies. However, no learning will take place when punishment is applied. Allowing employees to make, and learn from their own mistakes is one of the characteristics that is often pointed to as an attribute of World-Class organizations. Creating an environment where employees can learn from everyone's mistakes is the foundation for a system of continuously improving safety.

Discipline

Discipline is a form of negative reinforcement that does not have implied outcome of a worst-case scenario that is associated with punishment. When employees experience discipline they are given the opportunity to learn and grow from a potentially negative event. When discipline is used appropriately, the employee has the opportunity to preserve a great degree of self-respect. This is opposed to punishment where there is no positive outcome for the employee and no chance to grow and learn.

Indeed, the very root of the word discipline is disciple, one who learns and follows. To be self-disciplined is a virtuous attribute, but to de disciplined from an external source is to experience a form of corrective feedback that still contains the learning aspect. Discipline is forward looking and provides an opportunity to learn and grow. This is compared to punishment that has no opportunity to learn and amend behavior. The appropriate use of discipline can greatly enhance performance. However, if discipline is misapplied, it can be a very punishing experience.

When discipline is administered by another it does have an inherently retributory component. In its purest sense, discipline should reflect a call to attention for a serious transgression that involves inappropriate behavior. Although discipline administered by a supervisor or manager may have negative connotations, it is primarily intended to help employees learn and chart a new course of action. Indeed, *self-discipline* is a common term used both in business and society that reflects the idea of optimizing performance through rigorous adherence to a set of beliefs or principles.

Self-discipline describes the condition of following a rigorous system of beliefs and practices that are intended to lead to betterment of the individual. In most circles, self-discipline describes a fundamentally beneficial attribute and those who practice self-discipline are generally considered role models. Self-discipline is a condition that is actively sought after by employers and employees who practice self-discipline form the foundation of an optimal work system. On a societal level, self-discipline is a state of being that many people attempt to attain in an effort to reach tangible goals such as weight loss, smoking cessation, improved relationships and other positive outcomes that require effort and perseverance. In this regard, the concept of discipline is considered very forward thinking and is considered very positive. This creates an interesting paradox when the idea of discipline is considered within the workplace and is compared to this inherently noble attribute of self-discipline.

If discipline is applied in a work setting, it must include both discussion and problem solving components. This contrasts significantly with the ways that discipline is usually thought of, being almost synonymous with punishment. This attribute of discipline is critical for the employee to experience a productive interaction and to lead to an understanding of how their actions or behaviors must change. Keeping the ultimate goal in mind, of optimizing performance and engaging employees, it becomes self-evident that the importance of problem solving and providing a productive path forward is the most important part of workplace discipline. If this component is missing, it becomes more punitive and is more closely akin to punishment than to discipline.

According to Scott Stricoff Chief Operating Officer of BST, there are a number of important considerations that must be followed to ensure that discipline is applied consistently. These are, 1) that mixed messages must be avoided, 2) that the attribution of fault must be done objectively, 3) that realistic expectations of peoples performance going forward must be clear and 4) that there must be a narrow scope of focus when discipline is provided.[4] Following these principles will go a long way to establishing constructive discipline where an employee understands the causes that led to the admonishment and has a clear course of action to follow to meet expectations in the future.

So when should discipline be used? Discipline should be used whenever an employee has engaged in less than optimal behavior and their future behavior should be changed or improved. Discipline is appropriate any time a learning experience is needed for an employee, the violation or transgression did not involve the intentional action of putting oneself or others in harm's way, and when the employee requires an understanding of the seriousness of the situation.

Blame

Blame is a further extreme form of punishment. Blame is always implied when a person is experiencing punishment. However, blame can be used in situations where punishment and discipline are not even considered. Employees may find themselves in situations where they experience injuries, but where a formal application of discipline and punishment is absent. Even in these situations, employees may be blamed for being injured. Although there may be no formalized censure to the employee, the presence of blame is often all too common and very punishing and demotivating.

This informal application of blame can be rampant, insidious and highly demotivating. Informal blame can take many forms and may be applied by an individual or collectively. Individual application of blame occurs when a supervisor, manager, or even peer believes and articulates that the employee is likely to be at fault for some transgression. The situation can be further complicated if the person assigning the blame determines the causality for the blame. For example, if an employee is driving a company car and parks it in a legal parking space, only to return later and find the vehicle has been damaged in a hit and run accident. If the story is questioned or viewed with suspicion, the employee is likely being blamed for the event.

Most workplaces have numerous examples of people being blamed and even punished for actions that are not their doing. Collective blame is more subtle and often occurs as a result of a collective belief that people generally get what they deserve, or "*reap what they sow*" to coin a common term. If a younger employee has a manual material handling injury, the collective wisdom of blame may be that younger employees always view themselves as indestructible. As such, injuries to younger employees are to be expected because everyone knows that young employees work quickly, never ask for help, and put themselves at risk that other, more learned employees, would not. This means that a younger employee who experiences a manual material handling injury is almost always to blame because of the nature of the way they work.

Blame is rarely an appropriate response to an employee injury. No one intends to experience an injury and few people are cavalier enough to engage in actions that they realistically think are likely to lead to an accident. The conflict of assigning blame when it is not warranted is a long-standing challenge in the workplace, and in all social situations. The source of this conflict comes from two inherent psychological biases called the Fundamental Attribution Error and the Just World Phenomena.

In the Fundamental Attribution Error people tend to attribute beliefs and values to a person in light of their actions without an understanding of the true causes of the behaviors. For example, if an employee has a traffic accident, it is easy for management to attribute the accident to poor driving habits, a general lack of due care, or some other personal failing on the part of the person who experienced the accident. However, in many situations, the attribution of some failing on the part of the employee who had the accident may be misplaced. Consider an employee who sideswipes a car in traffic and causes an accident. If the employee did so to avoid hitting a small child who had just run into the street, then their actions would be considered completely laudable and beyond reproach.

The Just-World Phenomena is a bias that allows people to think that all good actions are ultimately rewarded and all bad actions eventually receive punishment. When this particular bias is in effect it allows people to blame those who have been injured. On some level, the injured person MUST have done something to cause the injury. This particular bias can be expressed in the concept of Karma or in the particular expression *"what goes around, comes around"*. Therefore an injured employee is clearly experiencing the results of some at risk act, poor belief system, or some other attribute for which they can be reasonably blamed for the incident.

Blame is never a helpful attribute of a safety process. The assignment of blame causes organizations to stop evaluating the true causes of risk and injuries. When a culture of blame is present, employees will not engage in the workplace, scapegoating becomes common, and any positive forward momentum will come to a stop. Safety professionals do well to work to remove the assignment of blame from any meaningful approaches to improving safety. As previously discussed, there will be times when discipline and even punishment are appropriate, but the same cannot be said for blame.

Accountability

Accountability is closely aligned with responsibility. When a person is accountable for something, they are responsible for ensuring that it occurs. When we claim to hold someone accountable we are essentially saying that this person is responsible for accomplishing something. Accountability is not a negative item, nor is being held accountable a form of discipline.

However, being accountable for something does not mean that it is necessarily something you are going to do. It simply means something you have to make sure gets done. According to Scott Eblin, Executive Coach, "If you're accountable, you answer for it. If you're responsible, you do it".[5] Therefore, it is possible to be accountable for the actions of others if it was up to you to make sure that something gets done.

Optimally, accountability should be a proactive means of managing performance. Assigning persons an amount of responsibility and then asking them to account for their actions can be done positively. In the most forward-thinking entities accountability is used frequently to recognize the achievement of accomplishments rather than simply being relegated to the use of pointing out failings. An example of positive accountability could involve an employee who takes personal responsibility for organizing, cataloging, and arranging a disposal plan for hazardous waste that has accumulated at a facility. Although this action might be outside the employee's normal responsibility, by taking ownership of the situation and creating a corrective action, the employee has created personal accountability. If the organization supports this effort, acknowledges the employee for making the workplace safer and more environmentally responsible, and reinforces the importance of contributions to the safety effort, then a positive shared accountability will be in existence.

Indeed, in the realm of safety, organizations should be striving for a shared accountability of performance. Expectations and responsibilities for employees, supervisors, managers, and senior leaders necessarily must align to the degree that all levels of the organization are responsible for controlling risk and reducing the chances of injury. Shared accountability implies that if a failure in the safety system occurs, that the failure is not the fault of any one individual, instead it represents an opportunity that the entire organization can embrace and work to improve. Organizations that practice true shared accountability rarely have to use discipline or punishment if all members of the organization fully accept their individual levels of responsibility. When members do not, then the application of discipline, or possibly punishment, *may* be warranted. That said, exploring the viability of other responses such as coaching, providing feedback, or disciplining should be explored first.

With all of this said about accountability, it is important to discuss the most common usage of the term in business. When someone is held accountable, they are disciplined for failing to achieve some objective or failing to avoid some negative outcome. Although this is the common usage of the term, it might be more correct to describe the interaction as "calling someone to account" rather than holding them accountable. Under both circumstances, the approach is very backward-looking and does not include the forward component found in discipline. Allowing persons to give an account of actions and outcomes proves much more fruitful than holding people accountable in the traditional sense.

When persons are given the opportunity to give an account of actions or inactions, it opens the door to problem solving and can quickly transition from a disciplinary event to a coaching or feedback event (see the following two sections). This becomes a much more proactive approach and can be developmentally very important for both the individual and the organization.

Optimally, a shared accountability for safety is developed. This shared accountability must be forward looking and ensures that when people are called to account, they have the ability to problem solve and move ahead in a constructive manner. A culture of accountability should be just that, everyone has responsibility for safety and all members of the organization are empowered to maintain safety as a core value. Doing so establishes an environment of facilitated safety success based on accomplishment and attribution instead of punitive model centered on failure avoidance and driven by blame or punishment.

Feedback

Feedback is the means of providing information to a person about his or her performance. Feedback is a form of consequence that is ubiquitous to social situations and is an outcome to behavior that people can find highly motivating. Feedback is typically verbal, but can also be visual. When done correctly, feedback will be directed and specific about the performance. Feedback can be both positively reinforcing for safe behaviors and actions, or it can be negatively reinforcing. When feedback is given positively, in order to reinforce a behavior and make sure it continues, it is termed appreciative feedback. When feedback is given to change behavior, it is called constructive feedback. Coaching is a form of constructive feedback that has been shown to yield the strongest results at changing behavior. Coaching will be discussed in the next section of this article.

Appreciative feedback is known to be one of the most powerful motivators of all. When desired behaviors are manifested by employees, the consequences to those actions will determine if they will be repeated or not. If the desired actions result in some type of positive outcome, such as appreciative feedback, then those behaviors will be likely to recur. The important attributes that make feedback such a powerful form of reinforcement are:

1. It can be given to anyone. Research supports that peer-to-peer feedback can be as powerful as supervisor-to-employee feedback. This is often because the feedback takes place in the form of a social reinforcer such as culture, along with the personal reinforcer of the feedback itself.
2. Feedback can be given specifically to reference the desired behavior. There will be no ambiguity about the association of feedback to the behavior as it is directly linked.
3. Feedback represents a very meaningful consequence. Feedback can be given immediately after a behavior and it can be certain to occur. Two of the most important characteristics of consequences that makes them meaningful are when they are presented immediately after the behavior and when they are certain to occur. Immediate and certain consequences are known to be the most important of all outcomes.
4. Feedback is a form of communication that can not only reinforce behavior, it can also strengthen relationships and improve the overall morale of the workplace.
5. Feedback allows standards of behavior to be established and continued. This allows feedback to link closely to the desired culture and the presentation of regular positive feedback becomes a self-sustaining normative practice within the culture.

Appreciative feedback is one of the best ways to sustain behavior and is even more effective at getting employees to do more than expected. In his revolutionary book *"Bringing Out the Best in People"* Aubrey Daniels termed the willingness of employees to do more than expected as **discretionary effort.** From a safety perspective, it is important to understand that when employees engage in safe work practices, they are by definition not using discretionary effort, as working safely is generally considered part of the daily routine of most workplaces. However, most Safety Professionals desire to have employees who take workplace safety personally, who take a high degree of ownership for safety, and who are interested in the safety and well-being of coworkers. In order to achieve a Safety Culture where employees are highly engaged and taking this level of ownership, organizations must use appreciative feedback when employees work safely and meet expectations. This is the best and most effective way to maximize discretionary effort and create a positive Safety Culture at the same time.

Coaching

Coaching is a form of feedback that is used when it is necessary to change or influence an undesired or an at risk behavior. Coaching is a form of negative feedback. However, the techniques used to present coaching require that it be done in a manner that is respectful and understanding, while obtaining a commitment to work safer or in a more appropriate manner in the future. Coaching can be used as an alternative to discipline if at risk behaviors result from employees making mistakes, engaging in poor behaviors that they have been positively reinforced for a long time and have now experienced an injury or other negative event, or when a legitimate knowledge gap exists and employees did not know the required work practices. The need for a coaching session arises when an at-risk behavior is observed.

When an at-risk action is noted, the use of coaching can be a powerful tool for beginning the process of positive behavior change. Coaching allows the employee to have a discussion about their current performance and the expected performance. In a coaching session, the person who observes the at risk behavior calls attention to the behavior and has a focused and specific discussion about how the at risk behavior is not acceptable and provides direction about what the appropriate behavior is. The coaching session should then conclude with the employee who was coached committing to be safe and agreeing to work in a more acceptable manner in the future. The biggest challenge with coaching is to make sure it is provided in a way that will facilitate positive behavioral change.

When a decision is made to coach an individual for an at risk behavior, there is one primary objective, beginning the process of behavior change. Unfortunately, many people view coaching not as part of the process, but rather as the singular tool needed to change behavior. Simply telling someone what to do or what not to do is not going to change behavior any more than telling a person when they have not performed appropriately. The best way to facilitate positive behavior change is to understand the process involved and to follow-it. Coaching is often the first step in a longer undertaking that includes providing positive feedback when the safe behavior is observed. Specifically, when an employee begins to change their behavior as a result of coaching, and then receives immediate and certain positive feedback for working safely, then the process of behavior change is well under way.

There are a number of important considerations to evaluate when coaching others that will help make a coaching session be accepted by the recipient. These items include:

1. Ensuring the coaching is provided in a respectful manner. Other people are much more open to hearing a message, even a potentially negative one, if the message is provided in a respectful manner. Showing respect means talking with people in an affirming tone, not making personal verbal assaults, and talking about the act or the behavior - not the individual. Finally, it is very respectful to coach a person in private whenever possible. Following these steps will help ensure the message is received and is valued rather than having the message be rejected.
2. Ask don't tell. No adult likes to be told what to do or what they may have done wrong. I have found that people are much more open to hearing a message and potentially receiving criticism when they are asked about a situation rather than told what they are doing is wrong. For example, if you observe an employee lifting unsafely, rather than approaching them and beginning to tell them what is wrong with their technique and how they should be lifting, begin by asking them if they need/would like some help. Then mention that what you observed seemed at risk and ask them if they thought so too, or if they felt like they were being safe. Having this kind of a dialog shows that you are concerned about their well-being and are much more interested in hearing their thoughts rather than just preaching to them about the right and wrong way to lift. People are much more open to a conversation if you ask them their thoughts and perspectives first rather than just diving into a message of what they did wrong.
3. Allow for personality characteristics when coaching. When coaching others, it is not unusual for some people to become defensive or withdrawn. By allowing for different personality characteristics it is easier to facilitate discussions and to help people become more receptive. With that, it is also possible to give feedback to people when they are not having a good day. Imagine how you would respond if a person wanted to talk with you about safety when you

just found out you have a serious illness. Allowing people to react to coaching, while still remaining resolute to deliver the message of the importance of working safely is a fine line. Successful coaches are able to do both by ensuring they allow people to respond to the coaching in a way that is natural to them, but still allows the dialog of the coaching session to continue.

4. Make certain the coaching session involves problem solving. A big part of being successful when you coach someone is to work with the person being coached to find a viable solution to ensure he or she works safely in the future. By working with their individual personality characteristics and asking them for solutions to working more safely, you will engage the person in the safety process. They will be likely to take a higher degree of ownership for the outcome and will take the solution much more personally. Problem solving can be as simple as asking an employee what do they need in order to work safely. If you are able to work with the employee to devise a strategy to remove barriers to safety, they will be much more open to following it and will likely encourage others to follow the solution as well.
5. Avoid the personal filters and biases on the part of the person providing the coaching. When coaching an employee it is critical to view behavior from their point of view. It is easy to sometimes believe that we can see the world through the eyes of others, or to assume that the way we would view something or act is the way everyone should also act. These kind of perceptions can filter our perspectives and can lead us to treating people the way we *think* they should be treated and not the way they need to be treated. For example, if a person has had challenges learning certain tasks and then is observed working at risk, it might be tempting to assume he or she has not learned the safe work procedures. However, it is very possible that they may know how to work safely, but are deciding to do so due to the probable consequences that they may save time and effort. Ensuring that personal biases remain in check and do not interfere in a coaching session is a crucial consideration.
6. Conduct the coaching in a comfortable and safe location. By selecting a location that puts the employee at ease during the coaching session, the employee will be much more likely to be open to hearing the message. People are often uncomfortable being involved in conversations where the discussions include a review of things they have done that are unacceptable, an overview of the right way to do things, and a request for an acknowledgement for them to work safely in the future. In an attempt to put them at ease, it is very important that the environment where the coaching takes place is neutral or one where the employee is comfortable. Considerations to putting the employee at ease include conducting coaching in private, providing it in the employee's work area, or giving it in a location such as a break room or training room. Whenever possible, avoid coaching an employee in your office, the supervisor's office, or some other area where they may feel insecure.
7. Obtain a commitment for future safe work practices. Whenever a coaching session concludes, it must be done so with a request for the person to work safely in the future. This will be a simple undertaking if the person coached was treated respectfully and solicited about how to work safely in the future. Getting the agreement from the person to work safely can be as simple as asking him or her to follow safe practices in the future. It can also be as complicated as verifying with the person that they know all the steps to Lockout/Tagout and obtaining a verbal agreement that the employee will follow the specific procedures as expected in the future.

A Call to Action...How to Motivate Employees to be Safe

What is needed is a redefinition of the fundamental paradigm of individual and collective motivation. Based on the experience of a diverse group of industries, a new framework where shared accountability in the absence of punitive measures for mistakes has been put in place. One excellent example is found in the airline industry. In the United States, in 2012 the Federal Aviation Administration (FAA) adopted a no-fault self-reporting process called the Air Traffic Safety Action Program (ATSAP)[8]. This program encouraged the reporting of incidents, near misses, and other events that could lead to airplane crashes. The program promises that if actions are reported within a certain time frame, that no punitive actions will be taken. This program has helped to increase the reporting of at risk events by over 50%. Air traffic controller participation in safety has increased considerably and has helped create a Safety Culture that is more positive and that facilitates reporting and investigation of at risk events.

The data clearly shows that when employees understand responsibilities and believe in a shared accountability, they are much more willing to be personally involved in safety efforts and to take safety very seriously. When punishment, discipline, and blame become normative values, they inhibit organizations from achieving greatness. As mentioned earlier, employees who are working vigorously to avoid failure are not working to accomplish anything. No person or company has ever been successful simply because it is excellent at avoiding failure.

Further, the misapplication of either discipline or blame can also become punishing for employees. When employees feel punished the changes in behavior may not be those that were expected. In the book *Removing Obstacles to Safety,* Agnew and Snyder state the following 5 short-comings when using punishment inappropriately. 1) Punishment has short-term effects and will not result in long-lasting or sustained behavior change. 2) Punishment may cause an at-risk behavior to stop, but it will not cause the at risk behavior to be automatically replaced by the safe behavior. Discipline is more likely to achieve this outcome with its forward looking component of education or coaching. 3) In order to punish effectively you must catch the perpetrator in the act. This can be difficult and a delay in the punishment will reduce its effectiveness. 4) Punishment will not work with lone workers and is impossible to administer. 5) The overuse of punishment can cause it to become an organizational value that then becomes the default response for all failures or mistakes. Although there remain viable reasons to administer discipline, and in some cases punishment, blame must be removed entirely from the safety equation.

So, if using blame creates the wrong outcomes and is representative of a poor approach to performance management, what is the best model to follow? Ideally, a measured and graded approach is generally the most viable. The following table outlines the various forms of responding to at risk employee behavior, the appropriate time to use them, and their likely outcomes.

Response	Time for Use	Anticipated Outcome
Punishment	When an employee violates a known rule with the intention of putting themselves or others in harm's way. Used when the exposure is serious and there is intentionality to not follow proscribed directives. An example is a company having a zero-tolerance policy for at risk working at heights. If employee understands the standard and chooses to work without	Removal of the Employee from the Organization

	the appropriate fall-arrest, then punishment is warranted.	
Discipline	Used when an employee violates a rule, but has done so due to a variety of reasons that may have nothing to do with having the intention of being at risk. An example may be, an employee who does not complete a pre-trip inspection because they have pressure to deliver an overdue order. An additional example is an employee fails to wear a seatbelt and is injured in a traffic crash. If the company had a written seat belt policy that was rarely or unevenly enforced with no feedback given, the employee may have not worn the seat belt for years and have never received any coaching.	Employees will find the experience somewhat punitive, but will have the ability to learn from the action and will have a plan to help them avoid the at risk behavior in the future.
Blame	Blame is never helpful in any setting. Even if an at-risk employee is removed, the application of blame ensures that no systemic review will take place. If the decision is that an employee is solely responsible for their actions that led to an injury, then the company must still ask, how did that employee get hired and how were they allowed to get themselves to that point. Blame is never a useful response to injuries.	No systemic review of the incident and a complete removal of the offending party.
Accountability	Because accountability is so closely tied with responsibility, its use as a form of performance management is somewhat limited. It is best applied when the discussion centers on giving an account of what has transpired. If there is true shared accountability for safety, then all discussions about possible failures are done in a blame-free environment. Indeed, the proper use of accountability is to make certain that those who are responsible for accomplishing a task safely are given the tools, support, and motivation to do so. If a failure occurs then having all the responsible parties provide an account of what happened is the best response. When a person is called to account, they are being asked to provide a self-report of a failure or shortcoming. There can be similar connotations to accountability as there are in discipline. However, when an organization is consistent and practices shared accountability these discussions will be held proactively rather than reactively.	When accountability is managed appropriately, the employees understand the area of responsibility for safety. If it is viewed as a collaboration between all members of the organization and one member fails in their area of responsibility, the accountability may feel like discipline but will have countermeasures that look at the problem systemically and not on an individual level.
Feedback	Feedback should be used at all times, with all employees, and under all circumstances. Feedback is a powerful social reinforcer that can be used to	Employees tend to be more successful and organizations tend to flourish in feedback rich

	encourage safe behaviors and then to coach at risk behaviors. Feedback must be given in a specific and directed manner. In the absence of feedback, people tend to give themselves their own feedback. It is not unusual for people to receive tacit approval for at risk behaviors. That is, for minor at risk behaviors to be ignored and not subjected to coaching. It is also common for people to receive positive reinforcement for working unsafely This can occur if an employee is praised for being productive, but they are productive by skipping safety rules. In that case, the praise for productivity comes at the expense of safety.	environments. When feedback is routine and regular, communication standards are higher, employees are more engaged in the workplace, and leadership is able to provide a vision that is seamlessly translated into actions.
Coaching	Coaching is feedback that is provided when at risk behavior is observed. Examples of situations when coaching is appropriate may include an employee who fails to maintain three points of contact on a ladder because no one else in the plant does, and falls from the bottom step. At risk behaviors occur regularly in most workplaces. In a model of safety that fosters continuous improvement, coaching all at risk behaviors is crucially important.	Coaching is often done on a peer-to-peer basis. This means there should never be any type of formal discipline associated with coaching. Coaching is a chance for an employee to dialogue about at risk behaviors and for problem solving to occur to ensure the employee can commit to working safely in the future.

The control of errors and the management of risk are critically important concepts for organizations. However, it has been repeatedly shown that the most effective way to control errors is to learn from them, not punish them. Further, when an error receives punishment the disengagement that employees will exhibit can drive such important traits as creativity, entrepreneurialism, and collaboration to extinction. As stated in a recent psychology publication, the use of discipline and punishment may result in a teachable moment is completely lost, and any attempt at making the organization better, is lost with it.[9] Without the ability to make and learn from errors, even in the safety perspective, the road to improvement and the achievement of Safety Excellence will never be possible.

Conclusion

This article has endeavored to provide an overview of the different responses that are possible when an employee works in an unsafe manner. Although the view in totality may seem complicated, the reality is actually very simple. Ensuring that employees who work safely receive positive reinforcement must be an organization's first priority. When employees work unsafely, it will be critical to carefully examine the situation and determine the appropriate response. In most cases, an examination of the systems of the workplace and a review of the past history of Safety Culture can provide insight into selecting the best response.

Generally, looking for opportunities to provide feedback, coach, develop a shared accountability for safety, and disciplining with a problem solving approach provide the most effective responses. Under all circumstances, blame is not an appropriate response. Punishment should be reserved for only those events where there is a blatant disregard for safety and wellbeing with the intention of putting the employee or others in harm's way. History and experience have demonstrated that employees almost universally want to avoid getting hurt. However, people will put themselves at risk when they believe the chance of injury is low and when the outcome for being at risk is savings in time and effort. Psychology tells us that the best way to change behaviors is to change consequences and give the desired behaviors positive outcomes all the time. The challenge of responding to at risk behavior becomes much simpler when these two points are understood and practiced.

References

Agnew, J. and G. Snyder. 2008. *Removing Obstacles to Safety.* Atlanta, GA: Performance Management Publications.

Daniels, A. 1999. *Bringing Out the Best in People.* Atlanta, GA: Performance Management Publications

Dekker, S. 2006. *The Field Guide to Understanding Human Error.* Burlington, VT: Ashgate.

Eblin, S. 2011. *Are you Accountable or Responsible?* The Government Excellence website, Retrieved July 16, 2014 http://www.govexec.com/excellence/executive-coach/2011/09/are-you-accountable-or-responsible/40709/

Federal Aviation Administration "Air Traffic Safety Action Program," 2012, Retrieved July 18, 2014 https://www.atsapsafety.com/atsap-home/

Johnson, Amy. 2008. *Practicing Good Ethics Gives Competitive Advantage. Denver Business Journal.* Denver, CO.

Margolis, Walsh & Kehmeyer. Business Roundtable… Institute for Corporate Ethics, Building the Business Case for Ethics, 2006. Retrieved July 15, 2014 from http://www.corporate-ethics.org/pdf/business_case.pdf

Skiba, R. "Zero Tolerance and Alternative Discipline Strategies," 2009, *National Association of School Psychologists Magazine.*

Stricoff, S. "Addressing the Safety Discipline Dilemma." webinar, 2009 Accessed July 17, 2014 http://vimeo.com/69326884

Session No. 752

Safety and Construction Cost Estimating: How to Increase your ROI to Your Company

Laura Casey CSP, CHST, EMT-P MSc
Principal
SafeCon Solutions, LLC
Beacon Falls, CT

Rosemarie Jesse
Manager Construction Cost Estimating
Hazen & Sawyer
New York, New York

Introduction

Return on investment or "ROI" is a term used most often in the world of finance regarding investments, although, it can be used by businesses and investors alike. The term can be applied to other things in addition to investments such as marketing ads, assets and even human capitol.

The Business Dictionary defines ROI as: "The earning power of assets measured as the ratio of the net income (profit less depreciation) to the average capital employed (or equity capital) in a company or project. Expressed usually as a percentage, return on investment is a measure of profitability that indicates whether or not a company is using its resources in an efficient manner.[1]

As a professional, how would you determine what you are worth to the company that you work for? Do you gauge your value by your salary? Do you gauge it by the fringe benefits the company provides you? Or do you use your title or a statement of "Great job!" to get a sense of your value to the company?

Return on Investment (ROI)

The return on an investment can be calculated and is typically expressed as a percentage and sometimes it is presented as a ratio. When companies consider the ROI of occupational safety, it most often is discussed in terms of cost of injuries, both realized and unrealized. It represents the monetary savings that a company reaps when incident and accident frequency and severity is decreased in the work place. Safety and Health Magazine reported in May of 2014 that the cost of

Occupational injuries and deaths for 2012 as a cost of $198.2 billion dollars or an injury impact per worker of $1,400 or an average combined cost per claim of $36,592.[2] Are these numbers correct or accurate? It is hard to say. As safety professionals we understand that statistics are only as good as the reporting and data collection, but we have embraced the concept that increased incident and accident frequency and severity costs money. The question that we are all trying to answer is how do we decrease the frequency and severity of incidents and accidents.

There are many schools of thought on the subject and most of the ideas have some element of prevention involved. We are challenging everyone in construction: owners, engineers, designers, architects, trades people, management teams and safety professionals to add another dimension to prevention: Start at the beginning".

Let's revisit the Business Dictionary's definition of ROI: "...return on investment is a measure of profitability that indicates whether or not a company is using its resources in an efficient manner". You, in your professional capacity are a resource. Are you being used in an efficient manner? We wanted to examine some of the non-traditional ways that you could get involved and integrate safety to decrease the frequency and severity of incidents and accidents, ultimately saving money. First, we should list them:

- Safety by design
- 30, 60 & 90% design reviews
- 30, 60 & 90% safety constructability reviews
- Safety consultation during the cost estimating process
- Bidders questions

At first glance, there does not seem to be anything earth shattering contained in this list. This list does, however contain some strategies not employed by most companies.

Lets take a brief look at each strategy on the list and see how each one can make a difference and connect cost estimating as an illustration.

Safety by Design

Safety by Design is not a new topic to the safety world but it still has not gained enough traction to become commonplace. It is often referred to as a "movement". OSHA has termed it "Design for Safety" and NIOSH has a national Prevention through Design initiative. The concept is to consider construction methodologies and workers as well as the end users (sometimes referred to as O&M or Operations and Maintenance) and make design or engineering changes to reduce or eliminate hazards. OSHA offers this quote:

"For example, an organization that spends money up front to design out a workplace safety hazard will often save money through reduced training costs, less need for personal protective equipment, and the savings associated with fewer workplace injuries and illnesses."[3]

For this concept to work, you need a design firm or architect that has a safety professional or consultant and is willing to allocate the cost for the work done by the safety professional. Another approach to the same idea is by having safety reviews at 30, 60 and 90% design completion stages.

30, 60, 90% Design Review

If you represent safety for the owner, the client or the engineer, discussing the potential costs savings with your management team can be the first step to engaging the positive step towards prevention. Designers and architects do not have any responsibility to consider OSHA or NFPA implications or a worker's safety and health when creating designs. The Life Safety Code is the code that is used during the design phase and is often mistakenly assumed to satisfy all safety requirements.

30, 60, 90% Safety Constructability Review

30, 60 & 90% safety constructability reviews take place during the construction phase. This initiative is particularly helpful if safety reviews during the design phase were not performed. In construction, safety is often viewed as a demon because work can be stopped for safety reasons. Work stoppage costs money and can effect the schedule. Schedule over runs can result in missing bonuses and even liquidated damages.

Time is money, so let's discuss the bonuses or liquidated damages and how they come into play in a construction contract. In some high profile projects, an owner can offer an incentive bonus to have components of the project, or even the entire project, completed by a specific date. Consider a high rise apartment building. The owner has a strong incentive to get it complete and units rented as early as possible. When initial schedules find that it will take four years to complete the job, a contract can state that for every floor or unit made available with full access, as much as $100,000 can be granted when the contractor completes the work early because ultimately, the owner benefits from having those units ready for lease signing.

Liquidated damages are more popular on public works projects where the public is impacted by non-completion of the project. A high profile job, or one that impedes public access to roadways or water, can be assessed as much as $10,000 per day.

Another popular form of liquidated damages is found in fines. These are assessed when a contractor has a limited time to perform the work, but they do not complete it in time. Very often, work within the public right of way is limited to active construction outside of rush hour. In high traffic corridors, fines as much as $1,000 can be charged by the hour or by the minute, depending on how sensitive the corridor is.

Constructability reviews have a different meaning to most engineers. Engineers review the plans and specifications to improve their effectiveness and usually perform the review during the pre-construction phase. A safety review when construction is 30, 60 and 90% complete can identify potential regulatory compliance issues, Federal, State and Local, that if not corrected or otherwise addressed could negatively impact cost, schedule and ultimate occupancy of your plant, building or facility.

When we look at a roof system on a building, are we taking into consideration the fact that in the future, maintenance engineers will need to access that roof. One way to incorporate safety into our design would be to install permanent anchors into that roof. In this example, we can clearly see that providing these anchors in the design of the roof is much less expensive than having to install it either under a change order or in the future.

Installed during construction:				
Install anchor points	12	ea	$ 48.26	**$ 579.11**
Installed after construction:				
Mobilization	1	ea	$ 368.50	$ 368.50
Install anchor points	12	ea	$ 286.41	$ 3,436.93
			$ 654.91	**$ 3,805.44**

Table 1.

Safety Consultation During Cost Estimating & Bidders Questions

The last two strategies go hand in hand: safety consultation during cost estimating and bidder's questions. These areas are where the safety professional can make a significant difference financially. You do not need to be part of the contractor's team for this to apply to you. Perhaps you work for the owner of the construction project, perhaps you work the engineering firm that was engaged by the owner or perhaps you do, in fact, work for the contractor, whichever category you belong to, you can increase your professional ROI by understanding and taking part in the cost estimating process.

From the owner's or engineering firm's perspective you should evaluate how clearly the safety requirements are spelled out in your technical specifications and contractual documents. Medium and smaller organizations like to include small notations that safety of the job, the work and the employees is the sole responsibility of the contractor. Medium to large organizations may have more prescriptive specifications that include the do's, the don'ts, and the musts of the safety requirements but after more than a combined 30 years in construction, these two professionals can assure you that it is a rare instance when safety is included as it's own lists of costs payable to the contractor.

As a bidding contractor, the cost of safety is typically carried the same way small tools are accounted for. A widely accepted rule of thumb is to carry 2% of the labor cost to account for shovels, hammers, wrenches and the like. Some larger companies may add another percentage on for safety integration but rarely do they look at the actual nuts and bolts of what they will need to perform the work. This can be a fatal flaw that often results in the idea that safety is expensive, or, as is most often the case, this is a place where the contractor may try to skimp. Or, if you are working for the engineer or the client, you hear "Change Order" or "Unforeseen Condition". This brings us back around to an earlier point. If you have clear, concise and detailed technical specifications for Safety and Health, the contractor will carry a more realistic number in their original bids and the owner is more likely to get what they expect.

Some special circumstances may be represented in your project and it is understood that there are additional costs associated with these circumstances such as presence of asbestos containing material, known hazardous substance, presence of poly-chlorinated bi-phenyls. How can you try to anticipate or identify the potentials for added safety costs during construction of a project if these circumstances are not directly called out in the project information?

A mock project example is the best way for us to illustrate the financial impact that safety can have during cost estimating. Our project will be ABC WWTP – WWTP = Wastewater Treatment

Plant. This is an existing facility in Podunkville, USA. The population has jumped dramatically in recent years and the municipality has determined they need to expand from XXX capacity to XXXX capacity. The technical specifications state that the contractors must comply with all Federal, State and Local regulations, laws and standards and safety is the responsibility of the contractor. You are the owner of the facility. You have a set of designs. You ask your engineer, what will this cost me? Lets start with the two main types of estimates.

Engineers who design public facilities are catching on to how important it is to have a good budget for the upgrades or new facilities required in a municipality. In fact, recently there is an engineer's estimate published with a notice for public letting. These estimates are developed differently depending on the phase of construction, or even by different engineers. Many firms look at past projects to come up with a budget. It is a quick way to have an idea of cost. For instance, if a pump station built in town A cost $30,000, then what are the similarities or differences if I build a similar pump station in town B and can I use a factor of comparison to come up with a cost.

In other instances, we may develop costs in a more detailed manner, understanding the earthwork, plumbing and electrical work associated with the specific facility we are pricing. This is very similar to what contractor's do when they are bidding on a project.

Now that you have a number, what this project should cost. The project goes out for competitive bid and the time comes to open the sealed bid submissions and you are shocked when you see a significant range of numbers. Below is a great example of a water main project in NYC. As you can see, the engineer had a good estimate, but they still range from $37M to $47M, and $10M is a lot of money.

Contract PIN No:	8502017WM0008C
Project Number:	BED798
Type of Work:	Construction/Regular/Water
Project Description:	REPLACEMENT OF TRUNK AND DISTRIBUTION WATER MAINS,SEWERS AND APPURTENANCES IN FLATBUSH AVE BET STERLING PLACE & 8TH AVENUE, ETC-BOROUGH OF BROOKLYN
Bid Date:	03/07/2017
DDC Engineer's Estimate:	$38,603,889.18

#	VENDOR NAME	BID AMOUNT
1	TRIUMPH CONSTRUCTION CORP.	$37,548,087.57
2	PERFETTO CONTRACTING CO. INC.	$41,923,967.72
3	JLJ IV ENTERPRISES INC.	$42,906,490.53
4	HALCYON CONSTRUCTION CORP.	$44,730,000.00
5	TULLY CONSTRUCTION CO. INC.	$45,493,938.28
6	E.E. CRUZ & COMPANY, INC.	$45,675,131.28
7	LAWS CONSTRUCTION CORP.	$46,659,000.00
8	JR CRUZ CORP	$47,237,287.79
9	RESTANI CONSTRUCTION CORP.	$47,784,500.00

Table 2.

You work for ACME Construction and your number is higher than all the other bidders. We need to understand why and figure out how to level the playing field. Let's take a look at another bid.

Project Number:	Description:	Engineer's Estimate:	Bid Date:
542B	Cleaning and Inspection, Queens and Brooklyn Tunnels	$ 32,000,000.00	3/2/2017
Bidders:			
1.	FRONTIER-KEMPER CONSTRUCTORS, INC	$ 22,148,600.00	
2.	TULLY CONSTRUCTION CO., INC	$ 30,919,500.00	
3.	JOHN P. PICONE INC	$ 58,802,620.00	

Table 3.

In both cases, how much do we think safety played a role in the bid? In both cases, there are significant hazards. Water main installations can require deep excavations, interference with other utilities in the roadway, or even something as simple as traffic control. The second project is the cleaning of a tunnel, where safety implications are more robust. Confined space, lighting, power in an area that is most likely wet. As a bidder, I may want to ask questions publicly to assure that every other bidder is aware of the circumstances in which the work will be done. Questions such as "how many access points are there currently available to the contractor?" are a great way to put every other bidder on a level playing field and will force the contractors to be aware of the means and methods required to perform the work. If additional safety considerations need to be made: training, equipment, 4 gas meters, fans, portable power to run it all; these items add up quickly. By helping your estimators look for things that could cost money, phrasing a question or two around it and having them submit, all contractors now have to address these costs in their bid.

Of course, during the design, these items can be incorporated into the project for the benefit and safety of all workers. Perhaps on the tunnel project, the engineer can specify where access points can be placed. We can also design to avoid a confined space on the water main projects by making water main vaults with two access points, easy entry points, and the list of options goes on and on. This not only makes it safer for the installers of the vault pipes and valves, but maintenance and operation of the facility eliminates dangerous hazards in the future.

Summary

There are unforeseen, positive ways to increase your company's return on investment in you as a safety professional. Each of the areas discussed above adds safety through various levels of a project and can save lives and yes, save money. If you understand how to have the conversation, and can clearly discuss the concepts and strategy's to save your company money, you can further managements objectives and your own simultaneously.

Endnotes

1. http://www.businessdictionary.com/definition/return-on-investment-ROI.html
2. http://www.safetyandhealthmagazine.com/articles/10414-the-roi-of-safety?page=2 May 23, 2014 Kyle W. Morrison
3. https://www.osha.gov/dcsp/products/topics/businesscase/designsafety.html

Rounding Up the Last Cowboys: Using Positive Reinforcement to Get Construction Workers Involved in Your Safety Effort

David M. Sims, Sr., Executive Vice President
The Bill Sims Company, Inc.
Irmo, South Carolina

Introduction

Construction workers have a reputation as independent, outspoken and self-reliant group. A hard group to get engaged in any corporate initiative, including safety.

Can transient construction workers be meaningfully engaged in safety efforts? And does their engagement produce tangible results? The evidence says "Yes!" to both question.

What's different about construction?

The construction industry is often separated into three broad categories or "sectors":

CONSTRUCTION OF BUILDINGS
The Construction of Buildings sector includes both residential building and nonresidential building. Think not just housing developments but apartment buildings and skyscrapers.

HEAVY AND CIVIL ENGINEERING CONSTRUCTION
Heavy and Civil Engineering Construction includes companies that design and manage entire projects, think interstate highways and dams; as well as specialty trade contractors that perform activities like excavation and grading.

SPECIALTY TRADE CONTRACTORS
Specialty Trade Contractors perform specific types of work, think plumbing, painting and electrical work.

While the types of companies and their size varies, the construction industry tends to draw workers who prefer to work with limited direct supervision.

Construction jobs remain some of the highest paying jobs for people with a high school education that there are. Last year the average union construction worker earned $55,000 and–in some industries–it is not unusual to see construction workers earning in the six figure range.

How are construction workers like cowboys?

Workers who prefer jobs with limited supervision like construction workers or long-haul truck drivers are often referred to as "cowboys" because of their independent nature. And the term is appropriate in more than one way when applied to our friends in construction.

Modern day construction workers work in conditions that are not dissimilar to the American cowboys of the 1800's.

Cowboys lived a hard life, revolving around two annual roundups, one in the spring and the other in the fall. Then a cattle drive to market.

Cattle drives could take as much as two months. A typical drive consisted of ten to fifteen cowboys herding between 1,500 and 2,000 head of cattle. Cowboys were isolated for much of the drive, with only each other to depend on. They had to cross treacherous rivers and travel through dangerous Indian territory to get the herd to the rail head on time and on budget.

Cowboys came from all walks of life. Following the American Civil War the cattle industry expanded and former soldiers from the Union and Confederacy came west seeking work. A significant number of African-American freedmen, Mexicans and American Indians also worked as cowboys, in part because there was less discrimination in the west as in other areas of American society at the time.

American writer Louis L'Amour wrote extensively about this era in American history and said this about the qualities that were admired in a cowboy:

"The term "riding for the brand" was an expression of loyalty to a man's employer or the particular outfit he rode for. It was considered a compliment of the highest order in an almost feudal society. If a man did not like a ranch or the way they conducted their affairs he was free to quit, and many did, but if he stayed on he gave loyalty and expected it."

"A man was rarely judged by his past, only by his actions. Many a man who came west left things behind him he would rather forget, so it was not the custom to ask questions. Much was forgiven if a man had courage and integrity and if he did his job. If a man gave *less than his best, somebody always had to take up the slack, and he was not admired."*

The environment construction workers face on projects is not much different than what cowboys faced in a cattle drive. The work often takes place at the mercy of the elements: heat, cold, rain and snow are common environment challenges. Groups in urban areas work in crime-ridden regions.

Like the cattle drive, construction crews tend to be small groups. Ten to fifteen in a workgroup, or smaller. The nature of the work and the size of the crew often leads to strong emotional bonds. It's common to see workgroups made up of individuals who have worked together for years.

Like cowboys, construction workers are judged by their actions. A worker is expected to "pull his weight" and loyally support the company who is writing his paycheck. This loyalty is especially important in the specialty trades such as plumbing, painting and electrical work. Frequently, specialty trade workers work under the supervision of the client, the "company man". They are expected to keep the company man happy while staying loyal to their employer.

What's Positive Reinforcement?

Positive reinforcement is a term commonly misunderstood and misused. The phrase is often viewed as the same as recognition or reward and is often confused with tangible awards given to a worker for achieving a safety milestone.

In fact, gifts such as plaques, trophies, jackets or other tangible items given upon reaching safety milestones may or may not be viewed as positive by the recipient. It is common for such items to show up at yard sales, much to the dismay of the Safety Director who so hard to source them.

Positive reinforcement is a term that has a much broader application. It is one of the key concepts in behavior analysis, a field within psychology. The father of behavior analysis is considered by many to be B.F. Skinner. Skinner received a PH.D. from

Harvard University in 1931 and was active in research and teaching until just before his death in 1990.

Skinner taught that human action depended on the consequences of previous actions. If the consequences of a person's actions are bad, there is a high chance the action will not be repeated. If the consequences are good, the actions that led to it being repeated become a lot more likely. He called this the principle of reinforcement.

Skinner held that reinforcement could be either *positive* or *negative.* Both types of reinforcement make behavior more likely to occur.

Positive reinforcement has occurred when three conditions are met:

1. A consequence is presented dependent on a behavior.
2. The behavior becomes more likely to occur.
3. The behavior becomes more likely to occur because and only because the consequence is presented dependent on the behavior.

As you can see, the definition of positive reinforcement is much more precise than just receiving an award.

Designing a Positive Reinforcement Delivery System

A construction client in Alaska had reached a plateau in their accident rates. Despite significant efforts in training and the expansion of the safety department their DART and TCIR rates did not budge. Their Experience Modification Rate (EMR), while below 1.0, was higher than their competitors, resulting in a competitive disadvantage when bidding projects.

Incidents were happening on job sites that didn't result in injuries, but that were nevertheless embarrassing. Material and equipment were damaged due to poor work practices. They had a reputation among their clients and their competitors for sloppy even reckless work.

They had an incentive program that rewarded workers for meeting lagging indicators such as no lost time accident and no recordable accident. But they weren't seeing any real effect with it, the accidents were still happening.

To design the system we reviewed three years' worth of observation reports, loss-run reports and OSHA 300 logs. We reviewed the findings with middle managers, veteran crew leaders and safety staff. We focused on what employees could do that would reduce risk.

The team struggled with identifying key behaviors that would reduce risk. Articulating what you DON'T want people to do is easy. Articulating what you WANT them to do can be surprisingly hard. Eventually key behaviors were identified for workers and two levels of supervision.

We designed an awareness program around those key behaviors and developed a tool and a methodology for crew leaders to focus on the behaviors, rewarding workers as the behaviors were observed and measuring the level of positive reinforcement by crew leader.

The client decided they would use the phrase "riding for the brand" from the book by Louie L'Amour as the "theme" for the effort. They emphasized the need for everyone to embrace the recognition and reinforcement of behaviors rather than the rewarding of lagging indicators such as working without an accident.

Results

Very quickly we began getting reports from the field of individual employees stopping jobs and pointing out hazards, two of the key behaviors we focused on at the start of the process.

One of the client's safety reps was onsite when their client's president came in by helicopter to the site. Stepping out of the helicopter the president put his hard hat on and began walking towards the work. One of our workers stopped him because he wasn't wearing a high visibility vest.

Our worker goes to the work trailer, gets a high visibility vest and brings it back to the president. The president toured the site wearing a vest with our client's logo on it.

After a month of focusing on housekeeping and proper hand tool usage we received a report from a senior manager remarking about the change at one location: "I couldn't help but notice the tool storage areas, equipment and materials were well kept and hazards eliminated. Everything from wheel chocks to personal PPE was in place. The employees were charged up and had great attitudes."

It did not take long before lagging indicators like the number of first aid cases, and the number of recordable cases began to drop. By the end of one year OSHA Recordable Injuries drop by more than 50%.

The client's senior health and safety manager congratulated supervisors and foremen at their annual meeting and commented: "I believe the biggest change we've had here is in the language-safety is now spoken not just 'understood'."

Key Factors in "Lassoing a Cowboy"

There are some key factors in getting meaningful engagement in safety efforts from construction workers. First and foremost is commitment from the top executive through the chain of command to support workers who do what is asked. For example: if "Stop the Job" is a critical behavior we ask workers to demonstrate then they should be supported when they exercise the behavior. If they're chastised for slowing down the project you can be sure they will not stop a job in the future.

During the design process hourly workers should be involved. The people closest to the risk are usual the ones who understand the hazards best, and how to mitigate them.

Focus communications to address behaviors and conditions that are within the control of the worker.

And last, persevere. Lasting behavior change that results in lasting results requires dedicated effort, day in and day out.

References

Athabasca University. 2017. "Positive Reinforcement Tutorial". (https://psych.athabascau.ca/open/prtut/definition.php).

Bureau of Labor Statistics (BLS). 2017. "About the Construction sector". (https://www.bls.gov/iag/tgs/iag23.htm).

______. 2017. "Industries by Supersector and NAICS Code". (https://www.bls.gov/iag/tgs/iag_index_naics.htm).

Hopkins, B. L. 1968. "Effects of candy and social reinforcement, instructions, and reinforcement schedule learning on the modification and maintenance of smiling." *Journal of Applied Behavior Analysis*, 1, 121-129.

L'Amour, Louis. 1986. *Riding for the Brand*. New York: Bantam Books

Schacter, Daniel, L., Daniel T. Gilbert, and Daniel M. Wegner. 2011. *Psychology,* Second Edition. New York: Worth Publishers.

Wikipedia contributors. 2017. "Cattle drives in the United States" Wikipedia, The Free Encyclopedia. (https://en.wikipedia.org/w/index.php?title=Cattle_drives_in_the_United_States&oldid=766875927)

Session No. 754

Violence at Work–
Shooting Your Mouth Off…or Other Things?

Melissa Black, MS, CSP, CIH, CEAS, CUSA
MsR3, LLC
Suwanee, GA.

Introduction

Pro-action for addressing risks and reducing them to ALARP (as low as reasonably practical/possible) is what we as Safety Professionals constantly aim to achieve, so why is Workplace Violence so different? We typically gather facts, analyze the data, identify areas of need and then put effort and action into risk reduction plans. We utilize a hierarchy of controls, implement the controls and strive to continually improve upon the plan of action. When it comes to Workplace Violence this sequence of action is no less effective. A team approach both internally and externally is necessary to reduce your risk to your employees, your customers, your organization's reputation, your liability and your peace of mind.

This author has worked with both governments and private industries on this topic. Having facts and implementing solid plans is the key to addressing this menacing issue. Getting organizations to remember and apply best practices in safety management is a challenging but necessary step for all involved to address this destructive emotional and economic drain on our society.

When the topic of Workplace Violence, or on a narrower but more fearful level Active Shooter situations, is discussed the emotional factors and individual reactions is quite interesting. The two topics are different in scope and will be discussed in the presentation. It is odd to me that one of the most common issues raised is the fear of the unpredictable behavior of human nature. The reality is that "no one can predict human behavior, and there is no specific "profile" of a potentially dangerous individual. However, studies indicate that incidents of violence are usually preceded by patterns of behavior or other activities that may serve as warning signs." (FEMA 2016) Like violent events in nature, human behavior can be unpredictable. However, just like earthquakes or hurricanes, we should plan and prepare to reduce the occurrences or effect of violent events. In safety management we encounter human behavior deviations from the norm on a daily basis. There are great ergonomic, engineering design solutions, administrative or training programs that result from responses to the root causes that human error contributes to accidents/incidents. When we relate root cause analysis to workplace violence I propose that we that are rational and typically act in a controlled manner are often at a quandary in understanding the logic or rational for extremes of anger, depression, predatory behavior or fanaticisms that often accompanies violent acts. It would seem that it is that lack of logical and rational understanding along with the unpredictability that unsettles most people regarding this topic.

When attempting to rationally discuss the issue of violence in a workplace setting for solutions, the unsettling effect is evident. This topic brings up defense reactions, protectionism for family and loved ones (even in a workplace discussion), gun control, freedom of speech, privacy issues including medical history and mental illnesses. Many of these topics are thought to be taboo discussions for workplaces yet they must be discussed and planned for to succeed regarding this topic. It is unfortunate that only in recent history we have begun to openly admit and discuss mental illness. Henriques, J. (2013) in her paper on Mass Shooting in American reported that "Shooters of the last thirty years, 65/67 shooters had mental health issues." This is not to imply that all with mental health illness will be violent. It should be noted that those that need assistance and demonstrated it in the workplace have historically lacked options or support. Laws and policies have been ill prepared for this level of personal sharing in the workplace although Employee Assistance Programs (EAP) are becoming more prevalent in businesses. According to Workforce writer Sipek (2015) "Employers are increasingly using employee assistance programs to help employees receive assistance for personal problems that might affect overall job performance." The report shows EAP usage rates are now at 7% in 2014 up from 6.2% in 2012. The slow shift to deal with mental illnesses in society and the workplace is evident when one reviews the data that workplace suicides have been on the rise (BLS 2014), yet this data is not well known, openly discussed or sensationalized in the media. Discussion on homicides and shooting occurs yet there has been a decrease in the workplace for those areas (BLS chart 2014).

Number of Workplace Deaths from Assaults and Violent Acts By a Person, 2003 – 2014

Year	Assaults & Violent Acts by Persons	Suicides/Self Inflicted Injuries	Homicides	Homicides		
				Shootings	Stabbing	Other Weapons
2003	854	218	632	487	58	87
2004	769	206	559	421	68	70
2005	750	180	567	441	60	66
2006	752	208	540	436	39	71
2007	828	196	628	503	45	80
2008	791	263	526	421	33	72
2009	837	263	542	434	49	59
2010	794	270	518	405	37	76
2011	754	242	458	358	42	58
2012	765	249	475	381	35	59
2013	738	282	404	322	38	44
2014	710	271	403	307	39	57

Source: U.S. Department of Labor, Bureau of Labor Statistics, Census of Fatal Occupational Injuries

When one considered the sheer number of high risk occupations: workers that handle cash, workers in convenience stores, workers in health care, workers alone or at night, workers in remote and high crime areas; there has been significant progress in proactive actions by many businesses in these high risk areas. More can and should be done however; critically look at the businesses you may frequent and notice improvements in large convenient stores, hotel chains and hospitals. Increased application of the principles of Crime Prevention Through Environmental Design (CPTED) is evident in large corporations that conduct these types of activities. Visibility, lighting, surveillance, communication, cash management, even room colours, landscape, layout, activity areas, egress and many more factors play into the improved general perception of a safe place, firm ownership/branding and the ultimate reduction of crime. CPTED is still a relatively new science, with origins in the 1960s it is a science still evolving with second generation disciplines such as conflict resolution, support and other concepts adding to solutions for crime reduction. It is a science that blends environmental factors with human behaviour and control of space and community. In 2006 I was fortunate to be a student of one of the pioneers in the field of CPTED, the late Timothy Crowe. CPTED is design and space management practices that assist with crime prevention and general quality of life. CPTED is just one proactive measure that has been implemented and has assisted in decreased workplace homicide rates.

However, do not be confused when hearing in the media that active shooter incidents are on the rise. Random shooter, mass killing events have increased in recent years. From the FBI study on Active Shooters Incidents in the US found that 64 incidents (40.0%) would have been categorized as falling within the new federal definition of "mass killing," which is defined as "three or more killings in a single incident." Some of the principles discussed in this presentation may also assist with reduction, response and recovery to these horrific events.

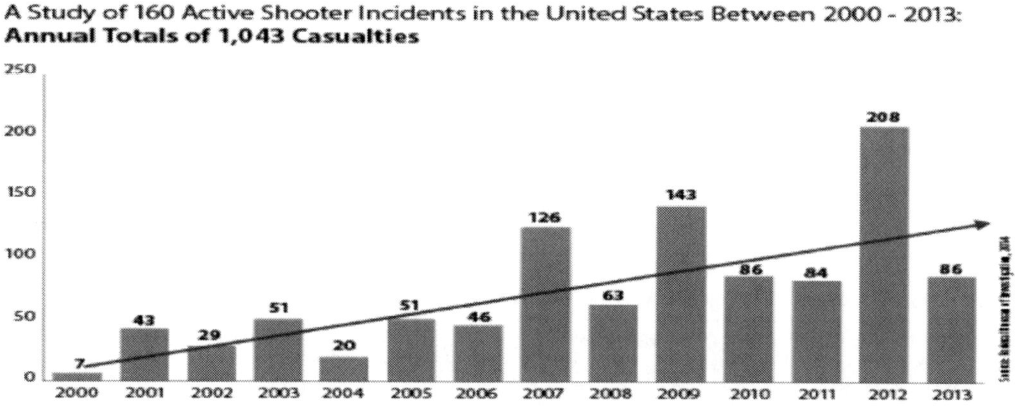

Brief Overview of History, Trends and Statistics

This presentation seeks to bring this emotionally charged topic of Workplace Violence Prevention back into reality and remind of actions and measures to reduce to ALARP. Data will be shared on history, trends, lessons learned, targets, and impact. We as Safety Professionals should know, there is a lag time to gathering reliable data from industry. It is worth reminding all that the Bureau of Labor Statics (BLS) has only been recording national data on fatal occupational injuries since 1992 (Kesy, Pegula 2014). Also when reviewing crime data validity capturing can also be called into question (i.e. was it a workplace violence issue or a drug or gang issue that occurred at a workplace site?). Bureau and Labor Statics (BLS), Federal Bureau of Investigation (FBI), Federal Emergency Management Agency (FEMA) data are utilized for this presentation. As in his article on incident information McManus (2014) writes "Report incompleteness is a common concern as is the small number of records available compared to the actual number of incidents that have likely occurred (OSHA, 1982a; 1982b; 1983; 1982; 1988; 1990). This is especially true when one considers the topic of bullying, threats, verbal assaults which are all the most prevalent forms of workplace violence.

From OSHA's fact sheet in 2002 estimate, roughly 2 million workers a year are affected by some form of workplace violence. The BLS chart below breaks down type of crime and victimization from 1993-'99. Comparing these organizations' data, it is disheartening to think that in just a 3 year period there was a 13% increase in estimates of the number of workers affected by workplace violence.

Average annual number, rate, and percent of workplace victimization by type of crime, 1993–99

Crime Category	Average annual workplace victimization	Rate per 1,000 persons in the workforce	Percent of workplace victimization
All Violent Crime	1,744,300	12.5	100%
Homicide	900	0.01	0.1
Rape/Sexual assault	36,500	0.3	2.1
Robbery	70,100	0.5	4.0
Aggravated assault	325,000	2.3	18.6
Simple assault	1,311,700	9.4	75.2

Sources: Homicide data are obtained from the Bureau of Labor Statistics Census of Fatal Occupational Injuries. Rape and sexual assault, robbery, aggravated assault, and simple assault data are from the NCVS.

NIOSH 2012 reports that Workplace Violence is the second leading cause of injury death for women in the workplace. From the FBI 2002 Crime National Center for the Analysis of Violent Crime (NCAVC) Workplace Violence "Although domestic violence and stalking are largely thought of as violence against women and thus as a "woman's problem," training and awareness programs should be directed at all employees, men and women alike." Do you have Domestic Violence literature and contacts posted to assist your employees? From the Huffington post Jeltsen (2015) "The majority of mass shootings in the U.S. take place in private. They occur in the home, and the victims are predominantly women and children." However, spillage of incidents into workplaces occurs at multiple levels with production impact, bullying transference, and even on site violence. Historical analysis of mass shooters events often exhibit pre-escalation of violence and warning signs. Jeltsen (2015) states "mass shootings that happen in the home are not unpredictable. They're frequently preceded by a flurry of red flags: 911 calls, hospitalizations, broken protective orders, repeat arrests, contentious custody battles, death threats and stalking". Awareness, planning and prevention are all needed at home and work.

The FBI made public statements after 911 that they needed a new way of doing business to achieve their mission of protection. However, even before 911 prevention verses reaction has been a part of law enforcement for a couple of decades now. From the U.S. Department of Justice, "FBI Workplace Violence Issues in Response Report (2002)":

> "Stephen Doherty, Chief of Police in Wakefield, Massachusetts, wrote about the shift—and its roots in domestic violence policing—in the April 2002 issue of Security Management Magazine. "Ten to fifteen years ago, the perception among police, businesses, and the public was that domestic violence was private. Over time, attitudes about domestic violence changed. Local police are now regularly provided domestic violence training. Ongoing partnerships with women's advocacy groups have raised awareness and incident reporting. …"Workplace violence," Doherty continued, "..is today where domestic violence was a decade ago...the more common but less dramatic lower-level incidents, such as threats and aggravated assaults, are still not viewed as an opportunity for early intervention. Companies tend to treat these situations internally-just as domestic violence was once treated as private. According to the Bureau of Justice Statistics, less than half (only 44.2 percent) of violent victimizations sustained at work are reported to the police."

In the workplace bullying is far more widespread than physical violence and yet though to be under reported. Per SIOP white paper "the consequences of workplace bullying have a range of costs including: (a) human costs, (b) organizational costs, and (c) spillover costs."…"workplace bullying does not occur in a social vacuum. Rather, it occurs in the context of an organizational environment that may aggravate or mitigate the incidence of workplace bullying." When we think of the safety domino theory do we apply it to workplace violence and bullying? When we think of shifting safety cultures do we factor in this important component, reducing bullying in the organization? From a business perspective the SIOP paper states "it should be unsurprising that workplace bullying in the United States costs organizations as much as $14,000 USD per employee in lost performance (Pearson & Porath, 2009)." This topic is not one that is just an American issue. From the Australian Human Rights Commission (2015) "A recent impact and cost assessment calculated that workplace bullying costs Australian employers between $6–$36 billion dollars every year when hidden and lost opportunity costs are considered." Bullying is being experienced in workplaces all over the world in various degrees. An unhealthy workplace breeds bullying. Per SIOP "Researchers have found that job insecurity and role stressors such as low job autonomy and high workload are associated with being both a perpetrator and a target of workplace bullying." Unfortunately it does not end with the boundaries of the workplace, it affects our society as a whole beyond economic impact SIOP report "research demonstrates the power of workplace mistreatment to extend beyond its relational context and beyond organizational borders." It is no surprise that beyond improved polices and communication, training for conflict resolution and improved workplace culture are all recommended to decrease the occurrence of bullying. Per SIOP "creating systems that can: (a) ensure role clarity and reasonable workloads, (b) offer leadership training that encourage supportive leadership styles, and (c) generate policies to ensure fair and just treatment, decisions, and out-comes may all help to reduce the prevalence of workplace aggression and bullying."

Response: Safety 101

The worst school mass murder in history: Bath, Michigan's 1927 school bombing has lessons we have still not learned and fully implemented. Biography (n.d.) regarding Andrew Kehoe who killed 37 children and that count of children would have been much more if the bomb on the other side of school house had functioned correctly. He used secondary devices and was known as angry about the schoolhouse due to taxes. He was known as a cruel man amongst those in the community. Yet he was let in to work in the basement alone. This historic event is discussed because of the lessons yet unlearned. . George Santayana wrote in *The Life of Reason*, 1905 "Those who cannot remember the past are condemned to repeat it.". When on review of the FBI 2002 Crime National Center for the Analysis of Violent Crime (NCAVC) Workplace Violence list, the list contains no extremely new or outlandish recommendations. Following is a summary from that report. As a Safety Professional, see if any of the list jogs your memory of Safety 101 concepts:

1. Support from the top…
2. There is no one-size-fits-all strategy...
4. Plan should be proactive, not reactive…
5. A plan should take into account the workplace culture…
6. Based on a multidisciplinary approach…
7. Communication-active at all levels…
8. Practice your plan!
9. Reevaluate, rethink and revise…
10. Plan for recovery.

Remember this is from a 2002 report. How many items on this list have you or your organization implemented with a focus on Workplace Violence Prevention?

Per a Society for Human Resource Management Report (SHRM) 2012 survey "Over one-third (36%) of organizations reported incidents of workplace violence" and only "47% responded they would respond with zero tolerance (immediate termination)". There obviously needs to be more done in prior preparation and policies. From my experience I have worked with many organizations that have drilled for fires but never for a workplace violence situation. Drills reduce panic and lead to lessons learned and plan improvements.

All too often I have seen Emergency Plans that weight heavily on Police and Emergency Responders as the answer for physical violence or active shooter situations. Per Blair, and Schweit 2014 report in conjunction with the Department of Justice, FBI "A Study of Active Shooter Incidents, 2000 - 2013 "many incidents ended before police arrived. Of the 160 incidents, at least 107 (66.9%) ended before police arrived and could engage the shooter, either because a citizen intervened, the shooter fled, or the shooter committed suicide or was killed by someone at the scene."…"(40.0%) of the shooters ultimately committed suicide." This is not to say that local law enforcement is not a valuable tool in Workplace Violence Prevention but companies need to look critically regarding traffic, location, plans and panic factors that may negatively impact them. One company I worked with was on the cusp of two jurisdictions and did not realize it. They had no idea who would actually respond or what response times would be in heavy metro traffic. From a FEMA interoperability meeting transcript held at Georgia Institute of Technology, Helens Mitchell (n.d.) with the Center for Advanced Communications Policy and the Rehabilitation Engineering Research Center on Wireless Technology had some great advice:

> "I just wanted to comment on the fact that one of the things we have found as far as getting in and out emergency information is that there's a disconnect between where information is posted by agencies and state, local and federal, and where the public goes to get the information. Then I just want to comment on Marcie's point because one thing we always try to tell people is a "yoyo" effect after an incident, which means, you're on your own for at least 78 hours. And if that's the case you have to make sure you have all of your equipment, everything And you were talking about not having technology as your only fall back…it's just really important that you carry more than one way of connecting in case of an emergency."

Redundancy applies to many systems, many plans and solutions and YOYO is something voiced by many. In a FEMA presentation on "Dangerous Disconnect: Public & Private Sectors Don't Understand their Partnership in Emergency Response and Readiness", Mitchell, B. 2012. Retired Police Commissioner's underlying theme throughout the presentation was "Denial is a problem and YOYO is a reality". His message is real world and should be heeded.

Should a violent or even near violent drama occur at your workplace, prompt implementation of a preconceived plan is necessary. As with fires, drills for egress or even "Run, Hide and Fight", a very well done City of Houston and FEMA video can be implemented. Which method will depend on circumstances of the violence and layout and operation particular to the business. There is no one route fits all solution.

Recovery

Following any event often constructive, recovery dialog is necessary. If facts are not given, employees will fill the void. Professional counselors are typically needed for worst case scenarios. Allowing some communication sharing in a non-judgmental, non-dramatic and focused sympathetic manner is healthy. However, with all levels of employees a cut off point for dialog and re-focusing on the reason for being in the workplace is advised.

Conclusion

It can seem that employers are in a quandary when it comes to Workplace Violence issues. The duty to provide a workplace "free from recognized hazards" is one we all know from OHSA's General Duty Clause. With all the laws on privacy such as the Health Information Protection Act (HIPA), antidiscrimination, gun rights and Americans with Disabilities Act (ADA) to name just a few seem to paralyze even the most honorable actions to assist. There are discussions taking place such as the Crime National Center for the Analysis of Violent Crime (NCAVC) and in other legal and government circles in response to roadblocks regarding privacy and information sharing. More rational and less restrictive rules for sharing of information are all real possibilities for the future of businesses. Until then, as with most safety efforts good faith, intent to protect and prevent as well as balance of risk and resources are all crucial components in this equation.

As with most other risks, prevention of workplace violence begins with the desire to improve. You are reading this article so your desire must be there. Plans of action and preparation will reduce vulnerability. Fear and vulnerability most often results from a feeling of a lack of control. You know how to control and reduce other risks; make your organization and yourself stronger and less of a target with proven Safety Management and Communication techniques.

Bibliography

Australian Human Rights Commission. 2015. "Fact Sheet: Workplace Bullying" (https://www.humanrights.gov.au/employers/good-practice-good-business-factsheets/workplace-discrimination-harassment-and-bullying)

Biography.com Editors (n.d.). "Andrew Kehoe Biography" (http://www.biography.com/people/andrew-kehoe-235986#public-response)

Blair, J. Pete and Katherine W Schweit. (2014). "A Study of Active Shooter Incidents, 2000 - 2013." Texas State University and Federal Bureau of Investigation, U.S. Department of Justice, Washington D.C. 2014.

Bureau of Labor Statistics (BLS). 2013. "Fact Sheet: Workplace Homicides from Shootings" (http://www.bls.gov/iifct/oshwc/cfoi/osar0016.htm).

______. 2015. "National Census of Fatal Occupational Injuries in 2014 (Preliminary Results)" (http://www.bls.gov/news.release/cfoi.t01.htm).

Crowe, T. "Crime Prevention Through Environmental Design (CPTED)." Five-day workshop presented in Buford, Georgia through Gwinnett County Parks and Recreation, May 2006.

Federal Emergency Management Agency (FEMA). 2016. "IS-106.16 Workplace Violence Awareness Training. 2016". (https://emilms.fema.gov/IS106.16/WV01.htm)

Hershcovis, M., T. Reich, and K. Niven. 2015. "Workplace Bullying: Causes, Consequences, and Intervention Strategies: White Paper for the Society for Industrial and Organizational Psychology (SIOP), Inc." (http://www.siop.org/WhitePapers/WorkplaceBullyingFINAL.pdf)

Jeltsen, M. (2015). "We're Missing The Big Picture On Mass Shootings." (http://www.huffingtonpost.com/entry/mass-shootings-domestic-violence-women_us_55d3806ce4b07addcb44542a)

Kesy L., Pegula S. (2014) "Census of Fatal Occupational Injuries commemorates 20 years of occupational safety and health data" (http://www.bls.gov/opub/btn/volume-3/census-of-fatal-occupational-injuries-commemorates-20-years.htm)

"Mass Shooting in America, A Historical Review." (http://www.globalresearch.ca/mass-shootings-in-america-a-historical-review/5355990)

Mitchell, H. (n.d.) FEMA and Georgia Institute of Technology interoperability meeting transcript (http://www.fema.gov/media-library-data/20130919-1030-5011-8953/transcripts20130919-5011-1hu6sqy.txt)

Mitchell, Bo. 2012. Police Commissioner (Ret.) FEMA Power Point Presentation on: "Dangerous Disconnect: Pub & Pvt Sectors Don't Understand their Partnership in Em Resp Readiness" (training.fema.gov/hiedu/12conf/presentations/mitchell%20-%20dangerous%20disconnect-the%20public%20and%20private%20sectors.ppt Dangerous Disconnect: Pub & Pvt Sectors Don't Understand their Partnership in Em Resp Readiness FEMA)

McManus, Neil. 2014. "Incident Records Understanding the Past to Prevent Future Hazardous Energy Incidents." *Professional Safety*. 58: 36.

National Institute of Occupational Safety and Health (NIOSH). 2012. "Total Worker Health." (www.cdc.gov/niosh/TWH/totalhealth.html/) .

National Institute of Occupational Safety and Health (NIOSH). 2012. "Women's Safety and Health Issues at Work." (http://www.cdc.gov/niosh/topics/women/serious-injury.html).

OSHwiki contributors, "Workplace violence," OSHwiki, http://oshwiki.eu/index.php?title=Workplace_violence&oldid=245186.

Ready Houston and Department of Homeland Security. (2012). "Video on Run, Hide and Fight" www.ReadyHoustonTX.gov

Santayana, G. 1905. *The Life of Reason*. New York: Scribner's.

SHRMA. 2012. "Society for Human Resource Survey Finding: Workplace Violence 2012" (http://www.shrm.org/research/surveyfindings/articles/pages/workplaceviolence.aspx)

Sipek, S. 2015. "Few Companies Taking a L(EAP) of Faith." (http://www.workforce.com/articles/21346-few-companies-taking-a-leap-of-faith)

U.S. Department of Justice, FBI. 2002. " Workplace Violence Issues in Response Critical Incident Response Group National Center for the Analysis of Violent Crime National Center for the Analysis of Violent Crime (NCAVC)" (https://www.fbi.gov/stats-services/publications/workplace-violence)

Zahm, D. 2007. "Using Crime Prevention in Environmental Design in Problem Solving" (http://www.popcenter.org/tools/pdfs/cpted.pdf)

Safety on College Campuses–Significant Challenges and Opportunities

Carol B. Schmeidler, MS, CSP
University at Buffalo
Buffalo, NY

Introduction

Approximately 20.5 million students were enrolled in colleges and universities in the United States in the fall of 2016. 12 million students are under the age of 25, which is 40% of the 18 to 24 year old age group, but also includes about 8.2 million students over 25 (2014 figures).[1] More than one million international students attend public and private institutions. Most are from China, India, Saudi Arabia, South Korea and Canada.[2]

A large network of people, and a vast assortment of properties and settings support students; trying to keep everyone safe poses significant and unique challenges, but also many opportunities.

Why College?

College graduates have better job opportunities and higher incomes than people with high school degrees. Graduate and professional degrees lead to higher incomes. It is estimated that, on average, someone with a Bachelor's Degree will earn about 66% more during a 40 year working life than someone with a high school degree. Having a professional degree will more than double that.[3]

Background

Large universities like my employer, the University at Buffalo (UB), are similar to cities. The University is the largest in the system of public colleges and universities in New York State. Approximately 14,000 full and part-time faculty and staff are employed for our population of more than 29,000 students. There are over 1300 acres on three campuses, in downtown Buffalo,

[1] National Center for Education Statistics (NCES). 2016. Fast Facts, Back to School Statistics (retrieved March 3, 2017). (http://www. https://nces.ed.gov/fastfacts/display.asp?id=372).

[2] Institute of International Education (IIE). International Students in the United States (retrieved March 3, 2017). (http://www.iie.org/Services/Project-Atlas/United-States/International-Students-In-US#.WNK78-lh3ow).

[3] The College Board. 2010. Trends in Higher Education, Lifetime Earnings by Education Level (retrieved March 3, 2017). (https://trends.collegeboard.org/education-pays/figures-tables/lifetime-earnings-education-level).

north Buffalo and suburban Amherst. There are more than 110 buildings, 38 miles of roads, even more miles of sidewalks and walkways, and 98 parking lots to maintain.

In the Fall of 2014, 17.4% of students at UB were international, from 106 countries.

Students live in dormitories and apartments on campus, and off-campus student housing can be found throughout the city and suburbs. Members of the public are on campus day and night, every day, in addition to faculty, staff and students. Contractors from numerous companies also work throughout the campuses daily. There is a police force. There are athletic facilities including a stadium and Olympic-sized pool, used by students and the public. National and local sporting events take place on campus, in addition to events for university teams. There are theaters and recital halls, and events occur in those all week and all year. National, state and local politicians stage events on campus; even the Dalai Lama appeared at UB. Food is served all day, every day, to students and the public. Libraries throughout the campuses are open from early morning until late at night, and many are open 24/7 for exams.

Some buildings are locked after hours due to security reasons and some buildings and rooms are locked all the time, but most areas are open and unlocked during the day.

Workers at a university include faculty, staff, from janitorial staff to departmental support people, both technical and administrative, to tradespeople, design/architects, to upper administration. Many are university employees, others contractors. There are multiple employers and multiple unions within the university, often within the same department.

Perspectives of College

What does going to college mean to students and parents?

Students (not an all-inclusive list)

- Learn and work hard when needed
- Grow up and become more independent from family
- Enjoy being a student!
- Live away from home
- Make friends to keep for a lifetime
- Learn about different ideas, different cultures

Parents (again, not an all-inclusive list)

- Worry
- Debt
- Pride in their children
- Hope that their children are getting a quality education that will enable them to get a good job
- Knowledge that their children will be more independent from their college experiences

Note the absence of the phrase "being safe". Surviving until graduation, or having your children survive to graduate, or even not being critically injured, is usually taken for granted.

Perspectives of College from Employees - Academics and Staff

Academics (not an all-inclusive list)

- Conduct research and make an impact
- Collaborate with colleagues
- Found a startup company
- Get a patent
- Teach and make an impact
- Publish findings
- Become tenured

Staff (again, not an all-inclusive list)

- Solid jobs, full or part time
- Union membership at some locations
- Opportunities for advancement and/or learning
- Ability to make a contribution to the overall good

While "being safe on the job" may be another factor, it is usually taken for granted.

Then, there are safety people.

Challenges

The hazards encountered at a college or university are similar to those in industry and construction, but the total and diversity of types of hazards can be daunting. Many institutions will have all of those hazards, and more.

Students are not employees (unless they are working, including teaching assistants, graduate assistants, or work study) and therefore are not regulated by employee safety and environmental standards. Nonetheless, they may be exposed to physical hazards in their environments, as well as health hazards in classes, studios or laboratories and must be protected. They are also exposed to dangers that are not "occupational hazards", including drugs, alcohol, stress, violence, fire safety and unsafe off-campus housing, for which there are standards and guidelines.

Members of the future workforce start their careers in colleges and universities; safety training and safe practices developed at this time can establish habits, patterns and concerns that will be carried on throughout a lifetime. Poor habits can be difficult to change, and can lead harmful to disastrous consequences.

Some of the challenges found on college and university campuses include:

Training

Undergraduate students rarely have any safety training or experience when they arrive for freshman year, and they might only have a brief safety/security talk during orientation, and a safety orientation in a first-year chemistry lab. While they are usually supervised during laboratories, the supervision often consists of graduate students who must deal with large numbers of students in full teaching labs.

Graduate students may also have limited safety training. If they are teaching or research assistants they are considered employees and fall under regulations requiring training. However, time to attend training is often affected by ongoing research, lab schedules, and professors' dictates.

Faculty and staff are required by federal and state regulations to attend training (Hazard Communication/RTK, hazardous waste, radiological safety, biosafety, respiratory protection, etc.). Compliance should be good; however, academia is not industry. In addition to tenured and tenure-track faculty, there are guest/visiting researchers who are not informed or not aware of campus training; special programs that require special hazard training, and new protocols and experiments that have never been done before. Contract workers may not have training or know that they should have training (depending on where they are from and their function on campus).

Safe habits and work practices cannot be developed if people do not know how to do so.

Cultural and Language Differences

Students, faculty and staff come from very diverse backgrounds and cultures. A new country with many differences in language, culture and traditions, combined with high pressure to succeed, can be daunting. Proficiency in English should be present; however, widely varying degrees of comprehension are found, and in different modes (speaking vs. reading, for example). This can pose challenges for comprehension of training and for reading procedures and instruction manuals.

Cultural norms can also vary widely; variations can occur between countries of origin, in different disciplines within an academic setting, and even within departments. Practices considered standard in one country may be totally unacceptable in others; different laws in other countries may allow actions not acceptable in this country. Despite our Globally Harmonized System of Hazard Communication, not all countries and cultures have adopted all or part.

Fire Safety

One of the foundations of the campus fire safety movement occurred in 2000, when a fire at a Seton Hall University dormitory in New Jersey killed 3 students and injured 67. The dormitory had no sprinklers; a state law was later enacted requiring dormitories to be protected by sprinklers.

The U. S Fire Administration, a division of the Federal Emergency Management Agency (FEMA), conducted an analysis of campus fire fatalities in residential buildings between 2000 and 2015. For this report, "campus" included both on and off-campus residential housing. The analysis concluded that on-campus housing is more controlled, i.e. regulated, than off-campus housing. Also, that students living on campus were somewhat more controlled by rules and regulations (such as bans on candles in dorm rooms) than students living in off-campus housing. The report included the following:

- More than 90% of fatal campus fires actually took place in off-campus housing
- There have been 85 fatal fires resulting in 118 fatalities, approximately 7 per school year

- Smoke alarms were inactive (disconnected or battery removed) or not there in 58% of the fires
- Fire sprinklers were not present in any of the 85 fires
- 70% of the fires occurred on Friday, Saturday and Sunday, and 73% occurred between midnight and 6AM
- Alcohol was a factor in 76% of the fatal fires, where at least one of the students was drinking and legally drunk
- Smoking was the leading cause of fatal fires in campus housing, followed by intentional actions, electrical and cooking
- April was the peak month; the least number of fires occurred in June, July and August
- There have been no fire fatalities in on-campus dormitories since April 2005[4]

Laboratory Safety

Safe work practices include attending training and applying that training, along with area/procedure-specific training, to work practices. Unfortunately, there have been many serious injuries and fatalities to show that that has not always been done.

New technology is developed on campuses and often widely used before potential hazards are assessed. What kinds of exposures and hazards are present during development? How are students/faculty/staff being protected? Some of the most hazardous chemicals, biological and radiological materials are used for research. Biological and radiological material use is more rigorously legislated and enforced than chemical use. How are the users protected, and how can practices and procedures used be shared?

Some of these incidents and their unfortunate/tragic consequences include:

- On December 29, 2008, graduate student Sheharbano Sangji was working in her laboratory at UCLA with an air-reactive (pyrophoric) chemical. She had been in the lab since October of that year and on that day, was not wearing a lab coat or goggles. Her closed system breached and ignited her polyester top and nitrile gloves. She suffered second and third degree burns over 40% of her body and died more than two weeks later. The University of California system was fined and Ms. Sangji's professor was charged with four felonies, which he settled without serving prison time.[5]
- On March 16, 2016, postdoctoral student Thea Elkins-Coward was working for the Hawaii Natural Energy Institute (HNEI) in her laboratory at the University of Hawaii's Manoa campus. She was using a combination of gases, moving them from high-pressure containers to a low pressure container when an explosion occurred, possibly from static electricity. She lost an arm in the explosion. She had received both general and lab-specific training, and the lab had been inspected earlier in the year. State officials charged the University with 15

---

[4] U. S. Fire Administration, Federal Emergency Management Agency. *Campus Fire Fatalities in Residential Buildings (2000-2015)* (retrieved February 23, 2017) (https://www.usfa.fema.gov/downloads/df/publications/campus_fire_fatalities_report.pdf)

[5] Van Noorden, Richard. "A death in the lab." *Nature*. April 2011:270-271.

serious violations of health and safety standards. Dr. Elkins-Coward is suing the University of Hawaii, her professor, and the Director of the HNEI. [6]

- On April 13, 2011, undergraduate physics and astronomy student Michelle Dufault was working along in a machine shop on the Yale Campus. Her hair became entangled in a lathe and Ms. Dufault died, apparently of asphyxiation. She was an experienced worker in the machine shop and was weeks away from graduating.[7]
- On January 7, 2010, chemistry graduate student Preston Brown was working in his lab at Texas Tech University in Lubbock when an explosion occurred, causing him to lose 3 fingers on his left hand, severely damaging his right hand, eyes and body. Due to the hazards of the material with which he was working, he should have been using small quantities at a time; subsequent investigation showed that he was using much larger quantities than he should have been using, was not wearing proper personal protective equipment at the time of the explosion, and was working on a benchtop rather than in a fume hood. Little attention was paid to the safe use of this highly hazardous material; procedures and labeling were found inadequate and housekeeping was poor. An internal investigation of the incident found numerous policy violations. This was the first academic chemical accident investigated by the Chemical Safety Board.[8]

Unfortunately, there are many more examples. While statistics show low rates of recordable injuries and illnesses in research and development, there are most likely far more occurrences of "near misses" and incidents than are reported.[9] We are sometimes ignored or treated as intrusive or interfering until something happens.

<u>Other Issues</u>

There are many other issues, including physical and health hazards, to be found on a typical campus. The hazards are not unlike those in construction or general industry, but can be as numerous as construction and general industry combined. These include:

- Vehicles – buses, cars, trucks, carts, scooters, bicycles, skateboards
    - Competing with students using phones
- Buildings of varying ages and conditions – maintenance issues, lead and asbestos
- Athletic facilities – injuries to athletes, public
- Dormitories and dining facilities – housing issues, food safety
- Construction, both new and remodeling
- Prevention through Design – limited budgets and competing priorities impact design/build
- Art facilities including foundry, ceramics, casting, painting

---

[6] Kemsley, Jyllian. "University of Hawaii fined $115,500 for lab explosion." *Chemical & Engineering News, American Chemical Society.* September 29, 2016 (retrieved March 5, 2017). (http://cen.acs.org/articles/94/web/2016/09/University-Hawaii-fined-115500-lab.html).

[7] Van Noorden

[8] Kemsley, Jyllian. "Texas Tech Lessons." Chemical & Engineering News, American Chemical Society. August 23, 2010: 34-37.

[9] Van Noorden

- Miles of roads, sidewalks and pathways
- Contractors unfamiliar with the campus
- Limited staff/budgets
- Constant stream of visitors and public in addition to employees and students

In addition, sources report declines in enrollment, with private, four-year undergraduate enrollment affected most.[10]Declining students mean declining revenues, leading to potentially smaller staffs and reduced maintenance.

Opportunities

A successful safety program reduces risk for our campus colleagues and eliminates serious injuries and fatalities, as well as destruction of property and experiments. As with safety practitioners in all disciplines, we can't do it alone. Our faculty, staff and students must work with us to make safety a priority.

Everyone has responsibilities for safety, top down and bottom up. The University at Buffalo has the "Commitment to Health and Safety in the Working and Learning Environment Policy," signed by the President of our University, outlining safety and health responsibilities.

Engaging Students

Providing safety information to incoming students that includes expectations as well as consequences will hopefully produce behaviors that last throughout a college career. These expectations and practices will also, hopefully, continue into the future, in industry, business and academia. The information has to be relevant, credible, and consistent.

Resident Assistants (RAs) in dormitories are provided with personal safety and fire safety information at the beginning of each college year. On our campus, this includes fire extinguisher use. RAs are the primary contacts with students. Student violators of fire safety rules (not exiting a building during drills, discharging fire extinguishers) face disciplinary actions.

Basic science classes and labs can include safety information; lab kits include safety glasses or goggles. Reading material for students can include specific lab information, as well as publications such as Safety in Academic Chemistry Laboratories from the American Chemical Society[11] and Prudent Practices in Laboratories[12].

---

[10] National Student Clearing House Research Center (NSCHRC). 2016. Current Term Enrollment Estimates – Fall 2016 (retrieved March 3, 2017) (https://nscresearchcenter.org/tag/enrollment-trends/).

[11] American Chemical Society Joint Board-Council Committee on Chemical Safety. 2003. Safety in Academic Chemistry Laboratories, Volume 2 (retrieved February 23, 2017) (https://www.acs.org/content/dam/acsorg/about/governance/committees/chemicalsafety/publications/safety-in-academic-chemistry-laboratories-faculty.pdf).

[12] National Research Council of the National Academies. Prudent Practices in Laboratories. Washington, DC: The National Academies Press, 2011.

Laboratory training for students, while not required by OSHA or the New York Department of Labor at public campuses in New York State, is often recommended or required by campus or department policy for specific experiments, specific chemicals and for advanced research. Summer research programs for visiting students and researchers frequently include safety training as part of their orientation; these students are often highly motivated, highly engaged, and very participative.

Undergraduates and graduate students attend chemical safety training that we try to make brief, relevant and engaging. Stories about incidents on our campus, as well as serious injuries and fatalities elsewhere are included. Handouts accompany lectures to aid comprehension. Quizzes are thorough but fair. Attendees receive training cards, valid for one year, which must be presented during laboratory inspections.

Engaging Faculty and Staff

Faculty and staff are required to attend training and to model safe behavior to peers and students. Professors are ultimately responsible for the activities and students in their laboratories. Unsafe laboratories can affect researchers, experiments and laboratories.

Working safely, knowing what should and should not be done are tasks that must be added to research methods, materials, procedures, grants, etc. Providing concise information and guidance to researchers can contribute greatly to safe laboratory operations. This is often the role of safety personnel.

Safety information is required in some grant applications and as part of some research protocols. Extreme incidents could affect funding. Safety officials sometimes have to approve documents; more opportunities for collaboration and assistance.

Recipients may not always appreciate communication about liability and risk, but the costs of doing things the "wrong way" can include death, injury, destruction of research and property, and legal consequences. Honest discussion and helpful information should be provided, along with clear definitions of responsibility and accountability.

Laboratory inspections, as well as facility audits, should be used to keep compliance on track. The people or areas being inspected/audited should understand what is being examined, and be present if possible. Feedback, and how to correct findings, should be provided promptly.

Employees with complaints sometimes contact unions and/or regulatory agencies; safety personnel work with all parties to investigate allegations/complaints and work towards resolutions. Conducting transparent investigations and providing prompt feedback and recommendations can often produce positive resolutions.

Sustainability and Technology

The intersection of safety and sustainability can lead to use of greener chemicals, recycling and energy savings. Sustainability practices can maintain or increase safety by using less toxic chemicals, smaller quantities of hazardous chemicals, more dilute chemicals, or modeling experiments rather than using hazardous chemicals.

Using recyclable materials and recycling batteries and other hazardous materials (fluorescent lights) will protect custodial staff from materials in garbage, reduce amounts of materials handled and landfilled, and prevent hazardous substances from polluting the environment in landfills. Energy costs can be reduced from utilizing heat recovery where possible, and utilizing energy-efficient lighting and building materials in new and remodeled buildings.

Technology developed on and for campuses can aid in observation, tracking and development of safe solutions for hazards.

Opportunities for Safety and Health Personnel

Large institutions may have numerous health and safety people, who specialize in one or two areas; smaller institutions, like small industrial settings, have fewer people who must be generalists, assisted by part-time employees, students and/or contractors.

There are a number of associations for safety and health professionals on college campuses to share information and practices. Most have listservs, and many have programs, webcasts and meetings. The group includes:

- SEHSA Environmental Health and Safety Association of New York, Inc. – for New York State (www.sehsa.org)
- CSHEMA, the Campus Safety, Health and Environmental Management Association – nation-wide (www.cshema.org)
- The Center for Campus Fire Safety – nationwide and international (www.campusfiresafety.org)
- The College and University Hazardous Material Management Conference (www.cuhmmc.org)
- The American Biological Safety Association (www.absa.org)
- ASSE is forming a Campus Safety Committee subgroup within the Management Practice Specialty

As safety and health professionals, we have significant responsibilities to keep people safe under almost any condition imaginable, but rewarding opportunities to work with people and to make a difference.

Bibliography

American Chemical Society Joint Board-Council Committee on Chemical Safety. 2003. *Safety in Academic Chemistry Laboratories, Volume 2* (retrieved February 23, 2017) (https://www.acs.org/content/dam/acsorg/about/governance/committees/chemicalsafety/publications/safety-in-academic-chemistry-laboratories-faculty.pdf).

The College Board. *Trends in Higher Education, Lifetime Earnings by Education Level.* 2010 (retrieved March 3, 2017). (https://trends.collegeboard.org/education-pays/figures-tables/lifetime-earnings-education-level).

Institute of International Education (IIE). *International Students in the United States*. (retrieved March 3, 2017). (http://www.iie.org/Services/Project-Atlas/United-States/International-Students-In-US#.WNK78-lh3ow).

Kemsley, Jyllian. "Texas Tech Lessons." *Chemical & Engineering News, American Chemical Society*. August 23, 2010: 34-37.

______. "University of Hawaii fined $115,500 for lab explosion." *Chemical & Engineering News, American Chemical Society*. September 29, 2016 (retrieved March 5, 2017).

(http://cen.acs.org/articles/94/web/2016/09/University-Hawaii-fined-115500-lab.html

National Center for Education Statistics (NCES). 2016. *Fast Facts, Back to School Statistics* (retrieved March 3, 2017). (http://www. https://nces.ed.gov/fastfacts/display.asp?id=372).

National Research Council of the National Academies. *Prudent Practices in Laboratories*. Washington, DC: The National Academies Press, 2011.

National Student Clearing House Research Center (NSCHRC). 2016. *Current Term Enrollment Estimates – Fall 2016* (retrieved March 3, 2017) (https://nscresearchcenter.org/tag/enrollment-trends/).

U. S. Fire Administration, Federal Emergency Management Agency. *Campus Fire Fatalities in Residential Buildings (2000-2015)* (retrieved February 23, 2017) (https://www.usfa.fema.gov/downloads/df/publications/campus_fire_fatalities_report.pdf)

Session No. 756

A Sports Medicine Approach to Ergonomics

Tony Kaczkowski, MHCA-MLS, LAT/ATC, CEES
InSite Health, LLC

Introduction: Sports Medicine and the Working Athlete

The term "Sports Medicine" is an everyday word in this day-and-age of hospital marketing campaigns, orthopedic surgeons, professional sports and even your child's soccer club. Do you really know what sports medicine is?

Our approach to sports medicine and the ***Working Athlete*** is very simple: the efficient, effective and concise delivery of services to a population that has an established relationship in place.

Here is a common scenario in today's traditional sports medicine world.

An NFL running back enters his first day of training camp, after the completion of an intensive conditioning program during the off-season. Prior to starting camp, he undergoes an assessment to determine any potential injury risks or weaknesses that need to be addressed by the sports medicine team. These are managed through daily treatment sessions in which the athletic trainer works to improve the function of the athlete by starting with the assessment results, collecting the athlete's feedback, developing a solution to the problem, implementing a plan, educating them on the situation, reassessing the progress and completing the stated tasks and goals. While using this process, the sports medicine team establishes the foundation on what will become a close relationship between them and the athlete.

This relationship is the key to managing the athlete's health once the season begins. Inevitably, that same running back will be breaking through the offensive line, making a cut to avoid a linebacker and suddenly feel a "pop" in their knee, falling to the ground in pain. This is where the relationship established between the sports medicine team and athlete is of great benefit. The athletic trainer will come onto the field, fully understanding the demands this athlete needs to be able to perform at. They will have the technical skill to assess the athlete to determine the extent of the problem. The relationship and trust established with the athlete and the baseline data to compare the new findings. They then will develop a plan, setting goals and tasks, to return the athlete to full function.

In this presentation, the reader will learn how to utilize a sports medicine approach for the ***Working Athlete***, to the delivery of ergonomic processes and injury prevention through the use of data gathering, project management and body mechanics.

The Sports Medicine Team

Why Athletic Trainers?

"What is an Athletic Trainer?" The most common question answered during our roll out process when first coming onsite with a client. Athletic trainers (ATs) are highly qualified, multi-skilled health care professionals who collaborate with physicians to provide preventative services, emergency care, clinical diagnosis, therapeutic intervention and rehabilitation of injuries and medical conditions. Athletic trainers work under the direction of a physician as prescribed by state licensure statutes. Athletic trainers are sometimes confused with personal trainers. There is, however, a large difference in the education, skillset, job duties and patients of an athletic trainer and a personal trainer. The athletic training academic curriculum and clinical training follows the medical model. Athletic trainers must graduate from an accredited baccalaureate or master's program, with 70% of ATs have a master's degree.[1]

Athletic trainers provide medical services to all types of patients, not just athletes participating in sports, and can work in a variety of job settings. Athletic trainers relieve widespread and future workforce shortages in primary care support and outpatient rehab professions and provide an unparalleled continuum of care for the patients.
Athletic trainers improve functional outcomes and specialize in patient education to prevent injury and re-injury. Preventative care provided by an athletic trainer has a positive return on investment for employers.[1] ATs are able to reduce injury and shorten rehabilitation time for their patients, which translates to lower absenteeism from work or school and reduced health care costs.

Athletic trainers are taught to ingrain themselves into the profession they are servicing from the beginning. This allows them to gain trust and thus help to increase continuity of care. The athletic trainer's skills are only as valuable as their ability to develop personal relationships. For this reason, athletic trainers are the obvious solution for our ergonomic interventions.

Another benefit to using athletic trainers in this health prevention and care model is their adaptability. Most ATs work with flexible schedules to fill the need of their team or

[1] http://www.nata.org/press-release/110414/athletic-trainers-provide-high-return-investment-today's-workplace. Accessed March 19, 2016.

client. This allows for the most efficient form of care and prevention. Schedule changes happen, but should not affect the level of care that is given.

Most of what athletic trainers learn occurs after graduation day. This is a valuable skill to have, the ability to learn on the job. This allows us to implement a system which they can independently navigate with minimal guidance. This allows the athletic trainer to be autonomous in nature but within a thought out environment. This is where athletic trainers succeed, having the ability to be creative with their services.

Having the right workforce in place is only half the battle. Teaching them the basics to get them off the ground is the other. As long as a systematic approach is taken the athletic trainer will be successful implementing his/her services within the organization's workforce.

On the Job Training

At the beginning of this discussion, two key principles were stated; relational care and the full understanding of the target client. To achieve the full understanding of the serviced clientele we need to learn on the job. This entails the most important aspect of complete integration, in market assessments. This is a term that relates to the athletic trainer going out to the job site, or on the driver's route, or into the market being serviced. Here is where they learn what the employee does on a regular basis. They will learn the lingo as well. Similar to being on a wrestling mat and understanding why it's important to have good head position, the athletic trainer needs to understand why multiple sets of gloves need to be worn before servicing a downed power line.

This is where the athletic trainer can relate previous experience to their current situation. Having serviced in the distribution and logistics market, I can advocate that one of the most important aspects of the job is communication. If communication breaks down, deliveries are delayed or delivered incorrectly or not at all. I can relate this to the football field where I spent most of my career. If the offense coordinator doesn't relay the play to the quarter back in the right manner, he will not be able to get the play to the other components which will ultimately end in a wasted or broken play. In football you get typically another play or another chance to correct but in the service industry one miscommunication could mean the termination of a contract and subsequent loss of revenue. This is why communication is a huge factor, and needs to be stressed to the athletic trainer onsite.

These soft skills that are learned on the job are just as important as the ones learned in the class room. In the next section, we will look at how everything is brought together to create an efficient systematic approach to ergonomics.

The Sports Medicine Approach

The sports medicine approach of efficient, effective and concise delivery of ergonomics and injury prevention needs to be delivered through a fully developed system. A true system allows for accuracy, accountability and consistency, no matter who uses the system. In this discussion, our system will consist of the following steps:

1. Assessment and data collection
2. Subjective interviewing and practical demonstration
3. Teamwork, data interpretation and solution identification
4. Goal setting, task assigning and time management
5. Education, Training and Coaching
6. Reassessment and continual improvement

Assessment and Data Collection

Just as in our traditional sports medicine example earlier, initial assessment and data needs to be collected to best understand the potential of injury risks. This data comes in two primary areas:

1. Physical assessment of the employee's musculoskeletal health.
2. Ergonomic assessment of the job they will be completing.

As in the early example of the NFL running back coming into camp, physical assessments on the employees aid in the prevention of injury due to poor physical conditioning or underlying musculoskeletal limits. This assessment is not designed to preclude employees from participating in work, but to proactively identify and address areas of opportunities to improve physical function.

This assessment can be related in a practical way to employees by use of general machine maintenance. As with any machine, regular inspection and maintenance is required to keep the machine operating properly and extend the longevity of the equipment. Generally, machines are inspected on a regular schedule, identifying issues and leading to addressing of the issues through corrective measures. The same philosophy should be taken in terms of the musculoskeletal health of the employee. A simple, and regular, musculoskeletal screen should be conducted an annual basis or with all new hires. This screening identifies areas of muscle or joint tightness as well as muscular weaknesses that may lead to a higher risk of injury. These risks are then addresses by the use of proactive stretching and strengthening exercises to improve the overall health and function of the employee. Not only has this been shown to reduce the incidence of injury, it more importantly greatly improved the health and comfort of the individual. By improving how an employee "feels" many positive gains are made both at work as well as at home.

The second area of data collection required is that of the ergonomic risks related to the employee performing a particular job. From a sports medicine point of view, this ergonomic assessment needs to be focused on the biomechanics and forces placed on the body, while conducting the job related tasks. The assessment conducted must include a general overview of the work being performed, including control over work, weights, heights, distances, duration, environment and tools used, or available to use, to complete this work.

Once the general overview is understood, the task is evaluated through the use of a standardized system or process that evaluates each region of the body and functions of the job. These regions include the following: hand/fingers, wrist, elbow, shoulder, neck, back, hip/knees/ankles/feet. To evaluate the functions, evaluators need to concentrate on: push/pull, lift/carry, repetitions, temperatures, general work environments and miscellaneous outside risks.

These assessment areas will need to be measured in a numeric scale that scores the overall job as well as to the individual body region with the highest risk potential. There are many different tools and assessments out there, but the key is to be fully understanding of the tool and consistent in its use and interpretation.

Once the initial assessment is completed and improvements made, it is also important to remember to reassess for the outcome data and determine if additional or new risks are present.

Subjective Interviewing and Practical Demonstration

During the ergonomic process, the evaluator must remember that the employee they are observing is most likely the "expert" in the process. This idea of "expert" is based off work conducted by Professor Anders Ericsson, at the University of Colorado. In Professor Ericsson's findings, he suggests that one becomes an "expert" at a specific task or activity by perform this activity for 10,000 hours (or 20 hours for 50 weeks a year for 10 years). With the average age of the American worker being 41.9 years, most employees will have achieved the status of "expert". (Lee and Mather, 2008)

With the philosophy that the employee is the "expert", the evaluator needs to collect subjective data, recommendations and elicit active participation in the process. Many times, the employee will have invaluable contributions to the assessment and the process of developing and implementing solutions.

Along with gathering subjective data (and when safe), the evaluator should perform a practical demonstration of the job or task. This practical experience allows for greater insight into the physical demands and nuances to completing the job and biomechanical risks that may not always be observed.

Teamwork, Data Interpretation and Solution Identification

Again, returning to the original description example of the traditional sports medicine setting in football, the athlete is cared for by a team. This Sports Medicine Team is comprised of athletic trainers, physicians, strength and conditioning coaches, massage therapists, physical therapist, sports psychologists, nutritionists and equipment managers. By involving a team, greater resources and perspectives are available to the athlete.

The same is true for the Ergonomics Team within an industrial setting. This team approach allows for improved data analysis, resource allocation and perspective in solution creations. Members of the team should include the athletic trainer, additional ergonomic evaluators, frontline employees, management representation, maintenance representation, supply chain/procurement, training department representation and any other key stakeholders needed to drive project completion.

The team will analyze the data collected and work towards an agreeable solution. Solutions may vary from basic training and education, to new tool selection to the full development of new processes and procedures. By having this team perspective, buy-in by the general employee population will be greater, solution implementation time reduced and management understanding improved. The quicker effective solutions can be implemented, the stronger the support for the ergonomics team will be.

Goal Setting, Task Assigning and Time Management

One of the most common pitfalls in ergonomic team success is the lack of goal setting, task assignment and time management. One of two problems tend to develop in this process today. The first is that teams become overwhelmed with the many needs within a site and the lack of resources, whether that be manpower, money or time, to accomplish the solutions. This can paralyze the team and significantly reduce its effectiveness and success. This also tends to follow the 80/20 rule being "80% of the work will be done by 20% of the people".

The second pitfall is the ability for the teams to develop a narrow focus and miss the big picture. One magnet for distraction is the desire to focus on only assessing and gathering data and not moving solutions forward. Many times it is easier for a team to determine what may be wrong with many job tasks, rather than implementing solutions that will greatly impact a few specific, high risk jobs. The second common distraction is the draw to the "Next Big Thing" in tools. Many teams will be sucked into the vortex of tool catalogs and vendors trying to convince them that all their problems will be eliminated through the purchase of their wares. A sports medicine approach to ergonomics must first start with improving the person's performance, effectively and efficiently, then utilize external aids to advance them to the next level.

To avoid these pitfalls, ergonomic teams need to work through these opportunities in a logical process. The first step is to annually conduct an overall assessment of the site and ergonomic team's strength's, weakness's, opportunities and threats. This process is commonly used in business and known as a SWOT analysis. The purpose of a SWOT is to look at what is going well (strength's), determine where there may be shortfalls (weakness's), evaluate areas of future success (opportunities) and forecast major problems today and into the future (threat's). Once the team gains greater insight into what they are working with and what needs to be addressed, a more focused approach can be taken to determine goals and projects for the year.

The next step in the process is to develop team goals that follow the SMART process. SMART goals are defined by goals that are: Specific, Measureable, Achievable, Realistic and Time Based. By utilizing the information gained in the SWOT and following this process for goal setting, the team will be better able to determine the most important projects to work on and truly layout a path to follow and measure its success in the process.

The third step in team success is to conquer and divide the project work. Teams need to continually refer to the 80/20 Rule and develop sub-teams or smaller groups to work on the individual projects. Once the projects are defined, groups / sub-teams created, tasks leading to the completion of the projects are identified and worked on. This is the way teams can break large mountains down to more manageable pieces. As smaller tasks are completed, overall projects are wrapped up, thus achieving the set goals of the team.

Education, Training and Coaching

Whether the ergonomic solutions are related to tools, processes or other interventions, a key to site success is education, training and coaching. The most effective tool ergonomics has is to train and coach the users of whatever solution is presented. This education and training may be in many different forms, verbal, written or in demonstration format. When providing this information, the presenter or format needs to provide an overall framework to what it is that is being accomplished, how it was developed, the background research and data used to come to this conclusion and why it will help. Employees or users of new ergonomic initiatives need to know that the solution is for their best interest and be taught how to use it.

Along with education and training, sports medicine professionals spend many hours coaching their athletes. This coaching, focused primarily of body mechanics, strength, flexibility or rehabilitation is another example of the merging of the traditional sports medicine setting with that of the industrial setting.

When approaching the employee's connection with the ergonomic solution, one must remember small specific changes lead to greater global change. A systematic approach to coaching all basis of healthcare promotion, prevention and intervention. This system needs to address the key factors to injury prevention as:

- Stage 1: Muscle tightness and fatigue
- Stage 2: Discomfort

- Stage 3: Pain
- Stage 4: Disability

With this all-encompassing paradigm athletic trainers are able to introduce stretching and strengthening to elicit change. These are a series of stretches and dynamic warm up movements tailored to address job specific injury prevention. The goal being preventing Stage 1, muscle tightness and fatigue, through the education of stretching and proper warm up techniques. Recommendations made to alleviate discomfort allows the individual to take control of their own healthcare.

Once the physical needs of the employee are addressed, the movement patterns used daily need to be observed and potentially improved. Movement coaching is a progressively building systematic approach to reducing discomforts and starts with the smallest change that exhibits the largest improvement, stance. With most sports where your feet start dictates the success of your movement or action. Staggered in nature similar to a wrestler grappling allows the individual to shift weight through each foot with the greatest proprioceptive ability. This allows for decrease in falls or coordination limitations. Once the feet are situated in a proper position allowing for optimal stability and safety the individuals body tends to reveal where there will be a breakdown in form.

Safe working zones need to be set and instructed to the employee. These are the safest statistically speaking areas to work in when looking at long term discomfort prevention. Each tackling the most common situations individuals will find themselves in on a given day with a given task.

Zone A: Working above the knees to below the shoulders. This allows the individual to have the greatest control over their body as well as the ability to move safely.

Zone B: Working within a forearms distance. Allowing leverages within the body system to be efficiently worked in a safe manor.

Zone C: Working within a 45-degree angle from the mid line of the body. Increases the body's awareness during any given task, as well as decreasing twisting motion and undue reaching that may lead to discomfort.

This is a small over view of the systematic approach to ergonomics that our athletic trainers use. The education on how to properly work the systems in any given workforce situation is where the first portion of this discussion is paramount. Proceeding to the final section, this system only works with a platform of ergonomic knowledge.

Reassessment and Continual Improvement
The final step in the sports medicine approach to ergonomics is the focus on continual improvement. As with the football example, or any athlete, one must strive to continually get better. Without the drive to improve, there is little to aim for.

This step of the process returns the team back to the beginning of the process by conducting an ergonomic assessment to determine how much of a reduction the implemented solution has provided as well as evaluate future opportunities to address additional risk factors. It is very important to be able to objectively measure the impact the team and solution has made for upper management support and the development of the team story.

Also, there may be the occasion where the solution implemented has led to unexpected risk factor being created. The team needs to make sure that this is not the case and if it does exist, address it through the same approach used throughout the described process.

Conclusion

With the ever increasing demands placed on an aging workforce, the risk of injury continues to rise and the need to efficiently and effectively manage this risk has never been greater. By tapping into the many years of sports medicine research, education and experience, one can implement a world-class ergonomics team that not only addresses ergonomic risk factors but improves the health and well-being of the entire workforce.

Bibliography

Lee, M. A., and M. Mather. 2008. U.S. Labor Force Trends. *Population Bulletin*, 5.

Session No. 757

Mindfulness and its Impact on Health and Safety

Kris Corbett, B.S. NSCA, ACSM
Wellness Coaches USA
Blue Bell, Pennsylvania

Introduction

The author has over 20 years of experience working in the Workplace Wellness and Injury Prevention industry, creating innovative and effective services and products for many Fortune 500 organizations. She and her associates have proven success in driving employee engagement without incentives, ultimately driving large scale health and safety risk improvements and risk reduction. Originally, injury and illness prevention industry standards provided training and education on hazards, common workplace injuries, and health promotion (OSHA 2012). However, the author and her team have discovered an additional concept and practice that effectively brings all of these learning points together: mindfulness.

Described by Time Magazine as the skill needed to survive the 21st century, mindfulness is more than just a buzzword (2016). It is a scientifically proven tool to change the way a person thinks, operates, and responds in situations. In addition to discussing how operating in mindlessness directly impacts health and safety, this paper explores how to incorporate mindfulness into everyday living. Lastly, the paper . In addition, it discusses how mindfulness was used to motivate and educate employees on the principles of worker readiness in an effort to support sustained achievement of injury free work performance.

What is Mindfulness?

Mindfulness is a practice that heightens our awareness, a tool that allows you to be responsive instead of reactive, and a method to help quiet constant brain chatter. Also known as staying in the present moment, mindfulness has applications at home, in relationships, and especially in the workplace. According to a recent study, employees that practice mindfulness are more productive, safe, and resilient, and have a more positive outlook on life than those who do not practice (Elder 2014). It can be utilized on a daily basis to impact nutrition, stress, injury prevention, and additional topics specific to a worker. Body mindfulness can also be applied to understanding how the body feels while performing tasks, promotes addressing body mechanics, and encourages proactive action on muscular skeletal disorders.

Mindfulness practices have been scientifically proven to increase grey matter in the hippocampus, an organ in the brain that regulates our emotions and manages learning and long-term memory (Pickut et al. 2013). Researchers have also done brain scans of people who regularly practiced mindfulness and found increased activity in their prefrontal cortex (Kozasa et al. 2012). This means the participants in the study had an enhanced ability to problem solve and process complex thoughts and emotions. In addition to these changes in the brain, Rinske Gotink and her team (2016) found decreased stimulation of the amygdala which may help reduce the symptoms of anxiety, depression, and pain from just a *couple* of weeks of regular mindfulness practice.

<u>Mindlessness and the Impact on Health and Safety</u>

Often, exploring mindlessness and its origins or triggers can provide greater understanding to mindfulness and mindful practice. Mindlessness is habitual, automatic, repetitive behavior generally brought on by multi-tasking. It can directly impact an individual's overall *wellness, stress level, attention, work and productivity, listening skills, injury risk and safety, and family relationships.* The next section of this paper will discuss a variety of examples of mindless living and how it directly impacts a worker's health and safety.

Highway Hypnosis

One of the most relatable forms of mindlessness is the concept coined as "Highway Hypnosis,". A driver's mind can be focused elsewhere while still able to fully operate a vehicle and upon reaching a destination, a driver in "Highway Hypnosis" may have no recollection of having consciously done the drive at all (Rothengatter & Huguein 2004). Highway Hypnosis is most commonly found in individuals that consider themselves experienced in the operation of machinery. For example, it is not often found in new drivers, but in populations that have been doing the task for a significant amount of time. Operating any type of machinery without full attention on the task clearly sets the stage for a safety hazard.

Mindless Eating

Often triggered by boredom, distractions (either technological or social), and overly emotional states, mindless eating is defined as not being fully aware of what and how much is eaten at any given time. It directly impacts health, because when not fully aware while consuming food or drink, individuals are more likely to overeat and make poor nutritional choices (Wasink 2011). Research shows that individuals with high Body Mass Index (BMI) or diagnosed as clinically obese have higher percentages of work impairment in addition to having severe disruptions of work, family, and social lives (Rodbard, Fox, & Grandy 2009).

Multi-Tasking

Multi-Tasking is the process of doing more than one task at a time. Although some multi-tasking has a place and purpose, most multi-tasking is a thief. Multi-tasking has been shown to steal time, take away from quality and accuracy, and even negatively impact an individual's mental health (Cravenshaw 2008).

Stress

Each time the body experiences any form of stressor (positive, negative, physical, or emotional), the sympathetic nervous system responds in the same way: increased breathing rate, elevated blood pressure, sweating, tense muscles, gastrointestinal dysfunction, dizziness, decreased immune function, and heightened sensitivity to light and sound. Without tools to appropriately respond and manage stress, chronic cortisol elevation can lead to calcium loss and overworked kidneys, increased fat accumulation, loss of cognitive function, and dysfunctional blood sugar metabolism (Mayo Clinic 2016). Constantly living with these physical conditions without appropriate ways to cope has been directly related to impacting both an individual's health and risk for injury (Williams and Anderson 2012).

Mindless Thinking

Judgmental and self-critical thoughts, projections of insecurity, and catastrophic thinking can occur in the face of social stressors without coping skills such as mindfulness. As human cognition, has evolved, the attitudes and perceptions of the body's physical reactions to stress have led to worry, negative self-talk, irrational thinking, and sometimes even depression, anxiety, or angry outbursts. When an individual is caught up in these emotions, their ability to think clearly and logically is directly impacted (Kozasa et al. 2012).

Origins of Mindlessness

Now that mindlessness has been thoroughly explored, this paper seeks to answer what has caused the significant increase of it in our society. Thanks to the introduction of smart phones, American workers have an increased amount of connection to their workplace. Due to the amount of time spent outside of the workplace checking emails, taking calls, and physically working from mobile devices, it is estimated that the average American's 40-hour work week has turned into 72 hours a week (Eblin 2014). Eblin also approximates that a total of 56 hours is spent on the basic necessities of life such as eating, sleeping, and bathing. With only 168 hours in a week, simple math equates to only 40 hours for everything else (family and relationships, hobbies and activities, chores and transportation, exercise and relaxation).

168 hours – (72 hours connected to work + 56 hours for necessities) = 40 hours

With only 40 hours available to do what gives individuals enjoyment, a sense of purpose, and happiness, American workers are rushed, over-worked, and overwhelmed. In order to keep up with unrealistic expectations, mindlessness, auto-pilot, and reactive responses to stimuli have taken over. It has been found, however, that this can be reversed. Simply being aware that one hasn't been aware, is very motivational to making sustainable and positive change.

Keys to Mindful Living

As shown in Figure 1, being mindful may simply be taking a purposeful approach to daily life and everyday tasks or practicing non-critical thinking.

Mindlessness	Mindful Living
• Auto-Pilot • Thought After Thought • Multi-Tasking • Self-Criticism • Habitual and Unconscious • Judgmental • Tense Muscles	• Voluntary; Purposeful • Quiets Brain Chatter • Being in the Moment • Non-Critical • Conscious Awareness • Non-Judgmental, Open, Curious

Figure 1. The differences between mindless and mindful living.

Mindfulness is referred to as a practice because it is a skill that requires regular maintenance and exercise. Practicing mindfulness doesn't require any fees, a significant time commitment, epiphanies, or even a major lifestyle change. Below are eight exercises to unlock mindful living.

Slow Down

As previously discussed, being pressed for time and rushing often leads to operating in auto-pilot. A common complaint among workers is that when late for work, they rush out of the house, only to realize they had forgotten something needed for the workday. Upon noticing the need to rush, an individual can elicit their own thinking by asking "What do I need to complete my tasks today?" As part of this example, utilizing this simple question is a form of mindful living that can directly impact a worker's safety by ensuring they have the tools necessary to complete the job.

Focus

Society places a constant pressure on an individual to multi-task. Making an effort to do one task at a time can improve a person's ability to stay in the present moment. Texting while driving is illegal in some states for a reason; doing so causes distraction and puts all drivers on the road at risk for injury. Reducing the amount of time spent multi-tasking increases the amount of time spent living mindfully. By focusing on one task, such as operating an automobile without distractions, the risk for injury decreases.

Engage the Senses

Sensory engagement is one of the simplest and quickest ways to get grounded and centered. After developing an awareness of operating in mindlessness, paying attention to the details of what can be heard, seen, felt, tasted, or smelled is a tool for switching the auto-pilot off. Practical application often begins with simply daily chores. For example, while washing the dishes, try paying attention to how the light reflects off of the wet utensils, the smell of the detergent, the texture of the suds on the sponge, and sounds of the water running. In addition to being fully present, this type of exercise also allows for practice of switching from mindlessness to mindfulness. With further practice, the skill is then applied in other areas of an individual's life that will directly impact the ability to be safe and work injury free.

Engaging in a Task As if For the First Time

As shown in Highway Hypnosis, experts are often prone to working and functioning in auto-pilot. What would be required of drivers to engage in the task as if for the first time? They might be more aware of all signs on the road, the speed at which they drive, and other drivers on the road. Being more mindful of these things can clearly lead to safer operation of the vehicle. Applying this exercise to other work tasks gives enough pause to examine what tools are needed, how an injury can occur, and planning and action for prevention.

Body Mindfulness

Body mindfulness is being aware of signals the body is communicating, exploring the origin, and taking action to address the causes. A study in 2007 showed that most workers described working through pain as "just part of the job" (Silverstein and Adams). This attitude and perception of body mindfulness creates a greater risk for workplace injury. Instead of ignoring or accepting discomfort and other physical abnormalities, body mindfulness allows for a proactive and preventative approach self-care management and injury free work performance.

Explore

Instead of reacting instantly in a situation, pausing and exploring the message of the emotion can allow for more appropriate responses. If experiencing sadness, ask "What have I lost?" With anxiety, ask, "What am I afraid of?" When it's anger, "How am I being threatened?" Lastly, follow those up with "Is this true? Is this the best way to be thinking right now?" Exploring emotions can help debunk irrational thinking with logic. Practicing mindfulness offers the option to turn down the volume of negative self-talk and irrational thinking to a manageable level. Once the thought is observed, mindfulness allows a shift of intentions to identifying the message of the emotion and then taking appropriate action.

Meditation

Meditation has many names from deep breathing, guided visualization, Shavasana, to progressive muscle relaxation. Contrary to what many believe, meditation does not involve "thinking about nothing" nor does it require an individual to come to some sort of epiphany. Instead, meditation is simply practicing the skill of allowing thoughts to enter and exit the mind without attaching judgment or reactions. Being able to respond wisely to things that happen instead of reacting blindly is a skill that requires practice. Through meditation, that skill is practice and then available for future applicable use. As a side effect, meditation also allows workers to become relaxed, calm, cleansed, and at a state full awareness with anywhere between 30 seconds to 60 minutes of practice.

Mindful Living Checklist

When used in the workplace, mindfulness is the skill to pause, analyze, respond, and take action. The very first step is teaching workers to first identify when they are operating in auto-pilot. This checklist allows workers to have a plan and approach to getting out of mindless living.

1. Identify auto-pilot?
2. Think about the task?
3. Is this the best way to operate?
4. Implement a mindfulness strategy?
5. Practice mindfulness?

Mindfulness as a Worker Readiness Tool

Upon evaluating the traditional injury prevention approach with a specific client, the author and her associates discovered that returns were diminishing. In environments where work assignments are modified, broadened, or increased, employee engagement in strictly hazard mitigation safety programs were not at the desired level. In addition, it was becoming more difficult to identify workplace stressors for a significant portion of workplace injuries; a majority of those injures appeared to be associated with low risk or routine tasks. In an effort to understand the shift, a survey and scale with which workers could rate their body mindfulness in regards to specific body parts were created. Using the survey, a total of 1129 workers were asked to identify their top two areas of concern.

Figure 2 shows the trends that were uncovered in the population. Not only did this provide direction for health promotion, training, and education, it also increased the mindfulness of all workers. Employees were educated on the trends within the company allowing them to act on increased prevention methods and behaviors.

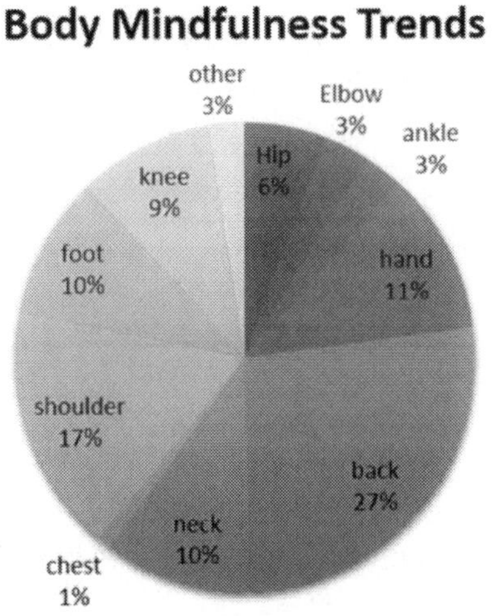

Figure 2. Body Mindfulness Trends

Through follow-up interviews, employees utilized the body mindfulness scale in Figure 3 to rate their concerns.

Body Mindfulness Rating	Example Verbal Description
0 None	I don't notice it at all; I don't think about it
1-3 Mild	Sometimes I feel it; Every once and a while
4-6 Moderate	I feel it often; It's intermittent
7-10 Persistent	I notice it all the time; It's nagging

Figure 3. Body Mindfulness Scale

By working with individuals on utilizing mindfulness techniques in their daily tasks, employees were also taught how to use body mindfulness: listening to, understanding, and acting on the body's warning signs. This lead to an increased amount of prevention behaviors in the workplace. With employees being more in tune with how their bodies feel, they are able to determine the root cause of their concerns in addition to seeking treatment for symptoms. The increase in overall mindfulness allowed employees to apply injury prevention activities, self-management techniques, and proper body mechanics in the workplace. After a year of education, training, and mindfulness practice, 48% of the workers self-reported improvement in these areas.

Conclusion

A comprehensive injury risk reduction program does require proactive health promotion, education on body mechanics, hazard reduction, and training. However, the effectiveness of these methods relies on adequate mindfulness training. Surveys have proved to be a valuable tool to understand employee concern and proactively identify issues that could present as injuries that also aligned with safety trends within the company.

Coaching can be a complement to employer safety strategies designed to reinforce, enhance, and boost employer efforts to prevent workplace injuries. Our Occupational Athlete model works through our unique, innovative application of evidenced-based coaching methodology to injury prevention. It is implemented by the professional, highly trained coaches we embed onsite, right in our customers' workplaces. They act as our "boots on the ground," coaching employees on injury prevention, one-on-one and face to face.

As described in this paper, mindfulness can have a direct positive impact on a worker's overall health and safety. When operating in auto-pilot or a mindless fashion, the likelihood of making mistakes and injury risk increases. There are many forms to practicing mindfulness and very few require little effort or change in everyday life. Research has found that mindfulness practice increases brain function, the ability to problem solve, and the management of pain. The methods of mindfulness discussed are applicable to all work environments, education levels, and populations.

Bibliography

Crenshaw, Dave. 2008. *The Myth of Multitasking: How "doing it all" gets nothing done.* San Francisco: Jossey-Bass.

Eblin, Scott. 2014. *Overworked and overwhelmed: The mindfulness alternative.* Hoboken, NJ: John Wiley & Sons.

Elder, C. "Effect of transcendental meditation on employee stress, depression, and burnout: A Randomized controlled study." 2014, *The Permanente Journal.*

Gotink, R. A. et al. "8-week Mindfulness based stress reduction induces brain changes similar to traditional long-term meditation practice – A systematic review." 2016, *Brain and Cognition, 108*, 32–41.

Kozasa, E. H. et al. "Meditation training increases brain efficiency in an attention task." 2012. *NeuroImage, 59*(1), 745–749. doi:10.1016/j.neuroimage.2011.06.088

Mayo Clinic. "Stress symptoms: Effects on your body and behavior". April 2016. http://www.mayoclinic.org/healthy-lifestyle/stress-management/in-depth/stress-symptoms/art-20050987.

"Mindfulness: The New Science of Health and Happiness". *TIME*. September 2016.

Occupational Safety & Health Administration (OSHA). 2012. "Injury and Illness Prevention Programs White Paper". (https://www.osha.gov/dsg/InjuryIllnessPreventionProgramsWhitePaper.html).

Pickut, B. A. et al. "Mindfulness based intervention in Parkinson's disease leads to structural brain changes on MRI." 2013. *Clinical Neurology and Neurosurgery*, 115(12), 2419–2425.

Rothengatter, Talib and Raphael D. Huguenin.2004. *Traffic and Transport Psychology: Theory and application.* Amsterdam: Elsevier.

Rodbard, Helena W., Kathleen M. Fox, and Susan Grandy. "Impact of Obesity on Work Productivity and Role Disability in Individuals With and at Risk for Diabetes Mellitus." American Journal of Health Promotion 23, no. 5 (2009): 353-60. doi:10.4278/ajhp.081010-quan-243.

Silverstein, Barbara, and D. Adams. "Injured at Work: What workers' compensation data reveal about work related musculoskeletal disorders (WMSDs)." (PDF). Olympia: National Institute for Occupational Safety and Health, April 16, 2007.

Wansink, Brian. 2011. *Mindless Eating: Why we eat more than we think.* London: Hay House.

Williams, Jean M., and Mark B. Andersen. "Psychosocial Antecedents of Sport Injury and Interventions for Risk Reduction." *Handbook of Sport Psychology*, 2007, Gershon Tenenbaum and Robert C. Eklund (Eds.), 379-403. Hoboken, NJ: John Wiley & Sons. (doi:10.1002/9781118270011.ch17.)

Hardwired Inhibitions: Hidden Forces that Keep Us Silent in the Face of Disaster

Phillip Ragain
The RAD Group
Houston, Texas

Overview

There is a common anecdote told by leaders and safety professionals who have been involved in incident investigations: *We found that people knew something was wrong prior to the incident, but no one said anything to prevent it.* Why do people so often stand by and say nothing as disasters unfold? This question confounds leaders, plagues industries and underscores the frustration felt by many who have investigated incidents.

After reviewing the results of a study of safety intervention that we conducted in 2010, this paper looks closely at four different factors that keep people silent. We draw from research in multiple disciplines to formulate a better understanding of low intervention rates during operational activity, and propose a preliminary step that organizations can take to increase the likelihood that their employees will speak up about unsafe activities.

The Significance of Employees' Silence

Employees' willingness and ability to talk about and stop unsafe operations is one of the most critical layers of protection against unwanted events. Managers and safety personnel cannot be everywhere at once. They cannot write rules and procedures for every possible situation, nor can they engineer the environment to remove every possible hazard.[1] As we have learned from detailed reports on recent, high-profile events, when catastrophic failures occur, it is often the result of a complex and unexpected interaction of many different elements.[2]

When people monitor the complex system in which they work — which includes equipment, processes, technologies, reporting structures and other people — and reliably stop both minor and

[1] Hudson (2001)
[2] See Dekker (2011), Hollnagel (2004), Hopkins (2009; 2012).

major unwanted events before they occur, they function as a kind of *agile* layer of protection. Much like the body's immune system, which attacks harmful foreign bodies automatically and without instruction from the brain's executive system, people throughout an organization are uniquely suited to stop unwanted events without direct oversight from leadership; but this agile layer of protection is not as robust as we would like, and many unwanted events occur as people stand by silently watching them unfold.

There is a second consequence of employees' silence — namely, its effect on the culture of the organization. People are inclined to behave in a way that they perceive to be congruent with the social values and expectations, or "norms," that constitute their organization's culture.[3] These behavioral norms are formed through social interaction and communication, as well as observing others' decisions and behaviors. When opinion leaders infrequently or inconsistently stop and address unsafe activities, it leads onlookers to believe that formal safety standards are not highly valued and employees are not genuinely expected to adhere to them.[4] In short, the low frequency of safety interventions in the workplace contributes to a culture in which employees are not positively influenced to work safely.

A Study of Safety Intervention

In 2010, my colleagues and I began studying safety intervention by surveying more than 2,600 employees across industries and in 14 countries. We found that, on average, employees intervene in only 39% of the unsafe operations that they recognize while at work.[5] To our surprise, there was little variation in the results among industries, countries or languages.

These numbers confirmed what most safety personnel knew from experience: Employees' silence is a critical gap; and it is a gap that needs to be honestly explored and resolved.

Stop Work Authority: Necessary and Insufficient

A preliminary effort to resolve the problem of low intervention rates is Stop Work Authority. While it has been clearly beneficial, it is insufficient. 97% of the people who participated in the 2010 study of intervention said that their company had explicitly given them the authority to stop unsafe operations. Stop Work Authority's value is in assuring employees that they will not be *formally* punished for insubordination or slowing productivity. While fear of formal punishment can discourage intervention, it is not the only inhibitor. There are other, perhaps more significant forces that keep people silent.

Many have reported that, among other things, the fear of *informal* punishment plays a significant role. While Stop Work Authority protects people from being formally disciplined for expressing their safety concerns, employees often explain that the potential for informal retaliation,

---

[3] Asch (1956) and Campbell et al. (1998)
[4] Schein (1992)
[5] Ragain et al. (2011)

such as alienation from the group, being labeled as a troublemaker or receiving undesirable job assignments, are more inhibitive than fear of formal punishment.

Additionally, as part of our 2010 study, my colleague and I asked people to give reasons for their silence. A quarter of respondents (24.6 percent) said that they usually choose not to intervene because the other person would become defensive or angry. We found that this is not an unwarranted concern, as the study also showed that employees react defensively in more than a quarter of interventions (27 percent) and angrily in one out of every six interventions (16 percent). Interestingly, respondents overwhelmingly indicated that they would welcome another employee's intervention in their own unsafe decision or behavior, which suggests that the prevalence of defensive and angry reactions has something to do with the manner in which interventions transpire.[6]

Why We Don't Speak Up

Many attribute this problem to employees' lack courage or motivation to speak up. This belief is unfortunately common among leadership, represented in a common refrain — "We communicated that it is their responsibility to intervene in unsafe operations; but they still don't do it. They don't take it seriously." This belief has driven the development of an array of training programs intended to persuade employees to be more "courageous" and "care" about their fellow workers. Our research indicates that these efforts are largely misguided.

Contrary to this common trend, we have conducted one-on-one interviews with more than two thousand frontline employees and nearly all of them, regardless of industry, culture, age or other demographic category, genuinely believe that they have the fundamental, even *moral* responsibility to watch out for and help to protect their coworkers. Their silence is not due to a simple lack of caring. People generally recognize their unique responsibility to help ensure the safety of others, and they genuinely believe that if they were to observe a situation in which they or someone else were at risk, they would speak up without hesitation. We know that when given the opportunity, though, people usually do not speak up, which presents us with a puzzling question: *How is it that people can take this responsibility seriously and believe that they would intervene, but, more often than not, remain silent when the opportunity presents itself?*

Context, Salience and Decisions

Behind this puzzling situation is something called the *Context Effect*, which is a psychological principle that helps us understand why people do all manner of things that they usually would consider to be inappropriate. It brief, it means that what people think about, remember and care about at any given moment is largely determined by the immediate context in which they find themselves. People perceive the world differently from one moment to the next as a result of the social, physical, mental and emotional factors that are most *salient* (obvious, pressing and significant) at that moment. As such, what people choose to do is subject to the peculiarities of their

[6] We have conducted additional research into this issue and found that two common phenomena play significant roles: *Reactance* and *The Fundamental Attribution Error*. The former affects how people interpret and respond to another person's intervention in what they are doing. The latter affects what people say when intervening. While they relate to the topic of this paper, they are not within the current scope.

specific situations. Clearly, the physical, social and mental context of operations is very different from that of safety meetings in which employees will heartily commit to watching each others' backs and stoping unsafe activities. With this concept as a backdrop, we arrive at the next question: *What factors in employees' production contexts play the most significant role in keeping people silent?*

Before answering this question, we must first define a term that was used in the previous paragraph: *Salience*. The factors that influence people's decisions can carry different weight, and the weight that they carry can change over time. Consider the commonly offered piece of advice, "Don't go to the grocery store on an empty stomach." Failing to heed this advice, of course, leads to buying more food than is actually needed, because the purchasing decisions that one makes on an empty stomach are heavily influenced by hunger. The pangs of hunger carry more weight than one's need to shop judiciously. In this context, hunger is more "salient" than perhaps we would like it to be.

Surprisingly Salient Factors

In the production context, there are certain factors that — like hunger at the grocery store — can be profoundly influential over people's decisions either to intervene or stay silent. They are more influential than we might think they should be, and this is because people's minds are "wired" in somewhat unexpected ways.

Production Pressure

It is well established that people tend to increase their speed when they feel pressure to produce. Personal experience and research both support this postulation; but production pressure can do more than make people rush. It literally changes the way that people perceive the world around them.

A study conducted by John Darley and Daniel Batson, which is commonly referred to as "The Good Samaritan Study," demonstrated the impact of production pressure on people's perception of their surroundings, as well as their willingness to help someone in need.[7] The subjects in this study were students at Princeton Theological Seminary. They were given the task of preparing a speech on the parable of the Good Samaritan — a story in which a man from Samaria voluntarily helps a stranger who was attacked by robbers and left beaten and half-dead along the road.

The participants were divided into different groups. In the experimental group, participants were rushed to complete this task. They were then sent from one building to another where, along the way, they ironically encountered a shabbily dressed confederate slumped over and appearing to need help. The researchers found that participants who were under pressure were much less likely to stop and help the person in need. People in the "High Pressure" group helped only 10 percent of the time while, by contrast, people in a "Low Pressure" group helped 63 percent of the time. Many participants in the high pressure group reported seeing the person but not recognizing that he needed help. The pressure that they were under changed the way they looked at the situation that they were in.

---

[7] Darley et al. (1973)

This study revealed that even people's deeply held moral convictions can be overridden by production pressure, not because it has eroded those convictions, but because it makes people see the world around them differently in that moment. The need to intervene was less salient than the need to get something done on time.

Groups

People are less likely to intervene when they are in a group. This phenomenon is known as the *Bystander Effect*. It came to be recognized by psychologists after an incident on March 13, 1964, when a young woman named Kitty Genovese was repeatedly stabbed and eventually murdered by an attacker outside of her home in Queens, New York. 37 people watched and listened from their windows for the half-hour that she screamed and attempted to escape. No one called the police or attempted to help, even when her attacker left on two occasions only to return and continue the attack.[8]

The onlookers' silence in this horrific event was popularly attributed to their "apathy," which, interestingly, is not unlike the prevailing belief today that people do not intervene at work because they lack "courage" or "caring." Social psychologists, including John Darley and Bibb Latané, suspected that there was more than simple apathy at play, and began conducting research into why people failed to intervene.[9] What they found time and again is that the mere presence of other people strongly inhibits individuals from intervening. In fact, the more people there are, the less likely it is that any one person will speak up. It is an unexpected oddity in how people's minds are wired.

In one study conducted by Latané, people's reactions were observed as they heard a woman fall and cry out in pain.[10] When subjects were alone, they were found to help 70 percent of the time; but when a second person was present, they helped only 7 percent of the time. Being in a group can profoundly inhibit people from intervening.

Tasks

It probably seems comically ill-advised to suggest that tasks — that is, work itself — inhibits people from intervening; but it does. Well-intentioned, caring and courageous individuals will often fail to intervene in unsafe activities because they are focused on a task. This is not only because their focus makes them blind to the risks around them, which is by now a well circulated phenomenon called "Inattentional Blindness,"[11] but also because people have as part of their cognitive functioning a certain distortion in how they make judgements. This distortion is called the *Unit Bias.*[12]

Think about a time when you were focused on something and realized that you should stop to deal with a different problem, but decided to complete the original task anyway? That is the *Unit Bias*. It is a distortion in the way that we view reality. In the moment, we conclude that complet-

---

[8] Martin (1964)
[9] Darley et al. (1968)
[10] Latané et al. (1964)
[11] Simmons et al. (1999)
[12] Geier et al. (2006)

ing the task at hand is more important than it really is, and so end up putting things off that, outside of the moment, we would recognize as far more important.

The role of the *Unit Bias* is perhaps most easily visible at the dinner table. Studies have shown, and many readers will recognize, that an effective way to control the amount of food one consumes is to change the size of the plate on which the food is served — that is, intentionally shrink the "unit" that someone feels compelled to complete.

Now imagine that an employee is focused on a task and sees a coworker working unsafely. "I'll get to it in a minute," he thinks to himself, "as soon as I'm done with this." The *Unit Bias* distorts judgement, and the importance of completing a task becomes more salient than protecting people's lives and wellbeing.

Authority Figures
In a 1963 study of human behavior, Stanley Milgram demonstrated that people would do things that are dramatically inconsistent with their own values when under instruction from a perceived authority figure. In his now famous "Behavioral Study of Obedience," 65 percent of participants willingly pressed a button that they believed was administering a massive 450-volt electrical shock to a man whom they had met only minutes before.[13] Following relatively minor petitions from a person in a white lab coat, and while showing clear emotional distress, these subjects were willing to kill another human.

This happened because the subjects "deferred" responsibility for their actions to the perceived authority figure, and this same deference of responsibility to authorities can happen when people observe an unsafe activity. Employees becomes significantly less likely to intervene in the presence of an authority figure because, often without realizing, they have surrendered that responsibility. The simple presence of a manager, supervisor or more experienced coworker can be a surprisingly salient force and a covert inhibitor of intervention on the job.

While our research has uncovered other factors that inhibit people and reduce intervention rates, including the social and cognitive forces behind people's assertion that they do not intervene because, "The other person will become defensive or angry," the purpose of this paper is to bring to light four significant but easily missed factors that keep people silent on the job.

Moreover, it is our hope that, by identifying these surprising influences on people's decisions to remain silent in the face of disaster, readers will move away from the common belief that their employees merely lack courage or caring. This "*apathy myth*" is simply inconsistent with the last six decades of research in intervention and other pro-social behaviors.

Overcoming the Inhibitors

Though a comprehensive strategy for increasing intervention rates is multifaceted in its structure, a preliminary step can be taken by organizations to move beyond Stop Work Authority and overcome the factors that suppress intervention. This involves educating employees about these com-

[13] Milgram (1963)

mon inhibiting forces and how working contexts are likely to shape their decisions in unanticipated ways.

With this first step as part of a comprehensive strategy for improving intervention rates, an offshore drilling contractor measured a 124 percent increase in "Stop Work" events during the three months following implementation. The correlation between this result and employees' understanding of the four "inhibiting forces" described here is supported by a study conducted by Arthur Beaman, in which people were given a lecture on the social and psychological factors that influence their decisions to help others in need — including the *Bystander Effect*. Two weeks later, in a different setting, subjects were presented with an emergency. In response, 67 percent of the subjects who had heard the lecture intervened. By comparison, only 27 percent of people who had not heard the lecture intervened.[14]

The need to improve intervention rates during operational activity carries with it a need to support those people who would intervene. This support must move beyond well intentioned cajoling to be more courageous or caring. It must seriously take into account the sometimes surprising, often hidden but nonetheless very real forces that keep people silent.

Bibliography

Asch, S.E. 1956. "Studies of Independence and Conformity: A majority of one against a unanimous majority." *Psychological Monographs.* 70(9): 1-70

Beaman, A. L. et al. 1978. "Increasing helping rates through information dissemination: Teaching pays." *Personality and Social Psychology Bulletin. 4*(3): 406-411.

Campbell, Jennifer D. and Patricia J. Fairey. 1989. "Informational and Normative Routes to Conformity: The effect of faction size as a function of norm extremity and attention to the stimulus." *Journal of Personality and Social Psychology.* 57(3): 457-468.

Darley, J. M. and C.D. Batson. 1973. "From Jerusalem to Jericho: A study of situational and dispositional variables in helping behavior." *Journal of Personality and Social Psychology.* 27: 100-108.

Darley, J.M., and B. Latané, B. 1968. "Bystander Intervention in Emergencies: Diffusion of responsibility." *Journal of Personality and Social Psychology.* 8(4): 377-383.

Dekker, S. 2011. *Drift into Failure: From hunting broken components to understanding complex systems.* Surrey: Ashgate Publishing, Ltd.

Geier, A.B.; P. Rozin, and G. Doros. 2006. "Unit bias. A new heuristic that helps explain the effect of portion size on food intake." Psychological Science. 17(6): 521-525.

Hollnagel, E. 2004. *Barriers and Accident Prevention.* Hampshire: Ashgate Publishing, Ltd.

[14] Beaman et al. (1978)

Hopkins, A. 2012. *Disastrous Decisions: The human and organisational causes of the Gulf of Mexico blowout*. Sydney: CCH Australia Limited.

______. 2009. *Failure to Learn: The BP Texas City Refinery disaster*. Sydney: CCH Australia Limited.

Hudson, P. T. W. 2001. "Safety Management and Safety Culture: The long, hard and winding road." *Occupational health and safety management systems*. 2001: 3-32.

Latane, B.; Rodin. J. 1964. "A lady in distress: Inhibiting effects of friends and strangers on by-stander intervention." *Journal of Personality and Social Psychology*. 5(2): 189-202.

Martin, G. 1964. "37 Who Saw Murder Didn't Call the Police; Apathy at Stabbing of Queens Woman Shocks Inspector." *The New York Times*. 27 March, 1964: 1.

Milgram, Stanley. 1963. "Behavioral Study of Obedience." *The Journal of abnormal and social psychology*. 67(4): 371.

Ragain, P.C. et al. 2011. "A Study of Safety Intervention: The causes and consequences of em-ployees' silent." *EHS Today*. (http://ehstoday.com/safety/management/study-safety-intervention-0701/).

Schein, E. 1992. *Organizational Culture and Leadership*. San Francisco: Jossey Bass.

Simons, D.J. and C.F. Chabris. 1999. "Gorillas in our midst: Sustained inattentional blindness for dynamic events." *Perception* 28(9): 1059-1074.

Session No. 762

Developing Leading Indicators for a FSI Program Using Risk Assessment Data

Richard D. Hilton, CSP
Manager, Corporate Safety

Paul Adams, Ph.D., CSP, PE, CPE
Safety Specialist

3M Company

Overview

An innovative process for managing hazards with higher potential to result in fatalities or serious injuries (FSI) is significantly reducing risk in a multinational company. After nearly four years of sustained implementation at over 200 manufacturing sites, successes and opportunities alike have been experienced. Data collected from over 60,000 risk assessments has provided a rich opportunity to tease out leading indicators of both performance and quality of risk reduction programs.

The strategy that our company has adopted to deal with infrequent yet severe incident outcomes integrates a common approach to risk assessment, prioritization and management. It embodies a systematic evaluation of safeguards meant to control those hazards that have a high potential to result in fatal or otherwise severe outcomes. This approach in 3M is called Risk Assessment and Prioritization (RAP). The RAP process is embedded in the company's Worldwide Incident Management System (WIMS).

Background

The RAP process is used as an investigative tool following actual incidents but it is most beneficial in risk reduction and incident prevention when used for analysis of near miss events and potential FSI events. The initial risk levels of hazards are systematically prioritized based upon how effectively (or not) the safeguards are found to be performing. The assessment results are then used to define the most effective and efficient mitigation strategy when further risk reduction is warranted.

A final hazard risk assessment is performed after mitigation is complete so that overall risk reduction can be quantified. When adequate hazard controls are in place, attention is directed at maintaining their effectiveness through a routine monitoring process.

This approach to risk management supplements rather than replaces existing hazard management tools. When used to its fullest potential, the approach can incorporate the results of other hazard-specific techniques into the determination of how effective the "system" actually is at controlling hazards with severe consequence potential.

Without question, the implementation of this tool has effectively lowered risk levels associated with many higher hazard tasks and activities to acceptable levels in a well-documented, defensible and quantifiable manner. The opportunity to understand performance and define the quality attributes of the risk management process has become a critical driver in the success of the program.

Developing Leading Indicators of Risk Management

WIMS was initially developed to capture data from incidents, near miss events and potential hazards. It includes results of investigations and the creation of corrective actions along with their tracking, completion and closure. Leading indicators that track the investigation process and closure of action items are well established. The WIMS database now delivers measures in risk management as part of the risk assessment steps which are required as a part of all investigations. The steps below highlight how performance measures in risk management have been developed.

Establish the Risk Assessment Process

The RAP process consists of the following key steps:

1. Identify high hazard activities (HHA) using analysis of facility incident data, standardized lists, and brainstorming by people familiar with plant operations.

2. Identify the specific tasks associated with the high hazard activities, along with existing safeguards and controls for safe and compliant operation.

3. Conduct team observations and interviews of people who perform the high hazard activities.

4. Evaluate existing safeguards and controls for strength and reliability during the observation and whether the potential need exists for improved or additional layers of protection.

6. Take appropriate prioritized action(s) to improve the strength, reliability and effectiveness of safeguards and controls for high hazard activities.

7. Establish regular monitoring of safeguards and controls for continued effectiveness.

Use a Data Management System to Conduct and Manage Risk Assessments

1. Information collected from risk assessments is maintained in a manner to monitor the progress and quality of HHA risk reduction.
2. Data in risk assessments is searchable and sortable including by types of HHA.

Define the Process Performance Criteria and Create Measures of Expected Performance

1. An evergreen list or "inventory" of HHA is maintained by individual departments or site.
2. A Risk Reduction Plan (RRP) is created by each site and sets objectives for completing HHA risk assessments during each year. Progress to plan is measured quarterly.
3. Risk Reduction achieved is measured.
4. Criteria for quality of risk assessments is built-in the WIMS database.

Develop Timely Reports to Monitor and Help Improve Risk Management Performance

1. Actual and potential significant events
2. Activities with high initial risk levels
3. Activities with high potential consequence
4. Time to closure of HHA assessments
5. Progress of risk level reduction by HHA
6. Risk assessment quality

Conclusion

To date, over 60,000 risk assessments have been completed which resulted in the identification and reduction of more than 17,000 high initial risk levels. These risk reduction activities have coincided with a major decrease in severe incidents over the past three years. This represents about 50 fewer significant injury/illness incidents per year in 3M or a rate reduction of over 40%.

True risk level reduction is highly dependent on understanding how well our safeguards and hazard controls are working. To reduce the frequency of high severity incidents, safety controls must be sufficiently robust so that the failure of any one or more elements will still not allow a fatality or severe event to occur. The RAP tool provides a disciplined and measurable approach to assess and monitor safety controls for high hazard activities.

Session No. 765

A Systems Approach to Occupational Safety

Paul Roberts, MS, CSHM
Mark Hargett, CHST

Trivela Safety LLC
Brighton, Colorado

Introduction

In order to consider occupational safety in the context of systems, one must first have some understanding of what a system is and how systems operate, interconnect, and adapt in light of systems of interest within an operating environment. While an in-depth analysis of systems would take several of these papers, Professor Derek Hitchins sums up systems as follows:

> Essentially, the [systems] approach considered a system-of-interest (SOI) to be open, dynamic, to exist in an environment, to interact with — and adapt to — other systems in that environment, and to form part of a larger, wider system. The systems could be of any kind, but are generally characterized as functional, i.e., the systems, subsystems, containing systems, etc., all perform functions and exhibit behavior (D. Hitchins, 2016).

In our application to Occupational Safety, this basic definition fits perfectly. The system is the organizational function and the systems of interests (SOI) are the specific operations of those organizational functions. For example, production would be a system, but breaking down production further would uncover SOIs, subsystems, and containing systems that are interconnected and must adapt to the overall system. Such systems could include purchasing, supply chain, human resources, accounting, and of course safety. However, in the organizational structure we know these SOIs are often free standing systems on their own with their own sets of SOIs and subsystems. Think of this in terms of interacting systems and SOIs contained within an operational environment.

The Problem

We see that many in the safety profession battle to find their place within this operational environment. We regularly hear safety professionals lamenting operational focus at the "expense of safety"; so they assert their authority from the periphery, often controversially, sometimes disruptively. One of our earliest memories of ASSE as new members was a first chapter meeting. Sitting silently, we listened to a group of five "seasoned" safety professional discussing the various processes they "shut down" that past week in response to management's "lack of regard for safety". Within the next 90 days, three of the five stood up at the beginning of the chapter meeting to announce they were "looking for their next opportunity". The three most vocal of that original group had been fired from their safety positions. One of those three has not held a consistent safety position for over five years now.

Conversely, we have heard frustration from executives and business owners who feel their safety people are out of touch with operational realities and simply don't understand business dynamics. They feel their safety departments cut into profits with no return on investment. They feel safety people are never happy, always want more, never feel supported, and can be very adversarial—often passive aggressively—and thereby disruptive to business operations. Some executives have gone as far to acknowledge the "necessary evil" of the safety manager, but more often the desire is a knowledgeable, responsible, practical, and cooperative approach that brings results, or in other words, true value to the organization.

Despite this—or maybe in support of it depending on your perspective—we have heard, observed, and worked with many organizations that state safety is a key focus, many times a #1 priority, while the safety director feels alone on his safety island and management is frustrated. Take these quotes from one of our clients who serves as a prominent and well known company's Corporate Director of Safety:

- CEO's speak safety for the benefit of the shareholders.
- Executives didn't care.
- Safety is a separate entity…until something happens.
- Safety is a nuisance.
- Safety is tolerated, but hated.
- Safety is truly a secondary consideration.

For some these quotes may be surprising, but we believe the majority of safety professionals can not only relate to these sentiments but may be mired in this same scenario upon returning to work after attending Safety 2017. However, consider this perspective from a client business owner:

- Safety people think I don't get it, but I do.
- Safety people aren't consistent.
- Safety looks for quick fixes, but doesn't look at long term impact.
- Safety doesn't account for ROI, they don't know how to think in terms of dollars and cents.
- Being profitable allows for more cost centered resources.
- Safety people don't understand the business operation as a whole.

So our questions are, how do we bring the sides together in a truly cooperative manner? How do we get everyone on the same page? How do we truly get safety into the company's consciousness? How do we get management buy-in at all levels? How do we provide return on the investment of safety? How do we add value to the organization? We often hear the answer is that a culture of safety is needed. In that context, we ask you to consider the following questions:

- What is safety?
- What is a safety culture?
- What is the role of the safety manager in safety?
- What is your role in safety?

- What is the role of your employees in safety?
- How do you define and accomplish a safe work place?
- Should you strive for safety culture?

While is in not our intent to answer these questions, per se, we do think there is reason to consider these questions in light of the topic at hand.

The Systems Concept

> Causation for successes or failures within any organization are rooted in the underlying systems that drive the organization. When weaknesses are present, the system is not fully implemented or the system is absent altogether, hazards and behaviors surface resulting in a myriad of exposures to the organization. At Trivela Safety, we believe safety systems are just one of the systems that drive organizational operations, but we believe safety systems are just as important to the success of your business as any other business system you utilize whether that system is your supply chain, production and operations management, inventory, or financial—just to name a few. As such, we believe safety systems should be used to measure success and determine the cause when safety failures occur (P. Roberts, 2013-2016).

The above statement was written to a client after spending more than 6 months assessing serious safety failures that manifested themselves in many types of incidents, some quite serious and catastrophic in nature. The goal in writing this statement was to define a fundamental approach to tackling occupational safety in the context of the existing organizational environment. The effect was profound in that it opened the door to serious and honest discussions on safety within the context of the existing organizational values of the subject client.

Since that time, this concept has been integrated into our firm's consciousness as a guiding directive on analyzing organizational safety as third party consultants. As we embraced this concept, the directive led our teams to a simple conclusion—we value people and we believe people matter. Protecting people quickly became our common purpose and our mission as a firm is now defined by this common purpose. But aside from that, it gave us the basis to tackle complex safety management issues in the context of the quotes and questions presented in *The Problem* section of this paper. You could say it gave us a different perspective.

The systems approach to occupational safety focuses on safety in context rather than trying to create a culture that may be counter to the organization's values. In saying this, we are not saying companies do not want to be safe. Rather we are saying safety as a priority is much different than safety as a value. After all, we know priorities change. But values, they are the driving force of who we are and what we do. Organizations operate fully engaged in their values whether these values are stated or not, and people who embrace the organization's values become the leaders of the organization. But in order to address failures within the organizational system, the organization and all who operate in it must be honest about these values. In its truest form, the systems approach works to identify failures equally if not better than it does to support proper functioning of the environment.

Is It Magic?

It has been said great values make for great organizations. It has also been said great values are what sets visionary organizations apart from their competitors and those who try to mimic their success (Collins, Porras, 2002). Such visionary organizations define, teach, promote, support, and are fully engaged in the values that define them. These organizations do not allow competing "values" or cultures. These values are deliberately nurtured by organizational leaders, and great leaders figure out how to get everyone working toward a common goal by harnessing the power of the defined values. We like this quote attributed to Walt Disney circa 1954 as he was building Disneyland: "Of all of the things I've done, the most vital is coordinating the talents of those who work for us and pointing them toward a certain goal" (Collins, Porras, 2002, p. 91).

If you ask a Disney cast member (employee) how they keep a park so clean or deliver such an amazing attraction, they will often answer "it's magic!" At a recent Disney Institute training we learned Disney is a company of storytellers. They tell stories to make emotional connections with their customers, who they call guests. Everyone in the organization values the guest experience. Everyone understands their role within the value system because Disney has designed four keys for delivering Disney's quality standards—the organizational value which they call the Common Purpose. Safety is not only one of these keys, but it is the number one key (Disney Institute, 2015).

Being skeptical, we questioned this directly to Disney management pointing out how it must be easy for a multi-billion company to make this level of commitment to safety integration. The response was strong and unequivocal, Disney is a multi-billion company because of this commitment. But it's not the commitment to safety that they were referring to, it was the commitment to the common purpose as defined and designed. It was the commitment to the system in action. Originally, we missed this because we were thinking one-dimensionally as safety professionals—something they seemingly picked up on as they answered multi-dimensionally as fully engaged leaders. But this does not mean Disney's culture is a "safety culture". Imagine walking through a Disney Resort that is defined by our traditional professional concept of safety culture. It would ruin the whole experience.

No, Disney is defined by the common purpose of "We create happiness…" (Disney Institute, 2011, p. 45) and everything they do is designed to give you, their guest, the best service experience possible. Why? So you are convinced the money you spent for the experience was worth it, was treasured. Why? Let's be honest, ultimately it is so you come back and spend more money. This is Disney's drive—the same drive as all for-profit companies—to make money, but, for Disney at least, not to make money by sacrificing your guest experience. This means, in part, safety is an ingrained and integral characteristic of the value system—after all, imagine the impact safety failures would have on the customer experience. However, safety is only one key and it works in conjunction with other key identifiables that all work together within all the operational systems which cause the entire organization to be viable. Here's the reality—Disney has defined how to fully integrate safety into their context of the organizational environment through defining the common purpose and designing the keys to achieving the common purpose. Simply put in terms of the systems approach, safety operates as a fully functioning SOI within all other Disney operating systems.

Does it work? Yes! I believe it. I have seen it in action. Not just at Disney, but at other organizations who understand safety as a system rather than safety as an entity or department or priority. Several exceptional organizations have integrated this systems approach—regardless of what they actually call it—successfully. They are purposed to deliver a service or product in a premeditated manner that takes into account the entire organizational environment. It is a universal or holistic approach to the entire business operation. It looks deeper than risk factors or compliance. It reaches into the very essence of the purpose of the organization's existence.

Safety as a System

I am certain many of us know the story of Paul O'Neill's first speech as the new CEO of Alcoa in 1987. He delivered a speech on safety that many say was the greatest CEO speech ever, and many credit this as the catalyst for Alcoa's monumental financial turnaround. In his book, *The Power of Habits* (2012), Charles Duhigg states that O'Neill "attacked one habit" that changed the entire organization. The result was a stunning $27 billion in increased market capitalization, effectively increasing the size of the company by a factor of 5 (p. 100). However, habits don't change with a simple declaration from a new CEO. There was much more involved in this process:

> [O'Neill] started with a powerful "why"...He then moved to "how" they were going to improve safety and then the "what". Other leaders communicate the "what", "how" and then the "why". It would have been easy (and expected) for him to have just talked about improving sales and reducing costs (what) and then "how: they would do it...Instead, he chose to talk about safety and become the champion for Alcoa workers (Ross, 2012).

When thinking about safety as a system, the idea of culture creation is scrapped and replaced with the development of a functioning system within the defined organizational value structure. This is not accomplished by leaving aspects of the operational environment to chance or to the winds of circumstances dictated by revenues, profits, employment challenges, collective bargaining, incident rates, regulatory mandates, or any other similar such factor. In the absence of clearly defined purpose or values, the organization becomes managed by priorities which are often in the moment, fleeting, sectionalized, and in conflict. Safety systems should function in the same context and with the same purpose/desired outcome as all other systems within the organization as defined by the common purpose of the organization. What is key is defining the common purpose and then designing how the common purpose is to be achieved.

Often, I hear many safety people lament the profit driven motives of their company while failing to understand their role in it. To be successful, safety people must understand and embrace the goal and responsibility of the for-profit company's profit driven operational environment. Simply put, it is what it is. Paul O'Neill did not roll out safety at Alcoa at the expense of profits, he did it with profits in full view. "With respect to safety, there seems to be near-perfect alignment between doing the right thing and enhancing profitability" (Basu, 2016). This fact has been proven over and over again, and while I think we know this, as safety people we too often times think one-dimensionally and miss the fact that there can be more than one positive outcome to our safety endeavors.

Today, safety must be part of the organizational environment; however, this does not mean it can be a culture within the organization. Rather, safety must be a defined element of the common purpose so that it can then become a valid SOI or subsystem of all the other systems functioning in the organization. This is not accomplished through the creation of safety departments with safety officers whose purpose is to enforce safety rules, although this is not a detriment to the process. Rather the idea is to integrate safety into the consciousness of the organization by supporting and promoting overall organizational values. This is accomplished through understanding the real purpose of the organization, and this is where true buy-in across all levels of the organization is achieved. This is where everyone comes to the realization they are all on the same team—the team that is working to promote the known defined values and common purpose of the organization.

Safety professionals should take strong notice of the fact it was the new CEO of Alcoa who is credited with instituting and implementing the successful integration of safety into Alcoa's consciousness—not a safety director. Many will argue safety was trying and never got the support until O'Neill came along, but that is selling other business leaders short. Business leaders are there to make money and they will listen to sound plans that increase revenues or lower costs with an eye toward increased profits. We must understand many business leaders want the same thing we want, and they are paying us to lead in this key financial area.

It is our position that as safety professionals, we must be more educated on the business aspects of the operation and more supportive of the common purpose of the organization. We must get out of our silos and seek to embrace a process that integrates safety into the organizational environment. We must develop sound safety business plans that seek to meet the company's common purpose and are at minimum fiscally responsible. But if you really want the ear of the owner, executive officers, management and even employees, you must seek to add value and add substantial value. Understanding how safety can work as an integrated system within the organizational context is how this can be achieved with long-term results.

Bibliography

Basu, Anirban. 2016. "Safety and profits: Increasing awareness, guidance and insight among construction firms." *Greater Baton Rouge Business Report.* (https://www.businessreport.com/business/safety-profits-increasing-awareness-guidance-insight-among-construction-firms).

Collins, James C. and Jerry I. Porras. 2002. *Built to Last: Successful Habits of Visionary Companies*. New York: HarperCollins Publishers, Inc.

Disney Institute. "*Disney's Approach to Quality Service.*" Four-day workshop presented in Anaheim, California by The Walt Disney Company., Anaheim, CA, November 2015.

Disney Institute and Theodore Kinni. 2001. *Be Our Guest: Perfecting the Art of Customer Service*. Los Angeles: Disney Editions.

Duhigg, Charles. 2012. *The Power of Habit: Why We Do What We Do In Life and Business*. New York: Random House LLC.

Hitchins, D.K. 2016. *"Systems World: Systems Approach."* (http://systems.hitchins.net/systems/systems-approach/)

Roberts, Paul. "*A Systems Approach*." (http://trivela-safety.squarespace.com/a-systems-approach/) (2013-2016).

Ross, Marie-Claire. 2012. "Is this the best CEO Safety Speech ever?" *Digicast Productions: Workplace Communicator Blog*. (http://www.digicast.com.au/blog/bid/90674/Is-this-the-best-CEO-Safety-Speech-ever).

Purple Project: The Innovative Use of Structural Engineering in a Safety Project

Daniel E. Agramonte, P.E.
Senior Managing Engineer
OBG
Atlanta, GA

Companies that manufacture retail products are frequently challenged to differentiate themselves from their competitors. This differentiation can take place at the Point of Sale (POS) and is quite acute in the food and beverage industry. One way to differentiate at the POS is to erect product displays which garner attention from consumers. This is a common industry practice and has led to larger and taller in-store displays. An unfortunate effect of the battle for attention has been an increase in hazards and risks associated with these in-store displays of merchandise. In some cases, this has led to injury to merchandisers, store personnel and consumers.

This paper will focus on these hazards and how a cross-functional team was employed to help address them. The unique team incorporated structural engineering and a process-driven approach to mitigate these risks and hazards. Similar to the United States Military which frequently mixes personnel from the various branches (Army, Air Force, Navy, etc.), resulting in "purple" teams, the complexity of the task at hand required the use of personnel from various disciplines to address these safety risks and hazards.

Background

In 2014, an incident occurred when an in-store product display collapsed onto one of their employees. This resulted in significant injury and a lost time accident for the employee. The ensuing post-review by the client revealed the extent of claims associated with in-store product displays and prompted them to review potential options to mitigate the risks and hazards associated with these displays. While the client was steadfast in the need to display their product using in-store displays, they nevertheless recognized that this practice could lead to potentially unsafe conditions. On the one hand, they didn't want to stifle the creativity of the personnel who take great pride in designing and erecting these in-store displays yet they also realized that reasonable steps could be helpful in mitigating the risks and hazards during the various phases of the in-store display lifecycle. This problem statement needed to be crisply articulated and used as a call to action.

Problem Statement

Based on a preliminary review, it was clear that a tradeoff existed between impactful and creative in-store displays that could be assembled, used and disassembled safely. In addition to the apparent tradeoff between creativity and safety, the post-incident review uncovered two high-level failure modes:

- Injury during assembly of the in-store display
- Injury during the operational lifespan (functional use) of the display

These two failure modes are distinct from each other in that the first failure mode deals with employees who are commonly referred to as "merchandisers" (*e.g.*, product distributor) and the second failure mode deals with retail customers or store employees where the display is located.

With this problem statement clearly understood, existing guidance was reviewed. Given the extent of the client's organization, which comprises hundreds of operational locations throughout North America, the displays are placed in thousands of stores. Based on the review of existing guidance, some conclusions were quickly able to be drawn:

- There was no current, centralized guidance document covering the assembly, deployment and disassembly of in-store display.
- Some areas of the country had taken the liberty of developing in-store guidance, though most of the available guidance documents focused on the marketing of the product and failed to address the risks and hazards associated with the displays themselves. Since it wasn't developed by the same parties, existing guidance appeared fragmented and sometimes even contradicted itself.
- Previous guidance about in-store display safety was limited to one slide in a PowerPoint training presentation. This training was now obsolete and the one slide was inadequate in addressing the specific issue of in-store display safety.

These absence or inconsistencies in guidance created a clear need that was not currently being met within the client's organization. The underlying question was where and how to begin to address this issue.

Starting Point

Following a review with the client, it became clear that any guidance to be prepared had to be credible and yet accessible to a wide range of employees. The number of employees who would need to understand this guidance would likely number in the thousands and they would cover a wide geographic range – North America. During these conversations it was suggested that the following factors would help drive the success of the effort:

- Creating a need for change
- Accessible guidance
- Credible reference materials

- A clear acknowledgment that personnel designing and assembling in-store displays take great pride in their work.

At this early point, key stakeholders were identified, which include in-store merchandising personnel who are typically responsible for in-store display design and assembly. These personnel are keenly aware of the in-store environment and work closely with their customers and retail store management on a daily basis. Their buy-in was essential to the success of the effort. In one of the subsequent meetings, the client identified key stakeholder personnel and a kick-off conference call was organized to discuss the issue. Other stakeholders included the safety department and merchandising management.

Most of the individuals on the kick-off conference call had heard of the incident where the merchandising team member had been injured by the collapsing display. They had also all experienced damaged displays that had been tampered with by store customers and agreed that a few reasonable measures might go a long way toward addressing the hazards and risks associated with the in-store displays.

Gameplan

The kick-off meeting with the merchandising team led to the formation of a small working group. This opportunity was also used to introduce two of the structural engineers to the merchandising team. This proved to be an eclectic team: Most of the merchandisers on the call had never met a structural engineer and the two structural engineers had never interacted with retail merchandising personnel. Surprisingly, most of the merchandising personnel had consistent guidelines they use while assembling in-store displays, and several "best practices" were discussed during the kick-off conference call. These included:

- The use of empty boxes above a certain height and limiting the ultimate height of the display
- Creating a base for the display
- Using certain methods to assemble the base (three cases followed by three cases rotated 90 degrees)
- Certain materials were especially well suited for securing the display base and their use greatly increased the stability of the display built on top of the base.

In a follow-on call, it was agreed that the two structural engineers would meet merchandising personnel at a client warehouse in Columbia, South Carolina. The purpose of the meeting would be to:

- Confirm the ultimate height
- Create guidance for the assembly of the display base, which would include: base assembly, base securing methods and base size/configuration
- Capture additional guidance that would be helpful in mitigation hazards and risks

It was also agreed among the small working group that the outcome of the meeting at the Columbia warehouse include draft guidance for merchandisers and that this guidance would be careful not to be overly prescriptive in order to avoid squelching the artistic flair of the display designers.

A draft version of the in-store display assembly guidance was prepared following the meeting at the Columbia warehouse and was reviewed by the small working group. After minor edits for clarity, this guidance was passed on to the broader merchandising organization team for their review. Below is an excerpt from the guidance that was reviewed by the merchandising organization:

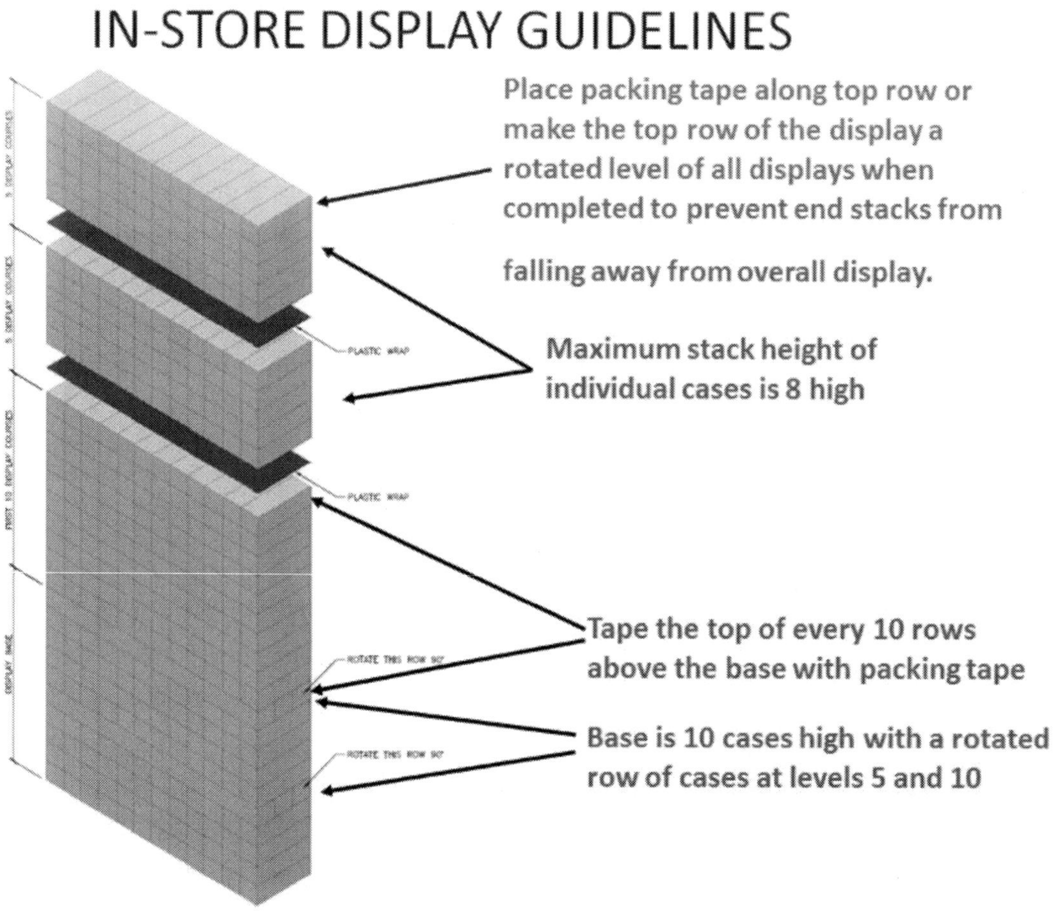

What Next?

On its own, the guidance presents a clear approach to be used when designing in-store displays. Unfortunately, the guidance by itself did not address the entirety of the risks and hazards that were originally noted in the post-incident review. Specifically, these included:

- Safe assembly and disassembly practices by merchandisers
- Protection of the in-store display

The lifecycle of in-store displays was mapped out as follows:

- **Planning** – the design of the in-store display includes the display dimensions and can also impact the method of construction. Most displays follow a theme (sporting event, movie release, etc.) that is typically incorporated into the planning phase.

- **Assembly** – the assembly of the display can begin in a non-store location such as a warehouse or back office. While many times the assembly takes place during non-store hours, assembly can also occur while store customers are present – especially in stores that are open 24 hours. The assembly of the display must take into account several factors such as: (1) safe assembly practices, (2) the surrounding areas of the display, and (3) access to saleable product by store customers.

- **Active** – this phase includes the active "life" of the display. During the active phase, the display must remain intact and cannot present a hazard to in-store shoppers.

- **Disassembly** – the final phase in the lifecycle of an in-store display includes the safe disassembly of the display. This phase included many of the same considerations noted in the assembly phase.

The impact of the display design guidance developed by the merchandisers and structural engineers was significant to the planning phase of the in-store display lifecycle. The limited focus on one portion of the lifecycle still allowed for significant risk.

In-Store Display Assembly

To better understand the assembly phase, the client filmed the assembly of a display in a "big box" grocery store. This video provided valuable insight into the various aspects of in-store display assembly. A safety specialists worked in conjunction with the client's safety team to highlight the following areas of significant concern:

- Isolation of the display assembly area during display assembly – several shoppers were observed passing less than 2 feet from the area where the display was being assembled.

- Potential interference with fire suppression system components (sprinklers, sensors, etc.) – a sprinkler head was noted above the display area and the display could easily have been erected to within 18 inches of the sprinkler head, impeding sprinkler performance.

- Safe lifting – during the video, several instances of potentially unsafe lifting of merchandise were observed. Merchandisers were also frequently seen working alone.

- Ladder safety – ladders were observed to be used on uneven surfaces (e.g., pallets) and the merchandisers were seen working alone on ladders.

- Interference with electrical panels – the designated "display area" is frequently determined by store personnel who may not be aware of the requirement to not block electrical panels.

- Proximity to store exits (in earthquake-prone areas) – the potential for placing displays near emergency exits could result in the prevention of egress. This is especially important in areas prone to earthquakes.

In order to help operationalize effective assembly requirements, an *ad hoc* team was formed to review the requirements above. The team identified key stakeholders within the client's merchandising and safety organizations and established roles and responsibilities for these personnel. Those who supervise the merchandisers were made responsible for the safe assembly of the display and were tasked with ensuring the merchandisers receive the appropriate training on how to safely assemble in-store displays. The client's existing safety guidance was reviewed for applicability to the high risk areas noted above and guidance was prepared to address the gaps that were identified. Since broader guidance already existed for ladder use and safe lifting, the new guidance focused on activities that are unique to the assembly of in-store displays such as:

- Isolation of the display – keeping in-store customers from being impacted by display assembly activities

- Locating the display – this includes proximity to fire suppression system components, electrical panels and store exists

Guidance and training materials were developed to address these assembly-related items. Remaining items were addressed next.

Active and Disassembly Phases

The risks and hazards associated with the disassembly of in-store displays were deemed to be very similar to those identified during the assembly phase. Accordingly, the team felt that the guidance provided during assembly, especially the guidance dealing with isolation of the display, was adequate to address the risks and hazards most likely to be encountered during disassembly. The risks and hazards noted during the "active phase" of the display were quite different.

While merchandising personnel frequently visit and are sometimes based at a store location, they rarely take proactive steps to inspect existing in-store displays. This was deemed a significant risk by the team and a guidance standard was established. Specifically, the guidance standard included:

- Clear roles and responsibilities for representatives from the merchandising and safety teams

- Weekly inspection of the in-store display during the active phase of the display

- Documentation of discrepancies or issues noted with the display

- Identification of a root cause in the event a discrepancy was noted

- Record retention requirements

The weekly inspection guidance had to be clear and be able to be understood by a wide range of merchandising employees:

Description	Yes	No	N/A
Are stacked display items protected by salable product?			
Does the display appear to be sturdy and otherwise structurally sound?			
Any missing elements of the display?			
Has store manager been trained on assessing display for potential hazards?			
Does the display appear to have been tampered with since the last inspection?			
Is the display at least 18" vertically from the nearest sprinkler?			
Are nearby fire alarm actuation/annunciator devices completely visually unobstructed and be least 36 inches away from product?			
Are there any signs of liquids near the display or indications that product from the display has leaked?			
Is the display at least 36" from the nearest exit?			
Is there 48" of aisle space in front of the display?			
Are there any issues that would force the display to be disassembled before the planned display removal date?			
Does the display conform with structural guidelines regarding maximum height?			
Does the display conform with structural guidelines regarding construction method for the base?			
Does the display conform with POS or customer display safety requirements?			

The development of the guidance standard addressing the active phase completed the analysis of the in-store display lifecycle.

Nearing Completion

With the development of guidance standards for the assembly, disassembly and active phases of the in-store display lifecycle, the project to make in-store displays was nearing completion. The documentation developed to date included:

- Design requirement for in-store displays
- Assembly and Disassembly guidance standard
- Active phase guidance standard (in-store display inspections)

Still, the team felt that they were missing the last element needed in order to operationalize the procedures. It was agreed that a PowerPoint training document would be needed to tie things together for the personnel responsible for carrying out the requirements set forth in the guidance standards. The training presentation was developed, included a script and is approximately 1-hour long. Representatives from the client's safety organization incorporated the training, together with the documentation noted above, into the client's health and safety management system and formally rolled out the training to the merchandising organization.

Recap

The scope and breadth of the problem that was identified during the unfortunate incident that resulted in injury to a merchandiser was significant. Addressing the main issues required a broad range of skills, which included:

- Project management
- Structural engineering
- Safety professionals
- Safety management systems expertise

These disparate skills were brought together in order to address the full scope of the issue in an effective manner. The rollout of the guidance began during the latter half of calendar year 2016 and the client will be monitoring claims associated with in-store displays as a trailing indicator of the success of the project.

Session No. 768

Boiler and Pressure Vessel Safety, Hazards in the Workplace!

Craig Bierl, ARM, AIS, AINS
Executive Field Specialist
Chubb Insurance Corporation
Whitehouse Station, NJ

Introduction

Boilers, water heaters, air tanks and other pressure retaining items are common to all occupancies. Many times they are hidden in mechanical rooms, basements and closets. These stationary devices can be hazardous to your facility in many ways, from water damage to carbon monoxide poisoning, to burns and explosions. As a Safety Professional, how do you know if your system is being properly monitored, operated, maintained and tested? Can you identify the basic hazards and conditions in your mechanical rooms?

As Safety Professionals, the immediate hazards we think about are the confined spaces, rotating machinery, electrical hazards and the other components of our safety program. There are many other hazards associated with boilers and pressure vessels that can be readily identified during your plants safety walk through.

Can automation can be hazardous?

Advances in automation and the need for efficiency gains have resulted in reduced maintenance resources in most facilities. Decades ago, boilers and pressure vessels were entrusted to experienced and often licensed operators who conducted routine inspections, tests and the required maintenance while monitoring the safe operation of these units. In today's facilities, it is the exception to have qualified and dedicated personnel monitoring the boiler and pressure vessel operations on a routine basis.

The hazards associated with stored energy have not changed. Boiler and pressure vessel accidents and injuries are still common and avoidable. In many cases, the design of modern boilers and pressure vessels are hidden under the cover of a common metal box. They are common hazards in the workplace and are overlooked on a frequent basis.

The advanced control systems found on modern boilers can give a false sense of security. They provide automatic operation with minimal operator intervention. These advances have created an increased reliability to overall operations yet they have created an apathetic boiler and pressure vessel operational culture.

Changes in boiler and pressure vessel laws in many cases have removed the requirement for a licensed operator. In some cases, there are requirements for jurisdictional inspections; however annual or longer jurisdictional inspection frequencies do not reveal the everyday hazards to the safety of your facility and you might be surprised at what they don't look for.

Boiler and Pressure Vessel Accidents and Injuries

Looking at OSHA.gov data and doing some key word searches you can find that boiler and pressure vessel accidents and injuries are still common. Many go unreported. The ones that are reported show some trends that we can use to identify both the source and the causes. They cross nearly every industry class, from medical facilities, schools, residential buildings, nursing homes, utilities, sports centers to larger industrial paper mills and chemical operations.

When you look at the accidents, injuries and fatalities since 2000, we can find some commonalities that are similar to vehicle accidents. We know that most vehicle accidents occur at intersections and when a vehicle is in reverse. Why? Because these are more complex tasks that require more attention to detail, yet they are routine. Two such areas also exist for boiler and pressure vessel accidents. Boiler and pressure vessel accidents have a much greater frequency during the more complex tasks that in many cases are not routine. These two areas are:

1. The startup and shut down of the systems
2. Maintenance and repair activities
   a. Interestingly, there are only a few "confined space" events recorded

These operations account for the majority of the accidents for both boilers and pressure vessels. Also interesting to note, while accidents during routine operations of boilers were minimal, pressure vessel accidents occur more frequently than boiler accidents during normal operations. Why is that? One possible explanation is that pressure vessels in many cases are largely ignored from an "internal inspection" standpoint. Many jurisdictions have no requirements for periodic pressure vessel inspections.

Looking at the type of injuries that occur and grouping them, we find that there are large concentrations in a few categories:

- Burns/Scalding injuries lead all others as a single type of injury.
- Contusions/cuts/fractures and other similar injuries

These next types occurred at a much lower rate than the top two:
- Electrical shock, however in each reported case it was a fatality
- Carbon monoxide poisoning
- Falls from height were almost always during a maintenance or repair activity
- Chemical exposure and burns

Hazards of Boilers and Pressure Vessels

Hazards we commonly find on boilers and pressure vessels include:

1. Stored Energy: This can be in the form of hot water, steam, compressed air or gases or hazardous substances such as ammonia. An uncontrolled release of the energy and/or contents can have devastating life threatening implications.
    - A standard 50 gallon air compressor at 125 psi contains 159,000 ft/lbs of energy. The threshold for injuries can be less than 1000 ft/lbs of energy.

2. Thermal hazards: Many of these systems contain piping, tanks and valves at a temperature capable of immediate damage to skin when contacted. The same thing can occur from piping and system leaks.
    - A third degree burn takes only five seconds of contact at 140'F. All steam boilers and most water heaters and hot water heating boilers operate above this temperature.

3. Carbon monoxide poisoning: Boiler and water heater combustion systems can produce CO and lead to an event, in most places CO monitoring is not a requirement.
    - There are 20,000 emergency room visits and 400 deaths each year from CO Poisoning.

4. Electrical hazards: Electrical energy can be present even when the system has been isolated form the "pressurized" content. There may be other hazards or systems that are not being properly de energized.

5. Confined Space Hazards: The maintenance of boilers and pressure vessels often requires an entry, yet this is done so infrequently by most plant personnel, are they adequately prepared for the task?
    - There is an average of 92 confined space fatalities each year.

6. Fall hazards: Many routine tasks on boilers and pressure vessels require the use of a ladder and a task at height. The actual task may take less time than the safety precautions required, this presents a challenge for the safety professional on site.

7. Mechanical hazards: Many of these systems have associated pumps and compressors that require mechanical guarding, yet because they are not a "production machine" they are often overlooked.

8. Inhalation Hazards: Some systems contain substances that are toxic or hazardous, yet, they are not always under the proper monitoring or control.

9. Chemical hazards: Boilers and systems are often treated with water treatment chemicals, from oxygen scavengers to polymer compounds. These create routine hazards. They can also be used in the "cleaning" or "maintenance" of the system and present an even greater danger to personnel.

Jurisdictional Inspection data

Data from the National Board of Boiler and Pressure Vessel Inspectors indicates that violations on pressure retaining items occur at a rate of 9.5% over all categories. Violations occur when a deficiency is found during the routine inspection of regulated objects. Examining the data collected from 2012-2016 provides additional insights into the types of issues that are commonly found on these objects.

The safety professional must also recognize that most jurisdictions only regulate and inspect a portion of the boilers, pressure vessels or water heaters you may have on site. There may not be a jurisdictional law on the object or it may be exempt from the law based on the type of object, size or location of the object. These objects would have a higher deficiency rate given that they are not subjected to routine regulatory inspections.

The safety professional should know what pressure retaining objects are regulated and inspected by the jurisdiction within their facility. It is also important that the non-regulated objects be part of an inspection program in order to control the hazards.

Jurisdictional inspections should not be relied upon to identify and prevent all hazards. In example, the State of Indiana does not regulate or inspect the fuel train components of a boiler, yet is a source of a violation in 15% of the objects inspected where it is regulated and 21% of the boiler and pressure vessel violations.

Controlling Energy

Control & Safety Device inspections

Boilers and pressure vessels have controls and safety devices that are required to be maintained, tested and inspected. Devices that are not in good condition and verifiable working order are significantly more likely to fail to perform the intended function at the exact moment when they are needed most. The generic categories are:

1. Pressure Relief devices –
2. Water level or Flow Control Devices
3. Pressure & Temperature control Devices
4. Combustion systems

Boilers and pressure vessels may remain in service for decades. They often remain in service without updates to the controls and safety devices. While some of these are simple single function devices, some, like the burner management systems (BMS) on a boiler conduct many complex operations that must be completed properly to prevent damages or injuries. Modern BMS controls have dozens to hundreds of safety mechanisms while the original model may only have less than a handful of protections built in. Think about this, in 1999 phone communication was completed primarily on land lines. How much has technology changed since then?

Pressure Relief Devices

Commonly known as pressure relief valves or safety valves (PRV's), however they can also be "rupture" or "burst" disks. PRV's account for 24% of the violations found during routine inspections. This device is used to control an unplanned over pressure condition. It is the last line of defense against an explosion and failure of the vessel. In many cases, it is one of the most ignored devices in the plant. A pressure relief device has to be the right type, rating and size, installed properly and it must be subjected to a periodic testing, inspection and maintenance program. We can look at some indications and verify by an external inspection if our PRV's are being properly maintained. PRV's will be found on all pressure retaining items, including water heaters!

- Are there any signs of corrosion? Replace the device
- Are there any signs of leakage? Replace the valve and make sure the system pressure is not causing the valve to leak.
- Are there any obstacles to proper and safe operation? Isolation valves, hoses, lack of a discharge line?
- Is the valve being tested? Does it have pull chains? A discharge pipe? Can it be safely tested?

PRV's must be maintained in a condition that is "like new" in order to ensure they will perform properly. The National Board of Boiler and Pressure Vessel Inspectors (NBBPV) recommends that all valves be tested and inspected routinely and that these devices be subjected to routine service (overhaul, inspection and testing by a qualified relief valve repair firm).

Water Level or Flow Control Devices

In the case of a fired vessel (boiler, water heater), the water keeps the metal from overheating. If the vessel should lose water, the metal will lose cooling and overheating will immediately begin. Did you know that they leading cause of boiler explosions and injuries have been low water failures for a number of decades? Most boilers will have at least two "low water fuel cut off" controls, yet this remains a leading cause of boiler failures and injuries and 9% of the boiler violations that are reported.

Like PRV's, water control devices must be inspected, tested and maintained to the manufacturer's specifications. Also like PRV's, this device is often overlooked in the routine maintenance of the plant. Water controls can be mechanical floats, flow switches or electrical probes. As with any control, the first rule is proper sizing and installation. Also just like the PRV there are many indications that we can find on an external inspection to determine if our devices will function when we need them most.

- Are the covers in place? Missing covers are a sign of adjustments that were probably not authorized and they expose the device to damage.
- Are there any signs of corrosion? It is an electrical device, a failure of the water seal will cause the electrical components to fail.
- Float devices: Are they being maintained, flushed, cleaned? Look for wrench marks and signs that the devices are routinely being serviced. If not, I would ask them when they plan to. Flushing them is NOT enough, they must be opened and cleaned.
- Operational tests – Are they logging operational testing of these devices? They should be, especially on steam boilers which have a lower margin of safety and are more prone to failures.
- Are they replacing the device per recommendations? One of the major level control manufacturers recommends replacement intervals of 3-15 years, yet it is very common to see devices that have never been replaced on boilers that are decades old.

Pressure and Tmperature Control

Boilers and pressure vessels have sensing devices that control the pressure and/or temperature at which they operate. There should be more than one control as a safety measure. One is known as the "operating control" as it turns the system on and off to regulate pressure or temperature. The second device is known as the "high limit" control and it strictly for safety. It trips the system when the vessel reaches the upper limit and normally requires a manual reset in order to bring the system back on line.

Approximately 12% of the jurisdictional violations occur on pressure and temperature devices. The nature of the violations includes improperly operating controls, missing controls and deteriorated controls. The issues that are readily identifiable are similar to water level controls.

- Are the covers in place? Missing covers are a sign of adjustments that were probably not authorized and they expose the device to damage.
- Are there any signs of corrosion? It is an electrical device, a failure of the water seal will cause the electrical components to fail.
- Operational tests – Are they logging operational testing of these devices? Are the parameters of the system being monitored, recorded and reviewed on a routine basis?
- Are the controls subjected to at least an annual formal service and calibration?

Combustion System Controls

Commonly known as "burner management systems" (BMS), they contain the fuel train components that deliver the fuel and combustion air to the boiler and control combustion. The BMS not only regulates the supply of fuel and air to the burner, it ensures that the sequence of operation is such that ignition only occurs when conditions are safe. Twenty one percent of boiler and water heater violations are found on BMS equipment.

Improperly operating combustion systems can result in furnace explosions and the overheating of controls. Many of the indicators of improper combustion are easy to identify.

- Are there hot spots (signs of overheating) on the casing?
- Are there indications of soot accumulation (improper combustion)?
- Is the fire being maintained within the boundaries of the boiler not only during routine operation, but during "light off"?

Useful Life Inspections

Boilers and pressure vessels can have extended life expectancies when they are properly maintained, however they still have finite life spans that require periodic evaluations to ensure continued fitness for service. As these objects approach their end of useful life, the frequency and severity of repairs tends to increase. Forty two percent of the violations found deficiencies with the vessel itself. Not all of the these are "end of life issues", however, many are.

The conditions that reduce a vessels life can be internal or external to the vessel. Corrosion and the operation of the vessels can reduce a vessels expected life. While this presentation won't detail all the aspects of useful life inspections, owners and users of pressure vessels must have the vessels periodically evaluated beyond routine inspections. Useful life inspections utilize corrosion rates and actual operations to predict the remaining life and necessary repairs to boilers and pressure vessels.

Safety professionals at facilities should ensure that boilers and pressure vessels are being replaced prior to experiencing "end of life" issues. As we have already noted, maintenance and repairs results in an increase frequency of accidents and injuries. This must be part of the decision making process when evaluating vessel life. Useful life inspections should be completed by experienced professionals prior to the expected end of the vessel useful life.

Summary

If your facility has a pressure retaining item, it has a potential hazard. Plant walk through inspections can identify many of the symptoms of an uncontrolled hazard. The safety professional can improve the safety and reliability of their facilities by taking an active role in ensuring that all pressure retaining items are being operated, inspected and maintained properly.

References

http://www.nationalboard.org/Index.aspx?pageID=164&ID=412

http://www.nationalboard.org/Index.aspx?pageID=385

http://www.pnl.gov/main/publications/external/technical_reports/PNNL-18696.pdf

http://www.ameriburn.org/Preven/2000Prevention/Scald2000PrevetionKit.pdf

https://www.cdc.gov/co/faqs.htm

https://www.cdc.gov/niosh/topics/confinedspace/default.html

http://www.nationalboard.org/index.aspx?pageID=164&ID=220

http://documentlibrary.xylemappliedwater.com/wp-content/blogs.dir/22/files/2012/09/MM-815D.pdf

Chubb is the marketing name used to refer to subsidiaries of Chubb Limited, providing insurance and related services. For a list of these subsidiaries, please visit our website, www.chubb.com

This document is necessarily general in content and intended to provide an overview of certain aspects of boiler and pressure vessel safety and has an edition date of 03/03/2017. This document is advisory in nature and is offered as a resource to be used together with your professional insurance advisors in maintaining a loss prevention program. The information provided should not be relied on as legal or insurance advice or a definitive statement of the law in any jurisdiction. For such advice, an applicant, insured, listener or reader should consult their own legal counsel or insurance consultant. No liability is assumed by reason of the information this document contains.

Session No. 769

Design for Safety

Gary Cunningham, CSP
Boston, MA

Jason Edic, CSP
Lee Kennedy Co., Inc.
Quincy, MA

Introduction

The authors have a combined experience of more than 50 years practicing safety, primarily in the construction industry. Over the course of this half century we have, in varied capacities, managed EHS programs for *billions* of dollars in construction and renovation projects. These statements are made not with intent of establishing bona fides or bragging – quite the opposite – they are self-deprecating. It has taken us decades of practice to understand and confirm what many of our colleagues in other industries have already figured out – an ounce of prevention is worth a pound of cure.

It is widely agreed in the safety profession that the hierarchy of controls represents a theory that translates effectively into sound practice – a theory that's been tried, tested and documented for centuries. In construction, we do not abide by the hierarchy of controls theory. In fact, we do not even come close – that's part two of the bad news. Working backwards on the cause and effect chain, part one of the bad news tells us that the vast majority of a structure's end users, facility maintainers and construction professionals are utterly powerless to impact EHS considerations in a truly meaningful way. Phrasing the latter statement another way, the groups that share in the greatest risks have the least influence over their safety and health.

Don't believe us? Walk into just about any building or structure and ask the facility manager to list the challenges he or she is pitted against on a regular basis. The list will surprise you – a filter bank situated atop an air handler with no access; a series of wall-washing light fixtures mounted in the building's lobby that is 28' above floor level, whose only means for changing the lamp is an extension ladder; a roof-top chiller unit located 4' from the roof's edge with no guardrails or parapets. You will find no shortage of examples when speaking to facilities professionals.

If you ask the trades people who are tasked with constructing our buildings, the answers you receive will be somewhat more convoluted. Trades people and their employers have succumbed to the absurdities that most engineers, designers and architects – albeit unintentionally – have placed before them and asked them to build. Trades people build what they're asked to build (i.e. whatever is on the drawings or in the specifications). Sometimes they do so reluctantly, but they still build it. For example, a pre-determined monitoring well that is situated directly beneath power transmission lines; or a 'double-connection' between two steel beams where the beams' end plates share common bolts, therefore requiring one of the beams to be disconnected in order to install the other beam. Again, you'll find no shortage of examples, in any trade or discipline, when speaking to constructors.

What do each of the five examples listed above have in common? Answer: each scenario resulted in the death of an individual. What else do they have in common? Answer: the employer was held ultimately responsible for the death – both in civil litigation and regulatory citations (we will not argue that employers must protect employees). And further still, what do they have in common? Answer: none of these scenarios employed the hierarchy of controls. In each of these cases, the hazard *could* have been **eliminated** through design. Thus we reach the million-dollar question and the purpose of this article.

These hazards *could* have been eliminated, but ***should*** these hazards have been eliminated? The answer is undoubtedly and resoundingly yes. So why were they not eliminated? In order to answer this question, we must look at several factors and barriers.

Misplaced Resources

Safety, like every other discipline in the construction process, is a limited resource. It's a resource that can be counted in hours, dollars, effort, ideas, innovation or any other quantitative or qualitative method but no matter how it's measured it is limited. Safety does not usually suffer from a lack of goodwill or good intentions (occupational safety benefits greatly when actual support matches the expressions of support it receives, but that doesn't always happen) but goodwill and good intentions on their own don't translate into effective safety management; it takes effort, and unlike sentiment there are limits to actual effort. So regardless of how the resources available for safety are measured it's clear that those resources are limited.

Let's assume that for every task on a construction project there are 100 units of safety effort available. Those 100 units can be used anywhere in the entire construction cycle to ensure a safe operation. Let's also assume that the 100 units exist from project inception until a particular task is completed. Envision also the hundreds of thousands or millions of individual tasks that must be completed on a project; envision also the innumerable potential injuries that each of those tasks might represent. You have to envision changing hundreds of blades on tools, you have to envision every task that human hands will perform, the contortions of every back, knee, shoulder and ankle; every lift, pull and push; every spark or splinter, every walking and working surface – all of them over the course of the project. And you have to imagine the millions or tens of millions of decisions that are made over the life of the project that are associated with each of those activities. A ladder, a lift, stairs, scaffold, a bucket, a milk crate, the rails on a lift, the preferred route or a shortcut – and on and on. Imagine it all.

We must accept that many tasks will be completed that use very few of the 100 units of safety that are available to them. The JHA or Pre-task Plan will not be written or not be reviewed, the toolbox talk will not be attended, orientation will be hurried, the operations manual will not be read – units of safety that were available will expire unused. Limited safety resources will be wasted; generated but not consumed. Those units of safety effort are perishable – the JHA must be completed before the task and it must be reviewed with the crew before the task. The knowledge of how to do a JHA makes up a portion of those 100 units and if the JHA is not performed properly, if the full value is not realized then those units of safety training on how to perform a JHA have been wasted and cannot be regained for that task that was just completed.

As with any resource, there is a general loss when it is wasted but with occupational safety the loss may be felt immediately and repeatedly in the form of injuries. Wasting natural resources can take generations before the loss is perceived but loss associated with wasting safety resources can be instantaneous. A safety conservationist might be someone that works to ensure that all safety resources that are available are consumed; not allowing them to perish unused.

Let's look at the inverse of tasks that are completed without the full benefit of available safety resources, tasks that take advantage of most of the units of safety effort that are available. It's nearly impossible to imagine all 100 units being consumed but certainly there are tasks that utilize more safety than others. Working backwards from a particular task we may observe good housekeeping and work that is being performed at a comfortable height with the best equipment. We might see PPE that's appropriate for the task and that fits well and is properly maintained. Good lighting and other environmental conditions can be observed, and as we move back in time, we can watch the crew as they review and discuss the work plan. How far back can we follow the trail of safety effort? How many of the 100 units can we trace back to project inception before the string breaks? If the crew has to work at height can we follow the decisions that were made related to access – was every conceivable method considered purely on the basis of worker safety or were there other considerations? Was there a cost implication or a schedule impact that broke the safety string? What decisions were made during buyout that the crew was forced to deal with in order to perform their task? Was a personnel hoist rejected due to cost? Was a job-made ladder rejected due to time? Was a lift not available? Was the installation of access equipment pushed further out in the schedule? Was an atrium designed without thought of construction workers? Was a site chosen that complicated every task that came after?

The 100 units are available to the owner as soon as they conceive a need to build. When the owner realizes that a new facility is necessary there are pieces that begin falling into place or being set in stone that will impact the life of the project and the life of the entire facility. Will a clean site be utilized? Is there demolition necessary? Support of excavation? Virgin soil? Many decisions will follow and every one will have an impact on safety. As the project moves forward decisions about the size, scope, features and elements of the site and structure accumulate more potential impacts. Decisions on foundations, underground structures, utilities, existing conditions, type of construction, structural elements and on and on go the decisions and considerations that will be dealt with down the road. Phasing, scheduling, manpower and staffing will come later but add to the ledger of plusses and minuses of occupational safety impacts. How many can be identified, analyzed and considered for just those safety impacts? And once identified and considered how many can be changed or modified to eliminate or mitigate a negative impact on occupational safety?

The further into the lifecycle of construction we go, the more challenging it becomes to make *any* type of change. Decisions made late in the design cycle will impact dozens if not hundreds of previously made decisions. For obvious reasons, making changes or adjustments to previously made choices will undoubtedly be met with resistance. Conversely, where informed decisions are made earlier in the process, the resistance is greatly lessened, and the positive impact on safety is maximized.

In summation, reason number one that hazards are not routinely considered, eliminated, substituted or engineered is that we as an industry do not apply our resources efficiently. This alone translates directly into a multitude of losses in the form of money, injuries and lives. While the direct losses are borne by employers, we submit that it is ultimately the "owners" who bear the monetary burden of these losses. In plain English, owners' monies are being wasted through misplaced resources – make no mistake, this is the owner's money.

Beast of Burden

On December 29, 1970, President Richard M. Nixon signed the Occupational Safety and Health Act into law. This law was the culmination of decades of both state and federal efforts to protect the American workforce, and for the first time a national regulation truly defined employers' responsibilities to protect its employees. While this Act laid the groundwork for sweeping changes to many, many industries in the U.S., it also inadvertently but severely limited direct involvement and responsibility for EHS considerations.

Section 2(b) of the OSH Act states "The Congress declares it to be its purpose and policy, through the exercise of its powers to regulate commerce among the several States and with foreign nations and to provide for the general welfare, to assure so far as possible every working man and woman in the Nation safe and healthful working conditions and to preserve our human resources by providing that employers and employees have separate but dependent responsibilities and rights with respect to achieving safe and healthful working conditions."[1]

Section 5(a)(1) of the Act (aka the General Duty Clause) further defines that each employer shall "furnish to each of his employees employment and a place of employment which are free from recognized hazards that are causing or are likely to cause death or serious physical harm to his employees."[2]

It is clear that employers hold the burden of safety and health for their employees, at least in the regulatory sense. It is equally clear that employers do not wield all of the power or hold all of the cards when it comes to influence. Go back to the hierarchy of controls – who truly has the power to eliminate a hazard? It is certainly rare in the construction industry that the employer of constructors has this influence. In fact, if you were to list out the entities that carry the biggest stick in terms of EHS sway and elimination of hazards, it would look like this (starting with the most powerful):

---

[1] United States Department of Labor Occupational Safety and Health Administration. "Occupational Safety and Health Act of 1970" (www.osha.gov/pls/oshaweb/owadisp.show_document?p_table=OSHACT&p_id=3356)

[2] United States Department of Labor Occupational Safety and Health Administration. "Occupational Safety and Health Act of 1970" (www.osha.gov/pls/oshaweb/owadisp.show_document?p_table=OSHACT&p_id=3359)

1. Owner
2. Design Team
3. Builder

Yet there is no regulation in existence in the U.S. that requires or even encourages the entities with the most power and influence over the hierarchy of controls to use that power to increase safety and decrease hazards.

So why has the design for safety idea not been adopted in construction? It certainly has been adopted for consumer protection, automobile safety, and in various sectors of manufacturing, but why not construction? We offer three overarching explanations:

Money

The answer to most questions these days – follow the money. Who is paying for design and construction? It is the owner, and the owner has a fiduciary responsibility to its shareholders to keep costs down. NOTE: The authors completely agree with this sentiment – it's part of what has kept us employed for so many years. If the company's profitable, we employees stay employed. First rule of business – remain profitable.

If "someone" was to identify and eliminate/substitute/engineer hazards from a construction design, this would cost money. This "someone" would need to be present starting with the conceptual phase of a project in order to provide the biggest impact. At the conceptual phase, it is not often that a safety professional (or a designer with enhanced knowledge of EHS) is involved, and rarer still to be involved to the level that impacts design decisions. Owners do not typically engage this type of service during design because (1) it costs money, (2) they do not understand the potential return on investment, and (3) there is no requirement to do so.

Let's go back to 100 units of safety effort and the limited resources for safety. Money and effort at the earliest phases of a project provides the largest return on investment. The fact that safety effort is not placed early and often on a project is penny wise and pound foolish. This leads us directly into the next barrier.

Education Level

In very simple terms, the majority of owners and the vast majority of designers, engineers and architects do not truly understand the impact(s) that their decisions play on the safety and health of constructors, end users and facility maintainers. It is not that they don't care – they simply do not understand enough of this topic to make pervasive changes in the industry. It may be the largest gap that we have in the industry – the disconnect between owners, the design team, constructors, users and maintainers in terms of needs and wants. This is a cavernous educational gap.

Design teams are rarely, if ever, asked to make EHS considerations as part of their design, hence there is no need to educate these teams to do so. Very few architecture/design/engineering curricula that the authors have seen include EHS considerations for constructors, end users and maintainers. There are many curricula that include code compliance and interpretation for a variety of disciplines. These are primarily geared towards life safety systems for occupants, public protection, or for end user critical safety systems (e.g. rails). There is minor mention in some codes (e.g. International Building Code) for safety during construction activities, but these codes (1) address life safety items for constructors (e.g. means of egress or fire protection), and (2) are typically contractual obligations of the constructor and *not* the design team.

Why are there no formal educational programs that address EHS considerations for designers, architects and engineers? The answer leads us to our next barrier.

Liability

There is a strong concern in the design and engineering community that if EHS considerations were to become a required part of the service, this would expose design professionals to increased liability should an injury occur. We believe this to be a completely valid, real and justified concern on the part of the design community.

While this last topic is summarized by only a few brief sentences, do not underestimate its force. The society that we live in today is overly litigious, and the construction industry is no exception. The fear of professional liability, errors & omissions and negligence-related claims is a powerful, motivating force – one that is capable of halting a movement.

The authors also believe that this concern will quickly be addressed through contract language and case precedence. We are willing to see the design community take this risk in order to *dramatically* increase protection of our greatest resource – our workforce. This would bring us full circle to barrier number one – money. Protection against liability costs money – direct and indirect money – and is a potent motivator.

Is it Possible?

Is the kind of change that we write about in this article possible? The answer again is "yes". While the U.S. is world's leader in many, many industries, this is *not* one of them. The U.K.'s Health and Safety Executive (Great Britain's version of OSHA) has led the way in this field with its Construction Design and Management (CDM) regulations. These regulations *require* that owners, designers and constructors partner during the construction lifecycle to identify, eliminate or control foreseeable risks both during construction and after the building is built.

Similarly, the European Union, Singapore and Australia have each initiated legislation and guidance to address this growing concern in the construction community. In short, if the U.S. were to address this issue, it would do so with a plethora of guidance and lessons learned. The authors believe that the time is now.

Conclusion

Construction is the equivalent of a dangerous toy or household appliance that is manufactured and sold in great quantity in the U.S. (somewhere north of a trillion dollars annually). This dangerous toy is sold to millions of consumers over the course of a single year. Consumers' use of this toy results in thousands of serious injuries and hundreds of deaths every single year in this country. Unfortunately, there is no consumer protection act in place to defend and protect the users of this dangerous toy. There is no national advocate that has the ability to declare this toy unsafe or unsuitable for use, and to demand that it be recalled and taken off of the shelf until it can be re-designed and made safe. Instead of making considerations for safe design of this dangerous toy, we as a nation expect each parent (employer) to require safeguards of their children. When the children are injured or killed using this toy, we penalize the parents – holding them *ultimately* responsible, when it is absolutely clear that the toy's designer is also complicit.

To the authors, it makes no sense whatsoever. Construction designs should come with their own version of a UL listing. **We believe that regulation and government intervention are going to be the most effective means to bring about this change.** We do not believe that the construction industry will perpetuate this initiative on its own in any meaningful way over the next decade. While we do see some positive changes in our industry (e.g. prefabrication), we believe that these changes are being lead primarily by the contractor base, and not due to an overarching concern for the safety and health of maintainers and constructors (i.e. follow the money).

In 2004, OSHA formed the Alliance Program Construction Roundtable. As part of this Roundtable, a Design for Safety workgroup was established. At the time it was very encouraging to know that the federal government was interested in this idea and was taking initial actions to research and investigate its potential. In the opinion of the authors, some very good work was produced by this group, and could have been the launching pad for a more sweeping set of changes. Unfortunately, this workgroup was discontinued in 2009.

This presentation is ultimately intended to be a call to action. If you agree with this concept, then do something to contribute to its propagation and the furthering of our profession and our industry. As the saying goes, it's time to fish or cut bait. We urge you to contact OSHA to request that they (re)initiate a Design for Safety program.

American Society of Safety Engineers
Professional Development Conference June 19-22, 2017

Session: 771

Fundamentals of SH&E: BASIC SAFETY MANAGEMENT

Joel N. Tietjens, CSP, CSHM
President
T-JENS & T-JENS, INC.

BASIC SAFETY MANAGEMENT ELEMENTS

7

2

BASIC ELEMENTS

The elements that make up an effective basic safety management plan

1. Management/Policy
2. Recordkeeping
3. Loss Analysis
4. Safety & Health Education/Training
5. Safety & Health Inspections/Surveys
6. Accident/Incident Reporting & Investigations
7. Plan & Programs Review

3

1. MANAGEMENT

4

1. MANAGEMENT

SAFETY 2017 ASSE

- The employer has the responsibility of providing a workplace free of any recognized hazards
- Safety management plan must rest on a solid foundation of management commitment and support

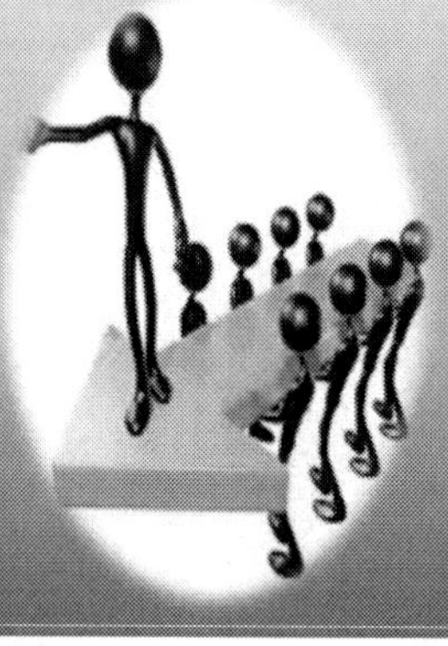

5

1. MANAGEMENT

SAFETY 2017 ASSE

- Commitment
  - To pledge or assign to some particular course or use
- Support
  - To provide resources
  - Uphold, advocate, champion

6

1. MANAGEMENT

SAFETY 2017
ASSE

- Management must thoughtfully and thoroughly develop a safety and health policy that can be understood, believed and sets the tone for action

7

1. MANAGEMENT

SAFETY 2017
ASSE

- Must understand what the policy means
  - Policy should be short and to the point (preferably approximately a half page)
- Must establish challenging (realistic) goals for that particular organization/industry

8

1. MANAGEMENT

- Must assign
  - Responsibility
  - Authority
  - Accountability to personnel for implementing the plan
- Must participate in safety forums, meetings, educational processes
- Must "walk the talk"

9

BASIC ELEMENTS

The elements

1. Management/Policy
2. **Recordkeeping**
3. Loss Analysis
4. Safety & Health Education/Training
5. Safety & Health Inspections/Surveys
6. Accident/Incident Reporting & Investigations
7. Plan & Programs Review

10

2. RECORDKEEPING

11

2. RECORDKEEPING

SAFETY 2017
ASSE

- Occupational Safety & Health Administration (OSHA) requires that employers with more than ten (10) employees maintain records

OSHA® Occupational Safety and Health Administration
www.osha.gov

12

2. RECORDKEEPING

SAFETY 2017 ASSE

- If the employer has ten (10) or less employees, they must maintain records if they have been requested to participate with the Bureau of Labor Statistics (BLS) in their annual survey of occupational illnesses and injuries

13

2. RECORDKEEPING

SAFETY 2017 ASSE

- A source of support for managing the safety management plan
- May be beneficial in strengthening other safety management programs (Loss/trend analysis)
- Support during legal or other evidentiary proceedings

14

2. RECORDKEEPING

- OSHA 300, Log of Work-Related Injuries and Illnesses
- Retained for a period of five (5) years, plus current year

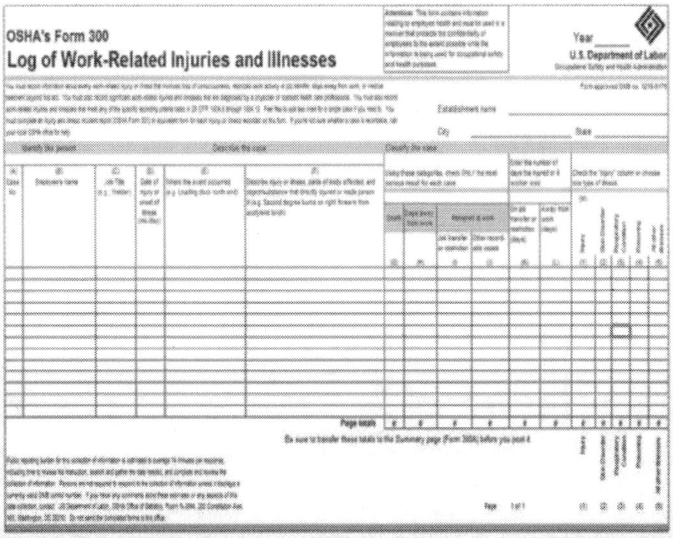

15

2. RECORDKEEPING

- OSHA 300A, Summary of Work-Related Injuries and Illnesses

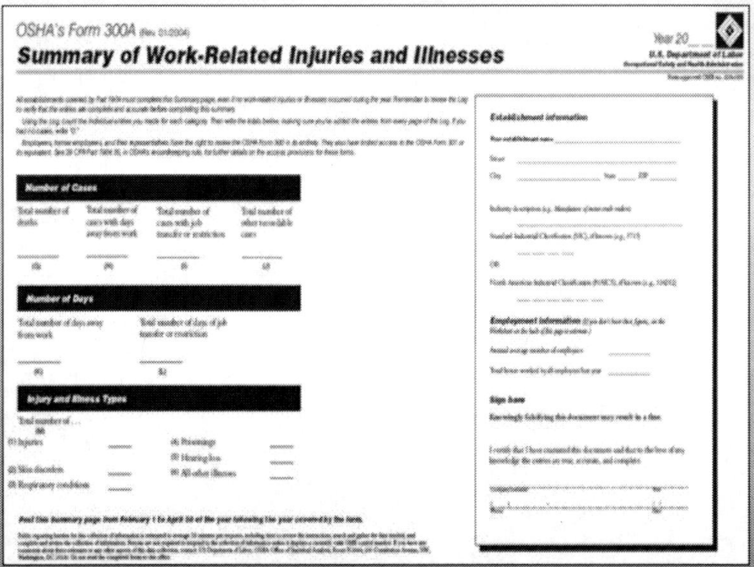

16

2. RECORDKEEPING

SAFETY 2017 ASSE

- OSHA 301, Injury and Illness Incident Report

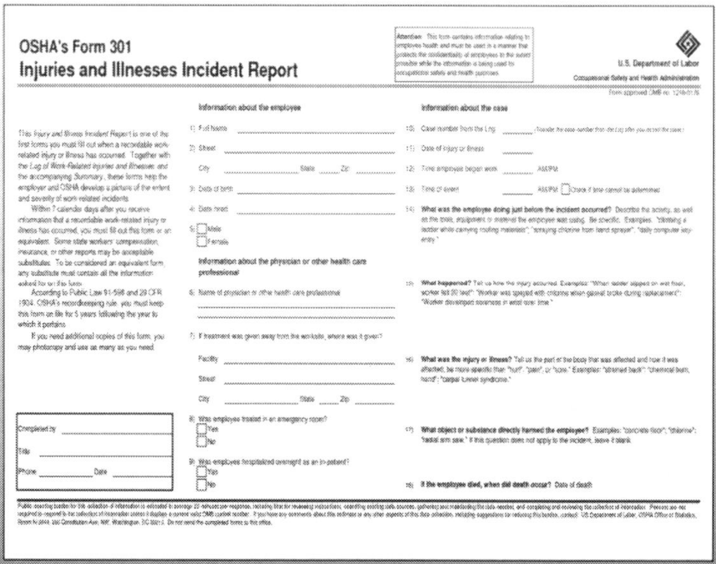
OSHA's Form 301
Injuries and Illnesses Incident Report

U.S. Department of Labor
Occupational Safety and Health Administration

Information about the employee

1) Full Name
2) Street
City State Zip
3) Date of birth
4) Date hired
5) Male / Female

Information about the physician or other health care professional

6) Name of physician or other health care professional
7) If treatment was given away from the worksite, where was it given?
Facility
Street
City State Zip
8) Was employee treated in an emergency room? Yes / No
9) Was employee hospitalized overnight as an in-patient? Yes / No

Completed by
Title
Phone Date

Information about the case

10) Case number from the Log
11) Date of injury or illness
12) Time employee began work AM/PM
13) Time of event AM/PM
14) What was the employee doing just before the incident occurred?
15) What happened?
16) What was the injury or illness?
17) What object or substance directly harmed the employee?
18) If the employee died, when did death occur? Date of death

17

2. RECORDKEEPING

SAFETY 2017 ASSE

- State reports
- First aid
- Property damage/losses
- Liability losses
- Motor vehicle liability/damage/losses/ maintenance
- Security losses
- Inspections/surveys
- Safety meetings or other related meetings

18

2. RECORDKEEPING

SAFETY 2017 ASSE

- Education/training records
- Equipment inspection/maintenance
- Records retention periods
- Other records

19

BASIC ELEMENTS

SAFETY 2017 ASSE

The elements

1. Management/Policy
2. Recordkeeping
3. **Loss Analysis**
4. Safety & Health Education/Training
5. Safety & Health Inspections/Surveys
6. Accident/Incident Reporting & Investigations
7. Plan & Programs Review

20

3. LOSS ANALYSIS

21

3. LOSS ANALYSIS

- The means of studying statistical data (favorable and/or unfavorable) to determine trends or identify problem areas
- Reason to perform loss analysis
  - Mistakes or errors result in damaged products, production delays, or employee/public accidents/incidents that effect profit

22

3. LOSS ANALYSIS

SAFETY 2017 ASSE

- Statistical Process Controls
  - Trend analysis

23

3. LOSS ANALYSIS

SAFETY 2017 ASSE

- Trending
  - Used to identify trends indicated by statistical data gathered from other program components
    - Accident/incident data
    - Inspection/survey data
  - Indicate where problem areas exist and where to allocate limited resources

24

3. LOSS ANALYSIS

SAFETY 2017 ASSE

- Statistical Process Control tools (SPC) include:
  - Pareto charts
  - Run (trend) charts
  - Histograms
  - Scatter diagrams
  - Tally charts
  - Control charts

25

3. LOSS ANALYSIS

SAFETY 2017 ASSE

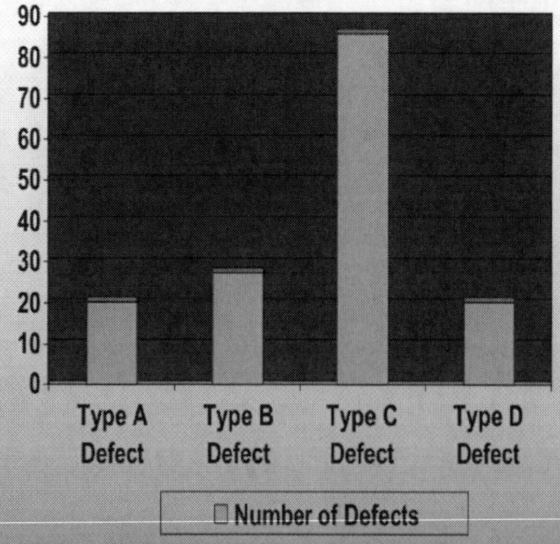

Pareto Chart

- Used to graphically summarize and display the relative importance of the differences between groups of data

26

3. LOSS ANALYSIS

SAFETY 2017 ASSE

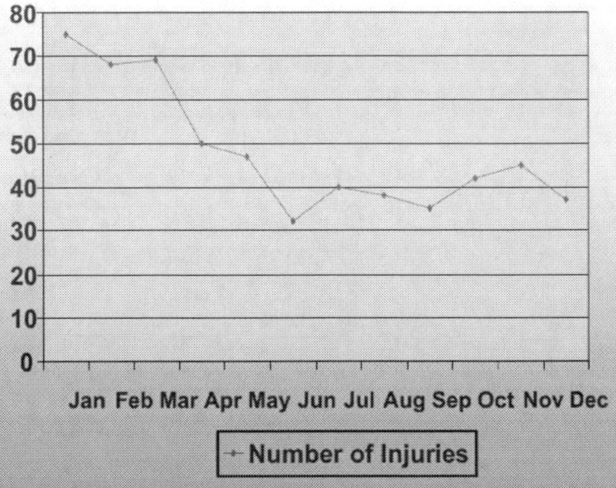

Run Chart

- Display process performance over time
- Upward and downward trends, cycles, and large aberrations may be spotted and investigated further
- Events, shown on the ***y*** axis, are graphed against a time period on the ***x*** axis

27

3. LOSS ANALYSIS

SAFETY 2017 ASSE

Histogram

- A specialized type of bar chart
- Individual data points are grouped together in classes, so that an idea of how frequently data in each class occur in the data set
- High bars indicate more points in a class, and low bars indicate less points

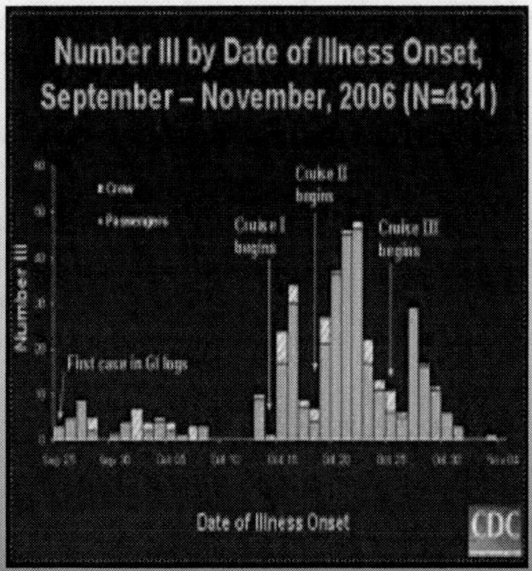

28

3. LOSS ANALYSIS

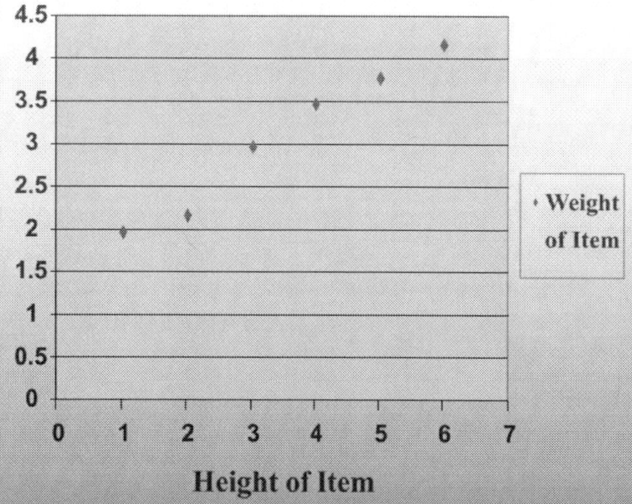

Scatter Diagram

- Used to investigate the possible relationship between two variables that both relate to the same "event"
- A straight line of best fit (using the least squares method) is often included

29

3. LOSS ANALYSIS

Tally Chart

Problem	Mon	Tue	Wed	Thu	Fri
Laceration	///	/		///	
Sprain	//		/	/	
Strain	/			/	
Contusion	/	/			////

- A grid used to help clearly show information as it is collected
- Has lines to show numbers and a total number to show the frequency
- Followed by a graph or a short sentence describing what the tally chart shows

30

3. LOSS ANALYSIS

SAFETY 2017 ASSE

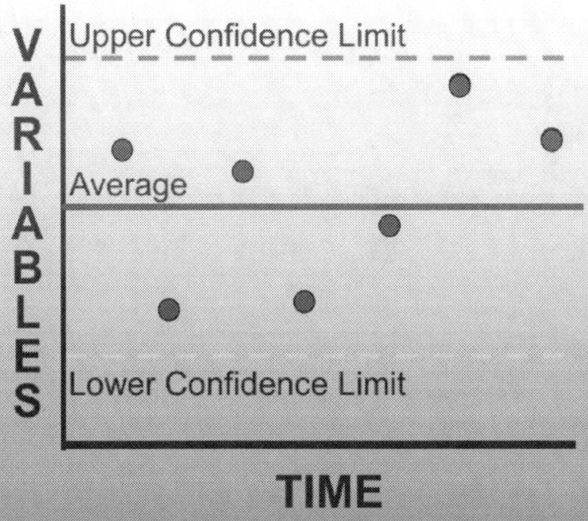

Control Chart

- Center line that represents the average value of the quality characteristic
- Two other horizontal lines
  - upper control limit (UCL)
  - lower control limit (LCL)
- If the process is in control, nearly all of the sample points will fall between them

31

3. LOSS ANALYSIS

SAFETY 2017 ASSE

- Conducting a Trend Analysis
  - Select the data to be analyzed
    - Injury, inspection/survey, etc. data
  - Determine time period
    - Monthly, quarterly, annually
  - Identify similarities
    - Injuries, job functions, etc.
  - Develop corrective actions

32

3. LOSS ANALYSIS

SAFETY 2017 ASSE

- Designate a person(s) responsible for conducting the analysis
- Including time frames and job tasks

33

BASIC ELEMENTS

SAFETY 2017 ASSE

The elements

1. Management/Policy
2. Recordkeeping
3. Loss Analysis
4. **Safety & Health Education/Training**
5. Safety & Health Inspections/Surveys
6. Accident/Incident Reporting & Investigations
7. Plan & Programs Review

34

4. SAFETY & HEALTH EDUCATION/TRAINING

35

4. EDUCATION/TRAINING

SAFETY 2017 ASSE

- Safety & health education/training is provided to minimize and/or eliminate unsafe acts (behaviors) while performing tasks
- Used to identify hazards involved in tasks, as well as, procedures to avoid them

36

4. EDUCATION/TRAINING

SAFETY 2017 ASSE

- All levels should receive education/training
  - Management
    - Line
    - Middle
    - Senior

37

4. EDUCATION/TRAINING

SAFETY 2017 ASSE

- Employees
  - New hires
  - Transferred
  - Continual
  - Temporary
- Contractors

38

4. EDUCATION/TRAINING

SAFETY 2017 ASSE

- How often should education/training be provided?
- Depending on the complexity of the subject and regulatory requirements
  - Quarterly
  - Semi-annually
  - Annually
  - Bi-annually
  - Tri-annually

39

4. EDUCATION/TRAINING

SAFETY 2017 ASSE

Seven (7) Steps to Education/Training

1. Determine if education/training is needed to solve the problem
2. Identify the education/training need
3. Identify the goals and objectives
4. Develop learning activities
5. Conduct the education/training
6. Evaluate education/training effectiveness
7. Improve the process

40

4. EDUCATION/TRAINING

- All training should be properly documented
- Selection of who conducts the education/training process

41

BASIC ELEMENTS

The elements

1. Management/Policy
2. Recordkeeping
3. Loss Analysis
4. Safety & Health Education/Training
5. **Safety & Health Inspections/Surveys**
6. Accident/Incident Reporting & Investigations
7. Plan & Programs Review

42

5. SAFETY & HEALTH INSPECTIONS/SURVEYS

43

5. INSPECTIONS/SURVEYS

SAFETY 2017 ASSE

- Inspections and surveys can be a proactive tool in identifying hazards and behaviors before the losses occur

- Inspections and surveys can also be a reactive tool, such as a thorough inspection after an accident or incident has occurred to determine what else could go wrong

44

5. INSPECTIONS/SURVEYS

- When to conduct inspections/surveys may be
  - Daily
  - Weekly
  - Monthly
  - Quarterly
  - Semi-annually
  - Annually
  - Special needs, i.e.,

- Plant turnarounds, non-routine activities, etc.
  - Follow-up

45

5. INSPECTIONS/SURVEYS

- Involvement in performing inspections/surveys
  - Employees
  - Line managers
  - Mid managers
  - Safety & health staff
  - Safety committee
  - Insurance carrier
  - Others

46

5. INSPECTIONS/SURVEYS

- Inspections/surveys should be properly documented
  - Checklists
  - Narratives
  - Follow-ups

47

BASIC ELEMENTS

The elements

1. Management/Policy
2. Recordkeeping
3. Loss Analysis
4. Safety & Health Education/Training
5. Safety & Health Inspections/Surveys
6. **Accident/Incident Reporting & Investigations**
7. Plan & Programs Review

48

6. ACCIDENT/INCIDENT REPORTING & INVESTIGATIONS

49

6. ACCIDENT/INCIDENT

SAFETY 2017 ASSE

- Why report and investigate?
  - Determine the cause
  - Uncover indirect incidents causes
  - Prevent similar incidents
  - Document facts
  - Cost information
  - Promote safety & health
  - Fact Find ***NOT*** Fault Find

50

6. ACCIDENT/INCIDENT

SAFETY 2017 ASSE

- Items to identify
  - Reporting requirements and procedures for both the employer and employees
    - Training?
  - Personnel responsible for investigations
    - Who should investigate?
  - Guidelines for how to conduct a thorough investigation
    - Training?

51

BASIC ELEMENTS

SAFETY 2017 ASSE

The elements

1. Management/Policy
2. Recordkeeping
3. Loss Analysis
4. Safety & Health Education/Training
5. Safety & Health Inspections/Surveys
6. Accident/Incident Reporting & Investigations
7. **Plan & Programs Review**

52

7. PLAN & PROGRAMS AUDIT/REVIEW

53

7. PLAN & PROGRAMS AUDIT/REVIEW

SAFETY 2017 ASSE

- The purpose of auditing or reviewing the safety management plan and its programs is to keep it working efficiently and effectively in preventing accidents, injuries, and other losses
- The plan and its programs must keep pace with the ever changing operating environment

54

7. PLAN & PROGRAMS AUDIT/REVIEW

- Determine who will conduct the audit/review
- Audits/reviews should be conducted
  - At least annually

55

7. PLAN & PROGRAMS AUDIT/REVIEW

- Documentation
  - All audits/reviews should be properly documented with the findings and the necessary actions to be taken

56

MODEL OCCUPATIONAL HEALTH & SAFETY MANAGEMENT SYSTEMS

57

MODELS

SAFETY 2017
ASSE

- **ANSI/AIHA/ASSE Z10-2012**
  Standard for Occupational Health & Safety Management Systems
- **OHSAS 18001**
  Occupational Health & Safety Assessment Series
- **ILO–OSH 2001**
  Guidelines for Occupational Safety& Health Management Systems
- **OSHA**
  Voluntary Protection Program (VPP)
- **ISO 45001 (Under Development)**
  Occupational Health & Safety Management System

58

IN SUMMARY

59

IN SUMMARY

To develop an effective safety management plan, the organization must have these elements

1. Management/Policy
2. Recordkeeping
3. Loss Analysis
4. Safety & Health Education/Training
5. Safety & Health Inspections/Surveys
6. Accident/Incident Reporting & Investigations
7. Plan & Programs Review

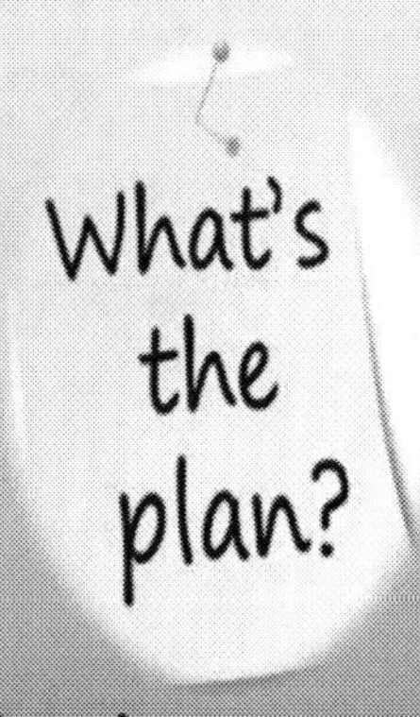

60

SAFETY 2017
ASSE

Questions & DISCUSSION

61

SAFETY 2017
ASSE

Joel N. Tietjens, CSP, CSHM
President - T-JENS & T-JENS, INC.

18803 Whitewood Drive, Spring, TX 77373-5613
Off: 281-353-3680
E-mail: Joel@tjens.com Web: www.tjens.com

T-JENS & T-JENS

62

Using Variable Teaching Modes to Engage Students and Promote Lifelong Learning

Monica Nevels, MS Occupational Safety Management, ASP
University of Central Missouri
Warrensburg, Missouri

Teaching Modes that Promote Content Retention and Develop Effective Self-learning Habits

Traditional methods of teaching, such as lecture, offer a means of quickly delivering a message but can be enhanced by following with activities that reinforce the material and promote the process of self-education. A number of approaches to active learning can be used to keep students on task and focused on the learning objectives. When students are engaged in the classroom, long-term retention of the material improves.

Traditional Methods: Do They Engage Students?

Traditional methods of teaching, such as lecture, offer a means of quickly delivering a message but can be better utilized if used in small amounts and by immediately following up with an interactive class activity that reinforces the material presented and promotes the process of active self-education. Several approaches to active learning can be used to help keep students on task and better focused on the learning objectives and outcomes for a course. When students are engaged in the classroom with a reinforcement activity, the retention of the lectured or read material, according to many studies, improves.

As the K-12 environment continues to evolve and incorporate the use of different teaching modes and technologies, students most likely arrive to the university setting with an expectation that the experience will be interactive in nature, with electronic devices integrated into their learning experience. Employers will most likely expect those emerging into their field of study to be proficient, not just with the current forms of technology, but with the ability to independently develop the skills needed to meet the expectations and goals of an organization. With this in mind, it is important that lesson plans are designed to incorporate the tools of active learning and the usage of electronic devices when they support or enhance the learning objectives. I firmly

believe that the decision to use different teaching modes should be based on the value they add to the lesson. It is also important that electronic devices support the learning practices of the students, and do not end up becoming a replacement for meaningful classroom discussions.
By utilizing different approaches to teaching, I attempt to tap into the different styles of learning and promote active and lively learning experiences in and out of the classroom. Incorporating some of the typical lean manufacturing tools into a lesson plan is one way of engaging students that also allows them to examine material and develop project management skills. I will identify a few key areas that, when addressed, have helped me draw students into the educational process and promote a lifestyle of learning and self-education.

Gaining Active Participation

The use of low stakes reinforcement quizzes, classroom preparation assignments, and other active learning tools, such as "brain-dump" exercises, are some of the ways I attempt to pull students into a topic and continue to develop their critical thinking skills. My overall goal is to use methods of teaching that require students to use the lectured or read material soon after it is delivered. When selecting an activity, I attempt to determine which would best serve the student in terms material retention and encourage critical thinking. I want to use the class time for learning, not just listening. With the increasing number of distractions to students, using a simple lean tool, such as a process map or a pick chart, allows them to physically move around the classroom, interact with the other students, and develop a skill set that will be used later in their careers. With instructor support present, it is easier for students to try out new approaches to learning and manage new information. Further research is needed to better determine the amount of impact a particular activity has on content retention, but when used in a manner that reinforces the topic, I believe there are positive and measurable improvements.

At the end of each summative exercise, I create a list of the most missed questions (marked incorrectly by more than 30% of the group), and compare that material to the delivery type. Frequently, this list contains topics not covered during the active learning sessions. I plan to continue to collect data on the possible connections between activity types and summative testing outcomes so I can better understand when and what to use in relation to the content type. I want to know what active learning approaches provide the best outcome in relation to the material covered.

I also noticed that students seemed to take more interest and ask more questions during traditional lecture periods that cover topics previously explored with a classroom activity. This response encouraged me to continue using various tools and to create lesson plans that included active learning strategies as often as time would allow. The reluctance to use active learning techniques sometimes stems from the time investment required. To avoid my temptation to opt out of using these activities, I construct lesson plans in a manner that does not require too much more preparation time than that of the traditional lectured material. Keeping the activity simple yet worthwhile is generally the challenge.

Some studies suggest that, when active learning models are used, students believe they are not learning as much as those in a traditional lecture setting, despite the evidence of higher testing scores and greater material retention. To avoid this negative perception, I combine short amounts of lecture with an active learning tool, and explain the reason and goal behind the activity. Many times, I sell the use of the tool as being a method of dividing and conquering large amounts of material in shorter amounts of time. Once the whole group is working on, for instance, a case study or supplementary text, the work associated seems to lessen, and the group is more likely to participate due to the end goal of lessening the overall workload for the individual. The goals for the professor and the student do to not necessarily have to be exactly the same for the active learning tools to be received and successfully used.

Classroom Preparation Assignments

The effectiveness of classroom preparation assignments (CPAs) seems to vary. When using these assignments to prepare for an active learning session, participation appears to improve when assigning a point value to the preparatory assignment, as well as the in-class activity. CPAs with no point value attached do not tend to promote participation, and students might then arrive without the information needed to participate in the planned group, team, or class activity. I have also noticed that CPAs graded in a formative manner seem to benefit students more than those graded for accuracy, which I attribute to the fact that it is a low-stakes type of exercise. My intent is to gain as much participation as possible in the learning of the material, and again help students explore different methods of learning. When the college experience has concluded, my hope is that each student will be familiar with a variety of teaching and learning methods, and be able to develop a learning approach that works for them long term. I encourage students to transition from assisted learners to self-learners. Regardless of the CPA used, the exercise encourages students to examine material on their own and reflect upon the content, knowing they will be expected to apply what they learned when they arrive to class. To encourage students to use CPAs more effectively, I explain that although the assignments by themselves do not significantly impact their grade in a course, due to the point value, the content is used in in-class activities and summative assessments, which will ultimately impact the final grade in the course. Since certain foundational concepts are needed in most courses in order for the rest of the material to be useful, CPAs are an excellent approach to reinforcing the underlying principles without a lot of academic risk. Once students understand how the CPAs and other active learning or self-learning approaches fit into the course objectives, participation is not normally problematic.

Electronic Devices

Rather than viewing the usage of electronic devices in the classroom as a distraction to the learning environment, I attempt to incorporate usage into the lesson plans. Approximately 1/3 of the points earned in my courses are the result of participation in activities that occur immediately after a short (20-minute) lecture that must be submitted before the end of the class period. This allows students to practice or reinforce what was presented by applying the information directly and immediately to a task or mini-project. These activities range from performing a quick topic search to using presentation applications that provide a means of assessment but require the use of an electronic device. Knowing when and how to supplement a lesson plan with an electronic device, I believe, is vital to successful usage.

One common way I use technology in the classroom is by dividing a class into several teams, and charging each with developing a particular piece of a topic. Once developed, I then ask the small groups or teams to present to their findings to the large group. This promotes not only the use of technology, but creativity, and learning accountability. Adding the presentation element to the exercise builds the students' communication skills necessary for a successful career. Some of the flaws with this approach can be that some students will not participate in the exercise and rely on other group members to do the bulk of the work. I address this by limiting group sizes and requiring that each member holds a designated function. The function of each must be documented for points to be awarded for the activity. Some parameters are set when assignments are made, such as the type of presentation application used, or the sources that can be used, but academic freedom is given as to the content the small groups want to emphasize, and how workload is distributed.

In one of my courses, Upton Sinclair's *The Jungle* is used as a supplementary text. In the first four weeks of the course, I usually give two short quizzes that are directly tied to the reading assignments. At the conclusion of the book, students spend two class periods developing a summary. The class is divided into four or five teams, with each small team assigned a different piece of the summary, along with a different mode of delivery. At the end of the exercise, the teams present their work, and the class then has a thorough examination of the text due to contributions of all of the teams. Summative assessments related to this exercise seem to yield better overall scores than the reading assignment/quiz approach used earlier in the semester. This type of engagement is much more enjoyable for me and the students seem to appreciate the opportunity to learn together. This process also allows me to correct any deficiencies in real time, so the students can focus on learning the content without the concern of making point-impacting errors, as with the quizzes.

Case Study Usage

The use of case studies is an excellent way to demonstrate direct, real-world outcomes in relation to key concepts. Since one of the courses I teach covers the legal liabilities associated with safety and health management, as well as workers' compensation, I use case studies as a method of showing the positive and negative impacts to employers in various compliance scenarios. Determining how to support the course objectives with the case studies is vital to their usefulness. For instance, when students are able to comprehend the impact of a particular statute, as well as the range of possible negative outcomes related to a failure to comply, the significance of that learning objective is then magnified. One tool that I used to help organize the events of the litigation process is a process map. Due to time limitations, I normally suggest that students use a very basic, level-two process map that shows the "what's" and the "who's" involved in a case. Since case summaries can include quite a lot of legal terminology, which can make understanding challenging, I suggest the use of a legal dictionary application during the exercise. During this event, students learn to break the case summary into steps, examine each step, document the order of events, and use their devices appropriately to support the exercise. Once presented to the larger group, the others get the benefit of understanding the case without weeding through the entire selection alone. I have also used an A3 project management tool in a similar manner to help organize events and arrive at a conclusion regarding the significance to safety professionals. The

A3 tool is particularly useful when relating the non-compliance conditions to a target, or desired state of compliance. Since the usefulness of lean manufacturing tools ress on the different methods of making information visual, they are a nice fit for the classroom.

Figure 1. A3 Project management tool used for a case study.

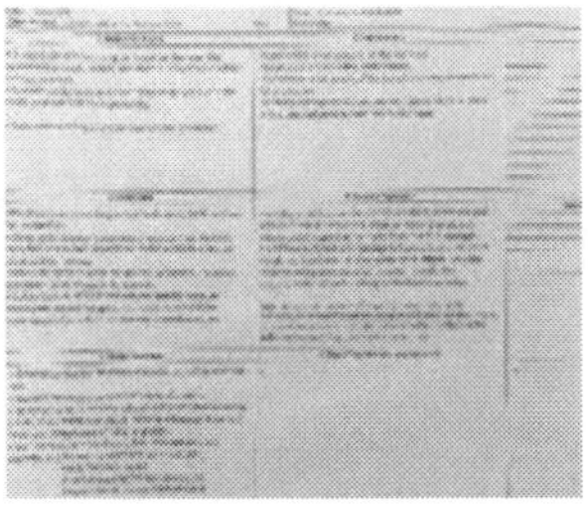

Figure 2. Process map showing criminal sanctions process.

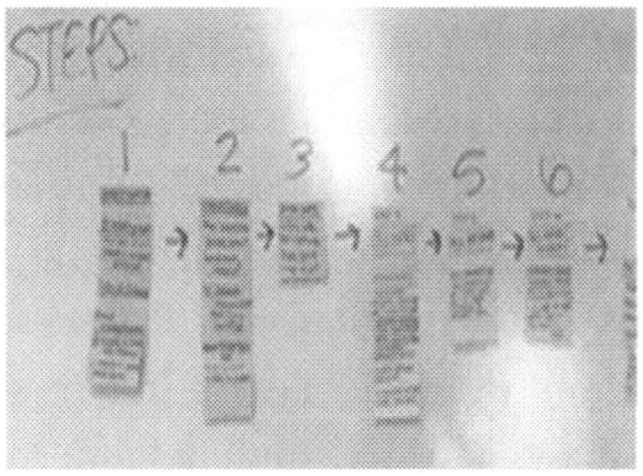

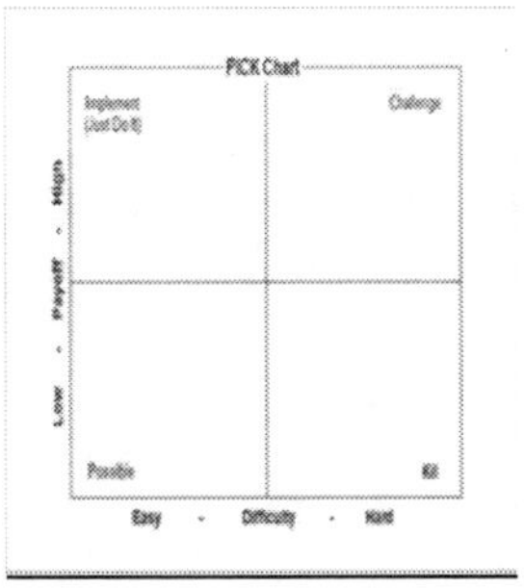

Figure 3. Pick chart used to help students categorize action items by difficulty level and impact.

A Matter of Choice

Despite our best efforts as educators, there are times when students are just not as engaged in the learning process as we would like. During my first semester teaching, I gave this issue quite a lot of thought, and started down the path of using various active learning tools in the classroom. It is important that students get as much out of the academic process as possible, and that the time spent in the university setting is truly value added. While trying out several strategies to address disengagement, I realized that an activity might appear to be very successful the first time it is used, but unsuccessful when used again with the same group. I am confident that there are many variables to consider when solving this problem, but aside from the many complexities, I often wonder if it can be reduced to the issue of choice. When students are allowed to have some choice, in either the construction of classroom activities or the selection of the material, engagement appears to improve. At times, I will have more than one method of delivery ready, then allow the students to vote on the lesson plan options for that day. The majority vote rules, and the lesson is then delivered in that chosen manner. Effectively meeting the same learning objectives, regardless of the method of delivery, can be somewhat challenging and requires more preparation time, when compared to the single delivery approach. For example, I might allow students to choose from an application-driven quiz, role play, or a traditional lecture. It should be noted that there are times when students will opt for traditional lecture over an active-learning-type lesson plan. I have noticed that this often happens at the end of a week, or when weeks are very busy with exams and other obligations. If I prepare an activity that does not end up getting used, I always have it in reserve for another group or another semester. Even though I prefer an active classroom, and enjoy deviating from traditional lecture, I think it is important to realize that there are times when that approach might not work as well, and that further study is needed so educators can identify when an how to apply active learning or student lead models. Allowing a choice in the delivery method of the material might possibly be one of the variables that warrants additional attention.

Successes, Lessons Learned, and End Goals

When students are participating in active discussions, motivated to find answers to questions, and the classroom is lively, I believe the lesson plan has been successful. Even though the typical definition of "success" includes the measurement of meaningful data that shows statistical evidence of improved learning and retention, evidence could also be present in the level of the overall satisfaction of the students and the positive tone of the classroom. When conversations about the material are lively, and extend beyond the classroom, I believe that is also evidence of success. When students emerge from the classroom motivated to learn more than what I have delivered, I believe the end goal is met.

Even though a number of studies indicate the existence of multiple variables when measuring the impact of active learning, and conflicts when relating the approach to long-term content retention, I believe it is important that educators continue to make student engagement activities part of the classroom experience. How much active learning models improve retention for all types of students seems to not yet be fully understood. Evidence suggests that a high performer might show marginal improvement in terms of material retention when an active learning model is used, and a low performer might show more noticeable improvement. Other studies indicate that group composition might also affect the degree of material retention when using these models. Since educators don't always have copious amounts of time to evaluate the many variables that could be present prior to development of a lesson plan, it might be difficult to determine in advance the best methodology to use with each activity.

Despite some of the challenges present in determining impact, I plan to continue to try different approaches to learning and use different tools to help engage my students. Aside from the short-term goal of classroom participation, my intent is to help students develop the skills needed that will support a lifestyle of learning and self-education. Exposing them to different learning techniques by the use of; process maps, project management tools, case studies, peer assessments, and CPAs, students begin to discover what helps them manage information and what helps them make connections between new concepts and those previously learned. When those connections are made successfully, my observation is that the motivation to continue to learn also improves. Knowing what to study is as important and knowing how to study. The more students are engaged and actively part of their own educational process, the more likely they are to continue the process post-graduation.

As I have attempted to incorporate a number of different interactive learning tools into the classroom, I have noticed that success does not appear to be consistent with any one approach. What seems to be well accepted and effective on one day, or with one group, will not necessarily yield the same results when repeated. I document my successes and record the content covered, along with the tool used. My goal is to learn more about how to select the most appropriate activities for the content type. With that documentation, I plan to improve my lesson plans and assessment techniques. Because students often express concern over cumulative exams, determining the effect of active learning on long-term retention will also be part of my focus when collecting data. Evaluating my own personal lessons learned in this process, and using that information to make the necessary adjustments, follows another a well-known continuous improvement model: Plan, Do, Check, Act (PDCA).

Even with careful planning, there are times when obvious signs of disengagement are present. When I discover that a well-developed activity seems to be going stagnant, I might, without warning, switch to a completely different method of delivery. This is a time when having an alternative lesson plan readily available becomes very useful. A change in delivery method seems to keep the students engaged, as long as it supports the overall learning objectives and outcomes. My daily goal is that students receive the education they need to be competitive and competent professionals. If the group appears to be digressing or detached from the learning process, it is my responsibility to attempt to bring them back into the content. If, for instance, we are presenting case studies, and the audience does not appear to be giving their attention to the presenters, the students might be asked to switch from large group presentation and divide into teams of two. The teams of two will use the content that was previously intended for a large group presentation, and create quiz questions based on just their own two cases. The questions from all the teams are compiled, and they become a study guide for an upcoming exam. My focus is to keep students engaged on a level that supports a habit of self-lead learning, which will end up becoming useful when managing complex safety and health issues. Since education and training is an important part of the safety and health profession, an additional objective in the process is to expose the students to techniques they can then use as professionals to address their own issues of classroom disengagement that might appear during their practice.

Studies will continue to be published regarding different approaches to learning and the different modes of delivery. Technology will continue to evolve to support various models. Regardless of the discipline, or the changes in curriculum educators will witness, students will need to emerge into their profession with the ability to engage themselves in a process of life-long learning. As an educator, my desire is to impart the message that learning is a part of everyday life, not just required to enter a profession, and not just a college experience. The excitement associated with the process fuels a desire to continue to learn. Excitement is fostered by exposing students to different methods of teaching and different ways of learning. When active learning models teach students not just the course content, but how to learn new material, the main objective is met and the process of lifelong learning continues.

Bibliography

Andrews, T. M., M. J. Leonard, C. A. Colgrove, and S. T. Kalinowski. "Active Learning Not Associated with Student Learning in a Random Sample of College Biology Courses." *Life Sciences Education.* Winter 2011.

Brazeal, Kathleen R., Tanya L. Brown, and Brian A. Couch. "Characterizing Student Perceptions of and Buy-In Toward Common Formative Assessment Techniques." *CBE—Life Sciences Education.* Winter 2016.

Connell, Georgianne L., Deborah A. Donovan, and Timothy G. Chambers. "Increasing the Use of Student-Centered Pedagogies from Moderate to High Improves Student Learning and Attitudes about Biology." *Life Sciences Education.* Spring 2016.

Lake, David A. "Student Performance and Perceptions of a Lecture-Based Course Compared with Same Course Utilizing Group Discussions." *Physical Therapy.* March 2001.

Michael, Joel. 2006. "Where's the Evidence That Active Learning Works?" *Advances in Physiology Education.* December 2006

Nunley, Kathie. "Active Research Leads to Active Classrooms." *Principal Leadership.* March 2002.

Piper, Katie. "Active Learning: What Do We Know About It?" *Australian Nursing and Midwife Federation ANMF.* September 2016.

Smart, Karl L. and Nancy Csapo. "Learning by Doing: Engaging Students Through Learner-Centered Activities." *Business Communication Quarterly.* December 2007.

Sobek, Smalley and Durward K., II. 2008. *Understanding A3 Thinking: Keys and Tools for PDCA Management.* Taylor & Francis.

Van Amburgh, Jenny A., John W. Devlin, Jennifer L. Kirwin, and Donna M. Qualters. "A Tool for Measuring Active Learning in the Classroom." *American Journal of Pharmaceutical Education.* October 15, 2007.

Session No. A2

Improving Academic Performance by Measuring and Improving Individual Learning Techniques

Todd William Loushine, Ph.D., P.E., CSP, CIH
University of Wisconsin-Whitewater
Whitewater, WI

Introduction

The purpose of a college education is to prepare students to enter their chosen profession with the knowledge, skills, and abilities required for early career success. The need for well-prepared occupational safety and health (OSH) graduates is increasing in the United States. A 2011 report from NIOSH (McAdams et al. 2011) predicted the current need in many industries for qualified OSH practitioners, with little indication of change in the near future. Credible four-year OSH degree programs are continuously assessing and improving curricular content, based on changing demands (or requests) by hiring institutions, new research literature, and changes in the attributes of students. Although publications from ASSE (Hertz and Ramsay 2017) and ABET (2017) on degree program curriculum and student outcome measures are very helpful and informative, there are no current OSH publications addressing the changing expectations or learning approach by college students. The result could be college graduates entering the OSH field ill-prepared to handling complex and continuous problem-solving requirements that protect people, the environment, and employer interests.

The traditional approach to assessing overall student performance, and predict future career success, is the grade point average (GPA). A student's GPA is assumed to be a reflection of their knowledge, skills, and abilities within their degree discipline, and the predicted ability to learn and develop over time. This approach to assessing student performance has validity, and makes logical sense. However, administrative controls over degree programs and ubiquitous use of online/web-delivery of course materials (such as lecture PowerPoint documents) have created a discrepancy in how students approach learning, and prepare course assignments/exams that contribute to a course grade. In the book entitled, *Academically Adrift,* it was shown that four-year college graduates were not developing, but actually declining, in their creative and problem-solving knowledge and skills (Aram & Roksa 2011). These skills are essential to OSH practitioners, particularly at the beginning of their careers.

Motivated by these results, a series of surveys (in 2012-2013) were conducted in two undergraduate occupational safety courses to investigate how students prepared for lectures, assignments, and exams. The survey results (and some informal classroom discussions) mostly agreed with the Aram & Roksa's (2011) conclusions, and found that a majority of students were not completing reading or homework assignments in a timely manner, but rather using the Internet search engines, like google.com, to gain a "gist" for a paper/article, and typically "copying and pasting" answers from websites to homework. In addition, students reported a preference of procrastinating and attempting to "cram" for exams and semester projects.

In an effort to combat procrastination and improve student-learning skills, a course-binder assignment was created. The plan required the development of course-binder assignment details and a grading rubric, which was presented to the class on the first day of a semester (lecture). The elements of the course binder attempted to replicate some of the attributes of highly effective people (Covey 2004).

The basis for this assignment and grading rubric is that successful students need to:

1. Organize course materials by keeping a well-labeled binder;
2. Keep a planner or detailed schedule, along with tracking activities;
3. Keep a journal by documenting reflections, ideas, and accomplishments;
4. Keep detailed, hand-written lecture notes, and notes from reading assignments; and
5. Complete homework assignments by responding to questions, and citing the source.

By requiring students to document (or create evidence) of their learning efforts, these artifacts can be scored by rubric, tracked over time, and compared to performance on other course-related work (such as exams). Feedback from students on their opinion of the binder assignment varies, but mostly they realize their importance, even though they do not like being forced to do the work. The primary focus of this study is whether students who demonstrate the "behaviors" of a successful learner perform better on exams that those students who do not. Three variables were used to study student performance: GPA, scoring on course-binder assignment, and scoring on examinations. The GPA is a collective or cumulative representation of a student's ability to learn and complete course assignments based on course expectations. The course-binder score represents a student's ability to document successful learning behaviors. The exam score represents the student's ability to recall course concepts or solve course-related problems based on course materials or course assignments.

The study objectives were:

1. Is there a relationship between GPA and course-binder score?
2. Is there a relationship between GPA and exam score?
3. Is there a relationship between binder score and exam score?
4. Is binder performance a better predictor of exam score versus GPA?

Methods

Data collected for this study came from grading records of two occupational safety courses over four semesters (Spring 2014-Fall 2015). One course is a mid-level, occupational safety and health standards course (mostly juniors, some seniors) and the other a high-level occupational safety course (mostly seniors). At the beginning of each semester, course registration lists were created in MS Excel, and student grade point averages (GPA) were added. Throughout the semester, students were required to build their course binders (planner/scheduler, journal, lecture notes, study sheets, and binder organization). The course binders were graded using a rubric (each of the five sections is scored out of 10 points, 8 representing "competent" work) mid-semester and at the end of the semester. The midterm and final exams were matched to the course binder scoring that preceded. Midterm and final exam scores were converted to percentages because exam length ranged from 35 to 100 questions (or items). The course-binder scores were out of 50 points (10 points for each of the five rubric items), and these scores were converted to percentages.

A total of 807 course binder and exam scores were collected over a two-year period. The first level of analysis will compare all entries against GPA and Binder Score Percentage (%) against Exam Score Percentage (%). A Pearson correlation analysis compared individual GPA against Binder Score% and Exam Score%, and then Binder Score% against Exam Score%. Linear regression (using SPSS) was also performed on these three variables to assess their relationships, and determine which represents a majority of variance in predicting exam performance.

The final step of analysis involved breaking down the dataset into four groups by GPA and Binder Score%. Reconfiguring the dataset into these distinct groups allows for within and between group comparisons, to assess how Binder Score% relates to Exam Score% within different GPA groups. Each entry (row) was coded to one of four groups based on GPA and binder score. The qualifications for each area and group are in Table 1. Table 2 shows the breakdown of number of observations per group/condition.

Table 1. Breakdown of GPA and Binder Groups

GPA group	GPA range	Count		Binder group	Binder Score range	Count
A	3.5-4.0	146		A	94-100	226
B	3.0-3.5	179		B	86-94	239
C	2.5-3.0	307		C	78-86	160
D	< 2.50	175		D	< 78	182

Table 2. Case Counts Assigned to GPA and Binder Groups

Group Assignments (N)	Binder Group A	Binder Group B	Binder Group C	Binder Group D
GPA Group A	92	37	13	4
GPA Group B	58	73	24	24
GPA Group C	54	87	80	86
GPA Group D	22	42	43	68

Results

The first three study objectives were to investigate the relationships between GPA, Binder Score%, and Exam Score%. Figures 1, 2, and 3 provide scatterplot and best-fit (linear) R2. All three comparison are significant at p<0.001 and appear to be related, but the question that needs investigation is whether Binder Score% is a mediator or moderator between GPA and Exam Score%.

Figure 1. Student Binder Score vs. GPA

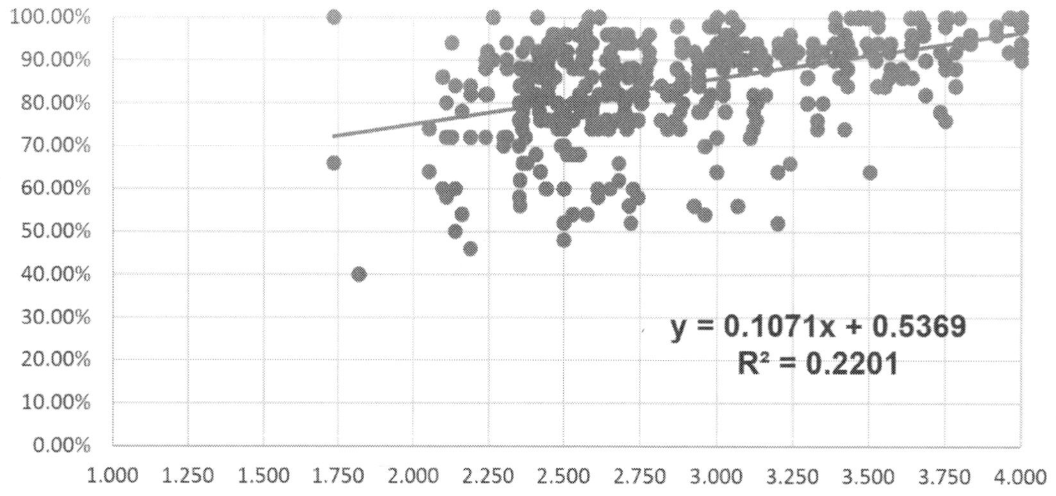

Figure 2. Student Exam Score vs. GPA

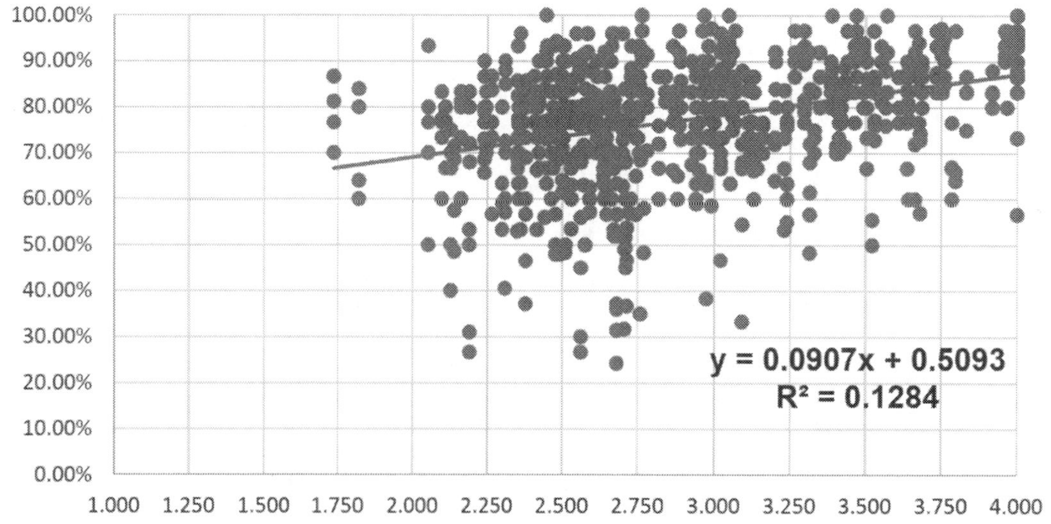

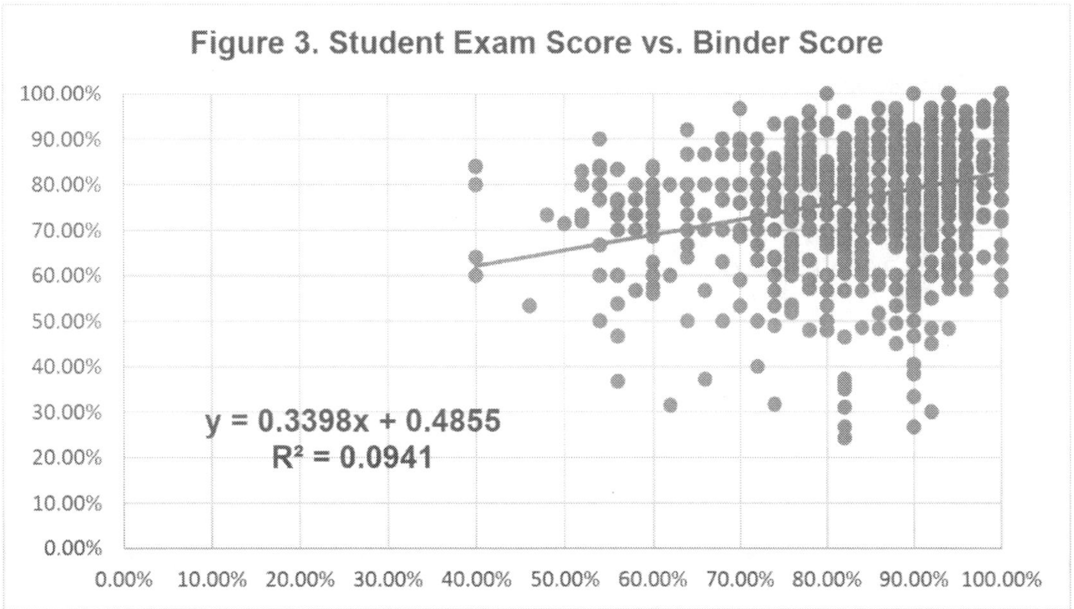

Tables 3, 4, 5, and 6 are the "output" of a linear regression analysis (using SPSS) between GPA, Binder Score% and Exam Score%. The results indicate that both GPA and Binder Score% are significantly related to the outcome measure of Exam Score%. It also shows that although GPA is the primary predictor of exam performance (Beta=0.27, p<0.000), that Binder Score% provides additional significant predictor power on exam performance (Beta=0.18, p<0.000). This is an interesting result, contending that the variance resulting in exam performance is shared by GPA and binder score.

Table 3. Descriptive Statistics

	Mean	Std. Deviation	N
Exam%	77.4%	13.3%	807
GPA	2.92	.526	807
Binder%	84.9%	12.0%	807

Table 4. Model Summary[b]

				Std. Error	Change Statistics				
Model	R	R Square	Adjusted R Square	of the Estimate	R Square Change	F Change	df1	df2	Sig. F Change
1	.391[a]	.153	.151	12.27%	.153	72.64	2	804	.000

a. Predictors: (Constant), Binder%, GPA

b. Dependent Variable: Exam%

Table 5. ANOVA[a]

Model		Sum of Squares	df	Mean Square	F	Sig.
1	Regression	21877.8	2	10938.9	72.6	.000[b]
	Residual	121073.3	804	150.6		
	Total	142951.1	806			

a. Dependent Variable: Exam%

b. Predictors: (Constant), Binder%, GPA

Table 6. Coefficients[a]

Model		Unstandardized Coefficients B	Unstandardized Coefficients Std. Error	Standardized Coefficients Beta	t	Sig.
1	(Constant)	40.36	3.27		12.33	.000
	GPA	6.95	.93	.27	7.48	.000
	Binder%	.197	.04	.18	4.84	.000

a. Dependent Variable: Exam%

The final step of analysis focused on sorting the dataset by GPA (into four groups) and Binder Score% (into four groups). The objective is to analyze whether binder performance can overcome the effect that GPA has on predicting exam performance. Table 7 shows the Binder Score% average and Exam Score% average for each GPA group. Table 8 provides the average Exam Score% for each combination of GPA and Binder Group.

Table 7. GPA Group Descriptive Statistics

GPA Group	GPA Average	Count	Binder% Average	Exam% Average
A	3.76	146	93.85%	85.38%
B	3.22	179	88.11%	80.19%
C	2.70	307	82.37%	74.54%
D	2.30	175	78.90%	73.02%
Average	*2.92*		*85.0%*	*77.4%*

Table 8. Exam Score Average per GPA and Binder Group

	Binder A	Binder B	Binder C	Binder D
Group A	86.75%	82.69%	84.22%	82.50%
Group B	84.67%	78.04%	78.25%	77.88%
Group C	81.20%	73.78%	74.51%	71.17%
Group D	76.69%	76.58%	71.01%	70.91%

Table 9. Pearson Correlation Results within GPA Groups

Variable Correlations	All	GPA Group A	GPA Group B	GPA Group C	GPA Group D
GPA v. Binder%	0.469***	0.433***	0.245**	0.080	0.275**
GPA v. Exam%	0.358***	0.339***	0.163*	0.096	0.062
Binder% v. Exam%	0.307***	0.205**	0.228**	0.182**	0.179

**p<0.05, **p<0.01, ***p<0.001*

Evidently, the predictability of exam performance varies based on GPA group. In Table 9, the two "higher" GPA groups demonstrated significant correlation between GPA, Binder Score% and Exam Score%, whereas the two "lower" GPA groups had limited corollary significance with the three performance variables.

Figure 4 provides a visual comparison of Exam Score% averages per GPA and Binder Group. Group A (average exam 85.4%) is the top GPA group, followed by B (average exam 80.2%), then C (average exam 74.5%), and finally, Group D (average exam 73.0%) was the lowest GPA group. Table 10 provides the results of T-tests, comparing exam averages within and between GPA groups to assess how binder performance can elevate student exam performance independent of GPA. The significance of binder performance on exam outcomes was determined in Tables 3-6, but not the extent or context of the effect.

Figure 4. Exam Avg vs. GPA and Binder Group

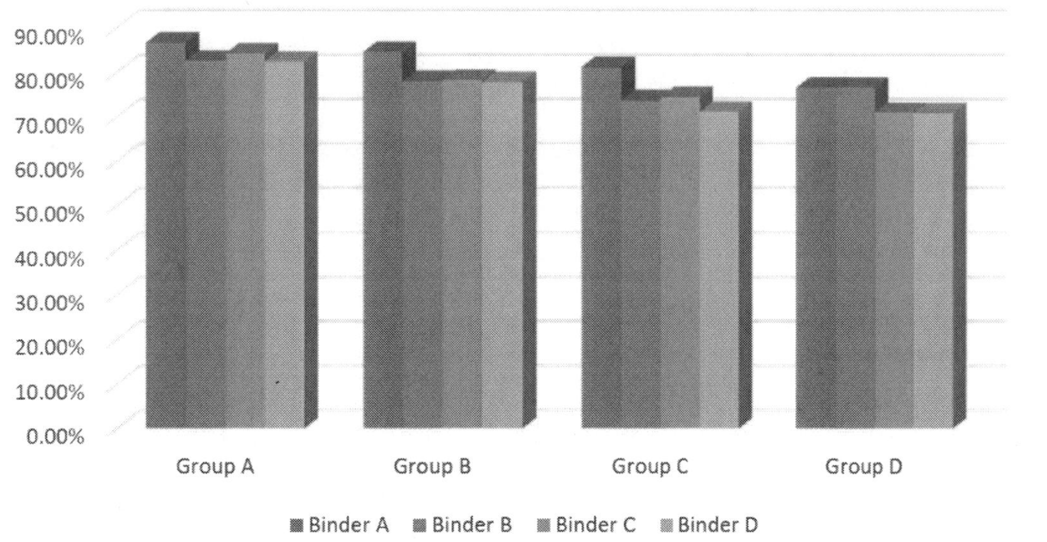

Table 10. T-Test Results, within and between GPA and Binder Groups

		Top Binder Score Group in Each GPA Group			
		AA	**BA**	**CA**	**DA**
Lowest Binder Score Group in Each GPA Group	**AD**	0.3882	0.6483	0.7830	0.2747
	BD	0.0000	0.0006	0.0881	0.6481
	CD	0.0000	0.0000	0.0000	0.0457
	DD	0.0000	0.0000	0.0000	0.0390

Table 10 is a matrix of the top (A) and bottom (D) Binder Group within each GPA group. T-test analyses were done within GPA groups (using Binder Group) and between groups (using both GPA and binder). For example, AA (GPA Group A and Binder Group A) is statistically different ($p<.000$) than DD (GPA Group D and Binder Group D). The comparison of exam score means is 86.75% to 70.91%. But that information is expected because the AA students are high-achievers and the DD students are under-achievers.

Within group analyses results from Table 10 indicate:

- No significant difference in exam scores within GPA group A ($p=0.39$)
- Significant difference in exam scores within GPA group B ($p<0.001$)
- Significant difference in exam scores within GPA group C ($p<0.000$)
- Modest significant difference in exam scores within GPA group D ($p<0.05$)

Between group analyses results from Table 10 indicate:

- No significant difference between the top Binder Score group in GPA groups B, C, or D and the lowest Binder Score group in Group A
- No significant difference between the top Binder Score group in GPA groups C or D and the lowest Binder Score group in Group B
- Modest significant difference ($p<0.05$) between the top Binder Score group in GPA group D and the lowest Binder Score group in Group C

The results of T-test analyses between GPA groups by top vs. lowest Binder Score groups demonstrate that, although GPA is a strong predictor of Binder Score and exam score for the high-achieving students, students with lower GPAs can significantly improve their exam scores (and learning outcomes) by performing well on the binder assignment.

Discussion

The relationships (or predictability) of pre-semester cumulative GPA and performance on binders and exams is statistically significant ($p<0.000$). The top GPA group had the least variance or "binder effect" of the four groups, which could be interpreted as high-performance students (high GPA) are more adaptable to course requirements/expectations or their learning technique is similar to the binder assignment and therefore, their higher exam scores reflected those achievements. Students in GPA groups below A showed a high binder performance, resulting in exam scores statistically similar to the GPA A group, although lower. This means that students who practice the elements of the course binder (document or demonstrate their learning techniques) tend to perform better on exams and therefore, should be able to earn a course grade higher than their pre-course GPA. This is a significant finding, considering that students are charged with determining how they develop and maintain the course binder. From a degree program perspective, the binder assignment provides artifacts and measures of academic performance that can be used for verification and/or continuous improvement.

The dataset used in this study covered courses taught from Spring 2014 to Fall 2015. In spring 2016, the course binder assignment was "optional" to observe student performance. Frankly, it was a disaster for the non-high performing students (i.e., GPA groups B, C, and D). Course reviews indicated that students wanted/needed the course-binder assignment to be required in order to take advantage of improved course performance. In fall 2016, the examinations and binder reviews were conducted one-on-one with the instructor (face-to-face). This new approach to assessment has uncovered a wealth of insight into how students perceive and manage (or not) their own learning experience, and are able to receive customized feedback on how to improve. This new face-to-face approach is very time consuming, but early feedback is indicating that the effects of practicing the course-binder technique is showing greater performance improvements than the results reported in this study.

The final section of this paper are personal reflections from a non-traditional, minority student who was academically dismissed in Fall 2015 but was able to develop a learning technique using the binder, and reached two consecutive semesters on the Dean's list. The following are excerpts from her reflective journal on her journey to graduation:

> *For the past 6 years I have always struggled to find the proper ways to learn material that I had not used for many years as well as learn and even more importantly, retain new material and earn grades higher than a C. For most of those years I was content with being an average student and there were even a few classes that I just hoped to pass even if it meant a D. The truth is I have painstakingly struggled to remember information moments before taking an exam and after I took the exam, I never looked back. By the time I was a senior in college it was clear to me that this method was extremely stressful and I was hurting myself both personally and academically, and after my GPA plummeted to a low 1.9/4.0, I was kicked out of college. When I was allowed to return (after a failed attempt of reconsideration), I took Dr. XXXXX's class in which he required a course binder and graded it twice during the semester. My initial feelings toward this requirement were negative and I reverted back to my old habits. I put the first binder together the night before it was due, partially completed the study sheets, it lacked handwritten notes, and I refused to include any journal entries. I received a 66% and I struggled on my first course midterm exam. However, in the margins of my binder, Dr. XXXXX provided detailed feedback with an explanation of why I received the grade and he invited me to come see him for further assistance in the course. This opened the door to comfortable communication between him and me and I was then able to understand the reasons for and benefits of each element of the binder requirements. The second time around the class binder became by "one stop" tool for the remainder of the semester and I received a score of 100% along with higher exam scores and a final grade of B. The following semester I used the*

binder for each of my four classes and college became fun and less stressful. I found that it was easier for me to truly work up to my potential. I always knew that I had so much more to offer, but all of my time had been spent juggling and trying not to drop the ball. I can honestly say that adapting to and changing my habits in the form of keeping a binder for my classes has changed my life for the better. The last semester I earned a 4.0 GPA which in turn has increased by cumulative GPA and put me back in good standing with the university and a spot on the Dean's List for my first time! I'm confident that I will complete my degree and start my career as a safety professional on the right foot.

This student's account of discovering a learning technique and using it to achieve academic excellence is not unique or novel. Many students have reported back after graduation that although they disliked the time and effort required for the course-binder assignment, that they now truly understand its utility and they continue to use those learning techniques in the current employment. In conclusion, students need to demonstrate a learning technique that parallels the problem-solving approach needed to be a successful safety professional. The results of this study (and the case study) demonstrate how student performance can be measured, and actual learning techniques developed and improved.

Bibliography

ABET 2017. *Criteria for Accrediting Applied Science Programs, 2017-2018 Accreditation Cycle.* ABET-Applied Science Accreditation Commission (ASAC). (Accessed 3/1/2017) (http://www.abet.org/wp-content/uploads/2016/12/R001-17-18-ASAC-Criteria-10-29-16-1.pdf).

Arum, R., and Roksa, J. 2011. *Academically adrift: Limited learning on college campuses.* Chicago: University of Chicago Press.

Covey, S.R. 2004. *The 7 habits of highly effective people: powerful lessons in personal change.* New York: Free Press

McAdams, M.T., Kerwin, J.J., Olivo, V., and Goksel, H.A. 2011. *National Assessment of the Occupational Safety and Health Workforce.* (Accessed 11/1/2016) (https://www.cdc.gov/niosh/oshworkforce/pdfs/nashw_final_report.pdf).

Ramsay, J. and Hartz, W. 2017. "Model Curriculum for OSH: Key to becoming a Profession." *Professional Safety*, vol. 62, no. 3, pp. 40-51.

Session No. A3

Activity-based Learning in the Safety Sciences: Panel Presentation

Helmut Paschold, Ph.D., CSP, CIH, PE
Tracey Cekada, ScD, CSP, CHSP
Jan Wachter, ScD, CSP, CIH
Lon Ferguson, EdD, CSP, CFPS
Indiana University of PA, Safety Sciences Department
Indiana, PA

Introduction

"Tell me and I forget, teach me and I may remember, involve me and I learn." – Benjamin Franklin.

Activity-based learning (ABL), or active learning, is not a new concept. It has received a significant amount of attention over recent years. However, it still continues to be an underutilized method of teaching in the academic environment. Active learning is commonly defined as "any instructional method that engages students in the learning process" (Prince, 2004, p. 223). In the college classroom, active learning involves "students in doing things and thinking about the things they are doing" (Bonwell, 1991, p.2). It goes beyond just watching, listening, or taking notes. It is a process where students become engaged in activities like writing, discussing, solving problems, or carrying out activities that promote the analyzing and processing of the content. Active learning is certainly based on activity–keep the body and mind moving–but it also embraces learning by hearing, visual input, touch, and the stimulation of intellectual processes concurrently. It involves learning by doing rather than learning by listening.

In this panel discussion, the theoretical framework drawing on published research and anecdotal experiences will be discussed first, along with a discussion/debate of the challenges and obstacles with incorporating active learning strategies into the classroom. For instance, how do the elements of the traditional Western culture educational practices stand at odds with the tenets of ABL? Basic strategies for effectively integrating ABL tactics into the classroom will be discussed, and will include how we learn from others (collaboration), discovery methods, and somatic demonstration. Generic applications of active learning, such as the use of question-and-answer formats, picture drawing, one-minute papers, think-pair-share approaches, and the use of responder cards will be discussed, as well as more specific implementation strategies that can be utilized for your specific course. Panel members will relate their personal experiences with resistance to active learning methods, and provide thoughtful actions to overcome opposition. Active learning examples for various classroom topics will be discussed, and some will be performed by the audience in attendance. The panel will conclude the presentation with an open question-and-answer session to draw upon the experiences of the audience.

Benefits and Challenges of Active Learning

Research shows student attention in lectures starts to fade after 10-15 minutes. In addition, research indicates that students remember 10% of what they read, 20% of what they hear, 30% of what they see, 50% of what they see and hear, 80% of what they say, and 90% of what they say and do (Thompson, 2000). Only 30% of students are listening at any given time in a lecture, with the remainder taking mental naps, thinking about their past, or planning their future (e.g., where to go to socialize that night). Because of this, traditional lecture-style format (e.g., reading and explaining slides to students) can be a relatively poor instructional tool for holding students' attention and imparting learning. However, incorporating active learning techniques regularly into lectures can aid in "bringing them back" or re-engaging students. An engaged student is a student who learns.

The benefits of engagement through active learning are essential to student success. Moreover, utilizing active learning strategies in the classroom allows us to provide more frequent feedback to our students and better gauge their learning because: we are interacting with them more regularly; we are listening to them; and we are watching them. If done effectively, active learning activities can feed off of students' varying learning styles, again making one more effective in the classroom. It can also help to build community in and outside of the classroom, and create a personal connection to the material.

Incorporating active learning strategies into the classroom brings its own challenges. Foremost is that it takes time. However, if the activities are well thought out and orchestrated, *valuable* use of the time can be made. Even though you may be able to cover more material in a traditional classroom, how much are students really retaining and learning? Active learning can focus on the *key* points or lessons that you want the student to take home, not everything.

Some also view active learning activities as a bunch of "fun and games," and are dismissive of the "frivolous" technique. This interpretation is especially prevalent in the private sector. ABL can be fun, but it must facilitate learning. Many active learning activities can present the student with challenges that requires a lot of thinking, dissecting, and effort, all important components to higher level learning.

A common way to incorporate active learning activities into the classroom is through the use of group work. The greatest benefit of group work is that students can learn from each other. There can be concerns with utilizing too much group work where students aren't able to learn individually or erroneous information gets passed around the group. In these instances, variety is critical to ensure that active learning activities don't always rely on groups to perform. Additionally, reviewing material with the entire class after the group work has ended can help to minimize the sharing of incorrect information.

Basic Strategies for Incorporating Active Learning Techniques Into the Classroom

If well prepared and orchestrated, active learning techniques can be regularly incorporated into the classroom with minimal disruption and waste of valuable class time. Some key suggestions for success include the following:

- *Prepare:* Plan for any out-of-class activities that students can do in advance so they are ready for the in-class activities. This may include activities such as finishing readings before class, watching videos in advance, or bringing specific items to class to discuss as part of the activity.

- *Start small*: Find a few simple active-learning strategies with which you are comfortable. These can be adapted for other classes instead of reinventing the wheel.

- *Be clear*: Explain the directions to the class clearly and make sure you are thinking about how you will control or facilitate this activity through each step. How will you introduce it? Do you need to put together written instructions or visuals to help them complete it? How do you quickly and effectively get the questions or instructions out them? How will you break the students into groups or organize them (if you are)? How much time do you allot? How will you wrap up everything at the end?

- *Explain the benefit*: You absolutely must explain the learning benefit of the activity and connect the benefits back to the class objectives and/or learning outcomes.

- *Establish time limits:* It is critical that you set and stick to time limits for your activities; otherwise, you could waste too much valuable class time. Make certain the students know that their time is limited so they understand the urgency in tackling their activity.

- *Debrief:* Always debrief or summarize major points at the end to ensure that everyone understands the lesson and to potentially clear up any misinformation passed around in groups, if applicable.

Active Learning in Practice

A variety of tactics can be used to promote active-learning strategies. The number is limited only by the imagination of the instructor. Some of the more common tactics include:

Question and Answer, Q&A. This method involves the learner because it causes the learner to develop an answer in her/his mind and respond.

Group discussion. This can be whole group or small group, depending on the task and number of students.

Individual problem solving. An example would be students who are given a mathematics or word problem to solve. They can be solved as a pair.

Case study. A "real life" situation is presented for class members to discuss and work through.

Role playing. Members of the class are assigned to play the part of someone else. For example, students could role play as a manager or an employee confronting an ethical dilemma.

Classroom demonstration. Presenter shows class members some skill, such as extinguishing a fire with a portable fire extinguisher. Learners then perform the task.

Classroom Examples

Below are some examples of how active learning strategies were incorporated into the safety science classrooms.

Use of Responder Cards

In Exhibit 1 below, students were presented with a problem that required critical thinking. In order to effectively gauge their understanding, responder cards were utilized in the classroom. These cards had a "Yes" response printed on one side and a "No" response printed on the other. This allowed the instructor to visually verify if all students were able to properly analyze the scenario, and come to the correct conclusion. In this particular situation, the instructor noted a student who initially indicated a "no" (incorrect) response, and then changed their mind to the correct "yes" response. The instructor was then able to inquire as to why the student changed their response. This helped to confirm that the student was able to critically analyze the scenario correctly and come to the correct decision.

Exhibit 1. Use of responder cards in the classroom.

In Exhibit 2 below, students were asked to visually depict the industrial hygiene process in a flow diagram. Students were asked to brainstorm on this activity. This activity enabled students to be introduced to the topic, share ideas, and gain new insights into the recognition, evaluation, and control of health hazards.

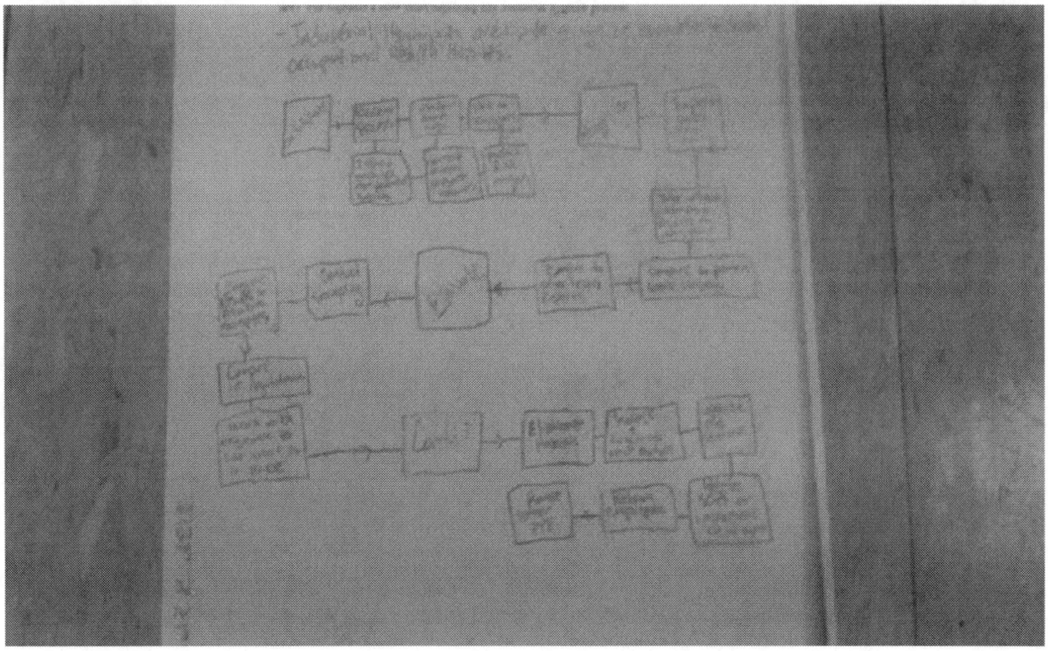

Exhibit 2. Example of brainstorming session to create visual flowcharts.

Conclusion

Incorporating active-learning strategies into the classroom can be an effective tool for enhancing learning. However, there can be associated challenges with ensuring its success. In order to make the most of your active-learning activities, it is crucial that you aptly plan your activities, and keep firm with established time limits.

Active-learning activities can be incorporated into virtually any classroom. The key is foresight and preparation. This panel discussion will be providing examples of actual activities conducted in the classroom, discussing their triumphs and challenges, and evoking ideas that can be incorporated across the curriculum.

Bibliography

Bonwell, C.C., and J.A. Eison. 1991. "Active Learning: Creating Excitement in the Classroom." *ASHEERIC Higher Education report No. 1.* George Washington University, Washington, DC.

Prince, M. 2004. "Does Active Learning Work? A Review of the Research." *Journal of Engineering Education*, 93(3), 223-231.

Thompson, M.R. 2000. "Planning, Writing and Producing Employee Education Programs." *Professional Safety.*

Risk Management for Road Safety Based on the ISO 31000

Juan Carlos Marín, MSc.
QHSE Consultant

Introduction

Worldwide, the problem of motor vehicle fatalities tends to surpass the prevention efforts from governments, institutions, and industry in general. Fatalities and injuries in traffic crashes do not seem to reduce significantly; on the contrary, they maintain their trend to increase year after year, becoming a social and economic problem burden.

The Community of Latin American and Caribbean States (CELAC) reported that the region presents one of the highest mortality rates due to traffic crashes. The burden of motor vehicle crashes has a significant impact on the society due to their high direct and indirect costs. Additionally, this type of event significantly targets working-age populations, with a subsequent impact on countries' competitiveness. In these countries, the estimated loss in gross domestic product oscillates between 1.6 and 3.0% per year.

According to the World Health Organization (WHO), approximately 1.25 million lives are lost each year because of traffic accidents, and between 20 and 50 million people suffer non-fatal injuries, including any form of disability (WHO, 2015). Although these numbers remain steady since 2013, the high number of fatalities associated with traffic risk is a matter of concern. Factors, such as the increase in population and in the use of the motor vehicle, compared to other forms of transport, make the population in general more and more vulnerable.

According to the scale of this problem, it is necessary to focus efforts in identifying new or improved methodologies able to effectively manage the risks associated with transit in all sectors. The objective of this paper is to present a model that manages the road-traffic accidents, through the implementation and development of an integrated risk assessment and management system that uses a logical and systematic process based on continuous improvement.

Data from Colombia and the Region

According to the National Institute of Forensic Medicine, in Colombia in 2015 were reported 52,690 cases from motor vehicle accidents. In these accidents, 6884 people died, representing 13.07% of the total number of accidents. As of December 2016, there were 6,800 fatalities in events associated with traffic accidents, which do not represented a significant reduction in comparison to previous years (Graph 1).

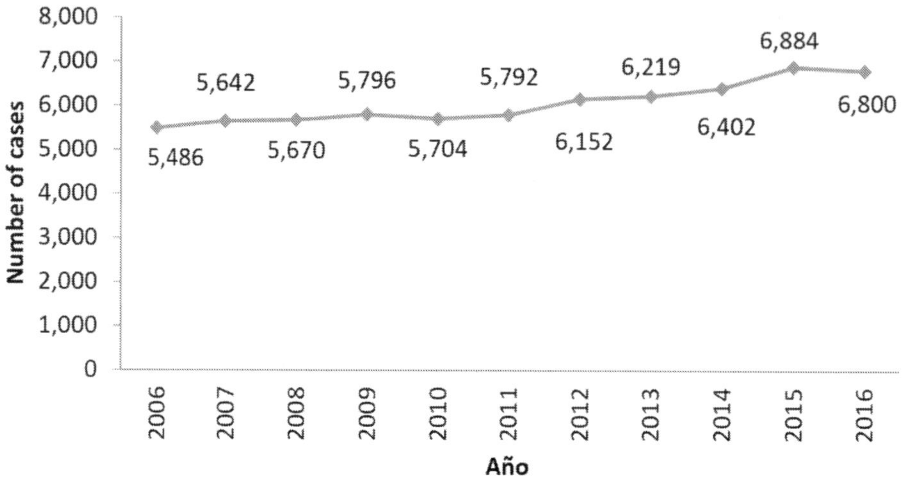

Graph 1. Deaths due to transportation accidents, cases per 100 thousand inhabitants. (*Source*: Author's elaboration with data from the Colombian National Institute of Legal Medicine and Forensic Sciences.)

From the perspective of the means of transportation affected, about 80% of the deaths and injuries in Colombia are concentrated in what the WHO calls vulnerable road users, such as: (1) pedestrians, who accounted for 26.57% of the total deaths and 20.93% of injuries; (2) bicycle users, with 5.53% of the deaths and 5.74% of injuries; and (3) motorcycle riders, with 47.36% of the deaths and 54.92% of the injuries (Forensis, 2015).

The condition of the victim of the transport accidents is the first variable of consideration in the analysis of road accidents (Vargas 2015). The statistical information determined that 66.74% of the deaths affected were vehicle occupants, while of the percentage of occupants of vehicles and pedestrian injured was 79.02% (Graph 2).

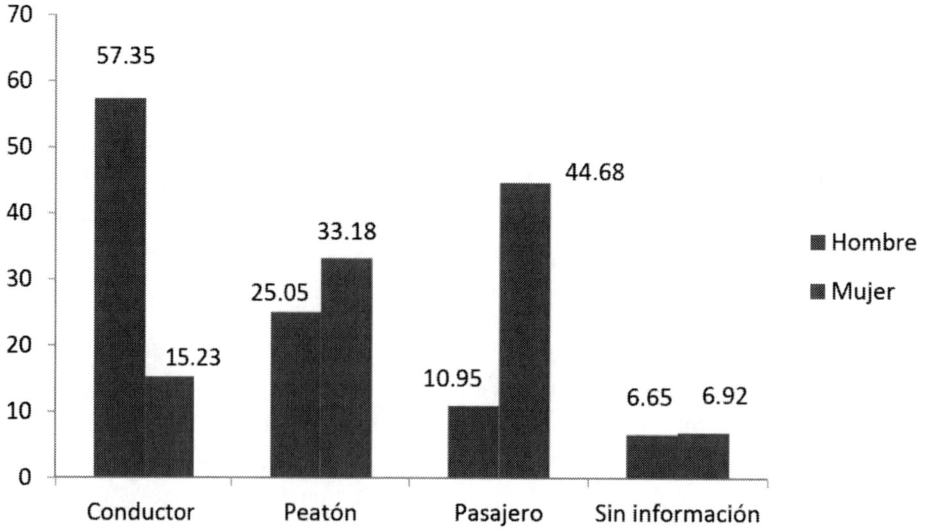

Graph 2. Users vulnerable due to transport accidents. (*Source:* Author's elaboration with data of the National Institute of Legal Medicine and Forensic Sciences.)

The regional scenario does not seem to be better, according to the United Nations the road safety management becomes a bigger problem. If they take into account the estimates in the countries of Latin America and the Caribbean between 2000 and 2020, the number of deaths due to traffic will grow by 48% (WHO, 2015).

Traffic Accidents in the Industry

The industrial sectors are not exempt from this problem, since in the development of their production processes, transportation takes on a significant importance due to the needs of moving raw materials, delivering finished products to the distribution centers or to the end customer. In order to fulfill their obligations, companies have opted for contracting different means of transport, with a significant increase in the use of motorcycles as a work tool. Although traffic accidents are not generated directly within the company's facilities, their severity is even greater than other work-related factors, since mortality rates are high and severe non-fatal cases are three times greater than common injuries.

Some jobs with a seemingly low risk may have a high mortality rate due to the use of a means of transport, without including independent professionals, who remain in the background as self-employed. Preventive actions of work-related traffic accidents have not been adequately implemented in organizations. However, with the new regulations that have been implemented in Colombia (Res. 1565 of 2014), it seems that special attention has been provided to this type of risk, moving efforts in the right direction. The framework in which the organizations should base their efforts lies in the implementation of policies and methodologies to prevent accidents, promoting prevention management of occupational risks related to traffic, with the purpose of generating social and economic benefits to both workers and companies. To achieve a healthier work environment, reduction of accidents and, therefore, of the direct and indirect costs derived from them, and the promotion of the good practices of safe driving, is a priority for any company.

Taking into account the great impact of road accidents in the society, the purpose of this paper is to present a model for the management of transit risk, applicable to any organization, independent of its economic activity. This model is based on the international standard ISO 31000 in order to manage risk through a systematic and logical process of identification, analysis, evaluation and monitoring, to provide effective responses in the treatment of this organizational risk.

Social Impact

The final result of accidents is often human losses (temporary or permanent), time, and damages in infrastructure or equipment, with the normal financial burden for organizations. According to CELAC (2005), the social and economic burden caused by traffic accidents accounts for the 1% of the gross national product of low-income countries, 1.5% in the middle-income countries and up to 5% in high-income countries. The deaths and injuries related to transportation accidents continue to be one of the situations that most affect our society, a fact that is reflected in the data where a decrease in road accidents is not visualized (Valbuena (2011).

Being a high-impact issue for society, it is necessary to address road safety management in a comprehensive way that integrates multiple public and private sectors. The World Health Organization (WHO) report on the global situation of road safety (2009) shows that road accidents have become a public health problem. In order to address this problem, a coordinated response is required, including the development and implementation of a strategy that integrates all the active sectors of the economy. This requires injury prevention strategies that include institutional policy and guarantee sufficient resources to plan activities to be carried out within a defined timeframe. It is estimated that by 2030, road traffic injuries will be the fifth leading cause of death in the world, making road accident prevention a must for all (Table 1).

Order	Main causes	%
1	Ischemic heart disease	14.2
2	Cerebrovascular disease	12.1
3	Obstructive pulmonary disease	8.6
4	Infections of the respiratory tract	3.8
5	Traumatic injuries due to traffic accidents	3.6

Table 1. Main causes of deaths in the world by the year 2030. (*Source:* Author's elaboration with data of the report on the world situation of the road safety, 2009.)

Legislation in Colombia

Colombia has understood the importance of road safety and has taken steps to improve the institutional frameworks needed to support road safety efforts. In this sense, it has promulgated measures such as Law 1503 of 2011, which promotes the formation of habits and behaviors in the way; The Decree 2851 of 2013 which regulates, among others, the Strategic Road Safety Plan, the Road Safety Map, the Safety Portal, and the Personal Commitment Card with road safety; and the most recent, the Resolution 1565 of 2014, which establishes the methodological guide for the elaboration of the strategic plan of road safety.

All these measures are aimed at preserving the integrity of people, since according to the statistics of the Institute of Legal Medicine in 2015, the working-age population between 18 and 39 years were the main victims of road accidents (Graph 3).

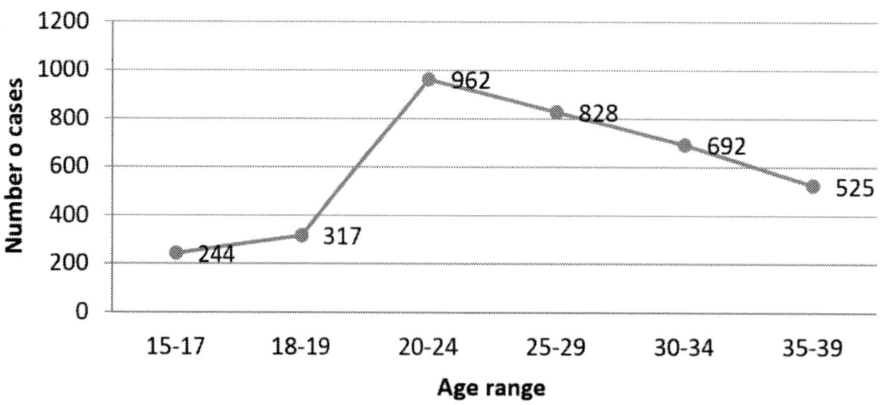

Graph 3. Age range most affected by road accidents in 2015. (*Source:* Report on the global situation of road safety, 2009.)

Technology plays a very important role in the prevention of traffic accidents, but it can often increase the likelihood of an accident on the road. Distractions due to the use of mobile equipment and state-of-the-art vehicles mean that the driver keeps his attention on other people's issues, and reduces his attention to the dangers of driving.

The technological improvements in vehicles have advanced considerably in the last several years. Innovations include adaptive speed control, in which the driver chooses the speed he wants the car to maintain, and the minimum safety distance. The intelligent speed assistant adds to the speed control, automatic detection of traffic signals, helps the car remain in the lane, and detects accidents, and is an undeniable improvement in road safety; however, there is a factor that is always present in the accident and that has been more difficult to control, the human factor.

Although the factors that influence the road accident are diverse, the human factor is considered as an important actor. The driver is responsible for acting appropriately or not, in the development of driving, receiving the information of the environment, valuing it, and executing the actions that it deems convenient. The person driving the vehicle is involved in about 80% of cases of traffic accidents. The most frequent causes of road accident are related specifically with unsafe acts that people perform when they are in front of the steering wheel (Table 2). Failing to maintain safe distance, disobeying or ignoring traffic signals, and driving in a state of intoxication, are the recurrent causes of transportation accidents.

Cause of traffic accidents	Number of accidents
Do not maintain safety distance between vehicles.	2. 551
Disobeying traffic signs.	1. 091
Overtake cars closing.	1.029
Lane change without directional.	430
Do not respect priority.	410
Reverse recklessly.	379
Forward invading via.	171
Speeding.	168
Start without precaution.	142
Driving in drunken or drugged state	139

Table 2. Range Causes of traffic accidents. (*Source:* Author's elaboration with data of the Secretary of mobility of Bogota.)

It is necessary, then, to apply a risk management model that serves as a conceptual framework in the prevention of road accidents, which provides private organizations or public entities with a frame of reference that goes beyond purely legal compliance and becomes a fundamental part of the company's performance, its strategic objectives and in general its risk prevention policy.

Risk Management

Just as the world political, social, financial and cultural environment has evolved towards a more globalized approach, nowadays the leading organizations focus more on risk identification, analysis and control. We can understand that there is not a human activity that does not generate a risk associated with its realization; for that reason, its identification and control becomes necessary to assume the results of its manifestation.

The business risk management began to take root in the late 1990s and, since then, has generally been recognized as an expectation of good and corporate governance. However, many companies still struggle with the implementation of a risk management system (Fraser, & Simkins, 2016). In the context of risk management, every day in organizations the need is even more evident to control threats that can affect their normal functioning and generate losses, ranging from the economic to the effect on people, through the deterioration of the environment or the corporate image (Mejía, 2013). Thus, over the years, different models of risk management have been created in different economic sectors such as banking, insurance, work environment, and industry in general.

In the search for a common model for systematic risk management, the ISO 31000 standard appeared in 2009, providing comprehensive principles and generic guidelines on risk management. This standard is not specific to any industry or sector so it can be used by any public, private or community company, association, group or individual. This standard can be applied throughout the timeline of an organization and to a wide range of activities, including strategies and decisions, operations, processes, functions, projects, products, services and assets (ISO, 2009). The ISO 31000 standard takes the fundamental principles of AS/NZS 460, which has been applied in Australia and New Zealand. The standard recommends that organizations develop a framework for fully integrating risk management, reporting, and accountability into the organization. This means that risk management is an active component in governance, strategy and planning, management, reporting processes, policies, values, and culture (Gjerdrum, 2010).

Risk management is understood as all "coordinated activities to direct and control an organization with respect to risk" (ISO, 2009). Taking into account that all organizations, regardless of their nature, size and rationale, are permanently exposed to different risks or events that may jeopardize their existence, it is necessary to manage their risks on a permanent basis, including internal and external risks, those own of its operation or acquired in the execution of specific projects.

Risk management in all organizations goes from being a good practice to becoming a necessity for the fulfillment of its strategic objectives. However, it is not enough to implement methodologies and norms in this sense, but the development of an organizational culture is required that leverages the achievement of the objectives outlined. People are one of the main sources of risk in organizations, although it is not the only one (Cori, 2014). In this sense, it is necessary to maintain a permanent synergy with all the collaborators and interested parties in order to direct improvement actions.

The analysis of internal and external risks that affect or may affect the organization must be developed in an environment that applies the basic concepts of risk management, through the application of a systematic process that allows the organization to make decisions based on the information, in order to minimize or mitigate the impact of the associated losses. Through this process, organizations communicate and consult with stakeholders, monitoring and review of the risk and controls that are modifying it, to ensure that no additional risk treatment is required. Although risk management practice has developed over time and in many sectors to meet a variety of needs, the adoption of consistent processes within a comprehensive framework can help to ensure that risk is effectively and efficiently managed throughout the organization (ISO, 2009).

The ISO 31000 of 2009 raises the monitoring and review as an indispensable stage to guarantee the effectiveness of the process of risk management in organizations. This stage is carried out through indicators, revisions of the plans, policies and framework of the risk management, as well as the detection of the changes present in the internal and external environment of the companies, and the identification of emerging risks in the implementation process, and the measurement of strategic risks (ISO, 2009).

In light of this situation, risk management appears to solve the needs that arise in the face of various threats, uncertainties and risk events to which all activities that are part of the development of a company or organization are exposed. According to ISO 31000, it defines risk as the "Effect of uncertainty on objectives, this effect is a deviation from what is expected, whether positive, negative or both." In the practice of risk management, we find the two scenarios possible; therefore, it is necessary to take measures on each of these effects:

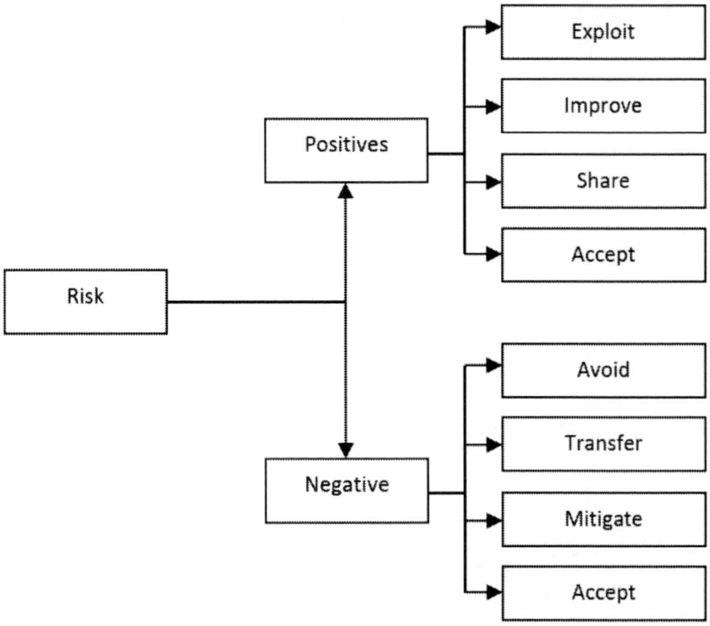

Figure 1. Characteristics of risks.

Each of the strategies designed to respond to the effects of risk has a direct influence on the condition:

Exploit: Tries to eliminate the uncertainty of the identified opportunity by making it concrete.
Improve: Makes the positive effect have a greater impact.
Share: Tries to involve another actor in order to have better results.
Avoid: Changes the original conditions of that event, to completely remove the identified risk.
Transfer: Transfers the negative impact of risk to a third party.
Mitigate: Carries out concrete actions that diminish or the probability of appearance of the threat, or its impact if it ends up materializing.
Accept: Does not change the original plan.

When approaching the process of integral risk management, and its planning stage, it is necessary to design and adjust each of the strategies that derive from the risks, whether positive or negative, so that its implementation engages with the corporate culture and promotes the achievement of objectives, in addition to establishing activities, responsibilities, competencies, criteria, operational controls and verification activities.

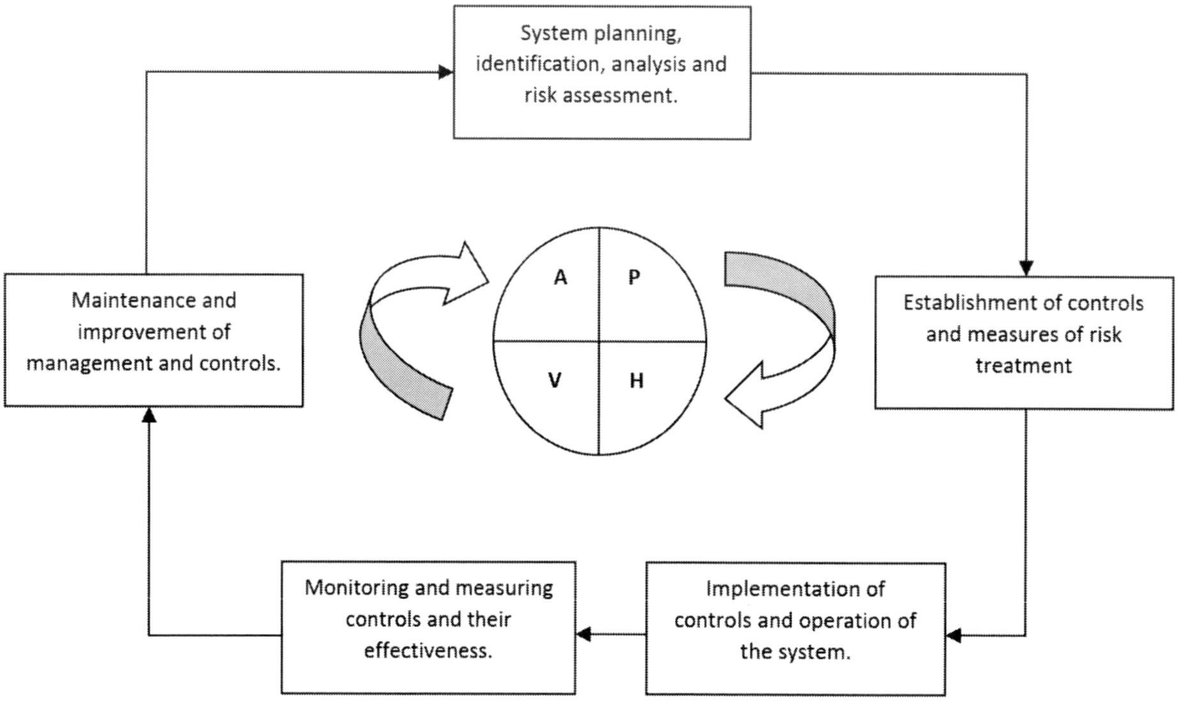

Figure 2. Planning with a risk management approach. (*Source:* Author's elaboration, with data of "The integrated management system," Francisco López, 2014.

Model for Traffic Risk Management

The first step to start a successful risk management depends on the definition of the reference framework in which the organization is going to move, since it is here that the bases and the dispositions that are implemented at all levels are established. ISO 31000 states that the framework ensures that the risk information derived from the process is adequately reported, and used as a basis for decision-making and accountability at all relevant organizational levels. All this is framed within an improvement process such as Plan-Do-Check-Act (PHVA). The establishment of the model of traffic risk management must be understood in the following steps, which are interrelated, as shown in Figure 3:

• Leadership and commitment
• Design of the reference framework
• Implementation of risk management
• Monitoring and review the reference framework
• Continuous improvement of the reference framework

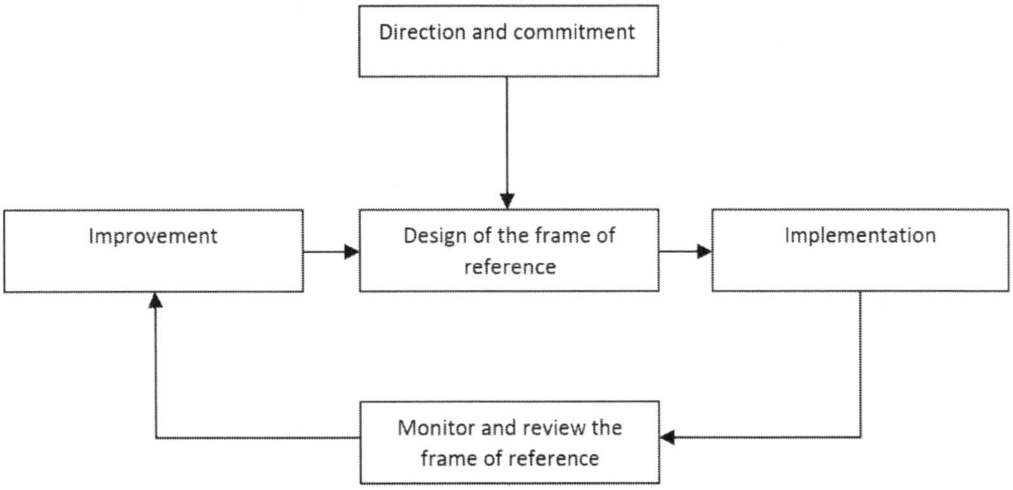

Figure 3. Interaction of the stages of the road risk management model. (*Source:* Author.)

Management commitment includes the introduction of risk management into the organizational structure, ensuring a sustainable management model and ongoing support from senior management. At this stage, senior management must lead the identification of the hazards associated with the transportation to their productive activities. This is achieved by carrying out an initial assessment to identify priorities for road safety, to define a risk management policy, and to ensure that the organization's culture and the risk management policy are aligned with the strategic objectives. Assigning obligations and responsibilities at the respective levels within the organization and establish the work plan ensure the sustainability of the system. The initial evaluation allows organizations to keep current security priorities within the corporate strategy, in line with changes in the conditions and work processes of the company and its environment (Figure 4).

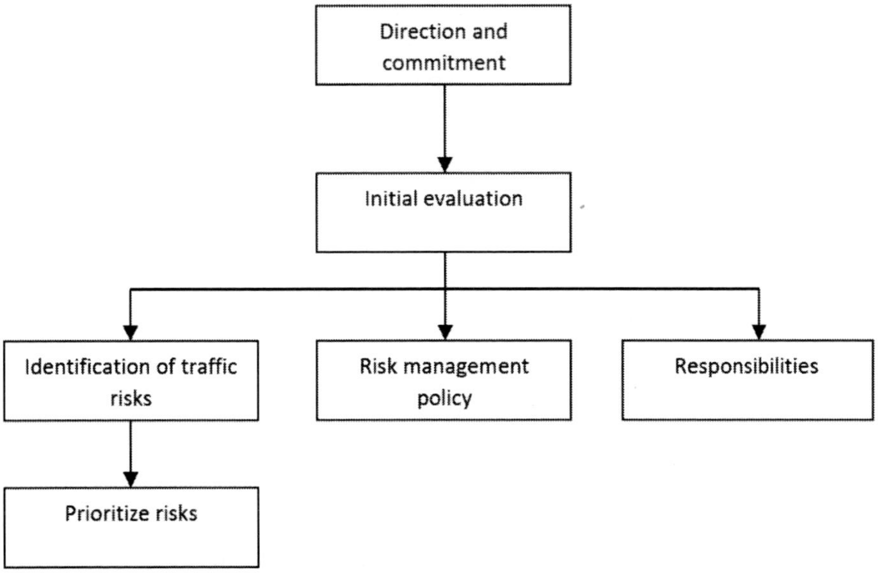

Figure 4. Stages of direction and commitment. (*Source:* Author.)

In the design of the reference framework, it is important to evaluate and understand the external and internal context of the organization implementing this risk management model. The evaluation of the external context of the organization may include, among others: the social and cultural environment, as well as political, legal, regulatory, financial, technological, economic, natural and competitive factors (ISO 31000). Taking into account the initial diagnosis and the identification of the hazards, it is necessary to review the information regarding the traffic accidents occurred in the period of study that is to be taken as a framework of reference; this to make a characterization that allows identification of different critical points in the operation.

The economic factor is essential in the design of the frame of reference. The ability to identify the leaks of capital to the organization represented by the traffic accidents allows decision making based on facts and real facts, thus adjusting its plans of action with respect to the cost benefit of its implementation.

Once the company's road safety issues are identified, an action plan is drawn up, describing the company's objectives, goals, strategies, activities, dates, indicators and managers. To solve the problems of higher priority, this action plan has to be adjusted to the reality of the company in terms of the available budget, and the process capacity to carry out the activities, so as to guarantee the achievement of the expected results. Another important component within the framework is the inclusion of all model mechanisms in the organization's practices and processes in a manner that is relevant, effective and efficient, from the definition of policies, the definition of strategic objectives, and the accountability for road risk management.

When implementing the organization's framework for road risk management, the appropriate time and strategy for its implementation should be defined. That is, to apply the process and policy for risk management to the organization's processes, comply with legal and regulatory requirements, ensuring that the decision-making arising from the implementation, including the development and establishment of objectives, is In line with the results of the processes for risk management (Figure 5).

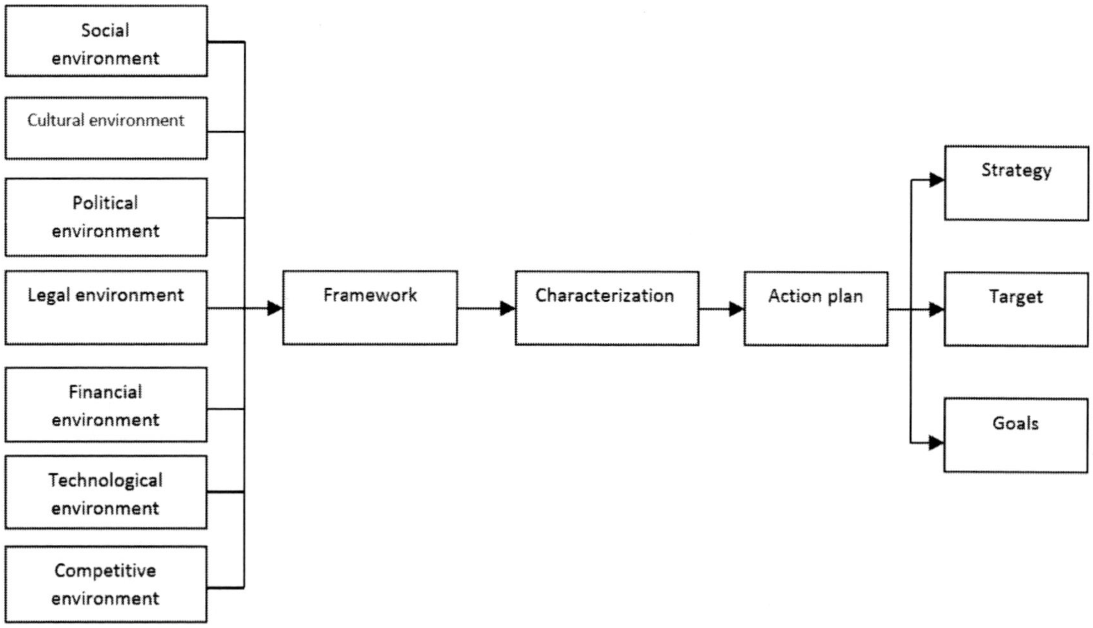

Figure 5. Inputs and outputs of the reference frame. (*Source:* Author.)

The action plans for the control of the risks identified must involve consideration of the causes and sources of risk, their positive and negative consequences, and the likelihood that such consequences may occur. The risk is analyzed by determining the consequences and their probability, and other attributes of the risk. An event can have multiple consequences and can affect multiple objectives. Existing controls and their effectiveness and efficiency should also be considered.

The action plan for risk management should consider the approach of ISO 31000, which are set out in Figure 6.

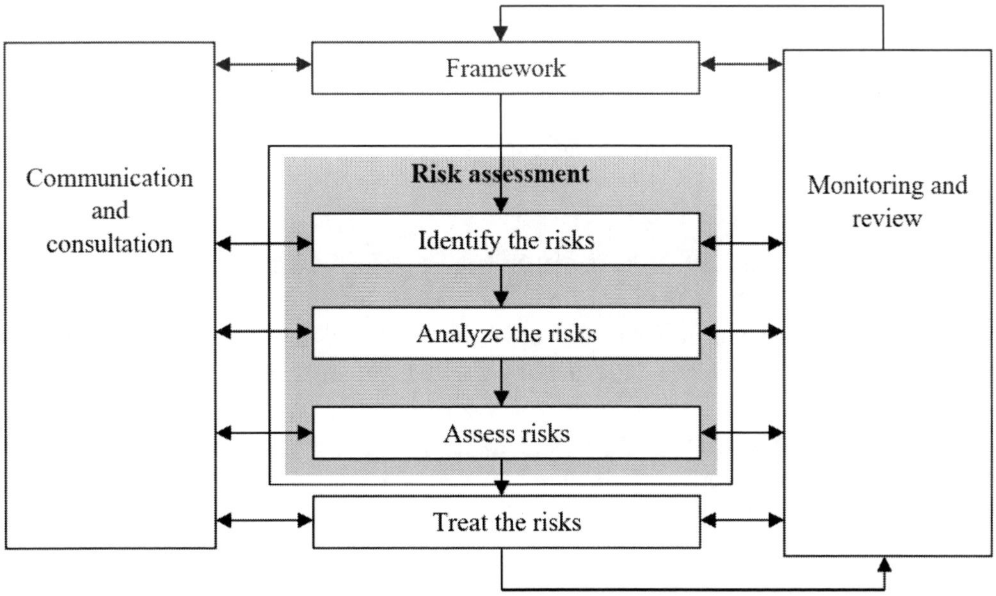

Figure 6. Risk analysis. (*Source:* Standard NTC, ISO 31000.)

According to ISO 31000, risk management should be implemented by ensuring that the process for risk management (Figure 6) is implemented through a risk management plan at all levels and relevant functions of the organization as part of their practices and processes. In the implementation phase, the organization should strengthen the competence of the personnel under its control, developing activities together towards the common goal of preventing road accidents. Once the strengths and weaknesses of the work team members are identified in relation to personal competencies, it is important to develop those who need it through a structured training program, defined for each of the job profiles.

The vehicle maintenance plan is another activity that is of great importance in the development of road risk management. The priority of this plan is to keep vehicles in optimum operating condition and thereby reduce the costs associated with unforeseen failures. It is necessary to carry out an inventory with the characteristics of each of the vehicles of the organization, keep a historical record of its useful life, maintain a resume, and implement the data sheet. The periodicity of maintenance will depend on the initial conditions identified in the resume; thus daily, weekly, monthly, or annual maintenance activities will be required (Figure 7).

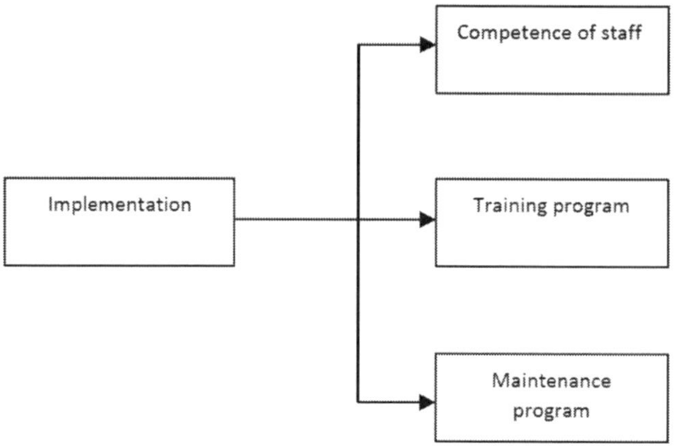

Figure 7. Typical outputs of the implementation. (*Source:* Author.)

In order to ensure that risk management is effective and continues to sustain the performance of the organization, the performance of risk management should be measured against the indicators prepared for that purpose. The objective of the evaluation is to measure and control the results achieved with respect to the actions carried out in a defined period. It also allows correcting errors and serving as a guide for decision making. Monitoring and review will periodically measure progress against the risk management plan and deviations from it, periodically review whether the risk management framework, policy and plan are still appropriate, as well as the external and internal context of the organization (ISO, 2009) (Figure 8).

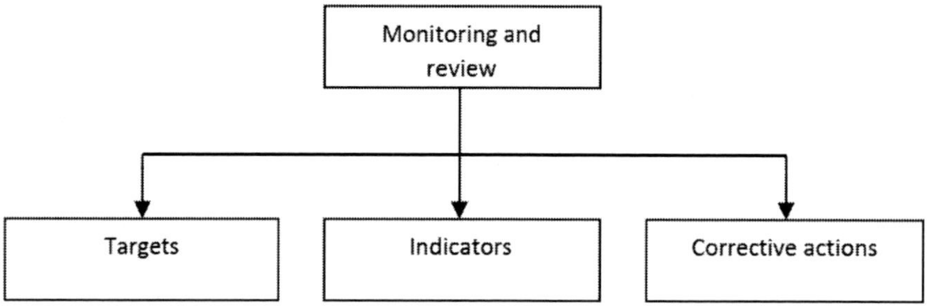

Figure 8. Monitoring and review outputs. (*Source:* Author.)

The improvement would be the final step in the implementation of a road risk management model. Excellence must be achieved through a continuous improvement process. Improvement in all areas of staff capacity, efficiency of resources, relationships between processes, the members of the organization, and society, translates into an improvement of the road accident prevention system. Achieving the best results is not a day's work. There can be no setbacks in a progressive process. The objectives of the organization must be met, preparing for the next challenges.

Continuous improvement implies the implementation of a system, the continuous learning of the organization, the monitoring of a management philosophy, and the active participation of all people.

Articulating all phases of the model, from the direction and commitment, to continuous improvement, is not an easy process. The culture of the organization and the deepening that is required in each one of the phases can become a process that takes a long time but will bring its fruits, since all the activities aimed at preserving human life will not be in vain.

Conclusions

Campaigns for prevention and promotion of road safety, courses, advertising, are all valid management tools in the management of road accidents. However, the main component, which should focus efforts, is in the awareness of the human factor. No strategy will be effective without the active participation of the people involved, and all will remain mere strategies without application, if the person does not appropriate them within their daily behavior and does not maintain them as a way of life. The efforts should be even greater in the prevention of road accidents, since people in productive ages are the main victims of this accident phenomenon.

It is necessary to understand that road accident is a global problem, affecting less or more countries depending on their level of development. For this reason, the public approach to road risk prevention must be based, among others, on the construction of an infrastructure, and the enactment of strong regulations in this regard. Private entities must also contribute to the construction of a safety culture regarding road risk, using fleets of safe vehicles, maintaining the competence of its personnel, and monitoring the measures established in order to improve the existing controls.

Again, it is important to reiterate that the preservation of human life should be the main objective in all strategies adopted in the sense of road accidents prevention.

References

Angulo Escrucería, D., & Ruiz Trujillo, M. L. 2016. Guía metodológica para el diseño e implementación de planes de seguridad vial.

Benavides, F. G., Delclos, J., Benach, J., & Serra, C. 2006. Lesiones por accidentes de trabajo, una prioridad en salud pública. Revista Española de Salud Pública.

Cabrera, G., Velásquez, N., & Valladares, M. 2009. Seguridad vial, un desafío de salud pública en la Colombia del siglo XXI. Facultad Nacional de Salud Pública: El escenario para la salud pública desde la ciencia.

Fraser, J. R., & Simkins, B. S. 2016. "The challenges of and solutions for implementing enterprise risk management." *Business Horizons*.

Gjerdrum, D. 2010. "The New ERM Gold Standard: ISO 31000:2009." *Professional Safety*.

Instituto Colombiano de Normas Técnicas y Certificación. 2012. Sistema de gestión del riesgo ISO 31000. Bogotá D.C., Colombia: ICONTEC.

Letayf, J. & González, C. 1994. Seguridad higiene y control ambiental 2d Ed. México D.F., México: McGraw Hill.

Martínez González, R., Bareño, O., & Ramiro, E. 2013. Metodología para la atención de puntos críticos para garantizar la seguridad vial en carreteras.

OMS. Departamento de Prevención de la Violencia y los Traumatismos y Discapacidad (VIP). 2015. Informe sobre la situación mundial de la seguridad vial. (www.who.int/violence_injury_prevention.)

Peden, M., ed. 2004. Informe mundial sobre prevención de los traumatismos causados por el tránsito (No. 599). Pan American Health Organization.

Planzer, R. 2005. La seguridad vial en la región de América Latina y el Caribe: Situación actual y desafíos. CEPAL.

Quijano, R. C. M. 2006. Administración de riesgos: un enfoque empresarial. Fondo Editorial Universidad EAFIT.

Valbuena, S. 2011. Muertes y lesiones no fatales por accidentes de transporte. Bogotá D.C.: Colombia. Instituto Nacional de Medicina Legal y Ciencias Forenses.

Vargas, D. 2015. Comportamiento de muertes y lesiones por accidentes de transporte. Colombia, Bogotá D.C.

Design and Implementation of SARES Seismic Emergency Plans

Gabriel Cortez Andrade, Master in Sciences, Master in Safety and Prevention of Occupational Hazards
Doctorate in Occupational Health
Universidad de las Américas UDLA
Quito, Pichincha, Ecuador

Introduction

Within the spectrum of natural disasters, earthquakes are those that generate the greatest number of stressors in the affected population, in terms of death and destruction. These phenomena of sudden jolts of the earth's crust, caused by the release of stored energy, which originate in geological faults by friction on the edge-of-plate tectonics, volcanic processes or even underground nuclear detonations, are those that test the responsiveness and resilience of populations, businesses and service teams government emergency and business.

During 2016 alone, 11,477 earthquakes have been reported worldwide, with magnitudes between 4.9 and 7.9 degrees on the Richter scale, and 5 to 27 km deep. Earthquakes have cost about 1209 lives and losses of between 1 to 3% of GDP in each country affected. If damage to the economic and productive infrastructure are considered, there have been about 100,000 jobs lost by the consequences of earthquakes, and about 20,000 businesses overall have had to close because of this. About 200,000 companies worldwide were toiling during the seismic period and tested business team's emergency plans.

The paper is based on the lessons learned during the earthquake of April 16, 2016, in the coastal area of Ecuador. It is presented as a study of protection and seismic reaction, then design and implement the emergency plan seismic, which allows not only response from infrastructure, but also from the socio-organizational sphere, and moving this functionality to the communities and populations affected, or potentially affected, by earthquakes.

Serving Suggestion

It is necessary that personnel in occupational safety and health are aware and manage definitions and specific concepts about earthquakes, intensities, magnitudes, scales, seismic measurement, return times, seismic propagation, and interpret efficiently the information sent to geophysical institutes on earthquakes in different magnitudes.

This helps to know, prevent, and evaluate the effects of earthquakes with its consequences within a population or a company. The information on the concepts of seismic cycle in their developmental stages helps prevent fatalities within the company or affected population and enable planning of seismic risk-triggering measures prediction, prevention and correction within companies in regard to infrastructure and production processes, as well as establishing new business locations or population.

It is essential that safety and occupational health staff learn and manage the cycle of disaster to prepare, auxiliary, restore, rebuild, and reduce risks and develop seismic emergency plans. Through this process, staff can understand the stages of the component cycle from the before the disaster until phase 2 of the post-disaster.

At the end of the presentation, participants will see a plan of seismic emergency, from the collection of business-relevant information, assessment of technical, operational, and administrative information, as well as information on threats, work support and rehabilitation, preparation of plan risks, and reports to maintain an efficient seismic emergency plan to help protect people and, at the same time, safeguarding the assets of the affected populations and the company.

Methodology

SARES was developed over four years, under the basic concepts of seismology and the methodology of risk and vulnerability analysis applied to the areas of emergencies, disasters, occupational safety and health. The method of investigation and data collection is based on the construction of matrices based on the characteristics of the hazard, the sources of seismic origin that result from the calculation of risk, probability, danger and vulnerability, separated in probable seismic scenarios. The methodology is based on the probabilistic and deterministic calculations, in order to estimate response spectra to maximum or expected earthquakes in each business or community scenario, following the line of disaggregation of the hazard (Benito, Molina, & Laín, 2015).

The calculation of probability gives the guidelines for the calculation of the magnitude of the seismic event, the calculation of the probability of the affect has on the location factors in the seismic zone studied with the archeosysmic or paleotosemic parameter, the geology, and the topography of the terrain of the seismic scenario. The magnitude, intensity and epicentral distance are used to calculate the performance and seismic behavior of the infrastructures, and are explained through the matrix of the seismic performance and the materials with which the buildings are built. The probabilistic analysis focuses on the calculation of the hazard in each infrastructure studied and inspected, with the contribution of simulated scenarios, which allows the construction of the logical tree of damages and seismic effects. The SARES analysis is characterized under the law of recurrence, maximum magnitude and minimum magnitude, intensity of infrastructure affectation, and obtains the danger curves for each of the defined scenarios and for each frequency of interest. For each defined scenario and for each frequency of interest, the response spectrum associated with the set probability is calculated, taking the values of the spectral indexes for that probability. Each seismic spectrum is represented by the expected movement in the indicated infrastructure, with the probability set, due to the seismic action of the sources encompassed in each scenario. This allows them to pick among themselves the spectra corresponding to the different movements given, according to the type of seismic wave, and it can be established which is more determinant for each frequency range that can evaluate which scenarios represent greater danger for a certain installation or construction. For the treatment of information produced by SARES, methodology of seismic engineering is used, which studies how the movement produced by an earthquake affects the constructions or other structures created by the human beings, taking into consideration displacement, speed, and acceleration of the seismic wave, thus verifying the seismic risk.

For the quantification of the macro scales, the ESI-07 scale is taken, compared to the Richter scale and the Mercalli scale (Silva & Rodriguez Pascua, 2014).

Seismic Emergency Action and Response System

The SARES system is a system for designing and implementing seismic emergency plans, based on three factors:
1. Diagnosis and analysis of vulnerability.
2. Seismic response action.
3. Recovery and return to activities.
The system is flexible in application and complements other general emergency plans that companies, families or communities may have.

Goals

• Know, analyze, and evaluate the presence of seismic phenomena and their effects on people, individuals, families, businesses and communities.
• Obtain estimates of the potential impact of earthquakes on daily activities and development.
• Include measures to reduce vulnerability and mitigate the effects of earthquakes.
• Program action and response operations to seismic emergency.

Factor 1. Diagnosis and Analysis of Vulnerabilities
The main objective of this factor is to collect objective and concise data, with the matrices provided; through this tool, the professional can carry out the efficient diagnoses and interpret appropriate information collected, and can establish a probability of action in case of earthquakes and their consequences on infrastructures, structures, and business or community processes.

The steps of this factor are as follows:
a. Qualification of professionals and inspectors.
b. Development of maps.
c. Development of matrices.
d. Monitoring of natural events and impacts.
e. Diagnostic report.

1. Qualification of professionals and inspectors
The occupational health and safety professional is the right professional to carry out the diagnosis and analysis of vulnerabilities from a complete vision; for this, it is necessary that they have the

necessary technical knowledge of evaluation, which includes a training process that allows you to look in detail at the assessment of structural safety of buildings and business and community processes.

2. Elaboration of instrumental maps

The use of maps to synthesize data on seismic events and combining them with socioeconomic data and productive processes facilitates the analysis and improvement of communication between participants in the seismic hazard management process between planners and managers. It is necessary to map multiple threats and critical facilities to structure the reaction plan and predict adverse effects on buildings after a seismic event.

The necessary maps are:
1. Geological map of the area of the company
2. Seismic hazard map
3. Map of seismic intensity
4. Map of instrumental or archaeological seismicity (previous 20 years)
5. Central map of buildings and buildings annex
6. Map of human seismic behaviors
7. Map of possible damages and restructuring of vital lines
8. Map of possible environmental impacts
9. Multiple threat map
10. Critical facilities maps with natural hazards

3. Development of instrumental matrices

The instrumental matrices are those that feed information to the SARES and allow structuring of the reaction plan and predicting harmful effects in the buildings after a seismic event.

The following matrices are qualified:
1. Qualification inspectors
2. Initial evaluation
3. Archeosysic catalog
4. Seismic catalog
5. Seismic information
6. Seismic affectation
7. Seismic performance
8. Damage assessment
9. Diagnostic report

4. Monitoring of natural events and impacts

For the business and community monitoring of natural events and impacts, it is necessary to create a situational crisis room that allows visualizing the deterioration or natural imbalance, especially after seismic factors. The evaluation of the monitoring is carried out through the analysis of the possible causes that emerge from an earthquake. The changes must be monitored according to the information obtained from the instrumental matrices, identified and quantified at the level of threat and vulnerability, where the probability of occurrence of the monitored phenomenon is essential in direct proportion to the degree of damage suffered by the structural and social elements of the company or community. The risk of impairment is measured by the expected impairment matrix in human and material losses. Monitoring, early-warning capacity indicators and structural performance help to establish the likelihood of damage, as well as to calculate human and material losses and how to deal with them from various environments.

5. Diagnostic report and post-earthquake report

The diagnostic report is part of the collection of all the information of the instrumental matrices and provides guidance on planning productive activities, reinforcing structures, hiring insurance for the company or the community, provides insight into the possible impacts on the business structure. The diagnostic report contains:

a. General information about the company
b. Inspector General Information
c. General information about buildings
d. Map of general affectation and risks (includes natural threats)
e. Impact of earthquakes on structures
f. Vulnerability of structures
g. Certification of habitability of the property
h. Early warning systems
i. Response system (human, business and socio-economic factor)
j. Recent drill reports

The post-seismic report should contain:

a. General information about the company
b. General data of the inspector and equipment to be used
c. Map of specific locations collapsed
d. Structural systems affected (by affected site)
e. Result of the general inspection
f. Structural Damage Cost Calculation Report
g. Structural damage (category)
h. Non-structural damage (category)
i. Affected corporate equity (costs)
j. Certification of habitability of the property
k. Initial recommendations
l. Urban risk report
m. Recovery, resilience, and business and community livelihoods

6. Training

SARES is based on training at different levels of the company or community, divided into the following factors:

Training on SARES responsibilities. The training is based on training the management, administrative, and operational staff in the following environments:

a. Executive staff. It is represented by the areas of the company that have the power to decide on the actions to be carried out before, during, and after an earthquake. The areas that can be trained are: legal Representative and/or general management, financial management, management resources human resources, production management, corporate image management, internal operative management, all the necessary managements:

i. Leadership training topics. Monitoring of natural events, damage assessment of buildings, damage and restoration of vital lines, environmental impacts, approval responsibilities, responsibilities of actions during emergencies, responsibilities for recovery of productive processes, and responsibilities for communication of the consequences of the earthquake in the company or community.

ii. Operational training topics. The administrative and operational personnel should be trained in the following subjects: emergency plans; incident command; handling construction information in reference to inherent risks; security zones; security zone per floor within buildings; conditions of seismic evacuation; external security of the building; evacuation routes between buildings; coastal security zones; signaling needs, alarms, emergency systems, and emergency response brigades; what to do during and after the earthquake; affect of the productive processes and return to labor or community activities; and about natural events annexed or produced by the earthquake.

b. Personal Security and Occupational Health. The occupational safety and health professional must be trained in the following topics: construction processes, structure analysis, topographic and geological analysis, cartographic analysis and map design, seismic risk analysis and natural disasters, trainer's trainer, emergency plans, evacuation and contingency plans, command of incidents, first aid, search and rescue, leadership in emergency, costs and finances of disasters, planning for recovery and reactivation of productive processes, natural disasters and consequences, human behavior in disasters, urban and community risks, and the INSARAG method.

c. Simulations and preparation for seismic emergencies. In order to comply with the first phase of SARES, it is necessary to carry out tests and simulations with the objective that the personnel of the company or community know how to react in case of a real adverse event. The mock plan must contain the way to act before, during, and after the drill. In the past, it is necessary to evaluate the company structures; identify hazards, safe areas, meeting points; identify escape routes and safety zones; check the exit route; and give responsibilities on closing or disconnection of critical areas. During the simulation, measure and evaluate reaction times: evacuation, the response of the first responders alerts, sound systems, and visual systems. After the drill, prepare the drill report, give recommendations, and take the necessary corrective measures.

7. Policies and procedures

SARES stipulates minimum policies and procedures for its functionality:

1. Emergency policy. It is mandatory to have this procedure; the body of the procedure must contain the responsibility of providing human, technical, technological, financial, and operational resources for the design, implementation, and execution of SARES or the seismic emergency plan. This policy must necessarily be known and understood by the staff of the company or community, there must be mandatory standards of action before, during, and after an earthquake.

2. SARES implementation procedures. This procedure establishes the need to implement SARES, and follow its parameters, in order to have a system of preparation, action, response, and systematic recovery.

3. Procedure on responsibilities in emergencies. This procedure includes the responsibilities of all managerial and managerial areas that influence the company or community.

4. Emergency procedure. This procedure includes training, enforcement of emergency regulations, and evacuation.
5. **Brigade procedure.** This includes selection, training, and activation of emergency brigades.
6. Procedure of request of activation of external support. This procedure includes the need tp, and the time in which it will, request help and support from external authorities.
7. Procedures on closures of buildings and stop production processes. This process includes the need to declare, within next 72 hours, whether a property is habitable or non-habitable, and to stop productive activities, and the costs that this suggests, for the operation of the company or community.
8. Procedure for reactivation and return to productive activities. This procedure includes post-inspection of buildings or production processes, the need to provide financial resources and human resources necessary to overcome the post-earthquake crisis.
9. Procedure for alert activities. It includes information about how to receive and issue seismic alerts and, during the seismic event, the activation of alarms and emergency communication systems, return to activities, recruitment of personnel, and necessary resources.
10. Needs and priorities procedures. This procedure includes assessing and reviewing priorities for areas of the business that may be affected.
11. Procedures on seismic damage scenarios. Depending on the seismic source, the magnitude of the event, the distance from the epicenter, and the soil conditions studied in the SARES seismic catalogs, it is possible to have multiple scenarios of seismic damage scenarios in the company or community:

a. A major earthquake will cause numerous fatalities and injuries, loss of business or community property, and disruption of public services, and will have a direct impact on the company or community. It should be taken into account which buildings may collapse and other structures of the company especially, old multi-story and unreinforced masonry that could represent problems. This can be predicted with the SARES analysis.

b. The extent of the number of casualties and damage will depend on factors such as the time of occurrence, severity of impact, climatic conditions, business population density, types of buildings, and the possibility of secondary events, such as landslides and fires, which may have potentially much greater impact than the earthquake's own vibrations.

12. Procedure for external damages. This should include dealing with damage to: public buildings; public roads and transportation system; utility systems; hospital capacity reduction; reduction of basic emergency response and safety capacity; damage to sanitation systems; direct impairment to external persons; problems of public order; problems in public information; damages to the system of city functions; as well as additional structural and functional damages, such as spills of toxic substances, the effect of aftershocks, the worsening social conditions of the population nearby to the company or community of low resources.

13. Procedure on the creation and activation of the crisis committee. This procedure includes the formation of the crisis committee, which should be chaired by the highest authority of the company or community and by the decision areas of the company. It should be stipulated in what moments the committee is activated, the duration of the same, and how to analyze the information that is received in order to make decisions.

14. Procedure for protection of critical areas. This procedure includes the protection of essential systems for the company or community, including generators, elevators and escalators, electric harrows, computer systems, drinking water systems, electrical systems, sewer systems, gas systems, and pumping emergency systems of water.

Factor 2. Action and Response

SARES occupies the INSARAG Guidelines to carry out action and response plans during and after the earthquake:

a. Administrative axis:

1. Seismic emergency policy, issued by the highest authority of the company, published and known by all staff of the company or community
2. Description of first responders
3. Institutional response capability matrix
4. Matrix of financial costs of action and response
5. Matrix of first responder necessary for the evacuation and rescue
6. Physical and mental requirements of brigade personnel
7. Selection of personnel for the brigade team
8. Roadmap for entrepreneurship in emergency
9. Checklist of personnel and equipment
10. Procedures for action, operational levels, standards, performance criteria

b. Operating axis:

1. Alert and activation of brigades
2. Mobilization of brigades
3. Operation of brigades
4. Cycle of response, preparation, mobilization of brigade leaders, operations, demobilization, post mission
5. Corporate emergency coordination structure (incident command)
6. Sectorization of brigade leaders
7. Areas of work of brigade leaders
8. Functions of brigade leaders
9. Codification and identification of brigades
10. Evaluation of the brigades
11. Classification (triage) of the work area
12. Information management during evacuation and post-earthquake inspection
13. Building marking system and habitability
14. Cleaning system
15. Victim classification system
16. Internal and external emergency communication system.
17. Situational internal report

c. External support:

1. External alert and activation
2. External mobilization
3. External operations
4. External demobilization
5. External inter-agency support (police, armed forces, prehospital care systems, emergency medical system and patient transfer)
6. Situational external report

Factor 3. Recovery and Return to Work or Community Activities

This factor displays the information that allows inspectors and managers to make decisions about the recovery and return-to-work activities. This factor includes:

1. Activation of the business crisis committee, which will receive reports and make decisions.
2. Inspection report of buildings, and safety measures to be taken by the company or community after the earthquake.
3. Matrix of specialized technical personnel to move to the area of the affected company or community.
4. Matrix of critical areas and essential systems affected.
5. Matrix of casualty personnel (injured and dead).
6. Evaluation of possible shelters and temporary accommodation for affected staff within the company or community.
7. Notification to the personnel of the company or community about closures of buildings and/or production processes.
8. Communicate if necessary to the external community about victims, partial or definitive closures of areas of the company or community.
9. Ask outside authorities for reports on the vital systems that were affected, to make decisions about the company or community.
10. Tentative budget on short-, medium- and long-term investment for the remediation of vital production areas and systems.
11. Make a decision to continue with work or community activities.

Concepts and Definitions

Earthquake: Sudden and transient strong movements of the soil, which propagate through waves in different directions. Sudden release of energy accumulated during variable times and from stresses and stresses in the upper part of the earth (Luis, 2012).

Microsismo: Small vibrations of the soil without beginning and definite end. These microsismos can increase their amplitude; they have times of ample appearance and usually disappear gradually; and they are contrasted with the sudden and transient of an earthquake. Micro-earthquakes have no tectonic origin (Luis, 2012).

Seismic origin: There are two classes of natural origin and those of anthropic origin. Natural are classified according to origin, tectonic, volcanic or impact. Anthropics have source controls like explosions or can be induced.

Seismic origin: Place in time and space where the concentration of energy occurs that, during an earthquake, will propagate in the form of seismic waves (Universidad Autonoma de México, 2014). The types of failure are by compression (reverse or cavalty failure), by combination of stresses (mixed failure), by shear stresses (transcurrent or bearing failure). The study of the seismic origin allows us to adequately control the structural performance of the buildings, through the knowledge of the mechanics of the seismic source and, in a specific way, the ductility and weakness of the foundations and grounds. A company, by means of the detection of the area of the crust and the constant microsismos or earthquakes produced in a fraction of time, can produce elastic deformation of the occupied land.

Seismic mechanics: The natural activity of the earth is based on the deforming force of the crust, when the ground is subjected to those forces and when the force ceases its original form (Universidad Autonoma de México, 2014). The elastic deformation generally occurs in a slow and gradual manner, producing stresses, shearing and accumulating enormous amounts of deformation energy in the material. When the resistance limit is reached or the frictional forces are exceeded, a bursting process is initiated in the weaker areas or in the zones with the greatest stress concentration. This fracturing is accompanied by an elastic rebound on both sides of the fault from the rupture start point, propagating along the fault plane, and causing the ground on both sides thereof to move in the opposite direction.

Radiation pattern: Seismic energy that is radiated in different directions, in different quantities, depending on the different types of waves and the geometry of the fracture (Universidad Autonoma de México, 2014).

Parameters of an earthquake: The parameters that determine a seismic point focus are: the geographical coordinates (latitude and longitude) related to the epicenter, and the depth to the interior of the earth from the epicenter to determine the hypocenter. The time of origin, i.e., the moment from which the release of energy began in the form of seismic waves. Hypocentral values and time of origin determine the travel times of the seismic waves, and from the propagation of the same, can know the physical properties of the earth. From this information can be determined the risk zones, building codes, seismic structural performance, seismic return time, seismic propagation, seismic swarm establishment (Luis, 2012).

Epicenter: Surface projection of the hypocenter, placed on the surface near the hypocenter.

Hypocentro: Focus of the earthquake or point in the depth of the earth where the rock is broken or collided, giving off seismic energy.

Earthquake: Earthquake whose depth is less than 30 km.

Premonitory earthquake and replica: Series of earthquakes that occurred in a limited area and time, the earthquakes previous to the main earthquake are called premonitors, and the later ones as replicas. Replicas follow known frequency-of-occurrence patterns, these being more variable for premonitors.

Seismic duration time: Usually, the major earthquake of greater magnitude and intensity has a variable duration time; however, in its strongest part and organoleptically human precept is usually around 1 to 20 seconds.

Seismic information: These are all epicentral location parameters, such as date, time, origin, and geographical coordinates, maximum seismic macro intensity, and momentum seismic magnitude where the earthquake occurred. The seismic quality parameters, historical period, and seismic data of the national seismic networks with pre-instrumental and instrumental periods are adhered to these concepts (Silva & Rodriguez Pascua, 2014).

Seismic information: Seismic information must be obtained from each earthquake and must always include parametric data and indices, such as data referring to epicentral depth, location errors, recorded magnitude, and associated events (replica or premonitors). In addition, general damage (destruction, deaths, injuries, machinery or production systems affected), time of reconstruction and return to activities, corporate seismotectonic zones, and environmental damages generated by each earthquake must be included in the file.

Seismic depth: Distance between the epicenter and the hypocenter in focal depth. There are three types: superficial, when they are smaller to 30 km; intermediate, between 30 and 200 km; and deep majors to 201 meters.

Seismic intensity: Effects to measure the severity of seismic shaking throughout the frequency range, including static deformations and the effects of ground vibration.

Magnitude: Amount of energy released during an earthquake with a single value.

Geographic location seismic: The effects on the ground are independent of the socio-economic conditions, urban development, and types of buildings present in the seismic zones. However, the area where a company is located if it assigns an intensity by amplification of the seismic wave.

Seismic waves: Propagation of the elastic energy released in the process of rupture by means of waves.

Primary Wave P.: The P wave propagates and behaves like sound waves, this wave alternately compresses and expands the rock, can propagate through solids and liquids. Such waves vibrate walls and windows, and do not usually cause damages to the structure. They have fast speeds according to the type of rock, can reach speeds of between 1100 and 8000 m/s; can be audible for animals; and, when the frequency is the suitable one, can be listened to by humans.

Second Wave S.: The S wave is a wave with vertical movement from top to bottom and vice versa and from side to side. For this reason, its speed is between 500 and 4400 m/s, and usually shakes the surface of the terrain vertically and horizontally. This is the movement responsible for damage to buildings, in areas close to the epicenter, and even at considerable distance

Surface waves (Love and Rayleig): Waves that are close to the surface show their maximum amplitude at the air-to-ground interface.

Seismic risk: Set of factors, such as the value of exposed goods (human life, buildings, roads, ports, pipelines, productive processes), vulnerability as an indicator of susceptibility to harm and danger as the probability of an earthquake occurring in a certain place. Intensity of a potentially harmful event. The degree of preparation determines the reduction of vulnerability and consequently the risk.

Seismic sensor: Electromechanical system, which can serve as a seismograph and detects P and S waves.

Seismic zoning: Classification of a territory according to different levels of seismic hazard. Geographic distribution of seismic sources, depth and magnitude ranges. Seismic zoning is used to guide the criteria for earthquake resistant construction.

Vulnerability analysis: The first SARES factor is the vulnerability analysis, where the risks and seismic hazards to which a company or community is exposed are identified and specified, as well as to verify the recurrent seismic periods in the time factor and the geological and damage of the infrastructures that were given in the passage of time. This factor is characterized by the development of research activities through observation and compilation and the creation of maps for evaluation and diagnosis

Seismic measurement scales and consequences: For SARES, the INQUA Environmental Seismic Intensity Scale is used as the measurement standard. It is based on the quantification of the environmental effects of earthquakes, divided into sweet and additional degrees, which are added descriptions of the diagnostic effects.

Seismic macro scale guide: The macro seismic scale is based on three types of effects produced on the earth's surface: human perception, effects and damages on buildings and constructions, and effects on the ground and alterations of the natural environment and those generated by the dynamics of the source. Seismic vibrations are measured by the displacement and deformations of the terrain, by the propagation of the superficial seismic waves given by the acceleration of the terrain and finally, by the amplification of the focal ground.

Bibliography

Aragon Cardenas, J., Flores , L., & López Batíz, Ó. 2011. *Manual de captura de datos para evaluación estructural.* Bogota: Publicaciones universales.

Benito, B., Molina, E., & Laín , L. 2015. *Metodología para estudio de amenaza sísmica, aplicación al diseño sismorresistente.* Madrid: Universidad Politécnica de Madrid.

Canter, D. 1985. *Studies of human behavior in Fire: Empirical results and their implications for education and desing. Building Research Establishmente Reporte: Fire Research Statio.* Surrey.

Diaz Paz, D. 2010. *Conducta humana en los incendios: el pánico.* Madrid: Mapfre.

FOPAE-AIS. 2009. *Guia Técnica para inspección de edificios despues de sismos.* Bogotá: FOPAE.

Gruipo Directivo INSARAG. 2015. *Guias INSARAG.* Washington: INSARAG.

Hurtado, E. 2014. *Fichas de lesiones típicas de daños por terremotos.* Santiago de Chile: Catalana.

Luis, E. 2012. *Apuntes de Ssimología.* Tucuma: Universitaria.

Mainstone, R. 2010. *Supplementary note on the stiffness and strength of infilled frames.* London: Build Res Establish.

National Fire Protection Association. 2010. *Manual de protección contra incendios.* Houston: Mapfre.

Pérez de Tudela, C. 2000. *Gestión informativa en una emergencia.* Sevilla: Mapfre.

Pérez de Tudela, C. 2013. *Comportamiento humano ante las catástrofes.* Sevilla: Mapfre.

Pérez de Tudela, C. 2014. *Importancia de la información de alerta.* Sevilla: Mapfre.

Silva, P. G., & Rodriguez Pascua, M. A. 2014. *Catálogo de los efectos geológicos de los terremotos.* Madrid: Instituto Geológico de España.

Suárez Miranda, C. 2000. *Conducta humana en los incendios: el miedo.* Madrid: Mapfre.

Timo, B., & Mingsheng, L. 2010. *Building damage detection using post seismic high resolution.* Washington: Satelite.

Universidad Autonoma de México. 2014. *Sismología, Conceptos básicos.* México D.F.: Universitas.

Wardlaw, D. 2009. *Human behavior in emergencies - is widesprad panic the myth?* Houston: International.

Wood, P. 2008. *El comportamiento de las personas en los incendios.* Barcelona: Mapfre.

Ergonomic Best Practices: Janitors, Custodians and Housekeepers

Manuel G. Samudio, MBA, ARM
Chubb – North America
New York, N.Y.

Introduction

Imagine a workplace without a janitor or a building cleaner to disinfect surfaces, repair fixtures or perform maintenance on building and HVAC systems that make workspaces a comfortable home away from home. The employee's workday would most likely be unclean, hazardous, and an unpleasant place to work for the average eight-hour day.

In 2015, there were 3,089,590 building cleaning workers employed in the U.S. with 2,146,000 classified as janitors/custodians. Many work for companies that provide these services to other entities. These employees work full time and part time, but many work nights and weekends and often are alone during their shift. Typically, all work is indoors but, on occasion, they may also spend time outdoors and, thus, are exposed to extremely hot, cold or otherwise inclement weather.

According to the Bureau of Labor Statistics (BLS), in 2015, janitors and cleaners had more than 42,500 injuries requiring days away from work and had the 13th highest total incidence rate of all occupations with 277.4 incidents; the incidence rates represent the number of injuries and illnesses per 10,000 full-time workers). Janitorial, custodial, and housekeeping tasks can cause a variety of musculoskeletal injuries. Sprain/strain injuries (approximately 40%) and overexertion injuries (approximately 13%) were the leading cause of musculoskeletal injuries requiring time off work for janitors and cleaners in 2015.

As a result, custodians/janitors have the third highest number of musculoskeletal injuries, according to BLS. Because of the physical and repetitive activities involved in housekeeping and custodial tasks, ergonomics is one of the hardest areas to implement, manage, and control since it's also difficult to totally eliminate the exposure. Custodial personnel may sustain injuries as a result of a lack of adequate material-handling equipment, poor condition of equipment, inexperience, unsafe acts, or inadequate training.

As a result of the rising population, obesity, as well as an aging workforce, the term "ergonomics" is now becoming more and more prevalent among employees while at work. There is a need for ergonomics practices to be reviewed, especially in the workplace where janitors and custodians are spending their eight-hour workday. Musculoskeletal disorders (MSDs) account for approximately 30% of the total number of workplace injuries and illnesses in the workplace (BLS 2015).

Ergonomics and the Custodial / Janitorial Industry

Ergonomics is the field of study concerned with finding ways to keep people safe, comfortable, and productive by accommodating human characteristics, capabilities, and limitations in the product or process, both at work and home. Rather than making the person adapt to the task, ergonomics is all about adapting the task to the person. It is simply "the science of designing jobs, equipment and workplaces to fit the person, and as a result, create a workplace in which the employees can minimize the wear and tear on their bodies."

This paper will provide guidance in order to help reduce the risks associated with common tasks, which include trash handling, bathroom cleaning, emptying garbage, snow shoveling, mopping and vacuuming. Tasks are commonly performed indoors but may also include outdoor work, potentially exposing this group to hot, cold or inclement weather. They often operate independently and after typical business hours.

When the task has been optimally fitted to the person, using ergonomic principles, a number of benefits can occur:

- Reduced lost work-time illnesses and injuries
- Reduced workers' compensation costs
- Reduced training/retraining costs
- Increased levels of productivity, efficiency, and quality
- Improved comfort of employees through proper equipment use
- Reduced long-term equipment costs by incorporating ergonomics specifications into purchasing
- Increased employee morale by providing a safer and healthier work environment

Janitorial and custodial work consists of many activities that require lifting, bending, and a variety of other tasks that may place a worker at risk for a back or a strain/sprain injury. The tasks range from mopping large areas to carrying garbage up and down stairs. These are the types of jobs that require force, repetition, and awkward postures that can place custodians/janitors at risk of an injury. Overexertion and sprain/strain injuries due to tasks that require lifting, bending, and reaching/pulling are considered so common that they are believed to be just "part of the job." This is based on the belief that the custodial work or task cannot be changed or modified, and therefore the discomfort, pain, and injuries are unavoidable. Changing work practices due to staff reduction, an aging workforce, and the lack of understanding of custodial exposures have contributed to high injury rates among custodians/housekeepers. This has led to challenges in developing safe work procedures and effective worker training aimed at preventing these types of injuries.

The best practices and procedures for the common custodial/janitorial tasks noted below will show ways of making custodial work safer and easier, so that the risks of exertion and sprain and strain injuries can be reduced. To achieve the above goals/benefits it requires employer commitment and employee training. Sometimes, new and better equipment may be needed, or a change in the task or procedure may also need to be done. In some cases, both equipment (tools) and procedures may need to be changed.

One of the objectives of this paper is to introduce procedures that can be used to identify and evaluate MSD exposures for janitors, custodians, and housekeepers, and provide best practices to minimize risk factors.

Some of the tasks that are performed by custodians while at work, such as lifting, reaching, and repeating the same movements, can strain their bodies. In some situations, these tasks and movements can result in injuries to the muscles, tendons, ligaments, and joints of the neck, shoulders, arms, wrists, legs, and back. Custodians work independently, and are required to maintain different areas of their assigned locations.

What Causes Injuries?

Certain tasks and actions can lead to fatigue, discomfort, or pain when you do them over and over without a break. Some examples of this include:

- Exerting force to perform a task or to use a tool (e.g., lifting tables, garbage bags or scrubbing with a brush)
- Working in awkward postures, such as bending or twisting the back, overhead reaching for dusting, or mopping with elbows away from the body
- Remaining in the same position for a long time with little or no movement
- Continuous pressure from a hard surface or edge on any part of the body (for example, kneeling or crawling)
- Working in hot or cold temperatures (for example, snow shoveling)
- Holding equipment that vibrates (for example, the handle of a pressure washer)

As previously noted, some janitorial tasks are often repetitive. As the workers become accustomed to the work routine, they may overlook the hazards and their consequences. A busy schedule, fatigue, or the past success of shortcuts may affect the worker's compliance to the safety rules and protocols. Other causes of unsafe behavior can be shortage of time, and the pressure to get the job done quickly. There may be some employees that will work fast, in a disorganized manner, and not pay attention to safety. Some employees may struggle to handle their duties and responsibilities, as well as urgent last-minute requests. Carelessness and rushing can also lead to injury.

The cleaning industry is intense, fast-paced, and physically demanding work. It is also repetitive, requiring the use of force, and awkward body postures. There are occasions when a supervisor expects greater productivity due to a purchase of new equipment, so he/she may add extra tasks or allow less time to complete the job. It is essential that the employer set reasonable workloads for their employees. It has been proven that productivity and work quality goes down as fatigue sets in. Always allow adequate rest periods for the staff.

Maintenance of custodial and janitorial equipment is an investment in better productivity. Employers that are purchasing the right equipment and materials, as well as training for their workers, will benefit from increased productivity and the potential for a reduction of their workplace injuries. By providing the proper supplies and equipment, employers are helping their staff to be as effective and efficient as possible, while promoting safety.

One way to protect the investment is to make a maintenance schedule, and adhere to it. Equipment that is in safe working order allows for positive work practices. Unexpected or forceful actions required to operate faulty or poorly maintained equipment can lead to strain and injuries.

Janitorial Tasks and Best Practices

Mopping

The traditional floor mopping task, using a mop-and-bucket system, have several MSD risk factors, especially given the fact that a bucket, when filled with water, can weigh as much as 40 pounds. Because of the way that most buckets are designed, filling or emptying the bucket requires the custodial/cleaning worker to lift the bucket from floor level to a waist-level sink. Even if the drain area is at floor level, the worker still must bend or squat to lift the bucket for draining, creating an opportunity for a strained back.

The actual mopping task is also perceived as a risk factor. It requires wringing out the mop head, lifting and carrying the mop while it is wet, and then using repetitive motions to perform the actual mopping. Additionally, many mop poles may be too short or too long for the worker, which can negatively impact shoulders and require reaching or bending to use the mop effectively. In addition, mopping a floor means the floor is now wet, making a slip-and-fall accident more likely. Some chemicals used to clean floors also can increase the likelihood of a slip-and-fall incident.

To reduce these risk factors, employers should consider having the custodial workers use lighter mop heads, such as microfiber mop heads. They also can consider purchasing bottom-draining buckets, buckets that can be filled with a hose system, and mop poles that are telescoping and can be adjusted to the appropriate length for each cleaning worker if within their operating budget.

Vacuuming

The risks for MSDs from vacuuming can impact janitorial workers because the vacuums are the one tool used by almost all cleaning workers regardless of industry. Most cleaning workers use upright vacuum cleaners, which require considerable hand and arm movement, as well as grip force to use. Awkward postures are also often required, and the worker must repeatedly lift and carry the machine, then plug it in and unplug it from the wall, all of which can put stress on the back, shoulders, arms, and wrists. Noise and soiled exhaust can further negatively impact the worker.

Newer vacuum cleaners have several features that make them easier to use, and help eliminate the risk for MSDs. Some vacuum cleaners now weigh as little as eight pounds, have 50-foot cords, are quieter, and have multi-stage or HEPA filtration systems that prevent dust from being released from the vacuum's exhaust to help protect indoor air quality.

Trash removal

Trash removal is one of the tasks that can create an awkward posture, as well as exert a lot of force as part of the daily duty of a custodial worker. The worker cannot assume that all trash cans or bags weigh the same each time they need to be emptied. Injuries can occur when lifting an unexpectedly heavy trash can.

Lifting heavy garbage bags repeatedly becomes even more difficult when the contents of the bag have been pushed down in order to avoid extra trips to a dumpster. A strong suction is generated by the vacuum of air when a filled garbage bag is lifted from a barrel. Extra force is then needed to pull the bag out. The force exerted by the body to pull the bag out comes from the shoulders and back, which can then be strained. Awkward postures are often involved.

Bags should be emptied before they become full. The dumping of heavy trash bags requires high force and exertion repeatedly.

Cleaning

Cleaning by hand while kneeling is forceful work, done again and again, with awkward arm and hand movements. The trunk of the body and neck are bent, and are often in fixed positions. Holding a duster and reaching up for long periods of time requires awkward and fixed positions of the arms, shoulders, and neck. This task can lead to pain and stiffness in the neck, shoulders, arms, and upper back. In addition, extending the shoulder and arm to wipe forcefully with the whole hand for long periods of time is also tiring. There is also twisting and bending of the wrist. Sweeping floors may involve awkward positions of wrists and prolonged contact pressure on hands. In addition, the back and neck are often in an awkward, forward-bent posture.

Workers should avoid bending or twisting of the hand and wrists, in addition to alternating between the right and left hands. Janitorial workers should also get closer to the work to avoid stretching and overreaching.

Carts

As more supplies and tools are loaded on the janitor's barrel/cart, the force needed to push it will be greater, especially on carpet. Materials unevenly distributed around the barrel or cart also creates instability. Moving the cart over uneven surfaces, such as elevator gaps or over thresholds, can further cause the barrel/cart to tip over. This task of moving the cart adds a force to the worker's back and shoulders, especially if the worker has to pull the cart. Workers should always push the cart and *not* pull it.

Conclusion

Employers, safety professionals, and workers should become more knowledgeable about the common hazards, controls, and best practices associated with the janitorial/custodial and housekeeping industries. Redesigning tasks and/or tools is not fully effective without changing employee behaviors at the workplace. Employee training is vital in ensuring that the workers are not engaged in tasks that may place them at risk of an injury. Training will need to be developed that will complement and reinforce the redesign of the tasks that are done by the worker. The workers will need to avoid awkward postures and not bend their backs as they are working, be alert as to how much force they are exerting while holding tools, and how often a task is done repeatedly, before they will use an alternate method in order to continue.

Bibliography

British Columbia School Safety Association/WorkSafety BC. "A Clean Sweep Safe Work Practices for Custodians." *Working safer and Easier – for Janitors, Custodians and Housekeepers*. June 2006.

Bukowski, Thomas. 2012. "Cleaning Up Safety - Janitors and Cleaners Face Multiple Hazards" *Safety & Health Magazine*.

Bureau of Labor Statistics (BLS). 2015. *Fatal Occupational injuries by industry and event or exposure, All U.S.* (www.bls.gov/iif/oshcdnew.pdf).

Cal/OSHA Consultation Service (DIR). *Working Safer and Easier – for Janitors, Custodians and Housekeepers*. (www.dir.ca.gov/dosh/dosh_publications/janitor.pdf).

University of Nebraska Lincoln. *Ergonomics and Back Safety for Custodial Workers - Safe Operating Procedure*. (http://ehs.unl.edu).

Four Elements to an Effective Safety Culture

Jose L. Valdez, CSP, CSHO
Assistant Safety Manager
KPost Company
Dallas, TX

Introduction

Since 2005, I have strived to honor and respect others during safety meetings and while enforcing safety policies. I have worked with a wide range of companies in the construction, agriculture, and general industries. However, when I started down this career path, my only qualification was being bilingual and trustworthy. Regardless of where you are on your career path, I want to encourage you to keep learning and keep growing. You are part of an industry that pays you to invest in yourself. Not only will this career path change your life; more importantly, you will be able to change the lives of many others. My personal vision is to empower others with practical ways that will help them understand, improve, and enjoy their lives. My mission is to lead by example, have a positive servant attitude, and remain accountable. I make every effort to leave a legacy of never giving up on any person or company. Safety is not about how much you know; it's about how you get along with people. The way you manage your safety and health program is a reflection of the way you manage life. Congratulations on accepting this challenge!

Unfortunately, I have found that there are unique challenges that safety professionals face. In this paper, we will identify four key elements required in a safety program. We will also discuss techniques on how to apply these elements. These methods will transform safety from a program into a culture. My goal is to help you understand these challenges and give you some solutions or ideas that will help you deal with these issues. Are you going to take the easy way out and ignore your company's culture, or will you step up to the challenge?

Perception vs. Reality

As safety professionals, you will be challenged by your employees to help solve a vast array of issues. Don't only focus on your job description and only doing "your job." In order to create transformation in your company's safety culture, you have to build genuine trust. To do so, you will need to learn to speak the employees' language. Spend time with them at the work site, in the shop, or at lunch. Talk with employees, listen to them, and they will begin to see that you truly care about them. You will begin to understand them better, and you will become relatable to them. This provides a bond where they can give you their honest and genuine feedback. Employee feedback is a priceless tool that directly affects your safety culture. You can use this feedback as a catalyst for change, and to strengthen your rapport.

Business owners and managers often ask me, "Why do employees fail to follow the rules even after being trained?" Many companies have safety policies and train their employees. The perception is that they are complying with the law and are doing great. However, the reality is that, in most cases, there is no difference on the production floor, even after implementing these safety policies and training employees. Training is just the first step of your safety program. Employees usually go back to business as usual when "the safety guy is not looking." On the production floor, safety professionals are considered the enemy who are slowing down their production. Top management expects safety professionals to fix all safety issues with training. If training doesn't correct the issues, there must be something wrong with the safety department. Training might correct some issues, but not every safety hazard is caused by a lack of training. Unfortunately, all the training and signatures in the world will not stop employees from getting injured or killed.

As safety professionals, it is your responsibility to do an accurate self-evaluation of your safety programs and leadership skills. It's also your responsibility to find the best use of the company's resources, and continually measure their effectiveness. How can you know if these processes or procedures are being effective? Some lagging indicators that help measure effectiveness are injuries, accidents, illnesses, fatalities, OSHA recordable, first-aid cases, and near misses. Leading indicators include employees' attitudes, feedback, cooperation, and observed safety compliance.

The company is the vehicle, and the safety culture is the transmission. The company's engine could be making a lot of noise but, if the transmission is stuck in first gear, you are wasting resources. These companies train, train, and re-train employees in hopes that all the safety issues will be fixed. They also feel protected because they think that they have transferred the liability from the company to the employees. In reality, the company's liability can never be fully transferred to the employees. You can't force safety or become complacent by thinking training fixes everything. Rather, continue improving and shifting gears to positively affect your safety culture.

You need all four gears of your safety culture's transmission actively participating in order to create an effective safety culture from within your organization. The four gears or elements we will focus on are adequate training, professional safety audits, compassionate enforcement, and encouraging reinforcement, which are the foundation. There are more gears or elements out there, but these four are the most important in getting your vehicle moving toward your vision. Later, as your safety culture accelerates to the next level, you can add a fifth or sixth gear in your transmission.

Adequate Training

Adequate safety training is the first gear of your safety culture's transmission. It includes company-wide monthly safety meetings, weekly toolbox talks, daily work hazard assessments, and on-the-job training during safety audits. There are some high-hazard tasks that should be taught at least once per year, for example: heat stress, fall protection, electrical, struck by, caught in/between. Safety training requires the most work, yet seems to produce the least return. You work hard, and give it your all, but sometimes it feels like you are going nowhere. Get creative; don't teach the exact same class year after year. During safety training, encourage employees to ask questions. Most employees won't ask questions until after the safety training is over. Answer their questions but explain how that question would have benefited the entire group. Even after conducting training on the same topic over and over again, there might be some employees who still do not seem to understand. Focus on those employees, one on one. Always document all of your training efforts, even if you don't get their signatures.

The first step in providing adequate safety training is that you must understand the technicalities of the topic. You must verify the information in your training program before you provide it. If you provide bad information, you will lose credibility. Rebuilding credibility is a tough mountain to climb. Research, attend specific training, and ask seek counsel from your peers. Make sure you understand where the requirements come from. Is it from OSHA, DOT, FMCSA, MSHA, DOL, or your company's safety policies? Explain to your employees the source of the requirement, and why it was created. When they begin to understand the why behind the regulations, they will be more likely to comply.

The second step to providing adequate safety training is knowing your audience. Your attire, appearance, and verbiage has to be appropriate for each type of audience. As you work with a company, you will find out how much information employees can handle in one session. You don't want to overload and overwhelm them. Each type of audience needs to have a balance between the information you provide, showing them how to apply it, and the benefit of compliance. If the safety training is for upper management personnel, you might need to add more detailed information and specific examples. Management personnel typically want to know which specific regulations require them to do what you are recommending. They also want to understand how to comply, external consequences for non-compliance, and how to remain under the radar. On the other hand, lower-level employees will want to know how to apply these new requirements, internal consequences for non-compliance, and how these extra efforts will help them. They need to understand the real daily value of compliance for their specific situation.

The third step is to make the information captivating. Just because employees are present physically does not mean they are there mentally. You have to find new and different ways to catch their attention, and draw them in. Variety is vital to properly engaging your audience. You can use PowerPoints, videos, YouTube clips, pictures, handouts, games, and/or quizzes. There is an art to professionally using each one of these tools. When showing videos, walk around the room and verify that the sound and visuals are at acceptable levels. This is also a great opportunity to observe employees' reactions. If needed, pause the video to highlight a certain point, answer questions, or correct something they said that doesn't apply to your situation. Make sure you preview all videos or clips before you show them to your employees. Videos should be short, and a review should be done after each. You can have questions prepared for the audience regarding what they saw. YouTube clips provide great entertainment, while providing an opportunity to conduct a real-life hazard assessment in the classroom. Another method of keeping employees' attention is to share about things they are interested in. For example, you can review the company's policies on raises, paid time off, paid vacation, etc. Also, provide rewards for classroom participation, showing up on time, answering questions, or sharing their experiences. Rewards can be small things like candy, chocolate bars, company promotional items, or even safety gear.

Lastly, adequate safety training must be relevant to the audience. Show them photographs of their actual work environment so they can see how to practically apply the things you are teaching them. They need to know how to apply this new knowledge to their daily work environment. Don't assume that they will automatically know how to apply what they are learning. On the other hand, don't give them too much to change all at once because it will become overwhelming and discouraging. Always end the training by challenging them with at least one or two practical methods of applying what they just learned. Help them understand that it might feel like it's taking more time at first. However, they are being compliant and safety is a requirement of their job. With time and practice of doing things safely, they will pick up speed because it will become second nature. Help employees understand that they are expected to follow the company's policies anytime they are on the company's premises or representing the company. Professionalism is not common knowledge, and must be taught. As safety professionals, you must remain positive, motivated, and continually looking for new ways to explain the same thing. Top speed in first gear is when you have personally trained and explained every safety requirement to all of your employees.

Professional Safety Audits

Once you are at top speed in first gear, you must switch more of your efforts into second gear. Second gear is conducting professional safety audits. Adequate safety training is part of your job in the office, but what you do during a safety audit shows employees your true character and values. These audits must be done on a more frequent basis and should remain, as much as possible, unannounced. Safety audits are when you validate to employees your true beliefs about their safety. Employees will see your passion, leadership, and integrity in action. This is when you set the tone for your safety culture.

The first step in conducting professional safety audits is to always respect your employees, regardless of the circumstances. Violating safety policies can seem disrespectful toward us because it's what we represent. However, employees are not intentionally trying to disrespect you or me. Don't take it personally. Safety is always our primary concern. Unfortunately, to employees, their own safety is not always top of mind. They ask themselves: Whom should I keep happy? My coworkers whom I spend all day with or the safety guy who may periodically show up. They typically tend to put more focus on getting along with their direct supervisor and coworkers. This results in them placing more value on production to impress those around them and build their value. Sadly, they risk their own safety for perceived job security.

The second step to conducting professional safety audits is remembering that safety is your suggestion on how they should do their job. To employees, safety rules are for ideal conditions if production allows. They don't see safety compliance as the law, the company, or a program. Safety is the person who enforces the rules. You cannot be the only person representing safety or you will burn out. Remember not to ignore safety hazards or show favoritism or you will ruin everything you have accomplished in first gear. During safety audits, if you find employees in immediately hazardous situations, remove them from that area, and help them address the safety hazards. Even if this means that you stop production to ensure everyone's safety. Make sure that you don't leave the worksite until all safety hazards have been properly addressed.

On the other hand, don't yell or threaten any employees. If disciplinary action is required, wait until after you have completed the safety audit and finished helping them resolve all the safety hazards. This will buy you some time to cool off, not overreact with a particular person, and remain professional. If you disrespect employees, your safety program and safety culture has just lost all of its credibility. Employees will not listen to advice from someone they don't respect. Don't damage your safety culture by losing your employees' respect.

The third step to conducting professional safety audits is to be able to prove what you observed. Employees will always have an excuse for why they violated a safety policy. They will even challenge you, and say that you must be mistaken in your observation. In order to help employees be honest and accountable, you must get creative about the ways you conduct your safety audits. For instance, you can use a large zoom camera and observe your employees for at least 15 minutes from a distance prior to your arrival. Try to find an inconspicuous location where you can see the most activity. Take pictures of the worksite, the surrounding area, safety innovations, safety compliance, and safety violations. This type of audit is more realistic to what an OSHA inspector would find on your worksite. The goal is to get a view of what really is happening when your employees think that nobody is watching. During the times when "nobody is watching" is when employees make errors in judgments that result in most injuries, accidents, and fatalities. In reality, anyone can call OSHA anonymously. Therefore, practicing this type of safety audit will help lead employees to the reality that someone is always watching. When they truly believe, they can peacefully accomplish their jobs, and you have achieved top speed for second gear and the beginning of a genuine culture change.

The final step to conducting professional safety audits is your official safety audit report. You should review the safety audit report with the supervisor at the worksite, and ask them to sign or initial the document. Your direct supervisor should also review your safety audit reports. Don't just archive these reports for the company's records. You must communicate the finding up the chain of command. My supervisor, the president of the company, saw every single safety audit report. I would print a copy for him, and he would initial it to show that he had reviewed it. After the safety audit report has been reviewed by your supervisor, you should archive it for record keeping.

There are several free apps that offer customizable safety reports. At first, I didn't like transferring my paper safety audits to a digital format. For many years, I used one double-sided sheet, and this app turned my reports into eight pages. On the other hand, I was able to customize my reports to our industry and insert pictures, comments, signatures, and time stamps. This app creates a PDF of the safety audit, which I can email, print, and archive. I was now able to do everything from an tablet, and everything was nicely organized in this app. I could easily refer to past safety audits while on the worksite, I never had to restock my paper safety audits, and I stopped looking for a pen because supervisors can sign the document with their finger. All this time and effort invested in customizing the app was worth it when my supervisor said, "Nice new fancy safety reports." The thoroughness and embedded visuals took my audit reports to the next level.

Compassionate Enforcement

Second- and third-gear work together, and you must shift back and forth between them until reaching top speed. Compassionate enforcement is understanding that your job is not to fix every safety issue. Your job is to teach and motivate employees, especially supervisors, to correct safety hazards immediately. All employees need to understand that. not only do they have the knowledge to recognize hazards and the authority to correct them, but they are also expected to correct or report these issues as part of their job description.

Helping employees recognize that compliance is mandatory not only keeps OSHA away, but it prevents injuries, keeps insurance rates low, makes bids more competitive, and results in more work for everyone. Safety compliance creates job security. Our company believes this so strongly that we created consequences for non-compliance. These consequences include warnings, write-ups, suspensions, and termination. Explain to employees that disciplinary action is not personal. You also must follow the company's policies and procedures to ensure their personal safety. As safety professionals, you must be consistent and fair regarding disciplinary action, which is what builds confidence in the safety culture. Employees will begin to trust that, regardless of ones' feelings, everyone must obey the company's policies and procedures. If you give one employee a pass, that would not be fair to all the other employees who have received disciplinary action in the past. It will also be a huge blow to employee morale and your rapport. Once you start deviating from the set policies and procedures, all employees will expect the same treatment. In the end, it will become very difficult to enforce any safety policies without losing all credibility.

When you visit the work sites, don't just complete your report in 15 minutes, and move on to the next one. The total time at the worksite should be no less than one hour. By now, you should be conducting as many safety audits as possible. During this time, you should be looking for opportunities to help the employees in the areas of safety or production. If it is close to break or lunch time, stick around and find out how the employees are doing. How is the job coming along? Is there anything you can do to help them? Can you replace any worn out or damaged personal protective equipment? Also, ask them about their personal lives. Are they married? Do they have children? How far away do they live? It is hard for employees to believe that you care about their safety if you don't even care enough to ask about their personal lives.

Use safety issues that are found as opportunities for continual on-the-job training. These are teaching opportunities to strengthen their understanding of how to apply safety. Have a short 10-15 minute safety talk to review any safety violations, how to correct them, and how to prevent them from recurring. You can also encourage them by pointing out all of the safety items they are doing well. Other methods of encouragement include reminding them of past successes when they had an excellent safety report, were recognized by other trades or peers, or received an award/reward for their safety efforts. Challenge them to live up to their potential, not just what they believe are the minimum expectations. If your company offers rewards, and they have not received any in the past, remind them of how they can be acknowledged. Encourage them to voice their opinions and concerns. Affirm that their feedback is very important, and they should not be afraid of any retaliation from their coworkers, supervisors, or management. It is your responsibility as a safety professional to follow up, and ensure that employees are protected against retaliation.

Encouraging Reinforcement

Encouraging reinforcement is the fourth gear, and it will help you gain the most speed with the least amount of effort. Remember, this is the fourth gear, not first or second. Only place the majority of your efforts and resources in fourth gear after reaching top speed in first, second, and third gears. Safety should not be a necessary evil of doing business. Safety compliance, when done right, is rewarding, fun, and exciting.

Encouraging reinforcement or reward programs must have various methods of being recognized; for example, implementing a new idea, completing all of the safety paperwork, or being engaged in safety training. Reward your employees in public, during safety meetings, special events, or scheduled production meetings. As safety professionals, you must find new and creative ways to reward your exemplary employees who take the initiative to be safety leaders. Encouraging reinforcement also includes gifts that should be given to all employees, for example, company promotional items or unique safety gear for extreme temperatures.

I am constantly looking for new and creative ways to help our employees beat extreme temperatures. Surprisingly, I received the most positive feedback from a commercial coffee maker I purchased for our field employees. It was the best morale booster and the least expensive. I also purchased an ice machine and water filtration system. Employees now have good-tasting water that they can keep cold throughout the day. We also provide Gatorade mix, Pedialyte, and electrolyte tablets to help prevent heat stress. Last but not least, we have an annual safety luncheon where everyone receives door prizes, the President of the company personally thanks everyone for their safety efforts, and we raffle off large-ticket items.

In Closing

Remember that there must be a balance between safety and production. Safety must be practical; don't just make new rules and policies to say you addressed a problem. Find the root cause of the goal problem; then find the best possible solution. Don't get caught in that trap of continually creating new rules. Lack of training or policies is generally not the issue. I have found that, in most cases, the employees know the correct thing to do but decide to ignore the proper processes. Investigate to find the core motivating factors that influenced that employee to make that poor decision on that day. Don't invest your time and efforts in treating the symptoms; instead, focus on properly diagnosing the condition.

To help change a culture, every employee, from the top down, must embrace and undertake the company's safety vision and goals, which should be published and visible to all employees. The expectations must be clear. Always keep reminding everyone that investing in safety provides a great return on their investment. Returns can be seen as insurance rate reductions, higher productivity, higher employee morale, employee longevity, etc. Using this approach toward your safety and health program will help prove to your employees that you care about their personal wellbeing. When all four gears are actively working together, your safety culture's transmission will run smoothly, effectively, and will drive your company to your desired destination.

Behavioral Observation Tool Designed In-house for Underground Mining

Ing. Juan Dumler, CSP
Hochschild Mining plc

Ing. Abel Del Carpio
Hochschild Mining plc

Introduction

When safety professionals investigate the unintended events occurring in the activities of our companies, regardless of the activity line, we find that, at our site, we have a replica of the statistics published by various governmental organizations of labor, and it is that most events are due to substandard acts.

In the same way, and in order to develop a preventive activity, we implemented a program of labor inspections, which is increasing every time an undesired event occurs, and we have the scheduled, unpublished, management, joint committee, the head of area, safety team, etc. All of them have another feature that does not add much to management: Most of them only detect substandard conditions. This is due to two main reasons: (1) We have the eye predisposed to detect errors; and (2) With so many activities assigned to supervision, the time to stay in a job is increasingly reduced.

With increasingly "lean" business environments, there is no possibility of permanent supervision on all work fronts that, acting as guardian angel, can warn the first line of collaborators of the dangers they have not identified and that can damage them.

In the case of certain industries, such as underground mining, this limitation becomes more critical due to the high risks and the distance between jobs that makes supervision very difficult. The question arises, "How does the first line behave when it is not being observed?" And the answer has a lot of uncertainty.

With this common background in all companies, we need to promote a tool that allows observation of the behaviors of workers, and time to detect deviations and correct them, while working, at the same time during these interventions, on behavior change and so progressively, if it were possible, not to continue, depending on supervision visits to the work.

It is common for the implementation of a tool of this type to use external specialists of renowned brands and famous products. However, in many cases, the tool does not allow adaptation or fine-tuning to our business, or its cost is so high it makes implementation unviable.

In the meantime, it is convenient to take action and try to implement a tool that will help us to identify the problem and align the behaviors, although there may be some criticism and, of course, continuous improvement.

Step 1. The Productive Process of the Company and Associated Risk Behaviors

The first step is to examine our production process to determine what acts/behaviors we should observe. We have three main sources of data to immediately involve:

A. Acts that appear in the investigations of accidents (fatal, lost time injured) and incidents (of high potential).

B. Acts as a result of field inspections.

C. Critical processes that can cause injuries of high severity, so no historical data is available that have occurred.

All these acts/behaviors will be the input that will serve to build the observation card.

If, in our industry, we have production processes that are very different, and trying to place all these acts can cause confusion, it is better to group them. In the specific case of mining, there are three consecutive processes that are totally different: (a) exploration and geology, (b) mining, and (c) process plant; to those processes is added the maintenance activity.

Given that, in the metric of the company where the case was implemented, the greater number of undesired events is given in the mining process, the tool was initiated in the mine with the acts/behaviors associated with them.

In order to organize/group these behaviors, and ensure that we covered all components that could be lost, we chose to follow the classic guidelines of the Professor Frank E. Bird Jr., who groups the components by the acronym PEMA:

- People
- Equipment
- Materials
- Work environment

We will see an example during the presentation.

Step 2. Improve Communication and Correct Approach

An observation of behavior requires, in principle, an observer, an observed, and a methodology to do it. As important as designing the tool is to train observers to do so without generating rejection, assuming that no one likes to be told that their work is not well done, nor does the worker like his supervisor to present himself as the "guru" who knows how to do things correctly.

To ensure that language is correct, a way of making observations should be designed, trained and retrained. In the case study to be presented, the observation methodology consists of five steps from a "canned" tool that was used as a pilot in the exploration division.

This approach tool, to be presented during the presentation, seeks to improve the language used by supervision, clearly oriented to production, and with very limited contact. Verticality distanced the relationship between supervisor and worker.

Care must be taken to ensure that, during training, people from the same hierarchy participate, so that their development is genuine, and not contaminated by the presence of their "bosses" or "subordinates."

As part of the training, observers must learn to listen because the natural tendency is to demonstrate that they know more and/or attract attention, while what is sought is for the observed to identify their own misalignment and correct themselves. This is only achieved if the observed becomes the protagonist of the intervention.

The five steps that were adopted are:

1. Securing the area, greeting and presentation

2. Observing of the task

3. Giving positive reinforcement

4. Having a dialogue on risks and controls

5. Obtaining the commitment

Step 3. Designing the Observation Tool

When having to make the observation in the workplace, it becomes necessary to have an observation card to help us keep track of everything we want to observe.

When the request was made to a major provider of behavioral tools to tailor its product to the type of our company activity, the company providing the tool replied that it was not possible because of the scientific studies behind its tool made it impossible.

The card designed in-house intentionally seeks to NOT look at conditions, but acts. A list of acts/behaviors was made as of the date of step 1, and the most significant of the most critical production process was chosen that, in the company of the presented case, is the mine.

These behaviors were grouped under the PEMA criterion, and each of them were listed next to a small box that indicated if it complied, did not comply, or if it did not apply,

Conceptually, it is sought that the observer does not have to write much on the card, and that it can be processed with ease type checklist.

Finally, on the card, the data of the place where the observation took place is recorded (this is because the workplaces inside the mine are geographically very distant and numerically more than 72) to process the data.

Step 4. The R +

Conceptually, R + (positive reinforcement) is indispensable during observation. In fact, in the five-step tool to make an observation, step 3 is the R +.

However, a verbal reinforcement, despite being potent, can be improved if there is also a tangible positive reinforcement, a prize; itis even better if this prize is awarded as recognition in front of the partners.

There is a postulate that tells us that when we have a winner, there are also several losers; thit is an interesting point to take into account. However, the culture of each company should be analyzed to see if it is more relevant to reward individually, with the potential consequences described above, or if it is better to make a collective award to the whole team, whether or not it is symbolic, like a lunch or a hat for all.

In the case presented, it was decided that the tangible prizes are individual and by weekly draw, thus maintaining the concept of opportunity by rewarding in the same week in which the behavior to be reinforced was performed. (Previously, it was rewarded for the accomplishment of the annual objective, and was done in the month of March of the following year, something that did not generate much enthusiasm among the observed ones, especially if, for some behaviors that resulted in accidents, the objective was not reached without matter that the majority of collaborators had exerted.)

Step 5. The Metrics to Generate Action Breakdowns

At the time of processing the data, the ratio of correct acts and incorrect acts is automatically obtained. The first is to continue reinforcing, and the second is to be corrected (this is the natural result that any typical tool would give). However, there are two (2) additional pieces of information that we need to process: (a) Where was the observation made? We need to know if we are seeing 100% of the work; and (b) In what subprocess was the observation made? This is to see if any activity that could be critical has not yet been observed. Both criteria are key so that the leader of the tool can direct the observers to where it is required.

Session No. S62

Perceptions of Safety in the Construction Industry in Colombia

Luz Stella Marín Sc.D.
Assistant Professor
Indiana University of Pennsylvania
Indiana, PA

Background

The construction industry is a substantial sector of the worldwide economy. It is one of the mainstays of the economy in most countries, contributing significantly to socio-economic development. However, it is also globally recognized as one of the most hazardous industrial sectors, with a long history of high injury and fatality rates. During 2015 in the United States, out of 4379 worker fatalities in private industry, 937, or 21.4%, were in the construction industry (OSHA 2015). In addition, the construction sector also ranked fourth among the sectors with high rates of fatal occupational injuries, with 10.1 per 100,000 full-time equivalent workers. In the United Kingdom, out of 247 fatalities during the period 2015/16, 45, or 18%, occurred in construction (HSE 2016). In Colombia, injury rates in the construction industry show an increasing trend, remaining higher than other industrial sectors (6.8 fatal injuries per 100,000 workers)(Fasecolda 2013).

Poor safety performance in the construction industry is very often attributed to the inevitably hazardous nature of the work, and/or to the poor behavioral choices of individual workers in that hazardous environment (Ringen et al. 1995; Gervais 2003). However, an inherent characteristic of the construction sector is the job fragmentation, which brings multiple contractors and subcontractors who perform tasks simultaneously and in close proximity, putting workers at risk not only due to their own performance but also due to other workers' practices and procedures (Gervais 2003; Lingard and Rowlinson 2005). The dynamic environment makes it difficult to anticipate, and thus control, hazards in a timely manner. This aspect of the construction sites brings the imperative need for approaches that are able to interpret the daily reality of the construction sites.

Perceptions of safety from those responsible for establishing policies and procedures, and for those promoting their execution at the worksite, must be integrated in order to address comprehensive safety interventions. While workers' perceptions are relevant to understand gaps occurring between formal policies and those in use, understanding perspectives from personnel positioned to implement the necessary changes to enhance workers' health and safety is a priority as well. Unlike injury rates and lost-time days reflecting workplace safety performance retrospectively, perceptions might be used as a leading indicator, which offers a proactive route for improving aspects affecting worksite safety.

Given the dynamic nature of the construction process, having accurate and opportune information about safety gaps in the work area can contribute to the design and implementation of more effective safety interventions. Risk factors are very often exacerbated by management decisions aimed at maximizing project profits by imposing short deadlines and budgetary constraints. In general, most injuries occur due to a combination of an unexpected event and ineffective safety measures, rather than a single cause (Hollnagel 2008). Extensive research has addressed causes of work-related injuries in order to improve safety outcomes in the construction industry. Interventions focused on single or multiple factors have been developed to implement safety measures and decrease injuries. Interventions, such as enforcing the implementation of engineering controls, safety campaigns and programs, housekeeping, worker training, and exercise programs, have been evaluated to corroborate their effectiveness in the complex work environment of construction sites (Rivara and Thompson 2000; Lehtola et al. 2008; Rinder et al. 2008).

Safety Perceptions

Recently, safety climate has been suggested as a key factor related to safety outcomes across different industries and environments. Several studies have focused on how to operationalize and measure safety climate, but to date, little research has explored its potential determinants. Such understanding would be relevant, since it may contribute to identification of safety interventions aimed at enhancing workplace safety climate, and therefore, lowering injury rates. Policies related to the implementation of safety standards, safety training, resources for safety, safety committees, and safety performance feedback have been considered as important elements of effective safety programs (i.e., Goldenhar 2001; Abudayyeh et al. 2006; Swuste et al. 2012; Cheng and Wu 2013). Safety climate emphasizes the perceptions held by employees regarding the importance of safety in their organization (DeJoy et al. 2004).

Construction Safety and Health in Colombia

As in many countries, the construction industry in Colombia is a cyclical business that responds to economic fluctuations and governmental policies. Housing and infrastructure have been identified by the 2010-2018 government as one of the five key sectors contributing to the economic growth. The government's medium-term goal in housing is to build 500,000 houses during the period of 2012-2018, and, in the infrastructure subsector, two large dams are currently being built. In addition, under the requirements of the approved free trade agreement with the United States, building and maintaining the road system is another priority of the current government.

Perceptions of Safety in the Construction Industry in Colombia

Perceptions of safety were evaluated among Colombian construction personnel from a sample of 30 commercial construction companies. The instrument chosen to assess safety and health perceptions for this study was the Nordic Occupational Safety Climate-NOSACQ-50, which was developed by the National Research Centre for the Working Environment (NRCWE) in Denmark (Kines et al. 2011). All responses were made on a four-point Likert scale, with strongly disagree, disagree, agree, and strongly agree options. The mathematical mean for the scale 1-2-3-4 is 2.5; thus, results over 2.5 represent positive perceptions of safety.

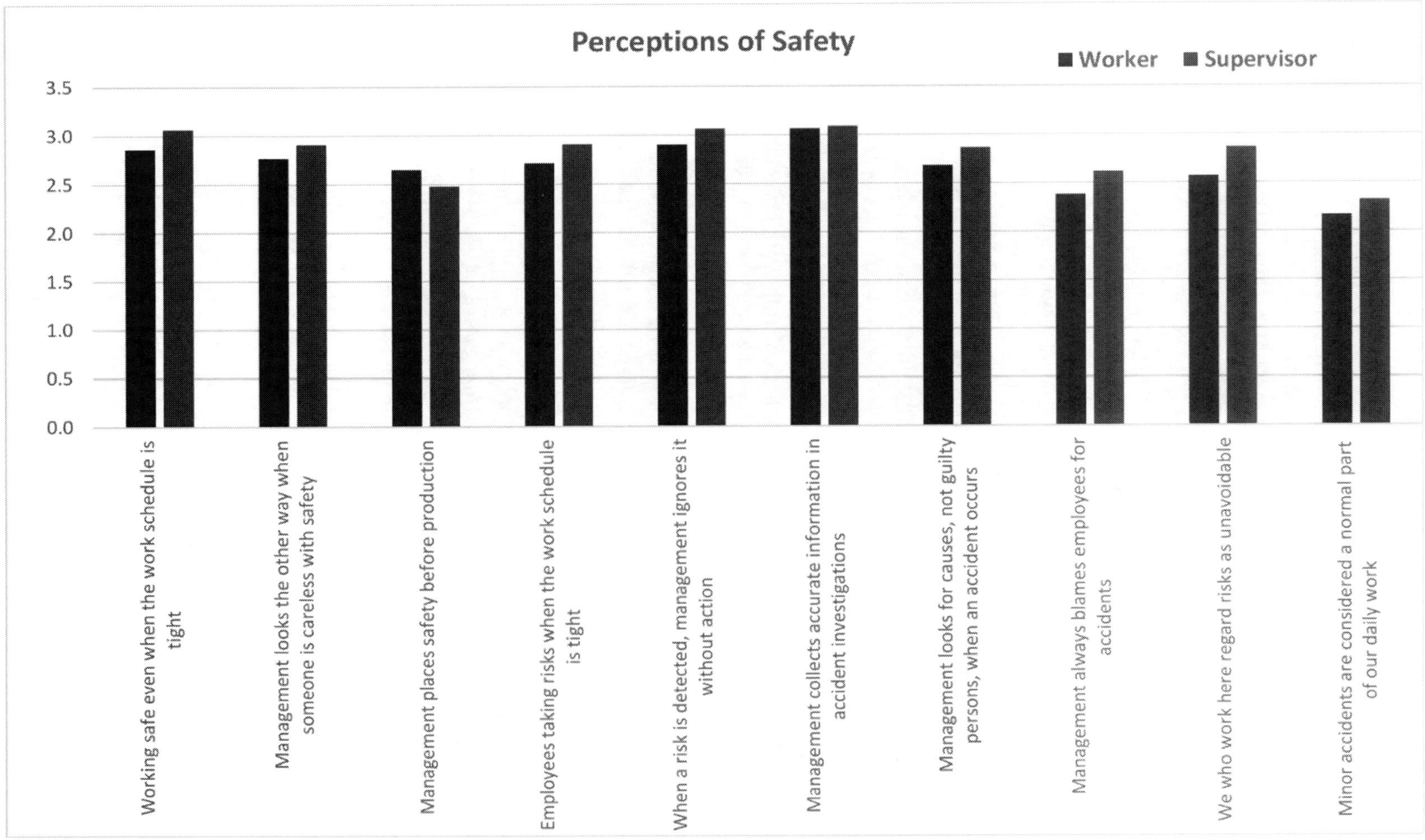

Exhibit 1. Perceptions of safety in the construction industry.

Perceptions of safety climate in this study, as well as in previous studies (Findley et al. 2007), varied between job groups (Exhibit 1). The total safety climate mean scores by job position ranged from 2.2 to 3.1. Workers had the lowest mean safety climate score both in the total mean safety climate scores as well as for each one of the NOSACQ-50 seven dimensions. Workers and supervisors had statistically significant differences in themes related to the importance of safety over production; management considers accidents as a normal part of the job; and perceptions that risk is unavoidable. In contrary, workers and supervisors shared perceptions regarding collecting accurate information regarding work-related accidents.

Conclusion

Differences in perceptions of safety may be explained by differences between the two groups studied in duties, responsibilities and characteristics, as well as educational background. The organizational hierarchy presents clear boundaries regarding factors such as tasks performed, educational requirements, experience and responsibilities. Wage labor in the construction industry attracts mainly unskilled manpower; it is very unlikely to find construction workers who hold a university degree or even a technology degree. Differences in safety perceptions across the groups may reflect dissimilarities in roles (Wu et al. 2007; Huang et al. 2011). For the purpose of achieving project objectives, a site manager's major role is to accomplish deadlines and budgeting. Thus, site supervisors perform mainly administrative tasks, whereas workers constantly experience the risky conditions of the worksite. Field supervisors combine low administrative tasks with more operational responsibilities that require their presence at the actual worksite, allowing them to closely experience the daily working conditions.

The construction industry presents basic differences in both work organization and process when compared to the manufacturing sector, where the safety climate concept was originally developed. The results agree with previous research that describes perceptions of safety as a shared group perception but not as a homogenous characteristic of the organization. Understanding differences in perception among hierarchical levels may contribute to designing comprehensive organizational safety interventions that incorporate different perspectives to improve the commitment to safety at all levels of the organization.

Bibliography

Abudayyeh, O., T. Fredericks, S. Butt and A. Shaar. 2006. "An investigation of management's commitment to construction safety." *Int J Proj Manag* **24**(2): 167.

Cheng, C. and T. Wu. 2013. "An investigation and analysis of major accidents involving foreign workers in Taiwan's manufacture and construction industries." *Saf Sci* **57**: 223–235.

DeJoy, D. M., B. S. Schaffer, M. G. Wilson, R. J. Vandenberg and M. M. Butts. 2004. "Creating safer workplaces: assessing the determinants and role of safety climate." *J Safety Res* **35**(1): 81-90.

Fasecolda, F. d. A. C. 2013. *Estadisticas del Sector.* (Retrieved February 20, 2017) (http://www.fasecolda.com/index.php/ramos/riesgos-laborales/estadisticas-del-ramo/.)

Findley, M., S. Smith, J. Gorski and M. O'neil. 2007. "Safety climate differences among job positions in a nuclear decommissioning and demolition industry: Employees' self-reported safety attitudes and perceptions." *Safety Science* **45**(8): 875-889.

Gervais, M. 2003. "Good management practice as a means of preventing back disorders in the construction sector." *Safety Science* **41**(1): 77-88.

Goldenhar, L. M. 2001. "Health and safety training in a sample of open-shop construction companies." **32:** 237-252.

Hollnagel, E. 2008. "Risk + barriers = safety." *Safety Science* **46**(2): 221.

HSE, H. a. S. E. 2016. *Statistics on fatal injuries in the workplace in Great Britain 2016.* (Retrieved March 10, 2017) (www.hse.gov.uk/statistics/fatals.htm.)

Huang, Y.-H., T. B. Leamon, T. K. Courtney, P. Y. Chen and S. DeArmond 2011. "A comparison of workplace safety perceptions among financial decision-makers of medium- vs. large-size companies." *Accident Analysis & Prevention* **43**(1): 1-10.

Kines, P., J. Lappalainen, K. L. Mikkelsen, E. Olsen, A. Pousette, J. Tharaldsen, K. Tómasson and M. Törner. 2011. "Nordic Safety Climate Questionnaire (NOSACQ-50): A new tool for diagnosing occupational safety climate." *Int J Ind Ergonom* **41**(6): 634-646.

Lehtola, M. M., H. F. van der Molen, J. Lappalainen, P. L. T. Hoonakker, H. Hsiao, R. A. Haslam, A. R. Hale and J. H. Verbeek. 2008. "*The Effectiveness of Interventions for Preventing Injuries in the Construction Industry. A Systematic Review.*" Am J Prev Med **35**(1): 77– 85.

Lingard, H. and S. Rowlinson. 2005. *Occupational Health and Safety in Construction Project Management.* New York: Spon Press.

OSHA, O. S. a. H. A. 2015. *Census of Fatal Occupational Injuries Summary, 2015.* (Retrieved February 20, 2017) (https://www.bls.gov/news.release/cfoi.nr0.htm.)

Rinder, M. M., A. Genaidy, S. Salem, R. Shell and W. Karwowski. 2008. "Interventions in the construction industry: A systematic review and critical appraisal." *Human Factors and Ergonomics in Manufacturing* **18**(2): 212-229.

Ringen, K., J. Seegal and A. Englund. 1995. "Safety and health in the construction industry." *Annu Rev Public Health* **16**: 165-188.

Rivara, F. P. and D. C. Thompson. 2000. "Prevention of falls in the construction industry: evidence for program effectiveness." *Am J Prev Med* **18**(4 Suppl): 23-26.

Swuste, P., A. Frijters and F. Guldenmund. 2012. "Is it possible to influence safety in the building sector?: A literature review extending from 1980 until the present." *Saf Sci* **50**(5): 1333-1343.

Wu, T.-C., C.-W. Liu and M.-C. Lu. 2007. "Safety climate in university and college laboratories: Impact of organizational and individual factors." *Journal of Safety Research* **38**(1): 91.

Seven Core Elements for Effective Safety and Health Programs: OSHA Guidelines and Employer Implementation

Diana Cortez, MS
Area Director, Department of Labor OSHA
Tarrytown, NY

Elisonia Valle, MS
EHS Consultant, Banda Group International LLC
Chandler, AZ

Every day, workplace injuries, illnesses and fatalities cause immeasurable pain and suffering to employees and their families. Recent estimates indicate that workplace injuries and illnesses cost billions of dollars to our nation's businesses. Implementing effective safety and health programs have proven to be a decisive factor in reducing the extent and severity of work-related injuries and illnesses.

According to the most recent data from the Bureau of Labor Statistics (BLS), 4836 workers were killed in 2015 while at work in various industries. The annual total number of worker fatalities in 2015 was the highest since 2008, with an overall fatality rate of 3.38 per 100,000 full-time equivalent (FTE) workers (See charts 1 and 2 below for more details) (BLS, 2016).

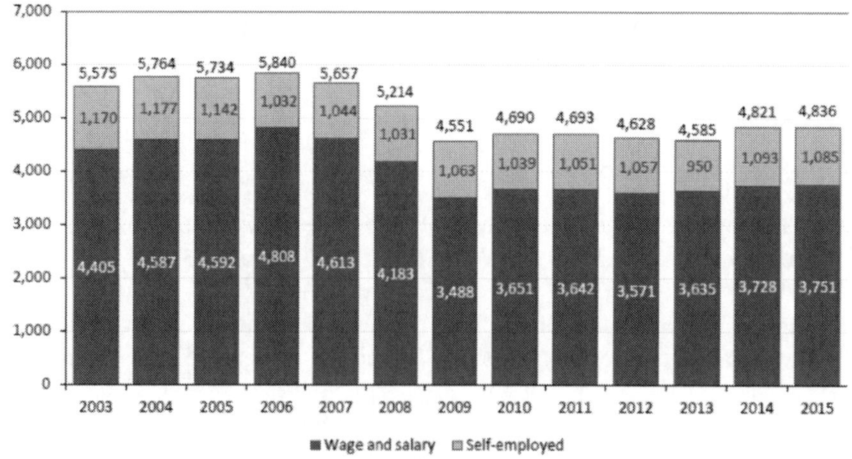

Chart 1 Number of fatal work injuries by employee status, 2003-15

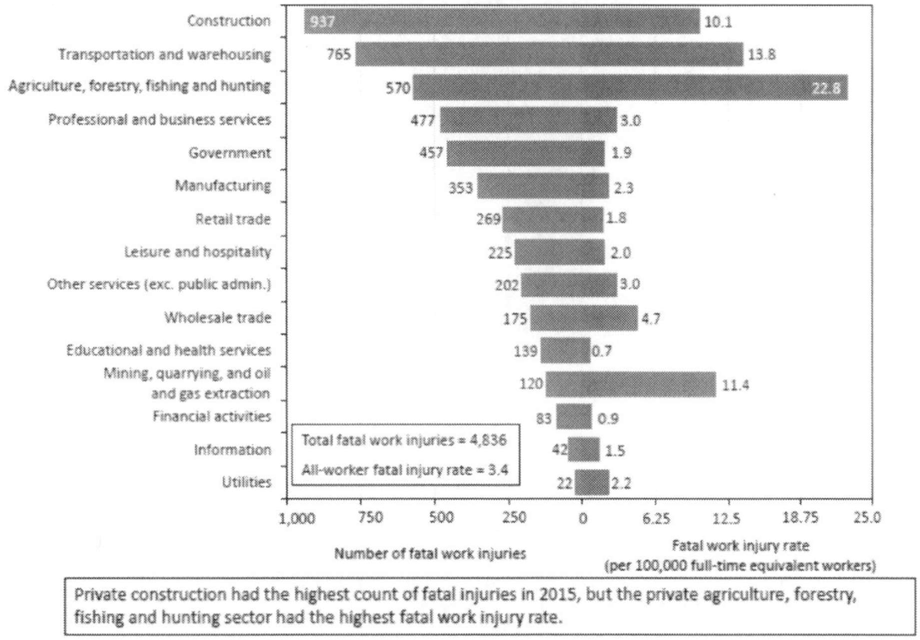

Chart 2. Number and Rate of Fatal Work Injuries by Industry Sector, 2015

Another 2.9 million workers suffered an injury or illness in 2015, of which 2.8 million suffered an injury. Charts 3 and 4 provide specific details including the injury and illness rates (BLS, 2016).

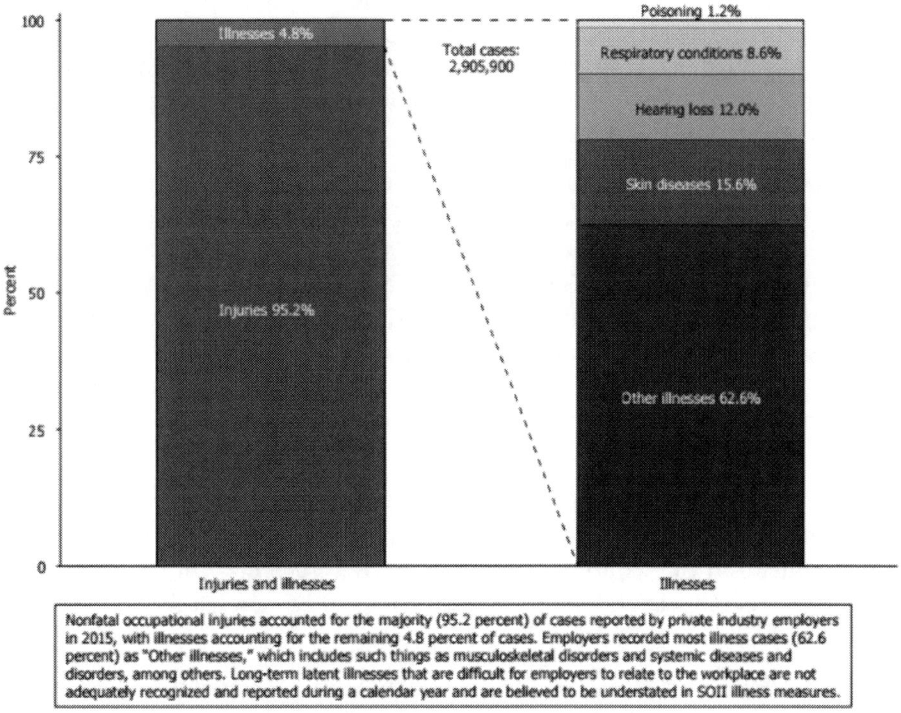

Chart 3. Distribution of nonfatal occupational injury and illness cases by category of illness, private industry, 2015

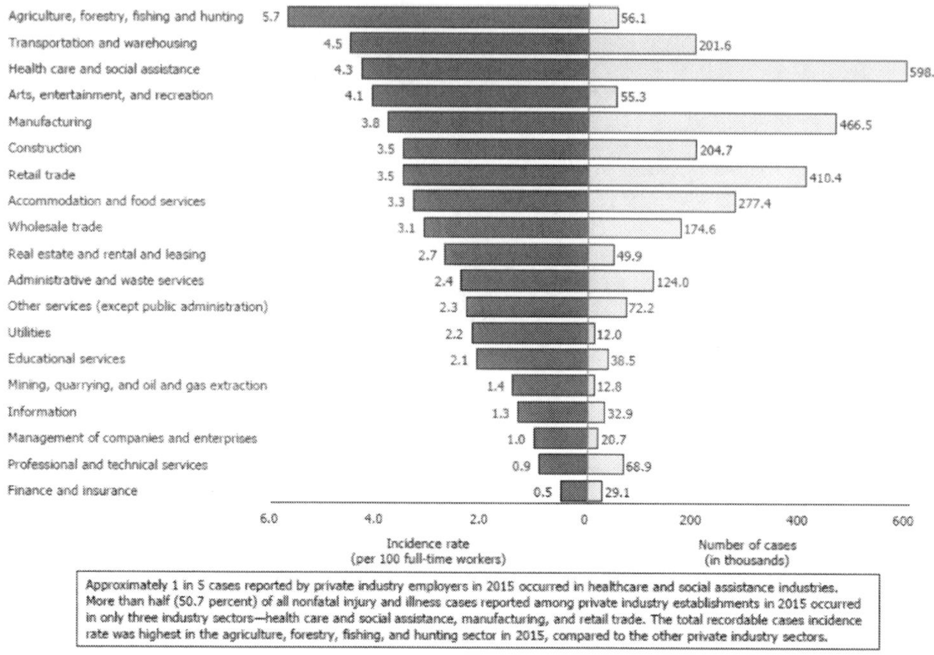

Chart 4. Incidence Rates and Numbers of Nonfatal Occupational Injuries and Illnesses by Private Industry, 2015

A recent study conducted by The Liberty Mutual Research Institute reports that costs associated with workplace injuries that cause employees to miss six or more days of work cost U.S. employers $59.9 billion in 2014 (businesswire.com, 2017). These incidents affect both the employees and their employers. Injured employees are affected not only by the physical harm, but also by the psychological and financial burden that they must shoulder long after the incident. In addition to direct costs, businesses also incur a variety of indirect costs when a worker is injured or becomes ill. These costs include wages paid to injured workers; time lost through work stoppages; administrative, such as in filing reports and claims; costs of training replacement employees; time spent by supervisors lost productivity; and replacement costs of damaged materials, machinery and property.

Despite the progress that has been made to protect workers, on average 12 workers are killed on the job every day, and nearly four million others suffer a serious injury or illness each year (BLS, 2016). Clearly, more needs to be done to prevent these incidents, most of which are known to be preventable. Implementing an effective safety and health program can help reduce the numbers of these workplace injuries and illnesses, which in turn improves efficiency and productivity. Injury and illness prevention programs are not new, as numerous studies have shown that such programs are effective in transforming workplace safety and health culture, leading to reductions in injuries, illnesses and fatalities; lowering worker's compensation and other costs; improving morale and communication; enhancing image and reputation; and improving processes, products and services.

The Occupational Safety and Health Administration (OSHA) has concluded a comprehensive review of studies looking at the effectiveness of these programs and examined the experience of the states that have implemented injury and illness prevention program requirements. On the basis of that review, OSHA estimates that these programs could reduce injuries by 15 to 35 percent for employers who do not now have safety and health programs.

Thousands of employers that are doing the right thing, protecting their workers, have implemented effective safety and health programs, including many that have been recognized under OSHA's Voluntary Protection Program (VPP) or Safety and Health Achievement Recognition Program

(SHARP) (OSHA.gov). These companies' fatality and injury/illness rates are significantly lower than those of others in their industry. Here's what some of them have to say:

- Omaha Steaks: "Employee involvement in improving ergonomics results in less incidence rates and best practices."
- Xerox: "Everyone contributes to health and safety solutions without fear of retribution."
- The Original Mattress: "Our daily morning warm-ups and stretches help reduce the number of repetitive motion and strain injuries that are typical of carrying heavy pieces of furniture, such as mattresses and box springs."

For more information on OSHA's VPP and SHARP programs, their success stories, best practices and resources, visit OSHA's website at www.osha.gov. (OSHA, 2017).

What is a safety and health program? It's a proactive way to manage hazards in the workplace to prevent injuries and illnesses. OSHA has recently updated its 1989 *Guidelines for Safety and Health Programs* to reflect changes in the economy, workplaces, and evolving safety and health issues. The new Recommended Practices for Safety and Health Programs are designed to be used in a wide variety of business settings and help employers establish a methodical approach to improving safety and health in their workplaces. (OSHA, 2016). Traditional approaches are often reactive; problems are usually addressed only after an incident happens where an employee is injured or suffers an illness. These recommended practices are proactive; that is, they are designed to find and fix hazards before they can cause an injury or illness, which is a far more effective approach. This proactive approach encourages all levels of employees to actively engage in identifying and correcting hazards, reducing complacency and embodying a mindset of continuous improvement. The recommendations include seven core elements: management leadership; worker participation; hazard identification and assessment; hazard prevention and control; education and training; program evaluation and improvement; and communication and coordination for host employers, contractors and staffing agencies.

Management leadership: Key to this process is the commitment and involvement of leaders at every level. The leadership must make worker safety and health a core organizational value by fully committing the necessary resources to implement and maintain the safety and health program, to eliminate hazards, to protect workers, and to continuously improve safety and health in the workplace. Additionally, a leader must set the example through their own actions, by walking the talk. Communication is critical, let your workers know that making sure they return home safe, sound and healthy is your top priority. Establish a written policy describing the company's commitment to safety and health and pledging to establish and maintain a safety and health program for all workers. Communicate this policy to all workers, including other relevant parties such as contractors, staffing agencies, temporary workers, suppliers and vendors, visitors and customers. Reinforce your commitment by incorporating safety and health in all business decisions, such as contractor and vendor selection and purchasing. While setting the program goals and objectives, keep in mind that these should be realistic and measurable, as well as contain plans on how these will be achieved, including specific roles and responsibilities. Provide for positive recognition for meeting or exceeding the safety and health goals aimed at preventing injury and illness. Some examples of these can include reporting a near miss, replacing a guard, and attending a training session. Sustaining a positive and encouraging attitude is paramount.

Worker Participation: In order for the program to be effective, engaging workers in the process is vital. Who better to provide input than those in the front lines actually performing the work? Your workers know the most about the process, the equipment, and hazards associated with their jobs. They can provide valuable input in identifying hazards and on how to remove them, and how to possibly introduce an equipment, tool or type of guard that will make the process safer while increasing production. Ensure that workers feel their input is valuable and welcomed. Give them the tools, resources, and empowerment that will provide for continuous involvement without fear of retaliation.

Hazard Identification and Assessment: Hazards are found and fixed. Working together, leadership and workers are able to inspect the workplace to identify hazards of existing and new operations, and correct these before they cause injuries or illnesses. The identified hazards must be evaluated, interim controls implemented when a quick fix is not feasible, and priority given to those hazards presenting the greatest risk to be addressed first. Other action items include the implementation of a reporting system where all workers can report a hazard as well as take action to stop any potential exposure; conduct regular worksite inspections to assess any new hazards or monitor the success of corrective actions taken; and conduct thorough incident investigations to identify root causes in order to prevent future occurrences.

Hazard Prevention and Control: Preventing and controlling the hazards identified and assessed will help employers provide workers with safe and healthful working conditions. Controls are selected, implemented, and continuously evaluated, ensuring their effectiveness. Employers should select the controls that are the most feasible, effective and permanent. Below is a graphic, Chart 5, provided by the National Institute of Occupational Safety and Health (NIOSH), describing the hierarchy of controls (NIOSH, 2017). The graphic illustrates that the control methods at the top "are potentially more effective and protective than those at the bottom."

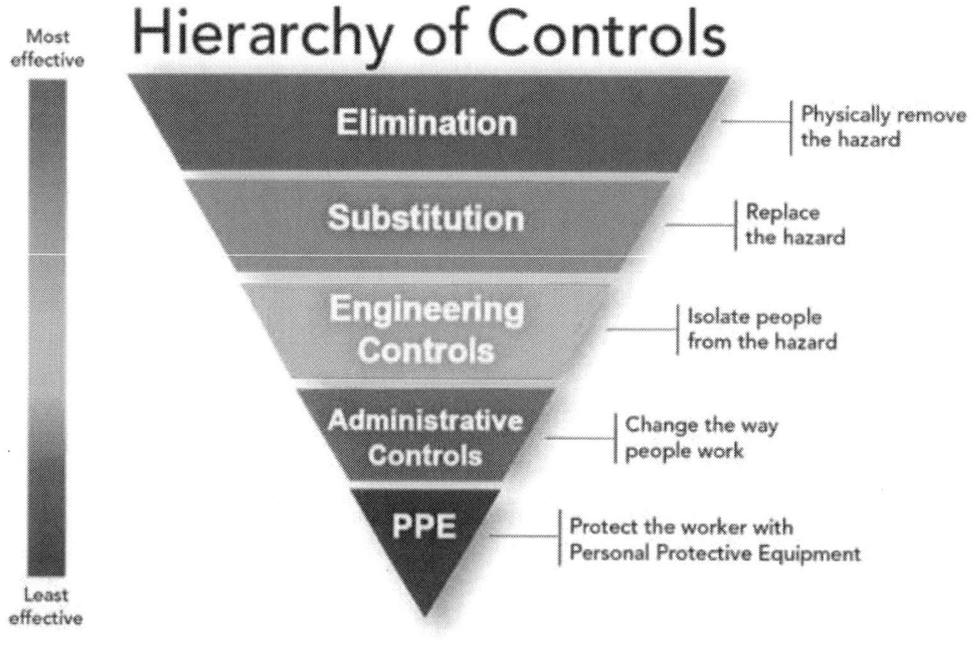

Chart 5. NIOSH's hierarchy of controls used as a means of determining how to implement feasible and effective control solutions

Education and Training: Workers and managers must be provided information and training about the program, awareness training to include an understanding of workplace hazards and how to identify them, report and control them; and training on their specific roles. The training should also include the hierarchy of controls, proper use of work practices, administrative controls, and when and how to use personal protective equipment (PPE). Training must also be conducted when a new process is introduced, and when workers are assigned to a new task. This training must be provided in the language(s) and at literacy level (methods) all workers can understand.

Program Evaluation and Improvement: Is the program on track to achieve its goals? Employers should periodically evaluate the program on whether the set goals are being met. To monitor the progress, employers can track indicators, such as the number and severity of injuries and illnesses, monitoring results of worker exposures, workers' compensation data, number of reported near misses, hazards identified and corrected, workers who have completed required training and whether corrective actions and preventive maintenance are completed timely. These results should be shared with all workers and managers, and engage them in providing input on how to correct program shortcomings, and how to further improve the performance of the program.

Communication and Coordination for Host Employers, Contractors, and Staffing Agencies: Host employers, staffing agencies, contractors and subcontractors have the shared responsibility to ensure workers are provided a safe and healthy workplace. As such, the host employer and staffing agency must clearly communicate and coordinate how they will provide and maintain a safe work environment for their workers. The same applies to contractors and subcontractors. Each employer and contractor needs to consider how its work and safety activities can affect the safety of other employers and workers at the site. In both the temporary worker and multiemployer situations, safety is enhanced if employers establish mechanisms to coordinate their efforts and communicate effectively to afford all workers equal protection against hazards.

OSHA's recommended seven core elements previously discussed offer a step-by-step approach to implementing successful safety and health programs. These practices provide a sound and flexible framework for addressing safety and health issues in a variety of workplaces. Instead of being reactive, management and workers collaborate to identify and correct hazards before a worker injury or illness occurs. These collaborative efforts build trust and improve communication, as well as have an impact on other business improvements. To help you get started or to improve your current programs, there is a plethora of free resources available that will help get you on your way. OSHA has developed guidelines and tools to assist employers in developing and implementing effective safety and health programs; these are on their website at www.osha.gov. The tools and resources available include how to get your safety and health program started, access tools for continual improvement, see state-specific requirements and other OSHA requirements, and see how others have implemented safety and health program in various industries.

Implementing safety and health programs will help employers improve compliance with existing laws and regulations, decrease the incidence of workplace injuries and illnesses, reduce costs and enhance their overall business operations.

Bibliography

Bureau of Labor Statistics (BLS). 2016. *Employer Reported Workplace Injuries and Illnesses – 2015.* (Retrieved from http://www.bls.gov/iif/oshsum.htm.)

_____. 2016. *2015 Survey of Occupational Injuries and Illnesses Summary Estimates Charts Package.* (Retrieved from https://www.bls.gov/iif/oshwc/osh/os/osch0057.pdf.)

_____. 2016. *Census of Fatal Occupational Injuries Charts, 1992-2015 (final data).* (Retrieved from http://www.bls.gov/iif/oshcfoi1.htm#charts.)

BusinessWire. 2017. *The Most Serious Workplace Injuries Cost U.S. Companies $59.9 Billion Per Year, According to 2017 Liberty Mutual Workplace Safety Index.* (Retrieved from http://www.businesswire.com/news/home/20170119005784/en.)

Occupational Health and Safety Administration (OSHA). 2016. *Recommended Practices for Safety and Health Programs.* (Retrieved from http://www.osha.gov/shpguidelines.)

_____. 2013. *OSHA Voluntary Protection Programs (VPP)/Success Stories.* (Retrieved from http://www.osha.gov/dcsp/vpp/success_stories.html.)

_____. 2013. *OSHA's Small Business/Success Stories.* (Retrieved from http://www.osha.gov/dcsp/smallbusiness/successes.html.)

Session No. S71

Safety Training Activities for the Ten Most Violated OSHA Violations

Linda M. Tapp, ALCM, CSP
SafetyFUNdamentals
Madison, NJ

Isabel Bravo, PHR, STS
Wonderful Citrus
Delano, CA

Introduction

The Occupational Safety & Health Administration (OSHA) is a part of the United States government that works to help employers and employees to reduce injuries, illnesses and deaths that occur on the job. One way OSHA does this is by issuing a list, updated every year, of the top ten standards that are most frequently cited during inspections. OSHA publishes this list so that employers can become aware of what the most common problems seem to be among all types of companies and not just in their own. By looking over the list and comparing it to activities and operations where you work, you can take steps to identify and address hazards that you may not have focused on in the past.

Much of the time, safety training is an integral part of ensuring that workers know about the hazards they face and how to protect themselves from injury. Making sure that your safety training classes are interesting and encourage interaction is key to getting trainees to pay attention and remember the information. Interactive safety training activities are a great way to get trainees involved and taking part in their own learning.

There are ten different training activities presented here, and each one corresponds to one of the standards on OSHA's Top Ten Citation List. Not every activity will be for every employer but most can be modified to make them easier or more difficult, shorter or longer or more specific to your workplace.

The Top Ten Most Frequently Cited OSHA Violations*

1. 1926.501 - Fall Protection
2. 1910.1200 - Hazard Communication
3. 1926.451 - Scaffolding
4. 1910.134 - Respiratory Protection
5. 1910.147 - Lockout/Tagout
6. 1910.178 - Powered Industrial Trucks
7. 1926.1053 - Ladders
8. 1910.305 - Electrical, Wiring Methods
9. 1910.212 - Machine Guarding
10. 1910.303 - Electrical, General Requirements

*This list and order reflects fiscal year 2015 and is accurate as of 1/5/16. Every year, the order may change slightly but generally, the same ten topic areas are found to be the "top citations."

A very brief description of each of these standards follows. For complete details and requirements under the standards, visit OSHA's website at www.OSHA.gov.

Fall Protection (1926.501)
OSHA's Fall Protection Standard requires fall protection (protection is required at different heights depending on the industry), various guards to protect from falls, clean and dry surfaces, proper personal protective equipment (PPE), training, and protection for those working below work being done overhead.

Fall Protection Safety Sequence is an activity that can be used during fall protection training to ensure that trainees understand the steps to properly don a fall protection harness. After a presentation has been made on the proper steps, divide the class up into teams, and give each team a set of the fall protection images shown below. (For a full-size set of these fall protection illustrations, visit http://bit.ly/toptenpacket. Use the password "safetyfundamentals" to download materials after clicking on the link provided.) Make sure the images in the set are mixed up and not in the correct order. Ask each team to work together to put the images in the correct order as quickly as possible, and tell them that the team to do so first will be declared winner. When you have a winning team, ask the members of that team to review their steps in the correct order. If the order is not correct, continue the activity until a team is able to present the correct order. It is important to note that, while the fall protection packet presented below contains eight steps, your company policy or specific fall protection procedures might include or not include other specific steps. Feel free to modify this card set as needed. Another very effective alternative, which takes a little more preparation, is to create your own fall protection illustrations. Before the training class, have a volunteer demonstrate the stages of donning a harness. Ask them to stop between each step so you can take their photo. Print out each photo separately, and run the activity with the photos instead of the images shown below.

The correct steps for putting on a harness may vary slightly in your organization but in general, the correct order should be:

1. Untangle and inspect harness.
2. Hold by D-ring and shake out straps.
3. Put arms through shoulder straps with D-ring positioned between shoulder blades.
4. Adjust sub-pelvic straps/place straps under buttocks.
5. Connect and adjust leg loops.
6. Adjust and buckle chest strap.
7. Check overall fit.
8. Buddy check.

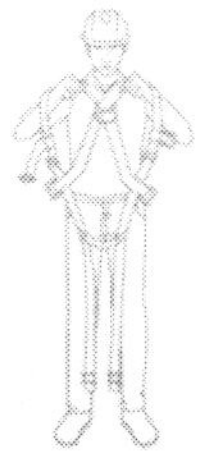

Untangle and inspect harness

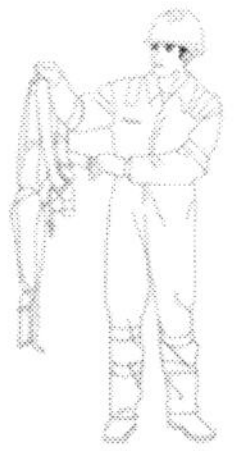

Hold by D-ring and shake out harness

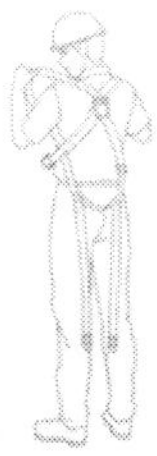

Put arms through shoulder straps with D-rings positioned between the shoulder blades

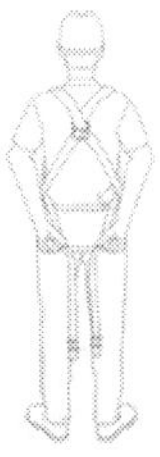

Adjust sub-pelvic straps/ Place straps under buttocks

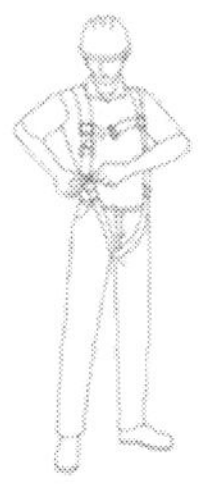

Connect and adjust leg loop

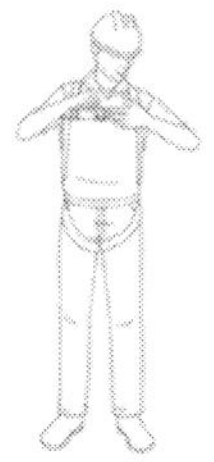

Adjust and buckle chest strap

Check overall fit

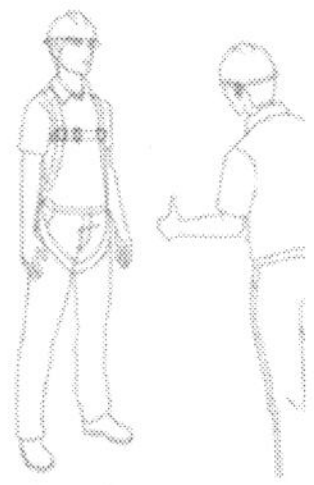

Buddy check

Exhibit 1. Fall Protection Safety Sequence

Hazard Communication (1910.1200)

Simply put, OSHA's Hazard Communication Standard provides instructions for classifying chemicals and communicating hazard information through labeling, safety data sheets, and training.

Pin the Pain is a training activity you can use as part of hazard communication training. This activity uses the body outline below. (For a full-size body outline for use in training, visit http://bit.ly/toptenpacket. Use the password "safetyfundamentals" to download materials after clicking on the link provided.) **Pin the Pain** can be used to identify the different ways a particular chemical or product can affect the body. Give each group of trainees a copy of the body outline and a copy of the safety data sheet or label for the material (or both). Tell each group they have five minutes to identify where the "pain" would be on the body outline if there was over-exposure to the substance. At the end of the five minutes, each team should be asked to show their body outline, and identify the places they marked as having "pain," and explain why.

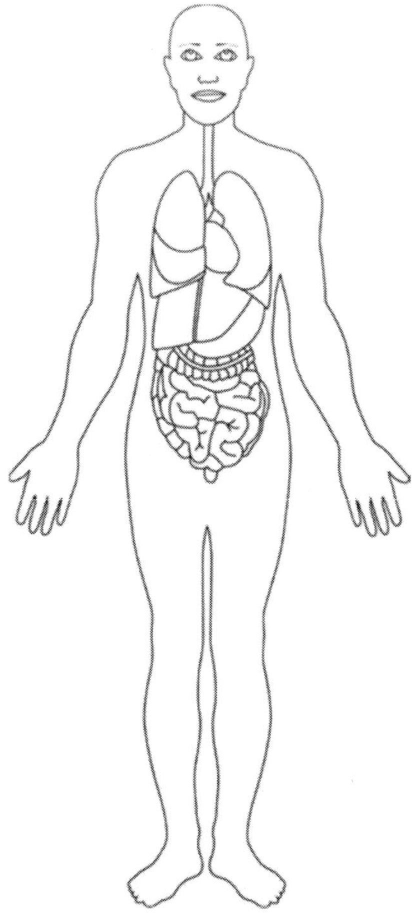

Exhibit 2. Body Outline

Scaffolding (1926.451)

OSHA's Scaffolding Standard addresses the safe use of specific types of scaffolds, including requirements for safe access, training, and fall protection.

Scaffolding Sort is a training activity that requires trainees to work with their teammates and discuss the different types of scaffolding hazards, and then attempt to prioritize the hazards in order of severity.

Give each team a set of ten scaffolding cards, and ask them to sort them in order, from most dangerous to least dangerous condition. At the end of a specified amount of time, ask each team to read aloud the order they placed their cards, from least dangerous to most dangerous. When they are finished, ask if any of the other teams had anything different. They most likely will have placed different conditions on different positions in their rankings. When a discrepancy is found, ask each team to explain why they placed a particular card where they did. The key to remember is that there is not a definitive order that is correct. The value of this activity is that each team will try to argue their way thinking, which is a great way to have the class lead their own discussion. Any time you can have trainees arguing about safety, it is a good thing! When the class works together to decide what may be a more dangerous condition than another, the concepts of probability and severity can also be discussed. At the end of the activity, the class may not have agreed upon a "most dangerous" and "least dangerous" condition, but remember, the value of this activity is in the discussion. (Note: For full-size scaffolding sort cards, visit http://bit.ly/toptenpacket. Use the password "safetyfundamentals" to download materials after clicking on the link provided.)

Employees are not wearing hard hats	Wheels are not locked
Scaffold is not level	Toeboards are missing
Ladder available for reaching top of scaffold is aluminum	It is lightly raining
The scaffold is 9.5 feet high, and no fall protection is provided	Bricks are used on one support leg to level the scaffold
The public is walking under the scaffold	Employees working on the scaffold are not wearing safety shoes

Exhibit 3. Scaffolding Safety Sort Cards

Respiratory Protection (1910.134)

The Respiratory Protection Standard provides guidance to employers when respirators must be worn in the workplace, including selection of appropriate respirators, appropriate maintenance of them, storage, medical evaluations for employees, and training, among other things.

Respiratory Protection Safety Bingo Safety is like regular Bingo except, instead of listening for a random number and letter to be called that you mark off on your Bingo sheet, you listen for clues, then look for that answer on the Safety Bingo card.

Each of the 25 squares on the Bingo sheet contain a different word related to respirator safety. There is a clue for each of these words (one for each word) on the clue sheet that can be found at http://bit.ly/toptenpacket. To start the game, the instructor should randomly read a clue out loud one at a time. If a trainee knows the answer to the clue, and that answer is on his or her safety Bingo card, the trainee should cross it out. The clues should continue to be called out until a trainee has been able to cross off five in a row (horizontal, vertical or on a diagonal). The first person to get five in a row should shout "Bingo;" after his or her answers are checked, they should be declared the winner. It is important not to only use one version of the Bingo sheet for the activity, since this would result in all of the trainees "winning" at the same time. An example of a Respirator Safety Bingo sheet follows. (For the full set of ten different respirator safety bingo cards, plus the call sheet containing the Respirator Safety Bingo clues, visit http://bit.ly/toptenpacket. Use the password "safetyfundamentals" to download materials after clicking on the link provided.)

Respirator Bingo

color	Self-Contained Breathing Apparatus	excessive exposure	16%	engineering controls
qualitative	facial hair	air-line respirators	dust mask	inspected
see or hear	training	FREE SQUARE	remove contaminant from the air	heat & sunlight
power air-purifying respirator	quantitative	fit test	tight seal	gas mask
air-purifying respirators	supplied-air respirator	monthly	confined spaces	emergency escape breathing apparatus

Exhibit 4. Sample Respirator Safety Bingo Card

1910.147 (Lockout/Tagout)

The Lockout/Tagout standard is in place to provide guidance on disabling machinery and equipment so that it can be worked on safely.

Lockout/Tagout Layers is an activity that stresses the proper steps to be taken when safely locking and tagging out a piece of equipment. The steps that need to be taken to properly lock out a piece of energized equipment may be slightly different at your facility and, if so, the steps listed here should be modified accordingly. The ten steps below are common in many places, and so are used in this example.

Each separate step should be written on one of ten Post-it® notes. A set of ten Post-it® notes like this should be given to each team. The Post-it® notes with the ten steps should be placed in a random order before being given to the teams of trainees. Ask each trainee to find a spot on the wall, and to place/stick their post-it notes in the correct order. It is fun to add a little competition to this activity by telling the teams that this is a race, and that the team that places their set of Post-it® notes in the correct order the fastest will be declared the winner. When there is a winner, have everyone return to their seats, and review the order of steps of the winning team. If the order is found to be incorrect, disqualify the team, and move on to the team that finished in second place, and repeat the process until a team has the correct order.

The steps to be written on the Post-it® notes are:

Turn Power Off
Verify Power Is Off
Clean up Work Area
Apply Lock & Tag
Remove Lock & Tag
Work on Equipment
Shut Down Equipment
Inform Others in Area
Inspect Work Area
Turn Power On

The following page shows the correct order of Post-it® notes. Remember, the Post-it® notes should not be given to each team in this order, but mixed up and out of sequence, so that the teams can work together to get the steps in the correct order.

Exhibit 5. Lockout/Tagout Layers

1910.178 (Powered Industrial Trucks)

The Powered Industrial Trucks Standard addresses fire protection, design, maintenance, and use, including training, for a variety of powered industrial trucks.

Forklift Safety Loteria is a version of a popular Mexican game called Lotería. If you have played Safety Bingo, you know that the game boards have words or short phrases related to the class topic. Safety Lotería is like Safety Bingo but, instead of words on the game boards, there are small pictures. The game can be played very simply, by the trainer randomly selecting a picture and then showing the card with the picture to the trainees. If a trainee has that picture on his or her card, they mark it off. Just like in regular Bingo, the first person to get all of the pictures in a line (across or down) marked off, wins. While this is an amusing game, it does not do a good job of teaching new skills related to safety. To make the activity more educational, read a clue that describes a picture on the Lotería card instead. If your audience is not English speaking, this clue will need to be spoken in the language of the trainees. By asking trainees to figure out which safety picture goes with each clue, you are asking them to remember and review what they just learned; this is necessary to be successful at the activity.

A sample Forklift Safety Lotería page follows. (For a complete set of Forklift Safety Lotería pages, along with a clue sheet for the activity, visit http://bit.ly/toptenpacket. Use the password "safetyfundamentals" to download materials after clicking on the link provided.) Like the Safety Bingo game described above, the activity is most effective as a game if different versions of the Forklift Safety Lotería sheet are used, so that not everyone wins at the same time.

Exhibit 6. Sample Forklift Safety Lotería

1926.1053 (Ladders)

OSHA's Ladder Safety standard provides for regulations for the proper use, maintenance, inspection, limits and placement of different types of ladders.

Step it Up is a training activity used you to review ladder safety terminology. Each team should be given a copy of the Step it Up Activity on the following page. (For a full-page version of the

Step It Up Activity, visit http://bit.ly/toptenpacket. Use the password "safetyfundamentals" to download materials after clicking on the link provided.) Explain to the teams that the object of the game is to work together to try to think of a fact related to the word on each step as it relates to ladder safety. Each step is worth a certain number of points (the points go up as the ladder gets higher). The teams can skip around to different steps but they only get the points for the steps they were able to write about successfully. The team with the highest number of points at the end of five minutes wins. At the end of five minutes, go around the room and ask each team to add up the points they have and tell you their score. If one team has the highest amount of points, ask them to tell the class what fact they listed for each step. If the fact they listed is correct, they get to keep their points for that step. If the fact is incorrect, they lose their points for that step, which may cause another team to take the lead. If so, switch to having the other team report on their facts and points until a winner is found.

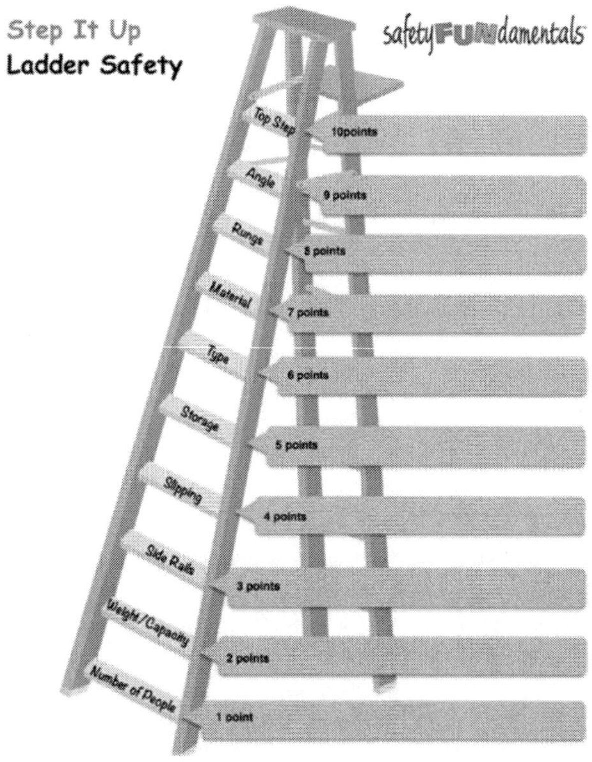

Exhibit 7. Step it Up Activity

Electrical Wiring (1910.305)

OSHA's Electrical Wiring standard covers rules pertaining to electrical wiring methods, components and equipment for general use.

Red Light Green Light is a sorting activity where trainees consider a variety of conditions and then place them into one of three categories: Red, Yellow or Green. Conditions that should result in an immediate stoppage of work should be placed in the Red group. Conditions that would permit work to be continued safely, but would still require a repair or other change, should be placed in the Yellow group. Findings that do not require anything should be placed in the Green group. You can easily make this activity specific to your workplace by listing and then printing out a set of real or imagined conditions that you may find in your facility's day-to-day operations or you can use the situations shown on the next page. You can print out a copy of the following page for each team, provide the teams with a red, green and yellow marker, and then ask the teams to work together to identify and mark the circles that would belong in each category. (For a full page version of Red Light/Green Light, visit http://bit.ly/toptenpacket. Use the password "safetyfundamentals" to download materials after clicking on the link provided.)

After each team has finished sorting the conditions into the the groups, ask one of the teams to announce to the class which items they placed in each. It is highly likely that different teams placed the items in different color categories. When this happens, you have a great opportunity to have discussion around the situation, what should be done when it occurs, and how quickly, and how to prevent it from happening again in the future.

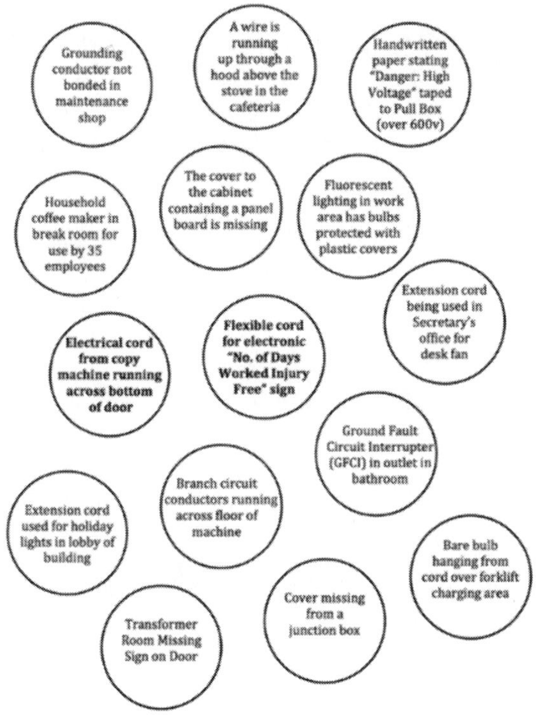

Exhibit 8. Red Light Green Light Activity

Machine Guarding (1910.212)
OSHA's Machine Guarding standard provides guidelines for different types of machine guarding for a variety of types of machines.

Race to Guard is a great way to wrap up a training class on machine guarding or as a way to open a refresher class. Divide the trainees into teams, and give each one a large piece of paper and a marker. A flip chart paper works great! Ask the teams to make two columns on their paper and write A-M down one column and N-Z down the other. Tell the teams that this is a race, and they need to try to write a word related to the class content that begins with each letter. They should work as a group, and can use their notes or other training materials. The teams should be told that the first team to finish will be declared the winner, so they should shout out when they are done. When one team has completed the game, have them read what they wrote for each letter and, if all of the words relate to the class topic, declare the team the winner. Often, teams will need to be creative in their responses (especially when thinking of words for certain letters like "X"), so you will need to be lenient and have fun with the activity. This activity is great to use in safety classes since it involves teamwork, collaboration with peers, and is highly interactive. It also causes the team members to think about what they just learned; this will help increase retention.

A sample A-Z worksheet is shown below, and a sample completed page can be found on the next page.

Exhibit 9. Race to Guard Activity

Race to Guard

A-	N-
B-	O-
C-	P-
D-	Q-
E-	R-
F-	S-
G-	T-
H-	U-
I-	V-
J-	W-
K-	X-
L-	Y-
M-	Z-

Example of a Completed Race Form

RACE TO GUARD

A – adjustable

B – band-saw

C – curtains

D – detects

E – enclosed

F – fixed

G – guard

H – hydraulic

I – interlock

J – jigsaws

K – kickback

L- light curtain

M - motion

N - nip

O - OSHA

P - protection

Q - quiet

R - reset

S – self-adjusting

T – training

U - unguarded

V - velocity

W- work rests

X - "X"pertise

Y -yellow

Z - zone

Electrical, General Requirements (1910.303)

This standard covers various areas concerning the safety of electrical equipment, including examination, installation, and the minimum working space required around electrical equipment. This standard also covers things like arcing, marking, and guarding.

Hazard Hunt activities are illustrations of hazards that could be found in the workplace. A Hazard Hunt activity with many electrical hazards is provided on the following page. The following hazards can be found in this illustration (there may be more than what is listed and not everything identified is strictly an electrical hazard):

1. Missing personal protective equipment
2. Jewelry worn
3. Storage in front of panel
4. Blocked exit door
5. Missing light cover
6. Tripping hazard
7. Metal ladder
8. Leaking ceiling
9. Liquid on floor/spill
10. Storage too close to equipment

For a copy of this Hazard Hunt plus answer key, visit http://bit.ly/toptenpacket. Use the password "safetyfundamentals" to download materials after clicking on the link provided:

Hazard Hunt activities should be completed by a small group working together to identify hazards in a supplied photo or drawing. If you are not using the illustration that follows, you can use your own by simply taking a photo of an area in your workplace that shows the types of hazards in which you are interested. A photo or drawing should have at least six hazards included, and the hazards should not all be obvious. If possible, the hazards should be related to class content, although this is a great game for general safety training tips, such as training on hazard recognition. With most safety training games and activities, it is recommended that a class of trainees be divided into small teams. After 10 minutes or so, stop the group discussions and ask each group to report how many hazards they identified. It is highly likely that the various groups will have identified a different number of hazards. Start with the groups with the longest list, and ask them to identify their hazards. If they name something that another team did not include, it will likely lead to lively discussion; that is the purpose of this exercise. To extend this activity, you could then ask each team to come up with a solution or corrective action for each item found.

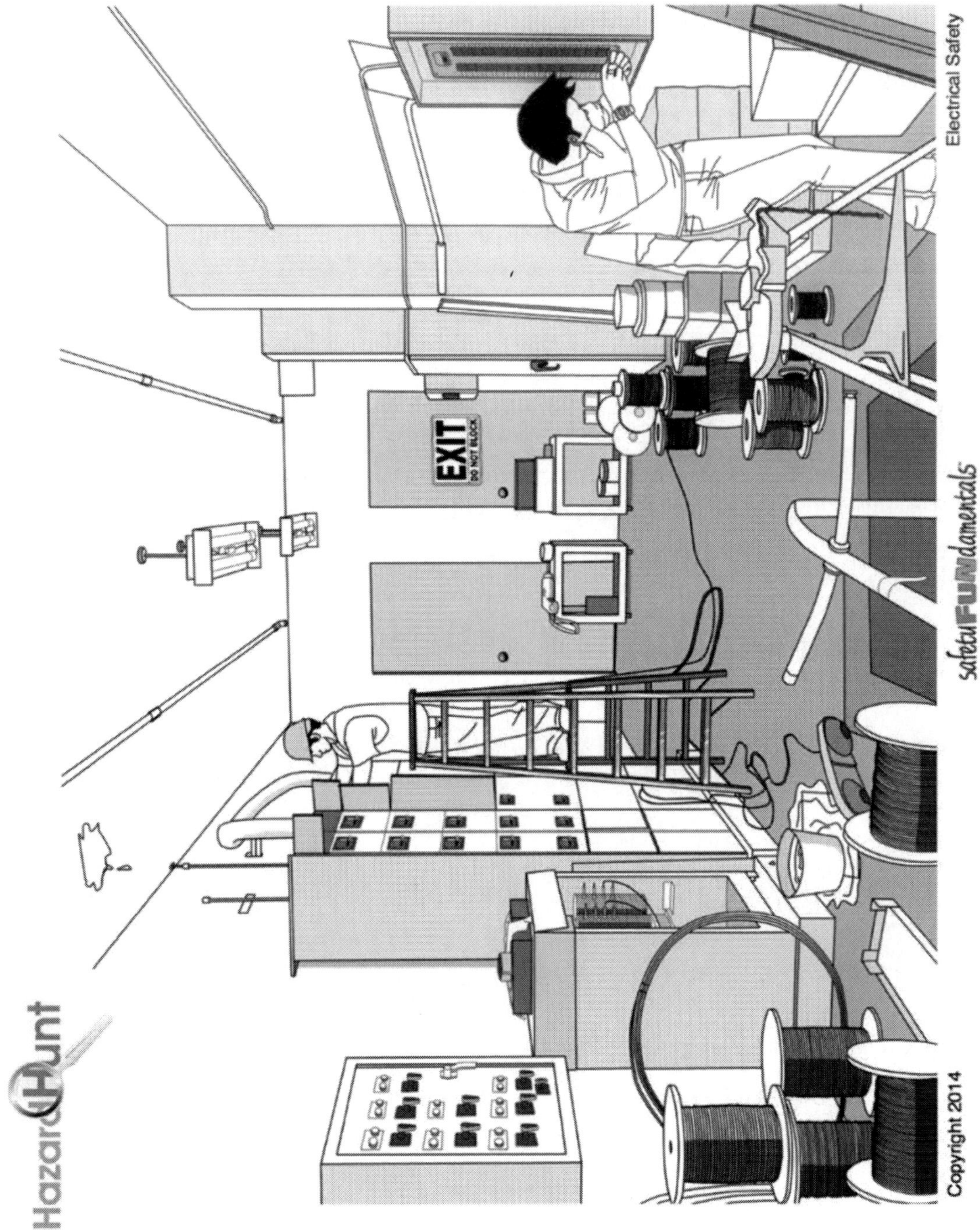

Exhibit 10. Hazard Hunt

Safety training, especially that required by OSHA Standards, is too important to be boring. Interactive training activities such as these increase class participation, trainee learning and retention. Most of these activities can be modified for classes of all sizes and all topics and most can also be incorporated into your training classes at very little expense. Most of the top ten OSHA citations involve a training component. Incorporate these interactive activities into these training classes to make this required training more effective.

Bibliography

Fairbanks, D. 1997. *Accelerated Learning.* Alexandra, VA: ASTD.

Furjanic, S. and L. Trotman.2000. *Turning Training into Learning.* New York: Amacom.

Meier, D. 2000. *The Accelerated Learning Handbook.* New York: McGraw Hill.

Occupational Safety and Health Administration. n.d. *Top Ten Standards.* (Retrieved Feb. 17, 2016.) (https://www.osha.gov/Top_Ten_Standards.html.)

Pike, R. 1994. *Creative Training Techniques Handbook.* Minneapolis: Lakewood Books.

Tapp, L. "Accelerated Learning Methods for Safety." *Safety 2005: Proceedings of the 44th Annual Professional Development Conference.* New Orleans, Louisiana, June 12-15, 2005.

Tapp, L. "It's All Fun and Games: Twenty Five Activities to Make Training Great." *Safety 2006: Proceedings of the 45th Annual Professional Development Conference.* Seattle, Washington, June 11-14, 2006.

Tapp, L. 2006. *SafetyFUNdamentals: 77 Games and Activities to Make Training Great!* Cherry Hill: Ennismore Publishing.

Tapp, L. 2015. *Training Activities for OSHA's Top Ten.* Madison: Ennismore Publishing.

Thiagarajan, S. 2003. *Design Your Own Games and Activities.* San Francisco: John Wiley & Sons.

Session No. S73

Let's Play Dominoes

C. Ezequiel Chalbaud, BA, OHST, HSE Advisor
Randall Tigert, BA, Alaska HSE Manager
CH2M
Anchorage, Alaska

Introduction

CH2M provides comprehensive solutions and services to the energy sector in the Alaska region. With over 1,200 employees, it is a private-sector, direct-hire business group that delivers operations management, project management, design-build construction management, and maintenance services for Alaska's upstream oil and gas industry. In the midst of one of the hardest years the oil industry has seen in the last 20 years, 2016 saw the lowest number of recordable injuries and resulting TRIR the company has ever achieved. This was largely the result of the implementation of a visionary leading indicator program that focused on developing leadership at all levels of the organization. The ROC (Risk Observation Card) Program is an observation program that is unlike traditional observation programs that focus on "safe behaviors;" the ROC program is designed to focus on the identification of at-risk conditions and behaviors and immediately intervening in those unsafe conditions/behaviors. It emphasizes near-miss reporting and leadership engagement with the workforce right where the action is: out in the field.

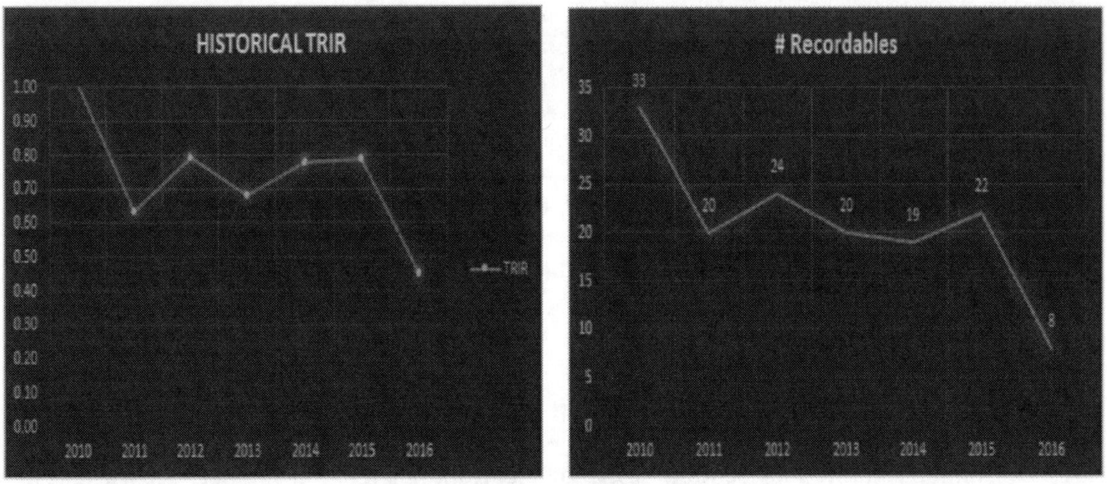

Figure 1. Lagging Indicators

Let's Play Dominoes takes a leadership-focused look at utilizing leading indicators in a proactive approach to risk management and prevention. It utilizes the Domino model as a simple, easy-to-understand incident causation model. It then analyzes people's and organization's safety commitment through the use of the SPIES model, which is a five-level model. Understanding of our moral commitment to safety allows us to explore our behaviors through the use of the accountability ladder. The ladder separates behavior into two main categories: "victim behaviors" and "accountability behaviors," each with four rungs or levels. Lastly, it explores the development and implementation of a leading indicators program to support a data-based focus for a proactive approach to safety.

The Domino Sequence

A Basic Model

As safety professionals, we could go back and forth all day on the difference between an incident and an accident, and whether or not one is preventable and the other is not. The definition often changes, depending on the context of the event. We will not debate on the topic; rather, we'll simply define it so that the reader can be on the same page as the writer throughout the extent of the paper. With that in mind, let's start by defining the terms. An incident is an individual occurrence or event, which is an umbrella term, including accidents, injuries, and illnesses that cannot be linked back to an event, intentional events, and criminal events. For the purpose of this paper, we will define accident as an unplanned, undesirable event that results in a loss. That loss is often an occupational injury or illness.

Let's begin our discussion talking about how accidents happen. Of all the undesired events that we are presented with in life and at work, 99.9% are preventable on one level or another. Sometimes we have the power to prevent them, and sometimes other people have the power to prevent them. But, at some point in the sequence of events leading to an accident, something could have been done differently to change the outcome, regardless of how complex the sequence might be. In safety, we understand that there is always a reason as to why an accident happens and therefore, they are always preventable. The reasons why accidents happen are simply called causes.

Through the incident investigation and root cause analysis process, we attempt to trace our steps back in time to determine what factors came together at the time of the incident that resulted in a loss. There is usually more than a single cause; therefore, we have different levels of causes. We could also argue about causation models for extended periods of time, as there is a wide range of causation models and root cause analysis methods. Some are more complex than others and better suited for certain applications. Keep in mind that our choice of causation models is based on simplicity, given that it is utilized as a training tool for our frontline leaders. The crew leads, foremen, and supervisors who are not safety professionals need a simple, easy-to-understand, one-size-fits-all causation model that they can utilize in their safety conversations with the hands out in the field.

The Domino causation model we use in our leadership training explains how incidents happen with a domino sequence. It is loosely based on the work presented by H.W. Heinrich, as it uses the domino series to represent the causation chain and the relationship between the loss, the incident, and the causes. Three of the dominos in our series relate to the causes. These are direct causes, indirect causes, and root causes. A model is a simplified representation of reality. The Domino model we use is a simplified, one-size-fits-all representation that most incidents are able to fit into. Logically, if all incidents have causes, and we eliminate those causes, then we effectively prevent the incident. The real challenge comes in being proactive and identifying those causes before the incident occurs.

Let's start with the last domino in the series, which is the loss. This is the undesired result we are left with after an incident. The loss resulting from an incident can vary from small stuff like spending resources to fix damaged property, to more serious, life-changing injuries with subsequent permanent disabilities, going as far as the ultimate loss that comes with the death of a loved one, a friend, or a coworker. The loss always affects more than just the person involved. It can have tremendous repercussions to our work groups, friends, families, and communities. The next-to-last domino in the sequence is the accident, which is an unplanned, undesired event resulting in a loss. The incident marks the point in time and space where the causes and contributing factors come together in a harmful way. The incident could be anything from a near miss or a small property damage or first aid, to a major catastrophic event with loss of life, major property damage and/or severe environmental impact.

Before the incidents come the causes. The first "cause" domino we find while going back through the domino sequence is the direct cause, which is an unsafe act and or an unsafe condition or, frequently, a combination of the two. Unsafe conditions are what we call hazards and are things and situations that have the potential to cause harm. They can vary from a missing handrail on an elevated platform; a missing guard on a piece of rotating equipment; or slick, icy roads or walking/working surfaces. An "unsafe act" is when people act in a manner that is unsafe. They come in as many varieties as there are mechanical hazards. They could range from simply not being aware of a hazard, not wearing the proper PPE, or taking a short cut to get something done in a hurry. They also include consciously choosing to take a risk by disregarding a safety rule, bypassing a safety system and thereby, putting ourselves and/or others in harm's way. Another way to think of unsafe conditions and unsafe acts is to think of them as mechanical and personal hazards respectively. They don't always result in incidents but anytime we ignore a hazard and take a chance, we are putting ourselves or our coworkers at risk. Personal and mechanical hazards exist through the fault of carelessness or poorly designed or improperly maintained equipment.

The next domino in the sequence is an indirect cause, which is a personal lack of training, resources, or belief. Indirect causes are basically how the social and work environment in which we live has shaped our perception of who we are and how we act in the face of hazards. Sometimes we may simply not have the proper training. Other times we may not have the needed resources for proper engineering and maintenance, or to get the materials and equipment that best suit the work. And other times we may not believe that the safety rules apply to us and have a high risk tolerance, or simply do not care. We may just be reckless with the belief, "Hey, it won't happen to me."

In order to understand why people do what they do, we must understand the basics of behavior so that we can identify what prompts a person to act a certain way, and what reinforces or discourages that action. There are three components of behavior: antecedent, behavior, and consequence. Consequences can be positive or negative, immediate or delayed, certain or uncertain. Consequences that are positive, immediate, and certain have the strongest effect on shaping future behavior. This is followed closely by negative, immediate and certain. Consequences that are negative, delayed, and uncertain have the weakest effect on shaping future behavior.

Lack of training leads a person to commit an unsafe act or create and unsafe condition because they don't know any better. Lack of training can mean: No training at all, training that was incomplete, training that was not repeated frequently enough, or training that was not comprehended or understood. Lack of resources can lead to unsafe acts/conditions because the resources (time, tools, equipment, materials, and people) necessary to do the job safely are either not provided, not proper (skills, size, abilities, and type), or not in safe operating condition. Lack of resources is linked to lack of planning in that we cannot properly resource a task until we have effectively planned to identify the resources needed. And we cannot complete planning until we have ensured that all resources needed are available. Lack of belief leads a person to commit an unsafe act or create an unsafe condition because they don't really believe that they or someone else will be hurt. There is a belief that a negative consequence will not result because of their action. Lack of belief is often caused by past experience. Cultural values also contribute significantly to a lack of belief. Other factors that can contribute to it are poor morale, peer pressure, productivity pressures, and inadequate resources.

The initial domino in the sequence is the root cause, which is the underlying cause, or the most basic or source cause that allows all other causes to exist. This is basically the social and work environment in which we live. In other words it is our safety culture, and the framework of rules and regulations under which we operate. In root cause analysis, it always relates back to an organizational failure to either plan, direct, organize, and/or control the work. Root causes generally fall under a few general categories that include procedures, training, quality control, communications, management systems, human engineering and work direction. Basic or root causes are where the accident causation sequence begins. Basic or root causes are as follows: organizational failure to properly and effectively plan work; organizational failure to properly and effectively organize work; organizational failure to properly and effectively direct (instruct/train) the people doing the work; and organizational failure to properly and effectively verify (monitor, control) work is being conducted as planned and directed.

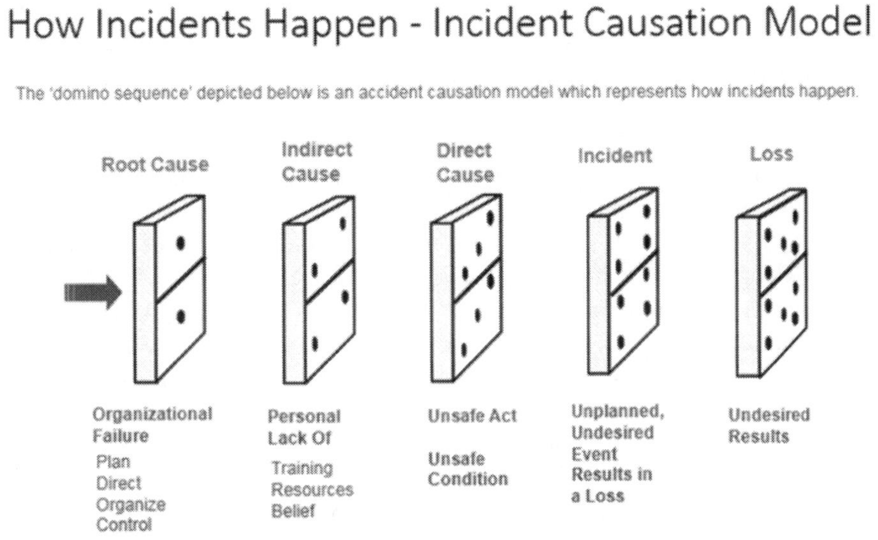

Figure 2. A Simple Causation Model

Let's Play

Now that we understand the Domino Theory, we can get into how we play the Domino Game. The purpose of the game is to keep the last two dominoes from ever being triggered. We do this by eliminating cause dominoes one at a time. Let's take a look at what we can do to keep the domino sequence from getting activated. The further back we go with our elimination, the better chances we have of not triggering an incident. Root causes have two distinct characteristics:

1. They have the greatest impact in the domino sequence; and
2. They are also the hardest ones to discover and effectively remove from the board.

This is not to say that the other two causal dominoes are not important. Any time we eliminate an indirect or direct "cause," we are effectively keeping the sequence from being activated, and keeping incidents and losses from occurring. Therefore, we want to seek to eliminate any cause that we can identify. There are a number of different ways we can make "cause" dominoes disappear. Let's explore the process.

It's About Commitment

We must start with ourselves, by taking a look at where we stand with safety in terms of our personal commitment and our behaviors. An easy way to measure our commitment to safety is to use the SPIES Model, which has five levels:

1. The first S stands for "Self," and it represents the lowest level of commitment. People at the Self level would say, "I do it to keep my job."
2. The next level up is the "Political" level, and it basically means the person is just playing politics. People a tthe Political level would say, "I do it satisfy my boss."
3. The next level is the "Intellectual," and it represents a breakthrough stage in embracing safety because it is here where people begin to have an understanding of the "why." People at the Intellectual level would say, "I understand its importance."

The next two levels of commitment come after we understand why it is important by embracing and accepting safety in our heart and mind because we realize that it is an intricate part of our values and beliefs.

4. The second highest level of commitment is the "Emotional" level. At this level, we start to care about safety, not because we want to keep our job, not because you want to make your boss happy, and not because you understand the importance, but because of what safety means to us, and because we care. People at this level of commitment would say, "I feel strongly about the need" to not hurt people or the company, which really are the same.
5. The final and highest level in commitment to safety is the "Spiritual." Here is where we are in it 100%. We live it not only at work but at home. We have seen the light, drank the Kool Aid, and we have a deep care and passion for doing things right and having our people going home safe and healthy every day. People at this level would say, "I believe it's the only way."

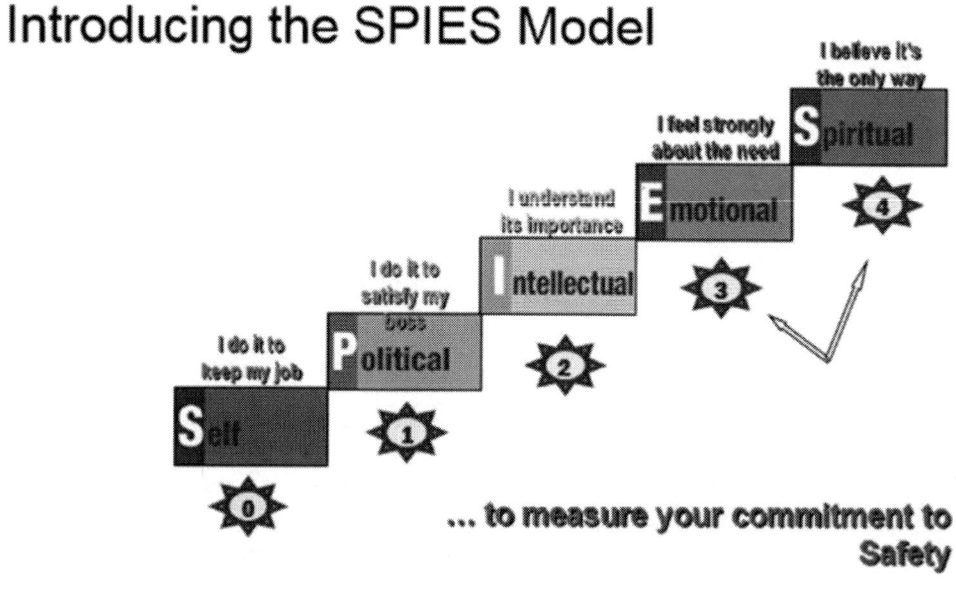

Figure 3. It's About Commitment

Taking Responsibility
Now that we have assessed our commitment, let's move on to our behaviors. After all, we are what we do, and we've all heard the quote by Ralph Waldo Emerson that says, "your actions speak so loudly I can't hear what you are saying." Actions are indeed louder than words, and if we don't walk the talk, no one is going to listen. For this exercise, we want to utilize the accountability ladder, which represents a progression in behaviors from being completely unaware to being fully accountable for our actions, and realizing that we have the power to make things happen. The bottom half of the ladder represents "Victim Behaviors," where things happen to us and we wash our hands of all responsibility. The upper half of the ladder represents

"accountable behaviors," where we become an active participant in the safety journey, and a contributor to our safety culture. So let's start from the bottom and climb the accountability ladder all the way to the top. The lowest level is "unaware or unconscious." Next, we move onto "blaming others," to later come up with "personal excuses." After we realize that these don't work, we might say it's just too hard, and "I can't," so we just sit, "wait and hope" for something to happen. Here is where we draw the line, and decide to make a change. That change starts with "acknowledging reality." We acknowledge the cards we have been dealt, and we "embrace it," and begin looking to "find solutions." Once we have envisioned a solution, we can begin to "make things happen." We can work our plan to truly make a difference in our life, the lives of others, the organization, and the community. Here is where we truly become leaders.

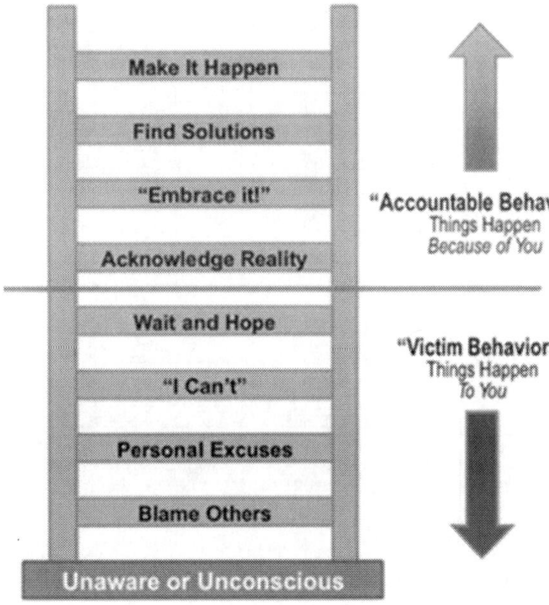

Figure 4. Taking Responsibility

Let's ROC

Once we have a good understanding of where we are and how we act, we'll see the path ahead in our personal growth towards becoming a safety-conscious leader. Employees then can provide valuable contributions by taking action every time they see a "cause" domino waiting to set off the incident sequence. One of the best tools we have in this great game of dominoes is the Risk Observation Card (ROC). The card is simply a way to document our actions in eliminating "cause" dominoes. It allows us to populate a database that is able to tell us what we need to work on as an organization and within our specific work groups. The ROC can be used to document every time we eliminate a "cause" domino by intervening. It doesn't matter if the domino is a root cause, an indirect cause, or a direct cause. Every time an employee identifies a possible cause to an incident, and we do something to help eliminate it, they can document it on a ROC.

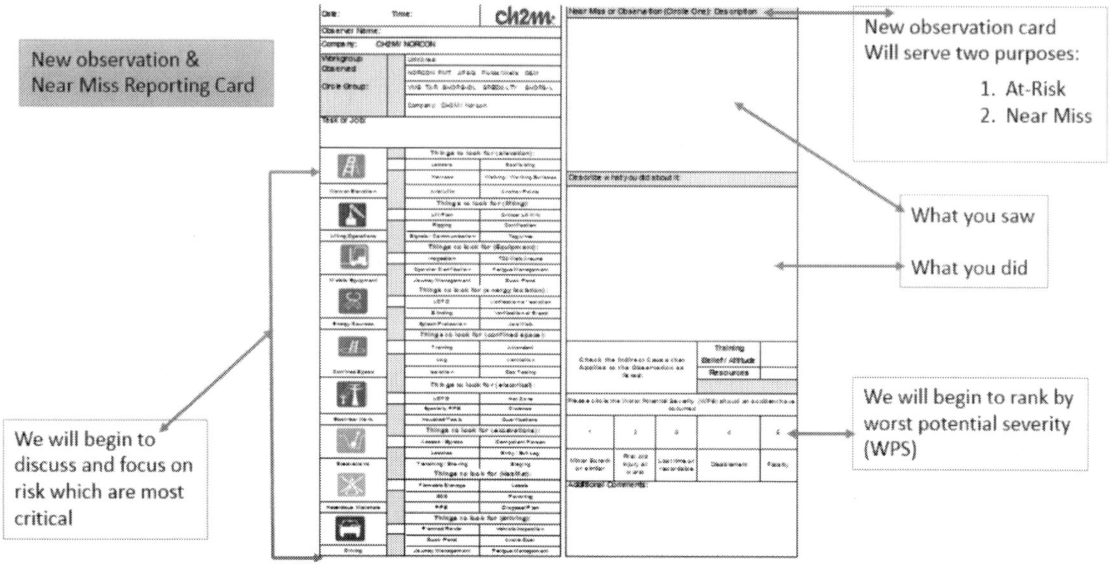

Figure 5. The ROC

An important data point we get from each card is the worst potential severity or WPS. This is a measure from one to five as follows:

(1) Minor scratch, small release contained onsite and no environmental impact, or a property damage less than $5K.
(2) First aid, release above the reportable limit with minor environmental impact, property damage $5K - $30K.
(3) Lost time or recordable injury, release significantly above reportable limit with some local impact, loss of $30K – 300K.
(4) Disablement, life-changing event, or intensive care, significant offsite environmental impact, some remediation required, loss of $300K - $3 million.
(5) Fatality, serious offsite environmental impact with significant remediation required, loss of more than $3 million.

It is essential to dedicate time coaching employees on how to arrive at the appropriate WPS taking into consideration only the factors actually present in the moment of the observation or near miss. In other words, what is the most likely outcome of the at-risk observation, condition, or near miss observed. Observations and near misses with a WPS of 4 or 5 require the supervisor to perform a brief investigation, with the goal of ascertaining action items that address the identified causes. The findings are reported in a one-page Leadership Intervention Report, which outlines the following: description of the event, direct cause, indirect cause, root cause, and corrective actions. The intent of an investigation is never to find blame. It is simply to identify the causes, so that we can learn from them, and address them with appropriate action items.

Let's say we: (a) Identify an organizational failure to effectively plan the work that puts people, environment, or property at risk; and (b) We act to eliminate this problem. Then we would document our observation/intervention in a ROC so that we can utilize that information to identify other opportunities for improvement that might be related to our finding. Let's say we identify an area where our employees don't have the proper training or resources, and we set a plan in motion to address the found deficiencies. We would write a ROC on it. Or maybe we are working with an individual that is a natural risk taker, and we realize that they are taking more risk than we would accept, and we have a conversation with that individual. Then we would write a card on it. Granted, we never write the name of the person other than ourselves (as an observer) on the card. But documenting these safety conversations is vital. Or maybe we find an at-risk condition that could get someone hurt, and we either fix it or get someone to fix it. Then we would write a card on it. Another great use for a ROC is to input near misses into the database. A near miss is when the domino sequence gets activated but, either by luck or quick reaction, we are able to avoid an incident. Something almost happens that nearly misses us that could have had negative consequences. Such near incidents give us a great opportunity to identify and address the causes associated with that specific domino sequence or chain of events.

Trend the Track

The idea behind documenting our observations is that it lets us to populate an ever-growing and evolving database that allows us to explore and analyze our interventions, near misses, and audits to find trends that point the direction in which we are currently headed. Knowing where we are headed helps us hone in on specific deficiencies steering us away from our target. We are predicting where we would end up if we were to keep doing what we are doing, and determine the areas of greater risk for our people and the organization.

First off, we track leadership participation. All leadership from crew leads and above are required to participate in the program. Everyone has been trained, and the expectations have been set that participation in the program is part of their responsibilities as leaders. Craft hands, on the other hand, are not required to participate; however, they are highly encouraged. Public recognitions are given to those individuals turning in quality risk observations and a variety of awards given out, from cash rewards to various items, such as hats and coolers.

We have developed a comprehensive leading indicator database where we log all our audits, near misses, and Risk Observation Cards. We utilize this information to produce detailed monthly reports for each of the 12 groups that encompass the CH2M Alaska region, as well as an overall regional report. This report is distributed to all levels of leadership to be shared and discussed with the crews in an effort to give employees a chance at presenting solutions for the major deficiencies identified. The monthly reports combine traditional lagging indicators, with the leading indicators derived from the information collected from our risk observation program.

The risk observations, audits, and near misses are grouped into nine critical risks. Categorizing our risk into these major groups helps us focus our efforts in those specific areas presenting the greatest risk to the organization and our people. The nine critical risks are the following: work at elevation, lifting operations, mobile equipment, energy sources, confined space, electrical work, excavations, hazardous materials, and driving. Furthermore, the nine critical risks can be further subdivided into 21 categories to pinpoint with precision where we need to focus our risk management efforts. These categories are as follows: fall protection, housekeeping, personal protective equipment (PPE), lifting and rigging, procedure violation, environmental, walking/working surfaces, mobile equipment, hot work, scaffolding, hazard recognition, driving hazards, job planning, energy isolation, body position, improper tool use, defective equipment, line of fire, lack of inspection, ladders, and barricades.

Following are a couple of examples of some of the data that is presented to the groups in the monthly reports. This specific snapshot is for the Wells group, which is part of our Wells and Fluids group, specializing in well intervention and support of BP's drilling operations in the North Slope of Alaska. The project is located in Prudhoe Bay, Alaska's northernmost region, 250 miles above the Arctic Circle. Due to geographical location, personnel are exposed to harsh environmental conditions. Winter temperatures frequently dip into minus 30 degrees Fahrenheit, and high winds are common. Well support intervention technicians work outside in the arctic environment, requiring specific tools and PPE in order to mitigate environmental hazards. Winter driving conditions exist for approximately nine months of the year. Slick, icy roads and drifting snow are common driving conditions, despite the continuous snow removal effort. These are only the environmental hazards. In addition, we find process hazards inherent to the drilling and well work required to keep the oil wells safe and operational. Some of the most common well work hazards include the following: caught-by, caught-in, caught-between, high pressure lines and

Wells

ROC Trends (Top 10)

Category	# Findings - YTD
Hazard Recognition	738
Personal Protective Equipment (PPE)	636
Driving Hazards	350
Procedure Violation	137
Job Planning	130
Defective Equipment	116
Environmental	106
Walking/Working Surfaces	99
Lifting and Rigging	86
Energy Isolation	77

Category	# Findings - December
Personal Protective Equipment (PPE)	89
Hazard Recognition	54
Driving Hazards	31
Energy Isolation	12
Ladders	11
Job Planning	11
Barricades	11
Walking/Working Surfaces	10
Lifting and Rigging	10
Environmental	10

Wells

Critical Risks & Worst Potential Severity

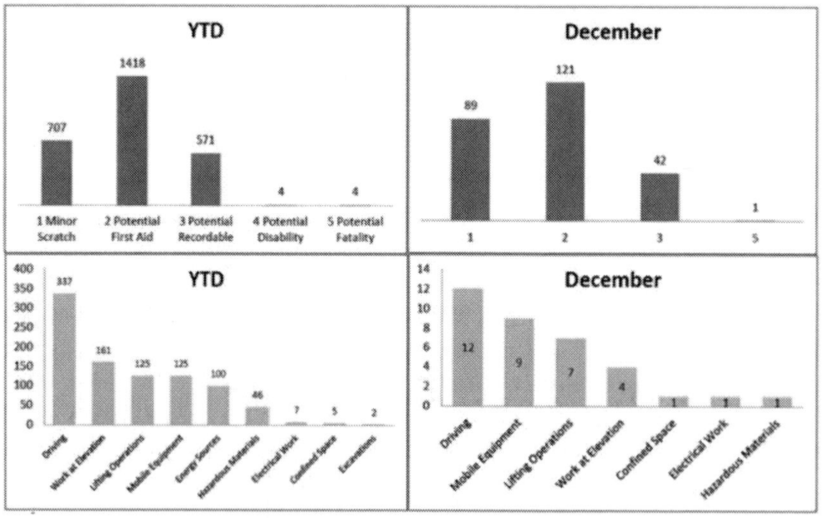

Figure 6. Reporting Snapshot

equipment, falls, confined spaces, explosions and fires, and ergonomic hazards.

In this great game of dominoes we want to be proactive and address those deficiencies before they become causes of incidents. We want to have an abundance of leading indicators, such as ROCs, audits, listening tours, effective training, pre-task safety plans and risk assessments, rather than lagging indicators, such as property damages, injuries, fluid releases, MVIs, etc. The more leading indicators we have the fewer incidents we'll have. There is a correlation between near-miss reporting and total incident ratio to Total Recordable Incident Rate (TRIR). Basically, the more near misses that are reported, the lower the TRIR will be. The more proactive we are, the less reactive we'll have to be. And when we are not being reactive, we can focus on the things that really matter, such as our people and the quality of the work that we perform. This proactive approach allows us to be the best we can be… not only locally, but throughout the industry… but more importantly in our lives. Now let's get out there and win at this game of dominoes!

Improving Safety and Health in Construction by Putting Research into Practice

Babak Memarian, Ph.D.
Linda M. Goldenhar, Ph.D.
Eileen Betit, B.A.
CPWR – The Center for Construction Research and Training
Silver Spring, MD

Introduction

CPWR – The Center for Construction Research and Training is a nonprofit organization that currently serves as the National Construction Center for the National Institute for Occupational Safety and Health (NIOSH). In cooperation with key federal and construction industry partners nationwide, CPWR's mission is to reduce occupational injuries, illnesses, and fatalities in the construction industry through research, training, and service programs. CPWR's research objective is to:

- Identify current and emerging hazards facing construction workers on the jobsite and conduct research to understand the safety and health implications of those hazards.
- Develop evidence-informed technologies and protective measures to prevent occupational injuries and illnesses.
- Publicize research findings and also translate them into practical tools and resources the industry can use to protect workers, also called Research to Practice (r2p).

To facilitate this last objective, researchers at CPWR, in close collaboration with academics and industry stakeholders, develop r2p resources including, but not limited to, informational guides and brochures, hazard-specific websites and online resources, toolkits, training programs, toolbox talks, and hazard alert cards. This paper will:

- Introduce some of our r2p resources currently available to stakeholders to advance safety and health in the construction industry.
- Describe the Construction Solutions Program, and how to navigate the solutions website.
- Describe our current r2p efforts to aid the industry in improving construction jobsite safety climate.

Building an Infrastructure to Advance r2p in Construction

Over the last twenty years, conceptual frameworks have been developed and operationalized and new strategies introduced to move research findings from peer-reviewed journals into practice by those the research is intended to benefit: research to practice (r2p) (Chambers, 2012). The National Academies' (Academies), in its 2008 review of the NIOSH Construction Research Program, emphasized the need to "keep the worker and contractor in mind as the ultimate destination" for r2p efforts, and identified two critical research questions to address: (1) "How can the program get vital information to the worker 'in the trench' or 'on the steel?'" and (2) "How does the program persuade contractors and workers to effectively use the interventions that are developed through research?" (Committee to Review the NIOSH Construction Research Program, National Research Council, 2008). The Academies' findings recognized that, while significant research exists on effective, evidence-informed safety and health interventions to reduce or eliminate many construction hazards and related injuries, illnesses and deaths, adoption by construction contractors and workers has been slow. In 2010, at the Academies' recommendation, CPWR began to develop a safety and health r2p infrastructure for the construction industry.

Integrating r2p into Research

Building a safety and health r2p infrastructure for construction is particularly challenging, given the decentralized nature of the industry, non-fixed worksites, prevalence of small, often isolated and under-resourced contractors, and a diverse workforce. The elements of the following three theoretical frameworks and models have influenced the approaches we are currently using to address these challenges:

- Wandersman et al.'s (2008) Interactive Systems Framework for Dissemination and Implementation's focus on the r2p process, including: the *synthesis and translation* of research findings into products (information and solutions) that target audiences can understand and act on; use of a variety of *dissemination* strategies to raise awareness and support use of research findings by target audiences; and *implementation,* the point where target audiences – construction contractors and workers – apply the research findings on jobsites.
- Kreuter and Bernhardt's (2009) findings, supporting the need to build an infrastructure for r2p, and the value of using private industry marketing and distribution systems to increase the use of research findings by end-users.
- The Institute for Work and Health's Knowledge Transfer and Exchange model's emphasis on the importance of involving "knowledge users" at both ends of the research spectrum, communicating research findings (push) and identifying research questions (pull) (Saunders, 2015, CPWR r2p Seminar; Reardon et al., 2006).

We are applying these elements to address the challenges listed above and to build a safety and health r2p infrastructure for developing tools to increase r2p capacity; creating and evaluating dissemination strategies and translational products; and expanding our reach and impact by establishing partnerships, networks, and communities of practice.

Building r2p Capacity

To build r2p capacity, we are developing tools researchers can use to integrate r2p throughout their research process. The following are selected examples developed to date:

- *CPWR's Research to Practice Roadmap* was developed to help researchers consider r2p activities from the inception of their research project, rather than waiting until the research is completed. This tool prompts researchers to consider key questions about the target audience(s) that will use their research findings, the best dissemination strategies and communications products to use, and the partners whose help will be needed for successful dissemination and implementation. This tool is now used by our research consortium members for their CPWR-funded projects, and training sessions on using the tool have been conducted with researchers at NIOSH and other organizations.
- To encourage the establishment of r2p partnerships with industry stakeholders, we developed the *Construction Research to Practice Partnership Toolkit.* The toolkit is designed as a resource for a range of audiences, from researchers and stakeholders interested in establishing a new r2p partnership to those interested in strengthening an existing one. It includes partnership best practices, lessons learned from our model partnerships, and other successful collaborations, and ways partnerships can facilitate research to practice. Housed in the partnership section of our website, this toolkit is a living document that we are updating as new lessons are learned and related resources developed.
- Building on Kreuters and Bernhard's (2009) findings and requests from our research consortium to learn more about how social marketing could be used to advance the adoption of their construction safety and health research findings, we developed a *Social Marketing Tool Kit for the Construction Industry* (Figure 1 in the Appendix at the end of the paper). This toolkit introduces researchers to social marketing principles, the steps involved in developing and implementing a social marketing program, and how to apply social marketing principles to research dissemination efforts.

These and other r2p tools and resources are housed in our online r2p Library (http://www.cpwr.com/research/research-practice-library).

<u>Developing and Testing Dissemination Strategies and Translational Products</u>

While the Internet, mobile devices, and other technological advances offer new platforms and mechanisms to share research findings, they also give people many more options for receiving and blocking information. To adapt, we have moved from finding the *best* approach to *finding the best mix of approaches* for responding to our target audiences' ever-changing preferences. The following are examples of our current dissemination strategies and translational products:

- *Websites*: Provide ready access to research findings and related translational products. Our family of websites includes ones that focus in depth on a particular hazard, group of related hazards, or safety and health topics, such as Stop Construction Falls (www.stopconstructionfalls.com), Choose Hand Safety.org (www.choosehandsafety.org), Nail Gun Safety: The Facts (www.nailgunfacts.org), Safety Climate Assessment Tool (www.safetyclimateassessment.org), and Work Safely with Silica (www.silica-safe.org) (Figure 2 in the Appendix). The silica website, for example, was designed with contractor and worker input as a one-stop source for the latest research on silica, regulatory requirements, and ways to protect workers from silica exposure. It includes a unique, free online tool contractors can use to develop the silica exposure control plan required under the new OSHA silica standard. We also have websites that provide end-users with access to information on a broader array of hazards, research findings, interventions, and training materials, such as CPWR.com, eLCOSH – The Electronic Library of Construction Occupational Safety and Health (www.elcosh.org), and CPWR Construction Solutions

(www.cpwrconstructionsolutions.org) and related Return on Investment calculator (www.safecalc.org). Collectively, these websites recorded more than 85,000 sessions on average each month during 2016.
- Our e-newsletter, *CPWR Update*, and social media outlets (e.g., LinkedIn, Facebook, and Twitter) are helping us rapidly reach a growing audience of researchers, contractors, trainers, government representatives, and other stakeholders with the latest research findings, new products and events, and safety and health reminders. The e-newsletter reached roughly 3700 direct subscribers in 2016, and our exposure through social media, as measured by tweets, retweets, impressions and likes, has increased significantly. For example, in 2016 our tweets were retweeted 685 times, for an increase of 325%, and the number of times our tweets were seen by others (measured by impressions) increased to 356,901, or by 340%.
- Our monthly webinars on timely safety and health topics, which can be viewed live or on-demand, are giving researchers a new platform to share their findings and receive direct feedback from stakeholders as their research progresses or when interventions are ready for use on jobsites. In 2016, more than 2300 individuals and organizations viewed the webinars, and some views led industry stakeholders to ask researchers to present their findings at conferences and meetings. For example, the silica webinar led an international equipment manufacturer to invite us to cohost a meeting on silica and develop a paper on the new silica standard, which was released at the 2017 World of Concrete event.
- Labor and management stakeholders' requests for training and educational programs on research findings (r2p) and emerging hazards (practice to research or p2r), such as the *Radiofrequency (RF) Radiation Awareness Program for the Construction Industry* and the *Foundations for Safety Leadership Course*, connect the dots between safety and health research and issues of immediate concern to contractors and workers. Integrating safety and health into training and educational programs is helping to reinforce best practices, respond to industry concerns, and ensure that research findings are reaching contractors and workers, who are the primary targets for safety and health research.

In addition to these dissemination strategies, we are also developing and testing traditional and new types of translational materials, including videos, infographics, hazard alert cards, and toolbox talks, many of which are available in both English and Spanish. We are finding that use of infographics, videos, and other visuals in our social media posts increases the likelihood that others will 'like' or 'retweet' the content to a broader audience. Similarly, hazard alert cards and toolbox talks, both of which focus on specific hazards and steps to reduce risks and prevent injuries and illnesses, provide other ways to share research findings in formats accessible to contractors and workers. Last year alone, we distributed more than 68,000 hard copies of the hazard alert cards, and thousands more were downloaded from the CPWR website. Our toolbox talk series recorded more than 260,000 downloads, and surveys of contractors represented by our r2p partnerships found that the vast majority (84%) consider toolbox talks to have a "somewhat" to "very" positive effect on a project's safety performance.

Establishing Networks and Partnerships

Building capacity and developing dissemination strategies and translational products are only parts of the formula required for effective r2p. Another equally critical element is generating the buy-in and active engagement from stakeholders.

To achieve this third element, we are establishing r2p partnerships to support both *push* and *pull* activities and ensure the sustainability of our r2p infrastructure. Examples include:

- The Masonry r2p Partnership that identified hand safety as an important issue, and was instrumental in developing the *ChooseHandSafety.org* website, and disseminating information on it to workers, contractors, and safety and health trainers. A baseline and follow-up survey of workers and contractors documented an overall increase in hand injury-related hazard awareness and the use of interventions, such as gloves, to prevent hand injuries and skin disorders.
- The Roofing r2p Partnership identified RF radiation as an emerging hazard. Their initial work on this topic led to a multi-trade, labor-management RF Radiation Work Group, and a new educational program, the *Radiofrequency (RF) Radiation Awareness Program for the Construction Industry*.

The information and products included in the *ChooseHandSafety.org* website and RF Awareness Program, including toolbox talks, hazard alert cards, and videos, are based on research findings. We are using the lessons learned from our work with these partnerships to shape our dissemination strategies, and encourage other industry segments to form r2p partnerships.

We are working to connect and better use networks of individuals and organizations that are interested in improving construction safety and health. One example is our Trainers & Researchers United Network, or TRU-Net, through which we are engaging thousands of trainers involved in CPWR's training program in our r2p activities and soliciting their input on emerging health and safety issues (p2r). We successfully used this network for a research project on noise. Hundreds of trainers provided input on awareness of noise hazards and solutions, training provided, and barriers to use of interventions. In addition, roughly 50 trainers served as trainer-researchers, fielding a noise survey to more than 4000 workers.

Finally, we are also exploring the establishment of communities of practice, which bring together researchers and industry stakeholders with a shared interest in addressing a specific hazard, as another mechanism to engage researchers and stakeholders in r2p activities. Our first community of practice, which includes researchers, insurance industry representatives, and construction stakeholders, is on ergonomics with a focus on identifying ways to reduce injuries due to manual materials handling.

Moving Forward

We are continuing to develop new r2p tools to aid researchers in their dissemination efforts, creating translational products to share research findings, such as the Roofing r2p Partnership's work on the development and pilot testing of the Safety Climate Assessment Tool (S-CAT) for Small Contractors, and identifying new dissemination strategies to reach target audiences.

We are also in the process of refining and co-branding our evidence-based series of English and Spanish toolbox talks with NIOSH. Following up on exploratory research on the use of this toolbox talk series by small employers and Spanish-speaking construction workers, we are planning to test how microgames, which are a new type of dissemination mechanism, can be incorporated into toolbox talks to reinforce safety practices. Microgames are two- to five-minute games played on mobile devices that can be accessed with a QR Code or URL printed on a toolbox talk. These games are designed to engage and encourage players to play multiple times, providing the opportunity to reinforce a good safety or health practice.

In addition, we are stepping up efforts to expand our reach through the development of a safety and health network, and to find ways to better measure the impact of our r2p activities.

CPWR Construction Solutions Program

As noted earlier, research has shown that, while evidence-based controls or "solutions" exist that could be used to reduce or eliminate many construction hazards, contractors and workers are often slow to adopt them for use on the jobsite (Glasgow et al., 2006 and Chen et al., 2010). CPWR defines a "solution" as an evidence-based practice or technology that, if implemented correctly, will reduce workers' exposure to occupational safety and health risks. The CPWR Solutions Program is another mechanism designed to address this by providing contractors and workers with information that not only encourages the use of a solution but also provides them with the data to identify and evaluate the "solution" likely to be most effective, given a particular hazard and situation.

The Construction Solutions database organizes and sorts evidence-based safety and health solutions by line of work (trade), task, and hazard, which are given priority according to the hierarchy of controls. Specifically, we: (1) identify the needs of stakeholders in the construction industry, (2) systematically summarize research findings in the field of construction safety and health in a format that is understandable to practitioners, (3) provide direction on how to translate research findings to applied solutions, and (4) continuously evaluate and enhance the content quality. This information is digitized and incorporated into an online database. Since 2008, the Construction Solutions database has grown steadily, with 268 solutions to date.

Construction Solutions is unique in that it is designed to be fully applied. It provides specific links to sources where "solutions" can be purchased, and gives guidance on user implementation. The Construction Solutions Program uses feedback from the industry to continuously improve content.

Solutions Development Process

The Construction Solutions Database uses a standardized format to compile information on solutions, and provides it to the industry in a way that allows contractors, workers, and other stakeholders to make informed decisions on the best methods for minimizing trade- and task-specific hazards. Most notably, the Construction Solutions Database is designed to be accessible to those with minimal computer experience and no professional training in occupational safety and health.

At the core of the Construction Solutions process is a system of selection, development, and peer review. Before a solution is chosen for development, it is assessed by the project team to determine if it is likely to be sufficiently relevant and effective. Development is then assigned to a researcher, known as the "author." Once completed, the solution is assigned to at least two external reviewers. After incorporating the reviewers' comments, the solution is posted to the website and becomes publicly available to all users.

Construction Solutions Website

Each solution included in the Construction Solutions website is linked to at least one trade, one task and one hazard. As shown in Figure 3 (in the Appendix), all three variables are required to initiate a search. Once all three variables are chosen, the user will be shown a list of all related solutions in the database and can then make a selection. In general, each Construction Solution provides a description, as well as information on the hazard(s) involved, hazard analysis, risk(s) involved, how to reduce risk(s), recommended solutions (tools, PPE, etc.) commercially available to mitigate the hazard, and manufacturers. Users are also provided with links to manufacturers' websites for further information. Figure 4 (see Appendix) shows a screenshot of the CPWR Construction Solutions website.

Improving Construction Jobsite Safety Climate

The term safety climate first appeared in the academic literature in 1980, when Zohar measured workers' perceptions of various aspects of job safety in manufacturing organizations with high- and low-accident rates, and found that the data could be used to proactively assess an organization's effectiveness in identifying and remediating work-related hazards, thereby reducing or preventing work-related ill health and injury (Zohar, 1980). Since then, others have examined the usefulness and accuracy of measuring safety climate, and whether such data can be used to understand or predict workplace safety and health performance across numerous industries. While the construction industry is a relative latecomer to the safety climate discussion (Schwatka et al, 2016), it is also a high-risk industry with the potential for catastrophic events. In 2012, CPWR began carrying out a research and research-to-practice agenda to learn about and begin to address the topic of safety climate in construction.

As a first step, we reviewed the academic and trade literature to gain a better understanding of the status of safety climate/safety culture (SC) in construction (Hecker and Goldenhar, 2012), and found a lack of agreement on three issues: (1) the definitions of safety culture and safety climate, (2) the indicators of safety culture and safety climate, and (3) ways to target the indicators to improve overall safety culture and safety climate. In an effort to come to some agreement on these three issues, we invited 70 construction industry stakeholders to attend a two-day workshop, entitled *Safety Climate and Safety Culture in Construction: Bridging the Gap from Research to Construction.*

2013: Stakeholder Workshop

Working in small groups first, and then coming together to vote as a group, attendees made great progress addressing all three issues. The group agreed on definitions of organizational safety culture, organizational safety climate and project/jobsite safety climate. The third definition is most relevant to this paper and states:

> Project/jobsite safety climate is perceptions of occupational safety and health on a particular construction project at a given point in time. It is a product of the multiple safety climates from the different organizations involved in the project including the project owner, construction manager/general contractor, and subcontractors. Project safety climate may be heavily influenced by local conditions such as project delivery method, schedule and planning, and incentives.

In the workshop attendees also identified what they perceived to be the eight leading indicators of safety climate (see Figure 5 in the Appendix).

For each indicator, we worked together to write rubric text a company could use to determine where they fell on a safety maturity scale, from inattentive to exemplary. Figure 6 (in the Appendix) contains text for the indicator demonstrating management commitment.

For the final workshop activity, the group brainstormed ideas that companies could use to improve each of the eight leading indicators and move them towards becoming exemplary on the safety climate maturity scale. All of the information generated at the workshop is presented in a final workshop report (Gillen et. al, 2013), and has been used as a starting point for our safety climate research to practice activities.

2014: r2p: Create a Practical Workbook

Our stakeholders told us that, while the workshop and report were a great start, companies needed practical tools they could use to better understand and improve their own jobsite safety climate. To address their request, we refined the workshop findings, solicited critical feedback from numerous industry stakeholders, and created a practical workbook called *Strengthening Jobsite Safety Climate by Using and Improving Leading Indicators* (Figure 7 in the Appendix). In it, there is a worksheet for each of the eight indicators that includes a five-level rubric scale for conducting a high-level safety climate maturity assessment and a list of ideas for strengthening that safety climate indicator. Over 5000 hard copies of the workbook have been distributed, and many more have been downloaded from CPWR's websites.

The feedback from our stakeholders was that, while they thought the workbook was very helpful, they would also like a way to measure their company's safety maturity in more detail than the high-level rubric.

2015: The Safety Climate Assessment Tool (S-CAT) and Updated Workbook

We were familiar with already developed safety climate surveys that ask respondents to answer questions, using a 1-5, strongly disagree to strongly agree scale. Rather than simply repeating what others had done, we partnered with researchers at Washington State University Vancouver to deconstruct each indicator's high-level rubric into its essential elements, develop rubric text for each element, and combine the resulting scales into a new 36-item Safety Climate Assessment Tool or S-CAT.

The S-CAT is a free tool available to anyone who wants tailored and actionable information to improve the safety of every worker at every jobsite. The pilot test of the S-CAT, with 350 individuals from numerous construction companies, showed high reliability and validity, with S-CAT scores predicting company injury rates.

We solicited stakeholder feedback on the workbook and used it, plus the new S-CAT, to create a second edition called *Worksheets and a Rating Tool to help you Strengthen Jobsite Safety Climate* (See Figure 8 in the Appendix).

2016: S-CAT Website

Our most recent r2p activity was to create and launch the S-CAT website (www.safetyclimateassessment.org) (Figure 9 in the Appendix). Once on the site, visitors can learn about construction safety climate, the eight leading indicators, improvement ideas, and most importantly, take the S-CAT. An individual can complete the S-CAT on their own or a company can ask their employees to answer the 36 questions. Either way, the information gathered is anonymous. A personalized feedback report is generated that contains a safety climate score for each indicator, as well as benchmarking and comparative information indicating current areas of success and ideas for making improvements. Companies can have their employees take the S-CAT periodically to track their progress at improving their jobsite safety climate. There is no charge for completing the S-CAT or receiving a report.

2017 and Beyond

We have three additional S-CAT-related activities planned. In terms of research, we will be conducting further analysis of the S-CAT database that, as of January 2017, had over 550 entries. We have developed and are pilot testing a *Workbook* and S-CAT specifically designed for smaller contractors, and will be translating the S-CAT into Spanish.

Bibliography

Chambers, D. 2012. *Foreword.* In R. C. Brownson, G. A. Colditz & E. K. Proctor, eds. *Dissemination and Implementation Research in Health: Translating Science to Practice.* New York: Oxford Press.

Chen, P.G., Diaz, N., Lucas, G., and Rosenthal, M.S. 2010. "Dissemination of results in community-based participatory research." *Am J Prev Med*, 39(4):372-8.

Committee to Review the NIOSH Construction Research Program, National Research Council. 2008. Construction Research at NIOSH: Reviews of Research Programs of the National Institute for Occupational Safety and Health, Section 4 - Future Research and Program Improvement.

Gillen, M., Goldenhar, L.M., Hecker, S., and Schneider, S. 2013. *Safety Culture and Safety Climate in Construction: Bridging the Gap between Research and Practice.* Workshop Report

Glasgow, R.E., Klesges, L.M., Dzewaltowski, D.A., Estabrooks, P.A., and Vogt, T.M. 2006. "Evaluating the impact of health promotion programs: using the RE-AIM framework to form summary measures for decision making involving complex issues." *Health Educ Res*, 21(5):688-94.

Hecker, S. and Goldenhar, L.M. 2014. *Safety Culture and Safety Climate in Construction: A Review of Current Literature.* CPWR Report. (Accessed January 2017) (http://www.cpwr.com/sites/default/files/publications/HeckerSafetyClimateSafetyCultureKF_2.pdf.); (accessed January 2017) (http://www.cpwr.com/safety-culture/workshop-safety-culture-and-climate-construction).

Kreuter, M.W. and Bernhardt, J.M. 2009. "Reframing the Dissemination Challenge: A Marketing and Distribution Perspective." *American Journal of Public Health*, 99(12), 2123–2127. http://www.ncbi.nlm.nih.gov/pmc/articles/PMC2775768/).

Reardon R., Lavis J., Gibson J. 2006. *From Research to Practice: A Knowledge Transfer Planning Guide.* (https://www.iwh.on.ca/from-research-to-practice.)

Saunders, R. 2015. "The approach to KTE at the Institute for Work & Health." *Presentation at the CPWR r2p Seminar & Partnership Workshop.* June, 2015, Silver Spring, MD.

Schwatka, N.V., Hecker, S., Goldenhar, L.M. 2016. "Defining and Measuring Safety Climate: A Review of the Construction Industry Literature." *Ann. Occup. Hyg,* Vol. 60, No. 5, 537–55.

Wandersman, A., Duffy, J., Flaspohler, P., Noonan, R., Lubell, K., Stillman, L., Blachman, M., Dunville, R., and Saul, J. 2008. "Bridging the gap between prevention research and practice: the interactive systems framework for dissemination and implementation." *American Journal of Community Psychology, 41*(3-4):171-81. doi: 10.1007/s10464-008-9174-z. (http://www.ncbi.nlm.nih.gov/pubmed/18302018.)

Zohar D. 1980. "Safety climate in industrial organizations: theoretical and applied implications." *J Appl Psychol;* 65: 96–102.

Appendix

Figure 1: Social Marketing Toolkit for the Construction Industry

Figure 2: Silica-Safe Website

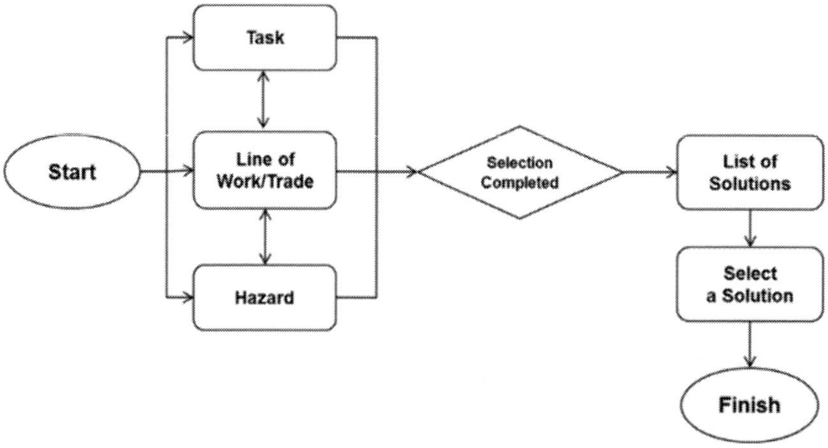

Figure 3: CPWR Construction Solutions Database

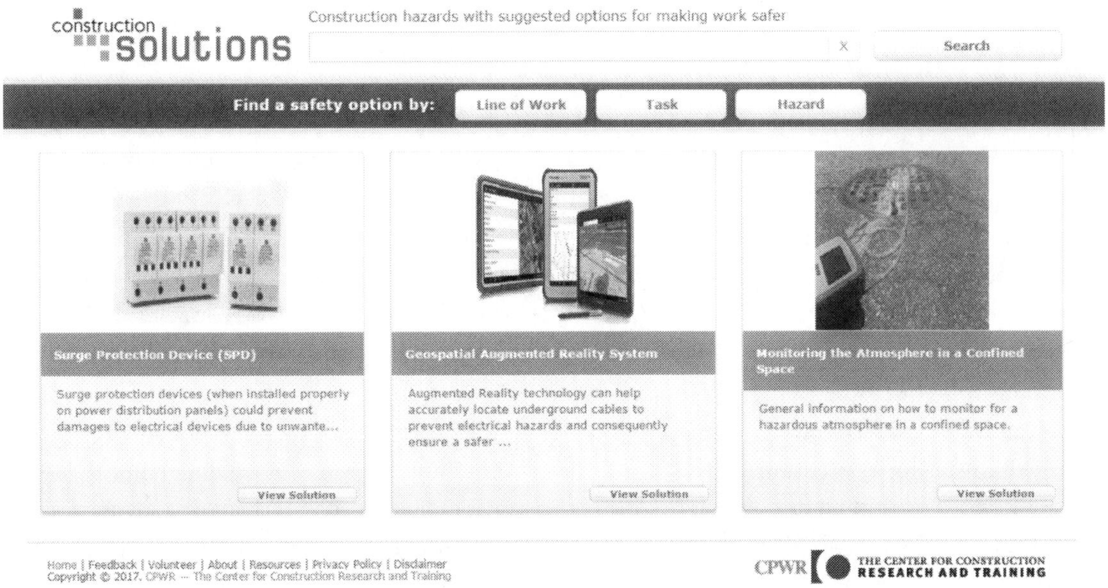

Figure 4: CPWR Construction Solutions Website

1. Demonstrating management commitment
2. Aligning and integrating safety as a value
3. Ensuring accountability at all levels
4. Improving site-safety leadership
5. Empowering and involving workers
6. Improving communication
7. Training at all levels
8. Encouraging owner/client involvement

Figure 5: Eight Leading Indicators of Safety Climate

Inattentive	*Reactive*	*Compliant*	*Proactive*	*Exemplary*
Management is mostly not visible onsite. They don't participate in safety audits. When they are onsite, they are oblivious to safety and don't follow safety rules or role-model good safety behavior.	Management gets involved after injury and may suspend or fire employees who get injured. Management enforces safety rules only after an incident or when audit results are negative.	Managers conform to OSHA regulations and participate in safety audits. Safety compliance based on owner or regulatory directives.	Management initiates and actively engages in participation with safety audits. Management meets with craft workers to ask for advice regarding hazard reduction. Management conducts spontaneous site visits and rewards safe behavior. Leadership participates in safety program development and provides safety resources.	Management addresses safety in every meeting and is constantly working to improve conditions and reduce hazards. External auditing of top management, management involvement, and analyzing safety trends and formulating corrective actions

Figure 6: Rubric Text for Indicator Demonstrating Management Commitment

Figure 7: Safety Climate Workbook (Edition 1)

Figure 8: Safety Climate Workbook (Edition 2)

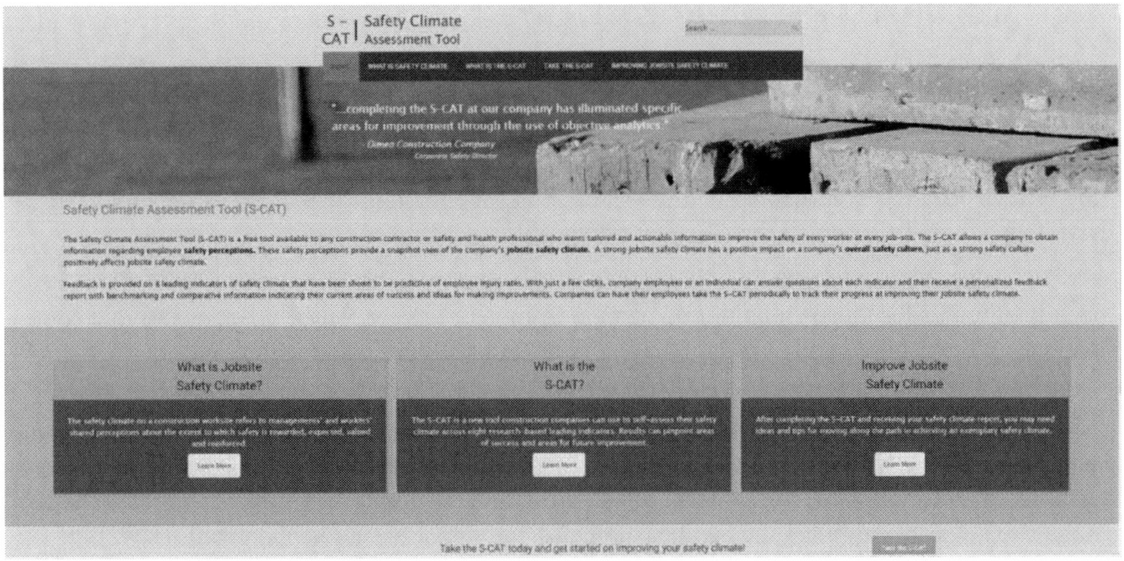

Figure 9: Safety Climate Assessment Tool (S-CAT) Website

And The Survey Says…

John M. Eliszewski, CSP
Technical Safety Specialist
W. W. Grainger, Inc.
Janesville, WI

Sally J. Smart, CSP
Technical Safety Specialist
W.W. Grainger, Inc.
Janesville, WI

Introduction

This is a discussion about the Occupational Safety and Health Administration's (OSHA's) annual Top 10 List of most frequently cited safety and health violations and its perceived value by fellow American Society of Safety Engineers (ASSE) peers.

Every fall the Occupational Safety and Health Administration (OHSA) releases a list of the 10 most frequently cited safety and health violations for the fiscal year (FY). It is compiled from more than 30,000 workplace inspections by Federal OSHA. The list rarely changes from year to year. In fact, the top three cited standards in fiscal year 2016 (October 1, 2015–September 30, 2016) have been the exact same top three since 2012, and in slightly different orders these three have topped the list for more than 10 years running.

However, the total number of inspections, the total number of violations cited and the penalty amounts for those violations do change from year to year. These changes provide some insights on how aggressively OSHA is citing their standards and what is their enforcement emphasis "top of mind".

OSHA publishes the Top 10 List to help employers improve the safety and health of their workers by reviewing their own operations for similar issues and taking steps to find and fix these hazards. In an October 2016 blog post, OSHA wrote: "If all employers simply corrected the top 10 hazards, we are confident the number of deaths, amputations and hospitalizations would drastically decline." (Galassi, Thomas, "Top 10 OSHA Citations of 2016: A Starting Point for Workplace Safety, U.S. Department of Labor Blog, https://blog.dol.gov/2016/10/18/top-10-osha-citations-of-2016-a-starting-point-for-workplace-safety/)

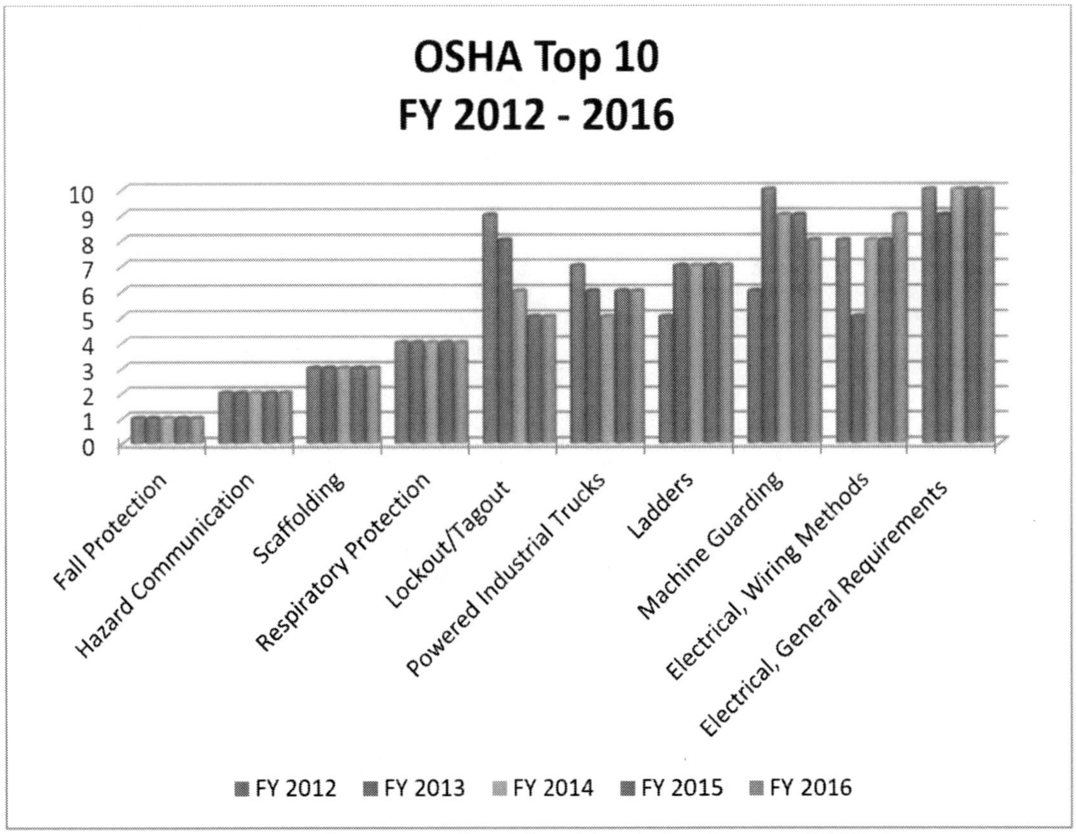

Source: www.osha.gov - Produced from historical and current OSHA data

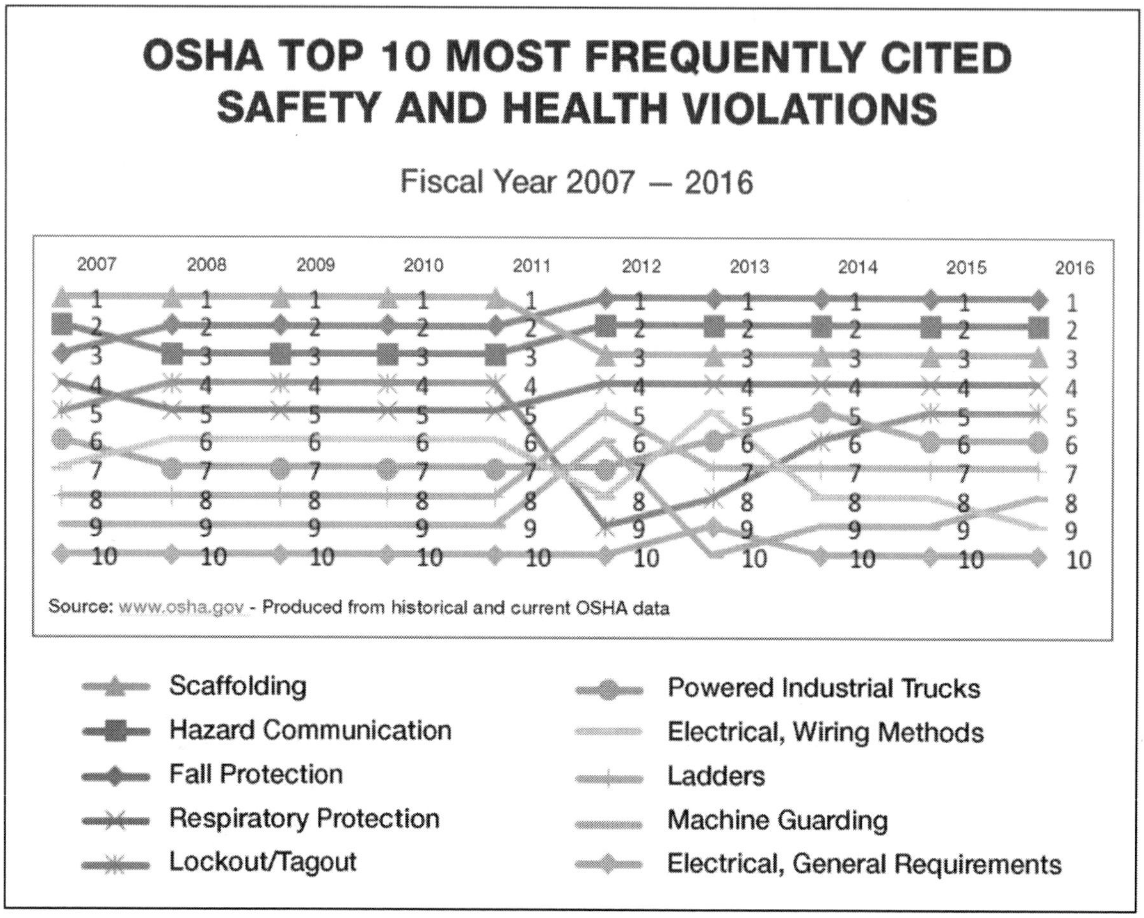

Even though the list is similar from year to year, there are things that change, such as the total number of inspections, the total number of violations cited, and the penalty amounts for those violations. These changes provide some insights on how aggressively OSHA is citing its standards and what its enforcement emphasis is. But what broader insights does the survey offer? Do safety professionals really use it? And are there ways that it could be improved? A recent survey sponsored by the American Society of Safety Engineers (ASSE) asked these and other questions?

Approach

In partnership with the ASSE, a pool of 13,553 members was asked for input on OSHA's annual Top 10 List. Questions addressed a range of issues, including the list's pertinence and perceived value. The response rate was 3.5% (477).

Surveys were sent to members representing the following industries:

- Agriculture/Forestry/Food
- Chemical/Allied Products
- Construction
- Governmental/Public Sector
- Healthcare
- Manufacturing
- Oil, Gas, Petrochemical
- Transportation
- Utilities

The majority of responders were either safety directors/managers or safety specialists/engineers who have been both in the safety profession and ASSE members for more than 15 years.

➤ What is your job title?

ANSWERS: **475** / 100% SKIPS: **2** / <1%

	COUNT	%
04 Director/Manager/Department Head/Chief—Safety	180	38
13 Safety Specialist Engineer	113	24
08 Branch/District/Division/Area/Regional Manager—Safety	56	12

➤ How many years have you been in the safety profession?

ANSWERS: **474** / 99% SKIPS: **3** / 1%

	COUNT	%
More than 15 years	313	66
6-10 years	62	13
11-15 years	59	12

➤ How long have you been a member of ASSE?

ANSWERS: **475** / 100% SKIPS: **2** / <1%

	COUNT	%
More than 15 years	180	38
6-10 years	79	17
11-15 years	66	14

Many industries were represented: 26% of respondents were in manufacturing, 25% were in construction, 10% were in oil, gas and petrochemicals, and 17% were in either utilities or government/public sector.

➤ Which of the following best describes your company's industry?
Please select one response.

ANSWERS: **477** / 100%

	0% — 13.5% — 27%	COUNT	%
A09 Manufacturing		125	26
A05 Construction		118	25
A12 Oil, Gas, Petrochemical		47	10
A06 Governmental/Public Sector		45	9
A17 Utilities		36	8

Findings

The majority of respondents were covered by federal OSHA (66%). The states of California, Minnesota and Washington were well represented by those respondents covered by a state plan.

➤ First, is your company covered by Federal OSHA or a state plan?

ANSWERS: **477** / 100%

	0% — 33% — 66%	COUNT	%
Federal OSHA		313	66
A state plan		164	34

➤ You said your company is covered by a state plan. From which state is your company covered?

ANSWERS: **164** / 34% SKIPS: **313** / 66%

	0% — 7.5% — 15%	COUNT	%
California		24	15
Minnesota		13	8
Washington		12	7

An overwhelming majority of respondents (98%) were either somewhat or very familiar with the annual Top 10 list and found it either somewhat or very valuable (89%). But only 54% watch for the list's annual release. This may be because, as one response suggested, the list does not have a recurring release date.

➤ How familiar are you with OSHA's annual Top 10 citations list?

ANSWERS: **477** / 100%

	0% 32% 64%	COUNT	%
Very familiar		303	64
Somewhat familiar		164	34
Somewhat unfamiliar		5	1
Not at all familiar		5	1

➤ How valuable do you find OSHA's Top 10 citations list to be?

ANSWERS: **477** / 100%

	0% 27.5% 55%	COUNT	%
Somewhat valuable		259	54
Very valuable		169	35
Not very valuable		43	9
Not at all valuable		6	1

➤ Do you watch for OSHA's annual release of the Top 10 citations?

ANSWERS: **477** / 100%

	0% 27% 54%	COUNT	%
Yes		256	54
No		221	46

Most respondents learned of the list's most recent release from the American Society of Safety Engineers (31%), OSHA's website (26%), or trade publications (20%).

➤ **How do you become aware of OSHA's most recent Top 10 citations list?** Please select all that apply.

ANSWERS: **477** / 100%

	COUNT	%
ASSE	329	31
OSHA's website	276	26
Trade publication	213	20

Surprisingly, only 26% of respondents recognized that the list reflects inspections conducted by Federal OSHA. The majority (49%) thought that the list uses data collected from both Federal and State OSHA, which is not the case. The list is compiled from over 30,000 inspections of workplaces by Federal OSHA staff.

➤ **Which of the following sources does OSHA use to develop the Top 10 citations list?**

ANSWERS: **477** / 100%

	COUNT	%
Both Federal and State inspections	236	49
Federal inspections	122	26
Don't know	95	20
Other option	19	4
State inspection	5	1

Respondents had a good sense of which standards have historically appeared on the list, correctly identifying nine of the ten. The most common answers, Fall Protection and Hazard Communication, have been the number 1 and 2 standards for the past five years. Excavation Requirements and Permit-Required Confined Spaces were mistakenly identified by a number of respondents.

➤ To the best of your knowledge, what OSHA standards have historically been cited on the Top 10 citations list?

ANSWERS: **477** / 100%

	COUNT	%
Fall Protection	438	11
Hazard Communications	419	11
Lockout/Tagout	411	11
Machine Guarding	351	9
Electrical, General Requirements	308	8
Scaffolding	288	7
Powered Industrial Trucks	251	6
Respiratory Protection	240	6
Ladders	197	5
Excavation Requirements	183	5
Electrical, Wiring	183	5
Permit-required Confined Spaces	180	5

The largest numbers of respondents were currently dealing with Fall Protection; Lockout/Tagout; Hazard Communication; Ladders; and Electrical, General Requirements. Before FY 2016, the top issues were Fall Protection; Hazard Communication; Machine Guarding; Lockout/Tagout; and Electrical, General Requirements. The issues identified as never being a problem included Scaffolding; Electrical, Wiring; and Respiratory Protection, while some respondents said that none of the Top 10 issues have ever been a problem in their facilities.

➤ **Within your facilities, which of the 2016 OSHA Top 10 citations, if any, is an issue you are dealing with currently? Which have you dealt with prior to 2016?**

ANSWERS: **477** / 100%

	TOPIC IS FY 2016 ISSUE	TOPIC WAS AN ISSUE PRIOR TO FY 2016	TOPIC HAS NEVER BEEN AN ISSUE
Fall Protection, 1926.501	190	227	131
Hazard Communications, 1910.1200	157	221	158
Scaffolding, 1926.451	98	117	304
Respiratory Protection, 1910.134	122	182	213
Lockout/Tagout, 1910.147	189	212	140
Powered Industrial Trucks, 1910.178	136	198	192
Ladders, 1926.1053	142	195	196
Machine Guarding, 1910.212	123	216	187
Electrical, Wiring, 1910.305	108	173	235
Electrical, General Requirements, 1910.303	139	202	189
None of the Top 10 citations are an issue for my facility	75	120	303

Respondents somewhat agreed that the Top 10 List does not change from year to year (48%), that it reflects trends in emerging safety issues (37%), and that it helps prioritize areas of focus for safety programs (40%). Respondents strongly agreed that a sector-specific list (like one for general industry, construction, maritime or agriculture) would be of greater value than the current Top 10 list (52%).

➤ How much do you agree or disagree with each of the following statements?

ANSWERS: **477** / 100%

	STRONGLY DISAGREE	SOMEWHAT DISAGREE	SOMEWHAT AGREE	STRONGLY AGREE
The OSHA Top 10 list reflects trends in emerging safety issues	61	118	175	123
A sector specific (e.g. General Industry, Construction, Maritime, Agriculture) Top 10 citations list would be of greater value than the current Top 10 citations list OSHA publishes	38	50	142	247
The OSHA Top 10 list does not change from year to year	36	148	231	62
The OSHA Top 10 list helps you prioritize areas of focus for your safety programs	66	75	190	146

The survey also asked three open-ended questions to get more insight into the perceived value of the list, how safety professionals use it, and what other sources they consult for safety information on their industries.

➤ How could OSHA make the Top 10 citations list more valuable for you?

Although 21% of respondents said that the list is fine or had no suggestions for improvement, others thought that it would be more valuable if it:

1. **Provided more details and specifics for each citation listed (20%)** — "More specifics on the exact problems."
2. **Offered solutions and training materials (19%)** — "Share successful compliance tools."
3. **Broke down the data by segments (construction versus general industry) or by industry (construction, healthcare, etc.) (18%)** — "Top 10 citations by industry would be much more valuable and relevant."

➤ How do you use the information included within OSHA's annual Top 10 citations list?

1 Safety professionals use the list to enhance training (33%)

"Concentrate more on our safety training in the areas of concern."

2 Safety professionals use the list to assess and benchmark compliance (29%)

"Performs audits of our programs."

3 Safety professionals use the list to raise awareness at all levels of their organization (23%)

"Committee, employee and management awareness."

➤ What resources or tools do you utilize to gather information on your industry's safety record?

Those participating in the survey were also asked what tools they use to gather information on their industry's safety record. Most are using the OSHA website (24%) and the Bureau of Labor Statistics (BLS) (23%), with trade associations and trade publications in third place (19%).

"ASSE, OSHA and NSC."

"The Bureau of Labor Statistics."

"BLS and trade magazines."

Conclusion

And the survey says:

- OSHA's Top 10 List is an annual recognized tradition within the safety profession, however the "source" of the data is not understood amongst even experienced safety professionals. When releasing the annual Top 10 List, OSHA should consider clearly stating that the information is derived from federal inspection data only.
- The Top 10 List has not changed significantly for more than 15 years and may have lost some of its impact.
- The Top 10 List could be made more valuable by:
    - Providing more details and specifics for each citation listed

    - Offering solutions and training materials
    - Breaking down the list by segments (Construction versus General Industry) or by industry type (Construction, Healthcare, North American Industry Classification System [NAICS], etc.)
- The Top 10 List is used primarily to enhance training, assess and benchmark compliance, and as an awareness tool for all levels of organizations.
- Safety professionals are turning from traditional resources such as OSHA and BLS to other resources such as trade associations and publications for industry-specific information and support. This indicates that the trade associations are doing a better job of providing industry specific and relevant safety information to its members.

Year after year, thousands of the same on-the-job hazards are identified and noted in OSHA's Top 10 List. Any one of these hazards could result in a fatality or severe injury. How can we work together to move the needle on this? As the survey indicated, there is an appetite for solutions and training. Perhaps safety organizations such as OSHA, ASSE, the National Safety Council (NSC) and others could collaborate to develop and organize training workshops to address one or two of the top-cited standards. Tackling the most commonly cited hazards is a good place to begin creating a safety culture that helps employers keep their people safe.

Five Buckets of Safety: Essential Considerations for the New Safety Manager

Valerie A. Stakes, M.A., CHST, COSS
Safety & Health Consultant
Zenith Insurance
San Diego, CA

Introduction

As a safety professional, I work with a wide variety of organizations and often serve as a coach and mentor to new safety coordinators and managers. These include those who have recently chosen safety as a vocation or have been promoted from within an organization. Many have taken on the role as safety manager in addition to their day-to-day responsibilities. These may include, for example, human resources professionals, office managers, or project superintendents.

As a new coordinator or manager, it can be challenging to know exactly where to start, not only to learn about safety in general, but how to build a basic program. For example, one may have to locate or piece together program materials from a former manager. Furthermore, there may be no safety program at all. In addition, new managers may not be thoroughly familiar with the processes involved at their organization, let alone the safety regulations that apply

The goal of the following discussion will be to emphasize how new safety managers do not have to build everything from the ground up. Instead, they can take advantage of existing and easily available resources, and take practical steps toward developing and implementing an effective safety program.

While this discussion is geared for new managers, more experienced safety professionals can also benefit from the concepts of planning for safety, hazard identification and control, training, communication, and ensuring the effectiveness of their safety activities. These topics or "buckets" frame basic safety considerations for the new safety manager. By becoming a sound practitioner in each, the safety manager will be setting the stage for program success, as well as their future development and capabilities.

Bucket One: Plan for Safety

The first "essential consideration" bucket to fill in order to help a new safety manager get their safety program in order is *planning*. Here, a blueprint is needed that has both strategic and tactical characteristics. Let's highlight three of those characteristics.

Address What Is New

Start by taking stock of what's "new" in the organization. This will help assess and address safety priorities and determine where to focus one's efforts. For example:

- *People:* Are there **new employees** or added positions? Both would trigger the need for new hire orientations and safety training.
- *Processes:* Has the organization expanded to include **new activities** and processes that require program updates or training?
- *Equipment:* Has **new equipment** been added or others being replaced? In addition to providing training, there may be a need to establish **new procedures** and/or specific programs, such as lockout/tagout.
- *Incidents*: What **new incidents** have occurred, and are there trends that need to be addressed?
- *Pertinent regulations*: Does the safety program reflect all applicable regulations, and are there any **new regulations** of which we should be aware?

By recognizing what is 'new' in an organization, appropriate priorities are established, and the new safety manager spends time on more needed and meaningful activities.

Identify Who, What, and Where

Next, it's critical to identify the resources at the safety manager's disposal, and determine what's needed. For example, take a look at the materials already in place, such as written programs and training materials. Once a program is analyzed as it currently exists, areas for improvement can be identified. Two fundamental questions to consider include:

- Is the safety program truly relevant and applies to what we do? Determine if the program is too general and should be structured to be more focused in areas, such as hazards, controls, and regulatory mandates.
- In addition, does the program match, where needed, the cultural and language needs of our workforce?

Engage Employees

By engaging employees, the safety manager is able to find out not only what they do in terms of job/tasks but what safety needs may exist. This includes, of course, the training and resources necessary to do their jobs in a safe manner.

Also, these employees are a valuable resource when it comes to safety program development and execution. They serve in many respects as subject matter experts who can help with augmenting your knowledge about operational procedures, and assist with program development and training.

Bucket Two: Hazard Recognition and Controls

How hazards are recognized constitutes the second bucket to fill. This bucket also includes how controls are implemented. It is important to incorporate hazards that currently exist in the workplace, and those that may develop based on future changes, such as new equipment, processes, materials, or to existing procedures.

What Is It That We Do Again?

In order to identify the hazards associated with an organization's work processes and the controls that need to be in place to prevent injuries, it's critical to understand the work being done. This includes learning as much as possible about the:

- **Equipment**: What machinery and tools do employees use? It's not necessary to be a machine operator or use tools to understand how work is done or the hazards involved. Talk to supervisors and experienced employees about their job tasks, what they perceive as hazards, and how equipment is operated as a starting point.

  In addition, there are owner's manuals with guidance on safe operation and guarding requirements. Many manufacturers also create videos on the features of their equipment, including safe operating guidelines. These videos can often be found on their websites or YouTube. These manuals and videos are not the only or sole source of resources to identify hazards, but they do speak to the concept that a new safety manager does not have to start from scratch; resources exist to get us all started.

- **Materials**: Become familiar with the materials used in work procedures, and understand their hazards. Labels and safety data sheets (SDSs) are valuable resources, and serve as a starting point.

- **People**: Learn what employees do by talking to them, watching them work, and reading their job descriptions. What an employee does and their job description should be in sync, but, unsafe acts that expose employees to hazards or create hazards need to be identified and controlled.

- **Processes**: Understand what the organization does, step by step, start to finish. There is some magic in this consideration. Encourage the organization to engage safety management as early as possible in the design, review, and acquisition of new processes, as safety management can often help build safety into processes.

The Hierarchy of Controls

In addition, keep in mind how hazard control focuses on a hierarchy, which includes:

- Engineering the problem away; always the first choice of control.
- Providing physical controls and barriers.
- Implementing administrative controls, such as job rotation.
- Providing personal protective equipment (PPE); always our last control approach.

Finally, training and education can serve as means to reinforce these methods of hazard control.

If It's Not Written: "It Never Happened"

Key to successful hazard identification and control is documentation. Documentation is extremely important when it comes to safety activities, such as hazard assessments and incident investigations. In addition, safety managers and employees need a tool to document findings, make corrections needed, and do follow-up activities.

One tool to consider is a checklist. Checklists, though basic, are a good starting point to assist managers and employees identify workplace hazards. As with written safety programs and employee training, these checklists should be developed to be site- and process-specific.

An additional method to identify and document workplace hazards is a job hazard analysis (JHA). JHAs are an excellent training tool for new employees, as well as for a general review or pre-task planning. In addition, they're extremely helpful when introducing a new procedure or piece of equipment, and when conducting incident reviews or investigations.

A JHA starts by listing job steps that a worker must perform to complete a task. The list can be created by watching an employee perform the operation or by using other task or production documents. From there, recording the steps and reviewing with a department manager and employees for completeness is essential. This is an excellent opportunity to involve employees in the hazard identification and controls process.

Next, write down the safety and health hazards that are, or may be, present during each phase of the activity.

Finally, document the controls that need to be in place in order to eliminate or reduce those hazards, calibrated to each step. Personal protective equipment (PPE) can be included in this or within in its own section.

Bucket Three: Training and Education

Our third bucket can be filled by identifying and providing the training that's needed and/or required within the organization:

- First, recognize the difference between education and training. Education is the "what and the why," and involves the rationale, awareness, and process overview.
- On the other hand, training is the "how" and includes understanding, coaching, mentoring, validation, and demonstration.
- Also, training establishes a baseline and expectations for employees, establishing that crucial foundation for success.
- In addition, initial and ongoing ***training*** should be task- and job-specific, not general. Don't forget to include what's new, which are all triggers for training, as discussed above.
- Keep in mind that once the initial training has been provided, a safety manager's job is not done. Successful training involves drill repetition, constant feedback, observation of demonstration of knowledge, and possibly retraining.

Bucket Four: Communication is Key

In addition to being aware of injury and illness exposures, employees must understand their organization's expectations of them, and what happens if safety procedures are not followed.

They also need to be able to express their concerns on safety issues and have a voice to say, "I don't feel well" or "I've been hurt." This includes having an avenue for anonymous notification of safety matters.

Finally, employees should have an active voice in resolving safety issues in their work environment. An easy, but critical, way to engage employees is through soliciting their input, such as with safety suggestions. Then, by meeting with those who have provided suggestions, and letting them know whether their suggestion will or may be considered for adoption, and why or why not. This should not be a one-time occurrence. Reach out periodically, and make this activity an ongoing part of your safety program.

Bucket Five: Ensuring the Success of Your Safety Activities

The fifth step or bucket needing to be filled is ensuring the success of our safety activities by stacking the odds in our favor. Achievement can be reached by:

- Put together a plan. Create a blueprint. It does not have to be complex, and it provides a place to start. Highlight what you need to accomplish, who you need to influence, timeframes, etc. What is important is to be able to measure if what you planned for has taken place, is on schedule, and is (over time), making a difference.

- Build a technical knowledge base. The new safety manager needs opportunities for ongoing safety education and training on the hazards in the workplace, and the controls that need to be in place. Find classes and other means to augment that knowledge.
- Don't' be afraid to practice safety.
- Engage and communicate with our workforce, which is essential. Do it often, use multiple distribution methods, and always follow-up on ideas and suggestions.
- Employees are a valuable resource; use them whenever possible.
- Establish formal ways to identify and suggest controls. From there, consistently document everything.
- Review your safety program and determine if it's old or forward-looking; this must be a consistent refrain.
- Examine your training program. Is it training or is it really educational and awareness-based? If it's the latter only, there's work to be done. Safety is an applied activity in many respects; it helps when training is kinesthetic, and the more opportunities employees have to demonstrate and validate safety skills, the better.
- Analyze employee safety communications. Are they only one-way? If so, determine how to open the 'communication highway,' such that information and dialogue goes in all directions.

Lastly, understand how a safety program is a constant work in progress, and needs to focus on progress, not perfection. It has to consistently worked on and enhanced, not allowed to go fallow.

Bibliography

Hopwood, Dan & Thompson, Steve, 2006. *Workplace Safety: A Guide for Small and Midsized Companies.* New Jersey: John Wiley & Sons.

AUTHOR INDEX

AUTHOR INDEX

AUTHOR INDEX